Geometry

(A = area, B = area of base, C = circumference, S = lateral area or surface area, V = volume)

1. Triangle

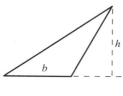

$$A = \frac{1}{2}bh$$

2. Similar Triangles

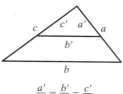

$$\frac{a'}{a} = \frac{b'}{b} = \frac{c'}{c}$$

3. Pythagorean Theorem

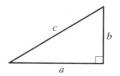

$$a^2 + b^2 = c^2$$

4. Parallelogram

$$A = bh$$

5. Trapezoid

$$A = \frac{1}{2}(a + b)h$$

6. Circle

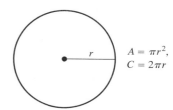

$$A = \pi r^2,$$
$$C = 2\pi r$$

7. Any Cylinder or Prism with Parallel Bases

$$V = Bh$$

8. Right Circular Cylinder

$$V = \pi r^2 h, \; S = 2\pi rh$$

9. Any Cone or Pyramid

$$V = \frac{1}{3}Bh$$

10. Right Circular Cone

$$V = \frac{1}{3}\pi r^2 h, \; S = \pi rs$$

11. Sphere

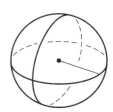

$$V = \frac{4}{3}\pi r^3, \; S = 4\pi r^2$$

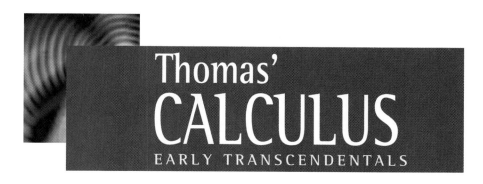

Thomas' CALCULUS
EARLY TRANSCENDENTALS

TENTH EDITION

PART I

Thomas'
CALCULUS
EARLY TRANSCENDENTALS

TENTH EDITION
PART I

Based on the original work by
George B. Thomas, Jr.
Massachusetts Institute of Technology

As revised by

Ross L. Finney
Maurice D. Weir
Naval Postgraduate School
and

Frank R. Giordano

Addison
Wesley

Boston San Francisco New York
London Toronto Sydney Tokyo Singapore Madrid
Mexico City Munich Paris Cape Town Hong Kong Montreal

Publisher	Greg Tobin
Senior Editor	Laurie Rosatone
Associate Project Editor	Ellen Keohane
Editorial Assistant	Susan Laferriere
Managing Editor	Karen Guardino
Senior Production Supervisor	Becky Malone
Editorial and Production Services	UG / GGS Information Services, Inc.
Art Editor	Geri Davis/The Davis Group, Inc.
Marketing Manager	Carter Fenton
Illustrators	Tech Graphics\cs
Technical Art Consultants	Scott Silva and Joe Vetere
Senior Prepress Supervisor	Caroline Fell
Compositor	UG / GGS Information Services, Inc.
Text Designer	Geri Davis/The Davis Group, Inc.
Cover Designer	Barbara T. Atkinson
Print Buyer	Evelyn Beaton

Photo Credits: 47, Courtesy of Agilent Technologies, Inc. **124,** John Elk III/Bruce Coleman, Inc. **139,** NASA/JPL **171, 245, 472, 631, 759, 864,** PSSC Physics 2/e, 1965; D.C. Heath & Co. with Educational Development Center, Inc., Newton, MA **236,** "Differentiation" by W.U. Walton et. Al., Project CALC, Educational Development Center, Inc., Newton, MA **239,** AP/Wide World Photos **323,** Scott Burns, Urbana, IL., www.designbyalgorithm.com **324,** COMAP, Inc. **466, 754,** Corbis **854,** Richard F. Voss/IBM Research **877,** Appalachian Mountain Club **911,** Department of History, US Military Academy, West Point, New York **929,** ND Roger-Viollet **1060, 1061,** Adapted from "NCFMF Book of Film Notes", 1974, MIT Press with Educational Development Center, Inc., Newton, MA **1061,** InterNetwork Media, Inc. and NASA/JPL

Library of Congress Cataloging-in-Publication Data
Finney, Ross L.
   Thomas' calculus : early transcendentals.—10th ed. / based on the original work by George B. Thomas, Jr., as revised by Ross L. Finney, Maurice D. Weir, and Frank R. Giordano.
     p. cm.
   Rev. ed. of Calculus and analytic geometry / George B. Thomas, Jr., Ross L. Finney, 9th ed. c1996.
   Includes bibliographical references and index.
   ISBN 0-201-66210-8
   1. Calculus. 2. Geometry, Analytic. I. Title: Calculus. II. Weir, Maurice D. III. Giordano, Frank R. IV. Thomas, George Brinton, 1914– Calculus and analytic geometry. V. Title.
   QA303.F48 2000
   515′.15—dc21   00-032790

Contents

 Preliminaries   **1**

 Limits and Continuity   **85**

Derivatives   **147**

3   Applications of Derivatives   241

4   Integration   333

5   Applications of Integrals   415

9   Vectors in the Plane and Polar Functions   717

Appendices   1143

Computer Algebra System (CAS) Exercises

To the Instructor

Throughout its illustrious history, *Thomas' Calculus* has been used to support a variety of courses and teaching methods, from traditional to experimental. This tenth edition is a substantial revision, yet it retains the traditional strengths of the text: sound mathematics, relevant and important applications to the sciences and engineering, and excellent exercises. This flexible and modern text contains all the elements needed to teach the many different kinds of courses that exist today.

A book does not make a course; the instructor and the students do. This text is a resource to support your course. With this in mind, we have added a number of features to the tenth edition making it even more flexible and useful, both for teaching and learning calculus.

Features of the Tenth Edition

- For the first time, this classic text is available in *both* standard and Early Transcendentals versions.
- The new *Annotated Instructor's Edition* contains suggestions for the incorporation of technology, highlighting how the Web site and CD-ROM can be used to enhance the presentation of chapter topics.
- As always, this text continues to be easy to read, conversational, and mathematically rich. Each new topic is motivated by clear, easy-to-understand examples and is then reinforced by its application to real-world problems of immediate interest to students.
- Each section now begins with a list of subsection headings, making key concepts readily apparent.
- Within the tenth edition is an increased emphasis on modeling and applications using real data. As a result, there is an improved balance of graphical, numerical, and analytic methods and techniques, accomplished without compromising the mathematical integrity of the book.
- Vectors and projectile motion in the plane are now covered separately from vectors in space, concluding the treatment of single-variable calculus. Three-dimensional vectors are then treated in conjunction with multivariable calculus.
- Exercise sets continue to be grouped within appropriate headings. Titles that indicate the content or application have been added for most word problems, and those requiring the use of a graphing utility are identified throughout the

text by the icon $\boxed{\text{T}}$. Computer Algebra System (CAS) exercises also appear in every chapter and are grouped in special subsections labeled "Computer Explorations."

- Together, the CD-ROM and Web site provide students and instructors with even more support:
  - A collection of Maple® and *Mathematica*® modules, videos, and Java applets are available to help students visualize key calculus concepts.
  - Interactive online tutorials help students review precalculus and textbook-specific material, take practice tests, and receive diagnostic feedback on their performance.
  - Chapter-by-chapter quizzes are also provided. These quizzes can be administered and graded online for skills-based mastery assessment.
  - Downloadable technology resources are provided for specific computer algebra systems and graphing calculators.
  - Expanded historical biographies are now on the Web site and CD-ROM, leaving more room in the margin of the text for notes, observations, and annotations and giving the book a more open look.

With all these changes, we have not compromised our belief that the fundamental goal of calculus is helping prepare students to enter the worlds of mathematics, science, and engineering.

Mastering Skills and Concepts

As always, this text continues to maintain a strong skill-building emphasis. Throughout this edition, we have included examples and discussions encouraging students to think visually, analytically, and numerically. Almost every exercise set contains problems requiring students to generate and interpret graphs as a tool for understanding mathematical or real-world relationships. Many sections also contain problems to extend the range of applications, mathematical ideas, and rigor.

Students are asked to explore and explain a variety of calculus concepts and applications in writing exercises placed throughout the text. In addition, each chapter ends with a list of questions to help students review and summarize what they have learned. Many of these review questions make great writing assignments.

Problem–Solving Strategies

We believe students learn best when procedural techniques are laid out as clearly and simply as possible. To this end, stepwise problem-solving summaries are included as appropriate, especially for the more difficult or complicated procedures. As always, we are especially careful that examples in the text illustrate the steps outlined by the summaries.

Exercises

Exercise sets have been carefully reviewed and revised in this new edition. They are grouped by topic, with special sections for computer explorations. These sections contain CAS explorations and projects.

Within exercise sets are practice and applied problems, critical thinking and challenging exercises (in subsections marked "Applications and Theory"), and exercises requiring students to write about important calculus concepts. Writing exercises appear throughout exercise sets. Exercises generally follow the order of presentation in the text, and those requiring a graphing utility (such as a graphing calculator) are identified throughout the text by the icon .

Chapter End Support Material

At the end of each chapter are three features summarizing the chapter contents:

"Questions to Guide Your Review" ask students to think about key chapter concepts and then verbalize their understanding of them and include illustrative examples. These questions are also suitable for writing exercises.

"Practice Exercises" provide a review of the techniques, computational and numerical skills, and key applications.

"Additional Exercises: Theory, Examples, and Applications" provide students with more theoretical or challenging applications and problems to further deepen their understanding of the mathematical ideas.

Applications and Examples

A hallmark of this book has been the application of calculus to science and engineering. These applied problems have been updated, improved, and extended continually over the last several editions. With this edition, we include more problems based on real data requiring graphical and numerical techniques for their solution. Throughout the text, we cite sources for the data or articles from which these applications are drawn, helping students understand that calculus is a current, dynamic field requiring a multiplicity of different techniques and approaches. Most of these applications are directed toward the physical sciences and engineering, but there are many from biology and the social sciences as well.

Technology: Graphing Utility and Computer Explorations

Virtually every section of the text contains exercises to explore numerical patterns or graphing utility exercises that ask students to generate and interpret graphs as a tool to understanding mathematical and real-world relationships. Many of the graphing utility exercises are suitable for classroom demonstration or for group work by students in or out of class. These exercises are identified throughout the text by the icon or the heading "Computer Explorations."

Computer Explorations

CD-ROM
WEBsite

Numbering more than 200, the computer explorations exercises have been solved using both *Mathematica* and Maple. In addition, *Mathematica* and Maple modules are available on the Web site and CD-ROM. These modules have been carefully designed to help students develop a geometric intuition and a deeper understanding and appreciation of calculus concepts, methodologies, and applications. CD/Web site icons mark the locations in the text where material related to these modules is covered.

Notes also appear throughout the text that encourage students to explore with graphing utilities and help them assess when the use of technology is helpful and when it may be misleading.

CD-ROM
WEBsite
Historical Biography

Expanded History and Biographies

Any student is enriched by seeing the human side of mathematics through its historical development. In previous editions, we featured history boxes describing the origins of ideas, conflicts concerning ownership of these ideas, and interesting sidelights into modern topics such as fractals and chaos. For the tenth edition, we have expanded and written more biographies and historical essays. These essays are now available on the CD-ROM and Web site; they are referenced by icons throughout the text, leaving more room in the margins for student notes, observations, and annotations.

The Many Faces of This Book

Mathematics Is a Formal and Beautiful Language

Calculus is one of the most powerful of human intellectual achievements. One goal of this book is to give students an appreciation of the beauty of calculus. As in previous editions, we have been careful to say only what is true and mathematically sound. Every definition, theorem, corollary, and proof has been reviewed for clarity and mathematical correctness.

Whether calculus is taught in a traditional lecture format or entirely in labs with individual and group learning focusing on numerical and graphical experimentation, its ideas and techniques need to be articulated clearly and accurately.

Students Will Learn from This Book for Many Years to Come

We intentionally provide far more material than any one instructor would want to teach. Students can continue to learn calculus from this book long after the class has ended. It provides an accessible review of the calculus a student has already studied and is a resource for the working engineer or scientist.

Highlights of New Content Features, by Chapter

Preliminaries

- All the familiar precalculus functions are covered completely.
- Parametric equations are introduced.
- Inverses of familiar functions, including inverse trigonometric functions, are also covered.
- Mathematical modeling, with modeling exercises, is introduced.
- New examples and exercises employ real data and regression analysis using a calculator.

Chapter 1   Limits and Continuity

- Limits are introduced by way of rates of change, with a concluding section on tangent lines to connect and complete the initial discussion.
- All the fundamental ideas on limits are now together in a single chapter, including finite limits, infinite limits, asymptotes, limit rules, and $\lim_{\theta \to 0} ((\sin \theta)/\theta)$.
- Both informal and precise definitions of the limit concept are given, but there is less emphasis on using the precise definition to prove theorems.

Chapter 2   Derivatives

- The derivative as a rate of change is presented earlier to stress its importance in studying motion along a line in modeling real-world phenomena.
- Differentiation rules are presented in two sections to enhance the clarity and flow of the presentation.
- First and second derivatives for parametric equations are included as an application of the Chain Rule.
- The Early Transcendentals version of this Chapter differs from the standard edition in that it has two additional setions, the first devoted to the derivatives of inverse trigonometric functions, and the second to derivatives of exponential and logarithmic functions. The number e is defined as the value of a that makes $\lim_{h \to 0} ((a^h - 1)/h)$ equal 1. The natural logarithm is defined as the inverse of e^x.
- With exponential and logarithmic functions now available, we study exponential change and verify the derivative power rule for arbitrary real powers.

Chapter 3   Applications of Derivatives

- The availability of the derivatives of inverse trigonometric, exponential, and logarithmic functions enriches the examples, exercises, and applications in this chapter.
- The treatment of using the first and second derivatives to determine the shape of a graph is more focused and streamlined.
- A new section on using the first and second derivative to produce graphical solutions to autonomous first-order differential equations acts as a graphical prelude to Chapters 4 and 6.
- The new section includes an introduction to population modeling.

Chapter 4   Integration

- As before, indefinite integrals are presented first, stressing their importance for solving elementary differential equations. The rules for antiderivatives and the substitution method follow next.
- The list of available integrable funtions now includes exponential functions and functions leading to logarithmic and inverse trigonometric functions. We no longer have to wait until Chapter 6, as in the standard edition, to integrate the tangent and cotangent functions.
- As in the previous edition, estimating with finite sums in a variety of application settings motivates the ideas of Riemann sums and definite integrals. Students see the definite integral early as more than just a tool for finding area.
- The section defining the definite integral as a limit of Riemann sums has been streamlined and now focuses on continuous functions. Piecewise-continuous functions are treated in the Additional Exercises at the end of the chapter.
- All the material on single integral area calculations (including areas between curves) is now treated in this chapter.

Chapter 5   Applications of Integrals

- The treatment of volumes has been combined from three into two sections.
- Arc length formulas are developed for both explicit function and parametric curves in the plane.
- Surface area has been moved to Chapter 13, where it is needed for surface integrals. There, it is treated in a unified fashion rather than as a special case of surfaces of revolution.
- The availability of exponential and logarithmic functions makes it possible to treat applications of separable first-order differential equations here instead of waiting until the next chapter as in the standard edition. Among the applications are modeling growth and decay, heat transfer, falling with resistance proportional to velocity, coasting to a stop, and torricelli's law.
- The important applications to springs, pumping and lifting, fluid forces, and moments have all been retained from the previous edition.

Chapter 6   Transcendental Functions and Differential Equations

- With so much of the standard material on transcendental functions now in earlier chapters, this chapter is considerably shorter than its standard-edition counterpart. It still begins with defining the natural logarithm as an integral, however, then defining the natural exponential function as its inverse and reconfirming the laws and differentiation rules derived at the end of Chapter 2.

- Linear first-order differential equations follow, modeling mixture problems and *RL* circuits.
- Euler's method and the improved Euler's method are combined with additional material on population models, illustrating graphical, numerical, and analytic solution methods.
- This chapter concludes with a short treatment of hyperbolic and inverse hyperbolic functions.

Chapter 7   Integration Techniques, L'Hôpital's Rule, and Improper Integrals

- Monte Carlo integration is now included with the use of integral tables or computer algebra systems (CAS) to find integrals.
- L'Hôpital's Rule is covered in this chapter just prior to its use for calculating some improper integrals and limits of sequences (in Chapter 8).

Chapter 8   Infinite Series

- The basic ideas concerning sequences of numbers and their limits are covered in the first section. The next section, which is optional, treats the more theoretical ideas involving subsequences and bounded monotonic sequences.
- Most of the important series convergence tests are presented together in a single, streamlined section.
- Two new optional sections at the end of the chapter cover the basics of Fourier series. This inclusion allows for an earlier introduction to these important concepts for students requiring their use right away in their applied science and engineering courses. Completing the elementary introduction to series, these sections illustrate important representations of functions by series other than power series.

Chapter 9   Vectors in the Plane and Polar Functions

- This is a new chapter on vectors and projectile motion in the plane, with two sections at the end covering polar coordinates and graphs and the calculus of polar curves to prepare students for their use in multivariable calculus. It permits an earlier self-contained treatment of planar vectors, if desired. The chapter can be covered any time after the coverage of the integral and the calculus of exponential and logarithmic functions.
- Chapters P through 9 now form a complete package treating the ideas of single variable calculus. Three-dimensional vectors are presented independently along with multivariable calculus, beginning in Chapter 10.
- Vector ideas are motivated by their application to studying paths, velocities, accelerations, and forces associated with bodies moving along planar paths.
- The detailed analytic geometry of conic sections and quadratic equations has been eliminated. These ideas are thoroughly covered in high school and precalculus courses, but we nevertheless review many of the basics throughout the text as needed.
- Parametrizations of plane curves has been moved to earlier chapters.

Chapter 10   Vectors and Motion in Space

- Three-dimensional vectors, the geometry of space, and vector-valued functions defining space curves are now organized together in this single chapter with fresh introductions and examples. This chapter now constitutes, and clearly delineates, the entry point for the multivariable calculus.
- Letters representing vectors have been changed from uppercase letters to the now more standard lowercase letters.
- Vectors in the plane are reviewed along with the development of the algebra and geometry of three-dimensional vectors to help students bridge any possible gap between Calculus II and Calculus III courses.
- The logical treatment and organization of motion along space curves and the **TNB** frame has been retained from the previous edition.

Chapter 11   Multivariable Functions and Their Derivatives

- The chapter has been reorganized to improve efficiency and flow. The treatment of partial derivatives with constrained variables has been moved toward the end of the chapter to follow the introduction to Lagrange multipliers. The treatment of linearization and differentials now follows the treatment of directional derivatives, gradient vectors, and tangent planes.

- The treatment of gradients and tangent planes is shorter and more direct.
- A new introduction to extreme values and saddle points compares and contrasts the multivariable case with the single-variable case.
- The exercise sets have been streamlined and all applications exercises labeled for quick identification.

Chapter 12   Multiple Integrals

- The treatment of the calculation of masses, moments, and centers of mass with multiple integrals is now self-contained. It no longer assumes previous exposure to the single-integral calculations in Chapter 5, which may now be bypassed entirely.
- Again, the practice of titling exercises makes them noticeably easier to select than before.

Chapter 13   Integration in Vector Fields

- In the treatment of Green's Theorem in the plane, circulation density at a point is introduced as the **k**-component of a more general circulation vector called the curl, which is treated in detail in the later section on Stokes' Theorem. This arrangement resolves the apparent inconsistency of having circulation in the plane represented by a scalar while circulation in space is represented by a vector.

Supplements for the Instructor

TestGen-EQ with QuizMaster-EQ

Windows and Macintosh CD (dual platform)
ISBN 0-201-70288-6
TestGen-EQ's friendly graphical interface enables instructors to view, edit, and add questions, transfer questions to tests, and print tests in a variety of fonts and forms easily. Search and sort features let the instructor quickly locate questions and arrange them in a preferred order. Six question formats are available, including short answer, true–false, multiple choice, essay, matching, and bimodal formats. A built-in question editor gives the user power to create graphs, import graphics, insert mathematical symbols and templates, and insert variable numbers or text. Computerized test banks include algorithmically defined problems organized according to each version of the textbook (standard and Early Transcendentals). An "Export to HTML" feature allows instructors to create practice tests for the Web.

QuizMaster-EQ enables instructors to create and save tests using *TestGen-EQ* so that students can take them for either practice or a grade on a computer network. Instructors can set preferences for how and when tests are administered. *QuizMaster-EQ* automatically grades exams, stores results on disk, and allows the instructor to view or print a variety of reports for individual students, classes, or courses.

This software is free to adopters of the text. Consult your Addison-Wesley representative for details.

Instructor's Solutions Manual

Volume I (Chapters P–9), ISBN 0-201-71010-2
Volume II (Chapters 8–13), ISBN 0-201-50404-9
The *Instructor's Solutions Manual* by Maurice D. Weir and John L. Scharf contains complete worked-out solutions to all the exercises in the text.

Answer Book

ISBN 0-201-71009-9
The *Answer Book* by Maurice D. Weir and John L. Scharf contains short answers to most of the exercises in the text.

Technology Resource Manuals for Computer Algebra Systems and Graphing Calculators

TI-Graphing Calculator Manual ISBN 0-201-72198-8
Maple Manual ISBN 0-201-72197-X
Mathematica Manual ISBN 0-201-72196-1
Each manual provides detailed guidance for integrating a specific software package or graphing calculator throughout the course, including syntax and commands.

Transparency Masters

Instructors may download from the CD-ROM a full set of color PowerPoint art transparencies featuring a number of the more complex figures from the text for use in the classroom.

Supplements for the Student

Student's Study Guide

Volume I (Chapters P–9), ISBN 0-201-71008-0
Volume II (Chapters 8–13), ISBN 0-201-50406-5
Organized to correspond with the text, the *Student's Study Guide* reinforces important concepts and provides study tips and additional practice problems.

Student's Solutions Manual

Volume I (Chapters P–9), ISBN 0-201-66211-6
Volume II (Chapters 8–13), ISBN 0-201-50402-2
The Student's Solutions Manual by Maurice D. Weir and John L. Scharf is designed for the student and contains carefully worked-out solutions to all the odd-numbered exercises in the text.

Just-in-Time Algebra and Trigonometry for Students of Calculus, Second Edition

ISBN 0-201-66974-9
Sharp algebra and trigonometry skills are critical to mastering calculus, and *Just-in-Time Algebra and Trigonometry for Students of Calculus, Second Edition* by Guntram Mueller and Ronald I. Brent is designed to bolster these skills while students study calculus. As students make their way through calculus, this text is with them every step of the way, showing them the necessary algebra or trigonometry topics and pointing out potential problem spots. The easy-to-use contents has algebra and trigonometry topics arranged in the order in which students will need them as they study calculus.

AWL Math Tutor Center

The AWL Math Tutor Center (www.awl.com/tutorcenter) provides assistance to students who take calculus and purchase a mathematics textbook published by Addison Wesley Longman. Help is provided via phone, fax, and e-mail. Students who use the service will be helped by tutors who are qualified mathematics instructors.

CD-ROM and Web Site

Maple and Mathematica Modules

Over 35 modules have been written by John L. Scharf and Marie M. Vanisko of Carroll College in Montana and Colonel D. Chris Arney of the U.S. Military Academy. These modules have been carefully designed to help students develop their geometric intuition and deepen their understanding of calculus concepts and methods. Based on real-world applications, they encourage students to visualize calculus and to discover its importance in everyday life. Users will need *Mathematica* or Maple to access these modules. Icons reference these modules throughout the text.

Interactive Calculus (Java Applets)

These unique interactive calculus Java applets are easy to use, with no syntax or special languages to learn. Students can manipulate equations and graphs in "real time." Topics span limits, projectile motion, slopes, tangents, derivatives, integrals, **TNB** frames, and the concept of curl. By bringing these applets into classroom demonstration and discussion, laboratory and homework assignments, or independent study, teachers and students can explore the mathematics of time and motion. These applets are designed to build a clear understanding of concepts when they are first encountered and to help students over the hurdles of abstraction that have often confused other students in the past.

Video Clips

Video clips of real-world situations provide motivation for learning and applying calculus. These videos have been developed specifically to accompany several of the calculus modules described above.

Expanded History and Biographies

Icons throughout the book refer to expanded historical biographies and notes on the Web site and CD-ROM. These materials have been written by Colonel D. Chris Arney of the U.S. Military Academy in collaboration with Joe B. Albree of Auburn University.

Just-in-Time Online Algebra and Trigonometry

Compiled by Ronald I. Brent and Guntram Mueller of the University of Massachusetts, Lowell, this interactive Web-based testing and tutorial system allows students to practice the algebra and trigonometry skills critical to mastering calculus. *Just-in-Time Online* tracks student progress and provides personalized study plans to help students succeed. The registration coupon at the back of this text provides access to this feature of the Web site.

Interactive Calculus Tutorial

Written by G. Donald Allen, Michael Stecher, and Philip B. Yasskin of Texas A&M University, this interactive online calculus tutorial lets students review textbook-specific material by chapter via practice quizzes and receive diagnostic feedback on their performance.

Skill Mastery Quizzes

A collection of chapter-by-chapter quizzes is also provided on the Web site. These quizzes can be administered and graded online for skills-based mastery assessment.

Transparency Masters

Instructors may download from the CD-ROM a full set of color PowerPoint art transparencies featuring a number of the more complex figures from the text for use in the classroom.

Downloadable Technology Resources for Specific Computer Algebra Systems and Graphing Calculators

Each manual provides detailed guidance for integrating a specific software package or graphing calculator throughout the course, including syntax and commands. These manuals are available on the Web site for downloading in PDF form.

Collaborative Network

The Collaborative Network is a suite of online communication tools, which includes message boards and i-chat. These tools can be used to deliver courses in a distance learning environment. Message boards allow users to post messages and check back periodically for responses. Students can also use message boards to obtain peer support for study guide activities, freeing up instructor time. I-chat is a perfect arena for instructor-led live discussions with groups of students. The i-chat auditorium allows instructors to post a series of slides in the upper part of the screen while fielding questions (text only) from students in the bottom portion. This feature is particularly useful either as a review of lectures or for class meetings of geographically dispersed students.

Syllabus Manager™

Syllabus Manager™ is a free online syllabus creation and management tool for instructors and students who use this text. It can be used by a nontechnical person to build and maintain one or more syllabi on the Web. Students may "turn on" an instructor's syllabus from the Web site.

Acknowledgments

We would like to express our thanks for the many valuable contributions of the people who reviewed this edition as it developed through its various stages:

Manuscript Reviewers

Tuncay Aktosun, North Dakota State University
Andrew G. Bennett, Kansas State University
Terri A. Bourdon, Virginia Polytechnic Institute and State University
Mark Brittenham, University of Nebraska, Lincoln
Bob Brown, Essex Community College
David A. Edwards, University of Delaware
Mark Farris, Midwestern State University
Kim Jongerius, Northwestern College
Jeff Knisley, East Tennessee State University
Slawomir Kwasik, Tulane University
Jeuel LaTorre, Clemson University
Daniel G. Martinez, California State University, Long Beach
Sandra E. McLaurin, University of North Carolina, Wilmington
Stephen J. Merrill, Marquette University
Shai Neumann, Brevard Community College
Linda Powers, Virginia Polytechnic Institute and State University
William L. Siegmann, Rensselaer Polytechnic Institute
Rick L. Smith, University of Florida
James W. Thomas, Colorado State University
Abraham Ungar, North Dakota State University
Harvey E. Wolff, University of Toledo

Technology Reviewers

Mark Brittenham, University of Nebraska, Lincoln
Warren J. Burch, Brevard Community College, Cocoa
Lyle Cochran, Whitworth College
Philip S. Crooke III, Vanderbilt University
Linda Powers, Virginia Polytechnic Institute and State University
David Ruch, Metropolitan State College of Denver
Paul Talaga, Weber State University
James W. Thomas, Colorado State University
Robert L. Wheeler, Virginia Polytechnic Institute and State University

Other Contributors

We especially thank Colonel D. Chris Arney, John L. Scharf, and Marie M. Vanisko for sharing their insights in using technology to help make calculus come alive for the student, and Colonel D. Chris Arney and Joe B. Albree for their contributions on the history of calculus. Their dedication, encouragement, and team efforts in working with us to conceive and create the technology modules and historical biographies and essays is deeply appreciated. Further thanks to John L. Scharf for his contributions to the solutions manuals.

To the Student

What Is Calculus?

Calculus is the mathematics of motion and change. Where there is motion or growth, where variable forces are at work producing acceleration, calculus is the mathematics to apply. This was true in the beginnings of the subject, and it is true today.

Calculus was first invented to meet the mathematical needs of the scientists of the sixteenth and seventeenth centuries, needs that were mainly mechanical in nature. Differential calculus dealt with the problem of calculating rates of change. It enabled people to define slopes of curves, to calculate velocities and accelerations of moving bodies, to find firing angles that would give cannons their greatest range, and to predict the times when planets would be closest together or farthest apart. Integral calculus dealt with the problem of determining a function from information about its rate of change. It enabled people to calculate the future location of a body from its present position and a knowledge of the forces acting on it, to find the areas of irregular regions in the plane, to measure the lengths of curves, and to find the volumes and masses of arbitrary solids.

Today, calculus and its extensions in mathematical analysis are far-reaching indeed, and the physicists, mathematicians, and astronomers who first invented the subject would surely be amazed and delighted, as we hope you will be, to see what a profusion of problems it solves and what a range of fields now use it in the mathematical models that bring understanding about the universe and the world around us. The goal of this edition is to present a modern view of calculus enhanced by the use of technology.

How to Learn Calculus

Learning calculus is not the same as learning arithmetic, algebra, and geometry. In those subjects, you learn primarily how to calculate with numbers; how to simplify algebraic expressions and calculate with variables; and how to reason about points, lines, and figures in the plane. Calculus involves those techniques and skills but develops others as well, with greater precision and at a deeper level. Calculus introduces so many new concepts and computational operations, in fact, that you will no longer be able to learn everything you need in class. You will have to learn a fair amount on your own or by working with other students. What should you do to learn?

1. *Read the text.* You will not be able to learn all the meanings and connections you need just by attempting the exercises. You will need to read relevant passages in

the book and work through examples step by step. Speed reading will not work here. You are reading and searching for detail in a step-by-step, logical fashion. This kind of reading, required by any deep and technical content, takes attention, patience, and practice.

2. *Do the homework*, keeping the following principles in mind.
   (a) *Sketch diagrams* whenever possible.
   (b) *Write your solutions in a connected step-by-step, logical fashion*, as if you were explaining it to someone else.
   (c) *Think about why* each exercise is there. Why was it assigned? How is it related to the other assigned exercises?

3. *Use your graphing calculator and computer* whenever possible. Complete as many grapher and Computer Exploration exercises as you can, even if they are not assigned. Graphs provide insight and visual representation of important concepts and relationships. Numbers can reveal important patterns. A graphing calculator or computer gives you the freedom to explore realistic problems and examples that involve calculations too difficult or lengthy to do by hand.

4. Try on your own to *write short descriptions of the key points* each time you complete a section of the text. If you succeed, you probably understand the material. If you do not, you will know where there is a gap in your understanding.

Learning calculus is a process; it does not come all at once. Be patient, persevere, ask questions, discuss ideas and work with classmates, and seek help when you need it, right away. The rewards of learning calculus will be very satisfying, both intellectually and professionally.

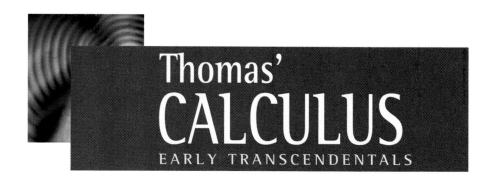

Thomas' CALCULUS
EARLY TRANSCENDENTALS

TENTH EDITION

PART I

P

Preliminaries

OVERVIEW   This chapter reviews the most important things you need to know to start learning calculus. It also introduces the use of a graphing utility as a tool to investigate mathematical ideas, to support analytical work, and to solve problems with numerical and graphical methods. The emphasis is on functions and graphs, the main building blocks of calculus.

Functions and parametric equations are the major tools for describing the real world in mathematical terms, from temperature variations to planetary motions, from brain waves to business cycles, and from heartbeat patterns to population growth. Many functions have particular importance because of the behavior they describe. Trigonometric functions describe cyclic, repetitive activity; exponential, logarithmic, and logistic functions describe growth and decay; and polynomial functions can approximate these and most other functions.

 Lines

Increments • Slope of a Line • Parallel and Perpendicular Lines • Equations of Lines • Applications • Regression Analysis with a Calculator

One reason calculus has proved to be so useful is that it is the right mathematics for relating the rate of change of a quantity to the graph of the quantity. Explaining that relationship is one goal of this book. It all begins with the slopes of lines.

Increments

When a particle in the plane moves from one point to another, the net changes or *increments* in its coordinates are found by subtracting the coordinates of its starting point from the coordinates of its stopping point.

The symbols Δx and Δy are read "delta x" and "delta y." The letter Δ is a Greek capital d for "difference." Neither Δx nor Δy denotes multiplication; Δx is not "delta times x" nor is Δy "delta times y."

Definition   Increments
If a particle moves from the point, (x_1, y_1) to the point (x_2, y_2), the
increments in its coordinates are

$$\Delta x = x_2 - x_1 \qquad \text{and} \qquad \Delta y = y_2 - y_1.$$

Increments can be positive, negative, or zero, as in Example 1.

Example 1   Finding Increments

The coordinate increments from $(4, -3)$ to $(2, 5)$ are

$$\Delta x = 2 - 4 = -2, \qquad \Delta y = 5 - (-3) = 8.$$

From $(5, 6)$ to $(5, 1)$, the increments are

$$\Delta x = 5 - 5 = 0, \qquad \Delta y = 1 - 6 = -5.$$

Slope of a Line

Each nonvertical line L has a *slope,* which we calculate as elevation change per unit of run in the following way. Let $P_1(x_1, y_1)$ and $P_2(x_2, y_2)$ be any two points on L (Figure 1). We call $\Delta y = y_2 - y_1$ the **rise** from P_1 to P_2, $\Delta x = x_2 - x_1$ the **run** from P_1 to P_2, and we define the slope of L to be $\Delta y / \Delta x$.

FIGURE 1 The slope of line L is

$$m = \frac{\text{rise}}{\text{run}} = \frac{\Delta y}{\Delta x}.$$

It is conventional to denote the slope by the letter m.

> **Definition**   Slope
>
> Let $P_1(x_1, y_1)$ and $P_2(x_2, y_2)$ be points on a nonvertical line, L. The **slope** of L is
>
> $$m = \frac{\text{rise}}{\text{run}} = \frac{\Delta y}{\Delta x} = \frac{y_2 - y_1}{x_2 - x_1}.$$

A line that goes uphill as x increases has a positive slope. A line that goes downhill as x increases has a negative slope. A horizontal line has slope zero since all its points have the same y-coordinate, making $\Delta y = 0$. For vertical lines, $\Delta x = 0$ and the ratio $\Delta y / \Delta x$ is undefined. We express this fact by saying *vertical lines have no slope.*

Parallel and Perpendicular Lines

Parallel lines form equal angles with the x-axis (Figure 2). Hence, nonvertical parallel lines have the same slope. Conversely, lines with equal slopes form equal angles with the x-axis and are therefore parallel.

If two nonvertical lines L_1 and L_2 are perpendicular, their slopes m_1 and m_2 satisfy $m_1 m_2 = -1$, so each slope is the *negative reciprocal* of the other:

$$m_1 = -\frac{1}{m_2}, \qquad m_2 = -\frac{1}{m_1}.$$

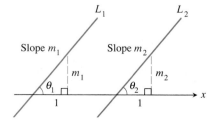

FIGURE 2 If $L_1 \parallel L_2$, then $\theta_1 = \theta_2$ and $m_1 = m_2$. Conversely, if $m_1 = m_2$, then $\theta_1 = \theta_2$ and $L_1 \parallel L_2$.

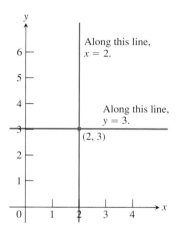

FIGURE 3 ΔADC is similar to ΔCDB. Hence, ϕ_1 is also the upper angle in ΔCDB, where $\tan \phi_1 = a/h$.

FIGURE 4 The standard equations for the vertical and horizontal lines through the point (2, 3) are $x = 2$ and $y = 3$. (Example 3)

The argument goes like this: In the notation of Figure 3, $m_1 = \tan \phi_1 = a/h$, whereas $m_2 = \tan \phi_2 = -h/a$. Hence, $m_1 m_2 = (a/h)(-h/a) = -1$.

Example 2   Determining Perpendicularity from Slope

If L is a line with slope $3/4$, any line with slope $-4/3$ is perpendicular to L.

Equations of Lines

The vertical line through the point (a, b) has equation $x = a$ since every x-coordinate on the line has the value a. Similarly, the horizontal line through (a, b) has equation $y = b$.

Example 3   Finding Equations for Vertical and Horizontal Lines

The vertical and horizontal lines through the point (2, 3) have equations $x = 2$ and $y = 3$, respectively (Figure 4).

We can write an equation for any nonvertical line if we know its slope m and the coordinates of one point $P_1(x_1, y_1)$ on it. For if $P(x, y)$ is *any* other point on the line then

$$\frac{y - y_1}{x - x_1} = m,$$

so that

$$y - y_1 = m(x - x_1) \qquad \text{or} \qquad y = m(x - x_1) + y_1.$$

Definition   Point–Slope Equation

The equation

$$y = m(x - x_1) + y_1$$

is the **point-slope equation** of the line through the point (x_1, y_1) with slope m.

Example 4   Using the Point-Slope Equation

Write an equation for the line through the point (2, 3) with slope $-3/2$.

Solution   We substitute $x_1 = 2$, $y_1 = 3$, and $m = -3/2$ into the point-slope equation and obtain

$$y = -\frac{3}{2}(x - 2) + 3 \qquad \text{or} \qquad y = -\frac{3}{2}x + 6.$$

Example 5   Using the Point-Slope Equation

Write an equation for the line through $(-2, -1)$ and $(3, 4)$.

Solution   The line's slope is

$$m = \frac{4 - (-1)}{3 - (-2)} = \frac{5}{5} = 1.$$

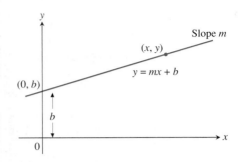

FIGURE 5 A line with slope m and y-intercept b.

We can use this slope with either of the two given points in the point-slope equation. Using $(x_1, y_1) = (-2, -1)$, we obtain

$$y = 1 \cdot (x - (-2)) + (-1)$$
$$y = x + 2 + (-1)$$
$$y = x + 1.$$

The y-coordinate of the point where a nonvertical line intersects the y-axis is the **y-intercept** of the line. Similarly, the x-coordinate of the point where a nonhorizontal line intersects the x-axis is the **x-intercept** of the line. A line with slope m and y-intercept b passes through $(0, b)$ (Figure 5), so

$$y = m(x - 0) + b, \qquad \text{or, more simply,} \qquad y = mx + b.$$

Definition   Slope–Intercept Equation

The equation

$$y = mx + b$$

is the **slope-intercept equation** of the line with slope m and y-intercept b.

Example 6   Writing Equations for Lines

Write an equation for the line through the point $(-1, 2)$ that is **(a)** parallel and **(b)** perpendicular to the line $L: y = 3x - 4$.

Solution   The line L, $y = 3x - 4$, has slope 3.

(a) The line $y = 3(x + 1) + 2$, or $y = 3x + 5$, passes through the point $(-1, 2)$ and is parallel to L because it has slope 3.

(b) The line $y = (-1/3)(x + 1) + 2$, or $y = (-1/3)x + 5/3$, passes through the point $(-1, 2)$ and is perpendicular to L because it has slope $-1/3$.

If A and B are not both zero, the graph of the equation $Ax + By = C$ is a line. Every line has an equation in this form, even lines with undefined slopes.

Definition   General Linear Equation

The equation

$$Ax + By = C \qquad (A \text{ and } B \text{ not both } 0)$$

is a **general linear equation** in x and y.

Although the general linear form helps in the quick identification of lines, the slope-intercept form is the one to enter into a calculator for graphing.

Example 7   Analyzing and Graphing a General Linear Equation

Find the slope and y-intercept of the line $8x + 5y = 20$. Graph the line.

$y = -\dfrac{8}{5}x + 4$

[-5, 7] by [-2, 6]

FIGURE 6 The line $8x + 5y = 20$.

Solution Solve the equation for y to put the equation in slope-intercept form.

$$8x + 5y = 20$$
$$5y = -8x + 20$$
$$y = -\frac{8}{5}x + 4.$$

This form reveals the slope ($m = -8/5$) and y-intercept ($b = 4$), and it puts the equation in a form suitable for graphing on a graphing calculator (Figure 6).

Applications

Many important variables are related by linear equations. For example, the relationship between Fahrenheit temperature and Celsius temperature is linear, a fact we use to advantage in the next example.

Example 8 Temperature Conversion

Find a formula relating Fahrenheit and Celsius temperature. Then find the Celsius equivalent of 90°F and the Fahrenheit equivalent of $-5°C$.

Solution Because the relationship between the two temperature scales is linear, it has the form $F = mC + b$. The freezing point of water is $F = 32°$ or $C = 0°$, whereas the boiling point is $F = 212°$ or $C = 100°$. Thus,

$$32 = m \cdot 0 + b \qquad \text{and} \qquad 212 = m \cdot 100 + b,$$

so $b = 32$ and $m = (212 - 32)/100 = 9/5$. Therefore,

$$F = \frac{9}{5}C + 32 \qquad \text{or} \qquad C = \frac{5}{9}(F - 32).$$

These relationships let us find equivalent temperatures. The Celsius equivalent of 90°F is

$$C = \frac{5}{9}(90 - 32) \approx 32.2°.$$

The Fahrenheit equivalent of $-5°C$ is

$$F = \frac{9}{5}(-5) + 32 = 23°.$$

Regression Analysis with a Calculator

It can be difficult to see patterns or trends in lists of paired numbers. For this reason, we sometimes begin by plotting the pairs (such a plot is called a **scatter plot**) to see whether the corresponding points have a pattern or trend of some kind. If they do, and if we can find an equation $y = f(x)$ for a curve that approximates the trend, then we have a formula that

1. summarizes the data with a simple expression, and

2. lets us predict values of y for other values of x.

Table 1 Price of a U.S. postage stamp	
Year *x*	**Cost** *y*
1885	0.02
1917	0.03
1919	0.02
1932	0.03
1958	0.04
1963	0.05
1968	0.06
1971	0.08
1974	0.10
1975	0.13
1977	0.15
1981	0.18
1981	0.20
1985	0.22
1987	0.25
1991	0.29
1995	0.32
1998	0.33

The process of finding a specific curve type to fit data is **regression analysis,** and the curve is a **regression curve.**

There are many useful types of regression curves, such as power, polynomial, exponential, logarithmic, and sinusoidal curves. In Example 9, we use a graphing calculator's linear regression feature to fit data from Table 1 with a linear equation. This process is equivalent to fitting the points in the scatter plot of these data with a line.

Example 9   Regression Analysis with a Calculator

Starting with the data in Table 1, build a model for the price of a postage stamp as a function of time. After verifying that the model is "reasonable," use it to predict the price in 2010.

Solution

Interpreting the Data

There is little change in the price of a U.S. stamp prior to 1968. Because we are really interested in the trend of the more recent data, we begin with that year. There were two increases in 1981, one of three cents followed by another of two cents. To make 1981 comparable with the other listed years, we lump them together as a single five-cent increase, giving the data in Table 2. Figure 7a gives the scatter plot for Table 2.

Table 2 Price of a U.S. postage stamp since 1968										
x	0	3	6	7	9	13	17	19	23	27	30
y	6	8	10	13	15	20	22	25	29	32	33

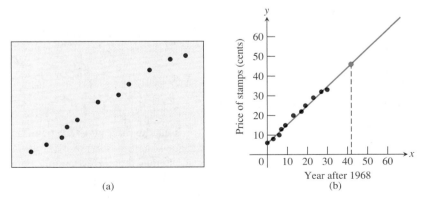

(a)   (b)

FIGURE 7 (a) Scatter plot of (x, y) data in Table 2. (b) Using the regression line to estimate the price of a stamp in 2010.

Model

Since the plot is fairly linear, we investigate a linear model. Upon entering the data into a graphing calculator (or spreadsheet) and selecting the linear regression option, we find the regression line to be

$$y = 0.96185x + 5.8978 .\tag{1}$$

Figure 7b shows the line and scatter plot together. The fit is remarkably good, so the model seems reasonable.

Solve Graphically

Our goal is to predict the price of a stamp in the year 2010. Reading from the graph in Figure 7b, we conclude that in 2010 ($x = 42$), the value of y is about 46.

Interpret

In the year 2010, a postage stamp will cost about 46 cents.

Confirm Algebraically

Evaluating Equation (1) for $x = 42$ gives

$$y = 0.96185(42) + 5.8978 \approx 46.3 .$$

Regression Analysis

Regression analysis has four steps:

Step 1. Plot the data (scatter plot).

Step 2. Find a regression equation. For a line, it has the form $y = mx + b$.

Step 3. Superimpose the graph of the regression equation on the scatter plot to see the fit.

Step 4. If the fit is satisfactory, use the regression equation to predict y-values for values of x not in the table.

EXERCISES 1

In Exercises 1 and 2, find the coordinate increments from A to B.

1. (a) $A(1, 2)$, $B(-1, -1)$ **(b)** $A(-3, 2)$, $B(-1, -2)$

2. (a) $A(-3, 1)$, $B(-8, 1)$ **(b)** $A(0, 4)$, $B(0, -2)$

In Exercises 3 and 4, let L be the line determined by points A and B.

  (i) Plot A and B.   **(ii)** Find the slope of L.

  (iii) Draw the graph of L.

3. (a) $A(1, -2)$, $B(2, 1)$ **(b)** $A(-2, -1)$, $B(1, -2)$

4. (a) $A(2, 3)$, $B(-1, 3)$ **(b)** $A(1, 2)$, $B(1, -3)$

In Exercises 5 and 6, write an equation for **(i)** the vertical line and **(ii)** the horizontal line through the point P.

5. (a) $P(2, 3)$ **(b)** $P(-1, 4/3)$

6. (a) $P(0, -\sqrt{2})$ **(b)** $P(-\pi, 0)$

In Exercises 7 and 8, write the point-slope equation for the line through the point P with slope m.

7. (a) $P(1, 1)$, $m = 1$ **(b)** $P(-1, 1)$, $m = -1$

8. (a) $P(0, 3)$, $m = 2$ **(b)** $P(-4, 0)$, $m = -2$

In Exercises 9 and 10, write a general linear equation for the line through the two points.

9. (a) $(0, 0)$, $(2, 3)$ **(b)** $(1, 1)$, $(2, 1)$

10. (a) $(-2, 0)$, $(-2, -2)$ **(b)** $(-2, 1)$, $(2, -2)$

In Exercises 11 and 12, write the slope-intercept equation for the line with slope m and y-intercept b.

11. (a) $m = 3$, $b = -2$ **(b)** $m = -1$, $b = 2$

12. (a) $m = -1/2$, $b = -3$ **(b)** $m = 1/3$, $b = -1$

In Exercises 13 and 14, the line contains the origin and the point in the upper right corner of the grapher screen. Write an equation for the line. In Exercise 13, a tick mark on the x-axis represents 1 unit, and on the y-axis, it represents 5 units. In Exercise 14, a tick mark on either axis represents 1 unit.

13.   **14.**

[−10, 10] by [−25, 25]   [−5, 5] by [−2, 2]

In Exercises 15 and 16, find the **(i)** slope and **(ii)** y-intercept and **(iii)** graph the line.

15. (a) $3x + 4y = 12$ **(b)** $x + y = 2$

16. (a) $\dfrac{x}{3} + \dfrac{y}{4} = 1$ **(b)** $y = 2x + 4$

In Exercises 17 and 18, write an equation for the line through P that is **(i)** parallel to L and **(ii)** perpendicular to L.

17. (a) $P(0, 0)$, $L: y = -x + 2$

   (b) $P(-2, 2)$, $L: 2x + y = 4$

18. (a) $P(-2, 4)$, $L: x = 5$

   (b) $P(-1, 1/2)$, $L: y = 3$

In Exercises 19 and 20, a table of values is given for the linear function $f(x) = mx + b$. Determine m and b.

19.

x	$f(x)$
1	2
3	9
5	16

20.

x	$f(x)$
2	−1
4	−4
6	−7

In Exercises 21 and 22, find the value of x or y for which the line through A and B has the given slope m.

21. $A(-2, 3)$, $B(4, y)$, $m = -2/3$

22. $A(-8, -2)$, $B(x, 2)$, $m = 2$

23. *Revisiting Example 5* Show that you get the same equation in Example 5 if you use the point $(3, 4)$ to write the equation.

24. *Writing to Learn: x-and y-intercepts*

   (a) Explain why c and d are the x-intercept and y-intercept, respectively, of the line

$$\frac{x}{c} + \frac{y}{d} = 1 \, .$$

   (b) How are the x-intercept and y-intercept related to c and d in the line

$$\frac{x}{c} + \frac{y}{d} = 2 \, ?$$

25. *Parallel and perpendicular lines* For what value of k are the two lines $2x + ky = 3$ and $x + y = 1$ **(a)** parallel? **(b)** perpendicular?

In Exercises 26–28, *work in groups of two or three* to solve the problem.

26. *Insulation* By measuring slopes in the figure, find the temperature change in degrees per inch for the following materials.

   (a) gypsum wallboard

   (b) fiberglass insulation

   (c) wood sheathing

   (d) *Writing to Learn* Which of the materials in parts **(a)** through **(c)** is the best insulator? The poorest? Explain.

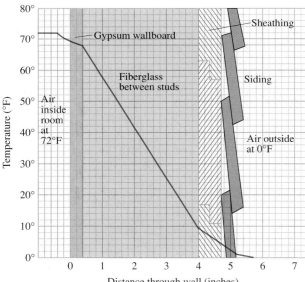

27. *Pressure under water* The pressure p experienced by a diver under water is related to the diver's depth d by an equation of the form $p = kd + 1$ (k a constant). When $d = 0$ meters, the pressure is 1 atmosphere. The pressure at 100 meters is 10.94 atmospheres. Find the pressure at 50 meters.

28. *Modeling distance traveled* A car starts from point P at time $t = 0$ and travels at 45 mph.

   (a) Write an expression $d(t)$ for the distance the car travels from P in t hours.

   (b) Graph $y = d(t)$.

(c) What is the slope of the graph in part (b)? What does it have to do with the car?

(d) *Writing to Learn* Create a scenario in which t could have negative values.

(e) *Writing to Learn* Create a scenario in which the y-intercept of $y = d(t)$ could be 30.

Extending the Ideas

29. *Fahrenheit versus Celsius* We found a relationship between Fahrenheit temperature and Celsius temperature in Example 8.

    (a) Is there a temperature at which a Fahrenheit thermometer and a Celsius thermometer give the same reading? If so, what is it?

T **(b)** *Writing to Learn* In the same viewing window graph $y_1 = (9/5)x + 32$, $y_2 = (5/9)(x - 32)$, and $y_3 = x$. Explain how this figure is related to the question in part (a).

30. *Parallelogram* Three different parallelograms have vertices at $(-1, 1)$, $(2, 0)$, and $(2, 3)$. Draw the three and give the coordinates of the missing vertices.

31. *Parallelogram* Show that if the midpoints of consecutive sides of any quadrilateral are connected, the result is a parallelogram.

32. *Tangent line* Consider the circle of radius 5 centered at $(0, 0)$. Find an equation of the line tangent to the circle at the point $(3, 4)$.

33. *Distance from a point to a line* This activity investigates how to find the distance from a point $P(a, b)$ to a line $L: Ax + By = C$. We suggest that students *work in groups of two or three*.

    (a) Write an equation for the line M through P perpendicular to L.

    (b) Find the coordinates of the point Q in which M and L intersect.

    (c) Find the distance from P to Q.

34. *Reflected light* A ray of light comes along the line $x + y = 1$ from the second quadrant and reflects off the x-axis. The angle of incidence is equal to the angle of reflection as measured from the perpendicular. Write an equation for the line along which the departing light travels.

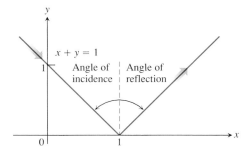

35. *The Mt. Washington cog railway* Civil engineers calculate the slope of roadbed as the ratio of the distance it rises or falls to the distance it runs horizontally. They call this ratio the **grade** of the roadbed, usually written as a percentage. Along the coast, commercial railroad grades are usually less than 2%. In the mountains, they may go as high as 4%. Highway grades are usually less than 5%.

The steepest part of the Mt. Washington cog railway in New Hampshire has an exceptional 37.1% grade. Along this part of the track, the seats in the front of the car are 14 ft above those in the rear. About how far apart are the front and rear rows of seats?

36. A 90° rotation counterclockwise about the origin takes $(2, 0)$ to $(0, 2)$, and $(0, 3)$ to $(-3, 0)$, as shown. Where does it take each of the following points?

    (a) $(4, 1)$ **(b)** $(-2, -3)$ **(c)** $(2, -5)$

    (d) $(x, 0)$ **(e)** $(0, y)$ **(f)** (x, y)

    (g) What point is taken to $(10, 3)$?

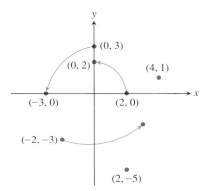

In Exercises 37 and 38, use linear regression analysis.

T **37.** Table 3 lists the ages and weights of nine girls.

Table 3 Girls' ages and weights	
Age (months)	**Weight (pounds)**
19	22
21	23
24	25
27	28
29	31
31	28
34	32
38	34
43	39

    (a) Find a linear regression equation for the data.

    (b) Find the slope of the regression line. What does the slope represent?

    (c) Superimpose the graph of the linear regression equation on a scatter plot of the data.

    (d) Use the regression equation to predict the approximate weight of a 30-month old girl.

T **38.** Table 4 shows the mean annual compensation of construction workers.

Table 4 Construction workers' average annual compensation	
Year	Annual compensation (dollars)
1980	22,033
1985	27,581
1988	30,466
1989	31,465
1990	32,836

Source: U.S. Bureau of Economic Analysis.

(a) Find a linear regression equation for the data.

(b) Find the slope of the regression line. What does the slope represent?

(c) Superimpose the graph of the linear regression equation on a scatter plot of the data.

(d) Use the regression equation to predict the construction workers' average annual compensation in 2000.

T **39.** The median price of existing single-family homes has increased consistently since 1970. The data in Table 5, however, show that there have been differences in various parts of the country.

(a) Find a linear regression equation for home cost in the Northeast.

(b) What does the slope of the regression line represent?

(c) Find a linear regression equation for home cost in the Midwest.

(d) Where is the median price increasing more rapidly, in the Northeast or the Midwest?

Table 5 Median price of single-family homes		
Year	Northeast (dollars)	Midwest (dollars)
1970	25,200	20,100
1975	39,300	30,100
1980	60,800	51,900
1985	88,900	58,900
1990	141,200	74,000

Source: National Association of Realtors®, *Home Sales Yearbook* (Washington DC, 1990).

2   Functions and Graphs

Functions • Domains and Ranges • Viewing and Interpreting Graphs • Increasing versus Decreasing Functions • Even Functions and Odd Functions: Symmetry • Functions Defined in Pieces • The Absolute Value Function • How to Shift a Graph • Composite Functions

Functions are the major tools for describing the real world in mathematical terms. This section discusses the basic ideas of functions, their graphs, and ways of shifting or combining them. Several important types of functions that arise in calculus are presented.

Functions

The values of one variable often depend on the values for another:

- The temperature at which water boils depends on elevation (the boiling point drops as you go up).

- The amount by which your savings will grow in a year depends on the interest rate offered by the bank.

In each case, the value of one variable quantity depends on the value of another. The boiling point of water, b, depends on the elevation, e; the amount of

FIGURE 8 A "machine" diagram for a function.

interest, I, depends on the interest rate, r. We call b and I **dependent variables** because they are determined by the values of the variables e and r on which they depend. The variables e and r are **independant variables.**

A rule that assigns to each element in one set a unique element in another set is called a *function*. The sets may be sets of any kind and do not have to be the same. A function is like a machine that assigns a unique output to every allowable input. The inputs make up the **domain** of the function; the outputs make up the **range** (Figure 8).

> **Definition**   Function
>
> A **function** from a set D to a set R is a rule that assigns a unique element in R to each element in D.

In this definition, D is the domain of the function and R is a set *containing* the range (Figure 9).

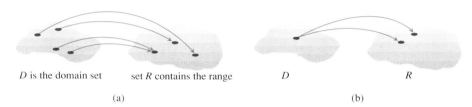

D is the domain set   set R contains the range D R

(a)   (b)

FIGURE 9 (a) A function from set D to set R. (b) *Not* a function. The assignment is not unique.

Years ago, Swiss mathematician Leonhard Euler invented a way to say "y is a function of x" symbolically:

$$y = f(x),$$

which we read as "y equals f of x." This notation enables us to give different functions different names by changing the letters we use. To say the boiling point of water is a function of elevation, we can write $b = f(e)$. To say the area of a circle is a function of the radius, we can write $A = A(r)$, giving the function the same name as the dependent variable.

The notation $y = f(x)$ also gives a way to denote specific values of a function. The value of f at a can be written as $f(a)$, read "f of a."

Example 1   The Circle-Area Function

The domain of the circle-area function $A(r) = \pi r^2$ is the set of all possible radii, the set of all positive real numbers. The range is also the set of all positive real numbers.

The value of A at $r = 2$ is

$$A(2) = \pi(2)^2 = 4\pi.$$

The area of a circle of radius 2 is 4π.

Domains and Ranges

In Example 1, the domain of the function is restricted by context: The independent variable is a radius and must be positive. When we define a function $y = f(x)$ with a formula and the domain is not stated explicitly or restricted by context, the domain is assumed to be the largest set of x-values for which the formula gives real y-values, the so-called **natural domain.** If we want to restrict the domain in some way, we must say so. The domain of $y = x^2$ is the entire set of real numbers. To restrict the function to, say, positive values of x, we would write "$y = x^2, x > 0$."

The domains and ranges of many real-valued functions of a real variable are intervals or combinations of intervals. The intervals may be open, closed, or half open (Figures 10 and 11) and finite or infinite (Figure 12).

Name: Open interval ab
Notation: $a < x < b$ or (a, b)

Closed at a and open at b
Notation: $a \le x < b$ or $[a, b)$

Name: Closed interval ab
Notation: $a \le x \le b$ or $[a, b]$

Open at a and closed at b
Notation: $a < x \le b$ or $(a, b]$

FIGURE 10 Open and closed finite intervals.   FIGURE 11 Half-open finite intervals.

The endpoints of an interval are called **boundary points.** They make up the interval's **boundary.** The remaining points are **interior points,** and they make up the interval's **interior.** Intervals containing all their boundary points are **closed.** Intervals containing no boundary points are **open.** Every point of an open interval is an interior point of the interval.

Example 2   Identifying Domain and Range

Verify the domains of these functions.

Function	Domain (x)	Range (y)
$y = x^2$	$(-\infty, \infty)$	$[0, \infty)$
$y = 1/x$	$(-\infty, 0) \cup (0, \infty)$	$(-\infty, 0) \cup (0, \infty)$
$y = \sqrt{x}$	$[0, \infty)$	$[0, \infty)$
$y = \sqrt{4 - x}$	$(-\infty, 4]$	$[0, \infty)$
$y = \sqrt{1 - x^2}$	$[-1, 1]$	$[0, 1]$

Solution   The formula $y = x^2$ gives a real y-value for any real number x, so the domain is $(-\infty, \infty)$.

The formula $y = 1/x$ gives a real y-value for every real x-value except $x = 0$. *We cannot divide any number by* 0.

The formula $y = \sqrt{x}$ gives a real y-value only when x is positive or zero.

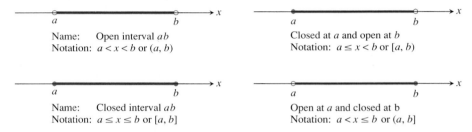

Name: The set of all real numbers
Notation: $-\infty < x < \infty$ or $(-\infty, \infty)$

Name: The set of numbers greater than a
Notation: $a < x$ or (a, ∞)

Name: The set of numbers greater than or equal to a
Notation: $a \le x$ or $[a, \infty)$

Name: The set of numbers less than b
Notation: $x < b$ or $(-\infty, b)$

Name: The set of numbers less than or equal to b
Notation: $x \le b$ or $(-\infty, b]$

FIGURE 12 Infinite intervals: rays on the number line and the number line itself. The symbol ∞ (infinity) is used merely for convenience; it does not mean that there is a number ∞.

The formula $y = \sqrt{4 - x}$ gives a real y-value only when $4 - x$ is greater than or equal to zero. So, $0 \le 4 - x$, or $x \le 4$.

The formula $y = \sqrt{1 - x^2}$ gives a real y-value for every value of x in the closed interval from -1 to 1. Outside this interval, $1 - x^2$ is negative and its square root is not a real number. The domain is $[-1, 1]$.

Viewing and Interpreting Graphs

The points (x, y) in the plane whose coordinates are the input-output pairs of a function $y = f(x)$ make up the function's **graph.** The graph of the function $y = x + 2$, for example, is the set of points with coordinates (x, y) for which $y = x + 2$.

Graphing with pencil and paper requires that you develop graph *drawing* skills. Graphing with a grapher requires that you develop graph *viewing* skills.

$y = \frac{1}{\sqrt{4 - x^2}}$

[−4, 4] by [−2, 4]

(a)

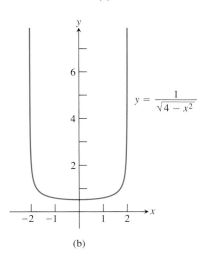

(b)

FIGURE 13 (a) Grapher failure. (b) A more accurate graph of $y = 1/\sqrt{4 - x^2}$. (Example 3)

Graph Viewing Skills

Step 1. Recognize that the graph is reasonable.

Step 2. See all the important characteristics of the graph.

Step 3. Interpret those characteristics.

Step 4. Recognize grapher failure.

Being able to recognize that a graph is reasonable comes with experience. You need to know the basic functions, their graphs, and how changes in their equations affect the graphs.

Grapher failure occurs when the graph produced by a grapher is less than precise—or even incorrect—which is usually due to the limitations of the screen resolution of the grapher.

Example 3 Recognizing Grapher Failure

Find the domain and range of $y = f(x) = 1/\sqrt{4 - x^2}$.

Solution The graph of f in Figure 13a seems to suggest that the domain of f is an interval between -2 and 2 and that the range is also a finite interval. The latter observation is the result of grapher failure; in this case, we can recognize the failure using algebra.

Solve Algebraically

The expression $4 - x^2$ must be greater than zero.

$$4 - x^2 > 0$$
$$x^2 < 4$$

Thus, $-2 < x < 2$, and the domain is $(-2, 2)$.

The smallest value of f is $1/2$ and occurs when $x = 0$. The values of f get very large as x approaches 2 from the left or -2 from the right, as suggested by the following table. (The values of f are rounded to three decimal places.)

x	± 1.99	± 1.999	± 1.9999	± 1.99999
$f(x)$	5.006	15.813	50.001	158.114

The range of f is $[0.5, \infty)$.

Figure 14 shows graphs of *power functions* that arise frequently in calculus. Knowing the general shapes of these graphs will help you recognize grapher failure. We will review other functions as the chapter continues.

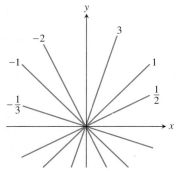

$y = mx$ for selected values of m
Domain: $-\infty < x < \infty$
Range: $-\infty < y < \infty$

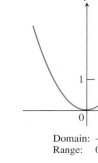

Domain: $-\infty < x < \infty$
Range: $0 \le y < \infty$

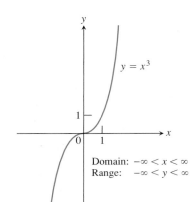

Domain: $-\infty < x < \infty$
Range: $-\infty < y < \infty$

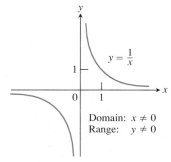

Domain: $x \ne 0$
Range: $y \ne 0$

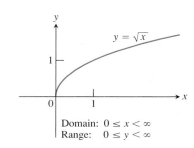

Domain: $0 \le x < \infty$
Range: $0 \le y < \infty$

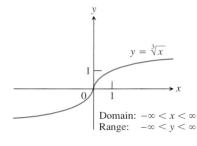

Domain: $-\infty < x < \infty$
Range: $-\infty < y < \infty$

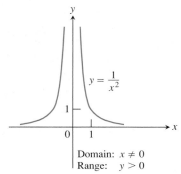

Domain: $x \ne 0$
Range: $y > 0$

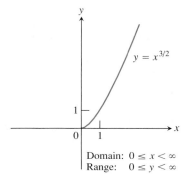

Domain: $0 \le x < \infty$
Range: $0 \le y < \infty$

Domain: $-\infty < x < \infty$
Range: $0 \le y < \infty$

FIGURE 14 Useful power functions.

Increasing versus Decreasing Functions

If the graph of a function *climbs* or *rises* as you move from left to right, we say that the function is *increasing*. If the graph *descends* or *falls* as you move from left to right, the function is *decreasing*. We give formal definitions of increasing functions and decreasing functions in Section 3.3. In that section, you will learn how to find the intervals over which a function is increasing and the intervals where it is decreasing. Here are examples from Figure 14.

Function	Where Increasing	Where Decreasing
$y = x^2$	$0 \le x < \infty$	$-\infty < x \le 0$
$y = x^3$	$-\infty < x < \infty$	Nowhere
$y = 1/x$	Nowhere	$-\infty < x < 0, 0 < x < \infty$
$y = 1/x^2$	$-\infty < x < 0$	$0 < x < \infty$
$y = \sqrt{x}$	$0 \le x < \infty$	Nowhere
$y = x^{2/3}$	$0 \le x < \infty$	$-\infty < x \le 0$

Even Functions and Odd Functions: Symmetry

The graphs of *even* and *odd* functions have characteristic symmetry properties.

> **Definitions** Even Function, Odd Function
>
> A function $y = f(x)$ is an
>
> $$\text{even function of } \boldsymbol{x} \text{ if } f(-x) = f(x),$$
> $$\text{odd function of } \boldsymbol{x} \text{ if } f(-x) = -f(x),$$
>
> for every x in the function's domain.

The names even and odd come from powers of x. If y is an even power of x, as in $y = x^2$ or $y = x^4$, it is an even function of x (because $(-x)^2 = x^2$ and $(-x)^4 = x^4$). If y is an odd power of x, as in $y = x$ or $y = x^3$, it is an odd function of x (because $(-x)^1 = -x$ and $(-x)^3 = -x^3$).

The graph of an even function is **symmetric about the *y*-axis.** Since $f(-x) = f(x)$, a point (x, y) lies on the graph if and only if the point $(-x, y)$ lies on the graph (Figure 15a).

The graph of an odd function is **symmetric about the origin.** Since $f(-x) = -f(x)$, a point (x, y) lies on the graph if and only if the point $(-x, -y)$ lies on the graph (Figure 15b). Equivalently, a graph is symmetric about the origin if a rotation of 180° about the origin leaves the graph unchanged.

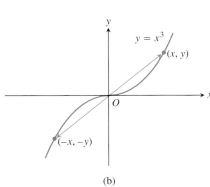

FIGURE 15 (a) The graph of $y = x^2$ (an even function) is symmetric about the *y*-axis. (b) The graph of $y = x^3$ (an odd function) is symmetric about the origin.

Example 4   Recognizing Even and Odd Functions

$f(x) = x^2$ Even function: $(-x)^2 = x^2$ for all x; symmetry about *y*-axis.

$f(x) = x^2 + 1$ Even function: $(-x)^2 + 1 = x^2 + 1$ for all x; symmetry about *y*-axis (Figure 16a).

$f(x) = x$ Odd function: $(-x) = -x$ for all x; symmetry about the origin.

$f(x) = x + 1$ Not odd: $f(-x) = -x + 1$, but $-f(x) = -x - 1$. The two are not equal.

Not even: $(-x) + 1 \neq x + 1$ for all $x \neq 0$ (Figure 16b).

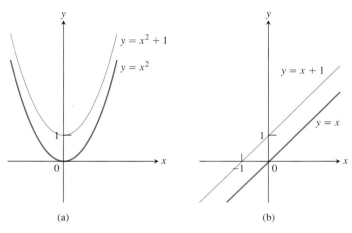

(a)   (b)

FIGURE 16 (a) When we add the constant term 1 to the function $y = x^2$, the resulting function $y = x^2 + 1$ is still even and its graph is still symmetric about the y-axis. (b) When we add the constant term 1 to the function $y = x$, the resulting function $y = x + 1$ is no longer odd. The symmetry about the origin is lost. (Example 4)

It is useful in graphing to recognize even and odd functions. Once we know the graph of either type of function on one side of the y-axis, we automatically know its graph on the other side.

Functions Defined in Pieces

Functions can be defined by applying different formulas to different parts of the domain.

Example 5   Graphing Piecewise-Defined Functions

Graph

$$y = f(x) = \begin{cases} -x, & x < 0 \\ x^2, & 0 \leq x \leq 1 \\ 1, & x > 1. \end{cases}$$

Solution   The values of f are given by three separate formulas: $y = -x$ when $x < 0$, $y = x^2$ when $0 \leq x \leq 1$, and $y = 1$ when $x > 1$. The function, however, is *just one function* whose domain is the entire set of real numbers (Figure 17).

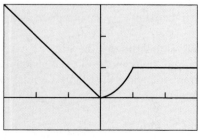

[−3, 3] by [−1, 3]

FIGURE 17 The graph of a piecewise defined function. (Example 5)

Example 6   Writing Formulas for Piecewise-Defined Functions

Write a formula for the function $y = f(x)$ whose graph consists of the two line segments in Figure 18.

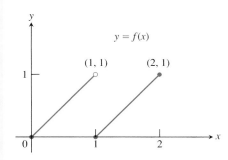

FIGURE 18 The segment on the left contains $(0, 0)$ but not $(1, 1)$. The segment on the right contains both of its endpoints. (Example 6)

Solution   We find formulas for the segments from $(0, 0)$ to $(1, 1)$ and from $(1, 0)$ to $(2, 1)$ and piece them together in the manner of Example 5.

Segment from (0, 0) to (1, 1)   The line through $(0, 0)$ and $(1, 1)$ has slope $m = (1 - 0)/(1 - 0) = 1$ and y-intercept $b = 0$. Its slope-intercept equation is $y = x$. The segment from $(0, 0)$ to $(1, 1)$ that includes the point $(0, 0)$ but not the point $(1, 1)$ is the graph of the function $y = x$ restricted to the half-open interval $0 \le x < 1$, namely,

$$y = x, \qquad 0 \le x < 1.$$

Segment from (1, 0) to (2, 1)   The line through $(1, 0)$ and $(2, 1)$ has slope $m = (1 - 0)/(2 - 1) = 1$ and passes through the point $(1, 0)$. The corresponding point-slope equation for the line is

$$y = 1(x - 1) + 0, \qquad \text{or} \qquad y = x - 1.$$

The segment from $(1, 0)$ to $(2, 1)$ that includes both endpoints is the graph of $y = x - 1$ restricted to the closed interval $1 \le x \le 2$, namely,

$$y = x - 1, \qquad 1 \le x \le 2.$$

Piecewise formula   Combining the formulas for the two pieces of the graph, we obtain

$$f(x) = \begin{cases} x, & 0 \le x < 1 \\ x - 1, & 1 \le x \le 2. \end{cases}$$

Remember that $\sqrt{a^2} = |a|$. Do not write $\sqrt{a^2} = a$ unless you already know that $a \ge 0$.

The Absolute Value Function

The absolute value function $y = |x|$ is defined piecewise by the formula

$$|x| = \begin{cases} -x, & x < 0 \\ x, & x \ge 0. \end{cases}$$

Absolute Value Properties

1. $|-a| = |a|$

2. $|ab| = |a||b|$

3. $\left|\dfrac{a}{b}\right| = \dfrac{|a|}{|b|}$

4. $|a + b| \le |a| + |b|$

The function is even, and its graph (Figure 19) is symmetric about the y-axis. Since the symbol $\sqrt{a}$ denotes the *nonnegative* square root of a, an alternate definition of $|x|$ is

$$|x| = \sqrt{x^2}.$$

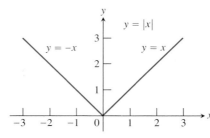

FIGURE 19 The absolute value function has domain $(-\infty, \infty)$ and range $[0, \infty)$.

How to Shift a Graph

To shift the graph of a function $y = f(x)$ straight up, add a positive constant to the right-hand side of the formula $y = f(x)$.

To shift the graph of a function $y = f(x)$ straight down, add a negative constant to the right-hand side of the formula $y = f(x)$.

Example 7   Shifting a Graph Vertically

Adding 1 to the right-hand side of the formula $y = x^2$ to get $y = x^2 + 1$ shifts the graph up 1 unit (Figure 20). Adding -2 to the right-hand side of the formula $y = x^2$ to get $y = x^2 - 2$ shifts the graph down 2 units (Figure 20).

To shift the graph of $y = f(x)$ to the left, add a positive constant to x. To shift the graph of $y = f(x)$ to the right, add a negative constant to x.

Example 8   Shifting a Graph Horizontally

Adding 3 to x in $y = x^2$ to get $y = (x + 3)^2$ shifts the graph 3 units to the left (Figure 21). Adding -2 to x in $y = x^2$ to get $y = (x - 2)^2$ shifts the graph 2 units to the right (Figure 21).

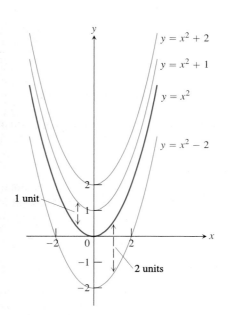

FIGURE 20 To shift the graph of $f(x) = x^2$ up (or down), we add positive (or negative) constants to the formula for f.

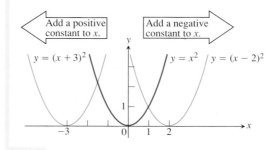

FIGURE 21 To shift the graph of $y = x^2$ to the left, we add a positive constant to x. To shift the graph to the right, we add a negative constant to x.

$y = |x-2| - 1$

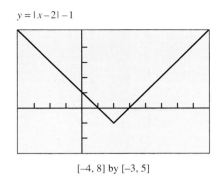

[-4, 8] by [-3, 5]

FIGURE 22   The lowest point of the graph of $f(x) = |x - 2| - 1$ is $(2, -1)$. (Example 9)

Shift Formulas

VERTICAL SHIFTS

$y = f(x) + k$ Shifts the graph *up* k units if $k > 0$

Shifts it *down* $|k|$ units if $k < 0$

HORIZONTAL SHIFTS

$y = f(x + h)$ Shifts the graph *left* h units if $h > 0$

Shifts it *right* $|h|$ units if $h < 0$

Example 9   Combining Shifts

Find the domain and range, and draw the graph of $f(x) = |x - 2| - 1$.

Solution   The graph of f is the graph of the absolute value function shifted 2 units horizontally to the right and 1 unit vertically downward (Figure 22). The domain of f is $(-\infty, \infty)$, and the range is $[-1, \infty)$.

Composite Functions

FIGURE 23   Two functions can be composed at x whenever the value of one at x lies in the domain of the other. The composite is denoted by $f \circ g$.

Suppose that some of the outputs of a function g can be used as inputs of a function f. We can then link g and f to form a new function whose inputs x are inputs of g and whose outputs are the numbers $f(g(x))$, as in Figure 23. We say the function $f(g(x))$ (read "f of g of x") is the **composite of g and f**. It is made by *composing g and f* in the order of first g, then f. The usual "stand-alone" notation for this composite is $f \circ g$, which is read as "f of g." The value of $f \circ g$ at x is $(f \circ g)(x) = f(g(x))$. Notice that in the notation $f \circ g$, we first apply g, and then the function f, to the input variable x.

Example 10   Viewing a Function as a Composite

The function $y = \sqrt{1 - x^2}$ in Example 2 can be thought of as first calculating $1 - x^2$ followed by taking the square root of the result. The function y is the composite of the function $g(x) = 1 - x^2$ and the function $f(x) = \sqrt{x}$. Notice that $1 - x^2$ cannot be negative. The domain of the composite is $[-1, 1]$.

Example 11   Finding a Formula for a Composite and Evaluating It

Find a formula for $f(g(x))$ if $g(x) = x^2$ and $f(x) = x - 7$. Then find $f(g(2))$.

Solution   To find $f(g(x))$, we replace x in the formula $f(x) = x - 7$ by the expression given for $g(x)$.

$$f(x) = x - 7$$
$$f(g(x)) = g(x) - 7 = x^2 - 7$$

We then find the value of $f(g(2))$ by substituting 2 for x.

$$f(g(2)) = (2)^2 - 7 = -3$$

EXERCISES 2

Finding Formulas for Functions

1. Express the area and perimeter of an equilateral triangle as a function of the triangle's side length x.

2. Express the side length of a square as a function of the length d of the square's diagonal. Then express the area as a function of the diagonal length.

3. Express the edge length of a cube as a function of the cube's diagonal length d. Then express the surface area and volume of the cube as a function of the diagonal length.

4. A point P in the first quadrant lies on the graph of the function $f(x) = \sqrt{x}$. Express the coordinates of P as functions of the slope of the line joining P to the origin.

In Exercises 5 and 6, which of the graphs are graphs of functions of x, and which are not? Give reasons for your answers.

5. (a)   **(b)**

6. (a)   **(b)**

    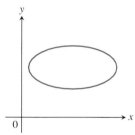

Domain and Range

In Exercises 7–10, find the domain and range of each function.

7. (a) $f(x) = 1 + x^2$ **(b)** $f(x) = 1 - \sqrt{x}$

8. (a) $F(t) = \dfrac{1}{\sqrt{t}}$ **(b)** $F(t) = \dfrac{1}{1 + \sqrt{t}}$

9. $g(z) = \sqrt{4 - z^2}$

10. $g(z) = \sqrt[3]{z - 3}$

Functions and Graphs

Graph the functions in Exercises 11 and 12. What symmetries, if any, do the graphs have?

11. (a) $y = -x^3$ **(b)** $y = -\dfrac{1}{x^2}$

12. (a) $y = \sqrt{|x|}$ **(b)** $y = -\dfrac{1}{x}$

13. Graph the following equations and explain why they are not graphs of functions of x.

    (a) $|y| = x$ **(b)** $y^2 = x^2$

14. Graph the following equations and explain why they are not graphs of functions of x.

    (a) $|x| + |y| = 1$ **(b)** $|x + y| = 1$

Even and Odd Functions

In Exercises 15–20, say whether the function is even, odd, or neither.

15. (a) $f(x) = 3$ **(b)** $f(x) = x^{-5}$

16. (a) $f(x) = x^2 + 1$ **(b)** $f(x) = x^2 + x$

17. (a) $g(x) = x^3 + x$ **(b)** $g(x) = x^4 + 3x^2 - 1$

18. (a) $g(x) = \dfrac{1}{x^2 - 1}$ **(b)** $g(x) = \dfrac{x}{x^2 - 1}$

19. (a) $h(t) = \dfrac{1}{t - 1}$ **(b)** $h(t) = |t^3|$

20. (a) $h(t) = \sqrt{t^2 + 3}$ **(b)** $h(t) = 2|t| + 1$

Piecewise-Defined Functions

In Exercises 21–24, **(a)** draw the graph of the function. Then find its **(b)** domain and **(c)** range.

21. (a) $f(x) = -|3 - x| + 2$ **(b)** $f(x) = 2|x + 4| - 3$

22. (a) $f(x) = \begin{cases} 3 - x, & x \le 1 \\ 2x, & 1 < x \end{cases}$ **(b)** $f(x) = \begin{cases} 1, & x < 0 \\ \sqrt{x}, & x \ge 0 \end{cases}$

23. $f(x) = \begin{cases} 4 - x^2, & x < 1 \\ (3/2)x + 3/2, & 1 \le x \le 3 \\ x + 3, & x > 3 \end{cases}$

24. $f(x) = \begin{cases} x^2, & x < 0 \\ x^3, & 0 \le x \le 1 \\ 2x - 1, & x > 1 \end{cases}$

25. *Writing to Learn* The *vertical line test* to determine whether a curve is the graph of a function states: If every vertical line in the *xy*-plane intersects a given curve in at most one point, then the curve is the graph of a function. Explain why this statement is true.

26. *Writing to Learn* For a curve to be *symmetric about the x-axis*, the point (x, y) must lie on the curve if and only if the point $(x, -y)$ lies on the curve. Explain why a curve that is symmetric about the *x*-axis is not the graph of a function, unless the function is $y = 0$.

In Exercises 27 and 28, write a piecewise formula for the function.

27. (a) **(b)**

(c) **(d)**

28. (a) **(b)**

(c) **(d)**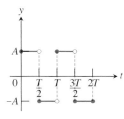

Shifting Graphs

29. Match the equations listed in **(a)** through **(d)** to the positions on the graphs in the accompanying figure.

(a) $y = (x - 1)^2 - 4$ **(b)** $y = (x - 2)^2 + 2$
(c) $y = (x + 2)^2 + 2$ **(d)** $y = (x + 3)^2 - 2$

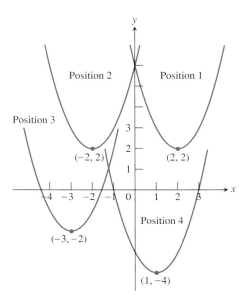

30. The figure shows the graph of $y = -x^2$ shifted to four new positions. Write an equation for each new graph.

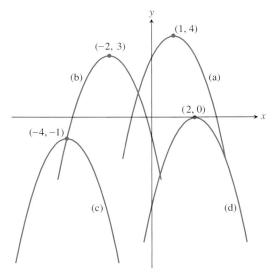

Exercises 31–36 tell how many units and in what directions the graphs of the given equations are to be shifted. Give an equation for the shifted graph. Then sketch the original and shifted graphs together, labeling each graph with its equation.

31. $x^2 + y^2 = 49$ Down 3, left 2

32. $y = x^3$ Left 1, down 1

33. $y = x^{2/3}$ Right 1, down 1

34. $y = -\sqrt{x}$ Right 3

35. $y = (1/2)(x + 1) + 5$ Down 5, right 1

36. $x = y^2$ Left 1

Composites of Functions

37. If $f(x) = x + 5$ and $g(x) = x^2 - 3$, find the following.

(a) $f(g(0))$ (b) $g(f(0))$

(c) $f(g(x))$ (d) $g(f(x))$

(e) $f(f(-5))$ (f) $g(g(2))$

(g) $f(f(x))$ (h) $g(g(x))$

38. If $f(x) = x - 1$ and $g(x) = 1/(x + 1)$, find the following.

(a) $f(g(1/2))$ (b) $g(f(1/2))$

(c) $f(g(x))$ (d) $g(f(x))$

(e) $f(f(2))$ (f) $g(g(2))$

(g) $f(f(x))$ (h) $g(g(x))$

39. If $u(x) = 4x - 5$, $v(x) = x^2$, and $f(x) = 1/x$, find formulas for the following.

(a) $u(v(f(x)))$ (b) $u(f(v(x)))$

(c) $v(u(f(x)))$ (d) $v(f(u(x)))$

(e) $f(u(v(x)))$ (f) $f(v(u(x)))$

40. If $f(x) = \sqrt{x}$, $g(x) = x/4$, and $h(x) = 4x - 8$, find formulas for the following.

(a) $h(g(f(x)))$ (b) $h(f(g(x)))$

(c) $g(h(f(x)))$ (d) $g(f(h(x)))$

(e) $f(g(h(x)))$ (f) $f(h(g(x)))$

Let $f(x) = x - 3$, $g(x) = \sqrt{x}$, $h(x) = x^3$, and $j(x) = 2x$. Express each of the functions in Exercises 41 and 42 as a composite involving one or more of f, g, h, and j.

41. (a) $y = \sqrt{x - 3}$ (b) $y = 2\sqrt{x}$

(c) $y = x^{1/4}$ (d) $y = 4x$

(e) $y = \sqrt{(x - 3)^3}$ (f) $y = (2x - 6)^3$

42. (a) $y = 2x - 3$ (b) $y = x^{3/2}$

(c) $y = x^9$ (d) $y = x - 6$

(e) $y = 2\sqrt{x - 3}$ (f) $y = \sqrt{x^3 - 3}$

43. Copy and complete the following table.

	$g(x)$	$f(x)$	$(f \circ g)(x)$
(a)	?	$\sqrt{x - 5}$	$\sqrt{x^2 - 5}$
(b)	?	$1 + 1/x$	x
(c)	$1/x$	?	x
(d)	$\sqrt{x}$	?	$|x|$

44. Copy and complete the following table.

	$g(x)$	$f(x)$	$(f \circ g)(x)$
(a)	$x - 7$	$\sqrt{x}$	?
(b)	$x + 2$	$3x$	?
(c)	?	$\sqrt{x - 5}$	$\sqrt{x^2 - 5}$
(d)	$\dfrac{x}{x - 1}$	$\dfrac{x}{x - 1}$	?
(e)	?	$1 + \dfrac{1}{x}$	x
(f)	$\dfrac{1}{x}$	?	x

45. The accompanying figure shows the graph of a function $f(x)$ with domain $[0, 2]$ and range $[0, 1]$. Find the domains and ranges of the following functions and sketch their graphs.

(a) $f(x) + 2$ (b) $f(x) - 1$

(c) $2f(x)$ (d) $-f(x)$

(e) $f(x + 2)$ (f) $f(x - 1)$

(g) $f(-x)$ (h) $-f(x + 1) + 1$

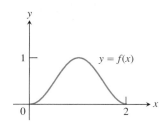

46. The accompanying figure shows the graph of a function $g(t)$ with domain $[-4, 0]$ and range $[-3, 0]$. Find the domains and ranges of the following functions and sketch their graphs.

(a) $g(-t)$ (b) $-g(t)$

(c) $g(t) + 3$ (d) $1 - g(t)$

(e) $g(-t + 2)$ (f) $g(t - 2)$

(g) $g(1 - t)$ (h) $-g(t - 4)$

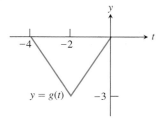

Theory and Examples

47. *The cone problem* Begin with a circular piece of paper with a 4 in. radius as shown in part (a). Cut out a sector with an arc length of x. Join the two edges of the remaining portion to form a cone with radius r and height h, as shown in part (b).

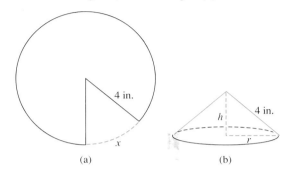

(a)   (b)

(a) Explain why the circumference of the base of the cone is $8\pi - x$.

(b) Express the radius r as a function of x.

(c) Express the height h as a function of x.

(d) Express the volume V of the cone as a function of x.

48. *Industrial costs* Dayton Power and Light, Inc., has a power plant on the Miami River where the river is 800 ft wide. To lay a new cable from the plant to a location in the city 2 mi downstream on the opposite side costs \$180 per foot across the river and \$100 per foot along the land.

(Not to scale)

(a) Suppose that the cable goes from the plant to a point Q on the opposite side that is x ft from the point P directly opposite the plant. Write a function $C(x)$ that gives the cost of laying the cable in terms of the distance x.

(b) Generate a table of values to determine if the least expensive location for point Q is less than 2000 ft or greater than 2000 ft from point P.

49. *Even and odd functions*

(a) Must the product of two even functions always be even? Give reasons for your answer.

(b) Can anything be said about the product of two odd functions? Give reasons for your answer.

(c) Can a function be both even and odd? Give reasons for your answer.

50. *A magic trick* You may have heard of a magic trick that goes like this: Take any number. Add 5. Double the result. Subtract 6. Divide by 2. Subtract 2. Now tell me your answer, and I'll tell you what you started with.

(a) Pick a number and try it.

(b) Why does it work with any number?

T **51.** Graph the functions $f(x) = \sqrt{x}$ and $g(x) = \sqrt{1 - x}$ together with their **(a)** sum, **(b)** product, **(c)** two differences, and **(d)** two quotients.

T **52.** Let $f(x) = x - 7$ and $g(x) = x^2$. Graph f and g together with $f \circ g$ and $g \circ f$.

Some graphers allow a function such as y_1 to be used as the independent variable of another function. With such a grapher, we can compose functions.

T **53. (a)** Enter the functions $y_1 = f(x) = 4 - x^2$, $y_2 = g(x) = \sqrt{x}$, $y_3 = y_2(y_1(x))$, and $y_4 = y_1(y_2(x))$. Which of y_3 and y_4 corresponds to $f \circ g$? To $g \circ f$?

(b) Graph y_1, y_2, and y_3 to make conjectures about the domain and range of y_3.

(c) Graph y_1, y_2, and y_4 to make conjectures about the domain and range of y_4.

(d) Confirm your conjectures algebraically by finding formulas for y_3 and y_4.

T **54.** Enter $y_1 = \sqrt{x}$, $y_2 = \sqrt{1 - x}$ and $y_3 = y_1 + y_2$ on your grapher.

(a) Graph y_3 in $[-3, 3]$ by $[-1, 3]$.

(b) Compare the domain of the graph of y_3 with the domains of the graphs of y_1 and y_2.

(c) Replace y_3 by

$$y_1 - y_2, \quad y_2 - y_1, \quad y_1 \cdot y_2, \quad y_1/y_2, \quad \text{and} \quad y_2/y_1,$$

in turn, and repeat the comparison of part (b).

(d) Based on your observations in parts (b) and (c), what would you conjecture about the domains of sums, differences, products, and quotients of functions?

Regression Analysis: Stern Waves and Stopping Distance

See page 5 for an introduction to regression analysis with a calculator.

T **55.** *Stern waves* Observations of the stern waves that follow a boat at right angles to its course have disclosed that the distance between the crests of these waves (their *wave length*) increases with the speed of the boat. Table 6 shows the relationship between wave length and the speed of the boat.

(a) Find a power regression equation $y = ax^b$ for the data in Table 6, where x is the wave length, and y the speed of the boat.

(b) Superimpose the graph of the power regression equation on a scatter plot of the data.

(c) Use the graph of the power regression equation to predict the speed of the boat when the wave length is 11 m. Confirm algebraically.

(d) Now use *linear* regression to predict the speed when the wave length is 11 m. Superimpose the regression line on a scatter plot of the data. Which gives the better fit, the line here or the curve in part (b)?

Table 6 Wave lengths

Wave length (m)	Speed (km/h)
0.20	1.8
0.65	3.6
1.13	5.4
2.55	7.2
4.00	9.0
5.75	10.8
7.80	12.6
10.20	14.4
12.90	16.2
16.00	18.0
18.40	19.8

 56. *Vehicular stopping distance* Table 7 shows the total stopping distance of a car as a function of its speed.

(a) Find the quadratic regression equation for the data in Table 7.

(b) Superimpose the graph of the quadratic regression equation on a scatter plot of the data.

(c) Use the graph of the quadratic regression equation to predict the average total stopping distance for speeds of 72 and 85 mph. Confirm algebraically.

(d) Now use *linear* regression to predict the average total stopping distance for speeds of 72 and 85 mph. Superimpose the regression line on a scatter plot of the data. Which gives the better fit, the line here or the graph in part (b)?

Table 7 Vehicular stopping distance

Speed (mph)	Average total stopping distance (ft)
20	42
25	56
30	73.5
35	91.5
40	116
45	142.5
50	173
55	209.5
60	248
65	292.5
70	343
75	401
80	464

Source: U.S. Bureau of Public Roads.

---

3   Exponential Functions

Exponential Growth • Population Growth • The Exponential Function e^x • What Happened to a^x?

Exponential functions are of particular importance in science and engineering applications. We review your experience with this class of functions in this section and discuss several important exponential models of growth and decay. The mathematics underlying the properties of these extraordinary functions and their relations to logarithms (next section) is beautiful and deep. We will investigate this in appropriate detail when we study the calculus of these functions in Chapters 3 and 6.

Exponential Growth

Table 8 shows the growth of $100 invested in 1996 at an interest rate of 5.5%, compounded annually. After the first year, the value of the account is always 1.055 times its value in the previous year. After n years, the value is $y = 100 \cdot (1.055)^n$.

Compound interest provides an example of *exponential growth* and is modeled by a function of the form $y = P \cdot a^x$, where P is the initial investment and a is equal to 1 plus the interest rate expressed as a decimal.

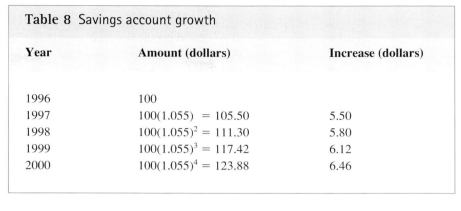

Table 8 Savings account growth

Year	Amount (dollars)	Increase (dollars)
1996	100	
1997	$100(1.055) = 105.50$	5.50
1998	$100(1.055)^2 = 111.30$	5.80
1999	$100(1.055)^3 = 117.42$	6.12
2000	$100(1.055)^4 = 123.88$	6.46

The equation $y = P \cdot a^x$, $a > 0$, $a \neq 1$, identifies a family of functions called *exponential functions*.

Example 1   Graphing $y = a^x$

Graph the functions $y = 2^x$, $y = 3^x$, and $y = 10^x$. For what values of x is it true that $2^x > 3^x > 10^x$?

Solution   From the graph in Figure 24, the functions are increasing for all values of x. For $x < 0$, we have $2^x > 3^x > 10^x$. At $x = 0$, we have $2^x = 3^x = 10^x = 1$. For $x > 0$, we have $2^x < 3^x < 10^x$.

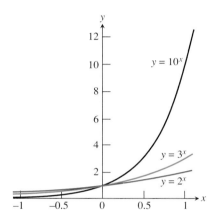

FIGURE 24 $y = 2^x$, $y = 3^x$, $y = 10^x$.

Definition   Exponential Function

Let a be a positive real number other than 1. The function

$$f(x) = a^x$$

is the **exponential function with base a**.

The domain of $f(x) = a^x$ is $(-\infty, \infty)$ and the range is $(0, \infty)$. If $a > 1$, the graph of f looks like the graph of $y = 2^x$ in Figure 25a. If $0 < a < 1$, the graph of f looks like the graph of $y = (1/2)^x = 2^{-x}$ in Figure 25b.

$y = 2^x$

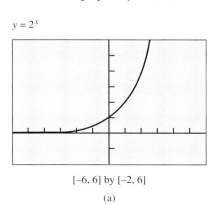

[–6, 6] by [–2, 6]

(a)

$y = 2^{-x}$

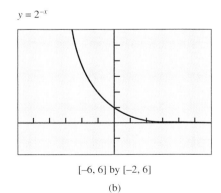

[–6, 6] by [–2, 6]

(b)

FIGURE 25 Graphs of (a) $y = 2^x$ and (b) $y = 2^{-x}$.

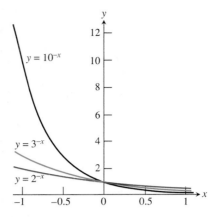

FIGURE 26 $y = 2^{-x}, y = 3^{-x}, y = 10^{-x}$.

Example 2   Graphing $y = a^{-x}$

Graph the functions $y = 2^{-x}$, $y = 3^{-x}$, and $y = 10^{-x}$. For what values of x is it true that $2^{-x} > 3^{-x} > 10^{-x}$?

Solution   From the graph in Figure 26, the functions are decreasing for all values of x. For $x < 0$, we have $2^{-x} < 3^{-x} < 10^{-x}$. At $x = 0$, we have $2^{-x} = 3^{-x} = 10^{-x} = 1$. For $x > 0$, we have $2^{-x} > 3^{-x} > 10^{-x}$.

Exponential functions obey the rules for exponents.

Rules for Exponents
If $a > 0$ and $b > 0$, the following hold true for all real numbers x and y.

1. $a^x \cdot a^y = a^{x+y}$

2. $\dfrac{a^x}{a^y} = a^{x-y}$

3. $(a^x)^y = (a^y)^x = a^{xy}$

4. $a^x \cdot b^x = (ab)^x$

5. $\dfrac{a^x}{b^x} = \left(\dfrac{a}{b}\right)^x$

Population Growth

Population growth can sometimes be modeled with an exponential function. In Table 9, we give some values for the population of the world. We have also divided the population in one year by the population in the previous year to get an idea of how the population is growing. These ratios are in the third column.

Example 3   Predicting World Population

Use the data in Table 9 and an exponential model to predict the population of the world in 2010.

Solution   Based on the third column in Table 9, although questionable given the variation in rates, we might be willing to conjecture that the population of the world in any year is about 1.018 times the population the year before. At any time t years after 1986, the world population would then be about $P(t) = 4936\,(1.018)^t$ million people. The population in 2010, which is $t = 24$ years after 1986, would be about

$$P(24) = 4936(1.018)^{24} \approx 7573.9$$

or 7.6 billion people.

The Exponential Function e^x

The most important exponential function for modeling natural, physical, and economic phenomena is the **natural exponential function,** whose base is the famous

$y = (1 + 1/x)^x$

[−10, 10] by [−5, 10]

X	Y1
1000	2.7169
2000	2.7176
3000	2.7178
4000	2.7179
5000	2.718
6000	2.7181
7000	2.7181

Y1 = (1+1/X)^X

FIGURE 27 A graph and table of values for $f(x) = (1 + 1/x)^x$ both suggest that as $x \to \infty$, $f(x) \to e \approx 2.718$.

Table 9 World population		
Year	**Population (millions)**	**Ratio**
1986	4936	
1987	5023	5023/4936 ≈ 1.0176
1988	5111	5111/5023 ≈ 1.0175
1989	5201	5201/5111 ≈ 1.0176
1990	5329	5329/5201 ≈ 1.0246
1991	5422	5422/5329 ≈ 1.0175

Source: Statistical Office of the United Nations, *Monthly Bulletin Statistics,* 1991.

number e, which is 2.718281828 to nine decimal places. We can define e to be the number that the function $f(x) = (1 + 1/x)^x$ approaches as x becomes large without bound. The graph and table in Figure 27 strongly suggest that such a number exists. You will learn more about the number e, and how it is obtained, in your study of calculus.

The exponential functions $y = e^{kx}$, where k is a nonzero constant, are frequently used as models of exponential growth or decay. For an example of exponential growth, interest **compounded continuously** uses the model $y = P \cdot e^{rt}$, where P is the initial investment, r is the interest rate as a decimal, and t is time in years. An example of exponential decay is the model $y = A \cdot e^{-1.2 \times 10^{-4}t}$, which represents how the radioactive element carbon-14 decays over time. Here A is the original amount of carbon-14 and t is the time in years. Carbon-14 decay is used to date the remains of dead organisms such as shells, seeds, and wooden artifacts.

Definitions Exponential Growth, Exponential Decay

The function $y = y_0 e^{kx}$ is a model for **exponential growth** if $k > 0$ and a model for **exponential decay** if $k < 0$.

Figure 28 shows graphs of exponential growth and exponential decay.

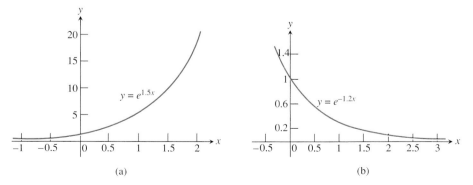

FIGURE 28 Graphs of (a) exponential growth, $k = 1.5 > 0$ and (b) exponential decay, $k = -1.2 < 0$.

Example 4   Savings Account Growth Revisited

Investment companies often use the model for continuous compounding in calculating the growth of an investment. Use this model to track the growth of $100 invested in 1996 at an annual interest rate of 5.5%, compounded continuously.

Solution

Model

Let $x = 0$ represent 1996, $x = 1$ represent 1997, and so on. Then the exponential growth model for continuous compounding is $y(x) = P \cdot e^{rx}$, where $P = 100$ (the initial investment), $r = 0.055$ (the annual interest rate expressed as a decimal), and x is time in years. To predict the amount in the account in 2000, for example, we take $x = 4$ and calculate

$$y(4) = 100 \cdot e^{0.055(4)}$$
$$= 100 \cdot e^{0.22}$$
$$= 124.61. \quad \text{\small Nearest cent}$$

Comparing this result with the $123.88 in the account when the interest is compounded annually (Table 10), we see that the investor earns more as the interest is compounded more frequently (in this case, *continuously*). In Table 10, we compare the values for the amount in the savings account for the years 1996 to 2000 when interest is compounded annually (Table 8) and continuously.

Table 10 Comparing savings account growth

Year	Amount (dollars), annual compounding	Amount (dollars), continuous compounding
1996	100.00	100.00
1997	105.50	105.65
1998	111.30	111.63
1999	117.42	117.94
2000	123.88	124.61

A bank might decide that it is worth the additional amount to be able to advertise, "We compound interest continuously," as a way to attract more customers.

Example 5   Modeling Radioactive Decay

Laboratory experiments indicate that some atoms emit a part of their mass as radiation, with the remainder of the atom re-forming to make an atom of some new element. For example, radioactive carbon-14 decays into nitrogen; radium eventually decays into lead. If y_0 is the number of radioactive nuclei present at time zero, the number still present at any later time t will be

$$y = y_0 e^{-rt}, \quad r > 0.$$

The number r is called the **decay rate** of the radioactive substance. For carbon-14, the decay rate has been determined experimentally to be about $r = 1.2 \times 10^{-4}$

when t is measured in years. Predict the percent of carbon-14 present after 866 years have elapsed.

Solution   If we start with an amount y_0 of carbon-14 nuclei, after 866 years we are left with the amount

$$y(866) = y_0 e^{(-1.2 \times 10^{-4})(866)}$$

$$\approx (0.901)y_0.$$

That is, after 866 years, we are left with about 90% of the original amount of carbon-14, so about 10% of the original nuclei have decayed. In Example 12 in the next section, you will see how to find the number of years required for half of the radioactive nuclei present in a sample to decay.

What Happened to a^x?

You may wonder why we use the family of functions $y = y_0 e^{kx}$ for different values of the constant k instead of the general exponential functions $y = Pa^x$. In the next section, we show that the exponential function a^x is the same as e^{kx} for an appropriate value of k. So the formula $y = y_0 e^{kx}$ covers the entire range of possibilities.

EXERCISES 3

In Exercises 1–6, match the function with the graphs from Figure 29. Try to do it without using your grapher.

1. $y = 2^x$ **2.** $y = 3^{-x}$

3. $y = -3^{-x}$ **4.** $y = -0.5^{-x}$

5. $y = 2^{-x} - 2$ **6.** $y = 1.5^x - 2$

(a)

(b)

(c)

(d)

(e)

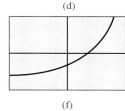

(f)

FIGURE 29 Graphs for Exercises 1–6.

In Exercises 7–10, graph the function. State its domain, range, and intercepts.

7. $y = -2^x + 3$ **8.** $y = e^x + 3$

9. $y = 3 \cdot e^{-x} - 2$ **10.** $y = -2^{-x} - 1$

In Exercises 11–14, rewrite the exponential expression to have the indicated base.

11. 9^{2x},   base 3   **12.** 16^{3x},   base 2

13. $(1/8)^{2x}$,   base 2   **14.** $(1/27)^x$,   base 3

In Exercises 15–18, copy and *work in groups of two or three* to complete the table for the function.

15. $y = 2x - 3$

x	y	Change (Δy)
1	?	
		?
2	?	
		?
3	?	
		?
4	?	

16. $y = -3x + 4$

x	y	Change (Δy)
1	?	
		?
2	?	
		?
3	?	
		?
4	?	

17. $y = x^2$

x	y	Change (Δy)
1	?	
2	?	?
3	?	?
4	?	?

18. $y = 3e^x$

x	y	Ratio (y_i/y_{i-1})
1	?	
2	?	?
3	?	?
4	?	?

19. *Writing to Learn* Explain how the change Δy is related to the slopes of the lines in Exercises 15 and 16. If the changes in x are constant for a linear function, what would you conclude about the corresponding changes in y?

20. *Writing to Learn* Describe how the change Δy from one x-value to the next in Exercise 17 is related to those x-values. What is the change Δy from $x = 1000$ to $x = 1001$? From $x = n$ to $x = n + 1$ for an arbitrary positive integer n?

Extending the Ideas

In Exercises 21 and 22, assume that the graph of the exponential function $f(x) = k \cdot a^x$ passes through the two points. Find the values of a and k.

21. $(1, 4.5), (-1, 0.5)$ **22.** $(1, 1.5), (-1, 6)$

Solving with Graphs

T In Exercises 23–26, use graphs to solve the equations.

23. $2^x = 5$ **24.** $e^x = 4$

25. $3^x - 0.5 = 0$ **26.** $3 - 2^{-x} = 0$

Theory and Examples

27. *World population* (*continuation of Example 3*) Use 1.018 and the population in 1991 to estimate the population of the world in 2010.

28. *Bacteria growth* The number of bacteria in a petri dish culture after t hours is

$$B = 100e^{0.693t}.$$

(a) What was the initial number of bacteria present?

(b) How many bacteria are present after 6 hours?

T (c) Approximately when will the number of bacteria be 200? Estimate the doubling time of the bacteria.

T In Exercises 29–40, use an exponential model and a graphing calculator to estimate the answer in each problem.

29. *Population growth* The population of Knoxville is 500,000 and is increasing at the rate of 3.75% each year. Approximately when will the population reach 1 million?

30. *Population growth* The population of Silver Run in the year 1890 was 6250. Assume the population increased at a rate of 2.75% per year.

(a) Estimate the population in 1915 and 1940.

(b) Approximately when did the population reach 50,000?

31. *Radioactive decay* The half-life of phosphorus-32 is about 14 days. There are 6.6 grams present initially.

(a) Express the amount of phosphorus-32 remaining as a function of time t.

(b) When will there be 1 gram remaining?

32. *Finding time* If John invests $2300 in a savings account with a 6% interest rate compounded annually, how long will it take until John's account has a balance of $4150?

33. *Doubling your money* Determine how much time is required for an investment to double in value if interest is earned at the rate of 6.25% compounded annually.

34. *Doubling your money* Determine how much time is required for an investment to double in value if interest is earned at the rate of 6.25% compounded monthly.

35. *Doubling your money* Determine how much time is required for an investment to double in value if interest is earned at the rate of 6.25% compounded continuously.

36. *Tripling your money* Determine how much time is required for an investment to triple in value if interest is earned at the rate of 5.75% compounded annually.

37. *Tripling your money* Determine how much time is required for an investment to triple in value if interest is earned at the rate of 5.75% compounded daily.

38. *Tripling your money* Determine how much time is required for an investment to triple in value if interest is earned at the rate of 5.75% compounded continuously.

39. *Cholera bacteria* Suppose that a colony of bacteria starts with 1 bacterium and doubles in number every half hour. How many bacteria will the colony contain at the end of 24 h?

40. *Eliminating a disease* Suppose that in any given year the number of cases of a disease is reduced by 20%. If there are 10,000 cases today, how many years will it take

(a) to reduce the number of cases to 1000?

(b) to eliminate the disease; that is, to reduce the number of cases to less than 1?

Regression Analysis: Exponential Models for Population

See page 5 for an introduction to regression analysis with a calculator.

T **41.** Table 11 gives data about the population of Mexico.

Table 11 Population of Mexico

Year	Population (millions)
1950	25.8
1960	34.9
1970	48.2
1980	66.8
1990	81.1

Source: The Statesman's Yearbook, 129th ed. (London: The Macmillan Press, Ltd., 1992).

(a) Let $x = 0$ represent 1900, $x = 1$ represent 1901, and so forth. Find an exponential regression equation for the data and superimpose its graph on a scatter plot of the data.

(b) Use the exponential regression equation to estimate the population of Mexico in 1900. How close is the estimate to the actual population in 1900 of 13,607,272?

(c) Use the exponential regression equation to estimate the annual rate of growth of the population of Mexico.

T **42.** Table 12 gives population data for South Africa.

Table 12 Population of South Africa

Year	Population (millions)
1904	5.2
1911	6.0
1921	6.9
1936	9.6
1946	11.4
1951	12.7
1960	16.0
1970	18.3
1980	20.6

Source: The Statesman's Yearbook, 129th ed. (London: The Macmillan Press, Ltd., 1992).

(a) Let $x = 0$ represent 1900, $x = 1$ represent 1901, and so forth. Find an exponential regression equation for the data and superimpose its graph on a scatter plot of the data.

(b) Use the exponential regression equation to estimate the population of South Africa in 1990.

(c) Use the exponential regression equation to estimate the annual rate of growth of the population of South Africa.

4 Inverse Functions and Logarithms

One-to-One Functions • Inverses • Finding Inverses •
Logarithmic Functions • Properties of Logarithms • Applications

In this section, we define what it means for functions to be inverses of one another and look at what this says about the formulas and graphs of function-inverse pairs. We then study a logarithm function as the inverse of an exponential function with appropriate base, and present several important applications of logarithm functions.

One-to-One Functions

As you know, a function is a rule that assigns a single value in its range to each point in its domain. Some functions assign the same output to more than one input. For example, $f(x) = x^2$ assigns the output 4 to both 2 and -2. Other functions

One-to-one: Graph meets each
horizontal line once.

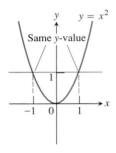

Not one-to-one: Graph meets some
horizontal lines more than once.

FIGURE 30 Using the horizontal line test,
we see that $y = x^3$ is one-to-one and
$y = x^2$ is not.

never output a given value more than once. For example, the cubes of different numbers are always different.

If each output value of a function is associated with exactly one input value, the function is *one-to-one*.

Definition   One-to-One Function

A function $f(x)$ is **one-to-one** on a domain D if $f(a) \neq f(b)$ whenever $a \neq b$.

The graph of a one-to-one function $y = f(x)$ can intersect any horizontal line at most once (the *horizontal line test*). If it intersects such a line more than once, it assumes the same y-value more than once and is not one-to-one (Figure 30).

Example 1   Using the Horizontal Line Test

Determine whether the functions are one-to-one.

$$\textbf{(a)}\ f(x) = x^{2/3} \qquad \textbf{(b)}\ g(x) = \sqrt{x}$$

Solution   As Figure 31a suggests, each horizontal line $y = c$, $c > 0$, intersects the graph of $f(x) = x^{2/3}$ twice so f is not one-to-one. As Figure 31b suggests, each horizontal line intersects the graph of $g(x) = \sqrt{x}$ either once or not at all. The function g is one-to-one.

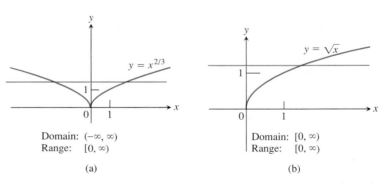

Domain: $(-\infty, \infty)$
Range: $[0, \infty)$

(a)

Domain: $[0, \infty)$
Range: $[0, \infty)$

(b)

FIGURE 31   (a) The graph of $f(x) = x^{2/3}$ and a horizontal line. (b) The graph of $g(x) = \sqrt{x}$ and a horizontal line. (Example 1)

Inverses

Since each output of a one-to-one function comes from just one input, a one-to-one function can be reversed to send outputs back to the inputs from which they came. The function defined by reversing a one-to-one function f is the **inverse of f**. The functions in Tables 13 and 14 are inverses of one another. The symbol for the inverse of f is f^{-1}, read "f inverse." The -1 in f^{-1} is not an exponent; $f^{-1}(x)$ does not mean $1/f(x)$.

As Tables 13 and 14 suggest, composing a function with its inverse in either order sends each output back to the input from which it came. In other words, the result of composing a function and its inverse in either order is the **identity func-**

Table 13 Rental charge versus time		Table 14 Time versus rental charge	
Time x (hours)	Charge y (dollars)	Charge x (dollars)	Time y (hours)
1	5.00	5.00	1
2	7.50	7.50	2
3	10.00	10.00	3
4	12.50	12.50	4
5	15.00	15.00	5
6	17.50	17.50	6

tion, the function that assigns each number to itself. Composing two functions f and g gives a way to test whether they are inverses of one another. Compute $f \circ g$ and $g \circ f$. If $(f \circ g)(x) = x$ and $(g \circ f)(x) = x$, then f and g are inverses of one another; otherwise, they are not. The functions $f(x) = x^3$ and $g(x) = x^{1/3}$ are inverses of one another because $(x^3)^{1/3} = x$ and $(x^{1/3})^3 = x$ for every number x.

A Test for Inverses

Functions f and g are an inverse pair if and only if

$$f(g(x)) = x \qquad \text{and} \qquad g(f(x)) = x.$$

In this case, $g = f^{-1}$ and $f = g^{-1}$.

Example 2   Testing for Inverses

(a) The functions

$$f(x) = 3x \qquad \text{and} \qquad g(x) = \frac{x}{3}$$

are an inverse pair because

$$f(g(x)) = f\left(\frac{x}{3}\right) = 3\left(\frac{x}{3}\right) = x \qquad \text{and} \qquad g(f(x)) = g(3x) = \frac{3x}{3} = x$$

for every number x.

(b) The functions

$$f(x) = x \qquad \text{and} \qquad g(x) = \frac{1}{x}$$

are *not* inverses because

$$f(g(x)) = f\left(\frac{1}{x}\right) = \frac{1}{x} \neq x.$$

Finding Inverses

How do we find the graph of the inverse of a function? Suppose, for example, that the function is the one pictured in Figure 32a. To read the graph, we start at the point x on the x-axis, go up to the graph, and then move over to the y-axis to read the value of y. If we start with y and want to find the x from which it came, we reverse the process (Figure 32b).

The graph of f is already the graph of f^{-1}, although the latter graph is not drawn in the usual way with the domain axis horizontal and the range axis vertical. For f^{-1}, the input-output pairs are reversed. To display the graph of f^{-1} in the usual way, we have to reverse the pairs by reflecting the graph in the 45° line $y = x$ (Figure 32c) and interchanging the letters x and y (Figure 32d). This step puts the independent variable, now called x, on the horizontal axis and the dependent variable, now called y, on the vertical axis.

That the graphs of f and f^{-1} are reflections of each other across the line $y = x$ is to be expected because the input-output pairs (a, b) of f have been reversed to produce the input-output pairs (b, a) of f^{-1}.

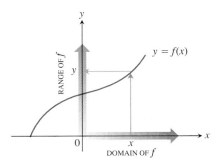

(a) To find the value of f at x, we start at x, go up to the curve, and then over to the y-axis.

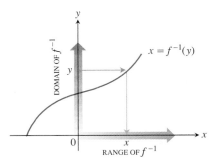

(b) The graph of f is already the graph of f^{-1}. To find the x that gave y, we start at y and go over to the curve and down to the x-axis. The domain of f^{-1} is the range of f. The range of f^{-1} is the domain of f.

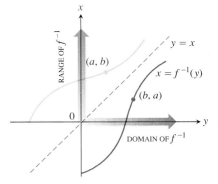

(c) To draw the graph of f^{-1} in the usual way, we reflect the system in the line $y = x$.

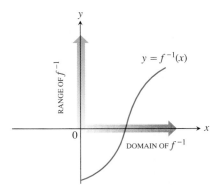

(d) Then we interchange the letters x and y. We now have a normal-looking graph of f^{-1} as a function of x.

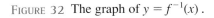

FIGURE 32 The graph of $y = f^{-1}(x)$.

The pictures in Figure 32 tell how to express f^{-1} as a function of x algebraically.

Writing f^{-1} as a Function of x

Step 1. Solve the equation $y = f(x)$ for x in terms of y.

Step 2. Interchange x and y. The resulting formula will be $y = f^{-1}(x)$.

Example 3   Finding the Inverse Function

Find the inverse of $y = \dfrac{1}{2}x + 1$, expressed as a function of x.

Solution

Step 1: Solve for x in terms of y:

$$y = \frac{1}{2}x + 1$$
$$2y = x + 2$$
$$x = 2y - 2.$$

Step 2: Interchange x and y:

$$y = 2x - 2.$$

The inverse of the function $f(x) = (1/2)x + 1$ is the function $f^{-1}(x) = 2x - 2$.

Check: To check, we verify that both composites give the identity function:

$$f^{-1}(f(x)) = 2\left(\frac{1}{2}x + 1\right) - 2 = x + 2 - 2 = x$$

$$f(f^{-1}(x)) = \frac{1}{2}(2x - 2) + 1 = x - 1 + 1 = x.$$

See Figure 33.

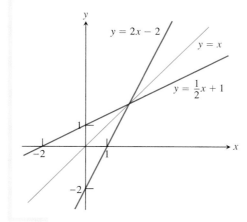

FIGURE 33 Graphing $f(x) = (1/2)x + 1$ and $f^{-1}(x) = 2x - 2$ together shows the graphs' symmetry with respect to the line $y = x$.

Example 4   Finding the Inverse Function

Find the inverse of the function $y = x^2, x \geq 0$, expressed as a function of x.

Solution

Step 1: Solve for x in terms of y:

$$y = x^2$$
$$\sqrt{y} = \sqrt{x^2} = |x| = x \qquad |x| = x \text{ because } x \geq 0$$

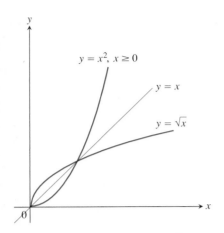

FIGURE 34 The functions $y = \sqrt{x}$ and $y = x^2$, $x \geq 0$, are inverses of one another. (Example 4)

Step 2: Interchange x and y:

$$y = \sqrt{x}.$$

The inverse of $y = x^2$, $x \geq 0$, is $y = \sqrt{x}$. See Figure 34.

Notice that, unlike the restricted function $y = x^2$, $x \geq 0$, the unrestricted function $y = x^2$ is not one-to-one and therefore has no inverse.

In Section 1.6, we show you an easy way to graph $y = f(x)$ and $y = f^{-1}(x)$ together on a graphing calculator or computer.

Logarithmic Functions

If a is any positive real number other than 1, the base a exponential function $f(x) = a^x$ is one-to-one. It therefore has an inverse. Its inverse is called the *base a logarithm function.*

> **Definition**   Base a Logarithm Function
>
> The **base a logarithm function** $y = \log_a x$ is the inverse of the base a exponential function $y = a^x$ $(a > 0, a \neq 1)$.

The domain of $\log_a x$ is $(0, \infty)$, the range of a^x. The range of $\log_a x$ is $(-\infty, \infty)$, the domain of a^x.

Because we have no technique for solving for x in terms of y in the equation $y = a^x$, we do not have an explicit formula for the logarithm function as a function of x. The graph of $y = \log_a x$, however, can be obtained by reflecting the graph of $y = a^x$ across the line $y = x$ (Figure 35).

Logarithms with base e and base 10 are so important in applications that calculators have special keys for them. They also have their own special notation and names:

$$\log_e x \quad \text{is written as} \quad \ln x.$$
$$\log_{10} x \quad \text{is written as} \quad \log x.$$

The function $y = \ln x$ is called the **natural logarithm function,** and $y = \log x$ is often called the **common logarithm function.**

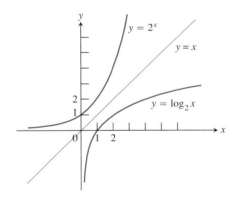

FIGURE 35 The graph of 2^x and its inverse, $\log_2 x$.

Properties of Logarithms

Because a^x and $\log_a x$ are inverses, composing them in either order gives the identity function.

> **Inverse Properties for a^x and $\log_a x$**
>
> **1.** Base a: $a^{\log_a x} = x$, $\log_a a^x = x$, $a > 0, a \neq 1, x > 0$
>
> **2.** Base e: $e^{\ln x} = x$, $\ln e^x = x$, $x > 0$

These properties help us solve equations that contain logarithms and exponential functions.

Example 5   Using the Inverse Properties

Solve for x: **(a)** $\ln x = 3t + 5$ **(b)** $e^{2x} = 10$

Solution

(a) $\ln x = 3t + 5$

$e^{\ln x} = e^{3t+5}$ Exponentiate both sides.

$x = e^{3t+5}$ Inverse Property

(b) $e^{2x} = 10$

$\ln e^{2x} = \ln 10$ Take logarithms of both sides.

$2x = \ln 10$ Inverse Property

$x = \dfrac{1}{2}\ln 10 \approx 1.15$

The logarithm function has the following arithmetic properties.

Properties of Logarithms

For any real numbers $x > 0$ and $y > 0$,

1. *Product Rule:* $\log_a xy = \log_a x + \log_a y$

2. *Quotient Rule:* $\log_a \dfrac{x}{y} = \log_a x - \log_a y$

3. *Power Rule:* $\log_a x^y = y \log_a x$

Substituting a^x for x in the equation $x = e^{\ln x}$ enables us to rewrite a^x as a power of e:

$a^x = e^{\ln(a^x)}$ Substitute a^x for x in $x = e^{\ln x}$.

$= e^{x \ln a}$ Power Rule for logs

$= e^{(\ln a)x}$.   Exponent rearranged

The exponential function a^x is the same as e^{kx} for $k = \ln a$. The usual formulation is the one without parentheses.

Every exponential function is a power of the natural exponential function.

$$a^x = e^{x \ln a}$$

That is, a^x is the same as e^x raised to the power $\ln a$.

Example 6   Writing Exponentials as Powers of e

$$2^x = e^{(\ln 2)x} = e^{x \ln 2}$$
$$5^{-3x} = e^{(\ln 5)(-3x)} = e^{-3x \ln 5}$$

Example 7   Solving Equations with Logarithms

Solve for x:

$$3^{\log_3 (7)} - 4^{\log_4 (2)} = 5^{(\log_5 x - \log_5 x^2)}$$

Solution:

$$3^{\log_3 (7)} - 4^{\log_4 (2)} = 5^{(\log_5 x - \log_5 x^2)}$$
$$3^{\log_3 (7)} - 4^{\log_4 (2)} = 5^{\log_5 (x/x^2)} \qquad \text{Quotient Rule}$$
$$7 - 2 = \frac{x}{x^2} \qquad \text{Inverse property}$$
$$5 = \frac{1}{x} \qquad \text{Cancellation, } x \neq 0$$
$$\frac{1}{5} = x$$

Returning once more to the properties of a^x and $\log_a x$, we have

$$\ln x = \ln a^{\log_a x} \qquad \text{Inverse Property for } a^x \text{ and } \log_a x$$
$$= (\log_a x)(\ln a). \qquad \text{Power Rule for logarithms, with } y = \log_a x$$

Rewriting this equation as $\log_a x = (\ln x)/(\ln a)$ shows that every logarithmic function is a constant multiple of $\ln x$.

Change of Base Formula
Every logarithmic function is a constant multiple of the natural logarithm.

$$\log_a x = \frac{\ln x}{\ln a} \qquad (a > 0, a \neq 1)$$

Example 8   Graphing a Base a Logarithm Function

Graph $f(x) = \log_2 x$.

Solution   We use the change of base formula to rewrite $f(x)$.

$$f(x) = \log_2 x = \frac{\ln x}{\ln 2}$$

Figure 36 gives the graph.

Applications

In Section 3, we used graphical methods to solve exponential growth and decay problems. Now we can use the properties of logarithms to solve the same problems algebraically.

$y = \dfrac{\ln x}{\ln 2} = \log_2 x$

[–6, 6] by [–4, 4]

FIGURE 36 The graph of $f(x) = \log_2 x$ using $f(x) = (\ln x)/(\ln 2)$. (Example 8)

Example 9   Finding Time

Sarah invests $1000 in an account that earns 5.25% interest compounded annually. How long will it take the account to reach $2500?

Solution

Model

The amount in the account at any time t in years is $1000(1.0525)^t$, so we need to solve the equation

$$1000(1.0525)^t = 2500 .$$

Solve Algebraically

$$(1.0525)^t = 2.5 \qquad \text{Divide by 1000.}$$
$$\ln (1.0525)^t = \ln 2.5 \qquad \text{Take logarithms of both sides.}$$
$$t \ln 1.0525 = \ln 2.5 \qquad \text{Power Rule}$$
$$t = \frac{\ln 2.5}{\ln 1.0525} \approx 17.9$$

Interpret

The amount in Sarah's account will be $2500 in about 17.9 years, or about 17 years and 11 months.

Example 10   Earthquake Intensity

Earthquake intensity is often reported on the logarithmic Richter scale. Here the formula is

$$\text{Magnitude } R = \log \left(\frac{a}{T} \right) + B ,$$

where a is the amplitude of the ground motion in microns at the receiving station, T is the period of the seismic wave in seconds, and B is an empirical factor that allows for the weakening of the seismic wave with increasing distance from the epicenter of the quake. For an earthquake 10,000 km from the receiving station, $B = 6.8$. If the recorded vertical ground motion is $a = 10$ microns and the period is $T = 1$ sec, the earthquake's magnitude is

$$R = \log \left(\frac{10}{1} \right) + 6.8 = 1 + 6.8 = 7.8 .$$

An earthquake of this magnitude does great damage near its epicenter.

Typical Sound Levels	
Threshold of hearing	0 db
Rustle of leaves	10 db
Average whisper	20 db
Quiet automobile	50 db
Ordinary conversation	65 db
Pneumatic drill 10 feet away	90 db
Threshold of pain	120 db

Example 11   Intensity of Sound

Another example of the use of common logarithms is the **decibel** or **db** ("dee bee") **scale** for measuring loudness. If I is the **intensity** of sound in watts per square meter, the decibel level of the sound is

$$\textbf{Sound level} = 10 \log (I \times 10^{12}) \text{ db} . \tag{1}$$

If you ever wondered why doubling the power of your audio amplifier increases the sound level by only a few decibels, Equation (1) provides the answer.

Doubling I in Equation (1) adds only about 3 db:

$$
\begin{aligned}
\text{Sound level with } I \text{ doubled} &= 10 \; \log\,(2I \times 10^{12}) \qquad \text{\small Eq. (1) with } 2I \text{ for } I \\
&= 10 \; \log\,(2 \cdot I \times 10^{12}) \\
&= 10 \; \log 2 \; + \; 10 \; \log\,(I \times 10^{12}) \\
&= \text{original sound level} \; + \; 10 \; \log 2 \\
&\approx \text{original sound level} \; + \; 3 \cdot \log 2 \approx 0.30
\end{aligned}
$$

Example 12   Half-Life of Polonium-210

The **half-life** of a radioactive element is the time required for half of the radioactive nuclei present in a sample to decay. It is a remarkable fact that the half-life is a constant that does not depend on the number of radioactive nuclei initially present in the sample, but only on the radioactive substance.

To see why, let y_0 be the number of radioactive nuclei initially present in the sample. Then the number y present at any later time t will be $y = y_0 \, e^{-kt}$. We seek the value of t at which the number of radioactive nuclei present equals half the original number:

$$
y_0 \, e^{-kt} = \frac{1}{2} \, y_0
$$

$$
e^{-kt} = \frac{1}{2}
$$

$$
-kt = \ln \frac{1}{2} = -\ln 2 \qquad \text{\small Reciprocal Rule for logarithms}
$$

$$
t = \frac{\ln 2}{k}. \tag{2}
$$

This value of t is the half-life of the element. It depends only on the value of k; the number y_0 does not enter in.

The effective radioactive lifetime of polonium-210 is so short that we measure it in days rather than years. The number of radioactive atoms remaining after t days in a sample that starts with y_0 radioactive atoms is

$$
y = y_0 \, e^{-5 \times 10^{-3} t}.
$$

The element's half-life is

$$
\begin{aligned}
\text{Half-life} &= \frac{\ln 2}{k} \qquad\qquad \text{\small Eq. (2)} \\
&= \frac{\ln 2}{5 \times 10^{-3}} \qquad \text{\small The k from polonium's decay equation} \\
&\approx 139 \text{ days.}
\end{aligned}
$$

EXERCISES 4

Identifying One-to-One Functions Graphically

Which of the functions graphed in Exercises 1–6 are one-to-one and which are not?

1.

$y = -3x^3$

2.

$y = x^4 - x^2$

3.

$y = 2|x|$

4.

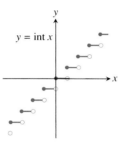

$y = \text{int } x$

5.

$y = \dfrac{1}{x}$

6.

$y = x^{1/3}$

Graphing Inverse Functions

Each of Exercises 7–10 shows the graph of a function $y = f(x)$. Copy the graph and draw in the line $y = x$. Then use symmetry with respect to the line $y = x$ to add the graph of f^{-1} to your sketch. (It is not necessary to find a formula for f^{-1}.) Identify the domain and range of f^{-1}.

7.

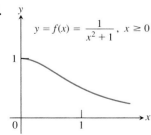

$y = f(x) = \dfrac{1}{x^2 + 1}, \quad x \geq 0$

8.

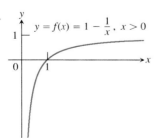

$y = f(x) = 1 - \dfrac{1}{x}, \quad x > 0$

9.

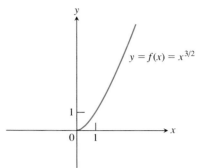

$y = f(x) = x^{3/2}$

10.

$y = f(x) = \sin x,$
$-\dfrac{\pi}{2} \leq x \leq \dfrac{\pi}{2}$

Formulas for Inverse Functions

Each of Exercises 11–16 gives a formula for a function $y = f(x)$ and shows the graphs of f and f^{-1}. Find a formula for f^{-1} in each case.

11. $f(x) = x^2 + 1, \quad x \geq 0$

12. $f(x) = x^2, \quad x \leq 0$

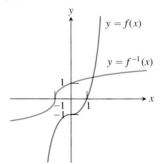

$y = f(x)$
$y = f^{-1}(x)$

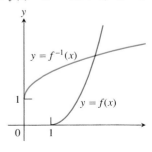

$y = f(x)$
$y = f^{-1}(x)$

13. $f(x) = x^3 - 1$

14. $f(x) = x^2 - 2x + 1, \quad x \geq 1$

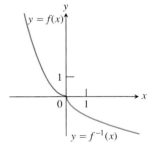

$y = f(x)$
$y = f^{-1}(x)$

15. $f(x) = (x + 1)^2$, $x \ge -1$ **16.** $f(x) = x^{2/3}$, $x \ge 0$

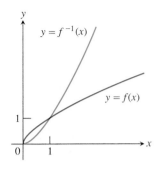

Finding Inverses

In Exercises 17–28, find f^{-1} and verify that

$$(f \circ f^{-1})(x) = (f^{-1} \circ f)(x) = x.$$

17. $f(x) = 2x + 3$ **18.** $f(x) = 5 - 4x$

19. $f(x) = x^3 - 1$ **20.** $f(x) = x^2 + 1$, $x \ge 0$

21. $f(x) = x^2$, $x \le 0$ **22.** $f(x) = x^{2/3}$, $x \ge 0$

23. $f(x) = -(x - 2)^2$, $x \le 2$

24. $f(x) = x^2 + 2x + 1$, $x \ge -1$

25. $f(x) = \dfrac{1}{x^2}$, $x > 0$ **26.** $f(x) = \dfrac{1}{x^3}$

27. $f(x) = \dfrac{2x + 1}{x + 3}$ **28.** $f(x) = \dfrac{x + 3}{x - 2}$

"Naturalizing" Exponential and Logarithmic Functions

In Exercises 29 and 30, express the exponential function as a power of e. Find the **(a)** domain and **(b)** range.

29. $y = 3^x - 1$ **30.** $y = 4^{x+1}$

In Exercises 31 and 32, express the function in terms of the natural logarithm. Find the **(a)** domain and **(b)** range. **(c)** Sketch the graph.

31. $y = 1 - (\ln 3) \log_3 x$ **32.** $y = (\ln 10) \log (x + 2)$

Solving Equations for Exponents

In Exercises 33–36, solve the equation algebraically. If you have a graphing calculator or computer grapher, support your solution graphically.

33. $(1.045)^t = 2$ **34.** $e^{0.05t} = 3$

35. $e^x + e^{-x} = 3$ **36.** $2^x + 2^{-x} = 5$

Solving Equations Containing Logarithmic Terms

In Exercises 37 and 38, solve for y.

37. $\ln y = 2t + 4$

38. $\ln (y - 1) - \ln 2 = x + \ln x$

Theory and Applications

39. Find a formula for f^{-1} and verify that $(f \circ f^{-1})(x) = (f^{-1} \circ f)(x) = x$.

(a) $f(x) = \dfrac{100}{1 + 2^{-x}}$ **(b)** $f(x) = \dfrac{50}{1 + 1.1^{-x}}$

40. *Inverse functions* We suggest that students *work in groups of two or three.*

$$\text{Let } y = f(x) = mx + b, \quad m \ne 0.$$

(a) *Writing to Learn* Give a convincing argument that f is a one-to-one function.

(b) Find a formula for the inverse of f. How are the slopes of the graphs of f and f^{-1} related?

(c) If the graphs of two functions are parallel lines with a nonzero slope, what can you say about the graphs of the inverses of the functions?

(d) If the graphs of two functions are perpendicular lines with a nonzero slope, what can you say about the graphs of the inverses of the functions?

41. *Radioactive decay* The half-life of a certain radioactive substance is 12 hours. There are 8 grams present initially.

(a) Express the amount of substance remaining as a function of time t.

(b) When will there be 1 gram remaining?

42. *Doubling your money* Determine how much time is required for a $500 investment to double in value if interest is earned at the rate of 4.75% compounded annually.

43. *Population growth* The population of Glenbrook is 375,000 and is increasing at the rate of 2.25% per year. Predict when the population will be 1 million.

44. *Audio amplifiers* By what factor k do you have to multiply the intensity of I of the sound from your audio amplifier to add 10 db to the sound level?

45. *Audio amplifiers* You multiplied the intensity of the sound of your audio system by a factor of 10. By how many decibels did this increase the sound level?

46. *Radon-222* The decay equation for radon-222 gas is known to be $y = y_0 e^{-0.18t}$, with t in days. About how long will it take the radon in a sealed sample of air to fall to 90% of its original value?

Solving Equations and Comparing Functions

T In Exercises 47–50, use a graphing calculator to find the points of intersection of the two curves. Round your answers to 2 decimal places.

47. $y = 2x - 3$, $y = 5$

48. $y = -3x + 5$, $y = -3$

49. (a) $y = 2^x$, $y = 3$ (b) $y = 2^x$, $y = -1$

50. (a) $y = e^{-x}$, $y = 4$ (b) $y = e^{-x}$, $y = -1$

T For each of the function pairs in Exercises 51–54:

   (a) Graph f and g together in a square window.

   (b) Graph $f \circ g$.

   (c) Graph $g \circ f$.

What can you conclude from the graphs?

51. $f(x) = x^3$, $g(x) = x^{1/3}$

52. $f(x) = x$, $g(x) = 1/x$

53. $f(x) = 3x$, $g(x) = x/3$

54. $f(x) = e^x$, $g(x) = \ln x$

T **55.** *Supporting the product rule* Let $y_1 = \ln (ax)$, $y_2 = \ln x$, and $y_3 = y_1 - y_2$.

   (a) Graph y_1 and y_2 for $a = 2, 3, 4$, and 5. How do the graphs of y_1 and y_2 appear to be related?

   (b) Support your finding by graphing y_3.

   (c) Confirm your finding algebraically.

T **56.** *Supporting the quotient rule* Let $y_1 = \ln (x/a)$, $y_2 = \ln x$, $y_3 = y_2 - y_1$, and $y_4 = e^{y_3}$.

   (a) Graph y_1 and y_2 for $a = 2, 3, 4$, and 5. How are the graphs of y_1 and y_2 related?

   (b) Graph y_3 for $a = 2, 3, 4$, and 5. Describe the graphs.

   (c) Graph y_4 for $a = 2, 3, 4$, and 5. Compare the graphs to the graph of $y = a$.

   (d) Use $e^{y_3} = e^{y_2 - y_1} = a$ to solve for y_1.

T **57.** The equation $x^2 = 2^x$ has three solutions: $x = 2$, $x = 4$, and one other. Estimate the third solution as accurately as you can by graphing.

T **58.** Could $x^{\ln 2}$ possibly be the same as $2^{\ln x}$ for $x > 0$? Graph the two functions and explain what you see.

Logarithmic Regression Analysis: Oil Production

See page 5 for an introduction to regression analysis with a calculator.

T **59.** *Oil production in Indonesia* Table 15 shows the number of metric tons of oil produced by Indonesia for three different years.

   (a) Use a calculator or computer to find the natural logarithm regression equation $y = a + b \ln x$ for the data in Table 15 and use it to estimate the number of metric tons of oil produced by Indonesia in 1982 and 2000. Let $x = 60$ represent 1960, $x = 70$ represent 1970, and so forth.

   (b) Superimpose the graph of the logarithmic regression equation on a scatter plot of the data.

   (c) Use the graph of the regression equation to predict the metric tons of oil production in 1982 and 2000.

Table 15 Indonesia's oil production	
Year	**Metric tons (millions)**
1960	20.56
1970	42.10
1990	70.10

Source: Statesman's Yearbook, 129th ed.
(London: The Macmillan Press, Ltd., 1992).

T **60.** *Oil production in Saudi Arabia*

   (a) Find a natural logarithm regression equation for the data in Table 16.

   (b) Estimate the number of metric tons of oil produced by Saudi Arabia in 1975.

   (c) Predict when Saudi Arabian oil production will reach 400 million metric tons.

Table 16 Saudi Arabian oil production	
Year	**Metric tons (millions)**
1960	61.09
1970	176.85
1990	321.93

Source: The Statesman's Yearbook, 129th ed.
(London: The Macmillan Press, Ltd., 1992).

5 Trigonometric Functions and Their Inverses

Radian Measure • Graphs of Trigonometric Functions • Values of Trigonometric Functions • Periodicity • Even and Odd Trigonometric Functions • Transformations of Trigonometric Graphs • Identities • The Law of Cosines • Inverse Trigonometric Functions • Identities Involving Arc Sine and Arc Cosine

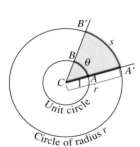

FIGURE 37 The radian measure of angle *ACB* is the length θ of arc *AB* on the unit circle centered at *C*. The value of θ can be found from any other circle, however, as the ratio s/r.

This section reviews the basic trigonometric functions and their inverses. The trigonometric functions are important because they are periodic, or repeating. They can therefore model many naturally occurring periodic processes such as daily temperature fluctuations in Earth's atmosphere, the wave behavior of musical notes, blood pressure in a heart, and the water level in a tidal basin.

Inverse trigonometric functions arise when we want to calculate angles from side measurements in triangles. You will see their usefulness in calculus in Chapters 6 and 7.

Radian Measure

The **radian measure** of the angle *ACB* at the center of the unit circle (Figure 37) equals the length of the arc that *ACB* cuts from the unit circle.

When an angle of measure θ is placed in *standard position* at the center of a circle of radius r (Figure 38), the six basic trigonometric functions of θ are defined as follows:

sine: $\sin \theta = \dfrac{y}{r}$ **cosecant:** $\csc \theta = \dfrac{r}{y}$

cosine: $\cos \theta = \dfrac{x}{r}$ **secant:** $\sec \theta = \dfrac{r}{x}$

tangent: $\tan \theta = \dfrac{y}{x}$ **cotangent:** $\cot \theta = \dfrac{x}{y}$

Conversion Formulas

1 degree $= \dfrac{\pi}{180}$ (≈ 0.02) radians

Degrees to radians: multiply by $\dfrac{\pi}{180}$

1 radian $= \dfrac{180}{\pi}$ (≈ 57) degrees

Radians to degrees: multiply by $\dfrac{180}{\pi}$

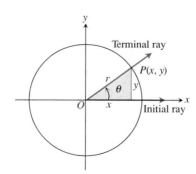

FIGURE 38 An angle θ in standard position.

Graphs of Trigonometric Functions

When we graph trigonometric functions in the coordinate plane, we usually denote the independent variable (radians) by x instead of θ (Figure 39).

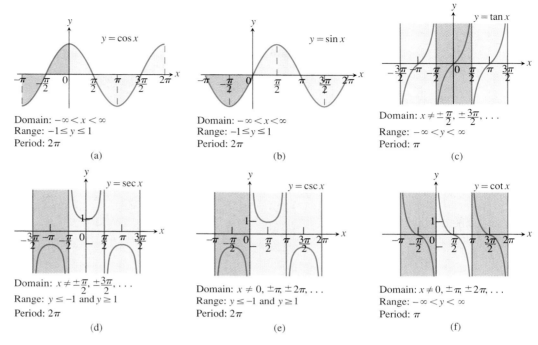

Domain: $-\infty < x < \infty$
Range: $-1 \le y \le 1$
Period: 2π

(a)

Domain: $-\infty < x < \infty$
Range: $-1 \le y \le 1$
Period: 2π

(b)

Domain: $x \ne \pm\frac{\pi}{2}, \pm\frac{3\pi}{2}, \ldots$
Range: $-\infty < y < \infty$
Period: π

(c)

Domain: $x \ne \pm\frac{\pi}{2}, \pm\frac{3\pi}{2}, \ldots$
Range: $y \le -1$ and $y \ge 1$
Period: 2π

(d)

Domain: $x \ne 0, \pm\pi, \pm 2\pi, \ldots$
Range: $y \le -1$ and $y \ge 1$
Period: 2π

(e)

Domain: $x \ne 0, \pm\pi, \pm 2\pi, \ldots$
Range: $-\infty < y < \infty$
Period: π

(f)

FIGURE 39 Graphs of the (a) cosine, (b) sine, (c) tangent, (d) secant, (e) cosecant, and
(f) cotangent functions using radian measure.

Values of Trigonometric Functions

If the circle in Figure 40 has radius $r = 1$, the equations defining $\sin\theta$ and $\cos\theta$
become

$$\cos\theta = x, \qquad \sin\theta = y.$$

We can then calculate the values of the cosine and sine directly from the coordinates of P, if we happen to know them, or indirectly from the acute reference triangle made by dropping a perpendicular from P to the x-axis (Figure 41). We read the

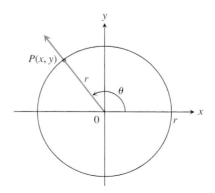

FIGURE 40 The trigonometric
functions of a general angle θ are
defined in terms of x, y, and r.

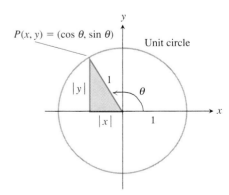

FIGURE 41 The acute reference triangle
for an angle θ.

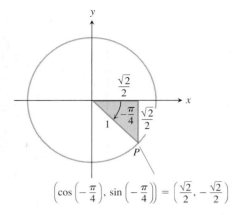

$$\left(\cos\left(-\frac{\pi}{4} \right), \sin\left(-\frac{\pi}{4} \right) \right) = \left(\frac{\sqrt{2}}{2}, -\frac{\sqrt{2}}{2} \right)$$

FIGURE 42 The triangle for calculating the sine and cosine of $-\pi/4$ radians. (Example 1)

magnitudes of x and y from the triangle's sides. The signs of x and y are determined by the quadrant in which the triangle lies.

Example 1   Finding Sine and Cosine Values

Find the sine and cosine of $-\pi/4$ radians.

Solution

Step 1: Draw the angle in standard position in the unit circle and write in the lengths of the sides of the reference triangle (Figure 42).

Step 2: Find the coordinates of the point P where the angle's terminal ray cuts the circle:

$$\cos\left(-\frac{\pi}{4} \right) = x\text{-coordinate of } P = \frac{\sqrt{2}}{2},$$

$$\sin\left(-\frac{\pi}{4} \right) = y\text{-coordinate of } P = -\frac{\sqrt{2}}{2}.$$

Calculations similar to those in Example 1 allow us to fill in Table 17. Most calculators and computers readily provide values of the trigonometric functions for angles given in either radians or degrees.

Table 17 Values of sin θ, cos θ, and tan θ for selected values of θ

Degrees	-180	-135	-90	-45	0	30	45	60	90	135	180
θ (radians)	$-\pi$	$-3\pi/4$	$-\pi/2$	$-\pi/4$	0	$\pi/6$	$\pi/4$	$\pi/3$	$\pi/2$	$3\pi/4$	π
sin θ	0	$-\sqrt{2}/2$	-1	$-\sqrt{2}/2$	0	$1/2$	$\sqrt{2}/2$	$\sqrt{3}/2$	1	$\sqrt{2}/2$	0
cos θ	-1	$-\sqrt{2}/2$	0	$\sqrt{2}/2$	1	$\sqrt{3}/2$	$\sqrt{2}/2$	$1/2$	0	$-\sqrt{2}/2$	-1
tan θ	0	1		-1	0	$\sqrt{3}/3$	1	$\sqrt{3}$		-1	0

Periodicity

When an angle of measure θ and an angle of measure $\theta + 2\pi$ are in standard position, their terminal rays coincide. The two angles therefore have the same trigonometric function values:

$$\cos(\theta + 2\pi) = \cos\theta \qquad \sin(\theta + 2\pi) = \sin\theta \qquad \tan(\theta + 2\pi) = \tan\theta$$
$$\sec(\theta + 2\pi) = \sec\theta \qquad \csc(\theta + 2\pi) = \csc\theta \qquad \cot(\theta + 2\pi) = \cot\theta \tag{1}$$

Similarly, $\cos(\theta - 2\pi) = \cos\theta$, $\sin(\theta - 2\pi) = \sin\theta$, and so on.

We see that the values of the trigonometric functions repeat at regular intervals. We describe this behavior by saying that the six basic trigonometric functions are *periodic*.

Definition   Periodic Function, Period

A function $f(x)$ is **periodic** if there is a positive number p such that $f(x + p) = f(x)$ for every value of x. The smallest such value of p is the **period** of f.

Periods of Trigonometric Functions

Period π: $\tan (x + \pi) = \tan x$
    $\cot (x + \pi) = \cot x$

Period 2π: $\sin (x + 2\pi) = \sin x$
    $\cos (x + 2\pi) = \cos x$
    $\sec (x + 2\pi) = \sec x$
    $\csc (x + 2\pi) = \csc x$

FIGURE 43 This compact patient monitor shows several periodic functions associated with the human body. This device dynamically monitors electrocardiogram (ECG) and respiration, and blood pressure.

As we can see in Figure 39, the functions $\cos x$, $\sin x$, $\sec x$, and $\csc x$ are periodic with period 2π. The functions $\tan x$ and $\cot x$ are periodic with period π.

Periodic functions are important because much of the behavior we study in science is periodic (Figure 43). Brain waves and heartbeats are periodic, as are household voltage and electric current. The electromagnetic field that heats food in a microwave oven is periodic, as are cash flows in seasonal businesses and the behavior of rotational machinery. The seasons are periodic, as is the weather. The phases of the moon are periodic, as are the motions of the planets. There is strong evidence that the ice ages are periodic, with a period of 90,000 to 100,000 years.

Why are trigonometric functions so important in the study of things periodic? The answer lies in a surprising and beautiful theorem from advanced calculus that says that every periodic function we want to use in mathematical modeling can be written as an algebraic combination of sines and cosines. Once we learn the calculus of sines and cosines, we can model the mathematical behavior of most periodic phenomena.

Even and Odd Trigonometric Functions

The graphs in Figure 39 suggest that $\cos x$ and $\sec x$ are even functions because their graphs are symmetric about the y-axis. The other four basic trigonometric functions are odd.

Example 2   Confirming Even and Odd

Show that the cosine is an even function and the sine is odd.

Solution   From Figure 44, it follows that

$$\cos (-\theta) = \frac{x}{r} = \cos \theta, \qquad \sin (-\theta) = \frac{-y}{r} = -\sin \theta,$$

so the cosine is an even function and the sine is odd.

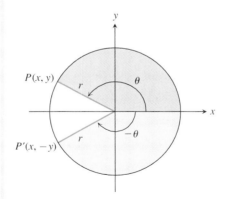

FIGURE 44 Angles of opposite sign. (Example 2)

We can use the results of Example 2 to establish the *parity* of the other four basic trigonometric functions. For example,

$$\sec (-\theta) = \frac{1}{\cos (-\theta)} = \frac{1}{\cos \theta} = \sec \theta,$$

$$\tan (-\theta) = \frac{\sin (-\theta)}{\cos (-\theta)} = \frac{-\sin \theta}{\cos \theta} = -\tan \theta,$$

so secant is an even function and tangent is odd. Similar steps will show that cosecant and cotangent are odd functions.

Transformations of Trigonometric Graphs

The rules for shifting, stretching, shrinking, and reflecting the graph of a function apply to the trigonometric functions. The following diagram will remind you of the controlling parameters.

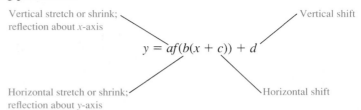

Vertical stretch or shrink; reflection about x-axis

Vertical shift

$$y = af(b(x + c)) + d$$

Horizontal stretch or shrink; reflection about y-axis

Horizontal shift

Example 3   Modeling Temperature in Alaska

The builders of the Trans-Alaska Pipeline used insulated pads to keep the pipeline heat from melting the permanently frozen soil beneath. To design the pads, it was necessary to take into account the variation in air temperature throughout the year. The variation was represented in the calculations by a general sine function or **sinusoid** of the form

$$f(x) = A \sin\left[\frac{2\pi}{B}(x - C)\right] + D, \tag{2}$$

where $|A|$ is the *amplitude*, $|B|$ is the *period*, C is the *horizontal shift*, and D is the *vertical shift* (Figure 45).

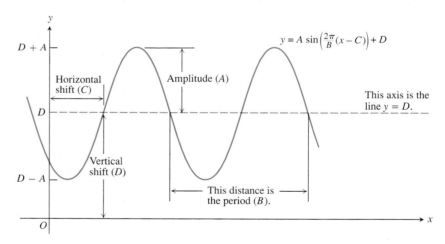

FIGURE 45 The general sine curve $y = A \sin[(2\pi/B)(x - C)] + D$, shown for A, B, C, and D positive. (Example 3)

Figure 46 shows how to use such a function to represent temperature data. The data points in the figure are plots of the mean daily air temperatures for Fairbanks, Alaska, based on records of the National Weather Service from 1941 to 1970. The sine function used to fit the data is

$$f(x) = 37 \sin\left[\frac{2\pi}{365}(x - 101)\right] + 25,$$

where f is temperature in degrees Fahrenheit and x is the number of the day counting from the beginning of the year. The fit is remarkably good.

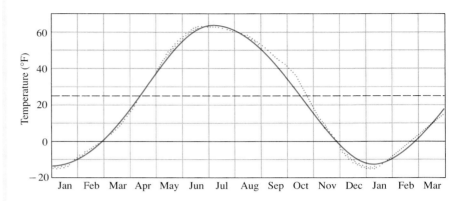

Source: "Is the Curve of Temperature Variation a Sine Curve?" by B. M. Lando and C. A. Lando, *The Mathematics Teacher*, Vol. 7, No. 6 (September 1977), Fig. 2, p. 53.

FIGURE 46 Normal mean air temperatures for Fairbanks, Alaska, plotted as data points (red). The approximating sine function (blue) is

$$f(x) = 37 \sin [(2\pi/365)\,(x - 101)] + 25 \,.$$

Identities

Applying the Pythagorean Theorem to the reference right triangle we obtain by dropping a perpendicular from the point $P(\cos \theta, \sin \theta)$ on the unit circle to the x-axis (Figure 47) gives

$$\cos^2 \theta + \sin^2 \theta = 1 \,. \tag{3}$$

This equation, true for all values of θ, is the most frequently used identity in trigonometry. Dividing this identity in turn by $\cos^2 \theta$ and $\sin^2 \theta$ gives

$$1 + \tan^2 \theta = \sec^2 \theta \,,$$
$$1 + \cot^2 \theta = \csc^2 \theta \,.$$

FIGURE 47 The reference triangle for a general angle θ.

The following formulas hold for all angles A and B.

All the trigonometric identities you will need in this book derive from Equations (3) and (4).

Angle Sum Formulas

$$\cos (A + B) = \cos A \cos B - \sin A \sin B$$
$$\sin (A + B) = \sin A \cos B + \cos A \sin B \tag{4}$$

Substituting θ for both A and B in the angle sum formulas gives two more useful identities.

Instead of memorizing Equations (5) you might find it helpful to remember Equations (4) and then recall where they came from.

Double-Angle Formulas

$$\cos 2\theta = \cos^2 \theta - \sin^2 \theta$$
$$\sin 2\theta = 2 \sin \theta \cos \theta$$

(5)

The Law of Cosines

If a, b, and c are sides of a triangle ABC and if θ is the angle opposite c, then

$$c^2 = a^2 + b^2 - 2ab \cos \theta .$$

(6)

This equation is called the **law of cosines**.

We can see why the law holds if we introduce coordinate axes with the origin at C and the positive x-axis along one side of the triangle, as in Figure 48. The coordinates of A are $(b, 0)$; the coordinates of B are $(a \cos \theta, a \sin \theta)$. The square of the distance between A and B is therefore

$$c^2 = (a \cos \theta - b)^2 + (a \sin \theta)^2$$
$$= a^2 \underbrace{(\cos^2 \theta + \sin^2 \theta)}_{1} + b^2 - 2ab \cos \theta$$
$$= a^2 + b^2 - 2ab \cos \theta .$$

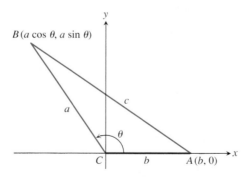

FIGURE 48 The square of the distance between A and B gives the law of cosines.

The law of cosines generalizes the Pythagorean Theorem. If $\theta = \pi/2$, then $\cos \theta = 0$ and $c^2 = a^2 + b^2$.

Inverse Trigonometric Functions

None of the six basic trigonometric functions graphed in Figure 39 is one-to-one. These functions do not have inverses. In each case, however, the domain can be restricted to produce a new function that does have an inverse, as illustrated in Example 4.

Example 4   Restricting the Domain of the Sine

Show that the function $y = \sin x$, $-\pi/2 \le x \le \pi/2$, is one-to-one, and graph its inverse.

Solution   Figure 49a shows the graph of this restricted sine function. This function is one-to-one because it does not repeat any output values. It therefore has an inverse, which we graph in Figure 49b by interchanging the ordered pairs as in Section 4.

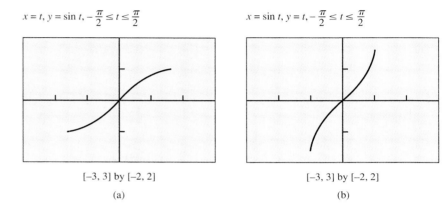

$$x = t, y = \sin t, -\frac{\pi}{2} \le t \le \frac{\pi}{2} \qquad\qquad x = \sin t, y = t, -\frac{\pi}{2} \le t \le \frac{\pi}{2}$$

[-3, 3] by [-2, 2]   [-3, 3] by [-2, 2]

(a)   (b)

FIGURE 49 (a) A restricted sine function and (b) its inverse. The graphs were generated by a graphing calculator in parametric mode. See Section 6 for a review of parametric equations. (Example 4)

The inverse of the restricted sine function of Example 4 is called the *inverse sine function*. The inverse sine of x is the angle in the interval $[-\pi/2 , \pi/2]$ whose sine is x. It is denoted by $\sin^{-1} x$ or arcsin x. Either notation is read "arcsine of x" or "the inverse sine of x."

The domains of the other basic trigonometric functions can also be restricted to produce a function with an inverse. The domains and ranges of the resulting inverse functions become parts of their definitions.

Definitions   Inverse Trigonometric Functions

Function	Domain	Range
$y = \cos^{-1} x$	$-1 \le x \le 1$	$0 \le y \le \pi$
$y = \sin^{-1} x$	$-1 \le x \le 1$	$-\dfrac{\pi}{2} \le y \le \dfrac{\pi}{2}$
$y = \tan^{-1} x$	$-\infty < x < \infty$	$-\dfrac{\pi}{2} < y < \dfrac{\pi}{2}$
$y = \sec^{-1} x$	$|x| \ge 1$	$0 \le y \le \pi, \quad y \neq \dfrac{\pi}{2}$
$y = \csc^{-1} x$	$|x| \ge 1$	$-\dfrac{\pi}{2} \le y \le \dfrac{\pi}{2}, \quad y \neq 0$
$y = \cot^{-1} x$	$-\infty < x < \infty$	$0 < y < \pi$

The graphs of the six inverse trigonometric functions are shown in Figure 50.

Domain: $-1 \leq x \leq 1$
Range: $0 \leq y \leq \pi$

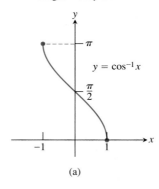

(a)

Domain: $-1 \leq x \leq 1$
Range: $-\dfrac{\pi}{2} \leq y \leq \dfrac{\pi}{2}$

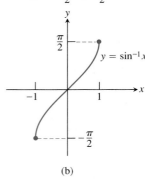

(b)

Domain: $-\infty < x < \infty$
Range: $-\dfrac{\pi}{2} < y < \dfrac{\pi}{2}$

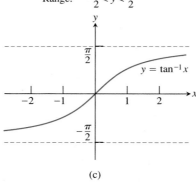

(c)

Domain: $x \leq -1$ or $x \geq 1$
Range: $0 \leq y \leq \pi, y \neq \dfrac{\pi}{2}$

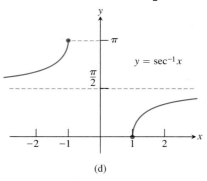

(d)

Domain: $x \leq -1$ or $x \geq 1$
Range: $-\dfrac{\pi}{2} \leq y \leq \dfrac{\pi}{2}, y \neq 0$

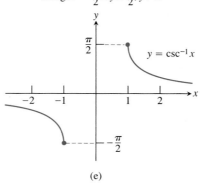

(e)

Domain: $-\infty < x < \infty$
Range: $0 < y < \pi$

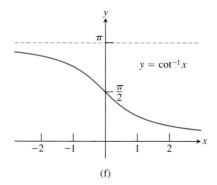

(f)

FIGURE 50 Graphs of (a) $y = \cos^{-1} x$, (b) $y = \sin^{-1} x$, (c) $y = \tan^{-1} x$, (d) $y = \sec^{-1} x$, (e) $y = \csc^{-1} x$, and (f) $y = \cot^{-1} x$.

The "Arc" in Arc Sine and Arc Cosine

The accompanying figure gives a geometric interpretation of $y = \sin^{-1} x$ and $y = \cos^{-1} x$ for radian angles in the first quadrant. For a unit circle, the equation $s = r\theta$ becomes $s = \theta$, so central angles and the arcs they subtend have the same measure. If $x = \sin y$, then, in addition to being the angle whose sine is x, y is also the length of arc on the unit circle that subtends an angle whose sine is x. So we call y "the arc whose sine is x."

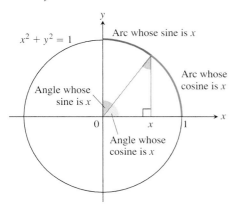

$x^2 + y^2 = 1$

Arc whose sine is x

Arc whose cosine is x

Angle whose sine is x

Angle whose cosine is x

The domain and ranges (where appropriate) of the inverse functions are chosen so that the functions will have the following relationships:

$$\sec^{-1} x = \cos^{-1}(1/x),$$
$$\csc^{-1} x = \sin^{-1}(1/x),$$
$$\cot^{-1} x = \pi/2 - \tan^{-1} x.$$

We use these relationships to find values of $\sec^{-1} x$, $\csc^{-1} x$, and $\cot^{-1} x$ on calculators that give only $\cos^{-1} x$, $\sin^{-1} x$, and $\tan^{-1} x$.

Example 5   Common Values of $Sin^{-1} x$

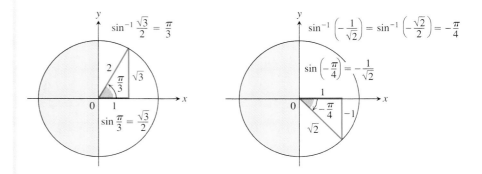

$\sin^{-1} \dfrac{\sqrt{3}}{2} = \dfrac{\pi}{3}$

$\sin \dfrac{\pi}{3} = \dfrac{\sqrt{3}}{2}$

$\sin^{-1}\left(-\dfrac{1}{\sqrt{2}}\right) = \sin^{-1}\left(-\dfrac{\sqrt{2}}{2}\right) = -\dfrac{\pi}{4}$

$\sin\left(-\dfrac{\pi}{4}\right) = -\dfrac{1}{\sqrt{2}}$

The angles come from the first and fourth quadrants because the range of $\sin^{-1} x$ is $[-\pi/2, \pi/2]$.

Example 6   Common Values of $Cos^{-1} x$

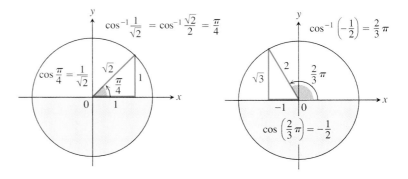

$\cos^{-1} \dfrac{1}{\sqrt{2}} = \cos^{-1} \dfrac{\sqrt{2}}{2} = \dfrac{\pi}{4}$

$\cos \dfrac{\pi}{4} = \dfrac{1}{\sqrt{2}}$

$\cos^{-1}\left(-\dfrac{1}{2}\right) = \dfrac{2}{3}\pi$

$\cos\left(\dfrac{2}{3}\pi\right) = -\dfrac{1}{2}$

The angles come from the first and second quadrants because the range of $\cos^{-1} x$ is $[0, \pi]$.

x	$\sin^{-1} x$
$\sqrt{3}/2$	$\pi/3$
$\sqrt{2}/2$	$\pi/4$
$1/2$	$\pi/6$
$-1/2$	$-\pi/6$
$-\sqrt{2}/2$	$-\pi/4$
$-\sqrt{3}/2$	$-\pi/3$

x	$\cos^{-1} x$
$\sqrt{3}/2$	$\pi/6$
$\sqrt{2}/2$	$\pi/4$
$1/2$	$\pi/3$
$-1/2$	$2\pi/3$
$-\sqrt{2}/2$	$3\pi/4$
$-\sqrt{3}/2$	$5\pi/6$

x	$\tan^{-1} x$
$\sqrt{3}$	$\pi/3$
1	$\pi/4$
$\sqrt{3}/3$	$\pi/6$
$-\sqrt{3}/3$	$-\pi/6$
1	$-\pi/4$
$-\sqrt{3}$	$-\pi/3$

Example 7   Common Values of Tan^{-1} x

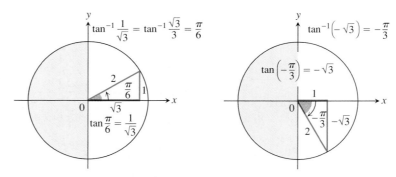

The angles come from the first and fourth quadrants because the range of $\tan^{-1} x$ is $(-\pi/2, \pi/2)$.

Example 8   Drift Correction

During an airplane flight from Chicago to St. Louis, the navigator determines that the plane is 12 mi off course, as shown in Figure 51. Find the angle a for a course parallel to the original, correct course, the angle b, and the correction angle $c = a + b$.

Solution

$$a = \sin^{-1} \frac{12}{180} \approx 0.067 \text{ radians} \approx 3.8°$$

$$b = \sin^{-1} \frac{12}{62} \approx 0.195 \text{ radians} \approx 11.2°$$

$$c = a + b \approx 15°.$$

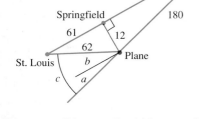

FIGURE 51 Diagram for drift correction (Example 8), with distances rounded to the nearest mile (drawing not to scale).

Identities Involving Arc Sine and Arc Cosine

The graph of $y = \sin^{-1} x$ is symmetric about the origin as shown in Figure 50b. The arc sine is therefore an odd function:

$$\sin^{-1} (-x) = -\sin^{-1} x. \tag{7}$$

The graph of $y = \cos^{-1} x$ has no such symmetry. Instead, we can see from Figure 52 that the arc cosine of x satisfies the identity

$$\cos^{-1} x + \cos^{-1} (-x) = \pi, \qquad (8)$$

or

$$\cos^{-1} (-x) = \pi - \cos^{-1} x. \qquad (9)$$

And we can see from the triangle in Figure 53 that for $x > 0$,

$$\sin^{-1} x + \cos^{-1} x = \pi/2. \qquad (10)$$

Equation (10) holds for the other values of x in $[-1, 1]$ as well.

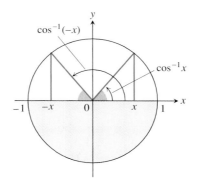

FIGURE 52 $\cos^{-1} x + \cos^{-1} (-x) = \pi$.

FIGURE 53 In this figure, $\sin^{-1} x + \cos^{-1} x = \pi/2$.

EXERCISES 5

Radians, Degrees, and Circular Arcs

1. On a circle of radius 10 m, how long is an arc that subtends a central angle of **(a)** $4\pi/5$ radians? **(b)** 110°?

2. A central angle in a circle of radius 8 is subtended by an arc of length 10π. Find the angle's radian and degree measures.

Evaluating Trigonometric Functions

3. Copy and complete the following table of function values. If the function is undefined at a given angle, enter "UND." Do not use a calculator or tables.

θ	$-\pi$	$-2\pi/3$	0	$\pi/2$	$3\pi/4$
$\sin \theta$					
$\cos \theta$					
$\tan \theta$					
$\cot \theta$					
$\sec \theta$					
$\csc \theta$					

4. Copy and complete the following table of function values. If the function is undefined at a given angle, enter "UND." Do not use a calculator or tables.

θ	$-3\pi/2$	$-\pi/3$	$-\pi/6$	$\pi/4$	$5\pi/6$
$\sin \theta$					
$\cos \theta$					
$\tan \theta$					
$\cot \theta$					
$\sec \theta$					
$\csc \theta$					

In Exercises 5 and 6, the value of one of $\sin x$, $\cos x$, and $\tan x$ is given. Find the values of the other two in the specified interval.

5. (a) $\sin x = \dfrac{3}{5}, \quad x \text{ in } \left[\dfrac{\pi}{2}, \pi\right]$

(b) $\cos x = \dfrac{1}{3}, \quad x \text{ in } \left[-\dfrac{\pi}{2}, 0\right]$

6. (a) $\tan x = \dfrac{1}{2}$, x in $\left[\pi, \dfrac{3\pi}{2}\right]$

(b) $\sin x = -\dfrac{1}{2}$, x in $\left[\pi, \dfrac{3\pi}{2}\right]$

Graphing Trigonometric Functions

Graph the functions in Exercises 7–10. What is the period of each function?

7. (a) $\sin 2x$ **(b)** $\cos \pi x$

8. (a) $-\sin \dfrac{\pi x}{3}$ **(b)** $-\cos 2\pi x$

9. (a) $\cos\left(x - \dfrac{\pi}{2}\right)$ **(b)** $\sin\left(x + \dfrac{\pi}{2}\right)$

10. (a) $\sin\left(x - \dfrac{\pi}{4}\right) + 1$ **(b)** $\cos\left(x + \dfrac{\pi}{4}\right) - 1$

Graph the functions in Exercises 11 and 12 in the ts-plane (t-axis horizontal, s-axis vertical). What is the period of each function? What symmetries do the graphs have?

11. $s = \cot 2t$ **12.** $s = \sec\left(\dfrac{\pi t}{2}\right)$

Using the Angle Sum Formulas

In Exercises 13 and 14, express the given quantity in terms of $\sin x$ and $\cos x$.

13. (a) $\cos(\pi + x)$ **(b)** $\sin(2\pi - x)$

14. (a) $\sin\left(\dfrac{3\pi}{2} - x\right)$ **(b)** $\cos\left(\dfrac{3\pi}{2} + x\right)$

Use the angle sum formulas to derive the identities in Exercises 15 and 16.

15. (a) $\cos\left(x - \dfrac{\pi}{2}\right) = \sin x$

(b) $\cos(A - B) = \cos A \cos B + \sin A \sin B$

16. (a) $\sin\left(x + \dfrac{\pi}{2}\right) = \cos x$

(b) $\sin(A - B) = \sin A \cos B - \cos A \sin B$

17. What happens if you take $B = A$ in the identity $\cos(A - B) = \cos A \cos B + \sin A \sin B$? Does the result agree with something you already know?

18. What happens if you take $B = 2\pi$ in the angle sum formulas? Do the results agree with something you already know?

General Sine Curves

Identify A, B, C, and D in Equation (2) for the sine functions in Exercises 19 and 20 and sketch their graphs.

19. (a) $y = 2\sin(x + \pi) - 1$

(b) $y = \dfrac{1}{2}\sin(\pi x - \pi) + \dfrac{1}{2}$

20. (a) $y = -\dfrac{2}{\pi}\sin\left(\dfrac{\pi}{-2}t\right) + \dfrac{1}{\pi}$

(b) $y = \dfrac{L}{2\pi}\sin\dfrac{2\pi t}{L}$, $L > 0$

21. *Temperature in Fairbanks, Alaska* Find the **(a)** amplitude, **(b)** period, **(c)** horizontal shift, and **(d)** vertical shift of the general sine function

$$f(x) = 37\sin\left(\dfrac{2\pi}{365}(x - 101)\right) + 25.$$

22. *Temperature in Fairbanks, Alaska* Use the equation in Exercise 21 to approximate the answers to the following questions about the temperature in Fairbanks, Alaska, shown in Figure 46. Assume that the year has 365 days.

(a) What are the highest and lowest mean daily temperatures shown?

(b) What is the average of the highest and lowest mean daily temperatures shown? Why is this average the vertical shift of the function?

Common Values of Inverse Trigonometric Functions

Use reference triangles like those in Examples 5–7 to find the angles in Exercises 23–26.

23. (a) $\tan^{-1} 1$ **(b)** $\tan^{-1}(-\sqrt{3})$ **(c)** $\tan^{-1}\left(\dfrac{1}{\sqrt{3}}\right)$

24. (a) $\sin^{-1}\left(\dfrac{-1}{2}\right)$ **(b)** $\sin^{-1}\left(\dfrac{1}{\sqrt{2}}\right)$ **(c)** $\sin^{-1}\left(\dfrac{-\sqrt{3}}{2}\right)$

25. (a) $\cos^{-1}\left(\dfrac{1}{2}\right)$ **(b)** $\cos^{-1}\left(\dfrac{-1}{\sqrt{2}}\right)$ **(c)** $\cos^{-1}\left(\dfrac{\sqrt{3}}{2}\right)$

26. (a) $\sec^{-1}(-\sqrt{2})$ **(b)** $\sec^{-1}\left(\dfrac{2}{\sqrt{3}}\right)$ **(c)** $\sec^{-1}(-2)$

Applications and Theory

27. You are sitting in a classroom next to the wall looking at the blackboard at the front of the room. The blackboard is 12 ft long and starts 3 ft from the wall you are sitting next to. Show that your viewing angle is

$$\alpha = \cot^{-1}\dfrac{x}{15} - \cot^{-1}\dfrac{x}{3}$$

if you are x ft from the front wall.

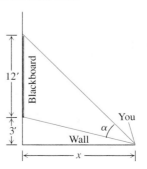

28. Find the angle α.

29. Apply the law of cosines to the triangle in the accompanying figure to derive a formula for $\cos (A - B)$.

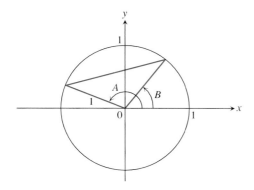

30. When applied to a figure similar to the one in Exercise 29, the law of cosines leads directly to the formula for $\cos (A + B)$. What is that figure, and how does the derivation go?

31. Here is an informal proof that $\tan^{-1} 1 + \tan^{-1} 2 + \tan^{-1} 3 = \pi$. Explain what is going on.

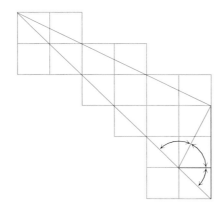

32. *Two derivations of the identity* $\sec^{-1} (-x) = \pi - \sec^{-1} x$

    (a) (*Geometric*) Here is a pictorial proof that $\sec^{-1} (-x) = \pi - \sec^{-1} x$. See if you can tell what is going on.

    (b) (*Algebraic*) Derive the identity $\sec^{-1} (-x) = \pi - \sec^{-1} x$ by combining the following two equations:

$$\cos^{-1} (-x) = \pi - \cos^{-1} x, \qquad \text{Eq. (9)}$$
$$\sec^{-1} x = \cos^{-1} (1/x).$$

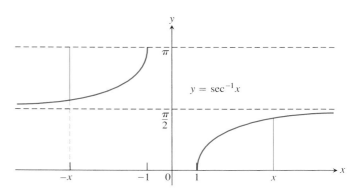

33. *The identity* $\sin^{-1} x + \cos^{-1} x = \pi/2$ Figure 53 establishes the identity for $0 < x < 1$. To establish it for the rest of $[-1, 1]$, verify by direct calculation that it holds for $x = 1, 0$, and -1. Then, for values of x in $(-1, 0)$, let $x = -a$, $a > 0$, and apply Equations (7) and (9) to the sum $\sin^{-1} (-a) + \cos^{-1} (-a)$.

34. Show that the sum $\tan^{-1} x + \tan^{-1} (1/x)$ is constant.

35. *The law of sines* The law of sines says that if a, b, and c are the sides opposite the angles A, B, and C in a triangle, then

$$\frac{\sin A}{a} = \frac{\sin B}{b} = \frac{\sin C}{c}.$$

Use the accompanying figures and the identity $\sin (\pi - \theta) = \sin \theta$, as required, to derive the law.

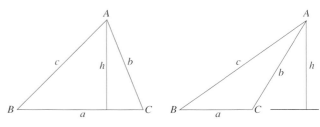

36. *The tangent sum formula* The standard formula for the tangent of the sum of two angles is

$$\tan (A + B) = \frac{\tan A + \tan B}{1 - \tan A \tan B}.$$

Derive the formula.

Solving Triangles and Comparing Functions

37. *Solving triangles*

    (a) A triangle has sides $a = 2$ and $b = 3$ and angle $C = 60°$. Find the length of side c.

    (b) A triangle has sides $a = 2$ and $b = 3$ and angle $C = 40°$. Find the length of side c.

38. *Solving triangles*

    (a) A triangle has sides $a = 2$ and $b = 3$ and angle $C = 60°$ (as in Exercise 37, part (a). Find the sine of angle B using the law of sines from Exercise 35.

(b) A triangle has side $c = 2$ and angles $A = \pi/4$ and $B = \pi/3$. Find the length a of the side opposite A.

T **39.** *The approximation* $\sin x \approx x$ It is often useful to know that, when x is measured in radians, $\sin x \approx x$ for numerically small values of x. In Section 3.6, we will see why the approximation holds. The approximation error is less than 1 in 500 if $|x| < 0.1$.

   (a) With your grapher in radian mode, graph $y = \sin x$ and $y = x$ together in a viewing window about the origin. What do you see happening as x nears the origin?

   (b) With your grapher in degree mode, graph $y = \sin x$ and $y = x$ together about the origin again. How is the picture different from the one obtained with radian mode?

   (c) *A quick radian mode check* Is your calculator in radian mode? Evaluate $\sin x$ at a value of x near the origin, say $x = 0.1$. If $\sin x \approx x$, the calculator is in radian mode; if not, it isn't. Try it.

T **40.** *Functions and their reciprocals*

   (a) Graph $y = \cos x$ and $y = \sec x$ together for $-3\pi/2 \le x \le 3\pi/2$. Comment on the behavior of $\sec x$ in relation to the signs and values of $\cos x$.

   (b) Graph $y = \sin x$ and $y = \csc x$ together for $-\pi \le x \le 2\pi$. Comment on the behavior of $\csc x$ in relation to the signs and values of $\sin x$.

T In Exercises 41 and 42, find the domain and range of each composite function. Then graph the composites on separate screens. Do the graphs make sense in each case? Give reasons for your answers. Comment on any differences you see.

41. **(a)** $y = \tan^{-1}(\tan x)$ **(b)** $y = \tan(\tan^{-1} x)$

42. **(a)** $y = \sin^{-1}(\sin x)$ **(b)** $y = \sin(\sin^{-1} x)$

In Exercises 43–46, solve the equation in the specified interval.

43. $\tan x = 2.5$, $\quad 0 \le x < 2\pi$

44. $\cos x = -0.7$, $\quad 2\pi \le x < 4\pi$

45. $\sec x = -3$, $\quad -\pi \le x < \pi$

46. $\sin x = -0.5$, $\quad -\infty < x < \infty$

T **47.** *Trigonometric identities* Let $f(x) = \sin x + \cos x$.

   (a) Graph $y = f(x)$. Describe the graph.

   (b) Use the graph to identify the amplitude, period, horizontal shift, and vertical shift.

   (c) Use the formula
$$\sin \alpha \cos \beta + \cos \alpha \sin \beta = \sin(\alpha + \beta)$$
   for the sine of the sum of two angles to confirm your answers.

T **48.** *Newton's serpentine* Graph Newton's serpentine, $y = 4x/(x^2 + 1)$. Then graph $y = 2 \sin(2 \tan^{-1} x)$ in the same graphing window. What do you see? Explain.

Sinusoidal Regression Analysis: Musical Notes and Temperature

See page 5 for an introduction to regression analysis with a calculator. A sinusoidal regression equation is a general sine curve; see Equation (2). Many calculators and computers produce these regression equations for given data sets.

T **49.** *Finding the frequency of a musical note* Musical notes are pressure waves in the air. The wave behavior can be modeled with great accuracy by general sine curves. Devices called Calculator Based Laboratory™ (CBL) systems can record these waves with a microphone. The data in Table 18 give pressure displacement versus time in seconds of a musical note produced by a tuning fork and recorded with a CBL system.

   (a) Find a sinusoidal regression equation (general sine curve) for the data and superimpose its graph on a scatter plot of the data.

   (b) The *frequency* of a musical note, or wave, is measured in cycles per second, or hertz (1 Hz = 1 cycle per second). The frequency is the reciprocal of the *period* of the wave, which is measured in seconds per cycle. Estimate the frequency of the note produced by the tuning fork.

Table 18 Tuning fork data

Time	Pressure	Time	Pressure
0.00091	−0.080	0.00362	0.217
0.00108	0.200	0.00379	0.480
0.00125	0.480	0.00398	0.681
0.00144	0.693	0.00416	0.810
0.00162	0.816	0.00435	0.827
0.00180	0.844	0.00453	0.749
0.00198	0.771	0.00471	0.581
0.00216	0.603	0.00489	0.346
0.00234	0.368	0.00507	0.077
0.00253	0.099	0.00525	−0.164
0.00271	−0.141	0.00543	−0.320
0.00289	−0.309	0.00562	−0.354
0.00307	−0.348	0.00579	−0.248
0.00325	−0.248	0.00598	−0.035
0.00344	−0.041		

T **50.** *Temperature data* Table 19 gives the average monthly temperatures for St. Louis for a 12-month period starting with January. Model the monthly temperature with an equation of the form
$$y = a \sin(b(t - h)) + k,$$
y in degrees Fahrenheit, t in months, as follows:

   (a) Find the value of b assuming that the period is 12 months.

   (b) How is the amplitude a related to the difference $80° - 30°$?

(c) Use the information in part (b) to find k.

(d) Find h and write an equation for y.

(e) Superimpose a graph of y on a scatter plot of the data.

Table 19 Temperature data for St. Louis

Time (months)	Temperature (°F)
1	34
2	30
3	39
4	44
5	58
6	67
7	78
8	80
9	72
10	63
11	51
12	40

T **51.** *Sinusoidal regression* Table 20 gives the values of the function

$$f(x) = a \sin (bx + c) + d$$

accurate to two decimals.

(a) Find a sinusoidal regression equation for the data.

(b) Rewrite the equation with a, b, c, and d rounded to the nearest integer.

Table 20 Values of a function

x	$f(x)$
1	3.42
2	0.73
3	0.12
4	2.16
5	4.97
6	5.97

T **52.** We suggest that students *work in groups of two or three*. A musical note like that produced with a tuning fork or pitch meter is a series of pressure waves. Table 21 gives frequencies (in hertz) of musical notes on the tempered scale. The pressure versus time tuning fork data in Table 22 were collected using a CBL system and a microphone.

(a) Find a sinusoidal regression equation for the data in Table 22 and superimpose its graph on a scatter plot of the data.

(b) Determine the frequency of and identify the musical note produced by the tuning fork.

Table 21 Frequencies of notes

Note	Frequency (Hz)
C	262
C♯ or D♭	277
D	294
D♯ or E♭	311
E	330
F	349
F♯ or G♭	370
G	392
G♯ or A♭	415
A	440
A♯ or B♭	466
B	494
C (next octave)	523

Source: CBL™ System Experimental Workbook, Texas Instruments, Inc., 1994.

Table 22 Tuning fork data

Time (s)	Pressure	Time (s)	Pressure
0.0002368	1.29021	0.0049024	−1.06632
0.0005664	1.50851	0.0051520	0.09235
0.0008256	1.51971	0.0054112	1.44694
0.0010752	1.51411	0.0056608	1.51411
0.0013344	1.47493	0.0059200	1.51971
0.0015840	0.45619	0.0061696	1.51411
0.0018432	−0.89280	0.0064288	1.43015
0.0020928	−1.51412	0.0066784	0.19871
0.0023520	−1.15588	0.0069408	−1.06072
0.0026016	−0.04758	0.0071904	−1.51412
0.0028640	1.36858	0.0074496	−0.97116
0.0031136	1.50851	0.0076992	0.23229
0.0033728	1.51971	0.0079584	1.46933
0.0036224	1.51411	0.0082080	1.51411
0.0038816	1.45813	0.0084672	1.51971
0.0041312	0.32185	0.0087168	1.50851
0.0043904	−0.97676	0.0089792	1.36298
0.0046400	−1.51971		

6   Parametric Equations

Parametrizations of Plane Curves • Lines and Other Curves •
Parametrizing Inverse Functions • An Application

When the path of a particle moving in the plane looks like the curve in Figure 54, we cannot describe it by an equation of the form $y = f(x)$ because there are vertical lines that intersect the curve more than once (see Exercise 25 in Section 2). Likewise, we cannot describe the curve by expressing x directly as a function of y. In this section, you learn another way to describe curves in terms of a third variable called a *parameter*. This powerful approach can also be used to describe ordinary function curves, like the ones we have been studying so far, and their inverses when they exist.

Parametrizations of Plane Curves

When the path of a particle moving in the plane looks like the curve in Figure 54, we express each of the particle's coordinates as a function of a third variable t and describe the path with a pair of equations, $x = f(t)$ and $y = g(t)$. For studying motion, t usually denotes time. Equations like these are better than a Cartesian formula because they tell us the particle's position $(x, y) = (f(t), g(t))$ at any time t.

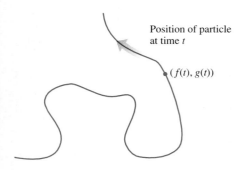

Position of particle at time t

$(f(t), g(t))$

FIGURE 54 The path traced by a particle moving in the xy-plane is not always the graph of a function of x or a function of y.

Example 1   Moving Along a Parabola

The position $P(x, y)$ of a particle moving in the xy-plane is given by the equations and parameter interval

$$x = \sqrt{t}, \qquad y = t, \qquad t \geq 0.$$

Identify the path traced by the particle and describe the motion.

Solution   We try to identify the path by eliminating t between the equations $x = \sqrt{t}$ and $y = t$. With any luck, this will produce a recognizable algebraic relation between x and y. We find that

$$y = t = (\sqrt{t})^2 = x^2.$$

Thus, the particle's position coordinates satisfy the equation $y = x^2$, so the particle moves along the parabola $y = x^2$.

It would be a mistake, however, to conclude that the particle's path is the entire parabola $y = x^2$; it is only half the parabola. The particle's x-coordinate is never negative. The particle starts at $(0, 0)$ when $t = 0$ and rises into the first quadrant as t increases (Figure 55).

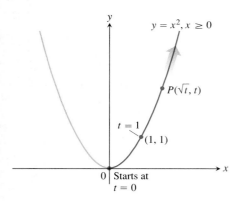

y

$y = x^2, x \geq 0$

$P(\sqrt{t}, t)$

$t = 1$

$(1, 1)$

x

0 Starts at
$t = 0$

FIGURE 55 The equations $x = \sqrt{t}$ and $y = t$ and the interval $t \geq 0$ describe the motion of a particle that traces the right-hand half of the parabola $y = x^2$. (Example 1)

Definitions   Parametric Curve, Parametric Equations

If x and y are given as functions

$$x = f(t), \qquad y = g(t)$$

over an interval of t-values, then the set of points $(x, y) = (f(t), g(t))$ defined by these equations is a **parametric curve**. The equations are **parametric equations** for the curve.

The variable t is a **parameter** for the curve, and its domain I is the **parameter interval.** If I is a closed interval, $a \le t \le b$, the point $(f(a), g(a))$ is the **initial point** of the curve. The point $(f(b), g(b))$ is the **terminal point.** When we give parametric equations and a parameter interval for a curve, we say that we have **parametrized** the curve. The equations and interval together constitute a **parametrization** of the curve.

In Example 1, the parameter interval is $[0, \infty)$, so $(0, 0)$ is the initial point. There is no terminal point.

A grapher can draw a parametrized curve only over a closed interval, so the portion it draws has endpoints even when the curve being graphed does not. Keep this in mind when you graph on a graphing calculator or computer.

Example 2   Moving Counterclockwise on a Circle

Graph the parametric curves

(a) $x = \cos t$, $y = \sin t$, $0 \le t \le 2\pi$

(b) $x = a \cos t$, $y = a \sin t$, $0 \le t \le 2\pi$.

Solution

(a) Since $x^2 + y^2 = \cos^2 t + \sin^2 t = 1$, the parametric curve lies along the unit circle $x^2 + y^2 = 1$. As t increases from 0 to 2π, the point $(x, y) = (\cos t, \sin t)$ starts at $(1, 0)$ and traces the entire circle once counterclockwise (Figure 56).

(b) For $x = a \cos t$, $y = a \sin t$, $0 \le t \le 2\pi$, we have $x^2 + y^2 = a^2 \cos^2 t + a^2 \sin^2 t = a^2$. The parametrization describes a motion that begins at the point $(a, 0)$ and traverses the circle $x^2 + y^2 = a^2$ once counterclockwise, returning to $(a, 0)$ at $t = 2\pi$.

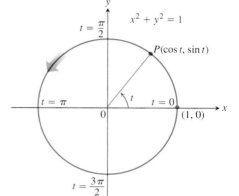

FIGURE 56 The equations $x = \cos t$ and $y = \sin t$ describe motion on the circle $x^2 + y^2 = 1$. The arrow shows the direction of increasing t. (Example 2)

Example 3   Traversing a Semicircle Clockwise

Graph the parametric curve

$$x = \cos t, \qquad y = -\sin t, \qquad 0 \le t \le \pi.$$

Find a Cartesian equation for a curve that contains the parametric curve. What portion of the graph of the Cartesian equation is traced by the parametric curve? Describe the motion.

Solution   The point $(x, y) = (\cos t, -\sin t)$ moves on the circle $x^2 + y^2 = 1$. In contrast to Example 2, the motion is now clockwise. As t increases from 0 toward π, y becomes negative and x decreases. The point (x, y) moves around the bottom half of the circle, first falling to $(0, -1)$ and then rising to $(-1, 0)$. The motion stops at $t = \pi$, leaving only the bottom half of the circle covered (Figure 57).

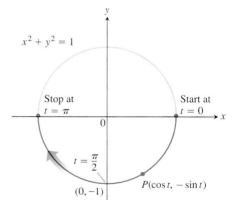

FIGURE 57 The point $P(\cos t, -\sin t)$ moves clockwise as t increases from 0 to π. (Example 3)

Lines and Other Curves

Many other curves, including lines and line segments, can be defined parametrically.

$x = 3t, y = 2-2t$

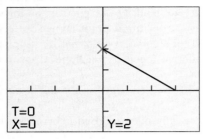

[–4, 4] by [–2, 4]

FIGURE 58 The graph of the line segment $x = 3t$, $y = 2 - 2t$, $0 \le t \le 1$, with trace on the initial point $(0, 2)$. (Example 4)

Example 4   Moving Along a Straight Line

Graph and identify the parametric curve

$$x = 3t, \qquad y = 2 - 2t, \qquad 0 \le t \le 1.$$

What happens if the restriction on t is removed?

Solution   When $t = 0$, the equations give $x = 0$ and $y = 2$. When $t = 1$, they give $x = 3$ and $y = 0$. If we substitute $t = x/3$ into the y equation, we obtain

$$y = 2 - 2\left(\frac{x}{3}\right) = -\frac{2}{3}x + 2.$$

Thus, the parametric curve traces the segment of the line $y = -(2/3)x + 2$ from $(0, 2)$ to $(3, 0)$ (Figure 58).

If we remove the restriction on t, changing the parameter interval from $[0, 1]$ to $(-\infty, \infty)$, the parametrization traces the entire line $y = -(2/3)x + 2$.

Example 5   Parametrizing a Line Segment

Find a parametrization for the line segment with endpoints $(-2, 1)$ and $(3, 5)$.

Solution   Using $(-2, 1)$ we create the parametric equations

$$x = -2 + at, \qquad y = 1 + bt.$$

These represent a line, as we can see by solving each equation for t and equating to obtain

$$\frac{x + 2}{a} = \frac{y - 1}{b}.$$

This line goes through the point $(-2, 1)$ when $t = 0$. We determine a and b so that the line goes through $(3, 5)$ when $t = 1$.

$$3 = -2 + a \implies a = 5 \qquad \text{\scriptsize $x = 3$ when $t = 1$.}$$
$$5 = 1 + b \implies b = 4 \qquad \text{\scriptsize $y = 5$ when $t = 1$.}$$

Therefore,

$$x = -2 + 5t, \qquad y = 1 + 4t, \qquad 0 \le t \le 1$$

is a parametrization of the line segment with initial point $(-2, 1)$ and terminal point $(3, 5)$.

Example 6   Moving Along the Ellipse $x^2/a^2 + y^2/b^2 = 1$

Describe the motion of a particle whose position $P(x, y)$ at time t is given by

$$x = a \cos t, \qquad y = b \sin t, \qquad 0 \le t \le 2\pi.$$

Solution   We find a Cartesian equation for the particle's coordinates by eliminating t between the equations

$$\cos t = \frac{x}{a}, \qquad \sin t = \frac{y}{b}.$$

$\dfrac{x^2}{a^2} + \dfrac{y^2}{b^2} = 1$

FIGURE 59 The ellipse in Example 6, drawn for $a > b > 0$. The coordinates of P are $x = a \cos t$, $y = b \sin t$.

The identity $\cos^2 t + \sin^2 t = 1$, yields

$$\left(\frac{x}{a}\right)^2 + \left(\frac{y}{b}\right)^2 = 1, \qquad \text{or} \qquad \frac{x^2}{a^2} + \frac{y^2}{b^2} = 1.$$

The particle's coordinates (x, y) satisfy the equation $(x^2/a^2) + (y^2/b^2) = 1$, so the particle moves along this ellipse. When $t = 0$, the particle's coordinates are

$$x = a \cos(0) = a, \qquad y = b \sin(0) = 0,$$

so the motion starts at $(a, 0)$. As t increases, the particle rises and moves toward the left, moving counterclockwise. It traverses the ellipse once, returning to its starting position $(a, 0)$ at $t = 2\pi$ (Figure 59).

Parametrizing Inverse Functions

We can graph or represent any function $y = f(x)$ parametrically as

$$x = t \qquad \text{and} \qquad y = f(t).$$

Interchanging t and $f(t)$ produces parametric equations for the inverse:

$$x = f(t) \qquad \text{and} \qquad y = t$$

(see Section 4).

For example, to graph the one-to-one function $f(x) = x^2$, $x \geq 0$, on a grapher together with its inverse and the line $y = x$, $x \geq 0$, use the parametric graphing option with

$$\begin{aligned} \text{Graph of } f: & \quad x_1 = t, & y_1 = t^2, & \quad t \geq 0 \\ \text{Graph of } f^{-1}: & \quad x_2 = t^2, & y_2 = t & \\ \text{Graph of } y = x: & \quad x_3 = t, & y_3 = t & \end{aligned}$$

Figure 60 shows the three graphs.

[−1.5, 3] by [−1, 2]

FIGURE 60 Parametric graphs of the function $f(x) = x^2$, $x \geq 0$, its inverse, and the line $y = x$.

An Application

Example 7   Dropping Emergency Supplies

A Red Cross aircraft is dropping emergency food and medical supplies into a disaster area. If the aircraft releases the supplies immediately above the edge of an open field 700 ft long and if the cargo moves along the path

$$x = 120t \qquad \text{and} \qquad y = -16t^2 + 500, \qquad t \geq 0$$

does the cargo land in the field? The coordinates x and y are measured in feet, and the parameter t (time since release) in seconds. Find a Cartesian equation for the path of the falling cargo (Figure 61).

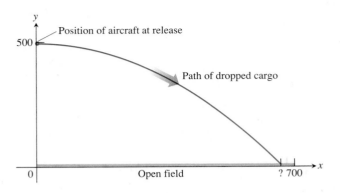

FIGURE 61 The path of the dropped cargo of supplies in Example 7.

Solution   The cargo hits the ground when $y = 0$, which occurs at time t when

$$-16t^2 + 500 = 0 \qquad \text{Set } y = 0.$$

$$t^2 = \frac{500}{16} \qquad \text{Solve for } t.$$

$$t = \frac{5\sqrt{5}}{2} \text{ sec}. \qquad t \geq 0$$

The x-coordinate at the time of the release is $x = 0$. At the time the cargo hits the ground, the x-coordinate is

$$x = 120t = 120\left(\frac{5\sqrt{5}}{2}\right) = 300\sqrt{5} \text{ ft}.$$

Since $300\sqrt{5} \approx 670.8 < 700$, the cargo does land in the field.

We find a Cartesian equation for the cargo's coordinates by eliminating t between the parametric equations:

$$y = -16t^2 + 500 \qquad \text{Parametric equation for } y$$

$$= -16\left(\frac{x}{120}\right)^2 + 500 \qquad \text{Substitute for } t \text{ from the equation } x = 120t.$$

$$= -\frac{16}{14400}x^2 + 500 \qquad \text{Simplify.}$$

or

$$y = -\frac{1}{900}x^2 + 500.$$

Thus, the cargo moves along the parabola

$$y = -\frac{x^2}{900} + 500.$$

Standard Parametrizations

CIRCLE $x^2 + y^2 = a^2$:   ELLIPSE $\dfrac{x^2}{a^2} + \dfrac{y^2}{b^2} = 1$:

$$x = a \cos t \qquad\qquad\qquad x = a \cos t$$
$$y = a \sin t \qquad\qquad\qquad y = b \sin t$$
$$0 \le t \le 2\pi \qquad\qquad\qquad 0 \le t \le 2\pi$$

FUNCTION $y = f(x)$:   INVERSE OF $y = f(x)$:

$$x = t \qquad\qquad\qquad\qquad x = f(t)$$
$$y = f(t) \qquad\qquad\qquad\qquad y = t$$

EXERCISES 6

Finding Cartesian Equations from Parametric Equations

Exercises 1–18 give parametric equations and parameter intervals for the motion of a particle in the xy-plane. Identify the particle's path by finding a Cartesian equation for it. Graph the Cartesian equation. (The graphs will vary with the equation used.) Indicate the portion of the graph traced by the particle and the direction of motion.

1. $x = \cos t, \qquad y = \sin t, \qquad 0 \le t \le \pi$

2. $x = \cos 2t, \qquad y = \sin 2t, \qquad 0 \le t \le \pi$

3. $x = \sin (2\pi t), \qquad y = \cos (2\pi t), \qquad 0 \le t \le 1$

4. $x = \cos (\pi - t), \qquad y = \sin (\pi - t), \qquad 0 \le t \le \pi$

5. $x = 4 \cos t, \qquad y = 2 \sin t, \qquad 0 \le t \le 2\pi$

6. $x = 4 \sin t, \qquad y = 5 \cos t, \qquad 0 \le t \le 2\pi$

7. $x = 3t, \qquad y = 9t^2, \qquad -\infty < t < \infty$

8. $x = -\sqrt{t}, \qquad y = t, \qquad t \ge 0$

9. $x = t, \qquad y = \sqrt{t}, \qquad t \ge 0$

10. $x = \sec^2 t - 1, \qquad y = \tan t, \qquad -\pi/2 < t < \pi/2$

11. $x = -\sec t, \qquad y = \tan t, \qquad -\pi/2 < t < \pi/2$

12. $x = 2t - 5, \qquad y = 4t - 7, \qquad -\infty < t < \infty$

13. $x = 1 - t, \qquad y = 1 + t, \qquad -\infty < t < \infty$

14. $x = 3 - 3t, \qquad y = 2t, \qquad 0 \le t \le 1$

15. $x = t, \qquad y = \sqrt{1 - t^2}, \qquad -1 \le t \le 0$

16. $x = \sqrt{t + 1}, \qquad y = \sqrt{t}, \qquad t \ge 0$

17. $x = e^t + e^{-t}, \qquad y = e^t - e^{-t}, \qquad -\infty < t < \infty$

18. $x = \cos (e^t), \qquad y = 2 \sin (e^t), \qquad -\infty < t < \infty$

Determining Parametric Equations

19. Find parametric equations and a parameter interval for the motion of a particle that starts at $(a, 0)$ and traces the circle $x^2 + y^2 = a^2$

   (a) once clockwise.   **(b)** once counterclockwise.

   (c) twice clockwise.   **(d)** twice counterclockwise.

   (There are many ways to do these, so your answers may not be the same as the ones in the back of the book.)

20. Find parametric equations and a parameter interval for the motion of a particle that starts at $(a, 0)$ and traces the ellipse $(x^2/a^2) + (y^2/b^2) = 1$

   (a) once clockwise.   **(b)** once counterclockwise.

   (c) twice clockwise.   **(d)** twice counterclockwise.

   (As in Exercise 19, there are many correct answers.)

In Exercises 21–26, find a parametrization for the curve.

21. the line segment with endpoints $(-1, -3)$ and $(4, 1)$

22. the line segment with endpoints $(-1, 3)$ and $(3, -2)$

23. the lower half of the parabola $x - 1 = y^2$

24. the left half of the parabola $y = x^2 + 2x$

25. the ray (half line) with initial point $(2, 3)$ that passes through the point $(-1, -1)$

26. the ray (half line) with initial point $(-1, 2)$ that passes through the point $(0, 0)$

Parametric Graphing

T In Exercises 27–30, match the parametric equations with their graph. State the approximate dimensions of the viewing window. Give a parameter interval that traces the curve exactly once.

27. $x = 3 \sin (2t)$, $\quad y = 1.5 \cos t$

28. $x = \sin^3 t$, $\quad y = \cos^3 t$

29. $x = 7 \sin t - \sin (7t)$, $\quad y = 7 \cos t - \cos (7t)$

30. $x = 12 \sin t - 3 \sin (6t)$, $\quad y = 12 \cos t + 3 \cos (6t)$

(a)

(b)

(c)

(d)

T In Exercises 31–38, use parametric graphing to graph f, f^{-1}, and $y = x$.

31. $f(x) = e^x$ **32.** $f(x) = 3^x$

33. $f(x) = 2^{-x}$ **34.** $f(x) = 3^{-x}$

35. $f(x) = \ln x$ **36.** $f(x) = \log x$

37. $f(x) = \sin^{-1} x$ **38.** $f(x) = \tan^{-1} x$

In Exercises 39–42, refer to the graph of

$$x = 3 - |t|, \qquad y = t - 1, \qquad -5 \le t \le 5,$$

shown in the accompanying figure. *Work in groups of two or three* to find the values of t that produce the graph in the given quadrant.

39. Quadrant I   **40.** Quadrant II

41. Quadrant III   **42.** Quadrant IV

[−6, 6] by [−8, 8]

T In Exercises 43–48, graph the equations over the given intervals.

43. *Ellipse* $x = 4 \cos t$, $y = 2 \sin t$, over

  (a) $0 \le t \le 2\pi$ **(b)** $0 \le t \le \pi$

  (c) $-\pi/2 \le t \le \pi/2$.

44. *Hyperbola branch* $x = \sec t$ (enter as $1/\cos (t)$), $y = \tan t$ (enter as $\sin (t)/\cos (t)$), over

  (a) $-1.5 \le t \le 1.5$ **(b)** $-0.5 \le t \le 0.5$

  (c) $-0.1 \le t \le 0.1$.

45. *Parabola* $x = 2t + 3$, $y = t^2 - 1$, $-2 \le t \le 2$

46. *A nice curve (a deltoid)*

  $x = 2 \cos t + \cos 2t$, $\quad y = 2 \sin t - \sin 2t$, $\quad 0 \le t \le 2\pi$

  What happens if you replace 2 with -2 in the equations for x and y? Graph the new equations to find out.

47. *An even nicer curve*

  $x = 3 \cos t + \cos 3t$, $\quad y = 3 \sin t - \sin 3t$, $\quad 0 \le t \le 2\pi$

  What happens if you replace 3 with -3 in the equations for x and y? Graph the new equations to find out.

48. *Cycloid* $x = t - \sin t$, $y = 1 - \cos t$, over

  (a) $0 \le t \le 2\pi$ **(b)** $0 \le t \le 4\pi$

  (c) $\pi \le t \le 3\pi$.

Extending the Ideas

49. *The witch of Agnesi* The bell-shaped witch of Agnesi can be constructed as follows. Start with the circle of radius 1, centered at the point $(0, 1)$, as shown in the accompanying figure.

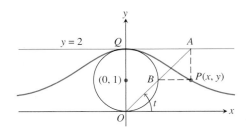

  Choose a point A on the line $y = 2$, and connect it to the origin with a line segment. Call the point where the segment crosses the circle B. Let P be the point where the vertical line through A crosses the horizontal line through B. The witch is the curve traced by P as A moves along the line $y = 2$.

  Find a parametrization for the witch by expressing the coordinates of P in terms of t, the radian measure of the angle that segment OA makes with the positive x-axis. The following equalities (which you may assume) will help:

  (i) $x = AQ$.

  (ii) $y = 2 - AB \sin t$.

  (iii) $AB \cdot AO = (AQ)^2$.

50. *Parametrizing lines and segments*

  (a) Show that the equations and parameter interval

  $$x = x_0 + (x_1 - x_0)t, \qquad y = y_0 + (y_1 - y_0)t, \qquad -\infty < t < \infty,$$

describe the line through the points (x_0, y_0) and (x_1, y_1) (Figure 62).

(b) Using the same parameter interval, write parametric equations for the line through a point (x_1, y_1) and the origin.

(c) Using the same parameter interval, write parametric equations for the line through $(-1, 0)$ and $(0, 1)$.

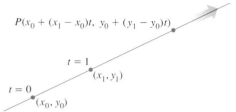

$$P(x_0 + (x_1 - x_0)t, \; y_0 + (y_1 - y_0)t)$$

$t = 1$

(x_1, y_1)

$t = 0$

(x_0, y_0)

FIGURE 62   The line in Exercise 50(a). The arrow shows the direction of increasing t.

 51. *Graphing the witch of Agnesi* The witch of Agnesi is the curve

$$x = 2 \cot t, \qquad y = 2 \sin^2 t, \qquad 0 < t < \pi.$$

(a) Draw the curve using the window in the accompanying figure. What did you choose as a closed parameter interval for your grapher? In what direction is the curve traced? How far to the left and right of the origin do you think the curve extends?

$x = 2 \cot t, \; y = 2 \sin^2 t$

$[-5, 5]$ by $[-2, 4]$

(b) Graph the same parametric equations using the parameter intervals $(-\pi/2, \pi/2)$, $(0, \pi/2)$, and $(\pi/2, \pi)$. In each case, describe the curve you see and the direction in which it is traced by your grapher.

(c) What happens if you replace $x = 2 \cot t$ by $x = -2 \cot t$ in the original parametrization? What happens if you use $x = 2 \cot (\pi - t)$?

 52. *Hyperboloids* Let $x = a \sec t$ and $y = b \tan t$.

(a) *Writing to Learn* Let $a = 1, 2,$ or 3, and $b = 1, 2,$ or 3, and graph using the parameter interval $(-\pi/2, \pi/2)$. Explain what you see and describe the role of a and b in these parametric equations. (Caution: If you get what appear to be asymptotes, try using the approximation $[-1.57, 1.57]$ for the parameter interval.)

(b) Let $a = 2$ and $b = 3$, and graph in the parameter interval $(\pi/2, 3\pi/2)$. Explain what you see.

(c) *Writing to Learn* Let $a = 2$ and $b = 3$, and graph using the parameter interval $(-\pi/2, 3\pi/2)$. Explain why you must be careful about graphing in this interval or any interval that contains $\pm \pi/2$.

(d) Use algebra to explain why

$$\left(\frac{x}{a}\right)^2 - \left(\frac{y}{b}\right)^2 = 1.$$

(e) Let $x = a \tan t$ and $y = b \sec t$. Repeat parts (a), (b), and (d) using an appropriate version of part (d).

---

7   Modeling Change

Mathematical Models   •   Simplification   •   Verifying a Model   •
A Model Construction Process   •   Empirical Modeling: Capturing the
Trend of Collected Data   •   Using Calculus in Modeling

CD-ROM
WEBsite

To help us better understand our world, we often describe a particular phenomenon mathematically (by means of a function or an equation, for instance). Such a **mathematical model** is an idealization of the real-world phenomenon and never a completely accurate representation. Although any model has its limitations, a good one can provide valuable results and conclusions. In this section, we look at the modeling process and several illustrative examples.

Mathematical Models

In modeling our world, we are often interested in predicting the value of a variable at some time in the future. Perhaps it is a population, a real estate value, or the number of people with a communicative disease. Often, a mathematical model can help us understand a behavior better or aid us in planning for the future. Let's think of a mathematical model as a mathematical construct designed to study a particular real-world system or behavior of interest. The model allows us to reach mathematical conclusions about the behavior, as illustrated in Figure 63. These conclusions can be interpreted to help a decision maker plan for the future.

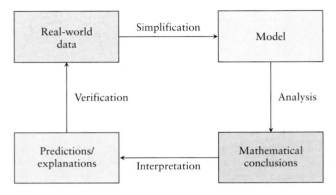

FIGURE 63 A flow of the modeling process beginning with an examination of real-world data.

Simplification

Most models simplify reality. Generally, models can only *approximate* real-world behavior. One very powerful simplifying relationship is **proportionality**.

Definition   Proportionality

Two variables y and x are **proportional** (to one another) if one is always a constant multiple of the other; that is, if

$$y = kx$$

for some nonzero constant k.

The definition means that the graph of y versus x lies along a straight line through the origin. This graphical observation is useful in testing whether a given data collection reasonably assumes a proportionality relationship. If a proportionality is reasonable, a plot of one variable against the other should approximate a straight line through the origin. Here is an example.

Example 1   Testing for Proportionality in Driver Reaction Distance

During a panic stop, the driver of a car must react to an emergency, apply the brakes, and bring the car to a stop. What is a *safe following distance* for motorists? To answer this question, it would be helpful to know how far a vehicle travels at a given speed before the brakes are applied (*driver reaction distance*). The U.S. Bureau of Public Roads collected data for reaction and braking distances for a large number of motorists. (*Braking distance* is how far a vehicle travels after the brakes are applied until the vehicle stops.) In Table 23, x is the speed of an automobile in miles per hour (mph) and y is the distance in feet (ft) the automobile travels before the brakes are applied.

Table 23 Driver reaction distance

x (mph)	20	25	30	35	40	45	50	55	60	65	70	75	80
y (ft)	22	28	33	39	44	50	55	61	66	72	77	83	88

For a collection of data representing many motorists, we might assume the time for a typical driver to react to an emergency to be approximately constant (independent of the speed). Then, the distance traveled while reacting is proportional to the speed. Let's test this proportionality assumption by plotting the distance traveled versus the speed. Figure 64 shows that the plotted points lie reasonably well along a straight line passing through the origin. Thus, the proportionality assumption appears valid.

We can even estimate the constant of proportionality from the graph. Using the first and last data points we find that a line approximating the data has a slope $rise/run = (88 - 22)/(80 - 20) = 1.1$. The proportionality model predicts driver reaction distance to be

$$y = 1.1x. \tag{1}$$

FIGURE 64 A plot of reaction distance versus speed.

Verifying a Model

We can test to see how well Equation (1) fits the data by superimposing its graph on the scatter plot. Another procedure is to examine the errors or *residuals* (Table 24):

$$\text{Residuals} = \text{observations} - \text{predictions}.$$

Table 24 Computing the residuals

Speed (mph) x	Observation (ft)	Prediction (ft) $y = 1.1x$	Residuals (ft)
20	22	22.0	0.0
25	28	27.5	0.5
30	33	33.0	0.0
35	39	38.5	0.5
40	44	44.0	0.0
45	50	49.5	0.5
50	55	55.0	0.0
55	61	60.5	0.5
60	66	66.0	0.0
65	72	71.5	0.5
70	77	77.0	0.0
75	83	82.5	0.5
80	88	88.0	0.0

The residuals in Table 26 are relatively small (the largest is 0.5 ft, compared with a range of 22 to 88 ft) and do not have a bothersome pattern. Notice that at 60 mph, or 88 ft/sec, the average driver travels 66 ft while reacting. A typical driver requires (66 ft)/(88 ft/sec) = 0.75 sec to stop. This reaction time seems reasonable for a large and varying population of drivers. Given the imprecise nature of the problem we are modeling, we would probably accept this simple model as suitable for predicting reaction distance. In the exercises, we ask you to analyze *braking distance* before suggesting a *safe following distance* and devising a simple rule for motorists to follow in estimating a safe distance.

A powerful technique for judging a model's adequacy, and for gaining insight on how to improve the model, is to plot the residuals versus the independent variable. Then observe the relative magnitude of the errors. A pattern indicates that the model could be improved by capturing and incorporating the trend of this pattern.

A Model Construction Process

In learning to build models, the following process has been found to be useful. In constructing Equation (1), we roughly completed the following steps:

Steps for Model Construction

Step 1. *Identify the problem.* To estimate a safe following distance, we first decided to estimate the driver reaction distance.

Step 2. *Make assumptions on which variables to include and the relationships among the included variables.* Reaction distance depends on many factors, including speed, visibility, weather, and the driver's age. For simplicity, we assumed that reaction distance depends only on speed. We further assumed that reaction distance is proportional to the speed.

Step 3. *Find a function or graph that satisfies the relationships.* We tested the proportionality assumption by determining if the plot of reaction distance versus speed lies approximately along a line through the origin. Since it does, we could calculate the slope, which is the constant of proportionality.

Step 4. *Verify the model.* Analyze the residuals for size and pattern.

Empirical Modeling: Capturing the Trend of Collected Data

In Example 1, we hypothesized a relationship between a dependent variable and an independent variable. Another method for constructing models is to collect data and find a model that captures the trend of the data. This empirical approach has both advantages and disadvantages.

Example 2   Finding a Curve to Predict Population Levels

We may want to predict the future size of a population, such as the number of trout or catfish living in a fish farm. Figure 65 shows a scatter plot of the data collected by R. Pearl for a collection of yeast cells (measured as **biomass**) growing over time (measured in hours) in a nutrient.

| Time (h) | Biomass |
x	y
0	9.6
1	18.3
2	29
3	47.2
4	71.1
5	119.1
6	174.6
7	257.3

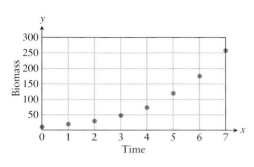

(Data from R. Pearl, "The Growth of Population," *Quart. Rev. Biol.*, Vol. 2 (1927), pp. 532–548.)

FIGURE 65 Biomass of a yeast culture versus elapsed time.

The plot of points appears to be reasonably smooth with an upward curving trend. We might attempt to capture this trend by fitting a polynomial (for example, a quadratic $y = ax^2 + bx + c$), a power curve ($y = ax^b$), or an exponential curve ($y = ae^{bx}$). Figure 66 shows the result of using a calculator to fit a quadratic model.

The quadratic model $y = 6.10x^2 - 9.28x + 16.43$ appears to fit the collected data reasonably well (Figure 66b). Using this model, we predict the population after 17 hours as $y(17) = 1622.65$. Let us examine more of Pearl's data to see if our quadratic model continues to be a good one.

(a)

(b)

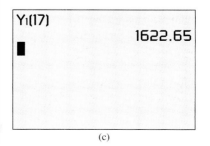

(c)

FIGURE 66 Using a calculator to (a) fit a quadratic; (b) overlay the data, model, and residuals; and (c) predict $y(17)$.

In Figure 67, we display all Pearl's data. Now you see that the prediction of $y(17) = 1622.65$ grossly overestimates the observed population of 659.6. Why did the quadratic model fail to predict a more accurate value?

Time (h)	Observed	Predicted
x	y	y
0	9.6	16.4
1	18.3	13.3
2	29.0	22.3
3	47.2	43.5
4	71.1	77.0
5	119.1	122.6
6	174.6	180.5
7	257.3	250.6
8	350.7	332.8
9	441.0	427.3
10	513.3	534.0
11	559.7	652.9
12	594.8	784.0
13	629.4	927.3
14	640.8	1082.9
15	651.1	1250.6
16	655.9	1430.5
17	659.6	1622.7
18	661.8	1827.0

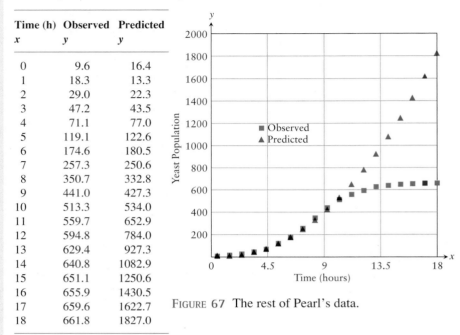

FIGURE 67 The rest of Pearl's data.

The problem lies in the danger of predicting beyond the range of data used to build the empirical model. (The range of data creating our model was $0 \le x \le 7$.) Such *extrapolation* is especially dangerous when the model selected is not supported by some underlying rationale suggesting the form of the model. In our yeast example, why would we expect a quadratic function as underlying population growth? Why not an exponential function? In the face of this, how then do we predict future values? Often, calculus can help.

Using Calculus in Modeling

The application of calculus involves the study of *change*. The origins of calculus lie in our curiosity about, and need to develop, a deeper understanding of motion. The search for laws governing planetary motion, the study of the pendulum and its application to clock making, and the laws governing the flight of a cannonball were the kinds of problems stimulating the minds of mathematicians and scientists in the sixteenth and seventeenth centuries. In many cases, we observe how change is taking place and hypothesize relationships among the variables, much as we did in Example 1. In Chapter 6, we model population growth with calculus. In the case of the growth of a yeast culture, you will see that the food source available to the yeast constrains its growth. That is, the environment can sustain only a limited population. As the population approaches this limiting value (called the *carrying capacity*), growth slows. The model underlying the growth of a yeast culture for Pearl's data will turn out to be the *logistic* function

$$P = \frac{665}{1 + 73.8e^{-0.55t}}. \tag{2}$$

The graph of Equation (2) superimposed on a scatter plot of Pearl's data is displayed in Figure 68. You will see how Equation (2) comes about in Chapter 6.

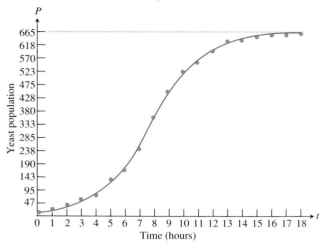

FIGURE 68 The logistic curve obtained from Equation (2) superimposed on a scatter plot of Pearl's observed data in Figure 67.

EXERCISES 7

1. *Constants of proportionality* Determine whether the following data support the stated proportionality assumption. If the assumption seems reasonable, estimate the constant of proportionality.

   (a) y is proportional to x

y	1	2	3	4	5	6	7	8
x	5.9	12.1	17.9	23.9	29.9	36.2	41.8	48.2

   (b) y is proportional to $x^{1/2}$

y	3.5	5	6	7	8
x	3	6	9	12	15

   (c) y is proportional to 3^x

y	5	15	45	135	405	1215	3645	10,935
x	0	1	2	3	4	5	6	7

   (d) y is proportional to $\ln x$

y	2	4.8	5.3	6.5	8.0	10.5	14.4	15.0
x	2.0	5.0	6.0	9.0	14.0	35.0	120.0	150.0

Constructing Models

2. *Spring elongation* The response of a spring to various loads must be modeled to design a vehicle such as a tank, dump truck, utility vehicle, or a luxury car that responds to road conditions in a desired way. We conducted an experiment to measure the stretch y of a spring in inches as a function of the number x of units of mass placed on the spring.

x (number of units of mass)	0	1	2	3	4	5
y (elongation in inches)	0	0.875	1.721	2.641	3.531	4.391

x (number of units of mass)	6	7	8	9	10
y (elongation in inches)	5.241	6.120	6.992	7.869	8.741

   (a) Build a model relating the elongation of a spring to the number of units of mass.

   (b) How well does your model fit the data?

   (b) Predict the elongation of the spring for 13 units of mass. How comfortable are you with this prediction?

3. *Braking distance* How far does an automobile travel once the brakes have been applied? Consider the data below, where x is the speed of an automobile in miles per hour and y is the distance in feet required to stop the automobile once the brakes have been applied.

x (mph)	20	25	30	35	40	45	50	55	60	65	70	75
y (ft)	32	47	65	87	112	140	171	204	241	282	325	376

   Build and test a model relating braking distance to speed.

4. *Safe following distance* Use Equation (1) for reaction distance and the model you built in Exercise 3 for braking distance to build a model for total stopping distance (reaction plus braking distances). A rule often given for safe following distance is to allow 2 seconds between your car and the car in front of you. Is this rule consistent with your model for total stopping distance? If not, suggest a better rule.

5. *Heart disease* Digoxin is used in the treatment of heart disease. Doctors must prescribe an amount that keeps the concentration of digoxin in the bloodstream above an *effective level* without exceeding a *safe level*. Begin by considering the rate of decay of digoxin in the bloodstream. Suppose that an initial dose of 0.5 mg is in the bloodstream. In the following table, x represents the number of days after taking the initial dose and y represents the amount of digoxin remaining in the bloodstream for a particular patient.

x	0	1	2	3	4	5	6	7	8
y	0.5000	0.345	0.238	0.164	0.113	0.078	0.054	0.037	0.026

(a) Build a model relating the amount of digoxin in the bloodstream to the number of days elapsed.

(b) How well does your model fit the data?

(c) Predict the amount of digoxin in the bloodstream after 12 days.

6. *Radioactivity* A radioactive dye is injected into a patient's veins to facilitate an X-ray procedure. Measuring the radioactivity in counts per minute (cpm) over the course of several minutes yielded the following table of values.

x time (min)	0	1	2	3	4	5
y radioactivity (cpm)	10,023	8174	6693	5500	4489	3683

x time (min)	6	7	8	9	10
y radioactivity (cpm)	3061	2479	2045	1645	1326

(a) Build a mathematical model relating radioactivity level to time elapsed.

(b) Compare the observations with the predictions.

(c) Use your model to predict when the radioactivity level will be below 500 cpm.

7. *Drug levels* As time passes, the blood concentration of a drug administered to laboratory animals decreases. The concentrations in parts per million (ppm) appear in the table below.

Concentration (ppm)	853	587	390	274	189	130
Time (days)	0	1	2	3	4	5

Concentration (ppm)	97	67	50	40	31
Time (days)	6	7	8	9	10

(a) Build a mathematical model relating the concentration level to the time elapsed.

(b) Compare the observations with the predictions.

(c) Use your model to predict when the concentration level will be below 10 parts per million (ppm).

8. *Ponderosa pines* In the table, x represents the girth (distance around) of a pine tree measured in inches (in.) at shoulder height; y represents the board feet (bf) of lumber finally obtained.

x (in.)	17	19	20	23	25	28	32	38	39	41
y (bf)	19	25	32	57	71	113	123	252	259	294

Formulate and test the following two models: that usable board feet is proportional to (a) the square of the girth and (b) the cube of the girth. Which is better? Does one model provide a better "explanation" than the other?

9. *Black bass* The following data represent the weight w in ounces of New York black bass for various lengths l in inches.

l (in.)	12.50	12.63	12.63	14.13	14.5	14.5	17.25	17.75
w (oz)	17	16	17	23	26	27	41	49

Build and test a model that assumes that weight is proportional to l^3. How well does the model fit?

10. *Heart rate of mammals* The following data relate the weight in grams (g) of some mammals to their heart rate in beats per minute (bpm). Plot the data. Is there a trend? If so, find a function that captures the trend of the data. (*Hint:* Try models of the form $y = x^{-(1/n)}$ for n an integer.)

Mammal	Body weight x (g)	Pulse rate y (bpm)
Vesperugo pipistrellus (a really small bat)	4	660
Mouse	25	670
Rat	200	420
Guinea pig	300	300
Rabbit	2,000	205
Little dog	5,000	120
Big dog	30,000	85
Sheep	50,000	70
Human	70,000	72
Horse	450,000	38
Ox	500,000	40
Elephant	3,000,000	48

Relating Graphs to Behaviors

For Exercises 11–14, select the graph (or suggest your own) that best describes the behavior qualitatively. Explain your choice. Possible answers are (a) line (b) concave up increasing (c) concave up decreasing (d) concave down increasing (e) concave down decreasing (f) logistic

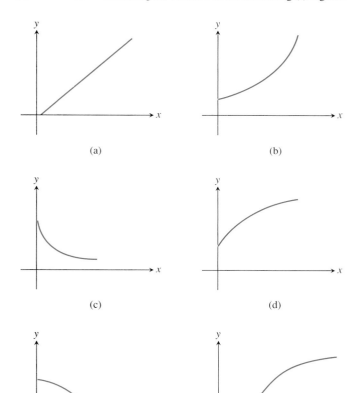

(a) (b)

(c) (d)

(e) (f)

11. The concentration of a prescription drug in the bloodstream versus time

12. Your proficiency in a subject versus time spent studying it

13. The amount of carbon-14 remaining in a work of art versus time

14. Water is draining from an outlet at the bottom of a tank. Relate

   (a) the velocity at which water is discharged with time elapsed.

   (b) the depth of the water in the tank to the time elapsed.

15. Suggest a behavior that is qualitatively described by each of the graphs above.

16. Layers of plastic tinting are used to reduce the intensity of light. Relate the intensity of light transmitted with the number of layers of tinting.

In Exercises 17–20, sketch a graph that qualitatively describes each of the behaviors.

17. A golf ball is dropped from a height of 6 feet onto a concrete walkway. Relate the ball's height to time elapsed.

18. A marble is dropped into a bucket of oil. Relate

   (a) the marble's speed with time.

   (b) distance fallen with time.

19. A skydiver jumps from an airplane. After 4 seconds of free fall, the chute opens. Relate

   (a) the skydiver's velocity with time.

   (b) distance fallen with time.

20. An animal preserve is capable of sustaining a population level of 500 deer. Relate the deer population to time if the preserve is initially stocked with

   (a) 300 deer.

   (b) 500 deer.

   (c) 600 deer.

21. Describe a behavior that is represented qualitatively by each of the following graphs.

(a)

(b)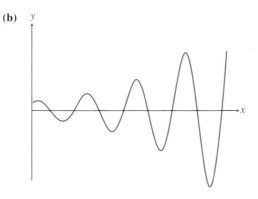

22. *Identifying a problem* From each of the following vaguely stated scenarios, identify a problem you would like to study. What

variables affect the behavior you have identified? Which variables are the most important?

(a) The population growth of a single species.

(b) An object is dropped from a great height. When and how hard will it hit the ground?

(c) How fast can a skier ski down a mountain slope?

(d) A physicist is interested in studying properties of light. She wants to understand the path of a ray of light as it travels

through the air into a smooth lake, particularly at the interface of the two different media.

(e) The U.S. Food and Drug Administration is interested in knowing whether a new drug is effective in the control of a certain disease in the population.

(f) A retail store intends to construct a new parking lot. How should the lot be illuminated?

Questions to Guide Your Review

1. How can you write an equation for a line if you know the coordinates of two points on the line? The line's slope and the coordinates of one point on the line? The line's slope and y-intercept? Give examples.

2. What are the standard equations for lines perpendicular to the coordinate axes?

3. How are the slopes of *mutually* perpendicular lines related? What about parallel lines? Give examples.

4. What is a function? Give examples. How do you graph a real valued function whose domain and range are real numbers? What is an increasing function? A decreasing function?

5. What is an even function? An odd function? What symmetry properties do the graphs of such functions have? What advantage can we make of this? Give an example of a function that is neither even nor odd.

6. What is a piecewise defined function? Give examples. Define the absolute value function and draw its graph.

7. When is it possible to compose one function with another? Give examples of composites and their values at various points. Does the order in which functions are composed ever matter?

8. How do you change the equation $y = f(x)$ to shift its graph up or down? To the left or right? Give examples.

9. What is an exponential function? Give examples. What laws of exponents does it obey? How does it differ from a simple power function like $f(x) = x^n$? What kind of real-world phenomena are modeled by exponential functions?

10. What is the number e, and how is it defined? What are the domain and range of $f(x) = e^x$? What does its graph look like? How do the values of e^x relate to x^2, x^3, and so on?

11. What functions have inverses? How do you know if two functions f and g are inverses of one another? Give examples of functions that are (are not) inverses of one another.

12. How are the domains, ranges, and graphs of functions and their inverses related? Give an example.

13. What procedure can you sometimes use to express the inverse of a function of x as a function of x? How can you graph a function

$y = f(x)$ together with its inverse $y = f^{-1}(x)$ parametrically on a graphing calculator or computer?

14. What is a logarithm function? What properties does it satisfy? What is the natural logarithm function? What are the domain and range of $y = \ln x$? What does its graph look like?

15. How is the graph of $\log_a x$ related to the graph of $\ln x$? What truth is there in the statement that there is really only one exponential function and one logarithmic function?

16. What is radian measure? How do you convert from radians to degrees? Degrees to radians?

17. Graph the six basic trigonometric functions. What symmetries do the graphs have?

18. How can you sometimes find values of trigonometric functions from triangles? Give examples.

19. What is a periodic function? Give examples. What are the periods of the six basic trigonometric functions?

20. How does the formula for the general sine function $f(x) = A \sin((2\pi/B)(x - C)) + D$ relate to the shifting, stretching, shrinking and reflection of its graph? Give examples. Graph the general sine curve and identify the constants A, B, C, and D.

21. Starting with the identity $\cos^2\theta + \sin^2\theta = 1$ and the formulas for $\cos(A + B)$ and $\sin(A + B)$, show how a variety of other trigonometric identities may be derived.

22. How are the inverse trigonometric functions defined? How can you sometimes use right triangles to find values of these functions? Give examples.

23. How can you find values of $\sec^{-1} x$, $\csc^{-1} x$, and $\cot^{-1} x$ using a calculator's keys for $\cos^{-1} x$, $\sin^{-1} x$, and $\tan^{-1} x$?

24. What is a parametrized curve in the xy plane? What is the initial point of the curve? The terminal point? If you find a Cartesian equation for the path of a particle whose motion in the plane is described parametrically, what kind of match can you expect between the Cartesian equation's graph and the path of motion? Give examples.

25. What is the standard parametrization for the circle $x^2 + y^2 = a^2$? For the ellipse $(x^2/a^2) + (y^2/b^2) = 1$? The graph of a function $y = f(x)$? The inverse of a function $y = f(x)$?

Practice Exercises

Lines

In Exercises 1–12, write an equation for the specified line.

1. through $(1, -6)$ with slope 3

2. through $(-1, 2)$ with slope $-1/2$

3. the vertical line through $(0, -3)$

4. through $(-3, 6)$ and $(1, -2)$

5. the horizontal line through $(0, 2)$

6. through $(3, 3)$ and $(-2, 5)$

7. with slope -3 and y-intercept 3

8. through $(3, 1)$ and parallel to $2x - y = -2$

9. through $(4, -12)$ and parallel to $4x + 3y = 12$

10. through $(-2, -3)$ and perpendicular to $3x - 5y = 1$

11. through $(-1, 2)$ and perpendicular to $(1/2)x + (1/3)y = 1$

12. with x-intercept 3 and y-intercept -5

Functions and Graphs

13. Express the area and circumference of a circle as functions of the circle's radius. Then express the area as a function of the circumference.

14. Express the radius of a sphere as a function of the sphere's surface area. Then express the surface area as a function of the volume.

15. A point P in the first quadrant lies on the parabola $y = x^2$. Express the coordinates of P as functions of the angle of inclination of the line joining P to the origin.

16. A hot-air balloon rising straight up from a level field is tracked by a range finder located 500 ft from the point of liftoff. Express the balloon's height as a function of the angle the line from the range finder to the balloon makes with the ground.

In Exercises 17–20, determine whether the graph of the function is symmetric about the y-axis, the origin, or neither.

17. $y = x^{1/5}$

18. $y = x^{2/5}$

19. $y = x^2 - 2x - 1$

20. $y = e^{-x^2}$

In Exercises 21–28, determine whether the function is even, odd, or neither.

21. $y = x^2 + 1$

22. $y = x^5 - x^3 - x$

23. $y = 1 - \cos x$

24. $y = \sec x \tan x$

25. $y = \dfrac{x^4 + 1}{x^3 - 2x}$

26. $y = 1 - \sin x$

27. $y = x + \cos x$

28. $y = \sqrt{x^4 - 1}$

In Exercises 29–38, find the (**a**) domain and (**b**) range.

29. $y = |x| - 2$

30. $y = -2 + \sqrt{1 - x}$

31. $y = \sqrt{16 - x^2}$

32. $y = 3^{2-x} + 1$

33. $y = 2e^{-x} - 3$

34. $y = \tan(2x - \pi)$

35. $y = 2\sin(3x + \pi) - 1$

36. $y = x^{2/5}$

37. $y = \ln(x - 3) + 1$

38. $y = -1 + \sqrt[3]{2 - x}$

Piecewise-Defined Functions

In Exercises 39 and 40, find the (**a**) domain and (**b**) range.

39. $y = \begin{cases} \sqrt{-x}, & -4 \le x \le 0 \\ \sqrt{x}, & 0 < x \le 4 \end{cases}$

40. $y = \begin{cases} -x - 2, & -2 \le x \le -1 \\ x, & -1 < x \le 1 \\ -x + 2, & 1 < x \le 2 \end{cases}$

In Exercises 41 and 42, write a piecewise formula for the function.

41.

42.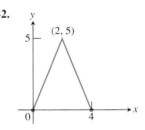

Composition of Functions

In Exercises 43 and 44, find

(**a**) $(f \circ g)(-1)$

(**b**) $(g \circ f)(2)$

(**c**) $(f \circ f)(x)$

(**d**) $(g \circ g)(x)$

43. $f(x) = \dfrac{1}{x}, \qquad g(x) = \dfrac{1}{\sqrt{x + 2}}$

44. $f(x) = 2 - x, \qquad g(x) = \sqrt[3]{x + 1}$

In Exercises 45 and 46, (**a**) write a formula for $f \circ g$ and $g \circ f$ and find the (**b**) domain and (**c**) range of each.

45. $f(x) = 2 - x^2, \qquad g(x) = \sqrt{x + 2}$

46. $f(x) = \sqrt{x}, \qquad g(x) = \sqrt{1 - x}$

Composition with absolute values In Exercises 47–52, graph f_1 and f_2 together. Then describe how applying the absolute value function before applying f_1 affects the graph.

$f_1(x)$	$f_2(x) = f_1(|x|)$
47. x	$|x|$
48. x^3	$|x|^3$
49. x^2	$|x|^2$
50. $\dfrac{1}{x}$	$\dfrac{1}{|x|}$
51. $\sqrt{x}$	$\sqrt{|x|}$
52. $\sin x$	$\sin|x|$

Composition with absolute values In Exercises 53–56, graph g_1 and g_2 together. Then describe how taking absolute values after applying g_1 affects the graph.

$g_1(x)$	$g_2(x) = |g_1(x)|$
53. x^3	$|x^3|$
54. $\sqrt{x}$	$|\sqrt{x}|$
55. $4 - x^2$	$|4 - x^2|$
56. $x^2 + x$	$|x^2 + x|$

Inverse Functions

57. (a) Graph the function $f(x) = \sqrt{1 - x^2}$, $0 \le x \le 1$. What symmetry does the graph have?

(b) Show that f is its own inverse. (Remember that $\sqrt{x^2} = x$ if $x \ge 0$.)

58. (a) Graph the function $f(x) = 1/x$. What symmetry does the graph have?

(b) Show that f is its own inverse.

In Exercises 59 and 60,

    (a) find f^{-1} and show that $(f \circ f^{-1})(x) = (f^{-1} \circ f)(x) = x$.

    (b) graph f and f^{-1} together.

59. $f(x) = 2 - 3x$

60. $f(x) = (x + 2)^2$, $x \ge -2$

61. (a) Show that $f(x) = x^3$ and $g(x) = \sqrt[3]{x}$ are inverses of one another.

T **(b)** Graph f and g over an x-interval large enough to show the graphs intersecting at $(1, 1)$ and $(-1, -1)$. Be sure the picture shows the required symmetry in the line $y = x$.

62. (a) Show that $h(x) = x^3/4$ and $k(x) = (4x)^{1/3}$ are inverses of one another.

T **(b)** Graph h and k over an x-interval large enough to show the graphs intersecting at $(2, 2)$ and $(-2, -2)$. Be sure the picture shows the required symmetry about the line $y = x$.

63. (a) Find the inverse of $f(x) = x + 1$. Graph f and its inverse together. Add the line $y = x$ to your sketch, drawing it with dashes or dots for contrast.

(b) Find the inverse of $f(x) = x + b$ (b constant). How is the graph of f^{-1} related to the graph of f?

(c) What can you conclude about the inverses of functions whose graphs are lines parallel to the line $y = x$?

64. (a) Find the inverse of $f(x) = -x + 1$. Graph the line $y = -x + 1$ together with the line $y = x$. At what angle do the lines intersect?

(b) Find the inverse of $f(x) = -x + b$ (b constant). What angle does the line $y = -x + b$ make with the line $y = x$?

(c) What can you conclude about the inverses of functions whose graphs are lines perpendicular to the line $y = x$?

T **65.** *A decimal representation of e* Find e to as many decimal places as your calculator allows by solving the equation $\ln x = 1$.

T **66.** *The inverse relation between e^x and $\ln x$* Find out how good your calculator is at evaluating the composites

$$e^{\ln x} \quad \text{and} \quad \ln(e^x).$$

Algebraic Calculations with the Exponential and Logarithm

Find simpler expressions for the quantities in Exercises 67–70.

67. (a) $e^{\ln 7.2}$ **(b)** $e^{-\ln x^2}$ **(c)** $e^{\ln x - \ln y}$

68. (a) $e^{\ln(x^2 + y^2)}$ **(b)** $e^{-\ln 0.3}$ **(c)** $e^{\ln \pi x - \ln 2}$

69. (a) $2 \ln \sqrt{e}$ **(b)** $\ln(\ln e^e)$ **(c)** $\ln(e^{-x^2 - y^2})$

70. (a) $\ln(e^{\sec \theta})$ **(b)** $\ln(e^{e^x})$ **(c)** $\ln(e^{2 \ln x})$

Trigonometry

In Exercises 71 and 72, find the measure of the angle in radians and degrees.

71. $\sin^{-1}(0.6)$ **72.** $\tan^{-1}(-2.3)$

73. Find the six trigonometric values of $\theta = \cos^{-1}(3/7)$. Give exact answers.

74. Solve the equation $\sin x = -0.2$ in the following intervals.

    (a) $0 \le x < 2\pi$ **(b)** $-\infty < x < \infty$

In Exercises 75 and 76, sketch the graph of the given function. What is the period of the function?

75. $y = \sin \dfrac{x}{2}$

76. $y = \cos \dfrac{\pi x}{2}$

77. Sketch the graph $y = 2 \cos \left(x - \dfrac{\pi}{3} \right)$.

78. Sketch the graph $y = 1 + \sin \left(x + \dfrac{\pi}{4} \right)$.

In Exercises 79–82, ABC is a right triangle with the right angle at C. The sides opposite angles A, B, and C are a, b, and c, respectively.

79. (a) Find a and b if $c = 2$, $B = \pi/3$.

    (b) Find a and c if $b = 2$, $B = \pi/3$.

80. (a) Express a in terms of A and c.

    (b) Express a in terms of A and b.

81. (a) Express a in terms of B and b.

    (b) Express c in terms of A and a.

82. (a) Express $\sin A$ in terms of a and c.

    (b) Express $\sin A$ in terms of b and c.

In Exercises 83 and 84, show that the function is periodic and find its period.

83. $y = \sin^3 x$ **84.** $y = |\tan x|$

Use the angle sum formulas to derive the identities in Exercises 85 and 86.

85. $\cos \left(x + \dfrac{\pi}{2} \right) = -\sin x$

86. $\sin \left(x - \dfrac{\pi}{2} \right) = -\cos x$

87. Evaluate $\sin \dfrac{7\pi}{12}$ as $\sin \left(\dfrac{\pi}{4} + \dfrac{\pi}{3} \right)$.

88. Evaluate $\cos \dfrac{11\pi}{12}$ as $\cos \left(\dfrac{\pi}{4} + \dfrac{2\pi}{3} \right)$.

Use reference triangles to find the angles in Exercises 89–92.

89. (a) $\sin^{-1} \left(\dfrac{1}{2} \right)$ **(b)** $\sin^{-1} \left(\dfrac{-1}{\sqrt{2}} \right)$ **(c)** $\sin^{-1} \left(\dfrac{\sqrt{3}}{2} \right)$

90. (a) $\cos^{-1} \left(\dfrac{-1}{2} \right)$ **(b)** $\cos^{-1} \left(\dfrac{1}{\sqrt{2}} \right)$ **(c)** $\cos^{-1} \left(\dfrac{-\sqrt{3}}{2} \right)$

91. (a) $\sec^{-1} \sqrt{2}$ **(b)** $\sec^{-1} \left(\dfrac{-2}{\sqrt{3}} \right)$ **(c)** $\sec^{-1} 2$

92. (a) $\cot^{-1} 1$ **(b)** $\cot^{-1}(-\sqrt{3})$ **(c)** $\cot^{-1} \left(\dfrac{1}{\sqrt{3}} \right)$

Evaluating Trigonometric and Inverse Trigonometric Functions

Find the values in Exercises 93–96.

93. $\sec \left(\cos^{-1} \dfrac{1}{2} \right)$

94. $\cot \left(\sin^{-1} \left(-\dfrac{\sqrt{3}}{2} \right) \right)$

95. $\tan (\sec^{-1} 1) + \sin (\csc^{-1} (-2))$

96. $\sec (\tan^{-1} 1 + \csc^{-1} 1)$

Evaluating Trigonometric Expressions

Evaluate the expressions in Exercises 97–100.

97. $\sec (\tan^{-1} 2x)$

98. $\tan \left(\sec^{-1} \dfrac{y}{5} \right)$

99. $\tan (\cos^{-1} x)$

100. $\sin \left(\tan^{-1} \dfrac{x}{\sqrt{x^2 + 1}} \right)$

Which of the expressions in Exercises 101–104 are defined and which are not? Give reasons for your answers.

101. (a) $\tan^{-1} 2$ **(b)** $\cos^{-1} 2$

102. (a) $\csc^{-1} \dfrac{1}{2}$ **(b)** $\csc^{-1} 2$

103. (a) $\sec^{-1} 0$ **(b)** $\sin^{-1} \sqrt{2}$

104. (a) $\cot^{-1} \left(\dfrac{1}{-2} \right)$ **(b)** $\cos^{-1} (-5)$

105. *Height of a pole* Two wires stretch from the top T of a vertical pole to points B and C on the ground, where C is 10 m closer to the base of the pole ,than is B. If wire BT makes an angle of $35°$ with the horizontal and wire CT makes an angle of $50°$ with the horizontal, how high is the pole?

106. *Height of a weather balloon* Observers at positions A and B 2 km apart simultaneously measure the angle of elevation of a weather balloon to be $40°$ and $70°$, respectively. If the balloon is directly above a point on the line segment between A and B, find the height of the balloon.

T 107. (a) Graph the function $f(x) = \sin x + \cos (x/2)$.

    (b) What appears to be the period of this function?

    (c) Confirm your finding in part (b) algebraically.

T 108. (a) Graph $f(x) = \sin (1/x)$.

    (b) What are the domain and range of f?

    (c) Is f periodic? Give reasons for your answer.

Parametrizations

In Exercises 109–112, a parametrization is given for a curve.

    (a) Find a Cartesian equation for a curve that contains the parametrized curve. What portion of the graph of the Cartesian equation is traced by the parametrized curve?

    (b) Graph the curve. Identify the initial and terminal points, if any. Indicate the direction in which the curve is traced.

109. $x = 5 \cos t$, $y = 2 \sin t$, $0 \le t \le 2\pi$

110. $x = 4 \cos t$, $y = 4 \sin t$, $\pi/2 \le t < 3\pi/2$

111. $x = 2 - t$, $y = 11 - 2t$, $-2 \le t \le 4$

112. $x = 1 + t$, $y = (t - 1)^2$, $t \le 1$

In Exercises 113–116, give a parametrization for the curve.

113. the line segment with endpoints $(-2, 5)$ and $(4, 3)$

114. the line through $(-3, -2)$ and $(4, -1)$

115. the ray with initial point $(2, 5)$ that passes through $(-1, 0)$

116. $y = x(x - 4)$, $x \le 2$

Additional Exercises: Theory, Examples, Applications

1. The graph of f is shown. Draw the graph of each function.

    (a) $y = f(-x)$ **(b)** $y = -f(x)$

    (c) $y = -2f(x + 1) + 1$ **(d)** $y = 3f(x - 2) - 2$

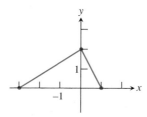

2. A portion of the graph of a function defined on $[-3, 3]$ is shown. Complete the graph assuming that the function is

    (a) even.   **(b)** odd.

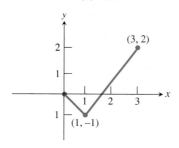

3. *Depreciation* Smith Hauling purchased an 18-wheel truck for $100,000. The truck depreciates at the constant rate of $10,000 per year for 10 years.

    (a) Write an expression that gives the value y after x years.

    (b) When is the value of the truck $55,000?

4. *Drug absorption* A drug is administered intravenously for pain. The function

$$f(t) = 90 - 52 \ln (1 + t), \qquad 0 \le t \le 4$$

gives the number of units of the drug remaining in the body after t hours.

    (a) What was the initial number of units of the drug administered?

    (b) How much is present after 2 hours?

    (c) Draw the graph of f.

5. *Finding time* If Juanita invests $1500 in a retirement account that earns 8% compounded annually, how long will it take this single payment to grow to $5000?

6. Explain the following "proof without words" of the law of cosines. (Source: Sidney H. Kung, "Proof without Words: The Law of Cosines," *Mathematics Magazine*, Vol. 63, No. 5, Dec. 1990, p. 342.)

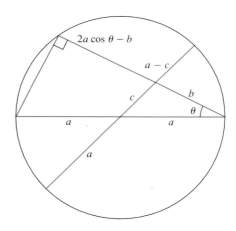

7. Show that the area of triangle ABC is given by $(1/2)ab \sin C = (1/2)bc \sin A = (1/2)ca \sin B$.

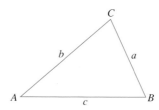

8. (a) Find the slope of the line from the origin to the midpoint P of side AB in the triangle in the accompanying figure $(a, b > 0)$.

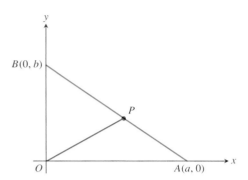

(b) When is OP perpendicular to AB?

9. The figure here shows an informal proof that

$$\tan^{-1} \frac{1}{2} + \tan^{-1} \frac{1}{3} = \frac{\pi}{4}.$$

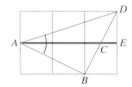

How does the argument go? (Source: Edward M. Harris, "Behold! Sums of Arctan," *College Mathematics Journal*, Vol. 18, No. 2, March 1987, p. 141.)

10. *Writing to Learn* For what $x > 0$, does $x^{(x^x)} = (x^x)^x$? Give reasons for your answer.

11. *Composition with an odd function*

(a) Let $h = g \circ f$ where g is an even function. Is h always an even function? Give reasons for your answer.

(b) Let $h = g \circ f$ where g is an odd function. Is h always an odd function? What if f is odd? What if f is even? Give reasons for your answer.

12. *The rule of 70* If you use the approximation $\ln 2 \approx 0.70$ (in place of $0.69314\ldots$), you can derive a rule of thumb that says, "To estimate how many years it will take an amount of money to double when invested at r percent compounded continuously, divide r into 70." For instance, an amount of money invested at 5% will double in about $70/5 = 14$ years. If you want it to double in 10 years instead, you have to invest it at $70/10 = 7\%$. Show how the rule of 70 is derived. (A similar "rule of 72" uses 72 instead of 70, because 72 has more integer factors.)

Functions and Graphs

13. Are there two functions f and g such that $f \circ g = g \circ f$? Give reasons for your answer.

14. Are there two functions f and g with the following property? The graphs of f and g are not straight lines but the graph of $f \circ g$ is a straight line. Give reasons for your answer.

15. If $f(x)$ is odd, can anything be said of $g(x) = f(x) - 2$? What if f is even instead? Give reasons for your answer.

16. If $g(x)$ is an odd function defined for all values of x, can anything be said about $g(0)$? Give reasons for your answer.

17. Graph the equation $|x| + |y| = 1 + x$.

18. Graph the equation $y + |y| = x + |x|$.

19. Show that if f is both even and odd, then $f(x) = 0$ for every x in the domain of f.

20. (a) *Even–odd decompositions* Let f be a function whose domain is symmetric about the origin; that is, $-x$ belongs to the domain whenever x does. Show that f is the sum of an even function and an odd function:

$$f(x) = E(x) + O(x),$$

where E is an even function and O is an odd function. (*Hint:* Let $E(x) = [f(x) + f(-x)]/2$. Show that $E(-x) = E(x)$, so that E is even. Then show that $O(x) = f(x) - E(x)$ is odd.)

(b) *Uniqueness* Show that there is only one way to write f as the sum of an even and an odd function. (*Hint:* One way is given in part (a). If also $f(x) = E_1(x) + O_1(x)$, where E_1 is even and O_1 is odd, show that $E - E_1 = O_1 - O$. Then use Exercise 19 to show that $E = E_1$ and $O = O_1$.)

21. *One-to-one functions* If f is a one-to-one function, prove that $g(x) = -f(x)$ is also one-to-one.

22. *One-to-one functions* If f is a one-to-one function and $f(x)$ is never zero, prove that $g(x) = 1/f(x)$ is also one-to-one.

23. *Domain and range* Suppose that $a \neq 0$, $b \neq 1$, and $b > 0$. Determine the domain and range of the function.

(a) $y = a(b^{c-x}) + d$ **(b)** $y = a \log_b (x - c) + d$

24. *Inverse functions* Let

$$f(x) = \frac{ax + b}{cx + d}, \qquad c \neq 0, \qquad ad - bc \neq 0.$$

(a) *Writing to Learn* Give a convincing argument that f is one-to-one.

(b) Find a formula for the inverse of f.

(c) Find the horizontal and vertical asymptotes of f.

(d) Find the horizontal and vertical asymptotes of f^{-1}. How are they related to those of f?

Modeling

25. *Constants of proportionality* Determine whether the following data support the stated proportionality assumption. If the assumption seems reasonable, estimate the constant of proportionality.

(a) y is proportional to x^2

y	6	13	24	39	58	81	108	139
x	0	1	2	3	4	5	6	7

(b) y is proportional to 4^x

y	0.6	2.4	9.6	38.4	153.6	614.4	2457.6	9830.4
x	0	1	2	3	4	5	6	7

26. *Cell count* A study of certain bacterial growth gives the cell counts shown.

x time (hr)	0	2	4	6	8	10
y cell count	597	893	1339	1995	2976	4433

x time (hr)	12	14	16	18	20
y cell count	6612	9865	14,719	21,956	32,763

(a) Build a mathematical model relating cell count to time elapsed.

(b) Compare the observations with the predictions.

(c) Use your model to predict when the cell count will reach 50,000.

27. *Spring elongation* The following table gives the elongation e in inches per inch (in./in.) for a given stress S on a steel wire measured in pounds per square in (lb/in.^{2}). Test the model $e = c_1 S$ by plotting the data. Estimate c_1 graphically.

$S \times 10^{-3}$	5	10	20	30	40	50
$e \times 10^5$	0	19	57	94	134	173

$S \times 10^{-3}$	60	70	80	90	100
$e \times 10^5$	216	256	297	343	390

(a) Build a model relating the elongation of a spring to the number of units of mass.

(b) How well does your model fit the data?

(c) Predict the elongation of the spring for a stress S of 200×10^{-3} lb/in.^{2}. How comfortable are you with this prediction?

28. *Vacuum pump* A mechanical vacuum pump, often called a "roughing pump," is used to evacuate a chamber of air. A pressure gauge measures the pressure in atmospheres (Pa). The data are recorded below.

Pressure (Pa)	100,000	36,788	13,537	4986	1837	671
Time (min)	0	1	2	3	4	5

(a) Build a model relating the pressure in the chamber to time elapsed.

(b) How well does your model fit the data?

(c) Predict when the pressure in the chamber will reach 200 Pa.

Regression Analysis

T **29.** *Doctoral degrees* Table 25 shows the number of doctoral degrees earned in the given academic year by Hispanic students. Let $x = 0$ represent 1970–71, $x = 1$ represent 1971–72, and so forth.

Table 25 Doctorates earned by Hispanic Americans	
Year	**Number of Degrees**
1976–77	520
1980–81	460
1984–85	680
1988–89	630
1990–91	730
1991–92	810
1992–93	830

Source: U.S. Department of Education, as reported in the *Chronicle of Higher Education*, April 28, 1995.

(a) Find a linear regression equation for the data and superimpose its graph on a scatter plot of the data.

(b) Use the regression equation to predict the number of doctoral degrees that will be earned by Hispanic Americans in the academic year 2000–01.

(c) *Writing to Learn* Find the slope of the regression line. What does the slope represent?

T **30.** *Estimating population growth* Use the data in Table 26 about the population of New York State. Let $x = 60$ represent 1960, $x = 70$ represent 1970, an so forth.

Table 26 Population of New York State	
Year	**Population (millions)**
1960	16.78
1980	17.56
1990	17.99

Source: The Statesman's Yearbook, 129th ed. (London: The Macmillan Press, Ltd., 1992).

(a) Find an exponential regression equation for the data.

(b) Use the regression equation to predict when the population will be 25 million.

(c) What annual rate of growth can we infer from the regression equation?

T **31.** *Sinusoidal regression* Table 27 gives the values of the function

$$f(x) = a \sin (bx + c) + d$$

accurate to two decimals.

(a) Find a sinusoidal regression equation for the data.

(b) Rewrite the equation with a, b, c, and d rounded to the nearest integer.

Table 27 Values of a function

x	$f(x)$
1	5.82
2	2.08
3	5.98
4	2.00
5	5.98
6	2.08

T **32.** *Oil production*

(a) Find a natural logarithm regression equation for the data in Table 28.

(b) Estimate the number of metric tons of oil produced by Canada in 1985.

(c) Predict when Canadian oil production will reach 120 metric tons.

Table 28 Canadian oil production

Year	Metric tons (millions)
1960	27.48
1970	69.95
1990	92.24

Source: The Statesman's Yearbook, 129th ed. (London: The Macmillan Press, Ltd., 1992).

T **33.** Table 29 gives some hypothetical data about energy consumption.

(a) Let $x = 0$ represent 1900, $x = 1$ represent 1910, and so forth. Find an exponential regression equation of the form $Q = ae^{bx}$ for the data and superimpose its graph on a scatter plot of the data.

(b) Use the exponential regression equation to estimate the energy consumption in 1996. What is the annual growth rate of energy consumption during the twentieth century?

Table 29 Energy consumption

Year	Consumption Q
1900	1.00
1910	2.01
1920	4.06
1930	8.17
1940	16.44
1950	33.12
1960	66.69
1970	134.29
1980	270.43
1990	544.57
2000	1096.63

1

Limits and Continuity

OVERVIEW   The concept of limit is one of the ideas that distinguish calculus from algebra and trigonometry. In this chapter, we show how to define and calculate limits of function values. The calculation rules are straightforward, and most of the limits we need can be found by substitution, graphical investigation, numerical approximation, algebra, or some combination of these.

The outputs of some functions vary continuously as their inputs vary—the smaller the change in the input, the smaller the change in the output. The values of other functions may jump or vary erratically no matter how carefully we control the inputs. The notion of limit gives a precise way to distinguish between these behaviors. We also use limits to define tangent lines to graphs of functions. This geometric application leads at once to the important concept of derivative of a function. The derivative, which we investigate thoroughly in Chapter 2, gives us a way to measure numerically the rate at which a function's values are changing at any instant.

1.1   Rates of Change and Limits

Average and Instantaneous Speed • Average Rates of Change and Secant Lines • Limits of Functions • Informal Definition of Limit • Precise Definition of Limit

CD-ROM
WEBsite

In this section, we introduce average and instantaneous rates of change. These lead to the main idea of the section, the idea of limit.

CD-ROM
WEBsite

Historical Biography

Zeno
(490BC — 430BC)

Average and Instantaneous Speed

A moving body's **average speed** during an interval of time is found by dividing the distance covered by the time elapsed. The unit of measure is length per unit time: kilometers per hour, feet per second, or whatever is appropriate to the problem at hand.

Example 1   Finding an Average Speed

A rock breaks loose from the top of a tall cliff. What is its average speed during the first 2 sec of fall?

Solution   Experiments show that a dense solid object dropped from rest to fall freely near the surface of the earth will fall

$$y = 16t^2$$

Free Fall

Near the surface of the earth, all bodies fall with the same constant acceleration. The distance a body falls after it is released from rest is a constant multiple of the square of the time fallen. At least, that is what happens when a body falls in a vacuum, where there is no air to slow it down. The square-of-time rule also holds for dense, heavy objects like rocks, ball bearings, and steel tools during the first few seconds of fall through air, before the velocity builds up to where air resistance begins to matter. When air resistance is absent or insignificant and the only force acting on a falling body is the force of gravity, we call the way the body falls *free fall*.

feet in the first t sec. The average speed of the rock over any given time interval is the distance traveled, Δy, divided by the length of the interval Δt. For the first 2 sec of fall, from $t = 0$ to $t = 2$, we have

$$\frac{\Delta y}{\Delta t} = \frac{16(2)^2 - 16(0)^2}{2 - 0} = 32 \text{ ft/sec .}$$

Example 2   Finding an Instantaneous Speed

Find the speed of the rock in Example 1 at the instant $t = 2$.

Solution

Solve Numerically

We can calculate the average speed of the rock over the interval from time $t = 2$ to any slightly later time $t = 2 + h$, $h > 0$, as

$$\frac{\Delta y}{\Delta t} = \frac{16(2 + h)^2 - 16(2)^2}{h}. \tag{1}$$

We cannot use this formula to calculate the speed at the exact instant $t = 2$ because that would require taking $h = 0$, and $0/0$ is undefined. We can, however, get a good idea of what is happening at $t = 2$ by evaluating the formula at values of h *close* to 0. When we do, we see a clear pattern (Table 1.1).

As h approaches 0, the average speed approaches the limiting value 64 ft/sec.

Confirm Algebraically

If we expand the numerator of Equation (1) and simplify, we find that

$$\frac{\Delta y}{\Delta t} = \frac{16(2 + h)^2 - 16(2)^2}{h} = \frac{16(4 + 4h + h^2) - 64}{h}$$

$$= \frac{64h + 16h^2}{h} = 64 + 16h .$$

For values of h different from 0, the expressions on the right and left are equivalent and the average speed is $64 + 16h$ ft/sec. We can now see why the average speed has the limiting value $64 + 16(0) = 64$ ft/sec as h approaches 0.

Average Rates of Change and Secant Lines

Given an arbitrary function $y = f(x)$, we calculate the average rate of change of y with respect to x over the interval $[x_1, x_2]$ by dividing the change in the value of y, $\Delta y = f(x_2) - f(x_1)$, by the length $\Delta x = x_2 - x_1 = h$ of the interval over which the change occurs.

Table 1.1 Average speeds over short time intervals starting at $t = 2$

$$\frac{\Delta y}{\Delta t} = \frac{16(2 + h)^2 - 16(2)^2}{h}$$

Length of time interval, h (sec)	Average speed for interval $\Delta y/\Delta t$ (ft/sec)
1	80
0.1	65.6
0.01	64.16
0.001	64.016
0.0001	64.0016
0.00001	64.00016

Definition   Average Rate of Change

The **average rate of change** of $y = f(x)$ with respect to x over the interval $[x_1, x_2]$ is

$$\frac{\Delta y}{\Delta x} = \frac{f(x_2) - f(x_1)}{x_2 - x_1} = \frac{f(x_1 + h) - f(x_1)}{h}, \qquad h \neq 0.$$

Geometrically, an average rate of change is a secant slope.

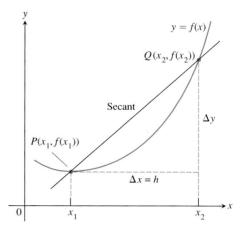

FIGURE 1.1 A secant to the graph $y = f(x)$. Its slope is $\Delta y / \Delta x$, the average rate of change of f over the interval $[x_1, x_2]$.

Notice that the rate of change of f over $[x_1, x_2]$ is the slope of the line through the points $P(x_1, f(x_1))$ and $Q(x_2, f(x_2))$ (Figure 1.1). In geometry, a line joining two points of a curve is a **secant** to the curve. Thus, the average rate of change of f from x_1 to x_2 is identical with the slope of secant PQ.

Engineers often want to know the rates at which temperatures change in materials to determine if cracks or other breaks will occur.

Example 3   The Change in Temperature of a Heat Shield

A mechanical engineer is designing a heat shield 1 in. thick for a reentry space vehicle. She has determined what the temperature u would be at each depth level x of the shield, as shown in Figure 1.2 (where the temperatures have been normalized to lie in the interval $0 \le u \le 1$). She must determine the maximum change in temperature per unit depth that the material would endure.

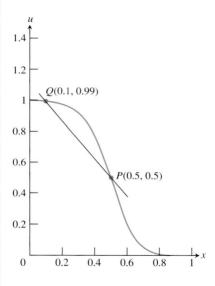

FIGURE 1.2 Temperature of a heat shield as a function of depth below the shield surface shortly after entering Earth's atmosphere.

Solution   From the graph of the temperature function (Figure 1.2), the engineer sees that the slope of the curve is steepest at the point P where the depth is 0.5 in. The average rate of change of the temperature from a depth of 0.1 in. at point Q to the depth of 0.5 in. is

$$\text{Average rate of change:} \quad \frac{\Delta u}{\Delta x} = \frac{0.99 - 0.5}{0.1 - 0.5} \approx -1.23 \text{ deg/in.}$$

This average is the slope of the secant through the points P and Q on the graph in Figure 1.2. This average, however, does not tell us how fast the temperature is changing at the point P itself. For that we need to examine the average rates of change over increasingly short depth intervals ending (or starting) at depth $x = 0.5$ in. In geometric terms, we find these rates by calculating the slopes of secants from P to Q, for a sequence of points Q approaching P along the curve (Figure 1.3).

The values in the table in Figure 1.3 show that the secant slopes go from -1.23 to -2.6 as the x-coordinate of Q increases from 0.1 to 0.4. Geometrically, the secants rotate clockwise about P and seem to approach a line that

Q	Slope of $PQ = \Delta u/\Delta x$
(0.1, 0.99)	$\dfrac{0.99 - 0.5}{0.1 - 0.5} \approx -1.23$
(0.2, 0.98)	$\dfrac{0.98 - 0.5}{0.2 - 0.5} \approx -1.60$
(0.3, 0.92)	$\dfrac{0.92 - 0.5}{0.3 - 0.5} \approx -2.10$
(0.4, 0.76)	$\dfrac{0.76 - 0.5}{0.4 - 0.5} \approx -2.60$

goes through P with the same steepness (slope) that the curve has at P. We will see that this line is called the *tangent* to the curve at P (Figure 1.4). Since the line appears to pass through the points $A(0.32, 1)$ and $B(0.68, 0)$, it has slope

$$\frac{1 - 0}{0.32 - 0.68} \approx -2.78 \text{ deg/in.}$$

At P, where the depth is 0.5 in., the temperature was changing at a rate of about -2.78 deg/in.

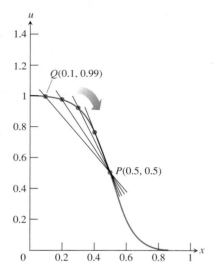

FIGURE 1.3 The positions and slopes of four secants through point P on the heat shield graph.

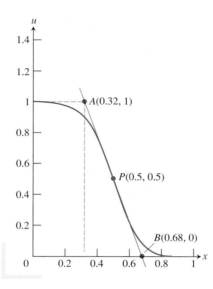

FIGURE 1.4 The tangent line at point P has the same steepness (slope) that the curve has at P.

The rate at which the rock in Example 2 was falling at the instant $t = 2$ and the rate at which the temperature in Example 3 was changing at depth 0.5 in. are called *instantaneous rates of change*. As the examples suggest, we find instantaneous rates as limiting values of average rates. In Example 3, we also pictured the tangent line to the temperature curve at depth 0.5 as a limiting position of secant lines. Instantaneous rates and tangent lines, intimately connected, appear in many other contexts. To talk about the two constructively and to understand the connection further, we need to investigate the process by which we determine limiting values, or *limits*, as we will soon call them.

Limits of Functions

Before we give a definition of limit, let us look at one more example.

CD-ROM
WEBsite

Example 4   Behavior of a Function Near a Point

How does the function

$$f(x) = \frac{x^2 - 1}{x - 1}$$

behave near $x = 1$?

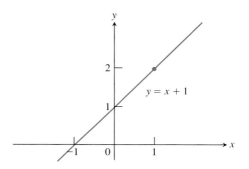

FIGURE 1.5 The graph of f is identical with the line $y = x + 1$ except at $x = 1$, where f is not defined.

Solution   The given formula defines f for all real numbers x except $x = 1$ (we cannot divide by zero). For any $x \neq 1$, we can simplify the formula by factoring the numerator and canceling common factors:

$$f(x) = \frac{(x-1)(x+1)}{x-1} = x + 1 \quad \text{for} \quad x \neq 1.$$

The graph of f is thus the line $y = x + 1$ with the point $(1, 2)$ *removed.* This removed point is shown as a "hole" in Figure 1.5. Even though $f(1)$ is not defined, it is clear that we can make the value of $f(x)$ *as close as we want* to 2 by choosing x *close enough* to 1 (Table 1.2).

Table 1.2 The closer x gets to 1, the closer $f(x) = (x^2 - 1)/(x - 1)$ seems to get to 2	
Values of x below and above 1	$f(x) = \dfrac{x^2 - 1}{x - 1} = x + 1, \quad x \neq 1$
0.9	1.9
1.1	2.1
0.99	1.99
1.01	2.01
0.999	1.999
1.001	2.001
0.999999	1.999999
1.000001	2.000001

We say that $f(x)$ approaches arbitrarily close to 2 as x approaches 1, or, more simply, that $f(x)$ approaches the *limit* 2 as x approaches 1. We write this as

$$\lim_{x \to 1} f(x) = 2, \quad \text{or} \quad \lim_{x \to 1} \frac{x^2 - 1}{x - 1} = 2.$$

Informal Definition of Limit

Let $f(x)$ be defined on an open interval about x_0, *except possibly at x_0 itself.* If $f(x)$ gets arbitrarily close to L for all x sufficiently close to x_0, we say that f approaches the **limit** L as x approaches x_0, and we write

$$\lim_{x \to x_0} f(x) = L.$$

This definition is "informal" because phrases like *arbitrarily close* and *sufficiently close* are imprecise; their meaning depends on the context. To a machinist manufacturing a piston, *close* may mean *within a few thousandths of an inch.* To an astronomer studying distant galaxies, *close* may mean *within a few thousand light-years*. The definition is clear enough, however, to enable us to recognize and evaluate limits of a number of specific functions.

Example 5   The Limit Value Does Not Depend On How the Function is Defined at x_0

The function f in Figure 1.6 has limit 2 as $x \to 1$ even though f is not defined at $x = 1$. The function g has limit 2 as $x \to 1$ even though $2 \neq g(1)$. The function h is the only one whose limit as $x \to 1$ equals its value at $x = 1$. For h, we have $\lim_{x\to 1} h(x) = h(1)$. This equality of limit and function value is special, and we return to it in Section 1.4.

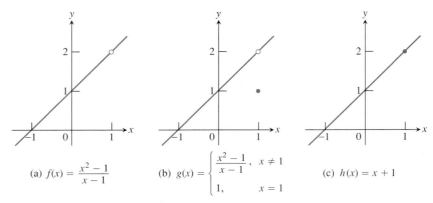

(a) $f(x) = \dfrac{x^2 - 1}{x - 1}$ (b) $g(x) = \begin{cases} \dfrac{x^2 - 1}{x - 1}, & x \neq 1 \\ 1, & x = 1 \end{cases}$ (c) $h(x) = x + 1$

FIGURE 1.6 $\lim_{x\to 1} f(x) = \lim_{x\to 1} g(x) = \lim_{x\to 1} h(x) = 2$.

Example 6   Two Functions That Have Limits at Every Point

(a) If f is the **identity function** $f(x) = x$, then for any value of x_0 (Figure 1.7a),

$$\lim_{x\to x_0} f(x) = \lim_{x\to x_0} x = x_0.$$

(b) If f is the **constant function** $f(x) = k$ (function with the constant value k), then for any value of x_0 (Figure 1.7b),

$$\lim_{x\to x_0} f(x) = \lim_{x\to x_0} k = k.$$

For instance,

$$\lim_{x\to 3} x = 3 \qquad \text{and} \qquad \lim_{x\to -7} (4) = \lim_{x\to 2} (4) = 4.$$

Some ways that limits can fail to exist are illustrated in Figure 1.8 and described in the next example.

(a) Identity function

(b) Constant function

FIGURE 1.7 The functions in Example 6.

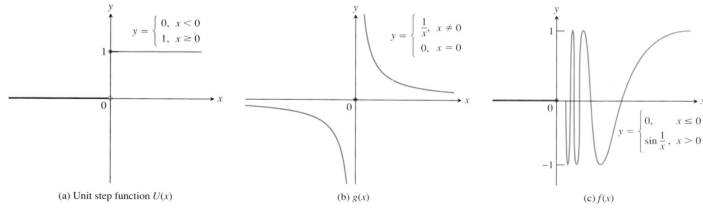

(a) Unit step function $U(x)$ (b) $g(x)$ (c) $f(x)$

FIGURE 1.8 The functions in Example 7.

CD-ROM
WEBsite

Historical Biography

Aristotle
(384BC — 322BC)

Example 7   Limits May Fail to Exist

Discuss the behavior of the following functions as $x \to 0$.

(a) $U(x) = \begin{cases} 0, & x < 0 \\ 1, & x \geq 0 \end{cases}$

(b) $g(x) = \begin{cases} \dfrac{1}{x}, & x \neq 0 \\ 0, & x = 0 \end{cases}$

(c) $f(x) = \begin{cases} 0, & x \leq 0 \\ \sin \dfrac{1}{x}, & x > 0 \end{cases}$

Solution

(a) It *jumps*: The **unit step function** $U(x)$ has no limit as $x \to 0$ because its values jump at $x = 0$. For negative values of x arbitrarily close to zero, $U(x) = 0$. For positive values of x arbitrarily close to zero, $U(x) = 1$. There is no *single* value L approached by $U(x)$ as $x \to 0$ (Figure 1.8a).

(b) It *grows too large to have a limit*: $g(x)$ has no limit as $x \to 0$ because the values of g grow arbitrarily large in absolute value as $x \to 0$ and do not stay close to *any* real number (Figure 1.8b).

(c) It *oscillates too much to have a limit*: $f(x)$ has no limit as $x \to 0$ because the function's values oscillate between $+1$ and -1 in every open interval containing 0. The values do not stay close to any one number as $x \to 0$ (Figure 1.8c).

Precise Definition of Limit

To show that the limit of $f(x)$ as $x \to x_0$ equals the number L, we need to show that the gap between $f(x)$ and L can be made "as small as we choose" if x is kept "close enough" to x_0. Let us see what this would require if we specified the size of the gap between $f(x)$ and L.

Example 8   Controlling a Linear Function

How close to $x_0 = 4$ must we hold the input x to be sure that the output $y = 2x - 1$ lies within 2 units of $y_0 = 7$?

Solution   We are asked: For what values of x is $|y - 7| < 2$? To find the answer we first express $|y - 7|$ in terms of x:

$$|y - 7| = |(2x - 1) - 7| = |2x - 8|.$$

The question then becomes: what values of x satisfy the inequality $|2x - 8| < 2$? To find out, we solve the inequality:

$$|2x - 8| < 2$$
$$-2 < 2x - 8 < 2$$
$$6 < 2x < 10$$
$$3 < x < 5$$
$$-1 < x - 4 < 1.$$

Keeping x within 1 unit of $x_0 = 4$ will keep y within 2 units of $y_0 = 7$ (Figure 1.9).

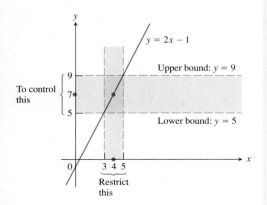

FIGURE 1.9 Keeping x within 1 unit of $x_0 = 4$ will keep y within 2 units of $y_0 = 7$.

In Example 8, we determined how close to hold a variable x to a particular value x_0 to ensure that the outputs $f(x)$ lie within a prescribed interval about the limit L. To show that the limit of $f(x)$ as $x \rightarrow x_0$ actually equals L, we must be able to show that the gap between $f(x)$ and L can be made less than *any prescribed error,* no matter how small, by holding x close enough to x_0.

Suppose that we are watching the values of a function $f(x)$ as x approaches x_0 (without taking on the value of x_0 itself). Certainly we want to be able to say that $f(x)$ stays within one-tenth of a unit of L as soon as x stays within some distance δ of x_0 (Figure 1.10). That in itself is not enough, however, because as x continues on its course toward x_0, what is to prevent $f(x)$ from jittering about within the interval from $L - 1/10$ to $L + 1/10$ without tending toward L? We can ensure that $f(x)$ is tending toward L by showing that no matter how tightly we narrow the gap between $f(x)$ and L, keeping x close enough to x_0 will keep $f(x)$ within that tolerance of L. This wording is the mathematical way to say that the closer x gets to x_0, the closer $y = f(x)$ gets to L.

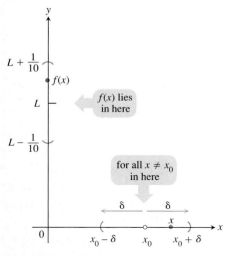

FIGURE 1.10 A preliminary stage in the development of the definition of limit.

Definition   Formal Definition of Limit

Let $f(x)$ be defined on an open interval about x_0, except possibly at x_0 itself. We say that $f(x)$ approaches the **limit** L as x approaches x_0 and write

$$\lim_{x \to x_0} f(x) = L,$$

if, for every number $\epsilon > 0$, there exists a corresponding number $\delta > 0$ such that for all x,

$$0 < |x - x_0| < \delta \implies |f(x) - L| < \epsilon.$$

The limit definition is illustrated in Figure 1.11.

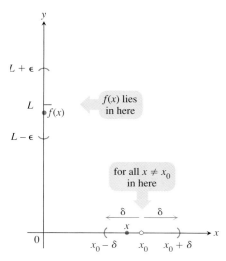

FIGURE 1.11 The relation of δ and ϵ in the definition of limit.

Example 9   Testing the Definition

Show that $\lim_{x \to 1} (5x - 3) = 2$.

Solution   Set $x_0 = 1$, $f(x) = 5x - 3$, and $L = 2$ in the definition of limit. For any given $\epsilon > 0$, we have to find a suitable $\delta > 0$ so that if $x \neq 1$ and x is within distance δ of $x_0 = 1$, that is, if

$$0 < |x - 1| < \delta,$$

then $f(x)$ is within distance ϵ of $L = 2$, that is

$$|f(x) - 2| < \epsilon.$$

We find δ by working backwards from the ϵ-inequality:

$$|(5x - 3) - 2| = |5x - 5| < \epsilon$$
$$5|x - 1| < \epsilon$$
$$|x - 1| < \epsilon/5.$$

Thus, we can take $\delta = \epsilon/5$ (Figure 1.12). If $0 < |x - 1| < \delta = \epsilon/5$, then

$$|(5x - 3) - 2| = |5x - 5| = 5|x - 1| < 5(\epsilon/5) = \epsilon,$$

which proves that $\lim_{x \to 1} (5x - 3) = 2$.

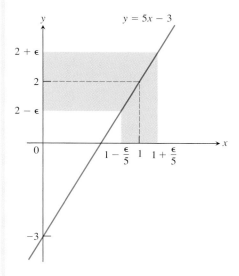

FIGURE 1.12 If $f(x) = 5x - 3$, then $0 < |x - 1| < \epsilon/5$ guarantees that $|f(x) - 2| < \epsilon$. (Example 9)

The value of $\delta = \epsilon/5$ is not the only value that will make $0 < |x - 1| < \delta$ imply $|5x - 5| < \epsilon$. Any smaller positive δ will do as well. The definition does not ask for a "best" positive δ, just one that will work.

In Example 9, the interval of values about x_0 for which $|f(x) - L|$ was less than ϵ was symmetric about x_0, and we could take δ to be half the length of the interval. When such symmetry is absent, as it usually is, we can take δ to be the distance from x_0 to the interval's nearer endpoint. This choice for δ is illustrated in the next example.

Example 10   Finding Delta Algebraically for a Given Epsilon

For the limit $\lim_{x \to 5} \sqrt{x - 1} = 2$, find a $\delta > 0$ that works for $\epsilon = 1$. That is, find a $\delta > 0$ such that for all x

$$0 < |x - 5| < \delta \;\Rightarrow\; |\sqrt{x - 1} - 2| < 1.$$

Solution   We organize the search into two steps. First, we solve the inequality $|\sqrt{x - 1} - 2| < 1$ to find an interval (a, b) about $x_0 = 5$ on which the inequality holds for all $x \neq x_0$. Then we find a value of $\delta > 0$ that places the interval $5 - \delta < x < 5 + \delta$ (centered at $x_0 = 5$) inside the interval (a, b).

Step 1: *Solve the inequality $|\sqrt{x - 1} - 2| < 1$ to find an interval about $x_0 = 5$ on which the inequality holds for all $x \neq x_0$.*

$$|\sqrt{x - 1} - 2| < 1$$
$$-1 < \sqrt{x - 1} - 2 < 1$$
$$1 < \sqrt{x - 1} < 3$$
$$1 < x - 1 < 9$$
$$2 < x < 10$$

The inequality holds for all x in the open interval $(2, 10)$, so it holds for all $x \neq 5$ in this interval as well.

Step 2: *Find a value of $\delta > 0$ to place the centered interval $5 - \delta < x < 5 + \delta$ inside the interval $(2, 10)$.* The distance from 5 to the nearer endpoint of $(2, 10)$ is 3 (Figure 1.13). If we take $\delta = 3$ or any smaller positive number, then the inequality $0 < |x - 5| < \delta$ will automatically place x between 2 and 10 to make $|\sqrt{x - 1} - 2| < 1$ (Figure 1.14):

$$0 < |x - 5| < 3 \;\Rightarrow\; |\sqrt{x - 1} - 2| < 1.$$

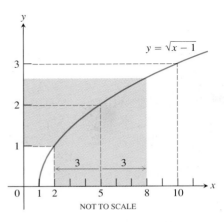

FIGURE 1.14 The function and intervals in Example 10.

FIGURE 1.13 An open interval of radius 3 about $x_0 = 5$ will lie inside the open interval $(2, 10)$.

EXERCISES 1.1

Average Rates of Change

In Exercises 1–4, find the average rate of change of the function over the given interval or intervals.

1. $f(x) = x^3 + 1$

    (a) $[2, 3]$ **(b)** $[-1, 1]$

2. $R(\theta) = \sqrt{4\theta + 1}$, $[0, 2]$

3. $h(t) = \cot t$

    (a) $[\pi/4, 3\pi/4]$ **(b)** $[\pi/6, \pi/2]$

4. $g(t) = 2 + \cos t$

    (a) $[0, \pi]$ **(b)** $[-\pi, \pi]$

5. *A Ford Mustang Cobra's speed* The accompanying figure shows the time-to-distance graph for a 1994 Ford Mustang Cobra accelerating from a standstill.

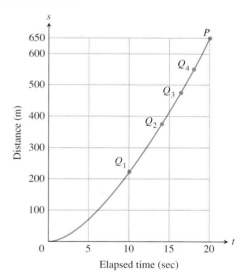

    (a) Estimate the slopes of secants PQ_1, PQ_2, PQ_3, and PQ_4, arranging them in order in a table. What are the appropriate units for these slopes?

    (b) Then estimate the Cobra's speed at time $t = 20$ sec.

6. *Speed of a falling wrench* The accompanying figure shows the plot of distance fallen versus time for a wrench that fell from the top of a communications mast on the moon to the station roof 80 m below.

    (a) Estimate the slopes of the secants PQ_1, PQ_2, PQ_3, and PQ_4, arranging them in a table like the one in Figure 1.3.

    (b) About how fast was the wrench going when it hit the roof?

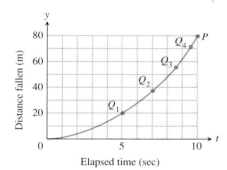

7. *Velocity of a ball* The accompanying data represent the distance a ball has traveled down an inclined plane. Estimate the instantaneous velocity at $t = 1$ by finding an upper and lower bound, and averaging them. That is, find $a \le v(1) \le b$ and estimate $v(1)$ as $(a + b)/2$.

Time t (sec)	Distance traveled (ft)
0	0
0.2	0.52
0.4	2.10
0.6	4.72
0.8	8.39
1.0	13.10
1.2	18.87
1.4	25.68

8. *Distance traveled by a train* A train accelerates from a standstill to a maximum cruising speed, then slows down to pass through a town at constant speed. After passing through the town the train accelerates to cruising speed. Finally, the train decelerates smoothly to a stop as it reaches its destination. Sketch a possible graph of the distance the train has traveled as a function of time.

Limits From Graphs

9. For the function $g(x)$ graphed here, find the following limits or explain why they do not exist.

   (a) $\lim_{x \to 1} g(x)$ (b) $\lim_{x \to 2} g(x)$ (c) $\lim_{x \to 3} g(x)$

10. For the function $f(t)$ graphed here, find the following limits or explain why they do not exist.

    (a) $\lim_{t \to -2} f(t)$ (b) $\lim_{t \to -1} f(t)$ (c) $\lim_{t \to 0} f(t)$

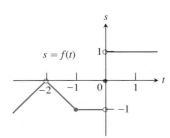

11. Which of the following statements about the function $y = f(x)$ graphed here are true, and which are false?

    (a) $\lim_{x \to 0} f(x)$ exists.

    (b) $\lim_{x \to 0} f(x) = 0$.

    (c) $\lim_{x \to 0} f(x) = 1$.

    (d) $\lim_{x \to 1} f(x) = 1$.

    (e) $\lim_{x \to 1} f(x) = 0$.

    (f) $\lim_{x \to x_0} f(x)$ exists at every point x_0 in $(-1, 1)$.

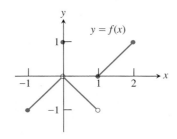

12. Which of the following statements about the function $y = f(x)$ graphed here are true, and which are false?

    (a) $\lim_{x \to 2} f(x)$ does not exist.

    (b) $\lim_{x \to 2} f(x) = 2$.

    (c) $\lim_{x \to 1} f(x)$ does not exist.

    (d) $\lim_{x \to x_0} f(x)$ exists at every point x_0 in $(-1, 1)$.

    (e) $\lim_{x \to x_0} f(x)$ exists at every point x_0 in $(1, 3)$.

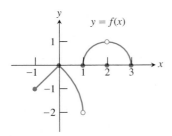

Existence of Limits

In Exercises 13 and 14, explain why the limits do not exist.

13. $\lim_{x \to 0} \dfrac{x}{|x|}$

14. $\lim_{x \to 1} \dfrac{1}{x - 1}$

15. *Writing to Learn* Suppose that a function $f(x)$ is defined for all real values of x except $x = x_0$. Can anything be said about the existence of $\lim_{x \to x_0} f(x)$? Give reasons for your answer.

16. *Writing to Learn* Suppose that a function $f(x)$ is defined for all x in $[-1, 1]$. Can anything be said about the existence of $\lim_{x \to 0} f(x)$? Give reasons for your answer.

17. *Writing to Learn* If $\lim_{x \to 1} f(x) = 5$, must f be defined at $x = 1$? If it is, must $f(1) = 5$? Can we conclude *anything* about the values of f at $x = 1$? Explain.

18. *Writing to Learn* If $f(1) = 5$, must $\lim_{x \to 1} f(x)$ exist? If it does, then must $\lim_{x \to 1} f(x) = 5$? Can we conclude *anything* about $\lim_{x \to 1} f(x)$? Explain.

Estimating Limits

T You will find a graphing calculator useful for Exercises 19–26.

19. Let $f(x) = (x^2 - 9)/(x + 3)$.

    (a) Make a table of the values of f at the points $x = -3.1$, -3.01, -3.001, and so on as far as your calculator can go. Then estimate $\lim_{x \to -3} f(x)$. What estimate do you arrive at if you evaluate f at $x = -2.9$, -2.99, -2.999, . . . instead?

    (b) Support your conclusions in part (a) by graphing f near $x_0 = -3$ and using Zoom and Trace to estimate y-values on the graph as $x \to -3$.

    (c) Find $\lim_{x \to -3} f(x)$ algebraically.

20. Let $g(x) = (x^2 - 2)/(x - \sqrt{2})$.

(a) Make a table of the values of g at the points $x = 1.4$, 1.41, 1.414, and so on through successive decimal approximations of $\sqrt{2}$. Estimate $\lim_{x \to \sqrt{2}} g(x)$.

(b) Support your conclusion in part (a) by graphing g near $x_0 = \sqrt{2}$ and using Zoom and Trace to estimate y-values on the graph as $x \to \sqrt{2}$.

(c) Find $\lim_{x \to \sqrt{2}} g(x)$ algebraically.

21. Let $G(x) = (x + 6)/(x^2 + 4x - 12)$.

(a) Make a table of the values of G at $x = -5.9$, -5.99, -5.999, and so on. Then estimate $\lim_{x \to -6} G(x)$. What estimate do you arrive at if you evaluate G at $x = -6.1$, -6.01, -6.001, . . . instead?

(b) Support your conclusions in part (a) by graphing G and using Zoom and Trace to estimate y-values on the graph as $x \to -6$.

(c) Find $\lim_{x \to -6} G(x)$ algebraically.

22. Let $h(x) = (x^2 - 2x - 3)/(x^2 - 4x + 3)$.

(a) Make a table of the values of h at $x = 2.9$, 2.99, 2.999, and so on. Then estimate $\lim_{x \to 3} h(x)$. What estimate do you arrive at if you evaluate h at $x = 3.1$, 3.01, 3.001, . . . instead?

(b) Support your conclusions in part (a) by graphing h near $x_0 = 3$ and using Zoom and Trace to estimate y-values on the graph as $x \to 3$.

(c) Find $\lim_{x \to 3} h(x)$ algebraically.

23. Let $g(\theta) = (\sin \theta)/\theta$.

(a) Make a table of the values of g at values of θ that approach $\theta_0 = 0$ from above and below. Then estimate $\lim_{\theta \to 0} g(\theta)$.

(b) Support your conclusion in part (a) by graphing g near $\theta_0 = 0$.

24. Let $G(t) = (1 - \cos t)/t^2$.

(a) Make tables of values of G at values of t that approach $t_0 = 0$ from above and below. Then estimate $\lim_{t \to 0} G(t)$.

(b) Support your conclusion in part (a) by graphing G near $t_0 = 0$.

25. Let $f(x) = x^{1/(1-x)}$.

(a) Make tables of values of f at values of x that approach $x_0 = 1$ from above and below. Does f appear to have a limit as $x \to 1$? If so, what is it? If not, why not?

(b) Support your conclusions in part (a) by graphing f near $x_0 = 1$.

26. Let $f(x) = (3^x - 1)/x$.

(a) Make tables of values of f at values of x that approach $x_0 = 0$ from above and below. Does f appear to have a limit as $x \to 0$? If so, what is it? If not, why not?

(b) Support your conclusions in part (a) by graphing f near $x_0 = 0$.

CD-ROM
WEBsite
Finding Deltas Graphically

In Exercises 27–30, use the graphs to find a $\delta > 0$ such that for all x,

$$0 < |x - x_0| < \delta \implies |f(x) - L| < \epsilon.$$

27.

28.

29.

30.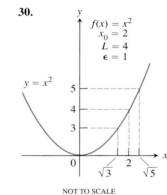

Finding Deltas Algebraically

Each of Exercises 31–36 gives a function $f(x)$ and numbers L, x_0, and $\epsilon > 0$. In each case, find an open interval about x_0 on which the inequality $|f(x) - L| < \epsilon$ holds. Then give a value for $\delta > 0$ such that for all x satisfying $0 < |x - x_0| < \delta$ the inequality $|f(x) - L| < \epsilon$ holds.

31. $f(x) = x + 1$, $L = 5$, $x_0 = 4$, $\epsilon = 0.01$

32. $f(x) = 2x - 2$, $L = -6$, $x_0 = -2$, $\epsilon = 0.02$

33. $f(x) = \sqrt{x + 1}$, $L = 1$, $x_0 = 0$, $\epsilon = 0.1$

34. $f(x) = \sqrt{19 - x}$, $L = 3$, $x_0 = 10$, $\epsilon = 1$

35. $f(x) = 1/x$, $L = 1/4$, $x_0 = 4$, $\epsilon = 0.05$

36. $f(x) = x^2$, $L = 3$, $x_0 = \sqrt{3}$, $\epsilon = 0.1$

Theory and Examples

37. *Grinding engine cylinders* Before contracting to grind engine cylinders to a cross-section area of 9 in.^{2}, you need to know how much deviation from the ideal cylinder diameter of $x_0 = 3.385$ in. you can allow and still have the area come within 0.01 in.^{2} of the required 9 in.^{2}. To find out, you let $A = \pi(x/2)^2$ and look for the interval in which you must hold x to make $|A - 9| \leq 0.01$. What interval do you find?

38. *Manufacturing electrical resistors* Ohm's law for electrical circuits like the one shown in the accompanying figure states that $V = RI$. In this equation, V is a constant voltage, I is the current in amperes, and R is the resistance in ohms. Your firm has been asked to supply the resistors for a circuit in which V will be 120 volts and I is to be 5 ± 0.1 amp. In what interval does R have to lie for I to be within 0.1 amp of the target value $I_0 = 5$?

T 39. *Controlling outputs* Let $f(x) = \sqrt{3x - 2}$.

(a) Show that $\lim_{x \to 2} f(x) = 2 = f(2)$.

(b) Use a graph to estimate values for a and b so that $1.8 < f(x) < 2.2$ provided $a < x < b$.

(c) Use a graph to estimate values for a and b so that $1.99 < f(x) < 2.01$ provided $a < x < b$.

T 40. *Controlling outputs* Let $f(x) = \sin x$.

(a) Find $f(\pi/6)$.

(b) Use a graph to estimate an interval (a, b) about $x = \pi/6$ so that $0.3 < f(x) < 0.7$ provided $a < x < b$.

(c) Use a graph to estimate an interval (a, b) about $x = \pi/6$ so that $0.49 < f(x) < 0.51$ provided $a < x < b$.

41. *Free fall* A water balloon dropped from a window high above the ground falls $y = 4.9t^2$ m in t sec. Find the balloon's

(a) average speed during the first 3 sec. of fall.

(b) speed at the instant $t = 3$.

42. *Free fall on a small airless planet* A rock released from rest to fall on a small airless planet falls $y = gt^2$ m in t sec, g a constant. Suppose that the rock falls to the bottom of a crevasse 20 m below the release point and reaches the bottom in 4 sec.

(a) Find the value of g.

(b) Find the average speed for the fall.

(c) With what speed did the rock hit the bottom?

In Exercises 43–46, complete the following tables and state what you believe $\lim_{x \to 0} f(x)$ to be.

(a)

x	−0.1	−0.01	−0.001	−0.0001	$\cdots$
$f(x)$	?	?	?	?	

(b)

x	0.1	0.01	0.001	0.0001	$\cdots$
$f(x)$	?	?	?	?	

43. $f(x) = x \sin \dfrac{1}{x}$

44. $f(x) = \sin \dfrac{1}{x}$

45. $f(x) = \dfrac{10^x - 1}{x}$

46. $f(x) = x \sin (\ln |x|)$

COMPUTER EXPLORATIONS

Graphical Estimates of Limits

In Exercises 47–50, use a CAS to perform the following steps:

(a) Plot the function near the point x_0 being approached.

(b) From your plot guess the value of the limit.

(c) Evaluate the limit symbolically. How close was your guess?

47. $\lim_{x \to 2} \dfrac{x^4 - 16}{x - 2}$

48. $\lim_{x \to 0} \dfrac{1 - \cos x}{x \sin x}$

49. $\lim_{x \to -1} \dfrac{x^3 - x^2 - 5x - 3}{(x + 1)^2}$

50. $\lim_{x \to 3} \dfrac{x^2 - 9}{\sqrt{x^2 + 7} - 4}$

Finding Deltas Graphically

In Exercises 51–54, you will further explore finding deltas graphically. Use a CAS to perform the following steps:

(a) Plot the function $y = f(x)$ near the point x_0 being approached.

(b) Guess the value of the limit L and then evaluate the limit symbolically to see if you guessed correctly.

(c) Using the value $\epsilon = 0.2$, graph the bounding lines $y_1 = L - \epsilon$ and $y_2 = L + \epsilon$ together with the function f near x_0.

(d) From your graph in part (c), estimate a $\delta > 0$ such that for all x

$$0 < |x - x_0| < \delta \quad \Rightarrow \quad |f(x) - L| < \epsilon.$$

Test your estimate by plotting f, y_1, and y_2 over the interval $0 < |x - x_0| < \delta$. For your viewing window, use

$$x_0 - 2\delta \leq x \leq x_0 + 2\delta \quad \text{and} \quad L - 2\epsilon \leq y \leq L + 2\epsilon.$$

If any function values lie outside the interval $[L - \epsilon, L + \epsilon]$, your choice of δ was too large. Try again with a smaller estimate.

(e) Repeat parts (c) and (d) successively for $\epsilon = 0.1, 0.05$, and 0.001.

51. $f(x) = \dfrac{x^4 - 81}{x - 3}$, $x_0 = 3$

52. $f(x) = \dfrac{5x^3 + 9x^2}{2x^5 + 3x^2}$, $x_0 = 0$

53. $f(x) = \dfrac{\sin 2x}{3x}$, $x_0 = 0$

CD-ROM
WEBsite **54.** $f(x) = \dfrac{x(1 - \cos x)}{x - \sin x}$, $x_0 = 0$

1.2 Finding Limits and One-Sided Limits

Properties of Limits • Eliminating Zero Denominators Algebraically •
Sandwich Theorem • One-Sided Limits • Limits Involving $(\sin \theta)/\theta$

CD-ROM
WEBsite

Historical Essay

Limits

In the preceding section, we evaluated limits by examining graphs and numerical patterns. In this section, we see that many limits can be evaluated algebraically, using arithmetic and a few basic rules.

Properties of Limits

The next theorem tells how to calculate limits of functions that are arithmetic combinations of functions whose limits we already know. The rules are proved in Appendix 2.

> **Theorem 1**   Limit Rules
> If L, M, c, and k are real numbers and
> $$\lim_{x \to c} f(x) = L \qquad \text{and} \qquad \lim_{x \to c} g(x) = M, \qquad \text{then}$$
>
> **1.** *Sum Rule:* $\qquad\qquad\qquad\qquad \displaystyle\lim_{x \to c} (f(x) + g(x)) = L + M$
>
> The limit of the sum of two functions is the sum of their limits.
>
> **2.** *Difference Rule:* $\qquad\qquad\qquad \displaystyle\lim_{x \to c} (f(x) - g(x)) = L - M$
>
> The limit of the difference of two functions is the difference of their limits.
>
> **3.** *Product Rule:* $\qquad\qquad\qquad\quad \displaystyle\lim_{x \to c} (f(x) \cdot g(x)) = L \cdot M$
>
> The limit of a product of two functions is the product of their limits.
>
> **4.** *Constant Multiple Rule:* $\qquad\quad \displaystyle\lim_{x \to c} (k \cdot f(x)) = k \cdot L$
>
> The limit of a constant times a function is the constant times the limit of the function.
>
> **5.** *Quotient Rule:* $\qquad\qquad\qquad\quad \displaystyle\lim_{x \to c} \frac{f(x)}{g(x)} = \frac{L}{M}$, $M \neq 0$
>
> The limit of a quotient of two functions is the quotient of their limits, provided the limit of the denominator is not zero.
>
> **6.** *Power rule:*   If r and s are integers, $s \neq 0$, then
> $$\lim_{x \to c} (f(x))^{r/s} = L^{r/s}$$
> provided that $L^{r/s}$ is a real number.
>
> The limit of a rational power of a function is that power of the limit of the function, provided the latter is a real number.

Here are some examples of how Theorem 1 can be used to find limits of polynomial and rational functions.

Example 1   Using Limit Rules

Use the observations $\lim_{x \to c} k = k$ and $\lim_{x \to c} x = c$ and the properties of limits to find the following limits.

(a) $\lim_{x \to c} (x^3 + 4x^2 - 3)$ (b) $\lim_{x \to c} \dfrac{x^4 + x^2 - 1}{x^2 + 5}$ (c) $\lim_{x \to -2} \sqrt{4x^2 - 3}$

Solution

(a) $\lim_{x \to c} (x^3 + 4x^2 - 3) = \lim_{x \to c} x^3 + \lim_{x \to c} 4x^2 - \lim_{x \to c} 3$ Sum and Difference Rules

$\qquad\qquad\qquad\qquad\quad = c^3 + 4c^2 - 3$ Product and Multiple Rules

(b) $\lim_{x \to c} \dfrac{x^4 + x^2 - 1}{x^2 + 5} = \dfrac{\lim_{x \to c} (x^4 + x^2 - 1)}{\lim_{x \to c} (x^2 + 5)}$ Quotient Rule

$\qquad\qquad\qquad = \dfrac{\lim_{x \to c} x^4 + \lim_{x \to c} x^2 - \lim_{x \to c} 1}{\lim_{x \to c} x^2 + \lim_{x \to c} 5}$ Sum and Difference Rules

$\qquad\qquad\qquad = \dfrac{c^4 + c^2 - 1}{c^2 + 5}$ Power or Product Rule

(c) $\lim_{x \to -2} \sqrt{4x^2 - 3} = \sqrt{\lim_{x \to -2} (4x^2 - 3)}$ Power Rule with $n = \frac{1}{2}$

$\qquad\qquad\qquad = \sqrt{\lim_{x \to -2} 4x^2 - \lim_{x \to -2} 3}$ Sum and Difference Rules

$\qquad\qquad\qquad = \sqrt{4(-2)^2 - 3}$ Product and Multiple Rules

$\qquad\qquad\qquad = \sqrt{16 - 3}$

$\qquad\qquad\qquad = \sqrt{13}$

As suggested by Example 1, the formulas in Theorem 1 lead us to conclude that limits of polynomial functions can be found by substitution. The same is true for rational functions if their denominators are different from zero at the point of evaluation.

Theorem 2   Limits of Polynomials Can Be Found by Substitution

If $P(x) = a_n x^n + a_{n-1} x^{n-1} + \cdots + a_0$, then

$$\lim_{x \to c} P(x) = P(c) = a_n c^n + a_{n-1} c^{n-1} + \cdots + a_0.$$

Theorem 3   Limits of Rational Functions Can Be Found by Substitution If the Limit of the Denominator Is Not Zero

If $P(x)$ and $Q(x)$ are polynomials and $Q(c) \neq 0$, then

$$\lim_{x \to c} \frac{P(x)}{Q(x)} = \frac{P(c)}{Q(c)}.$$

Example 2 Limit of a Rational Function

$$\lim_{x \to -1} \frac{x^3 + 4x^2 - 3}{x^2 + 5} = \frac{(-1)^3 + 4(-1)^2 - 3}{(-1)^2 + 5} = \frac{0}{6} = 0$$

This result is similar to the second limit in Example 1 with $c = -1$, now done in one step.

Eliminating Zero Denominators Algebraically

Theorem 3 applies only if the denominator of the rational function is not zero at the limit point c. If the denominator is zero, canceling common factors in the numerator and denominator may reduce the fraction to one whose denominator is no longer zero at c. If this happens, we can find the limit by substitution in the simplified fraction.

Example 3 Canceling a Common Factor

Evaluate

$$\lim_{x \to 1} \frac{x^2 + x - 2}{x^2 - x}.$$

Solution We cannot substitute $x = 1$ because it makes the denominator zero. We test the numerator to see if it, too, is zero at $x = 1$. It is, so it has a factor of $(x - 1)$ in common with the denominator. Canceling the $(x - 1)$'s gives a simpler fraction with the same values as the original for $x \neq 1$:

$$\frac{x^2 + x - 2}{x^2 - x} = \frac{(x - 1)(x + 2)}{x(x - 1)} = \frac{x + 2}{x}, \qquad \text{if } x \neq 1.$$

Using the simpler fraction, we find the limit of these values as $x \to 1$ by substitution:

$$\lim_{x \to 1} \frac{x^2 + x - 2}{x^2 - x} = \lim_{x \to 1} \frac{x + 2}{x} = \frac{1 + 2}{1} = 3.$$

See Figure 1.15.

Example 4 Creating and Canceling a Common Factor

Evaluate

$$\lim_{h \to 0} \frac{\sqrt{2 + h} - \sqrt{2}}{h}.$$

Solution We cannot substitute $h = 0$, and the numerator and denominator have no obvious factors. We can, however, create a common factor by multiplying the numerator and denominator by the *conjugate expression* $\sqrt{2 + h} + \sqrt{2}$, obtained by changing the sign between the square roots:

$$\frac{\sqrt{2 + h} - \sqrt{2}}{h} = \frac{\sqrt{2 + h} - \sqrt{2}}{h} \cdot \frac{\sqrt{2 + h} + \sqrt{2}}{\sqrt{2 + h} + \sqrt{2}}$$

$$= \frac{2 + h - 2}{h(\sqrt{2 + h} + \sqrt{2})}$$

$$= \frac{h}{h(\sqrt{2 + h} + \sqrt{2})} \qquad \text{Common factor of } h$$

$$= \frac{1}{\sqrt{2 + h} + \sqrt{2}}. \qquad \text{Cancel } h \text{ for } h \neq 0.$$

Identifying Common Factors

It can be shown that if $Q(x)$ is a polynomial and $Q(c) = 0$, then $(x - c)$ is a factor of $Q(x)$. Thus, if the numerator and denominator of a rational function of x are both zero at $x = c$, they have $(x - c)$ as a common factor.

(a)

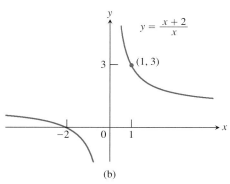

(b)

FIGURE 1.15 The graph of $f(x) = (x^2 + x - 2)/(x^2 - x)$ in (a) is the same as the graph of $g(x) = (x + 2)/x$ in (b) except at $x = 1$, where f is undefined. The functions have the same limit as $x \to 1$.

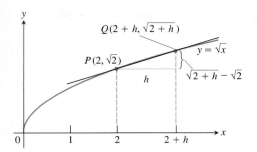

FIGURE 1.16 The limit of the slope $(\sqrt{2 + h} - \sqrt{2})/h$ of secant PQ as $Q \rightarrow P$ along the curve is $1/(2\sqrt{2})$. (Example 4)

Therefore,

$$\lim_{h \to 0} \frac{\sqrt{2 + h} - \sqrt{2}}{h} = \lim_{h \to 0} \frac{1}{\sqrt{2 + h} + \sqrt{2}}$$

$$= \frac{1}{\sqrt{2 + 0} + \sqrt{2}} \qquad \begin{array}{l} \text{Denominator not 0 at} \\ h = 0 \text{; substitute} \end{array}$$

$$= \frac{1}{2\sqrt{2}}.$$

Notice that $(\sqrt{2 + h} - \sqrt{2})/h$ is the slope of the secant through the points $P(2, \sqrt{2})$ and $Q(2 + h, \sqrt{2 + h})$ on the curve $y = \sqrt{x}$. Figure 1.16 shows the secant for $h > 0$. Our calculation shows that $1/(2\sqrt{2})$ is the limiting value of this slope as $Q \rightarrow P$ along the curve from either side.

Sandwich Theorem

If we cannot find a limit directly, we may be able to find it indirectly with the Sandwich Theorem. The theorem refers to a function f whose values are sandwiched between the values of two other functions, g and h. If g and h have the same limit as $x \rightarrow c$, then f has this limit, too (Figure 1.17).

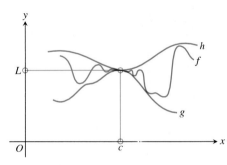

FIGURE 1.17 The graph of f is sandwiched between the graphs of g and h.

The Sandwich Theorem is sometimes called the Squeeze Theorem or the Pinching Theorem.

Theorem 4   Sandwich Theorem

Suppose that $g(x) \leq f(x) \leq h(x)$ for all x in some open interval containing c, except possibly at $x = c$ itself. Suppose also that

$$\lim_{x \to c} g(x) = \lim_{x \to c} h(x) = L.$$

Then $\lim_{x \to c} f(x) = L$.

You will find a proof of Theorem 4 in Appendix 2.

Example 5   Applying the Sandwich Theorem

Given that

$$1 - \frac{x^2}{4} \leq u(x) \leq 1 + \frac{x^2}{2} \qquad \text{for all } x \neq 0,$$

find $\lim_{x \to 0} u(x)$.

Solution Since

$$\lim_{x \to 0} (1 - (x^2/4)) = 1 \quad \text{and} \quad \lim_{x \to 0} (1 + (x^2/2)) = 1,$$

the Sandwich Theorem implies that $\lim_{x \to 0} u(x) = 1$ (Figure 1.18).

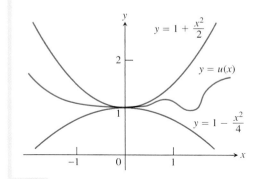

FIGURE 1.18 Any function $u(x)$ whose graph lies in the region between $y = 1 + (x^2/2)$ and $y = 1 - (x^2/4)$ has limit 1 as $x \to 0$.

Example 6 Another Application of the Sandwich Theorem

(a) (Figure 1.19a) Since $-|\theta| \le \sin \theta \le |\theta|$ for all θ and $\lim_{\theta \to 0} (-|\theta|) = \lim_{\theta \to 0} |\theta| = 0$, we have

$$\lim_{\theta \to 0} \sin \theta = 0.$$

(b) (Figure 1.19b) Since $0 \le 1 - \cos \theta \le |\theta|$ for all θ, we have $\lim_{\theta \to 0} (1 - \cos \theta) = 0$ or

$$\lim_{\theta \to 0} \cos \theta = 1.$$

(c) For any function $f(x)$, if $\lim_{x \to c} |f(x)| = 0$, then $\lim_{x \to c} f(x) = 0$. The argument: $-|f(x)| \le f(x) \le |f(x)|$ and $-|f(x)|$ and $|f(x)|$ have limit 0 as $x \to c$.

(a)

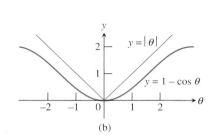

(b)

FIGURE 1.19 The Sandwich Theorem confirms that (a) $\lim_{\theta \to 0} \sin \theta = 0$ and (b) $\lim_{\theta \to 0} (1 - \cos \theta) = 0$.

One-Sided Limits

To have a limit L as x approaches a, a function f must be defined on *both sides* of a and its values $f(x)$ must approach L as x approaches a from either side. Because of this, ordinary limits are **two-sided.**

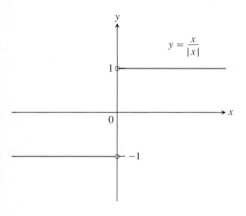

FIGURE 1.20 Different right-hand and left-hand limits at the origin.

___

The "+" and "−"

The significance of the signs in the notation for one-sided limits is that:

$x \rightarrow a^-$ means x approaches a from the negative side of a, through values less than a.

$x \rightarrow a^+$ means x approaches a from the positive side of a, through values greater than a.

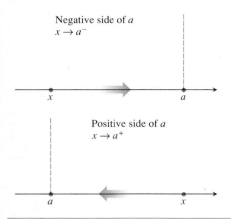

If f fails to have a two-sided limit at a, it may still have a one-sided limit, that is, a limit if the approach is only from one side. If the approach is from the right, the limit is a **right-hand** limit. From the left, it is a **left-hand** limit.

The function $f(x) = x/|x|$ (Figure 1.20) has limit 1 as x approaches 0 from the right, and limit -1 as x approaches 0 from the left.

> **Definitions**   Right–Hand and Left–Hand Limits
>
> Let $f(x)$ be defined on an interval (a, b), where $a < b$. If $f(x)$ approaches arbitrarily close to L as x approaches a from within that interval, then we say that f has **right-hand limit** L at a, and we write
>
> $$\lim_{x \to a^+} f(x) = L.$$
>
> Let $f(x)$ be defined on an interval (c, a), where $c < a$. If $f(x)$ approaches arbitrarily close to M as x approaches a from within the interval (c, a), then we say that f has **left-hand limit** M at a, and we write
>
> $$\lim_{x \to a^-} f(x) = M.$$

For the function $f(x) = x/|x|$ in Figure 1.20, we have

$$\lim_{x \to 0^+} f(x) = 1 \qquad \text{and} \qquad \lim_{x \to 0^-} f(x) = -1.$$

Example 7   One-Sided Limits for a Semicircle

The domain of $f(x) = \sqrt{4 - x^2}$ is $[-2, 2]$; its graph is the semicircle in Figure 1.21. We have

$$\lim_{x \to -2^+} \sqrt{4 - x^2} = 0 \qquad \text{and} \qquad \lim_{x \to 2^-} \sqrt{4 - x^2} = 0.$$

The function does not have a left-hand limit at $x = -2$ or a right-hand limit at $x = 2$. It does not have ordinary two-sided limits at either -2 or 2.

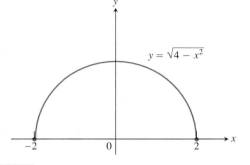

FIGURE 1.21 $\lim_{x \to 2^-} \sqrt{4 - x^2} = 0$ and $\lim_{x \to -2^+} \sqrt{4 - x^2} = 0$.

One-sided limits have all the properties listed in Theorem 1. The right-hand limit of the sum of two functions is the sum of their right-hand limits, and so on. The theorems for limits of polynomials and rational functions hold with one-sided limits, as does the Sandwich Theorem.

One-sided and two-sided limits are related in the following way:

The symbol ⇔

The symbol ⇔ is read "if and only if." It is a combination of the symbols ⇒ (implies) and ⇐ (is implied by).

Theorem 5   Relation Between One-Sided and Two–Sided Limits

A function $f(x)$ has a limit as x approaches c if and only if it has left-hand and right-hand limits there and these one-sided limits are equal:

$$\lim_{x \to c} f(x) = L \quad \Leftrightarrow \quad \lim_{x \to c^-} f(x) = L \quad \text{and} \quad \lim_{x \to c^+} f(x) = L.$$

Example 8   Limits of the Function Graphed in Figure 1.22

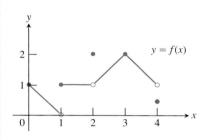

FIGURE 1.22 Graph of the function in Example 8.

At $x = 0$: $\lim_{x \to 0^+} f(x) = 1$,

$\lim_{x \to 0^-} f(x)$ and $\lim_{x \to 0} f(x)$ do not exist. The function is not defined to the left of $x = 0$.

At $x = 1$: $\lim_{x \to 1^-} f(x) = 0$ even though $f(1) = 1$,

$\lim_{x \to 1^+} f(x) = 1$,

$\lim_{x \to 1} f(x)$ does not exist. The right- and left-hand limits are not equal.

At $x = 2$: $\lim_{x \to 2^-} f(x) = 1$,

$\lim_{x \to 2^+} f(x) = 1$,

$\lim_{x \to 2} f(x) = 1$ even though $f(2) = 2$.

At $x = 3$: $\lim_{x \to 3^-} f(x) = \lim_{x \to 3^+} f(x) = \lim_{x \to 3} f(x) = f(3) = 2$

At $x = 4$: $\lim_{x \to 4^-} f(x) = 1$ even though $f(4) \neq 1$,

$\lim_{x \to 4^+} f(x)$ and $\lim_{x \to 4} f(x)$ do not exist. The function is not defined to the right of $x = 4$.

At every other point a in $[0, 4]$, $f(x)$ has limit $f(a)$.

The functions examined so far have had some kind of limit at each point of interest. In general, that need not be the case.

Example 9   A Function Oscillating Too Much

Show that $y = \sin (1/x)$ has no limit as x approaches zero from either side (Figure 1.23).

Solution   As x approaches zero, its reciprocal, $1/x$, grows without bound and the values of $\sin(1/x)$ cycle repeatedly from -1 to 1. There is no single number L that the function's values stay increasingly close to as x approaches zero, which is true even if we restrict x to positive values or to negative values. The function has neither a right-hand limit nor a left-hand limit at $x = 0$.

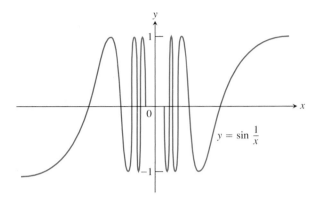

FIGURE 1.23 The function $y = \sin(1/x)$ has neither a right-hand nor a left-hand limit as x approaches zero. (Example 9)

Limits Involving $(\sin \theta)/\theta$

A central fact about $(\sin \theta)/\theta$ is that in radian measure its limit as $\theta \to 0$ is 1. We can see this in Figure 1.24 and confirm it algebraically using the Sandwich Theorem.

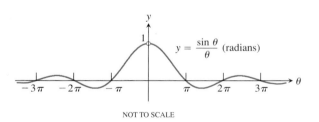

NOT TO SCALE

FIGURE 1.24 The graph of $f(\theta) = (\sin \theta)/\theta$.

Theorem 6

$$\lim_{\theta \to 0} \frac{\sin \theta}{\theta} = 1 \qquad (\theta \text{ in radians}) \qquad (1)$$

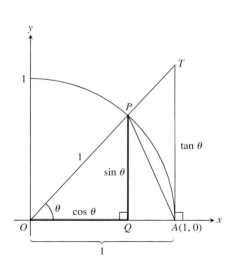

FIGURE 1.25 The figure for the proof of Theorem 6. $TA/OA = \tan \theta$, but $OA = 1$, so $TA = \tan \theta$.

Proof   The plan is to show that the right-hand and left-hand limits are both 1. Then we will know that the two-sided limit is 1 as well.

To show that the right-hand limit is 1, we begin with positive values of θ less than $\pi/2$ (Figure 1.25). Notice that

Area $\triangle OAP$ < area sector OAP < area $\triangle OAT$.

We can express these areas in terms of θ as follows:

Equation (2) is where radian measure comes in: The area of sector OAP is $\theta/2$ only if θ is measured in radians.

$$\text{Area } \Delta OAP = \frac{1}{2} \text{ base} \times \text{height} = \frac{1}{2}(1)(\sin \theta) = \frac{1}{2}\sin \theta$$

$$\text{Area sector } OAP = \frac{1}{2}r^2\theta = \frac{1}{2}(1)^2\theta = \frac{\theta}{2} \qquad (2)$$

$$\text{Area } \Delta OAT = \frac{1}{2} \text{ base} \times \text{height} = \frac{1}{2}(1)(\tan \theta) = \frac{1}{2}\tan \theta.$$

Thus,

$$\frac{1}{2}\sin \theta < \frac{1}{2}\theta < \frac{1}{2}\tan \theta.$$

This last inequality goes the same way if we divide all three terms by the positive number $(1/2)\sin \theta$:

$$1 < \frac{\theta}{\sin \theta} < \frac{1}{\cos \theta}.$$

Taking reciprocals reverses the inequalities:

$$1 > \frac{\sin \theta}{\theta} > \cos \theta.$$

Since $\lim_{\theta \to 0^+} \cos \theta = 1$, the Sandwich Theorem gives

$$\lim_{\theta \to 0^+} \frac{\sin \theta}{\theta} = 1.$$

Recall that $\sin \theta$ and θ are both *odd functions* (Preliminary Section 3). Therefore, $f(\theta) = (\sin \theta)/\theta$ is an *even function,* with a graph symmetric about the y-axis (see Figure 1.24). This symmetry implies that the left-hand limit at 0 exists and has the same value as the right-hand limit:

$$\lim_{\theta \to 0^-} \frac{\sin \theta}{\theta} = 1 = \lim_{\theta \to 0^+} \frac{\sin \theta}{\theta},$$

so $\lim_{\theta \to 0} (\sin \theta)/\theta = 1$ by Theorem 4.   ▬

CD-ROM
WEBsite

Example 10   Using $\lim\limits_{\theta \to 0} \dfrac{\sin \theta}{\theta} = 1$

Show that (a) $\lim\limits_{h \to 0} \dfrac{\cos h - 1}{h} = 0$ and (b) $\lim\limits_{x \to 0} \dfrac{\sin 2x}{5x} = \dfrac{2}{5}$.

Solution

(a) Using the half-angle formula $\cos h = 1 - 2\sin^2(h/2)$, we calculate

$$\lim_{h \to 0} \frac{\cos h - 1}{h} = \lim_{h \to 0} -\frac{2\sin^2(h/2)}{h}$$

$$= -\lim_{\theta \to 0} \frac{\sin \theta}{\theta} \sin \theta \qquad \text{Let } \theta = h/2.$$

$$= -(1)(0) = 0.$$

(b) Equation (1) does not apply to the original fraction. We need a $2x$ in the denominator, not a $5x$. We produce it by multiplying numerator and denominator by $2/5$:

$$\lim_{x \to 0} \frac{\sin 2x}{5x} = \lim_{x \to 0} \frac{(2/5) \cdot \sin 2x}{(2/5) \cdot 5x}$$

$$= \frac{2}{5} \lim_{x \to 0} \frac{\sin 2x}{2x} \qquad \begin{array}{l} \text{Now, Eq. (1) applies with} \\ \theta = 2x. \end{array}$$

$$= \frac{2}{5}(1) = \frac{2}{5}$$

EXERCISES 1.2

Estimating Limits Graphically

In Exercises 1–6, use the graph to estimate the limits and value of the function, or explain why the limits do not exist.

1.

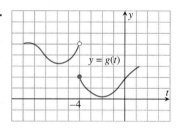

(a) $\lim_{x \to 3^-} f(x)$ **(b)** $\lim_{x \to 3^+} f(x)$ **(c)** $\lim_{x \to 3} f(x)$ **(d)** $f(3)$

2.

(a) $\lim_{t \to -4^-} g(t)$ **(b)** $\lim_{t \to -4^+} g(t)$ **(c)** $\lim_{t \to -4} g(t)$ **(d)** $g(-4)$

3.

(a) $\lim_{h \to 0^-} f(h)$ **(b)** $\lim_{h \to 0^+} f(h)$ **(c)** $\lim_{h \to 0} f(h)$ **(d)** $f(0)$

4.

(a) $\lim_{s \to -2^-} p(s)$ **(b)** $\lim_{s \to -2^+} p(s)$ **(c)** $\lim_{s \to -2} p(s)$ **(d)** $p(-2)$

5.

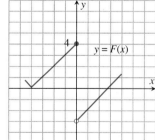

(a) $\lim_{x \to 0^-} F(x)$ **(b)** $\lim_{x \to 0^+} F(x)$ **(c)** $\lim_{x \to 0} F(x)$ **(d)** $F(0)$

6.

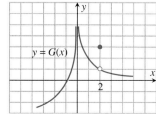

(a) $\lim_{x \to 2^-} G(x)$ **(b)** $\lim_{x \to 2^+} G(x)$ **(c)** $\lim_{x \to 2} G(x)$ **(d)** $G(2)$

Using Limit Rules

7. Suppose that $\lim_{x \to 0} f(x) = 1$ and $\lim_{x \to 0} g(x) = -5$. Name the rules in Theorem 1 that are used to accomplish steps (a), (b), and (c) of the following calculation.

$$\lim_{x \to 0} \frac{2f(x) - g(x)}{(f(x) + 7)^{2/3}}$$

$$= \frac{\lim_{x \to 0} (2f(x) - g(x))}{\lim_{x \to 0} (f(x) + 7)^{2/3}} \quad \text{(a)}$$

$$= \frac{\lim_{x \to 0} 2f(x) - \lim_{x \to 0} g(x)}{\left(\lim_{x \to 0} (f(x) + 7) \right)^{2/3}} \quad \text{(b)}$$

$$= \frac{2 \lim_{x \to 0} f(x) - \lim_{x \to 0} g(x)}{\left(\lim_{x \to 0} f(x) + \lim_{x \to 0} 7 \right)^{2/3}} \quad \text{(c)}$$

$$= \frac{(2)(1) - (-5)}{(1 + 7)^{2/3}} = \frac{7}{4}$$

8. Let $\lim_{x \to 1} h(x) = 5$, $\lim_{x \to 1} p(x) = 1$, and $\lim_{x \to 1} r(x) = -2$. Name the rules in Theorem 1 that are used to accomplish steps (a), (b), and (c) of the following calculation.

$$\lim_{x \to 1} \frac{\sqrt{5h(x)}}{p(x)(4 - r(x))}$$

$$= \frac{\lim_{x \to 1} \sqrt{5h(x)}}{\lim_{x \to 1} (p(x)(4 - r(x)))} \quad \text{(a)}$$

$$= \frac{\sqrt{\lim_{x \to 1} 5h(x)}}{\left(\lim_{x \to 1} p(x) \right)\left(\lim_{x \to 1} (4 - r(x)) \right)} \quad \text{(b)}$$

$$= \frac{\sqrt{5 \lim_{x \to 1} h(x)}}{\left(\lim_{x \to 1} p(x) \right)\left(\lim_{x \to 1} 4 - \lim_{x \to 1} r(x) \right)} \quad \text{(c)}$$

$$= \frac{\sqrt{(5)(5)}}{(1)(4 - 2)} = \frac{5}{2}$$

9. Suppose that $\lim_{x \to c} f(x) = 5$ and $\lim_{x \to c} g(x) = -2$. Find

(a) $\lim_{x \to c} f(x)g(x)$ **(b)** $\lim_{x \to c} 2f(x)g(x)$

(c) $\lim_{x \to c} (f(x) + 3g(x))$ **(d)** $\lim_{x \to c} \dfrac{f(x)}{f(x) - g(x)}$

10. Suppose that $\lim_{x \to 4} f(x) = 0$ and $\lim_{x \to 4} g(x) = -3$. Find

(a) $\lim_{x \to 4} (g(x) + 3)$ **(b)** $\lim_{x \to 4} xf(x)$

(c) $\lim_{x \to 4} (g(x))^2$ **(d)** $\lim_{x \to 4} \dfrac{g(x)}{f(x) - 1}$

Limit Calculations

Find the limits in Exercises 11–14.

11. (a) $\lim_{x \to -7} (2x + 5)$ **(b)** $\lim_{t \to 6} 8(t - 5)(t - 7)$

(c) $\lim_{y \to 2} \dfrac{y + 2}{y^2 + 5y + 6}$ **(d)** $\lim_{h \to 0} \dfrac{3}{\sqrt{3h + 1} + 1}$

12. (a) $\lim_{r \to -2} (r^3 - 2r^2 + 4r + 8)$ **(b)** $\lim_{x \to 2} \dfrac{x + 3}{x + 6}$

(c) $\lim_{y \to -3} (5 - y)^{4/3}$ **(d)** $\lim_{\theta \to 5} \dfrac{\theta - 5}{\theta^2 - 25}$

13. (a) $\lim_{t \to -5} \dfrac{t^2 + 3t - 10}{t + 5}$ **(b)** $\lim_{x \to -2} \dfrac{-2x - 4}{x^3 + 2x^2}$

(c) $\lim_{y \to 1} \dfrac{y - 1}{\sqrt{y + 3} - 2}$

14. (a) $\lim_{x \to -1} \dfrac{\sqrt{x^2 + 8} - 3}{x + 1}$ **(b)** $\lim_{\theta \to 1} \dfrac{\theta^4 - 1}{\theta^3 - 1}$

(c) $\lim_{t \to 9} \dfrac{3 - \sqrt{t}}{9 - t}$

Using the Sandwich Theorem

15. *Writing to Learn* **(a)** It can be shown that the inequalities

$$1 - \frac{x^2}{6} < \frac{x \sin x}{2 - 2 \cos x} < 1$$

hold for all values of x close to zero. What, if anything, does this tell you about

$$\lim_{x \to 0} \frac{x \sin x}{2 - 2 \cos x}$$

Give reasons for your answer.

T **(b)** Graph $y = 1 - (x^2/6)$, $y = (x \sin x)/(2 - 2 \cos x)$, and $y = 1$ together for $-2 \le x \le 2$. Comment on the behavior of the graphs as $x \to 0$.

16. *Writing to Learn* **(a)** The inequalities

$$\frac{1}{2} - \frac{x^2}{24} < \frac{1 - \cos x}{x^2} < \frac{1}{2}$$

hold for values of x close to zero. What, if anything, does this tell you about

$$\lim_{x \to 0} \frac{1 - \cos x}{x^2}?$$

Give reasons for your answer.

T **(b)** Graph $y = (1/2) - (x^2/24)$, $y = (1 - \cos x)/x^2$, and $y = 1/2$ together for $-2 \le x \le 2$. Comment on the behavior of the graphs as $x \to 0$.

Limits of Average Rates of Change

Because of their connections with secant lines, tangents, and instantaneous rates, limits of the form

$$\lim_{h \to 0} \frac{f(x_0 + h) - f(x_0)}{h}$$

occur frequently in calculus. In Exercises 17–20, evaluate this limit for the given x_0 and function f.

17. $f(x) = x^2$, $x_0 = 1$

18. $f(x) = 3x - 4$, $x_0 = 2$

19. $f(x) = 1/x$, $x_0 = -2$

20. $f(x) = \sqrt{x}$, $x_0 = 7$

Finding Limits Graphically

21. Which of the following statements about the function $y = f(x)$ graphed here are true and which are false? Give reasons for your answers.

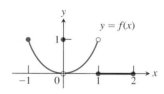

(a) $\lim\limits_{x \to -1^+} f(x) = 1$.

(b) $\lim\limits_{x \to 0^-} f(x) = 0$.

(c) $\lim\limits_{x \to 0} f(x) = 1$.

(d) $\lim\limits_{x \to 0^-} f(x) = \lim\limits_{x \to 0^+} f(x)$.

(e) $\lim\limits_{x \to 0} f(x)$ exists.

(f) $\lim\limits_{x \to 0} f(x) = 0$.

(g) $\lim\limits_{x \to 0} f(x) = 1$.

(h) $\lim\limits_{x \to 1} f(x) = 1$.

(i) $\lim\limits_{x \to 1} f(x) = 0$.

(j) $\lim\limits_{x \to 2^-} f(x) = 2$.

(k) $\lim\limits_{x \to -1^-} f(x)$ does not exist.

(l) $\lim\limits_{x \to 2^+} f(x) = 0$.

22. Let

$$f(x) = \begin{cases} 3 - x, & x < 2 \\ \dfrac{x}{2} + 1, & x > 2. \end{cases}$$

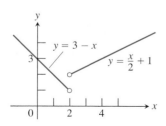

(a) Find $\lim\limits_{x \to 2^+} f(x)$ and $\lim\limits_{x \to 2^-} f(x)$.

(b) Does $\lim\limits_{x \to 2} f(x)$ exist? If so, what is it? If not, why not?

(c) Find $\lim\limits_{x \to 4^-} f(x)$ and $\lim\limits_{x \to 4^+} f(x)$.

(d) Does $\lim\limits_{x \to 4} f(x)$ exist? If so, what is it? If not, why not?

23. Let $f(x) = \begin{cases} 0, & x \le 0 \\ \sin \dfrac{1}{x}, & x > 0. \end{cases}$

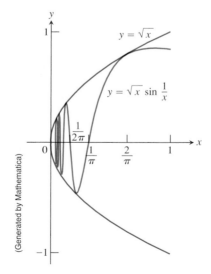

(a) Does $\lim\limits_{x \to 0^+} f(x)$ exist? If so, what is it? If not, why not?

(b) Does $\lim\limits_{x \to 0^-} f(x)$ exist? If so, what is it? If not, why not?

(c) Does $\lim\limits_{x \to 0} f(x)$ exist? If so, what is it? If not, why not?

24. Let $g(x) = \sqrt{x} \sin (1/x)$.

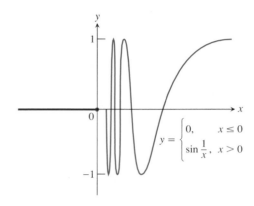

(a) Does $\lim\limits_{x \to 0^+} g(x)$ exist? If so, what is it? If not, why not?

(b) Does $\lim\limits_{x \to 0^-} g(x)$ exist? If so, what is it? If not, why not?

(c) Does $\lim\limits_{x \to 0} g(x)$ exist? If so, what is it? If not, why not?

Graph the functions in Exercises 25 and 26. Then answer these questions.

(a) What are the domain and range of f?

(b) At what point c, if any, does $\lim_{x \to c} f(x)$ exist?

(c) At what point does only the left-hand limit exist?

(d) At what point does only the right-hand limit exist?

25. $f(x) = \begin{cases} \sqrt{1 - x^2} & \text{if } 0 \le x < 1 \\ 1 & \text{if } 1 \le x < 2 \\ 2 & \text{if } x = 2 \end{cases}$

26. $f(x) = \begin{cases} x & \text{if } -1 \le x < 0 \quad \text{or} \quad 0 < x \le 1 \\ 1 & \text{if } x = 0 \\ 0 & \text{if } x < -1 \quad \text{or} \quad x > 1 \end{cases}$

Finding One-Sided Limits Algebraically

Find the limits in Exercises 27–32.

27. $\lim_{x \to -0.5^-} \sqrt{\dfrac{x + 2}{x + 1}}$

28. $\lim_{x \to -2^+} \left(\dfrac{x}{x + 1} \right) \left(\dfrac{2x + 5}{x^2 + x} \right)$

29. $\lim_{h \to 0^+} \dfrac{\sqrt{h^2 + 4h + 5} - \sqrt{5}}{h}$

30. $\lim_{h \to 0^-} \dfrac{\sqrt{6} - \sqrt{5h^2 + 11h + 6}}{h}$

31. (a) $\lim_{x \to -2^+} (x + 3) \dfrac{|x + 2|}{x + 2}$ **(b)** $\lim_{x \to -2^-} (x + 3) \dfrac{|x + 2|}{x + 2}$

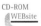

32. (a) $\lim_{x \to 1^+} \dfrac{\sqrt{2x}\,(x - 1)}{|x - 1|}$ **(b)** $\lim_{x \to 1^-} \dfrac{\sqrt{2x}\,(x - 1)}{|x - 1|}$

Theory and Examples

33. *Writing to Learn* If $x^4 \le f(x) \le x^2$ for x in $[-1, 1]$ and $x^2 \le f(x) \le x^4$ for $x < -1$ and $x > 1$, at what points c do you automatically know $\lim_{x \to c} f(x)$? What can you say about the value of the limit at these points?

34. *Writing to Learn* Suppose that $g(x) \le f(x) \le h(x)$ for all $x \ne 2$ and suppose that

$$\lim_{x \to 2} g(x) = \lim_{x \to 2} h(x) = -5.$$

Can we conclude anything about the values of f, g, and h at $x = 2$? Could $f(2) = 0$? Could $\lim_{x \to 2} f(x) = 0$? Give reasons for your answers.

35. *Deducing the value of a limit* If $\lim_{x \to -2} \dfrac{f(x)}{x^2} = 1$, find

(a) $\lim_{x \to -2} f(x)$ (b) $\lim_{x \to -2} \dfrac{f(x)}{x}$

36. *Deducing the value of a limit*

(a) If $\lim_{x \to 2} \dfrac{f(x) - 5}{x - 2} = 3$, find $\lim_{x \to 2} f(x)$.

(b) If $\lim_{x \to 2} \dfrac{f(x) - 5}{x - 2} = 4$, find $\lim_{x \to 2} f(x)$.

37. *Writing to Learn* Once you know $\lim_{x \to a^+} f(x)$ and $\lim_{x \to a^-} f(x)$ at an interior point of the domain of f, do you then know $\lim_{x \to a} f(x)$? Give reasons for your answer.

38. *Writing to Learn* If you know that $\lim_{x \to c} f(x)$ exists, can you find its value by calculating $\lim_{x \to c^+} f(x)$? Give reasons for your answer.

39. *Finding delta* Given $\epsilon > 0$, find an interval $I = (5, 5 + \delta)$, $\delta > 0$, such that if x lies in I, then $\sqrt{x - 5} < \epsilon$. What limit is being verified, and what is its value?

40. *Finding delta* Given $\epsilon > 0$, find an interval $I = (4 - \delta, 4)$, $\delta > 0$, such that if x lies in I, then $\sqrt{4 - x} < \epsilon$. What limit is being verified, and what is its value.

Even and Odd Functions

Recall that a function $y = f(x)$ defined on a domain D that is symmetric about the origin is **even** if $f(-x) = f(x)$ for all x in D and **odd** if $f(-x) = -f(x)$ for all x in D.

41. *Writing to Learn* Suppose that f is an odd function of x. Does knowing that $\lim_{x \to 0^+} f(x) = 3$ tell you anything about $\lim_{x \to 0^-} f(x)$? Give reasons for your answer.

42. *Writing to Learn* Suppose that f is an even function of x. Does knowing that $\lim_{x \to 2^-} f(x) = 7$ tell you anything about either $\lim_{x \to -2^-} f(x)$ or $\lim_{x \to -2^+} f(x)$? Give reasons for your answer.

COMPUTER EXPLORATIONS

43. (a) Graph $g(x) = x \sin(1/x)$ to estimate $\lim_{x \to 0} g(x)$, zooming in on the origin as necessary.

(b) *Writing to Learn* Now graph $k(x) = \sin(1/x)$. Compare the behaviors of g and k near the origin. What is the same? What is different?

44. (a) Graph $h(x) = x^2 \cos(1/x)$ to estimate $\lim_{x \to 0} h(x)$, zooming in on the origin as necessary.

(b) *Writing to Learn* Now graph $k(x) = \cos(1/x)$. Compare the behaviors of h and k near the origin. What is the same? What is different?

Limits Involving Infinity

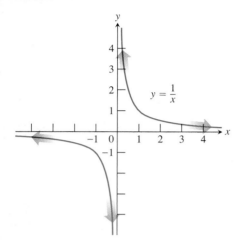

FIGURE 1.26 The graph of $y = 1/x$.

Finite Limits as $x \to \pm\infty$ • Limits of Rational Functions as $x \to \pm\infty$ • Horizontal and Vertical Asymptotes: Infinite Limits • Sandwich Theorem Revisited • Precise Definitions of Infinite Limits • End Behavior Models and Oblique Asymptotes

We analyze the graphs of rational functions (quotients of polynomial functions), as well as other functions with interesting limit behavior as $x \to \pm\infty$. Among the tools we use are horizontal and vertical asymptotes.

Finite Limits as $x \to \pm\infty$

The symbol for infinity (∞) does not represent a real number. We use ∞ to describe the behavior of a function when the values in its domain or range outgrow all finite bounds. For example, the function $f(x) = 1/x$ is defined for all $x \neq 0$ (Figure 1.26). When x is positive and becomes increasingly large, $1/x$ becomes increasingly small. When x is negative and its magnitude becomes increasingly large, $1/x$ again becomes small. We summarize these observations by saying that $f(x) = 1/x$ has limit 0 as $x \to \pm\infty$.

> **Definitions** Limits as $x \to \pm\infty$
>
> 1. We say that $f(x)$ has the **limit L as x approaches infinity** and write
>
> $$\lim_{x \to \infty} f(x) = L$$
>
> if, as x moves increasingly far from the origin in the positive direction, $f(x)$ gets arbitrarily close to L.
>
> 2. We say that $f(x)$ has the **limit L as x approaches minus infinity** and write
>
> $$\lim_{x \to -\infty} f(x) = L$$
>
> if, as x moves increasingly far from the origin in the negative direction, $f(x)$ gets arbitrarily close to L.

CD-ROM
WEBsite

The strategy for calculating limits of functions as $x \to \pm\infty$ is similar to the one for finite limits in Section 1.2. There we first found the limits of the constant and identity functions $y = k$ and $y = x$. We then extended these results to other functions by applying a theorem about limits of algebraic combinations. Here we do the same thing, except that the starting functions are $y = k$ and $y = 1/x$ instead of $y = k$ and $y = x$.

The basic facts to be verified as $x \to \pm\infty$ are given in the next example.

The Symbol Infinity (∞)

As always, the symbol ∞ does not represent a real number and we cannot use it in arithmetic in the usual way. Also, the symbol ∞ means $+\infty$. The two are used interchangeably.

Example 1   Limits of $1/x$ and k as $x \to \pm\infty$

Show that

(a) $\displaystyle\lim_{x \to \infty} \frac{1}{x} = \lim_{x \to -\infty} \frac{1}{x} = 0$

(b) $\displaystyle\lim_{x \to \infty} k = \lim_{x \to -\infty} k = k.$

Solution

(a) From Figure 1.26, we see that $y = 1/x$ gets closer and closer to zero as x moves increasingly far from the origin in either the positive or negative direction.

(b) No matter how far x is from the origin, the constant function $y = k$ always has exactly the value k.

Limits at infinity have properties similar to those of finite limits.

Theorem 7   Rules for Limits as x → ±∞

If L, M, and k, are real numbers and

$$\lim_{x \to \pm\infty} f(x) = L \quad\text{and}\quad \lim_{x \to \pm\infty} g(x) = M, \quad\text{then}$$

1. Sum Rule: $\qquad\qquad\qquad\qquad \lim_{x \to \pm\infty} (f(x) + g(x)) = L + M$

2. Difference Rule: $\qquad\qquad\quad \lim_{x \to \pm\infty} (f(x) - g(x)) = L - M$

3. Product Rule: $\qquad\qquad\qquad \lim_{x \to \pm\infty} (f(x) \cdot g(x)) = L \cdot M$

4. Constant Multiple Rule: $\qquad \lim_{x \to \pm\infty} (k \cdot f(x)) = k \cdot L$

5. Quotient Rule: $\qquad\qquad\qquad \lim_{x \to \pm\infty} \dfrac{f(x)}{g(x)} = \dfrac{L}{M}, \quad M \neq 0$

6. Power Rule: If r and s are integers, $s \neq 0$, then

$$\lim_{x \to \pm\infty} (f(x))^{r/s} = L^{r/s}$$

provided that $L^{r/s}$ is a real number.

These properties are just like the properties in Theorem 1, Section 1.2, and we use them the same way.

Example 2   Using Theorem 7

(a) $\displaystyle\lim_{x \to \infty} \left(5 + \frac{1}{x}\right) = \lim_{x \to \infty} 5 + \lim_{x \to \infty} \frac{1}{x}$ Sum Rule

$\qquad\qquad\qquad\quad = 5 + 0 = 5$ Known limits

(b) $\displaystyle\lim_{x \to -\infty} \frac{\pi\sqrt{3}}{x^2} = \lim_{x \to -\infty} \pi\sqrt{3} \cdot \frac{1}{x} \cdot \frac{1}{x}$

$\qquad\qquad\qquad = \lim_{x \to -\infty} \pi\sqrt{3} \cdot \lim_{x \to -\infty} \frac{1}{x} \cdot \lim_{x \to -\infty} \frac{1}{x}$ Product rule

$\qquad\qquad\qquad = \pi\sqrt{3} \cdot 0 \cdot 0 = 0$ Known limits

The **degree** of the polynomial $a_n x^n + a_{n-1}x^{n-1} + \cdots + a_1 x + a_0$, $a_n \neq 0$, is n, the largest exponent.

Limits of Rational Functions as x → ±∞

To determine the limit of a rational function as $x \to \pm\infty$, we can divide the numerator and denominator by the highest power of x in the denominator. What happens then depends on the degrees of the polynomials involved.

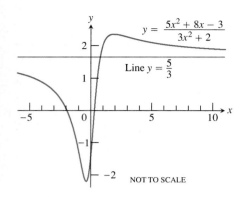

$$y = \frac{5x^2 + 8x - 3}{3x^2 + 2}$$

Line $y = \frac{5}{3}$

NOT TO SCALE

FIGURE 1.27 The function in Example 3.

Example 3 Numerator and Denominator of Same Degree

$$\lim_{x \to \infty} \frac{5x^2 + 8x - 3}{3x^2 + 2} = \lim_{x \to \infty} \frac{5 + (8/x) - (3/x^2)}{3 + (2/x^2)}$$

Divide numerator and denominator by x^2.

$$= \frac{5 + 0 - 0}{3 + 0} = \frac{5}{3}$$

See Fig. 1.27.

Example 4 Degree of Numerator Less Than Degree of Denominator

$$\lim_{x \to -\infty} \frac{11x + 2}{2x^3 - 1} = \lim_{x \to -\infty} \frac{(11/x^2) + (2/x^3)}{2 - (1/x^3)}$$

Divide numerator and denominator by x^3.

$$= \frac{0 + 0}{2 - 0} = 0$$

See Fig. 1.28.

Example 5 Degree of Numerator Greater Than Degree of Denominator

(a) $$\lim_{x \to -\infty} \frac{2x^2 - 3}{7x + 4} = \lim_{x \to -\infty} \frac{2x - (3/x)}{7 + (4/x)}$$

Divide numerator and denominator by x.

$$= -\infty$$

The numerator now approaches $-\infty$ while the denominator approaches 7, so the ratio $\to -\infty$. See Fig. 1.29.

(b) $$\lim_{x \to -\infty} \frac{-4x^3 + 7x}{2x^2 - 3x - 10} = \lim_{x \to -\infty} \frac{-4x + (7/x)}{2 - (3/x) - (10/x^2)}$$

Divide numerator and denominator by x^2.

$$= \infty$$

Numerator $\to \infty$, denominator $\to 2$; ratio $\to \infty$.

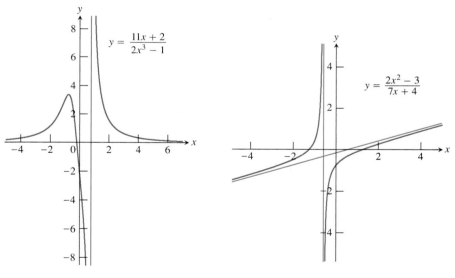

$$y = \frac{11x + 2}{2x^3 - 1}$$

$$y = \frac{2x^2 - 3}{7x + 4}$$

FIGURE 1.28 The graph of the function in Example 4. The graph approaches the x-axis as $|x|$ increases.

FIGURE 1.29 The function in Example 5(a).

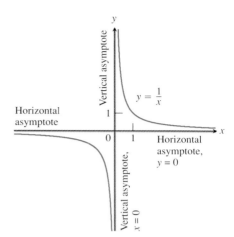

FIGURE 1.30 The coordinate axes are asymptotes of both branches of the hyperbola $y = 1/x$.

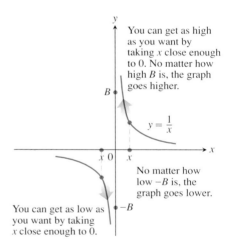

FIGURE 1.31 One-sided infinite limits:
$$\lim_{x \to 0^+} \frac{1}{x} = \infty \quad \text{and} \quad \lim_{x \to 0^-} \frac{1}{x} = -\infty.$$

Horizontal and Vertical Asymptotes: Infinite Limits

Looking at $f(x) = 1/x$ (Figure 1.30), we observe the following:

(a) As $x \to \infty$, $(1/x) \to 0$, and we write $\displaystyle\lim_{x \to \infty} (1/x) = 0$.

(b) As $x \to -\infty$, $(1/x) \to 0$, and we write $\displaystyle\lim_{x \to -\infty} (1/x) = 0$.

We say that the line $y = 0$ is a *horizontal asymptote* of the graph of f.

If the distance between the graph of a function and some fixed line approaches zero as the graph moves increasingly far from the origin, we say that the graph approaches the line asymptotically and that the line is an *asymptote* of the graph.

Let us look closely at the function $f(x) = 1/x$ in Figure 1.31. As $x \to 0^+$, the values of f grows without bound, eventually reaching and surpassing every positive real number. That is, given any positive real number B, however large, the values of f become larger still (Figure 1.31). Thus, f has no limit as $x \to 0^+$. It is nevertheless convenient to describe the behavior of f by saying that $f(x)$ approaches ∞ as $x \to 0^+$. We write

$$\lim_{x \to 0^+} f(x) = \lim_{x \to 0^+} \frac{1}{x} = \infty.$$

As $x \to 0^-$, the values of $f(x) = 1/x$ become arbitrarily large and negative. Given any negative real number $-B$, the values of f eventually lie below $-B$. (See Figure 1.31.) We write

$$\lim_{x \to 0^-} f(x) = \lim_{x \to 0^-} \frac{1}{x} = -\infty.$$

Notice that the denominator is zero at $x = 0$ and the function is undefined.

Definitions   Horizontal and Vertical Asymptotes

A line $y = b$ is a **horizontal asymptote** of the graph of a function $y = f(x)$ if either

$$\lim_{x \to \infty} f(x) = b \qquad \text{or} \qquad \lim_{x \to -\infty} f(x) = b.$$

A line $x = a$ is a **vertical asymptote** of the graph if either

$$\lim_{x \to a^+} f(x) = \pm\infty \qquad \text{or} \qquad \lim_{x \to a^-} f(x) = \pm\infty.$$

Example 6   Looking for Asymptotes

Find the asymptotes of the curve

$$y = \frac{x + 3}{x + 2}.$$

Solution   We are interested in the behavior as $x \to \pm\infty$ and as $x \to -2$, where the denominator is zero.

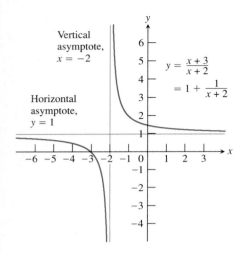

FIGURE 1.32 The lines $y = 1$ and $x = -2$ are asymptotes of the curve $y = (x + 3)/(x + 2)$. (Example 6)

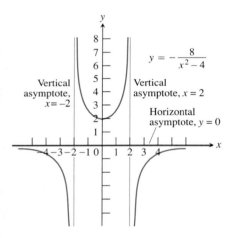

FIGURE 1.33 Graph of $y = -8/(x^2 - 4)$. Notice that the curve approaches the x-axis from only one side. Asymptotes do not have to be two-sided. (Example 7)

The asymptotes are quickly revealed if we recast the rational function as a polynomial with a remainder, by dividing $(x + 2)$ into $(x + 3)$.

$$
\begin{array}{r}
1 \\
x + 2 \overline{)x + 3} \\
\underline{x + 2} \\
1
\end{array}
$$

This result enables us to rewrite y:

$$y = 1 + \frac{1}{x + 2}.$$

We now see that the curve in question is the graph of $y = 1/x$ shifted 1 unit up and 2 units left (Figure 1.32). The asymptotes, instead of being the coordinate axes, are now the lines $y = 1$ and $x = -2$.

Example 7   Asymptotes Need Not Be Two-Sided

Find the asymptotes of the graph of

$$f(x) = -\frac{8}{x^2 - 4}.$$

Solution   We are interested in the behavior as $x \to \pm\infty$ and as $x \to \pm2$, where the denominator is zero. Notice that f is an even function of x, so its graph is symmetric with respect to the y-axis.

The behavior as $x \to \pm\infty$. Since $\lim_{x \to \infty} f(x) = 0$, the line $y = 0$ is an asymptote of the graph to the right. By symmetry it is an asymptote to the left as well (Figure 1.33).

The behavior as $x \to \pm2$. Since

$$\lim_{x \to 2^+} f(x) = -\infty \qquad \text{and} \qquad \lim_{x \to 2^-} f(x) = \infty,$$

the line $x = 2$ is a vertical asymptote both from the right and from the left. By symmetry, the same holds for the line $x = -2$.

There are no other asymptotes because f has a finite limit at every other point.

Example 8   Curves with Infinitely Many Asymptotes

The curves

$$y = \sec x = \frac{1}{\cos x} \qquad \text{and} \qquad y = \tan x = \frac{\sin x}{\cos x}$$

both have vertical asymptotes at odd-integer multiples of $\pi/2$, where $\cos x = 0$ (Figure 1.34).

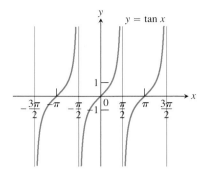

FIGURE 1.34 The graphs of sec x and tan x. (Example 8)

The graphs of

$$y = \csc x = \frac{1}{\sin x} \quad \text{and} \quad y = \cot x = \frac{\cos x}{\sin x}$$

have vertical asymptotes at integer multiples of π, where $\sin x = 0$ (Figure 1.35).

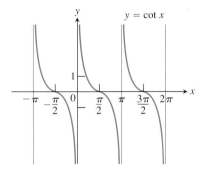

FIGURE 1.35 The graphs of csc x and cot x. (Example 8)

x	e^x
0	1.00000
-1	0.36788
-2	0.13534
-3	0.04979
-5	0.00674
-8	0.00034
-10	0.00005

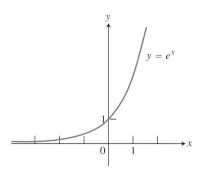

FIGURE 1.36 The line $y = 0$ is a horizontal asymptote of the graph $y = e^x$.

Example 9   Horizontal Asymptote of $y = e^x$

The curve

$$y = e^x$$

has the line $y = 0$ (the x-axis) as a horizontal asymptote. We see this from the graph in Figure 1.36 and its accompanying table of values. We write

$$\lim_{x \to -\infty} e^x = 0.$$

Notice that the values of e^x approach 0 quite rapidly.

We can investigate the behavior of $y = f(x)$ as $x \to \pm\infty$ by investigating the limit of $y = f(1/x)$ as $x \to 0$.

Example 10   Substituting a New Variable

Find $\lim\limits_{x\to\infty} \sin (1/x)$.

Solution   We introduce the new variable $t = 1/x$. From Figure 1.30, we know that $t \to 0^+$ as $x \to \infty$. Therefore,

$$\lim_{x\to\infty} \sin \frac{1}{x} = \lim_{t\to 0^+} \sin t = 0.$$

Likewise, we can investigate the behavior of $y = f(1/x)$ as $x \to 0$ by investigating $y = f(x)$ as $x \to \pm\infty$.

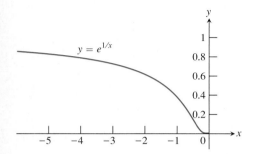

FIGURE 1.37 The graph of $y = e^{1/x}$ for $x < 0$ shows $\lim_{x\to 0^-} e^{1/x} = 0$. (Example 11)

Example 11   Using Substitution

Find $\lim\limits_{x\to 0^-} e^{1/x}$.

Solution   We let $t = 1/x$. From Figure 1.31, we know that $t \to -\infty$ as $x \to 0^-$. Therefore,

$$\lim_{x\to 0^-} e^{1/x} = \lim_{t\to -\infty} e^t = 0 \qquad \text{Example 9}$$

(Figure 1.37).

Sandwich Theorem Revisited

The Sandwich Theorem also holds for limits as $x \to \pm\infty$.

Example 12   Finding a Limit as x Approaches 0 or $\pm\infty$

Using the Sandwich Theorem, find the asymptotes of the curve

$$y = 2 + \frac{\sin x}{x}.$$

Solution   We are interested in the behavior as $x \to \pm\infty$ and as $x \to 0$, where the denominator is zero.

The behavior as $x \to 0$. We know that $\lim_{x\to 0} (\sin x)/x = 1$, so there is no asymptote at the origin.

The behavior as $x \to \pm\infty$. Since

$$0 \le \left| \frac{\sin x}{x} \right| \le \left| \frac{1}{x} \right|$$

and $\lim_{x\to\pm\infty} |1/x| = 0$, we have $\lim_{x\to\pm\infty} (\sin x)/x = 0$ by the Sandwich Theorem. Hence,

$$\lim_{x\to\pm\infty} \left(2 + \frac{\sin x}{x} \right) = 2 + 0 = 2,$$

and the line $y = 2$ is an asymptote of the curve on both left and right (Figure 1.38). This example illustrates that a curve may cross one of its horizontal asymptotes, perhaps many times.

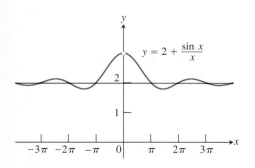

FIGURE 1.38 A curve may cross one of its asymptotes infinitely often. (Example 12)

Precise Definitions of Infinite Limits

Instead of requiring $f(x)$ to lie arbitrarily close to a finite number L for all x sufficiently close to x_0, the definitions of infinite limits require $f(x)$ to lie arbitrarily far from the origin. Except for this change, the language is identical with what we have seen before. Figures 1.39 and 1.40 accompany these definitions.

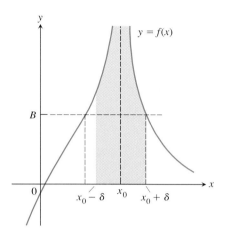

FIGURE 1.39 $\lim_{x \to x_0} f(x) = \infty$.

FIGURE 1.40 $\lim_{x \to x_0} f(x) = -\infty$.

Definition Infinite Limits

1. We say that $f(x)$ **approaches infinity as x approaches x_0**, and we write

$$\lim_{x \to x_0} f(x) = \infty,$$

   if for every positive real number B there exists a corresponding $\delta > 0$ such that for all x

$$0 < |x - x_0| < \delta \Rightarrow f(x) > B.$$

2. We say that $f(x)$ **approaches minus infinity as x approaches x_0**, and write

$$\lim_{x \to x_0} f(x) = -\infty,$$

   if for every negative real number $-B$ there exists a corresponding $\delta > 0$ such that for all x

$$0 < |x - x_0| < \delta \Rightarrow f(x) < -B.$$

The precise definitions of one-sided infinite limits at x_0 are similar.

End Behavior Models and Oblique Asymptotes

For numerically large values of x, we can sometimes model the behavior of a complicated function by a simpler one that acts virtually in the same way.

Example 13   Modeling Functions for $|x|$ Large

Let $f(x) = 3x^4 - 2x^3 + 3x^2 - 5x + 6$ and $g(x) = 3x^4$. Show that although f and g are quite different for numerically small values of x, they are virtually identical for $|x|$ large.

Solution

Solve Graphically

The graphs of f and g (Figure 1.41a), quite different near the origin, are virtually identical on a larger scale (Figure 1.41b).

$y = 3x^4 - 2x^3 + 3x^2 - 5x + 6$

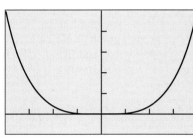

[−2, 2] by [−5, 20]   [−20, 20] by [−100,000, 500,000]

(a)   (b)

FIGURE 1.41   The graphs of f (upper curve) and g, (a) are distinct for $|x|$ small, and (b) nearly identical for $|x|$ large. (Example 13)

Confirm Analytically

We can test the claim that g models f for numerically large values of x by examining the ratio of the two functions as $x \to \pm\infty$. We find that

$$\lim_{x \to \pm\infty} \frac{f(x)}{g(x)} = \lim_{x \to \pm\infty} \frac{3x^4 - 2x^3 + 3x^2 - 5x + 6}{3x^4}$$

$$= \lim_{x \to \pm\infty} \left(1 - \frac{2}{3x} + \frac{1}{x^2} - \frac{5}{3x^3} + \frac{2}{x^4} \right)$$

$$= 1,$$

convincing evidence that f and g behave alike for $|x|$ large.

Definition   End Behavior Model

The function g is

(a) a **right end behavior model** for f if and only if

$$\lim_{x \to \infty} \frac{f(x)}{g(x)} = 1$$

(b) a **left end behavior model** for f if and only if

$$\lim_{x \to -\infty} \frac{f(x)}{g(x)} = 1 .$$

A function's right and left end behavior models need not be the same function.

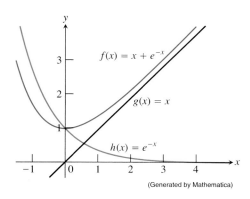

(Generated by Mathematica)

FIGURE 1.42 The graph of $f(x) = x + e^{-x}$ looks like the graph of $g(x) = x$ to the right of the y-axis and like the graph of $h(x) = e^{-x}$ to the left of the y-axis. (Example 14)

Example 14   Finding End Behavior Models

Let $f(x) = x + e^{-x}$. Show that $g(x) = x$ is a right end behavior model for f while $h(x) = e^{-x}$ is a left end behavior model for f.

Solution   On the right,

$$\lim_{x \to \infty} \frac{f(x)}{g(x)} = \lim_{x \to \infty} \frac{x + e^{-x}}{x} = \lim_{x \to \infty} \left(1 + \frac{e^{-x}}{x}\right) = 1 \text{ because } \lim_{x \to \infty} \frac{e^{-x}}{x} = 0.$$

On the left,

$$\lim_{x \to -\infty} \frac{f(x)}{h(x)} = \lim_{x \to -\infty} \frac{x + e^{-x}}{e^{-x}} = \lim_{x \to -\infty} \left(\frac{x}{e^{-x}} + 1\right) = 1 \text{ because } \lim_{x \to -\infty} \frac{x}{e^{-x}} = 0.$$

(See Exercise 51). The graph of f in Figure 1.42 supports these end behavior conclusions.

In some instances we can find end behavior models for rational functions. If the degree of the numerator is one greater than the degree of the denominator, the graph of the rational function $f(x)$ has an **oblique (slanted) asymptote,** as in Figure 1.29. We find an equation for the asymptote by dividing numerator by denominator to express f as a linear function plus a remainder that goes to zero as $x \to \pm\infty$. Here's an example.

Example 15   Finding an Oblique Asymptote

Find the oblique asymptote for the graph of

$$f(x) = \frac{2x^2 - 3}{7x + 4}$$

in Figure 1.29.

Solution   By long division, we find

$$f(x) = \frac{2x^2 - 3}{7x + 4}$$

$$= \underbrace{\left(\frac{2}{7}x - \frac{8}{49}\right)}_{\text{linear function } g(x)} + \underbrace{\frac{-115}{49(7x + 4)}}_{\text{remainder}}.$$

As $x \to \pm\infty$, the remainder, whose magnitude gives the vertical distance between the graphs of f and g, goes to zero, making the (slanted) line

$$g(x) = \frac{2}{7}x - \frac{8}{49}$$

an asymptote of the graph of f (Figure 1.29). The function g is both a right and a left hand behavior model for f.

EXERCISES 1.3

Calculating Limits as $x \to \pm\infty$

In Exercises 1–4, find the limit of each function (a) as $x \to \infty$ and (b) as $x \to -\infty$. (You may wish to visualize your answer with a grapher.)

1. $f(x) = \pi - \dfrac{2}{x^2}$

2. $g(x) = \dfrac{1}{2 + (1/x)}$

3. $h(x) = \dfrac{-5 + (7/x)}{3 - (1/x^2)}$

4. $h(x) = \dfrac{3 - (2/x)}{4 + (\sqrt{2}/x^2)}$

Find the limits in Exercises 5 and 6.

5. $\lim\limits_{x \to \infty} \dfrac{\sin 2x}{x}$

6. $\lim\limits_{t \to -\infty} \dfrac{2 - t + \sin t}{t + \cos t}$

Limits of Rational Functions

In Exercises 7–14, find the limit of each function (a) as $x \to \infty$ and (b) as $x \to -\infty$.

7. $f(x) = \dfrac{2x + 3}{5x + 7}$

8. $f(x) = \dfrac{x + 1}{x^2 + 3}$

9. $f(x) = \dfrac{1 - 12x^3}{4x^2 + 12}$

10. $h(x) = \dfrac{7x^3}{x^3 - 3x^2 + 6x}$

11. $g(x) = \dfrac{3x^2 - 6x}{4x - 8}$

12. $f(x) = \dfrac{2x^5 + 3}{-x^2 + x}$

13. $h(x) = \dfrac{-2x^3 - 2x + 3}{3x^3 + 3x^2 - 5x}$

14. $h(x) = \dfrac{-x^4}{x^4 - 7x^3 + 7x^2 + 9}$

Limits with Noninteger or Negative Powers

The process by which we determine limits of rational functions applies equally well to ratios containing noninteger or negative powers of x: Divide numerator and denominator by the highest power of x in the denominator and proceed from there. Find the limits in Exercises 15–20.

15. $\lim\limits_{x \to \infty} \dfrac{2\sqrt{x} + x^{-1}}{3x - 7}$

16. $\lim\limits_{x \to \infty} \dfrac{2 + \sqrt{x}}{2 - \sqrt{x}}$

17. $\lim\limits_{x \to -\infty} \dfrac{\sqrt[3]{x} - \sqrt[5]{x}}{\sqrt[3]{x} + \sqrt[5]{x}}$

18. $\lim\limits_{x \to \infty} \dfrac{x^{-1} + x^{-4}}{x^{-2} - x^{-3}}$

19. $\lim\limits_{x \to \infty} \dfrac{2x^{5/3} - x^{1/3} + 7}{x^{8/5} + 3x + \sqrt{x}}$

20. $\lim\limits_{x \to -\infty} \dfrac{\sqrt[3]{x} - 5x + 3}{2x + x^{2/3} - 4}$

Inventing Graphs from Values and Limits

In Exercises 21 and 22, sketch the graph of a function $y = f(x)$ that satisfies the given conditions. No formulas are required; just label the coordinate axes and sketch an appropriate graph. (The answers are not unique, so your graphs may not be exactly like those in the answer section.)

21. $f(0) = 0$, $f(1) = 2$, $f(-1) = -2$, $\lim\limits_{x \to -\infty} f(x) = -1$, and $\lim\limits_{x \to \infty} f(x) = 1$

22. $f(0) = 0$, $\lim\limits_{x \to \pm\infty} f(x) = 0$, $\lim\limits_{x \to 1^-} f(x) = \lim\limits_{x \to -1^+} f(x) = \infty$, $\lim\limits_{x \to 1^+} f(x) = -\infty$, and $\lim\limits_{x \to -1^-} f(x) = -\infty$

Inventing Functions

In Exercises 23 and 24, find a function that satisfies the given conditions and sketch its graph. (The answers here are not unique. Any function that satisfies the conditions is acceptable. Feel free to use formulas defined in pieces if that will help.)

23. $\lim\limits_{x \to \pm\infty} f(x) = 0$, $\lim\limits_{x \to 2^-} f(x) = \infty$, and $\lim\limits_{x \to 2^+} f(x) = \infty$

24. $\lim\limits_{x \to -\infty} h(x) = -1$, $\lim\limits_{x \to \infty} h(x) = 1$, $\lim\limits_{x \to 0^-} h(x) = -1$, and $\lim\limits_{x \to 0^+} h(x) = 1$

Graphing Rational Functions

T Graph the rational functions in Exercises 25–34. Include the graphs and equations of the asymptotes.

25. $y = \dfrac{1}{x - 1}$

26. $y = \dfrac{x + 1}{x + 2}$

27. $y = \dfrac{2x^2 + x - 1}{x^2 - 1}$

28. $y = \dfrac{x^2 - 1}{x}$

29. $y = \dfrac{x^4 + 1}{x^2}$

30. $y = \dfrac{x^2 - 4}{x - 1}$

31. $y = \dfrac{x^2 - x + 1}{x - 1}$

32. $y = \dfrac{x}{x^2 - 1}$

33. $y = \dfrac{8}{x^2 + 4}$ (Agnesi's witch)

34. $y = \dfrac{4x}{x^2 + 4}$ (Newton's serpentine)

End Behavior Models

In Exercises 35–38, match the function with the graph of its end behavior model.

35. $y = \dfrac{2x^3 - 3x^2 + 1}{x + 3}$

36. $y = \dfrac{x^5 - x^4 + x + 1}{2x^2 + x - 3}$

37. $y = \dfrac{2x^4 - x^3 + x^2 - 1}{2 - x}$

38. $y = \dfrac{x^4 - 3x^3 + x^2 - 1}{1 - x^2}$

(a)

(b)

(c)

(d)

In Exercises 39–42, find (**a**) a simple basic function as a right end be-havior model and (**b**) a simple basic function as a left end behavior model for the function.

39. $y = e^x - 2x$ **40.** $y = x^2 + e^{-x}$

41. $y = x + \ln|x|$ **42.** $y = x^2 + \sin x$

Theory and Examples

T **43.** (**a**) Estimate the value of

$$\lim_{x \to \infty} (\sqrt{x^2 + x + 1} - x)$$

by graphing the function

$$f(x) = \sqrt{x^2 + x + 1} - x.$$

(**b**) From a table of values of $f(x)$, guess the value of the limit in part (a). Then prove your guess is correct.

T **44.** Find $\lim_{x \to \infty} (\sqrt{x^2 + x} - \sqrt{x^2 - x})$ graphically and confirm it algebraically.

45. *Writing to Learn* How many horizontal asymptotes can the graph of a given rational function have? Give reasons for your answer.

46. *Writing to Learn* How many vertical asymptotes can the graph of a given rational function have? Give reasons for your answer.

COMPUTER EXPLORATIONS

Comparing Graphs With Formulas

Graph the curves in Exercises 47–50. Explain the relation between the curve's formula and what you see.

47. $y = \dfrac{x}{\sqrt{4 - x^2}}$ **48.** $y = \dfrac{-1}{\sqrt{4 - x^2}}$

49. $y = x^{2/3} + \dfrac{1}{x^{1/3}}$ **50.** $y = \sin\left(\dfrac{\pi}{x^2 + 1}\right)$

Substituting 1/x

In Exercises 51–54, use the graph of $y = f(1/x)$ to find $\lim_{x \to \infty} f(x)$ and $\lim_{x \to -\infty} f(x)$.

51. $f(x) = xe^x$ **52.** $f(x) = x^2 e^{-x}$

53. $f(x) = \dfrac{\ln|x|}{x}$ **54.** $f(x) = x \sin \dfrac{1}{x}$

55. $\lim_{x \to -\infty} \dfrac{\cos(1/x)}{1 + (1/x)}$ **56.** $\lim_{x \to \infty} \left(\dfrac{1}{x}\right)^{1/x}$

57. $\lim_{x \to \pm\infty} \left(3 + \dfrac{2}{x}\right)\left(\cos \dfrac{1}{x}\right)$

58. $\lim_{x \to \infty} \left(\dfrac{3}{x^2} - \cos \dfrac{1}{x}\right)\left(1 + \sin \dfrac{1}{x}\right)$

Finding Asymptotes

Graph the functions in Exercises 59–62. What asymptotes do the graphs have? Why are the asymptotes located where they are?

59. $y = -\dfrac{x^2 - 4}{x + 1}$ **60.** $y = \dfrac{x^3 - x^2 - 1}{x^2 - 1}$

61. $y = x^3 + \dfrac{3}{x}$ **62.** $y = 2 \sin x + \dfrac{1}{x}$

Graph the functions in Exercises 63 and 64. Then answer the follow-ing questions.

(**a**) How does the graph behave as $x \to 0^+$?

(**b**) How does the graph behave as $x \to \pm\infty$?

(**c**) How does the graph behave at $x = 1$ and $x = -1$?

Give reasons for your answers.

63. $y = \dfrac{3}{2}\left(x - \dfrac{1}{x}\right)^{2/3}$ **64.** $y = \dfrac{3}{2}\left(\dfrac{x}{x - 1}\right)^{2/3}$

1.4   Continuity

Continuity at a Point • Continuous Functions • Algebraic Combinations • Composites • Intermediate Value Theorem for Continuous Functions

CD-ROM
WEBsite

When we plot function values generated in the laboratory or collected in the field, we often connect the plotted points with an unbroken curve to show what the func-tion's values are likely to have been at the times we did not measure (Figure 1.43). In doing so, we are assuming that we are working with a *continuous function,* a function whose outputs vary continuously with the inputs and do not jump from one value to another without taking on the values in between.

Any function $y = f(x)$ whose graph can be sketched over its domain in one con-tinuous motion without lifting the pencil is an example of a continuous function. We study the idea of continuity in this section.

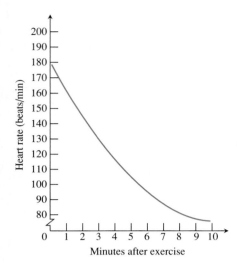

FIGURE 1.43 How the heart rate returns to a normal rate after running.

Continuity at a Point

Continuous functions are the functions we use to find a planet's closest point of approach to the sun or the peak concentration of antibodies in blood plasma. They are also the functions we use to describe how a body moves through space or how the speed of a chemical reaction changes with time. In fact, so many physical processes proceed continuously that throughout the eighteenth and nineteenth centuries it rarely occurred to anyone to look for any other kind of behavior. It came as a surprise when the physicists of the 1920's discovered that light comes in particles and that heated atoms emit light at discrete frequencies (Figure 1.44). As a result of these and other discoveries and because of the heavy use of discontinuous functions in computer science, statistics, and mathematical modeling, the issue of continuity has become one of practical as well as theoretical importance.

To understand continuity, we need to consider a function like the one in Figure 1.45, whose limits we investigated in Example 8, Section 1.2.

FIGURE 1.44 The laser was developed as a result of an understanding of the nature of the atom.

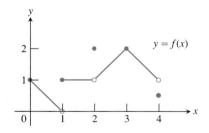

FIGURE 1.45 The function is continuous on $[0, 4]$ except at $x = 1$, $x = 2$, and $x = 4$. (Example 1)

Example 1   Investigating Continuity

Find the points at which the function f in Figure 1.45 is continuous and the points at which f is discontinuous.

Solution   The function f is continuous at every point in its domain $[0, 4]$ except at $x = 1$, $x = 2$, and $x = 4$. At these points, there are breaks in the graph. Note the relationship between the limit of f and the value of f at each point of the function's domain.

Points at which f is continuous:

At $x = 0$, $\lim\limits_{x \to 0^+} f(x) = f(0)$.

At $x = 3$, $\lim\limits_{x \to 3} f(x) = f(3)$.

At $0 < c < 4$, $c \neq 1, 2$, $\lim\limits_{x \to c} f(x) = f(c)$.

Points at which *f* is discontinuous:

At $x = 1$, $\lim\limits_{x \to 1} f(x)$ does not exist.

At $x = 2$, $\lim\limits_{x \to 2} f(x) = 1$, but $1 \neq f(2)$.

At $x = 4$, $\lim\limits_{x \to 4} f(x) = 1$, but $1 \neq f(4)$.

At $c < 0$, $c > 4$,   these points are not in the domain of *f*.

CD-ROM
WEBsite

Historical Biography

Johann van Waveren
Hudde
(1628 — 1704)

To define continuity at a point in a function's domain, we need to define continuity at an interior point (which involves a two-sided limit) and continuity at an endpoint (which involves a one-sided limit) (Figure 1.46).

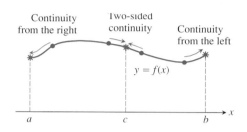

FIGURE 1.46 Continuity at points a, b, and c.

Definition   Continuity at a Point

Interior point: A function $y = f(x)$ is **continuous at an interior point *c*** of its domain if

$$\lim_{x \to c} f(x) = f(c).$$

Endpoint: A function $y = f(x)$ is **continuous at a left endpoint *a*** or is **continuous at a right endpoint *b*** of its domain if

$$\lim_{x \to a^+} f(x) = f(a) \qquad \text{or} \qquad \lim_{x \to b^-} f(x) = f(b), \qquad \text{respectively.}$$

If a function *f* is not continuous at a point *c*, we say that *f* is **discontinuous** at *c* and *c* is a **point of discontinuity** of *f*. Note that *c* need not be in the domain of *f*.

A function *f* is **right-continuous (continuous from the right)** at a point $x = c$ in its domain if $\lim_{x \to c^+} f(x) = f(c)$. It is **left-continuous (continuous from the left)** at *c* if $\lim_{x \to c^-} f(x) = f(c)$. Thus, a function is continuous at a left endpoint *a* of its domain if it is right-continuous at *a* and continuous at a right endpoint *b* of its domain if it is left-continuous at *b*. A function is continuous at an interior point *c* of its domain if and only if it is both right-continuous and left-continuous at *c* (Figure 1.46).

Example 2   A Function Continuous Throughout Its Domain

The function $f(x) = \sqrt{4 - x^2}$ is continuous at every point of its domain, $[-2, 2]$ (Figure 1.47), including $x = -2$, where *f* is right-continuous, and $x = 2$, where *f* is left-continuous.

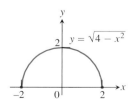

FIGURE 1.47 Continuous at every domain point.

FIGURE 1.48 Right-continuous at the origin.

Example 3   A Function with a Jump Discontinuity

The unit step function $U(x)$, graphed in (Figure 1.48), is right-continuous at $x = 0$, but is neither left-continuous nor continuous there. It has a jump discontinuity at $x = 0$.

We summarize continuity at a point in the form of a test.

Continuity Test

A function $f(x)$ is continuous at $x = c$ if and only if it meets the following three conditions.

1. $f(c)$ exists   (c lies in the domain of f)

2. $\lim_{x \to c} f(x)$ exists   (f has a limit as $x \to c$)

3. $\lim_{x \to c} f(x) = f(c)$ (the limit equals the function value)

For one-sided continuity and continuity at an endpoint, the limits in parts 2 and 3 of the test should be replaced by the appropriate one-sided limits.

Example 4   Finding Points of Continuity and Discontinuity

Find the points of continuity and the points of discontinuity of the greatest integer function $y = \text{int } x$ (Figure 1.49).

Solution   For the function to be continuous at $x = c$, the limit as $x \to c$ must exist and must equal the value of the function at $x = c$. The greatest integer function is discontinuous at every integer. For example,

$$\lim_{x \to 3^-} \text{int } x = 2 \quad \text{and} \quad \lim_{x \to 3^+} \text{int } x = 3,$$

so the limit as $x \to 3$ does not exist. Notice that $\text{int } 3 = 3$, so the greatest integer function is right-continuous at $x = 3$. In general, if n is any integer,

$$\lim_{x \to n^-} \text{int } x = n - 1 \quad \text{and} \quad \lim_{x \to n^+} \text{int } x = n,$$

so the limit as $x \to n$ does not exist. Since $\text{int } n = n$, the greatest integer function is right-continuous at every integer n (but not left-continuous).

The greatest integer function is continuous at every real number other than the integers. For example,

$$\lim_{x \to 1.5} \text{int } x = 1 = \text{int } 1.5 .$$

In general, if $n - 1 < c < n$, n an integer, then

$$\lim_{x \to c} \text{int } x = n - 1 = \text{int } c .$$

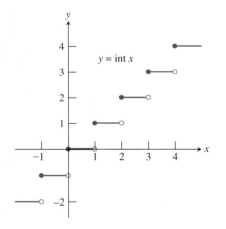

FIGURE 1.49 The function $\text{int } x$ is continuous at every noninteger point. It is right-continuous, but not left-continuous, at every integer point. (Example 4)

Figure 1.50 is a catalog of discontinuity types. The function in Figure 1.50a is continuous at $x = 0$. The function in Figure 1.50b would be continuous if it had

FIGURE 1.50 The function in (a) is continuous at $x = 0$; the functions in (b) through (f) are not.

$f(0) = 1$. The function in Figure 1.50c would be continuous if $f(0)$ were 1 instead of 2. The discontinuities in Figure 1.50b and c are **removable.** Each function has a limit as $x \to 0$, and we can remove the discontinuity by setting $f(0)$ equal to this limit.

The discontinuities in Figure 1.50d through f are more serious: $\lim_{x\to 0} f(x)$ does not exist, and there is no way to improve the situation by changing f at 0. The step function in Figure 1.50d has a **jump discontinuity:** The one-sided limits exist but have different values. The function $f(x) = 1/x^2$ in Figure 1.50e has an **infinite discontinuity.** The function in Figure 1.50f has an **oscillating discontinuity:** It oscillates too much to have a limit as $x \to 0$.

Continuous Functions

A function is **continuous on an interval** if and only if it is continuous at every point of the interval. A **continuous function** is one that is continuous at every point of its domain. A continuous function need not be continuous on every interval. For example, $y = 1/x$ is not continuous on $[-1, 1]$ (Figure 1.51).

Example 5   Identifying Continuous Functions

The function $y = 1/x$ (Figure 1.51) is a continuous function because it is continuous at every point of its domain. It has a point of discontinuity at $x = 0$, however, because it is not defined there.

The following types of functions are continuous at every point in their domains:

- polynomials
- rational functions

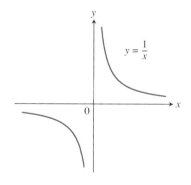

FIGURE 1.51 The function $y = 1/x$ is continuous at every value of x except $x = 0$. It has a point of discontinuity at $x = 0$. (Example 5)

- root functions ($y = \sqrt[n]{x}$, n a positive integer greater than 1)
- trigonometric functions
- inverse trigonometric functions
- exponential functions
- logarithmic functions.

Polynomial functions f are continuous at every number c because $\lim_{x \to c} f(x) = f(c)$. Rational functions are continuous at every point of their domains. They have points of discontinuity at the zeros of their denominators. From their graphs, we are not surprised that the sine and cosine functions are continuous.

The inverse function of any continuous function is continuous. We see this because the graph of a continuous function f has no breaks, and the graph of f^{-1} is obtained by reflecting the graph of f about the line $y = x$ (so the graph of f^{-1} has no breaks either).

The exponential function $y = a^x$ was defined to be continuous, and therefore its inverse $y = \log_a x$ is also continuous over its domain.

The function $f(x) = |x|$ is continuous at every value of x (Figure 1.52). If $x > 0$, we have $f(x) = x$, a polynomial. If $x < 0$, we have $f(x) = -x$, another polynomial. Finally, at the origin, $\lim_{x \to 0} |x| = 0 = |0|$.

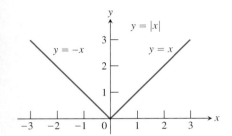

FIGURE 1.52 The sharp corner does not prevent the function from being continuous at the origin.

Algebraic Combinations

As you may have guessed, algebraic combinations of continuous functions are continuous wherever they are defined.

Theorem 8   Properties of Continuous Functions

If the functions f and g are continuous at $x = c$, then the following combinations are continuous at $x = c$.

1. *Sums*: $f + g$

2. *Differences*: $f - g$

3. *Products*: $f \cdot g$

4. *Constant multiples*: $k \cdot f$, for any number k

5. *Quotients*: f/g, provided $g(c) \neq 0$

Most of the results in Theorem 8 are easily proved from the limit rules in Theorem 1.

Composites

All composites of continuous functions are continuous. Thus, composites like

$$y = \sin(x^2) \qquad \text{and} \qquad y = |\cos x|$$

are continuous at every point at which they are defined. The idea is that if $f(x)$ is continuous at $x = c$ and $g(x)$ is continuous at $x = f(c)$, then $g \circ f$ is continuous at $x = c$ (Figure 1.53). In this case, the limit as $x \to c$ is $g(f(c))$.

Theorem 9   Composite of Continuous Functions

If f is continuous at c and g is continuous at $f(c)$, then the composite $g \circ f$ is continuous at c.

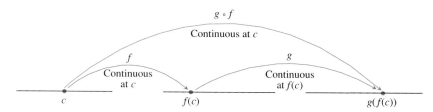

FIGURE 1.53 Composites of continuous functions are continuous.

Intuitively, Theorem 9 is reasonable because if x is close to c, then $f(x)$ is close to $f(c)$, and since g is continuous at $f(c)$, it follows that $g(f(x))$ is close to $g(f(c))$.

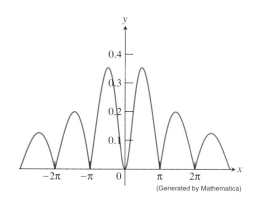

FIGURE 1.54 The graph suggests that $y = |(x \sin x)/(x^2 + 2)|$ is continuous. (Example 6)

Example 6   Using Theorem 9

Show that

$$y = \left| \frac{x \sin x}{x^2 + 2} \right|$$

is continuous.

Solution   The graph (Figure 1.54) of $y = |(x \sin x)/(x^2 + 2)|$ suggests that the function is continuous at every value of x. By letting

$$g(x) = |x| \qquad \text{and} \qquad f(x) = \frac{x \sin x}{x^2 + 2},$$

we see that y is the composite $g \circ f$.

We know that the absolute value function g is continuous. The function f is continuous by Theorem 8. Their composite is continuous by Theorem 9.

Intermediate Value Theorem for Continuous Functions

Functions that are continuous on intervals have properties that make them particularly useful in mathematics and its applications. One of these is the *Intermediate Value Property*. A function is said to have the **Intermediate Value Property** if it never takes on two values without taking on all the values in between.

Theorem 10 The Intermediate Value Theorem
for Continuous Functions

A function $y = f(x)$ that is continuous on a closed interval $[a, b]$ takes on every value between $f(a)$ and $f(b)$. In other words, if y_0 is any value between $f(a)$ and $f(b)$, then $y_0 = f(c)$ for some c in $[a, b]$.

Geometrically, the Intermediate Value Theorem says that any horizontal line $y = y_0$ crossing the y-axis between the numbers $f(a)$ and $f(b)$ will cross the curve $y = f(x)$ at least once over the interval $[a, b]$.

The continuity of f on the interval is essential to Theorem 10. If f is discontinuous at even one point of the interval, the theorem's conclusion may fail, as it does for the function graphed in Figure 1.55.

A Consequence for Graphing: Connectivity Theorem 10 is the reason the graph of a function continuous on an interval I cannot have any breaks over the interval. It will be **connected,** a single, unbroken curve, like the graph of $\sin x$. It will not have jumps like the graph of the greatest integer function int x or separate branches like the graph of $1/x$.

A Consequence for Root Finding We call a solution of the equation $f(x) = 0$ a **root** of the equation or **zero** of the function f. The Intermediate Value Theorem tells us that if f is continuous, then any interval on which f changes sign contains a zero of the function.

In practical terms, when we see the graph of a continuous function cross the horizontal axis on a computer screen, we know it is not stepping across. There really is a point where the function's value is zero. This consequence leads to a procedure for estimating the zeros of any continuous function we can graph:

1. Graph the function over a large interval to see roughly where the zeros are.

2. Zoom in on each zero to estimate its x-coordinate value.

You can practice this procedure on your graphing calculator or computer in some of the exercises.

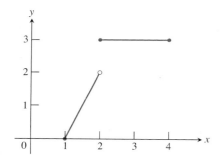

FIGURE 1.55 The function

$$f(x) = \begin{cases} 2x - 2, & 1 \le x < 2 \\ 3, & 2 \le x \le 4 \end{cases}$$

does not take on all values between $f(1) = 0$ and $f(4) = 3$; it misses all the values between 2 and 3.

A graphical procedure for root-finding.

USING TECHNOLOGY

Deceptive Pictures A graphing utility (calculator or Computer Algebra System) plots a graph much as you do when plotting by hand: by plotting points, or *pixels*, and then connecting them in succession. The resulting picture may be misleading when points on opposite sides of a point of discontinuity in the graph are incorrectly connected. To avoid incorrect connections, some systems allow you to use a "dot mode," which plots only the points. Dot mode, however, may not reveal enough information to portray the true behavior of the graph. Try the following four functions on your graphing device. If you can, plot them in both "connected" and "dot" modes.

$$y_1 = x \text{ int } x \qquad \text{at } x = 2 \qquad \text{jump discontinuity}$$

$$y_2 = \sin \frac{1}{x} \qquad \text{at } x = 0 \qquad \text{oscillating discontinuity}$$

$$y_3 = \frac{1}{x - 2} \qquad \text{at } x = 2 \qquad \text{infinite discontinuity}$$

$$y_4 = \frac{x^2 - 2}{x - \sqrt{2}} \qquad \text{at } x = \sqrt{2} \qquad \text{removable discontinuity}$$

(a)

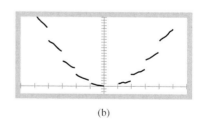

(b)

(a) $y_1 = x \times \text{int } x$ incorrectly graphed in connected mode. (b) $y_1 = x \times \text{int } x$ correctly graphed in dot mode.

Example 7 Using the Intermediate Value Theorem

Is any real number exactly 1 less than its cube?

Solution We answer this question by applying the Intermediate Value Theorem in the following way. Any such number must satisfy the equation $x = x^3 - 1$ or, equivalently, $x^3 - x - 1 = 0$. Hence, we are looking for a zero value of the continuous function $f(x) = x^3 - x - 1$ (Figure 1.56). The function changes sign between 1 and 2, so there must be a point c between 1 and 2 where $f(c) = 0$.

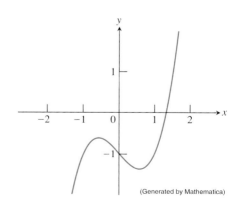

(Generated by Mathematica)

FIGURE 1.56 The graph of $f(x) = x^3 - x - 1$. (Example 7)

EXERCISES 1.4

Continuity from Graphs

In Exercises 1–4, say whether the function graphed is continuous on $[-1, 3]$. If not, where does it fail to be continuous and why?

1.

2.

3.

4.
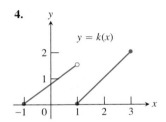

Exercises 5–10 are about the function

$$f(x) = \begin{cases} x^2 - 1, & -1 \le x < 0 \\ 2x, & 0 < x < 1 \\ 1, & x = 1 \\ -2x + 4, & 1 < x < 2 \\ 0, & 2 < x < 3 \end{cases}$$

graphed in Figure 1.57.

5. (a) Does $f(-1)$ exist?

   (b) Does $\lim_{x \to -1^+} f(x)$ exist?

   (c) Does $\lim_{x \to -1^+} f(x) = f(-1)$?

   (d) Is f continuous at $x = -1$?

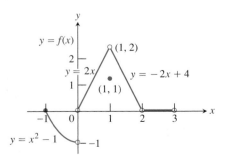

FIGURE 1.57 The graph for Exercises 5–10.

6. (a) Does $f(1)$ exist?

   (b) Does $\lim_{x \to 1} f(x)$ exist?

   (c) Does $\lim_{x \to 1} f(x) = f(1)$?

   (d) Is f continuous at $x = 1$?

7. (a) Is f defined at $x = 2$? (Look at the definition of f.)

   (b) Is f continuous at $x = 2$?

8. At what values of x is f continuous?

9. What value should be assigned to $f(2)$ to make the extended function continuous at $x = 2$?

10. To what new value should $f(1)$ be changed to remove the discontinuity?

Applying the Continuity Test

At which points do the functions in Exercises 11 and 12 fail to be continuous? At which points, if any, are the discontinuities removable? Not removable? Give reasons for your answers.

11. Exercise 11, Section 1.1

12. Exercise 12, Section 1.1

On what intervals are the functions in Exercises 13–20 continuous?

13. $y = \dfrac{1}{x - 2} - 3x$

14. $y = \dfrac{1}{(x + 2)^2} + 4$

15. $z = \dfrac{t + 1}{t^2 - 4t + 3}$

16. $u = \dfrac{1}{|t| + 1} - \dfrac{t^2}{2}$

17. $r = \dfrac{\cos \theta}{\theta}$

18. $y = \tan \dfrac{\pi \theta}{2}$

19. $s = \sqrt{2v + 3}$

20. $y = \sqrt[4]{3x - 1}$

Composite Functions

Find the limits in Exercises 21–24. Are the functions continuous at the point being approached?

21. $\lim\limits_{x \to \pi} \sin(x - \sin x)$

22. $\lim\limits_{t \to 0} \sin \left(\dfrac{\pi}{2} \cos(\tan t) \right)$

23. $\lim\limits_{y \to 1} \sec(y \sec^2 y - \tan^2 y - 1)$

24. $\lim\limits_{\theta \to 0} \tan \left(\dfrac{\pi}{4} \cos(\sin \theta^{1/3}) \right)$

Theory and Examples

25. *Writing to Learn* A function $y = f(x)$, continuous on $[0, 1]$, is known to be negative at $x = 0$ and positive at $x = 1$. What, if anything, does this say about the equation $f(x) = 0$? Illustrate with a sketch.

26. *Writing to Learn* Why does the equation $\cos x = x$ have at least one solution?

27. *Writing to Learn* Explain why the following five statements ask for the same information.

(a) Find the zeros of $f(x) = x^3 - 3x - 1$.

(b) Find the x-coordinate of the points where the curve $y = x^3$ crosses the line $y = 3x + 1$.

(c) Find all the values of x for which $x^3 - 3x = 1$.

(d) Find the x-coordinates of the points where the cubic curve $y = x^3 - 3x$ crosses the line $y = 1$.

(e) Solve the equation $x^3 - 3x - 1 = 0$.

28. *Solving an equation* If $f(x) = x^3 - 8x + 10$, show that there is at least one value of c for which $f(c)$ equals

(a) π

(b) $-\sqrt{3}$

(c) 5,000,000.

29. *Removable discontinuity* Give an example of a function $f(x)$ that is continuous for all values of x except $x = 2$, where it has a removable discontinuity. Explain how you know that f is discontinuous at $x = 2$ and how you know the discontinuity is removable.

30. *Nonremovable discontinuity* Give an example of a function $g(x)$ that is continuous for all values of x except $x = -1$, where it has a nonremovable discontinuity. Explain how you know that g is discontinuous there and why the discontinuity is not removable.

31. *Factoring a polynominal* Find rounded three-place values for r_1 through r_5 in the factorization

$$x^5 - x^4 - 5x^3 = (x - r_1)(x - r_2)(x - r_3)(x - r_4)(x - r_5).$$

T 32. *Factoring a polynomial* You want to rewrite the polynomial $x^3 - 3x - 1$ in the form $(x - r)q(x)$, where $q(x)$ is a quadratic polynomial. Rounded to three decimal places, what are your choices for r?

33. *A function discontinuous at every point*

(a) Use the fact that every nonempty interval of real numbers contains both rational and irrational numbers to show that the function

$$f(x) = \begin{cases} 1 & \text{if } x \text{ is rational} \\ 0 & \text{if } x \text{ is irrational} \end{cases}$$

is discontinuous at every point.

(b) Is f right-continuous or left-continuous at any point?

34. *Writing to Learn* If functions $f(x)$ and $g(x)$ are continuous for $0 \le x \le 1$, could $f(x)/g(x)$ possibly be discontinuous at a point of $[0, 1]$? Give reasons for your answer.

35. *Writing to Learn* Is it true that a continuous function that is never zero on an interval never changes sign on that interval? Give reasons for your answer.

36. *Stretching a rubber band* Is it true that if you stretch a rubber band by moving one end to the right and the other to the left, some point of the band will end up in its original position? Give reasons for your answer.

37. *A fixed point theorem* Suppose that a function f is continuous on the closed interval $[0, 1]$ and that $0 \le f(x) \le 1$ for every x in $[0, 1]$. Show that there must exist a number c in $[0, 1]$ such that $f(c) = c$ (c is called a **fixed point** of f).

38. *The sign-preserving property of continuous functions* Let f be defined on an interval (a, b) and suppose that $f(c) \ne 0$ at some c where f is continuous. Show that there is an interval $(c - \delta, c + \delta)$ about c where f has the same sign as $f(c)$. Notice how remarkable this conclusion is. Although f is defined throughout (a, b), it is not required to be continuous at any point except c. That and the condition $f(c) \ne 0$ are enough to make f different from zero (positive or negative) throughout an entire (small) interval.

T 39. *Salary negotiation* A welder's contract promises a 3.5% salary increase each year for 4 years, and Luisa has an initial salary of \$36,500.

(a) Show that Luisa's salary is given by

$$y = 36{,}500(1.035)^{\text{int } t},$$

where t is the time, measured in years, since Luisa signed the contract.

(b) Graph Luisa's salary function. At what values of t is it continuous?

T 40. *Airport parking* Valuepark charges \$1.10 per hour or fraction of an hour for airport parking. The maximum charge per day is \$7.25.

(a) Write a formula that gives the charge for x hours with $0 \le x \le 24$. (*Hint*: See Exercise 39.)

(b) Graph the function in part (a). At what values of x is it continuous?

COMPUTER EXPLORATIONS

Continuous Extension to a Point

As we saw in Section 1.2, a rational function may have a limit even at a point where its denominator is zero. If $f(c)$ is not defined but $\lim_{x \to c} f(x) = L$ exists, we can define a new function $F(x)$ by the rule

$$F(x) = \begin{cases} f(x) & \text{if } x \text{ is in the domain of } f \\ L & \text{if } x = c. \end{cases}$$

The function F is continuous at $x = c$. It is called the **continuous extension** of f to $x = c$. For rational functions f, continuous extensions are usually found by canceling common factors.

In Exercises 41–44, graph the function f to see whether it appears to have a continuous extension to the origin. If it does, use Trace and Zoom to find a good candidate for the extended function's value at $x = 0$. If the function does not appear to have a continuous extension, can it be extended to be continuous at the origin from the right or from the left? If so, what do you think the extended function's value(s) should be?

41. $f(x) = \dfrac{10^x - 1}{x}$

42. $f(x) = \dfrac{10^{|x|} - 1}{x}$

43. $f(x) = \dfrac{\sin x}{|x|}$

44. $f(x) = (1 + 2x)^{1/x}$

Solving Equations Graphically

Use a graphing calculator or computer grapher to solve the equations in Exercises 45–52. Round each solution to four decimal places.

45. $x^3 - 3x - 1 = 0$

46. $2x^3 - 2x^2 - 2x + 1 = 0$

47. $x(x - 1)^2 = 1$ (one root)

48. $x^x = 2$

49. $\sqrt{x} + \sqrt{1 + x} = 4$

50. $x^3 - 15x + 1 = 0$ (three roots)

51. $\cos x = x$ (one root). Be sure you are using radian mode.

52. $2 \sin x = x$ (three roots). Be sure you are using radian mode.

1.5   Tangent Lines

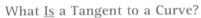

What *Is* a Tangent to a Curve?   •   Finding a Tangent to the Graph of a Function   •   Rates of Change: Derivative at a Point

This section continues the discussion of secants and tangents begun in Section 1.1. We calculate limits of secant slopes to find tangents to curves.

What *Is* a Tangent to a Curve?

For circles, tangency is straightforward. A line L is tangent to a circle at a point P if L passes through P perpendicular to the radius at P (Figure 1.58). Such a line just *touches* the circle. But what does it mean to say that a line L is tangent to some other curve C at a point P? Generalizing from the geometry of the circle, we might say that it means one of the following.

1. L passes through P perpendicular to the line from P to the center of C.

2.   L passes through only one point of C, namely P.

3.   L passes through P and lies on one side of C only.

Although these statements are valid if C is a circle, none of them works consistently for more general curves. Most curves do not have centers, and a line we may want to call tangent may intersect C at other points or cross C at the point of tangency (Figure 1.59).

FIGURE 1.58 L is tangent to the circle at P if it passes through P perpendicular to radius OP.

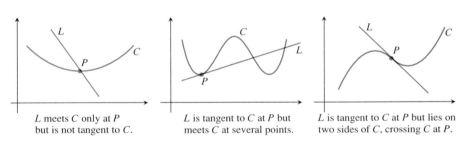

| L meets C only at P but is not tangent to C. | L is tangent to C at P but meets C at several points. | L is tangent to C at P but lies on two sides of C, crossing C at P. |

FIGURE 1.59 Exploding myths about tangent lines.

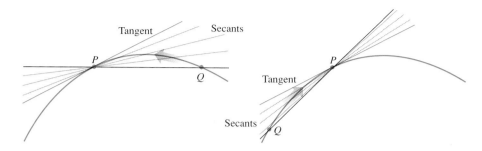

FIGURE 1.60 The dynamic approach to tangency. The tangent to the curve at P is the line through P whose slope is the limit of the secant slopes as $Q \to P$ from either side.

CD-ROM
WEBsite

Historical Biography

Pierre de Fermat
(1601 — 1665)

To define tangency for general curves, we need a *dynamic* approach that takes into account the behavior of the secants through P and nearby points Q as Q moves toward P along the curve (Figure 1.60). It goes like this:

1. We start with what we *can* calculate, namely the slope of the secant PQ.

2. Investigate the limit of the secant slope as Q approaches P along the curve.

3. If the limit exists, take it to be the slope of the curve at P and define the tangent to the curve at P to be the line through P with this slope.

This approach is what we were doing in the falling-rock and heat-shield examples in Section 1.1.

How do you find a tangent to a curve?

This problem was the dominant mathematical question of the early seventeenth century, and it is hard to overestimate how badly the scientists of the day wanted to know the answer. In optics, the tangent determined the angle at which a ray of light entered a curved lens. In mechanics, the tangent determined the direction of a body's motion at every point along its path. In geometry, the tangents to two curves at a point of intersection determined the angle at which the curves intersected. René Descartes went so far as to say that the problem of finding a tangent to a curve was "the most useful and most general problem not only that I know but even that I have any desire to know."

Example 1   Tangent Line to a Parabola

Find the slope of the parabola $y = x^2$ at the point $P(2, 4)$. Write an equation for the tangent to the parabola at this point.

Solution   We begin with a secant line through $P(2, 4)$ and $Q(2 + h, (2 + h)^2)$ nearby. We then write an expression for the slope of the secant PQ and investigate what happens to the slope as Q approaches P along the curve:

$$\text{Secant slope} = \frac{\Delta y}{\Delta x} = \frac{(2 + h)^2 - 2^2}{h} = \frac{h^2 + 4h + 4 - 4}{h}$$

$$= \frac{h^2 + 4h}{h} = h + 4.$$

If $h > 0$, then Q lies above and to the right of P, as in Figure 1.61. If $h < 0$, then Q lies to the left of P (not shown). In either case, as Q approaches P along the curve, h approaches zero and the secant slope approaches 4:

$$\lim_{h \to 0} (h + 4) = 4.$$

We take 4 to be the parabola's slope at P.

The tangent to the parabola at P is the line through P with slope 4:

$$y = 4 + 4(x - 2) \qquad \text{Point-slope equation}$$
$$y = 4x - 4.$$

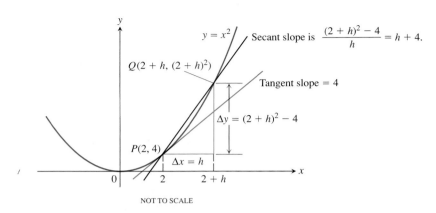

FIGURE 1.61 Diagram for finding the slope of the parabola $y = x^2$ at the point $P(2, 4)$. (Example 1)

Finding a Tangent to the Graph of a Function

To find a tangent to an arbitrary curve $y = f(x)$ at a point $P(x_0, f(x_0))$, we use the same dynamic procedure. We calculate the slope of the secant through P and a point $Q(x_0 + h, f(x_0 + h))$. We then investigate the limit of the slope as $h \to 0$ (Figure 1.62). If the limit exists, we call it the slope of the curve at P and define the tangent at P to be the line through P having this slope.

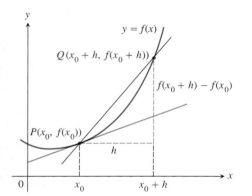

FIGURE 1.62 The tangent slope is
$$\lim_{h \to 0} \frac{f(x_0 + h) - f(x_0)}{h}$$

> **Definitions**   Slope and Tangent Line
> The **slope of the curve** $y = f(x)$ at the point $P(x_0, f(x_0))$ is the number
> $$m = \lim_{h \to 0} \frac{f(x_0 + h) - f(x_0)}{h} \qquad \text{(provided the limit exists).}$$
> The **tangent line** to the curve at P is the line through P with this slope.

CD-ROM
WEBsite

Whenever we make a new definition, it is a good idea to try it on familiar objects to be sure it gives the results we want in familiar cases. Example 2 shows that the new definition of slope agrees with the old definition when we apply it to non-vertical lines.

Example 2   Testing the Definition

Show that the line $y = mx + b$ is its own tangent at any point $(x_0, mx_0 + b)$.

Solution   We let $f(x) = mx + b$ and organize the work into three steps.

Step 1: *Find $f(x_0)$ and $f(x_0 + h)$.*

$$f(x_0) = mx_0 + b$$
$$f(x_0 + h) = m(x_0 + h) + b = mx_0 + mh + b$$

Step 2: *Find the slope* $\lim\limits_{h\to0} (f(x_0 + h) - f(x_0))/h$.

$$\lim_{h\to0} \frac{f(x_0 + h) - f(x_0)}{h} = \lim_{h\to0} \frac{(mx_0 + mh + b) - (mx_0 + b)}{h}$$

$$= \lim_{h\to0} \frac{mh}{h} = m$$

Step 3: *Find the tangent line using the point-slope equation.* The tangent line at the point $(x_0 , mx_0 + b)$ is

$$y = (mx_0 + b) + m(x - x_0)$$
$$y = mx_0 + b + mx - mx_0$$
$$y = mx + b .$$

How to Find the Tangent to the Curve
y = f(x) at (x₀ , y₀)

1. Calculate $f(x_0)$ and $f(x_0 + h)$.

2. Calculate the slope
$$m = \lim_{h\to0} \frac{f(x_0 + h) - f(x_0)}{h} .$$

3. If the limit exists, find the tangent line as
$$y = y_0 + m(x - x_0) .$$

Example 3   Slope and Tangent to $y = 1/x$

(a) Find the slope of the curve $y = 1/x$ at $x = a$.

(b) Where does the slope equal $-1/4$?

(c) What happens to the tangent to the curve at the point $(a , 1/a)$ as a changes?

Solution

(a) Here $f(x) = 1/x$. The slope at $(a , 1/a)$ is

$$\lim_{h\to0} \frac{f(a + h) - f(a)}{h} = \lim_{h\to0} \frac{\dfrac{1}{a + h} - \dfrac{1}{a}}{h}$$

$$= \lim_{h\to0} \frac{1}{h} \frac{a - (a + h)}{a(a + h)}$$

$$= \lim_{h\to0} \frac{-h}{ha(a + h)}$$

$$= \lim_{h\to0} \frac{-1}{a(a + h)} = -\frac{1}{a^2} .$$

Notice how we had to keep writing "$\lim_{h\to0}$" before each fraction until the stage where we could evaluate the limit by substituting $h = 0$.

(b) The slope of $y = 1/x$ at the point where $x = a$ is $-1/a^2$. It will be $-1/4$ provided that

$$-\frac{1}{a^2} = -\frac{1}{4} .$$

This equation is equivalent to $a^2 = 4$, so $a = 2$ or $a = -2$. The curve has slope $-1/4$ at the two points $(2, 1/2)$ and $(-2, -1/2)$ (Figure 1.63).

(c) Notice that the slope $-1/a^2$ is always negative. As $a \to 0^+$, the slope approaches $-\infty$ and the tangent becomes increasingly steep (Figure 1.64). We see this situation again as $a \to 0^-$. As a moves away from the origin in either direction, the slope approaches 0^- and the tangent levels off.

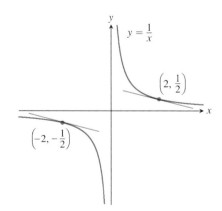

FIGURE 1.63   The two tangent lines to $y = 1/x$ having slope $-1/4$.

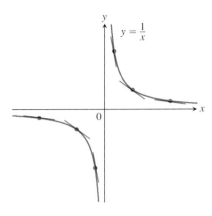

FIGURE 1.64 The tangent slopes, steep near the origin, become more gradual as the point of tangency moves away.

Rates of Change: Derivative at a Point

The expression

$$\frac{f(x_0 + h) - f(x_0)}{h}$$

is called the **difference quotient of f at x_0 with increment h.** If the difference quotient has a limit as h approaches zero, that limit is called the **derivative of f at x_0.** If we interpret the difference quotient as a secant slope, the derivative gives the slope of the curve and tangent at the point where $x = x_0$. If we interpret the difference quotient as an average rate of change, as we did in Section 1.1, the derivative gives the function's rate of change with respect to x at the point $x = x_0$. The derivative is one of the two most important mathematical objects considered in calculus. We begin a thorough study of it in Chapter 2. The other important object is the integral, and we initiate its study in Chapter 4.

All these refer to the same thing

1. The slope of $y = f(x)$ at $x = x_0$

2. The slope of the tangent to the curve $y = f(x)$ at $x = x_0$

3. The rate of change of $f(x)$ with respect to x at $x = x_0$

4. The derivative of f at $x = x_0$

5. $\displaystyle\lim_{h \to 0} \frac{f(x_0 + h) - f(x_0)}{h}$

Example 4   Instantaneous Speed (Continuation of Section 1.1, Examples 1 and 2)

In Examples 1 and 2 in Section 1.1, we studied the speed of a rock falling freely from rest near the surface of the earth. We knew that the rock fell $y = 16t^2$ feet during the first t sec, and we used a sequence of average rates over increasingly short intervals to estimate the rock's speed at the instant $t = 2$. Exactly what *was* the rock's speed at this time?

Solution   We let $f(t) = 16t^2$. The average speed of the rock over the interval between $t = 2$ and $t = 2 + h$ sec was

$$\frac{f(2 + h) - f(2)}{h} = \frac{16(2 + h)^2 - 16(2)^2}{h} = \frac{16(h^2 + 4h)}{h} = 16(h + 4).$$

The rock's speed at the instant $t = 2$ was

$$\lim_{h \to 0} 16(h + 4) = 16(0 + 4) = 64 \text{ ft/sec}.$$

Our original estimate of 64 ft/sec was right.

EXERCISES 1.5

Slopes and Tangent Lines

In Exercises 1–4, use the grid and a straight edge to make a rough estimate of the slope of the curve (in y-units per x-unit) at the points P_1 and P_2. Graphs can shift during a press run, so your estimates may be somewhat different from those in the back of the book.

1.

2.

3.

4.

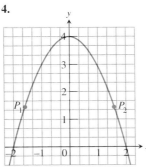

In Exercises 5–8, find an equation for the tangent to the curve at the given point. Sketch the curve and tangent together.

5. $y = 4 - x^2$, $(-1, 3)$

6. $y = 2\sqrt{x}$, $(1, 2)$

7. $y = x^3$, $(-2, -8)$

8. $y = \dfrac{1}{x^3}$, $(-2, -1/8)$

In Exercises 9–12, find the slope of the function's graph at the given point. Then find an equation for the tangent to the graph there.

9. $f(x) = x - 2x^2$, $(1, -1)$

10. $h(t) = t^3 + 3t$, $(1, 4)$

11. $g(u) = \dfrac{u}{u - 2}$, $(3, 3)$

12. $f(x) = \sqrt{x + 1}$, $(8, 3)$

In Exercises 13 and 14, find the slope of the curve at the indicated value of x.

13. $y = \dfrac{1}{x - 1}$, $x = 3$

14. $y = \dfrac{x - 1}{x + 1}$, $x = 0$

Tangent Lines with Specified Slopes

At what points do the graphs of the functions in Exercises 15 and 16 have horizontal tangents?

15. $f(x) = x^2 + 4x - 1$

16. $g(x) = x^3 - 3x$

17. Find equations for all tangents to the curve $y = 1/(x - 1)$ that have slope -1.

18. Find an equation for the tangent to the curve $y = \sqrt{x}$ that has slope $1/4$.

Rates of Change

19. *Object dropped from a tower* An object is dropped from the top of a 100-m-high tower. Its height above ground after t sec is $100 - 4.9t^2$ m. How fast is it falling 2 sec after it is dropped?

20. *Speed of a rocket* At t sec after liftoff, the height of a rocket is $3t^2$ ft. How fast is the rocket climbing 10 sec after liftoff?

21. *Circle's changing area* What is the rate of change of the area of a circle ($A = \pi r^2$) with respect to the radius when the radius is $r = 3$?

22. *Ball's changing volume* What is the rate of change of the volume of a ball ($V = (4/3)\pi r^3$) with respect to the radius when the radius is $r = 2$?

23. *Free fall on Mars* The equation for free fall at the surface of Mars is $s = 1.86t^2$ m, with t in seconds. Assume that a rock is dropped from the top of a 200-m cliff. Find the speed of the rock at $t = 1$ sec.

24. *Free fall on Jupiter* The equation for free fall at the surface of Jupiter is $s = 11.44t^2$ m with t in seconds. Assume that a rock is dropped from the top of a 500-m cliff. Find the speed of the rock at $t = 2$ sec.

Testing for Tangents

25. *Writing to Learn* Does the graph of

$$f(x) = \begin{cases} x^2 \sin (1/x), & x \neq 0 \\ 0, & x = 0 \end{cases}$$

have a tangent at the origin? Give reasons for your answer.

26. *Writing to Learn* Does the graph of

$$g(x) = \begin{cases} x \sin (1/x), & x \neq 0 \\ 0, & x = 0 \end{cases}$$

have a tangent at the origin? Give reasons for your answer.

Vertical Tangents

We say that the curve $y = f(x)$ has a **vertical tangent** at the point where $x = x_0$ if $\lim_{h \to 0} (f(x_0 + h) - f(x_0))/h = \infty$ or $-\infty$.

Vertical tangent at $x = 0$:

$$\lim_{h \to 0} \frac{f(0 + h) - f(0)}{h} = \lim_{h \to 0} \frac{h^{1/3} - 0}{h}$$

$$= \lim_{h \to 0} \frac{1}{h^{2/3}} = \infty$$

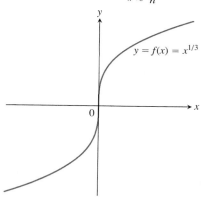

$$y = f(x) = x^{1/3}$$

VERTICAL TANGENT AT ORIGIN

No vertical tangent at $x = 0$:

$$\lim_{h \to 0} \frac{g(0 + h) - g(0)}{h} = \lim_{h \to 0} \frac{h^{2/3} - 0}{h}$$

$$= \lim_{h \to 0} \frac{1}{h^{1/3}}$$

does not exist, because the limit is ∞ from the right and $-\infty$ from the left.

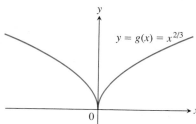

$$y = g(x) = x^{2/3}$$

NO VERTICAL TANGENT AT ORIGIN

27. *Writing to Learn* Does the graph of

$$f(x) = \begin{cases} -1, & x < 0 \\ 0, & x = 0 \\ 1, & x > 0 \end{cases}$$

have a vertical tangent at the origin? Give reasons for your answer.

28. *Writing to Learn* Does the graph of

$$U(x) = \begin{cases} 0, & x < 0 \\ 1, & x \geq 0 \end{cases}$$

have a vertical tangent at the point $(0, 1)$? Give reasons for your answer.

COMPUTER EXPLORATIONS

In Exercises 29–32, use a calculator or computer to find the average rate of change of the function over each interval.

29. $f(x) = e^x$

    (a) $[-2, 0]$ (b) $[1, 3]$

30. $f(x) = \ln x$

    (a) $[1, 4]$ (b) $[100, 103]$

31. $f(t) = \cot t$

    (a) $[\pi/4, 3\pi/4]$ (b) $[\pi/6, \pi/2]$

32. $f(t) = 2 + \cos t$

    (a) $[0, \pi]$ (b) $[-\pi, \pi]$

33. *INS Funding* Table 1.3 gives the amount of federal funding for the Immigration and Naturalization Service (INS) in the United States for several years.

Table 1.3 Federal INS funding	
Year	**Funding ($ billions)**
1993	1.5
1994	1.6
1995	2.1
1996	2.6
1997	3.1

Source: Immigration and Naturalization Service as reported by Bob Laird in *USA Today*, February 18, 1997.

(a) Find the average rate of change in funding from 1993 to 1995.

(b) Find the average rate of change from 1995 to 1997.

(c) Let $x = 0$ represent 1990, $x = 1$ represent 1991, and so forth. Find a quadratic regression equation for the data and superimpose its graph on a scatter plot of the data. (See page 5 for an introduction to regression analysis with a calculator.)

(d) Compute the average rates of change in parts (a) and (b) using the regression equation.

(e) Use the regression equation to find how fast the funding was growing in 1997.

34. *Academic funding by congress* Table 1.4 gives the amounts of money earmarked by the U.S. Congress for collegiate academic programs for several years.

Table 1.4 Congressional academic funding in the United States

Year	Funding ($ millions)
1988	225
1989	289
1990	270
1991	493
1992	684
1993	763
1994	651
1995	600
1996	296
1997	440

Source: The Chronicle of Higher Education, March 28, 1997.

(a) Let $x = 0$ represent 1980, $x = 1$ represent 1981, and so forth. Make a scatter plot of the data.

(b) Let P represent the point corresponding to 1997 and Q the point for any one of the previous years. Make a table of the slopes possible for the secant line PQ.

(c) *Writing to Learn* Based on the computations, explain why someone might be hesitant to make a prediction about the rate of change of congressional funding in 1997.

Grapher Explorations: Vertical Tangents

(a) Graph the curves in Exercises 35–44. Where do the graphs appear to have vertical tangents?

(b) Confirm your findings in part (a) with limit calculations. But before you do, read the introduction to Exercises 27 and 28.

35. $y = x^{2/5}$ **36.** $y = x^{4/5}$

37. $y = x^{1/5}$ **38.** $y = x^{3/5}$

39. $y = 4x^{2/5} - 2x$ **40.** $y = x^{5/3} - 5x^{2/3}$

41. $y = x^{2/3} - (x-1)^{1/3}$ **42.** $y = x^{1/3} + (x-1)^{1/3}$

43. $y = \begin{cases} -\sqrt{|x|}, & x \le 0 \\ \sqrt{x}, & x > 0 \end{cases}$ **44.** $y = \sqrt{|4-x|}$

Graphing Secant and Tangent Lines

Use a CAS to perform the following steps for the functions in Exercises 45–48.

(a) Plot $y = f(x)$ over the interval $x_0 - 1/2 \le x \le x_0 + 3$.

(b) Holding x_0 fixed, the difference quotient

$$q(h) = \frac{f(x_0 + h) - f(x_0)}{h}$$

at x_0 becomes a function of the step size h. Enter this function into your CAS workspace.

(c) Find the limit of q as $h \to 0$.

(d) Define the secant lines $y = f(x_0) + q*(x - x_0)$ for $h = 3, 2$, and 1. Graph them together with f and the tangent line over the interval in part (a).

45. $f(x) = x^3 + 2x, \quad x_0 = 0$

46. $f(x) = x + \frac{5}{x}, \quad x_0 = 1$

47. $f(x) = x + \sin(2x), \quad x_0 = \pi/2$

48. $f(x) = \cos x + 4\sin(2x), \quad x_0 = \pi$

Questions to Guide Your Review

1. What is the average rate of change of the function $g(t)$ over the interval from $t = a$ to $t = b$? How is it related to a secant line?

2. What limit must be calculated to find the rate of change of a function $g(t)$ at $t = t_0$?

3. What is an informal definition of the limit

$$\lim_{x \to x_0} f(x) = L?$$

Why is the definition "informal"? Give examples. What exactly does $\lim_{x \to x_0} f(x) = L$ mean?

4. Does the existence and value of the limit of a function $f(x)$ as x approaches x_0 ever depend on what happens at $x = x_0$? Explain and give examples.

5. What function behaviors might occur for which the limit may fail to exist? Give examples.

6. What theorems are available for calculating limits? Give examples of how the theorems are used.

7. How are one-sided limits related to limits? How can this relationship sometimes be used to calculate a limit or prove it does not exist? Give examples.

8. What is the value of $\lim_{\theta \to 0} ((\sin \theta)/\theta)$? Does it matter whether θ is measured in degrees or radians? Explain.

9. What do $\lim_{x \to \infty} f(x) = L$ and $\lim_{x \to -\infty} f(x) = L$ mean? Give examples.

10. What are $\lim_{x \to \pm \infty} k$ (k a constant) and $\lim_{x \to \pm \infty} (1/x)$? How do you extend these results to other functions? Give examples.

11. How do you find the limit of a rational function as $x \to \pm \infty$? Give examples.

12. What are horizontal, vertical, and oblique asymptotes? Give examples.

13. What conditions must be satisfied by a function if it is to be continuous at an interior point of its domain? At an endpoint?

14. How can looking at the graph of a function help you tell where the function is continuous?

15. What does it mean for a function to be right-continuous at a point? Left-continuous? How are continuity and one-sided continuity related?

16. What can be said about the continuity of polynomials? Of rational functions? Of trigonometric functions? Of exponential functions? Of logarithmic functions? Of rational powers and algebraic combinations of functions? Of composites of functions? Of absolute values of functions? Of inverses of functions?

17. What does it mean for a function to be continuous on an interval?

18. What does it mean for a function to be continuous? Give examples to illustrate that a function that is not continuous on its entire domain may still be continuous on selected intervals within the domain.

19. What are the basic types of discontinuity? Give an example of each. What is a removable discontinuity? Give an example.

20. What does it mean for a function to have the Intermediate Value Property? What conditions guarantee that a function has this property over an interval? What are the consequences for graphing and solving the equation $f(x) = 0$?

21. It is often said that a function is continuous if you can draw its graph without having to lift your pen from the paper. Why is that?

22. What does it mean for a line to be tangent to a curve C at a point P?

23. What is the significance of the formula
$$\lim_{h \to 0} \frac{f(x+h) - f(x)}{h} ?$$
Interpret the formula geometrically and physically.

24. How do you find the tangent to the curve $y = f(x)$ a point (x_0, y_0) on the curve?

25. How does the slope of the curve $y = f(x)$ at $x = x_0$ relate to the function's rate of change with respect to x at $x = x_0$? To the derivative of f at x_0?

Practice Exercises

Limits and Continuity

1. Graph the function
$$f(x) = \begin{cases} 1, & x \le -1 \\ -x, & -1 < x < 0 \\ 1, & x = 0 \\ -x, & 0 < x < 1 \\ 1, & x \ge 1. \end{cases}$$

Then discuss, in detail, limits, one-sided limits, continuity, and one-sided continuity of f at $x = -1, 0$, and 1. Are any of the discontinuities removable? Explain.

2. Repeat the instructions of Exercise 1 for
$$f(x) = \begin{cases} 0, & x \le -1 \\ 1/x, & 0 < |x| < 1 \\ 0, & x = 1 \\ 1, & x > 1. \end{cases}$$

3. Suppose that $f(t)$ and $g(t)$ are defined for all t and that $\lim_{t \to t_0} f(t) = -7$ and $\lim_{t \to t_0} g(t) = 0$. Find the limit as $t \to t_0$ of the following functions.

(a) $3f(t)$ (b) $(f(t))^2$

(c) $f(t) \cdot g(t)$ (d) $\dfrac{f(t)}{g(t) - 7}$

(e) $\cos (g(t))$ (f) $|f(t)|$

(g) $f(t) + g(t)$ (h) $1/f(t)$

4. Suppose that $f(x)$ and $g(x)$ are defined for all x and that $\lim_{x \to 0} f(x) = 1/2$ and $\lim_{x \to 0} g(x) = \sqrt{2}$. Find the limits as $x \to 0$ of the following functions.

(a) $-g(x)$ (b) $g(x) \cdot f(x)$

(c) $f(x) + g(x)$ (d) $1/f(x)$

(e) $x + f(x)$ (f) $\dfrac{f(x) \cdot \cos x}{x - 1}$

In Exercises 5 and 6, find the value that $\lim_{x\to 0} g(x)$ must have if the given limit statements hold.

5. $\lim_{x\to 0}\left(\dfrac{4-g(x)}{x}\right)=1$

6. $\lim_{x\to -4}\left(x\lim_{x\to 0} g(x)\right)=2$

7. On what intervals are the following functions continuous?

(a) $f(x)=x^{1/3}$ **(b)** $g(x)=x^{3/4}$

(c) $h(x)=x^{-2/3}$ **(d)** $k(x)=x^{-1/6}$

8. On what intervals are the following functions continuous?

(a) $f(x)=\tan x$ **(b)** $g(x)=\csc x$

(c) $h(x)=e^{-x}$ **(d)** $k(x)=\dfrac{\sin x}{x}$

Finding Limits

In Exercises 9–16, find the limit or explain why it does not exist.

9. $\lim \dfrac{x^2-4x+4}{x^3+5x^2-14x}$

(a) as $x\to 0$ **(b)** as $x\to 2$

10. $\lim \dfrac{x^2+x}{x^5+2x^4+x^3}$

(a) as $x\to 0$ **(b)** as $x\to -1$

11. $\lim_{x\to 1}\dfrac{1-\sqrt{x}}{1-x}$

12. $\lim_{x\to a}\dfrac{x^2-a^2}{x^4-a^4}$

13. $\lim_{h\to 0}\dfrac{(x+h)^2-x^2}{h}$

14. $\lim_{x\to 0}\dfrac{(x+h)^2-x^2}{h}$

15. $\lim_{x\to 0}\dfrac{\frac{1}{2+x}-\frac{1}{2}}{x}$

16. $\lim_{x\to 0}\dfrac{(2+x)^3-8}{x}$

Find the limits in Exercises 17–28.

17. $\lim_{x\to\infty}\dfrac{2x+3}{5x+7}$

18. $\lim_{x\to -\infty}\dfrac{2x^2+3}{5x^2+7}$

19. $\lim_{x\to -\infty}\dfrac{x^2-4x+8}{3x^3}$

20. $\lim_{x\to\infty}\dfrac{1}{x^2-7x+1}$

21. $\lim_{x\to -\infty}\dfrac{x^2-7x}{x+1}$

22. $\lim_{x\to\infty}\dfrac{x^4+x^3}{12x^3+128}$

23. $\lim_{x\to\infty}\dfrac{\sin x}{\text{int } x}$ (If you have a grapher, try graphing the function for $-5\le x\le 5$.)

24. $\lim_{\theta\to\infty}\dfrac{\cos\theta-1}{\theta}$ (If you have a grapher, try graphing $f(x)=x(\cos(1/x)-1)$ near the origin to "see" the limit at infinity.)

25. $\lim_{x\to\infty}\dfrac{x+\sin x+2\sqrt{x}}{x+\sin x}$

26. $\lim_{x\to\infty}\dfrac{x^{2/3}+x^{-1}}{x^{2/3}+\cos^2 x}$

27. $\lim_{x\to\infty} e^{-x^2}$

28. $\lim_{x\to -\infty} e^{1/x}$

T 29. Let $f(x)=x^3-x-1$.

(a) Show that f has a zero between -1 and 2.

(b) Solve the equation $f(x)=0$ graphically with an error of magnitude at most 10^{-8}.

(c) It can be shown that the exact value of the solution in part (b) is

$$\left(\frac{1}{2}+\frac{\sqrt{69}}{18}\right)^{1/3}+\left(\frac{1}{2}-\frac{\sqrt{69}}{18}\right)^{1/3}$$

Evaluate this exact answer and compare it with the value you found in part (b).

T 30. Let $f(\theta)=\theta^3-2\theta+2$.

(a) Show that f has a zero between -2 and 0.

(b) Solve the equation $f(\theta)=0$ graphically with an error of magnitude at most 10^{-4}.

(c) It can be shown that the exact value of the solution in part (b) is

$$\left(\sqrt{\frac{19}{27}}-1\right)^{1/3}-\left(\sqrt{\frac{19}{27}}+1\right)^{1/3}$$

Evaluate this exact answer and compare it with the value you found in part (b).

Additional Exercises: Theory, Examples, Applications

T 1. *Assigning a value to 0^0* The rules of exponents tell us that $a^0=1$ if a is any number different from zero. They also tell us that $0^n=0$ if n is any positive number.

If we tried to extend these rules to include the case 0^0, we would get conflicting results. The first rule would say $0^0=1$, whereas the second would say $0^0=0$.

We are not dealing with a question of right or wrong here. Neither rule applies as it stands, so there is no contradiction. We could, in fact, define 0^0 to have any value we wanted as long as we could persuade others to agree.

What value would you like 0^0 to have? Here is an example that might help you to decide. (See Exercise 2 below for another example.)

(a) Calculate x^x for $x=0.1$, 0.01, 0.001, and so on as far as your calculator can go. Record the values you get. What pattern do you see?

(b) Graph the function $y=x^x$ for $0<x\le 1$. Even though the function is not defined for $x\le 0$, the graph will approach the y-axis from the right. Toward what y-value does it seem to be headed? Zoom in to further support your idea.

T **2.** *A reason you might want 0^0 to be something other than 0 or 1* As the number x increases through positive values, the numbers $1/x$ and $1/(\ln x)$ both approach zero. What happens to the number

$$f(x) = \left(\frac{1}{x}\right)^{1/(\ln x)}$$

as x increases? Here are two ways to find out.

(a) Evaluate f for $x = 10, 100, 1000$, and so on as far as your calculator can reasonably go. What pattern do you see?

(b) Graph f in a variety of graphing windows, including windows that contain the origin. What do you see? Trace the y-values along the graph. What do you find?

3. *Lorentz contraction* In relativity theory, the length of an object, say a rocket, appears to an observer to depend on the speed at which the object is traveling with respect to the observer. If the observer measures the rocket's length as L_0 at rest, then at speed v the length will appear to be

$$L = L_0 \sqrt{1 - \frac{v^2}{c^2}}.$$

This equation is the Lorentz contraction formula. Here, c is the speed of light in a vacuum, about 3×10^8 m/sec. What happens to L as v increases? Find $\lim_{v \to c^-} L$. Why was the left-hand limit needed?

4. *Controlling the flow from a draining tank* Torricelli's law says that if you drain a tank like the one in the figure shown, the rate y at which water runs out is a constant times the square root of the water's depth x. The constant depends on the size and shape of the exit valve.

Exit rate y ft^3/min

Suppose that $y = \sqrt{x}/2$ for a certain tank. You are trying to maintain a fairly constant exit rate by adding water to the tank with a hose from time to time. How deep must you keep the water if you want to maintain the exit rate

(a) within 0.2 ft^3/min of the rate $y_0 = 1$ ft^3/min?

(b) within 0.1 ft^3/min of the rate $y_0 = 1$ ft^3/min?

5. *Thermal expansion in precise equipment* As you may know, most metals expand when heated and contract when cooled. The dimensions of a piece of laboratory equipment are sometimes so critical that

the shop where the equipment is made must be held at the same temperature as the laboratory where the equipment is to be used. A typical aluminum bar that is 10 cm wide at 70°F will be

$$y = 10 + (t - 70) \times 10^{-4}$$

centimeters wide at a nearby temperature t. Suppose that you are using a bar like this in a gravity wave detector, where its width must stay within 0.0005 cm of the ideal 10 cm. How close to $t_0 = 70°F$ must you maintain the temperature to ensure that this tolerance is not exceeded?

6. *Writing to Learn: Antipodal points* Is there any reason to believe that there is always a pair of antipodal (diametrically opposite) points on Earth's equator where the temperatures are the same? Explain.

T **7.** *Roots of a quadratic equation that is almost linear* The equation $ax^2 + 2x - 1 = 0$, where a is a constant, has two roots if $a > -1$ and $a \neq 0$, one positive and one negative:

$$r_+(a) = \frac{-1 + \sqrt{1 + a}}{a}, \qquad r_-(a) = \frac{-1 - \sqrt{1 + a}}{a}.$$

(a) What happens to $r_+(a)$ as $a \to 0$? As $a \to -1^+$?

(b) What happens to $r_-(a)$ as $a \to 0$? As $a \to -1^+$?

(c) Support your conclusions by graphing $r_+(a)$ and $r_-(a)$ as functions of a. Describe what you see.

(d) For added support, graph $f(x) = ax^2 + 2x - 1$ simultaneously for $a = 1, 0.5, 0.2, 0.1$, and 0.05.

8. *One-sided limits* If $\lim_{x \to 0^+} f(x) = A$ and $\lim_{x \to 0^-} f(x) = B$, find

(a) $\lim_{x \to 0^+} f(x^3 - x)$

(b) $\lim_{x \to 0^-} f(x^3 - x)$

(c) $\lim_{x \to 0^+} f(x^2 - x^4)$

(d) $\lim_{x \to 0^-} f(x^2 - x^4)$

9. *Limits and continuity* Which of the following statements are true, and which are false? If true, say why; if false, give a counterexample (that is, an example confirming the falsehood).

(a) If $\lim_{x \to a} f(x)$ exists but $\lim_{x \to a} g(x)$ does not exist, then $\lim_{x \to a} (f(x) + g(x))$ does not exist.

(b) If neither $\lim_{x \to a} (f(x)$ nor $\lim_{x \to a} g(x)$ exists, then $\lim_{x \to a} f(x) + g(x))$ does not exist.

(c) If f is continuous at x, then so is $|f|$.

(d) If $|f|$ is continuous at a, then so is f.

10. *Root of an equation* Show that the equation $x + 2 \cos x = 0$ has at least one solution.

Formal Definition of Limit

In Exercises 11–14, use the formal definition of limit to prove that the function is continuous at x_0.

11. $f(x) = x^2 - 7, \quad x_0 = 1$

12. $g(x) = 1/(2x), \quad x_0 = 1/4$

13. $h(x) = \sqrt{2x - 3}$, $\quad x_0 = 2$

14. $F(x) = \sqrt{9 - x}$, $\quad x_0 = 5$

15. *A function continuous at only one point* Let

$$f(x) = \begin{cases} x & \text{if } x \text{ is rational} \\ 0 & \text{if } x \text{ is irrational.} \end{cases}$$

(a) Show that f is continuous at $x = 0$.

(b) Use the fact that every nonempty open interval of real numbers contains both rational and irrational numbers to show that f is not continuous at any nonzero value of x.

16. *The Dirichlet ruler function* If x is a rational number, then x can be written in a unique way as a quotient of integers m/n where $n > 0$ and m and n have no common factors greater than 1. (We say that such a fraction is in *lowest terms*. For example, 6/4 written in lowest terms is 3/2.) Let $f(x)$ be defined for all x in the interval $[0, 1]$ by

$$f(x) = \begin{cases} 1/n & \text{if } x = m/n \text{ is a rational number in lowest terms} \\ 0 & \text{if } x \text{ is irrational.} \end{cases}$$

For instance, $f(0) = f(1) = 1$, $f(1/2) = 1/2$, $f(1/3) = f(2/3) = 1/3$, $f(1/4) = f(3/4) = 1/4$, and so on.

(a) Show that f is discontinuous at every rational number in $[0, 1]$.

(b) Show that f is continuous at every irrational number in $[0, 1]$. (*Hint*: If ϵ is a given positive number, show that there are only finitely many rational numbers r in $[0, 1]$ such that $f(r) \geq \epsilon$.)

(c) Sketch the graph of f. Why do you think f is called the "ruler function"?

2 Derivatives

OVERVIEW   In Chapter 1, we defined the slope of a curve at a point as the limit of secant slopes. This limit, called a derivative, measures the rate at which a function changes, and it is one of the most important ideas in calculus. Derivatives are used widely in engineering, science, economics, medicine, and computer science to calculate velocity and acceleration, to explain the behavior of machinery, to estimate the drop in water level as water is pumped out of a tank, and to predict the consequences of making errors in measurements. Finding derivatives by evaluating limits can be lengthy and difficult. In this chapter, we develop techniques to make calculating derivatives easier.

2.1 The Derivative as a Function

Definition of Derivative • Notation • Derivatives of Constants, Powers, Multiples, and Sums • Differentiable on an Interval; One-Sided Derivatives • Graphing f' from Estimated Values • Differentiable Functions are Continuous • Intermediate Value Property of Derivatives • Second- and Higher-Order Derivatives

CD-ROM
WEBsite

Historical Essay

The Derivative

CD-ROM
WEBsite

At the end of Chapter 1, we defined the slope of a curve $y = f(x)$ at the point where $x = x_0$ to be

$$\lim_{h \to 0} \frac{f(x_0 + h) - f(x_0)}{h}.$$

We called this limit, when it existed, the derivative of f at x_0. We now investigate the derivative as a *function* derived from f by considering the limit at each point of the domain of f.

Definition of Derivative

Definition   Derivative Function
The **derivative** of the function $f(x)$ with respect to the variable x is the function f' whose value at x is

$$f'(x) = \lim_{h \to 0} \frac{f(x + h) - f(x)}{h},$$

provided the limit exists.

Calculating $f'(x)$ from the Definition of Derivative

Step 1. Write expressions for $f(x)$ and $f(x + h)$.

Step 2. Expand and simplify the difference quotient

$$\frac{f(x + h) - f(x)}{h}.$$

Step 3. Using the simplified quotient, find $f'(x)$ by evaluating the limit

$$f'(x) = \lim_{h \to 0} \frac{f(x + h) - f(x)}{h}.$$

CD-ROM
WEBsite

The domain of f' is the set of points in the domain of f for which the limit exists. Its domain may be the same as the domain of f or it may be smaller. If f' exists at a particular x, we say that f is **differentiable (has a derivative)** at x. If f' exists at every point of the domain of f, we call f **differentiable.**

Example 1   Applying the Definition

(a) Find the derivative of $y = \sqrt{x}$ for $x > 0$.

(b) Find the tangent line to the curve $y = \sqrt{x}$ at $x = 4$.

Solution

(a)

Step 1: $f(x) = \sqrt{x}$ and $f(x + h) = \sqrt{x + h}$

Step 2: $\dfrac{f(x + h) - f(x)}{h} = \dfrac{\sqrt{x + h} - \sqrt{x}}{h}$

$$= \frac{(x + h) - x}{h(\sqrt{x + h} + \sqrt{x})} \qquad \text{Multiply by } \frac{\sqrt{x + h} + \sqrt{x}}{\sqrt{x + h} + \sqrt{x}}.$$

$$= \frac{1}{\sqrt{x + h} + \sqrt{x}}$$

Step 3: $f'(x) = \lim\limits_{h \to 0} \dfrac{1}{\sqrt{x + h} + \sqrt{x}} = \dfrac{1}{2\sqrt{x}}$

See Figure 2.1.

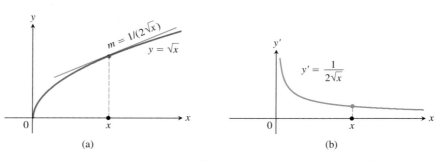

(a)   (b)

FIGURE 2.1 The graphs of (a) $y = \sqrt{x}$ and (b) $y' = 1/(2\sqrt{x})$, $x > 0$. The function is defined at $x = 0$, but its derivative is not. (Example 1)

(b) The slope of the curve at $x = 4$ is

$$f'(4) = \frac{1}{2\sqrt{4}} = \frac{1}{4}.$$

The tangent is the line through the point $(4, 2)$ with slope $1/4$ (Figure 2.2).

$$y = 2 + \frac{1}{4}(x - 4)$$

$$y = \frac{1}{4}x + 1$$

FIGURE 2.2 The curve $y = \sqrt{x}$ and its tangent at $(4, 2)$. The tangent's slope is found by evaluating y' at $x = 4$. (Example 1)

Notation

There are many ways to denote the derivative of a function $y = f(x)$. Besides $f'(x)$, the most common notations are these:

The "prime" notation comes from Newton's work, the d/dx notation from that of Leibniz.

y'	"y prime"	Nice and brief but does not name the independent variable
$\dfrac{dy}{dx}$	"$dy\,dx$"	Names the variables and uses d for derivative
$\dfrac{df}{dx}$	"$df\,dx$"	Emphasizes the function's name
$\dfrac{d}{dx} f(x)$	"ddx of $f(x)$"	Emphasizes the idea that differentiation is an operation performed on f (Figure 2.3)

We also read dy/dx as "the derivative of y with respect to x" and df/dx and $(d/dx)f(x)$ as "the derivative of f with respect to x."

The value

$$f'(a) = \lim_{h \to 0} \frac{f(a + h) - f(a)}{h}$$

of the derivative of $y = f(x)$ with respect to x at $x = a$ can also be denoted as

$$y'\big|_{x=a} \quad \text{or} \quad \frac{dy}{dx}\bigg|_{x=a} \quad \text{or} \quad \frac{d}{dx} f(x)\bigg|_{x=a}.$$

The symbol $\big|_{x=a}$, called an **evaluation symbol,** tells us to evaluate the expression to its left at $x = a$.

The process of calculating a derivative is called **differentiation.** Example 1 illustrates the process for the function $y = \sqrt{x}$. We now show how to differentiate functions without having to apply the definition each time.

FIGURE 2.3 Flow diagram for the operation of taking a derivative with respect to x.

Derivatives of Constants, Powers, Multiples, and Sums

The first rule is that the derivative of every constant function is the zero function.

Rule 1 Derivative of a Constant Function

If f has the constant value $f(x) = c$, then

$$\frac{df}{dx} = \frac{d}{dx}(c) = 0.$$

Example 2 Using Rule 1

If f has the constant value $f(x) = 8$, then

$$\frac{df}{dx} = \frac{d}{dx}(8) = 0.$$

Similarly,

$$\frac{d}{dx}\left(-\frac{\pi}{2}\right) = 0 \quad \text{and} \quad \frac{d}{dx}\left(\sqrt{3}\right) = 0.$$

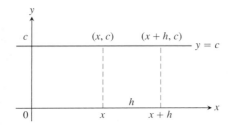

FIGURE 2.4 The rule $(d/dx)(c) = 0$ is another way to say that the values of constant functions never change and that the slope of a horizontal line is zero at every point.

Proof of Rule 1 We apply the definition of derivative to $f(x) = c$, the function whose outputs have the constant value c (Figure 2.4). At every value of x, we find that

$$f'(x) = \lim_{h \to 0} \frac{f(x+h) - f(x)}{h} = \lim_{h \to 0} \frac{c - c}{h} = \lim_{h \to 0} 0 = 0.$$

The second rule tells how to differentiate x^n if n is a positive integer.

Rule 2   Power Rule for Positive Integers

If n is a positive integer, then

$$\frac{d}{dx} x^n = nx^{n-1}.$$

To apply the Power Rule, we subtract 1 from the original exponent (n) and multiply the result by n.

Example 3   Interpreting Rule 2

f	x	x^2	x^3	x^4	$\cdots$
f'	1	$2x$	$3x^2$	$4x^3$	$\cdots$

Proof of Rule 2 If $f(x) = x^n$, then $f(x + h) = (x + h)^n$. Since n is a positive integer, we can use the fact that

$$a^n - b^n = (a - b)(a^{n-1} + a^{n-2}b + \cdots + ab^{n-2} + b^{n-1})$$

to simplify the difference quotient for f. Taking $x + h = a$ and $x = b$, we have $a - b = h$. Thus,

$$\frac{f(x+h) - f(x)}{h} = \frac{(x+h)^n - x^n}{h}$$

$$= \frac{(h)[(x+h)^{n-1} + (x+h)^{n-2}x + \cdots + (x+h)x^{n-2} + x^{n-1}]}{h}$$

$$= \underbrace{(x+h)^{n-1} + (x+h)^{n-2}x + \cdots + (x+h)x^{n-2} + x^{n-1}}_{n \text{ terms, each with limit } x^{n-1} \text{ as } h \to 0}.$$

Hence,

$$\frac{d}{dx} x^n = \lim_{h \to 0} \frac{f(x+h) - f(x)}{h} = nx^{n-1}.$$

The third rule says that when a differentiable function is multiplied by a constant, its derivative is multiplied by the same constant.

Rule 3   Constant Multiple Rule

If u is differentiable function of x, and c is a constant, then

$$\frac{d}{dx} (cu) = c \frac{du}{dx}.$$

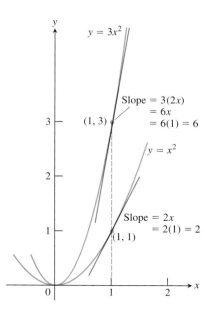

FIGURE 2.5 The graphs of $y = x^2$ and $y = 3x^2$. Tripling the y-coordinates triples the slope. (Example 4)

Denoting functions by u and v

The functions we are working with when we need a differentiation formula are likely to be denoted by letters like f and g. When we apply the formula, we do not want to find it using these same letters in some other way. To guard against this problem, we denote the functions in differentiation rules by letters like u and v that are not likely to be already in use.

Example 4   Using Rule 3

(a) $\dfrac{d}{dx}(3x^2) = 3 \cdot 2x = 6x$

Interpretation: Rescaling the graph of $y = x^2$ by multiplying each y-coordinate by 3 multiplies the slope at each point by 3 (Figure 2.5).

(b) *A useful special case:* The derivative of the negative of a differentiable function is the negative of the function's derivative. Rule 3 with $c = -1$ gives

$$\frac{d}{dx}(-u) = \frac{d}{dx}(-1 \cdot u) = -1 \cdot \frac{d}{dx}(u) = -\frac{du}{dx}.$$

Proof of Rule 3

$$\frac{d}{dx}cu = \lim_{h \to 0} \frac{cu(x + h) - cu(x)}{h} \qquad \text{Derivative definition with } f(x) = cu(x)$$

$$= c \lim_{h \to 0} \frac{u(x + h) - u(x)}{h} \qquad \text{Limit property}$$

$$= c \frac{du}{dx} \qquad u \text{ is differentiable.}$$

The next rule says that the derivative of the sum of two differentiable functions is the sum of their derivatives.

Rule 4   Derivative Sum Rule

If u and v are differentiable functions of x, then their sum $u + v$ is differentiable at every point where u and v are both differentiable. At such points,

$$\frac{d}{dx}(u + v) = \frac{du}{dx} + \frac{dv}{dx}.$$

Example 5   Derivative of a Sum

$$y = x^4 + 12x$$

$$\frac{dy}{dx} = \frac{d}{dx}(x^4) + \frac{d}{dx}(12x)$$

$$= 4x^3 + 12$$

Proof of Rule 4   We apply the definition of derivative to $f(x) = u(x) + v(x)$:

$$\frac{d}{dx}[u(x) + v(x)] = \lim_{h \to 0} \frac{[u(x + h) + v(x + h)] - [u(x) + v(x)]}{h}$$

$$= \lim_{h \to 0} \left[\frac{u(x + h) - u(x)}{h} + \frac{v(x + h) - v(x)}{h} \right]$$

$$= \lim_{h \to 0} \frac{u(x + h) - u(x)}{h} + \lim_{h \to 0} \frac{v(x + h) - v(x)}{h} = \frac{du}{dx} + \frac{dv}{dx}.$$

Combining the Sum Rule with the Constant Multiple rule gives the equivalent **Difference Rule,** which says that the derivative of a *difference* of differentiable functions is the difference of their derivatives.

$$\frac{d}{dx}(u - v) = \frac{d}{dx}[u + (-1)v] = \frac{du}{dx} + (-1)\frac{dv}{dx} = \frac{du}{dx} - \frac{dv}{dx}.$$

The Sum Rule also extends to sums of more than two functions, as long as there are only finitely many functions in the sum. If $u_1, u_2, \ldots, u_n$ are differentiable at x, then so is $u_1 + u_2 + \cdots + u_n$, and

$$\frac{d}{dx}(u_1 + u_2 + \cdots + u_n) = \frac{du_1}{dx} + \frac{du_2}{dx} + \cdots + \frac{du_n}{dx}.$$

Example 6   Derivative of a Polynomial

$$y = x^3 + \frac{4}{3}x^2 - 5x + 1$$

$$\frac{dy}{dx} = \frac{d}{dx}x^3 + \frac{d}{dx}\left(\frac{4}{3}x^2\right) - \frac{d}{dx}(5x) + \frac{d}{dx}(1)$$

$$= 3x^2 + \frac{4}{3}\cdot 2x - 5 + 0$$

$$= 3x^2 + \frac{8}{3}x - 5$$

All polynomials are differentiable.

Notice that we can differentiate any polynomial term by term, the way we differentiated the polynomials in Example 6.

Example 7   Finding Horizontal Tangents

Does the curve $y = x^4 - 2x^2 + 2$ have any horizontal tangents? If so, where?

Solution   The horizontal tangents, if any, occur where the slope dy/dx is zero. To find these points, we follow these steps.

1.   Calculate dy/dx:

$$\frac{dy}{dx} = \frac{d}{dx}(x^4 - 2x^2 + 2) = 4x^3 - 4x.$$

2.   Solve the equation $\dfrac{dy}{dx} = 0$ for x:

$$4x^3 - 4x = 0$$
$$4x(x^2 - 1) = 0$$
$$x = 0, 1, -1.$$

The curve $y = x^4 - 2x^2 + 2$ has horizontal tangents at $x = 0, 1,$ and -1. The corresponding points on the curve are $(0, 2)$, $(1, 1)$ and $(-1, 1)$. See Figure 2.6.

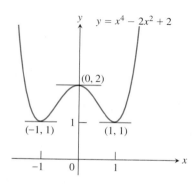

FIGURE 2.6 The curve $y = x^4 - 2x^2 + 2$ and its horizontal tangents. (Example 7)

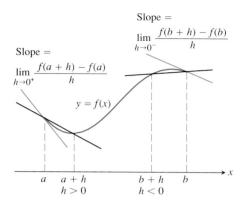

Slope =
$$\lim_{h \to 0^-} \frac{f(b + h) - f(b)}{h}$$

Slope =
$$\lim_{h \to 0^+} \frac{f(a + h) - f(a)}{h}$$

$y = f(x)$

| a | $a + h$ | | $b + h$ | b |
| | $h > 0$ | | $h < 0$ | |

FIGURE 2.7 Derivatives at endpoints are one-sided limits.

Differentiable on an Interval; One-Sided Derivatives

A function $y = f(x)$ is **differentiable** on an open interval (finite or infinite) if it has a derivative at each point of the interval. It is differentiable on a closed interval $[a, b]$ if it is differentiable on the interior (a, b) and if the limits

$$\lim_{h \to 0^+} \frac{f(a + h) - f(a)}{h} \qquad \textbf{Right-hand derivative at } a$$

$$\lim_{h \to 0^-} \frac{f(b + h) - f(b)}{h} \qquad \textbf{Left-hand derivative at } b$$

exist at the endpoints (Figure 2.7).

Right-hand and left-hand derivatives may be defined at any point of a function's domain. The usual relation between one-sided and two-sided limits holds for these derivatives. Because of Theorem 5, Section 1.2, a function has a derivative at a point if and only if it has left-hand and right-hand derivatives there, and these one-sided derivatives are equal.

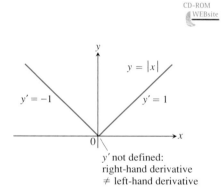

$y = |x|$

$y' = -1$ $y' = 1$

0

y' not defined:
right-hand derivative
$\neq$ left-hand derivative

FIGURE 2.8 The function $y = |x|$ is not differentiable at the origin where the graph has a "corner."

Example 8 $y = |x|$ is Not Differentiable at the Origin

Show that the function $y = |x|$ is differentiable on $(-\infty, 0)$ and $(0, \infty)$ but has no derivative at $x = 0$.

Solution   To the right of the origin,

$$\frac{d}{dx}(|x|) = \frac{d}{dx}(x) = \frac{d}{dx}(1 \cdot x) = 1. \qquad \frac{d}{dx}(mx + b) = m, \\ |x| = x$$

To the left,

$$\frac{d}{dx}(|x|) = \frac{d}{dx}(-x) = \frac{d}{dx}(-1 \cdot x) = -1 \qquad |x| = -x$$

(Figure 2.8). There can be no derivative at the origin because the one-sided derivatives differ there:

$$\text{Right-hand derivative of } |x| \text{ at zero} = \lim_{h \to 0^+} \frac{|0 + h| - |0|}{h} = \lim_{h \to 0^+} \frac{|h|}{h}$$

$$= \lim_{h \to 0^+} \frac{h}{h} \qquad |h| = h \text{ when } h > 0.$$

$$= \lim_{h \to 0^+} 1 = 1$$

$$\text{Left-hand derivative of } |x| \text{ at zero} = \lim_{h \to 0^-} \frac{|0 + h| - |0|}{h} = \lim_{h \to 0^-} \frac{|h|}{h}$$

$$= \lim_{h \to 0^-} \frac{-h}{h} \qquad |h| = -h \text{ when } h < 0.$$

$$= \lim_{h \to 0^-} -1 = -1.$$

In general, if the graph of a function has a "corner," then there is no tangent at this point and f is not differentiable there. Thus, differentiability is a "smoothness" condition.

Graphing f′ from Estimated Values

When we measure the values of a function $y = f(x)$ in the laboratory or in the field (pressure versus temperature, say, or population versus time), we usually connect the data points with lines or curves to picture the graph of f. We can often make a reasonable plot of $f′$ by estimating slopes on this graph. The following example shows how this is done and what can be learned from the process.

Example 9   Graphing a Derivative

Graph the derivative of the function $y = f(x)$ in Figure 2.9a.

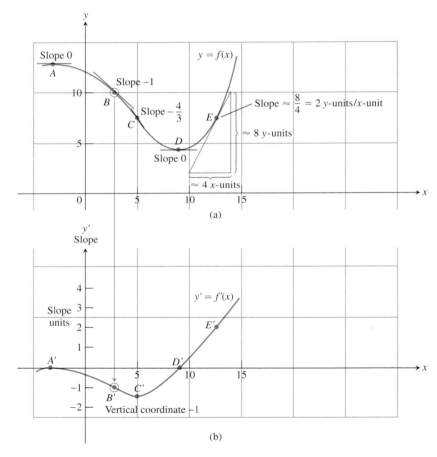

(a)

(b)

FIGURE 2.9 We made the graph of $y′ = f′(x)$ in (b) by plotting slopes from the graph of $y = f(x)$ in (a). The vertical coordinate of $B′$ is the slope at B and so on. The graph of $y′ = f′(x)$ is a visual record of how the slope of f changes with x.

Solution   We draw a pair of axes, marking the horizontal axis in x-units and the vertical axis in $y′$-units (Figure 2.9b). Next we sketch tangents to the graph of f at frequent intervals and use their slopes to estimate the values of $y′ = f′(x)$ at these points. We plot the corresponding $(x, y′)$ pairs and connect them with a smooth curve.

From the graph of $y' = f'(x)$ we see at a glance

1. where the rate of change of f is positive, negative, or zero

2. the rough size of the growth rate at any x and its size in relation to the size of $f(x)$

3. where the rate of change itself is increasing or decreasing.

You can experiment further with these ideas in Exercises 15 through 20.

Differentiable Functions Are Continuous

A function is continuous at every point where it has a derivative.

CAUTION   The converse of Theorem 1 is false. A function need not have a derivative at a point where it is continuous, as we saw in Example 8.

> **Theorem 1**   Differentiability Implies Continuity
> If f has a derivative at $x = c$, then f is continuous at $x = c$.

Proof   Given that $f'(c)$ exists, we must show that $\lim_{x \to c} f(x) = f(c)$, or equivalently, that $\lim_{h \to 0} f(c + h) = f(c)$. If $h \neq 0$, then

$$f(c + h) = f(c) + (f(c + h) - f(c))$$
$$= f(c) + \frac{f(c + h) - f(c)}{h} \cdot h.$$

Now take limits as $h \to 0$. By Theorem 1 of Section 1.2,

$$\lim_{h \to 0} f(c + h) = \lim_{h \to 0} f(c) + \lim_{h \to 0} \frac{f(c + h) - f(c)}{h} \cdot \lim_{h \to 0} h$$
$$= f(c) + f'(c) \cdot 0$$
$$= f(c) + 0$$
$$= f(c).$$

Similar arguments with one-sided limits show that if f has a derivative from one side (right or left) at $x = c$ then f is continuous from that side at $x = c$.

Theorem 1 gives another reason why a function may fail to have a derivative. If the function has a discontinuity at a point (for instance, a jump discontinuity), then it cannot be differentiable there. For instance, the greatest integer function $y = \text{int } x$ fails to be differentiable at every integer $x = n$ (Example 4, Section 1.4).

Intermediate Value Property of Derivatives

Not every function can be some function's derivative, as we see from the following theorem.

> **Theorem 2**   Intermediate Value Property of Derivatives
> If a and b are any two points in an interval on which f is differentiable, then f' takes on every value between $f'(a)$ and $f'(b)$.

Theorem 2 (which we will not prove) says that a function cannot *be* a derivative on an interval unless it has the Intermediate Value Property there (Figure 2.10). The question of when a function *is* a derivative is one of the central questions in all calculus, and Newton's and Leibniz's answer to this question revolutionized the world of mathematics. We see what their answer was in Chapter 4.

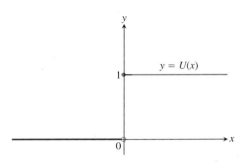

FIGURE 2.10 The unit step function does not have the Intermediate Value Property and cannot be the derivative of a function on the real line.

Second- and Higher-Order Derivatives

The derivative $y' = dy/dx$ is the **first (first-order) derivative** of y with respect to x. This derivative may itself be a differentiable function of x; if so, its derivative

$$y'' = \frac{dy'}{dx} = \frac{d}{dx}\left(\frac{dy}{dx}\right) = \frac{d^2y}{dx^2}$$

is called the **second (second-order) derivative** of y with respect to x.

If y'' is differentiable, its derivative, $y''' = dy''/dx = d^3y/dx^3$ is the **third (third-order) derivative** of y with respect to x. The names continue as you imagine, with

$$y^{(n)} = \frac{d}{dx}y^{(n-1)}$$

denoting the **nth (nth-order) derivative** of y with respect to x for any positive integer n.

We can interpret the second derivative as the rate of change of the slope of the tangent to the curve $y = f(x)$ at each point. You will see in the next chapter that the second derivative reveals whether a curve bends upward or downward from the tangent line as we move off the point of tangency. In the next section, we interpret both the second and third derivatives in terms of motion along a straight line.

Notice that

$$\frac{d}{dx}\left(\frac{dy}{dx}\right)$$

does not mean multiplication. It means "the derivative of the derivative."

How to read the symbols for derivatives

y'	"y prime"
y''	"y double prime"
$\dfrac{d^2y}{dx^2}$	"d squared y dx squared"
y'''	"y triple prime"
$y^{(n)}$	"y super n"
$\dfrac{d^ny}{dx^n}$	"d to the n of y by dx to the n"

Example 10   Finding Higher Derivatives

The first four derivatives of $y = x^3 - 3x^2 + 2$ are

First derivative:	$y' = 3x^2 - 6x$
Second derivative:	$y'' = 6x - 6$
Third derivative:	$y''' = 6$
Fourth derivative:	$y^{(4)} = 0.$

The function has derivatives of all orders, the fifth and later derivatives all being zero.

EXERCISES 2.1

Finding and Evaluating Derivative Functions

In Exercises 1–6, use the definition to find the function's derivative. Then evaluate the derivative at the indicated point or points.

1. $f(x) = 4 - x^2$, $f'(-3)$, $f'(0)$

2. $g(t) = 1/t^2$; $g'(-1)$, $g'(2)$

3. $\left.\dfrac{ds}{dt}\right|_{t=-1}$ if $s = t^3 - t^2$

4. $f(x) = x + \dfrac{9}{x}$, $x = -3$

5. $p(\theta) = \sqrt{3\theta}$, $\theta = 0.25$

6. $\left.\dfrac{dr}{d\theta}\right|_{\theta=0}$ if $r = \dfrac{2}{\sqrt{4 - \theta}}$

Derivative Calculations

In Exercises 7–10, find the first and second derivatives.

7. $y = x^2 + x + 8$

8. $s = 5t^3 - 3t^5$

9. $y = \dfrac{4x^3}{3} - 4$

10. $y = \dfrac{x^3 + 7}{x}$

Find the derivatives of all orders of the functions in Exercises 11 and 12.

11. $y = \dfrac{x^4}{2} - \dfrac{3}{2}x^2 - x$ **12.** $y = \dfrac{x^5}{120}$

Slopes and Tangents

13. (a) Find an equation for the tangent to the curve $y = x^3 - 4x + 1$ at the point (2, 1).

(b) What is the range of values of the curve's slope?

(c) Find equations for the tangents to the curve at the points where the slope of the curve is 8.

14. (a) Find an equation for the horizontal tangent to the curve $y = x - 3\sqrt{x}$.

(b) What is the range of values of the curve's slope?

Graphs

Match the functions graphed in Exercise 15–18 with the derivatives graphed in Figure 2.11.

15.

16.

17.

18.

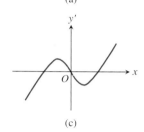

FIGURE 2.11 The derivative graphs for Exercises 15–18.

19. *Writing to Learn*

(a) The graph in Figure 2.12 is made of line segments joined end to end. At which points of the interval [−4, 6] is f' not defined? Give reasons for your answer.

(b) Graph the derivative of f. Call the vertical axis the y'-axis. The graph should show a step function.

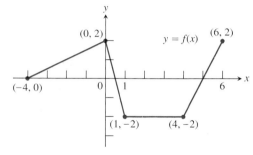

FIGURE 2.12 The graph for Exercise 19.

Recovering a Function from Its Derivative

20. (a) Use the following information to graph the function f over the closed interval $[-2, 5]$.

    i. The graph of f is made of line segments joined end to end.

    ii. The graph starts at the point $(-2, 3)$.

    iii. The derivative of f is the step function in Figure 2.13.

  (b) Repeat part (a) assuming that the graph starts at $(-2, 0)$ instead of $(-2, 3)$.

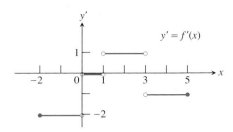

FIGURE 2.13 The derivative graph for Exercise 20.

One-Sided Derivatives

Compare the right-hand and left-hand derivatives to show that the functions in Exercises 21 and 22 are not differentiable at the point P.

21.   **22.**

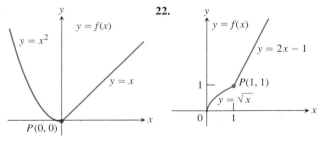

When Does a Function Not Have a Derivative at a Point?

A function has a derivative at a point x_0 if the slopes of the secant lines through $P(x_0, f(x_0))$ and a nearby point Q on the graph approach a limit as Q approaches P. Whenever the secants fail to take up a single limiting position or become vertical as Q approaches P, the derivative does not exist. A function whose graph is otherwise smooth will fail to have a derivative at a point where the graph has

(i) a *corner*, where the one-sided derivatives differ

(ii) a *cusp*, where the slope of PQ approaches ∞ from one side and $-\infty$ from the other

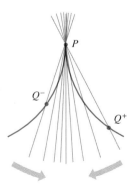

(iii) a *vertical tangent*, were the slope of PQ approaches ∞ from both sides or approaches $-\infty$ from both sides (here, $-\infty$)

(iv) a *discontinuity*.

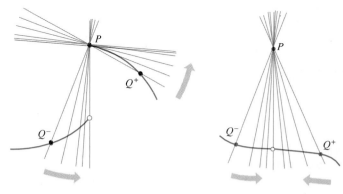

Each figure in Exercises 23–26 shows the graph of a function over a closed interval D. At what domain points does the function appear to be

    (a) differentiable?

    (b) continuous but not differentiable?

    (c) neither continuous nor differentiable?

Give reasons for your answers.

23.

24.

25.

26.

Interpreting Derivatives

In Exercises 27–30,

(a) Find the derivative $y' = f'(x)$ of the given function $y = f(x)$.

(b) Graph $y = f(x)$ and $y' = f'(x)$ side by side using separate sets of coordinate axes and answer the following questions.

(c) For what values of x, if any, is y' positive? Zero? Negative?

(d) Over what intervals of x-values, if any, does the function $y = f(x)$ increase as x increases? Decrease as x increases? How is this related to what you found in part (c)? (We say more about this relationship in Chapter 3.)

27. $y = -x^2$

28. $y = -1/x$

29. $y = x^3/3$

30. $y = x^4/4$

31. *Writing to Learn* Does the curve $y = x^3$ ever have a negative slope? If so, where? Give reasons for your answer.

32. *Writing to Learn* Does the curve $y = 2\sqrt{x}$ have any horizontal tangents? If so, where? Give reasons for your answer.

Theory and Examples

33. *Tangent to a parabola* Does the parabola $y = 2x^2 - 13x + 5$ have a tangent whose slope is -1? If so, find an equation for the line and the point of tangency. If not, why not?

34. *Tangent to* $y = \sqrt{x}$ Does any tangent to the curve $y = \sqrt{x}$ cross the x-axis at $x = -1$? If so, find an equation for the line and the point of tangency. If not, why not?

35. *Greatest integer in x* Does any function differentiable on $(-\infty, \infty)$ have $y = \text{int } x$, the greatest integer in x (see Figure 1.49) as its derivative? Give reasons for your answer.

36. *Derivative of* $y = |x|$ Graph the derivative of $f(x) = |x|$. Then graph $y = (|x| - 0)/(x - 0) = |x|/x$. What can you conclude?

37. *Derivative of* $-f$ Does knowing that a function $f(x)$ is differentiable at $x = x_0$ tell you anything about the differentiability of the function $-f$ at $x = x_0$? Give reasons for your answer.

38. *Derivative of multiples* Does knowing that a function $g(t)$ is differentiable at $t = 7$ tell you anything about the differentiability of the function $3g$ at $t = 7$? Give reasons for your answer.

39. *Limit of a quotient* Suppose that functions $g(t)$ and $h(t)$ are defined for all values of t and $g(0) = h(0) = 0$. Can $\lim_{t \to 0} (g(t))/(h(t))$ exist? If it does exist, must it equal zero? Give reasons for your answers.

40. (a) Let $f(x)$ be a function satisfying $|f(x)| \le x^2$ for $-1 \le x \le 1$. Show that f is differentiable at $x = 0$ and find $f'(0)$.

(b) Show that

$$f(x) = \begin{cases} x^2 \sin \dfrac{1}{x}, & x \ne 0 \\ 0, & x = 0 \end{cases}$$

is differentiable at $x = 0$ and find $f'(0)$.

T **41.** *Writing to Learn* Graph $y = 1/(2\sqrt{x})$ in a window that has $0 \le x \le 2$. Then, on the same screen, graph

$$y = \frac{\sqrt{x+h} - \sqrt{x}}{h}$$

for $h = 1, 0.5, 0.1$. Then try $h = -1, -0.5, -0.1$. Explain what is going on.

T **42.** *Writing to Learn* Graph $y = 3x^2$ in a window that has $-2 \le x \le 2$, $0 \le y \le 3$. Then, on the same screen, graph

$$y = \frac{(x+h)^3 - x^3}{h}$$

for $h = 2, 1, 0.2$. Then try $h = -2, -1, -0.2$. Explain what is going on.

T **43.** *Weierstrass's nowhere differentiable continuous function* The sum of the first eight terms of the Weierstrass function $f(x) = \sum_{n=0}^{\infty} (2/3)^n \cos(9^n \pi x)$ is

$$g(x) = \cos(\pi x) + (2/3)^1 \cos(9\pi x) + (2/3)^2 \cos(9^2 \pi x)$$
$$+ (2/3)^3 \cos(9^3 \pi x) + \cdots + (2/3)^7 \cos(9^7 \pi x).$$

Graph this sum. Zoom in several times. How wiggly and bumpy is this graph? Specify a viewing window in which the displayed portion of the graph is smooth.

T **44.** *Writing to Learn* Use a graphing utility to graph the two functions $f(x) = 1 + \ln(x+1)$ and $g(x) = |x| + 1$ in the same viewing rectangle. Use the Zoom and Trace features to analyze the graphs near the point $(0, 1)$. What do you observe? Which function appears to be differentiable at this point? Give reasons for your answers.

COMPUTER EXPLORATIONS

Use a CAS to perform the following steps for the functions in Exercises 45–50.

(a) Plot $y = f(x)$ to see that function's global behavior.

(b) Define the difference quotient q at a general point x, with general stepsize h.

(c) Take the limit as $h \to 0$. What formula does this give?

(d) Substitute the value $x = x_0$ and plot the function $y = f(x)$ together with its tangent line at that point.

(e) Substitute various values for x larger and smaller than x_0 into the formula obtained in part (c). Do the numbers make sense with your picture?

(f) *Writing to Learn* Graph the formula obtained in part (c). What does it mean when its values are negative? Zero? Positive? Does this make sense with your plot from part (a)? Give reasons for your answer.

45. $f(x) = x^3 + x^2 - x, \quad x_0 = 1$

46. $f(x) = x^{1/3} + x^{2/3}, \quad x_0 = 1$

47. $f(x) = \dfrac{4x}{x^2 + 1}, \quad x_0 = 2$

48. $f(x) = \dfrac{x - 1}{3x^2 + 1}, \quad x_0 = -1$

49. $f(x) = \sin 2x, \quad x_0 = \pi/2$

50. $f(x) = x^2 \cos x, \quad x_0 = \pi/4$

2.2 The Derivative as a Rate of Change

Instantaneous Rates of Change • Motion Along a Line: Displacement, Velocity, Speed, Acceleration, and Jerk • Sensitivity to Change • Derivatives in Economics

In Section 1.1, we initiated the study of average and instantaneous rates of change. In this section, we continue our investigations of applications in which derivatives are used to model the rates at which things change in the world around us. We revisit the study of motion along a line and examine other applications.

It is natural to think of change as change with respect to time, but other variables can be treated in the same way. For example, a physician may want to know how change in dosage affects the body's response to a drug. An economist may want to study how the cost of producing steel varies with the number of tons produced.

Instantaneous Rates of Change

If we interpret the difference quotient $(f(x + h) - f(x))/h$ as the average rate of change in f over the interval from x to $x + h$, we can interpret its limit as $h \to 0$ as the rate at which f is changing at the point x.

Instantaneous rates are limits of average rates.

Definition   Instantaneous Rate of Change

The **instantaneous rate of change** of f with respect to x at x_0 is the derivative

$$f'(x_0) = \lim_{h \to 0} \frac{f(x_0 + h) - f(x_0)}{h},$$

provided the limit exists.

It is conventional to use the word *instantaneous* even when x does not represent time. The word is, however, frequently omitted. When we say *rate of change*, we mean *instantaneous rate of change*.

Example 1   How a Circle's Area Changes with Its Diameter

The area A of a circle is related to its diameter by the equation

$$A = \frac{\pi}{4} D^2.$$

How fast does the area change with respect to the diameter when the diameter is 10 m?

Solution   The rate of change of the area with respect to the diameter is

$$\frac{dA}{dD} = \frac{\pi}{4} \cdot 2D = \frac{\pi D}{2}.$$

When $D = 10$ m, the area is changing at rate $(\pi/2)10 = 5\pi$ m²/m.

Motion Along a Line: Displacement, Velocity, Speed, Acceleration, and Jerk

Suppose that an object is moving along a coordinate line (say an s-axis) so that we know its position s on that line as a function of time t:

$$s = f(t).$$

Position at time t … and at time $t + \Delta t$

$s = f(t)$ $s + \Delta s = f(t + \Delta t)$

FIGURE 2.14 The positions of a body moving along a coordinate line at time t and shortly later at time $t + \Delta t$.

The **displacement** of the object over the time interval from t to $t + \Delta t$ (Figure 2.14) is

$$\Delta s = f(t + \Delta t) - f(t),$$

and the **average velocity** of the object over that time interval is

$$v_{av} = \frac{\text{displacement}}{\text{travel time}} = \frac{\Delta s}{\Delta t} = \frac{f(t + \Delta t) - f(t)}{\Delta t}.$$

To find the body's velocity at the exact instant t, we take the limit of the average velocity over the interval from t to $t + \Delta t$ as Δt shrinks to zero. This limit is the derivative of f with respect to t.

CD-ROM
WEBsite

Definition   (Instantaneous) Velocity
Velocity (instantaneous velocity) is the derivative of position with respect to time. If a body's position at time t is $s = f(t)$, then the body's velocity at time t is

$$v(t) = \frac{ds}{dt} = \lim_{\Delta t \to 0} \frac{f(t + \Delta t) - f(t)}{\Delta t}.$$

Example 2   Finding the Velocity of a Race Car

Figure 2.15 shows the time-to-distance graph of a 1996 Riley & Scott Mk III-Olds WSC race car. The slope of the secant PQ is the average velocity for the 3-sec interval from $t = 2$ to $t = 5$ sec; in this case, it is about 100 ft/sec or 68 mph.

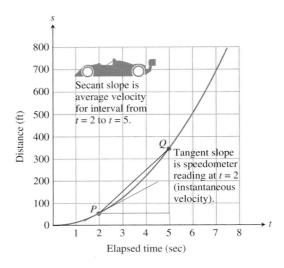

FIGURE 2.15 The time-to-distance graph for Example 2.

The slope of the tangent at P is the speedometer reading at $t = 2$ sec, about 57 ft/sec or 39 mph. The acceleration for the period shown is a nearly constant 28.5 ft/sec^2 during each second, which is about $0.89g$, where g is the acceleration due to gravity. The race car's top speed is an estimated 190 mph. (*Source: Road and Track*, March 1997.)

Besides telling how fast an object is moving, its velocity tells the direction of motion. When the object is moving forward (s increasing), the velocity is positive; when the body is moving backward (s decreasing), the velocity is negative.

If we drive to a friend's house and back at 30 mph, say, the speedometer will show 30 on the way over but it will not show -30 on the way back, even though our distance from home is decreasing. The speedometer always shows speed, which is the absolute value of velocity. Speed measures the rate of progress regardless of direction.

CD–ROM
WEBsite

Historical Biography

Galileo Galilei
(1564 — 1642)

Definition   Speed

Speed is the absolute value of velocity.

$$\text{Speed} = |v(t)| = \left| \frac{ds}{dt} \right|$$

Example 3   Horizontal Motion

Figure 2.16 shows the velocity $v = f'(t)$ of a particle moving on a coordinate line. The particle moves forward for the first 3 sec , moves backward for the next 2 sec, stands still for a second, and moves forward again. The particle achieves its greatest speed at time $t = 4$, while moving backward.

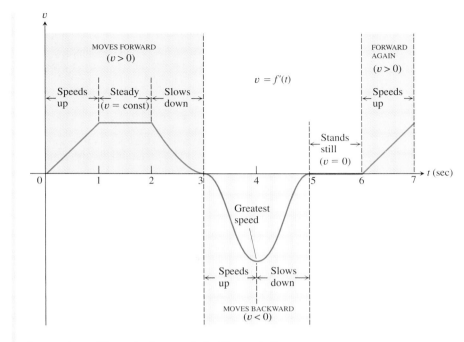

FIGURE 2.16 The velocity graph for Example 3.

The rate at which a body's velocity changes is the body's acceleration. The acceleration measures how quickly the body picks up or loses speed.

A sudden change in acceleration is called a "jerk." When a ride in a car or a bus is jerky, it is not that the accelerations involved are necessarily large but that the changes in acceleration are abrupt. Jerk is what spills our soft drink.

Definitions   Acceleration, Jerk

Acceleration is the derivative of velocity with respect to time. If a body's position at time t is $s = f(t)$, then the body's acceleration at time t is

$$a(t) = \frac{dv}{dt} = \frac{d^2s}{dt^2}.$$

Jerk is the derivative of acceleration with respect to time:

$$j(t) = \frac{da}{dt} = \frac{d^3s}{dt^3}.$$

Near the surface of Earth all bodies fall with the same constant acceleration. Galileo's experiments with free fall revealed that the distance a body released from rest falls in time t is proportional to the square of the amount of time it has fallen. Today, we express this by saying that

$$s = \frac{1}{2} gt^2,$$

where s is distance and g is the acceleration due to Earth's gravity. This equation holds in a vacuum, where there is no air resistance, and closely models the fall of dense, heavy objects, such as rocks or steel tools, for the first few seconds of their fall, before air resistance starts to slow them down.

The jerk of the constant acceleration of gravity ($g = 32$ ft/sec^2) is zero:

$$j = \frac{d}{dt}(g) = 0.$$

An object does not exhibit jerkiness during free fall.

The value of g in the equation $s = (1/2)gt^2$ depends on the units used to measure t and s. With t in seconds (the usual unit), we have the following values:

Abbreviations for Units of Measure

ft/sec^2 "feet per second squared" or "feet per second per second"

m/sec^2 "meters per second squared" or "meters per second per second"

Free-Fall Equations (Earth)

English units: $g = 32\ \dfrac{\text{ft}}{\text{sec}^2}$, $s = \dfrac{1}{2}(32)t^2 = 16t^2$ (s in feet)

Metric units: $g = 9.8\ \dfrac{\text{m}}{\text{sec}^2}$, $s = \dfrac{1}{2}(9.8)t^2 = 4.9t^2$ (s in meters)

Example 4   Modeling Free Fall

Figure 2.17 shows the free fall of a heavy ball bearing released from rest at time $t = 0$ sec.

(a) How many meters does the ball fall in the first 2 sec?

(b) What is its velocity, speed, and acceleration then?

Solution

(a) The metric free-fall equation is $s = 4.9t^2$. During the first 2 sec, the ball falls

$$s(2) = 4.9(2)^2 = 19.6 \text{ m}.$$

(b) At any time t, *velocity* is the derivative of position:

$$v(t) = s'(t) = \frac{d}{dt}(4.9t^2) = 9.8t.$$

At $t = 2$, the velocity is

$$v(2) = 19.6 \text{ m/sec}$$

in the downward (increasing s) direction. The *speed* at $t = 2$ is

$$\text{Speed} = |v(2)| = 19.6 \text{ m/sec}.$$

The *acceleration* at any time t is

$$a(t) = v'(t) = s''(t) = 9.8 \text{ m/sec}^2.$$

At $t = 2$, the acceleration is 9.8 m/sec^2.

FIGURE 2.17 A ball bearing falling from rest. (Example 4)

CD-ROM
WEBsite

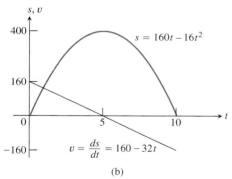

FIGURE 2.18 (a) The rock in Example 5.
(b) The graphs of s and v as functions of
time; s is largest when $v = ds/dt = 0$.
The graph of s is *not* the path of the rock:
It is a plot of height versus time. The
slope of the plot is the rock's velocity,
graphed here as a straight line.

Example 5   Modeling Vertical Motion

A dynamite blast blows a heavy rock straight up with a launch velocity of 160
ft/sec (about 109 mph) (Figure 2.18a). It reaches a height of $s = 160t - 16t^2$ ft
after t sec .

(a) How high does the rock go?

(b) What are the velocity and speed of the rock when it is 256 ft above the
ground on the way up? On the way down?

(c) What is the acceleration of the rock at any time t during its flight (after the
blast)?

(d) When does the rock hit the ground again?

Solution

(a) In the coordinate system we have chosen, s measures height from the ground
up, so the velocity is positive on the way up and negative on the way down.
The instant the rock is at its highest point is the one instant during the flight
when the velocity is 0. To find the maximum height, all we need to do is to
find when $v = 0$ and evaluate s at this time.

At any time t , the velocity is

$$v = \frac{ds}{dt} = \frac{d}{dt}(160t - 16t^2) = 160 - 32t \text{ ft/sec}.$$

The velocity is zero when

$$160 - 32t = 0 \qquad \text{or} \qquad t = 5 \text{ sec}.$$

The rock's height at $t = 5$ sec is

$$s_{\max} = s(5) = 160(5) - 16(5)^2 = 800 - 400 = 400 \text{ ft}.$$

See Figure 2.18b.

(b) To find the rock's velocity at 256 ft on the way up and again on the way
down, we first find the two values of t for which

$$s(t) = 160t - 16t^2 = 256.$$

To solve this equation, we write

$$16t^2 - 160t + 256 = 0$$
$$16(t^2 - 10t + 16) = 0$$
$$(t - 2)(t - 8) = 0$$
$$t = 2 \text{ sec} , t = 8 \text{ sec}.$$

The rock is 256 ft above the ground 2 sec after the explosion and again 8 sec
after the explosion. The rock's velocities at these times are

$$v(2) = 160 - 32(2) = 160 - 64 = 96 \text{ ft/sec}$$
$$v(8) = 160 - 32(8) = 160 - 256 = -96 \text{ ft/sec}.$$

At both instants, the rock's speed is 96 ft/sec.

(c) At any time during its flight following the explosion, the rock's acceleration is a constant

$$a = \frac{dv}{dt} = \frac{d}{dt}(160 - 32t) = -32 \text{ ft/sec}^2.$$

The acceleration is always downward. As the rock rises, it slows down; as it falls, it speeds up.

(d) The rock hits the ground at the positive time t for which $s = 0$. The equation $160t - 16t^2 = 0$ factors to give $16t(10 - t) = 0$, so it has solutions $t = 0$ and $t = 10$. At $t = 0$, the blast occurred and the rock was thrown upward. It returned to the ground 10 sec later.

> ## USING
> ### TECHNOLOGY
>
> *Simulation of motion on a vertical line* The parametric equations
>
> $$x(t) = c \qquad y(t) = f(t)$$
>
> will illuminate pixels along the vertical line $x = c$. If $f(t)$ denotes the height of a moving body at time t, graphing $(x(t), y(t)) = (c, f(t))$ will simulate the actual motion. Try it for the rock in Example 5 with $x(t) = 2$, say, and $y(t) = 160t - 16t^2$, in dot mode with t Step $= 0.1$. Why does the spacing of the dots vary? Why does the grapher seem to stop after it reaches the top? (Try the plots for $0 \leq t \leq 5$ and $5 \leq t \leq 10$ separately.)
>
> For a second experiment, plot the parametric equations
>
> $$x(t) = t, \qquad y(t) = 160t - 16t^2$$
>
> together with the vertical line simulation of the motion, again in dot mode. Use what you know about the behavior of the rock from the calculations of Example 5 to select a window size that will display all the interesting behavior.

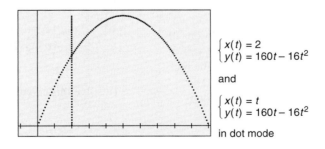

$$\begin{cases} x(t) = 2 \\ y(t) = 160t - 16t^2 \end{cases}$$

and

$$\begin{cases} x(t) = t \\ y(t) = 160t - 16t^2 \end{cases}$$

in dot mode

Sensitivity to Change

When a small change in x produces a large change in the value of a function $f(x)$, we say that the function is relatively **sensitive** to changes in x. The derivative $f'(x)$ is a measure of this sensitivity.

Example 6   Genetic Data and Sensitivity to Change

The Austrian monk Gregor Johann Mendel (1822–1884), working with garden peas and other plants, provided the first scientific explanation of hybridization.

His careful records showed that if p (a number between 0 and 1) is the frequency of the gene for smooth skin in peas (dominant) and $(1 - p)$ is the frequency of the gene for wrinkled skin in peas, then the proportion of smooth-skinned peas in the next generation will be

$$y = 2p(1 - p) + p^2 = 2p - p^2 .$$

The graph of y versus p in Figure 2.19a suggests that the value of y is more sensitive to a change in p when p is small than when p is large. Indeed, this fact is

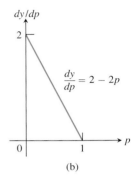

FIGURE 2.19   (a) The graph of $y = 2p - p^2$, describing the proportion of smooth-skinned peas. (b) The graph of dy/dp.

borne out by the derivative graph in Figure 2.19b, which shows that dy/dp is close to 2 when p is near 0 and close to 0 when p is near 1.

   The implication for genetics is that introducing a few more dominant genes into a highly recessive population (where the frequency of wrinkled skin peas is small) will have a more dramatic effect on later generations than will a similar increase in a highly dominant population.

Derivatives in Economics

Engineers use the terms *velocity* and *acceleration* to refer to the derivatives of functions describing motion. Economists, too, have a specialized vocabulary for rates of change and derivatives. They call them *marginals*.

   In a manufacturing operation, the *cost of production* $c(x)$ is a function of x, the number of units produced. The *marginal cost of production* is the rate of change of cost with respect to level of production, so it is dc/dx.

   Suppose that $c(x)$ represents the dollars needed to produce x tons of steel in one week. It costs more to produce $x + h$ units per week, and the cost difference, divided by h, is the average cost of producing each additional ton:

$$\frac{c(x + h) - c(x)}{h} = \frac{\text{average cost of each of the additional}}{h \text{ tons of steel produced}} .$$

   The limit of this ratio as $h \to 0$ is the *marginal cost* of producing more steel per week when the current weekly production is x tons (Figure 2.20).

$$\frac{dc}{dx} = \lim_{h \to 0} \frac{c(x + h) - c(x)}{h} = \text{marginal cost of production}.$$

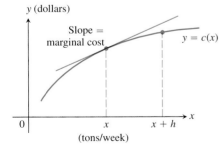

FIGURE 2.20   Weekly steel production: $c(x)$ is the cost of producing x tons per week. The cost of producing an additional h tons is $c(x + h) - c(x)$.

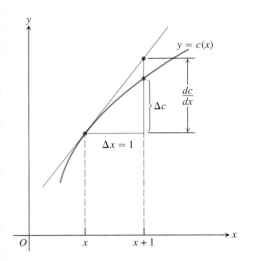

FIGURE 2.21 The marginal cost dc/dx is approximately the extra cost Δc of producing $\Delta x = 1$ more unit.

Choosing functions to illustrate economics

In case you are wondering why economists use polynomials of low degree to illustrate complicated phenomena like cost and revenue, here is the rationale: Although formulas for real phenomena are rarely available in any given instance, the theory of economics can still provide valuable guidance. The functions about which theory speaks can often be illustrated with low-degree polynomials on relevant intervals. Cubic polynomials provide a good balance between being easy to work with and being complicated enough to capture the behavior.

Sometimes the marginal cost of production is loosely defined to be the extra cost of producing one unit:

$$\frac{\Delta c}{\Delta x} = \frac{c(x + 1) - c(x)}{1},$$

which is approximated by the value of dc/dx at x. This approximation is acceptable if the slope of the graph of c does not change quickly near x. Then the difference quotient will be close to its limit dc/dx, which is the rise in the tangent line if $\Delta x = 1$ (Figure 2.21). The approximation works best for large values of x.

Example 7   Marginal Cost and Marginal Revenue

Suppose that it costs

$$c(x) = x^3 - 6x^2 + 15x$$

dollars to produce x radiators when 8 to 30 radiators are produced and that

$$r(x) = x^3 - 3x^2 + 12x$$

gives the dollar revenue from selling x radiators. Your shop currently produces 10 radiators a day. About how much extra will it cost to produce one more radiator a day, and what is your estimated increase in revenue for selling 11 radiators a day?

Solution   The cost of producing one more radiator a day when 10 are produced is about $c'(10)$:

$$c'(x) = \frac{d}{dx}(x^3 - 6x^2 + 15x) = 3x^2 - 12x + 15$$

$$c'(10) = 3(100) - 12(10) + 15 = 195.$$

The additional cost will be about \$195. The marginal revenue is

$$r'(x) = \frac{d}{dx}(x^3 - 3x^2 + 12x) = 3x^2 - 6x + 12.$$

The marginal revenue function estimates the increase in revenue that will result from selling one additional unit. If you currently sell 10 radiators a day, you can expect your revenue to increase by about

$$r'(10) = 3(100) - 6(10) + 12 = \$252$$

if you increase sales to 11 radiators a day.

Example 8   Marginal Tax Rate

To get some feel for the language of marginal rates, consider marginal tax rates. If your marginal income tax rate is 28% and your income increases by \$1000, you can expect to pay an extra \$280 in taxes. This does not mean that you pay 28% of your entire income in taxes. It just means that at your current income level I, the rate of increase of taxes T with respect to income is $dT/dI = 0.28$. You will pay \$0.28 out of every extra dollar you earn in taxes. Of course, if you earn a lot more, you may land in a higher tax bracket and your marginal rate will increase.

EXERCISES 2.2

Motion Along a Coordinate Line

Exercises 1–4 give the positions $s = f(t)$ of a body moving on a coordinate line, with s in meters and t in seconds.

(a) Find the body's displacement and average velocity for the given time interval.

(b) Find the body's speed and acceleration at the endpoints of the interval.

(c) When if ever during the interval does the body change direction?

1. $s = t^2 - 3t + 2, \quad 0 \le t \le 2$

2. $s = 6t - t^2, \quad 0 \le t \le 6$

3. $s = -t^3 + 3t^2 - 3t, \quad 0 \le t \le 3$

4. $s = (t^4/4) - t^3 + t^2, \quad 0 \le t \le 3$

5. *Particle motion* At time t, the position of a body moving along the s-axis is $s = t^3 - 6t^2 + 9t$ m.

(a) Find the body's acceleration each time the velocity is zero.

(b) Find the body's speed each time the acceleration is zero.

(c) Find the total distance traveled by the body from $t = 0$ to $t = 2$.

6. *Particle motion* At time $t \ge 0$, the velocity of a body moving along the s-axis is $v = t^2 - 4t + 3$.

(a) Find the body's acceleration each time the velocity is zero.

(b) When is the body moving forward? Backward?

(c) When is the body's velocity increasing? Decreasing?

Free-Fall Applications

7. *Free fall on Mars and Jupiter* The equations for free fall at the surfaces of Mars and Jupiter (s in meters, t in seconds) are $s = 1.86t^2$ on Mars and $s = 11.44t^2$ on Jupiter. How long does it take a rock falling from rest to reach a velocity of 27.8 m/sec (about 100 km/h) on each planet?

8. *Lunar projectile motion* A rock thrown vertically upward from the surface of the moon at a velocity of 24 m/sec (about 86 km/h) reaches a height of $s = 24t - 0.8t^2$ meters in t sec.

(a) Find the rock's velocity and acceleration at time t. (The acceleration in this case is the acceleration of gravity on the moon.)

(b) How long does it take the rock to reach its highest point?

(c) How high does the rock go?

(d) How long does it take the rock to reach half its maximum height?

(e) How long is the rock aloft?

9. *Finding g on a small airless planet* Explorers on a small airless planet used a spring gun to launch a ball bearing vertically upward from the surface at a launch velocity of 15 m/sec. Because the acceleration of gravity at the planet's surface was g_s m/sec^2, the explorers expected the ball bearing to reach a height of $s = 15t - (1/2)g_s\, t^2$ meters t sec later. The ball bearing reached its maximum height 20 sec after being launched. What was the value of g_s?

10. *Speeding bullet* A 45-caliber bullet fired straight up from the surface of the moon would reach a height of $s = 832t - 2.6t^2$ feet after t sec. On Earth, in the absence of air, its height would be $s = 832t - 16t^2$ ft after t sec. How long will the bullet be aloft in each case? How high will the bullet go?

11. *Free fall from the Tower of Pisa* Had Galileo dropped a cannonball from the Tower of Pisa, 179 ft above the ground, the ball's height above ground t sec into the fall would have been $s = 179 - 16t^2$.

(a) What would have been the ball's velocity, speed, and acceleration at time t?

(b) About how long would it have taken the ball to hit the ground?

(c) What would have been the ball's velocity at the moment of impact?

12. *Galileo's free-fall formula* Galileo developed a formula for a body's velocity during free fall by rolling balls from rest down increasingly steep inclined planks and looking for a limiting formula that would predict a ball's behavior when the plank was vertical and the ball fell freely; see part (a) of the accompanying figure. He found that, for any given angle of the plank, the ball's velocity t sec into motion was a constant multiple of t. That is, the velocity was given by a formula of the form $v = kt$. The value of the constant k depended on the inclination of the plank.

In modern notation—part (b) of the figure—with distance in meters and time in seconds, what Galileo determined by experiment was that, for any given angle θ, the ball's velocity t sec into the roll was

$$v = 9.8(\sin \theta)t \text{ m/sec}.$$

Free-fall position

(a)   (b)

(a) What is the equation for the ball's velocity during free fall?

(b) Building on your work in part (a), what constant acceleration does a freely falling body experience near the surface of Earth?

Conclusions About Motion from Graphs

13. The accompanying figure shows the velocity $v = ds/dt = f(t)$ (m/sec) of a body moving along a coordinate line.

v(m/sec)

$v = f(t)$

(a) When does the body reverse direction?

(b) When (approximately) is the body moving at a constant speed?

(c) Graph the body's speed for $0 \le t \le 10$.

(d) Graph the acceleration, where defined.

14. A particle P moves on the number line shown in part (a) of the accompanying figure. Part (b) shows the position of P as a function of time t.

(a)

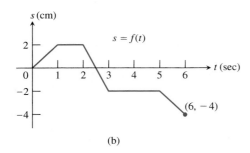

(b)

(a) When is P moving to the left? Moving to the right? Standing still?

(b) Graph the particle's velocity and speed (where defined).

15. *Launching a rocket* When a model rocket is launched, the propellant burns for a few seconds, accelerating the rocket upward. After burnout, the rocket coasts upward for a while and then begins to fall. A small explosive charge pops out a parachute shortly after the rocket starts down. The parachute slows the rocket to keep it from breaking when it lands.

   The figure here shows velocity data from the flight of the model rocket. Use the data to answer the following.

(a) How fast was the rocket climbing when the engine stopped?

(b) For how many seconds did the engine burn?

(c) When did the rocket reach its highest point? What was its velocity then?

(d) When did the parachute pop out? How fast was the rocket falling then?

(e) How long did the rocket fall before the parachute opened?

(f) When was the rocket's acceleration greatest?

(g) When was the acceleration constant? What was its value then (to the nearest integer)?

16. *A traveling truck* The accompanying graph shows the position s of a truck traveling on a highway. The truck starts at $t = 0$ and returns 15 h later at $t = 15$.

(a) Use the technique described in Section 2.1, Example 9, to graph the truck's velocity $v = ds/dt$ for $0 \le t \le 15$. Then repeat the process, with the velocity curve, to graph the truck's acceleration dv/dt.

(b) Suppose that $s = 15t^2 - t^3$. Graph ds/dt and d^2s/dt^2 and compare your graphs with those in part (a).

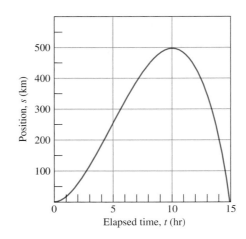

17. *Two falling balls* The multiflash photograph in the accompanying figure shows two balls falling from rest. The vertical rulers are marked in centimeters. Use the equation $s = 490t^2$ (the free-fall equation for s in centimeters and t in seconds) to answer the following questions.

(a) How long did it take the balls to fall the first 160 cm? What was their average velocity for the period?

(b) How fast were the balls falling when they reached the 160-cm mark? What was their acceleration then?

(c) About how fast was the light flashing (flashes per second)?

18. *Writing to Learn* The graphs in Figure 2.22 show the position s, velocity $v = ds/dt$, and acceleration $a = d^2s/dt^2$ of a body moving along a coordinate line as functions of time t. Which graph is which? Give reasons for your answers.

19. *Writing to Learn* The graphs in Figure 2.23 show the position s, the velocity $v = ds/dt$, and the acceleration $a = d^2s/dt^2$ of a body moving along the coordinate line as functions of time t. Which graph is which? Give reasons for your answers.

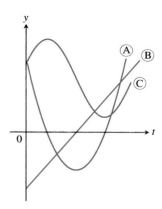

FIGURE 2.22 **The graphs for Exercise 18.**

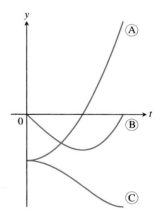

FIGURE 2.23 **The graphs for Exercise 19.**

Economics

20. *Marginal cost* Suppose that the dollar cost of producing x washing machines is $c(x) = 2000 + 100x - 0.1x^2$.

(a) Find the average cost per machine of producing the first 100 washing machines.

(b) Find the marginal cost when 100 washing machines are produced.

(c) Show that the marginal cost when 100 washing machines are produced is approximately the cost of producing one more washing machine after the first 100 have been made, by calculating the latter cost directly.

21. *Marginal revenue* Suppose that the revenue from selling x washing machines is

$$r(x) = 20{,}000 \left(1 - \frac{1}{x}\right)$$

dollars.

(a) Find the marginal revenue when 100 machines are produced.

(b) Use the function $r'(x)$ to estimate the increase in revenue that will result from increasing production from 100 machines a week to 101 machines a week.

(c) Find the limit of $r'(x)$ as $x \to \infty$. How would you interpret this number?

Additional Applications

22. *Bacterium population* When a bactericide was added to a nutrient broth in which bacteria were growing, the bacterium population continued to grow for a while, but then stopped growing and began to decline. The size of the population at time t (hours) was $b = 10^6 + 10^4 t - 10^3 t^2$. Find the growth rates at

(a) $t = 0$ h.

(b) $t = 5$ h.

(c) $t = 10$ h.

23. *Draining a tank* The number of gallons of water in a tank t minutes after the tank has started to drain is $Q(t) = 200(30 - t)^2$. How fast is the water running out at the end of 10 min? What is the average rate at which the water flows out during the first 10 min?

T **24.** *Draining a tank* It takes 12 h to drain a storage tank by opening the valve at the bottom. The depth y of fluid in the tank t h after the valve is opened is given by the formula

$$y = 6\left(1 - \frac{t}{12}\right)^2 \text{ m}.$$

(a) Find the rate dy/dt (m/h) at which the tank is draining at time t.

(b) When is the fluid level in the tank falling fastest? Slowest? What are the values of dy/dt at these times?

(c) Graph y and dy/dt together and discuss the behavior of y in relation to the signs and values of dy/dt.

25. *Inflating a balloon* The volume $V = (4/3)\pi r^3$ of a spherical balloon changes with the radius.

(a) At what rate (ft³/ft) does the volume change with respect to the radius when $r = 2$ ft?

(b) By approximately how much does the volume increase when the radius changes from 2 to 2.2 ft?

26. *Airplane takeoff* Suppose that the distance an aircraft travels along a runway before takeoff is given by $D = (10/9)t^2$, where D is measured in meters from the starting point and t is measured in seconds from the time the brakes are released. The aircraft will become airborne when its speed reaches 200 km/h. How long will it take to become airborne, and what distance will it travel in that time?

27. *Volcanic lava fountains* Although the November 1959 Kilauea Iki eruption on the island of Hawaii began with a line of fountains along the wall of the crater, activity was later confined to a single vent in the crater's floor, which at one point shot lava 1900 ft straight into the air (a world record). What was the lava's exit velocity in feet per second? In miles per hour? (*Hint*: If v_0 is the exit velocity of a particle of lava, its height t sec later will be $s = v_0 t - 16t^2$ ft. Begin by finding the time at which $ds/dt = 0$. Neglect air resistance.)

T Exercises 28–31 give the position function $s = f(t)$ of a body moving along the s-axis as a function of time t. Graph f together with the velocity function $v(t) = ds/dt = f'(t)$ and the acceleration function $a(t) = d^2 s/dt^2 = f''(t)$. Comment on the body's behavior in relation to the signs and values v and a. Include in your commentary such topics as the following:

(a) When is the body momentarily at rest?

(b) When does it move to the left (down) or to the right (up)?

(c) When does it change direction?

(d) When does it speed up and slow down?

(e) When is it moving fastest (highest speed)? Slowest?

(f) When is it farthest from the axis origin?

28. $s = 200t - 16t^2$, $0 \le t \le 12.5$ (a heavy object fired straight up from Earth's surface at 200 ft/sec)

29. $s = t^2 - 3t + 2$, $0 \le t \le 5$

30. $s = t^3 - 6t^2 + 7t$, $0 \le t \le 4$

31. $s = 4 - 7t + 6t^2 - t^3$, $0 \le t \le 4$

32. *Thoroughbred racing* A racehorse is running a 10-furlong race. (A furlong is 220 yards, although we will use furlongs and seconds as our units in this exercise.) As the horse passes each furlong marker (F), a steward records the time elapsed (t) since the beginning of the race, as shown in the table:

F	0	1	2	3	4	5	6	7	8	9	10
t	0	20	33	46	59	73	86	100	112	124	135

(a) How long does it take the horse to finish the race?

(b) What is the average speed of the horse over the first 5 furlongs?

(c) What is the approximate speed of the horse as it passes the 3-furlong marker?

(d) During which portion of the race is the horse running the fastest?

(e) During which portion of the race is the horse accelerating the fastest?

2.3 Derivatives of Products, Quotients, and Negative Powers

Products • Quotients • Negative Integer Powers of *x*

This section continues the discussion on how to differentiate functions without having to apply the definition each time.

Products

Although the derivative of the sum of two functions is the sum of their derivatives, the derivative of the product of two functions is *not* the product of their derivatives. For instance,

$$\frac{d}{dx}(x \cdot x) = \frac{d}{dx}(x^2) = 2x, \qquad \text{whereas} \qquad \frac{d}{dx}(x) \cdot \frac{d}{dx}(x) = 1 \cdot 1 = 1.$$

The derivative of a product of two functions is the sum of *two* products, as we now explain.

Rule 5 Derivative Product Rule

If u and v are differentiable at x, then so is their product uv, and

$$\frac{d}{dx}(uv) = u\frac{dv}{dx} + v\frac{du}{dx}.$$

The derivative of the product uv is u times the derivative of v plus v times the derivative of u. In prime notation, $(uv)' = uv' + vu'$.

Example 1 Using the Product Rule

Find the derivative of

$$y = \frac{1}{x}\left(x^2 + \frac{1}{x}\right).$$

Solution   We apply the product rule with $u = 1/x$ and $v = x^2 + (1/x)$:

$$\frac{d}{dx}\left[\frac{1}{x}\left(x^2 + \frac{1}{x}\right)\right] = \frac{1}{x}\left(2x - \frac{1}{x^2}\right) + \left(x^2 + \frac{1}{x}\right)\left(-\frac{1}{x^2}\right)$$

$$= 2 - \frac{1}{x^3} - 1 - \frac{1}{x^3} \qquad \frac{d}{dx}(uv) = u\frac{dv}{dx} + v\frac{du}{dx}, \text{ and}$$

$$= 1 - \frac{2}{x^3}. \qquad \frac{d}{dx}\left(\frac{1}{x}\right) = -\frac{1}{x^2} \text{ by}$$

Example 3, Section 1.5.

Proof of Rule 5

$$\frac{d}{dx}(uv) = \lim_{h \to 0} \frac{u(x+h)v(x+h) - u(x)v(x)}{h}$$

Picturing the product rule

If $u(x)$ and $v(x)$ are positive and increase when x increases, and if $h > 0$,

then the total shaded area in the picture is

$$u(x + h)v(x + h) - u(x)v(x)$$
$$= u(x + h)\,\Delta v + v(x + h)\,\Delta u$$
$$- \Delta u\, \Delta v \,.$$

Dividing both sides of this equation by h gives

$$\frac{u(x + h)v(x + h) - u(x)v(x)}{h}$$
$$= u(x + h)\frac{\Delta v}{h} + v(x + h)\frac{\Delta u}{h}$$
$$- \Delta u \frac{\Delta v}{h} \,.$$

As $h \to 0^+$,

$$\Delta u \cdot \frac{\Delta v}{h} \to 0 \cdot \frac{dv}{dx} = 0,$$

leaving

$$\frac{d}{dx}(uv) = u\frac{dv}{dx} + v\frac{du}{dx}\,.$$

To change this fraction into an equivalent one that contains difference quotients for the derivatives of u and v, we subtract and add $u(x + h)v(x)$ in the numerator:

$$\frac{d}{dx}(uv) = \lim_{h \to 0} \frac{u(x + h)v(x + h) - u(x + h)v(x) + u(x + h)v(x) - u(x)v(x)}{h}$$

$$= \lim_{h \to 0} \left[u(x + h)\frac{v(x + h) - v(x)}{h} + v(x)\frac{u(x + h) - u(x)}{h} \right]$$

$$= \lim_{h \to 0} u(x + h) \cdot \lim_{h \to 0} \frac{v(x + h) - v(x)}{h} + v(x) \cdot \lim_{h \to 0} \frac{u(x + h) - u(x)}{h}\,.$$

As h approaches zero, $u(x + h)$ approaches $u(x)$ because u, being differentiable at x, is continuous at x. The two fractions approach the values of dv/dx at x and du/dx at x. In short,

$$\frac{d}{dx}(uv) = u\frac{dv}{dx} + v\frac{du}{dx}\,. \qquad \rule{1em}{0.4pt}$$

In the following example, we have only numerical values with which to work.

Example 2   Derivative from Numerical Values

Let $y = uv$ be the product of the functions u and v. Find $y'(2)$ if

$$u(2) = 3, \qquad u'(2) = -4, \qquad v(2) = 1, \qquad \text{and} \qquad v'(2) = 2.$$

Solution   From the Product Rule, in the form

$$y' = (uv)' = uv' + vu',$$

we have

$$y'(2) = u(2)v'(2) + v(2)u'(2)$$
$$= (3)(2) + (1)(-4) = 6 - 4 = 2.$$

Quotients

Just as the derivative of the product of two differentiable functions is not the product of their derivatives, the derivative of the quotient of two functions is not the quotient of their derivatives. What happens instead is the Quotient Rule.

Rule 6   Derivative Quotient Rule

If u and v are differentiable at x and if $v(x) \neq 0$, then the quotient u/v is differentiable at x, and

$$\frac{d}{dx}\left(\frac{u}{v}\right) = \frac{v\dfrac{du}{dx} - u\dfrac{dv}{dx}}{v^2}\,.$$

Example 3   Using the Quotient Rule

Find the derivative of

$$y = \frac{t^2 - 1}{t^2 + 1}.$$

Solution   We apply the Quotient Rule with $u = t^2 - 1$ and $v = t^2 + 1$:

$$\frac{dy}{dt} = \frac{(t^2 + 1) \cdot 2t - (t^2 - 1) \cdot 2t}{(t^2 + 1)^2} \qquad \frac{d}{dt}\left(\frac{u}{v}\right) = \frac{v(du/dt) - u(dv/dt)}{v^2}$$

$$= \frac{2t^3 + 2t - 2t^3 + 2t}{(t^2 + 1)^2}$$

$$= \frac{4t}{(t^2 + 1)^2}.$$

Proof of Rule 6

$$\frac{d}{dx}\left(\frac{u}{v}\right) = \lim_{h \to 0} \frac{\dfrac{u(x+h)}{v(x+h)} - \dfrac{u(x)}{v(x)}}{h}$$

$$= \lim_{h \to 0} \frac{v(x)u(x+h) - u(x)v(x+h)}{h\,v(x+h)v(x)}$$

To change the last fraction into an equivalent one that contains the difference quotients for the derivatives of u and v, we subtract and add $v(x)u(x)$ in the numerator. We then get

$$\frac{d}{dx}\left(\frac{u}{v}\right) = \lim_{h \to 0} \frac{v(x)u(x+h) - v(x)u(x) + v(x)u(x) - u(x)v(x+h)}{h\,v(x+h)v(x)}$$

$$= \lim_{h \to 0} \frac{v(x)\dfrac{u(x+h) - u(x)}{h} - u(x)\dfrac{v(x+h) - v(x)}{h}}{v(x+h)v(x)}.$$

Taking the limit in the numerator and denominator now gives the Quotient Rule.

Negative Integer Powers of x

The Power Rule for negative integers is the same as the rule for positive integers.

Rule 7   Power Rule for Negative Integers

If n is a negative integer and $x \neq 0$, then

$$\frac{d}{dx}(x^n) = nx^{n-1}.$$

Example 4   Using Rule 7

(a) $\dfrac{d}{dx}\left(\dfrac{1}{x}\right) = \dfrac{d}{dx}(x^{-1}) = (-1)x^{-2} = -\dfrac{1}{x^2}$

(b) $\dfrac{d}{dx}\left(\dfrac{4}{x^3}\right) = 4\dfrac{d}{dx}(x^{-3}) = 4(-3)x^{-4} = -\dfrac{12}{x^4}$

Proof of Rule 7   The proof uses the Quotient Rule in a clever way. If n is a negative integer, then $n = -m$, where m is a positive integer. Hence, $x^n = x^{-m} = 1/x^m$, and

$$\dfrac{d}{dx}(x^n) = \dfrac{d}{dx}\left(\dfrac{1}{x^m}\right)$$

$$= \dfrac{x^m \cdot \dfrac{d}{dx}(1) - 1 \cdot \dfrac{d}{dx}(x^m)}{(x^m)^2} \qquad \text{Quotient Rule with } u = 1 \text{ and } v = x^m$$

$$= \dfrac{0 - mx^{m-1}}{x^{2m}} \qquad \text{Since } m > 0,\ \dfrac{d}{dx}(x^m) = mx^{m-1}.$$

$$= -mx^{-m-1}$$

$$= nx^{n-1}. \qquad \text{Since } -m = n$$

Example 5   Tangent to a Curve

Find an equation for the tangent to the curve

$$y = x + \dfrac{2}{x}$$

at the point $(1, 3)$ (Figure 2.24).

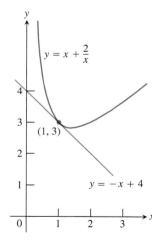

FIGURE 2.24 The tangent to the curve $y = x + (2/x)$ at $(1, 3)$. The curve has a third-quadrant portion not shown here. We see how to graph functions like this one in Chapter 3.

Solution   The slope of the curve is

$$\dfrac{dy}{dx} = \dfrac{d}{dx}(x) + 2\dfrac{d}{dx}\left(\dfrac{1}{x}\right) = 1 + 2\left(-\dfrac{1}{x^2}\right) = 1 - \dfrac{2}{x^2}.$$

The slope at $x = 1$ is

$$\frac{dy}{dx}\bigg|_{x=1} = \left[1 - \frac{2}{x^2}\right]_{x=1} = 1 - 2 = -1.$$

The line through $(1, 3)$ with slope $m = -1$ is

$$y - 3 = (-1)(x - 1) \qquad \text{Point-slope equation}$$
$$y = -x + 1 + 3$$
$$y = -x + 4.$$

The choice of which rules to use in solving a differentiation problem can make a difference in how much work you have to do. Here is an example.

Example 6   Choosing Which Rule to Use

Rather than using the Quotient Rule to find the derivative of

$$y = \frac{(x - 1)(x^2 - 2x)}{x^4},$$

expand the numerator and divide by x^4:

$$y = \frac{(x - 1)(x^2 - 2x)}{x^4} = \frac{x^3 - 3x^2 + 2x}{x^4} = x^{-1} - 3x^{-2} + 2x^{-3}.$$

Then use the Sum and Power Rules:

$$\frac{dy}{dx} = -x^{-2} - 3(-2)x^{-3} + 2(-3)x^{-4}$$

$$= -\frac{1}{x^2} + \frac{6}{x^3} - \frac{6}{x^4}.$$

Example 7   The Body's Reaction to Medicine

The reaction of the body to a dose of medicine is sometimes represented by an equation of the form

$$R = M^2\left(\frac{C}{2} - \frac{M}{3}\right),$$

where C is a positive constant and M is the amount of medicine absorbed in the blood. If the reaction is a change in blood pressure, R is measured in millimeters of mercury; if the reaction is a change in temperature, R is measured in degrees; and so on.

Find dR/dM. This derivative, as a function of M, is called the *sensitivity* of the body to the medicine.

Solution

$$\frac{dR}{dM} = 2M\left(\frac{C}{2} - \frac{M}{3}\right) + M^2(-1/3) \qquad \text{Product, Power, and Constant Multiple Rules}$$

$$= MC - M^2$$

EXERCISES 2.3

Derivative Calculations

In Exercises 1–4, find the first and second derivatives.

1. $y = 6x^2 - 10x - 5x^{-2}$

2. $w = 3z^{-3} - \dfrac{1}{z}$

3. $r = \dfrac{1}{3s^2} - \dfrac{5}{2s}$

4. $r = \dfrac{12}{\theta} - \dfrac{4}{\theta^3} + \dfrac{1}{\theta^4}$

In Exercises 5 and 6, find y' **(a)** by applying the Product Rule and **(b)** by multiplying the factors to produce a sum of simpler terms to differentiate.

5. $y = (3 - x^2)(x^3 - x + 1)$

6. $y = \left(x + \dfrac{1}{x}\right)\left(x - \dfrac{1}{x} + 1\right)$

Find the derivatives of the functions in Exercises 7–14.

7. $y = \dfrac{2x + 5}{3x - 2}$

8. $g(x) = \dfrac{x^2 - 4}{x + 0.5}$

9. $f(t) = \dfrac{t^2 - 1}{t^2 + t - 2}$

10. $v = (1 - t)(1 + t^2)^{-1}$

11. $f(s) = \dfrac{\sqrt{s} - 1}{\sqrt{s} + 1}$

12. $r = 2\left(\dfrac{1}{\sqrt{\theta}} + \sqrt{\theta}\right)$

13. $y = \dfrac{1}{(x^2 - 1)(x^2 + x + 1)}$

14. $y = \dfrac{(x + 1)(x + 2)}{(x - 1)(x - 2)}$

Find the first and second derivatives of the functions in Exercises 15–18.

15. $s = \dfrac{t^2 + 5t - 1}{t^2}$

16. $r = \dfrac{(\theta - 1)(\theta^2 + \theta + 1)}{\theta^3}$

17. $w = \left(\dfrac{1 + 3z}{3z}\right)(3 - z)$

18. $p = \left(\dfrac{q^2 + 3}{12q}\right)\left(\dfrac{q^4 - 1}{q^3}\right)$

Using Numerical Values

19. Suppose that u and v are functions of x that are differentiable at $x = 0$ and that

$$u(0) = 5, \qquad u'(0) = 3$$
$$v(0) = -1, \qquad v'(0) = 2.$$

Find the values of the following derivatives at $x = 0$.

(a) $\dfrac{d}{dx}(uv)$ **(b)** $\dfrac{d}{dx}\left(\dfrac{u}{v}\right)$

(c) $\dfrac{d}{dx}\left(\dfrac{v}{u}\right)$ **(d)** $\dfrac{d}{dx}(7v - 2u)$

20. Suppose that u and v are differentiable functions of x and that

$$u(1) = 2, \qquad u'(1) = 0$$
$$v(1) = 5, \qquad v'(1) = -1.$$

Find the values of the following derivatives at $x = 1$.

(a) $\dfrac{d}{dx}(uv)$ **(b)** $\dfrac{d}{dx}\left(\dfrac{u}{v}\right)$

(c) $\dfrac{d}{dx}\left(\dfrac{v}{u}\right)$ **(d)** $\dfrac{d}{dx}(7v - 2u)$

Slopes and Tangents

21. Find the tangents to *Newton's serpentine* (graphed here) at the origin and the point $(1, 2)$.

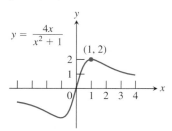

22. Find the tangent to the *witch of Agnesi* (graphed here) at the point $(2, 1)$.

23. The curve $y = ax^2 + bx + c$ passes through the point $(1, 2)$ and is tangent to the line $y = x$ at the origin. Find a, b, and c.

24. The curves $y = x^2 + ax + b$ and $y = cx - x^2$ have a common tangent at the point $(1, 0)$. Find a, b, and c.

Theory and Examples

25. *Writing to Learn* Suppose that the function v in the Product Rule has a constant value c. What does the Product Rule then say? What does this say about the Constant Multiple Rule?

26. *The Reciprocal Rule*

(a) The Reciprocal Rule says that at any point where the function $v(x)$ is differentiable and different from zero,

$$\frac{d}{dx}\left(\frac{1}{v}\right) = -\frac{1}{v^2}\frac{dv}{dx}.$$

Show that the Reciprocal Rule is a special case of the Quotient Rule.

(b) Show that the Reciprocal Rule and the Product Rule together imply the Quotient Rule.

27. *Generalizing the Product Rule* The Product Rule gives the formula

$$\frac{d}{dx}(uv) = u\frac{dv}{dx} + v\frac{du}{dx}$$

for the derivative of the product uv of two differentiable functions of x.

(a) What is the analogous formula for the derivative of the product uvw of *three* differentiable functions of x?

(b) What is the formula for the derivative of the product $u_1 u_2 u_3 u_4$ of *four* differentiable functions of x?

(c) What is the formula for the derivative of a product $u_1 u_2 u_3 \cdots u_n$ of a finite number n of differentiable functions of x?

28. *Rational powers*

(a) Find $\frac{d}{dx}(x^{3/2})$ by writing $x^{3/2}$ as $x \cdot x^{1/2}$ and using the Product Rule. Express your answer as a rational number times a rational power of x. Work parts (b) and (c) by a similar method.

(b) Find $\frac{d}{dx}(x^{5/2})$.

(c) Find $\frac{d}{dx}(x^{7/2})$.

(d) What patterns do you see in your answers to parts (a), (b), and (c)? Rational powers are explored in Section 2.6.

29. *Cylinder pressure* If gas in a cylinder is maintained at a constant temperature T, the pressure P is related to the volume V by a formula of the form

$$P = \frac{nRT}{V - nb} - \frac{an^2}{V^2},$$

in which a, b, n, and R are constants. Find dP/dV.

30. *The best quantity to order* One of the formulas for inventory management says that the average weekly cost of ordering, paying for, and holding merchandise is

$$A(q) = \frac{km}{q} + cm + \frac{hq}{2},$$

where q is the quantity you order when things run low (shoes, radios, brooms, or whatever the item might be); k is the cost of placing an order (the same, no matter how often you order); c is the cost of one item (a constant); m is the number of items sold each week (a constant); and h is the weekly holding cost per item (a constant that takes into account things such as space, utilities, insurance, and security). Find dA/dq and d^2A/dq^2.

2.4 Derivatives of Trigonometric Functions

Derivative of the Sine Function • Derivative of the Cosine Function • Simple Harmonic Motion • Derivatives of the Other Basic Trigonometric Functions • Continuity of Trigonometric Functions

Many of the phenomena we want information about are periodic (electromagnetic fields, heart rhythms, tides, weather). A surprising and beautiful theorem from advanced calculus says that every periodic function we are likely to use in mathematical modeling can be expressed in terms of sines and cosines. Thus, the derivatives of sines and cosines play a key role in describing periodic changes. This section shows how to differentiate the six basic trigonometric functions.

Derivative of the Sine Function

To calculate the derivative of $y = \sin x$, we combine the limits in Example 10(a) and Theorem 6 in Section 1.2 with the angle sum identity:

$$\sin (x + h) = \sin x \cos h + \cos x \sin h. \qquad (1)$$

We have

$$\frac{dy}{dx} = \lim_{h \to 0} \frac{\sin (x + h) - \sin x}{h} \qquad \text{Derivative definition}$$

$$= \lim_{h \to 0} \frac{(\sin x \cos h + \cos x \sin h) - \sin x}{h} \qquad \text{Eq. (1)}$$

$$= \lim_{h \to 0} \frac{\sin x (\cos h - 1) + \cos x \sin h}{h}$$

$$= \lim_{h \to 0} \left(\sin x \cdot \frac{\cos h - 1}{h} \right) + \lim_{h \to 0} \left(\cos x \cdot \frac{\sin h}{h} \right)$$

$$= \sin x \cdot \lim_{h \to 0} \frac{\cos h - 1}{h} + \cos x \cdot \lim_{h \to 0} \frac{\sin h}{h}$$

$$= \sin x \cdot 0 + \cos x \cdot 1 \qquad \text{Example 10a and}$$
$$\qquad\qquad\qquad\qquad \text{Theorem 6, Section 1.2}$$

$$= \cos x.$$

Radian measure in calculus

In case you are wondering why calculus uses radian measure when the rest of the world seems to use degrees, the answer lies in the argument that the derivative of the sine is the cosine. The derivative of $\sin x$ is $\cos x$ only if x is measured in radians, because

$$\lim_{h \to 0} \frac{\sin h}{h} = 1$$

only for radian measure.

The derivative of the sine function is the cosine function.

$$\frac{d}{dx} (\sin x) = \cos x$$

Example 1 Derivatives Involving the Sine

(a) $y = x^2 - \sin x$:

$$\frac{dy}{dx} = 2x - \frac{d}{dx} (\sin x) \qquad \text{Difference Rule}$$

$$= 2x - \cos x.$$

(b) $y = \dfrac{\sin x}{x}$:

$$\frac{dy}{dx} = \frac{x \cdot \dfrac{d}{dx} (\sin x) - \sin x \cdot 1}{x^2} \qquad \text{Quotient Rule}$$

$$= \frac{x \cos x - \sin x}{x^2}.$$

Derivative of the Cosine Function

With the help of the angle sum formula,

$$\cos (x + h) = \cos x \cos h - \sin x \sin h , \qquad (2)$$

we have

$$\frac{d}{dx} (\cos x) = \lim_{h\to 0} \frac{\cos (x + h) - \cos x}{h} \qquad \text{Derivative definition}$$

$$= \lim_{h\to 0} \frac{(\cos x \cos h - \sin x \sin h) - \cos x}{h} \qquad \text{Eq. (2)}$$

$$= \lim_{h\to 0} \frac{\cos x (\cos h - 1) - \sin x \sin h}{h}$$

$$= \lim_{h\to 0} \cos x \cdot \frac{\cos h - 1}{h} - \lim_{h\to 0} \sin x \cdot \frac{\sin h}{h}$$

$$= \cos x \cdot \lim_{h\to 0} \frac{\cos h - 1}{h} - \sin x \cdot \lim_{h\to 0} \frac{\sin h}{h}$$

$$= \cos x \cdot 0 - \sin x \cdot 1 \qquad \begin{array}{l}\text{Example 10(a) and}\\\text{Theorem 6, Section 1.2}\end{array}$$

$$= -\sin x .$$

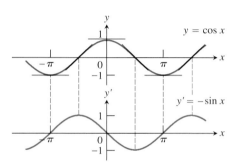

FIGURE 2.25 The curve $y' = -\sin x$ as the graph of the slopes of the tangents to the curve $y = \cos x$.

> **The derivative of the cosine function is the negative of the sine function.**
>
> $$\frac{d}{dx} (\cos x) = -\sin x$$

Figure 2.25 shows another way to visualize this result.

Example 2   Revisiting the Differentiation Rules

(a) $y = \sin x \cos x$:

$$\frac{dy}{dx} = \sin x \frac{d}{dx} (\cos x) + \cos x \frac{d}{dx} (\sin x) \qquad \text{Product Rule}$$

$$= \sin x (-\sin x) + \cos x (\cos x)$$

$$= \cos^2 x - \sin^2 x .$$

(b) $y = \dfrac{\cos x}{1 - \sin x}$:

$$\frac{dy}{dx} = \frac{(1 - \sin x) \dfrac{d}{dx} (\cos x) - \cos x \dfrac{d}{dx} (1 - \sin x)}{(1 - \sin x)^2} \qquad \text{Quotient Rule}$$

$$= \frac{(1 - \sin x)(-\sin x) - \cos x (0 - \cos x)}{(1 - \sin x)^2}$$

$$= \frac{1 - \sin x}{(1 - \sin x)^2} \qquad \sin^2 x + \cos^2 x = 1$$

$$= \frac{1}{1 - \sin x} .$$

Simple Harmonic Motion

The motion of a body bobbing freely up and down on the end of a spring or bungee cord is an example of *simple harmonic motion*. The next example describes a case in which there are no opposing forces such as friction or buoyancy to slow the motion down.

CD-ROM
WEBsite

Example 3   Motion on a Spring

A body hanging from a spring (Figure 2.26) is stretched 5 units beyond its rest position and released at time $t = 0$ to bob up and down. Its position at any later time t is

$$s = 5 \cos t.$$

What are its velocity and acceleration at time t?

Solution   We have

Position: $s = 5 \cos t$

Velocity: $v = \dfrac{ds}{dt} = \dfrac{d}{dt}(5 \cos t) = -5 \sin t$

Acceleration: $a = \dfrac{dv}{dt} = \dfrac{d}{dt}(-5 \sin t) = -5 \cos t.$

Notice how much we can learn from these equations:

1. As time passes, the weight moves down and up between $s = -5$ and $s = 5$ on the s-axis. The amplitude of the motion is 5. The period of the motion is 2π.

2. The velocity $v = -5 \sin t$ attains its greatest magnitude, 5, when $\cos t = 0$, as the graphs show in Figure 2.27. Hence, the speed of the weight, $|v| = 5 |\sin t|$, is greatest when $\cos t = 0$, that is, when $s = 0$ (the rest position). The speed of the weight is zero when $\sin t = 0$. This occurs when $s = 5 \cos t = \pm 5$, at the endpoints of the interval of motion.

3. The acceleration value is always the exact opposite of the position value. When the weight is above the rest position, gravity is pulling it back down; when the weight is below the rest position, the spring is pulling it back up.

4. The acceleration, $a = -5 \cos t$, is zero only at the rest position, where $\cos t = 0$ and the force of gravity and the force from the spring offset each other. When the weight is anywhere else, the two forces are unequal and acceleration is nonzero. The acceleration is greatest in magnitude at the points farthest from the rest position, where $\cos t = \pm 1$.

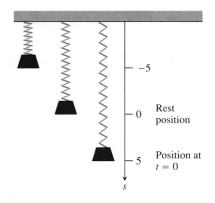

FIGURE 2.26 The body in Example 3.

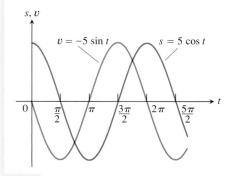

FIGURE 2.27 The graphs of the position and velocity of the body in Example 3.

Example 4 Jerk

The jerk of the simple harmonic motion in Example 3 is

$$j = \frac{da}{dt} = \frac{d}{dt}(-5\cos t) = 5\sin t .$$

It has its greatest magnitude when $\sin t = \pm 1$, not at the extremes of the displacement but at the rest position, where the acceleration changes direction and sign.

Derivatives of the Other Basic Trigonometric Functions

Because $\sin x$ and $\cos x$ are differentiable functions of x, the related functions

$$\tan x = \frac{\sin x}{\cos x}$$

$$\cot x = \frac{\cos x}{\sin x}$$

$$\sec x = \frac{1}{\cos x}$$

$$\csc x = \frac{1}{\sin x}$$

are differentiable at every value of x at which they are defined. Their derivatives, calculated from the Quotient Rule, are given by the following formulas.

$$\frac{d}{dx}(\tan x) = \sec^2 x \qquad \text{(3)}$$

$$\frac{d}{dx}(\sec x) = \sec x \tan x \qquad \text{(4)}$$

Notice the minus signs in the derivative formulas for the cofunctions.

$$\frac{d}{dx}(\cot x) = -\csc^2 x \qquad \text{(5)}$$

$$\frac{d}{dx}(\csc x) = -\csc x \cot x \qquad \text{(6)}$$

To show a typical calculation, we derive Equation (3). The other derivations are left to Exercise 44.

Example 5 Derivative of the Tangent Function

Find $d(\tan x)/dx$.

Solution

$$\frac{d}{dx}(\tan x) = \frac{d}{dx}\left(\frac{\sin x}{\cos x}\right) = \frac{\cos x \dfrac{d}{dx}(\sin x) - \sin x \dfrac{d}{dx}(\cos x)}{\cos^2 x} \qquad \text{Quotient Rule}$$

$$= \frac{\cos x \cos x - \sin x (-\sin x)}{\cos^2 x}$$

$$= \frac{\cos^2 x + \sin^2 x}{\cos^2 x}$$

$$= \frac{1}{\cos^2 x} = \sec^2 x$$

Example 6   A Trigonometric Second Derivative

Find y'' if $y = \sec x$.

Solution

$$y = \sec x$$

$$y' = \sec x \tan x \qquad \text{Eq. (4)}$$

$$y'' = \frac{d}{dx}(\sec x \tan x)$$

$$= \sec x \frac{d}{dx}(\tan x) + \tan x \frac{d}{dx}(\sec x) \qquad \text{Product Rule}$$

$$= \sec x (\sec^2 x) + \tan x (\sec x \tan x)$$

$$= \sec^3 x + \sec x \tan^2 x$$

Continuity of Trigonometric Functions

Since the six basic trigonometric functions are differentiable throughout their domains, they are continuous throughout their domains, by Theorem 1, Section 2.1. Hence, $\sin x$ and $\cos x$ are continuous for all x, supporting our observations in Section 1.4. Also, $\sec x$ and $\tan x$ are continuous except when x is a nonzero integer multiple of $\pi/2$, and $\csc x$ and $\cot x$ are continuous except at integer multiples of π. For each function, $\lim_{x \to c} f(x) = f(c)$ whenever $f(c)$ is defined. As a result, we can calculate the limits of many algebraic combinations and composites of trigonometric functions by direct substitution.

Example 7   Finding a Trigonometric Limit

$$\lim_{x \to 0} \frac{\sqrt{2 = \sec x}}{\cos(\pi - \tan x)} = \frac{\sqrt{2 + \sec 0}}{\cos(\pi - \tan 0)} = \frac{\sqrt{2 + 1}}{\cos(\pi - 0)} = \frac{\sqrt{3}}{-1} = -\sqrt{3}$$

EXERCISES 2.4

Derivatives

In Exercises 1–12, find dy/dx.

1. $y = -10x + 3 \cos x$

2. $y = \dfrac{3}{x} + 5 \sin x$

3. $y = \csc x - 4\sqrt{x} + 7$

4. $y = x^2 \cot x - \dfrac{1}{x^2}$

5. $y = (\sec x + \tan x)(\sec x - \tan x)$

6. $y = (\sin x + \cos x) \sec x$

7. $y = \dfrac{\cot x}{1 + \cot x}$

8. $y = \dfrac{\cos x}{1 + \sin x}$

9. $y = \dfrac{4}{\cos x} + \dfrac{1}{\tan x}$

10. $y = \dfrac{\cos x}{x} + \dfrac{x}{\cos x}$

11. $y = x^2 \sin x + 2x \cos x - 2 \sin x$

12. $y = x^2 \cos x - 2x \sin x - 2 \cos x$

In Exercises 13–16, find ds/dt.

13. $s = \tan t - t$

14. $s = t^2 - \sec t + 1$

15. $s = \dfrac{1 + \csc t}{1 - \csc t}$

16. $s = \dfrac{\sin t}{1 - \cos t}$

In Exercises 17–20, find $dr/d\theta$.

17. $r = 4 - \theta^2 \sin \theta$

18. $r = \theta \sin \theta + \cos \theta$

19. $r = \sec \theta \csc \theta$

20. $r = (1 + \sec \theta) \sin \theta$

In Exercises 21–24, find dp/dq.

21. $p = 5 + \dfrac{1}{\cot q}$

22. $p = (1 + \csc q) \cos q$

23. $p = \dfrac{\sin q + \cos q}{\cos q}$

24. $p = \dfrac{\tan q}{1 + \tan q}$

25. Find y'' if

    (a) $y = \csc x$.

    (b) $y = \sec x$.

26. Find $y^{(4)} = d^4 y/dx^4$ if

(a) $y = -2 \sin x$.   (b) $y = 9 \cos x$.

Tangent Lines

In Exercises 27–30, graph the curves over the given intervals, together with their tangents at the given values of x. Label each curve and tangent with its equation.

27. $y = \sin x$, $-3\pi/2 \le x \le 2\pi$

$x = -\pi, 0, 3\pi/2$

28. $y = \tan x$, $-\pi/2 < x < \pi/2$

$x = -\pi/3, 0, \pi/3$

29. $y = \sec x$, $-\pi/2 < x < \pi/2$

$x = -\pi/3, \pi/4$

30. $y = 1 + \cos x$, $-3\pi/2 \le x \le 2\pi$

$x = -\pi/3, 3\pi/2$

T Do the graphs of the functions in Exercises 31–34 have any horizontal tangents in the interval $0 \le x \le 2\pi$? If so, where? If not, why not? Visualize your findings by graphing the functions with a grapher.

31. $y = x + \sin x$ **32.** $y = 2x + \sin x$

33. $y = x - \cot x$ **34.** $y = x + 2\cos x$

35. Find all points on the curve $y = \tan x$, $-\pi/2 < x < \pi/2$, where the tangent line is parallel to the line $y = 2x$. Sketch the curve and tangent(s) together, labeling each with its equation.

36. Find all points on the curve $y = \cot x$, $0 < x < \pi$, where the tangent line is parallel to the line $y = -x$. Sketch the curve and tangent(s) together, labeling each with its equation.

In Exercises 37 and 38, find an equation for (a) the tangent to the curve at P and (b) the horizontal tangent to the curve at Q.

37.

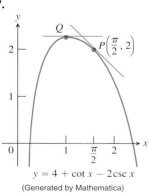

$y = 4 + \cot x - 2\csc x$

(Generated by Mathematica)

38.

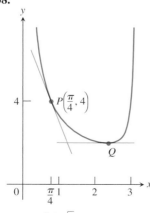

$y = 1 + \sqrt{2}\csc x + \cot x$

(Generated by Mathematica)

Simple Harmonic Motion

The equations in Exercises 39 and 40 give the position $s = f(t)$ of a body moving on a coordinate line (s in meters, t in seconds). Find the body's velocity, speed, acceleration, and jerk at time $t = \pi/4$ sec.

39. $s = 2 - 2\sin t$ **40.** $s = \sin t + \cos t$

Theory and Examples

41. *Writing to Learn* Is there a value of c that will make

$$f(x) = \begin{cases} \dfrac{\sin^2 3x}{x^2}, & x \neq 0 \\ c, & x = 0 \end{cases}$$

continuous at $x = 0$? Give reasons for your answer.

42. *Writing to Learn* Is there a value of b that will make

$$g(x) = \begin{cases} x + b, & x < 0 \\ \cos x, & x \geq 0 \end{cases}$$

continuous at $x = 0$? Differentiable at $x = 0$? Give reasons for your answers.

43. Find $d^{999}/dx^{999} (\cos x)$.

44. Derive the formula for the derivative with respect to x of

(a) $\sec x$.

(b) $\csc x$.

(c) $\cot x$.

T 45. Graph $y = \cos x$ for $-\pi \le x \le 2\pi$. On the same screen, graph

$$y = \frac{\sin(x + h) - \sin x}{h}$$

for $h = 1, 0.5, 0.3$, and 0.1. Then, in a new window, try $h = -1$, -0.5, and -0.3. What happens as $h \to 0^+$? As $h \to 0^-$? What phenomenon is being illustrated here?

T 46. Graph $y = -\sin x$ for $-\pi \le x \le 2\pi$. On the same screen, graph

$$y = \frac{\cos(x + h) - \cos x}{h}$$

for $h = 1, 0.5, 0.3$, and 0.1. Then, in a new window, try $h = -1, -0.5$, and -0.3. What happens as $h \to 0^+$? As $h \to 0^-$? What phenomenon is being illustrated here?

T 47. *Centered difference quotients* The **centered difference quotient**

$$\frac{f(x + h) - f(x - h)}{2h}$$

is used to approximate $f'(x)$ in numerical work because (1) its limit as $h \to 0$ equals $f'(x)$ when $f'(x)$ exists, and (2) it usually gives a better approximation of $f'(x)$ for a given value of h than Fermat's difference quotient

$$\frac{f(x+h) - f(x)}{h}.$$

See the accompanying figure.

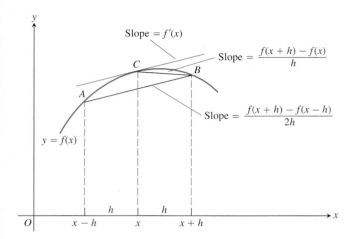

(a) To see how rapidly the centered difference quotient for $f(x) = \sin x$ converges to $f'(x) = \cos x$, graph $y = \cos x$ together with

$$y = \frac{\sin (x+h) - \sin (x-h)}{2h}$$

over the interval $[-\pi, 2\pi]$ for $h = 1, 0.5$, and 0.3. Compare the results with those obtained in Exercise 45 for the same values of h.

(b) To see how rapidly the centered difference quotient for $f(x) = \cos x$ converges to $f'(x) = -\sin x$, graph $y = -\sin x$ together with

$$y = \frac{\cos (x+h) - \cos (x-h)}{2h}$$

over the interval $[-\pi, 2\pi]$ for $h = 1, 0.5$, and 0.3. Compare the results with those obtained in Exercise 46 for the same values of h.

48. *A caution about centered difference quotients*   *(Continuation of Exercise 47)*   The quotient

$$\frac{f(x+h) - f(x-h)}{2h}$$

may have a limit as $h \to 0$ when f has no derivative at x. As a case in point, take $f(x) = |x|$ and calculate

$$\lim_{h \to 0} \frac{|0+h| - |0-h|}{2h}.$$

As you will see, the limit exists even though $f(x) = |x|$ has no derivative at $x = 0$. *Moral:* Before using a centered difference quotient, be sure the derivative exists.

T **49.** *Writing to Learn: Slopes on the graph of the tangent function* Graph $y = \tan x$ and its derivative together on $(-\pi/2, \pi/2)$. Does the graph of the tangent function appear to have a smallest slope? a largest slope? Is the slope ever negative? Give reasons for your answers.

T **50.** *Writing to Learn: Slopes on the graph of the cotangent function* Graph $y = \cot x$ and its derivative together for $0 < x < \pi$. Does the graph of the cotangent function appear to have a smallest slope? A largest slope? Is the slope ever positive? Give reasons for your answers.

T **51.** *Writing to Learn: Exploring (sin kx)/x*   Graph $y = (\sin x)/x$, $y = (\sin 2x)/x$, and $y = (\sin 4x)/x$ together over the interval $-2 \le x \le 2$. Where does each graph appear to cross the y-axis? Do the graphs really intersect the axis? What would you expect the graphs of $y = (\sin 5x)/x$ and $y = (\sin (-3x))/x$ to do as $x \to 0$? Why? What about the graph of $y = (\sin kx)/x$ for other values of k? Give reasons for your answers.

T **52.** *Radians versus degrees: degree mode derivatives* What happens to the derivatives of $\sin x$ and $\cos x$ if x is measured in degrees instead of radians? To find out, take the following steps.

(a) With your graphing calculator or computer grapher in *degree mode*, graph

$$f(h) = \frac{\sin h}{h}$$

and estimate $\lim_{h \to 0}\ f(h)$. Compare your estimate with $\pi/180$. Is there any reason to believe the limit *should* be $\pi/180$?

(b) With your grapher still in degree mode, estimate

$$\lim_{h \to 0} \frac{\cos h - 1}{h}.$$

(c) Now go back to the derivation of the formula for the derivative of $\sin x$ in the text and carry out the steps of the derivation using degree-mode limits. What formula do you obtain for the derivative?

(d) Work through the derivation of the formula for the derivative of $\cos x$ using degree-mode limits. What formula do you obtain for the derivative?

(e) The disadvantages of the degree-mode formulas become apparent as you start taking derivatives of higher order. Try it. What are the second and third degree-mode derivatives of $\sin x$ and $\cos x$?

2.5 The Chain Rule and Parametric Equations

Derivative of a Composite Function • "Outside–Inside" Rule •
Repeated Use of the Chain Rule • Slopes of Parametrized Curves •
Power Chain Rule • Melting Ice Cubes

As we see in Chapters 3 and 4, many engineering applications of calculus involve finding a function with a given derivative. Sometimes we identify such a function right away. We know that $\cos x$ is the derivative of $\sin x$, for example, that $2x$ is the derivative of x^2, and that the sum $\cos x + 2x$ is the derivative of $\sin x + x^2$. But what if we're starting with the *product* of two derivatives instead of their sum? We know it does not come from the product of the differentiated functions because the derivative of their product is not the product of their derivatives.

So, where does a product of derivatives come from? The answer lies in a rule for differentiating composite functions called the Chain Rule. This section describes the rule and how to use it.

Derivative of a Composite Function

We begin with examples.

Example 1 Relating Derivatives

The function $y = 6x - 10 = 2(3x - 5)$ is the composite of the functions $y = 2u$ and $u = 3x - 5$. How are the derivatives of these functions related?

Solution We have

$$\frac{dy}{dx} = 6, \qquad \frac{dy}{du} = 2, \qquad \text{and} \qquad \frac{du}{dx} = 3.$$

Since $6 = 2 \cdot 3$, we see that

$$\frac{dy}{dx} = \frac{dy}{du} \cdot \frac{du}{dx}.$$

Is it an accident that

$$\frac{dy}{dx} = \frac{dy}{du} \cdot \frac{du}{dx}?$$

If we think of the derivative as a rate of change, our intuition allows us to see that this relationship is reasonable. If $y = f(u)$ changes twice as fast as u and $u = g(x)$ changes three times as fast as x, then we expect y to change six times as fast as x. This effect is much like that of a multiple gear train (Figure 2.28).

C: y turns B: u turns A: x turns

FIGURE 2.28 When gear A makes x turns, gear B makes u turns and gear C makes y turns. By comparing circumferences or counting teeth, we see that $y = u/2$ and $u = 3x$, so $y = 3x/2$. Thus, $dy/du = 1/2$, $du/dx = 3$, and $dy/dx = 3/2 = (dy/du)(du/dx)$.

Example 2 Relating Derivatives

The function

$$y = 9x^4 + 6x^2 + 1 = (3x^2 + 1)^2$$

is the composite of $y = u^2$ and $u = 3x^2 + 1$. Calculating derivatives, we see that

$$\frac{dy}{du} \cdot \frac{du}{dx} = 2u \cdot 6x$$

$$= 2(3x^2 + 1) \cdot 6x \qquad u = 3x^2 + 1$$

$$= 36x^3 + 12x$$

and

$$\frac{dy}{dx} = \frac{d}{dx}(9x^4 + 6x^2 + 1) = 36x^3 + 12x.$$

Once again,

$$\frac{dy}{du} \cdot \frac{du}{dx} = \frac{dy}{dx}.$$

The derivative of the composite $f(g(x))$ at x is the derivative of f at $g(x)$ times the derivative of g at x. This observation is known as the Chain Rule (Figure 2.29).

FIGURE 2.29 Rates of change multiply: The derivative of $f \circ g$ at x is the derivative of f at the point $g(x)$ times the derivative of g at the point x.

Theorem 3   The Chain Rule

If $f(u)$ is differentiable at the point $u = g(x)$ and $g(x)$ is differentiable at x, then the composite function $(f \circ g)(x) = f(g(x))$ is differentiable at x, and

$$(f \circ g)'(x) = f'(g(x)) \cdot g'(x). \tag{1}$$

In Leibniz's notation, if $y = f(u)$ and $u = g(x)$, then

$$\frac{dy}{dx} = \frac{dy}{du} \cdot \frac{du}{dx}, \tag{2}$$

where dy/du is evaluated at $u = g(x)$.

It would be tempting to try to prove the Chain Rule by writing

$$\frac{\Delta y}{\Delta x} = \frac{\Delta y}{\Delta u} \cdot \frac{\Delta u}{\Delta x}$$

and taking the limit as $\Delta x \to 0$. This method would work if we knew that Δu, the change in u, was nonzero, but we do not know this. A small change in x could conceivably produce no change in u. The proof requires a different approach, using ideas in Section 3.6. (See Appendix 3.)

Example 3 Applying the Chain Rule

An object moves along the x-axis so that its position at any time $t \geq 0$ is given by $x(t) = \cos(t^2 + 1)$. Find the velocity of the object as a function of t.

Solution We know that the velocity is dx/dt. In this instance, x is a composite function: $x = \cos(u)$ and $u = t^2 + 1$. We have

$$\frac{dx}{du} = -\sin(u) \qquad x = \cos(u)$$

$$\frac{du}{dt} = 2t. \qquad u = t^2 + 1$$

By the Chain Rule,

$$\frac{dx}{dt} = \frac{dx}{du} \cdot \frac{du}{dt}$$

$$= -\sin(u) \cdot 2t$$

$$= -\sin(t^2 + 1) \cdot 2t$$

$$= -2t \sin(t^2 + 1).$$

"Outside–Inside" Rule

It sometimes helps to think about the Chain Rule this way: If $y = f(g(x))$, then

$$\frac{dy}{dx} = f'(g(x)) \cdot g'(x). \tag{3}$$

In words, differentiate the "outside" function f and evaluate it at the "inside" function $g(x)$ left alone; then multiply by the derivative of the "inside function."

Example 4 Differentiating from the Outside In

Differentiate $\sin(x^2 + x)$ with respect to x.

Solution

$$\frac{d}{dx} \sin \underbrace{(x^2 + x)}_{\text{inside}} = \cos \underbrace{(x^2 + x)}_{\substack{\text{inside} \\ \text{left alone}}} \cdot \underbrace{(2x + 1)}_{\substack{\text{derivative of} \\ \text{the inside}}}$$

Repeated Use of the Chain Rule

We sometimes have to use the Chain Rule two or more times to find a derivative. Here is an example.

Example 5   A Three-Link "Chain"

Find the derivative of $g(t) = \tan(5 - \sin 2t)$.

Solution   Notice here that the tangent is a function of $5 - \sin 2t$, whereas the sine is a function of $2t$, which is itself a function of t. Therefore, by the Chain Rule,

$$g'(t) = \frac{d}{dt}(\tan(5 - \sin 2t))$$

$$= \sec^2(5 - \sin 2t) \cdot \frac{d}{dt}(5 - \sin 2t) \qquad \text{Derivative of } \tan u \text{ with } u = 5 - \sin 2t$$

$$= \sec^2(5 - \sin 2t) \cdot \left(0 - \cos 2t \cdot \frac{d}{dt}(2t)\right) \qquad \text{Derivative of } 5 - \sin u \text{ with } u = 2t$$

$$= \sec^2(5 - \sin 2t) \cdot (-\cos 2t) \cdot 2$$

$$= -2(\cos 2t)\sec^2(5 - \sin 2t).$$

Slopes of Parametrized Curves

A parametrized curve $(x(t), y(t))$ is **differentiable** at t if x and y are differentiable at t. At a point on a differentiable parametrized curve where y is also a differentiable function of x, the derivatives dy/dt, dx/dt, and dy/dx are related by the Chain Rule:

$$\frac{dy}{dt} = \frac{dy}{dx} \cdot \frac{dx}{dt}.$$

If $dx/dt \neq 0$, we may divide both sides of this equation by dx/dt to solve for dy/dx.

Parametric Formula for dy/dx
If all three derivatives exist and $dx/dt \neq 0$,

$$\frac{dy}{dx} = \frac{dy/dt}{dx/dt}. \tag{4}$$

Example 6   Differentiating with a Parameter

Find the line tangent to the right-hand hyperbola branch defined parametrically by

$$x = \sec t, \qquad y = \tan t, \qquad -\frac{\pi}{2} < t < \frac{\pi}{2}$$

at the point $(\sqrt{2}, 1)$, where $t = \pi/4$ (Figure 2.30).

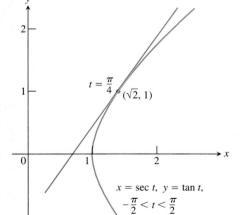

FIGURE 2.30 The hyperbola branch in Example 6. Equation (4) applies for every point on the graph except (1, 0).

Solution   All three of the derivatives in Equation (4) exist, and $dx/dt = \sec t \tan t \neq 0$ at the indicated point. Therefore, Equation (4) applies, and

$$\frac{dy}{dx} = \frac{dy/dt}{dx/dt}$$

$$= \frac{\sec^2 t}{\sec t \tan t}$$

$$= \frac{\sec t}{\tan t}$$

$$= \csc t.$$

Setting $t = \pi/4$ gives

$$\left. \frac{dy}{dx} \right|_{t=\pi/4} = \csc (\pi/4) = \sqrt{2}.$$

The equation of the tangent line is

$$y - 1 = \sqrt{2} (x - \sqrt{2})$$
$$y = \sqrt{2} x - 2 + 1$$
$$y = \sqrt{2} x - 1.$$

If parametric equations define y as a twice-differentiable function of x, we can apply Equation (4) to the function $dy/dx = y'$ to calculate d^2y/dx^2 as a function of t:

$$\frac{d^2y}{dx^2} = \frac{d}{dx} (y') = \frac{dy'/dt}{dx/dt} . \qquad \text{Eq. (4) with } y' \text{ in place of } y$$

Parametric Formula for d^2y/dx^2

If the equations $x = f(t)$, $y = g(t)$ define y as a twice-differentiable function of x, then at any point where $dx/dt \neq 0$,

$$\frac{d^2y}{dx^2} = \frac{dy'/dt}{dx/dt} .$$

Example 7   Finding d^2y/dx^2 for a Parametrized Curve

Find d^2y/dx^2 as a function of t if $x = t - t^2$, $y = t - t^3$.

Finding d^2y/dx^2 in terms of t

Step 1. Express $y' = dy/dx$ in terms of t.

Step 2. Find dy'/dt.

Step 3. Divide dy'/dt by dx/dt.

Solution

Step 1: Express $y' = dy/dx$ in terms of t.

$$y' = \frac{dy}{dx} = \frac{dy/dt}{dx/dt} = \frac{1 - 3t^2}{1 - 2t}$$

Step 2: Differentiate y' with respect to t.

$$\frac{dy'}{dt} = \frac{d}{dt} \left(\frac{1 - 3t^2}{1 - 2t} \right) = \frac{2 - 6t + 6t^2}{(1 - 2t)^2} . \qquad \text{Quotient Rule}$$

Step 3: Divide dy'/dt by dx/dt.

$$\frac{d^2y}{dx^2} = \frac{dy'/dt}{dx/dt} = \frac{(2 - 6t + 6t^2)/(1 - 2t)^2}{1 - 2t} = \frac{2 - 6t + 6t^2}{(1 - 2t)^3}$$

Power Chain Rule

If f is a differentiable function of u and if u is a differentiable function of x, then substituting $y = f(u)$ into the Chain Rule formula

$$\frac{dy}{dx} = \frac{dy}{du} \cdot \frac{du}{dx}$$

leads to the formula

$$\frac{d}{dx} f(u) = f'(u) \frac{du}{dx}.$$

Here's an example of how it works: If n is an integer and $f(u) = u^n$, the Power Rules (Rules 2 and 7) tell us that $f'(u) = nu^{n-1}$. If u is a differentiable function of x, then we can use the Chain Rule to extend this to the **Power Chain Rule:**

$$\frac{d}{dx} u^n = nu^{n-1} \frac{du}{dx}. \qquad \frac{d}{du}(u^n) = nu^{n-1} \tag{5}$$

Example 8 Finding Tangent Slopes

(a) Find the slope of the line tangent to the curve $y = \sin^5 x$ at the point where $x = \pi/3$.

(b) Show that the slope of every line tangent to the curve $y = 1/(1 - 2x)^3$ is positive.

Solution

(a) $\dfrac{dy}{dx} = 5 \sin^4 x \cdot \dfrac{d}{dx} \sin x$ ⎯⎯ Power Chain Rule with $u = \sin x$, $n = 5$

$\qquad\quad = 5 \sin^4 x \cos x$

The tangent line has slope

$$\left. \frac{dy}{dx} \right|_{x=\pi/3} = 5\left(\frac{\sqrt{3}}{2}\right)^4 \left(\frac{1}{2}\right) = \frac{45}{32}.$$

(b) $\dfrac{dy}{dx} = \dfrac{d}{dx}(1 - 2x)^{-3}$

$\qquad\quad = -3(1 - 2x)^{-4} \cdot \dfrac{d}{dx}(1 - 2x)$ ⎯⎯ Power Chain Rule with $u = (1 - 2x)$, $n = -3$

$\qquad\quad = -3(1 - 2x)^{-4} \cdot (-2)$

$\qquad\quad = \dfrac{6}{(1 - 2x)^4}$

At any point (x, y) on the curve, $x \neq 1/2$ and the slope of the tangent line is

$$\frac{dy}{dx} = \frac{6}{(1 - 2x)^4},$$

the quotient of two positive numbers.

$\sin^n x$ is short for $(\sin x)^n$, $n \neq -1$.

Example 9   Radians Versus Degrees

It is important to remember that the formulas for the derivatives of both sin x and cos x were obtained under the assumption that x is measured in radians, *not* degrees. The Chain Rule gives us new insight into the difference between the two. Since $180° = \pi$ radians, $x° = \pi x/180$ radians where $x°$ means the angle x measured in degrees.

By the Chain Rule,

$$\frac{d}{dx}\sin(x°) = \frac{d}{dx}\sin\left(\frac{\pi x}{180}\right) = \frac{\pi}{180}\cos\left(\frac{\pi x}{180}\right) = \frac{\pi}{180}\cos(x°).$$

See Figure 2.31. Similarly, the derivative of $\cos(x°)$ is $-(\pi/180)\sin(x°)$.

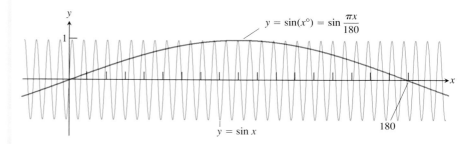

FIGURE 2.31 $\sin(x°)$ oscillates only $\pi/180$ times as often as $\sin x$ oscillates. Its maximum slope is $\pi/180$. (Example 9)

The factor $\pi/180$, annoying in the first derivative, would compound with repeated differentiation. We see at a glance the compelling reason for the use of radian measure.

Melting Ice Cubes

The state of California has serious droughts, and new sources of water are always under consideration. One of the proposals is to tow icebergs from polar waters to offshore locations near southern California, where the melting ice could provide fresh water. As a first approximation in analyzing this proposal, we might imagine the iceberg to be a large cube (or some other regularly shaped solid, like a rectangular solid or a pyramid).

Example 10   The Melting Ice Cube

How long will it take an ice cube to melt?

Solution   We start with a mathematical model. We assume that the cube retains its cubical shape as it melts. If we call its edge length s, its volume is $V = s^3$ and its surface area is $6s^2$. We assume that V and s are differentiable functions of time t. We assume also that the cube's volume decreases at a rate that is proportional to its surface area. This latter assumption seems reasonable enough when we think that the melting takes place at the surface: Changing the amount of surface changes the amount of ice exposed to melt. In mathematical terms,

$$\frac{dV}{dt} = -k(6s^2), \qquad k > 0.$$

The minus sign indicates that the volume is decreasing. We assume that the proportionality factor k is constant. It probably depends on many things, such as the relative humidity of the surrounding air, the air temperature, and the incidence or absence of sunlight, to name only a few.

Finally, we need at least one more piece of information: How long will it take a specific percentage of the ice cube to melt? We have nothing to guide us unless we make one or more observations, but now let us assume a particular set of conditions in which the cube lost 1/4 of its volume during the first hour. (You could use letters instead of particular numbers, say $n\%$ in r hours. Then your answer would be in terms of n and r.) Mathematically, we now have the following problem.

Given: $V = s^3$ and $\dfrac{dV}{dt} = -k(6s^2)$

$V = V_0$ when $t = 0$

$V = (3/4)V_0$ when $t = 1\,\text{h}.$

Find: The value of t when $V = 0$.

We apply the Chain Rule to differentiate $V = s^3$ with respect to t:

$$\frac{dV}{dt} = 3s^2 \frac{ds}{dt}.$$

We set this derivative equal to the given rate, $-k(6s^2)$, to get

$$3s^2 \frac{ds}{dt} = -6ks^2$$

$$\frac{ds}{dt} = -2k.$$

The side length is *decreasing* at the constant rate of $2k$ units per hour. Thus, if the initial length of the cube's side is s_0, the length of its side 1 hour later is $s_1 = s_0 - 2k$. This equation tells us that

$$2k = s_0 - s_1.$$

The melting time is the value of t that makes $2kt = s_0$. Hence,

$$t_{\text{melt}} = \frac{s_0}{2k} = \frac{s_0}{s_0 - s_1} = \frac{1}{1 - (s_1/s_0)},$$

but

$$\frac{s_1}{s_0} = \frac{\left(\frac{3}{4}V_0\right)^{1/3}}{(V_0)^{1/3}} = \left(\frac{3}{4}\right)^{1/3} \approx 0.91.$$

Therefore,

$$t_{\text{melt}} = \frac{1}{1 - 0.91} \approx 11\,\text{h}.$$

If 1/4 of the cube melts in 1 h, it will take about 10 h more for the rest of it to melt.

Of course the kinds of questions we ultimately need to answer are, How much of the ice will be lost in transit? And how long will it take the ice to turn into usable water? If we were interested in pursuing the matter further, our next step would be to test the model by running experiments and then to refine it on the basis of what we learned.

EXERCISES 2.5

Derivative Calculations

In Exercises 1–6, given $y = f(u)$ and $u = g(x)$, find $dy/dx = f'(g(x))g'(x)$.

1. $y = 6u - 9$, $u = (1/2)x^4$ **2.** $y = 2u^3$, $u = 8x - 1$

3. $y = \sin u$, $u = 3x + 1$ **4.** $y = \cos u$, $u = \sin x$

5. $y = \tan u$, $u = 10x - 5$ **6.** $y = -\sec u$, $u = x^2 + 7x$

In Exercises 7–12, write the function in the form $y = f(u)$ and $u = g(x)$. Then find dy/dx as a function of x.

7. $y = (4 - 3x)^9$ **8.** $y = \left(1 - \dfrac{x}{7}\right)^{-7}$

9. $y = \left(\dfrac{x^2}{8} + x - \dfrac{1}{x}\right)^4$ **10.** $y = \sec(\tan x)$

11. $y = \cot\left(\pi - \dfrac{1}{x}\right)$ **12.** $y = \sin^3 x$

Find the derivatives of the functions in Exercises 13–26.

13. $q = \sqrt{2r - r^2}$

14. $s = \sin\left(\dfrac{3\pi t}{2}\right) + \cos\left(\dfrac{3\pi t}{2}\right)$

15. $r = (\csc\theta + \cot\theta)^{-1}$ **16.** $r = -(\sec\theta + \tan\theta)^{-1}$

17. $y = x^2 \sin^4 x + x \cos^{-2} x$ **18.** $y = \dfrac{1}{x}\sin^{-5} x - \dfrac{x}{3}\cos^3 x$

19. $y = \dfrac{1}{21}(3x - 2)^7 + \left(4 - \dfrac{1}{2x^2}\right)^{-1}$

20. $y = (4x + 3)^4(x + 1)^{-3}$

21. $h(x) = x \tan(2\sqrt{x}) + 7$ **22.** $k(x) = x^2 \sec\left(\dfrac{1}{x}\right)$

23. $f(\theta) = \left(\dfrac{\sin\theta}{1 + \cos\theta}\right)^2$ **24.** $r = \sin(\theta^2)\cos(2\theta)$

25. $r = \sec\sqrt{\theta}\,\tan\left(\dfrac{1}{\theta}\right)$ **26.** $q = \sin\left(\dfrac{t}{\sqrt{t+1}}\right)$

In Exercises 27–32, find dy/dt.

27. $y = \sin^2(\pi t - 2)$ **28.** $y = (1 + \cos 2t)^{-4}$

29. $y = (1 + \cot(t/2))^{-2}$ **30.** $y = \sin(\cos(2t - 5))$

31. $y = \left(1 + \tan^4\left(\dfrac{t}{12}\right)\right)^3$

32. $y = \sqrt{1 + \cos(t^2)}$

Tangents to Parametrized Curves

In Exercises 33–40, find an equation for the line tangent to the curve at the point defined by the given value of t. Also, find the value of d^2y/dx^2 at this point.

33. $x = 2\cos t$, $y = 2\sin t$, $t = \pi/4$

34. $x = \cos t$, $y = \sqrt{3}\cos t$, $t = 2\pi/3$

35. $x = t$, $y = \sqrt{t}$, $t = 1/4$

36. $x = -\sqrt{t + 1}$, $y = \sqrt{3t}$, $t = 3$

37. $x = 2t^2 + 3$, $y = t^4$, $t = -1$

38. $x = t - \sin t$, $y = 1 - \cos t$, $t = \pi/3$

39. $x = \cos t$, $y = 1 + \sin t$, $t = \pi/2$

40. $x = \sec^2 t - 1$, $y = \tan t$, $t = -\pi/4$

Second Derivatives

Find y'' in Exercises 41–44.

41. $y = \left(1 + \dfrac{1}{x}\right)^3$ **42.** $y = (1 - \sqrt{x})^{-1}$

43. $y = \dfrac{1}{9}\cot(3x - 1)$ **44.** $y = 9\tan\left(\dfrac{x}{3}\right)$

Finding Numerical Values of Derivatives

In Exercises 45–50, find the value of $(f \circ g)'$ at the given value of x.

45. $f(u) = u^5 + 1$, $u = g(x) = \sqrt{x}$, $x = 1$

46. $f(u) = 1 - \dfrac{1}{u}$, $u = g(x) = \dfrac{1}{1 - x}$, $x = -1$

47. $f(u) = \cot\dfrac{\pi u}{10}$, $u = g(x) = 5\sqrt{x}$, $x = 1$

48. $f(u) = u + \dfrac{1}{\cos^2 u}$, $u = g(x) = \pi x$, $x = 1/4$

49. $f(u) = \dfrac{2u}{u^2 + 1}$, $\quad u = g(x) = 10x^2 + x + 1$, $\quad x = 0$

50. $f(u) = \left(\dfrac{u - 1}{u + 1}\right)^2$, $\quad u = g(x) = \dfrac{1}{x^2} - 1$, $\quad x = -1$

51. Suppose that functions f and g and their derivatives with respect to x have the following values at $x = 2$ and $x = 3$.

x	$f(x)$	$g(x)$	$f'(x)$	$g'(x)$
2	8	2	1/3	−3
3	3	−4	2π	5

Find the derivatives with respect to x of the following combinations at the given value of x.

(a) $2f(x)$, $\quad x = 2$ **(b)** $f(x) + g(x)$, $\quad x = 3$

(c) $f(x) \cdot g(x)$, $\quad x = 3$ **(d)** $f(x)/g(x)$, $\quad x = 2$

(e) $f(g(x))$, $\quad x = 2$ **(f)** $\sqrt{f(x)}$, $\quad x = 2$

(g) $1/g^2(x)$, $\quad x = 3$ **(h)** $\sqrt{f^2(x) + g^2(x)}$, $\quad x = 2$

52. Suppose that the functions f and g and their derivatives with respect to x have the following values at $x = 0$ and $x = 1$.

x	$f(x)$	$g(x)$	$f'(x)$	$g'(x)$
0	1	1	5	1/3
1	3	−4	−1/3	−8/3

Find the derivatives with respect to x of the following combinations at the given value of x.

(a) $5f(x) - g(x)$, $\quad x = 1$ **(b)** $f(x)g^3(x)$, $\quad x = 0$

(c) $\dfrac{f(x)}{g(x) + 1}$, $\quad x = 1$ **(d)** $f(g(x))$, $\quad x = 0$

(e) $g(f(x))$, $\quad x = 0$ **(f)** $(x^{11} + f(x))^{-2}$, $\quad x = 1$

(g) $f(x + g(x))$, $\quad x = 0$

53. Find ds/dt when $\theta = 3\pi/2$ if $s = \cos\theta$ and $d\theta/dt = 5$.

54. Find dy/dt when $x = 1$ if $y = x^2 + 7x - 5$ and $dx/dt = 1/3$.

Choices in Composition

What happens if you can write a function as a composite in different ways? Do you get the same derivative each time? The Chain Rule says you should. Try it with the functions in Exercises 55 and 56.

55. Find dy/dx if $y = x$ by using the Chain Rule with y as a composite of

(a) $y = (u/5) + 7$ and $u = 5x - 35$

(b) $y = 1 + (1/u)$ and $u = 1/(x - 1)$.

56. Find dy/dx if $y = x^{3/2}$ by using the Chain Rule with y as a composite of

(a) $y = u^3$ and $u = \sqrt{x}$ **(b)** $y = \sqrt{u}$ and $u = x^3$.

Tangents and Slopes

57. (a) Find the tangent to the curve $y = 2\tan(\pi x/4)$ at $x = 1$.

     (b) *Writing to Learn: Slopes on a tangent curve* What is the smallest value the slope of the curve can ever have on the interval $-2 < x < 2$? Give reasons for your answer.

58. *Writing to Learn: Slopes on sine curves*

     (a) Find equations for the tangents to the curves $y = \sin 2x$ and $y = -\sin(x/2)$ at the origin. Is there anything special about how the tangents are related? Give reasons for your answer.

     (b) Can anything be said about the tangents to the curves $y = \sin mx$ and $y = -\sin(x/m)$ at the origin (m a constant $\neq 0$)? Give reasons for your answer.

     (c) For a given m, what are the largest values the slopes of the curves $y = \sin mx$ and $y = -\sin(x/m)$ can ever have? Give reasons for your answer.

     (d) The function $y = \sin x$ completes one period on the interval $[0, 2\pi]$, the function $y = \sin 2x$ completes two periods, the function $y = \sin(x/2)$ completes half a period, and so on. Is there any relation between the number of periods $y = \sin mx$ completes on $[0, 2\pi]$ and the slope of the curve $y = \sin mx$ at the origin? Give reasons for your answer.

Theory, Examples, and Applications

59. *Running machinery too fast* Suppose that a piston is moving straight up and down and that its position at time t sec is

$$s = A\cos(2\pi bt),$$

with A and b positive. The value of A is the amplitude of the motion, and b is the frequency (number of times the piston moves up and down each second). What effect does doubling the frequency have on the piston's velocity, acceleration, and jerk? (Once you find out, you will know why machinery breaks when you run it too fast.) See Figure 2.32.

FIGURE 2.32 The internal forces in the engine get so large that they tear the engine apart when the velocity is too great.

60. *Temperatures in Fairbanks, Alaska* The graph in Figure 2.33 shows the average Fahrenheit temperature in Fairbanks, Alaska, during a typical 365-day year. The equation that approximates the temperature on day x is

$$y = 37 \sin\left[\frac{2\pi}{365}(x - 101)\right] + 25.$$

(a) On what day is the temperature increasing the fastest?

(b) About how many degrees per day is the temperature increasing when it is increasing at its fastest?

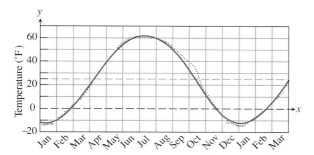

FIGURE 2.33 Normal mean air temperatures at Fairbanks, Alaska, plotted as data points, and the approximating sine function. (Exercise 60)

61. *Particle motion* The position of a particle moving along a coordinate line is $s = \sqrt{1 + 4t}$, with s in meters and t in seconds. Find the particle's velocity and acceleration at $t = 6$ sec.

62. *Constant acceleration* Suppose that the velocity of a falling body is $v = k\sqrt{s}$ m/sec (k a constant) at the instant the body has fallen s m from its starting point. Show that the body's acceleration is constant.

63. *Falling meteorite* The velocity of a heavy meteorite entering Earth's atmosphere is inversely proportional to $\sqrt{s}$ when it is s km from Earth's center. Show that the meteorite's acceleration is inversely proportional to s^2.

64. *Particle acceleration* A particle moves along the x-axis with velocity $dx/dt = f(x)$. Show that the particle's acceleration is $f(x)f'(x)$.

65. *Temperature and the period of a pendulum* For oscillations of small amplitude (short swings), we may safely model the relationship between the period T and the length L of a simple pendulum with the equation

$$T = 2\pi\sqrt{\frac{L}{g}},$$

where g is the constant acceleration of gravity at the pendulum's location. If we measure g in centimeters per second squared, we measure L in centimeters and T in seconds. If the pendulum is made of metal, its length will vary with temperature, either increas-

ing or decreasing at a rate that is roughly proportional to L. In symbols, with u being temperature and k the proportionality constant,

$$\frac{dL}{du} = kL.$$

Assuming this to be the case, show that the rate at which the period changes with respect to temperature is $kT/2$.

66. *Writing to Learn: Chain Rule* Suppose that $f(x) = x^2$ and $g(x) = |x|$. Then the composites

$$(f \circ g)(x) = |x|^2 = x^2 \quad \text{and} \quad (g \circ f)(x) = |x^2| = x^2$$

are both differentiable at $x = 0$ even though g itself is not differentiable at $x = 0$. Does this contradict the Chain Rule? Explain.

67. *Writing to Learn: Tangents* Suppose that $u = g(x)$ is differentiable at $x = 1$ and that $y = f(u)$ is differentiable at $u = g(1)$. If the graph of $y = f(g(x))$ has a horizontal tangent at $x = 1$, can we conclude anything about the tangent to the graph of g at $x = 1$ or the tangent to the graph of f at $u = g(1)$? Give reasons for your answer.

68. *Writing to Learn* Suppose that $u = g(x)$ is differentiable at $x = -5$, $y = f(u)$ is differentiable at $u = g(-5)$, and $(f \circ g)'(-5)$ is negative. What, if anything, can be said about the values of $g'(-5)$ and $f'(g(-5))$?

T **69.** *The derivative of sin 2x*   Graph the function $y = 2\cos 2x$ for $-2 \le x \le 3.5$. Then, on the same screen, graph

$$y = \frac{\sin 2(x + h) - \sin 2x}{h}$$

for $h = 1.0$, 0.5, and 0.2. Experiment with other values of h, including negative values. What do you see happening as $h \to 0$? Explain this behavior.

T **70.** *The derivative of cos (x²)* Graph $y = -2x\sin(x^2)$ for $-2 \le x \le 3$. Then, on the same screen, graph

$$y = \frac{\cos((x + h)^2) - \cos(x^2)}{h}$$

for $h = 1.0$, 0.7, and 0.3. Experiment with other values of h. What do you see happening as $h \to 0$? Explain this behavior.

T The curves in Exercises 71 and 72 are called *Bowditch curves* or *Lissajous figures*. In each case, find the point in the interior of the first quadrant where the tangent to the curve is horizontal, and find the equations of the two tangents at the origin.

71.

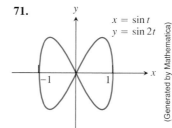

$x = \sin t$
$y = \sin 2t$

(Generated by Mathematica)

72.

$x = \sin 2t$
$y = \sin 3t$

(Generated by Mathematica)

COMPUTER EXPLORATIONS

Trigonometric Polynomials

73. As Figure 2.34 shows, the trigonometric "polynomial"

$$s = f(t) = 0.78540 - 0.63662 \cos 2t - 0.07074 \cos 6t$$
$$- 0.02546 \cos 10t - 0.01299 \cos 14t$$

gives a good approximation of the sawtooth function $s = g(t)$ on the interval $[-\pi, \pi]$. How well does the derivative of f approximate the derivative of g at the points where dg/dt is defined? To find out, carry out the following steps.

(a) Graph dg/dt (where defined) over $[-\pi, \pi]$.

(b) Find df/dt.

(c) Graph df/dt. Where does the approximation of dg/dt by df/dt seem to be best? Least good? Approximations by trigonometric polynomials are important in the theories of heat and oscillation, but we must not expect too much of them, as we see in the next exercise.

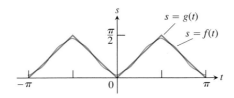

FIGURE 2.34 The approximation of a sawtooth function by a trigonometric "polynomial." (Exercise 73)

74. *(Continuation of Exercise 73)* In Exercise 73, the trigonometric polynomial $f(t)$ that approximated the sawtooth function $g(t)$ on $[-\pi, \pi]$ had a derivative that approximated the derivative of the sawtooth function. It is possible, however, for a trigonometric polynomial to approximate a function in a reasonable way without its derivative approximating the function's derivative at all well. As a case in point, the "polynomial"

$$s = h(t) = 1.2732 \sin 2t + 0.4244 \sin 6t + 0.25465 \sin 10t$$
$$+ 0.18189 \sin 14t + 0.14147 \sin 18t$$

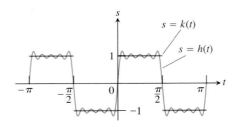

FIGURE 2.35 The approximation of a step function by a trigonometric "polynomial." (Exercise 74)

graphed in Figure 2.35 approximates the step function $s = k(t)$ shown there. Yet the derivative of h is nothing like the derivative of k.

(a) Graph dk/dt (where defined) over $[-\pi, \pi]$.

(b) Find dh/dt.

(c) Graph dh/dt to see how badly the graph fits the graph of dk/dt. Comment on what you see.

Parametrized Curves

Use a CAS to perform the following steps on the parametrized curves in Exercises 75–80.

(a) Plot the curve for the given interval of t values.

(b) Find dy/dx and d^2y/dx^2 at the point t_0.

(c) Find an equation for the tangent line to the curve at the point defined by the given value t_0. Plot the curve together with the tangent line on a single graph.

75. $x = \dfrac{1}{3}t^3$, $y = \dfrac{1}{2}t^2$, $0 \le t \le 1$, $t_0 = 1/2$

76. $x = 2t^3 - 16t^2 + 25t + 5$, $y = t^2 + t - 3$, $0 \le t \le 6$, $t_0 = 3/2$

77. $x = e^t - t^2$, $y = t + e^{-t}$, $-1 \le t \le 2$, $t_0 = 1$

78. $x = t - \cos t$, $y = 1 + \sin t$, $-\pi \le t \le \pi$, $t_0 = \pi/4$

79. $x = e^t + \sin 2t$, $y = e^t + \cos (t^2)$, $-\sqrt{2}\pi \le t \le \pi/4$, $t_0 = -\pi/4$

80. $x = e^t \cos t$, $y = e^t \sin t$, $0 \le t \le \pi$, $t_0 = \pi/2$

2.6 Implicit Differentiation

Implicitly Defined Functions • Derivatives of Higher Order • Rational Powers of Differentiable Functions

In the law that describes how light changes direction as it enters a lens, the important angles are the angles the light makes with the line perpendicular to the surface of the lens at the point of entry (angles A and B in Figure 2.36). This line is called the *normal* to the surface at the point of entry. In a profile view of a lens like the one

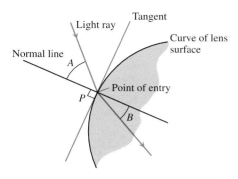

FIGURE 2.36 The profile of a lens, showing the bending (refraction) of a ray of light as it passes through the lens surface.

When Are the Functions Defined by F(x, y) = 0 Differentiable?

To justify implicit differentiation, we must know that the derivatives we want really do exist. That is, we need to know when we can rely on the functions defined by an expression $F(x, y) = 0$ to be differentiable. A theorem in advanced calculus guarantees this to be the case under certain conditions on the function F. All of the functions you will encounter in this section meet these conditions.

in Figure 2.36, the normal is the line perpendicular to the tangent to the profile curve at the point of entry.

Profiles of lenses are often described by equations of the form $F(x, y) = 0$. To find the normal to a point on the profile curve, we first need to find the slope of the tangent there by evaluating dy/dx. But what if we cannot put the equation in the form $y = f(x)$ to differentiate? In such cases, we may still be able to find dy/dx by a process called *implicit differentiation*. This section describes the technique and uses it to extend the Power Rule for differentiation to include rational exponents.

Implicitly Defined Functions

The graph of the equation $x^3 + y^3 - 9xy = 0$ (Figure 2.37) has a well-defined slope at nearly every point because it is the union of the graphs of the functions $y = f_1(x)$, $y = f_2(x)$, and $y = f_3(x)$, which are differentiable except at the origin and A. But how do we find the slope when we cannot conveniently solve the equation to find the functions? The answer is to treat y as a differentiable function of x and differentiate both sides of the equation with respect to x, using the differentiation rules for sums, products, and quotients and the Chain Rule. Then solve for dy/dx in terms of x and y *together* to obtain a formula that calculates the slope at any point (x, y) on the graph from the values of x and y. This process for finding dy/dx is called **implicit differentiation,** so named because the equation $x^3 + y^3 - 9xy = 0$ defines the functions f_1, f_2, and f_3 *implicitly* (i.e., hidden inside the equation), without giving us *explicit* formulas with which to work.

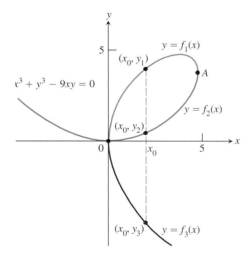

FIGURE 2.37 The curve $x^3 + y^3 - 9xy = 0$ is not the graph of any one function of x. The curve can, however, be divided into separate arcs that *are* the graphs of functions of x. This particular curve, called a *folium*, dates to Descartes in 1638.

Example 1 Differentiating Implicitly

Find dy/dx if $y^2 = x$.

Solution   The equation $y^2 = x$ defines two differentiable functions of x that we can actually find, namely $y_1 = \sqrt{x}$ and $y_2 = -\sqrt{x}$ (Figure 2.38). We know how to calculate the derivative of each of these for $x > 0$:

$$\frac{dy_1}{dx} = \frac{1}{2\sqrt{x}} \quad \text{and} \quad \frac{dy_2}{dx} = -\frac{1}{2\sqrt{x}}.$$

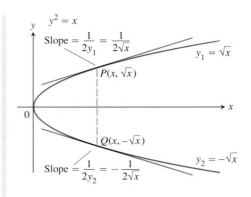

FIGURE 2.38 The equation $y^2 - x = 0$, or $y^2 = x$ as it is usually written, defines two differentiable functions of x on the interval $x \geq 0$. Example 1 shows how to find the derivatives of these functions without solving the equation $y^2 = x$ for y.

But suppose that we knew only that the equation $y^2 = x$ defined y as one or more differentiable functions of x for $x > 0$ without knowing exactly what these functions were. Could we still find dy/dx?

The answer is yes. To find dy/dx, we simply differentiate both sides of the equation $y^2 = x$ with respect to x, treating $y = f(x)$ as a differentiable function of x:

$$y^2 = x$$

$$2y \frac{dy}{dx} = 1 \qquad \text{The Chain Rule gives } \frac{d}{dx} y^2 = \frac{d}{dx} [f(x)]^2 = 2f(x) f'(x) = 2y \frac{dy}{dx}.$$

$$\frac{dy}{dx} = \frac{1}{2y}.$$

This one formula gives the derivatives we calculated for *both* explicit solutions $y_1 = \sqrt{x}$ and $y_2 = -\sqrt{x}$:

$$\frac{dy_1}{dx} = \frac{1}{2y_1} = \frac{1}{2\sqrt{x}} \qquad \text{and} \qquad \frac{dy_2}{dx} = \frac{1}{2y_2} = \frac{1}{2(-\sqrt{x})} = -\frac{1}{2\sqrt{x}}.$$

To calculate the derivatives of other implicitly defined functions, we proceed as in Example 1. We treat y as a differentiable implicit function of x and apply the usual rules to differentiate both sides of the defining equation.

Example 2   Differentiating Implicitly

Find dy/dx if $y^2 = x^2 + \sin xy$ (Figure 2.39).

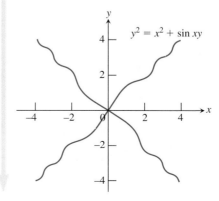

FIGURE 2.39 The graph of $y^2 = x^2 + \sin xy$ in Example 2. The example shows how to find slopes on this implicitly defined curve.

Implicit Differentiation Takes Four Steps

Step 1. Differentiate both sides of the equation with respect to x, treating y as a differentiable function of x.

Step 2. Collect the terms with dy/dx on one side of the equation.

Step 3. Factor out dy/dx.

Step 4. Solve for dy/dx.

Solution

$$y^2 = x^2 + \sin xy$$

$$\frac{d}{dx}(y^2) = \frac{d}{dx}(x^2) + \frac{d}{dx}(\sin xy) \qquad \text{Differentiate both sides with respect to } x \dots$$

$$2y\frac{dy}{dx} = 2x + (\cos xy)\frac{d}{dx}(xy) \qquad \text{... treating } y \text{ as a function of } x \text{ and using the Chain Rule.}$$

$$2y\frac{dy}{dx} = 2x + (\cos xy)\left(y + x\frac{dy}{dx}\right)$$

$$2y\frac{dy}{dx} - (\cos xy)\left(x\frac{dy}{dx}\right) = 2x + (\cos xy)\,y \qquad \text{Collect terms with } dy/dx \dots$$

$$(2y - x\cos xy)\frac{dy}{dx} = 2x + y\cos xy \qquad \text{... and factor out } dy/dx.$$

$$\frac{dy}{dx} = \frac{2x + y\cos xy}{2y - x\cos xy} \qquad \text{Solve for } dy/dx \text{ by dividing.}$$

Notice that the formula for dy/dx applies everywhere that the implicitly defined curve has a slope. Notice also that the derivative involves *both* variables x and y, not just the independent variable x.

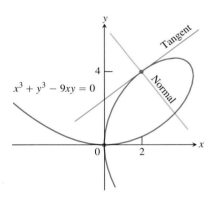

$$x^3 + y^3 - 9xy = 0$$

FIGURE 2.40 Example 3 shows how to find equations for the tangent and normal to the curve at $(2, 4)$.

Example 3 Tangent and Normal to the Folium of Descartes

Show that the point $(2, 4)$ lies on the curve $x^3 + y^3 - 9xy = 0$. Then find the tangent and normal to the curve there (Figure 2.40).

Solution The point $(2, 4)$ lies on the curve because its coordinates satisfy the equation given for the curve: $2^3 + 4^3 - 9(2)(4) = 8 + 64 - 72 = 0$.

To find the slope of the curve at $(2, 4)$, we first use implicit differentiation to find a formula for dy/dx:

$$x^3 + y^3 - 9xy = 0$$

$$\frac{d}{dx}(x^3) + \frac{d}{dx}(y^3) - \frac{d}{dx}(9xy) = \frac{d}{dx}(0) \qquad \text{Differentiate both sides with respect to } x.$$

$$3x^2 + 3y^2\frac{dy}{dx} - 9\left(x\frac{dy}{dx} + y\frac{dx}{dx}\right) = 0 \qquad \text{Treat } xy \text{ as a product and } y \text{ as a function of } x.$$

$$(3y^2 - 9x)\frac{dy}{dx} + 3x^2 - 9y = 0$$

$$3(y^2 - 3x)\frac{dy}{dx} = 9y - 3x^2 \qquad \text{Collect terms.}$$

$$\frac{dy}{dx} = \frac{3y - x^2}{y^2 - 3x}. \qquad \text{Solve for } dy/dx.$$

We then evaluate the derivative at $(x, y) = (2, 4)$:

$$\frac{dy}{dx}\bigg|_{(2,4)} = \frac{3y - x^2}{y^2 - 3x}\bigg|_{(2,4)} = \frac{3(4) - 2^2}{4^2 - 3(2)} = \frac{8}{10} = \frac{4}{5}.$$

The tangent at $(2, 4)$ is the line through $(2, 4)$ with slope $4/5$:

$$y = 4 + \frac{4}{5}(x - 2)$$

$$y = \frac{4}{5}x + \frac{12}{5}.$$

The normal to the curve at $(2, 4)$ is the line perpendicular to the tangent there, the line through $(2, 4)$ with slope $-5/4$:

$$y = 4 - \frac{5}{4}(x - 2)$$

$$y = -\frac{5}{4}x + \frac{13}{2}.$$

The Word Normal

When analytic geometry was developed in the seventeenth century, European scientists wrote about their work and ideas in Latin, a language that all educated Europeans could be counted on to read and understand. The word *normalis*, which scholars used for "perpendicular" in Latin, became *normal* when they discussed geometry in English.

Derivatives of Higher Order

Implicit differentiation can also be used to find derivatives of higher order. Here is an example.

Example 4   Finding a Second Derivative Implicitly

Find d^2y/dx^2 if $2x^3 - 3y^2 = 8$.

Solution   To start, we differentiate both sides of the equation with respect to x in order to find $y' = dy/dx$.

$$\frac{d}{dx}(2x^3 - 3y^2) = \frac{d}{dx}(8)$$

$$6x^2 - 6yy' = 0$$

$$x^2 - yy' = 0$$

$$y' = \frac{x^2}{y}, \qquad \text{when } y \neq 0$$

We now apply the Quotient Rule to find y''.

$$y'' = \frac{d}{dx}\left(\frac{x^2}{y}\right) = \frac{2xy - x^2y'}{y^2} = \frac{2x}{y} - \frac{x^2}{y^2} \cdot y'$$

Finally, we substitute $y' = x^2/y$ to express y'' in terms of x and y.

$$y'' = \frac{2x}{y} - \frac{x^2}{y^2}\left(\frac{x^2}{y}\right) = \frac{2x}{y} - \frac{x^4}{y^3}, \qquad \text{when } y \neq 0$$

Rational Powers of Differentiable Functions

We know that the rule

$$\frac{d}{dx}x^n = nx^{n-1}$$

holds when n is an integer. Using implicit differentiation we can show that it holds when n is any rational number.

Theorem 4   Power Rule for Rational Powers

If n is a rational number, then x^n is differentiable at every interior point of the domain of x^{n-1}, and

$$\frac{d}{dx}x^n = nx^{n-1}. \tag{1}$$

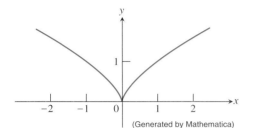

FIGURE 2.41 The graph of $y = x^{2/3}$ has a cusp at $x = 0$. (Example 5)

(Generated by Mathematica)

Example 5   Using the Rational Power Rule

(a) $\dfrac{d}{dx}\left(\sqrt{x}\right) = \dfrac{d}{dx}\left(x^{1/2}\right) = \dfrac{1}{2}x^{-1/2} = \dfrac{1}{2\sqrt{x}}$

Notice that $\sqrt{x}$ is defined at $x = 0$, but $1/(2\sqrt{x})$ is not.

(b) $\dfrac{d}{dx}\left(x^{2/3}\right) = \dfrac{2}{3}\left(x^{-1/3}\right) = \dfrac{2}{3x^{1/3}}$

The original function is defined for all real numbers, but the derivative is undefined at $x = 0$. Its graph has a *cusp* at $x = 0$ (Figure 2.41).

Proof of Theorem 4   Let p and q be integers with $q > 0$ and suppose that $y = \sqrt[q]{x^p} = x^{p/q}$. Then

$$y^q = x^p.$$

Since p and q are integers (for which we already have the Power Rule), we can differentiate both sides of the equation with respect to x and obtain

$$qy^{q-1}\frac{dy}{dx} = px^{p-1}.$$

If $y \neq 0$, we can divide both sides of the equation by qy^{q-1} to solve for dy/dx, obtaining

$$\frac{dy}{dx} = \frac{px^{p-1}}{qy^{q-1}}$$

$$= \frac{p}{q}\cdot\frac{x^{p-1}}{\left(x^{p/q}\right)^{q-1}} \qquad y = x^{p/q}$$

$$= \frac{p}{q}\cdot\frac{x^{p-1}}{x^{p-p/q}} \qquad \frac{p}{q}(q-1) = p - \frac{p}{q}$$

$$= \frac{p}{q}\cdot x^{(p-1)-(p-p/q)} \qquad \text{A law of exponents}$$

$$= \frac{p}{q}\cdot x^{(p/q)-1},$$

which proves the rule.   —

By combining this result with the Chain Rule, we get an extension of the Power Chain Rule to rational powers of u: If n is a rational number and u is a differentiable function of x, then u^n is a differentiable function of x and

$$\frac{d}{dx}u^n = nu^{n-1}\frac{du}{dx}, \tag{2}$$

provided that $u \neq 0$ if $n < 1$.

The restriction that $u \neq 0$ when $n < 1$ is necessary because 0 might be in the domain of u^n but not in the domain of u^{n-1}, as we see in the next example.

Example 6 Using the Rational Power and Chain Rules

(a) $\dfrac{d}{dx}\overbrace{(1 - x^2)^{1/4}}^{\text{function defined on } [-1,1]} = \dfrac{1}{4}(1 - x^2)^{-3/4}(-2x)$ Eq. (2) with $u = 1 - x^2$ and $n = 1/4$

$= \underbrace{\dfrac{-x}{2(1 - x^2)^{3/4}}}_{\text{derivative defined only on } (-1, 1)}$

(b) $\dfrac{d}{dx}(\cos x)^{-1/5} = -\dfrac{1}{5}(\cos x)^{-6/5}\dfrac{d}{dx}(\cos x)$

$= -\dfrac{1}{5}(\cos x)^{-6/5}(-\sin x)$

$= \dfrac{1}{5}(\sin x)(\cos x)^{-6/5}$

EXERCISES 2.6

Derivatives of Rational Powers

Find dy/dx in Exercises 1–6.

1. $y = x^{9/4}$

2. $y = \sqrt[3]{2x}$

3. $y = 7\sqrt{x + 6}$

4. $y = (1 - 6x)^{2/3}$

5. $y = x(x^2 + 1)^{1/2}$

6. $y = x(x^2 + 1)^{-1/2}$

Find the first derivatives of the functions in Exercises 7–12.

7. $s = \sqrt[7]{t^2}$

8. $r = \sqrt[4]{\theta^{-3}}$

9. $y = \sin((2t + 5)^{-2/3})$

10. $f(x) = \sqrt{1 - \sqrt{x}}$

11. $g(x) = 2(2x^{-1/2} + 1)^{-1/3}$

12. $h(\theta) = \sqrt[3]{1 + \cos(2\theta)}$

Differentiating Implicitly

Use implicit differentiation to find dy/dx in Exercises 13–22.

13. $x^2 y + x y^2 = 6$

14. $2xy + y^2 = x + y$

15. $x^3 - xy + y^3 = 1$

16. $x^2(x - y)^2 = x^2 - y^2$

17. $y^2 = \dfrac{x - 1}{x + 1}$

18. $x^2 = \dfrac{x - y}{x + y}$

19. $x = \tan y$

20. $x + \sin y = xy$

21. $y \sin\left(\dfrac{1}{y}\right) = 1 - xy$

22. $y^2 \cos\left(\dfrac{1}{y}\right) = 2x + 2y$

Find $dr/d\theta$ in Exercises 23–26.

23. $\theta^{1/2} + r^{1/2} = 1$

24. $r - 2\sqrt{\theta} = \dfrac{3}{2}\theta^{2/3} + \dfrac{4}{3}\theta^{3/4}$

25. $\sin(r\theta) = \dfrac{1}{2}$

26. $\cos r + \cos \theta = r\theta$

Higher Derivatives

In Exercises 27–30, use implicit differentiation to find dy/dx and then d^2y/dx^2.

27. $x^{2/3} + y^{2/3} = 1$

28. $y^2 = x^2 + 2x$

29. $2\sqrt{y} = x - y$

30. $xy + y^2 = 1$

31. If $x^3 + y^3 = 16$, find the value of d^2y/dx^2 at the point $(2, 2)$.

32. If $xy + y^2 = 1$, find the value of d^2y/dx^2 at the point $(0, -1)$.

Implicitly Defined Parametrizations

Assuming that the equations in Exercises 33–36 define x and y implicitly as differentiable functions $x = f(t)$, $y = g(t)$, find the slope of the curve $x = f(t)$, $y = g(t)$ at the given value of t.

33. $x^2 - 2tx + 2t^2 = 4$, $2y^3 - 3t^2 = 4$, $t = 2$

34. $x = \sqrt{5 - \sqrt{t}}$, $y(t - 1) = \ln y$, $t = 1$

35. $x + 2x^{3/2} = t^2 + t$, $y\sqrt{t + 1} + 2t\sqrt{y} = 4$, $t = 0$

36. $x \sin t + 2x = t$, $t \sin t - 2t = y$, $t = \pi$

Slopes, Tangents, and Normals

In Exercises 37 and 38, find the slope of the curve at the given points.

37. $y^2 + x^2 = y^4 - 2x$ at $(-2, 1)$ and $(-2, -1)$

38. $(x^2 + y^2)^2 = (x - y)^2$ at $(1, 0)$ and $(1, -1)$

In Exercises 39–46, verify that the given point is on the curve and find the lines that are **(a)** tangent and **(b)** normal to the curve at the given point.

39. $x^2 + xy - y^2 = 1$, $(2, 3)$

40. $x^2 y^2 = 9$, $(-1, 3)$

41. $y^2 - 2x - 4y - 1 = 0$, $(-2, 1)$

42. $6x^2 + 3xy + 2y^2 + 17y - 6 = 0$, $(-1, 0)$

43. $2xy + \pi \sin y = 2\pi$, $(1, \pi/2)$

44. $x \sin 2y = y \cos 2x$, $(\pi/4, \pi/2)$

45. $y = 2 \sin (\pi x - y)$, $(1, 0)$

46. $x^2 \cos^2 y - \sin y = 0$, $(0, \pi)$

47. *Parallel tangents* Find the two points where the curve $x^2 + xy + y^2 = 7$ crosses the x-axis and show that the tangents to the curve at these points are parallel. What is the common slope of these tangents?

48. *Tangents parallel to coordinate axes* Find points on the curve $x^2 + xy + y^2 = 7$ **(a)** where the tangent is parallel to the x-axis and **(b)** where the tangent is parallel to the y-axis. In the latter case, dy/dx is not defined, but dx/dy is. What value does dx/dy have at these points?

49. *The eight curve* Find the slopes of the curve $y^4 = y^2 - x^2$ at the two points shown here.

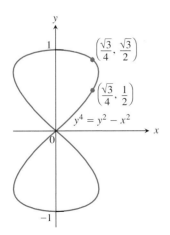

50. *The cissoid of Diocles (from about 200 B.C.)* Find equations for the tangent and normal to the cissoid of Diocles $y^2(2 - x) = x^3$ at $(1, 1)$.

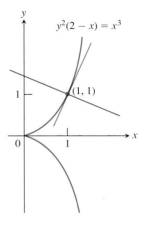

51. *The devil's curve (Gabriel Cramer, the Cramer of Cramer's rule, 1750)* Find the slopes of the devil's curve $y^4 - 4y^2 = x^4 - 9x^2$ at the four indicated points.

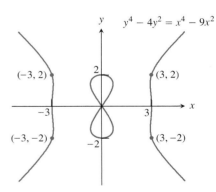

52. *The folium of Descartes* (See Figure 2.37)

(a) Find the slope of the folium of Descartes, $x^3 + y^3 - 9xy = 0$ at the points $(4, 2)$ and $(2, 4)$.

(b) At what point other than the origin does the folium have a horizontal tangent?

(c) Find the coordinates of the point A in Figure 2.37, where the folium has a vertical tangent.

Theory and Examples

53. Which of the following could be true if $f''(x) = x^{-1/3}$?

(a) $f(x) = \dfrac{3}{2} x^{2/3} - 3$

(b) $f(x) = \dfrac{9}{10} x^{5/3} - 7$

(c) $f'''(x) = -\dfrac{1}{3} x^{-4/3}$

(d) $f'(x) = \dfrac{3}{2} x^{2/3} + 6$

54. *Writing to Learn* Is there anything special about the tangents to the curves $2x^2 + 3y^2 = 5$ and $y^2 = x^3$ at the points $(1, \pm1)$? Give reasons for your answer.

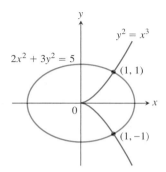

55. *Intersecting normal* The line that is normal to the curve $x^2 + 2xy - 3y^2 = 0$ at $(1, 1)$ intersects the curve at what other point?

56. *Normals parallel to a line* Find the normals to the curve $xy + 2x - y = 0$ that are parallel to the line $2x + y = 0$.

57. *Normals to a parabola* Show that if it is possible to draw three normals from the point $(a, 0)$ to the parabola $x = y^2$ shown here, then a must be greater than $1/2$. One of the normals is the x-axis. For what value of a are the other two normals perpendicular?

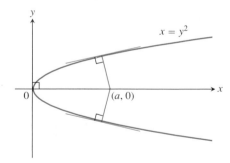

58. *Writing to Learn* What is the geometry behind the restrictions on the domains of the derivatives in Example 5 and Example 6(a)?

T In Exercises 59 and 60, find both dy/dx (treating y as a differential function of x) and dx/dy (treating x as a differentiable function of y). How do dy/dx and dx/dy seem to be related? Explain the relationship geometrically in terms of the graphs.

59. $xy^3 + x^2y = 6$ **60.** $x^3 + y^2 = \sin^2 y$

COMPUTER EXPLORATIONS

61. (a) Given that $x^4 + 4y^2 = 1$, find dy/dx two ways: (1) by solving for y and differentiating the resulting functions in the usual way and (2) by implicit differentiation. Do you get the same result each way?

(b) *Writing to Learn* Solve the equation $x^4 + 4y^2 = 1$ for y and graph the resulting functions together to produce a complete graph of the equation $x^4 + 4y^2 = 1$. Then add the graphs of the first derivatives of these functions to your display. Could you have predicted the general behavior of the derivative graphs from looking at the graph of $x^4 + 4y^2 = 1$? Could you have predicted the general behavior of the graph of $x^4 + 4y^2 = 1$ by looking at the derivative graphs? Give reasons for your answers.

62. (a) Given that $(x - 2)^2 + y^2 = 4$, find dy/dx two ways: (1) by solving for y and differentiating the resulting functions with respect to x and (2) by implicit differentiation. Do you get the same result each way?

(b) *Writing to Learn* Solve the equation $(x - 2)^2 + y^2 = 4$ for y and graph the resulting functions together to produce a complete graph of the equation $(x - 2)^2 + y^2 = 4$. Then add the graphs of the functions' first derivatives to your picture. Could you have predicted the general behavior of the derivative graphs from looking at the graph of $(x - 2)^2 + y^2 = 4$? Could you have predicted the general behavior of the graph of $(x - 2)^2 + y^2 = 4$ by looking at the derivative graphs? Give reasons for your answers.

Use a CAS to perform the following steps in Exercises 63–70.

(a) Plot the equation with the implicit plotter of a CAS. Check to see that the given point P satisfies the equation.

(b) Using implicit differentiation find a formula for the derivative dy/dx and evaluate it at the given point P.

(c) Use the slope found in part (b) to find an equation for the tangent line to the curve at P. Then plot the implicit curve and tangent line together on a single graph.

63. $x^3 - xy + y^3 = 7$, $P(2, 1)$

64. $x^5 + y^3x + yx^2 + y^4 = 4$, $P(1, 1)$

65. $y^2 + y = \dfrac{2 + x}{1 - x}$, $P(0, 1)$

66. $y^3 + \cos xy = x^2$, $P(1, 0)$

67. $x + \tan\left(\dfrac{y}{x}\right) = 2$, $P\left(1, \dfrac{\pi}{4}\right)$

68. $xy^3 + \tan(x + y) = 1$, $P\left(\dfrac{\pi}{4}, 0\right)$

69. $2y^2 + (xy)^{1/3} = x^2 + 2$, $P(1, 1)$

70. $x\sqrt{1 + 2y} + y = x^2$, $P(1, 0)$

2.7 Related Rates

Related Rate Equations • Solution Strategy

Suppose you want to measure the rate of ascent of a rocket during a vertical liftoff from the ground. With some elaborate and expensive precautions you could probably put an instrument of some kind on the ground under the rocket and take readings from it. But it would be safer and a lot cheaper just to stand some distance d away and measure the rate of change of the rocket's angle of elevation θ. With a little trigonometry you could then express the rocket's height h in terms of the angle and distance as $h = d \tan \theta$. Differentiating both sides of this equation with respect to time t would then express dh/dt, the rate you want, in terms of $d\theta/dt$, a rate you could easily measure. The problem of finding a rate you cannot measure easily from some other rate that you can is called a *related rate problem*. Related rate problems are the subject of this section.

Related Rate Equations

Suppose that a particle $P(x, y)$ is moving along a curve C in the plane so that its coordinates x and y are differentiable functions of time t. If D is the distance from the origin to P, then using the Chain Rule we can find an equation that relates dD/dt, dx/dt, and dy/dt.

$$D = \sqrt{x^2 + y^2}$$

$$\frac{dD}{dt} = \frac{1}{2}(x^2 + y^2)^{-1/2} \left(2x \frac{dx}{dt} + 2y \frac{dy}{dt} \right)$$

Any equation involving two or more variables that are differentiable functions of time t can be used to find an equation that relates their corresponding rates.

Example 1 Finding Related Rate Equations

Assume that the radius r and height h of a cone are differentiable functions of t and let V be the volume of the cone. Find an equation that relates dV/dt, dr/dt, and dh/dt.

Solution

$$V = \frac{\pi}{3} r^2 h \qquad \text{Cone volume formula}$$

$$\frac{dV}{dt} = \frac{\pi}{3} \left(r^2 \cdot \frac{dh}{dt} + 2r \frac{dr}{dt} \cdot h \right) = \frac{\pi}{3} \left(r^2 \frac{dh}{dt} + 2rh \frac{dr}{dt} \right)$$

Solution Strategy

How rapidly will the fluid level inside a vertical cylindrical storage tank drop if we pump the fluid out at some specified rate? A question like this one asks us to calculate a rate that we cannot measure directly from a rate that we can. To do so, we write an equation that relates the variables involved and differentiate it to get an equation that relates the rate we seek to the rate we know.

$$\frac{dh}{dt} = ?$$

h

$$\frac{dV}{dt} = -3000 \text{ L/min}$$

FIGURE 2.42 The cylindrical tank in Example 2.

Example 2   Pumping Out a Tank

How rapidly will the fluid level inside a vertical cylindrical tank drop if we pump the fluid out at the rate of 3000 L/min?

Solution   We draw a picture of a partially filled vertical cylindrical tank, calling its radius r and the height of the fluid h (Figure 2.42). Call the volume of the fluid V.

As time passes, the radius remains constant, but V and h change. We think of V and h as differentiable functions of time and use t to represent time. We are told that

$$\frac{dV}{dt} = -3000.$$ We pump out at the rate of 3000 L/min. The rate is negative because the volume is decreasing.

We are asked to find

$$\frac{dh}{dt}.$$ How fast will the fluid level drop?

To find dh/dt, we first write an equation that relates h to V. The equation depends on the units chosen for V, r, and h. With V in liters and r and h in meters, the appropriate equation for the cylinder's volume is

$$V = 1000\pi r^2 h$$

because a cubic meter contains 1000 L.

Since V and h are differentiable functions of t, we can differentiate both sides of the equation $V = 1000\pi r^2 h$ with respect to t to get an equation that relates dh/dt to dV/dt:

$$\frac{dV}{dt} = 1000\pi r^2 \frac{dh}{dt}.$$ r is a constant.

We substitute the known value $dV/dt = -3000$ and solve for dh/dt:

$$\frac{dh}{dt} = \frac{-3000}{1000\pi r^2} = -\frac{3}{\pi r^2}.$$

The fluid level will drop at the rate of $3/(\pi r^2)$ m/min.

Interpret

The equation $dh/dt = -3/\pi r^2$ shows how the rate at which the fluid level drops depends on the tank's radius. If r is small, dh/dt will be large; if r is large, dh/dt will be small.

If $r = 1$ m: $\dfrac{dh}{dt} = -\dfrac{3}{\pi} \approx -0.95$ m/min $= -95$ cm/min.

If $r = 10$ m: $\dfrac{dh}{dt} = -\dfrac{3}{100\pi} \approx -0.0095$ m/min $= -0.95$ cm/min.

Related Rate Problem Strategy

Step 1. *Draw a picture and name the variables and constants.* Use *t* for time. Assume that all variables are differentiable functions of *t*.

Step 2. *Write down the numerical information* (in terms of the symbols you have chosen).

Step 3. *Write down what you are asked to find* (usually a rate, expressed as a derivative).

Step 4. *Write an equation that relates the variables.* You may have to combine two or more equations to get a single equation that relates the variable whose rate you want to the variables whose rates you know.

Step 5. *Differentiate with respect to t.* Then express the rate you want in terms of the rate and variables whose values you know.

Step 6. *Evaluate.* Use known values to find the unknown rate.

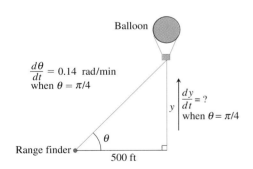

$$\frac{d\theta}{dt} = 0.14 \text{ rad/min}$$
when $\theta = \pi/4$

y | $\frac{dy}{dt} = ?$
when $\theta = \pi/4$

θ

Range finder

500 ft

FIGURE 2.43 The balloon in Example 3.

Example 3   A Rising Balloon

A hot-air balloon rising straight up from a level field is tracked by a range finder 500 ft from the liftoff point. At the moment the range finder's elevation angle is $\pi/4$, the angle is increasing at the rate of 0.14 rad/min . How fast is the balloon rising at that moment?

Solution   We answer the question in six steps.

Step 1: *Draw a picture and name the variables and constants* (Figure 2.43). The variables in the picture are

θ = the angle in radians the range finder makes with the ground

y = the height in feet of the balloon.

We let *t* represent time in minutes and assume that θ and *y* are differentiable functions of *t* .

The one constant in the picture is the distance from the range finder to the liftoff point (500 ft). There is no need to give it a special symbol.

Step 2: *Write down the additional numerical information.*

$$\frac{d\theta}{dt} = 0.14 \text{ rad/min} \qquad \text{when} \qquad \theta = \frac{\pi}{4}$$

Step 3: *Write down what we are to find.* We want dy/dt when $\theta = \pi/4$.

Step 4: *Write an equation that relates the variables y and θ .*

$$\frac{y}{500} = \tan \theta \qquad \text{or} \qquad y = 500 \tan \theta$$

Step 5: *Differentiate with respect to t using the Chain Rule.* The result tells how dy/dt (which we want) is related to $d\theta/dt$ (which we know).

$$\frac{dy}{dt} = 500 \, (\sec^2 \theta) \, \frac{d\theta}{dt}$$

Step 6: *Evaluate with* $\theta = \pi/4$ *and* $d\theta/dt = 0.14$ *to find* dy/dt.

$$\frac{dy}{dt} = 500(\sqrt{2})^2(0.14) = 140 \qquad \sec\frac{\pi}{4} = \sqrt{2}$$

Interpret

At the moment in question, the balloon is rising at the rate of 140 ft/min.

Example 4   A Highway Chase

A police cruiser, approaching a right-angled intersection from the north, is chasing a speeding car that has turned the corner and is now moving straight east. When the cruiser is 0.6 mi north of the intersection and the car is 0.8 mi to the east, the police determine with radar that the distance between them and the car is increasing at 20 mph. If the cruiser is moving at 60 mph at the instant of measurement, what is the speed of the car?

Solution   We carry out the steps of the strategy.

Step 1: *Picture and variables.* We picture the car and cruiser in the coordinate plane, using the positive x-axis as the eastbound highway and the positive y-axis as the southbound highway (Figure 2.44). We let t represent time and set

$$x = \text{position of car at time } t$$
$$y = \text{position of cruiser at time } t$$
$$s = \text{distance between car and cruiser at time } t.$$

We assume that x, y, and s are differentiable functions of t.

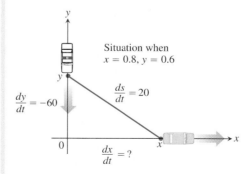

FIGURE 2.44 Figure for Example 4.

Step 2: *Numerical information.* At the instant in question,

$$x = 0.8 \text{ mi}, \qquad y = 0.6 \text{ mi}, \qquad \frac{dy}{dt} = -60 \text{ mph}, \qquad \frac{ds}{dt} = 20 \text{ mph}.$$

Note that dy/dt is negative because y is decreasing.

Step 3: *To find: dx/dt.*

Step 4: *How the variables are related:*

$$s^2 = x^2 + y^2$$

(We could also use $s = \sqrt{x^2 + y^2}$.)

Step 5: *Differentiate with respect to t.*

$$2s \frac{ds}{dt} = 2x \frac{dx}{dt} + 2y \frac{dy}{dt}$$

$$\frac{ds}{dt} = \frac{1}{s} \left(x \frac{dx}{dt} + y \frac{dy}{dt} \right)$$

$$= \frac{1}{\sqrt{x^2 + y^2}} \left(x \frac{dx}{dt} + y \frac{dy}{dt} \right)$$

Step 6: *Evaluate.* Use $x = 0.8$, $y = 0.6$, $dy/dt = -60$, $ds/dt = 20$, and solve for dx/dt.

$$20 = \frac{1}{\sqrt{(0.8)^2 + (0.6)^2}} \left(0.8 \frac{dx}{dt} + (0.6)(-60) \right)$$

$$\frac{dx}{dt} = \frac{20 \sqrt{(0.8)^2 + (0.6)^2} + (0.6)(60)}{0.8} = 70$$

Interpret

At the moment in question, the car's speed is 70 mph.

Example 5   Filling a Conical Tank

Water runs into a conical tank at the rate of 9 ft³/min. The tank stands point down and has a height of 10 ft and a base radius of 5 ft. How fast is the water level rising when the water is 6 ft deep?

Solution   We carry out the steps of the strategy.

Step 1: *Picture and variables.* Figure 2.45 shows a partially filled conical tank. The variables in the problem are

$V =$ volume (ft³) of the water in the tank at time t (min)

$x =$ radius (ft) of the surface of the water at time t

$y =$ depth (ft) of water in tank at time t.

We assume that V, x, and y are differentiable functions of t. The constants are the dimensions of the tank.

Step 2: *Numerical information.* At the time in question,

$$y = 6 \text{ ft} \qquad \text{and} \qquad \frac{dV}{dt} = 9 \text{ ft}^3/\text{min}.$$

Step 3: *To find: dy/dt.*

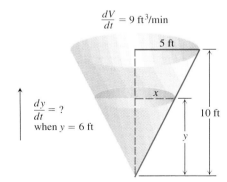

FIGURE 2.45 The conical tank in Example 5.

Step 4: *How the variables are related:* The water forms a cone with volume

$$V = \frac{1}{3}\pi x^2 y.$$

This equation involves x as well as V and y. Because no information is given about x and dx/dt at the time in question, we need to eliminate x. The similar triangles in Figure 2.45 give us a way to express x in terms of y:

$$\frac{x}{y} = \frac{5}{10} \qquad \text{or} \qquad x = \frac{y}{2}.$$

Therefore,

$$V = \frac{1}{3}\pi \left(\frac{y}{2}\right)^2 y = \frac{\pi}{12}y^3.$$

Step 5: *Differentiate with respect to* t.

$$\frac{dV}{dt} = \frac{\pi}{12}\cdot 3y^2\frac{dy}{dt} = \frac{\pi}{4}y^2\frac{dy}{dt}$$

Step 6: *Evaluate.* Use $y = 6$ and $dV/dt = 9$ to solve for dy/dt.

$$9 = \frac{\pi}{4}(6)^2\frac{dy}{dt}$$

$$\frac{dy}{dt} = \frac{1}{\pi} \approx 0.32 \text{ ft/min}$$

Interpret

At the moment in question, the water level is rising at about 0.32 ft/min.

EXERCISES 2.7

1. *Area* Suppose that the radius r and area $A = \pi r^2$ of a circle are differentiable functions of t. Write an equation that relates dA/dt to dr/dt.

2. *Surface area* Suppose that the radius r and surface area $S = 4\pi r^2$ of a sphere are differentiable functions of t. Write an equation that relates dS/dt to dr/dt.

3. *Volume* The radius r and height h of a right circular cylinder are related to the cylinder's volume V by the formula $V = \pi r^2 h$.

   (a) How is dV/dt related to dh/dt if r is constant?

   (b) How is dV/dt related to dr/dt if h is constant?

   (c) How is dV/dt related to dr/dt and dh/dt if neither r nor h is constant?

4. *Volume* The radius r and height h of a right circular cone are related to the cone's volume V by the equation $V = (1/3)\pi r^2 h$.

   (a) How is dV/dt related to dh/dt if r is constant?

   (b) How is dV/dt related to dr/dt if h is constant?

   (c) How is dV/dt related to dr/dt and dh/dt if neither r nor h is constant?

5. *Changing voltage* The voltage V (volts), current I (amperes), and resistance R (ohms) of an electric circuit like the one shown here are related by the equation $V = IR$. Suppose that V is increasing at the rate of 1 volt/sec while I is decreasing at the rate of 1/3 amp/sec. Let t denote time in seconds.

   (a) What is the value of dV/dt?

   (b) What is the value of dI/dt?

   (c) What equation relates dR/dt to dV/dt and dI/dt?

(d) Find the rate at which R is changing when $V = 12$ volts and $I = 2$ amp. Is R increasing, or decreasing?

6. *Electrical power* The power P (watts) of an electric circuit is related to the circuit's resistance R (ohms) and current i (amperes) by the equation $P = Ri^2$.

(a) How are dP/dt, dR/dt, and di/dt related if none of P, R, and i are constant?

(b) How is dR/dt related to di/dt if P is constant?

7. *Distance* Let x and y be differentiable functions of t and let $s = \sqrt{x^2 + y^2}$ be the distance between the points $(x, 0)$ and $(0, y)$ in the xy-plane.

(a) How is ds/dt related to dx/dt if y is constant?

(b) How is ds/dt related to dx/dt and dy/dt if neither x nor y is constant?

(c) How is dx/dt related to dy/dt if s is constant?

8. *Diagonals* If x, y, and z are lengths of the edges of a rectangular box, the common length of the box's diagonals is $s = \sqrt{x^2 + y^2 + z^2}$.

(a) Assuming that x, y, and z are differentiable functions of t, how is ds/dt related to dx/dt, dy/dt, and dz/dt?

(b) How is ds/dt related to dy/dt and dz/dt if x is constant?

(c) How are dx/dt, dy/dt, and dz/dt related if s is constant?

9. *Area* The area A of a triangle with sides of lengths a and b enclosing an angle of measure θ is

$$A = \frac{1}{2} ab \sin \theta.$$

(a) How is dA/dt related to $d\theta/dt$ if a and b are constant?

(b) How is dA/dt related to $d\theta/dt$ and da/dt if only b is constant?

(c) How is dA/dt related to $d\theta/dt$, da/dt, and db/dt if none of a, b, and θ are constant?

10. *Heating a plate* When a circular plate of metal is heated in an oven, its radius increases at the rate of 0.01 cm/min. At what rate is the plate's area increasing when the radius is 50 cm?

11. *Changing dimensions in a rectangle* The length l of a rectangle is decreasing at the rate of 2 cm/sec while the width w is increasing at the rate of 2 cm/sec. When $l = 12$ cm and $w = 5$ cm, find the rates of change of **(a)** the area, **(b)** the perimeter, and **(c)** the lengths of the diagonals of the rectangle. Which of these quantities are decreasing, and which are increasing?

12. *Changing dimensions in a rectangular box* Suppose that the edge lengths x, y, and z of a closed rectangular box are changing at the following rates:

$$\frac{dx}{dt} = 1 \text{ m/sec}, \qquad \frac{dy}{dt} = -2 \text{ m/sec}, \qquad \frac{dz}{dt} = 1 \text{ m/sec}.$$

Find the rates at which the box's **(a)** volume, **(b)** surface area, and **(c)** diagonal length $s = \sqrt{x^2 + y^2 + z^2}$ are changing at the instant when $x = 4$, $y = 3$, and $z = 2$.

13. *A sliding ladder* A 13-ft ladder is leaning against a house when its base starts to slide away. By the time the base is 12 ft from the house, the base is moving at the rate of 5 ft/sec.

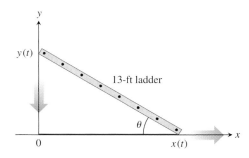

(a) How fast is the top of the ladder sliding down the wall then?

(b) At what rate is the area of the triangle formed by the ladder, wall, and ground changing then?

(c) At what rate is the angle θ between the ladder and the ground changing then?

14. *Commercial air traffic* Two commercial airplanes are flying at 40,000 ft along straight-line courses that intersect at right angles. Plane A is approaching the intersection point at a speed of 442 knots (nautical miles per hour; a nautical mile is 2000 yd). Plane B is approaching the intersection at 481 knots. At what rate is the distance between the planes changing when A is 5 nautical miles from the intersection point and B is 12 nautical miles from the intersection point?

15. *Flying a kite* A girl flies a kite at a height of 300 ft, the wind carrying the kite horizontally away from her at a rate of 25 ft/sec. How fast must she let out the string when the kite is 500 ft away from her?

16. *Boring a cylinder* The mechanics at Lincoln Automotive are reboring a 6-in.-deep cylinder to fit a new piston. The machine they are using increases the cylinder's radius one-thousandth of an inch every 3 min. How rapidly is the cylinder volume increasing when the bore (diameter) is 3.800 in.?

17. *A growing sand pile* Sand falls from a conveyor belt at the rate of 10 m³/min onto the top of a conical pile. The height of the pile is always three-eighths of the base diameter. How fast are the **(a)** height and **(b)** radius changing when the pile is 4 m high? Answer in cm/min.

18. *A draining conical reservoir* Water is flowing at the rate of 50 m³/min from a shallow concrete conical reservoir (vertex down) of base radius 45 m and height 6 m.

(a) How fast (cm/min) is the water level falling when the water is 5 m deep?

(b) How fast is the radius of the water's surface changing then? Answer in centimeters per minute.

19. *A draining hemispherical reservoir* Water is flowing at the rate of 6 m³/min from a reservoir shaped like a hemispherical bowl of radius 13 m, shown here in profile. Answer the following questions, given that the volume of water in a hemispherical bowl of radius R is $V = (\pi/3)y^2(3R - y)$ when the water is y meters deep.

(a) At what rate is the water level changing when the water is 8 m deep?

(b) What is the radius r of the water's surface when the water is y m deep?

(c) At what rate is the radius r changing when the water is 8 m deep?

20. *A growing raindrop* Suppose that a drop of mist is a perfect sphere and that, through condensation, the drop picks up moisture at a rate proportional to its surface area. Show that under these circumstances the drop's radius increases at a constant rate.

21. *The radius of an inflating balloon* A spherical balloon is inflated with helium at the rate of 100π ft³/min. How fast is the balloon's radius increasing at the instant the radius is 5 ft? How fast is the surface area increasing?

22. *Hauling in a dinghy* A dinghy is pulled toward a dock by a rope from the bow through a ring on the dock 6 ft above the bow. The rope is hauled in at the rate of 2 ft/sec.

(a) How fast is the boat approaching the dock when 10 ft of rope are out?

(b) At what rate is the angle θ changing then (see the figure)?

Ring at edge of dock

23. *A balloon and a bicycle* A balloon is rising vertically above a level, straight road at a constant rate of 1 ft/sec. Just when the balloon is 65 ft above the ground, a bicycle moving at a constant rate of 17 ft/sec passes under it. How fast is the distance $s(t)$ between the bicycle and balloon increasing 3 sec later?

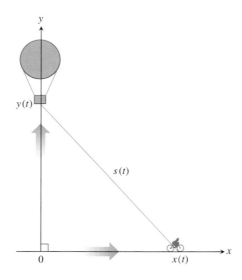

24. *Making coffee* Coffee is draining from a conical filter into a cylindrical coffeepot at the rate of 10 in³/min.

(a) How fast is the level in the pot rising when the coffee in the cone is 5 in. deep?

(b) How fast is the level in the cone falling then?

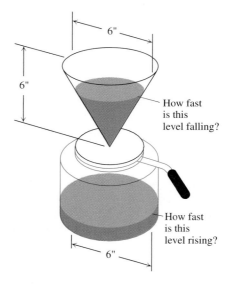

How fast is this level falling?

How fast is this level rising?

25. *Cardiac output* In the late 1860s, Adolf Fick, a professor of physiology in the Faculty of Medicine in Würtzberg, Germany, developed one of the methods we use today for measuring how much blood your heart pumps in a minute. Your cardiac output as you read this sentence is probably about 7 L/min. At rest it is likely to be a bit under 6 L/min. If you are a trained marathon runner running a marathon, your cardiac output can be as high as 30 L/min.

Your cardiac output can be calculated with the formula

$$y = \frac{Q}{D},$$

where Q is the number of milliliters of CO_2 you exhale in a minute and D is the difference between the CO_2 concentration (ml/L) in the blood pumped to the lungs and the CO_2 concentration in the blood returning from the lungs. With $Q = 233$ ml/min and $D = 97 - 56 = 41$ ml/L,

$$y = \frac{233 \text{ ml/min}}{41 \text{ ml/L}} \approx 5.68 \text{ L/min},$$

fairly close to the 6 L/min that most people have at basal (resting) conditions. (Data courtesy of J. Kenneth Herd, M.D., Quillan College of Medicine, East Tennessee State University.)

Suppose that when $Q = 233$ and $D = 41$, we also know that D is decreasing at the rate of 2 units a minute but that Q remains unchanged. What is happening to the cardiac output?

26. *Cost, revenue, and profit* A company can manufacture x items at a cost of $c(x)$ thousand dollars, a sales revenue of $r(x)$ thousand dollars, and a profit of $p(x) = r(x) - c(x)$ thousand dollars. Find dc/dt, dr/dt, and dp/dt for the following values of x and dx/dt.

(a) $r(x) = 9x$, $c(x) = x^3 - 6x^2 + 15x$, and $dx/dt = 0.1$ when $x = 2$

(b) $r(x) = 70x$, $c(x) = x^3 - 6x^2 + 45/x$, and $dx/dt = 0.05$ when $x = 1.5$

27. *Moving along a parabola* A particle moves along the parabola $y = x^2$ in the first quadrant in such a way that its x-coordinate (measured in meters) increases at a steady 10 m/sec. How fast is the angle of inclination θ of the line joining the particle to the origin changing when $x = 3$ m?

28. *Moving along another parabola* A particle moves from right to left along the parabolic curve $y = \sqrt{-x}$ in such a way that its x-coordinate (measured in meters) decreases at the rate of 8 m/sec. How fast is the angle of inclination θ of the line joining the particle to the origin changing when $x = -4$?

29. *Motion in the plane* The coordinates of a particle in the metric xy-plane are differentiable functions of time t with $dx/dt = -1$ m/sec and $dy/dt = -5$ m/sec. How fast is the particle's distance from the origin changing as it passes through the point $(5, 12)$?

30. *A moving shadow* A man 6 ft tall walks at the rate of 5 ft/sec toward a streetlight that is 16 ft above the ground. At what rate is the tip of his shadow moving? At what rate is the length of his shadow changing when he is 10 ft from the base of the light?

31. *Another moving shadow* A light shines from the top of a pole 50 ft high. A ball is dropped from the same height from a point 30 ft away from the light. (See accompanying figure.) How fast is the shadow of the ball moving along the ground 1/2 sec later? (Assume the ball falls a distance $s = 16t^2$ ft in t sec.)

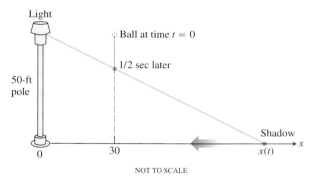

NOT TO SCALE

32. *Videotaping a moving car* You are videotaping a race from a stand 132 ft from the track, following a car that is moving at 180 mi/h (264 ft/sec). How fast will your camera angle θ be changing when the car is right in front of you? A half second later?

33. *A melting ice layer* A spherical iron ball 8 in. in diameter is coated with a layer of ice of uniform thickness. If the ice melts at the rate of 10 in^3/min, how fast is the thickness of the ice decreasing when it is 2 in. thick? How fast is the outer surface area of ice decreasing?

34. *Highway patrol* A highway patrol plane flies 3 mi above a level, straight road at a steady 120 mi/h. The pilot sees an oncoming car and with radar determines that at the instant the line-of-sight distance from plane to car is 5 mi, the line-of-sight distance is decreasing at the rate of 160 mi/h. Find the car's speed along the highway.

35. *A building's shadow* On a morning of a day when the sun will pass directly overhead, the shadow of an 80-ft building on level

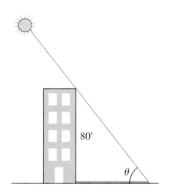

ground is 60 ft long. At the moment in question, the angle θ the sun makes with the ground is increasing at the rate of 0.27°/min. At what rate is the shadow decreasing? (Remember to use radians. Express your answer in inches per minute, to the nearest tenth.)

36. *Walkers* A and B are walking on straight streets that meet at right angles. A approaches the intersection at 2 m/sec ; B moves away from the intersection 1 m/sec . At what rate is the angle θ changing when A is 10 m from the intersection and B is 20 m from the intersection? Express your answer in degrees per second to the nearest degree.

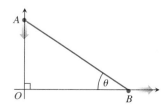

37. *Baseball players* A baseball diamond is a square 90 ft on a side. A player runs from first base to second at a rate of 16 ft/sec .

(a) At what rate is the player's distance from third base changing when the player is 30 ft from first base?

(b) At what rates are angles θ_1 and θ_2 (see the figure) changing at that time?

(c) The player slides into second base at the rate of 15 ft/sec. At what rates are angles θ_1 and θ_2 changing as the player touches base?

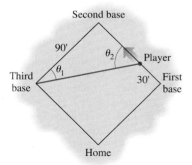

38. *Ships* Two ships are steaming straight away from a point O along routes that make a 120° angle. Ship A moves at 14 knots (nautical miles per hour; a nautical mile is 2000 yd). Ship B moves at 21 knots. How fast are the ships moving apart when $OA = 5$ and $OB = 3$ nautical miles?

2.8   Derivatives of Inverse Trigonometric Functions

Derivatives of Inverse Functions • Derivative of the Arcsine •
Derivative of the Arctangent • Derivative of the Arcsecant •
Derivatives of the Other Three

In this section, we learn when a differentiable function has an inverse function that is also differentiable. We then use that result (Theorem 5) to find formulas for the derivatives of the inverse trigonometric functions reviewed in Preliminary Section 5.

Derivatives of Inverse Functions

In Preliminary Section 4, we learned that the graph of the inverse of a function f can be obtained by reflecting the graph of f across the line $y = x$. If we combine that with our understanding of what makes a function differentiable, we can gain insight into the differentiability of inverse functions.

Example 1   Relating Slopes of Reflected Lines

If we calculate the derivatives of $f(x) = (1/2)x + 1$ and its inverse $f^{-1}(x) = 2x - 2$ from Example 3 in Preliminary Section 4, we see that

$$\frac{d}{dx} f(x) = \frac{d}{dx}\left(\frac{1}{2}x + 1\right) = \frac{1}{2}$$

$$\frac{d}{dx} f^{-1}(x) = \frac{d}{dx}(2x - 2) = 2.$$

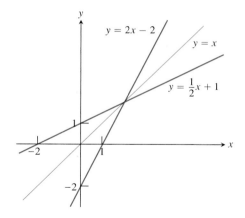

FIGURE 2.46 Graphing $f(x) = (1/2)x + 1$ and $f^{-1}(x) = 2x - 2$ together shows the graphs' symmetry with respect to the line $y = x$. Their slopes are reciprocals of one another.

The derivatives are reciprocals of one another. The graph of f is the line $y = (1/2)x + 1$, and the graph of f^{-1} is the line $y = 2x - 2$ (Figure 2.46). Their slopes are reciprocals of one another.

This case is not special. Reflecting any nonhorizontal or nonvertical line across the line $y = x$ inverts the line's slope. If the original line has slope $m \neq 0$ (Figure 2.47), the reflected line has slope $1/m$ (Exercise 38).

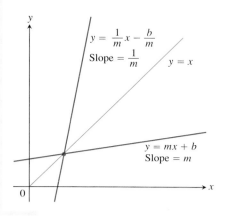

FIGURE 2.47 The slopes of nonvertical lines reflected across the line $y = x$ are reciprocals of one another.

As Figure 2.48 suggests, the reflection of a continuous curve with no cusps or corners will be another continuous curve with no cusps or corners. Indeed, if there is a tangent line to the graph of f at the point $(a, f(a))$, that line will reflect across $y = x$ to become a line tangent to the graph of f^{-1} at the point $(f(a), a)$. And, as we have just seen, the *slope* of the reflected tangent line, when it exists, will be the *reciprocal* of the slope of the original tangent line.

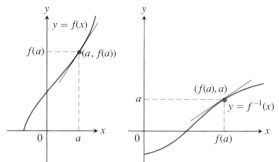

The slopes are reciprocal: $\left. \dfrac{df^{-1}}{dx} \right|_{f(a)} = \dfrac{1}{\left. \dfrac{df}{dx} \right|_a}$.

FIGURE 2.48 The graphs of inverse functions have reciprocal slopes at corresponding points.

All this serves as an introduction to the following theorem, which we will assume as we proceed to find derivatives of inverse functions. Although the essentials of the proof are illustrated in the geometry of Figure 2.48, a careful analytic proof is more appropriate for an advanced calculus text and is omitted here.

> **Theorem 5**   Derivatives of Inverse Functions
>
> If f is differentiable at every point of an interval I and df/dx is never zero on I, then f has an inverse and f^{-1} is differentiable at every point of the interval $f(I)$.

Derivative of the Arcsine

We know that the function $x = \sin y$ is differentiable in the open interval $-\pi/2 < y < \pi/2$ and that its derivative, the cosine, is positive there. Theorem 5 therefore assures us that the inverse function $y = \sin^{-1}(x)$ (the *arcsine* of x) is differentiable throughout the interval $-1 < x < 1$. We cannot expect the inverse to be differentiable at $x = -1$ or $x = 1$, however, because the tangents to the graph are vertical at these points (Figure 2.49).

We find the derivative of $y = \sin^{-1}(x)$ as follows:

$$y = \sin^{-1}x$$

$$\sin y = x \qquad \text{Inverse function relationship}$$

$$\frac{d}{dx}(\sin y) = \frac{d}{dx}x \qquad \text{Differentiate both sides.}$$

$$\cos y \frac{dy}{dx} = 1 \qquad \text{Implicit differentiation.}$$

$$\frac{dy}{dx} = \frac{1}{\cos y}.$$

The division in the last step is safe because $\cos y \neq 0$ for $-\pi/2 < y < \pi/2$. In fact, $\cos y$ is *positive* for $-\pi/2 < y < \pi/2$, so we can replace $\cos y$ with $\sqrt{1 - (\sin y)^2}$, which is $\sqrt{1 - x^2}$. Thus,

$$\frac{d}{dx}(\sin^{-1}x) = \frac{1}{\sqrt{1 - x^2}}.$$

If u is a differentiable function of x with $|u| < 1$, we apply the Chain Rule to get

$$\frac{d}{dx}\sin^{-1}u = \frac{1}{\sqrt{1 - u^2}}\frac{du}{dx}, \qquad |u| < 1.$$

Example 2   Applying the Formula

$$\frac{d}{dx}(\sin^{-1}x^2) = \frac{1}{\sqrt{1 - (x^2)^2}} \cdot \frac{d}{dx}(x^2) = \frac{2x}{\sqrt{1 - x^4}}$$

Derivative of the Arctangent

Although the function $y = \sin^{-1}(x)$ has a rather narrow domain of $[-1, 1]$, the function $y = \tan^{-1}x$ is defined for all real numbers. It is also differentiable for all

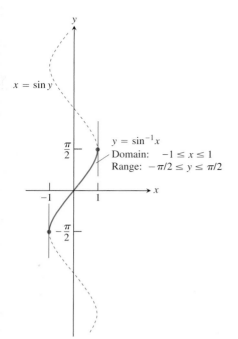

$x = \sin y$

$y = \sin^{-1}x$
Domain: $-1 \le x \le 1$
Range: $-\pi/2 \le y \le \pi/2$

FIGURE 2.49 The graph of $y = \sin^{-1}x$ has vertical tangents $x = -1$ and $x = 1$.

real numbers, as we will now see. The differentiation proceeds exactly as with the arcsine function above.

$$y = \tan^{-1} x$$

$$\tan y = x \qquad \text{Inverse function relationship}$$

$$\frac{d}{dx}(\tan y) = \frac{d}{dx} x$$

$$\sec^2 y \frac{dy}{dx} = 1 \qquad \text{Implicit differentiation}$$

$$\frac{dy}{dx} = \frac{1}{\sec^2 y}$$

$$= \frac{1}{1 + (\tan y)^2} \qquad \text{Trig identity: } \sec^2 y = 1 + \tan^2 y$$

$$= \frac{1}{1 + x^2}$$

The derivative is defined for all real numbers. If u is a differentiable function of x, we get the Chain Rule form:

$$\frac{d}{dx} \tan^{-1} u = \frac{1}{1 + u^2} \frac{du}{dx}.$$

Example 3 Finding the Velocity of a Moving Particle

A particle moves along the x-axis so that its position at any time $t \geq 0$ is $x(t) = \tan^{-1} \sqrt{t}$. What is the velocity of the particle when $t = 16$?

Solution

$$v(t) = \frac{d}{dt} \tan^{-1} \sqrt{t} = \frac{1}{1 + (\sqrt{t})^2} \cdot \frac{d}{dt} \sqrt{t} = \frac{1}{1 + t} \cdot \frac{1}{2\sqrt{t}}$$

When $t = 16$, the velocity is

$$v(16) = \frac{1}{1 + 16} \cdot \frac{1}{2\sqrt{16}} = \frac{1}{136}.$$

Derivative of the Arcsecant

We find the derivative of $y = \sec^{-1} x$, $|x| > 1$, beginning as we did with the other inverse trigonometric functions.

$$y = \sec^{-1} x$$

$$\sec y = x \qquad \text{Inverse function relationship}$$

$$\frac{d}{dx}(\sec y) = \frac{d}{dx} x$$

$$\sec y \tan y \frac{dy}{dx} = 1$$

$$\frac{dy}{dx} = \frac{1}{\sec y \tan y} \qquad \begin{array}{l} \text{Since } |x| > 1, \, y \text{ lies in} \\ (0, \pi/2) \cup (\pi/2, \pi) \text{ and} \\ \sec y \tan y \neq 0. \end{array}$$

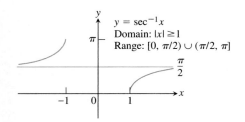

FIGURE 2.50 The slope of the curve $y = \sec^{-1} x$ is positive for both $x < -1$ and $x > 1$.

To express the result in terms of x, we use the relationships

$$\sec y = x \quad \text{and} \quad \tan y = \pm \sqrt{\sec^2 y - 1} = \pm \sqrt{x^2 - 1}$$

to get

$$\frac{dy}{dx} = \pm \frac{1}{x\sqrt{x^2 - 1}}.$$

Can we do anything about the $\pm$ sign? A glance at Figure 2.50 shows that the slope of the graph $y = \sec^{-1} x$ is always positive. That means that

$$\frac{d}{dx} \sec^{-1} x = \begin{cases} + \dfrac{1}{x\sqrt{x^2 - 1}} & \text{if } x > 1 \\[3mm] - \dfrac{1}{x\sqrt{x^2 - 1}} & \text{if } x < -1. \end{cases}$$

With the absolute value symbol we can write

$$\frac{d}{dx} \sec^{-1} x = \frac{1}{|x|\sqrt{x^2 - 1}}.$$

No $\pm$ is required in this new formula. If u is a differentiable function of x with $|u| > 1$, we have

$$\frac{d}{dx} \sec^{-1} u = \frac{1}{|u|\sqrt{u^2 - 1}} \frac{du}{dx}, \qquad |u| > 1$$

Example 4 Using the Formula

$$\frac{d}{dx} \sec^{-1} (5x^4) = \frac{1}{|5x^4|\sqrt{(5x^4)^2 - 1}} \frac{d}{dx}(5x^4)$$

$$= \frac{1}{5x^4\sqrt{25x^8 - 1}} (20x^3)$$

$$= \frac{4}{x\sqrt{25x^8 - 1}}$$

Derivatives of the Other Three

We could use the same technique to find the derivatives of the other three inverse trigonometric functions: arccosine, arccotangent, and arccosecant, but that is unnecessary, thanks to the following identities.

Inverse Function—Inverse Cofunction Identities

$$\cos^{-1} x = \pi/2 - \sin^{-1} x$$
$$\cot^{-1} x = \pi/2 - \tan^{-1} x$$
$$\csc^{-1} x = \pi/2 - \sec^{-1} x$$

It follows from these identities that the derivatives of the inverse cofunctions are the negatives of the derivatives of the corresponding inverse functions (see Exercises 23–25).

Example 5   A Tangent Line to the Arccotangent Curve

Find an equation for the line tangent to the graph of $y = \cot^{-1} x$ at $x = -1$.

Solution   First, we note that

$$\cot^{-1}(-1) = \pi/2 - \tan^{-1}(-1) = \pi/2 - (-\pi/4) = 3\pi/4.$$

The slope of the tangent line is

$$\left.\frac{dy}{dx}\right|_{x=-1} = \left.-\frac{1}{1 + x^2}\right|_{x=-1} = -\frac{1}{1 + (-1)^2} = -\frac{1}{2},$$

so the tangent line has equation $y - 3\pi/4 = (-1/2)(x + 1)$.

EXERCISES 2.8

Finding Derviatives

In Exercises 1–18, find the derivative of y with respect to the appropriate variable.

1. $y = \cos^{-1}(x^2)$

2. $y = \cos^{-1}(1/x)$

3. $y = \sin^{-1}\sqrt{2t}$

4. $y = \sin^{-1}(1 - t)$

5. $y = \sec^{-1}(2s + 1)$

6. $y = \sec^{-1} 5s$

7. $y = \csc^{-1}(x^2 + 1), \quad x > 0$

8. $y = \csc^{-1} x/2$

9. $y = \sec^{-1}\dfrac{1}{t}, \quad 0 < t < 1$

10. $y = \sin^{-1}\dfrac{3}{t^2}$

11. $y = \cot^{-1}\sqrt{t}$

12. $y = \cot^{-1}\sqrt{t - 1}$

13. $y = s\sqrt{1 - s^2} + \cos^{-1} s$

14. $y = \sqrt{s^2 - 1} - \sec^{-1} s$

15. $y = \tan^{-1}\sqrt{x^2 - 1} + \csc^{-1} x, \quad x > 1$

16. $y = \cot^{-1}\dfrac{1}{x} - \tan^{-1} x$

17. $y = x \sin^{-1} x + \sqrt{1 - x^2}$

18. $y = \dfrac{1}{\sin^{-1}(2x)}$

Theory and Examples

19. **(a)** Find an equation for the line tangent to the graph of $y = \tan x$ at the point $(\pi/4, 1)$.

(b) Find an equation for the line tangent to the graph of $y = \tan^{-1} x$ at the point $(1, \pi/4)$.

20. Let $f(x) = x^5 + 2x^3 + x - 1$.

(a) Find $f(1)$ and $f'(1)$.

(b) Find $f^{-1}(3)$ and $(f^{-1})'(3)$.

21. Let $f(x) = \cos x + 3x$.

(a) Show that f has a differentiable inverse.

(b) Find $f(0)$ and $f'(0)$.

(c) Find $f^{-1}(1)$ and $(f^{-1})'(1)$.

22. *Motion of a particle* A particle moves along the x-axis so that its position at any time $t \geq 0$ is $x = \arctan t$.

(a) Prove that the particle is always moving to the right.

(b) Prove that the particle is always decelerating.

(c) What is the limiting position of the particle as t approaches infinity?

In Exercises 23–25, use the inverse function–inverse cofunction identities to derive the formula for the derivative of the function.

23. arccosine

24. arccotangent

25. arccosecant

26. *Identities* Confirm the following identities for $x > 0$.

(a) $\cos^{-1} x + \sin^{-1} x = \pi/2$

(b) $\tan^{-1} x + \cot^{-1} x = \pi/2$

(c) $\sec^{-1} x + \csc^{-1} x = \pi/2$

End Behavior Models

In Exercises 27–30, find **(a)** a right end behavior model, **(b)** a left end behavior model, and **(c)** any horizontal tangents for the function if they exist.

27. $y = \tan^{-1} x$

28. $y = \cot^{-1} x$

29. $y = \sec^{-1} x$

30. $y = \csc^{-1} x$

Derivatives of Inverse Functions

In Exercises 31–34:

(a) Find $f^{-1}(x)$.

(b) Graph f and f^{-1} together.

(c) Evaluate df/dx at $x = a$ and df^{-1}/dx at $x = f(a)$ to show that at these points $df^{-1}/dx = 1/(df/dx)$.

31. $f(x) = 2x + 3, \quad a = -1$

32. $f(x) = (1/5)x + 7, \quad a = -1$

33. $f(x) = 5 - 4x$, $a = 1/2$

34. $f(x) = 2x^2$, $x \geq 0$, $a = 5$

35. (a) Show that $f(x) = x^3$ and $g(x) = \sqrt[3]{x}$ are inverses of one another.

$\boxed{\text{T}}$ **(b)** Graph f and g over an x-interval large enough to show the graphs intersecting at $(1, 1)$ and $(-1, -1)$. Be sure the picture shows the required symmetry in the line $y = x$.

(c) Find the slopes of the tangents to the graphs of f and g at $(1, 1)$ and $(-1, -1)$ (four tangents in all).

(d) What lines are tangent to the curves at the origin?

36. (a) Show that $h(x) = x^3/4$ and $k(x) = (4x)^{1/3}$ are inverses of one another.

$\boxed{\text{T}}$ **(b)** Graph h and k over an x-interval large enough to show the graphs intersecting at $(2, 2)$ and $(-2, -2)$. Be sure the picture shows the required symmetry about the line $y = x$.

(c) Find the slopes of the tangents to the graphs at h and k at $(2, 2)$ and $(-2, -2)$.

(d) What lines are tangent to the curves at the origin?

37. (a) Find the inverse of the function $f(x) = mx$, where m is a constant different from zero.

(b) What can you conclude about the inverse of a function $y = f(x)$ whose graph is a line through the origin with a nonzero slope m?

38. Show that the graph of the inverse of $f(x) = mx + b$, where m and b are constants and $m \neq 0$, is a line with slope $1/m$ and y-intercept $-b/m$.

COMPUTER EXPLORATIONS

Inverses and Derivatives

In Exercises 39–46, you will explore some functions and their inverses together with their derivatives and tangent lines at specified points. Perform the following steps using a CAS:

(a) Plot the function $y = f(x)$ together with its derivative over the given interval. Explain why you know that f is one-to-one over the interval.

(b) Solve the equation $y = f(x)$ for x as a function of y and name the resulting inverse function g.

(c) Find the equation for the line tangent to the graph of f at the specified point $(x_0, f(x_0))$.

(d) If we write $g(x)$ for the inverse $f^{-1}(x)$, the derivative of the inverse can be written as

$$g'(f(a)) = \frac{1}{f'(a)}, \quad \text{or} \quad g'(f(a)) \cdot f'(a) = 1.$$

Find the equation for the line tangent to the graph of g at the point $(f(x_0), x_0)$ located symmetrically across the $45°$ line $y = x$ (which is the graph of the identity function). Use the formula above to find the slope of this tangent line.

(e) Plot the functions f and g, the identity, the two tangent lines, and the line segment joining the points $(x_0, f(x_0))$ and $(f(x_0), x_0)$. Discuss the symmetries you see across the main diagonal.

39. $y = \sqrt{3x - 2}$, $2/3 \leq x \leq 4$, $x_0 = 3$

40. $y = \dfrac{3x + 2}{2x - 11}$, $-2 \leq x \leq 2$, $x_0 = 1/2$

41. $y = \dfrac{4x}{x^2 + 1}$, $-1 \leq x \leq 1$, $x_0 = 1/2$

42. $y = \dfrac{x^3}{x^2 + 1}$, $-1 \leq x \leq 1$, $x_0 = 1/2$

43. $y = x^3 - 3x^2 - 1$, $2 \leq x \leq 5$, $x_0 = 27/10$

44. $y = 2 - x - x^3$, $-2 \leq x \leq 2$, $x_0 = 3/2$

45. $y = e^x$, $-3 \leq x \leq 5$, $x_0 = 1$

46. $y = \sin x$, $-\pi/2 \leq x \leq \pi/2$, $x_0 = 1$

In Exercises 47 and 48, repeat the steps above to solve for the functions $y = f(x)$ and $x = f^{-1}(y)$ defined implicitly by the given equations over the interval.

47. $y^{1/3} - 1 = (x + 2)^3$, $-5 \leq x \leq 5$, $x_0 = -3/2$

48. $\cos y = x^{1/5}$, $0 \leq x \leq 1$, $x_0 = 1/2$

2.9 Derivatives of Exponential and Logarithmic Functions

Derivative of e^x • Exponential Change • Derivative of a^x •

Geometric Significance of $\lim\limits_{h \to 0} \dfrac{a^h - 1}{h}$ • Derivative of $\ln x$ •

Derivative of $\log_a x$ • Power Rule for Arbitrary Real Powers

The most important function–inverse pair in mathematics and science is the pair consisting of the natural logarithm function $\ln x$ and the exponential function e^x. In this section, we learn how to differentiate these functions. We also begin to explore the properties that account for the amazing frequency with which the exponential function appears in science and engineering applications. We look at many of these applications in later chapters.

Derivative of e^x

At the end of the brief review of exponential functions in Preliminary Section 3, we mentioned that the function $y = e^{kx}$ was a particularly important function for modeling exponential growth. The number e was defined in that section to be the limit of $(1 + 1/x)^x$ as $x \to \infty$. This intriguing number shows up in many interesting applications, because the exponential function $y = e^x$ has the property that it is its own derivative. Let's see why this is true.

When we apply the definition of the derivative to $f(x) = a^x$, we see that the derivative is a constant multiple of a^x itself:

$$\frac{d}{dx}(a^x) = \lim_{h \to 0} \frac{a^{x+h} - a^x}{h} \qquad \text{Derivative definition}$$

$$= \lim_{h \to 0} \frac{a^x \cdot a^h - a^x}{h} \qquad a^{x+h} = a^x \cdot a^h$$

$$= \lim_{h \to 0} a^x \cdot \frac{a^h - 1}{h} \qquad \text{Factoring out } a^x$$

$$= a^x \cdot \lim_{h \to 0} \frac{a^h - 1}{h} \qquad a^x \text{ is constant as } h \to 0.$$

$$= \underbrace{\left(\lim_{h \to 0} \frac{a^h - 1}{h} \right)}_{\text{a fixed number } L} \cdot a^x. \qquad (1)$$

It can be proved that the limit L exists. To estimate its value, we graph $y = (a^h - 1)/h$ as a function of h to see where the graph approaches the y-axis as $h \to 0$. Figure 2.51 shows the graphs for $a = 2$, 2.5, and 3 and for one other value we will discuss in a moment. The value of L is approximately 0.69 if $a = 2$, 0.92 if $a = 2.5$, and 1.10 if $a = 3$. Accordingly,

$$\frac{d}{dx}(2^x) \approx (0.69)2^x$$

$$\frac{d}{dx}(2.5^x) \approx (0.92)2.5^x$$

$$\frac{d}{dx}(3^x) \approx (1.10)3^x.$$

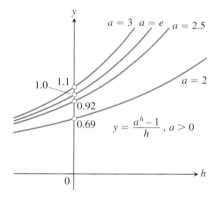

FIGURE 2.51 The position of the curve $y = (a^h - 1)/h$, $a > 0$, varies continuously with a.

The good news is that we now have some idea of the values of these derivatives. The bad news is that the multipliers accumulate if we continue to differentiate. The calculations get messy. For instance,

$$\frac{d}{dx}(2^x) \approx (0.69)2^x$$

$$\frac{d^2}{dx^2}(2^x) \approx (0.69)\frac{d}{dx}(2^x) \approx (0.69)^2 \cdot 2^x$$

$$\vdots$$

$$\frac{d^n}{dx^n}(2^x) \approx (0.69)^n \cdot 2^x.$$

Another look at Figure 2.51 suggests a way out. For $a = 2.5$, the graph of $y = (a^h - 1)/h$ steps across the y-axis at about 0.92. For $a = 3$, the graph steps across at about 1.1. Thus, at some value of a between 2.5 and 3, the graph should step across the axis at $y = 1$. The particular value of a for which this is true is the number e. That is, the number e gives the limit

$$\lim_{h \to 0} \frac{e^h - 1}{h} = 1. \tag{2}$$

That the limit is 1 creates a remarkable relationship between the function e^x and its derivative:

$$\frac{d}{dx}(e^x) = \lim_{h \to 0} \left(\frac{e^h - 1}{h} \right) \cdot e^x \qquad \text{Eq. (1) with } a = e$$

$$= 1 \cdot e^x \qquad \text{Eq. (2)}$$

$$= e^x.$$

In other words, the derivative of this particular function is itself!

$$\frac{d}{dx}(e^x) = e^x$$

If u is a differentiable function of x, then we have

$$\frac{d}{dx} e^u = e^u \frac{du}{dx}. \tag{3}$$

Is any other function its own derivative?

The zero function is also its own derivative, but this hardly seems worth mentioning. (Its value is always 0 and its slope is always 0.) In addition to e^x, however, we can also see that any constant *multiple* of e^x is its own derivative:

$$\frac{d}{dx}(c \cdot e^x) = c \cdot e^x.$$

The next obvious question is whether there are still *other* functions that are their own derivatives, and this time the answer is no. The only functions that satisfy the condition $dy/dx = y$ are functions of the form $y = ke^x$ (and notice that the zero function can be included in this category). We prove this significant fact in Chapter 4.

We make extensive use of this formula when we study exponential growth and decay in Chapter 4.

Example 1   Differentiating Exponentials

(a) $\dfrac{d}{dx}(e^{5x}) = e^{5x} \cdot \dfrac{d}{dx}(5x) = 5e^{5x}$ Eq. (3) with $u = 5x$

(b) $\dfrac{d}{dx}(e^{kx}) = e^{kx} \cdot \dfrac{d}{dx}(kx) = ke^{kx}$ Eq. (3) with $u = kx$, any number k

(c) $\dfrac{d}{dx}(e^{-x}) = -e^{-x}$ Eq. (3) with $u = -1 \cdot x$

(d) $\dfrac{d}{dx}(e^{x^2}) = e^{x^2} \cdot \dfrac{d}{dx}(x^2) = 2xe^{x^2}$ Eq. (3) with $u = x^2$

(e) $\dfrac{d}{dx} e^{\sin x} = e^{\sin x} \cdot \dfrac{d}{dx}(\sin x) = e^{\sin x} \cdot \cos x$ Eq. (3) with $u = \sin x$

Exponential Change

In many instances in science, some positive quantity increases or decreases at a rate that at any given time t is proportional to the amount that is present at time t. In Section 5.4, we show that in these instances the amount can be represented by an equation of the form $y = y_0 e^{kt}$, where y_0 is the amount initially present at time $t = 0$.

The equation

$$y = y_0 e^{kt} \qquad (4)$$

is called the **law of exponential change.**

Example 2   Forecasting the Incidence of a Disease

One model for the way diseases spread assumes that the rate at which the number y of infected people changes is proportional to y itself. The more infected people there are, the faster the disease will spread. The fewer there are, the slower it will spread. If y_0 is the number of infected people at time $t = 0$, then the number of infected people at any time in the near future will be about

$$y = y_0 e^{kt}.$$

Let's use this equation to answer questions about the course of a particular disease. To be specific, suppose that a worldwide eradication program is reducing the number y of cases at the rate of 20% a year. There are 10,000 recorded cases today, and we want to know how many years it will take to lower the number to 1000.

Solution

Starting with the equation $y = y_0 e^{kt}$, there are three things to find:

1. the value of y_0

2. the value of k

3. the value of t that makes $y = 1000$.

The value of y_0

If we start counting time from today, then $y = 10,000$ when $t = 0$, so $y_0 = 10,000$. Our equation is

$$y = 10,000\, e^{kt}. \qquad (5)$$

The value of k

The available information tells us that when $t = 1$ (i.e., when one year has passed), the number of cases will be 80% of its present value, or 8000. Hence,

$$10,000\, e^{k(1)} = 8000 \qquad \text{Eq. (5) with } t = 1, y = 8000$$

$$e^k = 0.8$$

$$\ln e^k = \ln 0.8 \qquad \text{Logarithm of both sides}$$

$$k \approx -0.22 \qquad \text{Calculator result}$$

The value of t that makes $y = 1000$

We find t by solving the following equation for t:

$$10{,}000\, e^{-0.22t} = 1000 \qquad \text{Eq. (5) with } k = -0.22,\, y = 1000$$

$$e^{-0.22t} = 0.1$$

$$\ln e^{-0.22t} = \ln 0.1 \qquad \text{Logarithm of both sides}$$

$$-0.22t \approx -2.3$$

$$t \approx 10.5.$$

The number of cases will drop from the original 10,000 to 1000 in about 10.5 years.

Derivative of a^x

What about an exponential function with a base other than e? We will assume that the base is positive and different from 1, since negative numbers to arbitrary real powers are not always real numbers, and $y = 1^x$ is a constant function.

If $a > 0$ and $a \neq 1$, we can write a^x in terms of e^x. The formula for doing so is

$$a^x = e^{x \ln a},$$

reviewed in Preliminary Section 4. We can then find the derivative of a^x with the Chain Rule:

$$\frac{d}{dx}\, a^x = \frac{d}{dx}\, e^{x \ln a} = e^{x \ln a} \cdot \frac{d}{dx}\,(x \ln a) = e^{x \ln a} \cdot \ln a = a^x \ln a.$$

Thus, if u is a differentiable function of x, we get the following rule.

For $a > 0$ and $a \neq 1$,

$$\frac{d}{dx}\,(a^u) = a^u \ln a\, \frac{du}{dx}. \tag{6}$$

Geometric Significance of $\displaystyle\lim_{h \to 0} \frac{a^h - 1}{h}$

You may have noticed that the graph of each of the functions $f(x) = a^x$ passes through the point $(0, 1)$ with a different slope. What is this slope? It is none other than

$$f'(0) = \lim_{h \to 0} \frac{a^{0+h} - a^0}{h} = \lim_{h \to 0} \frac{a^h - 1}{h}. \tag{7}$$

Since $f'(x) = a^x \ln a$, we see that $f'(0) = a^0 \ln a = \ln a$. Equation (7) then gives

$$\lim_{h \to 0} \frac{a^h - 1}{h} = \ln a \tag{8}$$

Example 3 Finding a Point on a Curve with a Specified Slope

At what point on the graph of the function $y = 2^t - 3$ does the tangent line have slope 21?

Solution The slope is the derivative:

$$\frac{d}{dt}(2^t - 3) = 2^t \cdot \ln 2 - 0 = 2^t \ln 2.$$

We want the value of t for which $2^t \ln 2 = 21$:

$$2^t \ln 2 = 21$$

$$2^t = \frac{21}{\ln 2}$$

$$\ln 2^t = \ln\left(\frac{21}{\ln 2}\right) \qquad \text{Logarithm of both sides}$$

$$t \cdot \ln 2 = \ln 21 - \ln(\ln 2) \qquad \text{Properties of logarithms}$$

$$t = \frac{\ln 21 - \ln(\ln 2)}{\ln 2}$$

$$t \approx 4.921 \qquad \text{Calculator result}$$

$$y = 2^t - 3 \approx 27.297. \qquad \text{Using the value of } t$$

The point is approximately (4.9, 27.3).

Derivative of ln x

Now that we know the derivative of e^x, it is relatively easy to find the derivative of its inverse function, $\ln x$.

$$y = \ln x$$

$$e^y = x \qquad \text{Inverse function relationship}$$

$$\frac{d}{dx}(e^y) = \frac{d}{dx}(x) \qquad \text{Differentiate implicitly.}$$

$$e^y \frac{dy}{dx} = 1$$

$$\frac{dy}{dx} = \frac{1}{e^y} = \frac{1}{x}$$

This equation answers what was once a perplexing problem: Is there a function with derivative x^{-1}? All the other power functions follow the Power Rule,

$$\frac{d}{dx}x^n = nx^{n-1}.$$

This formula, however, is not much help if one is looking for a function with x^{-1} as its derivative! Now we know why: the function we should be looking for is not a power function at all; it is the natural logarithm function.

If u is a differentiable function of x and $u > 0$,

$$\frac{d}{dx}\ln u = \frac{1}{u}\frac{du}{dx}. \qquad (9)$$

Example 4 Finding a Tangent Line Through the Origin

A line with slope m passes through the origin and is tangent to the graph of $y = \ln x$. What is the value of m?

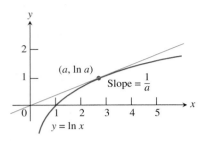

FIGURE 2.52 The tangent line intersects the curve at some point $(a, \ln a)$, where the slope of the curve is $1/a$. (Example 4)

Solution   This problem is a little harder than it looks, since we do not know the point of tangency. We do, however, know two important facts about that point:

1. It has coordinates $(a, \ln a)$ for some positive a, and

2. The tangent line there has slope $m = 1/a$ (Figure 2.52).

Since the tangent line passes through the origin, its slope is

$$m = \frac{\ln a - 0}{a - 0} = \frac{\ln a}{a}.$$

Setting these two formulas for m equal to each other, we have

$$\frac{\ln a}{a} = \frac{1}{a}$$
$$\ln a = 1$$
$$e^{\ln a} = e^1$$
$$a = e$$
$$m = \frac{1}{e}.$$

Derivative of $\log_a x$

To find the derivative of $\log_a x$ for an arbitrary base $(a > 0, a \neq 1)$, we use the change-of-base formula for logarithms (reviewed in Preliminary Section 4) to express $\log_a x$ in terms of natural logarithms, as follows:

$$\log_a x = \frac{\ln x}{\ln a}.$$

The rest is easy:

$$\begin{aligned}
\frac{d}{dx} \log_a x &= \frac{d}{dx}\left(\frac{\ln x}{\ln a}\right) \\
&= \frac{1}{\ln a} \cdot \frac{d}{dx} \ln x \qquad \text{Since } \ln a \text{ is a constant} \\
&= \frac{1}{\ln a} \cdot \frac{1}{x} \\
&= \frac{1}{x \ln a}.
\end{aligned}$$

So, if u is a differentiable function of x and $u > 0$, the formula is as follows.

For $a > 0$ and $a \neq 1$,

$$\frac{d}{dx} \log_a u = \frac{1}{u \ln a}\frac{du}{dx}. \tag{10}$$

Example 5   Going the Long Way with the Chain Rule

Find dy/dx if $y = \log_a a^{\sin x}$.

Solution   Carefully working from the outside in, we apply the Chain Rule to get:

$$\frac{d}{dx}(\log_a a^{\sin x}) = \frac{1}{a^{\sin x}\ln a} \cdot \frac{d}{dx}(a^{\sin x}) \qquad \log_a u, \quad u = a^{\sin x}$$

$$= \frac{1}{a^{\sin x}\ln a} \cdot a^{\sin x}\ln a \cdot \frac{d}{dx}(\sin x) \qquad a^u, \quad u = \sin x$$

$$= \frac{a^{\sin x}\ln a}{a^{\sin x}\ln a} \cdot \cos x$$

$$= \cos x.$$

We could have saved ourselves a lot of work in Example 5 if we had noticed at the beginning that $\log_a a^{\sin x}$, being the composite of inverse functions, is equal to $\sin x$. It is always a good idea to simplify functions *before* differentiating, wherever possible. On the other hand, it is comforting to know that all these rules do work if applied correctly.

Power Rule for Arbitrary Real Powers

We are now ready to prove the Power Rule in its final form. As long as $x > 0$, we can write any real power of x as a power of e, specifically

$$x^n = e^{n\ln x}.$$

Doing so enables us to differentiate x^n for any real power n, as follows:

$$\frac{d}{dx}(x^n) = \frac{d}{dx}(e^{n\ln x})$$

$$= e^{n\ln x} \cdot \frac{d}{dx}(n\ln x) \qquad e^u, \quad u = n\ln x$$

$$= e^{n\ln x} \cdot \frac{n}{x}$$

$$= x^n \cdot \frac{n}{x}$$

$$= nx^{n-1}.$$

The Chain Rule extends this result to the Power Rule's final form.

Rule 8   Power Rule for Arbitrary Real Powers

If u is a positive differentiable function of x and n is any real number, then u^n is a differentiable function of x, and

$$\frac{d}{dx}u^n = nu^{n-1}\frac{du}{dx}. \tag{11}$$

Example 6   Using the Power Rule in All Its Power

(a) If $y = x^{\sqrt{2}}$, then

$$\frac{dy}{dx} = \sqrt{2}x^{(\sqrt{2}-1)}.$$

(b) If $y = (2 + \sin 3x)^{\pi}$, then

$$\frac{d}{dx}(2 + \sin 3x)^{\pi} = \pi(2 + \sin 3x)^{\pi-1}(\cos 3x) \cdot 3$$

$$= 3\pi(2 + \sin 3x)^{\pi-1}(\cos 3x).$$

Example 7   Finding a Derivative's Domain

If $f(x) = \ln(x - 3)$, find $f'(x)$. Identify the domain of f'.

Solution   The domain of f is $(3, \infty)$, and

$$f'(x) = \frac{1}{x - 3}.$$

The domain of f' appears to be all $x \neq 3$. Since f is not defined for $x < 3$, however, neither is f'. Thus,

$$f'(x) = \frac{1}{x - 3}, \qquad x > 3.$$

That is, the domain of f' is $(3, \infty)$.

Sometimes the properties of logarithms can be used to simplify the differentiation process, even if we must introduce the logarithms ourselves as a step in the process. Example 8 shows a clever way to differentiate $y = x^x$ for $x > 0$.

Example 8   Logarithmic Differentiation

Find dy/dx for $y = x^x$, $x > 0$.

Solution

$$y = x^x$$

$$\ln y = \ln x^x \qquad \text{Logarithm of both sides}$$

$$\ln y = x \ln x \qquad \text{Property of logs}$$

$$\frac{d}{dx}(\ln y) = \frac{d}{dx}(x \ln x) \qquad \text{Differentiate implicitly}$$

$$\frac{1}{y}\frac{dy}{dx} = 1 \cdot \ln x + x \cdot \frac{1}{x}$$

$$\frac{dy}{dx} = y(\ln x + 1)$$

$$\frac{dy}{dx} = x^x(\ln x + 1)$$

EXERCISES 2.9

Differentiating Exponentials

In Exercises 1–20, find dy/dx.

1. $y = 2e^x$

2. $y = e^{x+\sqrt{2}}$

3. $y = e^{-3x/2}$

4. $y = e^{-5x}$

5. $y = e^{2x/3}$

6. $y = e^{-x/4}$

7. $y = xe^x - e^x$

8. $y = x^2 e^x - xe^x$

9. $y = e^{\sqrt{x}}$

10. $y = e^{(x^3)}$

11. $y = x^\pi$

12. $y = x^{1+\sqrt{2}}$

13. $y = x^{-\sqrt{2}}$

14. $y = x^{1-e}$

15. $y = 8^x$

16. $y = 9^{-x}$

17. $y = 3^{\csc x}$

18. $y = 3^{\cot x}$

19. $y = \dfrac{e^x}{e^{-x} + 1}$

20. $y = \dfrac{e^{-x}}{e^x + 1}$

Differentiating Logarithms

In Exercises 21–40, find dy/dx.

21. $y = \ln (x^2)$

22. $y = (\ln x)^2$

23. $y = \ln (1/x)$

24. $y = \ln (10/x)$

25. $y = \ln (x + 2)$

26. $y = \ln (2x + 2)$

27. $y = \ln (2 - \cos x)$

28. $y = \ln (x^2 + 1)$

29. $y = \ln (\ln x)$

30. $y = x \ln x - x$

31. $y = \log_4 x^2$

32. $y = \log_5 \sqrt{x}$

33. $y = \log_2 (3x + 1)$

34. $y = \log_{10} \sqrt{x + 1}$

35. $y = \log_2 (1/x)$

36. $y = 1/\log_2 x$

37. $y = \ln 2 \cdot \log_2 x$

38. $y = \log_3 (1 + x \ln 3)$

39. $y = \log_{10} e^x$

40. $y = \ln 10^x$

Logarithmic Differentiation

In Exercises 41–46, use the technique of logarithmic differentiation to find dy/dx.

41. $y = x^{\ln x}$

42. $y = x^{(1/\ln x)}$

43. $y = (\sin x)^x, \quad 0 < x < \pi/2$

44. $y = x^{\tan x}, \quad x > 0$

45. $y = \sqrt[5]{\dfrac{(x-3)^4(x^2+1)}{(2x+5)^3}}$

46. $y = \dfrac{x\sqrt{x^2+1}}{(x+1)^{2/3}}$

Theory and Applications

47. *Tangent line* Find an equation for a line that is tangent to the graph of $y = e^x$ and goes through the origin.

48. *Normal line* Find an equation for a line that is normal to the graph of $y = xe^x$ and goes through the origin.

49. *Radioactive decay* The amount A (in grams) of radioactive plutonium remaining in a 20-g sample after t days is given by the formula

$$A = 20 \cdot (1/2)^{t/140}.$$

At what rate is the plutonium decaying when $t = 2$ days? Answer in appropriate units.

50. *Human evolution continues* The analysis of tooth shrinkage by C. Loring Brace and colleagues at the University of Michigan's Museum of Anthropology indicates that human tooth size is continuing to decrease and that the evolutionary process did not come to a halt some 30,000 years ago as many scientists contend. In northern Europeans, for example, tooth size reduction now has a rate of 1% per 1000 years.

(a) If t represents time in years and y represents tooth size, use the condition that $y = 0.99y_0$ when $t = 1000$ to find the value of k in the equation $y = y_0 e^{kt}$. Then use this value of k to answer the following questions.

(b) In about how many years will human teeth be 90% of their present size?

(c) What will be our descendants' tooth size 20,000 years from now (as a percentage of our present tooth size)?

(*Source: LSA Magazine*, Vol. 12, No. 2 (Spring 1989), p. 19.)

51. *The U.S. population* The Museum of Science in Boston displays a running total of the U.S. population. On May 11, 1993, the total was increasing at the rate of 1 person every 14 sec. The displayed population figure for 3:45 P.M. that day was 257,313,431.

(a) Assuming exponential growth at a constant rate, find the rate constant for the population's growth (people per 365-day year).

(b) At this rate, what will the U.S. population be at 3:45 P.M. Boston time on May 11, 2001?

52. *The mean life of a radioactive nucleus* Physicists using the radioactivity equation $y = y_0 e^{-kt}$ call the number $1/k$ the *mean life* of a radioactive nucleus. The mean life of a radon nucleus is about $1/0.18 \approx 5.6$ days. The mean life of a carbon-14 nucleus is more than 8000 years. Show that 95% of the radioactive nuclei originally present in a sample will disintegrate within three mean lifetimes, that is, by time $t = 3/k$. Thus, the mean life of a nucleus gives a quick way to estimate how long the radioactivity of a sample will last.

53. *Writing to Learn: The derivative of* $g(x) = (0.5)^x$

(a) How are the derivative of $g(x) = (0.5)^x$ and the number

$$L = \lim_{h \to 0} \frac{(0.5)^h - 1}{h}$$

related? Give reasons for your answer.

(b) What is the approximate value of L?

54. *Orthogonal families of curves* Prove that all curves in the family

$$y = -\frac{1}{2}x^2 + k$$

(k any constant) are perpendicular to all curves in the family $y = \ln x + c$ (c any constant) at their points of intersection. (See the accompanying figure.)

[−3, 6] by [−3, 3]

T 55. *Writing to Learn: Why the graph of $y = \ln x$ has no horizontal asymptote* The graph of $y = \ln x$ looks as though it might be approaching a horizontal asymptote. Try it. Write an argument based on the graph of $y = e^x$ to explain why it does not.

[−3, 6] by [−3, 3]

T 56. *Solving $x^2 = 2^x$* How many solutions does the equation $x^2 = 2^x$ have? What are they?

57. *Which is bigger, π^e or e^π?* Calculators have taken some of the mystery out of this once-challenging question. (Go ahead and check; you will see that it is a surprisingly close call.) You can answer the question without a calculator, though, by using the result from Example 4 of this section.

Recall from that example that the line through the origin tangent to the graph of $y = \ln x$ has slope $1/e$.

[−3, 6] by [−3, 3]

(a) Find an equation for this tangent line.

(b) Give an argument based on the graphs of $y = \ln x$ and the tangent line to explain why $\ln x < x/e$ for all positive $x \neq e$.

(c) Show that $\ln (x^e) < x$ for all positive $x \neq e$.

(d) Conclude that $x^e < e^x$ for all positive $x \neq e$.

(e) So which is bigger, π^e or e^π?

T 58. *$x^{\ln 2}$ versus $2^{\ln x}$* Could $x^{\ln 2}$ possibly be the same as $2^{\ln x}$ for $x > 0$? Graph the two functions and explain what you see.

Questions To Guide Your Review

1. What is the derivative of a function f? How is its domain related to the domain of f? Give examples.

2. What role does the derivative play in defining slopes, tangents, and rates of change?

3. How can you sometimes graph the derivative of a function when all you have is a table of the function's values?

4. What does it mean for a function to be differentiable on an open interval? On a closed interval?

5. How are derivatives and one-sided derivatives related?

6. Describe geometrically when a function typically does *not* have a derivative at a point.

7. How is a function's differentiability at a point related to its continuity there if at all?

8. Could the unit step function

$$U(x) = \begin{cases} 0, & x < 0 \\ 1, & x \geq 0 \end{cases}$$

possibly be the derivative of some other function on $[-1, 1]$? Explain.

9. What rules do you know for calculating derivatives? Give some examples.

10. Explain how the three formulas

(a) $\dfrac{d}{dx}(x^n) = nx^{n-1}$

(b) $\dfrac{d}{dx}(cu) = c\dfrac{du}{dx}$

(c) $\dfrac{d}{dx}(u_1 + u_2 + \cdots + u_n) = \dfrac{du_1}{dx} + \dfrac{du_2}{dx} + \cdots + \dfrac{du_n}{dx}$

enable us to differentiate any polynomial.

11. What formula do we need, in addition to the three listed in Question 10, to differentiate rational functions?

12. What is a second derivative? A third derivative? How many derivatives do the functions you know have? Give examples.

13. What is the relationship between a function's average and instantaneous rates of change? Give an example.

14. How do derivatives arise in the study of motion? What can you learn about a body's motion along a line by examining the derivatives of the body's position function? Give examples.

15. How can derivatives arise in economics?

16. Give examples of still other applications of derivatives.

17. What do the limits $\lim_{h\to 0} (\sin h)/h$ and $\lim_{h\to 0} ((\cos h - 1)/h)$ have to do with the derivatives of the sine and cosine functions? What *are* the derivatives of these functions?

18. Once you know the derivatives of $\sin x$ and $\cos x$, how can you find the derivatives of $\tan x$, $\cot x$, $\sec x$, and $\csc x$? What *are* the derivatives of these functions?

19. At what points are the six basic trigonometric functions continuous? How do you know?

20. What is the rule for calculating the derivative of a composite of two differentiable functions? How is such a derivative evaluated? Give examples.

21. What is the formula for the slope dy/dx of a parametrized curve $x = f(t)$, $y = g(t)$? When does the formula apply? When can you expect to be able to find d^2y/dx^2 as well? Give examples.

22. What is implicit differentiation? When do you need it? Give examples.

23. How do related rate problems arise? Give examples.

24. Outline a strategy for solving related rate problems. Illustrate with an example.

25. What are the derivatives of the inverse trigonometric functions? How do the domains of the derivatives compare with the domains of the functions?

26. What are the derivatives of the exponential function e^x and natural logarithm function $\ln x$? How do the domains of the derivatives compare with the domains of the functions?

27. What is the law of exponential change? What are some applications of the law?

28. What is the derivative of the exponential function a^x, $a > 0$ and $a \neq 1$? What is the geometric significance of the limit of $(a^h - 1)/h$ as $h \to 0$? What is the limit when a is the number e?

29. What is the derivative of $\log_a x$? Are there any restrictions on a?

30. What is logarithmic differentiation? Give an example.

31. How can you write any real power of x as a power of e? Are there any restrictions on x? How does this lead to the power rule for differentiating arbitrary real powers?

Practice Exercises

Derivatives of Functions

Find the derivatives of the functions in Exercises 1–42.

1. $y = x^5 - 0.125x^2 + 0.25x$

2. $y = x^3 - 3(x^2 + \pi^2)$

3. $y = x^7 + \sqrt{7}x - \dfrac{1}{\pi + 1}$

4. $y = (2x - 5)(4 - x)^{-1}$

5. $y = (\theta^2 + \sec \theta + 1)^3$

6. $s = \dfrac{\sqrt{t}}{1 + \sqrt{t}}$

7. $s = \dfrac{1}{\sqrt{t} - 1}$

8. $y = 2 \tan^2 x - \sec^2 x$

9. $y = \dfrac{1}{\sin^2 x} - \dfrac{2}{\sin x}$

10. $s = \cos^4 (1 - 2t)$

11. $s = \cot^3 \left(\dfrac{2}{t}\right)$

12. $s = (\sec t + \tan t)^5$

13. $r = \sqrt{2\theta \sin \theta}$

14. $r = \sin (\theta + \sqrt{\theta + 1})$

15. $y = \dfrac{1}{2}x^2 \csc \dfrac{2}{x}$

16. $y = x^{-1/2} \sec (2x)^2$

17. $y = 5 \cot x^2$

18. $y = x^2 \sin^2 (2x^2)$

19. $s = \left(\dfrac{4t}{t + 1}\right)^{-2}$

20. $y = \left(\dfrac{\sqrt{x}}{1 + x}\right)^2$

21. $y = 4x\sqrt{x + \sqrt{x}}$

22. $r = \left(\dfrac{\sin \theta}{\cos \theta - 1}\right)^2$

23. $y = 20(3x - 4)^{1/4}(3x - 4)^{-1/5}$

24. $y = \dfrac{3}{(5x^2 + \sin 2x)^{3/2}}$

25. $y = \dfrac{1}{4}xe^{4x} - \dfrac{1}{16}e^{4x}$

26. $y = x^2 e^{-2/x}$

27. $y = \ln (\sin^2 \theta)$

28. $y = \log_2 (x^2/2)$

29. $y = \log_5 (3x - 7)$

30. $y = 8^{-t}$

31. $y = 5x^{3.6}$

32. $y = \sqrt{2}x^{-\sqrt{2}}$

33. $y = (x + 2)^{x+2}$

34. $y = 2(\ln x)^{x/2}$

35. $y = \sin^{-1} \sqrt{1 - u^2}, \quad 0 < u < 1$

36. $y = \ln \cos^{-1} x$

37. $y = z \cos^{-1} z - \sqrt{1 - z^2}$

38. $y = t \tan^{-1} t - \dfrac{1}{2} \ln t$

39. $y = (1 + t^2) \cot^{-1} 2t$

40. $y = z \sec^{-1} z - \sqrt{z^2 - 1}, \quad z > 1$

41. $y = \csc^{-1} (\sec \theta), \quad 0 < \theta < \pi/2$

42. $y = (1 + x^2) e^{\tan^{-1} x}$

Implicit Differentiation

In Exercises 43–54, find dy/dx.

43. $xy + 2x + 3y = 1$

44. $x^2 + xy + y^2 - 5x = 2$

45. $x^3 + 4xy - 3y^{4/3} = 2x$

46. $5x^{4/5} + 10y^{6/5} = 15$

47. $\sqrt{xy} = 1$

48. $x^2y^2 = 1$

49. $e^{x+2y} = 1$

50. $y^2 = 2e^{-1/x}$

51. $\ln(x/y) = 1$

52. $x\sin^{-1} y = 1 + x^2$

53. $ye^{\tan^{-1}x} = 2$

54. $x^y = \sqrt{2}$

In Exercises 55 and 56, find dr/ds.

55. $r\cos 2s + \sin^2 s = \pi$

56. $2rs - r - s + s^2 = -3$

57. Find d^2y/dx^2 by implicit differentiation:

(a) $x^3 + y^3 = 1$

(b) $y^2 = 1 - \dfrac{2}{x}$

58. (a) By differentiating $x^2 - y^2 = 1$ implicitly, show that $dy/dx = x/y$.

(b) Then show that $d^2y/dx^2 = -1/y^3$.

Numerical Values of Derivatives

59. Suppose that functions $f(x)$ and $g(x)$ and their first derivatives have the following values at $x = 0$ and $x = 1$.

x	$f(x)$	$g(x)$	$f'(x)$	$g'(x)$
0	1	1	-3	$1/2$
1	3	5	$1/2$	-4

Find the first derivatives of the following combinations at the given value of x.

(a) $6f(x) - g(x)$, $x = 1$

(b) $f(x)g^2(x)$, $x = 0$

(c) $\dfrac{f(x)}{g(x)+1}$, $x = 1$

(d) $f(g(x))$, $x = 0$

(e) $g(f(x))$, $x = 0$

(f) $(x + f(x))^{3/2}$, $x = 1$

(g) $f(x + g(x))$, $x = 0$

60. Suppose that the function $f(x)$ and its first derivative have the following values at $x = 0$ and $x = 1$.

x	$f(x)$	$f'(x)$
0	9	-2
1	-3	$1/5$

Find the first derivatives of the following combinations at the given value of x.

(a) $\sqrt{x}f(x)$, $x = 1$

(b) $\sqrt{f(x)}$, $x = 0$

(c) $f(\sqrt{x})$, $x = 1$

(d) $f(1 - 5\tan x)$, $x = 0$

(e) $\dfrac{f(x)}{2 + \cos x}$, $x = 0$

(f) $10\sin\left(\dfrac{\pi x}{2}\right)f^2(x)$, $x = 1$

61. Find the value of dy/dt at $t = 0$ if $y = 3\sin 2x$ and $x = t^2 + \pi$.

62. Find the value of ds/du at $u = 2$ if $s = t^2 + 5t$ and $t = (u^2 + 2u)^{1/3}$.

63. Find the value of dw/ds at $s = 0$ if $w = \sin(e^{\sqrt{r}})$ and $r = 3\sin(s + \pi/6)$.

64. Find the value of dr/dt at $t = 0$ if $r = (\theta^2 + 7)^{1/3}$ and $\theta^2 e^t + \theta = 1$.

65. If $y^3 + y = 2\cos x$, find the value of d^2y/dx^2 at the point $(0, 1)$.

66. If $x^{1/3} + y^{1/3} = 4$, find d^2y/dx^2 at the point $(8, 8)$.

Derivative Definition

In Exercises 67 and 68, find the derivative using the definition.

67. $f(t) = \dfrac{1}{2t + 1}$

68. $g(x) = 2x^2 + 1$

69. *Writing to Learn*

(a) Graph the function

$$f(x) = \begin{cases} x^2, & -1 \le x < 0 \\ -x^2, & 0 \le x \le 1. \end{cases}$$

(b) Is f continuous at $x = 0$?

(c) Is f differentiable at $x = 0$?

Give reasons for your answers.

70. *Writing to Learn*

(a) Graph the function

$$f(x) = \begin{cases} x, & -1 \le x < 0 \\ \tan x, & 0 \le x \le \pi/4. \end{cases}$$

(b) Is f continuous at $x = 0$?

(c) Is f differentiable at $x = 0$?

Give reasons for your answers.

71. *Writing to Learn*

(a) Graph the function

$$f(x) = \begin{cases} x, & 0 \le x \le 1 \\ 2 - x, & 1 < x \le 2. \end{cases}$$

(b) Is f continuous at $x = 1$?

(c) Is f differentiable at $x = 1$?

Give reasons for your answers.

72. *Writing to Learn* For what value or values of the constant m, if any, is

$$f(x) = \begin{cases} \sin 2x, & x \le 0 \\ mx, & x > 0 \end{cases}$$

(a) continuous at $x = 0$?

(b) differentiable at $x = 0$?

Give reasons for your answers.

Slopes, Tangents, and Normals

73. *Tangents with specified slope* Are there any points on the curve $y = (x/2) + 1/(2x - 4)$ where the slope is $-3/2$? If so, find them.

74. *Tangents with specified slope* Are there any points on the curve $y = x - e^{-x}$ where the slope is 2? If so, find them.

75. *Tangents intercepts* Find the x- and y-intercepts of the line that is tangent to the curve $y = x^3$ at the point $(-2, -8)$.

76. *Tangents perpendicular or parallel to lines* Find the points on the curve $y = 2x^3 - 3x^2 - 12x + 20$ where the tangent is

(a) perpendicular to the line $y = 1 - (x/24)$

(b) parallel to the line $y = \sqrt{2} - 12x$.

77. *Intersecting tangents* Show that the tangents to the curve $y = (\pi \sin x)/x$ at $x = \pi$ and $x = -\pi$ intersect at right angles.

78. *Tangent and normal lines* Find equations for the tangent and normal to the curve $y = 1 + \cos x$ at the point $(\pi/2, 1)$. Sketch the curve, tangent, and normal together, labeling each with its equation.

79. *Tangent parabola* The parabola $y = x^2 + C$ is to be tangent to the line $y = x$. Find C.

80. *Normal to a circle* Show that the normal line at any point of the circle $x^2 + y^2 = a^2$ passes through the origin.

Tangents and Normals to Implicitly Defined Curves

In Exercises 81–86, find equations for the lines that are tangent and normal to the curve at the given point.

81. $x^2 + 2y^2 = 9$, $(1, 2)$ **82.** $e^x + y^2 = 2$, $(0, 1)$

83. $xy + 2x - 5y = 2$, $(3, 2)$ **84.** $(y - x)^2 = 2x + 4$, $(6, 2)$

85. $x + \sqrt{xy} = 6$, $(4, 1)$ **86.** $x^{3/2} + 2y^{3/2} = 17$, $(1, 4)$

87. Find the slope of the curve $x^3 y^3 + y^2 = x + y$ at the points $(1, 1)$ and $(1, -1)$.

88. *Writing to Learn* The graph shown suggests that the curve $y = \sin (x - \sin x)$ might have horizontal tangents at the x-axis. Does it? Give reasons for your answer.

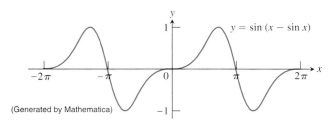

(Generated by Mathematica)

Analyzing Graphs

Each of the figures in Exercises 89 and 90 shows two graphs, the graph of a function $y = f(x)$ together with the graph of its derivative $f'(x)$. Which graph is which? How do you know?

89.

90.

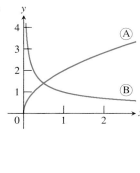

91. Use the following information to graph the function $y = f(x)$ for $-1 \le x \le 6$.

i. The graph of f is made of line segments joined end to end.

ii. The graph starts at the point $(-1, 2)$.

iii. The derivative of f, where defined, agrees with the step function shown here.

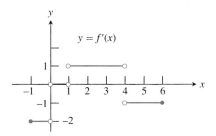

92. Repeat Exercise 91, supposing that the graph starts at $(-1, 0)$ instead of $(-1, 2)$.

Exercises 93 and 94 are about the graphs in Figure 2.53. The graphs in part (a) show the numbers of rabbits and foxes in a small arctic population. They are plotted as functions of time for 200 days. The number of rabbits increases at first, as the rabbits reproduce. But the foxes prey on the rabbits and, as the number of foxes increases, the rabbit population levels off and then drops. Figure 2.53b shows the graph of the derivative of the rabbit population. We made it by plotting slopes.

93. (a) What is the value of the derivative of the rabbit population in Figure 2.53 when the number of rabbits is largest? Smallest?

(b) What is the size of the rabbit population in Figure 2.53 when its derivative is largest? Smallest?

FIGURE 2.53 Rabbits and foxes in an arctic predator-prey food chain.

94. In what units should the slopes of the rabbit and fox population curves be measured?

Related Rates

95. *Right circular cylinder* The total surface area S of a right circular cylinder is related to the base radius r and height h by the equation $S = 2\pi r^2 + 2\pi rh$.

(a) How is dS/dt related to dr/dt if h is constant?

(b) How is dS/dt related to dh/dt if r is constant?

(c) How is dS/dt related to dr/dt and dh/dt if neither r nor h is constant?

(d) How is dr/dt related to dh/dt if S is constant?

96. *Right circular cone* The lateral surface area S of a right circular cone is related to the base radius r and height h by the equation $S = \pi r\sqrt{r^2 + h^2}$.

(a) How is dS/dt related to dr/dt if h is constant?

(b) How is dS/dt related to dh/dt if r is constant?

(c) How is dS/dt related to dr/dt and dh/dt if neither r nor h is constant?

97. *Circle's changing area* The radius of a circle is changing at the rate of $-2/\pi$ m/sec. At what rate is the circle's area changing when $r = 10$ m?

98. *Cube's changing edges* The volume of a cube is increasing at the rate of 1200 cm³/min at the instant its edges are 20 cm long. At what rate are the edges changing at that instant?

99. *Resistors connected in parallel* If two resistors of R_1 and R_2 ohms are connected in parallel in an electric circuit to make an R-ohm resistor, the value of R can be found from the equation

$$\frac{1}{R} = \frac{1}{R_1} + \frac{1}{R_2}.$$

If R_1 is decreasing at the rate of 1 ohm/sec and R_2 is increasing at the rate of 0.5 ohm/sec, at what rate is R changing when $R_1 = 75$ ohms and $R_2 = 50$ ohms?

100. *Impedance in a series circuit* The impedance Z (ohms) in a series circuit is related to the resistance R (ohms) and reactance X (ohms) by the equation $Z = \sqrt{R^2 + X^2}$. If R is increasing at 3 ohms/sec and X is decreasing at 2 ohms/sec, at what rate is Z changing when $R = 10$ ohms and $X = 20$ ohms?

101. *Speed of moving particle* The coordinates of a particle moving in the metric xy-plane are differentiable functions of time t with $dx/dt = 10$ m/sec and $dy/dt = 5$ m/sec. How fast is the particle moving away from the origin as it passes through the point $(3, -4)$?

102. *Motion of a particle* A particle moves along the curve $y = x^{3/2}$ in the first quadrant in such a way that its distance from the origin increases at the rate of 11 units per second. Find dx/dt when $x = 3$.

103. *Draining a tank* Water drains from the conical tank shown in Figure 2.54 at the rate of 5 ft³/min.

(a) What is the relation between the variables h and r in the figure?

(b) How fast is the water level dropping when $h = 6$ ft?

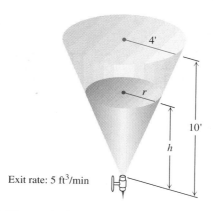

Exit rate: 5 ft³/min

FIGURE 2.54 The conical tank in Exercise 103.

104. *Rotating spool* As television cable is pulled from a large spool to be strung from the telephone poles along a street, it unwinds from the spool in layers of constant radius (see Figure 2.55). If the truck pulling the cable moves at a steady 6 ft/sec (a touch over 4 mph), use the equation $s = r\theta$ to find how fast (in radians per second) the spool is turning when the layer of radius 1.2 ft is being unwound.

FIGURE 2.55 The television cable in Exercise 104.

105. *Moving searchlight beam* The figure shows a boat 1 km offshore, sweeping the shore with a searchlight. The light turns at a constant rate, $d\theta/dt = -0.6$ rad/sec.

(a) How fast is the light moving along the shore when it reaches point A?

(b) How many revolutions per minute is 0.6 rad/sec?

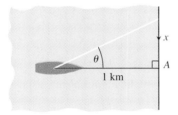

106. *Points moving on coordinate axes* Points A and B move along the x- and y-axes, respectively, in such a way that the distance r (meters) along the perpendicular from the origin to line AB remains constant. How fast is OA changing, and is it increasing, or decreasing, when $OB = 2r$ and B is moving toward O at the rate of $0.3r$ m/sec?

107. *Motion of a particle* A particle is traveling upward and to the right along the curve $y = \ln x$. Its x-coordinate is increasing at the rate $(dx/dt) = \sqrt{x}$ m/sec. At what rate is the y-coordinate changing at the point $(e^2, 2)$?

108. *Sliding down a slide* A girl is sliding down a slide shaped like the curve $y = 9e^{-x/3}$. Her y-coordinate is changing at the rate $dy/dt = (-1/4)\sqrt{9 - y}$ ft/sec. At approximately what rate is her x-coordinate changing when she reaches the bottom of the slide at $x = 9$ ft? (Take e^3 to be 20 and round your answer to the nearest foot per second.)

Logarithms

109. *Writing to Learn* The functions $f(x) = \ln 5x$ and $g(x) = \ln 3x$ differ by a constant. What constant? Give reasons for your answer.

110. *Writing to Learn*

(a) If $(\ln x)/x = (\ln 2)/2$, must $x = 2$?

(b) If $(\ln x)/x = -2 \ln 2$, must $x = 1/2$?

Give reasons for your answers.

111. *Writing to Learn* The quotient $(\log_4 x)/(\log_2 x)$ has a constant value. What value?
Give reasons for your answer.

112. $\log_x (2)$ *versus* $\log_2 (x)$ How does $f(x) = \log_x (2)$ compare with $g(x) = \log_2 (x)$? Here is one way to find out.

(a) Use the equation $\log_a b = (\ln b)/(\ln a)$ to express $f(x)$ and $g(x)$ in terms of natural logarithms.

T **(b)** Graph f and g together. Comment on the behavior of f in relation to the signs and values of g.

113. *Fundamental frequency of a vibrating piano string* We measure the frequencies at which wires vibrate in cycles (trips back and forth) per second. The unit of measure is a *hertz*:1 cycle/sec. Middle A on a piano has a frequency 440 hertz. For any given wire, the fundamental frequency y is a function of four variables:

r: the radius of the wire

l: the length

d: the density of the wire

T: the tension (force) holding the wire taut.

With r and l in centimeters, d in grams per centimeter, and T in dynes (it takes about 100,000 dynes to lift an apple), the fundamental frequency of the wire is

$$y = \frac{1}{2rl}\sqrt{\frac{T}{\pi d}}.$$

If we keep all the variables fixed except one, then y can be alternatively thought of as four different functions of one variable, $y(r)$, $y(l)$, $y(d)$, and $y(T)$. How would changing each variable affect the string's fundamental frequency? To find out, calculate $y'(r)$, $y'(l)$, $y'(d)$, and $y'(T)$.

114. *Spread of measles* The spread of measles in a certain school is given by

$$P(t) = \frac{200}{1 + e^{5-t}},$$

where t is the number of days since the measles first appeared and $P(t)$ is the total number of students who have caught the measles to date.

(a) Estimate the initial number of students infected with measles.

(b) About how many students in all will get the measles?

(c) When will the rate of spread of measles be greatest? What is the rate?

115. Graph the function $f(x) = \tan^{-1}(\tan 2x)$ in the window $[-\pi, \pi]$
⊤ by $[-4, 4]$. Then answer the following questions.

(a) What is the domain of f?

(b) What is the range of f?

(c) At which points is f not differentiable?

(d) Describe the graph of f'.

116. If $x^2 - y^2 = 1$, find d^2y/dx^2 at the point $(2, \sqrt{3})$.

Additional Exercises: Theory, Examples, Applications

1. An equation like $\sin^2 \theta + \cos^2 \theta = 1$ is called an **identity** because it holds for all values of θ. An equation like $\theta = 0.5$ is not an identity because it holds only for selected values of θ, not all. If you differentiate both sides of a trigonometric identity in θ with respect to θ, the resulting new equation will also be an identity.

Differentiate the following to show that the resulting equations hold for all θ.

(a) $\sin 2\theta = 2 \sin \theta \cos \theta$

(b) $\cos 2\theta = \cos^2 \theta - \sin^2 \theta$

2. *Writing to Learn* If the identity $\sin(x + a) = \sin x \cos a + \cos x \sin a$ is differentiated with respect to x, is the resulting equation also an identity? Does this principle apply to the equation $x^2 - 2x - 8 = 0$? Explain.

3. (a) Find values for the constants $a, b,$ and c that will make

$$f(x) = \cos x \quad \text{and} \quad g(x) = a + bx + cx^2$$

satisfy the conditions

$$f(0) = g(0), \quad f'(0) = g'(0), \quad \text{and} \quad f''(0) = g''(0).$$

(b) Find values for b and c that will make

$$f(x) = \sin(x + a) \quad \text{and} \quad g(x) = b \sin x + c \cos x$$

satisfy the conditions

$$f(0) = g(0) \quad \text{and} \quad f'(0) = g'(0).$$

(c) For the determined values of $a, b,$ and $c,$ what happens for the third and fourth derivatives of f and g in each of parts (a) and (b)?

4. *Solutions to differential equations*
(a) Show that $y = \sin x$, $y = \cos x$, and $y = a \cos x + b \sin x$ (a and b constants) all satisfy the equation

$$y'' + y = 0.$$

(b) How would you modify the functions in part (a) to satisfy the equation

$$y'' + 4y = 0?$$

Generalize this result.

5. *An osculating circle* Find the values of h, k, and a that make the circle $(x - h)^2 + (y - k)^2 = a^2$ tangent to the parabola $y = x^2 + 1$ at the point $(1, 2)$ and that also make the second derivatives d^2y/dx^2 have the same value on both curves there. Circles like this one that are tangent to a curve and have the same second derivative as the curve at the point of tangency are called *osculating circles* (from the Latin *osculari*, meaning "to kiss"). We encounter them again in Chapter 10.

6. *Marginal revenue* A bus will hold 60 people. The number x of people per trip who use the bus is related to the fare charged (p dollars) by the law $p = [3 - (x/40)]^2$. Write an expression for the total revenue $r(x)$ per trip received by the bus company. What number of people per trip will make the marginal revenue dr/dx equal to zero? What is the corresponding fare? (This fare is the one that maximizes the revenue, so the bus company should probably rethink its fare policy.)

7. *Industrial production*
(a) Economists often use the expression "rate of growth" in relative rather than absolute terms. For example, let $u = f(t)$ be the number of people in the labor force at time t in a given industry. (We treat this function as though it were differentiable even though it is an integer-valued step function.)

Let $v = g(t)$ be the average production per person in the labor force at time t. The total production is then $y = uv$. If the labor force is growing at the rate of 4% per year ($du/dt = 0.04u$) and the production per worker is growing at the rate of 5% per year ($dv/dt = 0.05v$), find the rate of growth of the total production, y.

(b) Suppose that the labor force in part (a) is decreasing at the rate of 2% per year while the production per person is increasing at the rate of 3% per year. Is the total production increasing, or is it decreasing, and at what rate?

8. *Designing a gondola* The designer of a 30-ft-diameter spherical hot-air balloon wants to suspend the gondola 8 ft below the bottom of the balloon with cables tangent to the surface of the balloon, as shown in the accompanying figure. Two of the cables are shown running from the top edges of the gondola to their points of tangency, $(-12, -9)$ and $(12, -9)$. How wide should the gondola be?

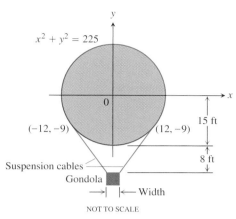

$x^2 + y^2 = 225$

$(-12, -9)$ $(12, -9)$

15 ft

8 ft

Suspension cables

Gondola

←— Width

NOT TO SCALE

9. *Pisa by parachute* The photograph shows Mike McCarthy parachuting from the top of the Tower of Pisa on August 6, 1988. Make a rough sketch to show the shape of the graph of his speed during the jump.

Mike McCarthy of London jumped from the Tower of Pisa and then opened his parachute in what he said was a world record low-level parachute jump of 179 ft. (*Source: Boston Globe*, Aug. 6, 1988.)

10. *Motion of a particle* The position at time $t \geq 0$ of a particle moving along a coordinate line is

$$s = 10 \cos (t + \pi/4).$$

(a) What is the particle's starting position ($t = 0$)?

(b) What are the points farthest to the left and right of the origin reached by the particle?

(c) Find the particle's velocity and acceleration at the points in part (b).

(d) When does the particle first reach the origin? What are its velocity, speed, and acceleration then?

11. *Shooting a paper clip* On Earth, you can easily shoot a paper clip 64 ft straight up into the air with a rubber band. In t seconds after firing, the paper clip is $s = 64t - 16t^2$ ft above your hand.

(a) How long does it take the paper clip to reach its maximum height? With what velocity does it leave your hand?

(b) On the moon, the same acceleration will send the paper clip to a height of $s = 64t - 2.6t^2$ ft in t sec. About how long will it take the paper clip to reach its maximum height and how high will it go?

12. *Velocities of two particles* At time t sec, the positions of two particles on a coordinate line are $s_1 = 3t^3 - 12t^2 + 18t + 5$ m and $s_2 = -t^3 + 9t^2 - 12t$ m. When do the particles have the same velocities?

13. *Velocity of a particle* A particle of constant mass m moves along the x-axis. Its velocity v and position x satisfy the equation

$$\frac{1}{2} m (v^2 - v_0^2) = \frac{1}{2} k (x_0^2 - x^2),$$

where k, v_0, and x_0 are constants. Show that whenever $v \neq 0$,

$$m \frac{dv}{dt} = -kx.$$

14. *Average and instantaneous velocity*

(a) Show that if the position x of a moving point is given by a quadratic function of time t, $x = At^2 + Bt + C$, then the average velocity over any time interval $[t_1, t_2]$ is equal to the instantaneous velocity at the midpoint of the time interval.

(b) What is the geometric significance of the result in part (a)?

15. Find all values of the constants m and b for which the function

$$y = \begin{cases} \sin x & \text{for } x < \pi \\ mx + b & \text{for } x \geq \pi \end{cases}$$

is

(a) continuous at $x = \pi$.

(b) differentiable at $x = \pi$.

16. *Writing to Learn* Does the function

$$f(x) = \begin{cases} \dfrac{1 - \cos x}{x} & \text{for } x \neq 0 \\ 0 & \text{for } x = 0 \end{cases}$$

have a derivative at $x = 0$? Explain.

17. **(a)** For what values of a and b will

$$f(x) = \begin{cases} ax, & x < 2 \\ ax^2 - bx + 3, & x \geq 2 \end{cases}$$

be differentiable for all values of x?

(b) *Writing to Learn* Discuss the geometry of the resulting graph of f.

18. (a) For what values of a and b will

$$g(x) = \begin{cases} ax + b, & x \le -1 \\ ax^3 + x + 2b, & x > -1 \end{cases}$$

be differentiable for all values of x?

(b) *Writing to Learn* Discuss the geometry of the resulting graph of g.

19. *Odd differentiable functions* Is there anything special about the derivative of an odd differentiable function of x? Give reasons for your answer.

20. *Even differentiable functions* Is there anything special about the derivative of an even differentiable function of x? Give reasons for your answer.

21. *A surprising result* Suppose that the functions f and g are defined throughout an open interval containing the point x_0, that f is differentiable at x_0, that $f(x_0) = 0$, and that g is continuous at x_0. Show that the product fg is differentiable at x_0. This process shows, for example, that although $|x|$ is not differentiable at $x = 0$, the product $x|x|$ is differentiable at $x = 0$.

22. (*Continuation of Exercise 21*) Use the result of Exercise 21 to show that the following functions are differentiable at $x = 0$.

(a) $|x| \sin x$ **(b)** $x^{2/3} \sin x$ **(c)** $\sqrt[3]{x}(1 - \cos x)$

(d) $h(x) = \begin{cases} x^2 \sin (1/x), & x \ne 0 \\ 0, & x = 0 \end{cases}$

23. *Writing to Learn* Is the derivative of

$$h(x) = \begin{cases} x^2 \sin (1/x), & x \ne 0 \\ 0, & x = 0 \end{cases}$$

continuous at $x = 0$? How about the derivative of $k(x) = xh(x)$? Give reasons for your answers.

24. Suppose that a function f satisfies the following conditions for all real values of x and y:

i. $f(x + y) = f(x) \cdot f(y)$.

ii. $f(x) = 1 + xg(x)$, where $\lim_{x \to 0} g(x) = 1$.

Show that the derivative $f'(x)$ exists at every value of x and that $f'(x) = f(x)$.

25. *The generalized product rule* Use mathematical induction (Appendix 1) to prove that if $y = u_1 u_2 \cdots u_n$ is a finite product of differentiable functions, then y is differentiable on their common domain and

$$\frac{dy}{dx} = \frac{du_1}{dx} u_2 \cdots u_n + u_1 \frac{du_2}{dx} \cdots u_n + \cdots + u_1 u_2 \cdots u_{n-1} \frac{du_n}{dx}.$$

3

Applications of Derivatives

OVERVIEW In this chapter, we show how to use derivatives to find maximum and minimum values of functions, to predict and analyze the shapes of graphs, and to draw conclusions about the behavior of functions that solve differential equations. We also see how a tangent line captures the shape of a curve near the point of tangency and how it can be used to find the zeros of functions numerically. The key to many of these accomplishments is the Mean Value Theorem, a theorem whose corollaries provide the gateway to integral calculus in Chapter 4.

Extreme Values of Functions

The Drilling-Rig Problem • Absolute (Global) Extreme Values • Local (Relative) Extreme Values • Finding Extreme Values

One of the most useful things to learn from a function's derivative is whether the function assumes any maximum or minimum values on a given interval and where these values are located if it does. Once we can do this, we can solve problems like the drilling-rig problem.

The Drilling–Rig Problem

Example 1   Piping Oil from a Drilling Rig to a Refinery

A drilling rig 12 mi offshore is to be connected by pipe to a refinery onshore, 20 mi straight down the coast from the rig. If underwater pipe costs $50,000 per mile and land-based pipe costs $30,000 per mile, what combination of the two will give the least expensive connection?

Preliminary analysis:

We try a few possibilities to get a feel for the problem:

(a) *Smallest amount of underwater pipe*

Underwater pipe is the more expensive, so we use as little as we can. We run straight to shore (12 mi) and use land pipe for the 20 mi to the refinery.

$$\text{Dollar cost} = 12(50,000) + 20(30,000)$$
$$= 1,200,000.$$

(b) *All pipe under water (most direct route)*

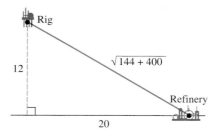

We go straight to the refinery under water.

$$\text{Dollar cost} = \sqrt{544}\,(50,000)$$
$$= 1,166,190$$

This is less expensive than plan (a).

(c) *Something in between*

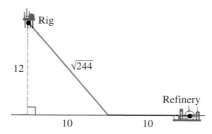

We run pipe underwater to the halfway point 10 mi from the refinery and go by land from there.

$$\text{Dollar cost} = \sqrt{244}\,(50,000) + 10(30,000)$$
$$\approx 1,081,025$$

Neither extreme (shortest underwater segment or all pipe underwater) gives the best solution. Something in between is better.

The 10 mi point was an arbitrary choice. Would another choice be even better? If so, how do we find it? What's the best we can do? We find out using the mathematics we are about to develop, and we return to solve this problem at the end of the section.

Absolute (Global) Extreme Values

The largest and smallest values a function takes on, both locally and globally (see Figure 3.1), are always of great interest.

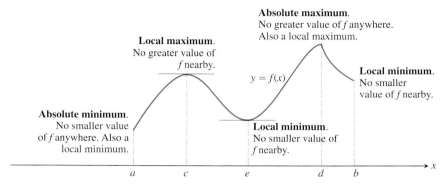

FIGURE 3.1 How to classify maxima and minima.

Definition   Absolute Extreme Values

Let f be a function with domain D. Then $f(c)$ is the

(a) absolute maximum value on D if and only if $f(x) \leq f(c)$ for all x in D

(b) absolute minimum value on D if and only if $f(x) \geq f(c)$ for all x in D.

Absolute (or **global**) maximum and minimum values are also called **absolute extrema** (plural of the Latin *extremum*). We often omit the term "absolute" or "global" and just say maximum and minimum.

Example 2 shows that extreme values can occur at interior points or endpoints of intervals.

Example 2   Exploring Extreme Values

On $[-\pi/2, \pi/2]$, $f(x) = \cos x$ takes on a maximum value of 1 (once) and a minimum value of 0 (twice). The function $g(x) = \sin x$ takes on a maximum value of 1 and a minimum value of -1 (Figure 3.2).

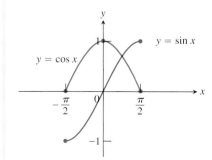

FIGURE 3.2 The graphs for Example 2.

Functions with the same defining rule can have different extrema, depending on the domain.

Example 3   Exploring Absolute Extrema

The absolute extrema of the following functions on their domains can be seen in Figure 3.3.

	Function rule	Domain D	Absolute extrema on D
(a)	$y = x^2$	$(-\infty, \infty)$	No absolute maximum. Absolute minimum of 0 at $x = 0$.
(b)	$y = x^2$	$[0, 2]$	Absolute maximum of 4 at $x = 2$. Absolute minimum of 0 at $x = 0$.
(c)	$y = x^2$	$(0, 2]$	Absolute maximum of 4 at $x = 2$. No absolute minimum.
(d)	$y = x^2$	$(0, 2)$	No absolute extrema.

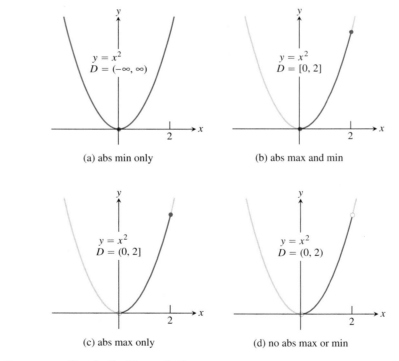

FIGURE 3.3 Graphs for Example 3.

Example 3 shows that a function may fail to have a maximum or minimum value. This cannot happen, however, with a continuous function on a finite closed interval.

Theorem 1 The Extreme-Value Theorem for Continuous Functions

If f is continuous at every point of a closed interval I, then f assumes both an absolute maximum value M and an absolute minimum value m somewhere in I. That is, there are numbers x_1 and x_2 in I with $f(x_1) = m$, $f(x_2) = M$, and $m \leq f(x) \leq M$ for every other x in I (Figure 3.4).

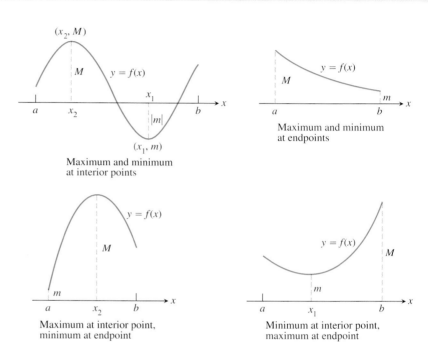

FIGURE 3.4 Some possibilities for a continuous function's maximum and minimum on a closed interval $[a, b]$.

The proof of Theorem 1 requires a detailed knowledge of the real number system, and we do not give it here.

Example 4 Extrema Depending on Continuity

As Figure 3.5 shows, the requirements in Theorem 1 that the interval be closed and the function continuous are key ingredients. Without them, the conclusion of the theorem need not hold.

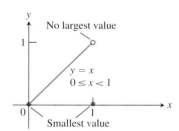

FIGURE 3.5 Even a single point of discontinuity can keep a function from having either a maximum or minimum value on a closed interval. The function

$$y = \begin{cases} x, & 0 \leq x < 1 \\ 0, & x = 1 \end{cases}$$

is continuous at every point of $[0, 1]$ except $x = 1$, yet its graph over $[0, 1]$ does not have a highest point.

Local (Relative) Extreme Values

Figure 3.1 shows a graph with five points where a function has extreme values on its domain $[a, b]$. The function's absolute minimum occurs at a even though at e the function's value is smaller than at any other point *nearby*. The curve rises to the left and falls to the right around c, making $f(c)$ a maximum locally. The function attains its absolute maximum at d.

> **Definition**   Local Extreme Values
>
> Let c be an interior point of the domain of the function f. Then $f(c)$ is a
>
> (a) **local maximum value** at c if and only if $f(x) \leq f(c)$ for all x in some open interval containing c
>
> (b) **local minimum value** at c if and only if $f(x) \geq f(c)$ for all x in some open interval containing c.

Local extrema are also called **relative extrema.** We can extend the definitions of local extrema to endpoints of intervals. A function f has a local maximum or local minimum *at an endpoint c* if the appropriate inequality holds for all x in some half-open interval containing c.

An absolute extremum is also a local extremum, because being an extreme value overall makes it an extreme value in its immediate neighborhood. Hence, *a list of all local extrema will automatically include the absolute extrema if there are any.*

Finding Extreme Values

The interior domain points where the function in Figure 3.1 has local extreme values are points where either f' is zero or f' does not exist. This is generally the case, as we see from the following theorem.

> **Theorem 2**   Local Extreme Values
>
> If a function f has a local maximum value or a local minimum value at an interior point c of its domain, and if f' exists at c, then
>
> $$f'(c) = 0.$$

Theorem 2 says that a function's first derivative is always zero at an interior point where the function has a local extreme value and the derivative is defined. Hence the only places where a function f can possibly have an extreme value (local or global) are

1. interior points where $f' = 0$

2. interior points where f' is undefined

3. endpoints of the domain of f.

Most quests for extreme values call for finding the absolute extrema of a continuous function on a closed interval. Theorem 1 assures us that such values exist; Theorem 2 tells us where to look. The following definition helps us summarize these findings.

Definition Critical Point

A point in the domain of a function f at which $f' = 0$ or f' does not exist is a **critical point** of f.

Thus, in summary, extreme values occur only at critical points and endpoints.

Example 5 Finding Absolute Extrema on a Closed Interval

Find the absolute maximum and minimum values of $f(x) = 10x(2 - \ln x)$ on the interval $[1, e^2]$.

Solution Figure 3.6 suggests that f has its absolute maximum value near $x = 3$ and its absolute minimum value of 0 at $x = e^2$.

CD-ROM
WEBsite

FIGURE 3.6 The extreme values of $f(x) = 10x(2 - \ln x)$ occur at $x = e$ and $x = e^2$. (Example 5)

We evaluate the function at the critical points and endpoints and take the largest and smallest of the resulting values.

The first derivative

$$f'(x) = 10(2 - \ln x) - 10x\left(\frac{1}{x}\right) = 10(1 - \ln x)$$

The only critical point in the domain $[1, e^2]$ is the point $x = e$, where $\ln x = 1$. The values of f at this one critical point and at the endpoints are

Critical point value: $\quad f(e) = 10e$

Endpoint values: $\quad f(1) = 10(2 - \ln 1) = 20$

$\qquad\qquad\qquad\qquad f(e^2) = 10e^2(2 - 2 \ln e) = 0.$

We can see from this list that the function's absolute maximum value is $10e \approx 27.2$; it occurs at the critical interior point $x = e$. The absolute minimum value is 0 and occurs at the right endpoint $x = e^2$.

How to Find the Absolute Extrema of a Continuous Function f on a Closed Interval

Step 1. Evalute f at all critical points and endpoints.

Step 2. Take the largest and smallest of these values.

Proof of Theorem 2 To show that $f'(c)$ is zero at an interior local extremum, we show first that $f'(c)$ cannot be positive and second that $f'(c)$ cannot be negative. The only number that is neither positive nor negative is zero, so that is what $f'(c)$ must be.

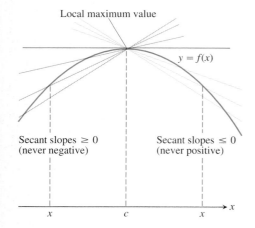

Local maximum value

$y = f(x)$

Secant slopes ≥ 0
(never negative)

Secant slopes ≤ 0
(never positive)

x c x

FIGURE 3.7 A curve with a local maximum value. The slope at c, simultaneously the limit of nonpositive numbers and nonnegative numbers, is zero.

To begin, suppose that f has a local maximum value at $x = c$ (Figure 3.7) so that $f(x) - f(c) \leq 0$ for all values of x near enough to c. Since c is an interior point of the domain of f, $f'(c)$ is defined by the two-sided limit

$$\lim_{x \to c} \frac{f(x) - f(c)}{x - c}.$$

Hence, the right-hand and left-hand limits both exist at $x = c$ and equal $f'(c)$. When we examine these limits separately, we find that

$$f'(c) = \lim_{x \to c^+} \frac{f(x) - f(c)}{x - c} \leq 0. \qquad \text{Because } (x - c) > 0 \atop \text{and } f(x) \leq f(c) \qquad (1)$$

Similarly,

$$f'(c) = \lim_{x \to c^-} \frac{f(x) - f(c)}{x - c} \geq 0. \qquad \text{Because } (x - c) < 0 \atop \text{and } f(x) \leq f(c) \qquad (2)$$

Together, (1) and (2) imply $f'(c) = 0$.

This proves the theorem for local maximum values. To prove it for local minimum values, we simply use $f(x) \geq f(c)$, which reverses the inequalities in (1) and (2). ▬

In Example 6, we investigate the function whose graph was drawn in Example 3 of Preliminary Section 2.

Example 6   Finding Extreme Values

Find the extreme values of

$$f(x) = \frac{1}{\sqrt{4 - x^2}}.$$

Solution   Figure 3.8 suggests that f has an absolute minimum of about 0.5 at $x = 0$. There also appear to be local maxima at $x = -2$ and $x = 2$. At these points, however, f is not defined, and there do not appear to be maxima anywhere else.

[-4, 4] by [-2, 4]

FIGURE 3.8 The graph of

$$f(x) = \frac{1}{\sqrt{4 - x^2}}.$$

(Example 6)

Let's confirm these graphical observations. The function f is defined only for $4 - x^2 > 0$, so its domain is the open interval $(-2, 2)$. The domain has no endpoints, so all the extreme values must occur at critical points. We rewrite the formula for f to find f':

$$f(x) = \frac{1}{\sqrt{4 - x^2}} = (4 - x^2)^{-1/2}.$$

Thus,

$$f'(x) = -\frac{1}{2}(4-x^2)^{-3/2}(-2x) = \frac{x}{(4-x^2)^{3/2}}.$$

The only critical point in the domain $(-2, 2)$ is $x = 0$. The value

$$f(0) = \frac{1}{\sqrt{4-0^2}} = \frac{1}{2}$$

is therefore the sole candidate for an extreme value.

To determine whether $1/2$ is an extreme value of f, we examine the formula

$$f(x) = \frac{1}{\sqrt{4-x^2}}.$$

As x moves away from 0 on either side, the denominator gets smaller, the values of f increase, and the graph rises. We have a minimum value at $x = 0$, and the minimum is absolute.

The function has no maxima, either local or absolute. This does not violate Theorem 1 (the Extreme Value Theorem) because here f is defined on an *open* interval. To guarantee extreme points, Theorem 1 requires the interval be closed.

Example 7   Critical Points Need Not Give Extreme Values

Although a function's extrema can occur only at critical points and endpoints, not every critical point or endpoint signals the presence of an extreme value. Figure 3.9 illustrates this for interior points. Exercise 62 describes a function that fails to assume an extreme value at an endpoint of its domain.

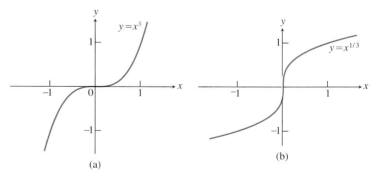

(a)

(b)

FIGURE 3.9 Critical points without extreme values. (a) $y' = 3x^2$ is 0 at $x = 0$, but $y = x^3$ has no extremum there. (b) $y' = (1/3)x^{-2/3}$ is undefined at $x = 0$, but $y = x^{1/3}$ has no extremum there.

FIGURE 3.10 The figure for solving the drilling-rig problem. (Example 8)

Example 8   Solution to the Drilling-Rig Problem

As in the preliminary analysis at the beginning of the section, we draw a picture and sketch the relevant dimensions. This time, though, we also add the length x of underwater pipe and the length y of land-based pipe as variables (Figure 3.10).

The right angle opposite the rig is the key to expressing the relationship between x and y, for the Pythagorean Theorem gives

$$x^2 = 12^2 + (20 - y)^2$$
$$x = \sqrt{144 + (20 - y)^2}. \qquad (3)$$

Only the positive root has meaning in this model.

The dollar cost of the pipeline is

$$c = 50{,}000x + 30{,}000y.$$

To express c as a function of a single variable, we can substitute for x, using Equation (3):

$$c = 50{,}000\sqrt{144 + (20 - y)^2} + 30{,}000y.$$

Our goal now is to find the minimum value of $c(y)$ on the interval $0 \le y \le 20$. The first derivative of c with respect to y is

$$c' = 50{,}000 \cdot \frac{1}{2} \cdot \frac{2(20 - y)(-1)}{\sqrt{144 + (20 - y)^2}} + 30{,}000$$

$$= -50{,}000\,\frac{20 - y}{\sqrt{144 + (20 - y)^2}} + 30{,}000.$$

Setting c' equal to zero gives

$$50{,}000\,(20 - y) = 30{,}000\,\sqrt{144 + (20 - y)^2}$$

$$\frac{5}{3}\,(20 - y) = \sqrt{144 + (20 - y)^2}$$

$$\frac{25}{9}\,(20 - y)^2 = 144 + (20 - y)^2$$

$$\frac{16}{9}\,(20 - y)^2 = 144$$

$$(20 - y) = \pm\frac{3}{4} \cdot 12 = \pm 9$$

$$y = 20 \pm 9$$

$$y = 11 \qquad \text{or} \qquad y = 29.$$

Only $y = 11$ lies in the interval of interest. The values of c at this one critical point and at the endpoints are

$$c(11) = 1{,}080{,}000$$
$$c(0) = 1{,}166{,}190$$
$$c(20) = 1{,}200{,}000.$$

The least expensive connection costs \$1,080,000, and we achieve it by running the line underwater to the point on shore 11 mi from the refinery.

EXERCISES 3.1

Finding Extrema from Graphs

In Exercises 1–6, determine from the graph whether the function has any absolute extreme values on [a, b]. Then explain how your answer is consistent with Theorem 1.

1.   **2.**

3.   **4.**

5.   **6.**

In Exercises 7–10, find the extreme values and where they occur.

7.   **8.**

9.   **10.**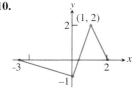

In Exercises 11–14, match the table with a graph.

11.

x	$f'(x)$
a	0
b	0
c	5

12.

x	$f'(x)$
a	0
b	0
c	-5

13.

x	$f'(x)$
a	does not exist
b	0
c	-2

14.

x	$f'(x)$
a	does not exist
b	does not exist
c	-1.7

(a)   (b)

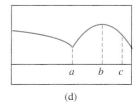
(c)   (d)

Absolute Extrema on Intervals

In Exercises 15–26, find the absolute extreme values of each function on the interval. Then graph the function. Identify the points on the graph where the absolute extrema occur and include their coordinates.

15. $f(x) = \frac{2}{3}x - 5, \quad -2 \le x \le 3$

16. $f(x) = x^2 - 6x + 9, \quad 0 \le x \le 5$

17. $f(x) = 4 - x^2, \quad -3 \le x \le 1$

18. $f(x) = \sqrt{3 + 2x - x^2}, \quad -0.5 \le x \le 3$

    (*Hint:* Sketch the parabola first, then take the square root.)

19. $g(x) = \sin\left(x + \frac{\pi}{4}\right), \quad 0 \le x \le \frac{7\pi}{4}$

20. $g(x) = \sec x, \quad -\frac{\pi}{2} < x < \frac{3\pi}{2}$

21. $F(x) = -\frac{1}{x^2}, \quad 0.5 \le x \le 2$

22. $h(x) = \sqrt[3]{x}, \quad -1 \le x \le 8$

23. $f(x) = \frac{1}{x} + \ln x, \quad 0.5 \le x \le 4$

24. $g(x) = e^{-x}, \quad -1 \le x \le 1$

25. $h(x) = \ln(x + 1), \quad 0 \le x \le 3$

T **26.** $k(x) = e^{-x^2}, \quad -\infty < x < \infty$

Finding Extreme Values

In Exercises 27–36, find the extreme values of the function and where they occur.

27. $y = 2x^2 - 8x + 9$

28. $y = |3 + 2x - x^2|$

    (*Hint:* Sketch the parabola first, then take absolute values.)

29. $y = \dfrac{1}{\sqrt{1 - x^2}}$

30. $y = \dfrac{1}{x^2 - 1}$

T **31.** $y = \dfrac{\ln x}{x}$

T **32.** $y = 4 + ex^2 - e^{-x^2}$

T **33.** $y = \dfrac{x}{x^2 + 1}$

T **34.** $y = \dfrac{x + 1}{x^2 + 2x + 2}$

35. $y = x^3 + x^2 - 8x + 5$

T **36.** $y = xe^{2x}$

Local Extrema and Critical Points

In Exercises 37–44, find the derivative at each critical point and determine the local extreme values.

37. $y = x^{2/3}(x + 2)$

38. $y = x^{2/3}(x^2 - 4)$

39. $y = x\sqrt{4 - x^2}$

40. $y = x^2\sqrt{3 - x}$

41. $y = \begin{cases} 4 - 2x, & x \le 1 \\ x + 1, & x > 1 \end{cases}$

42. $y = \begin{cases} 3 - x, & x < 0 \\ 3 + 2x - x^2, & x \ge 0 \end{cases}$

43. $y = \begin{cases} -x^2 - 2x + 4, & x \le 1 \\ -x^2 + 6x - 4, & x > 1 \end{cases}$

44. $y = \begin{cases} -\frac{1}{4}x^2 - \frac{1}{2}x + \frac{15}{4}, & x \le 1 \\ x^3 - 6x^2 + 8x, & x > 1 \end{cases}$

In Exercises 45 and 46, give reasons for your answers.

45. *Writing to Learn* Let $f(x) = (x - 2)^{2/3}$.

    (a) Does $f'(2)$ exist?

    (b) Show that the only local extreme value of f occurs at $x = 2$.

    (c) Does the result in part (b) contradict the Extreme Value Theorem?

    (d) Repeat parts (a) and (b) for $f(x) = (x - a)^{2/3}$, replacing 2 by a.

46. *Writing to Learn* Let $f(x) = |x^3 - 9x|$.

    (a) Does $f'(0)$ exist?

    (b) Does $f'(3)$ exist?

    (c) Does $f'(-3)$ exist?

    (d) Determine all extrema of f.

Optimization Applications

Whenever you are maximizing or minimizing a function of a single variable, we urge you to graph the function over the domain that is appropriate to the problem you are solving. The graph will provide insight before you begin to calculate and will furnish a visual context for understanding your answer.

47. *Constructing a pipeline* Supertankers off-load oil at a docking facility 4 mi offshore. The nearest refinery is 9 mi east of the shore point nearest the docking facility. A pipeline must be constructed connecting the docking facility with the refinery. The pipeline costs $300,000 per mile if constructed underwater and $200,000 per mile if overland.

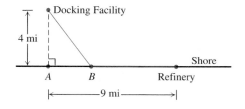

(a) Locate Point *B* to minimize the cost of the construction.

(b) The cost of underwater construction is expected to increase, whereas the cost of overland construction is expected to stay constant. At what cost does it become optimal to construct the pipeline directly to Point *A*?

48. *Upgrading a highway* A highway must be constructed to connect Village *A* with Village *B*. There is a rudimentary roadway that can be upgraded 50 mi south of the line connecting the two villages. The cost of upgrading the existing roadway is $300,000 per mile, whereas the cost of constructing a new highway is $500,000 per mile. Find the combination of upgrading and new construction that minimizes the cost of connecting the two villages. Clearly define the location of the proposed highway.

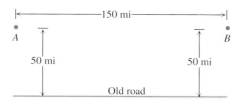

49. *Locating a pumping station* Two towns exist on the south side of a river. A pumping station is to be located to serve the two towns. A pipeline will be constructed from the pumping station to each of the towns along the line connecting the town and the pumping station. Locate the pumping station to minimize the amount of pipeline that must be constructed.

50. *Length of a guy wire* One tower is 50 ft high and another tower is 30 ft high. The towers are 150 ft apart. A guy wire is to run from Point *A* to the top of each tower.

(a) Locate Point *A* so that the total length of guy wire is minimal.

(b) Show in general that regardless of the height of the tower, the length of guy wire is minimized if the angles at *A* are equal.

51. *Writing to Learn* The function

$$V(x) = x(10 - 2x)(16 - 2x), \qquad 0 < x < 5,$$

models the volume of a box.

(a) Find the extreme values of *V*.

(b) Interpret any values found in part (a) in terms of volume of the box.

52. *Writing to Learn* The function

$$P(x) = 2x + \frac{200}{x}, \qquad 0 < x < \infty,$$

models the perimeter of a rectangle of dimensions *x* by $100/x$.

(a) Find any extreme values of *P*.

(b) Give an interpretation in terms of perimeter of the rectangle for any values found in part (a).

53. *Area of a right triangle* What is the largest possible area for a right triangle whose hypotenuse is 5 cm long?

54. *Area of an athletic field* An athletic field is to be built in the shape of a rectangle *x* units long capped by semicircular regions of radius *r* at the two ends. The field is to be bounded by a 400 m racetrack.

(a) Express the area of the rectangular portion of the field as a function of *x* alone or *r* alone (your choice).

(b) What values of *x* and *r* give the rectangular portion the largest possible area?

55. *Maximum height of a vertically moving body* The height of a body moving vertically is given by

$$s = -\frac{1}{2} gt^2 + v_0 t + s_0, \qquad g > 0,$$

with *s* in meters and *t* in seconds. Find the body's maximum height.

56. *Peak alternating current* Suppose that at any given time *t* (in seconds) the current *i* (in amperes) in an alternating current circuit is $i = 2 \cos t + 2 \sin t$. What is the peak current for this circuit (largest magnitude)?

Theory and Examples

57. *A minimum with no derivatives* The function $f(x) = |x|$ has an absolute minimum value at $x = 0$ even though *f* is not differentiable at $x = 0$. Is this consistent with Theorem 2? Give reasons for your answer.

58. *Even functions* If an even function $f(x)$ has a local maximum value at $x = c$, can anything be said about the value of *f* at $x = -c$? Give reasons for your answer.

59. *Odd functions* If an odd function $g(x)$ has a local minimum value at $x = c$, can anything be said about the value of *g* at $x = -c$? Give reasons for your answer.

60. *Writing to Learn* We know how to find the extreme values of a continuous function $f(x)$ by investigating its values at critical points and endpoints. But what if there *are* no critical points or endpoints? What happens then? Do such functions really exist? Give reasons for your answers.

61. *Cubic functions* Consider the cubic function

$$f(x) = ax^3 + bx^2 + cx + d.$$

(a) Show that f can have 0, 1, or 2 critical points. Give examples and graphs to support your argument.

(b) How many local extreme values can f have?

T 62. *Functions with no extreme values at endpoints*

(a) Graph the function

$$f(x) = \begin{cases} \sin \dfrac{1}{x}, & x > 0 \\ 0, & x = 0. \end{cases}$$

Explain why $f(0) = 0$ is not a local extreme value of f.

(b) Construct a function of your own that fails to have an extreme value at a domain endpoint.

T Graph the functions in Exercises 63–66. Then find the extreme values of the function on the interval and where they occur.

63. $f(x) = |x - 2| + |x + 3|, \quad -5 \le x \le 5$

64. $g(x) = |x - 1| - |x - 5|, \quad -2 \le x \le 7$

65. $h(x) = |x + 2| - |x - 3|, \quad -\infty < x < \infty$

66. $k(x) = |x + 1| + |x - 3|, \quad -\infty < x < \infty$

COMPUTER EXPLORATIONS

In Exercises 67–74, you will use a CAS to help find the absolute extrema of the given function over the specified closed interval. Perform the following steps.

(a) Plot the function over the interval to see its general behavior there.

(b) Find the interior points where $f' = 0$. (In some exercises, you may have to use the numerical equation solver to approximate a solution.) You may want to plot f' as well.

(c) Find the interior points where f' does not exist.

(d) Evaluate the function at all points found in parts (b) and (c) and at the endpoints of the interval.

(e) Find the function's absolute extreme values on the interval and identify where they occur.

67. $f(x) = x^4 - 8x^2 + 4x + 2, \quad [-20/25, 64/25]$

68. $f(x) = -x^4 + 4x^3 - 4x + 1, \quad [-3/4, 3]$

69. $f(x) = x^{2/3}(3 - x), \quad [-2, 2]$

70. $f(x) = 2 + 2x - 3x^{2/3}, \quad [-1, 10/3]$

71. $f(x) = \sqrt{x} + \cos x, \quad [0, 2\pi]$

72. $f(x) = x^{3/4} - \sin x + \dfrac{1}{2}, \quad [0, 2\pi]$

73. $f(x) = \pi x^2 e^{-3x/2}, \quad [0, 5]$

74. $f(x) = \ln(2x + x \sin x), \quad [1, 15]$

3.2 The Mean Value Theorem and Differential Equations

Rolle's Theorem • Mean Value Theorem • A Physical Interpretation • Mathematical Consequences • Finding Velocity and Position from Acceleration • Differential Equations and the Height of a Projectile

We have seen how to find the position of a body falling freely from rest as a function of time and from it how to derive the body's velocity and acceleration functions. But suppose that we had started with knowing only the body's acceleration, that is, with knowing the effect of gravity on the body and nothing else. Could we have worked backward to find the body's velocity and position functions?

The underlying mathematical question here is, What functions can have another function as derivative? What velocity functions can have a given acceleration function? What position functions can have a given velocity function? The corollaries of the Mean Value Theorem provide the answers.

The Mean Value Theorem itself connects the average rate of change of a function over an interval with the instantaneous rate of change of the function at a point within the interval.

Rolle's Theorem

There is strong geometric evidence that between any two points where a differentiable curve crosses the *x*-axis there is a point on the curve where the tangent is horizontal. A 300-year-old theorem of Michel Rolle assures us this is always the case.

> **Theorem 3** Rolle's Theorem
>
> Suppose that $y = f(x)$ is continuous at every point of $[a, b]$ and differentiable at every point of (a, b). If
>
> $$f(a) = f(b) = 0,$$
>
> then there is at least one number c in (a, b) at which $f'(c) = 0$ (Figure 3.11).

FIGURE 3.11 Rolle's Theorem says that a differentiable curve has at least one horizontal tangent between any two points where it crosses the *x*-axis. The curve here has three.

Proof Being continuous, f assumes absolute maximum and minimum values on $[a, b]$. These can occur only

1. at interior points where f' is zero
2. at interior points where f' does not exist
3. at the endpoints of the function's domain, in this case, a and b.

By hypothesis, f has a derivative at every interior point of $[a, b]$. That rules out option 2, leaving us with interior points where $f' = 0$ and with the two endpoints a and b.

If either the maximum or the minimum occurs at a point c inside the interval, then $f'(c) = 0$ by Theorem 2 in Section 3.1, and we have found a point for Rolle's Theorem.

If both maximum and minimum are at a or b, then the maximum and minimum values of f are both 0. Thus, f has the constant value 0, so $f' = 0$ throughout (a, b) and c can be taken anywhere in the interval. This now completes the proof. ▬

The hypotheses of Theorem 3 are essential. If they fail at even one point, the graph may not have a horizontal tangent (Figure 3.12).

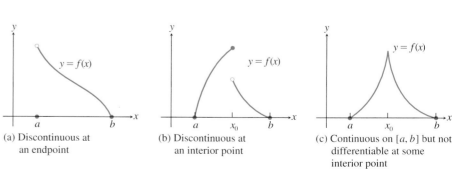

(a) Discontinuous at an endpoint

(b) Discontinuous at an interior point

(c) Continuous on $[a, b]$ but not differentiable at some interior point

FIGURE 3.12 No horizontal tangent.

Mean Value Theorem

The Mean Value Theorem is Rolle's Theorem on a slant.

Theorem 4   The Mean Value Theorem

Suppose that $y = f(x)$ is continuous on a closed interval $[a, b]$ and differentiable on the interval's interior (a, b). Then there is at least one point c in (a, b) at which

$$\frac{f(b) - f(a)}{b - a} = f'(c). \tag{1}$$

Proof   We picture the graph of f as a curve in the plane and draw a line through the points $A(a, f(a))$ and $B(b, f(b))$ (see Figure 3.13). The line is the graph of the function

$$g(x) = f(a) + \frac{f(b) - f(a)}{b - a}(x - a) \tag{2}$$

(point-slope equation). The vertical difference between the graphs of f and g at x is

$$h(x) = f(x) - g(x)$$

$$= f(x) - f(a) - \frac{f(b) - f(a)}{b - a}(x - a). \tag{3}$$

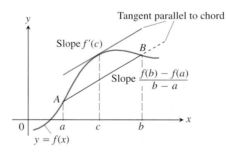

FIGURE 3.13 Geometrically, the Mean Value Theorem says that somewhere between A and B the curve has at least one tangent parallel to chord AB.

Figure 3.14 shows the graphs of f, g, and h together.

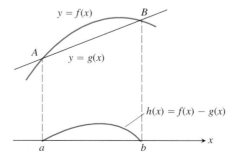

FIGURE 3.14 The chord AB is the graph of the function $g(x)$. The function $h(x) = f(x) - g(x)$ gives the vertical distance between the graphs of f and g at x.

The function h satisfies the hypotheses of Rolle's Theorem on $[a, b]$. It is continuous on $[a, b]$ and differentiable on (a, b) because f and g are. Also, $h(a) = h(b) = 0$ because the graphs of f and g both pass through A and B. Therefore, $h' = 0$ at some point c in (a, b). This is the point we want for Equation (1).

To verify Equation (1), we differentiate both sides of Equation (3) with respect to x and then set $x = c$:

$$h'(x) = f'(x) - \frac{f(b) - f(a)}{b - a} \qquad \text{Derivative of Eq. (3)} \ldots$$

$$h'(c) = f'(c) - \frac{f(b) - f(a)}{b - a} \qquad \ldots \text{with } x = c$$

$$0 = f'(c) - \frac{f(b) - f(a)}{b - a} \qquad h'(c) = 0$$

$$f'(c) = \frac{f(b) - f(a)}{b - a}, \qquad \text{Rearranged}$$

which is what we set out to prove. ▬

The hypotheses of the Mean Value Theorem do not require f to be differentiable at either a or b. Continuity at a and b is enough (Figure 3.15).

We usually do not know any more about c than the theorem says, which is that c exists. In a few cases, we can satisfy our curiosity about the identity of c, as in the next example. Our ability to identify c is the exception rather than the rule, however, and the importance of the theorem lies elsewhere.

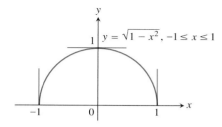

FIGURE 3.15 The function $f(x) = \sqrt{1 - x^2}$ satisfies the hypotheses (and conclusion) of the Mean Value Theorem on $[-1, 1]$ even though f is not differentiable at -1 and 1.

Example 1   Exploring the Mean Value Theorem

The function $f(x) = x^2$ (Figure 3.16) is continuous for $0 \le x \le 2$ and differentiable for $0 < x < 2$. Since $f(0) = 0$ and $f(2) = 4$, the Mean Value Theorem says that at some point c in the interval the derivative $f'(x) = 2x$ must have the value $(4 - 0)/(2 - 0) = 2$. In this (exceptional) case, we can identify c by solving the equation $2c = 2$ to get $c = 1$.

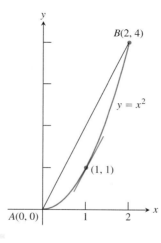

FIGURE 3.16 As we find in Example 1, $c = 1$ is where the tangent is parallel to the chord.

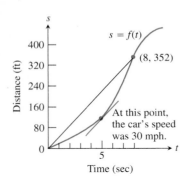

FIGURE 3.17 Distance versus elapsed
time for the car in Example 2.

Historical Biography

Bernard le Bouyer
Fontenelle
(1657 — 1757)

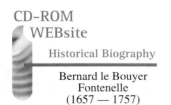

A Physical Interpretation

Think of the number $(f(b) - f(a))/(b - a)$ as the average change in f over $[a, b]$ and $f'(c)$ as an instantaneous change. The Mean Value Theorem says that the instantaneous change at some interior point must equal the average change over the entire interval.

Example 2   Interpreting the Mean Value Theorem

If a car accelerating from zero takes 8 sec to go 352 ft, its average velocity for the 8-sec interval is $352/8 = 44$ ft/sec. At some point during the acceleration, the speedometer must read exactly 30 mph (44 ft/sec) (Figure 3.17).

Mathematical Consequences

The first corollary of the Mean Value Theorem tells us what kind of function has a zero derivative.

Corollary 1   Functions with Zero Derivatives Are Constant Functions

If $f'(x) = 0$ at each point of an interval I, then $f(x) = C$ for all x in I, where C is a constant.

We know that if a function f has a constant value on an interval I, then f is differentiable on I and $f'(x) = 0$ for all x in I. Corollary 1 provides the converse.

Proof of Corollary 1   We want to show f has a constant value on I. We do so by showing that if x_1 and x_2 are any two points in I, then $f(x_1) = f(x_2)$.

Suppose that x_1 and x_2 are two points in I, numbered from left to right so that $x_1 < x_2$. Then f satisfies the hypotheses of the Mean Value Theorem on $[x_1, x_2]$: It is differentiable at every point of $[x_1, x_2]$ and hence continuous at every point as well. Therefore,

$$\frac{f(x_2) - f(x_1)}{x_2 - x_1} = f'(c)$$

at some point c between x_1 and x_2. Since $f' = 0$ throughout I, this equation translates successively into

$$\frac{f(x_2) - f(x_1)}{x_2 - x_1} = 0, \qquad f(x_2) - f(x_1) = 0, \qquad \text{and} \qquad f(x_1) = f(x_2). \quad \rule[0.5ex]{1.5em}{0.4pt}$$

At the beginning of the section, we asked if we could work backward from the acceleration of a body falling freely from rest to find the body's velocity and position functions. The answer is yes, as a consequence of the next corollary.

**Corollary 2   Functions with the Same Derivative Function on an
Interval Differ by a Constant**

If $f'(x) = g'(x)$ at each point of an interval I, then there exists a constant C such that $f(x) = g(x) + C$ for all x in I.

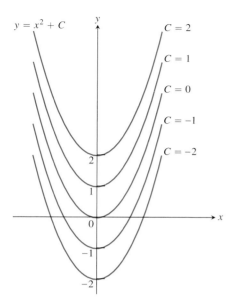

FIGURE 3.18 From a geometric point of view, Corollary 2 of the Mean Value Theorem says that the graphs of functions with identical derivatives on an interval can differ only by a vertical shift there. The graphs of the functions with derivative $2x$ are the parabolas $y = x^2 + C$, shown here for selected values of C.

Proof   At each point x in I, the derivative of the difference function $h = f - g$ is

$$h'(x) = f'(x) - g'(x) = 0.$$

Thus, $h(x) = C$ on I (Corollary 1). That is, $f(x) - g(x) = C$ on I, so $f(x) = g(x) + C$.   ▬

Corollary 2 says that functions have identical derivatives on an interval only if their values on the interval have a constant difference. We know, for instance, that the derivative of $f(x) = x^2$ on $(-\infty, \infty)$ is $2x$. Any other function with derivative $2x$ on $(-\infty, \infty)$ must equal $x^2 + C$ for some value of C (Figure 3.18).

Example 3   Applying Corollary 2

Find the function $f(x)$ whose derivative is $\sin x$ and whose graph passes through the point $(0, 2)$.

Solution   Since f has the same derivative as $g(x) = -\cos x$, we know that $f(x) = -\cos x + C$ for some constant C. The value of C can be determined from the condition that $f(0) = 2$ (the graph of f passes through $(0, 2)$):

$$f(0) = -\cos (0) + C = 2, \qquad \text{so } C = 3.$$

The formula for f is $f(x) = -\cos x + 3$.

Finding Velocity and Position from Acceleration

Here is how to find the velocity $v(t)$ and position $s(t)$ of a body falling freely from rest with acceleration 9.8 m/sec^2, the metric equivalent of 32 ft/sec^2.

We know $v(t)$ is a function whose derivative is 9.8. We also know the derivative of $g(t) = 9.8t$ is 9.8. By Corollary 2,

$$v(t) = 9.8t + C$$

for some constant C. Since the body falls from rest, $v(0) = 0$. This determines C:

$$9.8(0) + C = 0, \qquad \text{so } C = 0.$$

The velocity function must be $v(t) = 9.8t$. How about the position function $s(t)$?

We know $s(t)$ is a function whose derivative is $9.8t$. We also know the derivative of $h(t) = 4.9t^2$ is $9.8t$. By Corollary 2,

$$s(t) = 4.9t^2 + C$$

for some constant C. Since $s(0) = 0$,

$$4.9(0)^2 + C = 0 \qquad \text{and} \qquad C = 0.$$

The position function must be $s(t) = 4.9t^2$.

Differential Equations and the Height of a Projectile

A **differential equation** is an equation relating an unknown function and one or more of its derivatives. A function whose derivatives satisfy a differential equation is called a **solution** to the differential equation.

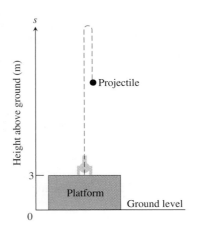

FIGURE 3.19 The sketch for modeling the projectile motion in Example 5.

Example 4   Solutions to Differential Equations

(a) The function $s(t) = 4.9t^2$ is a solution of the differential equation $d^2s/dt^2 = 9.8$ m/sec^2.

(b) The function $y = -\cos x + 3$ solves the differential equation $dy/dx = \sin x$.

Example 5   Finding a Projectile's Height from Its Acceleration, Initial Velocity, and Initial Position

A heavy projectile is fired straight up from a platform 3 m above the ground, with an initial velocity of 160 m/sec. Assume that the only force affecting the projectile during its flight is from gravity, which produces a downward acceleration of 9.8 m/sec^2. Find an equation for the projectile's height above the ground as a function of time t if $t = 0$ when the projectile is fired. How high above the ground is the projectile 3 sec after firing?

Solution   To model the motion, we draw a figure (Figure 3.19) and let s denote the projectile's height above the ground at time t. We assume s to be a twice-differentiable function of t and represent the projectile's velocity and acceleration with the derivatives

$$v = \frac{ds}{dt} \quad \text{and} \quad a = \frac{dv}{dt} = \frac{d^2s}{dt^2}.$$

Since gravity acts in the direction of *decreasing* s in our model, the problem is to solve the differential equation

$$\frac{d^2s}{dt^2} = -9.8$$

knowing also that

$$v(0) = 160 \quad \text{and} \quad s(0) = 3.$$

Since the derivative of $g(t) = -9.8t$ is -9.8, Corollary 2 gives

$$v(t) = -9.8t + C$$

for some constant C. We can find the value of C from the initial condition $v(0) = 160$:

$$v(0) = 160$$
$$-9.8(0) + C = 160$$
$$C = 160.$$

This completes the formula for ds/dt:

$$\frac{ds}{dt} = -9.8t + 160.$$

We know $s(t)$ is a function whose derivative is $-9.8t + 160$. We also know the derivative of $h(t) = -4.9t^2 + 160t$ is $-9.8t + 160$. By Corollary 2,

$$s = -4.9t^2 + 160t + C.$$

We apply the second condition to find this new C:

$$s(0) = 3$$
$$-4.9(0)^2 + 160(0) + C = 3$$
$$C = 3.$$

This completes the formula for s as a function of t:

$$s = -4.9t^2 + 160t + 3.$$

To find the projectile's height 3 sec into the flight, we set $t = 3$ in the formula for s. The height is

$$s = -4.9(3)^2 + 160(3) + 3 = 438.9 \text{ m.}$$

We say more about solving differential equations in Chapters 4 and 6.

EXERCISES 3.2

Checking and Using Hypotheses

In Exercises 1–4, (a) show that the function f satisfies the hypotheses of the Mean Value Theorem on the given interval $[a, b]$ and (b) find each value of c in (a, b) that satisfies the equation

$$f'(c) = \frac{f(b) - f(a)}{b - a}.$$

1. $f(x) = x^2 + 2x - 1$, $[0, 1]$
2. $f(x) = x^{2/3}$, $[0, 1]$
3. $f(x) = \sin^{-1} x$, $[-1, 1]$
4. $f(x) = \ln(x - 1)$, $[2, 4]$
5. *Writing to Learn* The function

$$f(x) = \begin{cases} x, & 0 \le x < 1 \\ 0, & x = 1 \end{cases}$$

is zero at $x = 0$ and $x = 1$ and differentiable on $(0, 1)$, but its derivative on $(0, 1)$ is never zero. How can this be? Doesn't Rolle's Theorem say the derivative has to be zero somewhere in $(0, 1)$? Give reasons for your answer.

6. For what values of a, m, and b does the function

$$f(x) = \begin{cases} 3, & x = 0 \\ -x^2 + 3x + a, & 0 < x < 1 \\ mx + b, & 1 \le x \le 2 \end{cases}$$

satisfy the hypotheses of the Mean Value Theorem on the interval $[0, 2]$?

Differential Equations

7. *Writing to Learn* Suppose that $f(-1) = 3$ and that $f'(x) = 0$ for all x. Must $f(x) = 3$ for all x? Give reasons for your answer.

8. *Writing to Learn* Suppose that $g(0) = 5$ and that $g'(t) = 2$ for all t. Must $g(t) = 2t + 5$ for all t? Give reasons for your answer.

In Exercises 9–12, find all possible functions with the given derivative.

9. (a) $y' = x$
   (b) $y' = x^2$
   (c) $y' = x^3$

10. (a) $y' = 2x$
    (b) $y' = 2x - 1$
    (c) $y' = 3x^2 + 2x - 1$

11. (a) $r' = \frac{1}{\theta}$
    (b) $r' = 1 - \frac{1}{\theta}$
    (c) $r' = 5 + \frac{1}{\theta}$

12. (a) $y' = \frac{1}{2\sqrt{t}}$
    (b) $y' = \frac{1}{\sqrt{t}}$
    (c) $y' = 4t - \frac{1}{\sqrt{t}}$

In Exercises 13–16, find the function with the given derivative whose graph passes through the point P.

13. $f'(x) = 2x - 1$, $P(0, 0)$
14. $g'(x) = \frac{1}{x} + 2x$, $P(1, -1)$
15. $f'(x) = e^{2x}$, $P\left(0, \frac{3}{2}\right)$
16. $r'(t) = \sec t \tan t - 1$, $P(0, 0)$

Finding Position from Velocity

Exercises 17–20 give the velocity $v = ds/dt$ and initial position of a body moving along a coordinate line. Find the body's position at time t.

17. $v = 9.8t + 5, \quad s(0) = 10$

18. $v = 32t - 2, \quad s(0.5) = 4$

19. $v = \sin \pi t, \quad s(0) = 0$

20. $v = \dfrac{1}{t + 2}, \quad t > -2, \quad s(-1) = \dfrac{1}{2}$

Finding Position from Acceleration

Exercises 21–24 give the acceleration $a = d^2s/dt^2$, initial velocity, and initial position of a body moving on a coordinate line. Find the body's position at time t.

21. $a = e^x; \quad v(0) = 20, \quad s(0) = 5$

22. $a = 9.8; \quad v(0) = -3, \quad s(0) = 0$

23. $a = -4 \sin 2t; \quad v(0) = 2, \quad s(0) = -3$

24. $a = \dfrac{9}{\pi^2} \cos \dfrac{3t}{\pi}; \quad v(0) = 0, \quad s(0) = -1$

Gravitational Acceleration

25. *Free fall on the moon* On our moon, the acceleration of gravity is 1.6 m /sec^2. If a rock is dropped into a crevasse, how fast will it be going just before it hits bottom 30 sec later?

26. *Speed of a rocket* A rocket lifts off the surface of Earth with a constant acceleration of 20 m/sec^2. How fast will the rocket be going 1 min later?

27. *Diving off a platform* With approximately what velocity do you enter the water if you dive from a 10 m platform? (Use $g = 9.8$ m/sec^2.)

28. *Projectile height on Mars* The acceleration of gravity near the surface of Mars is 3.72 m/sec^2. If a rock is blasted straight up from the surface with an initial velocity of 93 m/sec (about 208 mph), how high does it go? (*Hint:* When is the velocity zero?)

Linear Motion

29. *Motion along a coordinate line* A particle moves on a coordinate line with acceleration $a = d^2s/dt^2 = 15\sqrt{t} - (3/\sqrt{t})$, subject to the conditions that $ds/dt = 4$ and $s = 0$ when $t = 1$. Find

(a) the velocity $v = ds/dt$ in terms of t

(b) the position s in terms of t.

30. *Finding displacement from velocity* **(a)** Suppose that the velocity of a body moving along the s-axis is

$$\frac{ds}{dt} = v = 9.8t - 3.$$

i. Find the body's displacement over the time interval from $t = 1$ to $t = 3$ given that $s = 5$ when $t = 0$.

ii. Find the body's displacement from $t = 1$ to $t = 3$ given that $s = -2$ when $t = 0$.

iii. Now find the body's displacement from $t = 1$ to $t = 3$ given that $s = s_0$ when $t = 0$.

(b) *Writing to Learn* Suppose that the position s of a body moving along a coordinate line is a differentiable function of time t. Is it true that once you know a function whose derivative is the velocity function ds/dt you can find the body's displacement from $t = a$ to $t = b$ even if you do not know the body's exact position at either of those times? Give reasons for your answer.

Applications

31. *Temperature change* It took 14 sec for a mercury thermometer to rise from -19°C to 100°C when it was taken from a freezer and placed in boiling water. Show that somewhere along the way the mercury was rising at the rate of 8.5°C/sec.

32. *Speeding* A trucker handed in a ticket at a toll booth showing that in 2 h she had covered 159 mi on a toll road with speed limit 65 mph. The trucker was cited for speeding. Why?

33. *Triremes* Classical accounts tell us that a 170-oar trireme (ancient Greek or Roman warship) once covered 184 sea miles in 24 h. Explain why at some point during this feat the trireme's speed exceeded 7.5 knots (sea miles per hour).

34. *Running a marathon* A marathoner ran the 26.2 mi New York City Marathon in 2.2 h. Show that at least twice the marathoner was running at exactly 11 mph.

Theory and Examples

35. *The geometric mean of a and b* The **geometric mean** of two positive numbers a and b is the number $\sqrt{ab}$. Show that the value of c in the conclusion of the Mean Value Theorem for $f(x) = 1/x$ on an interval of positive numbers $[a, b]$ is $c = \sqrt{ab}$.

36. *The arithmetic mean of a and b* The **arithmetic mean** of two numbers a and b is the number $(a + b)/2$. Show that the value of c in the conclusion of the Mean Value Theorem for $f(x) = x^2$ on any interval $[a, b]$ is $c = (a + b)/2$.

T **37.** *Writing to Learn: A surprising graph* Graph the function

$$f(x) = \sin x \sin (x + 2) - \sin^2 (x + 1).$$

What does the graph do? Why does the function behave this way? Give reasons for your answers.

T **38.** *Rolle's Theorem*

(a) Construct a polynomial $f(x)$ that has zeros at $x = -2, -1, 0, 1$, and 2.

(b) Graph f and its derivative f' together. How is what you see related to Rolle's theorem?

(c) Do $g(x) = \sin x$ and its derivative g' illustrate the same phenomenon?

39. *Unique solution* Assume that f is continuous on $[a, b]$ and differentiable on (a, b). Also assume that $f(a)$ and $f(b)$ have opposite signs and that $f' \neq 0$ between a and b. Show that $f(x) = 0$ exactly once between a and b.

40. *Parallel tangents* Assume that f and g are differentiable on $[a, b]$ and that $f(a) = g(a)$ and $f(b) = g(b)$. Show that there is at least one point between a and b where the tangents to the graphs of f and g are parallel or the same line. Illustrate with a sketch.

41. *Writing to Learn: Identical graphs* If the graphs of two differentiable functions $f(x)$ and $g(x)$ start at the same point in the plane and the functions have the same rate of change at every point, do the graphs have to be identical? Give reasons for your answer.

42. *Upper bounds* Show that for any numbers a and b, the inequality $|\sin b - \sin a| \leq |b - a|$ is true.

43. *Sign of f'* Assume that f is differentiable on $a \leq x \leq b$ and that $f(b) < f(a)$. Show that f' is negative at some point between a and b.

44. Let f be a function defined on an interval $[a, b]$. What conditions could you place on f to guarantee that

$$\min f' \leq \frac{f(b) - f(a)}{b - a} \leq \max f',$$

where $\min f'$ and $\max f'$ refer to the minimum and maximum values of f' on $[a, b]$? Give reasons for your answers.

45. Use the inequalities in Exercise 44 to estimate $f(0.1)$ if $f'(x) = 1/(1 + x^4 \cos x)$ for $0 \leq x \leq 0.1$ and $f(0) = 1$.

46. Use the inequalities in Exercise 44 to estimate $f(0.1)$ if $f'(x) = 1/(1 - x^4)$ for $0 \leq x \leq 0.1$ and $f(0) = 2$.

COMPUTER EXPLORATIONS

Graphing Solutions of Differential Equations

Use a CAS to explore graphically solutions to each of the differential equations in Exercises 47–50. Perform the following steps to help with your explorations.

(a) Find the solution containing the arbitrary constant C using your CAS differential equation solver.

(b) Graph the solutions for $C = -2, -1, 0, 1, 2$ together.

(c) Find and graph the solution that passes through the specified point $P(x_0, y_0)$ over the given interval $[a, b]$.

47. $y' = x\sqrt{1 - x}, \quad [0, 1], \quad P(1/2, 1)$

48. $y' = \frac{1}{x}, \quad [1, 4], \quad P(2, -1)$

49. $y' = x \sin x, \quad [-4, 4], \quad P(\pi, -1)$

50. $y' = \frac{1}{1 + \sin x}, \quad [-\pi/2, \pi/2], \quad P(0, 1)$

3.3 The Shape of a Graph

First Derivative Test for Increasing Functions and Decreasing Functions • The First Derivative Test for Local Extrema • Concavity • Points of Inflection • Second Derivative Test for Local Extrema • Learning about Functions from Derivatives

What do we need to know to determine the shape of a graph? We need to know how it rises or falls as it goes along and how it bends. These distinguishing features are displayed in Figure 3.20. In this section, we see how the first and second derivatives of a function provide the information we need to determine a graph's shape. We begin by defining formally what it means for a function to be increasing or decreasing over an interval.

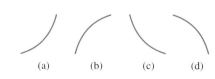

(a)   (b)   (c)   (d)

FIGURE 3.20 The graph in (a) rises and bends upward. The graph in (b) rises and bends downward. The graph in (c) falls and bends upward. The graph in (d) falls and bends downward.

First Derivative Test for Increasing Functions and Decreasing Functions

What kinds of functions have positive derivatives or negative derivatives? The answer, provided by the Mean Value Theorem's third corollary, is this: The only functions with positive derivatives are increasing functions; the only functions with negative derivatives are decreasing functions.

Definitions   Increasing Function, Decreasing Function

Let f be a function defined on an interval I. Then

1. f **increases** on I if for all points x_1 and x_2 in I, $x_1 < x_2 \Rightarrow f(x_1) < f(x_2)$.

2. f **decreases** on I if for all points x_1 and x_2 in I, $x_1 < x_2 \Rightarrow f(x_2) < f(x_1)$.

Corollary 3   The First Derivative Test for Increasing and Decreasing

Suppose that f is continuous on $[a, b]$ and differentiable on (a, b).

If $f' > 0$ at each point of (a, b), then f increases on $[a, b]$.

If $f' < 0$ at each point of (a, b), then f decreases on $[a, b]$.

Proof   Let x_1 and x_2 be two points in $[a, b]$ with $x_1 < x_2$. The Mean Value Theorem applied to f on $[x_1, x_2]$ says that

$$f(x_2) - f(x_1) = f'(c)(x_2 - x_1)$$

for some c between x_1 and x_2. The sign of the right-hand side of this equation is the same as the sign of $f'(c)$ because $x_2 - x_1$ is positive. Therefore, $f(x_2) > f(x_1)$ if f' is positive on (a, b) and $f(x_2) < f(x_1)$ if f' is negative on (a, b). �---

Here is how we apply the first derivative test to find where a function is increasing and decreasing. The critical points of a function f partition the x-axis into intervals on which f' is either positive or negative. We determine the sign of f' in each interval by evaluating f' for one value of x in the interval. Then we apply Corollary 3.

Example 1   Using the First Derivative Test for Increasing and Decreasing

Find the critical points of $f(x) = x^3 - 12x - 5$ and identify the intervals on which f is increasing and decreasing.

Solution   Figure 3.21 suggests that f has two critical points. Since f is continuous and differentiable for all real numbers, the critical points occur only at the zeros of f'.

$$f'(x) = 3x^2 - 12 = 3(x^2 - 4)$$
$$= 3(x + 2)(x - 2)$$

The zeros of f' are $x = -2$ and $x = 2$. They partition the x-axis into intervals as follows.

Intervals	$-\infty < x < -2$	$-2 < x < 2$	$2 < x < \infty$
Sign of f'	$+$	$-$	$+$
Behavior of f	increasing	decreasing	increasing

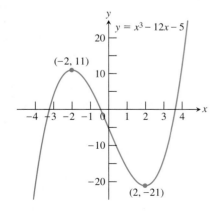

FIGURE 3.21 The graph of $f(x) = x^3 - 12x - 5$. (Example 1)

To determine the sign of f' on each interval, we found the sign of each factor on the interval and "multiplied" the signs of the factors to get the sign of f'. We then applied Corollary 3 to see that f increases on $(-\infty, -2)$, decreases on $(-2, 2)$, and increases on $(2, \infty)$.

Knowing where a function increases and decreases tells us how to test for the nature of local extreme values.

The First Derivative Test for Local Extrema

In Figure 3.22, at the points where f has a minimum value, $f' < 0$ immediately to the left and $f' > 0$ immediately to the right. (If the point is an endpoint, there is only one side to consider.) Thus, the curve is falling (values decreasing) on the left of the minimum value and rising (values increasing) on its right. Similarly, at the points where f has a maximum value, $f' > 0$ immediately to the left and $f' < 0$ immediately to the right. Thus, the curve is rising (values increasing) on the left of the maximum value and falling (values decreasing) on its right.

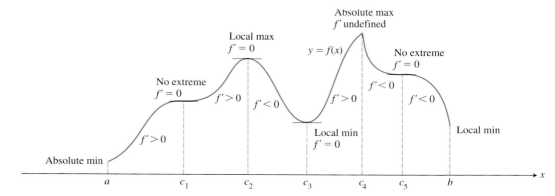

FIGURE 3.22 A function's first derivative tells how the graph rises and falls.

These observations lead to a test for the presence and nature of local extreme values of differentiable functions.

First Derivative Test for Local Extrema

At a critical point $x = c$,

1. f has a *local minimum* if f' changes from negative to positive at c

2. f has a *local maximum* if f' changes from positive to negative at c

3. f has *no local extreme* if f' has the same sign on both sides of c.

The test for local extrema at endpoints is similar but there is only one side to consider.

Example 2   Using the First Derivative Test for Local Extrema

Find the critical points of

$$f(x) = (x^2 - 3)\, e^x.$$

Identify the intervals on which f is increasing and decreasing. Find the function's local and absolute extreme values.

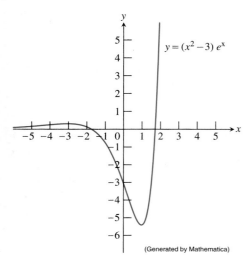

FIGURE 3.23 The graph of $f(x) = (x^2 - 3)e^x$. (Example 2)

Solution This time it is a little harder to see one of the extrema (Figure 3.23). The function f is continuous and differentiable for all real numbers, so the critical points occur only at the zeros of f'.

Using the Product Rule, we find

$$f'(x) = (x^2 - 3) \cdot \frac{d}{dx} e^x + \frac{d}{dx}(x^2 - 3) \cdot e^x$$

$$= (x^2 - 3) \cdot e^x + (2x) \cdot e^x$$

$$= (x^2 + 2x - 3)e^x.$$

Since e^x is never zero, the first derivative is zero if and only if

$$x^2 + 2x - 3 = 0$$

$$(x + 3)(x - 1) = 0.$$

The zeros $x = -3$ and $x = 1$ partition the x-axis into intervals as follows.

Intervals	$x < -3$	$-3 < x < 1$	$1 < x$
Sign of f'	$+$	$-$	$+$
Behavior of f	increasing	decreasing	increasing

We can see from the table that there is a local maximum (about 0.299) at $x = -3$ and a local minimum (about -5.437) at $x = 1$. The local minimum value is also an absolute minimum because $f(x) > 0$ for $|x| > \sqrt{3}$. There is no absolute maximum. The function increases on $(-\infty, -3)$ and $(1, \infty)$ and decreases on $(-3, 1)$.

We now show how to determine the way the graph of a function $y = f(x)$ bends. We know the information must be contained in y', but how do we find it? The answer, for functions that are twice differentiable except perhaps at isolated points, is to differentiate y'. Together y' and y'' tell us the shape of the function's graph. We see in the next section how this enables us to sketch solutions of certain differential equations.

Concavity

As you can see in Figure 3.24, the function $y = x^3$ rises as x increases, but the portions defined on the intervals $(-\infty, 0)$ and $(0, \infty)$ *turn* in different ways. Looking at tangents as we scan from left to right, we see that the slope y' of the curve decreases on the interval $(-\infty, 0)$ and then increases on the interval $(0, \infty)$. The curve $y = x^3$ is *concave down* on $(-\infty, 0)$ and *concave up* on $(0, \infty)$. The curve lies below the tangents where it is concave down and above the tangents where it is concave up.

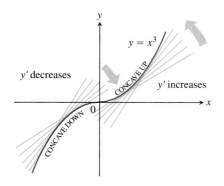

FIGURE 3.24 The graph of $f(x) = x^3$ is concave down on $(-\infty, 0)$ and concave up on $(0, \infty)$.

Definition Concavity

The graph of a differentiable function $y = f(x)$ is

(a) concave up on an open interval I if y' is increasing on I

(b) concave down on an open interval I if y' is decreasing on I.

If a function $y = f(x)$ has a second derivative, then we can conclude that y' increases if $y'' > 0$ and y' decreases if $y'' < 0$.

Second Derivative Test for Concavity
The graph of a twice-differentiable function $y = f(x)$ is

(a) concave up on any interval where $y'' > 0$

(b) concave down on any interval where $y'' < 0$.

Example 3   Applying the Concavity Test

The curve $y = x^2$ (Figure 3.25) is concave up on $(-\infty, \infty)$ because its second derivative $y'' = 2$ is always positive.

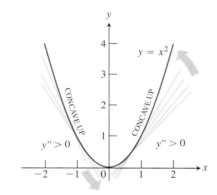

FIGURE 3.25 The graph of $f(x) = x^2$ is concave up on every interval.

Example 4   Determining Concavity

Determine the concavity of $y = 3 + \sin x$ on $[0, 2\pi]$.

Solution   The graph of $y = 3 + \sin x$ is concave down on $(0, \pi)$, where $y'' = -\sin x$ is negative. It is concave up on $(\pi, 2\pi)$, where $y'' = -\sin x$ is positive (Figure 3.26).

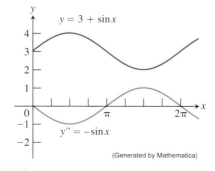

(Generated by Mathematica)

FIGURE 3.26 Using the graph of y'' to determine the concavity of y. (Example 4)

Points of Inflection

The curve $y = 3 + \sin x$ in Example 4 changes concavity at the point $(\pi, 3)$. We call $(\pi, 3)$ a *point of inflection* of the curve.

> **Definition** Point of Inflection
> A point where the graph of a function has a tangent line and where the concavity changes is a **point of inflection.**

A point on a curve where y'' is positive on one side and negative on the other is a point of inflection. At such a point, y'' is either zero (because derivatives have the Intermediate Value Property) or undefined. If y is a twice-differentiable function, $y'' = 0$ at a point of inflection and y' has a local maximum or minimum.

To study the motion of a body moving along a line as a function of time, we often are interested in knowing when the body's acceleration, given by the second derivative, is positive or negative. The points of inflection on the graph of the body's position function reveal where the acceleration changes sign.

Example 5 Studying Motion Along a Line

A particle is moving along a horizontal line with position function

$$s(t) = 2t^3 - 14t^2 + 22t - 5, \qquad t \geq 0.$$

Find the velocity and acceleration, and describe the motion of the particle.

Solution The velocity is

$$v(t) = s'(t) = 6t^2 - 28t + 22 = 2(t - 1)(3t - 11),$$

and the acceleration is

$$a(t) = v'(t) = s''(t) = 12t - 28 = 4(3t - 7).$$

When the function $s(t)$ is increasing, the particle is moving to the right; when $s(t)$ is decreasing, the particle is moving to the left.

Notice that the first derivative $(v = s')$ is zero when $t = 1$ and $t = 11/3$.

Intervals	$0 < t < 1$	$1 < t < 11/3$	$11/3 < t$
Sign of $v = s'$	+	−	+
Behavior of s	increasing	decreasing	increasing
Particle motion	right	left	right

The particle is moving to the right in the time intervals $[0, 1)$ and $(11/3, \infty)$, and moving to the left in $(1, 11/3)$.

The acceleration $a(t) = s''(t) = 4(3t - 7)$ is zero when $t = 7/3$.

Intervals	$0 < t < 7/3$	$7/3 < t$
Sign of $a = s''$	−	+
Graph of s	concave down	concave up

The accelerating force is directed toward the left during the time interval $[0, 7/3)$, is momentarily zero at $t = 7/3$, and is directed toward the right thereafter.

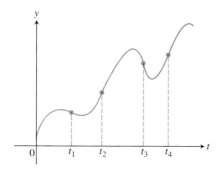

FIGURE 3.27 A hypothetical version of the Dow Jones Industrial Average for Example 6.

Example 6   Inflection Points and the Stock Market

The graph in Figure 3.27 shows a hypothetical version of the Dow Jones Industrial Average. The Dow Jones Industrial Average is a stock market *index*, capturing the overall increase of the stock market along with the local dips and rises.

One way to invest in the stock market is to buy shares of an index fund, which in turn buys a number of different stocks with the goal of tracking the index. The goal of an index fund director would certainly be to buy low (at local minima) and sell high (at local maxima). Such market timing is elusive, however, since it is impossible to predict the extreme values of the market. By the time an investor realizes that the market is truly on the upswing, the minimum has already passed.

Inflection points provide a way for an investor to predict a reversing trend before it happens, since inflection points signal fundamental changes in the *growth rate* of a function. Buying at (or close to) inflection points allows an investor to stay with the historical upward trend (the inflection points warn of a change in trend). This way an investor dampens the fluctuations of the market and, over the long run, captures the upward trend.

Example 7   Finding No Inflection Point Where $y'' = 0$

The curve $y = x^4$ has no inflection point at $x = 0$ (Figure 3.28). Even though $y'' = 12x^2$ is zero there, it does not change sign.

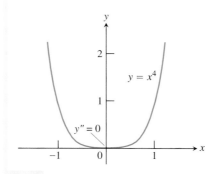

FIGURE 3.28 The graph of $y = x^4$ has no inflection point at the origin, even though $y'' = 0$ there.

Example 8   Finding an Inflection Point Where y'' Does Not Exist

The curve $y = x^{1/3}$ has a point of inflection at $x = 0$ (Figure 3.29), but y'' does not exist there.

$$y'' = \frac{d^2}{dx^2}\left(x^{1/3}\right) = \frac{d}{dx}\left(\frac{1}{3}x^{-2/3}\right) = -\frac{2}{9}x^{-5/3}$$

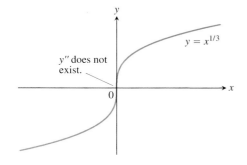

FIGURE 3.29 A point where y'' fails to exist can be a point of inflection.

We see from Example 7 that a zero second derivative does not always produce a point of inflection. From Example 8, we see that inflection points can also occur where there *is no* second derivative.

Second Derivative Test for Local Extrema

Instead of looking for sign changes in y' at critical points, we can sometimes use the following test to determine the presence of local extrema.

> **Theorem 5** Second Derivative Test for Local Extrema
>
> 1. If $f'(c) = 0$ and $f''(c) < 0$, then f has a local maximum at $x = c$.
>
> 2. If $f'(c) = 0$ and $f''(c) > 0$, then f has a local minimum at $x = c$.

This test requires us to know f'' *only at c itself* and not in an interval about c. This makes the test easy to apply. That's the good news. The bad news is that the test fails if $f''(c) = 0$ or if $f''(c)$ fails to exist. When this happens, go back to the first derivative test for local extreme values.

In Example 9, we apply the second derivative test to the function in Example 1.

Example 9   Using the Second Derivative Test

Find the extreme values of $f(x) = x^3 - 12x - 5$.

Solution   We have

$$f'(x) = 3x^2 - 12 = 3(x^2 - 4)$$
$$f''(x) = 6x.$$

Testing the critical points $x = \pm 2$ (there are no endpoints), we find

$$f''(-2) = -12 < 0 \Rightarrow f \text{ has a local maximum at } x = -2$$

and

$$f''(2) = 12 > 0 \Rightarrow f \text{ has a local minimum at } x = 2.$$

Example 10   Using f' and f'' to Graph f

Sketch a graph of the function

$$f(x) = x^4 - 4x^3 + 10$$

using the following steps.

(a) Identify where the extrema of f occur.

(b) Find the intervals on which f is increasing and the intervals on which f is decreasing.

(c) Find where the graph of f is concave up and where it is concave down.

(d) Sketch a possible graph for f.

Solution f is continuous since $f'(x) = 4x^3 - 12x^2$ exists. The domain of f' is $(-\infty, \infty)$, so the domain of f is also $(-\infty, \infty)$. Thus, the critical points of f occur only at the zeros of f'. Since

$$f'(x) = 4x^3 - 12x^2 = 4x^2(x - 3)$$

the first derivative is zero at $x = 0$ and $x = 3$.

Intervals	$x < 0$	$0 < x < 3$	$3 < x$
Sign of f'	$-$	$-$	$+$
Behavior of f	decreasing	decreasing	increasing

(a) Using the first derivative test for local extrema and the table above, we see that there is no extremum at $x = 0$ and a local minimum at $x = 3$.

(b) Using the table above, we see that f is decreasing in $(-\infty, 0]$ and $[0, 3]$, and increasing in $[3, \infty)$.

(c) $f''(x) = 12x^2 - 24x = 12x(x - 2)$ is zero at $x = 0$ and $x = 2$.

Intervals	$x < 0$	$0 < x < 2$	$2 < x$
Sign of f''	$+$	$-$	$+$
Behavior of f	concave up	concave down	concave up

We see that f is concave up on the intervals $(-\infty, 0)$ and $(2, \infty)$, and concave down on $(0, 2)$.

(d) Summarizing the information in the two tables above, we obtain

$x < 0$	$0 < x < 2$	$2 < x < 3$	$3 < x$
decreasing	decreasing	decreasing	increasing
concave up	concave down	concave up	concave up

Figure 3.30 shows the graph of f.

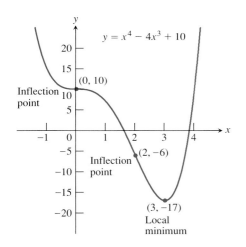

FIGURE 3.30 The graph of $f(x) = x^4 - 4x^3 + 10$. (Example 10)

The steps in Example 10 give a general procedure for graphing by hand.

Learning about Functions from Derivatives

As we saw in Example 10, we can learn almost everything we need to know about a twice-differentiable function $y = f(x)$ by examining its first derivative. We can find where the function's graph rises and falls and where any local extrema are assumed. We can differentiate y' to learn how the graph bends as it passes over the intervals of rise and fall. We can determine the shape of the function's graph. The only information we cannot get from the derivative is how to place the graph in the xy-plane. But, as we discovered in Section 3.2, the only additional information we need to position the graph is the value of f at one point.

Procedure for Graphing y = f(x) by Hand

Step 1. Find y' and y''.

Step 2. Find the rise and fall of the curve.

Step 3. Determine the concavity of the curve.

Step 4. Make a summary and show the curve's general shape.

Step 5. Plot specific points and sketch the curve.

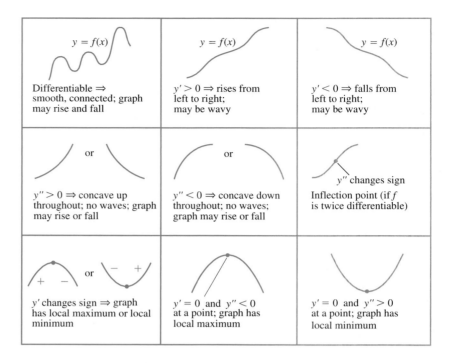

$y = f(x)$	$y = f(x)$	$y = f(x)$
Differentiable $\Rightarrow$ smooth, connected; graph may rise and fall	$y' > 0 \Rightarrow$ rises from left to right; may be wavy	$y' < 0 \Rightarrow$ falls from left to right; may be wavy
or	or	y'' changes sign
$y'' > 0 \Rightarrow$ concave up throughout; no waves; graph may rise or fall	$y'' < 0 \Rightarrow$ concave down throughout; no waves; graph may rise or fall	Inflection point (if f is twice differentiable)
or		
y' changes sign $\Rightarrow$ graph has local maximum or local minimum	$y' = 0$ and $y'' < 0$ at a point; graph has local maximum	$y' = 0$ and $y'' > 0$ at a point; graph has local minimum

EXERCISES 3.3

Sketching y from Graphs of y′ and y″

Each of Exercises 1–4 shows the graphs of the first and second derivatives of a function $y = f(x)$. Copy the picture and add to it a sketch of the approximate graph of f, given that the graph passes through the point P.

1.

2.

3.

4.

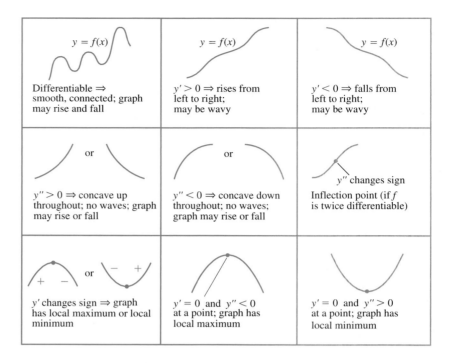

Sketching y from Signs of y′ and y″

5. Sketch the graph of a twice-differentiable function $y = f(x)$ with the following properties. Label coordinates where possible.

x	y	Derivatives
$x < 2$		$y' < 0, \quad y'' < 0$
2	1	$y' = 0, \quad y'' < 0$
$2 < x < 4$		$y' < 0, \quad y'' < 0$
4	4	$y' > 0, \quad y'' = 0$
$4 < x < 6$		$y' > 0, \quad y'' < 0$
6	7	$y' = 0, \quad y'' < 0$
$x > 6$		$y' < 0, \quad y'' < 0$

6. Sketch the graph of a twice-differentiable function $y = f(x)$ that passes through the points $(-2, 2)$, $(-1, 1)$, $(0, 0)$, $(1, 1)$ and $(2, 2)$ and whose first two derivatives have the following sign patterns:

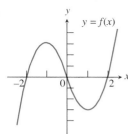

Using Graphs to Analyze Functions

In Exercises 7 and 8, use the graph of the function f to estimate where (a) f' and (b) f'' are 0, positive, and negative.

7.

8.

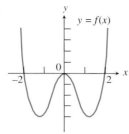

In Exercises 9–12, use the graph of f' to estimate the intervals on which the function f is (a) increasing or (b) decreasing. (c) Estimate where f has local extreme values.

9.

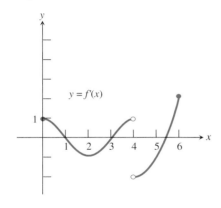

10.

11. The domain of f' is $[0, 4) \cup (4, 6]$.

12. The domain of f' is $[0, 1) \cup (1, 2) \cup (2, 3]$.

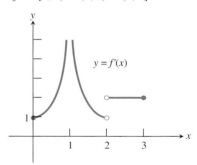

Analyzing f Given f'

Answer the following questions about the functions whose derivatives are given in Exercises 13–16.

  (a) What are the critical points of f?

  (b) On what intervals is f increasing or decreasing?

  (c) At what points, if any, does f assume local maximum and minimum values?

13. $f'(x) = (x - 1)(x + 2)$

14. $f'(x) = (x - 1)^2(x + 2)$

15. $f'(x) = (x - 1)e^{-x}$

16. $f'(x) = x^{-1/3}(x + 2)$

Shape of a Graph

In Exercises 17–22, use analytic methods to find the intervals on which the function is

  (a) increasing

  (b) decreasing

  (c) concave up

  (d) concave down.

Then locate and identify any

  (e) local extreme values

  (f) inflection points.

17. $y = x^2 - x - 1$ **18.** $y = -2x^3 + 6x^2 - 3$

19. $y = 2x^4 - 4x^2 + 1$ **20.** $y = xe^{1/x}$

21. $y = x\sqrt{8 - x^2}$ **22.** $y = \begin{cases} 3 - x^2, & x < 0 \\ x^2 + 1, & x \geq 0 \end{cases}$

T In Exercises 23–32, graph the function to find the intervals on which the function is

  (a) increasing

  (b) decreasing

  (c) concave up

  (d) concave down.

Then locate and identify any

(e) local extreme values

(f) inflection points.

23. $y = 4x^3 + 21x^2 + 36x - 20$ **24.** $y = -x^4 + 4x^3 - 4x + 1$

25. $y = 2x^{1/5} + 3$ **26.** $y = 5 - x^{1/3}$

27. $y = x^{1/3}(x - 4)$ **28.** $y = x^2\sqrt{9 - x^2}$

29. $y = xe^{1/x^2}$ **30.** $y = \dfrac{\ln x}{x}$

31. $y = x^{1/4}(x + 3)$ **32.** $y = \dfrac{x}{x^2 + 1}$

Extrema and Points of Inflection

In Exercises 33 and 34, use the derivative of the function $y = f(x)$ to find the points at which f has a

(a) local maximum

(b) local minimum

(c) point of inflection.

33. $y' = (x - 1)^2(x - 2)$ **34.** $y' = (x - 1)^2(x - 2)(x - 4)$

In Exercises 35 and 36, *work in groups of two or three.*

(a) Find the absolute extrema of f and where they occur.

(b) Find any points of inflection.

(c) Sketch a possible graph of f.

35. f is continuous on $[0, 3]$ and satisfies the following.

x	0	1	2	3
f	0	2	0	-2
f'	3	0	does not exist	-3
f''	0	-1	does not exist	0

x	$0 < x < 1$	$1 < x < 2$	$2 < x < 3$
f	$+$	$+$	$-$
f'	$+$	$-$	$-$
f''	$-$	$-$	$-$

36. f is an even function, continuous on $[-3, 3]$, and satisfies the following.

x	0	1	2
f	2	0	-1
f'	does not exist	0	does not exist
f''	does not exist	0	does not exist

x	$0 < x < 1$	$1 < x < 2$	$2 < x < 3$
f	$+$	$-$	$-$
f'	$-$	$-$	$+$
f''	$+$	$-$	$-$

T In Exercises 37–40, find the inflection points (if any) on the graph of the function and the coordinates of the points on the graph where the function has a local maximum or local minimum value. Then graph the function in a region large enough to show all these points simultaneously. Add to your picture the graphs of the function's first and second derivatives. How are the values at which these graphs intersect the x-axis related to the graph of the function? In what other ways are the graphs of the derivatives related to the graph of the function?

37. $y = x^5 - 5x^4 - 240$

38. $y = x^3 - 12x^2$

39. $y = \dfrac{4}{5}x^5 + 16x^2 - 25$

40. $y = \dfrac{x^4}{4} - \dfrac{x^3}{3} - 4x^2 + 12x + 20$

 41. Graph $f(x) = 2x^4 - 4x^2 + 1$ and its first two derivatives together. Comment on the behavior of f in relation to the signs and values of f' and f''.

 42. Graph $f(x) = x \cos x$ and its second derivative together for $0 \le x \le 2\pi$. Comment on the behavior of the graph of f in relation to the signs and values of f''.

Motion Along a Line

In Exercises 43–46, a particle is moving along a line with position function $s(t)$. Find the (a) velocity and (b) acceleration, and (c) describe the motion of the particle for $t \ge 0$.

43. $s(t) = t^2 - 4t + 3$ **44.** $s(t) = 6 - 2t - t^2$

45. $s(t) = t^3 - 3t + 3$ **46.** $s(t) = 3t^2 - 2t^3$

In Exercises 47 and 48, the graph of the position function $y = s(t)$ of a particle moving along a line is given. At approximately what times is the particle's

(a) velocity equal to zero?

(b) acceleration equal to zero?

47.

48.

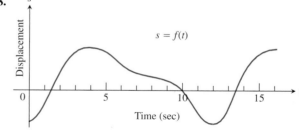

Theory and Examples

49. *Writing to Learn* If $f(x)$ is a differentiable function and $f'(c) = 0$ at an interior point c of f's domain, must f have a local maximum or minimum at $x = c$? Explain.

50. *Writing to Learn* If $f(x)$ is a twice-differentiable function and $f''(c) = 0$ at an interior point c of f's domain, must f have an inflection point at $x = c$? Explain.

51. *Connecting f and f'* Sketch a smooth curve $y = f(x)$ through the origin with the properties that $f'(x) < 0$ for $x < 0$ and $f'(x) > 0$ for $x > 0$.

52. *Connecting f and f''* Sketch a smooth curve $y = f(x)$ through the origin with the properties that $f''(x) < 0$ for $x < 0$ and $f''(x) > 0$ for $x > 0$.

53. *Connecting f, f', and f''* Sketch a continuous curve $y = f(x)$ with the following properties. Label coordinates where possible.

$f(-2) = 8$	$f'(x) > 0$ for $|x| > 2$
$f(0) = 4$	$f'(x) < 0$ for $|x| < 2$
$f(2) = 0$	$f''(x) < 0$ for $x < 0$
$f'(2) = f'(-2) = 0$	$f''(x) > 0$ for $x > 0$

54. *Writing to Learn: Horizontal tangents* Does the curve $y = x^2 + 3\sin 2x$ have a horizontal tangent near $x = -3$? Give reasons for your answer.

55. *Writing to Learn* For $x > 0$, sketch a curve $y = f(x)$ that has $f(1) = 0$ and $f'(x) = 1/x$. Can anything be said about the concavity of such a curve? Give reasons for your answer.

56. *Writing to Learn* Can anything be said about the graph of a function $y = f(x)$ that has a continuous second derivative that is never zero? Give reasons for your answer.

57. *Quadratic curves* What can you say about the inflection points of a quadratic curve $y = ax^2 + bx + c$, $a \ne 0$? Give reasons for your answer.

58. *Cubic curves* What can you say about the inflection points of a cubic curve $y = ax^3 + bx^2 + cx + d$, $a \ne 0$? Give reasons for your answer.

Counting Zeros

When we solve an equation $f(x) = 0$ numerically, we usually want to know beforehand how many solutions to look for in a given interval. With the help of Corollary 3, we can sometimes find out.

Suppose that

    1. f is continuous on $[a, b]$ and differentiable on (a, b)

    2. $f(a)$ and $f(b)$ have opposite signs

    3. $f' > 0$ on (a, b) or $f' < 0$ on (a, b).

Then f has exactly one zero between a and b: It cannot have more than one because it is either increasing on $[a, b]$ or decreasing on $[a, b]$. Yet it has at least one, by the Intermediate Value Theorem (Section 1.4). For example, $f(x) = x^3 + 3x + 1$ has exactly one zero

on $[-1, 1]$ because f is differentiable on $[-1, 1]$, $f(-1) = -3$ and $f(1) = 5$ have opposite signs, and $f'(x) = 3x^2 + 3 > 0$ for all x (see the accompanying figure.)

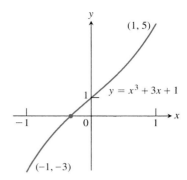

Show that the functions in Exercises 59–62 have exactly one zero in the given interval.

59. $f(x) = x^4 + 3x + 1$, $\quad [-2, -1]$

60. $g(t) = \sqrt{t} + \sqrt{1 + t} - 4$, $\quad (0, \infty)$

61. $r(\theta) = \theta + \sin^2\left(\dfrac{\theta}{3}\right) - 8$, $\quad (-\infty, \infty)$

62. $r(\theta) = \tan\theta - \cot\theta - \theta$, $\quad (0, \pi/2)$

COMPUTER EXPLORATIONS

63. *Writing to Learn: A family of cubic functions*

    (a) On a common screen, graph $f(x) = x^3 + kx$ for $k = 0$ and nearby positive and negative values of k. How does the value of k seem to affect the shape of the graph?

    (b) Find $f'(x)$. As you will see, $f'(x)$ is a quadratic function of x. Find the discriminant of the quadratic (the discriminant of $ax^2 + bx + c$ is $b^2 - 4ac$). For what values of k is the discriminant positive? Zero? Negative? For what values of k does f' have two zeros? One or no zeros? Now explain what the value of k has to do with the shape of the graph of f.

    (c) Experiment with other values of k. What appears to happen as $k \to -\infty$? As $k \to \infty$?

64. *Writing to Learn: A family of quartic functions*

    (a) On a common screen, graph $f(x) = x^4 + kx^3 + 6x^2$, $-1 \le x \le 4$ for $k = -4$, and some nearby values of k. How does the value of k seem to affect the shape of the graph?

    (b) Find $f''(x)$. As you will see, $f''(x)$ is a quadratic function of x. What is the discriminant of this quadratic (see Exercise 63b)? For what values of k is the discriminant positive? Zero? Negative? For what values of k does $f''(x)$ have two zeros? One or no zeros? Now explain what the value of k has to do with the shape of the graph of f.

In Exercises 65 and 66, use a CAS to solve the problem.

65. *Logistic functions* Let $f(x) = c / (1 + ae^{-bx})$ with $a > 0$, $abc \neq 0$.

    (a) Show that f is increasing on the interval $(-\infty, \infty)$ if $abc > 0$ and decreasing if $abc < 0$.

    (b) Show that the point of inflection of f occurs at $x = (\ln a)/b$.

66. *Quartic polynomial functions* Let $f(x) = ax^4 + bx^3 + cx^2 + dx + e$ with $a \neq 0$.

    (a) Show that the graph of f has 0 or 2 points of inflection.

    (b) Write a condition that must be satisfied by the coefficients if the graph of f has 0 or 2 points of inflection.

3.4 Graphical Solutions of Autonomous Differential Equations

Equilibrium Values and Phase Lines   •   Stable and Unstable Equilibria   •
Cooling, a Falling Body Encountering Resistance, and Logistic Growth

We can build on our knowledge of how derivatives determine the shape of a graph to solve differential equations graphically. The starting points for doing so are the notions of *phase line* and *equilibrium value*. We arrive at these notions by investigating what happens when the derivative of a differentiable function is zero from a new point of view.

Equilibrium Values and Phase Lines

We have seen the important role the critical points play in determining how a function behaves and in finding its extreme points. Let's investigate what happens when the derivative of a function is zero from a slightly different point of view. In this case, the derivative dy/dx will be a function of y only (the dependent variable). For example, differentiating the equation

$$y^2 = x + 1$$

implicitly gives

$$2y \frac{dy}{dx} = 1 \qquad \text{or} \qquad \frac{dy}{dx} = \frac{1}{2y}.$$

A differential equation for which dy/dx is a function of y is called an **autonomous** differential equation.

> **Definition**   Equilibrium Values or Rest Points
>
> If $dy/dx = g(y)$ is an autonomous differential equation, then the values of y for which $dy/dx = 0$ are called **equilibrium values** or **rest points.**

Thus, equilibrium values are those for which no change occurs in the dependent variable, so y is at *rest*. The emphasis is on the value of y where $dy/dx = 0$, not the value of x as in the preceding section.

Example 1   Finding Equilibrium Values

The equilibrium values for the autonomous differential equation

$$\frac{dy}{dx} = (y + 1)(y - 2)$$

are $y = -1$ and $y = 2$.

To construct a graphical solution to an autonomous differential equation like the one in Example 1, we first make a **phase line** for the equation, a plot on the y-axis that shows the equation's equilibrium values along the intervals where dy/dx and d^2y/dx^2 are positive and negative. Then we know where the solutions are increasing and decreasing, and the concavity of the solution curves. These are the essential features we found in the previous section, so we can determine the shapes of the solution curves without having to find formulas for them.

Example 2   Drawing a Phase Line and Sketching Solution Curves

Draw a phase line for the differential equation

$$\frac{dy}{dx} = (y + 1)(y - 2)$$

and use it to sketch solutions to the equation.

Solution

Step 1. *Draw a number line for y and mark the equilibrium values $y = -1$ and $y = 2$, where $dy/dx = 0$.*

Step 2. *Identify and label the intervals where $y' > 0$ and $y' < 0$.* This step resembles what we did in the preceding section, only now we are marking the y-axis instead of the x-axis.

We can encapsulate the information about the sign of y' on the phase line itself. Since $y' > 0$ on the interval to the left of $y = -1$, a solution of the differential equation with a y-value less than -1 will increase from there toward $y = -1$. We display this information by drawing an arrow on the interval pointing to -1.

Similarly, $y' < 0$ between $y = -1$ and $y = 2$, so any solution with a value in this interval will decrease toward $y = -1$.

For $y > 2$, we have $y' > 0$, so a solution with a y-value greater than 2 will increase from there without bound.

In short, solution curves below the horizontal line $y = -1$ in the xy-plane rise toward $y = -1$. Solution curves between the lines $y = -1$ and $y = 2$ fall away from $y = 2$ toward $y = -1$. Solution curves above $y = 2$ rise away from $y = 2$ and keep going up.

Step 3. *Calculate y'' and mark the intervals where $y'' > 0$ and $y'' < 0$.* To find y'', we differentiate y' *with respect to x*, using implicit differentiation.

$$y' = (y + 1)(y - 2) = y^2 - y - 2 \qquad \text{Formula for } y' \ldots$$

$$y'' = \frac{d}{dx}(y') = \frac{d}{dx}(y^2 - y - 2)$$

$$= 2yy' - y' \qquad \text{\ldots differentiated implicitly with respect to } x.$$

$$= (2y - 1)y'$$

$$= (2y - 1)(y + 1)(y - 2).$$

From this formula, we see that y'' changes sign at $y = -1$, $y = 1/2$, and $y = 2$. We add the sign information to the phase line.

Step 4. *Sketch an assortment of solution curves in the xy-plane.* The horizontal lines $y = -1$, $y = 1/2$, and $y = 2$ partition the plane into horizontal bands in which we know the signs of y' and y''. In each band, this information tells us whether the solution curves rise or fall and how they bend as x increases (Figure 3.31).

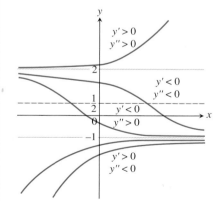

FIGURE 3.31 Graphical solutions from Example 2.

The "equilibrium lines" $y = -1$ and $y = 2$ are also solution curves. (The constant functions $y = -1$ and $y = 2$ satisfy the differential equation.) Solution curves that cross the line $y = 1/2$ have an inflection point there. The concavity changes from concave down (above the line) to concave up (below the line).

As predicted in step 2, solutions in the middle and lower bands approach the equilibrium value $y = -1$ as x increases. Solutions in the upper band rise steadily away from the value $y = 2$.

Stable and Unstable Equilibria

Look at Figure 3.31 once more, in particular at the behavior of the solution curves near the equilibrium values. Once a solution curve has a value near $y = -1$, it tends steadily toward that value; $y = -1$ is a **stable equilibrium.** The behavior near $y = 2$ is just the opposite: all solutions except the equilibrium solution $y = 2$ itself move *away* from it as x increases. We call $y = 2$ an **unstable equilibrium.** If the solution is *at* that value, it stays, but if it is off by any amount, no matter how small, it moves away. (Sometimes an equilibrium value is unstable because a solution moves away from it only on one side of the point.)

Now that we know what to look for, we can already see this behavior on the initial phase line. The arrows lead away from $y = 2$ and, once to the left of $y = 2$, toward $y = -1$.

Cooling, a Falling Body Encountering Resistance, and Logistic Growth

In the next example, we use a phase line to understand the behavior of a physical model. Isaac Newton postulated that the rate of change in the temperature of a cooled or heated object is proportional to the difference in temperature between the object and its surrounding medium. We can use this idea to describe how the object's temperature will change over time.

This postulate is called *Newton's Law of Cooling* (even though it applies equally well to heating).

Example 3   Cooling Soup

What happens to the temperature of the soup when a cup of hot soup is placed on a table in a room? We know the soup cools down, but what does a typical temperature curve look like as a function of time?

Solution   We assume that the soup's Celsius temperature H is a differentiable function of time t, choose a suitable unit for t—minutes, say—and start measuring time at $t = 0$. We also assume that the volume of the surrounding medium is large enough so that the heat of the soup has a negligible effect on its surrounding temperature.

Suppose that the surrounding medium has a constant temperature of 15°C. We can then express the difference in temperature as $H(t) - 15$. According to Newton's Law of Cooling, there is a constant of proportionality $k > 0$ such that

$$\frac{dH}{dt} = -k(H - 15) \tag{1}$$

(*minus k* to give a negative derivative when $H > 15$).

Since $dH/dt = 0$ at $H = 15$, the temperature 15°C is an equilibrium value. If $H > 15$, Equation (1) tells us that $(H - 15) > 0$ and $dH/dt < 0$. If the object is hotter than the room, it will get cooler. Similarly, if $H < 15$, then $(H - 15) < 0$ and $dH/dt > 0$. An object cooler than the room will warm up. Thus, the behavior described by Equation (1) agrees with our intuition of how temperature should behave. These observations are captured in the initial phase line diagram in Figure 3.32.

FIGURE 3.32 First step in constructing the phase line for Newton's Law of Cooling in Example 3. The temperature tends towards the equilibrium (surrounding-medium) value in the long run.

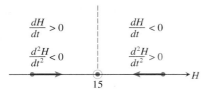

$$\frac{dH}{dt} > 0 \qquad \frac{dH}{dt} < 0$$

$$\frac{d^2H}{dt^2} < 0 \qquad \frac{d^2H}{dt^2} > 0$$

FIGURE 3.33 The complete phase line for Example 3.

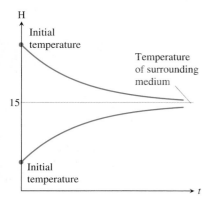

FIGURE 3.34 Temperature versus time. Regardless of initial temperature, the object's temperature $H(t)$ tends toward 15°C, the temperature of the surrounding medium.

We determine the concavity of the solution curves by differentiating both sides of Equation (1) with respect to t:

$$\frac{d}{dt}\left(\frac{dH}{dt}\right) = \frac{d}{dt}\left(-k(H-15)\right)$$

$$\frac{d^2H}{dt^2} = -k\,\frac{dH}{dt}.$$

Since $-k$ is negative, we see that d^2H/dt^2 is positive when $dH/dt < 0$ and negative when $dH/dt > 0$. Figure 3.33 adds this information to the phase line.

The completed phase line shows that if the temperature of the object is above the equilibrium value of 15°C, the graph of $H(t)$ will be decreasing and concave upward. If the temperature is below 15°C (the temperature of the surrounding medium), the graph of $H(t)$ will be increasing and concave downward. We use this information to sketch typical solution curves (Figure 3.34).

From the upper solution curve in Figure 3.34, we see that as the object cools down, the rate at which it cools slows down because dH/dt approaches zero. This observation is implicit in Newton's Law of Cooling and contained in the differential equation, but the flattening of the graph as time advances gives an immediate visual representation of the phenomenon. The ability to discern physical behavior from graphs is a powerful tool in understanding real-world systems.

Example 4 Analyzing the Fall of a Body Encountering a Resistive Force

Galileo and Newton both observed that the rate of change in momentum encountered by a moving object is equal to the net force applied to it. In mathematical terms,

$$F = \frac{d}{dt}(mv), \tag{2}$$

where F is the force and m and v the object's mass and velocity. If m varies with time, as it will if the object is a rocket burning fuel, the right-hand side of Equation (2) expands to

$$m\frac{dv}{dt} + v\frac{dm}{dt}$$

using the product rule. In many situations, however, m is constant, $dm/dt = 0$, and Equation (2) takes the simpler form

$$F = m\frac{dv}{dt} \qquad \text{or} \qquad F = ma, \tag{3}$$

known as *Newton's second law of motion*.

In free fall, the constant acceleration due to gravity is denoted by g and the one force acting downward on the falling body is

$$F_p = mg,$$

the propulsion due to gravity. If, however, we think of a real body falling through the air—say, a penny from a great height or a parachutist from an even greater height—we know that at some point air resistance is a factor in the speed of the fall. A more realistic model of free fall would include air resistance, shown as a force F_r in the schematic diagram in Figure 3.35.

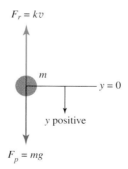

FIGURE 3.35 An object falling under the influence of gravity with a resistive force assumed to be proportional to the velocity.

For speeds well below the speed of sound, physical experiments have shown that F_r is approximately proportional to the body's velocity. The net force on the falling body is therefore

$$F = F_p - F_r$$

giving

$$ma = mg - kv$$

$$\frac{dv}{dt} = g - \frac{k}{m} v. \tag{4}$$

We can use a phase line to analyze the velocity functions that solve this differential equation.

The equilibrium point, obtained by setting the right-hand side of Equation (4) equal to zero, is

$$v = \frac{mg}{k}.$$

If the body is initially moving faster than this, dv/dt is negative and the body slows down. If the body is moving at a velocity below mg/k, $dv/dt > 0$ and the body speeds up. These observations are captured in the initial phase line diagram in Figure 3.36.

We determine the concavity of the solution curves by differentiating both sides of Equation (4) with respect to t:

$$\frac{d^2v}{dt^2} = \frac{d}{dt}\left(g - \frac{k}{m} v\right) = -\frac{k}{m}\frac{dv}{dt}.$$

We see that $d^2v/dt^2 < 0$ when $v < mg/k$ and $d^2v/dt^2 > 0$ when $v > mg/k$. Figure 3.37 adds this information to the phase line. Notice the similarity to the phase

FIGURE 3.36 Initial phase line for Example 4.

FIGURE 3.37 The completed phase line for Example 4.

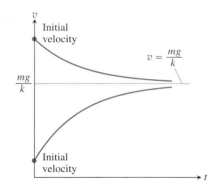

FIGURE 3.38 Typical velocity curves in Example 4. The value $v = mg/k$ is the terminal velocity.

Skydivers can vary their terminal velocity from 95 mph to 180 mph by changing the amount of body area opposing the fall.

line for Newton's Law of Cooling (Figure 3.33). The solution curves are similar as well (Figure 3.38).

Figure 3.38 shows two typical solution curves. Regardless of the initial velocity, we see the body's velocity tending toward the limiting value $v = mg/k$. This value, a stable equilibrium point, is called the body's **terminal velocity.**

Example 5   Analyzing Population Growth in a Limiting Environment

Suppose that $P = P(t)$ represents the number of individuals in a particular population at time t. For instance, P might be the number of yeast cells in a nutrient fluid or the number of white owls in the western United States. Assume that during a small increment of time Δt, a certain percentage of the population is born and another percentage dies. Then the average rate of change of P over the interval $[t, t + \Delta t]$ is

$$\frac{\Delta P}{\Delta t} = kP(t), \tag{5}$$

where $k > 0$ is the birth rate minus the death rate per individual per unit time.

Because the natural environment has only a limited number of resources to sustain life, it is reasonable to assume that only a maximum population M can be accommodated. As the population approaches this **limiting population** or **carrying capacity,** resources become less abundant and the growth rate k decreases. A simple relationship exhibiting this behavior is

$$k = r(M - P),$$

where $r > 0$ is a constant. Notice that k decreases as P increases toward M and that k is negative if P is greater than M. Substituting $r(M - P)$ for k in Equation (5) and taking the limit as $\Delta t \to 0$ gives the differential equation

$$\frac{dP}{dt} = r(M - P)P = rMP - rP^2. \tag{6}$$

The model (6) is referred to as **logistic growth.**

We can forecast the behavior of the population over time by analyzing the phase line for Equation (6). The equilibrium values are $P = M$ and $P = 0$, and we can see that $dP/dt > 0$ if $0 < P < M$ and $dP/dt < 0$ if $P > M$. These observations are recorded on the phase line in Figure 3.39.

$$\frac{dP}{dt} > 0 \qquad \frac{dP}{dt} < 0$$

FIGURE 3.39 The initial phase line for Equation 6.

We determine the concavity of the population curves by differentiating both sides of Equation (6) with respect to t:

$$\frac{d^2P}{dt^2} = \frac{d}{dt}(rMP - rP^2)$$

$$= rM\frac{dP}{dt} - 2rP\frac{dP}{dt}$$

$$= r(M - 2P)\frac{dP}{dt}. \tag{7}$$

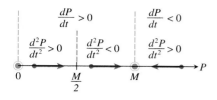

FIGURE 3.40 The completed phase line for logistic growth (Equation 6).

If $P = M/2$, then $d^2P/dt^2 = 0$. If $P < M/2$, then $(M - 2P)$ and dP/dt are positive and $d^2P/dt^2 > 0$. If $M/2 < P < M$, then $(M - 2P) < 0$, $dP/dt > 0$, and $d^2P/dt^2 < 0$. If $P > M$, then $(M - 2P)$ and dP/dt are both negative and $d^2P/dt^2 > 0$. We add this information to the phase line (Figure 3.40).

The lines $P = M/2$ and $P = M$ divide the first quadrant of the tP-plane into horizontal bands in which we know the signs of both dP/dt and d^2P/dt^2. In each band, we know how the solution curves rise and fall and how the bend behaves as time passes. The equilibrium lines $P = 0$ and $P = M$ are both population curves. Population curves crossing the line $P = M/2$ have an inflection point there, giving them a **sigmoid** shape (curved in two directions like a letter S). Figure 3.41 displays typical population curves. Figure 3.42 shows the growth of a laboratory yeast culture based on experimental data.

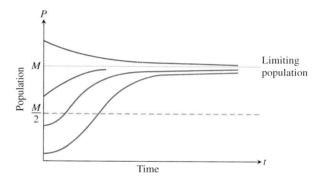

FIGURE 3.41 Population curves in Example 5.

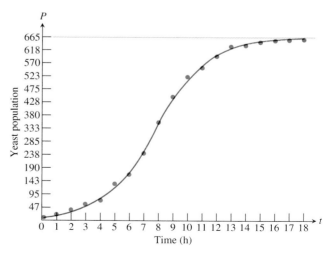

FIGURE 3.42 Logistic curve showing the growth of yeast in a culture. The dots indicate observed values. (Data from R. Pearl, "Growth of Population," *Quart. Rev. Biol.* 2 (1927): 532–548.)

EXERCISES 3.4

Phase Lines and Solution Curves

In Exercises 1–8,

    (a) Identify the equilibrium values. Which are stable and which are unstable?

    (b) Construct a phase line. Identify the signs of y' and y''.

    (c) Sketch several solution curves.

1. $\dfrac{dy}{dx} = (y + 2)(y - 3)$ **2.** $\dfrac{dy}{dx} = y^2 - 4$

3. $\dfrac{dy}{dx} = y^3 - y$ **4.** $\dfrac{dy}{dx} = y^2 - 2y$

5. $y' = \sqrt{y}, \quad y > 0$ **6.** $y' = y - \sqrt{y}, \quad y > 0$

7. $y' = (y - 1)(y - 2)(y - 3)$ **8.** $y' = y^3 - y^2$

Models of Population Growth

The autonomous differential equations in Exercises 9–12 represent models for population growth. For each exercise, use a phase line analysis to sketch solution curves for $P(t)$ selecting different starting values $P(0)$ (as in Example 5). Which equilibria are stable, and which are unstable?

9. $\dfrac{dP}{dt} = 1 - 2P$ **10.** $\dfrac{dP}{dt} = P(1 - 2P)$

11. $\dfrac{dP}{dt} = 2P(P - 3)$ **12.** $\dfrac{dP}{dt} = 3P(1 - P)\left(P - \dfrac{1}{2}\right)$

13. *Catastrophic continuation of Example 5* Suppose that a healthy population of some species is growing in a limited environment and that the current population P_0 is fairly close to the carrying capacity M_0. You might imagine a population of fish living in a freshwater lake in a wilderness area. Suddenly a catastrophe such as the Mount St. Helens volcanic eruption contaminates the lake and destroys a significant part of the food and oxygen on which the fish depend. The result is a new environment with a carrying capacity M_1 considerably less than M_0 and, in fact, less than the current population P_0. Starting at some time before the catastrophe, sketch a "before-and-after" curve that shows how the fish population responds to the change in environment.

14. *Controlling a population* The fish and game department in a certain state is planning to issue hunting permits to control the deer population (one deer per permit). It is known that if the deer population falls below a certain level m, the deer will become extinct. It is also known that if the deer population goes above the maximum carrying capacity M, the population will decrease back to M through disease and malnutrition.

    (a) Discuss the reasonableness of the following model for the growth rate of the deer population as a function of time:

$$\frac{dP}{dt} = rP(M - P)(P - m),$$

    where P is the population of the deer and r is a positive constant of proportionality. Include a phase line.

    (b) Explain how this model differs from the logistic model $dP/dt = rP(M - P)$. Is it better or worse than the logistic model?

    (c) Show that if $P > M$ for all t, then $\lim_{t \to \infty} P(t) = M$.

    (d) What happens if $P < m$ for all t?

    (e) Discuss the solutions to the differential equation. What are the equilibrium points of the model? Explain the dependence of the steady-state value of P on the initial values of P. About how many permits should be issued?

Applications and Examples

15. *Skydiving* If a body of mass m falling from rest under the action of gravity encounters an air resistance proportional to the square of velocity, then the body's velocity t seconds into the fall satisfies the equation

$$m\frac{dv}{dt} = mg - kv^2, \quad k > 0$$

where k is a constant that depends on the body's aerodynamic properties and the density of the air. (We assume that the fall is too short to be affected by changes in the air's density.)

    (a) Draw a phase line for the equation.

    (b) Sketch a typical velocity curve.

    (c) For a 160-lb skydiver ($mg = 160$) and with time in seconds and distance in feet, a typical value of k is 0.005. What is the diver's terminal velocity?

16. *Resistance proportional to $\sqrt{v}$* A body of mass m is projected vertically downward with initial velocity v_0. Assume that the resisting force is proportional to the square root of the velocity and find the terminal velocity from a graphical analysis.

17. *Sailing* A sailboat is running along a straight course with the wind providing a constant forward force of 50 lb. The only other force acting on the boat is resistance as the boat moves through the water. The resisting force is numerically equal to five times the boat's speed, and the initial velocity is 1 ft/sec. What is the maximum velocity in feet per second of the boat under this wind?

18. *The spread of information* Sociologists recognize a phenomenon called **social diffusion,** which is the spreading of a piece of information, technological innovation, or cultural fad among a population. The members of the population can be divided into two classes: those who have the information and those who do not. In a fixed population whose size is known, it is reasonable to assume that the rate of diffusion is proportional to the number who have the information times the number yet to receive it. If X denotes the number of individuals who have the information in a population of N people, then a mathematical model for social diffusion is given by

$$\frac{dX}{dt} = kX(N - X),$$

where t represents time and k is a positive constant.

(a) Discuss the reasonableness of the model.

(b) Construct a phase line identifying the sign of X' and X''.

(c) Sketch representative solution curves.

(d) Predict the value of X for which the information is spreading most rapidly. How many people eventually receive the information?

19. *Current in a circuit* The accompanying diagram represents an electrical circuit whose total resistance is a constant R ohms and whose self-inductance, shown as a coil, is L henries, also a constant. There is a switch whose terminals at a and b can be closed to connect a constant electrical source of V volts.
Ohm's law, $V = Ri$, has to be modified for such a circuit. The modified form is

$$L\frac{di}{dt} + Ri = V,$$

where i is the intensity of the current in amperes and t is the time in seconds. By solving this equation, we can predict how the current will flow after the switch is closed.

Use a phase line analysis to sketch the solution curve assuming that the switch in the RL circuit is closed at time $t = 0$. What happens to the current as $t \to \infty$? This value is called the **steady-state solution.**

20. *A pearl in shampoo* Suppose that a pearl is sinking in a thick fluid, like shampoo, subject to a frictional force opposing its fall and proportional to its velocity. Suppose that there is also a resistive buoyant force exerted by the shampoo. According to **Archimedes' principle,** the buoyant force equals the weight of the fluid displaced by the pearl. If m is the mass of the pearl and P is the mass of the shampoo displaced by the pearl as it descends, complete the following steps.

(a) Draw a schematic diagram showing the forces acting on the pearl as it sinks, as in Figure 3.35.

(b) Using $v(t)$ for the pearl's velocity as a function of time t, write a differential equation modeling the velocity of the pearl as a falling body.

(c) Construct a phase line identifying the signs of v' and v''.

(d) Sketch typical solution curves.

(e) What is the terminal velocity of the pearl?

3.5 Modeling and Optimization

Examples from Business and Industry • Examples from Mathematics and Physics • Fermat's Principle and Snell's Law • Examples from Economics • Modeling Discrete Phenomena with Differentiable Functions

To optimize something means to maximize or minimize some aspect of it. What is the size of the most profitable production run? What is the least expensive shape for an oil can? What is the stiffest beam we can cut from a 12 in. log? In the mathematical models in which we use functions to describe the things that interest us, we usually answer such questions by finding the greatest or smallest value of a differentiable function.

To illustrate, suppose that a cabinet maker uses exotic materials to produce custom furnishings. She has a contract to produce 5 units per day, which is also her capacity. For each of the raw materials she uses, she wants to determine *how much* and *how often* to have the material delivered. She is charged a delivery cost independent of the amount delivered, and she can rent as much storage as she likes. She reasons that if the delivery cost of a particular material is high and the storage cost low, she should deliver infrequently and store much. On the other hand, if the delivery cost is low and the storage cost high, she should deliver frequently and store little. But what is the minimum combined total cost of delivery and storage for each item? We answer this question later in this section.

Examples from Business and Industry

Example 1   Fabricating a Box

An open-top box is to be made by cutting small congruent squares from the corners of a 12-in.-by-12-in. sheet of tin and bending up the sides. How large should the squares cut from the corners be to make the box hold as much as possible?

Solution   We start with a picture (Figure 3.43). In the figure, the corner squares are x in. on a side. The volume of the box is a function of this variable:

$$V(x) = x(12 - 2x)^2 = 144x - 48x^2 + 4x^3. \qquad V = h\,/\,w$$

Since the sides of the sheet of tin are only 12 in. long, $x \leq 6$ and the domain of V is the interval $0 \leq x \leq 6$.

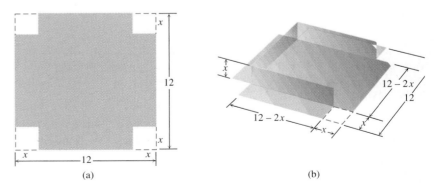

(a)   (b)

FIGURE 3.43 An open box made by cutting the corners from a square sheet of tin. (Example 1)

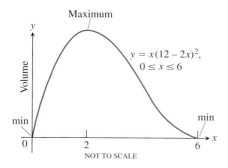

Maximum

$y = x(12 - 2x)^2,$
$0 \le x \le 6$

min

min

0   2   6

NOT TO SCALE

FIGURE 3.44 The volume of the box in Figure 3.43 graphed as a function of x.

A graph of V (Figure 3.44) suggests a minimum value of 0 at $x = 0$ and $x = 6$ and a maximum near $x = 2$. To learn more, we examine the first derivative of V with respect to x:

$$\frac{dV}{dx} = 144 - 96x + 12x^2 = 12(12 - 8x + x^2) = 12(2 - x)(6 - x).$$

Of the two zeros, $x = 2$ and $x = 6$, only $x = 2$ lies in the interior of the function's domain and makes the critical-point list. The values of V at this one critical point and two endpoints are

Critical-point value: $V(2) = 128$

Endpoint values: $V(0) = 0,$ $V(6) = 0.$

The maximum volume is 128 in^3. The cutout squares should be 2 in. on a side.

Example 2   Designing an Efficient Oil Can

You have been asked to design a 1 L oil can shaped like a right circular cylinder (Figure 3.45). What dimensions will use the least material?

Solution

Volume of can: If r and h are measured in centimeters, then the volume of the can in cubic centimeters is

$$\pi r^2 h = 1000. \qquad \text{1 liter} = 1000 \text{ cm}^3$$

Surface area of can: $A = \underbrace{2\pi r^2}_{\substack{\text{circular} \\ \text{ends}}} + \underbrace{2\pi rh}_{\substack{\text{cylinder} \\ \text{wall}}}$

FIGURE 3.45 This 1 L can uses the least material when $h = 2r$. (Example 2)

How can we interpret the phrase "least material"? One possibility is to ignore the thickness of the material and the waste in manufacturing. Then we ask for dimensions r and h that make the total surface area as small as possible while satisfying the constraint $\pi r^2 h = 1000$. (Exercise 15 describes one way to take waste into account.)

Model

To express the surface area as a function of one variable, we solve for one of the variables in $\pi r^2 h = 1000$ and substitute that expression into the surface area formula. Solving for h is easier:

$$h = \frac{1000}{\pi r^2}.$$

Thus,

$$A = 2\pi r^2 + 2\pi rh$$
$$= 2\pi r^2 + 2\pi r \left(\frac{1000}{\pi r^2} \right)$$
$$= 2\pi r^2 + \frac{2000}{r}.$$

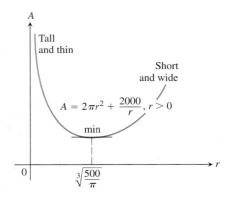

FIGURE 3.46 The graph of $A = 2\pi r^2 + 2000/r$ is concave up.

Solve Analytically

Our goal is to find a value of $r > 0$ that minimizes the value of A. Figure 3.46 suggests that such a value exists.

Notice from the graph that for small r (a tall thin container, like a piece of pipe), the term $2000/r$ dominates and A is large. For large r (a short wide container, like a pizza pan), the term $2\pi r^2$ dominates and A again is large. Since A is differentiable on $r > 0$, an interval with no endpoints, it can have a minimum value only where its first derivative is zero.

$$\frac{dA}{dr} = 4\pi r - \frac{2000}{r^2}$$

$$0 = 4\pi r - \frac{2000}{r^2} \qquad \text{Set } dA/dr = 0.$$

$$4\pi r^3 = 2000 \qquad \text{Multiply by } r^2.$$

$$r = \sqrt[3]{\frac{500}{\pi}} \approx 5.42 \qquad \text{Solve for } r.$$

Something happens at $r = \sqrt[3]{500/\pi}$, but what?

If the domain of A were a closed interval, we could find out by evaluating A at this critical point and the endpoints and comparing the results. But the domain is an open interval, so we must learn what is happening at $r = \sqrt[3]{500/\pi}$ by referring to the shape of A's graph. The second derivative

$$\frac{d^2A}{dr^2} = 4\pi + \frac{4000}{r^3}$$

is positive throughout the domain of A. The graph is therefore concave up and the value of A at $r = \sqrt[3]{500/\pi}$ an absolute minimum. The corresponding value of h (after a little algebra) is

$$h = \frac{1000}{\pi r^2} = 2\sqrt[3]{\frac{500}{\pi}} = 2r.$$

Interpret

The 1 L can that uses the least material has height equal to the diameter, with $r \approx 5.42$ cm and $h \approx 10.84$ cm.

CD-ROM
WEBsite
Historical Biography
Marin Mersenne
(1588 — 1648)

Strategy for Solving Max-Min Problems

Step 1: *Understand the Problem* Read the problem carefully. Identify the information you need to solve the problem. What is unknown? What is given? What is sought?

Step 2:. *Develop a Mathematical Model of the Problem* Draw pictures and label the parts that are important to the problem. Introduce a variable to represent the quantity to be maximized or minimized. Using the variable, write a function whose extreme value gives the information sought.

Step 3: *Find the Domain of the Function* Determine what values of the variable make sense in the problem. Graph the function if possible.

Step 4: *Identify the Critical Points and Endpoints* Find where the derivative is zero or fails to exist. Use what you know about the shape of the function's graph and the physics of the problem. Use the first and second derivatives to identify and classify critical points (where $f' = 0$ or does not exist).

Step 5: *Solve the Mathematical Model* If unsure of the result, support or confirm your solution with another method.

Step 6: *Interpret the Solution* Translate your mathematical result into the problem setting and decide whether the result makes sense.

Examples from Mathematics and Physics

Example 3   Inscribing Rectangles

A rectangle is to be inscribed in a semicircle of radius 2. What is the largest area the rectangle can have, and what are its dimensions?

Solution

Model

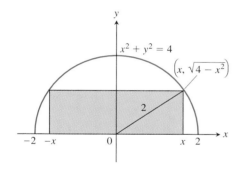

FIGURE 3.47 The rectangle and semicircle in Example 3.

Let $(x, \sqrt{4 - x^2})$ be the coordinates of the corner of the rectangle obtained by placing the circle and rectangle in the coordinate plane (Figure 3.47). The length, height, and area of the rectangle can then be expressed in terms of the position x of the lower right-hand corner:

$$\text{Length: } 2x, \qquad \text{height: } \sqrt{4 - x^2}, \qquad \text{area: } 2x \cdot \sqrt{4 - x^2}.$$

Notice that the values of x are to be found in the interval $0 \le x \le 2$, where the selected corner of the rectangle lies.

Our mathematical goal is now to find the absolute maximum value of the continuous function

$$A(x) = 2x\sqrt{4 - x^2}$$

on the domain $[0, 2]$.

Identify the Critical and Endpoints

The derivative

$$\frac{dA}{dx} = \frac{-2x^2}{\sqrt{4 - x^2}} + 2\sqrt{4 - x^2}$$

is not defined when $x = 2$ and is equal to zero when

$$\frac{-2x^2}{\sqrt{4 - x^2}} + 2\sqrt{4 - x^2} = 0$$

$$-2x^2 + 2(4 - x^2) = 0 \qquad \text{Multiply both sides by } \sqrt{4 - x^2}.$$

$$8 - 4x^2 = 0$$

$$x^2 = 2$$

$$x = \pm\sqrt{2}$$

Of the two zeros, $x = \sqrt{2}$ and $x = -\sqrt{2}$, only $x = \sqrt{2}$ lies in the interior of A's domain and makes the critical-point list. The values of A at the endpoints and at this one critical point are

Critical-point value: $A(\sqrt{2}) = 2\sqrt{2}\sqrt{4-2} = 4$

Endpoint values: $A(0) = 0, \qquad A(2) = 0.$

Interpret

The area has a maximum value of 4 when the rectangle is $\sqrt{4-x^2} = \sqrt{2}$ units high and $2x = 2\sqrt{2}$ units long.

Fermat's Principle and Snell's Law

The speed of light depends on the medium through which it travels and tends to be slower in denser media. In a vacuum, it travels at the speed $c = 3 \times 10^8$ m/sec, but in Earth's atmosphere it travels slightly slower than that, and in glass slower still (about two-thirds as fast).

Fermat's principle in optics states that light always travels from one point to another along the quickest route (shortest time). This observation enables us to predict the path light will take when it travels from a point in one medium (air, say) to a point in another medium (say, glass or water).

Example 4   Finding the Path of a Light Ray

Find the path that a ray of light will follow in going from a point A in a medium where the speed of light is c_1 across a straight boundary to a point B in a medium where the speed of light is c_2.

Solution   Since light traveling from A to B will do so by the quickest route, we look for a path that will minimize the travel time.

Model

We assume that A and B lie in the xy-plane and that the line separating the two media is the x-axis (Figure 3.48).

In a uniform medium, where the speed of light remains constant, "shortest time" means "shortest path," and the ray of light will follow a straight line. Hence, the path from A to B will consist of a line segment from A to a boundary point P, followed by another line segment from P to B. From the formula distance equals rate times time, we have

$$\text{Time} = \frac{\text{distance}}{\text{rate}}.$$

The time required for light to travel from A to P is therefore

$$t_1 = \frac{AP}{c_1} = \frac{\sqrt{a^2 + x^2}}{c_1}.$$

From P to B, the time is

$$t_2 = \frac{PB}{c_2} = \frac{\sqrt{b^2 + (d-x)^2}}{c_2}.$$

The time from A to B is the sum of these:

$$t = t_1 + t_2 = \frac{\sqrt{a^2 + x^2}}{c_1} + \frac{\sqrt{b^2 + (d-x)^2}}{c_2}.$$

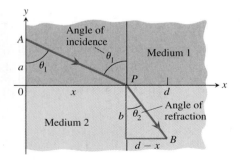

FIGURE 3.48 A light ray refracted (deflected from its path) as it passes from one medium to another. (Example 4)

This equation expresses t as a differentiable function of x whose domain is $[0, d]$, and we want to find the absolute minimum value of t on this closed interval.

Identify the Critical and Endpoints

We find

$$\frac{dt}{dx} = \frac{x}{c_1\sqrt{a^2 + x^2}} - \frac{(d - x)}{c_2\sqrt{b^2 + (d - x)^2}}. \tag{1}$$

In terms of the angles θ_1 and θ_2 in Figure 3.48,

$$\frac{dt}{dx} = \frac{\sin\theta_1}{c_1} - \frac{\sin\theta_2}{c_2}.$$

We can see from Equation (1) that $dt/dx < 0$ at $x = 0$ and $dt/dx > 0$ at $x = d$. Hence, $dt/dx = 0$ at some point x_0 in between (Figure 3.49). There is only one such point because dt/dx is an increasing function of x (Exercise 58). At this point,

$$\frac{\sin\theta_1}{c_1} = \frac{\sin\theta_2}{c_2}.$$

This equation is **Snell's Law** or the **Law of Refraction.**

Interpret

We conclude that the path the ray of light follows is the one described by Snell's Law. Figure 3.50 shows how this works for air and water.

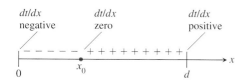

FIGURE 3.49 The sign pattern of dt/dx in Example 4.

FIGURE 3.50 For air and water at room temperature, the light velocity ratio is 1.33, and Snell's Law becomes $\sin\theta_1 = 1.33 \sin\theta_2$. In this laboratory photograph, $\theta_1 = 35.5°$, $\theta_2 = 26°$ and $(\sin 35.5°/\sin 26°) \approx 0.581/0.438 \approx 1.33$, as predicted.

This photograph also illustrates that angle of reflection = angle of incidence. (Exercise 42)

Examples from Economics

Here we want to point out two more places where calculus makes a contribution to economic theory. The first has to do with maximizing profit. The second has to do with minimizing average cost.

Suppose that

$$r(x) = \text{the revenue from selling } x \text{ items}$$
$$c(x) = \text{the cost of producing the } x \text{ items}$$
$$p(x) = r(x) - c(x) = \text{the profit from selling } x \text{ items.}$$

The **marginal revenue, marginal cost,** and **marginal profit** at this production level (x items) are

$$\frac{dr}{dx} = \text{marginal revenue}$$

$$\frac{dc}{dx} = \text{marginal cost}$$

$$\frac{dp}{dx} = \text{marginal profit.}$$

The first observation is about the relationship of p to these derivatives.

Theorem 6   Maximum Profit
At a production level yielding maximum profit, marginal revenue equals marginal cost.

Proof   We assume that $r(x)$ and $c(x)$ are differentiable for all $x > 0$, so if $p(x) = r(x) - c(x)$ has a maximum value, it occurs at a production level at which $p'(x) = 0$. Since $p'(x) = r'(x) - c'(x)$, $p'(x) = 0$ implies that

$$r'(x) - c'(x) = 0 \qquad \text{or} \qquad r'(x) = c'(x).$$

Figure 3.51 gives more information about this situation.

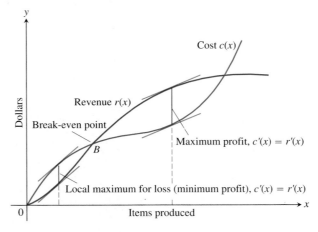

FIGURE 3.51 The graph of a typical cost function starts concave down and later turns concave up. It crosses the revenue curve at the break-even point B. To the left of B, the company operates at a loss. To the right, the company operates at a profit, with the maximum profit occurring where $c'(x) = r'(x)$. Farther to the right, cost exceeds revenue (perhaps because of a combination of rising labor and material costs and market saturation), and production levels become unprofitable again.

What guidance do we get from this observation? We know that a production level at which $p'(x) = 0$ need not be a level of maximum profit. It might be a level of minimum profit, for example. If we are making financial projections for our company, however, we should look for production levels at which marginal cost seems to equal marginal revenue. If there is a most profitable production level, it will be one of these.

Example 5   Maximizing Profit

Suppose that $r(x) = 9x$ and $c(x) = x^3 - 6x^2 + 15x$, where x represents thousands of units. Is there a production level that maximizes profit? If so, what is it?

Solution   Notice that $r'(x) = 9$ and $c'(x) = 3x^2 - 12x + 15$.

$$3x^2 - 12x + 15 = 9 \qquad \text{Set } c'(x) = r'(x).$$
$$3x^2 - 12x + 6 = 0$$

The two solutions of the quadratic equation are

$$x_1 = \frac{12 - \sqrt{72}}{6} = 2 - \sqrt{2} \approx 0.586 \qquad \text{and}$$

$$x_2 = \frac{12 + \sqrt{72}}{6} = 2 + \sqrt{2} \approx 3.414.$$

The possible production levels for maximum profit are $x \approx 0.586$ thousand units or $x \approx 3.414$ thousand units. The graphs in Figure 3.52 show that maximum profit occurs at about $x = 3.414$ (where revenue exceeds cost) and maximum loss occurs at about $x = 0.586$.

FIGURE 3.52 The cost and revenue curves for Example 5.

Example 6   Minimizing Costs

In the introduction to the chapter, we considered a cabinet maker using raw materials to produce 5 furnishings each day. Suppose that each delivery of a particular exotic material is $5000, whereas the storage of that material is $10 per day per unit stored, where a unit is the amount of material needed by her to produce one furnishing. How much material should be ordered each time and how often should the material be delivered to minimize her average daily cost in the production cycle between deliveries?

Solution

Model

If she asks for a delivery every x days, then she must order $5x$ units to have enough material for that delivery cycle. The *average* amount in storage is approximately one-half of the delivery amount, or $5x/2$. Thus, the cost of delivery and storage for each cycle is approximately

$$\text{Cost per cycle} = \text{delivery cost} + \text{storage costs}$$

$$\text{Cost per cycle} = \underbrace{5000}_{\substack{\text{delivery} \\ \text{cost}}} + \underbrace{\left(\frac{5x}{2}\right)}_{\substack{\text{average amount} \\ \text{stored}}} \cdot \underbrace{x}_{\substack{\text{number of} \\ \text{days stored}}} \cdot \underbrace{10}_{\substack{\text{storage cost} \\ \text{per day}}}$$

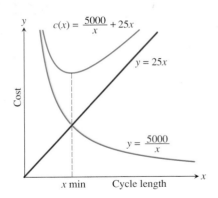

FIGURE 3.53 The average daily cost $c(x)$ is the sum of a hyperbola and a linear function. (Example 6)

We compute the *average daily cost* $c(x)$ by dividing the cost per cycle by the number of days x in the cycle (see Figure 3.53)

$$c(x) = \frac{5000}{x} + 25x, \quad x > 0.$$

As $x \rightarrow 0$ and as $x \rightarrow \infty$, the average daily cost becomes large. So we expect a minimum to exist, but where? Our goal is to determine the number of days x between deliveries that provides the absolute minimum cost.

Identify the Critical Points

We find the critical points by determining where the derivative is equal to zero:

$$c'(x) = -\frac{5000}{x^2} + 25 = 0$$

$$x = \pm\sqrt{200} \approx \pm 14.14.$$

Of the two critical points, only $\sqrt{200}$ lies in the domain of $c(x)$. The critical-point value of the average daily cost is

$$c(\sqrt{200}) = \frac{5000}{\sqrt{200}} + 25\sqrt{200} = 500\sqrt{2} \approx \$707.11.$$

We note that $c(x)$ is defined over the open interval $(0, \infty)$ with $c''(x) = 10000/x^3 > 0$. Thus, an absolute minimum exists at $x = \sqrt{200} \approx 14.14$ days.

Interpret

The cabinet maker should schedule a delivery of $5(14) = 70$ units of the exotic wood every 14 days.

Modeling Discrete Phenomena with Differentiable Functions

In case you are wondering how we can use differentiable functions $c(x)$ and $r(x)$ to describe the cost and revenue that comes from producing a number of items x that can only be an integer, here is the rationale.

When x is large, we can reasonably fit the cost and revenue data with smooth curves $c(x)$ and $r(x)$ that are defined not only at integer values of x but at the values in between. Once we have these differentiable functions, which are supposed to behave like the real cost and revenue when x is an integer, we can apply calculus to draw conclusions about their values. We then translate these mathematical conclusions into inferences about the real world that we hope will have predictive value. When they do, as is the case with the economic theory here, we say that the functions give a good model of reality.

What do we do when our calculus tells us that the best solution is a value of x that isn't an integer? In Example 6, if the number of days between deliveries must be an integer and certain materials come in lot sizes, we must round our answers. Should we round up or round down? That is, in the vicinity of the optimal time between deliveries, how *sensitive* is the change in cost to an increase or decrease in the time between deliveries?

Example 7   Sensitivity of the Minimum Cost

Should we round the number of days between deliveries up or down for the best solution in Example 6?

Solution
The average daily cost will increase by about $0.03 if we round down from 14.14 to 14 days:

$$c(14) = \frac{5000}{14} + 25(14) = \$707.14$$

and

$$c(14) - c(14.14) = \$707.14 - \$707.11 = \$0.03.$$

On the other hand, $c(15) = \$708.33$, and our cost would increase by $\$708.33 - \$707.11 = \$1.22$ if we round up. Thus, it is better that we round x down to 14 days. In Exercise 51, you are asked to derive a general formula to obtain the optimal time between deliveries given the delivery and storage costs for any material.

EXERCISES 3.5

Whenever you are maximizing or minimizing a function of a single variable, we urge you to graph it over the domain that is appropriate to the problem you are solving. The graph will provide insight before you calculate and will furnish a visual context for understanding your answer.

Applications in Geometry

1. *Minimizing perimeter* What is the smallest perimeter possible for a rectangle whose area is 16 in.^{2}, and what are its dimensions?

2. *Finding area* Show that among all rectangles with an 8-m perimeter, the one with largest area is a square.

3. *Inscribing rectangles* The figure at right shows a rectangle inscribed in an isosceles right triangle whose hypotenuse is 2 units long.

    (a) Express the y-coordinate of P in terms of x. (*Hint:* Write an equation for the line AB.)

    (b) Express the area of the rectangle in terms of x.

    (c) What is the largest area the rectangle can have, and what are its dimensions?

4. *Largest rectangle* A rectangle has its base on the x-axis and its upper two vertices on the parabola $y = 12 - x^2$. What is the largest area the rectangle can have, and what are its dimensions?

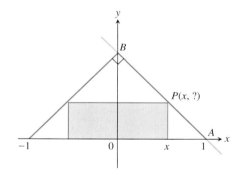

5. *Optimal dimensions* You are planning to make an open rectangular box from an 8-in.-by-15-in. piece of cardboard by cutting congruent squares from the corners and folding up the sides. What are the dimensions of the box of largest volume you can make this way, and what is its volume?

6. *Closing off the first quadrant* You are planning to close off a corner of the first quadrant with a line segment 20 units long running from $(a, 0)$ to $(0, b)$. Show that the area of the triangle enclosed by the segment is largest when $a = b$.

7. *The best fencing plan* A rectangular plot of farmland will be bounded on one side by a river and on the other three sides by a single-strand electric fence. With 800 m of wire at your disposal, what is the largest area you can enclose, and what are its dimensions?

8. *The shortest fence* A 216 m² rectangular pea patch is to be enclosed by a fence and divided into two equal parts by another fence parallel to one of the sides. What dimensions for the outer rectangle will require the smallest total length of fence? How much fence will be needed?

9. *Designing a tank* Your iron works has contracted to design and build a 500 ft³, square-based, open-top, rectangular steel holding tank for a paper company. The tank is to be made by welding thin stainless steel plates together along their edges. As the production engineer, your job is to find dimensions for the base and height that will make the tank weigh as little as possible.

   (a) What dimensions do you tell the shop to use?

   (b) *Writing to Learn* Briefly describe how you took weight into account.

10. *Catching rainwater* A 1125 ft³ open-top rectangular tank with a square base x ft on a side and y ft deep is to be built with its top flush with the ground to catch runoff water. The costs associated with the tank involve not only the material from which the tank is made but also an excavation charge proportional to the product xy.

   (a) If the total cost is

   $$c = 5(x^2 + 4xy) + 10xy,$$

   what values of x and y will minimize it?

   (b) *Writing to Learn* Give a possible scenario for the cost function in part (a).

11. *Designing a poster* You are designing a rectangular poster to contain 50 in.² of printing with a 4-in. margin at the top and bottom and a 2-in. margin at each side. What overall dimensions will minimize the amount of paper used?

12. *Inscribing a cone* Find the volume of the largest right circular cone that can be inscribed in a sphere of radius 3.

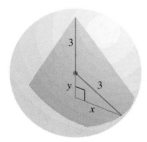

13. *Finding an angle* Two sides of a triangle have lengths a and b, and the angle between them is θ. What value of θ will maximize the triangle's area? (*Hint: $A = (1/2)ab \sin \theta$.*)

14. *Designing a can* What are the dimensions of the lightest open-top right circular cylindrical can that will hold a volume of 1000 cm³? Compare the result here with the result in Example 2.

15. *Designing a can* You are designing a 1000 cm³ right circular cylindrical can whose manufacture will take waste into account. There is no waste in cutting the aluminum for the side, but the top and bottom of radius r will be cut from squares that measure $2r$ units on a side. The total amount of aluminum used up by the can will therefore be

$$A = 8r^2 + 2\pi rh$$

rather than the $A = 2\pi r^2 + 2\pi rh$ in Example 2. In Example 2, the ratio of h to r for the most economical can was 2 to 1. What is the ratio now?

16. *Designing a box with a lid* A piece of cardboard measures 10 in. by 15 in. Two equal squares are removed from the corners of a 10 in. side as shown in the figure. Two equal rectangles are removed from the other corners so that the tabs can be folded to form a rectangular box with lid.

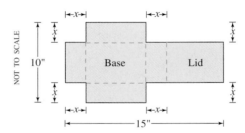

   (a) Write a formula $V(x)$ for the volume of the box.

   (b) Find the domain of V for the problem situation and graph V over this domain.

   (c) Use a graphical method to find the maximum volume and the value of x that gives it.

   (d) Confirm your result in part (c) analytically.

T 17. *Designing a suitcase* A 24-in.-by-36-in. sheet of cardboard is folded in half to form a 24-in-by-18-in. rectangle as shown in the accompanying figure. Then four congruent squares of side length x are cut from the corners of the folded rectangle. The sheet is unfolded, and the six tabs are folded up to form a box with sides and a lid.

   (a) Write a formula $V(x)$ for the volume of the box.

   (b) Find the domain of V for the problem situation and graph V over this domain.

   (c) Use a graphical method to find the maximum volume and the value of x that gives it.

   (d) Confirm your result in (c) analytically.

   (e) Find a value of x that yields a volume of 1120 in.³.

   (f) *Writing to Learn* Write a paragraph describing the issues that arise in part (b).

The sheet is then unfolded.

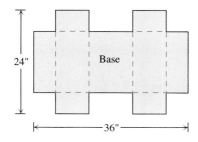

18. *Inscribing rectangles* A rectangle is to be inscribed under the arch of the curve $y = 4 \cos (0.5x)$ from $x = -\pi$ to $x = \pi$. What are the dimensions of the rectangle with largest area, and what is the largest area?

19. *Maximizing volume* Find the dimensions of a right circular cylinder of maximum volume that can be inscribed in a sphere of radius 10 cm. What is the maximum volume?

20. (a) The U.S. Postal Service will accept a box for domestic shipment only if the sum of its length and girth (distance around) does not exceed 108-in. What dimensions will give a box with a square end the largest possible volume?

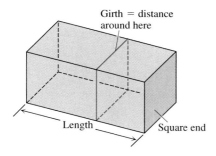

Girth = distance around here

Length   Square end

T **(b)** Graph the volume of a 108-in. box (length plus girth equals 108 in.) as a function of its length and compare what you see with your answer in part (a).

21. *(Continuation of Exercise 20.)*

(a) Suppose that instead of having a box with square ends you have a box with square sides so that its dimensions are h by h by w and the girth is $2h + 2w$. What dimensions will give the box its largest volume now?

T **(b)** Graph the volume as a function of h and compare what you see with your answer in part (a).

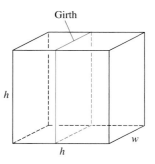

22. *Designing a window* A window is in the form of a rectangle surmounted by a semicircle. The rectangle is of clear glass, whereas the semicircle is of tinted glass that transmits only half as much light per unit area as clear glass does. The total perimeter is fixed. Find the proportions of the window that will admit the most light. Neglect the thickness of the frame.

23. *Constructing a silo* A silo (base not included) is to be constructed in the form of a cylinder surmounted by a hemisphere. The cost of construction per square unit of surface area is twice as great for the hemisphere as it is for the cylindrical sidewall. Determine the dimensions to be used if the volume is fixed and the cost of construction is to be kept to a minimum. Neglect the thickness of the silo and waste in construction.

24. *Finding an angle* The trough in the figure is to be made to the dimensions shown. Only the angle θ can be varied. What value of θ will maximize the trough's volume?

25. *Paper folding (Work in groups of two or three.)* A rectangular sheet of 8.5-in.-by-11-in. paper is placed on a flat surface. One of the corners is placed on the opposite longer edge, as shown in the figure, and held there as the paper is smoothed flat. The problem is to make the length of the crease as small as possible. Call the length L. Try it with paper.

(a) Show that $L^2 = 2x^3/(2x - 8.5)$.

(b) What value of x minimizes L^2?

(c) What is the minimum value of L?

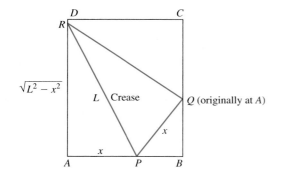

26. *Constructing cylinders* Compare the answers to the following two construction problems.

(a) A rectangular sheet of perimeter 36 cm and dimensions x cm by y cm is to be rolled into a cylinder as shown in part (a) of the figure. What values of x and y give the largest volume?

(b) The same sheet is to be revolved about one of the sides of length y to sweep out the cylinder as shown in part (b) of the figure. What values of x and y give the largest volume?

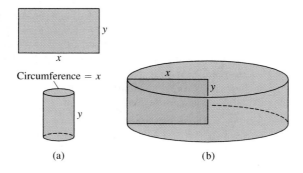

Circumference $= x$

(a)   (b)

27. *Constructing cones* A right triangle whose hypotenuse is $\sqrt{3}$ m long is revolved about one of its legs to generate a right circular cone. Find the radius, height, and volume of the cone of greatest volume that can be made this way.

28. *Finding parameter values* What value of a makes $f(x) = x^2 + (a/x)$ have

(a) a local minimum at $x = 2$?

(b) a point of inflection at $x = 1$?

29. *Inscribing rectangles* The rectangle shown here has one side on the positive y-axis, one side on the positive x-axis, and its upper right-hand vertex on the curve $y = e^{-x^2}$. What dimensions give the rectangle its largest area, and what is that area?

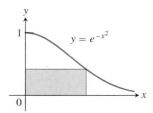

30. *Inscribing rectangles* The rectangle shown here has one side on the positive y-axis, one side on the positive x-axis, and its upper right-hand vertex on the curve $y = (\ln x)/x^2$. What dimensions give the rectangle its largest area, and what is that area?

Physical Applications

31. *Vertical motion* The height of an object moving vertically is given by

$$s = -16t^2 + 96t + 112,$$

with s in feet and t in seconds. Find

(a) the object's velocity when $t = 0$

(b) its maximum height and when it occurs

(c) its velocity when $s = 0$.

32. *Quickest route* Jane is 2 mi offshore in a boat and wishes to reach a coastal village 6 mi down a straight shoreline from the point nearest the boat. She can row 2 mph and can walk 5 mph. Where should she land her boat to reach the village in the least amount of time?

33. *Shortest beam* The 8 ft wall shown in the accompanying figure stands 27 ft from the building. Find the length of the shortest straight beam that will reach to the side of the building from the ground outside the wall.

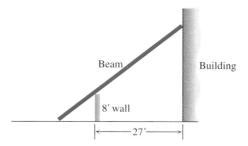

Beam Building

8' wall

|←——27'——→|

T **34.** *Strength of a beam* The strength S of a rectangular wooden beam
is proportional to its width times the square of its depth. (See the
accompanying figure.)

(a) Find the dimensions of the strongest beam that can be cut
from a 12-in.-diameter cylindrical log.

(b) *Writing to Learn* Graph S as a function of the beam's width w,
assuming the proportionality constant to be $k = 1$. Reconcile
what you see with your answer in part (a).

(c) *Writing to Learn* On the same screen, graph S as a function of
the beam's depth d, again taking $k = 1$. Compare the graphs
with one another and with your answer in part (a). What
would be the effect of changing to some other value of k?
Try it.

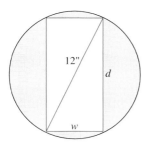

12"

d

w

T **35.** *Stiffness of a beam* The stiffness S of a rectangular beam is proportional to its width times the cube of its depth.

(a) Find the dimensions of the stiffest beam that can be cut from
a 12-in.-diameter cylindrical log.

(b) *Writing to Learn* Graph S as a function of the beam's width w,
assuming the proportionality constant to be $k = 1$. Reconcile
what you see with your answer in part (a).

(c) *Writing to Learn* On the same screen, graph S as a function of
the beam's depth d, again taking $k = 1$. Compare the graphs
with one another and with your answer in part (a). What
would be the effect of changing to some other value of k?
Try it.

36. *Locating a solar station* You are under contract to build a solar station at ground level on the east–west line between the two buildings shown here. How far from the taller building should you
place the station to maximize the number of hours it will be in the

sun on a day when the sun passes directly overhead? Begin by
observing that

$$\theta = \pi - \cot^{-1} \frac{x}{60} - \cot^{-1} \frac{50 - x}{30}.$$

Then find the value of x that maximizes θ.

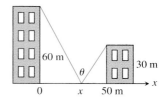

60 m

30 m

θ

0   x 50 m x

37. *Motion on a line* The positions of two particles on the s-axis are
$s_1 = \sin t$ and $s_2 = \sin (t + \pi/3)$, with s_1 and s_2 in meters and t in
seconds.

(a) At what time(s) in the interval $0 \le t \le 2\pi$ do the particles
meet?

(b) What is the farthest apart that the particles ever get?

(c) When in the interval $0 \le t \le 2\pi$ is the distance between the
particles changing the fastest?

38. *Electrical current* Suppose that at any time t(seconds) the current
i (amp) in an alternating current circuit is $i = 2 \cos t + 2 \sin t$.
What is the peak (largest magnitude) current for this circuit?

39. *Frictionless cart* A small frictionless cart, attached to the wall by a
spring, is pulled 10 cm from its rest position and released at time
$t = 0$ to roll back and forth for 4 sec. Its position at time t is
$s = 10 \cos \pi t$.

(a) What is the cart's maximum speed? When is the cart moving
that fast? Where is it then? What is the magnitude of the acceleration then?

(b) Where is the cart when the magnitude of the acceleration is
greatest? What is the cart's speed then?

0   10 s

40. *Side-by-side hanging masses* Two masses hanging side by side from springs have positions $s_1 = 2 \sin t$ and $s_2 = \sin 2t$, respectively.

   (a) At what times in the interval $0 < t$ do the masses pass each other? (*Hint:* $\sin 2t = 2 \sin t \cos t$.)

   (b) When in the interval $0 \le t \le 2\pi$ is the vertical distance between the masses the greatest? What is this distance? (*Hint:* $\cos 2t = 2 \cos^2 t - 1$.)

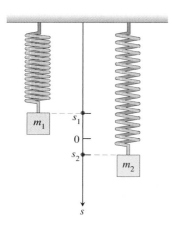

41. *Distance between two ships* At noon, ship A was 12 nautical miles due north of a ship B. Ship A was sailing south at 12 knots (nautical miles per hour; a nautical mile is 2000 yd) and continued to do so all day. Ship B was sailing east at 8 knots and continued to do so all day.

   (a) Start counting time with $t = 0$ at noon and express the distance s between the ships as a function of t.

   (b) How rapidly was the distance between the ships changing at noon? One hour later?

   (c) The visibility that day was 5 nautical miles. Did the ships ever sight each other?

   T **(d)** Graph s and ds/dt together as functions of t for $-1 \le t \le 3$, using different colors if possible. Compare the graphs and reconcile what you see with your answers in (b) and (c).

   (e) The graph of ds/dt looks as if it might have a horizontal asymptote in the first quadrant. This in turn suggests that ds/dt approaches a limiting value as $t \to \infty$. What is this value? What is its relation to the ships' individual speeds?

42. *Fermat's principle in optics* Fermat's principle in optics states that light always travels from one point to another along a path that minimizes the travel time. Light from a source A is reflected by a plane mirror to a receiver at point B, as shown in the figure. Show that for the light to obey Fermat's principle, the angle of incidence must equal the angle of reflection, both measured from the line normal to the reflecting surface. (This result can also be derived without calculus. There is a purely geometric argument, which you may prefer.)

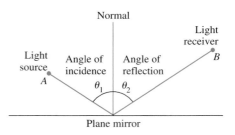

43. *Tin pest* When metallic tin is kept below 13.2°C, it slowly becomes brittle and crumbles to a gray powder. Tin objects eventually crumble to this gray powder spontaneously if kept in a cold climate for years. The Europeans who saw tin organ pipes in their churches crumble away years ago called the change *tin pest* because it seemed to be contagious, and indeed it was, for the gray powder is a catalyst for its own formation.

   A *catalyst* for a chemical reaction is a substance that controls the rate of reaction without undergoing any permanent change in itself. An *autocatalytic reaction* is one whose product is a catalyst for its own formation. Such a reaction may proceed slowly at first if the amount of catalyst present is small and slowly again at the end, when most of the original substance is used up. But in between, when both the substance and its catalyst product are abundant, the reaction proceeds at a faster pace.

   In some cases, it is reasonable to assume that the rate $v = dx/dt$ of the reaction is proportional both to the amount of the original substance present and to the amount of product. That is, v may be considered to be a function of x alone, and

$$v = kx(a - x) = kax - kx^2,$$

where

      $x = $ the amount of product

      $a = $ the amount of substance at the beginning

      $k = $ a positive constant.

At what value of x does the rate v have a maximum? What is the maximum value of v?

44. *Airplane landing path* An airplane is flying at altitude H when it begins its descent to an airport runway that is at horizontal ground distance L from the airplane, as shown in the figure. Assume that the landing path of the airplane is the graph of a cubic polynomial function $y = ax^3 + bx^2 + cx + d$, where $y(-L) = H$ and $y(0) = 0$.

   (a) What is dy/dx at $x = 0$?

   (b) What is dy/dx at $x = -L$?

   (c) Use the values for dy/dx at $x = 0$ and $x = -L$ together with $y(0) = 0$ and $y(-L) = H$ to show that

$$y(x) = H\left[2\left(\frac{x}{L}\right)^3 + 3\left(\frac{x}{L}\right)^2 \right].$$

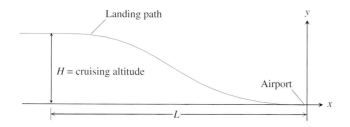

Business and Economics

45. *Selling backpacks* It costs you c dollars each to manufacture and distribute backpacks. If the backpacks sell at x dollars each, the number sold is given by

$$n = \frac{a}{x - c} + b(100 - x),$$

where a and b are certain positive constants. What selling price will bring a maximum profit?

46. *Tour service* You operate a tour service that offers the following rates:

> $200 per person if 50 people (the minimum number to book the tour) go on the tour

> For each additional person, up to a maximum of 80 people total, the rate per person is reduced by $2.

It costs $6000 (a fixed cost) plus $32 per person to conduct the tour. How many people does it take to maximize your profit?

47. *Wilson lot size formula* One of the formulas for inventory management says that the average weekly cost of ordering, paying for, and holding merchandise is

$$A(q) = \frac{km}{q} + cm + \frac{hq}{2},$$

where q is the quantity you order when things run low (shoes, radios, brooms, or whatever the item might be), k is the cost of placing an order (the same, no matter how often you order), c is the cost of one item (a constant), m is the number of items sold each week (a constant), and h is the weekly holding cost per item (a constant that takes into account things such as space, utilities, insurance, and security).

(a) Your job, as the inventory manager for your store, is to find the quantity that will minimize $A(q)$. What is it? (The formula you get for the answer is called the *Wilson lot size formula*.)

(b) Shipping costs sometimes depend on order size. When they do, it is more realistic to replace k by $k + bq$, the sum of k and a constant multiple of q. What is the most economical quantity to order now?

48. *Production level* Prove that the production level (if any) at which average cost is smallest is a level at which the average cost equals marginal cost.

49. *Production level* Show that if $r(x) = 6x$ and $c(x) = x^3 - 6x^2 + 15x$ are your revenue and cost functions, then the best you can do is break even (have revenue equal cost).

50. *Production level* Suppose that $c(x) = x^3 - 20x^2 + 20{,}000x$ is the cost of manufacturing x items. Find a production level that will minimize the average cost of making x items.

51. *Average daily cost* In Example 6, assume for any material that a cost of d is incurred per delivery, the storage cost is s per unit stored per day, and the production rate is p units per day.

(a) How much should be delivered every x days?

(b) Show that

$$\text{cost per cycle} = d + \frac{px}{2} sx.$$

(c) Find the time between deliveries x^* and the amount to deliver that minimizes the *average daily cost* of delivery and storage.

(d) Show that x^* occurs at the intersection of the hyperbola $y = d/x$ and the line $y = psx/2$.

52. *Minimizing average cost* Suppose that $c(x) = 2000 + 96x + 4x^{3/2}$ where x represents thousands of units. Is there a production level that minimizes average cost? If so, what is it?

Medicine

53. *Sensitivity to medicine* (Continuation of Example 7, Section 2.3) Find the amount of medicine to which the body is most sensitive by finding the value of M that maximizes the derivative dR/dM, where

$$R = M^2 \left(\frac{C}{2} - \frac{M}{3} \right)$$

and C is a constant.

54. *How we cough*

(a) When we cough, the trachea (windpipe) contracts to increase the velocity of the air going out. This raises the questions of how much it should contract to maximize the velocity and whether it really contracts that much when we cough.

Under reasonable assumptions about the elasticity of the tracheal wall and about how the air near the wall is slowed by friction, the average flow velocity v can be modeled by the equation

$$v = c(r_0 - r)r^2 \text{ cm/sec}, \qquad \frac{r_0}{2} \le r \le r_0,$$

where r_0 is the rest radius of the trachea in centimeters and c is a positive constant whose value depends in part on the length of the trachea.

Show that v is greatest when $r = (2/3)r_0$, that is, when the trachea is about 33% contracted. The remarkable fact is that X-ray photographs confirm that the trachea contracts about this much during a cough.

⊤ **(b)** Take r_0 to be 0.5 and c to be 1 and graph v over the interval $0 \le r \le 0.5$. Compare what you see with the claim that v is at a maximum when $r = (2/3)r_0$.

Theory and Examples

55. *An inequality for positive integers* Show that if a, b, c, and d are positive integers, then

$$\frac{(a^2 + 1)(b^2 + 1)(c^2 + 1)(d^2 + 1)}{abcd} \ge 16.$$

56. *The derivative dt /dx in Example 4*

(a) Show that

$$f(x) = \frac{x}{\sqrt{a^2 + x^2}}$$

is an increasing function of x.

(b) Show that

$$g(x) = \frac{d - x}{\sqrt{b^2 + (d - x)^2}}$$

is a decreasing function of x.

(c) Show that

$$\frac{dt}{dx} = \frac{x}{c_1\sqrt{a^2 + x^2}} - \frac{d - x}{c_2\sqrt{b^2 + (d - x)^2}}$$

is an increasing function of x.

57. *Writing to Learn* Let $f(x)$ and $g(x)$ be the differentiable functions graphed here. Point c is the point where the vertical distance between the curves is the greatest. Is there anything special about the tangents to the two curves at c? Give reasons for your answer.

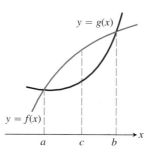

58. *Writing to Learn* You have been asked to determine whether the function $f(x) = 3 + 4 \cos x + \cos 2x$ is ever negative.

(a) Explain why you need consider values of x only in the interval $[0, 2\pi]$.

(b) Is f ever negative? Explain.

59. *Absolute maximum*

(a) The function $y = \cot x - \sqrt{2} \csc x$ has an absolute maximum value on the interval $0 < x < \pi$. Find it.

⊤ **(b)** Graph the function and compare what you see with your answer in part (a).

60. *Absolute minimum*

(a) The function $y = \tan x + 3 \cot x$ has an absolute minimum value on the interval $0 < x < \pi/2$. Find it.

⊤ **(b)** Graph the function and compare what you see with your answer in part (a).

61. *Calculus and geometry*

(a) How close does the curve $y = \sqrt{x}$ come to the point $(3/2, 0)$? (*Hint:* If you minimize the *square* of the distance, you can avoid square roots.)

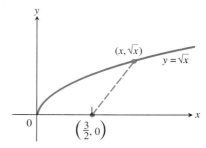

⊤ **(b)** Graph the distance function and $y = \sqrt{x}$ together and reconcile what you see with your answer in part (a).

62. *Calculus and geometry*

(a) How close does the semicircle $y = \sqrt{16 - x^2}$ come to the point $(1, \sqrt{3})$?

⊤ **(b)** Graph the distance function and $y = \sqrt{16 - x^2}$ together and reconcile what you see with your answer in part (a).

COMPUTER EXPLORATIONS

In Exercises 63 and 64, you may find it helpful to use a CAS.

63. *Generalized cone problem* A cone of height h and radius r is constructed from a flat, circular disk of radius a in. by removing a sector AOC of arc length x in. and then connecting the edges OA and OC.

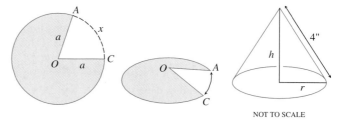

NOT TO SCALE

(a) Find a formula for the volume V of the cone in terms of x and a.

(b) Find r and h in the cone of maximum volume for $a = 4, 5, 6, 8$.

(c) *Writing to Learn* Find a simple relationship between r and h that is independent of a for the cone of maximum volume. Explain how you arrived at your relationship.

64. *Circumscribing an ellipse* Let $P(x, a)$ and $Q(-x, a)$ be two points on the upper half of the ellipse

$$\frac{x^2}{100} + \frac{(y - 5)^2}{25} = 1$$

centered at $(0, 5)$. A triangle RST is formed by using the tangent lines to the ellipse at Q and P as shown in the figure.

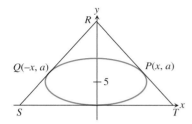

(a) Show that the area of the triangle is

$$A(x) = -f'(x) \left[x - \frac{f(x)}{f'(x)} \right]^2,$$

where $y = f(x)$ is the function representing the upper half of the ellipse.

(b) What is the domain of A? Draw the graph of A. How are the asymptotes of the graph related to the problem situation?

(c) Determine the height of the triangle with minimum area. How is it related to the y-coordinate of the center of the ellipse?

(d) Repeat parts (a) through (c) for the ellipse

$$\frac{x^2}{C^2} + \frac{(y - B)^2}{B^2} = 1$$

centered at $(0, B)$. Show that the triangle has minimum area when its height is $3B$.

3.6 Linearization and Differentials

Linearization • Differentials • Estimating Change with Differentials • Absolute, Relative, and Percentage Change • Sensitivity to Change • Error in Differential Approximation • Converting Mass to Energy

Sometimes we can approximate complicated functions with simpler ones that give the accuracy we want for specific applications and are easier to work with. The approximating functions discussed in this section are called *linearizations*, and they are based on tangent lines. Other approximating functions are discussed in Chapter 8.

We introduce new variables dx and dy and define them in a way that gives new meaning to the Leibniz notation dy/dx. We use dy to estimate error in measurement and sensitivity of a function to change.

CD-ROM
WEBsite

Linearization

As you can see in Figure 3.54, the tangent to the curve $y = x^2$ lies close to the curve near the point of tangency. For a brief interval to either side, the y-values along the tangent line give a good approximation to the y-values on the curve. We observe this phenomenon by zooming in on the two graphs at the point of tangency, or by looking at tables of values for the difference between $f(x)$ and its tangent line near the x-coordinate of the point of tangency. Locally, every curve behaves like a straight line.

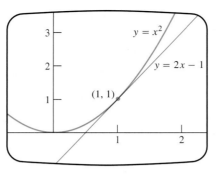

$y = x^2$ and its tangent $y = 2x - 1$ at $(1, 1)$.

Tangent and curve very close near $(1, 1)$.

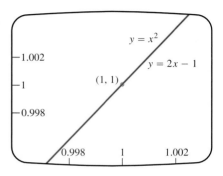

Tangent and curve very close throughout entire x-interval shown.

Tangent and curve closer still. Computer screen cannot distinguish tangent from curve on this x-interval.

FIGURE 3.54 The more we magnify the graph of a function near a point where the function is differentiable, the flatter the graph becomes and the more it resembles its tangent.

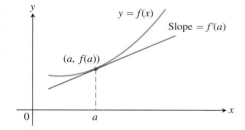

FIGURE 3.55 The tangent to the curve $y = f(x)$ at $x = a$ is the line $y = f(a) + f'(a)(x - a)$.

CD-ROM
WEBsite

In general, the tangent to $y = f(x)$ at a point $x = a$, where f is differentiable (Figure 3.55) passes through the point $(a, f(a))$, so its point-slope equation is

$$y = f(a) + f'(a)(x - a).$$

Thus, the tangent line is the graph of the linear function

$$L(x) = f(a) + f'(a)(x - a).$$

For as long as the line remains close to the graph of f, $L(x)$ gives a good approximation to $f(x)$.

Definition   Linearization

If f is differentiable at $x = a$, then the approximating function

$$L(x) = f(a) + f'(a)(x - a) \qquad (1)$$

is the **linearization** of f at a.

CD-ROM
WEBsite

The approximation $f(x) \approx L(x)$ is the **standard linear approximation** of f at a. The point $x = a$ is the **center** of the approximation.

Example 1   Finding a Linearization

Find the linearization of $f(x) = \sqrt{1 + x}$ at $x = 0$ (Figure 3.56).

Solution   Since

$$f'(x) = \frac{1}{2} (1 + x)^{-1/2},$$

we have $f(0) = 1, f'(0) = 1/2$, and

$$L(x) = f(a) + f'(a)(x - a) = 1 + \frac{1}{2} (x - 0) = 1 + \frac{x}{2}.$$

See Figure 3.56.

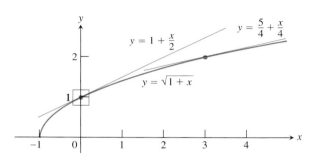

FIGURE 3.56 The graph of $y = \sqrt{1 + x}$ and its linearizations at $x = 0$ and $x = 3$. Figure 3.57 shows a magnified view of the small window about 1 on the y-axis.

Look at how accurate the approximation $\sqrt{1 + x} \approx 1 + (x/2)$ is for values of x near 0.

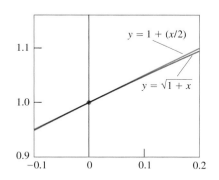

FIGURE 3.57 Magnified view of the window in Figure 3.56.

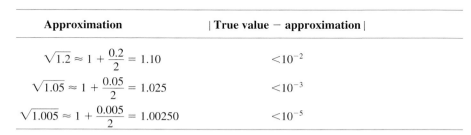

Approximation	|True value − approximation|
$\sqrt{1.2} \approx 1 + \dfrac{0.2}{2} = 1.10$	$< 10^{-2}$
$\sqrt{1.05} \approx 1 + \dfrac{0.05}{2} = 1.025$	$< 10^{-3}$
$\sqrt{1.005} \approx 1 + \dfrac{0.005}{2} = 1.00250$	$< 10^{-5}$

As we move away from zero, we lose accuracy. For example, for $x = 2$, the linearization gives 2 as the approximation for $\sqrt{3}$, which is not even accurate to one decimal place.

Do not be misled by the preceding calculations into thinking that whatever we do with a linearization is better done with a calculator. In practice, we would never use a linearization to find a particular square root. The utility of a linearization is its ability to replace a complicated formula by a simpler one over an entire interval of values. If we have to work with $\sqrt{1 + x}$ for x close to 0 and can tolerate the small amount of error involved, we can work with $1 + (x/2)$ instead. Of course, we then need to know how much error there is. We have a full story on error in Chapter 8.

A linear approximation normally loses accuracy away from its center. As Figure 3.56 suggests, the approximation $\sqrt{1 + x} \approx 1 + (x/2)$ will probably be too crude to be useful near $x = 3$. There, we need the linearization at $x = 3$.

CD-ROM
WEBsite
Historical Biography
François Viete
(1540 — 1603)

Example 2   Finding a Second Linearization

Find the linearization of $f(x) = \sqrt{1 + x}$ at $x = 3$.

Solution   We evaluate Equation (1) for f at $a = 3$. With

$$f(3) = 2, \qquad f'(3) = \frac{1}{2}(1 + x)^{-1/2}\Big|_{x=3} = \frac{1}{4},$$

we have

$$L(x) = 2 + \frac{1}{4}(x - 3) = \frac{5}{4} + \frac{x}{4}.$$

At $x = 3.2$, the linearization in Example 2 gives

$$\sqrt{1 + x} = \sqrt{1 + 3.2} \approx \frac{5}{4} + \frac{3.2}{4} = 1.250 + 0.800 = 2.050,$$

which differs from the true value $\sqrt{4.2} \approx 2.04939$ by less than one one-thousandth. The linearization in Example 1 gives

$$\sqrt{1 + x} = \sqrt{1 + 3.2} \approx 1 + \frac{3.2}{2} = 1 + 1.6 = 2.6,$$

a result that is off by more than 25%.

Example 3   Finding Roots and Powers

The most important linear approximation for roots and powers is

$$(1 + x)^k \approx 1 + kx \qquad (x \text{ near } 0; \text{ any number } k) \tag{2}$$

(Exercise 7). This approximation, good for values of x sufficiently close to zero, has broad application.

Common Linear Approximations, x ≈ 0

$$\sin x \approx x$$
$$\cos x \approx 1$$
$$\tan x \approx x$$
$$e^x \approx 1 + x$$
$$\ln(1 + x) \approx x$$
$$(1 + x)^k \approx 1 + kx$$

(Exercises 6 and 7)

Example 4   Applying Example 3

The following approximations are consequences of Example 3.

$$\sqrt{1 + x} \approx 1 + \frac{1}{2}x \qquad\qquad k = 1/2$$

$$\frac{1}{1 - x} = (1 - x)^{-1} \approx 1 + (-1)(-x) = 1 + x \qquad \begin{array}{l} k = -1; \text{ replace} \\ x \text{ by } - x. \end{array}$$

$$\sqrt[3]{1 + 5x^4} = (1 + 5x^4)^{1/3} \approx 1 + \frac{1}{3}(5x^4) = 1 + \frac{5}{3}x^4 \qquad \begin{array}{l} k = 1/3; \text{ replace} \\ x \text{ by } 5x^4. \end{array}$$

$$\frac{1}{\sqrt{1 - x^2}} = (1 - x^2)^{-1/2} \approx 1 + \left(-\frac{1}{2}\right)(-x^2) = 1 + \frac{1}{2}x^2 \qquad \begin{array}{l} k = -1/2; \\ \text{replace } x \text{ by } -x^2. \end{array}$$

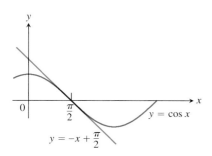

FIGURE 3.58 The graph of $f(x) = \cos x$ and its linearization at $x = \pi/2$. Near $x = \pi/2$, $\cos x \approx -x + (\pi/2)$. (Example 5)

Example 5   Finding a Linearization

Find the linearization of $f(x) = \cos x$ at $x = \pi/2$ (Figure 3.58).

Solution   Since $f(\pi/2) = \cos(\pi/2) = 0$, $f'(x) = -\sin x$, and $f'(\pi/2) = -\sin(\pi/2) = -1$, we have

$$L(x) = f(a) + f'(a)(x - a)$$
$$= 0 + (-1)\left(x - \frac{\pi}{2}\right)$$
$$= -x + \frac{\pi}{2}.$$

Differentials

We sometimes use the notation dy/dx to represent the derivative y' of y with respect to x. Contrary to its appearance, it is not a ratio. We now introduce two new variables dx and dy with the property that if their ratio exists, it will be equal to the derivative.

> **Definition   Differentials**
>
> Let $y = f(x)$ be a differentiable function. The **differential dx** is an independent variable. The **differential dy** is
>
> $$dy = f'(x)\, dx.$$

Unlike the independent variable dx, the variable dy is always a dependent variable. It depends on both x and dx.

Example 6   Finding the Differential dy

Find dy if

(a) $y = x^5 + 37x$

(b) $y = \sin 3x$.

Solution

(a) $dy = (5x^4 + 37)\, dx$ **(b)** $dy = (3 \cos 3x)\, dx$

If $dx \neq 0$, then the quotient of the differential dy by the differential dx is equal to the derivative $f'(x)$ because

$$\frac{dy}{dx} = \frac{f'(x)\, dx}{dx} = f'(x).$$

We sometimes write

$$df = f'(x)\, dx$$

The Meaning of dx and dy

In most contexts, the differential dx of the independent variable is its change Δx, but we do not impose this restriction on the definition.

Unlike the independent variable dx, the variable dy is always a dependent variable. It depends on both x and dx.

---

in place of $dy = f'(x)\,dx$, calling df the **differential of f.** For instance, if $f(x) = 3x^2 - 6$, then

$$df = d(3x^2 - 6) = 6x\,dx.$$

Every differentiation formula like

$$\frac{d(u + v)}{dx} = \frac{du}{dx} + \frac{dv}{dx} \qquad \text{or} \qquad \frac{d(\sin u)}{dx} = \cos u \frac{du}{dx}$$

has a corresponding differential form like

$$d(u + v) = du + dv \qquad \text{or} \qquad d(\sin u) = \cos u\,du.$$

Example 7 Finding Differentials of Functions

(a) $d(\tan 2x) = \sec^2(2x)\,d(2x) = 2\sec^2 2x\,dx$

(b) $d\left(\dfrac{x}{x+1}\right) = \dfrac{(x+1)\,dx - x\,d(x+1)}{(x+1)^2} = \dfrac{x\,dx + dx - x\,dx}{(x+1)^2} = \dfrac{dx}{(x+1)^2}$

Estimating Change with Differentials

Suppose that we know the value of a differentiable function $f(x)$ at a point a and we want to predict how much this value will change if we move to a nearby point $a + dx$. If dx is small, f and its linearization L at a will change by nearly the same amount (Figure 3.59). Since the values of L are simple to calculate, calculating the change in L offers a practical way to estimate the change in f.

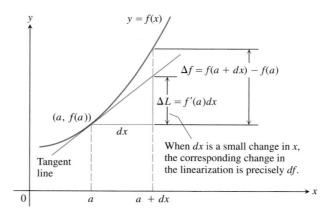

FIGURE 3.59 Approximating the change in the function f by the change in the linearization of f.

In the notation of Figure 3.59, the change in f is

$$\Delta f = f(a + dx) - f(a).$$

The corresponding change in L is

$$\begin{aligned}\Delta L &= L(a + dx) - L(a)\\ &= \underbrace{f(a) + f'(a)[(a + dx) - a]}_{L(a+dx)} - \underbrace{f(a)}_{L(a)}\\ &= f'(a)\,dx.\end{aligned}$$

Thus, the differential $df = f'(x)\,dx$ has a geometric interpretation: The value of df at $x = a$ is ΔL, the change in the linearization of f corresponding to the change dx.

> **Differential Estimate of Change**
>
> Let $f(x)$ be differentiable at $x = a$. The approximate change in the value of f when x changes from a to $a + dx$ is
>
> $$df = f'(a)\,dx.$$

FIGURE 3.60 When dr is small compared with a, as it is when $dr = 0.1$ and $a = 10$, the differential $dA = 2\pi a\,dr$ gives a good estimate of ΔA. (Example 8)

Example 8   Estimating Change with Differentials

The radius r of a circle increases from $a = 10$ m to 10.1 m (Figure 3.60). Use dA to estimate the increase in the circle's area A. Compare this estimate with the true change ΔA.

Solution   Since $A = \pi r^2$, the estimated increase is

$$dA = A'(a)\,dr = 2\pi a\,dr = 2\pi(10)(0.1) = 2\pi \text{ m}^2.$$

The true change is

$$\Delta A = \pi(10.1)^2 - \pi(10)^2 = (102.01 - 100)\pi = (\underbrace{2\pi}_{dA} + \underbrace{0.01\pi}_{\text{error}}) \text{ m}^2.$$

Absolute, Relative, and Percentage Change

As we move from a to a nearby point $a + dx$, we can describe the change in f in three ways:

	True	**Estimated**
Absolute change	$\Delta f = f(a + dx) - f(a)$	$df = f'(a)\,dx$
Relative change	$\dfrac{\Delta f}{f(a)}$	$\dfrac{df}{f(a)}$
Percentage change	$\dfrac{\Delta f}{f(a)} \times 100$	$\dfrac{df}{f(a)} \times 100$

Example 9   Computing Percentage Change

The estimated percentage change in the area of the circle in Example 8 is

$$\frac{dA}{A(a)} \times 100 = \frac{2\pi}{100\pi} \times 100 = 2\%.$$

The true percentage change is

$$\frac{\Delta A}{A(a)} \times 100 = \frac{2.01\pi}{100\pi} \times 100 = 2.01\%.$$

Example 10   Unclogging Arteries

In the late 1830s, French physiologist Jean Poiseuille ("pwa-ZOY") discovered the formula we use today to predict how much the radius of a partially clogged artery has to be expanded to restore normal flow. His formula,

$$V = kr^4,$$

says that the volume V of fluid flowing through a small pipe or tube in a unit of time at a fixed pressure is a constant times the fourth power of the tube's radius r. How will a 10% increase in r affect V?

Angiography

An opaque dye is injected into a partially blocked artery to make the inside visible under X-rays. This reveals the location and severity of the blockage.

Angioplasty

A balloon-tipped catheter is inflated inside the artery to widen it at the blockage site.

Solution   The differentials of r and V are related by the equation

$$dV = \frac{dV}{dr}\, dr = 4kr^3\, dr.$$

The relative change in V is

$$\frac{dV}{V} = \frac{4kr^3\, dr}{kr^4} = 4\frac{dr}{r}.$$

The relative change in V is 4 times the relative change in r, so a 10% increase in r will produce a 40% increase in the flow.

Sensitivity to Change

The equation $df = f'(x)\, dx$ tells how *sensitive* the output of f is to a change in input at different values of x. The larger the value of f' at x, the greater the effect of a given change dx.

Example 11   Finding the Depth of a Well

You want to calculate the depth of a well from the equation $s = 16t^2$ by timing how long it takes a heavy stone you drop to splash into the water below. How sensitive will your calculations be to a 0.1 sec error in measuring the time?

Solution   The size of ds in the equation

$$ds = 32t\, dt$$

depends on how big t is. If $t = 2$ sec, the error caused by $dt = 0.1$ is only

$$ds = 32(2)(0.1) = 6.4 \text{ ft}.$$

Three seconds later at $t = 5$ sec, the error caused by the same dt is

$$ds = 32(5)(0.1) = 16 \text{ ft}.$$

Error in Differential Approximation

Let $f(x)$ be differentiable at $x = a$ and suppose that Δx is an increment of x. We have two ways to describe the change in f as x changes from a to $a + \Delta x$:

The true change:	$\Delta f = f(a + \Delta x) - f(a)$
The differential estimate:	$df = f'(a)\Delta x.$

How well does df approximate Δf?

We measure the approximation error by subtracting df from Δf:

$$
\begin{aligned}
\text{Approximation error} &= \Delta f - df \\
&= \Delta f - f'(a)\,\Delta x \\
&= \underbrace{f(a + \Delta x) - f(a)}_{\Delta f} - f'(a)\Delta x \\
&= \underbrace{\left(\frac{f(a + \Delta x) - f(a)}{\Delta x} - f'(a) \right)}_{\text{Call this part } \epsilon} \Delta x \\
&= \epsilon \cdot \Delta x.
\end{aligned}
$$

As $\Delta x \to 0$, the difference quotient

$$\frac{f(a + \Delta x) - f(a)}{\Delta x}$$

approaches $f'(a)$ (remember the definition of $f'(a)$), so the quantity in parentheses becomes a very small number (which is why we called it ϵ). In fact, $\epsilon \to 0$ as $\Delta x \to 0$. When Δx is small, the approximation error $\epsilon\,\Delta x$ is smaller still.

$$\underbrace{\Delta f}_{\substack{\text{true} \\ \text{change}}} = \underbrace{f'(a)\,\Delta x}_{\substack{\text{estimated} \\ \text{change}}} + \underbrace{\epsilon\,\Delta x}_{\text{error}}$$

Although we do not know exactly how small the error is and will not be able to make much progress on this front until Chapter 8, there is something worth noting here, namely the *form* taken by the equation.

Change in $y = f(x)$ near $x = a$

If $y = f(x)$ is differentiable at $x = a$, and x changes from a to $a + \Delta x$, the change Δy in f is given by an equation of the form

$$\Delta y = f'(a)\,\Delta x + \epsilon\,\Delta x \qquad (3)$$

in which $\epsilon \to 0$ as $\Delta x \to 0$.

Converting Mass to Energy

Example 12   Using Approximations in Einstein's Physics

Newton's second law,

$$F = \frac{d}{dt}(mv) = m\frac{dv}{dt} = ma,$$

is stated with the assumption that mass is constant, but we know this is not strictly true because the mass of a body increases with velocity. In Einstein's corrected formula, mass has the value

$$m = \frac{m_0}{\sqrt{1 - v^2/c^2}},$$

where the "rest mass" m_0 represents the mass of a body that is not moving and c is the speed of light, which is about 300,000 km/sec. Use the approximation

$$\frac{1}{\sqrt{1 - x^2}} \approx 1 + \frac{1}{2}x^2 \qquad (4)$$

from Example 4 to estimate the increase Δm in mass resulting from the added velocity v.

Solution
When v is very small compared with c, v^2/c^2 is close to zero and it is safe to use the approximation

$$\frac{1}{\sqrt{1 - v^2/c^2}} \approx 1 + \frac{1}{2}\left(\frac{v^2}{c^2}\right)$$

(Equation (4) with $x = v/c$) to write

$$m = \frac{m_0}{\sqrt{1 - v^2/c^2}} \approx m_0\left[1 + \frac{1}{2}\left(\frac{v^2}{c^2}\right)\right] = m_0 + \frac{1}{2}m_0v^2\left(\frac{1}{c^2}\right).$$

or

$$m \approx m_0 + \frac{1}{2}m_0v^2\left(\frac{1}{c^2}\right). \qquad (5)$$

Equation (5) expresses the increase in mass that results from the added velocity v.

Energy Interpretation
In Newtonian physics, $(1/2)\,m_0v^2$ is the kinetic energy (KE) of the body, and if we rewrite Equation (5) in the form

$$(m - m_0)c^2 \approx \frac{1}{2}m_0v^2,$$

we see that

$$(m - m_0)c^2 \approx \frac{1}{2}m_0v^2 = \frac{1}{2}m_0v^2 - \frac{1}{2}m_0(0)^2 = \Delta(\text{KE}),$$

or

$$(\Delta m)c^2 \approx \Delta(\text{KE}). \qquad (6)$$

In other words, the change in kinetic energy $\Delta(KE)$ in going from velocity 0 to velocity v is approximately equal to $(\Delta m)c^2$.

With c equal to 3×10^8 m/sec, Equation (6) becomes

$$\Delta(KE) \approx 90{,}000{,}000{,}000{,}000{,}000\,\Delta m \text{ joules} \qquad \text{Mass in kilograms}$$

and we see that a small change in mass can create a large change in energy. The energy released by exploding a 20-kiloton atomic bomb, for instance, is the result of converting only 1 g of mass to energy. The products of the explosion weigh only 1 g less than the material exploded. A U.S. penny weighs about 3 g.

EXERCISES 3.6

Finding Linearizations

CD-ROM WEBsite

In Exercises 1–5, find the linearization $L(x)$ of $f(x)$ at $x = a$.

1. $f(x) = x^3 - 2x + 3, \quad a = 2$

2. $f(x) = \sqrt{x^2 + 9}, \quad a = -4$

3. $f(x) = x + \dfrac{1}{x}, \quad a = 1$

4. $f(x) = \sqrt[3]{x}, \quad a = -8$

5. $f(x) = \tan x, \quad a = \pi$

6. *Common linear approximations at x = 0* Find the linearizations of the following functions at $x = 0$.

   (a) $\sin x$ **(b)** $\cos x$ **(c)** $\tan x$

   (d) e^x **(e)** $\ln(1 + x)$

Linearizations for Powers and Roots

CD-ROM WEBsite

7. Show that the linearization of $f(x) = (1 + x)^k$ at $x = 0$ is $L(x) = 1 + kx$.

CD-ROM WEBsite

8. Use the linear approximation $(1 + x)^k \approx 1 + kx$ to find an approximation for the function $f(x)$ for values of x near zero.

   (a) $f(x) = (1 - x)^6$

   (b) $f(x) = \dfrac{2}{1 - x}$

   (c) $f(x) = \dfrac{1}{\sqrt{1 + x}}$

   (d) $f(x) = \sqrt{2 + x^2}$

   (e) $f(x) = (4 + 3x)^{1/3}$

   (f) $f(x) = \sqrt[3]{\left(1 - \dfrac{1}{2 + x}\right)^2}$

Linearization for Approximation

In Exercises 9–12, choose a linearization with center not at $x = a$ but at a nearby value at which the function and its derivative are easy to evaluate. State the linearization and the center.

9. $f(x) = 2x^2 + 4x - 3, \quad a = -0.9$

10. $f(x) = \sqrt[3]{x}, \quad a = 8.5$

11. $f(x) = \dfrac{x}{x + 1}, \quad a = 1.3$

12. $f(x) = \cos x, \quad a = 1.7$

13. *Faster than a calculator* Use the approximation $(1 + x)^k \approx 1 + kx$ to estimate the following.

   (a) $(1.0002)^{50}$ **(b)** $\sqrt[3]{1.009}$.

14. *Writing to Learn* Find the linearization of $f(x) = \sqrt{x + 1} + \sin x$ at $x = 0$. How is it related to the individual linearizations of $\sqrt{x + 1}$ and $\sin x$ at $x = 0$?

Derivatives in Differential Form

In Exercises 15–24, find dy.

15. $y = x^3 - 3\sqrt{x}$

16. $y = x\sqrt{1 - x^2}$

17. $y = x^2 \ln x$

18. $y = \dfrac{2\sqrt{x}}{3(1 + \sqrt{x})}$

19. $2y^{3/2} + xy - x = 0$

20. $xy^2 - 4x^{3/2} - y = 0$

21. $y = e^{\sin x}$

22. $y = \cos(x^2)$

23. $y = xe^x$

24. $y = \sec(x^2 - 1)$

Approximation Error

In Exercises 25–28, the function f changes value when x changes from a to $a + dx$. Find

   (a) the absolute change $\Delta f = f(a + dx) - f(a)$

(b) the estimated change $df = f'(a)\, dx$

(c) the approximation error $|\Delta f - df|$.

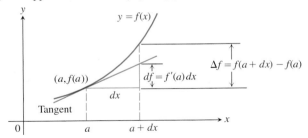

25. $f(x) = x^2 + 2x, \quad a = 0, \quad dx = 0.1$

26. $f(x) = x^3 - x, \quad a = 1, \quad dx = 0.1$

27. $f(x) = x^{-1}, \quad a = 0.5, \quad dx = 0.05$

28. $f(x) = x^4, \quad a = 1, \quad dx = 0.01$

Differential Estimates of Change

In Exercises 29–32, write a differential formula that estimates the given change in volume or surface area.

29. *Volume* The change in the volume $V = (4/3)\pi r^3$ of a sphere when the radius changes from a to $a + dr$

$V = \frac{4}{3}\pi r^3, \quad S = 4\pi r^2$

$V = x^3, \quad S = 6x^2$

30. *Surface area* The change in the surface area $S = 4\pi r^2$ of a sphere when the radius changes from a to $a + dr$

31. *Volume* The change in the volume $V = x^3$ of a cube when the edge lengths change from a to $a + dx$

32. *Surface area* The change in the surface area $S = 6x^2$ of a cube when the edge lengths change from a to $a + dx$

Theory and Examples

33. *Expanding circle* The radius of a circle is increased from 2.00 to 2.02 m.

   (a) Estimate the resulting change in area.

   (b) Express the estimate as a percentage of the circle's original area.

34. *Growing tree* The diameter of a tree was 10 in. During the following year, the circumference increased 2 in. About how much did the tree's diameter increase? The tree's cross-section area?

35. *Estimating volume* Estimate the volume of material in a cylindrical shell with height 30 in., radius 6 in., and shell thickness 0.5 in.

0.5 in.

30 in.

6 in.

36. *Estimating height* A surveyor, standing 30 ft from the base of a building, measures the angle of elevation to the top of the building to be 75°. How accurately must the angle be measured for the percentage error in estimating the height of the building to be less than 4%?

37. *Tolerance* The height and radius of a right circular cylinder are equal, so the cylinder's volume is $V = \pi h^3$. The volume is to be calculated with an error of no more than 1% of the true value. Find approximately the greatest error that can be tolerated in the measurement of h, expressed as a percentage of h.

38. *Tolerance*

   (a) About how accurately must the interior diameter of a 10-m high cylindrical storage tank be measured to calculate the tank's volume to within 1% of its true value?

   (b) About how accurately must the tank's exterior diameter be measured to calculate the amount of paint it will take to paint the side of the tank to within 5% of the true amount?

39. *Minting coins* A manufacturer contracts to mint coins for the federal government. How much variation dr in the radius of the coins can be tolerated if the coins are to weigh within $1/1000$ of their ideal weight? Assume that the thickness does not vary.

40. *Profit* The profit P for a certain manufacturer selling x items is

$$P(x) = 200xe^{-x/400}.$$

Estimate the change and percent change in P as sales change from $x = 145$ to $x = 150$ items.

41. *The effect of flight maneuvers on the heart* The amount of work done by the heart's main pumping chamber, the left ventricle, is given by the equation

$$W = PV + \frac{V\delta v^2}{2g},$$

where W is the work per unit time, P is the average blood pressure, V is the volume of blood pumped out during the unit of time, δ ("delta") is the weight density of the blood, v is the average velocity of the exiting blood, and g is the acceleration of gravity.

When P, V, δ, and v remain constant, W becomes a function of g, and the equation takes the form

$$W = a + \frac{b}{g} \qquad (a, b \text{ constant}).$$

As a member of NASA's medical team, you want to know how sensitive W is to apparent changes in g caused by flight maneuvers, and this depends on the initial value of g. As part of your investigation, you decide to compare the effect on W of a given change dg on the moon, where $g = 5.2$ ft/sec^2, with the effect the same change dg would have on Earth, where $g = 32$ ft/sec^2. Use the simplified equation above to find the ratio of dW_{moon} to dW_{Earth}.

42. *Measuring acceleration of gravity* When the length L of a clock pendulum is held constant by controlling its temperature, the pendulum's period T depends on the acceleration of gravity g. The period will therefore vary slightly as the clock is moved from place to place on the earth's surface, depending on the change in g. By keeping track of ΔT, we can estimate the variation in g from the equation $T = 2\pi(L/g)^{1/2}$ that relates T, g, and L.

(a) With L held constant and g as the independent variable, calculate dT and use it to answer parts (b) and (c).

(b) *Writing to Learn* If g increases, will T increase or decrease? Will a pendulum clock speed up or slow down? Explain.

(c) A clock with a 100 cm pendulum is moved from a location where $g = 980$ cm/sec^2 to a new location. This increases the period by $dT = 0.001$ sec. Find dg and estimate the value of g at the new location.

T 43. *Zooming in to "see" differentiability* Is either of these functions differentiable at $x = 0$?

$$f(x) = |x| + 1, \qquad g(x) = \sqrt{x^2 + 0.0001} + 0.99$$

(a) We already know that f is not differentiable at $x = 0$; its graph has a corner there. Graph f and zoom in at the point $(0, 1)$ several times. Does the corner show signs of straightening out?

(b) Now do the same thing with g. Does the graph of g show signs of straightening out? We know g *is* differentiable at $x = 0$ and, in fact, has a horizontal tangent there.

(c) How many zooms does it take before the graph of g looks exactly like a horizontal line?

(d) Now graph f and g *together* in a standard square viewing window. They appear to be identical until you start zooming in. The differentiable function eventually straightens out, whereas the nondifferentiable function remains impressively unchanged.

T 44. *Reading derivatives from graphs* The idea that differentiable curves flatten out when magnified can be used to estimate the values of the derivatives of functions at particular points. We magnify the curve until the portion we see looks like a straight line through the point in question, and then we use the screen's coordinate grid to read the slope of the curve as the slope of the line it resembles.

(a) To see how the process works, try it first with the function $y = x^2$ at $x = 1$. The slope you read should be 2.

(b) Then try it with the curve $y = e^x$ at $x = 1$, $x = 0$, and $x = -1$. In each case, compare your estimate of the derivative with the value of e^x at the point. What pattern do you see? Test it with other values of x.

45. *The linearization is the best linear approximation* (This is why we use the linearization.) Suppose that $y = f(x)$ is differentiable at $x = a$ and that $g(x) = m(x - a) + c$ is a linear function in which m and c are constants. If the error $E(x) = f(x) - g(x)$ were small enough near $x = a$, we might think of using g as a linear approximation of f instead of the linearization $L(x) = f(a) + f'(a)(x - a)$. Show that if we impose on g the conditions

1. $E(a) = 0$ — The approximation error is zero at $x = a$.

2. $\displaystyle \lim_{x \to a} \frac{E(x)}{x - a} = 0$ — The error is negligible when compared with $x - a$.

then $g(x) = f(a) + f'(a)(x - a)$. Thus, the linearization $L(x)$ gives the only linear approximation whose error is both zero at $x = a$ and negligible in comparison with $x - a$.

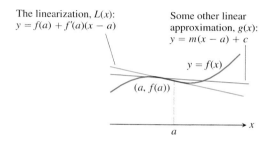

The linearization, $L(x)$:
$y = f(a) + f'(a)(x - a)$

Some other linear approximation, $g(x)$:
$y = m(x - a) + c$

$y = f(x)$

$(a, f(a))$

46. *Quadratic approximations*

(a) Let $Q(x) = b_0 + b_1(x - a) + b_2(x - a)^2$ be a quadratic approximation to $f(x)$ at $x = a$ with the properties:

   i. $Q(a) = f(a)$

   ii. $Q'(a) = f'(a)$

   iii. $Q''(a) = f''(a)$.

   Determine the coefficients b_0, b_1, and b_2.

(b) Find the quadratic approximation to $f(x) = 1/(1 - x)$ at $x = 0$.

T (c) Graph $f(x) = 1/(1 - x)$ and its quadratic approximation at $x = 0$. Then zoom in on the two graphs at the point $(0, 1)$. Comment on what you see.

T (d) Find the quadratic approximation to $g(x) = 1/x$ at $x = 1$. Graph g and its quadratic approximation together. Comment on what you see.

T (e) Find the quadratic approximation to $h(x) = \sqrt{1 + x}$ at $x = 0$. Graph h and its quadratic approximation together. Comment on what you see.

(f) What are the linearizations of f, g, and h at the respective points in parts (b), (d), and (e)?

47. *The linearization of 2^x*

(a) Find the linearization of $f(x) = 2^x$ at $x = 0$. Then round its coefficients to two decimal places.

T (b) Graph the linearization and function together for $-3 \leq x \leq 3$ and $-1 \leq x \leq 1$.

48. *The linearization of $\log_3 x$*

(a) Find the linearization of $f(x) = \log_3 x$ at $x = 3$. Then round its coefficients to two decimal places.

T (b) Graph the linearization and function together in the window $0 \leq x \leq 8$ and $2 \leq x \leq 4$.

49. *Linearizations at inflection points* As Figure 3.58 suggests, linearizations fit particularly well at inflection points. You will understand why in Chapter 8. As another example, graph *Newton's serpentine*, $f(x) = 4x/(x^2 + 1)$, together with its linearizations at $x = 0$ and $x = \sqrt{3}$.

T **50.** *Writing to Learn: Repeated root-taking*

(a) Enter 2 in your calculator and take successive square roots by pressing the square root key repeatedly (or raising the displayed number repeatedly to the 0.5 power). What pattern do you see emerging? Explain what is going on. What happens if you take successive tenth roots instead?

(b) Repeat the procedure with 0.5 in place of 2 as the original entry. What happens now? Can you use any positive number x in place of 2? Explain what is going on.

COMPUTER EXPLORATIONS

In Exercises 51–54, you will use a CAS to estimate the magnitude of the error in using the linearization in place of the function over a specified interval I. Perform the following steps.

(a) Plot the function f over I.

(b) Find the linearization L of the function at the point a.

(c) Plot f and L together on a single graph.

(d) Plot the absolute error $|f(x) - L(x)|$ over I and find its maximum value.

(e) From your graph in part (d), estimate as large a $\delta > 0$ as you can, satisfying

$$|x - a| < \delta \implies |f(x) - L(x)| < \epsilon$$

for $\epsilon = 0.5$, 0.1, and 0.01. Then check graphically to see if your δ-estimate holds true.

51. $f(x) = x^3 + x^2 - 2x$, $\quad [-1, 2]$, $\quad a = 1$

52. $f(x) = \dfrac{x - 1}{4x^2 + 1}$, $\quad [-3/4, 1]$, $\quad a = \dfrac{1}{2}$

53. $f(x) = x^{2/3}(x - 2)$, $\quad [-2, 3]$, $\quad a = 2$

54. $f(x) = \sqrt{x} - \sin x$, $\quad [0, 2\pi]$, $\quad a = 2$

3.7 Newton's Method

Procedure for Newton's Method • The Practice • Convergence Is Usually Assured • But Things Can Go Wrong • Fractal Basins and Newton's Method

CD-ROM
WEBsite
Historical Biography
Neils Henrik Abel
(1802 — 1829)

We know simple formulas for solving linear and quadratic equations, and there are somewhat more complicated formulas for cubic and quartic equations (equations of degree three and four). At one time it was hoped that similar formulas might be found for quintic and higher-degree equations, but Norwegian mathematician Neils Henrik Abel showed that no formulas like these are possible for polynomial equations of degree greater than four.

When exact formulas for solving an equation $f(x) = 0$ are not available, we can turn to numerical techniques from calculus to approximate the solutions we seek. One of these techniques is *Newton's method* or, as it is more accurately called, the *Newton–Raphson method*. It is based on the idea of using tangent lines to replace the graph of $y = f(x)$ near the points where f is zero. Once again, linearization is the key to solving a practical problem.

Procedure for Newton's Method

Newton's method is a numerical technique for approximating a zero of a function with zeros of its linearizations. Under favorable circumstances, the zeros of the linearizations *converge* rapidly to an accurate approximation. Moreover the method applies to a wide range of functions and usually gets results in only a few steps. Here is how it works.

The initial estimate, x_0, may be found by graphing or just plain guessing. The method then uses the tangent to the curve $y = f(x)$ at $(x_0, f(x_0))$ to approximate the curve, calling the point where the tangent meets the *x*-axis x_1 (Figure 3.61). The number x_1 is usually a better approximation to the solution than is x_0. The point x_2 where the tangent to the curve at $(x_1, f(x_1))$ crosses the *x*-axis is the next approximation in the sequence. We continue on, using each approximation to generate the next, until we are close enough to the root to stop.

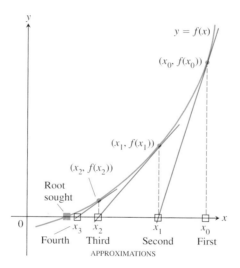

FIGURE 3.61 Newton's method starts with an initial guess x_0 and (under favorable circumstances) improves the guess one step at a time.

There is a formula for finding the $(n + 1)$st approximation x_{n+1} from the *n*th approximation x_n. The point-slope equation for the tangent to the curve at $(x_n, f(x_n))$ is

$$y - f(x_n) = f'(x_n)(x - x_n).$$

We can find where it crosses the *x*-axis by setting $y = 0$ (Figure 3.62).

$$0 - f(x_n) = f'(x_n)(x - x_n)$$
$$-f(x_n) = f'(x_n) \cdot x - f'(x_n) \cdot x_n$$
$$f'(x_n) \cdot x = f'(x_n) \cdot x_n - f(x_n)$$
$$x = x_n - \frac{f(x_n)}{f'(x_n)} \qquad \text{If } f'(x_n) \neq 0$$

This value of *x* is the next approximation x_{n+1}. Here is a summary of Newton's method.

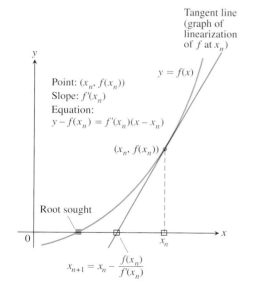

FIGURE 3.62 The geometry of the successive steps of Newton's method. From x_n we go up to the curve and follow the tangent line down to find x_{n+1}.

> **Procedure for Newton's Method**
> 1. Guess a first approximation to a solution of the equation $f(x) = 0$. A graph of $y = f(x)$ may help.
> 2. Use the first approximation to get a second, the second to get a third, and so on, using the formula
>
> $$x_{n+1} = x_n - \frac{f(x_n)}{f'(x_n)}. \qquad (1)$$

CD-ROM
WEBsite

Algorithm and Iteration

It is customary to call a specified sequence of computational steps like the one in Newton's method an *algorithm*. When an algorithm proceeds by repeating a given set of steps over and over, using the answer from the previous step as the input for the next, the algorithm is called *iterative* and each repetition is called an *iteration*. Newton's method is one of the really fast iterative techniques for finding roots.

The Practice

In our first example, we find decimal approximations to $\sqrt{2}$ by estimating the positive root of the equation $f(x) = x^2 - 2 = 0$.

Example 1   Finding the Square Root of 2

Find the positive root of the equation

$$f(x) = x^2 - 2 = 0.$$

Solution   With $f(x) = x^2 - 2$ and $f'(x) = 2x$, Equation (1) becomes

$$x_{n+1} = x_n - \frac{x_n^2 - 2}{2x_n}.$$

To use our calculator efficiently, we rewrite this equation in a form that uses fewer arithmetic operations:

$$x_{n+1} = x_n - \frac{x_n}{2} + \frac{1}{x_n}$$

$$= \frac{x_n}{2} + \frac{1}{x_n}.$$

The equation

$$x_{n+1} = \frac{x_n}{2} + \frac{1}{x_n}.$$

enables us to go from each approximation to the next with just a few keystrokes. With the starting value $x_0 = 1$, we get the results in the first column of the following table. (To five decimal places, $\sqrt{2} = 1.41421$.)

	Error	Number of correct figures
$x_0 = 1$	−0.41421	1
$x_1 = 1.5$	0.08579	1
$x_2 = 1.41667$	0.00246	3
$x_3 = 1.41422$	0.00001	5

Newton's method is the method used by most calculators to calculate roots because it converges so fast (more about this later). If the arithmetic in the table in Example 1 had been carried to 13 decimal places instead of 5, then going one step further would have given $\sqrt{2}$ correctly to more than 10 decimal places.

Example 2   Using Newton's Method

Find the x-coordinate of the point where the curve $y = x^3 - x$ crosses the horizontal line $y = 1$.

Solution   The curve crosses the line when $x^3 - x = 1$ or $x^3 - x - 1 = 0$. When does $f(x) = x^3 - x - 1$ equal zero? The graph of f (Figure 3.63) shows a single root, located between $x = 1$ and $x = 2$. We apply Newton's method to f with the starting value $x_0 = 1$. The results are displayed in Table 3.1 and Figure 3.64.

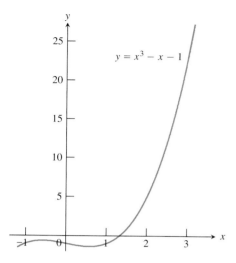

FIGURE 3.63 The graph of $f(x) = x^3 - x - 1$. (Example 2)

Table 3.1 The result of applying Newton's method to $f(x) = x^3 - x - 1$ with $x_0 = 1$

n	x_n	$f(x_n)$	$f'(x_n)$	$x_{n+1} = x_n - \dfrac{f(x_n)}{f'(x_n)}$
0	1	-1	2	1.5
1	1.5	0.875	5.75	1.3478 26087
2	1.3478 26087	0.1006 82173	4.4499 05482	1.3252 00399
3	1.3252 00399	0.0020 58362	4.2684 68292	1.3247 18174
4	1.3247 18174	0.0000 00924	4.2646 34722	1.3247 17957
5	1.3247 17957	$1.8672\text{E}{-13}$	4.2646 32999	1.3247 17957

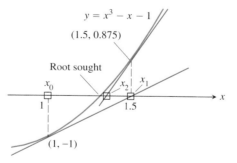

FIGURE 3.64 The first three x-values in Table 3.1.

At $n = 5$, we come to the result $x_6 = x_5 = 1.3247\ 17957$. When $x_{n+1} = x_n$, Equation (1) shows that $f(x_n) = 0$. We have found a solution of $f(x) = 0$ to nine decimals.

In Figure 3.65, we have indicated that the process in Example 2 might have started at the point $B_0(3, 23)$ on the curve, with $x_0 = 3$. Point B_0 is quite far from the x-axis, but the tangent at B_0 crosses the x-axis at about $(2.12, 0)$, so x_1 is still an improvement over x_0. If we use Equation (1) repeatedly as before, with $f(x) = x^3 - x - 1$ and $f'(x) = 3x^2 - 1$, we confirm the nine-place solution $x_7 = x_6 = 1.3247\ 17957$ in seven steps.

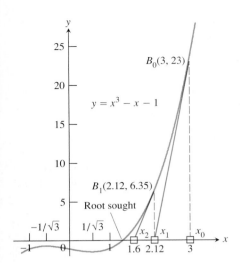

FIGURE 3.65 Any starting value x_0 to the right of $x = 1/\sqrt{3}$ will lead to the root.

The curve in Figure 3.65 has a local maximum at $x = -1/\sqrt{3}$ and a local minimum at $x = +1/\sqrt{3}$. We would not expect good results from Newton's method if we were to start with x_0 between these points, but we can start any place to the right of $x = 1/\sqrt{3}$ and get the answer. It would not be very clever to do so, but we could even begin far to the right of B_0, for example with $x_0 = 10$. It takes a bit longer, but the process still converges to the same answer as before.

Convergence Is Usually Assured

In practice, Newton's method usually converges with impressive speed, but this is not guaranteed. One way to test convergence is to begin by graphing the function to estimate a good starting value for x_0. You can test that you are getting closer to a zero of the function by evaluating $|f(x_n)|$ and check that the method is converging by evaluating $|x_n - x_{n+1}|$.

Theory does provide some help. A theorem from advanced calculus says that if

$$\left| \frac{f(x)f''(x)}{[f'(x)]^2} \right| < 1 \qquad (2)$$

for all x in an interval about a root r, then the method will converge to r for any starting value x_0 in that interval.

Newton's method always converges if the curve $y = f(x)$ is convex ("bulges") toward the x-axis in the interval between x_0 and the root sought. (See Figure 3.66.)

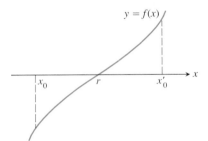

FIGURE 3.66 Newton's method will converge to r from either starting point.

Under favorable circumstances, the speed with which Newton's method converges to r is expressed by the advanced calculus formula

$$\underbrace{|x_{n+1} - r|}_{\text{error } e_{n+1}} \leq \frac{\max |f''|}{2 \min |f'|} |x_n - r|^2 = \text{constant} \cdot \underbrace{|x_n - r|^2}_{\text{error } e_n}, \qquad (3)$$

where max and min refer to the maximum and minimum values in an interval surrounding r. The formula says that the error in step $n + 1$ is no greater than a constant times the square of the error in step n. This may not seem like much, but think of what it says. If the constant is less than or equal to 1 and $|x_n - r| < 10^{-3}$, then $|x_{n+1} - r| < 10^{-6}$. *In a single step,* the method moves from three decimal places of accuracy to six!

The results in Equations (2) and (3) both assume that f is "nice." Hence, in the case of Equation (3), this means that f has only a single root at r, so that $f'(r) \neq 0$. If f has a multiple root at r, the convergence may be slower.

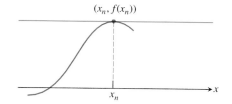

FIGURE 3.67 If $f'(x_n) = 0$, there is no intersection point to define x_{n+1}.

But Things Can Go Wrong

Newton's method stops if $f'(x_n) = 0$ (Figure 3.67). In that case, try a new starting point. Of course, f and f' may have a common root. To detect whether this is so, you could first find the solutions of $f'(x) = 0$ and check f at those values, or you can graph f and f' together.

Newton's method does not always converge. For instance, if

$$f(x) = \begin{cases} -\sqrt{r - x}, & x < r \\ \sqrt{x - r}, & x \geq r, \end{cases}$$

the graph will be like the one in Figure 3.68. If we begin with $x_0 = r - h$, we get $x_1 = r + h$, and successive approximations go back and forth between these two values. No amount of iteration brings us closer to the root than our first guess.

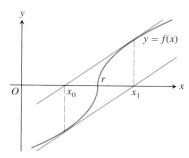

FIGURE 3.68 Newton's method fails to converge. You go from x_0 to x_1 and back to x_0, never getting any closer to r.

If Newton's method does converge, it converges to a root. Be careful, however. There are situations in which the method appears to converge but there is no root there. Fortunately, such situations are rare.

When Newton's method converges to a root, it may not be the root you have in mind. Figure 3.69 shows two ways this can happen.

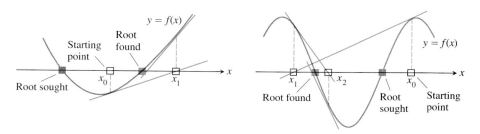

FIGURE 3.69 If you start too far away, Newton's method may miss the root you want.

Fractal Basins and Newton's Method

The process of finding roots by Newton's method can be uncertain in the sense that for some equations the final outcome can be extremely sensitive to the starting value's location.

The equation $4x^4 - 4x^2 = 0$ is a case in point (Figure 3.70a). Starting values in the blue zone on the x-axis lead to root A. Starting values in the black lead to root B, and starting values in the red zone lead to root C. The points $\pm\sqrt{2}/2$ give

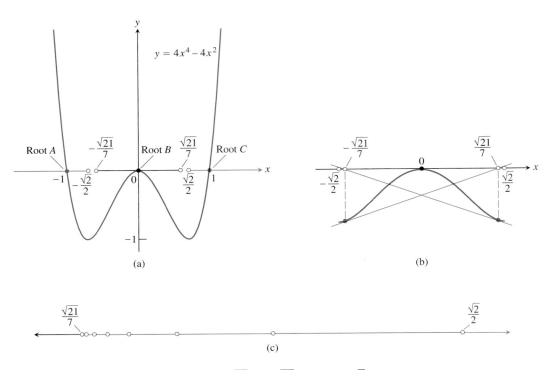

FIGURE 3.70 (a) Starting values in $(-\infty, -\sqrt{2}/2)$, $(-\sqrt{21}/7, \sqrt{21}/7)$, and $(\sqrt{2}/2, \infty)$ lead respectively to roots A, B, and C. (b) The values $x = \pm\sqrt{21}/7$ lead only to each other. (c) Between $\sqrt{21}/7$ and $\sqrt{2}/2$, there are infinitely many open intervals of points attracted to A alternating with open intervals of points attracted to C. This behavior is mirrored in the interval $(-\sqrt{2}/2, -\sqrt{21}/7)$.

horizontal tangents. The points $\pm\sqrt{21}/7$ "cycle," each leading to the other, and back (Figure 3.70b).

The interval between $\sqrt{21}/7$ and $\sqrt{2}/2$ contains infinitely many open intervals of points leading to root A, alternating with intervals of points leading to root C (Figure 3.70c). The boundary points separating consecutive intervals (there are infinitely many) do not lead to roots, but cycle back and forth from one to another. Moreover, as we select points that approach $\sqrt{21}/7$ from the right, it becomes increasingly difficult to distinguish which lead to root A and which to root C. On the same side of $\sqrt{21}/7$, we find arbitrarily close together points whose ultimate destinations are far apart.

If we think of the roots as "attractors" of other points, the coloring in Figure 3.70 shows the intervals of the points they attract (the "intervals of attraction"). You might think that points between roots A and B would be attracted to either A or B, but, as we see, that is not the case. Between A and B there are infinitely many intervals of points attracted to C. Similarly, between B and C lie infinitely many intervals of points attracted to A.

We encounter an even more dramatic example of such behavior when we apply Newton's method to solve the complex-number equation $z^6 - 1 = 0$. It has six solutions: 1, -1, and the four numbers $\pm(1/2) \pm (\sqrt{3}/2)i$. As Figure 3.71 suggests, each of the six roots has infinitely many "basins" of attraction in the complex plane (Appendix 3). Starting points in red basins are attracted to the root 1, those in the green basin to the root $(1/2) + (\sqrt{3}/2)i$, and so on. Each basin has a boundary whose complicated pattern repeats without end under successive magnifications. These basins are called **fractal basins.**

FIGURE 3.71 This computer-generated initial value portrait uses color to show where different points in the complex plane end up when they are used as starting values in applying Newton's method to solve the equation $z^6 - 1 = 0$. Red points go to 1, green points to $(1/2) + (\sqrt{3}/2)i$, dark blue points to $(-1/2) + (\sqrt{3}/2)i$, and so on. Starting values that generate sequences that do not arrive within 0.1 units of a root after 32 steps are colored black.

EXERCISES 3.7

Root-Finding
CD-ROM
WEBsite

1. Use Newton's method to estimate the solutions of the equation $x^2 + x - 1 = 0$. Start with $x_0 = -1$ for the left-hand solution and with $x_0 = 1$ for the solution on the right. Then, in each case, find x_2.

2. Use Newton's method to estimate the one real solution of $x^3 + 3x + 1 = 0$. Start with $x_0 = 0$ and then find x_2.

3. Use Newton's method to estimate the two zeros of the function $f(x) = x^4 + x - 3$. Start with $x_0 = -1$ for the left-hand zero and with $x_0 = 1$ for the zero on the right. Then, in each case, find x_2.

4. Use Newton's method to estimate the two zeros of the function $f(x) = 2x - x^2 + 1$. Start with $x_0 = 0$ for the left-hand zero and with $x_0 = 2$ for the zero on the right. Then, in each case, find x_2.

In Exercises 5 and 6, use Newton's method to find all roots of the equation correct to six decimal places.

5. $e^{-x} = 2x + 1$

6. $\tan^{-1} x = 1 - 2x$

Theory, Examples, and Applications

7. *Guessing a root* Suppose that your first guess is lucky, in the sense that x_0 is a root of $f(x) = 0$. Assuming that $f'(x_0)$ is defined and not 0, what happens to x_1 and later approximations?

8. *Writing to Learn: Estimating pi* You plan to estimate $\pi/2$ to five decimal places by using Newton's method to solve the equation $\cos x = 0$. Does it matter what your starting value is? Give reasons for your answer.

9. *Oscillation* Show that if $h > 0$, applying Newton's method to

$$f(x) = \begin{cases} \sqrt{x}, & x \geq 0 \\ \sqrt{-x}, & x < 0 \end{cases}$$

leads to $x_1 = -h$ if $x_0 = h$ and to $x_1 = h$ if $x_0 = -h$. Draw a picture that shows what is going on.

10. *Approximations that get worse and worse* Apply Newton's method to $f(x) = x^{1/3}$ with $x_0 = 1$ and calculate $x_1, x_2, x_3,$ and x_4. Find a formula for $|x_n|$. What happens to $|x_n|$ as $n \to \infty$? Draw a picture that shows what is going on.

11. *Writing to Learn* Explain why the following four statements ask for the same information:

   i. Find the roots of $f(x) = x^3 - 3x - 1$.

   ii. Find the x-coordinates of the intersections of the curve $y = x^3$ with the line $y = 3x + 1$.

   iii. Find the x-coordinates of the points where the curve $y = x^3 - 3x$ crosses the horizontal line $y = 1$.

   iv. Find the values of x where the derivative of $g(x) = (1/4)x^4 - (3/2)x^2 - x + 5$ equals zero.

12. *Locating a planet* To calculate a planet's space coordinates, we have to solve equations like $x = 1 + 0.5 \sin x$. Graphing the function $f(x) = x - 1 - 0.5 \sin x$ suggests that the function has a root near $x = 1.5$. Use one application of Newton's method to improve this estimate. That is, start with $x_0 = 1.5$ and find x_1. (The value of the root is 1.49870 to five decimal places.) Remember to use radians.

13. *A program for using Newton's method on a grapher* Let $f(x) = x^3 + 3x + 1$. Here is a home screen program to perform the computations in Newton's method.

(a) Let $y_0 = f(x)$ and $y_1 = \text{NDER } f(x)$.

(b) Store $x_0 = -0.3$ into x.

(c) Then store $x - (y_0/y_1)$ into x and press the "Enter" key over and over. Watch as the numbers converge to the zero of f.

(d) Use different values for x_0 and repeat steps (b) and (c).

(e) Write your own equation and use this approach to solve it using Newton's method. Compare your answer with the answer given by the built-in feature of your calculator that gives zeros of functions.

14. *(Continuation of Exercise 11)*

(a) Use Newton's method to find the two negative zeros of $f(x) = x^3 - 3x - 1$ to five decimal places.

(b) Graph $f(x) = x^3 - 3x - 1$ for $-2 \le x \le 2.5$. Use the zoom and trace features to estimate the zeros of f to five decimal places.

(c) Graph $g(x) = 0.25x^4 - 1.5x^2 - x + 5$. Use the zoom and trace features with appropriate rescaling to find, to five decimal places, the values of x where the graph has horizontal tangents.

15. *Intersecting curves* The curve $y = \tan x$ crosses the line $y = 2x$ between $x = 0$ and $x = \pi/2$. Use Newton's method to find where.

16. *Real solutions of a quartic* Use Newton's method to find the two real solutions of the equation $x^4 - 2x^3 - x^2 - 2x + 2 = 0$.

17. *Finding solutions*

(a) How many solutions does the equation $\sin 3x = 0.99 - x^2$ have?

(b) Use Newton's method to find them.

18. *Intersection of curves*

(a) Does $\cos 3x$ ever equal x? Give reasons for your answer.

(b) Use Newton's method to find where.

19. *Multiple zeros* Find the four real zeros of the function $f(x) = 2x^4 - 4x^2 + 1$.

20. *Estimating pi* Estimate π to as many decimal places as your calculator will display by using Newton's method to solve the equation $\tan x = 0$ with $x_0 = 3$.

21. *Intersection of curves* At what value(s) of x does $e^{-x^2} = x^2 - x + 1$?

22. *Intersection of curves* At what value(s) of x does $\ln(1 - x^2) = x - 1$?

23. *Finding a root* Use the Intermediate Value Theorem from Section 1.4 to show that $f(x) = x^3 + 2x - 4$ has a root between $x = 1$ and $x = 2$. Then find the root to five decimal places.

24. *Factoring a quartic* Find the approximate values of r_1 through r_4 in the factorization

$$8x^4 - 14x^3 - 9x^2 + 11x - 1 = 8(x - r_1)(x - r_2)(x - r_3)(x - r_4).$$

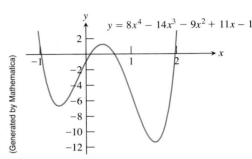

25. *Converging to different zeros* Use Newton's method to find the zeros of $f(x) = 4x^4 - 4x^2$ using the given starting values (Figure 3.70).

(a) $x_0 = -2$ and $x_0 = -0.8$, lying in $(-\infty, -\sqrt{2}/2)$

(b) $x_0 = -0.5$ and $x_0 = 0.25$, lying in $(-\sqrt{21}/7, \sqrt{21}/7)$

(c) $x_0 = 0.8$ and $x_0 = 2$, lying in $(\sqrt{2}/2, \infty)$

(d) $x_0 = -\sqrt{21}/7$ and $x_0 = \sqrt{21}/7$

26. *The sonobuoy problem* In submarine location problems, it is often necessary to find a submarine's closest point of approach (CPA) to a sonobuoy (sound detector) in the water. Suppose that the submarine travels on a parabolic path $y = x^2$ and that the buoy is located at the point $(2, -1/2)$.

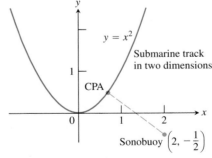

(Source: The Contraction Mapping Principle, by C. O. Wilde, UMAP Unit 326, Arlington, MA, COMAP, Inc.)

(a) Show that the value of x that minimizes the distance between the submarine and the buoy is a solution of the equation $x = 1/(x^2 + 1)$.

(b) Solve the equation $x = 1/(x^2 + 1)$ with Newton's method.

27. *Curves that are nearly flat at the root* Some curves are so flat that, in practice, Newton's method stops too far from the root to give a useful estimate. Try Newton's method on $f(x) = (x - 1)^{40}$ with a starting value of $x_0 = 2$ to see how close your machine comes to the root $x = 1$.

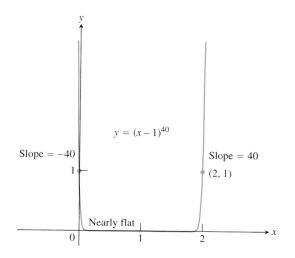

$y = (x - 1)^{40}$

Slope $= -40$

Slope $= 40$

(2, 1)

Nearly flat

28. *Finding a root different from the one sought* All three roots of $f(x) = 4x^4 - 4x^2$ can be found by starting Newton's method near $x = \sqrt{21}/7$. Try it. See Figure 3.70.

29. *Finding an ion concentration* While trying to find the acidity of a saturated solution of magnesium hydroxide in hydrochloric acid, you derive the equation

$$\frac{3.64 \times 10^{-11}}{[\mathrm{H_3O^+}]^2} = [\mathrm{H_3O^+}] + 3.6 \times 10^{-4}$$

for the hydronium ion concentration $[\mathrm{H_3O^+}]$. To find the value of $[\mathrm{H_3O^+}]$, you set $x = 10^4[\mathrm{H_3O^+}]$ and convert the equation to

$$x^3 + 3.6x^2 - 36.4 = 0.$$

You then solve this by Newton's method. What do you get for x? (Make it good to two decimal places.) For $[\mathrm{H_3O^+}]$?

T 30. *Complex roots* If you have a computer or a calculator that can be programmed to do complex-number arithmetic, experiment with Newton's method to solve the equation $z^6 - 1 = 0$. The recursion relation to use is

$$z_{n+1} = z_n - \frac{z_n^6 - 1}{6z_n^5} \quad \text{or} \quad z_{n+1} = \frac{5}{6}z_n + \frac{1}{6z_n^5}.$$

Try these starting values (among others): $2, i, \sqrt{3} + i$.

Questions to Guide Your Review

1. What can be said about the values of a function that is continuous on a closed interval?

2. What does it mean for a function to have a local extreme value on its domain? An absolute extreme value? How are local and absolute extreme values related, if at all? Give examples.

3. What is true of f' at an interior point where a local extremum occurs? How does this fact lead to a procedure for finding a function's local extreme values?

4. How do you find the absolute extrema of a continuous function on a closed interval? Give examples.

5. What are the hypotheses and conclusion of Rolle's Theorem? Are the hypotheses really necessary? Explain.

6. What are the hypotheses and conclusion of the Mean Value Theorem? What physical interpretations might the theorem have?

7. State the Mean Value Theorem's three corollaries.

8. How can you sometimes identify a function $f(x)$ by knowing f' and knowing the value of f at a point $x = x_0$? Give an example.

9. What is a differential equation? What is a solution to a differential equation? Give examples.

10. What is the First Derivative Test for Increasing and Decreasing? How can it be used to test for local extrema?

11. How do you test a twice-differentiable function to determine where its graph is concave up or concave down? Give examples.

12. What is an inflection point? Give an example. What physical significance do inflection points sometimes have?

13. What is the Second Derivative Test for Local Extrema? Give examples of how it is applied.

14. What do the derivatives of a function tell you about the shape of its graph?

15. What is an autonomous differential equation? What are its equilibrium values? How do they differ from critical points? What is a stable equilibrium value? Unstable?

16. How do you construct the phase line for an autonomous differential equation? How does the phase line help you produce a graph which qualitatively depicts a solution to the differential equation?

17. Outline a general strategy for solving max-min problems. Give examples.

18. What is the linearization $L(x)$ of a function $f(x)$ at a point $x = a$? What is required of f at a for the linearization to exist? How are linearizations used? Give examples.

19. If x moves from a to a nearby value $a + dx$, how do you estimate the corresponding change in the value of a differentiable function $f(x)$? How do you estimate the relative change? The percentage change? Give an example.

20. Describe Newton's method for solving equations. Give an example. What are some of the things to watch out for when you use the method?

Practice Exercises

Conclusions from Graphs

In Exercises 1–4, use the graph to answer the questions.

1. Identify any global extreme values of f and the values of x at which they occur.

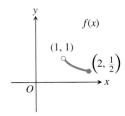

2. At which of the five points on the graph of $y = f(x)$ shown here

   (a) are y' and y'' both negative?

   (b) is y' negative and y'' positive?

3. Estimate the intervals on which the function $y = f(x)$ is

   (a) increasing

   (b) decreasing.

   (c) Use the given graph of f' to indicate where any local extreme values of the function occur, and whether each extreme is a relative maximum or minimum.

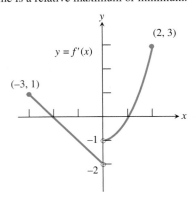

4. Here is the graph of a fruit fly population. On approximately what day did the population's growth rate change from increasing to decreasing?

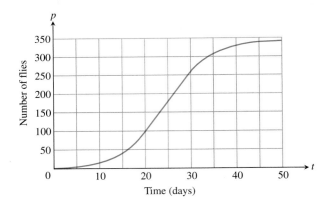

Existence of Extreme Values

5. *Writing to Learn: Local extrema* Does $f(x) = x^3 + 2x + \tan x$ have any local maximum or minimum values? Gives reasons for your answer.

6. *Writing to Learn: Local maxima* Does $g(x) = \csc x + 2 \cot x$ have any local maximum values? Give reasons for your answer.

7. *Writing to Learn: Extreme values* Does $f(x) = (7 + x)(11 - 3x)^{1/3}$ have an absolute minimum value? An absolute maximum? If so, find them or give reasons why they fail to exist. List all critical points of f.

8. *Writing to Learn: Local extrema* Find values of a and b such that the function

$$f(x) = \frac{ax + b}{x^2 - 1}$$

has a local extreme value of 1 at $x = 3$. Is this extreme value a local maximum, or a local minimum? Give reasons for your answer.

9. *Writing to Learn* The greatest integer function $f(x) = \mathrm{int}\, x$, defined for all values of x, assumes a local maximum value of 0 at each point of $[0, 1)$. Could any of these local maximum values also be local minimum values of f? Give reasons for your answer.

10. (a) Give an example of a differentiable function f whose first derivative is zero at some point c even though f has neither a local maximum nor a local minimum at c.

   (b) *Writing to Learn* How is this consistent with Theorem 2 in Section 3.1? Give reasons for your answer.

11. *Absolute extrema* The function $y = 1/x$ does not take on either a maximum or a minimum on the interval $0 < x < 1$ even though the function is continuous on this interval. Does this contradict the Extreme Value Theorem for Continuous Functions? Why?

12. *Absolute extrema* What are the maximum and minimum values of the function $y = |x|$ on the interval $-1 \le x < 1$? Notice that the interval is not closed. Is this consistent with the Extreme Value Theorem for Continuous Functions? Why?

The Mean Value Theorem

13. (a) Show that $g(t) = \sin^2 t - 3t$ decreases on every interval in its domain.

(b) *Writing to Learn* How many solutions does the equation $\sin^2 t - 3t = 5$ have? Give reasons for your answer.

14. (a) Show that $y = \tan \theta$ increases on every interval in its domain.

(b) *Writing to Learn* If the conclusion in (a) is really correct, how do you explain the fact that $\tan \pi = 0$ is less than $\tan (\pi/4) = 1$?

15. (a) Show that the equation $x^4 + 2x^2 - 2 = 0$ has exactly one solution on $[0, 1]$.

T **(b)** Using a calculator, find the solution to as many decimal places as you can.

16. (a) *An increasing function* Show that $f(x) = x/(x + 1)$ increases on every interval in its domain.

(b) *A function with no local extrema* Show that $f(x) = x^3 + 2x$ has no local maximum or minimum values.

17. *Water in a reservoir* As a result of a heavy rain, the volume of water in a reservoir increased by 1400 acre-ft in 24 h. Show that at some instant during that period, the reservoir's volume was increasing at a rate in excess of 225,000 gal /min. (An acre-foot is $43,560 \text{ ft}^3$, the volume that would cover one acre to the depth of one foot. A cubic foot holds 7.48 gal.)

18. *Writing to Learn* The formula $F(x) = 3x + C$ gives a different function for each value of C. All these functions, however, have the same derivative with respect to x, namely $F'(x) = 3$. Are these the only differentiable functions whose derivative is 3? Could there be any others? Give reasons for your answers.

19. *Writing to Learn* Show that

$$\frac{d}{dx}\left(\frac{x}{x+1}\right) = \frac{d}{dx}\left(-\frac{1}{x+1}\right)$$

even though

$$\frac{x}{x+1} \ne -\frac{1}{x+1}.$$

Doesn't this contradict Corollary 2 of the Mean Value Theorem? Give reasons for your answer.

20. *Comparing derivatives* Calculate the first derivatives of $f(x) = x^2/(x^2 + 1)$ and $g(x) = -1/(x^2 + 1)$. What can you conclude about the graphs of these functions?

Graphs and Sketching

Sketch the curves in Exercises 21–26.

21. $y = x^2 - (x^3/6)$

22. $y = -x^3 + 6x^2 - 9x + 3$

23. $y = (1/8)(x^3 + 3x^2 - 9x - 27)$

24. $y = x^3(8 - x)$

25. $y = (x - 3)^2 e^x$

26. $y = x\sqrt{3 - x}$

Each of Exercises 27 and 28 gives the first derivative of a function $y = f(x)$.

(a) At what points, if any, does the graph of f have a local maximum, local minimum, or inflection point?

(b) Sketch the general shape of the graph.

27. $y' = 16 - x^2$

28. $y' = 6x(x + 1)(x - 2)$

Finding Extreme Values

In Exercises 29 and 30, find the absolute maxima and minima of the functions and say where they are assumed.

29. $f(x) = e^{x/\sqrt{x^4+1}}$

30. $g(x) = e^{\sqrt{3-2x-x^2}}$

Motion

Drawing conclusions about motion from graphs Each of the graphs in Exercises 31 and 32 is the graph of the position function $s = f(t)$ of a body moving on a coordinate line (t represents time). At approximately what times (if any) is each body's

(a) velocity equal to zero?

(b) acceleration equal to zero?

During approximately what time intervals does the body move

(c) forward?

(d) backward?

31.

32.

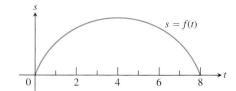

33. *Motion along a line* A particle is moving along a line with position function $s(t) = 3 + 4t - 3t^2 - t^3$. Find the

(a) velocity

(b) acceleration.

(c) Describe the motion of the particle for $t \geq 0$.

34. *Motion along a line* A particle is moving along a line with position function $s(t) = (1/2)t^4 - 4t^3 + 6t^2, t \geq 0$. During what time intervals does the particle move forward? Move backward?

Differential Equations

In Exercises 35–38, find all possible functions with the given derivative.

35. $f'(x) = x^{-5} + e^{-x}$　　　**36.** $f'(x) = \sec x \tan x$

37. $f'(x) = \dfrac{2}{x^2} + x^2 + 1, x > 0$　　**38.** $f'(x) = \sqrt{x} + \dfrac{1}{\sqrt{x}}$

In Exercises 39 and 40, the velocity v or acceleration a of a particle is given along with its initial position. Find the particle's position s at time t.

39. $v = 9.8t + 5, \quad s = 10 \quad \text{when} \quad t = 0$

40. $a = 32, \quad v = 20 \quad \text{and} \quad s = 5 \quad \text{when} \quad t = 0$

Autonomous Differential Equations and Phase Lines

In Exercises 41 and 42,

(a) Identify the equilibrium values. Which are stable and which are unstable?

(b) Construct a phase line. Identify the signs of y' and y''.

(c) Sketch a representative selection of solution curves.

41. $\dfrac{dy}{dx} = y^2 - 1$　　　　**42.** $\dfrac{dy}{dx} = y - y^2$

Optimization

43. *Area of sector* If the perimeter of the circular sector shown here is fixed at 100 ft, what values of r and s will give the sector the greatest area?

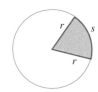

44. *Area of triangle* An isosceles triangle has its vertex at the origin and its base parallel to the x-axis with the vertices above the axis on the curve $y = 27 - x^2$. Find the largest area the triangle can have.

45. *Inscribing a cylinder* Find the height and radius of the largest right circular cylinder that can be put into a sphere of radius $\sqrt{3}$ as described in the figure.

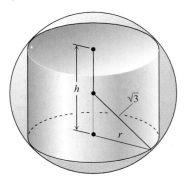

46. *Cone in a cone* The figure here shows two right circular cones, one upside down inside the other. The two bases are parallel, and the vertex of the smaller cone lies at the center of the larger cone's base. What values of r and h will give the smaller cone the largest possible volume?

47. *Manufacturing tires* Your company can manufacture x hundred grade A tires and y hundred grade B tires a day, where $0 \leq x \leq 4$ and

$$y = \frac{40 - 10x}{5 - x}.$$

Your profit on a grade A tire is twice your profit on a grade B tire. What is the most profitable number of each kind to make?

48. *Particle motion* The positions of two particles on the s-axis are $s_1 = \cos t$ and $s_2 = \cos(t + \pi/4)$.

(a) What is the farthest apart the particles ever get?

(b) When do the particles collide?

T 49. *Open-top box* An open-top rectangular box is constructed from a 10-in.-by-16.-in piece of cardboard by cutting squares of equal side length from the corners and folding up the sides. Find analytically the dimensions of the box of largest volume and the maximum volume. Support your answers graphically.

50. *Designing a vat* You are to design an open-top rectangular stainless-steel vat. It is to have a square base and a volume of 32 ft³, to be welded from quarter-inch plate, and weigh no more than necessary. What dimensions do you recommend?

Linearization

51. Find the linearizations of

(a) $\tan x$ at $x = -\pi/4$ (b) $\sec x$ at $x = -\pi/4$.

Graph the curves and linearizations together.

52. We can obtain a useful linear approximation of the function $f(x) = 1/(1 + \tan x)$ at $x = 0$ by combining the approximations

$$\frac{1}{1 + x} \approx 1 - x \qquad \text{and} \qquad \tan x \approx x$$

to get

$$\frac{1}{1 + \tan x} \approx 1 - x.$$

Show that this result is the standard linear approximation of $1/(1 + \tan x)$ at $x = 0$.

53. Find the linearization of $f(x) = e^x + \sin x - 0.5$ at $x = 0$.

54. Find the linearization of $f(x) = 2/(1 - x) + \sqrt{1 + x} - 3.1$ at $x = 0$.

Differential Estimates of Change

55. *Volume of a cone* Write a formula that estimates the change that occurs in the volume of a right circular cone when the radius changes from r_0 to $r_0 + dr$ and the height does not change.

$V = \frac{1}{3}\pi r^2 h$

$S = \pi r \sqrt{r^2 + h^2}$

(Lateral surface area)

56. *Controlling error*

(a) How accurately should you measure the edge of a cube to be reasonably sure of calculating the cube's surface area with an error of no more than 2%?

(b) Suppose that the edge is measured with the accuracy required in part (a). About how accurately can the cube's volume be calculated from the edge measurement? To find out, estimate the percentage error in the volume calculation that might result from using the edge measurement.

57. *Compounding error* The circumference of the equator of a sphere is measured as 10 cm with a possible error of 0.4 cm. This measurement is then used to calculate the radius. The radius is then used to calculate the surface area and volume of the sphere. Estimate the percentage errors in the calculated values of

(a) the radius

(b) the surface area

(c) the volume.

58. *Finding height* To find the height of a lamppost (see figure), you stand a 6 ft pole 20 ft from the lamp and measure the length a of its shadow, finding it to be 15 ft, give or take an inch. Calculate the height of the lamppost using the value $a = 15$ and estimate the possible error in the result.

Newton's Method

[T] In Exercises 59–62, use Newton's method to estimate the zeros of the given function. Use a calculator and state your answers accurate to six decimal places.

59. $f(x) = 3x - x^3, \quad 1 \le x \le 2$

60. $f(x) = x^3 + \dfrac{4}{x^2} + 7, \quad x < 0$

61. $g(t) = 2 \cos t - \sqrt{1 - t}, \quad -\infty < t < \infty$

62. $g(t) = \sqrt{t} + \sqrt{1 + t} - 4, \quad t > 0$

Additional Exercises: Theory, Examples, Applications

1. *Writing to Learn* What can you say about a function whose maximum and minimum values on an interval are equal? Give reasons for your answer.

2. *Writing to Learn* Is it true that a discontinuous function cannot have both an absolute maximum and an absolute minimum value on a closed interval? Give reasons for your answer.

3. *Writing to Learn* Can you conclude anything about the extreme values of a continuous function on an open interval? On a half-open interval? Give reasons for your answer.

4. *Local extrema* Use the sign pattern for the derivative

$$\frac{df}{dx} = 6(x-1)(x-2)^2(x-3)^3(x-4)^4$$

to identify the points where f has local maximum and minimum values.

5. *Local extrema*

(a) Suppose that the first derivative of $y = f(x)$ is

$$y' = 6(x+1)(x-2)^2.$$

At what points, if any, does the graph of f have a local maximum, local minimum, or point of inflection?

(b) Suppose that the first derivative of $y = f(x)$ is

$$y' = 6x(x+1)(x-2).$$

At what points, if any, does the graph of f have a local maximum, local minimum, or point of inflection?

6. *Writing to Learn: Bounding a function* If $f'(x) \leq 2$ for all x, what is the most the values of f can increase on $[0, 6]$? Give reasons for your answer.

7. *Bounding a function* Suppose that f is continuous on $[a, b]$ and that c is an interior point of the interval. Show that if $f'(x) \leq 0$ on $[a, c)$ and $f'(x) \geq 0$ on $(c, b]$, then $f(x)$ is never less than $f(c)$ on $[a, b]$.

8. *An inequality*

(a) Show that $-1/2 \leq x/(1 + x^2) \leq 1/2$ for every value of x.

(b) Suppose that f is a function whose derivative is $f'(x) = x/(1 + x^2)$. Use the result in part (a) to show that

$$|f(b) - f(a)| \leq \frac{1}{2}|b - a|$$

for any a and b.

9. *Writing to Learn* The derivative of $f(x) = x^2$ is zero at $x = 0$, but f is not a constant function. Doesn't this contradict the corollary of

the Mean Value Theorem that says that functions with zero derivatives are constant? Give reasons for your answer.

10. *Extrema and inflection points* Let $h = fg$ be the product of two differentiable functions of x.

(a) If f and g are positive, with local maxima at $x = a$, and if f' and g' change sign at a, does h have a local maximum at a?

(b) If the graphs of f and g have inflection points at $x = a$, does the graph of h have an inflection point at a?

In either case, if the answer is yes, give a proof. If the answer is no, give a counterexample.

11. *Finding a function* Use the following information to find the values of a, b, and c in the formula $f(x) = (x + a)/(bx^2 + cx + 2)$.

i. The values of a, b, and c are either 0 or 1.

ii. The graph of f passes through the point $(-1, 0)$.

iii. The line $y = 1$ is an asymptote of the graph of f.

12. *Horizontal tangent* For what value or values of the constant k will the curve $y = x^3 + kx^2 + 3x - 4$ have exactly one horizontal tangent?

13. *Largest inscribed triangle* Points A and B lie at the ends of a diameter of a unit circle and point C lies on the circumference. Is it true that the perimeter of triangle ABC is largest when the triangle is isosceles? How do you know?

14. *The ladder problem* What is the approximate length (in feet) of the longest ladder you can carry horizontally around the corner of the corridor shown here? Round your answer down to the nearest foot.

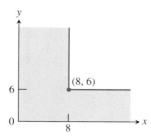

15. *Hole in a water tank* You want to bore a hole in the side of the tank shown here at a height that will make the stream of water coming out hit the ground as far from the tank as possible. If you drill the hole near the top, where the pressure is low, the water will exit slowly but spend a relatively long time in the air. If you drill the hole near the bottom, the water will exit at a higher velocity but have only a short time to fall. Where is the best place, if any, for the hole? (*Hint:* How long will it take an exiting particle of water to fall from height y to the ground?)

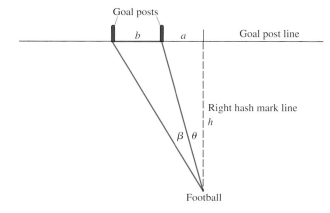

16. *Kicking a field goal* An American football player wants to kick a field goal with the ball being on a right hash mark. Assume that the goal posts are b feet apart and that the hash mark line is a distance $a > 0$ feet from the right goal post. (See the accompanying figure.) Find the distance h from the goal post line that gives the kicker his largest angle β. Assume that the football field is flat.

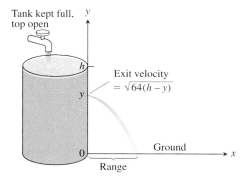

17. *A max-min problem with a variable answer* Sometimes the solution of a max-min problem depends on the proportions of the shapes involved. As a case in point, suppose that a right circular cylinder of radius r and height h is inscribed in a right circular cone of radius R and height H, as shown here. Find the value of r (in terms of R and H) that maximizes the total surface area of the cylinder (including top and bottom). As you will see, the solution depends on whether $H \le 2R$ or $H > 2R$.

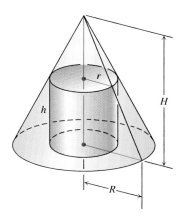

18. *Minimizing a parameter* Find the smallest value of the positive constant m that will make $mx - 1 + (1/x)$ greater than or equal to zero for all positive values of x.

19. *The second derivative test* The second derivative test for local maxima and minima (Section 3.3) says:

(a) f has a local maximum value at $x = c$ if $f'(c) = 0$ and $f''(c) < 0$

(b) f has a local minimum value at $x = c$ if $f'(c) = 0$ and $f''(c) > 0$.

To prove statement (a), let $\epsilon = (1/2)|f''(c)|$. Then use the fact that

$$f''(c) = \lim_{h \to 0} \frac{f'(c + h) - f'(c)}{h} = \lim_{h \to 0} \frac{f'(c + h)}{h}$$

to conclude that for some $\delta > 0$,

$$0 < |h| < \delta \;\Rightarrow\; \frac{f'(c + h)}{h} < f''(c) + \epsilon < 0.$$

Thus, $f'(c + h)$ is positive for $-\delta < h < 0$ and negative for $0 < h < \delta$. Prove statement (b) in a similar way.

20. *Schwarz's inequality*

(a) Show that if $a > 0$, then $f(x) = ax^2 + 2bx + c \ge 0$ for all (real) x if and only if $b^2 \le ac$.

(b) Derive **Schwarz's inequality,**

$$(a_1 b_1 + a_2 b_2 + \cdots + a_n b_n)^2 \le$$
$$(a_1^2 + a_2^2 + \cdots + a_n^2)(b_1^2 + b_2^2 + \cdots + b_n^2),$$

by applying what you learned in part (a) to the sum

$$(a_1 x + b_1)^2 + (a_2 x + b_2)^2 + \cdots + (a_n x + b_n)^2.$$

(c) Show that equality holds in Schwarz's inequality only if there exists a real number x that makes $a_i x$ equal $-b_i$ for every value of i from 1 to n.

21. *The period of a clock pendulum* The period T of a clock pendulum (time for one full swing and back) is given by the formula $T^2 = 4\pi^2 L/g$, where T is measured in seconds, $g = 32.2 \text{ ft/sec}^2$, and L, the length of the pendulum, is measured in feet. Find approximately

(a) the length of a clock pendulum whose period is $T = 1$ sec

(b) the change dT in T if the pendulum in part (a) is lengthened 0.01 ft

(c) the amount the clock gains or loses in a day as a result of the period's changing by the amount dT found in part (b).

22. *Estimating reciprocals without division* You can estimate the value of the reciprocal of a number a without ever dividing by a if you apply Newton's method to the function $f(x) = (1/x) - a$. For example, if $a = 3$, the function involved is $f(x) = (1/x) - 3$.

(a) Graph $y = (1/x) - 3$. Where does the graph cross the x-axis?

(b) Show that the recursion formula in this case is

$$x_{n+1} = x_n(2 - 3x_n),$$

so there is no need for division.

T 23. *Free fall in the fourteenth century* In the middle of the fourteenth century, Albert of Saxony (1316–1390) proposed a model of free fall that assumed that the velocity of a falling body was proportional to the distance fallen. It seemed reasonable to think that a body that had fallen 20 ft might be moving twice as fast as a body that had fallen 10 ft. And besides, none of the instruments in use at the time were accurate enough to prove otherwise. Today, we can see just how far off Albert of Saxony's model was by solving the initial value problem implicit in his model. Solve the problem and compare your solution graphically with the equation $s = 16t^2$. You will see that it describes a motion that starts too slowly at first and then becomes too fast too soon to be realistic.

24. *Group blood testing* During World War II, it was necessary to administer blood tests to large numbers of recruits. There are two standard ways to administer a blood test to N people. In method 1, each person is tested separately. In method 2, the blood samples of x people are pooled and tested as one large sample. If the test is negative, this one test is enough for all x people. If the test is positive, then each of the x people is tested separately, requiring a total of $x + 1$ tests. Using the second method and some probability theory, it can be shown that, on the average, the total number of tests y will be

$$y = N\left(1 - q^x + \frac{1}{x}\right).$$

With $q = 0.99$ and $N = 1000$, find the integer value of x that minimizes y. Also find the integer value of x that maximizes y. (This second result is not important to the real-life situation.) The group testing method was used in World War II with a savings of

80% over the individual testing method, but not with the given value of q.

25. *The best branching angles for blood vessels and pipes* When a smaller pipe branches off from a larger one in a flow system, we may want it to run off at an angle that is best from some energy-saving point of view. We might require, for instance, that energy loss due to friction be minimized along the section AOB shown in the following figure. In this diagram, B is a given point to be reached by the smaller pipe, A is a point in the larger pipe upstream from B, and O is the point where the branching occurs. A law due to Poiseuille states that the loss of energy due to friction in nonturbulent flow is proportional to the length of the path and inversely proportional to the fourth power of the radius. Thus, the loss along AO is $(kd_1)/R^4$ and along OB is $(kd_2)/r^4$, where k is a constant, d_1 is the length of AO, d_2 is the length of OB, R is the radius of the larger pipe, and r is the radius of the smaller pipe. The angle θ is to be chosen to minimize the sum of these two losses:

$$L = k\frac{d_1}{R^4} + k\frac{d_2}{r^4}.$$

In our model, we assume that $AC = a$ and $BC = b$ are fixed. Thus, we have the relations

$$d_1 + d_2 \cos\theta = a \qquad d_2 \sin\theta = b$$

so that

$$d_2 = b \csc\theta, \qquad d_1 = a - d_2 \cos\theta = a - b\cot\theta.$$

We can express the total loss L as a function of θ:

$$L = k\left(\frac{a - b\cot\theta}{R^4} + \frac{b\csc\theta}{r^4}\right).$$

(a) Show that the critical value of θ for which $dL/d\theta$ equals zero is

$$\theta_c = \cos^{-1}\frac{r^4}{R^4}.$$

(b) If the ratio of the pipe radii is $r/R = 5/6$, estimate to the nearest degree the optimal branching angle given in part (a).

The mathematical analysis described here is also used to explain the angles at which arteries branch in an animal's body. See *Introduction to Mathematics for Life Scientists*, 2nd ed., by E. Batschelet (New York: Springer-Verlag, 1976).

4 Integration

OVERVIEW   We have seen how the need to calculate instantaneous rates of change led the discoverers of calculus to an investigation of the slopes of tangent lines and, ultimately, to the derivative, to what we call *differential* calculus. But they knew that derivatives revealed only half the story. In addition to a calculation method (a "calculus") to describe how functions were changing at a given instant, they also needed a method to describe how those instantaneous changes could accumulate over an interval to produce the function. That is, by studying how a behavior *changed*, they wanted to learn about the behavior itself. For example, from knowing the velocity of a moving object, they wanted to be able to determine its position as a function of time. This is why they were also investigating *areas under curves,* an investigation that ultimately led to the second main branch of calculus, called *integral* calculus.

Once they had the calculus for finding slopes of tangent lines and the calculus for finding areas under curves—two geometric operations that would seem to have nothing at all to do with each other—the challenge for Newton and Leibniz was to prove the connection that they know intuitively had to be there. The discovery of this connection (called the Fundamental Theorem of Calculus) brought differential and integral calculus together to become the single most powerful tool mathematicians ever acquired for understanding the universe.

4.1   Indefinite Integrals, Differential Equations, and Modeling

Finding Antiderivatives: Indefinite Integrals   •   Initial Value Problems   •
Mathematical Modeling

CD-ROM
WEBsite

The process of determining a function $f(x)$ from one of its known values and its derivative $f'(x)$ has two steps. The first is to find a formula that gives all the functions that could possibly have f as a derivative. These functions are called the *antiderivatives* of f, and the formula that gives them all is called the *indefinite integral* of f. The second step is to use the known value to select the particular antiderivative we want from those in the indefinite integral.

Finding a formula that gives all a function's antiderivatives might seem like an impossible task, or at least to require a little magic, but this is not the case at all. If we can find even one antiderivative, we can find them all, because of the first two corollaries of the Mean Value Theorem of Section 3.2.

333

CD-ROM
WEBsite

Historical Essay

The Integral

Finding Antiderivatives: Indefinite Integrals

We begin with a definition.

Definition   Antiderivative of a Function

A function $F(x)$ is an **antiderivative** of a function $f(x)$ if

$$F'(x) = f(x)$$

for all x in the domain of f. The set of all antiderivatives of f is the **indefinite integral** of f with respect to x, denoted by

$$\int f(x)\, dx.$$

The symbol $\int$ is an **integral sign**. The function f is the **integrand** of the integral and x is the **variable of integration.**

According to Corollary 2 of the Mean Value Theorem (Section 3.2), once we have found one antiderivative F of a function f, the other antiderivatives differ from it by a constant. We indicate this in integral notation in the following way:

$$\int f(x)\, dx = F(x) + C \qquad (1)$$

The constant C is the **constant of integration** or **arbitrary constant.** Equation (1) is read, "The indefinite integral of f with respect to x is $F(x) + C$." When we find $F(x) + C$, we say that we have **integrated** f and **evaluated** the integral.

Example 1   Finding An Indefinite Integral

Evaluate $\int e^{2x}\, dx$.

Solution

$$\int e^{2x}\, dx = \frac{1}{2} e^{2x} + C$$

an antiderivative of e^{2x}

the arbitrary constant

The formula $(1/2)e^{2x} + C$ generates all the antiderivatives of the function e^{2x}. The functions $(1/2)e^{2x} + 1$, $(1/2)e^{2x} - \pi$, and $(1/2)e^{2x} + \sqrt{2}$ are all antiderivatives of the function e^{2x}, as you can check by differentiation.

Many of the indefinite integrals needed in scientific work are found by reversing derivative formulas. You will see what we mean if you look at Table 4.1, which lists a number of standard integral forms side by side with their derivative-formula sources.

In case you are wondering why the integrals of the tangent, cotangent, secant, and cosecant do not appear in the table, the answer is that the usual formulas for them require logarithms. In Section 4.2, we find their antiderivatives.

Table 4.1 Integral formulas

Indefinite integral	Reversed derivative formula

1. $\displaystyle\int x^n \, dx = \frac{x^{n+1}}{n+1} + C,$ $\qquad\qquad \dfrac{d}{dx}\left(\dfrac{x^{n+1}}{n+1}\right) = x^n$

   $n \neq -1,\ n$ rational

$\displaystyle\int dx = \int 1 \, dx = x + C$ (special case) $\quad \dfrac{d}{dx}(x) = 1$

2. $\displaystyle\int \sin kx \, dx = -\frac{\cos kx}{k} + C$ $\qquad \dfrac{d}{dx}\left(-\dfrac{\cos kx}{k}\right) = \sin kx$

3. $\displaystyle\int \cos kx \, dx = \frac{\sin kx}{k} + C$ $\qquad \dfrac{d}{dx}\left(\dfrac{\sin kx}{k}\right) = \cos kx$

4. $\displaystyle\int \sec^2 x \, dx = \tan x + C$ $\qquad \dfrac{d}{dx}\tan x = \sec^2 x$

5. $\displaystyle\int \csc^2 x \, dx = -\cot x + C$ $\qquad \dfrac{d}{dx}(-\cot x) = \csc^2 x$

6. $\displaystyle\int \sec x \tan x \, dx = \sec x + C$ $\qquad \dfrac{d}{dx}\sec x = \sec x \tan x$

7. $\displaystyle\int \csc x \cot x \, dx = -\csc x + C$ $\qquad \dfrac{d}{dx}(-\csc x) = \csc x \cot x$

8. $\displaystyle\int e^{kx} \, dx = \frac{1}{k} e^{kx} + C$ $\qquad \dfrac{d}{dx}\left(\dfrac{1}{k} e^{kx}\right) = e^{kx}$

9. $\displaystyle\int \frac{1}{x} \, dx = \ln x + C, \quad x > 0$ $\qquad \dfrac{d}{dx}(\ln x) = \dfrac{1}{x}, \quad x > 0$

10. $\displaystyle\int \frac{1}{\sqrt{1-x^2}} \, dx = \sin^{-1} x + C$ $\qquad \dfrac{d}{dx}(\sin^{-1} x) = \dfrac{1}{\sqrt{1-x^2}}$

11. $\displaystyle\int \frac{1}{1+x^2} \, dx = \tan^{-1} x + C$ $\qquad \dfrac{d}{dx}(\tan^{-1} x) = \dfrac{1}{1+x^2}$

12. $\displaystyle\int \frac{1}{x\sqrt{x^2-1}} \, dx = \sec^{-1} x + C,$ $\qquad \dfrac{d}{dx}(\sec^{-1} x) = \dfrac{1}{x\sqrt{x^2-1}}, \quad x > 1$

   $x > 1$

13. $\displaystyle\int a^x \, dx = \left(\frac{1}{\ln a}\right) a^x,$ $\qquad \dfrac{d}{dx}\left(\dfrac{1}{\ln a} a^x\right) = a^x, \quad a > 0, a \neq 1$

   $a > 0, \quad a \neq 1$

Example 2   Selected Integrals from Table 4.1

(a) $\displaystyle\int x^5 \, dx = \frac{x^6}{6} + C$ $\qquad$ Formula 1 with $n = 5$

(b) $\displaystyle\int \frac{1}{\sqrt{x}} \, dx = \int x^{-1/2} \, dx = 2x^{1/2} + C = 2\sqrt{x} + C$ $\qquad$ Formula 1 with $n = -1/2$

(c) $\displaystyle\int \sin 2x \, dx = -\frac{\cos 2x}{2} + C$ Formula 2 with $k = 2$

(d) $\displaystyle\int e^{-3x} \, dx = -\frac{1}{3} e^{-3x} + C$ Formula 8 with $k = -3$

(e) $\displaystyle\int 2^x \, dx = \left(\frac{1}{\ln 2}\right) 2^x + C$ Formula 13 with $a = 2$

(f) $\displaystyle\int \cos \frac{x}{2} \, dx = \int \cos\left(\frac{1}{2}x\right) dx$

$\displaystyle = \frac{\sin (1/2)x}{1/2} + C = 2 \sin \frac{x}{2} + C$ Formula 3 with $k = 1/2$

Finding an integral formula can sometimes be difficult, but checking it, once found, is relatively easy: differentiate the right-hand side. The derivative should be the integrand.

Example 3   Checking Correctness of an Indefinite Integral

Right: $\displaystyle\int x \cos x \, dx = x \sin x + \cos x + C$

Reason: The derivative of the right-hand side is the integrand:

$$\frac{d}{dx}(x \sin x + \cos x + C) = x \cos x + \sin x - \sin x + 0 = x \cos x.$$

Wrong: $\displaystyle\int x \cos x \, dx = x \sin x + C$

Reason: The derivative of the right-hand side is not the integrand:

$$\frac{d}{dx}(x \sin x + C) = x \cos x + \sin x + 0 \neq x \cos x.$$

Do not worry about how to derive the correct integral formula in Example 3. We present a technique for doing so in Chapter 7.

Initial Value Problems

The problem of finding a function y of x when we know its derivative and its value y_0 at a particular point x_0 is called an **initial value problem.** We solve such problems in two steps, as demonstrated in Example 4.

Example 4   Finding a Curve from its Slope Function and a Point

Find the curve whose slope at the point (x, y) is $3x^2$ if the curve is required to pass through the point $(1, -1)$.

Solution   In mathematical language, we are asked to solve the initial value problem that consists of the following.

The differential equation: $\dfrac{dy}{dx} = 3x^2$ The curve's slope is $3x^2$.

The initial condition: $y(1) = -1$

1. Solve the differential equation:

$$\frac{dy}{dx} = 3x^2$$

$$\int \frac{dy}{dx}\, dx = \int 3x^2\, dx$$

$$y + C_1 = x^3 + C_2$$

$$y = x^3 + C \qquad \text{Constants of integration combined, giving the general solution}$$

This result tells us that y eqauls $x^3 + C$ for some value of C. We find that value from the condition $y(1) = -1$.

2. Evaluate C:

$$y = x^3 + C$$

$$-1 = (1)^3 + C \qquad \text{Initial condition } y(1) = -1$$

$$C = -2.$$

The curve we want is $y = x^3 - 2$ (Figure 4.1).

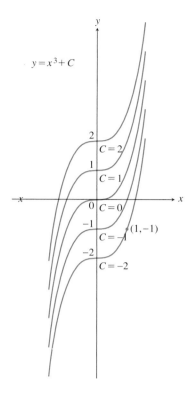

FIGURE 4.1 The curves $y = x^3 + C$ fill the coordinate plane without overlapping. In Example 4, we identify the curve $y = x^3 - 2$ as the one that passes through the given point $(1, -1)$.

The indefinite integral $F(x) + C$ of the function $f(x)$ gives the **general solution** $y = F(x) + C$ of the differential equation $dy/dx = f(x)$. The general solution gives all the solutions of the equation (there are infinitely many, one for each value of C). We **solve** the differential equation by finding its general solution. We then solve the initial value problem by finding the **particular solution** that satisfies the initial condition $y(x_0) = y_0$ (y has the value y_0 when $x = x_0$).

Solving initial value problems is important in mathematical modeling, the process by which we, as scientists and engineers, use mathematics to learn about the real world.

The Modeling Process

Step 1. Observe real-world behavior.

Step 2. Make assumptions to identify variables and their relationships, creating a model.

Step 3. Solve the model to obtain mathematical solutions.

Step 4. Interpret the model and verify it is consistent with real-world observations.

Mathematical Modeling

The development of a mathematical model usually takes four steps: First we observe something in the real world (for example, a ball bearing falling from rest or the trachea contracting during a cough) and construct a system of mathematical variables and relationships that imitate some of its important features. We build a mathematical metaphor for what we see. Next we apply mathematics to the variables and relationships to solve the model and draw conclusions about the variables. After that we translate the mathematical conclusions into information about the system under study. Finally, we check the information against observation to see if the model has predictive value. We also investigate the possibility that the model applies to other systems. The really good models are the ones that lead to conclusions that are consistent with observation, that have predictive value and broad application, and that are not too hard to use.

The natural cycle of mathematical imitation, deduction, interpretation, and confirmation is shown in the diagram for free fall.

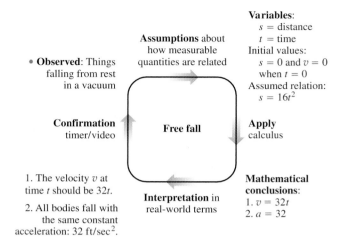

Variables:
s = distance
t = time
Initial values:
$s = 0$ and $v = 0$
when $t = 0$
Assumed relation:
$s = 16t^2$

• **Observed:** Things falling from rest in a vacuum

Assumptions about how measurable quantities are related

Free fall

Confirmation timer/video

Apply calculus

1. The velocity v at time t should be $32t$.

2. All bodies fall with the same constant acceleration: 32 ft/sec².

Interpretation in real-world terms

Mathematical conclusions:
1. $v = 32t$
2. $a = 32$

Example 5   Dropping a Package from an Ascending Balloon

A balloon ascending at the rate of 12 ft/sec is at a height 80 ft above the ground when a package is dropped. How long does it take the package to reach the ground?

Solution   Let $v(t)$ denote the velocity of the package at time t, and let $s(t)$ denote its height above the ground. The acceleration of gravity near the surface of the earth is 32 ft/sec². Assuming no other forces act on the dropped package, we have

$$\frac{dv}{dt} = -32. \qquad \text{\small Negative because gravity acts in the direction of decreasing } s.$$

This leads to the initial value problem

Differential equation: $\qquad \dfrac{dv}{dt} = -32$

Initial condition: $\qquad v(0) = 12,$

which is our mathematical model for the package's motion. We solve the initial value problem to obtain the velocity of the package.

1. Solve the differential equation:

$$\frac{dv}{dt} = -32$$

$$\int \frac{dv}{dt}\, dt = \int -32\, dt$$

$$v = -32t + C. \qquad \text{\small Constants combined as one}$$

Having found the general solution of the differential equation, we use the initial condition to find the particular solution that solves our problem.

2. Evaluate C:

$$12 = -32(0) + C \qquad \text{Initial condition } v(0) = 12$$
$$C = 12.$$

The solution of the initial value problem is

$$v = -32t + 12.$$

Since velocity is the derivative of height and the height of the package is 80 ft at the time $t = 0$ when it is dropped, we now have a second initial value problem.

Differential equation: $\qquad \dfrac{ds}{dt} = -32t + 12 \qquad$ Set $v = ds/dt$ in the last equation.

Initial condition: $\qquad s(0) = 80$

We solve this initial value problem to find the height as a function of t.
Solve the differential equation:

$$\frac{ds}{dt} = -32t + 12$$

$$\int \frac{ds}{dt}\,dt = \int (-32t + 12)\,dt$$

$$s = -16t^2 + 12t + C. \qquad \begin{array}{l}\text{Constants of integration}\\\text{combined, giving the}\\\text{general solution.}\end{array}$$

Evaluate C:

$$80 = -16(0)^2 + 12(0) + C \qquad \text{Initial condition } s(0) = 80$$
$$C = 80.$$

The package's height above ground at time t is

$$s = -16t^2 + 12t + 80.$$

Use the solution: To find how long it takes the package to reach the ground, we set s equal to 0 and solve for t:

$$-16t^2 + 12t + 80 = 0$$
$$-4t^2 + 3t + 20 = 0$$
$$t = \frac{-3 \pm \sqrt{329}}{-8} \qquad \text{Quadratic formula}$$
$$t \approx -1.89, \qquad t \approx 2.64.$$

The package hits the ground about 2.64 sec after it is dropped from the balloon. (The negative root has no physical meaning.)

EXERCISES 4.1

Finding Antiderivatives

In Exercises 1–8, find an antiderivative for each function. Do as many as you can mentally. Check your answers by differentiation.

1. (a) $6x$ (b) x^7 (c) $x^7 - 6x + 8$

2. (a) $-3x^{-4}$ (b) x^{-4} (c) $x^{-4} + 2x + 3$

3. (a) $-\dfrac{2}{x^3}$ (b) $\dfrac{1}{2x^3}$ (c) $x^3 - \dfrac{1}{x^3}$

4. (a) $\dfrac{3}{2}\sqrt{x}$ (b) $\dfrac{1}{2\sqrt{x}}$ (c) $\sqrt{x} + \dfrac{1}{\sqrt{x}}$

5. (a) $\dfrac{2}{3}x^{-1/3}$ (b) $\dfrac{1}{3}x^{-2/3}$ (c) $-\dfrac{1}{3}x^{-4/3}$

6. (a) $-\pi \sin \pi x$ (b) $3 \sin x$ (c) $\sin \pi x - 3 \sin 3x$

7. (a) $\sec^2 x$ (b) $\dfrac{2}{3}\sec^2 \dfrac{x}{3}$ (c) $-\sec^2 \dfrac{3x}{2}$

8. (a) $\sec x \tan x$ (b) $4 \sec 3x \tan 3x$ (c) $\sec \dfrac{\pi x}{2} \tan \dfrac{\pi x}{2}$

Evaluating Integrals

Evaluate the integrals in Exercises 9–26. Check your answers by differentiation.

9. $\displaystyle\int (x + 1)\, dx$

10. $\displaystyle\int \left(3t^2 + \dfrac{t}{2}\right) dt$

11. $\displaystyle\int \left(\dfrac{1}{x} - \dfrac{5}{x^2 + 1}\right) dx$

12. $\displaystyle\int \left(\dfrac{1}{x^2} - x^2 - \dfrac{1}{3}\right) dx$

13. $\displaystyle\int \left(e^{-x} + 4^x\right) dx$

14. $\displaystyle\int \left(\sqrt{x} + \sqrt[3]{x}\right) dx$

15. $\displaystyle\int \left(\dfrac{2}{\sqrt{1 - y^2}} - \dfrac{1}{y^{1/4}}\right) dy$

16. $\displaystyle\int \left(\dfrac{\sqrt{x}}{2} + \dfrac{2}{\sqrt{x}}\right) dx$

17. $\displaystyle\int \left(\dfrac{1}{7} - \dfrac{1}{y^{5/4}}\right) dy$

18. $\displaystyle\int 2x\,(1 - x^{-3})\, dx$

19. $\displaystyle\int \dfrac{t\sqrt{t} + \sqrt{t}}{t^2}\, dt$

20. $\displaystyle\int (-2 \cos t)\, dt$

21. $\displaystyle\int 7 \sin \dfrac{\theta}{3}\, d\theta$

22. $\displaystyle\int (-3 \csc^2 x)\, dx$

23. $\displaystyle\int (1 + \tan^2 \theta)\, d\theta$ (*Hint:* $1 + \tan^2 \theta = \sec^2 \theta$)

24. $\displaystyle\int \cot^2 x\, dx$ (*Hint:* $1 + \cot^2 x = \csc^2 x$)

25. $\displaystyle\int \cos \theta\, (\tan \theta + \sec \theta)\, d\theta$

26. $\displaystyle\int \dfrac{\csc \theta}{\csc \theta - \sin \theta}\, d\theta$

Checking Integration Formulas

Verify the integral formulas in Exercises 27–30 by differentiation. In Section 4.2, we will see where formulas like these come from.

27. $\displaystyle\int (7x - 2)^3\, dx = \dfrac{(7x - 2)^4}{28} + C$

28. $\displaystyle\int (3x + 5)^{-2}\, dx = -\dfrac{(3x + 5)^{-1}}{3} + C$

29. $\displaystyle\int \csc^2 \left(\dfrac{x - 1}{3}\right) dx = -3 \cot \left(\dfrac{x - 1}{3}\right) + C$

30. $\displaystyle\int \dfrac{1}{x + 1}\, dx = \ln (x + 1) + C, \qquad x > -1$

31. Right or wrong? Say which for each formula and give a brief reason for each answer.

(a) $\displaystyle\int x \sin x\, dx = \dfrac{x^2}{2} \sin x + C$

(b) $\displaystyle\int x \sin x\, dx = -x \cos x + C$

(c) $\displaystyle\int x \sin x\, dx = -x \cos x + \sin x + C$

32. Right or wrong? Say which for each formula and give a brief reason for each answer.

(a) $\displaystyle\int (2x + 1)^2\, dx = \dfrac{(2x + 1)^3}{3} + C$

(b) $\displaystyle\int 3(2x + 1)^2\, dx = (2x + 1)^3 + C$

(c) $\displaystyle\int 6(2x + 1)^2\, dx = (2x + 1)^3 + C$

Initial Value Problems

33. *Writing to Learn* Which of the following graphs show the solution of the initial value problem

$$\dfrac{dy}{dx} = 2x, \qquad y = 4 \text{ when } x = 1?$$

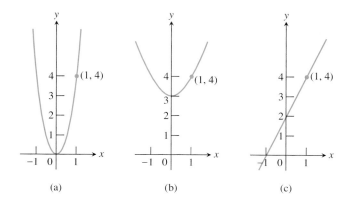

(a)   (b)   (c)

Give reasons for your answer.

34. *Writing to Learn* Which of the following graphs shows the solution of the initial value problem

$$\frac{dy}{dx} = -x, \qquad y = 1 \text{ when } x = -1?$$

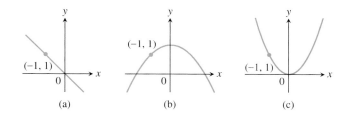

(a)   (b)   (c)

Give reasons for your answer.

Solve the initial value problems in Exercises 35–46.

35. $\frac{dy}{dx} = 2x - 7, \quad y(2) = 0$

36. $\frac{dy}{dx} = \frac{1}{x^2} + x, \quad x > 0; \quad y(2) = 1$

37. $\frac{dy}{dx} = 3x^{-2/3}, \quad y(-1) = -5$

38. $\frac{dy}{dx} = \frac{1}{2x}, \quad y(1) = -1$

39. $\frac{ds}{dt} = \cos t + \sin t, \quad s(\pi) = 1$

40. $\frac{dr}{d\theta} = -\pi \sin \pi\theta, \quad r(0) = 0$

41. $\frac{dv}{dt} = \frac{3}{t\sqrt{t^2 - 1}}, \quad t > 1, v(2) = 0$

42. $\frac{dv}{dt} = \frac{8}{1 + t^2} + \sec^2 t, \quad v(0) = 1$

43. $\frac{d^2y}{dx^2} = 2 - 6x; \quad y'(0) = 4, \quad y(0) = 1$

44. $\frac{d^2r}{dt^2} = \frac{2}{t^3}; \quad \frac{dr}{dt}\Big|_{t=1} = 1, \quad r(1) = 1$

45. $\frac{d^3y}{dx^3} = 6; \quad y''(0) = -8, \quad y'(0) = 0, \quad y(0) = 5$

46. $y^{(4)} = -\sin t + \cos t; \quad y'''(0) = 7, \quad y''(0) = y'(0) = -1, \quad y(0) = 0$

Finding Position from Velocity

Exercises 47 and 48 give the velocity $v = ds/dt$ and initial position of a body moving along a coordinate line. Find the body's position at time t.

47. $v = 9.8t + 5, \quad s(0) = 10$

48. $v = \frac{2}{\pi} \cos \frac{2t}{\pi}, \quad s(\pi^2) = 1$

Finding Position from Acceleration

Exercises 49 and 50 give the acceleration $a = d^2s/dt^2$, initial velocity, and initial position of a body moving on a coordinate line. Find the body's position at time t.

49. $a = 32; \quad v(0) = 20, \quad s(0) = 5$

50. $a = -4 \sin 2t; \quad v(0) = 2, \quad s(0) = -3$

Finding Curves

51. Find the curve $y = f(x)$ in the xy-plane that passes through the point $(9, 4)$ and whose slope at each point is $3\sqrt{x}$.

52. **(a)** Find a curve $y = f(x)$ with the following properties:

i. $\frac{d^2y}{dx^2} = 6x$

ii. Its graph passes through the point $(0, 1)$ and has a horizontal tangent there.

(b) *Writing to Learn* How many curves like this are there? How do you know?

Solution (Integral) Curves

Exercises 53–56 show solution curves of differential equations. In each exercise, find an equation for the curve through the labeled point.

53.   **54.**

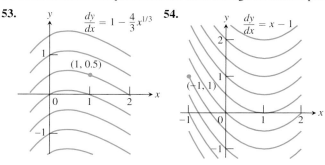

55.

$\frac{dy}{dx} = \sin x - \cos x$

$(-\pi, -1)$

56.

$\frac{dy}{dx} = \frac{1}{2\sqrt{x}} + \pi \sin \pi x$

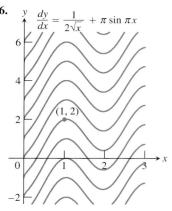

$(1, 2)$

Applications

57. *Falling on the moon* On the moon, the acceleration due to gravity is 1.6 m/sec². If a rock is dropped into a crevasse, how fast will it be going just before it hits the bottom 30 sec later?

58. *Liftoff from Earth* A rocket lifts off the surface of Earth with a constant acceleration of 20 m/sec². How fast will the rocket be going 1 min later?

59. *Stopping a car in time* You are driving along a highway at a steady 60 mph (88 ft/sec) when you see an accident ahead and slam on the brakes. What constant deceleration is required to stop your car in 242 ft? To find out, carry out the following steps.

Step 1: Solve the initial value problem

Differential equation: $\frac{d^2s}{dt^2} = -k$ (k constant)

Initial conditions: $\frac{ds}{dt} = 88$ and $s = 0$ when $t = 0$.

<small>Measuring time and distance from when the brakes are applied</small>

Step 2: Find the value of t that makes $ds/dt = 0$. (The answer will involve k.)

Step 3: Find the value of k that makes $s = 242$ for the value of t you found in step 2.

60. *Stopping a motorcycle* The State of Illinois Cycle Rider Safety Program requires riders to be able to brake from 30 mph (44 ft/sec) to 0 in 45 ft. What constant deceleration does it take to do that?

61. *Motion along a coordinate line* A particle moves on a coordinate line with acceleration $a = d^2s/dt^2 = 15\sqrt{t} - (3/\sqrt{t})$, subject to the conditions that $ds/dt = 4$ and $s = 0$ when $t = 1$. Find

(a) the velocity $v = ds/dt$ in terms of t

(b) the position s in terms of t.

T **62.** *The hammer and the feather* When *Apollo 15* astronaut David Scott dropped a hammer and a feather on the moon to demonstrate that

in a vacuum all bodies fall with the same (constant) acceleration, he dropped them from about 4 ft above the ground. The television footage of the event shows the hammer and feather falling more slowly than on Earth, where, in a vacuum, they would have taken only half a second to fall the 4 ft. How long did it take the hammer and feather to fall 4 ft on the moon? To find out, solve the following initial value problem for s as a function of t. Then find the value of t that makes s equal to 0.

Differential equation: $\frac{d^2s}{dt^2} = -5.2$ ft/sec²

Initial conditions: $\frac{ds}{dt} = 0$ and $s = 4$ when $t = 0$

63. *Motion with constant acceleration* The standard equation for the position s of a body moving with a constant acceleration a along a coordinate line is

$$s = \frac{a}{2}t^2 + v_0t + s_0, \tag{2}$$

where v_0 and s_0 are the body's velocity and position at time $t = 0$. Derive this equation by solving the initial value problem

Differential equation: $\frac{d^2s}{dt^2} = a$

Initial conditions: $\frac{ds}{dt} = v_0$ and $s = s_0$ when $t = 0$.

64. *Writing to Learn: Free fall near the surface of a planet* (*Continuation of Exercise 63*) For free fall near the surface of a planet where the acceleration due to gravity has a constant magnitude of g length-units/sec², Equation (2) in Exercise 63 takes the form

$$s = -\frac{1}{2}gt^2 + v_0t + s_0, \tag{3}$$

where s is the body's height above the surface. The equation has a minus sign because the acceleration acts downward, in the direction of decreasing s. The velocity v_0 is positive if the object is rising at time $t = 0$ and negative if the object is falling.

Instead of using the result of Exercise 63, you can derive Equation (3) directly by solving an appropriate initial value problem. What initial value problem? Solve it to be sure you have the right one, explaining the solution steps as you go along.

Theory and Examples

65. *Finding displacement from an antiderivative of velocity*

(a) Suppose that the velocity of a body moving along the s-axis is

$$\frac{ds}{dt} = v = 9.8t - 3.$$

i. Find the body's displacement over the time interval from $t = 1$ to $t = 3$ given that $s = 5$ when $t = 0$.

ii. Find the body's displacement from $t = 1$ to $t = 3$ given that $s = -2$ when $t = 0$.

iii. Now find the body's displacement from $t = 1$ to $t = 3$ given that $s = s_0$ when $t = 0$.

(b) Suppose that the position s of a body moving along a coordinate line is a differentiable function of time t. Is it true that once you know an antiderivative of the velocity function ds/dt you can find the body's displacement from $t = a$ to $t = b$ even if you do not know the body's exact position at either of those times? Give reasons for your answer.

66. *Uniqueness of solutions* If differentiable functions $y = F(x)$ and $y = G(x)$ both solve the initial value problem

$$\frac{dy}{dx} = f(x), \qquad y(x_0) = y_0,$$

on an interval I, must $F(x) = G(x)$ for every x in I? Give reasons for your answer.

COMPUTER EXPLORATIONS

Use a CAS to solve the initial value problems in Exercises 67–72. Plot the solution curves.

67. $y' = \cos^2 x + \sin x, \quad y(\pi) = 1$

68. $y' = xe^x, \quad y(\ln 2) = 0$

69. $y' = \ln x, \quad y(1) = 2$

70. $y' = \dfrac{1}{\sqrt{4 - x^2}}, \quad y(0) = 2$

71. $y'' = 3e^{x/2} + x, \quad y(0) = -1, \quad y'(0) = 4$

72. $y'' = \dfrac{2}{x} + \sqrt{x}, \quad y(1) = 0, \quad y'(1) = 0$

4.2 Integral Rules; Integration by Substitution

Rules of Algebra for Antiderivatives • The Integrals of $\sin^2 x$ and $\cos^2 x$ • The Power Rule in Integral Form • Substitution: Running the Chain Rule Backwards • The Integral $\int (1/u)\, du$ • The Integrals of $\tan x$ and $\cot x$

Just as limits and derivatives obey algebraic rules, so do antiderivatives and indefinite integrals. In this section, we present and apply those rules to find antiderivatives for a variety of functions.

Rules of Algebra for Antiderivatives

We know the following from our previous study of the derivative.

1. A function is an antiderivative of a constant multiple kf of a function f if and only if it is k times an antiderivative of f.

2. In particular, a function is an antiderivative of $-f$ if and only if it is the negative of an antiderivative of f.

3. A function is an antiderivative of a sum or difference $f \pm g$ if and only if it is the sum or difference of an antiderivative of f and an antiderivative of g.

When we express these observations in integral notation, we get the standard arithmetic rules for indefinite integration (Table 4.2).

CD-ROM
WEBsite

Historical Biography

Jakob Bernoulli
(1654 — 1705)

Table 4.2 Rules for indefinite integration

1. *Constant Multiple Rule:* $\displaystyle\int kf(x)\, dx = k \int f(x)\, dx$
(Does not work if k varies with x.)

2. *Rule for Negatives:* $\displaystyle\int -f(x)\, dx = -\int f(x)\, dx$
(Rule 1 with $k = -1$)

3. *Sum and Difference Rule:* $\displaystyle\int [f(x) \pm g(x)]\, dx = \int f(x)\, dx \pm \int g(x)\, dx$

Example 1   Rewriting the Constant of Integration

$$\int 5 \sec x \, \tan x \, dx = 5 \int \sec x \, \tan x \, dx \qquad \text{Table 4.2, Rule 1}$$
$$= 5(\sec x + C) \qquad \text{Table 4.1, Formula 6}$$
$$= 5 \sec x + 5C \qquad \text{First form}$$
$$= 5 \sec x + C' \qquad \text{Shorter form, where } C' \text{ is } 5C$$
$$= 5 \sec x + C. \qquad \text{Usual form, no prime. Since 5 times an arbitrary constant is an arbitrary constant, we rename } C'.$$

What about the three different forms in Example 1? Each one gives all the antiderivatives of $f(x) = 5 \sec x \tan x$, so each answer is correct, but the least complicated of the three, and the usual choice, is the last one

$$\int 5 \sec x \, \tan x \, dx = 5 \sec x + C.$$

The Sum and Difference Rule for integration enables us to integrate expressions term by term. When we do so, we combine the individual constants of integration into a single arbitrary constant at the end.

Example 2   Term-by-Term Integration

Evaluate

$$\int (x^2 - 2x + 5) \, dx.$$

Solution   If we recognize that $(x^3/3) - x^2 + 5x$ is an antiderivative of $x^2 - 2x + 5$, we can evaluate the integral as

$$\int (x^2 - 2x + 5) \, dx = \overbrace{\frac{x^3}{3} - x^2 + 5x}^{\text{antiderivative}} \underbrace{+ C.}_{\text{arbitrary constant}}$$

If we do not recognize the antiderivative right away, we can generate it term by term with the Sum and Difference Rule:

$$\int (x^2 - 2x + 5) \, dx = \int x^2 \, dx - \int 2x \, dx + \int 5 \, dx$$
$$= \frac{x^3}{3} + C_1 - x^2 + C_2 + 5x + C_3.$$

This formula is more complicated than it needs to be. If we combine C_1, C_2, and C_3 into a single constant $C = C_1 + C_2 + C_3$, the formula simplifies to

$$\frac{x^3}{3} - x^2 + 5x + C$$

and *still* gives all the antiderivatives there are. For this reason we recommend that you go right to the final form even if you elect to integrate term by term. Write

$$\int (x^2 - 2x + 5) \, dx = \int x^2 \, dx - \int 2x \, dx + \int 5 \, dx$$
$$= \frac{x^3}{3} - x^2 + 5x + C.$$

Find the simplest antiderivative you can for each part and add the constant of integration at the end.

The Integrals of $\sin^2 x$ and $\cos^2 x$

We can sometimes use trigonometric identities to transform integrals we do not know how to evaluate into integrals we do know how to evaluate. The integral formulas for $\sin^2 x$ and $\cos^2 x$ arise frequently in applications.

Example 3 Integrating $\sin^2 x$ and $\cos^2 x$

(a) $\displaystyle \int \sin^2 x \, dx = \int \frac{1 - \cos 2x}{2} \, dx \qquad \sin^2 x = \frac{1 - \cos 2x}{2}$

$$= \frac{1}{2} \int (1 - \cos 2x) \, dx = \frac{1}{2} \int dx - \frac{1}{2} \int \cos 2x \, dx$$

$$= \frac{1}{2} x - \frac{1}{2} \frac{\sin 2x}{2} + C = \frac{x}{2} - \frac{\sin 2x}{4} + C$$

(b) $\displaystyle \int \cos^2 x \, dx = \int \frac{1 + \cos 2x}{2} \, dx \qquad \cos^2 x = \frac{1 + \cos 2x}{2}$

$$= \frac{x}{2} + \frac{\sin 2x}{4} + C \qquad \text{As in part (a), but with a sign change}$$

CD-ROM
WEBsite

Historical Biography

Hippocrates of Chios
(ca. 440 BC)

The Power Rule in Integral Form

When u is a differentiable function of x and n is a rational number different from -1, the Chain Rule tells us that

$$\frac{d}{dx} \left(\frac{u^{n+1}}{n + 1} \right) = u^n \frac{du}{dx}.$$

This same equation, from another point of view, says that $u^{n+1}/(n + 1)$ is one of the antiderivatives of the function $u^n (du/dx)$. Therefore,

$$\int \left(u^n \frac{du}{dx} \right) dx = \frac{u^{n+1}}{n + 1} + C.$$

The integral on the left-hand side of this equation is usually written in the simpler "differential" form,

$$\int u^n \, du,$$

obtained by treating the dx's as differentials that cancel. We are thus led to the following rule.

Equation (1) actually holds for any real exponent $n \neq -1$, as we see in Chapter 6.

If u is any differentiable function, then

$$\int u^n \, du = \frac{u^{n+1}}{n + 1} + C \qquad (n \neq -1, n \text{ rational}). \tag{1}$$

In deriving Equation (1), we assumed u to be a differentiable function of the variable x, but the name of the variable does not matter and does not appear in the final formula. We could have represented the variable with θ, t, y, or any other letter. Equation (1) says that whenever we can cast an integral in the form

$$\int u^n \, du \qquad (n \neq -1),$$

with u a differentiable function and du its differential, we can evaluate the integral as $[u^{n+1}/(n+1)] + C$.

Example 4   Using the Power Rule

$$\int \sqrt{1 + y^2} \cdot 2y \, dy = \int u^{1/2} \, du \qquad \begin{array}{l}\text{Let } u = 1 + y^2, \\ du = 2y \, dy.\end{array}$$

$$= \frac{u^{(1/2)+1}}{(1/2) + 1} + C \qquad \begin{array}{l}\text{Integrate, using} \\ \text{Eq. (1) with } n = 1/2.\end{array}$$

$$= \frac{2}{3} u^{3/2} + C \qquad \text{Simpler form}$$

$$= \frac{2}{3} (1 + y^2)^{3/2} + C \qquad \text{Replace } u \text{ by } 1 + y^2.$$

Example 5   Adjusting the Integrand by a Constant

$$\int \sqrt{4t - 1} \, dt = \int u^{1/2} \cdot \frac{1}{4} \, du \qquad \begin{array}{l}\text{Let } u = 4t - 1, \, du = 4 \, dt, \\ (1/4) \, du = dt.\end{array}$$

$$= \frac{1}{4} \int u^{1/2} \, du \qquad \begin{array}{l}\text{With the } 1/4 \text{ out front, the integral is} \\ \text{now in standard form.}\end{array}$$

$$= \frac{1}{4} \cdot \frac{u^{3/2}}{3/2} + C \qquad \begin{array}{l}\text{Integrate, using Eq. (1)} \\ \text{with } n = 1/2.\end{array}$$

$$= \frac{1}{6} u^{3/2} + C \qquad \text{Simpler form}$$

$$= \frac{1}{6} (4t - 1)^{3/2} + C \qquad \text{Replace } u \text{ by } 4t - 1.$$

The Substitution Method of Integration

Take these steps to evaluate the integral

$$\int f(g(x)) g'(x) \, dx,$$

when f and g' are continuous functions:

Step 1. Substitute $u = g(x)$ and $du = g'(x) \, dx$ to obtain the integral

$$\int f(u) \, du.$$

Step 2. Integrate with respect to u.

Step 3. Replace u by $g(x)$ in the result.

Substitution: Running the Chain Rule Backwards

The substitutions in Examples 4 and 5 are all instances of the following general rule.

$$\int f(g(x)) \cdot g'(x) \, dx = \int f(u) \, du \qquad \text{1. Substitute } u = g(x), \, du = g'(x) \, dx.$$

$$= F(u) + C \qquad \begin{array}{l}\text{2. Evaluate by finding an antiderivative} \\ F(u) \text{ of } f(u). \text{ (Any one will do.)}\end{array}$$

$$= F(g(x)) + C \qquad \text{3. Replace } u \text{ by } g(x).$$

These three steps are the steps of the substitution method of integration. The method works because $F(g(x))$ is an antiderivative of $f(g(x)) \cdot g'(x)$ whenever F is an antiderivative of f:

$$\frac{d}{dx} F(g(x)) = F'(g(x)) \cdot g'(x) \qquad \text{Chain Rule}$$

$$= f(g(x)) \cdot g'(x). \qquad \text{Because } F' = f$$

Example 6 Using Substitution

$$\int \cos(7\theta + 5)\, d\theta = \int \cos u \cdot \frac{1}{7}\, du$$

Let $u = 7\theta + 5$, $du = 7\, d\theta$, $(1/7)\, du = d\theta$.

$$= \frac{1}{7} \int \cos u\, du$$

With the 1/7 out front, the integral is now in standard form.

$$= \frac{1}{7} \sin u + C$$

Integrate with respect to u.

$$= \frac{1}{7} \sin(7\theta + 5) + C$$

Replace u by $7\theta + 5$.

Example 7 Using Substitution

$$\int x^2 \sin(x^3)\, dx = \int \sin(x^3) \cdot x^2\, dx$$

$$= \int \sin u \cdot \frac{1}{3}\, du$$

Let $u = x^3$, $du = 3x^2\, dx$, $(1/3)\, du = x^2\, dx$.

$$= \frac{1}{3} \int \sin u\, du$$

$$= \frac{1}{3}(-\cos u) + C$$

Integrate with respect to u.

$$= -\frac{1}{3} \cos(x^3) + C$$

Replace u by x^3.

Example 8 Using Identities and Substitution

Evaluate

$$\int \frac{e^t\, dt}{\cos^2(e^t - 2)}.$$

Solution

$$\int \frac{e^t\, dt}{\cos^2(e^t - 2)} = \int \frac{du}{\cos^2 u}$$

$u = e^t - 2$, $du = e^t\, dt$.

$$= \int \sec^2 u\, du$$

$\frac{1}{\cos u} = \sec u$

$$= \tan u + C$$

$\frac{d}{du} \tan u = \sec^2 u$

$$= \tan(e^t - 2) + C$$

The success of the substitution method depends on finding a substitution that will change an integral we cannot evaluate directly into one that we can. If the first substitution fails, we can try to simplify the integrand further with an additional substitution or two. (You will see what we mean if you do Exercises 41 and 42). Alternatively, we can start afresh. There can be more than one good way to start, as in the next example.

Example 9 Using Different Substitutions

Evaluate

$$\int \frac{2z\, dz}{\sqrt[3]{z^2 + 1}}.$$

Solution   We can use the substitution method of integration as an exploratory tool: Substitute for the most troublesome part of the integrand and see how things work out. For the integral here, we might try $u = z^2 + 1$ or we might even press our luck and take u to be the entire cube root. Here is what happens in each case.

Solution 1   Substitute $u = z^2 + 1$.

$$\int \frac{2z\,dz}{\sqrt[3]{z^2+1}} = \int \frac{du}{u^{1/3}} \qquad \text{Let } u = z^2+1, \; du = 2z\,dz.$$

$$= \int u^{-1/3}\,du \qquad \text{In the form } \int u^n\,du$$

$$= \frac{u^{2/3}}{2/3} + C$$

$$= \frac{3}{2}u^{2/3} + C \qquad \text{Integrate with respect to } u.$$

$$= \frac{3}{2}(z^2+1)^{2/3} + C \qquad \text{Replace } u \text{ by } z^2+1.$$

Solution 2   Substitute $u = \sqrt[3]{z^2+1}$ instead.

$$\int \frac{2z\,dz}{\sqrt[3]{z^2+1}} = \int \frac{3u^2\,du}{u} \qquad \begin{array}{l}\text{Let } u = \sqrt[3]{z^2+1}, \\ u^3 = z^2+1, \\ 3u^2\,du = 2z\,dz.\end{array}$$

$$= 3\int u\,du$$

$$= 3\cdot\frac{u^2}{2} + C \qquad \text{Integrate with respect to } u.$$

$$= \frac{3}{2}(z^2+1)^{2/3} + C \qquad \text{Replace } u \text{ by } (z^2+1)^{1/3}.$$

The Integral $\int (1/u)\,du$

The equation

$$\frac{d}{dx}\ln u = \frac{1}{u}\frac{du}{dx}, \qquad u > 0$$

leads to the integral formula

$$\int \frac{1}{u}\,du = \ln u + C \qquad\qquad (2)$$

when u is a positive differentiable function, but what if u is negative? If u is negative, then $-u$ is positive and

$$\int \frac{1}{u}\,du = \int \frac{1}{(-u)}\,d(-u)$$

$$= \ln(-u) + C. \qquad \text{Eq. (2) with } u \text{ replaced by } -u \;\; (3)$$

We can combine Equations (2) and (3) into a single formula by noticing that in each case the expression on the right is $\ln|u| + C$. In Equation (2), $\ln u = \ln|u|$ because $u > 0$; in Equation (3), $\ln(-u) = \ln|u|$ because $u < 0$. Whether u is positive or negative, the integral of $(1/u)\,du$ is $\ln|u| + C$.

$$\int \frac{1}{u} \, du = \ln |u| + C. \qquad (4)$$

Formulas like Equation (4) with a single constant of integration are assumed to hold only over domains that are intervals.

Example 10   Using Equation 4

$$\int \frac{2x}{x^2 - 5} \, dx = \int \frac{du}{u} \qquad u = x^2 - 5, \, du = 2x \, dx$$

$$= \ln |u| + C \qquad \text{Eq. (4)}$$

$$= \ln |x^2 - 5| + C \qquad u = x^2 - 5$$

The Integrals of tan x and cot x

Equation (4) tells us at last how to integrate the tangent and cotangent functions. For the tangent,

$$\int \tan x \, dx = \int \frac{\sin x}{\cos x} \, dx = \int \frac{-du}{u} \qquad \begin{matrix} u = \cos x, \\ du = -\sin x \, dx \end{matrix}$$

$$= -\int \frac{du}{u} = -\ln |u| + C \qquad \text{Eq. (4)}$$

$$= -\ln |\cos x| + C = \ln \frac{1}{|\cos x|} + C \qquad \text{Reciprocal Rule}$$

$$= \ln |\sec x| + C.$$

For the cotangent,

$$\int \cot x \, dx = \int \frac{\cos x \, dx}{\sin x} = \int \frac{du}{u} \qquad \begin{matrix} u = \sin x, \\ du = \cos x \, dx \end{matrix}$$

$$= \ln |u| + C = \ln |\sin x| + C = -\ln |\csc x| + C.$$

$$\int \tan u \, du = -\ln |\cos u| + C = \ln |\sec u| + C$$

$$\int \cot u \, du = \ln |\sin u| + C = -\ln |\csc x| + C$$

EXERCISES 4.2

Evaluating Integrals

Evaluate the indefinite integrals in Exercises 1–10 by using the given substitutions to reduce the integrals to standard form.

1. $\int x \sin (2x^2) \, dx, \quad u = 2x^2$

2. $\int \left(1 - \cos \frac{t}{2}\right)^2 \sin \frac{t}{2} \, dt, \quad u = 1 - \cos \frac{t}{2}$

3. $\int 28(7x - 2)^{-5} \, dx, \quad u = 7x - 2$

4. $\int x^3 (x^4 - 1)^2 \, dx, \quad u = x^4 - 1$

5. $\int \frac{9r^2 \, dr}{\sqrt{1 - r^3}}, \quad u = 1 - r^3$

6. $\int 12(y^4 + 4y^2 + 1)^2 (y^3 + 2y) \, dy, \quad u = y^4 + 4y^2 + 1$

7. $\int \sqrt{x} \sin^2 (x^{3/2} - 1) \, dx, \quad u = x^{3/2} - 1$

8. $\int \frac{1}{x^2} \cos^2 \left(\frac{1}{x}\right) dx, \quad u = -\frac{1}{x}$

9. $\int \csc^2 2\theta \cot 2\theta \, d\theta$
   (a) Using $u = \cot 2\theta$ (b) Using $u = \csc 2\theta$

10. $\int \frac{dx}{\sqrt{5x + 8}}$
   (a) Using $u = 5x + 8$ (b) Using $u = \sqrt{5x + 8}$

Evaluate the integrals in Exercises 11–40.

11. $\int \sqrt{3 - 2s} \, ds$

12. $\int \frac{1}{\sqrt{5s + 4}} \, ds$

13. $\int \frac{3 \, dx}{(2 - x)^2}$

14. $\int \theta \sqrt[4]{1 - \theta^2} \, d\theta$

15. $\int 3y \sqrt{7 - 3y^2} \, dy$

16. $\int \frac{1}{\sqrt{x} (1 + \sqrt{x})^2} \, dx$

17. $\int \frac{2e^{\sqrt{x}}}{\sqrt{x}} \, dx$

18. $\int \cos (3z + 4) \, dz$

19. $\int \sec^2 (3x + 2) \, dx$

20. $\int \sin^5 \frac{x}{3} \cos \frac{x}{3} \, dx$

21. $\int \frac{1}{t^2} e^{1/t} \, dt$

22. $\int r^2 \left(\frac{r^3}{18} - 1\right)^5 dr$

23. $\int x^{1/2} \sin (x^{3/2} + 1) \, dx$

24. $\int \sec \left(v + \frac{\pi}{2}\right) \tan \left(v + \frac{\pi}{2}\right) dv$

25. $\int \frac{\sin (2t + 1)}{\cos^2 (2t + 1)} \, dt$

26. $\int \frac{6 \cos t}{(2 + \sin t)^3} \, dt$

27. $\int \sqrt{\cot y} \csc^2 y \, dy$

28. $\int \frac{1}{t^2} \cos \left(\frac{1}{t} - 1\right) dt$

29. $\int \frac{dx}{2x - 1}$

30. $\int \frac{x \, dx}{x^2 + 4}$

31. $\int \frac{\sin t}{2 - \cos t} \, dt$

32. $\int \frac{1}{\sin t \cos t} \, dt$

33. $\int \frac{dx}{\sqrt{1 - 4x^2}}$

34. $\int \frac{dx}{9 + 3x^2}$

35. $\int \frac{dx}{x \sqrt{25x^2 - 2}}$

36. $\int \frac{3 \, dr}{\sqrt{1 - 4(r - 1)^2}}$

37. $\int \frac{dx}{1 + (3x + 1)^2}$

38. $\int \frac{y \, dy}{\sqrt{1 - y^4}}$

39. $\int \frac{e^x \, dx}{1 + e^{2x}}$

40. $\int \frac{4 \, dt}{t(1 + \ln^2 t)}$

Simplifying Integrals Step by Step

If you do not know what substitution to make, try reducing the integral step by step, using a trial substitution to simplify the integral a bit and then another to simplify it some more. You will see what we mean if you try the sequences of substitutions in Exercises 41 and 42.

41. $\int \frac{18 \tan^2 x \sec^2 x}{(2 + \tan^3 x)^2} \, dx$
   (a) $u = \tan x$, followed by $v = u^3$, then by $w = 2 + v$
   (b) $u = \tan^3 x$, followed by $v = 2 + u$
   (c) $u = 2 + \tan^3 x$

42. $\int \sqrt{1 + \sin^2 (x - 1)} \sin (x - 1) \cos (x - 1) \, dx$
   (a) $u = x - 1$, followed by $v = \sin u$, then by $w = 1 + v^2$
   (b) $u = \sin (x - 1)$, followed by $v = 1 + u^2$
   (c) $u = 1 + \sin^2 (x - 1)$

Evaluate the integrals in Exercises 43 and 44.

43. $\int \frac{(2r - 1) \cos \sqrt{3(2r - 1)^2 + 6}}{\sqrt{3(2r - 1)^2 + 6}} \, dr$

44. $\int \frac{\sin \sqrt{\theta}}{\sqrt{\theta} \cos^3 \sqrt{\theta}} \, d\theta$

Initial Value Problems

Solve the initial value problems in Exercises 45–58.

45. $\frac{ds}{dt} = 12t(3t^2 - 1)^3, \quad s(1) = 3$

46. $\frac{dy}{dx} = 4x(x^2 + 8)^{-1/3}, \quad y(0) = 0$

47. $\frac{ds}{dt} = 8 \sin^2 \left(t + \frac{\pi}{12}\right), \quad s(0) = 8$

48. $\frac{dy}{dx} = 1 + \frac{1}{x}, \quad y(1) = 3$

49. $\frac{dy}{dt} = e^t \sin (e^t - 2), \quad y(\ln 2) = 0$

50. $\frac{dy}{dt} = e^{-t} \sec^2 (\pi e^{-t}), \quad y(\ln 4) = 2/\pi$

51. $\frac{d^2s}{dt^2} = -4 \sin \left(2t - \frac{\pi}{2}\right), \quad s'(0) = 100, \quad s(0) = 0$

52. $\frac{d^2y}{dx^2} = \sec^2 x, \quad y(0) = (0) \quad \text{and} \quad y'(0) = 1$

53. $\dfrac{d^2y}{dx^2} = 2e^{-x}$, $y(0) = 1$ and $y'(0) = 0$

54. $\dfrac{d^2y}{dt^2} = 1 - e^{2t}$, $y(1) = -1$ and $y'(1) = 0$

55. $\dfrac{dy}{dx} = \dfrac{1}{\sqrt{1-x^2}}$, $y(0) = 0$

56. $\dfrac{dy}{dx} = \dfrac{1}{x^2+1} - 1$, $y(0) = 1$

57. $\dfrac{dy}{dx} = \dfrac{1}{x\sqrt{x^2-1}}$, $x > 1$; $y(2) = \pi$

58. $\dfrac{dy}{dx} = \dfrac{1}{1+x^2} - \dfrac{2}{\sqrt{1-x^2}}$, $y(0) = 2$

59. *Particle motion* The velocity of a particle moving back and forth on a line is $v = ds/dt = 6\sin 2t$ m/sec for all t. If $s = 0$ when $t = 0$, find the value of s when $t = \pi/2$ sec.

60. *Particle motion* The acceleration of a particle moving back and forth on a line is $a = d^2s/dt^2 = \pi^2 \cos \pi t$ m/sec² for all t. If $s = 0$ and $v = 8$ m/sec when $t = 0$, find s when $t = 1$ sec.

Generalizing Integration Formulas

Formulas 10 and 11 in Table 4.1, Section 4.1, generalize as follows:

10. $\displaystyle\int \dfrac{1}{\sqrt{a^2 - x^2}}\, dx = \sin^{-1}\dfrac{x}{a} + C$

11. $\displaystyle\int \dfrac{1}{\sqrt{a^2 - x^2}}\, dx = \dfrac{1}{a}\tan^{-1}\dfrac{x}{a} + C$

61. Verify formula 10.

62. Verify formula 11.

63. Evaluate

 (a) $\displaystyle\int \dfrac{1}{\sqrt{9 - x^2}}\, dx$

 (b) $\displaystyle\int \dfrac{1}{3 + x^2}\, dx$

64. Evaluate

 (a) $\displaystyle\int \dfrac{1}{\sqrt{16 - 25x^2}}\, dx$

 (b) $\displaystyle\int \dfrac{8}{1 + 4x^2}\, dx$

Theory and Examples

65. *Adjusting the integrand: Multiply it by a form of 1* Evaluate $\int \sec x\, dx$ (*Hint:* Multiply the integrand by

$$\frac{\sec x + \tan x}{\sec x + \tan x}$$

and then use a substitution to integrate the result.)

66. *Writing to Learn: Using different substitutions* It looks as if we can integrate $2\sin x \cos x$ with respect to x in three different ways:

 (a) $\displaystyle\int 2\sin x \cos x\, dx = \int 2u\, du \qquad u = \sin x$

 $\qquad\qquad = u^2 + C_1 = \sin^2 x + C_1$

 (b) $\displaystyle\int 2\sin x \cos x\, dx = \int -2u\, du \qquad u = \cos x$

 $\qquad\qquad = -u^2 + C_2 = -\cos^2 x + C_2$

 (c) $\displaystyle\int 2\sin x \cos x\, dx = \int \sin 2x\, dx \qquad 2\sin x \cos x = \sin 2x$

 $\qquad\qquad = -\dfrac{\cos 2x}{2} + C_3.$

Can all three integrations be correct? Give reasons for your answer.

4.3 Estimating With Finite Sums

Area and Cardiac Output • Distance Traveled • Displacement versus Distance Traveled • Volume of a Sphere • Average Value of a Nonnegative Function • Conclusion

CD-ROM
WEBsite

This section shows how practical questions can lead in natural ways to approximations by finite sums.

Area and Cardiac Output

The number of liters of blood your heart pumps in a fixed time interval is called your *cardiac output*. For a person at rest, the rate might be 5 or 6 L per minute. During strenuous exercise the rate might be as high as 30 L per minute. It might also be altered significantly by disease. How can a physician measure a patient's cardiac output without interrupting the flow of blood?

Table 4.3 Dye concentration data	
Seconds after injection t	Dye concentration (adjusted for recirculation) c
5	0
7	3.8
9	8.0
11	6.1
13	3.6
15	2.3
17	1.45
19	0.91
21	0.57
23	0.36
25	0.23
27	0.14
29	0.09
31	0

One technique is to inject a dye into a main vein near the heart. The dye is drawn into the right side of the heart and pumped through the lungs and out the left side of the heart into the aorta, where its concentration can be measured every few seconds as the blood flows past. The data in Table 4.3 and the plot in Figure 4.2 (obtained from the data) show the response of a healthy, resting patient to an injection of 5.6 mg of dye.

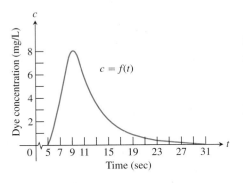

FIGURE 4.2 The dye concentrations from Table 4.3, plotted and fitted with a smooth curve. Time is measured with $t = 0$ at the time of injection. The dye concentrations are zero at the beginning, while the dye passes through the lungs. They then rise to a maximum at about $t = 9$ sec and taper to zero by $t = 31$ sec.

The graph shows dye concentration (measured in milligrams of dye per liter of blood) as a function of time (in seconds). How can we use this graph to obtain the cardiac output (measured in liters of blood per second)? The trick is to divide the *number of milligrams of dye* by the *area under the dye concentration curve*. You can see why this works if you consider what happens to the units:

$$\frac{\text{mg of dye}}{\text{units of area under curve}} = \frac{\text{mg of dye}}{\dfrac{\text{mg of dye}}{\text{L of blood}} \cdot \text{sec}}$$

$$= \frac{\text{mg of dye}}{\text{sec}} \cdot \frac{\text{L of blood}}{\text{mg of dye}}$$

$$= \frac{\text{L of blood}}{\text{sec}}.$$

So you are now ready to compute like a cardiologist.

Example 1   Computing Cardiac Output From Dye Concentration

Estimate the cardiac output of the patient whose data appear in Table 4.3 and Figure 4.2. Give the estimate in liters per minute.

Solution   We have seen that we can obtain the cardiac output by dividing the amount of dye (5.6 mg for our patient) by the area under the curve in Figure 4.2. Now we need to find the area. None of the area formulas we know can be used for this irregularly shaped region, but we can get a good estimate of this area by approximating the region between the curve and the t-axis with rectangles and adding the areas of the rectangles (Figure 4.3). Each rectangle omits some of the area under the curve but includes area from outside the curve, which compensates. In Figure 4.3, each rectangle has a base 2 units long and a height that is equal to the height of the curve above the midpoint of the base. The rectangle's height acts as a sort of average value of the function over the time interval on which the rectangle stands. After reading rectangle heights from the curve, we

FIGURE 4.3 The region under the concentration curve of Figure 4.2 is approximated with rectangles. We ignore the portion from $t = 29$ to $t = 31$; its contribution is negligible.

multiply each rectangle's height and base to find its area and then get the following estimate:

Area under curve ≈ sum of rectangle areas

$$\approx f(6) \cdot 2 + f(8) \cdot 2 + f(10) \cdot 2 + \cdots + f(28) \cdot 2$$
$$\approx 2 \cdot (1.4 + 6.3 + 7.5 + 4.8 + 2.8 + 1.9 + 1.1$$
$$+ 0.7 + 0.5 + 0.3 + 0.2 + 0.1)$$
$$\approx 2 \cdot (27.6) = 55.2 \text{ (mg/L)} \cdot \sec.$$

Dividing 5.6 mg by this figure gives an estimate for cardiac output in liters per second. Multiplying by 60 converts the estimate to liters per minute:

$$\frac{5.6 \text{ mg}}{55.2 \text{ mg} \cdot \sec /L} \cdot \frac{60 \sec}{1 \min} \approx 6.09 \text{ L/min.}$$

USING TECHNOLOGY *Using a Grapher to Calculate Finite Sums* If your graphing utility has a method for evaluating sums, you might want to use it in this section. It is useful for approximating "definite" integrals later in the chapter. There will be other uses still later in your study of calculus.

Distance Traveled

Suppose that we know the velocity function $v = ds/dt = f(t)$ m/sec of a car moving down a highway and want to know how far the car will travel in the time interval $a \le t \le b$. If we know an antiderivative F of f, we can find the car's position function $s = F(t) + C$ and calculate the distance traveled as the difference between the car's positions at times $t = a$ and $t = b$ (as in Section 4.1, Exercise 65).

If we do not know an antiderivative of $v = f(t)$, we can approximate the answer with a sum in the following way. We partition $[a, b]$ into short time intervals *on each of which v is fairly constant*. Since velocity is the rate at which the car is traveling, we approximate the distance traveled on each time interval with the formula

$$\text{Distance} = \text{rate} \times \text{time} = f(t) \cdot \Delta t$$

and add the results across $[a, b]$. To be specific, suppose that the partitioned interval looks like

with the subintervals all of length Δt. Let t_1 be a point in the first subinterval. If the interval is short enough so that the rate is almost constant, the car will move about $f(t_1) \Delta t$ m during that interval. If t_2 is a point in the second interval, the car will move an additional $f(t_2) \Delta t$ m during that interval, and so on. The sum of these products approximates the total distance D traveled from $t = a$ to $t = b$. If we use n subintervals, then

$$D \approx f(t_1) \Delta t + f(t_2) \Delta t + \cdots + f(t_n) \Delta t. \qquad (1)$$

Let's try this on the projectile in Example 5, Section 3.2. The projectile was fired straight into the air. Its velocity t sec into the flight was $v = f(t) = 160 - 9.8t$, and it rose 435.9 m from a height of 3 m to a height of 438.9 m during the first 3 sec of flight.

Example 2   Estimating the Height of a Projectile

The velocity function of a projectile fired straight into the air is $f(t) = 160 - 9.8t$. Use the summation techniques just described to estimate how far the projectile rises during the first 3 sec. How close do the sums come to the exact figure of 435.9 m?

Solution   We explore the results for different numbers of intervals and different choices of evaluation points.

Three subintervals of length 1, with f evaluated at left-hand endpoints:

With f evaluated at $t = 0$, 1, and 2, we have

$$D \approx f(t_1)\,\Delta t + f(t_2)\,\Delta t + f(t_3)\,\Delta t \qquad \text{Eq. (1)}$$
$$\approx [160 - 9.8(0)](1) + [160 - 9.8(1)](1) + [160 - 9.8(2)](1)$$
$$\approx 450.6.$$

Three subintervals of length 1, with f evaluated at right-hand endpoints:

With f evaluated at $t = 1$, 2, and 3, we have

$$D \approx f(t_1)\,\Delta t + f(t_2)\,\Delta t + f(t_3)\,\Delta t \qquad \text{Eq. (1)}$$
$$\approx [160 - 9.8(1)](1) + [160 - 9.8(2)](1) + [160 - 9.8(3)](1)$$
$$\approx 421.2.$$

With six subintervals of length 1/2, we get

Using left-hand endpoints: $D \approx 443.25.$
Using right-hand endpoints: $D \approx 428.55.$

These six-interval estimates are somewhat closer than the three-interval estimates. The results improve as the subintervals get shorter.

As we can see in Table 4.4, the left-endpoint sums approach the true value 435.9 from above, whereas the right-endpoint sums approach it from below. The

Table 4.4 Travel-distance estimates

Number of subintervals	Length of each subinterval	Left-endpoint sum	Right-endpoint sum
3	1	450.6	421.2
6	0.5	443.25	428.55
12	0.25	439.58	432.23
24	0.125	437.74	434.06
48	0.0625	436.82	434.98
96	0.03125	436.36	435.44
192	0.015625	436.13	435.67

Error magnitude = | true value − calculated value |

true value lies between these upper and lower sums. The magnitude of the error in the closest entries is 0.23, a small percentage of the true value.

$$\text{Error percentage} = \frac{0.23}{435.9} \approx 0.05\%.$$

It would be safe to conclude from the table's last entities that the projectile rose about 436 m during its first 3 sec of flight.

Displacement versus Distance Traveled

If a body with position function $s(t)$ moves along a coordinate line without changing direction, we can calculate the total distance it travels from $t = a$ to $t = b$ by summing the distance traveled over small intervals, as in Example 2. If the body changes direction one or more times during the trip, then we need to use the body's *speed* $|v(t)|$, which is the absolute value of its velocity function, $v(t)$, to find the total distance traveled. Using the velocity itself, as in Example 2, only gives an estimate to the body's **displacement,** $s(b) - s(a)$, the difference between its initial and final positions.

To see why, partition the time interval $[a, b]$ into small enough equal subintervals Δt so that the body's velocity does not change very much from time t_{k-1} to t_k. Then $v(t_k)$ will give a good approximation of the velocity throughout the interval. Accordingly, the change in the body's position coordinate during the time interval will be about

$$v(t_k)\,\Delta t.$$

The change will be positive if $v(t_k)$ is positive and negative if $v(t_k)$ is negative.

In either case, the distance traveled during the subinterval will be about

$$|v(t_k)|\,\Delta t.$$

The **total trip distance** will be approximately the sum

$$|v(t_1)|\,\Delta t + |v(t_2)|\,\Delta t + \cdots + |v(t_n)|\,\Delta t. \tag{2}$$

Volume of a Sphere

Notice the mathematical similarity between Examples 1 and 2. In each case, we have a function f defined on a closed interval and estimate what we want to know with a sum of function values multiplied by interval lengths. We can use similar sums to estimate volumes.

Example 3 Estimating the Volume of a Sphere

CD-ROM
WEBsite

Estimate the volume of a solid sphere of radius 4.

Solution We picture the sphere as if its surface were generated by revolving the graph of the function $f(x) = \sqrt{16 - x^2}$ about the x-axis (Figure 4.4a). We partition the interval $-4 \le x \le 4$ into n subintervals of equal length $\Delta x = 8/n$. We then slice the sphere with planes perpendicular to the x-axis at the partition points, cutting it like a round loaf of bread into n parallel slices of width Δx. When n is large, each slice can be approximated by a cylinder, a familiar geometric shape of known volume, $\pi r^2 h$. In our case, the cylinders lie on their sides and h is Δx, whereas r varies according to where we are on the x-axis. Let's choose $n = 8$ cylinders and pick the radius of each as the height $f(c_i) = \sqrt{16 - c_i^2}$ at its left-hand endpoint $x = c_i$ of the subinterval (Figure 4.4b). (The cylinder at $x = -4$ is degenerate because the cross section there is just a point.) We can then approximate the volume of the sphere by summing the cylinder volumes,

$$\pi r^2 h = \pi \left(\sqrt{16 - c_i^2}\right)^2 \Delta x.$$

The sum of the eight cylinders' volumes is

$$S_8 = \pi \left[\sqrt{16 - c_1^2}\right]^2 \Delta x + \pi \left[\sqrt{16 - c_2^2}\right]^2 \Delta x + \pi \left[\sqrt{16 - c_3^2}\right]^2 \Delta x$$

$$+ \cdots + \pi \left[\sqrt{16 - c_8^2}\right]^2 \Delta x \qquad \Delta x = \dfrac{8}{n} = 1$$

$$= \pi[(16 - (-4)^2) + (16 - (-3)^2) + (16 - (-2)^2) + \cdots + (16 - (3)^2)]$$

$$= \pi[0 + 7 + 12 + 15 + 16 + 15 + 12 + 7]$$

$$= 84\pi.$$

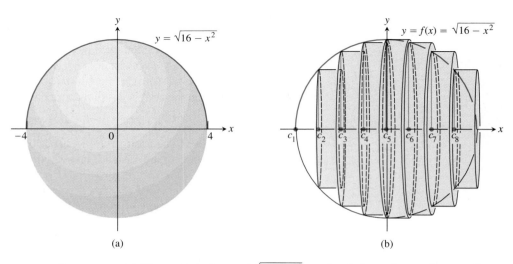

(a)   (b)

FIGURE 4.4 (a) The semicircle $y = \sqrt{16 - x^2}$ revolved about the x-axis to outline a sphere. (b) The solid sphere approximated with cross-section-based cylinders.

This result compares favorably with the sphere's true volume,

$$V = \frac{4}{3}\pi r^3 = \frac{4}{3}\pi(4)^3 = \frac{256\pi}{3}.$$

The difference between S_8 and V is a small percentage of V:

$$\text{Error percentage} = \frac{|V - S_8|}{V} = \frac{(256/3)\pi - 84\pi}{(256/3)\pi}$$

$$= \frac{256 - 252}{256} = \frac{1}{64} \approx 1.6\%.$$

With a finer partition (more subintervals), the approximation would be even better.

Average Value of a Nonnegative Function

To find the average of a finite set of values, we add them and divide by the number of values added. But what happens if we want to find the average of an infinite number of values? For example, what is the average value of the function $f(x) = x^2$ on the interval $[-1, 1]$? To see what this kind of "continuous" average might mean, imagine that we are pollsters sampling the function. We pick random x's between -1 and 1, square them, and average the squares. As we take larger samples, we expect this average to approach some number, which seems reasonable to call the *average of f over* $[-1, 1]$.

The graph in Figure 4.5a suggests that the average square should be less than 1/2, because numbers with squares less than 1/2 make up more than 70% of the interval $[-1, 1]$. If we had a computer to generate random numbers, we could carry

(a)

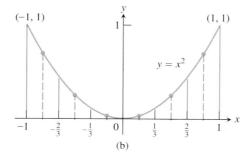

(b)

FIGURE 4.5 (a) The graph of $f(x) = x^2$, $-1 \le x \le 1$. (b) Values of f sampled at regular intervals.

out the sampling experiment described above, but it is much easier to estimate the average value with a finite sum.

Example 4   Estimating Average Value

Estimate the average value of the function $f(x) = x^2$ on the interval $[-1, 1]$.

Solution   We look at the graph of $y = x^2$ and partition the interval $[-1, 1]$ into 6 subintervals of length $\Delta x = 1/3$ (Figure 4.5b).

It appears that a good estimate for the average square on each subinterval is the square of the midpoint of the subinterval. Since the subintervals have the same length, we can average these six estimates to get a final estimate for the average value over $[-1, 1]$.

$$\text{Average value} \approx \frac{\left(-\frac{5}{6}\right)^2 + \left(-\frac{3}{6}\right)^2 + \left(-\frac{1}{6}\right)^2 + \left(\frac{1}{6}\right)^2 + \left(\frac{3}{6}\right)^2 + \left(\frac{5}{6}\right)^2}{6}$$

$$\approx \frac{1}{6} \cdot \frac{25 + 9 + 1 + 1 + 9 + 25}{36} = \frac{70}{216} \approx 0.324$$

We will be able to show later that the average value is $1/3$.

Notice that

$$\frac{\left(-\frac{5}{6}\right)^2 + \left(-\frac{3}{6}\right)^2 + \left(-\frac{1}{6}\right)^2 + \left(\frac{1}{6}\right)^2 + \left(\frac{3}{6}\right)^2 + \left(\frac{5}{6}\right)^2}{6}$$

$$= \frac{1}{2}\left[\left(-\frac{5}{6}\right)^2 \cdot \frac{1}{3} + \left(-\frac{3}{6}\right)^2 \cdot \frac{1}{3} + \cdots + \left(\frac{5}{6}\right)^2 \cdot \frac{1}{3}\right]$$

$$= \frac{1}{\text{length of } [-1, 1]} \cdot \left[f\left(-\frac{5}{6}\right) \cdot \frac{1}{3} + f\left(-\frac{3}{6}\right) \cdot \frac{1}{3} + \cdots + f\left(\frac{5}{6}\right) \cdot \frac{1}{3}\right]$$

$$= \frac{1}{\text{length of } [-1, 1]} \cdot \left[\begin{matrix}\text{a sum of function values} \\ \text{multiplied by interval lengths}\end{matrix}\right].$$

Once again our estimate has been achieved by multiplying function values by interval lengths and summing the results for all the intervals.

Conclusion

The examples in this section describe instances in which sums of function values multiplied by interval lengths provide approximations that are good enough to answer practical questions. You will find additional examples in the exercises.

The distance approximations in Example 2 improved as the intervals involved became shorter and more numerous. We knew this because we had already found the exact answer with antiderivatives in Section 3.2. If we had made our partitions of the time interval still finer, would the sums have approached the exact answer as a limit? Is the connection between the sums and the antiderivative in this case just a coincidence? Could we have calculated the area in Example 1, the volume in Example 3, and the average value in Example 4 with antiderivatives as well? As we will see, the answers are "Yes, they would have," "No, it is not a coincidence," and "Yes, we could have."

EXERCISES 4.3

Cardiac Output

1. The accompanying table gives dye concentrations for a dye-dilution cardiac-output determination like the one in Example 1. The amount of dye injected in this case was 5 mg instead of 5.6 mg. Use rectangles to estimate the area under the dye concentration curve and then go on to estimate the patient's cardiac output.

Seconds after injection	Dye concentration (adjusted for recirculation)
2	0
4	0.6
6	1.4
8	2.7
10	3.7
12	4.1
14	3.8
16	2.9
18	1.7
20	1.0
22	0.5
24	0

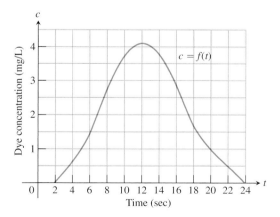

2. The accompanying table gives dye concentrations for a cardiac-output determination like the one in Example 1. The amount of dye injected in this case was 10 mg. Plot the data and connect the data points with a smooth curve. Estimate the area under the curve and calculate the cardiac output from this estimate.

Seconds after injection t	Dye concentration (adjusted for recirculation) c	Seconds after injection t	Dye concentration (adjusted for recirculation) c
0	0	16	7.9
2	0	18	7.8
4	0.1	20	6.1
6	0.6	22	4.7
8	2.0	24	3.5
10	4.2	26	2.1
12	6.3	28	0.7
14	7.5	30	0

Distance

3. *Distance traveled* The accompanying table shows the velocity of a model train engine moving along a track for 10 sec. Estimate the distance traveled by the engine using 10 subintervals of length 1 with

(a) left-endpoint values

(b) right-endpoint values.

Time (sec)	Velocity (in./sec)	Time (sec)	Velocity (in./sec)
0	0	6	11
1	12	7	6
2	22	8	2
3	10	9	6
4	5	10	0
5	13		

4. *Distance traveled upstream* You are sitting on the bank of a tidal river watching the incoming tide carry a bottle upstream. You record the velocity of the flow every five minutes for an hour, with the results shown in the table on the following page. About how far upstream did the bottle travel during that hour? Find an estimate using 12 subintervals of length 5 with

(a) left-endpoint values

(b) right-endpoint values.

Time (min)	Velocity (m/sec)	Time (min)	Velocity (m/sec)
0	1	35	1.2
5	1.2	40	1.0
10	1.7	45	1.8
15	2.0	50	1.5
20	1.8	55	1.2
25	1.6	60	0
30	1.4		

5. *Length of a road* You and a companion are about to drive a twisty stretch of dirt road in a car whose speedometer works but whose odometer (mileage counter) is broken. To find out how long this particular stretch of road is, you record the car's velocity at 10 sec intervals, with the results shown in the accompanying table. Estimate the length of the road using

(a) left-endpoint values

(b) right-endpoint values.

Time (sec)	Velocity (converted to ft/sec) (30 mi/h = 44 ft/sec)	Time (sec)	Velocity (converted to ft/sec) (30 mi/h = 44 ft/sec)
0	0	70	15
10	44	80	22
20	15	90	35
30	35	100	44
40	30	110	30
50	44	120	35
60	35		

6. *Distance from velocity data* The accompanying table gives data for the velocity of a vintage sports car accelerating from 0 to 142 mi/h in 36 sec (10 thousandths of an hour).

Time (h)	Velocity (mi/h)	Time (h)	Velocity (mi/h)
0.0	0	0.006	116
0.001	40	0.007	125
0.002	62	0.008	132
0.003	82	0.009	137
0.004	96	0.010	142
0.005	108		

(a) Use rectangles to estimate how far the car traveled during the 36 sec it took to reach 142 mi/h.

(b) Roughly how many seconds did it take the car to reach the halfway point? About how fast was the car going then?

Volume

7. *Volume of a sphere* (*Continuation of Example 3*) Suppose that we approximate the volume V of the sphere in Example 3 by partitioning the interval $-4 \le x \le 4$ into four subintervals of length 2 and using cylinders based on the cross sections at the subintervals' left-hand endpoints. (As in Example 3, the leftmost cylinder will have a zero radius.)

(a) Find the sum S_4 of the volumes of the cylinders.

(b) Express $|V - S_4|$ as a percentage of V to the nearest percent.

8. *Volume of a solid sphere* To estimate the volume V of a solid sphere of radius 5, you partition its diameter into five subintervals of length 2. You then slice the sphere with planes perpendicular to the diameter at the subintervals' left-hand endpoints and add the volumes of cylinders of height 2 based on the cross sections of the sphere determined by these planes.

(a) Find the sum S_5 of the volumes of the cylinders.

(b) Express $|V - S_5|$ as a percentage of V to the nearest percent.

9. *Volume of a solid hemisphere* To estimate the volume V of a solid hemisphere of radius 4, imagine its axis of symmetry to be the interval [0, 4] on the x-axis. Partition [0, 4] into eight subintervals of equal length and approximate the solid with cylinders based on the circular cross sections of the hemisphere perpendicular to the x-axis at the subintervals' left-hand endpoints. (See the accompanying profile view.)

(a) *Writing to Learn* Find the sum S_8 of the volumes of the cylinders. Do you expect S_8 to overestimate V, or to underestimate V? Give reasons for your answer.

(b) Express $|V - S_8|$ as a percentage of V to the nearest percent.

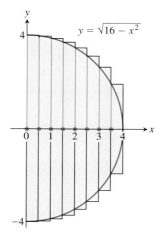

construct cylinders of height 1 based on cross sections at these points. (See the accompanying figure.)

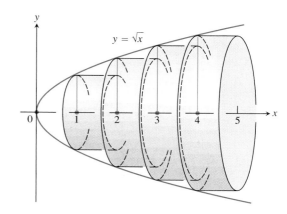

10. *Volume of a solid hemisphere* Repeat Exercise 9 using cylinders based on cross sections at the *right-hand* endpoints of the subintervals.

11. *Volume of water in a reservoir* A reservoir shaped like a hemispherical bowl of radius 8 m is filled with water to a depth of 4 m.

   (a) Find an estimate S of the water's volume by approximating the water with eight circumscribed solid cylinders.

   (b) As you will see in Section 4.5, Exercise 43, the water's volume is $V = 320\pi/3$ m^3. Find the error $|V - S|$ as a percentage of V to the nearest percent.

12. *Volume of water in a swimming pool* A rectangular swimming pool is 30 ft wide and 50 ft long. The accompanying table shows the depth $h(x)$ of the water at 5 ft intervals from one end of the pool to the other. Estimate the volume of water in the pool using

   (a) left-endpoint values of h

   (b) right-endpoint values of h.

Position x ft	Depth $h(x)$ ft	Position x ft	Depth $h(x)$ ft
0	6.0	30	11.5
5	8.2	35	11.9
10	9.1	40	12.3
15	9.9	45	12.7
20	10.5	50	13.0
25	11.0		

13. *Volume of a nose cone* The nose "cone" of a rocket is a paraboloid obtained by revolving the curve $y = \sqrt{x}$, $0 \le x \le 5$, about the x-axis, where x is measured in feet. Estimate the volume V of the nose cone by partitioning the closed interval [0, 5] into five subintervals of equal length, slicing the cone with planes perpendicular to the x-axis at the subintervals' left-hand endpoints, and

   (a) *Writing to Learn* Find the sum S_5 of the volumes of the cylinders. Do you expect S_5 to overestimate V or to underestimate V? Give reasons for your answer.

   (b) As you will see in Section 4.5, Exercise 44, the volume of the nose cone is $V = 25\pi/2$ ft^3. Express $|V - S_5|$ as a percentage of V to the nearest percent.

14. *Volume of a nose cone* Repeat Exercise 13 using cylinders based on cross sections at the *right-hand* endpoints of the subintervals.

Velocity and Distance

15. *Free fall with air resistance* An object is dropped straight down from a helicopter. The object falls faster and faster but its acceleration (rate of change of its velocity) decreases over time because of air resistance. The acceleration is measured in feet per seconds squared and recorded every second after the drop for 5 sec, as shown in the table below.

t	0	1	2	3	4	5
a	32.00	19.41	11.77	7.14	4.33	2.63

   (a) Find an upper estimate for the speed when $t = 5$.

   (b) Find a lower estimate for the speed when $t = 5$.

   (c) Find an upper estimate for the distance fallen when $t = 3$.

16. *Distance traveled by a projectile* An object is shot straight upward from sea level with an initial velocity of 400 ft/sec.

   (a) Assuming that gravity is the only force acting on the object, give an upper estimate for its velocity after 5 sec have elapsed. Use $g = 32$ ft/sec^2 for the gravitational constant.

   (b) Find a lower estimate for the height attained after 5 sec.

Average Value of a Function

In Exercises 17–20, use a finite sum to estimate the average value of f on the given interval by partitioning the interval into four subintervals of equal length and evaluating f at the subinterval midpoints.

17. $f(x) = x^3$ on $[0, 2]$

18. $f(x) = 1/x$ on $[1, 9]$

19. $f(t) = (1/2) + \sin^2 \pi t$ on $[0, 2]$

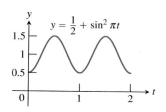

20. $f(t) = 1 - \left(\cos \dfrac{\pi t}{4}\right)^4$ on $[0, 4]$

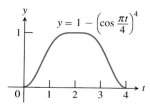

Pollution Control

21. *Water pollution* Oil is leaking out of a tanker damaged at sea. The damage to the tanker is worsening as evidenced by the increased leakage each hour, recorded in the following table.

Time (hours)	0	1	2	3	4
Leakage (gal/hr)	50	70	97	136	190

Time (hours)	5	6	7	8
Leakage (gal/hr)	265	369	516	720

(a) Give an upper and a lower estimate of the total quantity of oil that has escaped after 5 hours.

(b) Repeat part (a) for the quantity of oil that has escaped after 8 hours.

(c) The tanker continues to leak 720 gal/h after the first 8 hours. If the tanker originally contained 25,000 gal of oil, approximately how many more hours will elapse in the worst case before all the oil has spilled. In the best case?

22. *Air pollution* A power plant generates electricity by burning oil. Pollutants produced as a result of the burning process are removed by scrubbers in the smokestacks. Over time, the scrubbers become less efficient and eventually they must be replaced when the amount of pollution released exceeds government standards. Measurements are taken at the end of each month determining the rate at which pollutants are released into the atmosphere, recorded as follows.

Month	Jan	Feb	Mar	Apr	May	Jun
Pollutant Release rate (tons/day)	0.20	0.25	0.27	0.34	0.45	0.52

Month	Jul	Aug	Sep	Oct	Nov	Dec
Pollutant Release rate (tons/day)	0.63	0.70	0.81	0.85	0.89	0.95

(a) Assuming a 30-day month and that new scrubbers allow only 0.05 ton/day released, give an upper estimate of the total tonnage of pollutants released by the end of June. What is a lower estimate?

(b) In the best case, approximately when will a total of 125 tons of pollutants have been released into the atmosphere?

Area

23. *Area of a circle* Inscribe a regular n-sided polygon inside a circle of radius 1 and compute the area of the polygon for the following values of n.

(a) 4 (square)   **(b)** 8 (octagon)   **(c)** 16

(d) Compare the areas in parts (a), (b), and (c) with the area of the circle.

24. *(Continuation of Exercise 23)*

(a) Inscribe a regular n-sided polygon inside a circle of radius 1 and compute the area of one of the n congruent triangles formed by drawing radii to the vertices of the polygon.

(b) Compute the limit of the area of the inscribed polygon as $n \to \infty$.

(c) Repeat the computations in parts (a) and (b) for a circle of radius r.

COMPUTER EXPLORATIONS

Estimating Average Value

In Exercises 25–28, use a CAS to perform the following steps.

(a) Plot the functions over the given interval.

(b) Partition the interval into $n = 100, 200,$ and 1000 subintervals of equal length, and evaluate the function at the midpoint of each subinterval.

(c) Compute the average value of the function values generated in part (b).

(d) Solve the equation $f(x) =$ (average value) for x using the average value calculated in part (c) for the $n = 1000$ partitioning.

25. $f(x) = \sin x$ on $[0, \pi]$

26. $f(x) = \sin^2 x$ on $[0, \pi]$

27. $f(x) = x \sin \dfrac{1}{x}$ on $\left[\dfrac{\pi}{4}, \pi\right]$

28. $f(x) = x \sin^2 \dfrac{1}{x}$ on $\left[\dfrac{\pi}{4}, \pi\right]$

4.4   Riemann Sums and Definite Integrals

Riemann Sums   •   Terminology and Notation of Integration   •   Area Under the Graph of a Nonnegative Function   •   Average Value of an Arbitrary Continuous Function   •   Properties of Definite Integrals

In the preceding section, we estimated distances, area, volumes, and average values with finite sums. The terms in the sums were obtained by multiplying selected function values by the lengths of intervals. In this section, we move beyond finite sums to see what happens in the limit, as the lengths of the intervals become infinitely small and their number infinitely large.

Riemann Sums

Sigma notation enables us to express a large sum in compact form:

$$\sum_{k=1}^{n} a_k = a_1 + a_2 + a_3 + \cdots + a_{n-1} + a_n.$$

The Greek capital letter Σ (sigma) stands for "sum." The index k tells us where to begin the sum (at the number below the Σ) and where to end (at the number above). If the symbol ∞ appears above the Σ, it indicates that the terms go on indefinitely.

Example 1   Using Sigma Notation

The sum in sigma notation	The sum written out, one term for each value of k	The value of the sum
$\displaystyle\sum_{k=1}^{5} k$	$1 + 2 + 3 + 4 + 5$	15
$\displaystyle\sum_{k=1}^{3} (-1)^k k$	$(-1)^1(1) + (-1)^2(2) + (-1)^3(3)$	$-1 + 2 - 3 = -2$
$\displaystyle\sum_{k=1}^{2} \dfrac{k}{k+1}$	$\dfrac{1}{1+1} + \dfrac{2}{2+1}$	$\dfrac{1}{2} + \dfrac{2}{3} = \dfrac{7}{6}$
$\displaystyle\sum_{k=4}^{5} \dfrac{k^2}{k-1}$	$\dfrac{4^2}{4-1} + \dfrac{5^2}{5-1}$	$\dfrac{16}{3} + \dfrac{25}{4} = \dfrac{139}{12}$

The lower limit of summation does not have to be 1; it can be any integer.

The sums in which we will be interested are called *Riemann* ("*ree*-mahn") *sums*, after Georg Friedrich Bernhard Riemann. Riemann sums are constructed in a particular way. We now describe that construction formally, in a more general context that does not confine us to nonnegative functions.

We begin with an arbitrary continuous function $f(x)$ defined on a closed interval $[a, b]$. Like the function graphed in Figure 4.6, it may have negative values as well as positive values.

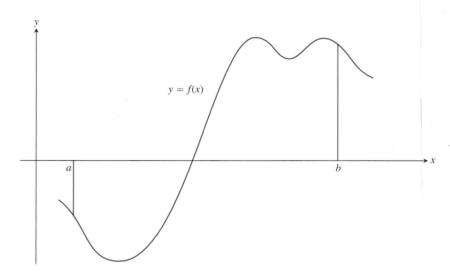

$y = f(x)$

FIGURE 4.6 A typical continuous function $y = f(x)$ over a closed interval $[a, b]$.

We then partition the interval $[a, b]$ into n subintervals by choosing $n - 1$ points, say $x_1, x_2, \cdots, x_{n-1}$, between a and b subject only to the condition that

$$a < x_1 < x_2 < \cdots < x_{n-1} < b.$$

To make the notation consistent, we denote a by x_0 and b by x_n. The set

$$P = \{x_0, x_1, x_2, \cdots, x_n\}$$

is called a **partition** of $[a, b]$.

The partition P defines n closed **subintervals**

$$[x_0, x_1], [x_1, x_2], \ldots, [x_{n-1}, x_n].$$

The typical closed subinterval $[x_{k-1}, x_k]$ is called the **kth subinterval** of P.

kth subinterval

$x_0 = a$ x_1 x_2 $\cdots$ x_{k-1} x_k $\cdots$ x_{n-1} $x_n = b$

The length of the kth subinterval is $\Delta x_k = x_k - x_{k-1}$.

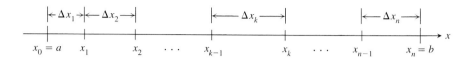

In each subinterval we select some number. Denote the number chosen from the kth subinterval by c_k.

Then, on each subinterval we stand a vertical rectangle that reaches from the x-axis to touch the curve at $(c_k, f(c_k))$. These rectangles could lie either above or below the x-axis (Figure 4.7).

(a)

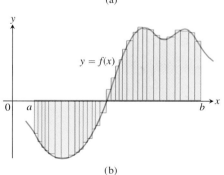

(b)

FIGURE 4.8 The curve of Figure 4.7 with rectangles from finer partitions of $[a, b]$. Finer partitions create more rectangles with shorter bases.

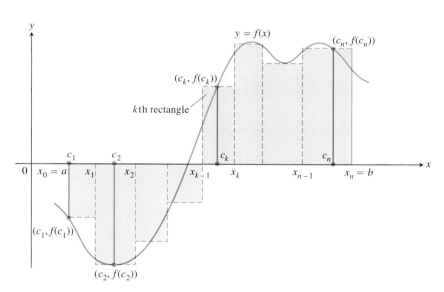

FIGURE 4.7 The rectangles approximate the region between the graph of the function $y = f(x)$ and the x-axis.

On each subinterval, we form the product $f(c_k) \cdot \Delta x_k$. This product can be positive, negative, or zero, depending on $f(c_k)$.

Finally, we take the sum of these products:

$$S_n = \sum_{k=1}^{n} f(c_k) \cdot \Delta x_k.$$

This sum, which depends on the partition P and the choice of the numbers c_k, is a **Riemann sum for f on the interval $[a, b]$.**

As the partitions of $[a, b]$ become finer and finer, we would expect the rectangles defined by the partitions to approximate the region between the x-axis and the graph of f with increasing accuracy (Figure 4.8). So we expect the associated Riemann sums to have a limiting value. Theorem 1 below assures us that they do, as long as the lengths of the subintervals all tend to zero. This latter condition is assured by requiring the longest subinterval length, called the **norm** of the partition and denoted by $\|P\|$, to tend to zero.

Despite the potential for variety in the sums $\Sigma f(c_k) \, \Delta x_k$ as the partitions change

---

Definition   The Definite Integral as a Limit of Riemann Sums

Let f be a function defined on a closed interval $[a, b]$. For any partition P of $[a, b]$, let the numbers c_k be chosen arbitrarily in the subintervals $[x_{k-1}, x_k]$.

If there exists a number I such that

$$\lim_{\|P\| \to 0} \sum_{k=1}^{n} f(c_k) \, \Delta x_k = I$$

no matter how P and the c_k's are chosen, then f is **integrable** on $[a, b]$ and I is the **definite integral** of f over $[a, b]$.

---

and as the c_k's are chosen arbitrarily in the intervals of each partition, the sums always have the same limit as $\|P\| \to 0$ as long as f is *continuous* on $[a, b]$.

Theorem 1   The Existence of Definite Integrals

All continuous functions are integrable. That is, if a function f is continuous on an interval $[a, b]$, then its definite integral over $[a, b]$ exists.

Terminology and Notation of Integration

Leibniz's clever choice of notation for the derivative, dy/dx, had the advantage of retaining an identity as a "fraction" even though both numerator and denominator had tended to zero. Although not really fractions, derivatives can *behave* like fractions, so the notation makes profound results like the Chain Rule

$$\frac{dy}{dx} = \frac{dy}{du} \cdot \frac{du}{dx}$$

seem almost simple.

The notation that Leibniz introduced for the definite integral was equally inspired. In his derivative notation, the Greek letters ("Δ" for "difference") switch to Roman letters ("d" for "differential") in the limit,

$$\lim_{\Delta x \to 0} \frac{\Delta y}{\Delta x} = \frac{dy}{dx}.$$

In his definite integral notation, the Greek letters again become Roman letters in the limit,

$$\lim_{n \to \infty} \sum_{k=1}^{n} f(c_k) \, \Delta x = \int_{a}^{b} f(x) \, dx.$$

Notice that the difference Δx has again tended to zero, becoming a differential dx. The Greek "Σ" has become an elongated Roman "S," so that the integral can retain its identity as a "sum." The c_k's have become so crowded together in the limit that we no longer think of a choppy selection of x values between a and b, but rather of a continuous, unbroken sampling of x values from a to b. It is as if we were summing *all* products of the form $f(x)\,dx$ as x goes from a to b, so we can abandon the k and the n used in the finite sum expression.

The symbol

$$\int_a^b f(x)\,dx$$

is read as "the integral from a to b of f of x dee x" or sometimes as "the integral from a to b of f of x with respect to x." The component parts also have names:

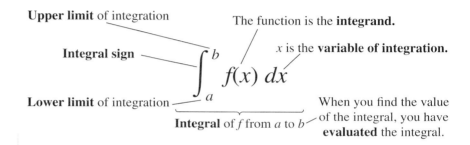

The value of the definite integral of a function over any particular interval depends on the function and not on the letter we choose to represent its independent variable. If we decide to use t or u instead of x, we simply write the integral as

$$\int_a^b f(t)\,dt \qquad \text{or} \qquad \int_a^b f(u)\,du \qquad \text{instead of} \qquad \int_a^b f(x)\,dx.$$

No matter how we represent the integral, it is the same *number*, defined as a limit of Riemann sums. Since it does not matter what letter we use to run from a to b, the variable of integration is called a **dummy variable.**

Example 2   Using the Notation

The interval $[-1, 3]$ is partitioned into n subintervals of equal length $\Delta x = 4/n$. Let m_k denote the midpoint of the kth subinterval. Express the limit

$$\lim_{n \to \infty} \sum_{k=1}^{n} (3(m_k)^2 - 2m_k + 5)\,\Delta x$$

as an integral.

Solution   Since the midpoints m_k have been chosen from the subintervals of the partition, this expression is indeed a limit of Riemann sums. (The points chosen did not have to be midpoints; they could have been chosen from the subintervals in any arbitrary fashion.) The function being integrated is $f(x) = 3x^2 - 2x + 5$ over the interval $[-1, 3]$. Therefore,

$$\lim_{n \to \infty} \sum_{k=1}^{n} (3(m_k)^2 - 2m_k + 5)\,\Delta x = \int_{-1}^{3} (3x^2 - 2x + 5)\,dx.$$

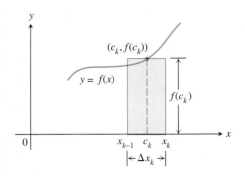

FIGURE 4.9 A term of a Riemann sum $\sum f(c_k)\,\Delta x_k$ for a nonnegative function f is either zero or the area of a rectangle such as the one shown.

Area Under the Graph of a Nonnegative Function

In Example 1, Section 4.3, we saw that we could approximate the area under the graph of a nonnegative continuous function $y = f(x)$ by summing the areas of finitely many rectangles with height equal to the height of the curve above the midpoint of the base subinterval. We now know why this is true. If an integrable function $y = f(x)$ is nonnegative throughout an interval $[a, b]$, each nonzero term $f(c_k)\,\Delta x_k$ is the area of a rectangle reaching from the x-axis up to the curve $y = f(x)$. (See Figure 4.9.)

The Riemann sum

$$\sum f(c_k)\,\Delta x_k,$$

which is the sum of the areas of these rectangles, gives an estimate of the area of the region between the curve and the x-axis from a to b. Since the rectangles give an increasingly good approximation of the region as we use partitions with smaller and smaller norms, we call the limiting value the area under the curve.

> **Definition** Area Under a Curve (as a Definite Integral)
> If $y = f(x)$ is nonnegative and integrable over a closed interval $[a, b]$, then the **area under the curve $y = f(x)$ from a to b** is the integral of f from a to b,
> $$A = \int_a^b f(x)\,dx.$$

This definition works both ways: We can use integrals to calculate areas *and* we can use areas to calculate integrals.

Example 3 Area Under the Curve $f(x) = x$

Evaluate

$$\int_a^b x\,dx, \qquad 0 < a < b.$$

Solution We sketch the region under the curve $y = x$, $a \le x \le b$ (Figure 4.10), and see that it is a trapezoid with height $(b - a)$ and bases a and b. The value of the integral is the area of this trapezoid:

$$\int_a^b x\,dx = (b - a)\cdot\frac{a + b}{2} = \frac{b^2}{2} - \frac{a^2}{2}.$$

Thus,

$$\int_1^{\sqrt{5}} x\,dx = \frac{(\sqrt{5})^2}{2} - \frac{(1)^2}{2} = 2$$

and so on.

Notice that $x^2/2$ is an antiderivative of x, further evidence of a connection between antiderivatives and summation.

FIGURE 4.10 The region in Example 3.

Average Value of an Arbitrary Continuous Function

In Section 4.3, Example 4, we discussed the average value of a nonnegative continuous function. We are now ready to define average value without requiring f to be nonnegative and to show that every continuous function assumes its average value at least once.

We start once again with the idea from arithmetic that the average of n numbers is the sum of the numbers divided by n. For a continuous function f on a closed interval $[a, b]$, there may be infinitely many values to consider, but we can sample them in an orderly way. We partition $[a, b]$ into n subintervals of equal length (the length is $\Delta x = (b - a)/n$) and evaluate f at a point c_k in each subinterval (Figure 4.11). The average of the n sampled values is

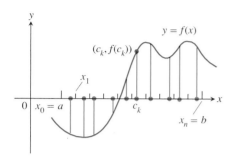

$$\frac{f(c_1) + f(c_2) + \cdots + f(c_n)}{n} = \frac{1}{n} \cdot \sum_{k=1}^{n} f(c_k) \qquad \text{the sum in sigma notation}$$

$$= \frac{\Delta x}{b - a} \cdot \sum_{k=1}^{n} f(c_k) \qquad \Delta x = \frac{b - a}{n}$$

$$= \frac{1}{b - a} \cdot \underbrace{\sum_{k=1}^{n} f(c_k)\, \Delta x}_{\text{a Riemann sum for } f \text{ on } [a, b]}.$$

FIGURE 4.11 A sample of values of a function on an interval $[a, b]$.

Thus, the average of the sampled values is always $1/(b - a)$ times a Riemann sum for f on $[a, b]$. As we increase the size of the sample and let the norm of the partition approach zero, the average must approach $(1/(b - a)) \int_a^b f(x)\, dx$. We are led by this remarkable fact to the following definition.

Definition   Average (Mean) Value
If f is integrable on $[a, b]$, then its **average (mean) value** on $[a, b]$ is

$$\text{av}\,(f) = \frac{1}{b - a} \int_a^b f(x)\, dx.$$

CD-ROM
WEBsite

Example 4   Finding an Average Value

Find the average value of $f(x) = \sqrt{4 - x^2}$ on $[-2, 2]$.

Solution   We recognize $f(x) = \sqrt{4 - x^2}$ as a function whose graph is the upper semicircle of radius 2 centered at the origin.

The area between the semicircle and the x-axis from -2 to 2 can be computed using the geometry formula

$$\text{Area} = \frac{1}{2} \cdot \pi r^2 = \frac{1}{2} \cdot \pi(2)^2 = 2\pi.$$

Because the area is also the value of the integral of f from -2 to 2,

$$\int_{-2}^{2} \sqrt{4 - x^2}\, dx = 2\pi.$$

Therefore, the average value of f is

$$\operatorname{av}(f) = \frac{1}{2 - (-2)} \int_{-2}^{2} \sqrt{4 - x^2}\, dx = \frac{1}{4}(2\pi) = \frac{\pi}{2}.$$

Properties of Definite Integrals

In defining $\int_a^b f(x)\, dx$ as a limit of sums $\Sigma f(c_k)\, \Delta x_k$, we moved from left to right across the interval $[a, b]$. What would happen if we integrated in the *opposite* direction? The integral would become $\int_b^a f(x)\, dx$—again a limit of sums of the form $\Sigma f(c_k)\, \Delta x_k$—but this time each of the Δx_k's would be negative as the x-values *decreased* from b to a. This would change the signs of all the terms in each Riemann sum and ultimately, the sign of the definite integral. This suggests the rule

$$\int_b^a f(x)\, dx = -\int_a^b f(x)\, dx.$$

Since the original definition did not apply to integrating backwards over an interval, we can treat this rule as a logical extension of the definition.

Although $[a, a]$ is technically not an interval, another logical extension of the definition is that $\int_a^a f(x)\, dx = 0$.

These are the first two rules in Table 4.5. The others are inherited from rules that hold for Riemann sums.

Table 4.5 Rules for definite integrals

1. *Order of Integration*:	$\displaystyle\int_b^a f(x)\, dx = -\int_a^b f(x)\, dx$	A definition
2. *Zero*:	$\displaystyle\int_a^a f(x)\, dx = 0$	Also a definition
3. *Constant Multiple*:	$\displaystyle\int_a^b k f(x)\, dx = k \int_a^b f(x)\, dx$	Any number k
	$\displaystyle\int_a^b -f(x)\, dx = -\int_a^b f(x)\, dx$	$k = -1$
4. *Sum and Difference*:	$\displaystyle\int_a^b (f(x) \pm g(x))\, dx = \int_a^b f(x)\, dx \pm \int_a^b g(x)\, dx$	
5. *Additivity*:	$\displaystyle\int_a^b f(x)\, dx + \int_b^c f(x)\, dx = \int_a^c f(x)\, dx$	

6. *Max-Min Inequality*: If max f and min f are the maximum and minimum values of f on $[a, b]$, then

$$\min f \cdot (b - a) \le \int_a^b f(x)\, dx \le \max f \cdot (b - a).$$

7. *Domination*: $f(x) \ge g(x)$ on $[a, b] \Rightarrow \displaystyle\int_a^b f(x)\, dx \ge \int_a^b g(x)\, dx$

$$f(x) \ge 0 \text{ on } [a, b] \Rightarrow \int_a^b f(x)\, dx \ge 0 \qquad \text{(special case)}$$

Example 5 Using the Rules for Definite Integrals

Suppose that

$$\int_{-1}^{1} f(x)\,dx = 5, \qquad \int_{1}^{4} f(x)\,dx = -2, \qquad \text{and} \qquad \int_{-1}^{1} h(x)\,dx = 7.$$

Then

1. $\displaystyle\int_{4}^{1} f(x)\,dx = -\int_{1}^{4} f(x)\,dx = -(-2) = 2$ Rule 2

2. $\displaystyle\int_{-1}^{1} [2f(x) + 3h(x)]\,dx = 2\int_{-1}^{1} f(x)\,dx + 3\int_{-1}^{1} h(x)\,dx$

$$= 2(5) + 3(7) = 31 \qquad \text{Rules 3 and 4}$$

3. $\displaystyle\int_{-1}^{4} f(x)\,dx = \int_{-1}^{1} f(x)\,dx + \int_{1}^{4} f(x)\,dx = 5 + (-2) = 3.$ Rule 5

Proof of Rule 3 Rule 3 says that the integral of k times a function is k times the integral of the function. This is true because

$$\int_{a}^{b} k f(x)\,dx = \lim_{\|P\| \to 0} \sum_{i=1}^{n} k f(c_i)\,\Delta x_i$$

$$= \lim_{\|P\| \to 0} k \sum_{i=1}^{n} f(c_i)\,\Delta x_i$$

$$= k \lim_{\|P\| \to 0} \sum_{i=1}^{n} f(c_i)\,\Delta x_i = k \int_{a}^{b} f(x)\,dx.$$

Figure 4.12 illustrates Rule 5 with a positive function, but the rule applies to any integrable function.

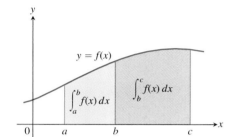

FIGURE 4.12 Additivity for definite integrals:

$$\int_{a}^{b} f(x)\,dx + \int_{b}^{c} f(x)\,dx = \int_{a}^{c} f(x)\,dx$$

$$\int_{b}^{c} f(x)\,dx = \int_{a}^{c} f(x)\,dx - \int_{a}^{b} f(x)\,dx.$$

Proof of Rule 6 Rule 6 says that the integral of f over $[a, b]$ is never smaller than the minimum value of f times the length of the interval and never larger than the maximum value of f times the length of the interval. The reason is that for every partition of $[a, b]$ and for every choice of the points c_k,

$$\min f \cdot (b - a) = \min f \cdot \sum_{k=1}^{n} \Delta x_k \qquad \sum_{k=1}^{n} \Delta x_k = b - a$$

$$= \sum_{k=1}^{n} \min f \cdot \Delta x_k$$

$$\leq \sum_{k=1}^{n} f(c_k)\,\Delta x_k \qquad \min f \leq f(c_k)$$

$$\leq \sum_{k=1}^{n} \max f \cdot \Delta x_k \qquad f(c_k) \leq \max f$$

$$= \max f \cdot \sum_{k=1}^{n} \Delta x_k$$

$$= \max f \cdot (b - a).$$

In short, all Riemann sums for f on $[a, b]$ satisfy the inequality

$$\min f \cdot (b - a) \le \sum_{k=1}^{n} f(c_k) \, \Delta x_k \le \max f \cdot (b - a).$$

Hence, their limit, the integral, does too.

Example 6   Finding Bounds for an Integral

Show that the value of $\int_{0}^{1} \sqrt{1 + \cos x} \, dx$ is less than $3/2$.

Solution   The Max-Min Inequality for definite integrals (Rule 6) says that $\min f \cdot (b - a)$ is a *lower bound* for the value of $\int_{a}^{b} f(x) \, dx$ and that $\max f \cdot (b - a)$ is an *upper bound*. The maximum value of $\sqrt{1 + \cos x}$ on $[0, 1]$ is $\sqrt{1 + 1} = \sqrt{2}$, so

$$\int_{0}^{1} \sqrt{1 + \cos x} \, dx \le \sqrt{2} \cdot (1 - 0) = \sqrt{2}.$$

Since $\int_{0}^{1} \sqrt{1 + \cos x} \, dx$ is bounded from above by $\sqrt{2}$ (which is $1.414 \ldots$), the integral is less than 3.2.

EXERCISES 4.4

Sigma Notation

Write the sums in Exercises 1–6 without sigma notation. Then evaluate them.

1. $\sum_{k=1}^{2} \dfrac{6k}{k + 1}$

2. $\sum_{k=1}^{3} \dfrac{k - 1}{k}$

3. $\sum_{k=1}^{4} \cos k\pi$

4. $\sum_{k=1}^{5} \sin k\pi$

5. $\sum_{k=1}^{3} (-1)^{k+1} \sin \dfrac{\pi}{4}$

6. $\sum_{k=1}^{4} (-1)^{k} \cos k\pi$

Rectangles for Riemann Sums

In Exercises 7–10, graph each function $f(x)$ over the given interval. Partition the interval into four subintervals of equal length. Then add to your sketch the rectangles associated with the Riemann sum $\sum_{k=1}^{4} f(c_k) \, \Delta x_k$, given that c_k is the **(a)** left-hand endpoint, **(b)** right-hand endpoint, and **(c)** midpoint of the kth subinterval. (Make a separate sketch for each set of rectangles.)

7. $f(x) = x^2 - 1, \quad [0, 2]$

8. $f(x) = -x^2, \quad [0, 1]$

9. $f(x) = \sin x, \quad [-\pi, \pi]$

10. $f(x) = \sin x + 1, \quad [-\pi, \pi]$

Expressing Limits as Integrals

Express the limits in Exercises 11–16 as definite integrals.

11. $\lim_{\|P\| \to 0} \sum_{k=1}^{n} c_k^2 \, \Delta x_k$, where P is a partition of $[0, 2]$

12. $\lim_{\|P\| \to 0} \sum_{k=1}^{n} 2c_k^3 \, \Delta x_k$, where P is a partition of $[-1, 0]$

13. $\lim_{\|P\| \to 0} \sum_{k=1}^{n} (c_k^2 - 3c_k) \, \Delta x_k$, where P is a partition of $[-7, 5]$

14. $\lim_{\|P\| \to 0} \sum_{k=1}^{n} \dfrac{1}{1 - c_k} \, \Delta x_k$, where P is a partition of $[2, 3]$

15. $\lim_{\|P\| \to 0} \sum_{k=1}^{n} \sqrt{4 - c_k^2} \, \Delta x_k$, where P is a partition of $[0, 1]$

16. $\lim_{\|P\| \to 0} \sum_{k=1}^{n} (\sec c_k) \, \Delta x_k$, where P is a partition of $[-\pi/4, 0]$

Using Area to Evaluate Integrals

In Exercises 17–22, graph the integrands and use areas to evaluate the integrals.

17. $\int_{-2}^{4} \left(\dfrac{x}{2} + 3 \right) dx$

18. $\int_{-3}^{3} \sqrt{9 - x^2} \, dx$

19. $\int_{-2}^{1} |x|\, dx$

20. $\int_{-1}^{1} (2 - |x|)\, dx$

21. $\int_{0}^{b} x\, dx, \quad b > 0$

22. $\int_{a}^{b} 2s\, ds, \quad 0 < a < b$

Average Value

In Exercises 23–26, find the average value of the function over the given interval, using the geometric method of Example 4.

23. $f(x) = 1 - x$ on $[0, 1]$

24. $f(x) = |x|$ on $[-1, 1]$

25. $f(x) = \sqrt{1 - x^2}$ on $[0, 1]$

26. $f(x) = \sqrt{1 - (x - 2)^2}$ on $[1, 2]$

Using Properties and Known Values to Find Other Integrals

27. Suppose that f and g are continuous and that

$$\int_{1}^{2} f(x)\, dx = -4, \qquad \int_{1}^{5} f(x)\, dx = 6, \qquad \int_{1}^{5} g(x)\, dx = 8.$$

Use the rules in Table 4.5 to find the following.

(a) $\int_{2}^{2} g(x)\, dx$

(b) $\int_{5}^{1} g(x)\, dx$

(c) $\int_{1}^{2} 3f(x)\, dx$

(d) $\int_{2}^{5} f(x)\, dx$

(e) $\int_{1}^{5} [f(x) - g(x)]\, dx$

(f) $\int_{1}^{5} [4f(x) - g(x)]\, dx$

28. Suppose that f and h are continuous and that

$$\int_{1}^{9} f(x)\, dx = -1, \qquad \int_{7}^{9} f(x)\, dx = 5, \qquad \int_{7}^{9} h(x)\, dx = 4.$$

Use the rules in Table 4.5 to find the following.

(a) $\int_{1}^{9} -2 f(x)\, dx$

(b) $\int_{7}^{9} [f(x) + h(x)]\, dx$

(c) $\int_{7}^{9} [2f(x) - 3h(x)]\, dx$

(d) $\int_{9}^{1} f(x)\, dx$

(e) $\int_{1}^{7} f(x)\, dx$

(f) $\int_{9}^{7} [h(x) - f(x)]\, dx$

29. Suppose that $\int_{1}^{2} f(x)\, dx = 5$. Find the following.

(a) $\int_{1}^{2} f(u)\, du$

(b) $\int_{1}^{2} \sqrt{3}\, f(z)\, dz$

(c) $\int_{2}^{1} f(t)\, dt$

(d) $\int_{1}^{2} [-f(x)]\, dx$

30. Suppose that $\int_{-3}^{0} g(t)\, dt = \sqrt{2}$. Find the following.

(a) $\int_{0}^{-3} g(t)\, dt$

(b) $\int_{-3}^{0} g(u)\, du$

(c) $\int_{-3}^{0} [-g(x)]\, dx$

(d) $\int_{-3}^{0} \frac{g(r)}{\sqrt{2}}\, dr$

31. Suppose that f is continuous and that $\int_{0}^{3} f(z)\, dz = 3$ and $\int_{0}^{4} f(z)\, dz = 7$. Find the following.

(a) $\int_{3}^{4} f(z)\, dz$

(b) $\int_{4}^{3} f(t)\, dt$

32. Suppose that h is continuous and that $\int_{-1}^{1} h(r)\, dr = 0$ and $\int_{-1}^{3} h(r)\, dr = 6$. Find the following.

(a) $\int_{1}^{3} h(r)\, dr$

(b) $-\int_{3}^{1} h(u)\, du$

Theory and Examples

33. *Maximizing an integral* What values of a and b maximize the value of

$$\int_{a}^{b} (x - x^2)\, dx\,?$$

34. *Minimizing an integral* What values of a and b minimize the value of

$$\int_{a}^{b} (x^4 - 2x^2)\, dx\,?$$

35. *Writing to Learn* Explain why the rule

$$\int_{a}^{b} k\, dx = k(b - a)$$

holds for *any* constant k.

36. *Integrals of nonnegative functions* Use the max-min inequality to show that if f is integrable then

$$f(x) \geq 0 \quad \text{on} \quad [a, b] \Rightarrow \int_{a}^{b} f(x)\, dx \geq 0.$$

37. *Upper and lower bounds* Use the max-min inequality to find upper and lower bounds for the value of

$$\int_{0}^{1} \frac{1}{1 + x^2}\, dx.$$

38. *Upper and lower bounds* (*Continuation of Exercise 37*) Use the max-min inequality to find upper and lower bounds for

$$\int_{0}^{0.5} \frac{1}{1 + x^2}\, dx \qquad \text{and} \qquad \int_{0.5}^{1} \frac{1}{1 + x^2}\, dx.$$

Add these to arrive at an improved estimate of

$$\int_{0}^{1} \frac{1}{1 + x^2}\, dx.$$

39. *Average speed on a trip* If you average 30 mph on a 150 mi trip and then return over the same 150 mi at the rate of 50 mph, what is your average speed for the round trip? Give reasons for your answer. (*Source*: David H. Pleacher, *The Mathematics Teacher*, Vol. 85, No. 6 (September 1992), pp 445–446.)

40. *Average rate of water release* A dam released 1000 m³ of water at 10 m³/min and then released another 1000 m³ at 20 m³/min. What was the average rate at which the water was released? Give reasons for your answer.

COMPUTER EXPLORATIONS

Finding Riemann Sums

If your CAS can draw rectangles associated with Riemann sums, use it to draw rectangles associated with Riemann sums that converge to

the integrals in Exercises 41–46. Use $n = 4, 10, 20,$ and 50 subintervals of equal length in each case.

41. $\int_0^1 (1 - x)\, dx = \frac{1}{2}$

42. $\int_0^1 (x^2 + 1)\, dx = \frac{4}{3}$

43. $\int_{-\pi}^{\pi} \cos x\, dx = 0$

44. $\int_0^{\pi/4} \sec^2 x\, dx = 1$

45. $\int_{-1}^{1} |x|\, dx = 1$

46. $\int_1^2 \frac{1}{x}\, dx$ (The integral's value is ln 2.)

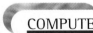 **4.5** ## The Mean Value and Fundamental Theorems

Mean Value Theorem for Definite Integrals • Fundamental Theorem, Part 1 • A Geometric Interpretation • Fundamental Theorem, Part 2 • Area Connection

CD-ROM
WEBsite

Historical Biography

Sir Isaac Newton
(1642—1727)

This section presents two of the most important theorems in integral calculus. The Mean Value Theorem for Definite Integrals asserts that a continuous function on a closed interval assumes its average value at least once in the interval. The Fundamental Theorem connects integration and differentiation and comes in two parts. Its independent discovery by Leibniz and Newton started the mathematical developments that fueled the scientific revolution for the next 200 years and constitutes what is still regarded as the most important computational discovery in the history of the world.

Mean Value Theorem for Definite Integrals

In the previous section, we defined the average value of a continuous function over a closed interval $[a, b]$ as the definite integral $\int_a^b f(x)\, dx$ divided by the length $b - a$ of the interval. The Mean Value Theorem for Definite Integrals asserts that this average value is *always* taken on at least once in the interval. This is no mere coincidence. Look at the graph in Figure 4.13 and imagine rectangles with base $(b - a)$ and heights ranging from the minimum of f (a rectangle too small to give the integral) to the maximum of f (a rectangle too large). Somewhere in between there is a "just right" rectangle, and its topside will intersect the graph of f if f is continuous.

FIGURE 4.13 The value $f(c)$ in the Mean Value Theorem is, in a sense, the average (or *mean*) height of f on $[a, b]$. When $f \geq 0$, the area of the shaded rectangle is the area under the graph of f from a to b,

$$f(c)(b - a) = \int_a^b f(x)\, dx.$$

> **Theorem 2** The Mean Value Theorem for Definite Integrals
>
> If f is continuous on $[a, b]$, then at some point c in $[a, b]$,
>
> $$f(c) = \frac{1}{b - a} \int_a^b f(x)\, dx.$$

FIGURE 4.14 The rectangle with base [0, 3] and with height equal to 5/2 (the average value of the function $f(x) = 4 - x$) has area equal to the area between the graph of f and the x-axis from 0 to 3. (Example 1)

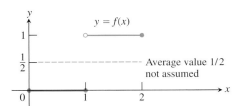

FIGURE 4.15 A discontinuous function need not assume its average value.

Example 1   Applying Theorem 2

Find the average value of $f(x) = 4 - x$ on [0, 3] and where f actually takes on this value at some point in the given domain.

Solution

$$\mathrm{av}\,(f) = \frac{1}{b - a} \int_a^b f(x)\, dx$$

$$= \frac{1}{3 - 0} \int_0^3 (4 - x)\, dx = \frac{1}{3}\left(\int_0^3 4\, dx - \int_0^3 x\, dx \right)$$

$$= \frac{1}{3}\left(4(3 - 0) - \left(\frac{3^2}{2} - \frac{0^2}{2} \right) \right) \qquad \text{Section 4.4, Example 3}$$

$$= 4 - \frac{3}{2} = \frac{5}{2}.$$

The average value of $f(x) = 4 - x$ over [0, 3] is 5/2. The function assumes this value when $4 - x = 5/2$ or $x = 3/2$ (Figure 4.14).

In Example 1, we found a point where f assumed its average value by setting $f(x)$ equal to the calculated average value and solving for x, but this does not prove that such a point will always exist. It proves only that it existed in Example 1. To prove Theorem 2, we need a more general argument.

Proof of Theorem 2   If we divide both sides of the max-min inequality (Table 4.5, Rule 6) by $(b - a)$, we obtain

$$\min f \leq \frac{1}{b - a} \int_a^b f(x)\, dx \leq \max f.$$

Since f is continuous, the Intermediate Value Theorem for Continuous Functions (Section 1.4) says that f must assume every value between $\min f$ and $\max f$. It must therefore assume the value $(1/(b - a)) \int_a^b f(x)\, dx$ at some point c in $[a, b]$.   ▬

The continuity of f is important here. A discontinuous function can step over its average value (Figure 4.15).

What else can we learn from Theorem 2? Here is an example.

Example 2   Average Value of Zero

Show that if f is continuous on $[a, b]$, $a \neq b$, and if

$$\int_a^b f(x)\, dx = 0,$$

then $f(x) = 0$ at least once in $[a, b]$.

Solution   The average value of f on $[a, b]$ is

$$\mathrm{av}\,(f) = \frac{1}{b - a} \int_a^b f(x)\, dx = \frac{1}{b - a} \cdot 0 = 0.$$

By Theorem 2, f assumes this value at some point c in $[a, b]$.

Fundamental Theorem, Part 1

If $f(t)$ is an integrable function, the integral from any fixed number a to another number x defines a function F whose value at x is

$$F(x) = \int_a^x f(t)\, dt. \tag{1}$$

For example, if f is nonnegative and x lies to the right of a, $F(x)$ is the area under the graph from a to x. The variable x is the upper limit of integration of an integral, but F is just like any other real-valued function of a real variable. For each value of the input x, there is a well-defined numerical output, in this case the integral of f from a to x.

Equation (1) gives an important way to define new functions and to describe solutions of differential equations (more about this later). The reason for mentioning Equation (1) now, however, is the connection it makes between integrals and derivatives. If f is any continuous function whatever, then F is a differentiable function of x whose derivative is f itself. At every value of x,

$$\frac{d}{dx} F(x) = \frac{d}{dx} \int_a^x f(t)\, dt = f(x).$$

This idea is so important that it is the first part of the Fundamental Theorem of Calculus.

Theorem 3   The Fundamental Theorem of Calculus, Part 1

If f is continuous on $[a, b]$, then the function

$$F(x) = \int_a^x f(t)\, dt$$

has a derivative at every point x in $[a, b]$, and

$$\frac{dF}{dx} = \frac{d}{dx} \int_a^x f(t)\, dt = f(x). \tag{2}$$

This conclusion is beautiful, powerful, deep, and surprising, and Equation (2) may well be one of the most important equations in mathematics. It says that the differential equation $dF/dx = f$ has a solution for every continuous function f. It says that every continuous function f is the derivative of some other function, namely $\int_a^x f(t)\, dt$. It says that every continuous function has an antiderivative. And it says that the process of integration and differentiation are inverses of one another.

CD-ROM
WEBsite

Example 3   Applying the Fundamental Theorem

Find

$$\frac{d}{dx} \int_{-\pi}^x \cos t\, dt \quad \text{and} \quad \frac{d}{dx} \int_0^x \frac{1}{1+t^2}\, dt$$

by using the Fundamental Theorem.

Solution

$$\frac{d}{dx}\int_{-\pi}^{x}\cos t\,dt = \cos x \qquad\qquad \text{Eq. 2 with } f(t) = \cos t$$

$$\frac{d}{dx}\int_{0}^{x}\frac{1}{1+t^{2}}\,dt = \frac{1}{1+x^{2}}. \qquad \text{Eq. 2 with } f(t) = \frac{1}{1+t^{2}}.$$

Example 4   Applying the Fundamental Theorem With the Chain Rule

Find dy/dx if $y = \int_{1}^{x^{2}}\cos t\,dt$.

Solution   The upper limit of integration is not x but x^{2}. This makes y a composite of

$$y = \int_{1}^{u}\cos t\,dt \qquad \text{and} \qquad u = x^{2}.$$

We must therefore apply the Chain Rule when finding dy/dx.

$$\frac{dy}{dx} = \frac{dy}{du}\cdot\frac{du}{dx}$$

$$= \left(\frac{d}{du}\int_{1}^{u}\cos t\,dt\right)\cdot\frac{du}{dx}$$

$$= \cos u\cdot\frac{du}{dx}$$

$$= \cos(x^{2})\cdot 2x$$

$$= 2x\,\cos x^{2}$$

Example 5   Variable Lower Limits of Integration

Find dy/dx.

(a) $y = \int_{x}^{5} 3t\sin t\,dt$ **(b)** $y = \int_{1+3x^{2}}^{4}\frac{1}{2+e^{t}}\,dt$

Solution   Rule 1 for integrals in Section 4.4 sets these up for the Fundamental Theorem.

(a) $\quad \dfrac{d}{dx}\displaystyle\int_{x}^{5} 3t\sin t\,dt = \dfrac{d}{dx}\left(-\int_{5}^{x} 3t\sin t\,dt\right)$ Rule 1

$$= -\frac{d}{dx}\int_{5}^{x} 3t\sin t\,dt$$

$$= -3x\,\sin x$$

(b) $\dfrac{d}{dx}\displaystyle\int_{1+3x^{2}}^{4}\frac{1}{2+e^{t}}\,dt = \dfrac{d}{dx}\left(-\int_{4}^{1+3x^{2}}\frac{1}{2+e^{t}}\,dt\right)$ Rule 1

$$= -\frac{d}{dx}\int_{4}^{1+3x^{2}}\frac{1}{2+e^{t}}\,dt$$

$$= -\frac{1}{2+e^{(1+3x^{2})}}\frac{d}{dx}(1+3x^{2}) \qquad\quad \begin{array}{l}\text{Eq. (2) and the Chain}\\ \text{Rule}\end{array}$$

$$= -\frac{6x}{2+e^{(1+3x^{2})}}$$

Example 6   Constructing a Function With A Given Derivative and Value

Find a function $y = f(x)$ with derivative

$$\frac{dy}{dx} = \tan x$$

that satisfies the condition $f(3) = 5$.

Solution   The Fundamental Theorem makes it easy to construct a function with derivative $\tan x$:

$$y = \int_{3}^{x} \tan t\, dt.$$

Since $y(3) = 0$, we have only to add 5 to this function to construct one with derivative $\tan x$ whose value at $x = 3$ is 5:

$$f(x) = \int_{3}^{x} \tan t\, dt + 5.$$

Although the solution to the problem in Example 6 satisfies the two required conditions, you might question whether it is in a useful form. Not many years ago, this form might have posed a computation problem. Indeed, for such problems much effort has been expended over the centuries trying to find solutions that do not involve integrals. Since $\ln |\cos t|$ is an antiderivative of $\tan t$, you will see momentarily that we can write the solution in Example 6 as

$$y = \ln \left| \frac{\cos 3}{\cos x} \right| + 5.$$

Now that computers and calculators are capable of evaluating integrals, however, the form given in Example 6 is not only useful, but in some ways preferable. It is certainly easier to find and is always available.

A Geometric Interpretation

If the values of f are positive, the equation

$$\frac{d}{dx} \int_{a}^{x} f(t)\, dt = f(x)$$

has a nice geometric interpretation: The integral of f from a to x is the area $A(x)$ of the region between the graph of f and the x-axis from a to x. Imagine the area swept out by a wiper blade clearing the raindrops from the windshield on a bus. As the blade moves past x, the rate at which the cleared area is being swept out is precisely the height of the vertical blade $f(x)$ (Figure 4.16).

Proof of Theorem 3   We prove Theorem 3 by applying the definition of derivative directly to the function $F(x)$. This means writing out the difference quotient

$$\frac{F(x + h) - F(x)}{h} \tag{3}$$

and showing that its limit as $h \to 0$ is the number $f(x)$.

FIGURE 4.16 The rate at which the wiper blade on a bus clears the windshield of rain as the blade moves past x is the height of the blade. In symbols, $dA/dx = f(x)$.

When we replace $F(x + h)$ and $F(x)$ by their defining integrals, the numerator in Equation (3) becomes

$$F(x + h) - F(x) = \int_a^{x+h} f(t)\, dt - \int_a^x f(t)\, dt.$$

The Additivity Rule for integrals (Table 4.5, Rule 5) simplifies the right-hand side to

$$\int_x^{x+h} f(t)\, dt,$$

so that Equation (3) becomes

$$\frac{F(x + h) - F(x)}{h} = \frac{1}{h}[F(x + h) - F(x)]$$

$$= \frac{1}{h}\int_x^{x+h} f(t)\, dt. \qquad (4)$$

According to the Mean Value Theorem for Definite Integrals, the value of the last expression in Equation (4) is one of the values taken on by f in the interval joining x and $x + h$. That is, for some number c in this interval,

$$\frac{1}{h}\int_x^{x+h} f(t)\, dt = f(c). \qquad (5)$$

We can therefore find out what happens to $(1/h)$ times the integral as $h \to 0$ by watching what happens to $f(c)$ as $h \to 0$.

What does happen to $f(c)$ as $h \to 0$? As $h \to 0$, the endpoint $x + h$ approaches x, pushing c ahead of it like a bead on a wire.

So c approaches x, and, since f is continuous at x, $f(c)$ approaches $f(x)$:

$$\lim_{h \to 0} f(c) = f(x). \qquad (6)$$

Going back to the beginning, then, we have

$$\frac{dF}{dx} = \lim_{h \to 0} \frac{F(x + h) - F(x)}{h} \qquad \text{Definition of derivative}$$

$$= \lim_{h \to 0} \frac{1}{h}\int_x^{x+h} f(t)\, dt \qquad \text{Eq. (4)}$$

$$= \lim_{h \to 0} f(c) \qquad \text{Eq. (5)}$$

$$= f(x). \qquad \text{Eq. (6)}$$

This concludes the proof.

Fundamental Theorem, Part 2

The second part of the Fundamental Theorem of Calculus shows how to evaluate definite integrals directly from antiderivatives.

CD-ROM
WEBsite

Theorem 3 (continued)   The Fundamental Theorem of Calculus, Part 2

If f is continuous at every point of $[a, b]$ and if F is any antiderivative of f on $[a, b]$, then

$$\int_a^b f(x)\,dx = F(b) - F(a).$$

This part of the Fundamental Theorem is also called the **Integral Evaluation Theorem.**

Proof   Part 1 of the Fundamental Theorem tells us that an antiderivative of f exists, namely

$$G(x) = \int_a^x f(t)\,dt.$$

Thus, if F is *any* antiderivative of f, then $F(x) = G(x) + C$ for some constant C (by Corollary 2 of the Mean Value Theorem for Derivatives, Section 3.2).

Evaluating $F(b) - F(a)$, we have

$$F(b) - F(a) = [G(b) + C] - [G(a) + C]$$

$$= G(b) - G(a)$$

$$= \int_a^b f(t)\,dt - \int_a^a f(t)\,dt$$

$$= \int_a^b f(t)\,dt - 0$$

$$= \int_a^b f(t)\,dt.$$

At the risk of repeating ourselves: It is difficult to overestimate the power of the simple equation

How to Evaluate $\int_a^b f(x)\,dx$

Step 1. Find an antiderivative F of f. Any antiderivative will do, so pick the simplest one you can.

Step 2. Calculate the number $F(b) - F(a)$.

This number will be $\int_a^b f(x)\,dx$.

$$\int_a^b f(x)\,dx = F(b) - F(a).$$

It says that any definite integral of any continuous function f can be calculated without taking limits, without calculating Riemann sums, and often without effort, as long as an antiderivative of f can be found. If you can imagine what it was like before this theorem (and before computing machines), when approximations by tedious sums were the only alternative for solving many real-world problems, then you can imagine what a miracle calculus was thought to be. If any equation deserves to be called the Fundamental Theorem of Calculus, this equation is surely the (second) one.

Integral Evaluation Notation

The usual notation for $F(b) - F(a)$ is

$$F(x)\Big]_a^b \quad \text{or} \quad \Big[F(x)\Big]_a^b,$$

depending on whether F has one or more terms. This notation provides a compact "recipe" for the evaluation, allowing us to show the antiderivative in an intermediate step.

Example 7   Evaluating an Integral

Evaluate $\int_{-1}^{3} (x^3 + 1)\, dx$ using an antiderivative.

Solution

A simple antiderivative of $x^3 + 1$ is $(x^4/4) + x$. Therefore,

$$\int_{-1}^{3} (x^3 + 1)\, dx = \left[\frac{x^4}{4} + x \right]_{-1}^{3}$$

$$= \left(\frac{81}{4} + 3 \right) - \left(\frac{1}{4} - 1 \right)$$

$$= 24.$$

Area Connection

We can now compute areas using antiderivatives, but we must be careful to distinguish net area (in which area below the x-axis is counted as negative) from total area. The unmodified word "area" will be taken to mean *total area*.

Example 8   Finding Area Using Antiderivatives

Find the area of the region between the x-axis and the graph of $f(x) = x^3 - x^2 - 2x$, $-1 \le x \le 2$.

Solution   First find the zeros of f. Since

$$f(x) = x^3 - x^2 - 2x = x(x^2 - x - 2) = x(x + 1)(x - 2),$$

the zeros are $x = 0, -1,$ and 2 (Figure 4.17). The zeros partition $[-1, 2]$ into two subintervals: $[-1, 0]$, on which $f \ge 0$ and $[0, 2]$, on which $f \le 0$. We integrate f over each subinterval and add the absolute values of the calculated values.

Integral over $[-1, 0]$: $\displaystyle\int_{-1}^{0} (x^3 - x^2 - 2x)\, dx = \left[\frac{x^4}{4} - \frac{x^3}{3} - x^2 \right]_{-1}^{0}$

$$= 0 - \left[\frac{1}{4} + \frac{1}{3} - 1 \right] = \frac{5}{12}$$

Integral over $[0, 2]$: $\displaystyle\int_{0}^{2} (x^3 - x^2 - 2x)\, dx = \left[\frac{x^4}{4} - \frac{x^3}{3} - x^2 \right]_{0}^{2}$

$$= \left[4 - \frac{8}{3} - 4 \right] - 0 = -\frac{8}{3}$$

Enclosed area: Total enclosed area $= \dfrac{5}{12} + \left| -\dfrac{8}{3} \right| = \dfrac{37}{12}$

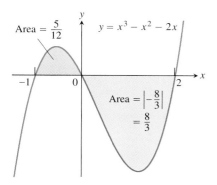

FIGURE 4.17 The region between the curve $y = x^3 - x^2 - 2x$ and the x-axis. (Example 8)

How to Find Total Area

To find the area between the graph of $y = f(x)$ and the x-axis over the interval $[a, b]$ analytically, do the following.

Step 1. Partition $[a, b]$ with the zeros of f.

Step 2. Integrate f over each subinterval.

Step 3. Add the absolute values of the integrals.

Example 9   Household Electricity

We model the voltage in our home wiring with the sine function

$$V = V_{\max} \sin 120\pi t,$$

which expresses the voltage V in volts as a function of time t in seconds. The function runs through 60 cycles each second (its frequency is 60 hertz, or 60 Hz). The positive constant $V_{\max}$ ("vee max") is the **peak voltage.**

The average value of V over the half-cycle from 0 to $1/120$ sec (see Figure 4.18) is

$$V_{\mathrm{av}} = \frac{1}{(1/120) - 0} \int_{0}^{1/120} V_{\max} \sin 120\pi t \; dt$$

$$= 120 \, V_{\max} \left[-\frac{1}{120\pi} \cos 120\pi t \right]_{0}^{1/120}$$

$$= \frac{V_{\max}}{\pi} [-\cos \pi + \cos 0]$$

$$= \frac{2V_{\max}}{\pi}.$$

The average value of the voltage over a full cycle, as we can see from Figure 4.18, is zero. (Also see Exercise 52.) If we measured the voltage with a standard moving-coil galvanometer, the meter would read zero.

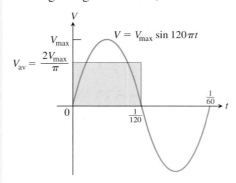

FIGURE 4.18 The graph of the voltage $V = V_{\max} \sin 120\pi t$ over a full cycle. Its average value over a half-cycle is $2V_{\max}/\pi$. Its average value over a full cycle is zero. (Example 9)

To measure the voltage effectively, we use an instrument that measures the square root of the average value of the square of the voltage, namely

$$V_{\mathrm{rms}} = \sqrt{(V^2)_{\mathrm{av}}} \; .$$

The subscript "rms" (read the letters separately) stands for "root mean square." Since the average value of $V^2 = (V_{max})^2 \sin^2 120\pi t$ over a cycle is

$$(V^2)_{av} = \frac{1}{(1/60) - 0} \int_0^{1/60} (V_{max})^2 \sin^2 120\pi t \, dt = \frac{(V_{max})^2}{2} \qquad (7)$$

(Exercise 52, part (c)), the rms voltage is

$$V_{rms} = \sqrt{\frac{(V_{max})^2}{2}} = \frac{V_{max}}{\sqrt{2}}. \qquad (8)$$

The values given for household currents and voltages are always rms values. Thus, "115 volts ac" means that the rms voltage is 115. The peak voltage,

$$V_{max} = \sqrt{2} \, V_{rms} = \sqrt{2} \cdot 115 \approx 163 \text{ volts},$$

obtained from Equation (8), is considerably higher.

EXERCISES 4.5

Evaluating Integrals

Evaluate the integrals in Exercises 1–14.

1. $\displaystyle\int_{-2}^{0} (2x + 5) \, dx$

2. $\displaystyle\int_{0}^{4} \left(3x - \frac{x^3}{4} \right) dx$

3. $\displaystyle\int_{0}^{1} (x^2 + \sqrt{x}) \, dx$

4. $\displaystyle\int_{-2}^{-1} \frac{2}{x^2} \, dx$

5. $\displaystyle\int_{0}^{\pi} (1 + \cos x) \, dx$

6. $\displaystyle\int_{0}^{\pi/3} 2 \sec^2 x \, dx$

7. $\displaystyle\int_{\pi/4}^{3\pi/4} \csc \theta \cot \theta \, d\theta$

8. $\displaystyle\int_{0}^{\pi/2} \frac{1 + \cos 2t}{2} \, dt$

9. $\displaystyle\int_{-\pi/2}^{\pi/2} (8y^2 + \sin y) \, dy$

10. $\displaystyle\int_{-1}^{1} (r + 1)^2 \, dr$

11. $\displaystyle\int_{1}^{\sqrt{2}} \left(\frac{u^2}{2} - \frac{1}{u^5} \right) du$

12. $\displaystyle\int_{4}^{9} \frac{1 - \sqrt{u}}{\sqrt{u}} \, du$

13. $\displaystyle\int_{0}^{1} x e^{x^2} \, dx$

14. $\displaystyle\int_{0}^{\ln 2} e^{3x} \, dx$

Derivatives of Integrals

Find the derivatives in Exercises 15–18

(a) by evaluating the integral and differentiating the result.

(b) by differentiating the integral directly.

15. $\displaystyle\frac{d}{dx} \int_{0}^{\sqrt{x}} \cos t \, dt$

16. $\displaystyle\frac{d}{dx} \int_{1}^{\sin x} 3t^2 \, dt$

17. $\displaystyle\frac{d}{dt} \int_{0}^{t^4} \sqrt{u} \, du$

18. $\displaystyle\frac{d}{d\theta} \int_{0}^{\tan \theta} \sec^2 y \, dy$

Find dy/dx in Exercises 19–24.

19. $\displaystyle y = \int_{0}^{x} \sqrt{1 + t^2} \, dt$

20. $\displaystyle y = \int_{1}^{x} \frac{1}{t} \, dt, \quad x > 0$

21. $\displaystyle y = \int_{\sqrt{x}}^{0} \sin (t^2) \, dt$

22. $\displaystyle y = \int_{0}^{x^2} \cos \sqrt{t} \, dt$

23. $\displaystyle y = \int_{1}^{x^{1/3}} e^{(t^3 + 1)} \, dt$

24. $\displaystyle y = \int_{e^x}^{e} \ln t \, dt, \quad x > 1$

Evaluating Integrals Using Substitutions

In Exercises 25–28, use a substitution to find an antiderivative and then apply the Fundamental Theorem to evaluate the integral.

25. $\displaystyle\int_{0}^{1} (1 - 2x)^3 \, dx$

26. $\displaystyle\int_{0}^{1} t \sqrt{t^2 + 1} \, dt$

27. $\displaystyle\int_{0}^{\pi} \sin^2 \left(1 + \frac{\theta}{2} \right) d\theta$

28. $\displaystyle\int_{0}^{\pi} \sin^2 \frac{x}{4} \cos \frac{x}{4} \, dx$

Initial Value Problems

Solve the initial value problems in Exercises 29–32.

29. $\displaystyle\frac{dy}{dx} = \sec x, \quad y(2) = 3$

30. $\dfrac{dy}{dx} = x\sqrt{1 + x^2}, \quad y(1) = -2$

31. $\dfrac{dy}{dt} = e^t \sin (e^t - 2), \quad y(\ln 2) = 0$

32. $\dfrac{d^2y}{dt^2} = 1 - e^{2t}, \quad y(1) = -1 \text{ and } y'(1) = 0$

Area

In Exercises 33–36, find the total area of the region between the curve and the x-axis.

33. $y = -x^2 - 2x, \quad -3 \le x \le 2$

34. $y = x^3 - 3x^2 + 2x, \quad 0 \le x \le 2$

35. $y = x^3 - 4x, \quad -2 \le x \le 2$

36. $y = x^{1/3} - x, \quad -1 \le x \le 8$

Find the areas of the shaded regions in Exercises 37 and 38.

37.

38.

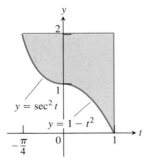

Applications

39. *Cost from marginal cost* The marginal cost of printing a poster when x posters have been printed is

$$\frac{dc}{dx} = \frac{1}{2\sqrt{x}}$$

dollars. Find

(a) $c(100) - c(1)$, the cost of printing posters 2–100

(b) $c(400) - c(100)$, the cost of printing posters 101–400.

40. *Revenue from marginal revenue* Suppose that a company's marginal revenue from the manufacture and sale of egg beaters is

$$\frac{dr}{dx} = 2 - \frac{2}{(x + 1)^2},$$

where r is measured in thousands of dollars and x in thousands of units. How much money should the company expect from a production run of $x = 3$ thousand egg beaters? To find out, integrate the marginal revenue from $x = 0$ to $x = 3$.

Drawing Conclusions about Motion from Graphs

41. *Writing to Learn* Suppose that f is the differentiable function shown in the accompanying graph and that the position at time t (seconds) of a particle moving along a coordinate axis is

$$s(t) = \int_0^t f(x) \, dx$$

meters. Use the graph to answer the following questions. Give reasons for your answers.

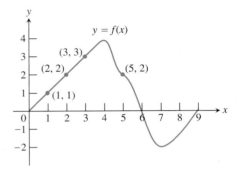

(a) What is the particle's velocity at time $t = 5$?

(b) Is the acceleration of the particle at time $t = 5$ positive or negative?

(c) What is the particle's position at time $t = 3$?

(d) At what time during the first 9 sec does s have its largest value?

(e) Approximately when is the acceleration zero?

(f) When is the particle moving toward the origin? Away from the origin?

(g) On which side of the origin does the particle lie at time $t = 9$?

42. *Writing to Learn* Suppose that g is the differentiable function shown in the accompanying graph and that the position at time t (sec) of a particle moving along a coordinate axis is

$$s(t) = \int_0^t g(x) \, dx$$

meters. Use the graph to answer the following questions. Give reasons for your answers.

(a) What is the particle's velocity at $t = 3$?

(b) Is the acceleration at time $t = 3$ positive or negative?

(c) What is the particle's position at time $t = 3$?

(d) When does the particle pass through the origin?

(e) When is the acceleration zero?

(f) When is the particle moving away from the origin? Toward the origin?

(g) On which side of the origin does the particle lie at $t = 9$?

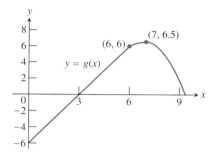

Volumes from Section 4.3

43. (*Continuation of Section 4.3, Exercise 11*) The approximating sums for the volume of water in Exercise 11, Section 4.3, are Riemann sums for an integral. What integral? Evaluate it to find the volume.

44. (*Continuation of Section 4.3, Exercise 13*) The approximating sums for the volume of the rocket nose cone in Exercise 13, Section 4.3, is a Riemann sum for an integral. What integral? Evaluate it to find the volume.

Theory and Examples

45. Suppose that $\int_1^x f(t)\,dt = x^2 - 2x + 1$. Find $f(x)$.

46. Find $f(4)$ if $\int_0^x f(t)\,dt = x \cos \pi x$.

47. *Linearization* Find the linearization of

$$f(x) = 2 - \int_2^{x+1} \frac{9}{1+t}\,dt$$

at $x = 1$.

48. *Linearization* Find the linearization of

$$g(x) = 3 + \int_1^{x^2} \sec(t - 1)\,dt$$

at $x = -1$.

49. *Writing to Learn* Suppose that f has a positive derivative for all values of x and that $f(1) = 0$. Which of the following statements must be true of the function

$$g(x) = \int_0^x f(t)\,dt?$$

Give reasons for your answers.

(a) g is a differentiable function of x.

(b) g is a continuous function of x.

(c) The graph of g has a horizontal tangent at $x = 1$.

(d) g has a local maximum at $x = 1$.

(e) g has a local minimum at $x = 1$.

(f) The graph of g has an inflection point at $x = 1$.

(g) The graph of dg/dx crosses the x-axis at $x = 1$.

50. *Writing to Learn* Suppose that f has a negative derivative for all values of x and that $f(1) = 0$. Which of the following statements must be true of the function

$$h(x) = \int_0^x f(t)\,dt?$$

Give reasons for your answers.

(a) h is a twice-differentiable function of x.

(b) h and dh/dx are both continuous.

(c) The graph of h has a horizontal tangent at $x = 1$.

(d) h has a local maximum at $x = 1$.

(e) h has a local minimum at $x = 1$.

(f) The graph of h has an inflection point at $x = 1$.

(g) The graph of dh/dx crosses the x-axis at $x = 1$.

51. *Archimedes' area formula for parabolas* Archimedes (287–212 B.C.), inventor, military engineer, physicist, and the greatest mathematician of classical times in the western world, discovered that the area under a parabolic arch is two-thirds the base times the height.

(a) Use an integral to find the area under the arch

$$y = 6 - x - x^2, \quad -3 \le x \le 2.$$

(b) Find the height of the arch.

(c) Show that the area is two-thirds the base b times the height h.

(d) Sketch the parabolic arch $y = h - (4h/b^2)x^2$, $-b/2 \le x \le b/2$, assuming that h and b are positive. Then use calculus to find the area of the region enclosed between the arch and the x-axis.

52. (*Continuation of Example 9*) *Household electricity*

(a) Show by evaluating the integral in the expression

$$\frac{1}{(1/60) - 0} \int_0^{1/60} V_{max} \sin 120\pi t\,dt$$

that the average value of $V = V_{max} \sin 120\pi t$ over a full cycle is zero.

(b) The circuit that runs your electric stove is rated 240 volts rms. What is the peak value of the allowable voltage?

(c) Show that

$$\int_0^{1/60} (V_{max})^2 \sin^2 120\pi t\,dt = \frac{(V_{max})^2}{120}.$$

T **53.** *The Fundamental Theorem* If f is continuous, we expect

$$\lim_{h \to 0} \frac{1}{h} \int_x^{x+h} f(t)\, dt$$

to equal $f(x)$, as in the proof of Part 1 of the Fundamental Theorem. For instance, if $f(t) = \cos t$, then

$$\frac{1}{h} \int_x^{x+h} \cos t\, dt = \frac{\sin(x+h) - \sin x}{h}. \qquad (9)$$

The right-hand side of Equation (9) is the difference quotient for the derivative of the sine, and we expect its limit as $h \to 0$ to be $\cos x$.

Graph $\cos x$ for $-\pi \le x \le 2\pi$. Then, in a different color if possible, graph the right-hand side of Equation (9) as a function of x for $h = 2, 1, 0.5,$ and 0.1. Watch how the latter curves converge to the graph of the cosine as $h \to 0$.

T **54.** Repeat Exercise 53 for $f(t) = 3t^2$. What is

$$\lim_{h \to 0} \frac{1}{h} \int_x^{x+h} 3t^2\, dt = \lim_{h \to 0} \frac{(x+h)^3 - x^3}{h}?$$

Graph $f(x) = 3x^2$ for $-1 \le x \le 1$. Then graph the quotient $((x+h)^3 - x^3)/h$ as a function of x for $h = 1, 0.5, 0.2,$ and 0.1. Watch how the latter curves converge to the graph of $3x^2$ as $h \to 0$.

COMPUTER EXPLORATIONS

In Exercises 55–58, let $F(x) = \int_a^x f(t)\, dt$ for the specified function f and interval $[a, b]$. Use a CAS to perform the following steps and answer the questions posed.

(a) Plot the functions f and F together over $[a, b]$.

(b) Solve the equation $F'(x) = 0$. What can you see to be true about the graphs of f and F at points where $F'(x) = 0$? Is your observation borne out by Part 1 of the Fundamental Theorem coupled with information provided by the derivative? Explain your answer.

(c) Over what intervals (approximately) is the function F increasing and decreasing? What is true about f over those intervals?

(d) Calculate the derivative f' and plot it together with F. What can you see to be true about the graph of F at points where $f'(x) = 0$? Is your observation borne out by Part 1 of the Fundamental Theorem? Explain your answer.

55. $f(x) = x^3 - 4x^2 + 3x, \quad [0, 4]$

56. $f(x) = 2x^4 - 17x^3 + 46x^2 - 43x + 12, \quad [0, 9/2]$

57. $f(x) = \sin 2x \cos \dfrac{x}{3}, \quad [0, 2\pi]$

58. $f(x) = x \cos \pi x, \quad [0, 2\pi]$

In Exercises 59–64, let $F(x) = \int_a^{u(x)} f(t)\, dt$ for the specified a, u, and f. Use a CAS to perform the following steps and answer the questions posed.

(a) Find the domain of F.

(b) Calculate $F'(x)$ and determine its zeros. Over what intervals is F increasing? decreasing?

(c) Calculate $F''(x)$ and determine its zero. Identify the local extrema and the points of inflection of F.

(d) Using the information from parts (a) through (c), draw a rough sketch of $y = F(x)$ over its domain. Then graph $F(x)$ on your CAS to support your sketch.

59. $a = 1, \quad u(x) = x^2, \quad f(x) = \sqrt{1 - x^2}$

60. $a = 0, \quad u(x) = x^2, \quad f(x) = \sqrt{1 - x^2}$

61. $a = 0, \quad u(x) = 1 - x, \quad f(x) = x^2 - 2x - 3$

62. $a = 0, \quad u(x) = 1 - x^2, \quad f(x) = x^2 - 2x - 3$

63. Calculate $(d/dx) \displaystyle\int_a^{u(x)} f(t)\, dt$ and check your answer using a CAS.

64. Calculate $(d^2/dx^2) \displaystyle\int_a^{u(x)} f(t)\, dt$ and check your answer using a CAS.

4.6 Substitution in Definite Integrals

Substitution Formula • Area Between Curves • Boundaries with Changing Formulas

There are two methods for evaluating a definite integral by substitution, and they both work well. One is to find the corresponding indefinite integral by substitution and use one of the resulting antiderivatives to evaluate the definite integral by the Fundamental Theorem. The other is to use a formula we now study.

Substitution Formula

CD-ROM
WEBsite

Historical Biography

Isaac Barrow
(1630—1677)

Substitution in Definite Integrals

THE FORMULA

$$\int_a^b f(g(x)) \cdot g'(x)\, dx = \int_{g(a)}^{g(b)} f(u)\, du \qquad (1)$$

HOW TO USE IT

Substitute $u = g(x)$, $du = g'(x)\, dx$, and integrate from $g(a)$ to $g(b)$.

To use the formula, make the same u-substitution you would use to evaluate the corresponding indefinite integral. Then integrate with respect to u from the value u has at $x = a$ to the value u has at $x = b$.

To see why Equation (1) is true, we let F denote any antiderivative of f. Then

$$\int_a^b f(g(x)) \cdot g'(x)\, dx = F(g(x)) \Big]_{x=a}^{x=b} \qquad \begin{aligned} &\frac{d}{dx}F(g(x)) \\ &= F'(g(x))\, g'(x) \\ &= f(g(x))g'(x) \end{aligned}$$

$$= F(g(b)) - F(g(a))$$

$$= F(u) \Big]_{u=g(a)}^{u=g(b)}$$

$$= \int_{g(a)}^{g(b)} f(u)\, du. \qquad \text{Fundamental Theorem, Part 2}$$

Example 1   Using the Substitution Formula

Use Equation (1) to evaluate

$$\int_{-1}^{1} 3x^2 \sqrt{x^3 + 1}\, dx$$

Solution
Transform the integral and evaluate the transformed integral with the transformed limits given by Equation (1).

$$\int_{-1}^{1} 3x^2 \sqrt{x^3 + 1}\, dx$$

$$= \int_0^2 \sqrt{u}\, du \qquad \begin{aligned} &\text{Let } u = x^3 + 1,\, du = 3x^2\, dx. \\ &\text{When } x = -1,\, u = (-1)^3 + 1 = 0. \\ &\text{When } x = 1,\, u = (1)^3 + 1 = 2. \end{aligned}$$

$$= \frac{2}{3}\, u^{2/3} \Big]_0^2 \qquad \text{Evaluate the new definite integral.}$$

$$= \frac{2}{3} [2^{3/2} - 0^{3/2}] = \frac{2}{3} [2\sqrt{2}] = \frac{4\sqrt{2}}{3}$$

Instead of using The Substitution Formula we could use antiderivatives and the Fundamental Theorem.

Example 2   Evaluation without Using the Formula

Evaluate

$$\int_{-1}^{1} 3x^2 \sqrt{x^3 + 1}\; dx$$

by transforming the integral as an indefinite integral, integrating, changing back to x, and using the original x-limits.

Solution

$$\int 3x^2 \sqrt{x^3 + 1}\; dx = \int \sqrt{u}\; du \qquad \text{Let } u = x^3 + 1,\; du = 3x^2\; dx.$$

$$= \frac{2}{3} u^{3/2} + C \qquad \text{Integrate with respect to } u.$$

$$= \frac{2}{3} (x^3 + 1)^{3/2} + C \qquad \text{Replace } u \text{ by } x^3 + 1.$$

$$\int_{-1}^{1} 3x^2 \sqrt{x^3 + 1}\; dx = \frac{2}{3} (x^3 + 1)^{3/2} \bigg]_{-1}^{1} \qquad \begin{array}{l}\text{Use the integral just found, with}\\ \text{limits of integration for } x.\end{array}$$

$$= \frac{2}{3} [((1)^3 + 1)^{3/2} - ((-1)^3 + 1)^{3/2}]$$

$$= \frac{2}{3} [2^{3/2} - 0^{3/2}] = \frac{2}{3} [2\sqrt{2}] = \frac{4\sqrt{2}}{3}$$

Which method is better, transforming the integral, integrating, and transforming back to use the original limits of integration or evaluating the transformed integral with transformed limits? For the integrand $3x^2 \sqrt{x^3 + 1}$, using the Substitution Formula in Example 1 seems easier, but that is not always the case. As a rule, it is best to know both methods and to use whichever one seems better at the time.

Here is another example of evaluating a transformed integral with transformed limits.

Example 3   Using the Substitution Formula

$$\int_{\pi/4}^{\pi/2} \cot \theta \csc^2 \theta\; d\theta = \int_{1}^{0} u \cdot (-du) \qquad \begin{array}{l}\text{Let } u = \cot \theta,\; du = -\csc^2 \theta\; d\theta,\\ -du = \csc^2 \theta\; d\theta.\\ \text{When } \theta = \pi/4,\; u = \cot (\pi/4) = 1.\\ \text{When } \theta = \pi/2,\; u = \cot (\pi/2) = 0.\end{array}$$

$$= -\int_{1}^{0} u\; du$$

$$= -\frac{u^2}{2} \bigg]_{1}^{0}$$

$$= -\left[\frac{(0)^2}{2} - \frac{(1)^2}{2} \right] = \frac{1}{2}$$

USING TECHNOLOGY   *Visualizing Integrals with Elusive Antiderivatives*
Many integrable functions, such as the important

$$f(x) = e^{-x^2}$$

from probability theory, *do not* have antiderivatives that can be expressed in terms of elementary functions. Nevertheless, we know the antiderivative of f exists by Part 1 of the Fundamental Theorem of Calculus. Use your graphing utility to visualize the integral function

$$F(x) = \int_0^x e^{-t^2}\, dt.$$

What can you say about $F(x)$? Where is it increasing and decreasing? Where are its extreme values, if any? What can you say about the concavity of its graph?

Area Between Curves

We next determine how to find the areas of regions in the coordinate plane by integrating the functions that define the regions' boundaries.

Suppose that we want to find the area of a region that is bounded above by the curve $y = f(x)$, below by the curve $y = g(x)$, and on the left and right by the lines $x = a$ and $x = b$ (Figure 4.19). The region might accidentally have a shape whose area we could find with geometry, but if f and g are arbitrary continuous functions we usually have to find the area with an integral.

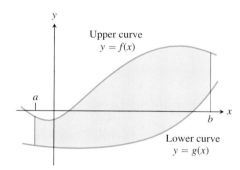

FIGURE 4.19 The region between the curves $y = f(x)$ and $y = g(x)$ and the lines $x = a$ and $x = b$.

To see what the integral should be, we first approximate the region with n vertical rectangles based on a partition $P = \{x_0, x_1, \ldots, x_n\}$ of $[a, b]$ (Figure 4.20). The area of the kth rectangle (Figure 4.21) is

$$\Delta A_k = \text{height} \times \text{width} = [f(c_k) - g(c_k)]\, \Delta x_k.$$

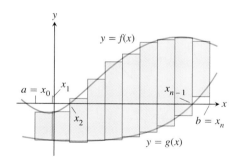

FIGURE 4.20 We approximate the region with rectangles perpendicular to the x-axis.

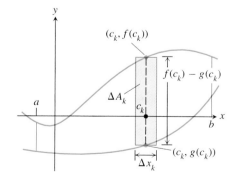

FIGURE 4.21 $\Delta A_k = $ area of kth rectangle, $f(c_k) - g(c_k) = $ height, and $\Delta x_k = $ width.

We then approximate the area of the region by adding the areas of the n rectangles:

$$A \approx \sum_{k=1}^n \Delta A_k = \sum_{k=1}^n [f(c_k) - g(c_k)]\, \Delta x_k. \quad \text{Riemann sum}$$

As $\| P \| \to 0$ the sums on the right approach the limit $\int_a^b [f(x) - g(x)]\, dx$ because f and g are continuous. We take the area of the region to be the value of this integral. That is,

$$A = \lim_{\|P\|\to 0} \sum_{k=1}^n [f(c_k) - g(c_k)]\, \Delta x_k = \int_a^b [f(x) - g(x)]\, dx.$$

CD-ROM
WEBsite

> **Definition**   Area Between Curves
>
> If f and g are continuous with $f(x) \geq g(x)$ throughout $[a, b]$, then the area of the region between the curves $y = f(x)$ and $y = g(x)$ from a to b is the integral of $[f - g]$ from a to b:
>
> $$A = \int_a^b [f(x) - g(x)]\, dx. \tag{2}$$

To apply Equation (2) we take the following steps.

How to Find the Area Between Two Curves

Step 1. *Graph the curves and draw a representative rectangle.* This reveals which curve is f (upper curve) and which is g (lower curve). It also helps find the limits of integration if you do not already know them.

Step 2. *Find the limits of integration.*

Step 3. *Write a formula for $f(x) - g(x)$.* Simplify it if you can.

Step 4. *Integrate $[f(x) - g(x)]$ from a to b.* The number you get is the area.

Example 4   Area Between Intersecting Curves

Find the area of the region enclosed by the parabola $y = 2 - x^2$ and the line $y = -x$.

Solution

Step 1. *Sketch the curves and a vertical rectangle* (Figure 4.22). Identifying the upper and the lower curves, we take $f(x) = 2 - x^2$ and $g(x) = -x$. The x-coordinates of the intersection points are the limits of integration.

Step 2. *Find the limits of integration.* We find the limits of integration by solving $y = 2 - x^2$ and $y = -x$ simultaneously for x:

$$2 - x^2 = -x \qquad \text{Equate } f(x) \text{ and } g(x).$$
$$x^2 - x - 2 = 0 \qquad \text{Rewrite.}$$
$$(x + 1)(x - 2) = 0 \qquad \text{Factor.}$$
$$x = -1, \qquad x = 2. \qquad \text{Solve.}$$

The region runs from $x = -1$ to $x = 2$. The limits of integration are $a = -1$, $b = 2$.

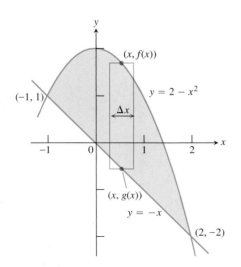

FIGURE 4.22 The region in Example 4 with a typical approximating rectangle.

Step 3. *Simplify the formula for $f(x) - g(x)$.*

$$f(x) - g(x) = (2 - x^2) - (-x) = 2 - x^2 + x \qquad \text{Rearrangement}$$
$$= 2 + x - x^2 \qquad \text{a matter of taste}$$

Step 4. *Integrate $[f(x) - g(x)]$ from a to b.*

$$A = \int_a^b [f(x) - g(x)]\, dx = \int_{-1}^2 (2 + x - x^2)\, dx = \left[2x + \frac{x^2}{2} - \frac{x^3}{3} \right]_{-1}^2$$

$$= \left(4 + \frac{4}{2} - \frac{8}{3} \right) - \left(-2 + \frac{1}{2} + \frac{1}{3} \right)$$

$$= 6 + \frac{3}{2} - \frac{9}{3} = \frac{9}{2}$$

Boundaries with Changing Formulas

If the formula for a bounding curve changes at one or more points, we partition the region into subregions that correspond to the formula changes and apply Equation (2) to each subregion.

Example 5   Changing the Integral to Match a Boundary Change

Find the area of the region in the first quadrant that is bounded above by $y = \sqrt{x}$ and below by the x-axis and the line $y = x - 2$.

Solution

Step 1. The sketch (Figure 4.23) shows that the region's upper boundary is the graph of $f(x) = \sqrt{x}$. The lower boundary changes from $g(x) = 0$ for $0 \le x \le 2$ to $g(x) = x - 2$ for $2 \le x \le 4$ (there is agreement at $x = 2$). We partition the region at $x = 2$ into subregions A and B and sketch a representative rectangle for each subregion.

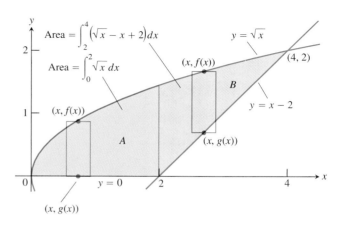

FIGURE 4.23 When the formula for a bounding curve changes, the area integral changes to match. (Example 5)

Step 2. The limits of integration for region A are $a = 0$ and $b = 2$. The left-hand limit for region B is $a = 2$. To find the right-hand limit, we solve the equations $y = \sqrt{x}$ and $y = x - 2$ simultaneously for x:

$$\sqrt{x} = x - 2 \qquad \text{Equate } f(x) \text{ and } g(x).$$
$$x = (x - 2)^2 = x^2 - 4x + 4 \qquad \text{Square both sides.}$$
$$x^2 - 5x + 4 = 0 \qquad \text{Rewrite.}$$
$$(x - 1)(x - 4) = 0 \qquad \text{Factor.}$$
$$x = 1, \qquad x = 4. \qquad \text{Solve.}$$

Only the value $x = 4$ satisfies the equation $\sqrt{x} = x - 2$. The value $x = 1$ is an extraneous root introduced by squaring. The right-hand limit is $b = 4$.

Step 3. For $0 \le x \le 2$: $\qquad f(x) - g(x) = \sqrt{x} - 0 = \sqrt{x}$

For $2 \le x \le 4$: $\qquad f(x) - g(x) = \sqrt{x} - (x - 2) = \sqrt{x} - x + 2$

Step 4. We add the area of subregions A and B to find the total area:

$$\text{Total area} = \underbrace{\int_0^2 \sqrt{x}\, dx}_{\text{area of } A} + \underbrace{\int_2^4 (\sqrt{x} - x + 2)\, dx}_{\text{area of } B}$$

$$= \left[\frac{2}{3} x^{3/2} \right]_0^2 + \left[\frac{2}{3} x^{3/2} - \frac{x^2}{2} + 2x \right]_2^4$$

$$= \frac{2}{3} (2)^{3/2} - 0 + \left(\frac{2}{3} (4)^{3/2} - 8 + 8 \right) - \left(\frac{2}{3} (2)^{3/2} - 2 + 4 \right)$$

$$= \frac{2}{3} (8) - 2 = \frac{10}{3}.$$

USING TECHNOLOGY

The Intersection of Two Graphs One of the difficult and sometimes frustrating parts of integration applications is finding the limits of integration. To do this, you often have to find the zeroes of a function or the intersection points of two curves.

To solve the equation $f(x) = g(x)$ using a graphing utility, you enter

$$y_1 = f(x) \qquad \text{and} \qquad y_2 = g(x)$$

and use the grapher routine to find the points of intersection. Alternatively, you can solve the equation $f(x) - g(x) = 0$ with a root finder. Try both procedures with

$$f(x) = \ln x \qquad \text{and} \qquad g(x) = 3 - x.$$

When points of intersection are not clearly revealed or you suspect hidden behavior, additional work with the graphing utility or further use of calculus may be necessary.

(a) The intersecting curves $y_1 = \ln x$ and $y_2 = 3 - x$, using a built-in function to find the intersection

(b) Using a built-in root finder to find the zero of $f(x) = \ln x - 3 + x$

EXERCISES 4.6

Use the substitution formula to evaluate the integrals in Exercises 1–16.

1. (a) $\displaystyle\int_0^3 \sqrt{y+1}\, dy$

  (b) $\displaystyle\int_{-1}^0 \sqrt{y+1}\, dy$

2. (a) $\displaystyle\int_0^{\pi/4} \tan x\ \sec^2 x\, dx$

  (b) $\displaystyle\int_{-\pi/4}^0 \tan x\ \sec^2 x\, dx$

3. (a) $\displaystyle\int_0^{\pi} 3\cos^2 x\ \sin x\, dx$

  (b) $\displaystyle\int_{2\pi}^{3\pi} 3\cos^2 x\ \sin x\, dx$

4. (a) $\displaystyle\int_0^{\sqrt 7} t\,(t^2+1)^{1/3}\, dt$

  (b) $\displaystyle\int_{-\sqrt 7}^0 t\,(t^2+1)^{1/3}\, dt$

5. (a) $\displaystyle\int_{-1}^1 \frac{5r}{(4+r^2)^2}\, dr$

  (b) $\displaystyle\int_0^1 \frac{5r}{(4+r^2)^2}\, dr$

6. (a) $\displaystyle\int_0^{\sqrt 3} \frac{4x}{\sqrt{x^2+1}}\, dx$

  (b) $\displaystyle\int_{-\sqrt 3}^{\sqrt 3} \frac{4x}{\sqrt{x^2+1}}\, dx$

7. (a) $\displaystyle\int_0^{2\pi} \frac{\cos z}{\sqrt{4+3\sin z}}\, dz$

  (b) $\displaystyle\int_{-\pi}^{\pi} \frac{\cos z}{\sqrt{4+3\sin z}}\, dz$

8. $\displaystyle\int_0^1 \sqrt{t^5+2t}\,(5t^4+2)\, dt$

9. $\displaystyle\int_1^4 \frac{dy}{2\sqrt y\,(1+\sqrt y)^2}$

10. $\displaystyle\int_0^{\pi/6} \cos^{-3} 2\theta\ \sin 2\theta\, d\theta$

11. $\displaystyle\int_0^{\pi/4} (1-\sin 2t)^{3/2}\cos 2t\, dt$

12. $\displaystyle\int_0^1 (4y-y^2+4y^3+1)^{-2/3}(12y^2-2y+4)\, dy$

13. $\displaystyle\int_0^{\pi/2} e^{\sin x}\cos x\, dx$

14. $\displaystyle\int_0^{\pi/4} (1+e^{\tan\theta})\sec^2\theta\, d\theta$

15. $\displaystyle\int_{\ln(\pi/6)}^{\ln(\pi/2)} 2e^v\cos e^v\, dv$

16. $\displaystyle\int_0^{\sqrt{\ln \pi}} 2xe^{x^2}\cos(e^{x^2})\, dx$

Initial Value Problems

Solve the initial value problems in Exercises 17 and 18.

17. $\dfrac{dy}{dt} = e^{-t}\sec^2(\pi e^{-t}),\quad y(\ln 4) = 2/\pi$

18. $\dfrac{d^2 y}{dx^2} = 2e^{-x},\quad y(0)=1 \text{ and } y'(0)=0$

Area

Find the areas of the shaded regions in Exercises 19–24.

19.

20.

21.

NOT TO SCALE

22.

23.

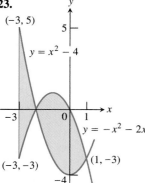

24.

In Exercises 25–28, graph the function over the given interval. Then

    (a) integrate the function over the interval

    (b) find the area of the region between the graph and the x-axis.

25. $y = x^2 - 6x + 8$, $[0, 3]$

26. $y = -x^2 + 5x - 4$, $[0, 2]$

27. $y = 2x - x^2$, $[0, 3]$

28. $y = x^2 - 4x$, $[0, 5]$

Find the areas of the regions enclosed by the lines and curves in Exercises 29–38.

29. $y = x^2 - 2$ and $y = 2$

30. $y = -x^2 - 2x$ and $y = x$

31. $y = x^2$ and $y = -x^2 + 4x$

32. $y = 7 - 2x^2$ and $y = x^2 + 4$

33. $y = x^4 - 4x^2 + 4$ and $y = x^2$

34. $y = |x^2 - 4|$ and $y = (x^2/2) + 4$

35. $y = 2 \sin x$ and $y = \sin 2x$, $0 \leq x \leq \pi$

36. $y = 8 \cos x$ and $y = \sec^2 x$, $-\pi/3 \leq x \leq \pi/3$

37. $y = \sin(\pi x/2)$ and $y = x$

38. $y = \sec^2 x$, $y = \tan^2 x$, $x = -\pi/4$, and $x = \pi/4$

39. Find the area of the region in the first quadrant bounded by the line $y = x$, the line $x = 2$, the curve $y = 1/x^2$, and the x-axis.

40. Find the area of the "triangular" region in the first quadrant bounded on the left by the y-axis and on the right by the curves $y = \sin x$ and $y = \cos x$.

41. Find the area of the region between the curve $y = 3 - x^2$ and the line $y = -1$.

42. Find the area of the region in the first quadrant bounded on the left by the y-axis, below by the line $y = x/4$, above left by the curve $y = 1 + \sqrt{x}$, and above right by the curve $y = 2/\sqrt{x}$.

COMPUTER EXPLORATIONS

In Exercises 43–46, you will find the area between curves in the plane when you cannot find their points of intersection using simple algebra. Use a CAS to perform the following steps.

    (a) Plot the curves together to see what they look like and how many points of intersection they have.

    (b) Use the numerical equation solver in your CAS to find all the points of intersection.

    (c) Integrate $|f(x) - g(x)|$ over consecutive pairs of intersection values.

    (d) Sum together the integrals found in part (c).

43. $f(x) = \dfrac{x^3}{3} - \dfrac{x^2}{2} - 2x + \dfrac{1}{3}$, $g(x) = x - 1$

44. $f(x) = \dfrac{x^4}{2} - 3x^3 + 10$, $g(x) = 8 - 12x$

45. $f(x) = x + \sin(2x)$, $g(x) = x^3$

46. $f(x) = x^2 \cos x$, $g(x) = x^3 - x$

4.7 Numerical Integration

Trapezoidal Approximations • Error in the Trapezoidal Approximation
• Approximations Using Parabolas • Error in Simpson's Rule •
Which Rule Gives Better Results? • Round-off Errors

As we have seen, the ideal way to evaluate a definite integral $\int_a^b f(x)\, dx$ is to find a formula $F(x)$ for one of the antiderivatives of $f(x)$ and calculate the number $F(b) - F(a)$. But some antiderivatives are hard to find, and still others, like the antiderivatives of $(\sin x)/x$ and $\sqrt{1 + x^4}$, have no elementary formulas. We do not mean merely that no one has yet succeeded in finding elementary formulas for the antiderivatives of $(\sin x)/x$ and $\sqrt{1 + x^4}$. We mean it has been proved that no such formulas exist.

Whatever the reason, when we cannot evaluate a definite integral with an antiderivative, we turn to numerical methods such as the Trapezoidal Rule and Simpson's Rule, described in this section.

CD-ROM
WEBsite

Trapezoidal Approximations

When we cannot find a workable antiderivative for a function f that we have to integrate, we partition the interval of integration, replace f by a closely fitting polynomial on each subinterval, integrate the polynomials, and add the results to approximate the integral of f. We start with straight-line segments giving trapezoids.

As shown in Figure 4.24, if $[a, b]$ is partitioned into n subintervals of equal length $h = (b - a)/n$, the graph of f on $[a, b]$ can be approximated by a straight-line segment over each subinterval.

The region between the curve and the x-axis is then approximated by the trapezoids, the area of each trapezoid being the length of its horizontal "altitude" times

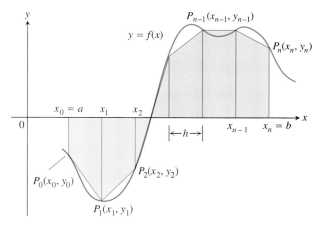

The length $h = (b - a)/n$ is called the **step size.** It is conventional to use h in this context instead of Δx.

FIGURE 4.24 The Trapezoidal Rule approximates short stretches of the curve $y = f(x)$ with line segments. To approximate the integral of f from a to b, we add the "signed" areas of the trapezoids made by joining the ends of the segments to the x-axis.

the average of its two vertical "bases." We add the areas of the trapezoids, counting area above the x-axis as positive and area below the axis as negative:

$$T = \frac{1}{2}(y_0 + y_1)h + \frac{1}{2}(y_1 + y_2)h + \cdots + \frac{1}{2}(y_{n-2} + y_{n-1})h + \frac{1}{2}(y_{n-1} + y_n)h$$

$$= h\left(\frac{1}{2}y_0 + y_1 + y_2 + \cdots + y_{n-1} + \frac{1}{2}y_n\right)$$

$$= \frac{h}{2}(y_0 + 2y_1 + 2y_2 + \cdots + 2y_{n-1} + y_n),$$

where

$$y_0 = f(a), \quad y_1 = f(x_1), \quad \ldots, \quad y_{n-1} = f(x_{n-1}), \quad y_n = f(b).$$

The Trapezoidal Rule says: Use T to estimate the integral of f from a to b.

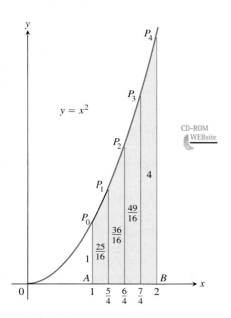

FIGURE 4.25 The trapezoidal approximation of the area under the graph of $y = x^2$ from $x = 1$ to $x = 2$ is a slight overestimate.

The Trapezoidal Rule

To approximate $\int_a^b f(x)\, dx$, use

$$T = \frac{h}{2}(y_0 + 2y_1 + 2y_2 + \cdots + 2y_{n-1} + y_n),$$

The y's are the values of f at the partition points

$$x_0 = a, \quad x_1 = a + h, \quad x_2 = a + 2h, \quad \ldots, \quad x_{n-1} = a + (n-1)h, \quad x_n = b,$$

where $h = (b - a)/n$.

Example 1   Applying the Trapezoidal Rule

Use the Trapezoidal Rule with $n = 4$ to estimate $\int_1^2 x^2\, dx$. Compare the estimate with the exact value.

Solution   Partition $[1, 2]$ into four subintervals of equal length (Figure 4.25). Then evaluate $y = x^2$ at each partition point (Table 4.6).

Using these y values, $n = 4$, and $h = (2 - 1)/4 = 1/4$ in the Trapezoidal Rule, we have

$$T = \frac{h}{2}(y_0 + 2y_1 + 2y_2 + 2y_3 + y_4)$$

$$= \frac{1}{8}\left(1 + 2\left(\frac{25}{16}\right) + 2\left(\frac{36}{16}\right) + 2\left(\frac{49}{16}\right) + 4\right)$$

$$= \frac{75}{32} = 2.34375.$$

The exact value of the integral is

$$\int_1^2 x^2\, dx = \left.\frac{x^3}{3}\right]_1^2 = \frac{8}{3} - \frac{1}{3} = \frac{7}{3}.$$

Table 4.6

x	$y = x^2$
1	1
$\frac{5}{4}$	$\frac{25}{16}$
$\frac{6}{4}$	$\frac{36}{16}$
$\frac{7}{4}$	$\frac{49}{16}$
2	4

The T approximation overestimates the integral by about half a percent of its true value of 7/3. The percentage error is $(2.34375 - 7/3)/(7/3) \approx 0.00446$, or 0.446%.

We could have predicted that the Trapezoidal Rule would overestimate the integral in Example 1 by considering the geometry of the graph in Figure 4.25. Since the parabola is concave *up*, the approximating segments lie above the curve, giving each trapezoid slightly more area than the corresponding strip under the curve. In Figure 4.24, we see that the straight segments lie *under* the curve on those intervals where the curve is concave *down*, causing the Trapezoidal Rule to *underestimate* the integral on those intervals. The interpretation of "area" changes where the curve lies below the x-axis, but it is still the case that the higher y-values give the greater signed area. Thus, we can always say that T overestimates the integral where the graph is concave up and underestimates the integral where the graph is concave down.

Example 2   Averaging Temperatures

An observer measures the outside temperature every hour from noon until midnight, recording the temperatures in the following table.

Time	N	1	2	3	4	5	6	7	8	9	10	11	M
Temp	63	65	66	68	70	69	68	68	65	64	62	58	55

What was the average temperature for the 12-hour period?

Solution   We are looking for the average value of a continuous function (temperature) for which we know values at discrete times that are one unit apart. We need to find

$$\text{av}\,(f) = \frac{1}{b-a} \int_a^b f(x)\,dx,$$

without having a formula for $f(x)$. The integral, however, can be approximated by the Trapezoid Rule, using the temperatures in the table as function values at the points of a 12-subinterval partition of the 12-hour interval (making $h = 1$).

$$T = \frac{h}{2}(y_0 + 2y_1 + 2y_2 + \cdots + 2y_{11} + y_{12})$$

$$= \frac{1}{2}(63 + 2\cdot65 + 2\cdot66 + \cdots + 2\cdot58 + 55)$$

$$= 782$$

Using T to approximate $\int_a^b f(x)\,dx$, we have

$$\text{av}\,(f) \approx \frac{1}{b-a}\cdot T = \frac{1}{12}\cdot782 \approx 65.17.$$

Rounding to be consistent with the data given, we estimate the average temperature as 65 degrees.

Error in the Trapezoidal Approximation

Pictures suggest that the magnitude of the error

$$E_T = \int_a^b f(x)\,dx - T$$

in the trapezoidal approximation will decrease as the **step size** h decreases, because the trapezoids fit the curve better as their number increases. A theorem from advanced calculus assures us that this will be the case if f has a continuous second derivative.

Error Estimate for the Trapezoidal Rule

If f'' is continuous and M is any upper bound for the values of $|f''|$ on $[a, b]$, then

$$|E_T| \le \frac{b-a}{12}\,h^2 M, \qquad\qquad (1)$$

where $h = (b - a)/n$.

Although theory tells us there will always be a smallest safe value of M, in practice we can hardly ever find it. Instead, we find the best value we can and go on from there to estimate $|E_T|$. This may seem sloppy, but it works. To make $|E_T|$ small for a given M, we make h small.

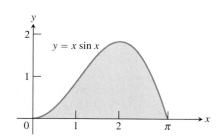

FIGURE 4.26 Graph of the integrand in Example 3.

Example 3   Bounding the Trapezoidal Rule Error

Find an upper bound for the error incurred in estimating

$$\int_0^\pi x\,\sin\,x\,dx$$

with the trapezoidal rule with $n = 10$ steps (Figure 4.26).

Solution   With $a = 0$, $b = \pi$, and $h = (b - a)/n = \pi/10$, Equation (1) gives

$$|E_T| \le \frac{b-a}{12}\,h^2 M = \frac{\pi}{12}\left(\frac{\pi}{10}\right)^2 M = \frac{\pi^3}{1200}\,M.$$

The number M can be any upper bound for the magnitude of the second derivative of $f(x) = x \sin x$ on $[0, \pi]$. A routine calculation gives

$$f''(x) = 2\cos x - x\sin x,$$

so

$$|f''(x)| = |2\,\cos x - x\sin\,x|$$

$$\le 2\,|\cos x| + |x\|\sin x|$$

$$\le 2\cdot 1 + \pi\cdot 1 = 2 + \pi.$$

Triangle inequality:
$|a + b| \le |a| + |b|$

$|\cos x|$ and $|\sin x|$ never exceed 1, and $0 \le x \le \pi$.

We can safely take $M = 2 + \pi$. Therefore,

$$|E_T| \le \frac{\pi^3}{1200} M = \frac{\pi^3(2 + \pi)}{1200} < 0.133. \qquad \text{Rounded up to be safe}$$

The absolute error is no greater than 0.133.

For greater accuracy, we would not try to improve M but would take more steps. With $n = 100$ steps, for example, $h = \pi/100$ and

$$|E_T| \le \frac{\pi}{12}\left(\frac{\pi}{100}\right)^2 M = \frac{\pi^3(2 + \pi)}{120,000} < 0.00133 = 1.33 \times 10^{-3}.$$

Approximations Using Parabolas

Riemann sums and the Trapezoidal Rule both give reasonable approximations to the integral of a continuous function over a closed interval. The Trapezoidal Rule is more efficient, giving a better approximation for small values of n, which makes it a faster algorithm for numerical integration.

Indeed, the only shortcoming of the Trapezoidal Rule seems to be that it depends on approximating curved arcs with straight segments. You might think that an algorithm that approximates the curve with *curved* pieces would be even more efficient (and hence faster for machines), and you would be right. One such algorithm uses parabolas and is known as *Simpson's Rule*. Simpson's Rule for approximating $\int_a^b f(x)\,dx$ is based on approximating f with quadratic polynomials instead of linear polynomials. We approximate the graph with parabolic arcs instead of line segments (Figure 4.27).

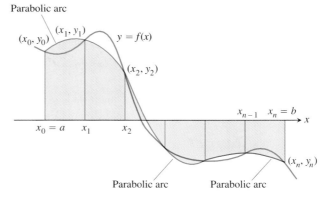

FIGURE 4.27 Simpson's Rule approximates short stretches of a curve with parabolic arcs.

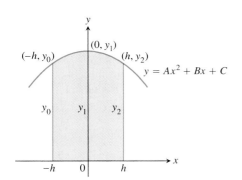

FIGURE 4.28 By integrating from $-h$ to h, we find the shaded area to be $\frac{h}{3}(y_0 + 4y_1 + y_2)$.

The integral of the quadratic polynomial $y = Ax^2 + Bx + C$ in Figure 4.28 from $x = -h$ to $x = h$ is

$$\int_{-h}^{h}(Ax^2 + Bx + C)\,dx = \frac{h}{3}(y_0 + 4y_1 + y_2) \qquad (2)$$

(Appendix 4). Simpson's Rule follows from partitioning $[a, b]$ into an even number of subintervals of equal length h, applying Equation (2) to successive interval pairs, and adding the results.

Simpson's Rule

To approximate $\int_a^b f(x)\, dx$, use

$$S = \frac{h}{3}(y_0 + 4y_1 + 2y_2 + 4y_3 + \cdots + 2y_{n-2} + 4y_{n-1} + y_n).$$

The y's are the values of f at the partition points

$$x_0 = a, \quad x_1 = a + h, \quad x_2 = a + 2h, \quad \ldots, \quad x_{n-1} = a + (n-1)h, \quad x_n = b.$$

The number n is even, and $h = (b - a)/n$.

Table 4.7

x	$y = 5x^4$
0	0
$\frac{1}{2}$	$\frac{5}{16}$
1	5
$\frac{3}{2}$	$\frac{405}{16}$
2	80

CD-ROM
WEBsite

Example 4   Applying Simpson's Rule

Use Simpson's Rule with $n = 4$ to approximate $\int_0^2 5x^4\, dx$.

Solution   Partition $[0, 2]$ into four subintervals and evaluate $y = 5x^4$ at the partition points (Table 4.7). Then apply Simpson's Rule with $n = 4$ and $h = 1/2$:

$$S = \frac{h}{3}(y_0 + 4y_1 + 2y_2 + 4y_3 + y_4)$$

$$= \frac{1}{6}\left(0 + 4\left(\frac{5}{16}\right) + 2(5) + 4\left(\frac{405}{16}\right) + 80\right)$$

$$= 32\frac{1}{12}.$$

This estimate differs from the exact value (32) by only $1/12$, a percentage error of less than three-tenths of one percent, and this was with just four subintervals.

Error in Simpson's Rule

The magnitude of the Simpson's Rule error,

$$E_S = \int_a^b f(x)\, dx - S,$$

decreases with the step size, as we would expect from our experience with the Trapezoidal Rule. The inequality for controlling the Simpson's Rule error, however, assumes f to have a continuous fourth derivative instead of merely a continuous second derivative. The formula, once again from advanced calculus, is this.

Error Estimate for Simpson's Rule

If $f^{(4)}$ is continuous and M is an upper bound for the values of $|f^{(4)}|$ on $[a, b]$, then

$$|E_S| \leq \frac{b - a}{180} h^4 M, \tag{3}$$

where $h = (b - a)/n$.

As with the Trapezoidal Rule, we can almost never find the smallest possible value of M. We just find the best value we can and go on from there.

Example 5   Bounding the Error in Simpson's Rule

What estimate does Equation (3) give for the error in Simpson's approximation in Example 4?

Solution   To estimate the error, we first find an upper bound M for the magnitude of the fourth derivative of $f(x) = 5x^4$ on the interval $0 \le x \le 2$. Since the fourth derivative has the constant value $f^{(4)}(x) = 120$, we may safely take $M = 120$. With $b - a = 2$ and $h = 1/2$, Equation (3) gives

$$|E_S| \le \frac{b-a}{180} h^4 M = \frac{2}{180} \left(\frac{1}{2}\right)^4 (120) = \frac{1}{12}.$$

Which Rule Gives Better Results?

The answer lies in the error-control formulas

$$|E_T| \le \frac{b-a}{12} h^2 M, \qquad |E_S| \le \frac{b-a}{180} h^4 M.$$

The M's, of course, mean different things, the first being an upper bound on $|f''|$ and the second an upper bound on $|f^{(4)}|$. But there is more. The factor $(b - a)/180$ in the Simpson formula is one-fifteenth of the factor $(b - a)/12$ in the trapezoidal formula. More important still, the Simpson formula has an h^4, whereas the trapezoidal formula has only an h^2. If h is one-tenth, then h^2 is one-hundredth but h^4 is only one ten-thousandth. If both M's are 1, for example, and $b - a = 1$, then, with $h = 1/10$,

$$|E_T| \le \frac{1}{12} \left(\frac{1}{10}\right)^2 \cdot 1 = \frac{1}{1200},$$

whereas

$$|E_S| \le \frac{1}{180} \left(\frac{1}{10}\right)^4 \cdot 1 = \frac{1}{1,800,000} = \frac{1}{1500} \cdot \frac{1}{1200}.$$

For roughly the same amount of computational effort, we get better accuracy with Simpson's Rule, at least in this case.

The h^2 versus h^4 is the key. If h is less than 1, then h^4 can be significantly smaller than h^2. On the other hand, if h equals 1, there is no difference between h^2 and h^4. If h is greater than 1, the value of h^4 may be significantly larger than the value of h^2. In the latter two cases, the error formulas offer little help. We have to go back to the geometry of the curve $y = f(x)$ to see whether trapezoids or parabolas, if either, are going to give the results we want.

Example 6   Comparing the Trapezoidal Rule and Simpson's Rule Approximations

We know from the Fundamental Theorem that the value of ln 2 can be calculated from the integral

$$\ln 2 = \int_1^2 \frac{1}{x} \, dx.$$

Trapezoidal versus Simpson

If Simpson's Rule is more accurate, why bother with the Trapezoidal Rule? There are two reasons. First, the Trapezoidal Rule is useful in a number of specific applications because it leads to much simpler expressions. Second, the Trapezoidal Rule is the basis for *Rhomberg integration,* one of the most satisfactory machine methods when high precision is required.

Table 4.8 shows T and S values for approximations of $\int_1^2 (1/x)\, dx$ using various values of n. Notice how Simpson's Rule dramatically improves over the Trapezoidal Rule. In particular, notice that when we double the value of n (thereby halving the value of h), the T error is divided by 2 *squared*, whereas the S error is divided by 2 *to the fourth*.

Table 4.8 Trapezoidal Rule approximations (T_n) and Simpson's Rule approximations (S_n) of $\ln 2 = \int_1^2 (1/x)\, dx$

n	T_n	|Error| less than ...	S_n	|Error| less than ...
10	0.6937714032	0.0006242227	0.6931502307	0.0000030502
20	0.6933033818	0.0001562013	0.6931473747	0.0000001942
30	0.6932166154	0.0000694349	0.6931472190	0.0000000385
40	0.6931862400	0.0000390595	0.6931471927	0.0000000122
50	0.6931721793	0.0000249988	0.6931471856	0.0000000050
100	0.6931534305	0.0000062500	0.6931471809	0.0000000004

This has a dramatic effect as h gets very small. The Simpson approximation for $n = 50$ rounds accurately to seven places and for $n = 100$ agrees to nine decimal places (billionths)!

Example 7   Draining A Swamp

A town wants to drain and fill a small polluted swamp (Figure 4.29). The swamp averages 5 ft deep. About how many cubic yards of dirt will it take to fill the area after the swamp is drained?

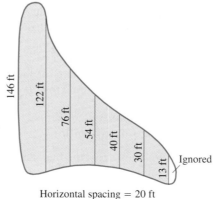

Horizontal spacing = 20 ft

FIGURE 4.29 The swamp in Example 7.

Solution To calculate the volume of the swamp, we estimate the surface area and multiply by 5. To estimate the area, we use Simpson's Rule with $h = 20$ ft and the y's equal to the distances measured across the swamp, as shown in Figure 4.29.

$$S = \frac{h}{3}(y_0 + 4y_1 + 2y_2 + 4y_3 + 2y_4 + 4y_5 + y_6)$$

$$= \frac{20}{3}(146 + 488 + 152 + 216 + 80 + 120 + 13) = 8100$$

The volume is about $(8100)(5) = 40{,}500$ ft^3 or 1500 yd^3.

Round-off Errors

Although decreasing the step size h reduces the error in the Simpson and trapezoidal approximations in theory, it may fail to do so in practice. When h is very small, say $h = 10^{-5}$, computer or calculator round-off errors in the arithmetic required to evaluate S and T may accumulate to such an extent that the error formulas no longer describe what is going on. Shrinking h below a certain size can actually make things worse. Although this is not an issue in this book, you should consult a text on numerical analysis for alternative methods if you are having problems with round-off.

EXERCISES 4.7

Estimating Integrals

The instructions for the integrals in Exercises 1–10 have two parts, one for the Trapezoidal Rule and one for Simpson's Rule.

I. *Using the Trapezoidal Rule*

(a) Estimate the integral with $n = 4$ steps and use Equation (1) to find an upper bound for $|E_T|$.

(b) Evaluate the integral directly and find $|E_T|$.

(c) Use the formula $(|E_T|/(\text{true value})) \times 100$ to express $|E_T|$ as a percentage of the integral's true value.

II. *Using Simpson's Rule*

(a) Estimate the integral with $n = 4$ steps and use Equation (3) to find an upper bound for $|E_S|$.

(b) Evaluate the integral directly, and find $|E_S|$.

(c) Use the formula $(|E_S|/(\text{true value})) \times 100$ to express $|E_S|$ as a percentage of the integral's true value.

1. $\int_1^2 x\,dx$

2. $\int_1^3 (2x - 1)\,dx$

3. $\int_{-1}^1 (x^2 + 1)\,dx$

4. $\int_{-2}^0 (x^2 - 1)\,dx$

5. $\int_0^2 (t^3 + t)\,dt$

6. $\int_{-1}^1 (t^3 + 1)\,dt$

7. $\int_1^2 \frac{1}{s^2}\,ds$

8. $\int_2^4 \frac{1}{(s-1)^2}\,ds$

9. $\int_0^\pi \sin t\,dt$

10. $\int_0^1 \sin \pi t\,dt$

In Exercises 11–14, use the tabulated values of the integrand to estimate the integral with **(a)** the Trapezoidal Rule and **(b)** Simpson's Rule with $n = 8$ steps. Round your answers to five decimal places. Then **(c)** find the integral's exact value and the approximation error E_T or E_S, as appropriate.

11. $\int_0^1 x\sqrt{1 - x^2}\,dx$

x	$x\sqrt{1-x^2}$
0	0.0
0.125	0.12402
0.25	0.24206
0.375	0.34763
0.5	0.43301
0.625	0.48789
0.75	0.49608
0.875	0.42361
1.0	0

12. $\int_0^3 \dfrac{\theta}{\sqrt{16 + \theta^2}}\, d\theta$

θ	$\theta/\sqrt{16 + \theta^2}$
0	0.0
0.375	0.09334
0.75	0.18429
1.125	0.27075
1.5	0.35112
1.875	0.42443
2.25	0.49026
2.625	0.58466
3.0	0.6

13. $\int_{-\pi/2}^{\pi/2} \dfrac{3 \cos t}{(2 + \sin t)^2}\, dt$

t	$(3 \cos t)/(2 + \sin t)^2$
−1.57080	0.0
−1.17810	0.99138
−0.78540	1.26906
−0.39270	1.05961
0	0.75
0.39270	0.48821
0.78540	0.28946
1.17810	0.13429
1.57080	0

14. $\int_{\pi/4}^{\pi/2} (\csc^2 y)\sqrt{\cot y}\, dy$

y	$(\csc^2 y)\sqrt{\cot y}$
0.78540	2.0
0.88357	1.51606
0.98175	1.18237
1.07992	0.93998
1.17810	0.75402
1.27627	0.60145
1.37445	0.46364
1.47262	0.31688
1.57080	0

Applications

15. *Volume of water in a swimming pool* A rectangular swimming pool is 30 ft wide and 50 ft long. The table below shows the depth $h(x)$ of the water at 5 ft intervals from one end of the pool to the other. Estimate the volume of water in the pool using the Trapezoidal Rule with $n = 10$, applied to the integral.

$$V = \int_0^{50} 30 \cdot h(x)\, dx.$$

Position (ft) x	Depth (ft) $h(x)$	Position (ft) x	Depth (ft) $h(x)$
0	6.0	30	11.5
5	8.2	35	11.9
10	9.1	40	12.3
15	9.9	45	12.7
20	10.5	50	13.0
25	11.0		

16. *Stocking a fish pond* As the fish and game warden of your township, you are responsible for stocking the town pond with fish before the fishing season. The average depth of the pond is 20 ft. Using a scaled map, you measure distance across the pond at 200 ft intervals, as shown in the accompanying diagram.

(a) Use the Trapezoidal Rule to estimate the volume of the pond.

(b) You plan to start the season with one fish per 1000 cubic feet. You intend to have at least 25% of the opening day's fish population left at the end of the season. What is the maximum number of licenses the town can sell if the average seasonal catch is 20 fish per license?

Vertical spacing = 200 ft

17. *Ford® Mustang Cobra™* The accompanying table shows time-to-speed data for a 1994 Ford Mustang Cobra accelerating from rest to 130 mph. How far had the Mustang traveled by the time it reached this speed? (Use trapezoids to estimate the area under the velocity curve, but be careful: The time intervals may vary in length.)

Speed change	Time (sec)
Zero to 30 mph	2.2
40 mph	3.2
50 mph	4.5
60 mph	5.9
70 mph	7.8
80 mph	10.2
90 mph	12.7
100 mph	16.0
110 mph	20.6
120 mph	26.2
130 mph	37.1

Source: Car and Driver, April 1994.

18. *Aerodynamic drag* A vehicle's aerodynamic drag is determined in part by its cross-section area, so, all other things being equal, engineers try to make this area as small as possible. Use Simpson's Rule to estimate the cross-section area of the body of James Worden's solar-powered Solectria® automobile at MIT from the diagram.

19. *Wing design* The design of a new airplane requires a gasoline tank of constant cross-section area in each wing. A scale drawing of a cross section is shown here. The tank must hold 5000 lb of gasoline, which has a density of 42 lb/ft³. Estimate the length of the tank.

$y_0 = 1.5$ ft, $y_1 = 1.6$ ft, $y_2 = 1.8$ ft, $y_3 = 1.9$ ft,
$y_4 = 2.0$ ft, $y_5 = y_6 = 2.1$ ft   Horizontal spacing = 1 ft

20. *Oil Consumption on Pathfinder Island* A diesel generator runs continuously, consuming oil at a gradually increasing rate until it must be temporarily shut down to have the filters replaced.

Use the Trapezoidal Rule to estimate the amount of oil consumed by the generator during that week.

Day	Oil consumption rate (liters/hour)
Sun	0.019
Mon	0.020
Tue	0.021
Wed	0.023
Thu	0.025
Fri	0.028
Sat	0.031
Sun	0.035

Theory and Examples

21. *Polynomials of low degree* The magnitude of the error in the trapezoidal approximation of $\int_a^b f(x)\,dx$ is

$$|E_T| = \frac{b-a}{12} h^2 |f''(c)|,$$

where c is some point (usually unidentified) in (a, b). If f is a linear function of x, then $f''(c) = 0$, so $E_T = 0$ and T gives the exact value of the integral for any value of h. This is no surprise, really, for if f is linear, the line segments approximating the graph of f fit the graph exactly. The surprise comes with Simpson's Rule. The magnitude of the error in Simpson's Rule is

$$|E_S| = \frac{b-a}{180} h^4 |f^{(4)}(c)|,$$

where once again c lies in (a, b). If f is a polynomial of degree less than 4, then $f^{(4)} = 0$ no matter what c is, so $E_S = 0$ and S

gives the integral's exact value, even if we use only two steps. As a case in point, use Simpson's Rule with $n = 2$ to estimate

$$\int_0^2 x^3 \, dx.$$

Compare your answer with the integral's exact value.

22. *Usable values of the sine-integral function* **The sine-integral function,**

$$Si\,(x) = \int_0^x \frac{\sin t}{t} \, dt, \qquad \text{"Sine integral of } x\text{"}$$

is one of the many functions in engineering whose formulas cannot be simplified. There is no elementary formula for the antiderivative of $(\sin t)/t$. The values of $Si\,(x)$, however, are readily estimated by numerical integration.

Although the notation does not show it explicitly, the function being integrated is

$$f(t) = \begin{cases} \dfrac{\sin t}{t}, & t \neq 0 \\ 1, & t = 0, \end{cases}$$

the continuous extension of $(\sin t)/t$ to the interval $[0, x]$. The function has derivatives of all orders at every point of its domain. Its graph is smooth and you can expect good results from Simpson's Rule.

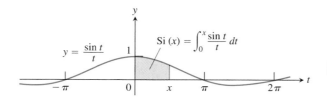

(a) Use the fact that $\left| f^{(4)} \right| \leq 1$ on $[0, \pi/2]$ to give an upper bound for the error that will occur if

$$Si\left(\frac{\pi}{2}\right) = \int_0^{\pi/2} \frac{\sin t}{t} \, dt$$

is estimated by Simpson's Rule with $n = 4$.

(b) Estimate $Si\,(\pi/2)$ by Simpson's Rule with $n = 4$.

(c) Express the error bound you found in part (a) as a percentage of the value you found in part (b).

23. *The error function* **The error function,**

$$erf\,(x) = \frac{2}{\sqrt{\pi}} \int_0^x e^{-t^2} \, dt,$$

important in probability and in the theories of heat flow and signal transmission, must be evaluated numerically because there is no elementary expression for the antiderivative of e^{-t^2}.

(a) Use Simpson's Rule with $n = 10$ to estimate $erf\,(1)$.

(b) In $[0, 1]$,

$$\left| \frac{d^4}{dt^4}\left(e^{-t^2}\right) \right| \leq 12.$$

Give an upper bound for the magnitude of the error of the estimate in part (a).

24. *Writing to Learn* In Example 2 (before rounding), we found the average temperature to be 65.17 degrees when we used the integral approximation, yet the average of the 13 discrete temperatures is only 64.69 degrees. Considering the shape of the temperature curve, explain why you would expect the average of the 13 discrete temperatures to be less than the average value of the temperature function on the entire interval.

In Exercises 25 and 26, *work in groups of two or three.*

T **25.** Consider the integral $\int_{-1}^{1} \sin\,(x^2) \, dx$.

(a) Find f'' for $f(x) = \sin\,(x^2)$.

(b) Graph $y = f''(x)$ in the viewing window $[-1, 1]$ by $[-3, 3]$.

(c) Explain why the graph in part (b) suggests that $\left| f''(x) \right| \leq 3$ for $-1 \leq x \leq 1$.

(d) Show that the error estimate for the Trapezoidal Rule in this case becomes

$$\left| E_T \right| \leq \frac{h^2}{2}.$$

(e) Show that the Trapezoidal Rule error will be less than or equal to 0.01 in magnitude if $h \leq 0.1$.

(f) How large must n be for $h \leq 0.1$?

T **26.** Consider the integral $\int_{-1}^{1} \sin\,(x^2) \, dx$.

(a) Find $f^{(4)}$ for $f(x) = \sin\,(x^2)$.

(b) Graph $y = f^{(4)}(x)$ in the viewing window $[-1, 1]$ by $[-30, 10]$.

(c) Explain why the graph in part (b) suggests that $\left| f^{(4)}(x) \right| \leq 30$ for $-1 \leq x \leq 1$.

(d) Show that the error estimate for Simpson's Rule in this case becomes

$$\left| E_S \right| \leq \frac{h^4}{3}.$$

(e) Show that the Simpson's Rule error will be less than or equal to 0.01 in magnitude if $h \leq 0.4$.

(f) How large must n be for $h \leq 0.4$?

COMPUTER EXPLORATIONS

As we mentioned at the beginning of the section, the definite integrals of many continuous functions cannot be evaluated with the Fundamental Theorem of Calculus because their antiderivatives lack elementary formulas. Numerical integration offers a practical way to estimate the values of these so-called *nonelementary integrals*. If your calculator or computer has a numerical integration routine, try it on the integrals in Exercises 27–30.

27. $\int_{-1}^{1} 2\sqrt{1 - x^2}\, dx$ The exact value is π.

28. $\int_{0}^{1} \sqrt{1 + x^4}\, dx$ An integral that came up in Newton's research

29. $\int_{0}^{\pi/2} \frac{\sin x}{x}\, dx$

30. $\int_{0}^{\pi/2} \sin (x^2)\, dx$ An integral associated with the diffraction of light

31. Consider the integral $\int_{0}^{\pi} \sin x\, dx$.

  (a) Find the Trapezoidal Rule approximations for $n = 10$, 100, and 1000.

  (b) Record the errors with as many decimal places of accuracy as you can.

  (c) What pattern do you see?

  (d) *Writing to Learn* Explain how the error bound for E_T accounts for the pattern in part (c).

32. (*Continuation of Exercise 31*) Repeat Exercise 31 with Simpson's Rule and E_S.

Questions To Guide Your Review

1. Can a function have more than one antiderivative? If so, how are the antiderivatives related? Explain.

2. What is an indefinite integral? How do you evaluate one? What general formulas do you know for evaluating indefinite integrals?

3. How can you sometimes solve a differential equation of the form $dy/dx = f(x)$?

4. What is an initial value problem? How do you solve one? Give an example.

5. How can you sometimes use a trigonometric identity to transform an unfamiliar integral into one you know how to evaluate?

6. If you know the acceleration of a body moving along a coordinate line as a function of time, what more do you need to know to find the body's position function? Give an example.

7. How is integration by substitution related to the Chain Rule?

8. How can you sometimes evaluate indefinite integrals by substitution? Give examples.

9. How can you sometimes estimate quantities like distance traveled, area, volume, and average value with finite sums? Why might you want to do so?

10. What is sigma notation? What advantage does it offer? Give examples.

11. What is a Riemann Sum? Why might you want to consider such a sum?

12. What is the norm of a partition of a closed interval?

13. What is the definite integral of a function f over a closed interval $[a, b]$? When can you be sure it exists?

14. What is the relation between definite integrals and area? Describe some other interpretations of definite integrals.

15. What is the average value of an integrable function over a closed interval? Must the function assume its average value? Explain.

16. What does a function's average value have to do with sampling a function's values?

17. Describe the rules for working with definite integrals (Table 4.5). Give examples.

18. What is the Fundamental Theorem of Calculus? Why is it so important? Illustrate each part of the theorem with an example.

19. How does the Fundamental Theorem provide a solution to the initial value problem $dy/dx = f(x)$, $y(x_0) = y_0$, when f is continuous?

20. How does the method of substitution work for definite integrals? Give examples.

21. How do you define and calculate the area of the region between the graphs of two continuous functions? Give an example.

22. You are collaborating to produce a short "how-to" manual for numerical integration, and you are writing about the Trapezoidal Rule.

  (a) What would you say about the rule itself and how to use it? How to achieve accuracy?

  (b) What would you say if you were writing about Simpson's Rule instead?

23. How would you compare the relative merits of Simpson's Rule and the Trapezoidal Rule?

Practice Exercises

Evaluating Indefinite Integrals

Evaluate the integrals in Exercises 1–20.

1. $\displaystyle\int (x^3 + 5x - 7)\, dx$

2. $\displaystyle\int \left(8t^3 - \frac{t^2}{2} + t\right) dt$

3. $\displaystyle\int \left(3\sqrt{t} + \frac{4}{t^2}\right) dt$

4. $\displaystyle\int \left(\frac{1}{2\sqrt{t}} - \frac{3}{t^4}\right) dt$

5. $\displaystyle\int \frac{r\, dr}{(r^2 + 5)^2}$

6. $\displaystyle\int \frac{6r^2\, dr}{(r^3 - \sqrt{2})^3}$

7. $\displaystyle\int 3\theta\sqrt{2 - \theta^2}\, d\theta$

8. $\displaystyle\int \frac{\theta^2}{9\sqrt{73 + \theta^3}}\, d\theta$

9. $\displaystyle\int e^x \sin\,(e^x)\, dx$

10. $\displaystyle\int e^t \cos\,(3e^t - 2)\, dt$

11. $\displaystyle\int e^x \sec^2 (e^x - 7)\, dx$

12. $\displaystyle\int e^y \csc\,(e^y + 1)\, \cot\,(e^y + 1)\, dy$

13. $\displaystyle\int \frac{\tan\,(\ln v)}{v}\, dv$

14. $\displaystyle\int \sec \frac{\theta}{3}\, \tan \frac{\theta}{3}\, d\theta$

15. $\displaystyle\int \sin^2 \frac{x}{4}\, dx$

16. $\displaystyle\int \frac{(\ln x)^{-3}}{x}\, dx$

17. $\displaystyle\int \frac{1}{r} \csc^2 (1 + \ln r)\, dr$

18. $\displaystyle\int \frac{\cos\,(1 - \ln v)}{v}\, dv$

19. $\displaystyle\int x 3^{x^2}\, dx$

20. $\displaystyle\int 2^{\tan x} \sec^2 x\, dx$

Finite Sums and Estimates

21. *Flight of a model rocket* The accompanying figure shows the graph of the velocity (ft/sec) of a model rocket for the first 8 sec after launch. The rocket accelerated straight up for the first 2 sec and then coasted to reach its maximum height at $t = 8$ sec.

Time after launch (sec)

(a) Assuming that the rocket was launched from ground level, about how high did it go? (This is the rocket in Section 2.2, Exercise 15, but you do not need to do Exercise 15 to do the exercise here.)

(b) Sketch a graph of the rocket's height aboveground as a function of time for $0 \le t \le 8$.

22. *Analyzing linear motion*

(a) The accompanying figure shows the velocity (m/sec) of a body moving along the s-axis during the time interval from $t = 0$ to $t = 10$ sec. About how far did the body travel during those 10 sec?

(b) Sketch a graph of the body's position s as a function of t for $0 \le t \le 10$ assuming $s(0) = 0$.

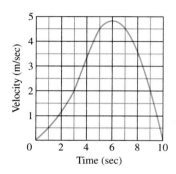

Time (sec)

Definite Integrals

In Exercises 23–26, express each limit as a definite integral. Then evaluate the integral to find the value of the limit. In each case, P is a partition of the given interval and the numbers c_k are chosen from the subintervals of P.

23. $\displaystyle\lim_{\|P\|\to 0} \sum_{k=1}^{n} (2c_k - 1)^{-1/2}\, \Delta x_k$, where P is a partition of $[1, 5]$

24. $\displaystyle\lim_{\|P\|\to 0} \sum_{k=1}^{n} c_k (c_k^2 - 1)^{1/3}\, \Delta x_k$, where P is a partition of $[1, 3]$

25. $\displaystyle\lim_{\|P\|\to 0} \sum_{k=1}^{n} \left(\cos\left(\frac{c_k}{2}\right)\right) \Delta x_k$, where P is a partition of $[-\pi, 0]$

26. $\displaystyle\lim_{\|P\|\to 0} \sum_{k=1}^{n} (\sin c_k)(\cos c_k)\, \Delta x_k$, where P is a partition of $[0, \pi/2]$

Using Properties and Known Values to Find Other Integrals

27. If $\int_{-2}^{2} 3f(x)\, dx = 12$, $\int_{-2}^{5} f(x)\, dx = 6$, and $\int_{-2}^{5} g(x)\, dx = 2$, find the values of the following.

(a) $\displaystyle\int_{-2}^{2} f(x)\, dx$

(b) $\displaystyle\int_{2}^{5} f(x)\, dx$

(c) $\displaystyle\int_{5}^{-2} g(x)\, dx$

(d) $\displaystyle\int_{-2}^{5} (-\pi g(x))\, dx$

(e) $\displaystyle\int_{-2}^{5} \left(\frac{f(x) + g(x)}{5}\right) dx$

28. If $\int_0^2 f(x)\,dx = \pi$, $\int_0^2 7g(x)\,dx = 7$, and $\int_0^1 g(x)\,dx = 2$, find the values of the following.

(a) $\displaystyle\int_0^2 g(x)\,dx$

(b) $\displaystyle\int_1^2 g(x)\,dx$

(c) $\displaystyle\int_2^0 f(x)\,dx$

(d) $\displaystyle\int_0^2 \sqrt{2}\,f(x)\,dx$

(e) $\displaystyle\int_0^2 (g(x) - 3f(x))\,dx$

Evaluating Definite Integrals

Evaluate the integrals in Exercises 29–52.

29. $\displaystyle\int_{-1}^1 (3x^2 - 4x + 7)\,dx$

30. $\displaystyle\int_0^1 (8s^3 - 12s^2 + 5)\,ds$

31. $\displaystyle\int_1^4 \left(\frac{x}{8} + \frac{1}{2x}\right) dx$

32. $\displaystyle\int_1^8 \left(\frac{2}{3x} - \frac{8}{x^2}\right) dx$

33. $\displaystyle\int_1^4 \frac{dt}{t\sqrt{t}}$

34. $\displaystyle\int_1^4 \frac{(1 + \sqrt{u})^{1/2}}{\sqrt{u}}\,du$

35. $\displaystyle\int_0^1 \frac{36\,dx}{(2x + 1)^3}$

36. $\displaystyle\int_0^1 \frac{dr}{\sqrt[3]{(7 - 5r)^2}}$

37. $\displaystyle\int_0^{\ln 5} e^r (3e^r + 1)^{-3/2}\,dr$

38. $\displaystyle\int_0^{\ln 9} e^\theta (e^\theta - 1)^{1/2}\,d\theta$

39. $\displaystyle\int_e^{e^2} \frac{1}{x\sqrt{\ln x}}\,dx$

40. $\displaystyle\int_0^{\pi/4} \cos^2\left(4t - \frac{\pi}{4}\right) dt$

41. $\displaystyle\int_1^e \frac{1}{x}(1 + 7\ln x)^{-1/3}\,dx$

42. $\displaystyle\int_{\pi/4}^{3\pi/4} \csc^2 x\,dx$

43. $\displaystyle\int_{-2}^2 \frac{3\,dt}{4 + 3t^2}$

44. $\displaystyle\int_1^3 \frac{(\ln (v + 1))^2}{v + 1}\,dv$

45. $\displaystyle\int_{-\pi/3}^0 \sec x \tan x\,dx$

46. $\displaystyle\int_{\pi/4}^{3\pi/4} \csc z \cot z\,dz$

47. $\displaystyle\int_{\sqrt{2}/3}^{2/3} \frac{dy}{|y|\sqrt{9y^2 - 1}}$

48. $\displaystyle\int_{-1}^1 2x \sin (1 - x^2)\,dx$

49. $\displaystyle\int_0^{\pi/2} \frac{3 \sin x \cos x}{\sqrt{1 + 3 \sin^2 x}}\,dx$

50. $\displaystyle\int_{-2}^{-1} \frac{2\,dv}{v^2 + 4v + 5}$

51. $\displaystyle\int_{-1}^1 \frac{3\,dv}{4v^2 + 4v + 4}$

52. $\displaystyle\int_{\pi^2/36}^{\pi^2/4} \frac{\cos \sqrt{t}}{\sqrt{t} \sin \sqrt{t}}\,dt$

Area

In Exercises 53–56, find the total area of the region between the graph of f and the x-axis.

53. $f(x) = x^2 - 4x + 3$, $\quad 0 \le x \le 3$

54. $f(x) = 1 - (x^2/4)$, $\quad -2 \le x \le 3$

55. $f(x) = 5 - 5x^{2/3}$, $\quad -1 \le x \le 8$

56. $f(x) = 1 - \sqrt{x}$, $\quad 0 \le x \le 4$

Find the areas of the regions enclosed by the curves and lines in Exercises 57–64.

57. $y = x$, $\quad y = 1/x^2$, $\quad x = 2$

58. $y = x$, $\quad y = 1/\sqrt{x}$, $\quad x = 2$

59. $\sqrt{x} + \sqrt{y} = 1$, $\quad x = 0$, $\quad y = 0$

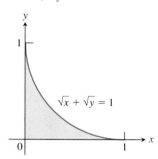

60. $x^3 + \sqrt{y} = 1$, $\quad x = 0$, $\quad y = 0$, $\quad$ for $\quad 0 \le x \le 1$

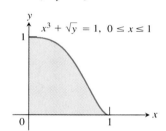

61. $y = \sin x$, $\quad y = x$, $\quad 0 \le x \le \pi/4$

62. $y = |\sin x|$, $\quad y = 1$, $\quad -\pi/2 \le x \le \pi/2$

63. $y = 2 \sin x$, $\quad y = \sin 2x$, $\quad 0 \le x \le \pi$

64. $y = 8 \cos x$, $\quad y = \sec^2 x$, $\quad -\pi/3 \le x \le \pi/3$

65. Find the extreme values of $f(x) = x^3 - 3x^2$ and find the area of the region enclosed by the graph of f and the x-axis.

66. Find the area of the region cut from the first quadrant by the curve $x^{1/3} + y^{1/3} = 1$.

Initial Value Problems

Solve the initial value problems in Exercises 67–70.

67. $\dfrac{dy}{dx} = \dfrac{x^2 + 1}{x^2}$, $\quad y(1) = -1$

68. $\dfrac{dy}{dx} = e^{-x-y-2}$, $\quad y(0) = -2$

69. $\dfrac{d^2 r}{dt^2} = 15\sqrt{t} + \dfrac{3}{\sqrt{t}}$; $\quad r'(1) = 8$, $\quad r(1) = 0$

70. $\dfrac{d^3r}{dt^3} = -\cos t; \quad r''(0) = r'(0) = 0, \quad r(0) = -1$

71. Show that $y = x^2 + \int_1^x (1/t)\, dt$ solves the initial value problem

$$\frac{d^2y}{dx^2} = 2 - \frac{1}{x^2}; \qquad y'(1) = 3, \qquad y(1) = 1.$$

72. Show that $y = \int_0^x (1 + 2\sqrt{\sec t})\, dt$ solves the initial value problem

$$\frac{d^2y}{dx^2} = \sqrt{\sec x}\,\tan x; \qquad y'(0) = 3, \qquad y(0) = 0.$$

Express the solutions of the initial value problems in Exercises 73 and 74 in terms of integrals.

73. $\dfrac{dy}{dx} = \dfrac{\sin x}{x}, \quad y(5) = -3$

74. $\dfrac{dy}{dx} = \sqrt{2 - \sin^2 x}, \quad y(-1) = 2$

Average Values

75. Find the average value of $f(x) = mx + b$

  (a) over $[-1, 1]$

  (b) over $[-k, k]$.

76. Find the average value of

  (a) $y = \sqrt{3x}$ over $[0, 3]$

  (b) $y = \sqrt{ax}$ over $[0, a]$.

77. *Writing to Learn: Average and instantaneous rates of change* Let f be a function that is differentiable on $[a, b]$. In Chapter 1, we defined the average rate of change of f over $[a, b]$ to be

$$\frac{f(b) - f(a)}{b - a}$$

and the instantaneous rate of change of f at x to be $f'(x)$. In this chapter, we defined the average value of a function. For the new definition of average to be consistent with the old one, we should have

$$\frac{f(b) - f(a)}{b - a} = \text{average value of } f' \text{ on } [a, b].$$

Is this the case? Give reasons for your answer.

78. *Writing to Learn: Average value* Is it true that the average value of an integrable function over an interval of length 2 is half the function's integral over the interval? Give reasons for your answer.

Differentiating Integrals

In Exercises 79–82, find dy/dx.

79. $y = \displaystyle\int_2^x \sqrt{2 + \cos^3 t}\, dt$

80. $y = \displaystyle\int_2^{7x^2} \sqrt{2 + \cos^3 t}\, dt$

81. $y = \displaystyle\int_x^1 \frac{6}{3 + t^4}\, dt$

82. $y = \displaystyle\int_{\sec x}^2 \frac{1}{t^2 + 1}\, dt$

Numerical Integration

83. A direct calculation from Example 3, Section 4.2 shows that

$$\int_0^\pi 2\,\sin^2 x\, dx = \pi.$$

How close do you come to this value by using the Trapezoidal Rule with $n = 6$? Simpson's Rule with $n = 6$? Try them and find out.

84. *Fuel efficiency* An automobile computer gives a digital readout of fuel consumption in gallons per hour. During a trip, a passenger recorded the fuel consumption every 5 min for a full hour of travel.

Time	Gal/h	Time	Gal/h
0	2.5	35	2.5
5	2.4	40	2.4
10	2.3	45	2.3
15	2.4	50	2.4
20	2.4	55	2.4
25	2.5	60	2.3
30	2.6		

  (a) Use the Trapezoidal Rule to approximate the total fuel consumption during the hour.

  (b) If the automobile covered 60 mi in the hour, what was its fuel efficiency (in miles per gallon) for that portion of the trip?

85. *Mean temperature* Compute the average value of the temperature function

$$f(x) = 37\,\sin\left(\frac{2\pi}{365}(x - 101)\right) + 25$$

for a 365-day year. This is one way to estimate the annual mean air temperature in Fairbanks, Alaska. The National Weather Service's official figure, a numerical average of the daily normal mean air temperatures for the year, is 25.7°F, which is slightly higher than the average value of $f(x)$. Figure 2.34 shows why.

86. *Specific heat of a gas* Heat capacity C_v is the amount of heat required to raise the temperature of a given mass of gas with constant volume by 1°C, measured in units of cal/deg-mol (calories per degree gram molecular weight). The heat capacity of oxygen depends on its temperature T and satisfies the formula

$$C_v = 8.27 + 10^{-5}(26T - 1.87T^2).$$

Find the average value of C_v for $20° \leq T \leq 675°C$ and the temperature at which it is attained.

87. *A new parking lot* To meet the demand for parking, your town has allocated the area shown here. As the town engineer, you have been asked by the town council to find out if the lot can be built for $11,000. The cost to clear the land will be $0.10 a square foot, and the lot will cost $2.00 a square foot to pave. Can the job be done for $11,000?

0 ft

36 ft

54 ft

51 ft

49.5 ft

54 ft

64.4 ft

67.5 ft

42 ft

Ignored

Vertical spacing = 15 ft

88. *Rubber-band-powered sled* A sled powered by a wound rubber band moves along a track until friction and the unwinding of the rubber band gradually slow it to a stop. A speedometer in the sled monitors its speed, which is recorded at 3 sec intervals during the 27 sec run.

Time (sec)	Speed (ft/sec)
0	5.30
3	5.25
6	5.04
9	4.71
12	4.25
15	3.66
18	2.94
21	2.09
24	1.11
27	0

(a) Give an upper estimate and a lower estimate for the distance traveled by the sled.

(b) Use the Trapezoidal Rule to estimate the distance traveled by the sled.

Theory and Examples

89. *Writing to Learn* Is it true that every function $y = f(x)$ that is differentiable on $[a, b]$ is itself the derivative of some function on $[a, b]$? Give reasons for your answer.

90. *Writing to Learn* Suppose that $F(x)$ is an antiderivative of $f(x) = \sqrt{1 + x^4}$. Express $\int_0^1 \sqrt{1 + x^4}\, dx$ in terms of F and give a reason for your answer.

91. *Expressing a solution as a definite integral* Express the function $y(x)$ with

$$\frac{dy}{dx} = \frac{\sin x}{x} \quad \text{and} \quad y(5) = 3$$

as a definite integral.

92. *A differential equation* Show that $y = \sin x + \int_x^\pi \cos 2t\, dt + 1$ satisfies both of the following conditions:

   i. $y'' = -\sin x + 2 \sin 2x$

   ii $y = 1$ and $y' = -2$ when $x = \pi$.

93. *A function defined by an integral* The graph of a function f consists of a semicircle and two line segments as shown below.

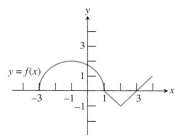

Let $g(x) = \int_1^x f(t)\, dt$.

(a) Find $g(1)$.   (b) Find $g(3)$.   (c) Find $g(-1)$.

(d) Find all the values of x on the open interval $(-3, 4)$ at which g has a relative maximum.

(e) Write an equation for the line tangent to the graph of g at $x = -1$.

(f) Find the x-coordinate of each point of inflection of the graph of g on the open interval $(-3, 4)$.

(g) Find the range of g.

94. *Skydiving* Skydivers A and B are in a helicopter hovering at 6400 ft. Skydiver A jumps and descends for 4 sec before opening her parachute. The helicopter then climbs to 7000 ft and hovers there. Forty-five seconds after A leaves the aircraft, B jumps and descends for 13 sec before opening her parachute. Both skydivers descend at 16 ft/sec with parachutes open. Assume that the skydivers fall freely (with acceleration -32 ft/sec^2) before their parachutes open.

(a) At what altitude does A's parachute open?

(b) At what altitude does B's parachute open?

(c) Which skydiver lands first?

Average Daily Inventory

Average value is used in economics to study such things as average daily inventory. If $I(t)$ is the number of radios, tires, shoes, or whatever product a firm has on hand on day t (we call I an **inventory function**), the average value of I over a time period $[0, T]$ is called the firm's average daily inventory for the period.

$$\textbf{Average daily inventory} = \text{av}\,(I) = \frac{1}{T}\int_0^T I(t)\,dt.$$

If h is the dollar cost of holding one item per day, the product av $(I) \cdot h$ is the **average daily holding cost** for the period.

95. As a wholesaler, Tracey Burr Distributors (TBD) receives a shipment of 1200 cases of chocolate bars every 30 days. TBD sells the chocolate to retailers at a steady rate, and t days after a shipment arrives, its inventory of cases on hand is $I(t) = 1200 - 40t$, $0 \le t \le 30$. What is TBD's average daily inventory for the 30-day period? What is its average daily holding cost if the cost of holding one case is 3¢ a day?

96. Rich Wholesale Foods, a manufacturer of cookies, stores its cases of cookies in an air-conditioned warehouse for shipment every 14 days. Rich tries to keep 600 cases on reserve to meet occasional peaks in demand, so a typical 14-day inventory function is $I(t) = 600 + 600t$, $0 \le t \le 14$. The daily holding cost for each case is 4¢ per day. Find Rich's average daily inventory and average daily holding cost.

97. Solon Container receives 450 drums of plastic pellets every 30 days. The inventory function (drums on hand as a function of days) is $I(t) = 450 - t^2/2$. Find the average daily inventory. If the holding cost for one drum is 2¢ per day, find the average daily holding cost.

98. Mitchell Mailorder receives a shipment of 600 cases of athletic socks every 60 days. The number of cases on hand t days after the shipment arrives is $I(t) = 600 - 20\sqrt{15t}$. Find the average daily inventory. If the holding cost for one case is 1/2¢ per day, find the average daily holding cost.

Additional Exercises: Theory, Examples, Applications

Theory and Examples

1. (a) If $\int_0^1 7 f(x)\,dx = 7$, does $\int_0^1 f(x)\,dx = 1$?

   (b) If $\int_0^1 f(x)\,dx = 4$ and $f(x) \ge 0$, does $\int_0^1 \sqrt{f(x)}\,dx = \sqrt{4} = 2$?

   Give reasons for your answers.

2. Suppose that $\int_{-2}^2 f(x)\,dx = 4$, $\int_2^5 f(x)\,dx = 3$, $\int_{-2}^5 g(x)\,dx = 2$. Which, if any, of the following statements are true?

   (a) $\int_5^2 f(x)\,dx = -3$ **(b)** $\int_{-2}^5 (f(x) + g(x))\,dx = 9$

   (c) $f(x) \le g(x)$ on the interval $-2 \le x \le 5$

3. *Initial value problem* Show that

$$y = \frac{1}{a}\int_0^x f(t)\,\sin\,a(x-t)\,dt$$

solves the initial value problem

$$\frac{d^2y}{dx^2} + a^2 y = f(x), \quad \frac{dy}{dx} = 0 \quad \text{and} \quad y = 0 \quad \text{when} \quad x = 0.$$

(*Hint*: $\sin\,(ax - at) = \sin ax \cos at - \cos ax \sin at$.)

4. *Proportionality* Suppose that x and y are related by the equation

$$x = \int_0^y \frac{1}{\sqrt{1 + 4t^2}}\,dt.$$

Show that d^2y/dx^2 is proportional to y and find the constant of proportionality.

5. Find $f(4)$ if

   (a) $\int_0^{x^2} f(t)\,dt = x \cos \pi x$

   (b) $\int_0^{f(x)} t^2\,dt = x \cos \pi x$.

6. Find $f(\pi/2)$ from the following information.

   i. f is positive and continuous.

   ii. The area under the curve $y = f(x)$ from $x = 0$ to $x = a$ is

$$\frac{a^2}{2} + \frac{a}{2}\sin a + \frac{\pi}{2}\cos a.$$

7. The area of the region in the xy-plane enclosed by the x-axis, the curve $y = f(x)$, $f(x) \ge 0$, and the lines $x = 1$ and $x = b$ is equal to $\sqrt{b^2 + 1} - \sqrt{2}$ for all $b > 1$. Find $f(x)$.

8. Prove that

$$\int_0^x \left(\int_0^u f(t)\,dt\right)du = \int_0^x f(u)(x - u)\,du.$$

(*Hint*: Express the integral on the right-hand side as the difference of two integrals. Then show that both sides of the equation have the same derivative with respect to x.)

9. *Finding a curve* Find the equation for the curve in the xy-plane that passes through the point $(1, -1)$ if its slope at x is always $3x^2 + 2$.

10. *Shoveling dirt* You sling a shovelfull of dirt up from the bottom of a hole with an initial velocity of 32 ft/sec. The dirt must rise 17 ft above the release point to clear the edge of the hole. Is that enough speed to get the dirt out, or had you better duck?

Piecewise Continuous Functions

Although we are mainly interested in continuous functions, many functions in applications are piecewise continuous. A function $f(x)$ is **piecewise continuous on a closed interval** I if f has only finitely many discontinuities in I, the limits

$$\lim_{x \to c^-} f(x) \quad \text{and} \quad \lim_{x \to c^+} f(x)$$

exist and are finite at every interior point of I, and the appropriate one-sided limits exist and are finite at the endpoints of I. All piecewise continuous functions are integrable. The points of discontinuity partition I into open and half-open subintervals on which f is continuous, and the limit criteria above guarantee that f has a continuous extension to the closure of each subinterval. To integrate a piecewise continuous function, we integrate the individual extensions and add the results. The integral of

$$f(x) = \begin{cases} 1 - x, & -1 \le x < 0 \\ x^2, & 0 \le x < 2 \\ -1, & 2 \le x \le 3 \end{cases}$$

(Figure 4.30) over $[-1, 3]$ is

$$\int_{-1}^{3} f(x)\, dx = \int_{-1}^{0} (1 - x)\, dx + \int_{0}^{2} x^2\, dx + \int_{2}^{3} (-1)\, dx$$

$$= \left[x - \frac{x^2}{2} \right]_{-1}^{0} + \left[\frac{x^3}{3} \right]_{0}^{2} + \left[-x \right]_{2}^{3}$$

$$= \frac{3}{2} + \frac{8}{3} - 1 = \frac{19}{6}.$$

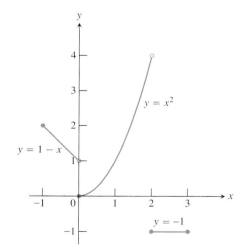

FIGURE 4.30 Piecewise continuous functions like this are integrated piece by piece.

The Fundamental Theorem applies to piecewise continuous functions with the restriction that $(d/dx) \int_a^x f(t)\, dt$ is expected to equal $f(x)$ only at values of x at which f is continuous. There is a similar restriction on Leibniz's Rule below.

Graph the functions in Exercises 11–16 and integrate them over their domains.

11. $f(x) = \begin{cases} x^{2/3}, & -8 \le x < 0 \\ -4, & 0 \le x \le 3, \end{cases}$

12. $f(x) = \begin{cases} \sqrt{-x}, & -4 \le x < 0 \\ x^2 - 4, & 0 \le x \le 3 \end{cases}$

13. $g(t) = \begin{cases} t, & 0 \le t < 1 \\ \sin \pi t, & 1 \le t \le 2 \end{cases}$

14. $h(z) = \begin{cases} \sqrt{1 - z}, & 0 \le z < 1 \\ (7z - 6)^{-1/3}, & 1 \le z \le 2 \end{cases}$

15. $f(x) = \begin{cases} 1, & -2 \le x < -1 \\ 1 - x^2, & -1 \le x < 1 \\ 2, & 1 \le x \le 2 \end{cases}$

16. $h(r) = \begin{cases} r, & -1 \le r < 0 \\ 1 - r^2, & 0 \le r < 1 \\ 1, & 1 \le r \le 2 \end{cases}$

17. Find the average value of the function graphed in Figure 4.31a.

18. Find the average value of the function graphed in Figure 4.31b.

(a)

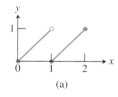

(b)

FIGURE 4.31 The graphs for Exercises 17 and 18.

Leibniz's Rule

In applications, we sometimes encounter functions like

$$f(x) = \int_{\sin x}^{x^2} (1 + t)\, dt \quad \text{and} \quad g(x) = \int_{\sqrt{x}}^{2\sqrt{x}} \sin t^2\, dt,$$

defined by integrals that have variable upper limits of integration and variable lower limits of integration at the same time. The first integral can be evaluated directly, but the second cannot. We may find the derivative of either integral, however, by a formula called **Leibniz's Rule.**

Leibniz's Rule

If f is continuous on $[a, b]$ and if $u(x)$ and $v(x)$ are differentiable functions of x whose values lie in $[a, b]$, then

$$\frac{d}{dx} \int_{u(x)}^{v(x)} f(t)\, dt = f(v(x)) \frac{dv}{dx} - f(u(x)) \frac{du}{dx}.$$

Figure 4.32 gives a geometric interpretation of Leibniz's Rule. It shows a carpet of variable width $f(t)$ that is being rolled up at the left at the same time x as it is being unrolled at the right. (In this interpretation, time is x, not t.) At time x, the floor is covered from $u(x)$ to $v(x)$. The rate du/dx at which the carpet is being rolled up need not be the same as the rate dv/dx at which the carpet is being laid down. At any given time x, the area covered by carpet is

$$A(x) = \int_{u(x)}^{v(x)} f(t)\, dt.$$

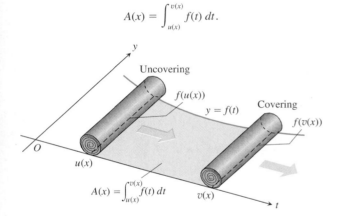

y

Uncovering

$f(u(x))$

$y = f(t)$ Covering

$f(v(x))$

O

$u(x)$

$A(x) = \int_{u(x)}^{v(x)} f(t)\, dt$ $v(x)$ t

FIGURE 4.32 Rolling and unrolling a carpet; a geometric interpretation of Leibniz's Rule:

$$\frac{dA}{dx} = f(v(x)) \frac{dv}{dx} - f(u(x)) \frac{du}{dx}.$$

At what rate is the covered area changing? At the instant x, $A(x)$ is increasing by the width $f(v(x))$ of the unrolling carpet times the rate dv/dx at which the carpet is being unrolled. That is, $A(x)$ is being increased at the rate

$$f(v(x)) \frac{dv}{dx}.$$

At the same time, A is being decreased at the rate

$$f(u(x)) \frac{du}{dx},$$

the width at the end that is being rolled up times the rate du/dx. The net rate of change in A is

$$\frac{dA}{dx} = f(v(x)) \frac{dv}{dx} - f(u(x)) \frac{du}{dx},$$

which is precisely Leibniz's Rule.

To prove the rule, let F be an antiderivative of f on $[a, b]$. Then

$$\int_{u(x)}^{v(x)} f(t)\, dt = F(v(x)) - F(u(x)).$$

Differentiating both sides of this equation with respect to x gives the equation we want:

$$\frac{d}{dx} \int_{u(x)}^{v(x)} f(t)\, dt = \frac{d}{dx} [F(v(x)) - F(u(x))]$$

$$= F'(v(x)) \frac{dv}{dx} - F'(u(x)) \frac{du}{dx}$$

$$= f(v(x)) \frac{dv}{dx} - f(u(x)) \frac{du}{dx}.$$

You will see another way to derive the rule in Chapter 11, additional Exercise 3.

Use Leibniz's Rule to find the derivatives of the functions in Exercises 19–21.

19. $f(x) = \displaystyle\int_{1/x}^{x} \frac{1}{t}\, dt$

20. $f(x) = \displaystyle\int_{\cos x}^{\sin x} \frac{1}{1 - t^2}\, dt$

21. $g(y) = \displaystyle\int_{\sqrt{y}}^{2\sqrt{y}} \sin t^2\, dt$

22. Use Leibniz's Rule to find the value of x that maximizes the value of the integral

$$\int_{x}^{x+3} t(5 - t)\, dt.$$

Problems like this arise in the mathematical theory of political elections. See "The Entry Problem in a Political Race" by Steven J. Brams and Philip D. Straffin Jr., in *Political Equilibrium*, edited by Peter Ordeshook and Kenneth Shepfle (Boston: Kluwer-Nijhoff, 1982), pp. 181–195.

5 Applications of Integrals

OVERVIEW   Many things we want to know can be calculated with integrals: the volumes of solids, the lengths of curves, the amount of work it takes to pump liquids from below ground, the forces against floodgates, the coordinates of the points where solid objects will balance. We define all these as limits of Riemann sums of continuous functions on closed intervals—that is, as integrals—and evaluate these limits with calculus.

5.1 Volumes by Slicing and Rotation About an Axis

Volumes by Slicing   •   Solids of Revolution: Circular Cross Sections   •
Solids of Revolution: Washer Cross Sections

In Section 4.3, Example 3, we estimated the volume of a sphere by partitioning it into thin slices that were nearly cylindrical and summing the cylinders' volumes in what turned out later to be a Riemann sum. Had we known how at the time, we could have continued on to express the volume of the sphere as a definite integral.

Starting the same way, we can now find the volumes of a great many solids by integration.

Volumes by Slicing

Suppose that we want to find the volume of a solid like the one in Figure 5.1. The cross section of the solid at each point x in the interval $[a, b]$ is a region $R(x)$ of area $A(x)$. If A is a continuous function of x, we can use it to define and calculate the volume of the solid as an integral in the following way.

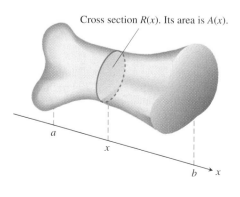

Cross section $R(x)$. Its area is $A(x)$.

FIGURE 5.1 If the area $A(x)$ of the cross section $R(x)$ is a continuous function of x, we can find the volume of the solid by integrating $A(x)$ from a to b.

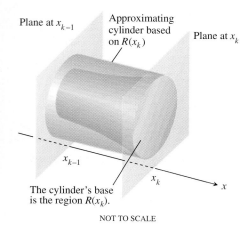

Plane at x_{k-1}

Approximating cylinder based on $R(x_k)$

Plane at x_k

x_{k-1}

x_k

The cylinder's base is the region $R(x_k)$.

NOT TO SCALE

FIGURE 5.2 Enlarged view of the slice of the solid between the planes at x_{k-1} and x_k and its approximating cylinder.

We partition $[a, b]$ into subintervals of length Δx and slice the solid, as we would a loaf of bread, by planes perpendicular to the x-axis at the partition points. The kth slice, the one between the planes at x_{k-1} and x_k, has approximately the same volume as the cylinder between the two planes based on the region $R(x_k)$ (Figure 5.2).

The volume of the cylinder is

$$V_k = \text{base area} \times \text{height} = A(x_k) \times \Delta x.$$

The sum

$$\sum V_k = \sum A(x_k) \times \Delta x$$

approximates the volume of the solid.

This is a Riemann sum for $A(x)$ on $[a, b]$. We expect the approximations to improve as the norms of the partitions go to zero, so we define their limiting integral to be the *volume of the solid.*

Definition   Volume of a Solid

The **volume** of a solid of known integrable cross-section area $A(x)$ from $x = a$ to $x = b$ is the integral of A from a to b,

$$V = \int_a^b A(x)\, dx.$$

To apply this formula, we proceed as follows.

CD-ROM
WEBsite

How to Find Volume by the Method of Slicing

Step 1. *Sketch the solid and a typical cross section.*

Step 2. *Find a formula for $A(x)$.*

Step 3. *Find the limits of integration.*

Step 4. *Integrate $A(x)$ to find the volume.*

Typical cross section

x

0

3

x

x

3

x (m)

3

FIGURE 5.3 The cross sections of the pyramid in Example 1 are squares.

Example 1   Volume of a Pyramid

A pyramid 3 m high has a square base that is 3 m on a side. The cross section of the pyramid perpendicular to the altitude x m down from the vertex is a square x m on a side. Find the volume of the pyramid.

Solution

Step 1: *A sketch.* We draw the pyramid with its altitude along the x-axis and its vertex at the origin and include a typical cross section (Figure 5.3).

Step 2: *A formula for $A(x)$.* The cross section at x is a square x meters on a side, so its area is

$$A(x) = x^2.$$

Step 3: *The limits of integration.* The squares go from $x = 0$ to $x = 3$.

Step 4: *Integrate to find the volume.*

$$V = \int_0^3 A(x)\, dx = \int_0^3 x^2\, dx = \frac{x^3}{2}\Big]_0^3 = 9 \text{ m}^3$$

Example 2   Cavalieri's Volume Theorem

Cavalieri's volume theorem says that solids with equal altitudes and identical cross-section areas at each height have the same volume (Figure 5.4). This follows immediately from the definition of volume, because the cross-section area function $A(x)$ and the interval $[a, b]$ are the same for both solids.

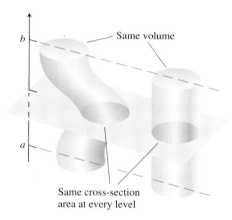

FIGURE 5.4 Cavalieri's Theorem: These solids have the same volume. You can illustrate this yourself with stacks of coins.

FIGURE 5.5 The wedge of Example 3, sliced perpendicular to the x-axis. The cross sections are rectangles.

Example 3   Volume of a Wedge

A curved wedge is cut from a cylinder of radius 3 by two planes. One plane is perpendicular to the axis of the cylinder. The second plane crosses the first plane at a 45° angle at the center of the cylinder. Find the volume of the wedge.

Solution

Step 1: *A sketch.* We draw the wedge and sketch a typical cross section perpendicular to the x-axis (Figure 5.5).

Step 2: *The formula for $A(x)$.* The cross section at x is a rectangle of area

$$A(x) = (\text{height})(\text{width}) = (x)\,(2\sqrt{9 - x^2})$$

$$= 2x\sqrt{9 - x^2}, \text{ units squared.}$$

Step 3: *The limits of integration.* The rectangles run from $x = 0$ to $x = 3$.

Step 4: *Integrate to find the volume.*

$$V = \int_a^b A(x)\, dx = \int_0^3 2x\sqrt{9 - x^2}\, dx$$

$$= -\frac{2}{3}(9 - x^2)^{3/2}\Big]_0^3$$

$$= 0 + \frac{2}{3}(9)^{3/2}$$

Let $u = 9 - x^2$, $du = -2x\, dx$, integrate, and substitute back.

$$= 18, \text{ units cubed.}$$

Solids of Revolution: Circular Cross Sections

The most common application of the method of slicing is to solids of revolution. **Solids of revolution** are solids whose shapes can be generated by revolving plane regions about axes. The only thing that changes when the cross sections are circular is the formula for the area $A(x)$.

The typical cross section of the solid perpendicular to the axis of revolution is a disk of radius $R(x)$ and area

$$A(x) = \pi(\text{radius})^2 = \pi[R(x)]^2.$$

For this reason, the method is often called the **disk method.** Here are several examples.

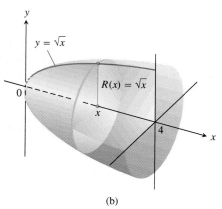

FIGURE 5.6 The region (a) and solid (b) in Example 4.

Example 4 A Solid of Revolution (Rotation About the x-axis)

The region between the curve $y = \sqrt{x}$, $0 \le x \le 4$, and the x-axis is revolved around the x-axis to generate a solid. Find its volume.

Solution We draw figures showing the region, a typical radius, and the generated solid (Figure 5.6). The volume is

$$V = \int_a^b \pi[R(x)]^2\, dx$$

$$= \int_0^4 \pi[\sqrt{x}]^2\, dx \qquad R(x) = \sqrt{x}$$

$$= \pi\int_0^4 x\, dx = \pi\,\frac{x^2}{2}\Big]_0^4 = \pi\,\frac{(4)^2}{2} = 8\pi \text{ units cubed.}$$

The axis of revolution in the next example is not the x-axis, but the rule for calculating the volume is the same: Integrate $\pi(\text{radius})^2$ between appropriate limits.

Example 5 A Solid of Revolution (Rotation About the Line $y = 1$)

Find the volume of the solid generated by revolving the region bounded by $y = \sqrt{x}$ and the lines $y = 1$, $x = 4$ about the line $y = 1$.

Solution   We draw figures showing the region, a typical radius, and the generated solid (Figure 5.7). The volume is

$$V = \int_1^4 \pi [R(x)]^2 \, dx$$

$$= \int_1^4 \pi [\sqrt{x} - 1]^2 \, dx \qquad R(x) = \sqrt{x} - 1$$

$$= \pi \int_1^4 [x - 2\sqrt{x} + 1] \, dx$$

$$= \pi \left[\frac{x^2}{2} - 2 \cdot \frac{2}{3} x^{3/2} + x \right]_1^4 = \frac{7\pi}{6} \text{ units cubed.}$$

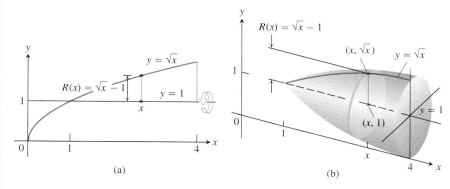

FIGURE 5.7 The region (a) and solid (b) in Example 5.

CD-ROM
WEBsite

How to Find Volumes for Circular Cross Sections (Disk Method)

Step 1. *Draw the region and identify the radius function $R(x)$.*

Step 2. *Square $R(x)$ and multiply by π.*

Step 3. *Integrate to find the volume.*

To find the volume of a solid generated by revolving a region between the y-axis and a curve $x = R(y)$, $c \le y \le d$, about the y-axis, we use the same method with x replaced by y. In this case, the circular cross section is

$$A(y) = \pi \, [\text{radius}]^2 = \pi \, [R(y)]^2.$$

Example 6   Revolution About the y-axis

Find the volume of the solid generated by revolving the region between the y-axis and the curve $x = 2/y$, $1 \le y \le 4$, about the y-axis.

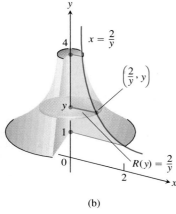

(b)

FIGURE 5.8 The region (a) and solid (b) in Example 6.

Solution We draw figures showing the region, a typical radius, and the generated solid (Figure 5.8). The volume is

$$V = \int_1^4 \pi [R(y)]^2 \, dy$$

$$= \int_1^4 \pi \left(\frac{2}{y} \right)^2 dy \qquad R(y) = \frac{2}{y}$$

$$= \pi \int_1^4 \frac{4}{y^2} \, dy = 4\pi \left[-\frac{1}{y} \right]_1^4 = 4\pi \left[\frac{3}{4} \right]$$

$$= 3\pi \text{ units cubed.}$$

Example 7 Revolution About a Vertical Axis

Find the volume of the solid generated by revolving the region between the parabola $x = y^2 + 1$ and the line $x = 3$ about the line $x = 3$.

Solution We draw figures showing the region, a typical radius, and the generated solid (Figure 5.9). The volume is

$$V = \int_{-\sqrt{2}}^{\sqrt{2}} \pi \, [R(y)]^2 \, dy$$

$$= \int_{-\sqrt{2}}^{\sqrt{2}} \pi \, [2 - y^2]^2 \, dy \qquad \begin{aligned} R(y) &= 3 - (y^2 + 1) \\ &= 2 - y^2 \end{aligned}$$

$$= \pi \int_{-\sqrt{2}}^{\sqrt{2}} [4 - 4y^2 + y^4] \, dy$$

$$= \pi \left[4y - \frac{4}{3} y^3 + \frac{y^5}{5} \right]_{-\sqrt{2}}^{\sqrt{2}}$$

$$= \frac{64\pi\sqrt{2}}{15} \text{ units cubed.}$$

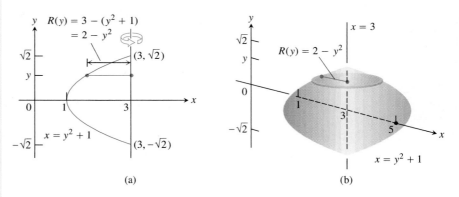

(a)   (b)

FIGURE 5.9 The region (a) and solid (b) in Example 7.

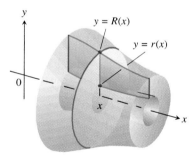

FIGURE 5.10   The cross sections of the solid of revolution generated here are washers, not disks, so the integral $\int_a^b A(x)\,dx$ leads to a slightly different formula.

Solids of Revolution: Washer Cross Sections

If the region we revolve to generate a solid does not border on or cross the axis of revolution, the solid has a hole in it (Figure 5.10). The cross-sections perpendicular to the axis of revolution are washers instead of disks. The dimensions of a typical washer are

Outer radius: $R(x)$

Inner radius: $r(x)$

The washer's area is

$$A(x) = \pi[R(x)]^2 - \pi[r(x)]^2 = \pi([R(x)]^2 - [r(x)]^2).$$

Example 8   A Washer Cross Section (Rotation About the x-axis)

The region bounded by the curve $y = x^2 + 1$ and the line $y = -x + 3$ is revolved about the x-axis to generate a solid. Find the volume of the solid.

Solution

Step 1: Draw the region and sketch a line segment across it perpendicular to the axis of revolution (the red segment in Figure 5.11).

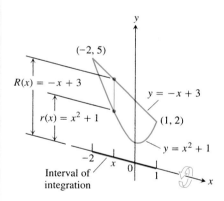

FIGURE 5.11.   The region in Example 8 spanned by a line segment perpendicular to the axis of revolution. When the region is revolved about the x-axis, the line segment will generate a washer.

Step 2: Find the limits of integration by finding the x-coordinates of the intersection points of the curve and line in Figure 5.11.

$$x^2 + 1 = -x + 3$$
$$x^2 + x - 2 = 0$$
$$(x + 2)(x - 1) = 0$$
$$x = -2, \qquad x = 1$$

Step 3: Find the outer and inner radii of the washer that would be swept out by the line segment if it were revolved about the x-axis along with the region. (We

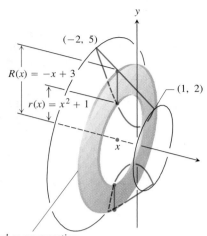

$R(x) = -x + 3$

$(-2, 5)$

$r(x) = x^2 + 1$

$(1, 2)$

Washer cross section
Outer radius: $R(x) = -x + 3$
Inner radius: $r(x) = x^2 + 1$

FIGURE 5.12 The inner and outer radii of the washer swept out by the line segment in Figure 5.11.

The area of a washer is $\pi R^2 - \pi r^2$.

drew the washer in Figure 5.12, but in your own work you need not do that.) These radii are the distances of the ends of the line segment from the axis of revolution.

Outer radius: $R(x) = -x + 3$

Inner radius: $r(x) = x^2 + 1$

Step 4: Evaluate the volume integral.

$$V = \int_a^b \pi([R(x)]^2 - [r(x)]^2)\,dx$$

$$= \int_{-2}^1 \pi((-x + 3)^2 - (x^2 + 1)^2)\,dx \qquad \text{Values from steps 2 and 3}$$

$$= \int_{-2}^1 \pi(8 - 6x - x^2 - x^4)\,dx \qquad \begin{array}{l}\text{Expressions squared and}\\\text{combined}\end{array}$$

$$= \pi\left[8x - 3x^2 - \frac{x^3}{3} - \frac{x^5}{5}\right]_{-2}^1 = \frac{117\pi}{5}\ \text{units cubed.}$$

How to Find Volumes for Washer Cross Sections

Step 1. *Draw the region and sketch a line segment across it perpendicular to the axis of revolution.* When the region is revolved, this segment will generate a typical washer cross section of the generated solid.

Step 2. *Find the limits of integration.*

Step 3. *Find the outer and inner radii* of the washer swept out by the line segment.

Step 4. *Integrate to find the volume.*

To find the volume of a solid generated by revolving a region about the y-axis, we use the steps listed above but integrate with respect to y instead of x.

Example 9   A Washer Cross Section (Rotation About the y-axis)

The region bounded by the parabola $y = x^2$ and the line $y = 2x$ in the first quadrant is revolved about the y-axis to generate a solid. Find the volume of a solid.

Solution

Step 1: Draw the region and sketch a line segment across it perpendicular to the axis of revolution, in this case the y-axis (Figure 5.13).

Step 2. The line and parabola intersect at $y = 0$ and $y = 4$, so the limits of integration are $c = 0$ and $d = 4$.

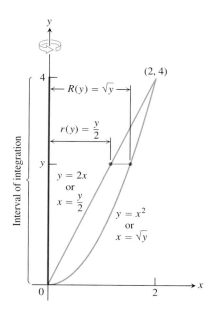

FIGURE 5.13 The region, limits of integration, and radii in Example 9.

Step 3. The radii of the washer swept out by the line segment are $R(y) = \sqrt{y}$, $r(y) = y/2$ (Figures 5.13 and 5.14).

Step 4. Integrate to find the volume:

$$V = \int_c^d \pi([R(y)]^2 - [r(y)]^2)\, dy$$

$$= \int_0^4 \pi\left(\left[\sqrt{y}\right]^2 - \left[\frac{y}{2}\right]^2\right) dy \qquad \text{Values from steps 2 and 3}$$

$$= \pi \int_0^4 \left(y - \frac{y^2}{4}\right) dy = \pi\left[\frac{y^2}{2} - \frac{y^3}{12}\right]_0^4 = \frac{8}{3}\pi \text{ units cubed.}$$

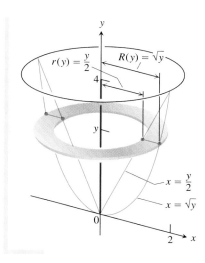

FIGURE 5.14 The washer swept out by the line segment in Figure 5.13.

EXERCISES 5.1

Cross-Section Areas

In Exercises 1 and 2, find a formula for the area $A(x)$ of the cross sections of the solid perpendicular to the x-axis.

1. The solid lies between planes perpendicular to the x-axis at $x = -1$ and $x = 1$. In each case, the cross sections perpendicular to the x-axis between these planes run from the semicircle $y = -\sqrt{1 - x^2}$ to the semicircle $y = \sqrt{1 - x^2}$.

   (a) The cross sections are circular disks with diameters in the xy-plane.

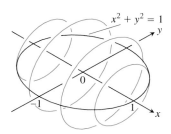

(b) The cross sections are squares with bases in the xy-plane.

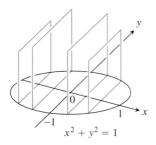

$x^2 + y^2 = 1$

(c) The cross sections are squares with diagonals in the xy-plane. (The length of a square's diagonal is $\sqrt{2}$ times the length of its sides.)

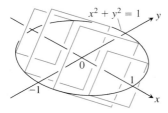

$x^2 + y^2 = 1$

(d) The cross sections are equilateral triangles with bases in the xy-plane.

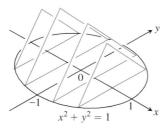

$x^2 + y^2 = 1$

2. The solid lies between planes perpendicular to the x-axis at $x = 0$ and $x = 4$. The cross sections perpendicular to the x-axis between these planes run from the parabola $y = -\sqrt{x}$ to the parabola $y = \sqrt{x}$.

(a) The cross sections are circular disks with diameters in the xy-plane.

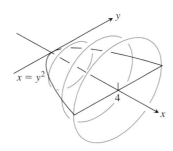

$x = y^2$

4

(b) The cross sections are squares with bases in the xy-plane.

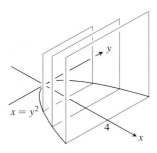

$x = y^2$

4

(c) The cross sections are squares with diagonals in the xy-plane.

(d) The cross sections are equilateral triangles with bases in the xy-plane.

Volumes by Slicing

Find the volumes of the solids in Exercises 3–10.

3. The solid lies between planes perpendicular to the x-axis at $x = 0$ and $x = 4$. The cross sections perpendicular to the x-axis on the interval $0 \le x \le 4$ are squares whose diagonals run from the parabola $y = -\sqrt{x}$ to the parabola $y = \sqrt{x}$.

4. The solid lies between planes perpendicular to the x-axis at $x = -1$ and $x = 1$. The cross sections perpendicular to the x-axis are circular disks whose diameters run from the parabola $y = x^2$ to the parabola $y = 2 - x^2$.

2

$y = x^2$

$y = 2 - x^2$

5. The solid lies between planes perpendicular to the x-axis at $x = -1$ and $x = 1$. The cross sections perpendicular to the x-axis are

(a) circles whose diameters stretch from the curve $y = -1/\sqrt{1 + x^2}$ to the curve $y = 1/\sqrt{1 + x^2}$

(b) vertical squares whose base edges run from the curve $y = -\sqrt{1 - x^2}$ to the curve $y = 1/\sqrt{1 + x^2}$.

6. The solid lies between planes perpendicular to the x-axis at $x = -\sqrt{2}/2$ and $x = \sqrt{2}/2$. The cross sections are

(a) circles whose diameters stretch from the x-axis to the curve $y = 2/\sqrt[4]{1 - x^2}$

(b) squares whose diagonals stretch from the x-axis to the curve $y = 2/\sqrt[4]{1 - x^2}$.

7. The base of a solid is the region between the curve $y = 2\sqrt{\sin x}$ and the interval $[0, \pi]$ on the x-axis. The cross sections perpendicular to the x-axis are

(a) equilateral triangles with bases running from the x-axis to the curve as shown in the figure

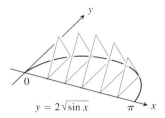

$$y = 2\sqrt{\sin x}$$

(b) squares with bases running from the x-axis to the curve.

8. The solid lies between planes perpendicular to the x-axis at $x = -\pi/3$ and $x = \pi/3$. The cross sections perpendicular to the x-axis are

(a) circular disks with diameters running from the curve $y = \tan x$ to the curve $y = \sec x$

(b) squares whose bases run from the curve $y = \tan x$ to the curve $y = \sec x$.

9. The solid lies between planes perpendicular to the y-axis at $y = 0$ and $y = 2$. The cross sections perpendicular to the y-axis are circular disks with diameters running from the y-axis to the parabola $x = \sqrt{5}y^2$.

10. The base of the solid is the disk $x^2 + y^2 \le 1$. The cross sections by planes perpendicular to the y-axis between $y = -1$ and $y = 1$ are isosceles right triangles with one leg in the disk.

$$x^2 + y^2 = 1$$

11. *A twisted solid* A square of side length s lies in a plane perpendicular to a line L. One vertex of the square lies on L. As this square moves a distance h along L, the square turns one revolution about L to generate a corkscrew-like column with square cross sections.

(a) Find the volume of the column.

(b) *Writing to Learn* What will the volume be if the square turns twice instead of once? Give reasons for your answer.

12. *Writing to Learn* A solid lies between planes perpendicular to the x-axis at $x = 0$ and $x = 12$. The cross sections by planes perpendicular to the x-axis are circular disks whose diameters run from the line $y = x/2$ to the line $y = x$ as shown in the accompanying

figure. Explain why the solid has the same volume as a right circular cone with base radius 3 and height 12.

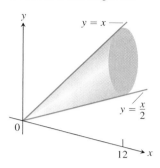

Solids of Revolution: Circular Cross Sections

In Exercises 13–16, find the volume of the solid generated by revolving the shaded region about the given axis.

13. About the x-axis

14. About the y-axis

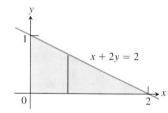

$$x + 2y = 2$$

$$x = 3y/2$$

15. About the y-axis

16. About the x-axis

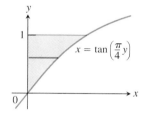

$$x = \tan\left(\frac{\pi}{4}y\right)$$

$$y = \sin x \cos x$$

Find the volumes of the solids generated by revolving the regions bounded by the lines and curves in Exercises 17–22, about the x-axis.

17. $y = x^2$, $\quad y = 0$, $\quad x = 2$ **18.** $y = x^3$, $\quad y = 0$, $\quad x = 2$

19. $y = \sqrt{9 - x^2}$, $\quad y = 0$ **20.** $y = e^{-x}$, $\quad y = 0$, $\quad x = 0$, $\quad x = 1$

21. $y = \sqrt{\cos x}$, $\quad 0 \le x \le \pi/2$, $\quad y = 0$, $\quad x = 0$

22. $y = \sqrt{\cot x}$, $\quad y = 0$, $\quad x = \pi/6$, $\quad x = \pi/2$

In Exercises 23 and 24, find the volume of the solid generated by revolving the region about the given line.

23. The region in the first quadrant bounded above by the line $y = \sqrt{2}$, below by the curve $y = \sec x \tan x$, and on the left by the y-axis, about the line $y = \sqrt{2}$

24. The region in the first quadrant bounded above by the line $y = 2$, below by the curve $y = 2 \sin x$, $0 \le x \le \pi/2$, and on the left by the y-axis, about the line $y = 2$

CD-ROM
WEBsite

Find the volumes of the solids generated by revolving the regions bounded by the lines and curves in Exercises 25–30 about the y-axis.

25. The region enclosed by $x = \sqrt{5}y^2$, $x = 0$, $y = -1$, $y = 1$

26. The region enclosed by $x = y^{3/2}$, $x = 0$, $y = 2$

27. The region enclosed by $x = \sqrt{2}\sin 2y$, $0 \le y \le \pi/2$, $x = 0$

28. The region enclosed by $x = 2/\sqrt{y + 1}$, $y = 3$, $x = 0$, $y = 0$

29. $x = 2/(y + 1)$, $x = 0$, $y = 0$, $y = 3$

30. $x = \sqrt{2y}/(y^2 + 1)$, $x = 0$, $y = 1$

Solids of Revolution: Washer Cross Sections

Find the volumes of the solids generated by revolving the shaded regions in Exercises 31 and 32 about the indicated axes.

31. The x-axis   **32.** The y-axis

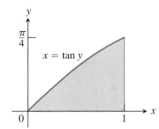

Find the volumes of the solids generated by revolving the regions bounded by the lines and curves in Exercises 33–38 about the x-axis.

33. $y = x$, $y = 1$, $x = 0$

34. $y = 2\sqrt{x}$, $y = 2$, $x = 0$

35. $y = x^2 + 1$, $y = x + 3$

36. $y = 4 - x^2$, $y = 2 - x$

37. $y = \sec x$, $y = \sqrt{2}$, $-\pi/4 \le x \le \pi/4$

38. $y = \sec x$, $y = \tan x$, $x = 0$, $x = 1$

In Exercises 39–42, find the volume of the solid generated by revolving each region about the y-axis.

39. The region enclosed by the triangle with vertices $(1, 0)$, $(2, 1)$, and $(1, 1)$

40. The region enclosed by the triangle with vertices $(0, 1)$, $(1, 0)$, and $(1, 1)$

41. The region in the first quadrant bounded above by the parabola $y = x^2$, below by the x-axis, and on the right by the line $x = 2$

42. The region in the first quadrant bounded on the left by the circle $x^2 + y^2 = 3$, on the right by the line $x = \sqrt{3}$, and above by the line $y = \sqrt{3}$

In Exercises 43 and 44, find the volume of the solid generated by revolving each region about the given axis.

43. The region in the first quadrant bounded above by the curve $y = x^2$, below by the x-axis, and on the right by the line $x = 1$, about the line $x = -1$

44. The region in the second quadrant bounded above by the curve $y = -x^3$, below by the x-axis, and on the left by the line $x = -1$, about the line $x = -2$

Volumes of Solids of Revolution

45. Find the volume of the solid generated by revolving the region bounded by $y = \sqrt{x}$ and the lines $y = 2$ and $x = 0$ about

   (a) the x-axis

   (b) the y-axis

   (c) the line $y = 2$

   (d) the line $x = 4$.

46. Find the volume of the solid generated by revolving the triangular region bounded by the lines $y = 2x$, $y = 0$, and $x = 1$ about

   (a) the line $x = 1$

   (b) the line $x = 2$.

47. Find the volume of the solid generated by revolving the region bounded by the parabola $y = x^2$ and the line $y = 1$ about

   (a) the line $y = 1$

   (b) the line $y = 2$

   (c) the line $y = -1$.

48. By integration, find the volume of the solid generated by revolving the triangular region with vertices $(0, 0)$, $(b, 0)$, $(0, h)$ about

   (a) the x-axis

   (b) the y-axis.

Theory and Applications

49. *The volume of a torus* The disk $x^2 + y^2 \le a^2$ is revolved about the line $x = b$ $(b > a)$ to generate a solid shaped like a doughnut and called a *torus*. Find its volume. (*Hint:* $\int_{-a}^{a} \sqrt{a^2 - y^2}\, dy = \pi a^2/2$, since it is the area of a semicircle of radius a.)

50. *Volume of a bowl* A bowl has a shape that can be generated by revolving the graph of $y = x^2/2$ between $y = 0$ and $y = 5$ about the y-axis.

   (a) Find the volume of the bowl.

   (b) *Related rates* If we fill the bowl with water at a constant rate of 3 cubic units per second, how fast will the water level in the bowl be rising when the water is 4 units deep?

51. *Volume of a bowl*

   (a) A hemispherical bowl of radius a contains water to a depth h. Find the volume of water in the bowl.

   (b) *Related rates* Water runs into a sunken concrete hemispherical bowl of radius 5 m at the rate of 0.2 m^3/sec. How fast is the water level in the bowl rising when the water is 4 m deep?

52. *Writing to Learn* Explain how you could estimate the volume of a solid of revolution by measuring the shadow cast on a table parallel to its axis of revolution by a light shining directly above it.

53. *Volume of a hemisphere* Derive the formula $V = (2/3)\,\pi R^3$ for the volume of a hemisphere of radius R by comparing its cross sections with the cross sections of a solid right circular cylinder of radius R and height R from which a solid right circular cone of base radius R and height R has been removed as suggested by the accompanying figure.

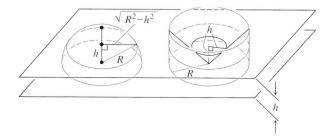

54. *Consistency of volume definitions* The volume formulas in calculus are consistent with the standard formulas from geometry in the sense that they agree on objects to which both apply.

(a) As a case in point, show that if you revolve the region enclosed by the semicircle $y = \sqrt{a^2 - x^2}$ and the x-axis about the x-axis to generate a solid sphere, the calculus formula for volume at the beginning of the section will give $(4/3)\pi a^3$ for the volume just as it should.

(b) Use calculus to the find the volume of a right circular cone of height h and base radius r.

55. *Designing a wok* You are designing a wok frying pan that will be shaped like a spherical bowl with handles. A bit of experimentation at home persuades you that you can get one that holds about 3 L if you make it 9 cm deep and give the sphere a radius of 16 cm. To be sure, you picture the wok as a solid of revolution, as shown here, and calculate its volume with an integral. To the nearest cubic centimeter, what volume do you really get? ($1\,L = 1000\,cm^3$.)

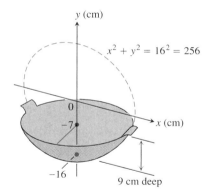

56. *Designing a plumb bob* Having been asked to design a brass plumb bob that will weigh in the neighborhood of 190 g, you decide to shape it like the solid of revolution shown here. Find the plumb bob's volume. If you specify a brass that weighs $8.5\,g/cm^3$, how much will the plumb bob weigh (to the nearest gram)?

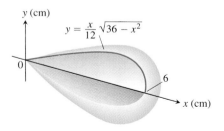

57. *Max-Min* The arch $y = \sin x$, $0 \leq x \leq \pi$, is revolved about the line $y = c$, $0 \leq c \leq 1$, to generate the solid in the figure.

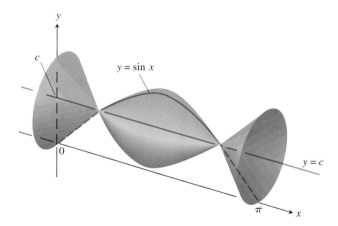

(a) Find the value of c that minimizes the volume of the solid. What is the minimum volume?

(b) What value of c in $[0, 1]$ maximizes the volume of the solid?

(c) *Writing to Learn* Graph the solid's volume as a function of c, first for $0 \leq c \leq 1$ and then on a larger domain. What happens to the volume of the solid as c moves away from $[0, 1]$? Does this make sense physically? Give reasons for your answers.

58. *An auxiliary fuel tank* You are designing an auxiliary fuel tank that will fit under a helicopter's fuselage to extend its range. After some experimentation at your drawing board, you decide to shape the tank like the surface generated by revolving the curve $y = 1 - (x^2/16)$, $-4 \leq x \leq 4$, about the x-axis (dimensions in feet).

(a) How many cubic feet of fuel will the tank hold (to the nearest cubic foot)?

(b) A cubic foot holds 7.481 gal. If the helicopter gets 2 mi to the gallon, how many additional miles will the helicopter be able to fly once the tank is installed (to the nearest mile)?

59. *A vase* We wish to estimate the volume of a flower vase using only a calculator, a string, and a ruler. We measure the height of the vase to be 6 in. We then use the string and the ruler to find circumferences of the vase (in inches) at half-inch intervals. (We list them from the top down to correspond with the picture of the vase.)

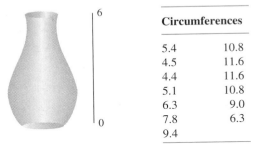

Circumferences	
5.4	10.8
4.5	11.6
4.4	11.6
5.1	10.8
6.3	9.0
7.8	6.3
9.4	

(a) Find the areas of the cross sections that correspond to the given circumferences.

(b) Express the volume of the vase as an integral with respect to y over the interval $[0, 6]$.

(c) Approximate the integral using the Trapezoidal Rule with $n = 12$.

(d) *Writing to Learn* Approximate the integral using Simpson's Rule with $n = 12$. Which result do you think is more accurate? Give reasons for your answer.

60. *A sailboat's displacement* To find the volume of water displaced by a sailboat, the common practice is to partition the waterline into 10 subintervals of equal length, measure the cross-section area $A(x)$ of the submerged portion of the hull at each partition point, and then use Simpson's Rule to estimate the integral of $A(x)$ from one end of the waterline to the other. The table here lists the area measurements at "Stations" 0 through 10, as the partition points are called, for the cruising sloop *Pipedream*, shown here. The common subinterval length (distance between consecutive stations) is $h = 2.54$ ft (about 2 ft 6-1/2 in., chosen for the convenience of the builder).

(a) Estimate *Pipedream*'s displacement volume to the nearest cubic foot.

Station	Submerged area (ft²)
0	0
1	1.07
2	3.84
3	7.82
4	12.20
5	15.18
6	16.14
7	14.00
8	9.21
9	3.24
10	0

(b) The figures in the table are for seawater, which weighs 64 lb/ft³. How many pounds of water does *Pipedream* displace? (Displacement is given in pounds for small craft and in long tons (1 long ton = 2240 lb) for larger vessels.)

(Data from *Skene's Elements of Yacht Design* by Francis S. Kinney, (Dodd, Mead, 1962.))

(c) *Prismatic coefficients* A boat's prismatic coefficient is the ratio of the displacement volume to the volume of a prism whose height equals the boat's waterline length and whose base equals the area of the boat's largest submerged cross section. The best sailboats have prismatic coefficients between 0.51 and 0.54. Find *Pipedream*'s prismatic coefficient, given a waterline length of 25.4 ft and a largest submerged cross section area of 16.14 ft² (at Station 6).

5.2   Modeling Volume Using Cylindrical Shells

Volume by Cylindrical Shells • The Shell Formula

There is another way to find volumes of solids of rotation that can be useful when the axis of revolution is perpendicular to the axis containing the natural interval of integration. Instead of summing volumes of thin slices, we sum volumes of thin cylindrical shells that grow outward from the axis of revolution like tree rings.

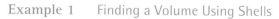

FIGURE 5.15 The graph of the region in Example 1, before revolution.

Volume by Cylindrical Shells

Here is an example of using thin cylindrical shells to find the volume of a solid.

Example 1   Finding a Volume Using Shells

The region enclosed by the x-axis and the parabola $y = f(x) = 3x - x^2$ is revolved about the line $x = -1$ to generate the shape of a solid (Figures 5.15 and 5.16). What is the volume of the solid?

Solution   Integrating with respect to y would be awkward here, as it is not easy to get the original parabola in terms of y. (Try finding the volume by washers and you will soon see what we mean.) To integrate with respect to x, you can do the problem by *cylindrical shells*, which requires that you cut the solid in a rather unusual way.

FIGURE 5.16 The region in Figure 5.15 is revolved about the line $x = -1$ to form a solid cake. The natural interval of integration is along the x-axis, perpendicular to the axis of revolution. (Example 1)

Step 1: Instead of cutting a wedge shape, cut a *cylindrical* slice by cutting straight down (parallel to the axis of revolution) all the way around close to the inside hole. Then cut another cylindrical slice around the enlarged hole, then another, and so on. The radii of the cylinders gradually increase, and the heights of the cylinders follow the contour of the parabola: smaller to larger, then back to smaller (Figure 5.17). Each slice is sitting over a subinterval of the x-axis of length Δx. Its radius is approximately $(1 + x_k)$, and its height is approximately $3x_k - x_k^2$.

FIGURE 5.17 Cutting the solid into thin cylindrical slices, working from the inside out. Each slice occurs at some x_k between 0 and 3 and has thickness Δx. (Example 1)

Step 2: If you unroll the cylinder at x_k and flatten it out, it becomes (essentially) a rectangular slab with thickness Δx (Figure 5.18). The inner circumference of the

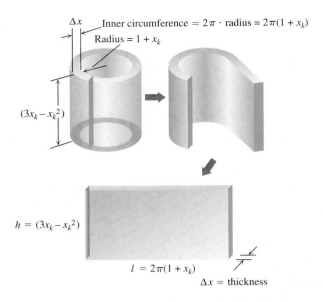

FIGURE 5.18 Imagine cutting and unrolling a cylindrical shell to get a flat (nearly) rectangular solid. (Example 1)

cylinder is $2\pi \cdot$ radius $= 2\pi(1 + x_k)$, and this is the length of the rolled out rectangular slab. Therefore, the volume of the flat (nearly) rectangular solid is

$$\Delta V \approx \text{length} \times \text{height} \times \text{thickness}$$
$$\approx 2\pi(1 + x_k) \cdot (3x_k - x_k^2) \cdot \Delta x$$

Step 3: Summing together the volumes of the individual cylindrical shells over the interval $0 \le x \le 3$ gives the Riemann sum $\Sigma \, 2\pi(x_k + 1)(3x_k - x_k^2) \, \Delta x$. Taking the limit as the thickness $\Delta x \to 0$ gives the volume integral

$$V = \int_0^3 2\pi(x + 1)(3x - x^2) \, dx$$

$$= \int_0^3 2\pi(3x^2 + 3x - x^3 - x^2) \, dx$$

$$= 2\pi \int_0^3 (2x^2 + 3x - x^3) \, dx$$

$$= 2\pi \left[\frac{2}{3}x^3 + \frac{3}{2}x^2 - \frac{1}{4}x^4 \right]_0^3$$

$$= \frac{45\pi}{2} \text{ units cubed.}$$

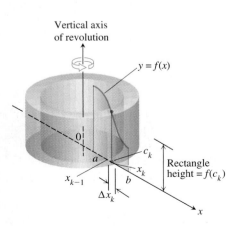

Vertical axis
of revolution

$y = f(x)$

c_k

Rectangle
height $= f(c_k)$

a

x_{k-1} x_k b

Δx_k

x

FIGURE 5.19 The shell swept out by the kth rectangle.

The Shell Formula

Suppose that we revolve the tinted region in Figure 5.19 about a vertical line to generate a solid. To estimate the volume of the solid, we can approximate the region with rectangles based on a partition P of the interval $[a, b]$ over which the region stands. The typical approximating rectangle is Δx_k units wide by $f(c_k)$ units high, where c_k is the midpoint of the rectangle's base. A formula from geometry tells us that the volume of the shell swept out by the rectangle is

$$\Delta V_k = 2\pi \times \text{average shell radius} \times \text{shell height} \times \text{thickness}.$$

We approximate the volume of the solid by adding the volumes of the shells swept out by the n rectangles based on P:

$$V \approx \sum_{k=1}^{n} \Delta V_k.$$

The limit of this sum as $\| P \| \rightarrow 0$ gives the volume of the solid:

$$V = \lim_{\| P \| \to 0} \sum \Delta V_k$$

$$= \int_a^b 2\pi \left(\begin{array}{c} \text{shell} \\ \text{radius} \end{array} \right) \left(\begin{array}{c} \text{shell} \\ \text{height} \end{array} \right) dx.$$

> **Shell Formula for Revolution About a Vertical Line**
> The volume of the solid generated by revolving the region between the x-axis and the graph of a continuous function $y = f(x) \geq 0, 0 \leq a \leq x \leq b,$ about a vertical line is
>
> $$V = \int_a^b 2\pi \left(\begin{array}{c} \text{shell} \\ \text{radius} \end{array} \right) \left(\begin{array}{c} \text{shell} \\ \text{height} \end{array} \right) dx$$

Example 2   Cylindrical Shells Revolving About the y-axis

The region bounded by the curve $y = \sqrt{x}$, the x-axis, and the line $x = 4$ is revolved about the y-axis to generate a solid. Find the volume of the solid.

Solution

Step 1: Sketch the region and draw a line segment across it *parallel* to the axis of revolution (Figure 5.20). Label the segment's height (shell height) and distance from the axis of revolution (shell radius). The width of the segment is the shell thickness dx. (We drew the shell in Figure 5.21, but you need not do that.)

FIGURE 5.20 The region, shell dimensions, and interval of integration in Example 2.

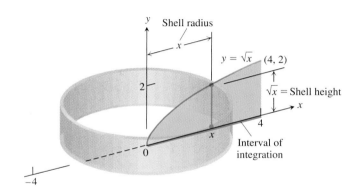

FIGURE 5.21 The shell swept out by the line segment in Figure 5.20.

Step 2: Find the limits of integration for the thickness variable (x runs from $a = 0$ to $b = 4$) and write the volume integral using the Shell Formula:

$$V = \int_a^b 2\pi \left(\begin{array}{c} \text{shell} \\ \text{radius} \end{array}\right)\left(\begin{array}{c} \text{shell} \\ \text{height} \end{array}\right) dx$$

$$= \int_0^4 2\pi(x)(\sqrt{x})\, dx .$$

Step 3: Integrate to find the volume:

$$V = \int_0^4 2\pi(x)(\sqrt{x})\, dx$$

$$= 2\pi \int_0^4 x^{3/2}\, dx = 2\pi \left[\frac{2}{5}x^{5/2}\right]_0^4 = \frac{128\pi}{5} \quad \text{units cubed.}$$

So far, we have used vertical axes of revolution. For horizontal axes, we replace the x's with y's.

Example 3   Cylindrical Shells Revolving About the x-axis

The region bounded by the curve $y = \sqrt{x}$, the x-axis, and the line $x = 4$ is revolved about the x-axis to generate a solid. Find the volume of the solid.

Solution

Step 1: Sketch the region and draw a line segment across it parallel to the axis of revolution (Figure 5.22). Label the segment's length (shell height) and distance from the axis of revolution (shell radius). The width of the segment is the shell thickness dy. (We drew the shell in Figure 5.23, but you need not do that.)

Step 2: Find the limits of integration for the thickness variable (y runs from 0 to 2) and write the volume integral using the Shell Formula:

$$V = \int_0^2 2\pi \left(\begin{array}{c} \text{shell} \\ \text{radius} \end{array}\right)\left(\begin{array}{c} \text{shell} \\ \text{height} \end{array}\right) dy$$

$$= \int_0^2 2\pi(y)(4 - y^2)\, dy .$$

Step 3: Integrate to find the volume:

$$V = \int_0^2 2\pi(y)(4 - y^2)\, dy$$

$$= 2\pi \left[2y^2 - \frac{y^4}{4}\right]_0^2 = 8\pi \quad \text{units cubed.}$$

FIGURE 5.22 The region, shell dimensions, and interval of integration in Example 3.

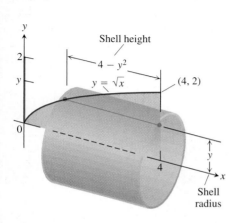

FIGURE 5.23 The shell swept out by the line segment in Figure 5.22.

How to Use the Shell Method

Regardless of the position of the axis of revolution (horizontal or vertical), the steps for implementing the shell method are these.

Step 1. *Draw the region and sketch a line segment* across it *parallel* to the axis of revolution. *Label* the segment's height or length (shell height), distance from the axis of revolution (shell radius), and width (shell thickness).

Step 2. *Find the limits of integration* for the thickness variable and write the volume integral.

Step 3. *Integrate* the product 2π (shell radius) (shell height) with respect to the thickness variable (x or y) to find the volume.

EXERCISES 5.2

In Exercises 1–6, use the shell method to find the volumes of the solids generated by revolving the shaded region about the indicated axis.

1.

$y = 1 + \dfrac{x^2}{4}$

2.

$y = 2 - \dfrac{x^2}{4}$

3.

$y = \sqrt{2}$
$x = y^2$

4.
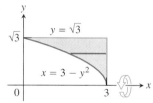
$y = \sqrt{3}$
$x = 3 - y^2$

5. The y-axis
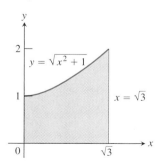
$y = \sqrt{x^2 + 1}$
$x = \sqrt{3}$

6. The y-axis
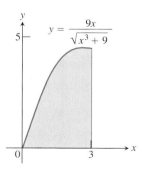
$y = \dfrac{9x}{\sqrt{x^3 + 9}}$

Revolution About the y-axis

Use the shell method to find the volumes of the solids generated by revolving the regions bounded by the curves and lines in Exercises 7–14 about the y-axis.

7. $y = x, \quad y = -x/2, \quad x = 2$

8. $y = 2x, \quad y = x/2, \quad x = 1$

9. $y = x^2, \quad y = 2 - x, \quad x = 0$, for $x \geq 0$

10. $y = 2 - x^2, \quad y = x^2, \quad x = 0$

11. $y = e^{-x^2}, \quad y = 0, \quad x = 0, \quad x = 1$

12. $y = 3/(2\sqrt{x}), \quad y = 0, \quad x = 1, \quad x = 4$

13. Let $f(x) = \begin{cases} (\sin x)/x, & 0 < x \leq \pi \\ 1, & x = 0 \end{cases}$

  (a) Show that $xf(x) = \sin x, 0 \leq x \leq \pi$.

  (b) Find the volume of the solid generated by revolving the shaded region about the y-axis.

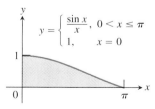
$y = \begin{cases} \dfrac{\sin x}{x}, & 0 < x \leq \pi \\ 1, & x = 0 \end{cases}$

14. Let $g(x) = \begin{cases} (\tan x)^2/x, & 0 < x \leq \pi/4 \\ 0, & x = 0 \end{cases}$

  (a) Show that $xg(x) = (\tan x)^2, 0 \leq x \leq \pi/4$.

(b) Find the volume of the solid generated by revolving the shaded region about the *y*-axis.

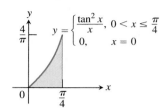

$$y = \begin{cases} \dfrac{\tan^2 x}{x}, & 0 < x \le \dfrac{\pi}{4} \\ 0, & x = 0 \end{cases}$$

Revolution About the x-axis

Use the shell method to find the volumes of the solids generated by revolving the regions bounded by the curves and lines in Exercises 15–22 about the *x*-axis.

15. $x = \sqrt{y}$, $x = -y$, $y = 2$

16. $x = y^2$, $x = -y$, $y = 2$, $y \ge 0$

17. $x = 2y - y^2$, $x = 0$

18. $x = 2y - y^2$, $x = y$

19. $y = |x|$, $y = 1$

20. $x = e^{y^2}$, $y = 0$, $x = 0$, $y = 1$

21. $y = \sqrt{x}$, $y = 0$, $y = x - 2$

22. $y = \sqrt{x}$, $y = 0$, $y = 2 - x$

Revolution About Horizontal Lines

In Exercises 23 and 24, use the shell method to find the volumes of the solids generated by revolving the shaded regions about the indicated axes.

23. **(a)** The *x*-axis

(b) The line $y = 1$

(c) The line $y = 8/5$

(d) The line $y = -2/5$

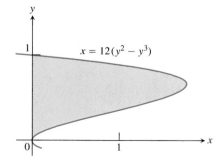

$x = 12(y^2 - y^3)$

24. **(a)** The *x*-axis

(b) The line $y = 2$

(c) The line $y = 5$

(d) The line $y = -5/8$

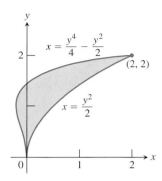

$$x = \frac{y^4}{4} - \frac{y^2}{2}$$

$$x = \frac{y^2}{2}$$

(2, 2)

Comparing the Washer and Shell Models

For some regions, both the washer and shell methods work well for the solid generated by revolving the region about the coordinate axes, but this is not always the case. When a region is revolved about the *y*-axis, for example, and washers are used, we must integrate with respect to *y*. It may not be possible, however, to express the integrand in terms of *y*. In such a case, the shell method allows us to integrate with respect to *x* instead, Exercises 25 and 26 provide some insight.

25. Compute the volume of the solid generated by revolving the region bounded by $y = x$ and $y = x^2$ about each coordinate axis using

(a) the shell method

(b) the washer method.

26. Compute the volume of the solid generated by revolving the triangular region bounded by the lines $2y = x + 4$, $y = x$, and $x = 0$ about

(a) the *x*-axis using the washer method

(b) the *y*-axis using the shell method

(c) the line $x = 4$ using the shell method

(d) the line $y = 8$ using the washer method.

Choosing Shells or Washers

In Exercises 27–34, find the volumes of the solids generated by revolving the regions about the given axes. If you think it would be better to use washers in any given instance, feel free to do so.

27. The triangle with vertices (1, 1), (1, 2), and (2, 2) about

(a) the *x*-axis

(b) the *y*-axis

(c) the line $x = 10/3$

(d) the line $y = 1$

28. The region bounded by $y = \sqrt{x}$, $y = 2$, $x = 0$ about

(a) the *x*-axis

(b) the *y*-axis

(c) the line $x = 4$

(d) the line $y = 2$

29. The region in the first quadrant bounded by the curve $x = y - y^3$ and the y-axis about

   (a) the x-axis

   (b) the line $y = 1$

30. The region in the first quadrant bounded by $x = y - y^3$, $x = 1$, and $y = 1$ about

   (a) the x-axis

   (b) the y-axis

   (c) the line $x = 1$

   (d) the line $y = 1$

31. The region bounded by $y = \sqrt{x}$ and $y = x^2/8$ about

   (a) the x-axis

   (b) the y-axis

32. The region bounded by $y = 2x - x^2$ and $y = x$ about

   (a) the y-axis

   (b) the line $x = 1$

33. The region in the first quadrant that is bounded above by the curve $y = 1/x^{1/4}$, on the left by the line $x = 1/16$, and below by the line $y = 1$ is revolved about the x-axis to generate a solid. Find the volume of the solid by

   (a) the washer method

   (b) the shell method.

34. The region in the first quadrant that is bounded above by the curve $y = 1/\sqrt{x}$, on the left by the line $y = 1/4$, and below by the line $y = 1$ is revolved about the y-axis to generate a solid. Find the volume of the solid by

   (a) the washer method   **(b)** the shell method.

Choosing Disks, Washers, or Shells

35. The region shown here is to be revolved about the x-axis to generate a solid. Which of the methods (disk, washer, shell) could you use to find the volume of the solid? How many integrals would be required in each case? Explain.

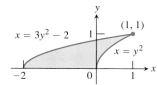

36. The region shown here is to be revolved about the y-axis to generate a solid. Which of the methods (disk, washer, shell) could you use to find the volume of the solid? How many integrals would be required in each case? Give reasons for your answers.

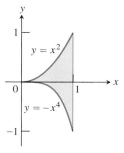

5.3 Lengths of Plane Curves

A Sine Wave • Length of a Smooth Curve • Dealing with Discontinuities in dy/dx • The Short Differential Formula • Parametric Arc Length Formula

If you are hiking along the rim of the Grand Canyon, how many miles do you walk? As a highway engineer, how do you estimate the cost of paving a curving mountainous road based on its total length? To answer questions like this, you need to know how to calculate the length of a curve.

A Sine Wave

How long is a sine wave (Figure 5.24)?

   The usual meaning of *wavelength* refers to the fundamental period, which for $y = \sin x$ is 2π. But how long is the curve itself? If you straightened it out like a piece of string along the positive x-axis with one end at 0, where would the other end be?

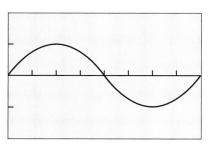

[0, 2π] by [−2, 2]

FIGURE 5.24 One wave of a sine curve has to be longer than 2π.

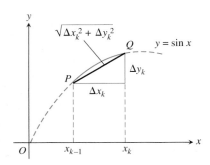

FIGURE 5.25 The line segment approximating the arc PQ of the sine curve above the subinterval $[x_{k-1}, x_k]$. (Example 1)

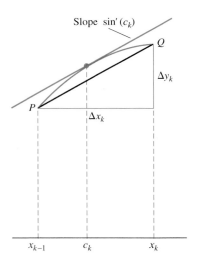

FIGURE 5.26 The portion of the sine curve above $[x_{k-1}, x_k]$. At some c_k in the interval, $\sin'(c_k) = \Delta y_k / \Delta x_k$, the slope of segment PQ. (Example 1)

Example 1   The Length of a Sine Wave

What is the length of the curve $y = \sin x$ from $x = 0$ to $x = 2\pi$?

Solution   We answer this question with integration, following our usual plan of breaking the whole into measurable parts. We partition $[0, 2\pi]$ into intervals so short that the pieces of curve (call them "arcs") lying directly above the intervals are nearly straight. That way, each arc is nearly the same as the line segment joining its two ends, and we can take the length of the segment as an approximation to the length of the arc.

Figure 5.25 shows the segment approximating the arc above the subinterval $[x_{k-1}, x_k]$. The length of the segment is $\sqrt{\Delta x_k^2 + \Delta y_k^2}$. The sum

$$\sum \sqrt{\Delta x_k^2 + \Delta y_k^2}$$

over the entire partition approximates the length of the curve. All we need now is to find the limit of this sum as the norms of the partitions go to zero. That's the usual plan, but this time there is a problem. Do you see it?

The problem is that the sums as written are not Riemann sums. They do not have the form $\sum f(c_k) \, \Delta x$. We can rewrite them as Riemann sums if we multiply and divide each square root by Δx_k.

$$\sum \sqrt{\Delta x_k^2 + \Delta y_k^2} = \sum \frac{\sqrt{(\Delta x_k)^2 + (\Delta y_k)^2}}{\Delta x_k} \, \Delta x_k$$

$$= \sum \sqrt{1 + \left(\frac{\Delta y_k}{\Delta x_k}\right)^2} \, \Delta x_k$$

This is better, but we still need to write the last square root as a function evaluated at some c_k in the kth subinterval. For this, we call on the Mean Value Theorem for differentiable functions (Section 3.2), which says that since $\sin x$ is continuous on $[x_{k-1}, x_k]$ and is differentiable on (x_{k-1}, x_k), there is a point c_k in $[x_{k-1}, x_k]$ at which $\Delta y_k / \Delta x_k = \sin' c_k$ (Figure 5.26). That gives us

$$\sum \sqrt{1 + (\sin' c_k)^2} \, \Delta x_k,$$

which *is* a Riemann sum.

Now we take the limit as the norms of the subdivisions go to zero and find that the length of one wave of the sine function is

$$\int_0^{2\pi} \sqrt{1 + (\sin' x)^2} \, dx = \int_0^{2\pi} \sqrt{1 + \cos^2 x} \, dx \approx 7.64.$$

We used a numerical integrator on a calculator to obtain the (approximate) value of the integral.

Length of a Smooth Curve

We are almost ready to define the length of a curve as a definite integral, using the procedure of Example 1. We first call attention to two properties of the sine function that came into play along the way.

We obviously used *differentiability* when we invoked the Mean Value Theorem to replace $\Delta y_k / \Delta x_k$ by $\sin'(c_k)$ for some c_k in the interval $[x_{k-1}, x_k]$. Less obviously, we used the continuity of the derivative of sine in passing from $\sum \sqrt{1 + (\sin' c_k)^2} \, \Delta x_k$ to the Riemann integral. The requirement for finding the

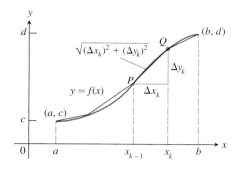

FIGURE 5.27 The graph of f, approximated by line segments.

length of a curve by this method, then, is that the function have a continuous first derivative. We call this property **smoothness.** A function with a continuous first derivative is **smooth** and its graph is a **smooth curve.**

Let us review the process, this time with a general smooth function $f(x)$. Suppose that the graph of f begins at the point (a, c) and ends at (b, d), as shown in Figure 5.27. We partition the interval $a \leq x \leq b$ into subintervals so short that the arcs of the curve above them are nearly straight. The length of the segment approximating the arc above the subinterval $[x_{k-1}, x_k]$ is $\sqrt{\Delta x_k^2 + \Delta y_k^2}$. The sum $\Sigma \sqrt{\Delta x_k^2 + \Delta y_k^2}$ approximates the length of the curve. We apply the Mean Value Theorem to f on each subinterval to rewrite the sum as a Riemann sum,

$$\sum \sqrt{\Delta x_k^2 + \Delta y_k^2} = \sum \sqrt{1 + \left(\frac{\Delta y_k}{\Delta x_k}\right)^2} \, \Delta x_k$$

$$= \sum \sqrt{1 + (f'(c_k))^2} \, \Delta x_k. \qquad \text{For some point } c_k \text{ in } (x_{k-1}, x_k)$$

Passing to the limit as the norms of the subdivisions go to zero gives the length of the curve as

$$L = \int_a^b \sqrt{1 + (f'(x))^2} \, dx = \int_a^b \sqrt{1 + \left(\frac{dy}{dx}\right)^2} \, dx.$$

We could as easily have transformed $\Sigma \sqrt{\Delta x_k^2 + \Delta y_k^2}$ into a Riemann sum by dividing and multiplying by Δy_k, giving a formula that involves x as a function of y (say, $x = g(y)$) on the interval $[c, d]$:

$$L \approx \sum \frac{\sqrt{(\Delta x_k)^2 + (\Delta y_k)^2}}{\Delta y_k} \, \Delta y_k = \sum \sqrt{1 + \left(\frac{\Delta x_k}{\Delta y_k}\right)^2} \, \Delta y_k$$

$$= \sum \sqrt{1 + (g'(c_k))^2} \, \Delta y_k. \qquad \text{For some } c_k \text{ in } (y_{k-1}, y_k)$$

The limit of these sums, as the norms of the subdivisions go to zero, gives another reasonable way to calculate the curve's length,

$$L = \int_c^d \sqrt{1 + (g'(y))^2} \, dy = \int_c^d \sqrt{1 + \left(\frac{dx}{dy}\right)^2} \, dy.$$

Putting these two formulas together, we have the following definition for the length of a smooth curve.

Arc Length Formulas for Length of a Smooth Curve
If f is smooth on $[a, b]$, the **length** of the curve $y = f(x)$ from a to b is the number

$$L = \int_a^b \sqrt{1 + \left(\frac{dy}{dx}\right)^2} \, dx. \qquad (1)$$

If g is smooth on $[c, d]$, the **length** of the curve $x = g(y)$ from c to d is the number

$$L = \int_c^d \sqrt{1 + \left(\frac{dx}{dy}\right)^2} \, dy \qquad (2)$$

Example 2   Applying the Arc Length Formula

Find the *exact* length of the curve

$$y = \frac{x^2}{2} - \frac{\ln x}{4} \quad \text{for} \quad 2 \le x \le 4.$$

Solution

$$\frac{dy}{dx} = x - \frac{1}{4x},$$

which is continuous on $[2, 4]$. Therefore,

$$L = \int_2^4 \sqrt{1 + \left(\frac{dy}{dx}\right)^2}\, dx$$

$$= \int_2^4 \sqrt{1 + \left(x^2 - \frac{1}{2} + \frac{1}{16x^2}\right)}\, dx \qquad \text{Square } x - \frac{1}{4x} \text{ for } dy/dx$$

$$= \int_2^4 \sqrt{x^2 + \frac{1}{2} + \frac{1}{16x^2}}\, dx \qquad \text{Combine terms}$$

$$= \int_2^4 \sqrt{\left(x + \frac{1}{4x}\right)^2}\, dx \qquad \text{Factor}$$

$$= \int_2^4 \left(x + \frac{1}{4x}\right) dx \qquad x + (1/4x) \ge 0 \text{ on } [2, 4]$$

$$= \left(\frac{x^2}{2} + \frac{\ln x}{4}\right)\Bigg]_2^4 \qquad \text{Integrate}$$

$$= 6 + \frac{\ln 4}{4} - \frac{\ln 2}{4}$$

$$= 6 + \frac{\ln 2}{4} \approx 6.173.$$

Dealing with Discontinuities in dy/dx

At a point on a curve where dy/dx fails to exist, dx/dy may exist and we may be able to find the curve's length by expressing x as a function of y and applying Equation (2).

Example 3   Applying Equation (2)

Find the length of the curve $y = (x/2)^{2/3}$ from $x = 0$ to $x = 2$.

Solution   The derivative

$$\frac{dy}{dx} = \frac{2}{3}\left(\frac{x}{2}\right)^{-1/3}\left(\frac{1}{2}\right) = \frac{1}{3}\left(\frac{2}{x}\right)^{1/3}$$

is not defined at $x = 0$, so we cannot find the curve's length with Equation (1).

We therefore rewrite the equation to express x in terms of y:

$$y = \left(\frac{x}{2}\right)^{2/3}$$

$$y^{3/2} = \frac{x}{2} \qquad \text{Raise both sides to the power } 3/2.$$

$$x = 2y^{3/2}. \qquad \text{Solve for } x.$$

From this, we see that the curve whose length we want is also the graph of $x = 2y^{3/2}$ from $y = 0$ to $y = 1$ (Figure 5.28).

The derivative

$$\frac{dx}{dy} = 2\left(\frac{3}{2}\right)y^{1/2} = 3y^{1/2}$$

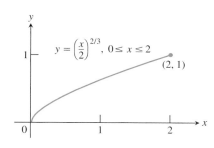

FIGURE 5.28 The graph of $y = (x/2)^{2/3}$ from $x = 0$ to $x = 2$ is also the graph of $x = 2y^{3/2}$ from $y = 0$ to $y = 1$.

is continuous on $[0, 1]$. We may therefore use Equation (2) to find the curve's length:

$$L = \int_c^d \sqrt{1 + \left(\frac{dx}{dy}\right)^2}\, dy = \int_0^1 \sqrt{1 + 9y}\, dy \qquad \begin{array}{l}\text{Eq. (2) with } c = 0, \\ d = 1\end{array}$$

$$= \frac{1}{9} \cdot \frac{2}{3}(1 + 9y)^{3/2}\Big]_0^1 \qquad \begin{array}{l}\text{Let } u = 1 + 9y \\ \text{and } du/9 = dy, \\ \text{integrate, and} \\ \text{substitute back.}\end{array}$$

$$= \frac{2}{27}(10\sqrt{10} - 1) \approx 2.27.$$

The Short Differential Formula

Equations (1) and (2),

$$L = \int_a^b \sqrt{1 + \left(\frac{dy}{dx}\right)^2}\, dx \qquad \text{and} \qquad L = \int_c^d \sqrt{1 + \left(\frac{dx}{dy}\right)^2}\, dy,$$

are often written with differentials instead of derivatives. This is done formally by thinking of the derivatives as quotients of differentials and bringing the dx and dy inside the radicals to cancel the denominators. In the first integral, we have

$$\sqrt{1 + \left(\frac{dy}{dx}\right)^2}\, dx = \sqrt{1 + \frac{dy^2}{dx^2}}\, dx = \sqrt{dx^2 + \frac{dy^2}{dx^2}\, dx^2} = \sqrt{dx^2 + dy^2}.$$

In the second integral, we have

$$\sqrt{1 + \left(\frac{dx}{dy}\right)^2}\, dy = \sqrt{1 + \frac{dx^2}{dy^2}}\, dy = \sqrt{dy^2 + \frac{dx^2}{dy^2}\, dy^2} = \sqrt{dx^2 + dy^2}.$$

In either case, we end up with the same differential formula:

$$L = \int_\alpha^\beta \sqrt{dx^2 + dy^2}. \tag{3}$$

Of course, dx and dy must be expressed in terms of a common variable, and appropriate limits of integration α and β must be found before the integration in Equation (3) is performed.

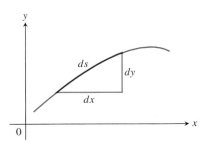

FIGURE 5.29 Think of dx and dy as two sides of a small triangle whose "hypotenuse" is $ds = \sqrt{dx^2 + dy^2}$.

We can shorten Equation (3) still further. Think of dx and dy as two sides of a small triangle whose "hypotenuse" is $ds = \sqrt{dx^2 + dy^2}$ (Figure 5.29). The differential ds is then regarded as a differential of arc length that can be integrated between appropriate limits to give the length of the curve. With $\sqrt{dx^2 + dy^2}$ set equal to ds, the integral in Equation (3) simply becomes the integral of ds.

Definition   The Arc Length Differential and the Differential Formula for Arc Length

$$ds = \sqrt{dx^2 + dy^2} \qquad\qquad L = \int ds$$

arc length differential   differential formula for arc length

---

Parametric Arc Length Formula

Another way to deal with discontinuities in dy/dx is to use a parametrization for the plane curve.

Suppose that a curve C can be described by the parametric equations $x = f(t)$ and $y = g(t)$, $\alpha \le t \le \beta$. The curve is **smooth** if f and g have continuous first derivatives that are not simultaneously zero. We find an integral for the length of a smooth curve $x = f(t)$, $y = g(t)$, $\alpha \le t \le \beta$, by rewriting the integral $L = \int \sqrt{dx^2 + dy^2}$ from Equation (3) in the following way:

$$L = \int_{t=\alpha}^{t=\beta} \sqrt{dx^2 + dy^2}$$

$$= \int_{\alpha}^{\beta} \sqrt{\left(\frac{(dx)^2}{(dt)^2} + \frac{(dy)^2}{(dt)^2}\right) dt^2} = \int_{\alpha}^{\beta} \sqrt{\left(\frac{dx}{dt}\right)^2 + \left(\frac{dy}{dt}\right)^2} \, dt.$$

The only requirement besides the continuity of the integrand is that the point $P(x, y) = P(f(t), g(t))$ not trace any portion of the curve more than once as t moves from α to β.

CD-ROM
WEBsite

Historical Biography

James Gregory
(1638 — 1675)

Parametric Formula for Arc Length
If a curve C is described by the parametric equations $x = f(t)$, $y = g(t)$, $\alpha \le t \le \beta$, where f' and g' are continuous and not simultaneously zero on $[\alpha, \beta]$, and if C is traversed exactly once as t increases from α to β, then the length of C is

$$L = \int_{a}^{\beta} \sqrt{\left(\frac{dx}{dt}\right)^2 + \left(\frac{dy}{dt}\right)^2} \, dt. \qquad (4)$$

What if there are two different parametrizations for a curve whose length we want to find; does it matter which one we use? The answer, from advanced calculus, is no, as long as the parametrization we choose meets the conditions preceding Equation (4).

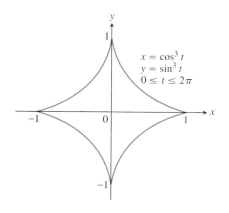

$x = \cos^3 t$
$y = \sin^3 t$
$0 \le t \le 2\pi$

FIGURE 5.30 The astroid in Example 4.

Example 4   Applying the Parametric Formula

Find the length of the astroid (Figure 5.30)

$$x = \cos^3 t, \qquad y = \sin^3 t, \qquad 0 \le t \le 2\pi.$$

Solution   Because of the curve's symmetry with respect to the coordinate axes, its length is four times the length of the first-quadrant portion. We have

$$x = \cos^3 t, \qquad y = \sin^3 t$$

$$\left(\frac{dx}{dt}\right)^2 = [3 \cos^2 t \,(-\sin t)]^2 = 9 \cos^4 t \sin^2 t$$

$$\left(\frac{dy}{dt}\right)^2 = [3 \sin^2 t \,(\cos t)]^2 = 9 \sin^4 t \cos^2 t$$

$$\sqrt{\left(\frac{dx}{dt}\right)^2 + \left(\frac{dy}{dt}\right)^2} = \sqrt{9 \cos^2 t \sin^2 t \,\underbrace{(\cos^2 t + \sin^2 t)}_{1}}$$

$$= \sqrt{9 \cos^2 t \sin^2 t}$$

$$= 3\,|\cos t\ \sin t\,|$$

$$= 3\ \cos t\ \sin t. \qquad \text{\small $\cos t \sin t \ge 0$ for $0 \le t \le \pi/2$}$$

Therefore,

$$\text{Length of first-quadrant portion} = \int_0^{\pi/2} 3\ \cos t\ \sin t\ dt$$

$$= \frac{3}{2} \int_0^{\pi/2} \sin 2t\ dt \qquad \text{\small $\begin{array}{l}\cos t \sin t = \\ (1/2) \sin 2t\end{array}$}$$

$$= -\frac{3}{4} \cos 2t \,\Big]_0^{\pi/2} = \frac{3}{2}.$$

The length of the astroid is four times this: $4(3/2) = 6$.

EXERCISES 5.3

Finding Lengths of Curves

Find the lengths of the curves in Exercises 1–10. If you have a grapher, you may want to graph these curves to see what they look like.

1. $y = (1/3)(x^2 + 2)^{3/2}$ from $x = 0$ to $x = 3$

2. $y = x^{3/2}$ from $x = 0$ to $x = 4$

3. $x = (y^3/3) + 1/(4y)$ from $y = 1$ to $y = 3$

(*Hint:* $1 + (dx/dy)^2$ is a perfect square.)

4. $x = (y^{3/2}/3) - y^{1/2}$ from $y = 1$ to $y = 9$

(*Hint:* $1 + (dx/dy)^2$ is a perfect square.)

5. $x = (y^4/4) + 1/(8y^2)$ from $y = 1$ to $y = 2$

(*Hint:* $1 + (dx/dy)^2$ is a perfect square.)

6. $x = (y^3/6) + 1/(2y)$ from $y = 2$ to $y = 3$

(*Hint:* $1 + (dx/dy)^2$ is a perfect square.)

7. $y = (3/4)x^{4/3} - (3/8)x^{2/3} + 5$, $1 \le x \le 8$

8. $y = (x^3/3) + x^2 + x + 1/(4x + 4)$, $0 \le x \le 2$

9. $x = \int_0^y \sqrt{\sec^4 t - 1}\ dt$, $-\pi/4 \le y \le \pi/4$

10. $y = \int_{-2}^x \sqrt{3t^4 - 1}\ dt$, $-2 \le x \le -1$

Lengths of Parametrized Curves

Find the lengths of the curves in Exercises 11–16.

11. $x = a \cos t$, $y = a \sin t$, $0 \le t \le 2\pi$

12. $x = \cos t$, $y = t + \sin t$, $0 \le t \le \pi$

13. $x = e^t - t$, $y = 4e^{t/2}$, $0 \le t \le 3$

14. $x = t^2/2$, $y = (2t + 1)^{3/2}/3$, $0 \le t \le 4$

15. $x = e^t \cos t$, $y = e^t \sin t$, $0 \le t \le \pi$

16. $x = 8 \cos t + 8t \sin t$, $y = 8 \sin t - 8t \cos t$, $0 \le t \le \pi/2$

Theory and Examples

17. *Writing to Learn* Is there a smooth curve $y = f(x)$ whose length over the interval $0 \le x \le a$ is always $\sqrt{2}a$? Give reasons for your answer.

18. *Using tangent fins to derive the length formula for curves* Assume that f is smooth on $[a, b]$ and partition the interval $[a, b]$ in the usual way. In each subinterval $[x_{k-1}, x_k]$, construct the *tangent fin* at the point $(x_{k-1}, f(x_{k-1}))$, as shown in the figure.

(a) Show that the length of the kth tangent fin over the interval $[x_{k-1}, x_k]$ equals $\sqrt{(\Delta x_k)^2 + (f'(x_{k-1}) \, \Delta x_k)^2}$.

(b) Show that

$$\lim_{n \to \infty} \sum_{k=1}^{n} (\text{length of } k\text{th tangent fin}) = \int_a^b \sqrt{1 + (f'(x))^2} \, dx,$$

which is the length L of the curve $y = f(x)$ from a to b.

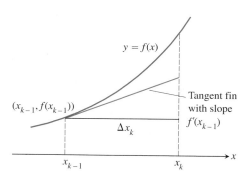

19. **(a)** Find a curve through the point $(1, 1)$ whose length integral is

$$L = \int_1^4 \sqrt{1 + \frac{1}{4x}} \, dx.$$

(b) *Writing to Learn* How many such curves are there? Give reasons for your answer.

20. **(a)** Find a curve through the point $(0, 1)$ whose length integral is

$$L = \int_1^2 \sqrt{1 + \frac{1}{y^4}} \, dy.$$

(b) *Writing to Learn* How many such curves are there? Give reasons for your answer.

Finding Integrals for Lengths of Curves

In Exercises 21–28, do the following.

    (a) Set up an integral for the length of the curve.

T **(b)** Graph the curve to see what it looks like.

T **(c)** Use your grapher's or computer's integral evaluator to find the curve's length numerically.

21. $y = x^2$, $-1 \le x \le 2$

22. $y = \tan x$, $-\pi/3 \le x \le 0$

23. $x = \sin y$, $0 \le y \le \pi$

24. $x = \sqrt{1 - y^2}$, $-1/2 \le y \le 1/2$

25. $y^2 + 2y = 2x + 1$ from $(-1, -1)$ to $(7, 3)$

26. $y = \sin x - x \cos x$, $0 \le x \le \pi$

27. $y = \int_0^x \tan t \, dt$, $0 \le x \le \pi/6$

28. $x = \int_0^y \sqrt{\sec^2 t - 1} \, dt$, $-\pi/3 \le y \le \pi/4$

29. *Fabricating metal sheets* Your metal fabrication company is bidding for a contract to make sheets of corrugated steel roofing like the one shown here. The cross sections of the corrugated sheets are to conform to the curve

$$y = \sin\left(\frac{3\pi}{20} x\right),   \qquad 0 \le x \le 20 \text{ in.}$$

If the roofing is to be stamped from flat sheets by a process that does not stretch the material, how wide should the original material be? Give your answer to two decimal places.

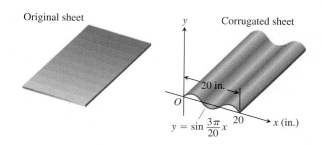

30. *Tunnel construction* Your engineering firm is bidding for the contract to construct the tunnel shown in the accompanying figure. The tunnel is 300 ft long and 50 ft wide at the base. The cross section is shaped like one arch of the curve $y = 25 \cos(\pi x/50)$. Upon completion, the tunnel's inside surface (excluding the roadway) will be treated with a waterproof sealer that costs $1.75 per square foot to apply. How much will it cost to apply the sealer?

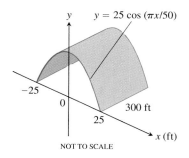

y = 25 cos (πx/50)

NOT TO SCALE

(b) Find the corresponding approximation to the length of the curve by summing the lengths of the line segments.

(c) Evaluate the length of the curve using an integral. Compare your approximations for $n = 2, 4, 8$ with the actual length given by the integral. How does the actual length compare with the approximations as n increases? Explain your answer.

31. $f(x) = \sqrt{1 - x^2}, \quad -1 \le x \le 1$

32. $f(x) = x^{1/3} + x^{2/3}, \quad 0 \le x \le 2$

33. $f(x) = \ln(1 - x^2), \quad 0 \le x \le 1/2$

34. $f(x) = x^2 \cos x, \quad 0 \le x \le \pi$

35. $f(x) = \dfrac{x - 1}{4x^2 + 1}, \quad -\dfrac{1}{2} \le x \le 1$

36. $f(x) = \ln\left(\dfrac{e^x + 1}{e^x - 1}\right), \quad 1 \le x \le 2$

COMPUTER EXPLORATIONS

Polygonal Approximations

In Exercises 31–36, use a CAS to perform the following steps for the given curve over the closed interval.

(a) Plot the curve together with the polygonal path approximations for $n = 2, 4, 8$ partition points over the interval. (See Figure 5.27).

5.4 First-Order Separable Differential Equations

General First-Order Differential Equations and Solutions • Separable Equations • Slope Fields: Viewing Solution Curves • Solving the Initial Value Problem $dy/dt = ky$, $y(0) = y_0$ • Heat Transfer: Newton's Law of Cooling Revisited • Resistance Proportional to Velocity • A Moving Body Coasting to a Stop • Torricelli's Law

In calculating derivatives by implicit differentiation (Section 2.6), we found that the expression for the derivative dy/dx often contained both variables x and y, not just the independent variable x. In this section, we study initial value problems in which the derivative has the form $dy/dx = f(x, y)$.

General First-Order Differential Equations and Solutions

A **first-order differential equation** is a relation

$$\frac{dy}{dx} = f(x, y) \tag{1}$$

in which $f(x, y)$ is a function of two variables defined on a region in the xy-plane. A **solution** of Equation (1) is a differentiable function $y = y(x)$ defined on an interval of x-values (perhaps infinite) such that

$$\frac{d}{dx} y(x) = f(x, y(x))$$

on that interval. The initial condition that $y(x_0) = y_0$ amounts to requiring the solution curve $y = y(x)$ to pass through the point (x_0, y_0).

Example 1   Verifying that a Function is a Solution

Show that the function

$$y = \frac{1}{x} + \frac{x}{2}$$

solves the first-order initial value problem

$$\frac{dy}{dx} = 1 - \frac{y}{x}, \qquad y(2) = \frac{3}{2}.$$

Solution   The equation

$$\frac{dy}{dx} = 1 - \frac{y}{x}$$

is a first-order differential equation in which $f(x, y) = 1 - (y/x)$.

The function $y = (1/x) + (x/2)$ solves the differential equation because the two sides of the equation agree when we substitute $(1/x) + (x/2)$ for y:

On the left: $\dfrac{dy}{dx} = \dfrac{d}{dx}\left(\dfrac{1}{x} + \dfrac{x}{2}\right) = -\dfrac{1}{x^2} + \dfrac{1}{2}.$

On the right: $1 - \dfrac{y}{x} = 1 - \dfrac{1}{x}\left(\dfrac{1}{x} + \dfrac{x}{2}\right)$

$$= 1 - \frac{1}{x^2} - \frac{1}{2} = -\frac{1}{x^2} + \frac{1}{2}.$$

The function satisfies the initial condition because

$$y(2) = \left(\frac{1}{x} + \frac{x}{2}\right)_{x=2} = \frac{1}{2} + \frac{2}{2} = \frac{3}{2}.$$

Separable Equations

The equation $y' = f(x, y)$ is **separable** if f can be expressed as a product of a function of x and a function of y. The differential equation then has the form

$$\frac{dy}{dx} = g(x)\, h(y).$$

We sometimes write $y' = f(x, y)$ for $dy/dx = f(x, y)$.

If $h(y) \neq 0$, we can **separate the variables** by dividing both sides by h, obtaining, in succession

$$\frac{1}{h(y)}\frac{dy}{dx} = g(x)$$

$$\int \frac{1}{h(y)}\frac{dy}{dx}\, dx = \int g(x)\, dx \qquad \text{Integrate with respect to } x.$$

$$\int \frac{1}{h(y)}\, dy = \int g(x)\, dx.$$

With x and y now separated, we simply integrate each side to get the solutions we seek by expressing y either explicitly or implicitly as a function of x, up to an arbitrary constant.

Example 2   Solving a Separable Equation

Solve the differential equation

$$\frac{dy}{dx} = (1 + y^2)e^x.$$

Solution   Since $1 + y^2$ is never zero, we can solve the equation by separating the variables.

$$\frac{dy}{dx} = (1 + y^2)e^x$$

Treat dy/dx as a quotient of differentials and multiply both sides by dx.

$$dy = (1 + y^2)e^x\,dx$$

$$\frac{dy}{1 + y^2} = e^x\,dx$$

Divide by $(1 + y^2)$.

$$\int \frac{dy}{1 + y^2} = \int e^x\,dx$$

Integrate both sides.

$$\tan^{-1} y = e^x + C$$

C represents the combined constants of integration.

The equation $\tan^{-1} y = e^x + C$ gives y as an implicit function of x. When $-\pi/2 < e^x + C < \pi/2$, we can solve for y as an explicit function of x by taking the tangent of both sides:

$$\tan\,(\tan^{-1} y) = \tan\,(e^x + C)$$

$$y = \tan\,(e^x + C).$$

Slope Fields: Viewing Solution Curves

CD-ROM
WEBsite

Each time we specify an initial condition $y(x_0) = y_0$ for the solution of a differential equation $y' = f(x, y)$, the **solution curve** (graph of the solution) is required to pass through the point (x_0, y_0) and to have slope $f(x_0, y_0)$ there. We can picture these slopes graphically by drawing short line segments of slope $f(x, y)$ at selected points (x, y) in the region of the xy-plane that constitutes the domain of f. Each segment has the same slope as the solution curve through (x, y) and so is tangent to the curve there. We see how the curves behave by following these tangents (Figure 5.31).

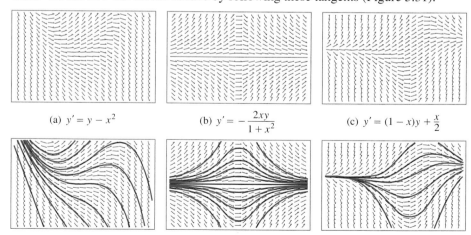

(a) $y' = y - x^2$ (b) $y' = -\dfrac{2xy}{1 + x^2}$ (c) $y' = (1 - x)y + \dfrac{x}{2}$

FIGURE 5.31 Slope fields (top row) and selected solution curves (bottom row). In computer renditions, slope segments are sometimes portrayed with arrows, as they are here. This is not to be taken as an indication that slopes have directions, however, because they do not.

Constructing a slope field with pencil and paper can be quite tedious. All our examples were generated by a computer.

Solving the Initial Value Problem $dy/dt = ky$, $y(0) = y_0$

We know that all the functions $y = Ae^{kt}$, A and k constant, satisfy the differential equation $dy/dt = ky$ because

$$\frac{d}{dt}(Ae^{kt}) = A\frac{d}{dt}(e^{kt}) = A(ke^{kt}) = ky.$$

Using the fact that $\int (1/y)\,dy = \ln|y| + C$, we can show that these are the *only* functions satisfying the equation.

We see right away that the constant function $y = 0$ is a solution of $dy/dt = ky$. To find the nonzero solutions we separate the variables and integrate with respect to t:

$$\frac{dy}{dt} = ky$$

$$\frac{1}{y}\frac{dy}{dt} = k \qquad\qquad y \neq 0$$

$$\int \frac{1}{y}\frac{dy}{dt}\,dt = \int k\,dt$$

$$\int \frac{1}{y}\,dy = \int k\,dt \qquad dy = (dy/dt)\,dt$$

$$\ln|y| = kt + C \qquad \text{Constants of integration combined}$$

$$|y| = e^{kt+C} \qquad \text{Exponentiate}$$

$$|y| = e^{C} \cdot e^{kt} \qquad \text{Law of exponents}$$

$$y = \pm e^{C}e^{kt} \qquad \text{If } |y| = r, \text{ then } y = \pm r.$$

$$y = Ae^{kt}. \qquad \text{Writing } A \text{ for } \pm e^{C}$$

Thus, all solutions of the equation $dy/dt = ky$ have the form $y = Ae^{kt}$. If, further, we have the initial condition $y = y_0$ when $t = 0$, then

$$y_0 = Ae^{k(0)} = A$$

and the solution of the initial value problem is $y = y_0 e^{kt}$.

We can summarize our work with this problem as follows.

The Initial Value Problem $dy/dt = ky$, $y(0) = y_0$
The unique solution of the problem is

$$y = y_0 e^{kt}.$$

This equation is the **law of exponential change.** It describes growth if $k > 0$ and decay if $k < 0$. The number k is the **rate constant** of the equation.

Heat Transfer: Newton's Law of Cooling Revisited

In Section 3.4, we obtained the shape of the temperature curve of a heated object immersed in a cooler surrounding medium, such as hot soup in a tin cup cooling to the temperature of the surrounding air. The differential equation describing this cooling process is based on the principle that the rate at which an object's temperature is changing at any given time is roughly proportional to the difference between its temperature and the temperature of the surrounding medium (*Newton's Law of Cooling*, which applies to warming as well).

If H is the temperature of the object at time t and if H_s is the constant surrounding temperature, then the differential equation is

$$\frac{dH}{dt} = -k(H - H_s). \tag{2}$$

If we substitute y for $(H - H_s)$, then

$$\frac{dy}{dt} = \frac{d}{dt}(H - H_s) = \frac{dH}{dt} - \frac{d}{dt}(H_s)$$

$$= \frac{dH}{dt} - 0 \qquad \text{\small{H_s is a constant.}}$$

$$= \frac{dH}{dt}$$

$$= -k(H - H_s) \qquad \text{\small{Eq. (2)}}$$

$$= -ky. \qquad \text{\small{$H - H_s = y$}}$$

Now we know that the solution of $dy/dt = -ky$ is $y = y_0 e^{-kt}$, where $y(0) = y_0$. Substituting $(H - H_s)$ for y gives

$$H - H_s = (H_0 - H_s)e^{-kt}, \tag{3}$$

where H_0 is the temperature at $t = 0$. This is the equation of Newton's Law of Cooling.

Example 3   Cooling a Hard-Boiled Egg

A hard-boiled egg at 98°C is put in a sink of 18°C water. After 5 min, the egg's temperature is 38°C. Assuming that the water has not warmed appreciably, how much longer will it take the egg to reach 20°C?

Solution   We find how long it would take the egg to cool from 98°C to 20°C and subtract the 5 min that have already elapsed.

Step 1: *Solve the initial value problem.* Using Equation (3) with $H_s = 18$ and $H_0 = 98$, the egg's temperature t min after it is put in the sink is

$$H = 18 + (98 - 18)e^{-kt} = 18 + 80e^{-kt}.$$

To find k, we use the information that $H = 38$ when $t = 5$:

$$38 = 18 + 80e^{-5k}$$

$$e^{-5k} = \frac{1}{4}$$

$$\ln(e^{-5k}) = \ln\left(\frac{1}{4}\right) \qquad \text{\small{Natural logarithm of both sides}}$$

$$-5k = \ln\frac{1}{4} = -\ln 4$$

$$k = \frac{1}{5}\ln 4 = 0.2 \ln 4.$$

The egg's temperature at time t is $H = 18 + 80e^{-(0.2 \ln 4)t}$.

Step 2: *Find the time t when H = 20.*

$$20 = 18 + 80e^{-(0.2 \ln 4)t}$$

$$80e^{-(0.2 \ln 4)t} = 2$$

$$e^{-(0.2 \ln 4)t} = \frac{1}{40}$$

$$-(0.2 \ln 4)t = \ln \frac{1}{40} = -\ln 40 \qquad \text{Logarithms of both sides}$$

$$t = \frac{\ln 40}{0.2 \ln 4} \approx 13 \text{ min}$$

Step 3: *Interpretation of the time required to reach H = 20.* The egg's temperature will reach 20°C about 13 min after it is put in water to cool. Since it took 5 min to reach 38°C, it will take about 8 min more to reach 20°C.

Resistance Proportional to Velocity

In some cases, it is reasonable to assume that, other forces being absent, the resistance encountered by a moving object, such as a car coasting to a stop, is proportional to the object's velocity. The slower the object moves, the less its forward progress is resisted by the air through which it passes. To describe this in mathematical terms, we picture the object as a mass m moving along a coordinate line with position function s and velocity v at time t. The resisting force opposing the motion is

$$\text{Force} = \text{mass} \times \text{acceleration} = m\frac{dv}{dt}.$$

We can express the assumption that the resisting force is proportional to velocity by writing

$$m\frac{dv}{dt} = -kv \qquad \text{or} \qquad \frac{dv}{dt} = -\frac{k}{m}v \qquad (k > 0).$$

This is a differential equation of exponential change. The solution to the differential equation with initial condition $v = v_0$ at $t = 0$ is

$$v = v_0 e^{-(k/m)t}. \tag{4}$$

A Moving Body Coasting to a Stop

What can we learn from Equation (4)? For one thing, we can see that if m is something large, like the mass of a 20,000-ton ore boat in Lake Erie, it will take a long time for the velocity to approach zero. For another, we can integrate the equation to find s as a function of t.

Suppose that a body is coasting to a stop and the only force acting on it is a resistance proportional to its speed. How far will it coast? To find out, we start with Equation (4) and solve the initial value problem

$$\frac{ds}{dt} = v_0 e^{-(k/m)t}, \qquad s(0) = 0.$$

Integrating with respect to t gives

$$s = -\frac{v_0 m}{k} e^{-(k/m)t} + C.$$

Substituting $s = 0$ when $t = 0$ gives

$$0 = -\frac{v_0 m}{k} + C \qquad \text{and} \qquad C = \frac{v_0 m}{k}.$$

The body's position at time t is therefore

$$s(t) = -\frac{v_0 m}{k} e^{-(k/m)t} + \frac{v_0 m}{k} = \frac{v_0 m}{k}(1 - e^{-(k/m)t}). \tag{5}$$

To find how far the body will coast, we find the limit of $s(t)$ as $t \to \infty$. Since $-(k/m) < 0$, we know that $e^{-(k/m)t} \to 0$ as $t \to \infty$, so that

$$\lim_{t \to \infty} s(t) = \lim_{t \to \infty} \frac{v_0 m}{k}(1 - e^{-(k/m)t})$$

$$= \frac{v_0 m}{k}(1 - 0) = \frac{v_0 m}{k}.$$

Thus,

$$\text{Distance coasted} = \frac{v_0 m}{k}. \tag{6}$$

This is an ideal figure, of course. Only in mathematics can time stretch to infinity. The number $v_0 m/k$ is only an upper bound (albeit a useful one). It is true to life in one respect, at least: if m is large it will take a lot of energy to stop the body. That is why ocean liners have to be docked by tugboats. Any liner of conventional design entering a slip with enough speed to steer would smash into the pier before it could stop.

Weight versus Mass

Weight is the force that results from gravity pulling on a mass. The two are related by the equation in Newton's second law,

Weight = mass × acceleration.

To convert mass to weight, multiply by the acceleration of gravity. To convert weight to mass, divide by the acceleration of gravity. In the metric system,

Newtons = kilograms × 9.8

and

Newtons/9.8 = kilograms.

In the English system, where weight is measured in pounds, mass is measured in slugs. Thus,

Pounds = slugs × 32

and

Pounds/32 = slugs.

A skater weighing 192 lb has a mass of

192/32 = 6 slugs.

Example 4 A Coasting Ice Skater

For a 192 lb skater, the k in Equation (4) is about $1/3$ slug/sec and $m = 192/32 = 6$ slugs. How long will it take the skater to coast from 11 ft/sec (7.5 mph) to 1 ft/sec? How far will the skater coast before coming to a complete stop?

Solution We answer the first question by solving Equation (4) for t:

$$11e^{-t/18} = 1 \qquad \text{Eq. (4) with } k = 1/3,$$
$$\qquad\qquad\qquad\qquad m = 6, v_0 = 11, v = 1$$
$$e^{-t/18} = 1/11$$
$$-t/18 = \ln(1/11) = -\ln 11$$
$$t = 18 \ln 11 \approx 43 \text{ sec.}$$

We answer the second question with Equation (6):

$$\text{Distance coasted} = \frac{v_0 m}{k} = \frac{11 \cdot 6}{1/3}$$
$$= 198 \text{ ft.}$$

FIGURE 5.32 The rate at which water runs out is $0.5\sqrt{x}$ ft³/min. (Example 5)

CD-ROM
WEBsite

Historical Biography

Evangelista Torricelli
(1608 — 1647)

Torricelli's Law

Torricelli's law says that if you drain a tank like the one in Figure 5.32, the rate at which the water runs out is a constant times the square root of the water's depth x. The constant depends on the size of the exit value. In Example 5, we assume that the constant is $1/2$.

Example 5   Draining a Tank

A right circular cylindrical tank with radius 5 ft and height 16 ft that was initially full of water is being drained at the rate of $0.5\sqrt{x}$ ft³/min. Find a formula for the depth and the amount of water in the tank at any time t. How long will it take the tank to empty?

Solution   The volume of a right circular cylinder with radius r and height h is $V = \pi r^2 h$.

Model

The volume of water in the tank (Figure 5.32) is

$$V = \pi r^2 h = \pi (5)^2 x = 25 \pi x.$$

Differential equation: $\dfrac{dV}{dt} = 25\pi \dfrac{dx}{dt}$

$-0.5\sqrt{x} = 25\pi \dfrac{dx}{dt}$ Negative because V is decreasing and $dx/dt < 0$

$\dfrac{dx}{dt} = -\dfrac{\sqrt{x}}{50\pi}$ Torricelli's law

Initial condition: $x(0) = 16$ The water is 16 ft deep when $t = 0$.

Solve Analytically

We first solve the differential equation by separating the variables.

$$x^{-1/2}\, dx = -\frac{1}{50\pi}\, dt$$

$$\int x^{-1/2}\, dx = -\int \frac{1}{50\pi}\, dt \qquad \text{Integrate both sides.}$$

$$2x^{1/2} = -\frac{1}{50\pi}\, t + C \qquad \text{Constants combined}$$

The initial condition $x(0) = 16$ determines the value of C.

$$2(16)^{1/2} = -\frac{1}{50\pi}(0) + C$$

$$C = 8$$

With $C = 8$, we have

$$2x^{1/2} = -\frac{1}{50\pi}\, t + 8 \qquad \text{or} \qquad x^{1/2} = 4 - \frac{t}{100\pi}.$$

The formulas we seek are

$$x = \left(4 - \frac{t}{100\pi}\right)^2 \qquad \text{and} \qquad V = 25\pi x = 25\pi \left(4 - \frac{t}{100\pi}\right)^2.$$

Interpret

At any time t, the water in the tank is $(4 - t/(100\pi))^2$ ft deep and the amount of water is $25\pi(4 - t/(100\pi))^2$ ft^3. At $t = 0$, we have $x = 16$ ft and $V = 400\pi$ ft^3, as required. The tank will empty ($V = 0$) in $t = 400\pi$ minutes, which is about 21 h.

EXERCISES 5.4

Verifying Solutions

In Exercises 1 and 2, show that each function $y = f(x)$ is a solution of the accompanying differential equation.

1. $2y' + 3y = e^{-x}$

   (a) $y = e^{-x}$

   (b) $y = e^{-x} + e^{-(3/2)x}$

   (c) $y = e^{-x} + Ce^{-(3/2)x}$

2. $y' = y^2$

   (a) $y = -\dfrac{1}{x}$ **(b)** $y = -\dfrac{1}{x + 3}$ **(c)** $y = -\dfrac{1}{x + C}$

In Exercises 3 and 4, show that each function is a solution of the given initial value problem.

3. Differential equation: $y' = e^{-x^2} - 2xy$

   Initial condition: $y(2) = 0$

   Solution candidate: $y = (x - 2)e^{-x^2}$

4. Differential equation: $xy' + y = -\sin x, \quad x > 0$

   Initial condition: $y\left(\dfrac{\pi}{2}\right) = 0$

   Solution candidate: $y = \dfrac{\cos x}{x}$

Separable Equations

Solve the differential equation in Exercises 5–14.

5. $2\sqrt{xy}\,\dfrac{dy}{dx} = 1, \quad x, y > 0$

6. $\dfrac{dy}{dx} = x^2\sqrt{y}, \quad y > 0$

7. $\dfrac{dy}{dx} = e^{x-y}$

8. $\dfrac{dy}{dx} = 3x^2 e^{-y}$

9. $\dfrac{dy}{dx} = \sqrt{y}\cos^2\sqrt{y}$

10. $\sqrt{2xy}\,\dfrac{dy}{dx} = 1$

11. $\sqrt{x}\,\dfrac{dy}{dx} = e^{y+\sqrt{x}}, \quad x > 0$

12. $(\sec x)\,\dfrac{dy}{dx} = e^{y+\sin x}$

13. $\dfrac{dy}{dx} = 2x\sqrt{1 - y^2}, \quad -1 < y < 1$

14. $\dfrac{dy}{dx} = \dfrac{e^{2x-y}}{e^{x+y}}$

Applications

15. *Atmospheric pressure* Earth's atmospheric pressure p is often modeled by assuming that the rate dp/dh at which p changes with the altitude h above sea level is proportional to p. Suppose that the pressure at sea level is 1013 millibars (about 14.7 pounds per square inch) and that the pressure at an altitude of 20 km is 90 millibars.

   (a) Solve the initial value problem:

        Differential equation: $dp/dh = kp$ (k a constant)

        Initial condition: $p = p_0$ when $h = 0$

        to express p in terms of h. Determine the values of p_0 and k from the given altitude-pressure data.

   (b) What is the atmospheric pressure at $h = 50$ km?

   (c) At what altitude does the pressure equal 900 millibars?

16. *First-order chemical reactions* In some chemical reactions, the rate at which the amount of a substance changes with time is proportional to the amount present. For the change of δ-glucono lactone into gluconic acid, for example,

$$\frac{dy}{dt} = -0.6y$$

when t is measured in hours. If there are 100 g of δ-glucono lactone present when $t = 0$, how many grams will be left after the first hour?

17. *The inversion of sugar* The processing of raw sugar has a step called "inversion" that changes the sugar's molecular structure. Once the process has begun, the rate of change of the amount of raw sugar is proportional to the amount of raw sugar remaining. If 1000 kg of raw sugar reduces to 800 kg of raw sugar during the first 10 h, how much raw sugar will remain after another 14 h?

18. *Working underwater* The intensity $L(x)$ of light x ft beneath the surface of the ocean satisfies the differential equation

$$\frac{dL}{dx} = -kL.$$

As a diver, you know from experience that diving to 18 ft in the Caribbean Sea cuts the intensity in half. You cannot work without artificial light when the intensity falls below one-tenth of the surface value. About how deep can you expect to work without artificial light?

19. *Voltage in a discharging capacitor* Suppose that electricity is draining from a capacitor at a rate that is proportional to the voltage V across its terminals and that, if t is measured in seconds,

$$\frac{dV}{dt} = -\frac{1}{40} V.$$

Solve this equation for V, using V_0 to denote the value of V when $t = 0$. How long will it take the voltage to drop to 10% of its original value?

20. *Oil depletion* Suppose that the amount of oil pumped from one of the canyon wells in Whittier, California decreases at the continuous rate of 10% per year. When will the well's output fall to one-fifth of its present value?

21. *Glucose fed intravenously* A hospital patient is fed glucose intravenously (directly to the bloodstream) at a rate of r units per minute. The body removes glucose from the bloodstream at a rate proportional to the amount $Q(t)$ present in the bloodstream at time t.

(a) Write a differential equation modeling the change of glucose in the bloodstream over time.

(b) Solve the differential equation in part (a) with the initial condition $Q(0) = Q_0$.

(c) Find the limit of $Q(t)$ as $t \to \infty$.

22. *Cooling soup* Suppose that a cup of soup cooled from 90°C to 60°C after 10 min in a room whose temperature was 20°C. Use Newton's Law of Cooling to answer the following questions.

(a) How much longer would it take the soup to cool to 35°C?

(b) Instead of being left to stand in the room, the cup of 90°C soup is put in a freezer whose temperature is -15°C. How long will it take the soup to cool from 90°C to 35°C?

23. *A beam of unknown temperature* An aluminum beam was brought from the outside cold into a machine shop where the temperature was held at 65°F. After 10 min, the beam warmed to 35°F, and after another 10 min, it was 50°F. Use Newton's Law of Cooling to estimate the beam's initial temperature.

24. *Surrounding medium of unknown temperature* A pan of warm water (46°C) was put in a refrigerator. Ten minutes later, the water's temperature was 39°C; 10 min after that, it was 33°C. Use Newton's Law of Cooling to estimate how cold the refrigerator was.

25. *Continuous price discounting* To encourage buyers to place 100-unit orders, your firm's sales department applies a continuous discount that makes the unit price a function $p(x)$ of the number of units x ordered. The discount decreases the price at the rate of $0.01 per unit ordered. The price per unit for a 100-unit order is $p(100) = \$20.09$.

(a) Find $p(x)$ by solving the following initial value problem:

   Differential equation: $\dfrac{dp}{dx} = -\dfrac{1}{100} p$

   Initial condition: $p(100) = 20.09$.

(b) Find the unit price $p(10)$ for a 10-unit order and the unit price $p(90)$ for a 90-unit order.

(c) The sales department has asked you to find out if it is discounting so much that the firm's revenue, $r(x) = x \cdot p(x)$, will actually be less for a 100-unit order than, say, for a 90-unit order. Reassure them by showing that r has its maximum value at $x = 100$.

26. *World's population* The table gives estimates of the world population, in millions, over two centuries:

Year	1750	1800	1850	1900	1950
Population	728	906	1171	1608	2517

(a) Use the exponential model $dP/dt = kP$, $k > 0$, and the population figures for 1750 and 1800 to predict the world population in 1900 and 1950. Compare with the actual figures.

(b) Use the exponential model and the population figures for 1850 and 1900 to predict the world population in 1950. Compare with the actual population.

(c) *Writing to Learn* Use the exponential model and the population figures for 1900 and 1950 to predict the world population in 1999. Compare with the actual 1999 population of 6 billion and try to explain the discrepancy.

27. *Radioactive decay* Experiments have shown that the rate at which a radioactive element decays (as measured by the number of nuclei that change per unit time) is approximately proportional to the number $y(t)$ of radioactive nuclei present at time t. The constant of proportionality is called the **decay constant.** Write and solve a differential equation modeling radioactive decay assuming the initial condition $y(0) = y_0$.

28. *Half-life (Continuation of Exercise 27)* The **half-life** of a radioactive element is the time required for half of the radioactive nuclei present in a sample to decay (see Exercise 27).

(a) Show that the half-life of a radioactive element with decay constant k is

$$\text{Half-life} = \frac{\ln 2}{k}.$$

(b) *Polonium-210* The effective radioactive lifetime of polonium-210 is so short we measure it in days rather than years. Find the half-life of polonium-210 assuming that its (per day) decay constant is $k = 5 \times 10^{-3}$.

29. *Carbon-14 dating* People who do carbon-14 dating use a figure of 5700 years for its half-life (see Exercises 27 and 28). Find the age of a sample in which 10% of the radioactive nuclei originally present have decayed.

30. *The age of Crater Lake* The charcoal from a tree killed in the volcanic eruption that formed Crater Lake in Oregon contained 44.5% of the carbon-14 found in living matter. About how old is Crater Lake?

31. *The sensitivity of carbon-14 dating to measurement* To see the effect of a relatively small error in the estimate of the amount of carbon-14 in a sample being dated, consider this hypothetical situation.

(a) A fossilized bone found in central Illinois in the year A.D. 2000 contains 17% of its original carbon-14 content. Estimate the year the animal died.

(b) Repeat part (a) assuming 18% instead of 17%.

(c) Repeat part (a) assuming 16% instead of 17%.

32. *Art forgery* A painting attributed to Vermeer (1632–1675), which should contain no more than 96.2% of its original carbon-14, contains 99.5% instead. About how old is the forgery?

Resistance Proportional to Velocity

33. *A coasting cyclist* For a 160 lb cyclist on a 15 lb bicycle on level ground, the k in Equation (4) is about 1/5 slug/sec and $m =$ $160/32 = 5$ slugs. The cyclist starts coasting at 22 ft/sec (15 mph).

(a) About how far will the cyclist coast before reaching a complete stop?

(b) To the nearest second, about how long will it take the cyclist's speed to drop to 1 ft/sec?

34. *A coasting battleship* For a 56,000-ton Iowa class battleship, $m = 1,750,000$ slugs and the k in Equation (4) might be 3000 slugs/sec. Suppose that the battleship loses power when it is moving at a speed of 22 ft/sec (13.0 knots).

(a) About how far will the ship coast before it stops?

(b) About how long will it take the ship's speed to drop to 1 ft/sec?

COMPUTER EXPLORATIONS

Slope Fields and Solution Curves

In Exercises 35–40, obtain a slope field and graph the solution curves passing through the given points.

35. $y' = y$ with
CD-ROM
WEBsite **(a)** $(0, 1)$
(b) $(0, 2)$
(c) $(0, -1)$

36. $y' = 2(y - 4)$ with
CD-ROM
WEBsite **(a)** $(0, 1)$
(b) $(0, 4)$
(c) $(0, 5)$

37. $y' = y(2 - y)$ with
CD-ROM
WEBsite **(a)** $(0, 1/2)$
(b) $(0, 3/2)$
(c) $(0, 2)$
(d) $(0, 3)$

38. $y' = y^2$ with
CD-ROM
WEBsite **(a)** $(0, 1)$
(b) $(0, 2)$
(c) $(0, -1)$
(d) $(0, 0)$

39. $y' = \dfrac{3y}{x}$ with
CD-ROM
WEBsite **(a)** $(-3, 2)$
(b) $(1, 1)$
(c) $(2, 4)$

40. $y' = \dfrac{xy}{x^2 + 4}$ with
CD-ROM
WEBsite **(a)** $(0, 2)$
(b) $(0, -6)$
(c) $(-2\sqrt{3}, -4)$

5.5 Springs, Pumping and Lifting

CD-ROM
WEBsite

Work Done by a Constant Force • Work Done by a Variable Force Along a Line • Hooke's Law for Springs: $F = kx$ • Pumping Liquids from Containers

Most dams are built with an overflow device known as a "glory hole" to provide an outlet for the water behind the dam when the water level exceeds a certain height. Unfortunately, the glory hole can get clogged by debris that can be cleared only after the water in the hole is pumped out. To acquire powerful enough pumps to do the job, it is necessary to find the amount of work required to pump the water from the hole.

In everyday life, *work* means an activity that requires muscular or mental effort. In science, the term refers specifically to a force acting on a body and the body's subsequent displacement. This section shows how to calculate work.

Work Done by a Constant Force

When a body moves a distance d along a straight line as a result of being acted on by a force of constant magnitude F in the direction of motion, we calculate the **work** W done by the force on the body with the formula

$$W = Fd \qquad \text{(constant-force formula for work)}. \tag{1}$$

Right away, we can see a considerable difference between what we are used to calling work and what this formula says work is. If you push a car down the street, you will be doing work on the car, both by your own reckoning and by Equation (1). But if you push against the car and the car does not move, Equation (1) says you will do no work on the car, even if you push for an hour.

From Equation (1), we see that the unit of work in any system is the unit of force multiplied by the unit of distance. In SI units (SI stands for *Système International*, or International System), the unit of force is a newton, the unit of distance is a meter, and the unit of work is a newton-meter (N · m). This combination appears so often it has a special name, the **joule.** In the British system, the unit of work is the foot-pound, a unit frequently used by engineers.

Joules

The joule, abbreviated **J** and pronounced "jewel," is named after the English physicist James Prescott Joule (1818–1889). The defining equation is

1 joule = (1 newton)(1 meter).

In symbols, 1 J = 1 N · m.

It takes a force of about 1 N to lift an apple from a table. If you lift it 1 m, you have done about 1 J of work on the apple.

Example 1   Jacking Up a Car

If you jack up the side of a 2000 lb car 1.25 ft to change a tire (you have to apply a constant vertical force of about 1000 lb), you will perform $1000 \times 1.25 = 1250$ ft · lb of work on the car. In SI units, you have applied a force of 4448 N through a distance of 0.381 m to do $4448 \times 0.381 \approx 1695$ J of work.

Work Done by a Variable Force Along a Line

If the force you apply varies along the way, as it will if you are lifting a leaking bucket or compressing a spring, the formula $W = Fd$ has to be replaced by an integral formula that takes the variation in F into account.

Suppose that the force performing the work acts along a line that we can model with the x-axis and that its magnitude F is a continuous function of the position. We want to find the work done over the interval from $x = a$ to $x = b$. We partition $[a, b]$ in the usual way and choose an arbitrary point c_k in each subinterval $[x_{k-1}, x_k]$. If the subinterval is short enough, F, being continuous, will not vary much from x_{k-1} to x_k. The amount of work done across the interval will be about $F(c_k)$ times the distance Δx_k, the same as it would be if F were constant and we could apply Equation (1). The total work done from a to b is therefore approximated by the Riemann sum

$$\sum_{k=1}^{n} F(c_k)\, \Delta x_k.$$

We expect the approximation to improve as the norm of the partition goes to zero, so we define the work done by the force from a to b to be the integral of F from a to b.

CD-ROM
WEBsite

Definition Work

The **work** done by a variable force $F(x)$ directed along the x-axis from $x = a$ to $x = b$ is

$$W = \int_a^b F(x)\,dx. \qquad (2)$$

The units of the integral are joules if F is in newtons and x is in meters, and foot-pounds if F is in pounds and x in feet.

Example 2   Applying the Definition of Work

The work done by a force of $F(x) = 1/x^2$ N along the x-axis from $x = 1$ m to $x = 10$ m is

$$W = \int_1^{10} \frac{1}{x^2}\,dx = -\frac{1}{x}\Big]_1^{10} = -\frac{1}{10} + 1 = 0.9\ \text{J}.$$

Example 3   Lifting a Leaky Bucket

A leaky 5 lb bucket is lifted from the ground into the air by pulling in 20 ft of rope at a constant speed (Figure 5.33). The rope weighs 0.08 lb/ft. The bucket starts with 2 gal of water (16 lb) and leaks at a constant rate. It finishes draining just as it reaches the top. How much work was spent

(a) lifting the water alone

(b) lifting the water and bucket together

(c) lifting the water, bucket, and rope?

Solution

(a) *The water alone.* The force required to lift the water is equal to the water's weight, which varies steadily from 16 to 0 lb over the 20-ft lift. When the bucket is x ft off the ground, the water weighs

$$F(x) = \underbrace{16}_{\substack{\text{original weight} \\ \text{of water}}} \cdot \underbrace{((20 - x)/20)}_{\substack{\text{proportion left} \\ \text{at elevation } x}} = 16\left(1 - \frac{x}{20}\right) = 16 - \frac{4x}{5}\ \text{lb}.$$

The work done is

$$W = \int_a^b F(x)\,dx \qquad \text{Use Eq. (2) for variable forces.}$$

$$= \int_0^{20} \left(16 - \frac{4x}{5}\right)dx$$

$$= \left[16x - \frac{2x^2}{5}\right]_0^{20}$$

$$= 320 - 160 = 160\ \text{ft} \cdot \text{lb}$$

FIGURE 5.33 The leaky bucket in Example 3.

(b) *The water and bucket together.* According to Equation (1), it takes $5 \times 20 = 100$ ft·lb to lift a 5 lb weight 20 ft. Therefore,

$$160 + 100 = 260 \text{ ft} \cdot \text{lb}$$

of work were spent lifting the water and bucket together.

(c) *The water, bucket, and rope.* Now the total weight at level x is

$$F(x) = \left(16 - \frac{4x}{5}\right) + 5 + \overbrace{(0.08)}^{\text{lb/ft}}\,\overbrace{(20 - x)}^{\text{ft}}.$$

$$\underbrace{\phantom{\left(16 - \frac{4x}{5}\right)}}_{\substack{\text{variable weight}\\\text{of water}}} \quad \underbrace{}_{\substack{\text{constant weight}\\\text{of bucket}}} \quad \underbrace{}_{\substack{\text{weight of rope}\\\text{paid out at}\\\text{elevation } x}}$$

The work lifting the rope is

$$\text{Work on rope} = \int_0^{20} (0.08)(20 - x)\, dx = \int_0^{20} (1.6 - 0.08x)\, dx$$

$$= \Big[1.6x - 0.04x^2\Big]_0^{20} = 32 - 16 = 16 \text{ ft} \cdot \text{lb}.$$

The total work for the water, bucket, and rope combined is

$$160 + 100 + 16 = 276 \text{ ft} \cdot \text{lb}.$$

Hooke's Law for Springs: F = kx

Hooke's law says that the force it takes to stretch or compress a spring x length units from its natural (unstressed) length is proportional to x. In symbols,

$$F = kx. \tag{3}$$

The constant k, measured in force units per unit length, is a characteristic of the spring, called the **force constant** (or spring constant) of the spring. Hooke's law Equation (3), gives good results as long as the force doesn't distort the metal in the spring. We assume that the forces in this section are too small to do that.

Example 4   Compressing a Spring

Find the work required to compress a spring from its natural length of 1 ft to a length of 0.75 ft if the force constant is $k = 16$ lb/ft.

Solution   We picture the uncompressed spring laid out along the x-axis with its movable end at the origin and its fixed end at $x = 1$ ft (Figure 5.34). This enables us to describe the force required to compress the spring from 0 to x with the formula $F = 16x$. To compress the spring from 0 to 0.25 ft, the force must increase from

$$F(0) = 16 \cdot 0 = 0 \text{ lb} \qquad \text{to} \qquad F(0.25) = 16 \cdot 0.25 = 4 \text{ lb}.$$

The work done by F over this interval is

$$W = \int_0^{0.25} 16x\, dx = 8x^2 \Big]_0^{0.25} = 0.5 \text{ ft} \cdot \text{lb}. \qquad \begin{array}{l}\text{Eq. (2) with } a = 0,\\ b = 0.25, F(x) = 16x\end{array}$$

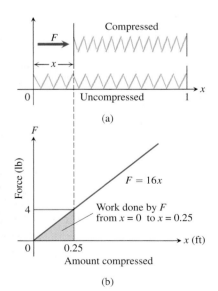

FIGURE 5.34 The force F needed to hold a spring under compression increases linearly as the spring is compressed.

Example 5   Stretching a Spring

A spring has a natural length of 1 m. A force of 24 N stretches the spring to a length of 1.8 m.

(a) Find the force constant k.

(b) How much work will it take to stretch the spring 2 m beyond its natural length?

(c) How far will a 45 N force stretch the spring?

Solution

(a) *The force constant.* We find the force constant from Equation (3). A force of 24 N stretches the spring 0.8 m, so

$$24 = k(0.8) \qquad \text{Eq. (3) with } F = 24, x = 0.8$$
$$k = 24/0.8 = 30 \text{ N/m.}$$

(b) *The work to stretch the spring* 2 m. We imagine the unstressed spring hanging along the x-axis with its free end at $x = 0$ (Figure 5.35). The force required to stretch the spring x m beyond its natural length is the force required to pull the free end of the spring x units from the origin. Hooke's law with $k = 30$ says that this force is

$$F(x) = 30x.$$

The work done by F on the spring from $x = 0$ m to $x = 2$ m is

$$W = \int_0^2 30x\,dx = 15x^2 \bigg]_0^2 = 60 \text{ J.}$$

(c) *How far will a 45 N force stretch the spring?* We substitute $F = 45$ in the equation $F = 30x$ to find

$$45 = 30x, \qquad \text{or} \qquad x = 1.5 \text{ m.}$$

A 45 N force will stretch the spring 1.5 m. No calculus is required to find this.

FIGURE 5.35 A 24 N weight stretches this spring 0.8 m beyond its unstressed length.

$x = 0$

0.8

1 24 N

x (m)

Pumping Liquids from Containers

How much work does it take to pump all or part of the liquid from a container? To find out, we imagine lifting the liquid out one thin horizontal slab at a time and applying the equation $W = Fd$ to each slab. We then evaluate the integral this leads to as the slabs become thinner and more numerous. The integral we get each time depends on the weight of the liquid and the dimensions of the container, but the way we find the integral is always the same. The next examples show what to do.

Example 6   Pumping Water from a Cylindrical Tank

How much work does it take to pump the water from a full upright circular cylindrical tank of radius 5 m and height 10 m to a level of 4 m above the top of the tank?

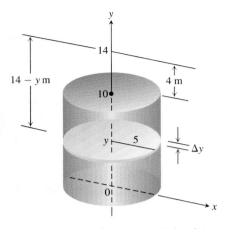

FIGURE 5.36 To find the work it takes to pump the water from a tank, think of lifting the water one thin slab at a time.

Solution   We draw the tank (Figure 5.36), add coordinate axes, and imagine the water divided into thin horizontal slabs by planes perpendicular to the y-axis at the points of a partition P of the interval $[0, 10]$.

The typical slab between the planes at y and $y + \Delta y$ has a volume of

$$\Delta V = \pi (\text{radius})^2 (\text{thickness}) = \pi (5)^2 \, \Delta y = 25\pi \, \Delta y \ \text{m}^3.$$

The force $F(y)$ required to lift the slab is equal to its weight,

$$F(y) = 9800 \, \Delta V \qquad \text{Water weighs 9800 N /m}^3.$$
$$= 9800(25\pi \, \Delta y) = 245{,}000\pi \, \Delta y \ \text{N}.$$

The distance through which F must act is about $(14 - y)$ m, so the work done lifting the slab is about

$$\Delta W = \text{force} \times \text{distance} = 245{,}000\pi (14 - y) \, \Delta y \ \text{J}.$$

The work it takes to lift all the water is approximately

$$W \approx \sum_{0}^{10} \Delta W = \sum_{0}^{10} 245{,}000\pi (14 - y) \, \Delta y \ \text{J}.$$

This is a Riemann sum for the function $245{,}000\pi (14 - y)$ over the interval $0 \leq y \leq 10$. The work of pumping the tank dry is the limit of these sums as $\| P \| \to 0$:

$$W = \int_{0}^{10} 245{,}000\pi (14 - y) \, dy = 245{,}000\pi \int_{0}^{10} (14 - y) \, dy$$

$$= 245{,}000\pi \left[14y - \frac{y^2}{2} \right]_{0}^{10} = 245{,}000\pi[90]$$

$$\approx 69{,}272{,}118 \approx 69.3 \times 10^6 \ \text{J}.$$

A 1-horsepower output motor rated at 746 J/sec could empty the tank in a little less than 26 h.

Example 7   Pumping Oil from a Conical Tank

The conical tank in Figure 5.37 is filled to within 2 ft of the top with olive oil weighing 57 lb/ft^3. How much work does it take to pump the oil to the rim of the tank?

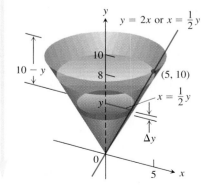

FIGURE 5.37 The olive oil in Example 7.

Solution   We imagine the oil divided into thin slabs by planes perpendicular to the y-axis at the points of a partition of the interval $[0, 8]$.

The typical slab between the planes at y and $y + \Delta y$ has a volume of about

$$\Delta V = \pi(\text{radians})^2(\text{thickness}) = \pi \left(\frac{1}{2}y\right)^2 \Delta y = \frac{\pi}{4}y^2 \, \Delta y \text{ ft}^3.$$

The force $F(y)$ required to lift this slab is equal to its weight,

$$F(y) = 57 \, \Delta V = \frac{57\pi}{4}y^2 \, \Delta y \text{ lb}$$

Weight = weight per unit volume × volume

The distance through which $F(y)$ must act to lift this slab to the level of the rim of the cone is about $(10 - y)$ ft, so the work done lifting the slab is about

$$\Delta W = \frac{57\pi}{4}(10 - y)y^2 \, \Delta y \text{ ft} \cdot \text{lb}.$$

The work done lifting all the slabs from $y = 0$ to $y = 8$ to the rim is approximately

$$W \approx \sum_{0}^{8} \frac{57\pi}{4}(10 - y)y^2 \, \Delta y \text{ ft} \cdot \text{lb}.$$

This is a Riemann sum for the function $(57\pi/4)(10 - y)y^2$ on the interval from $y = 0$ to $y = 8$. The work of pumping the oil to the rim is the limit of these sums as the norm of the partition goes to zero.

$$W = \int_{0}^{8} \frac{57\pi}{4}(10 - y)y^2 \, dy$$

$$= \frac{57\pi}{4} \int_{0}^{8} (10y^2 - y^3) \, dy$$

$$= \frac{57\pi}{4}\left[\frac{10y^3}{3} - \frac{y^4}{4}\right]_{0}^{8} \approx 30,561 \text{ ft} \cdot \text{lb}$$

Example 8   Pumping Water from a Glory Hole

A glory hole is a vertical drain pipe that keeps the water behind a dam from getting too high. The top of the glory hole for a dam is 14 ft below the top of the dam and 375 ft above the bottom (Figure 5.38). The hole needs to be pumped out from time to time to permit the removal of seasonal debris as mentioned in the opening remarks to this section.

From the cross section in Figure 5.38a, we see that the glory hole is a funnel-shaped drain. The throat of the funnel is 20 ft wide and the head is 120 ft across. The outside boundary of the head cross section are quarter circles formed with 50 ft radii, shown in Figure 5.38b. The glory hole is formed by rotating a cross section around its center. Consequently, all horizontal cross sections are

389 ft

|← 120 ft →| 375 ft above bottom of dam

325 ft above bottom of dam

← Throat 20 ft wide

(a)

r

Quarter circle of radius 50 ft

(b)

FIGURE 5.38 (a) Cross section of the glory hole for a dam and (b) the top of the glory hole.

circular disks throughout the entire glory hole. We calculate the work required to pump water from

(a) the throat of the hole

(b) the funnel portion.

Solution

(a) *Pumping from the throat.* A typical slab in the throat between the planes at y and $y + \Delta y$ has a volume of about

$$\Delta V = \pi (\text{radius})^2 (\text{thickness}) = \pi (10)^2 \, \Delta y \ \text{ft}^3.$$

The force $F(y)$ required to lift this slab is equal to its weight (about 62.4 lbs/ft^3 for water),

$$F(y) = 62.4 \, \Delta V = 6{,}240 \pi \, \Delta y \ \text{lb}.$$

The distance through which $F(y)$ must act to lift this slab to the top of the hole is $(375 - y)$ ft, so the work done lifting the slab is

$$\Delta W = 6240 \pi (375 - y) \, \Delta y \ \text{ft} \cdot \text{lb}.$$

The work done in pumping the water from the throat is the integral of a differential slab element from $y = 0$ to $y = 325$,

$$W = \int_0^{325} 6240 \pi (375 - y) \, dy$$

$$= 6240 \pi \left[375 y - \frac{y^2}{2} \right]_0^{325}$$

$$\approx 1{,}353{,}869{,}354 \ \text{ft} \cdot \text{lb}$$

(b) *Pumping from the funnel.* To compute the work necessary to pump water from the funnel portion of the glory hole, from $y = 325$ to $y = 375$, we need to compute ΔV for approximating elements in the funnel as shown in Figure 5.39. As can be seen from the figure, the radii of the slabs vary with the height y.

In Exercises 33 and 34, you are asked to complete the analysis to determine the total work required to pump the water and power of the pumps necessary to pump out the glory hole.

FIGURE 5.39 The glory hole funnel portion.

EXERCISES 5.5

Work Done by a Variable Force

1. *Leaky bucket* The workers in Example 3 changed to a larger bucket that held 5 gal (40 lb) of water, but the new bucket had an even larger leak so that it, too, was empty by the time it reached the top. Assuming that the water leaked out at a steady rate, how much work was done lifting the water? (Do not include the rope and bucket.)

2. *Leaky bucket* The bucket in Example 3 is hauled up twice as fast so that there is still 1 gal (8 lb) of water left when the bucket reaches the top. How much work is done lifting the water this time? (Do not include the rope and bucket.)

3. *Lifting a rope* A mountain climber is about to haul up a 50 m length of hanging rope. How much work will it take if the rope weighs 0.624 N/m?

4. *Leaky sand bag* A bag of sand originally weighing 144 lb was lifted at a constant rate. As it rose, sand also leaked out at a constant rate. The sand was half gone by the time the bag had been lifted 18 ft. How much work was done lifting the sand this far? (Neglect the weight of the bag and lifting equipment.)

5. *Lifting an elevator cable* An electric elevator with a motor at the top has a multistrand cable weighing 4.5 lb/ft. When the car is at the first floor, 180 ft of cable are paid out, and effectively 0 ft are out when the car is at the top floor. How much work does the motor do just lifting the cable when it takes the car from the first floor to the top?

6. *Force of attraction* When a particle of mass m is at $(x, 0)$, it is attracted toward the origin with a force whose magnitude is k/x^2. If the particle starts from rest at $x = b$ and is acted on by no other forces, find the work done on it by the time it reaches $x = a$, $0 < a < b$.

7. *Compressing gas* Suppose that the gas in a circular cylinder of cross-section area A is being compressed by a piston. If p is the pressure of the gas in pounds per square inch and V is the volume in cubic inches, show that the work done in compressing the gas from state (p_1, V_1) to state (p_2, V_2) is given by the equation

$$\text{Work} = \int_{(p_1, V_1)}^{(p_2, V_2)} p \, dV.$$

(*Hint:* In the coordinates suggested in the figure here, $dV = A \, dx$. The force against the piston is pA.)

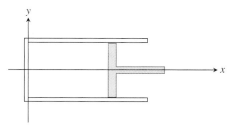

8. (*Continuation of Exercise 7*) Use the integral in Exercise 7 to find the work done in compressing the gas from $V_1 = 243$ in.^{3} to $V_2 = 32$ in.^{3} if $p_1 = 50$ lb/in^3 and p and V obey the gas law $pV^{1.4} =$ constant (for adiabatic processes).

Springs

9. *Spring constant* It took 1800 J of work to stretch a spring from its natural length of 2 m to a length of 5 m. Find the spring's force constant.

10. *Stretching a spring* A spring has a natural length of 10 in. An 800 lb force stretches the spring to 14 in.

(a) Find the force constant.

(b) How much work is done in stretching the spring from 10 in. to 12 in.?

(c) How far beyond its natural length will a 1600 lb force stretch the spring?

11. *Stretching a rubber band* A force of 2 N will stretch a rubber band 2 cm (0.02 m). Assuming Hooke's law applies, how far will a 4 N force stretch the rubber band? How much work does it take to stretch the rubber band this far?

12. *Stretching a spring* If a force of 90 N stretches a spring 1 m beyond its natural length, how much work does it take to stretch the spring 5 m beyond its natural length?

13. *Subway car springs* It takes a force of 21,714 lb to compress a coil spring assembly on a New York City Transit Authority subway car from its free height of 8 in. to its fully compressed height of 5 in.

(a) What is the assembly's force constant?

(b) How much work does it take to compress the assembly the first half inch? the second half inch? Answer to the nearest in. · lb.

(Data courtesy of Bombardier, Inc., Mass Transit Division, for spring assemblies in subway cars delivered to the New York City Transit Authority from 1985 to 1987.)

14. *Bathroom scale* A bathroom scale is compressed 1/16 in. when a 150 lb person stands on it. Assuming that the scale behaves like a spring that obeys Hooke's law, how much does someone who compresses the scale 1/8 in. weigh? How much work is done compressing the scale 1/8 in.?

Pumping Liquids from Containers

The Weight of Water
Because of Earth's rotation and variations in its gravitational field, the weight of a cubic foot of water at sea level can vary from about 62.26 lb at the equator to as much as 62.59 lb near the poles, a variation of about 0.5%. A cubic foot that weighs about 62.4 lb in Melbourne and New York City will weigh 62.5 lb in Juneau and Stockholm. Although 62.4 is a typical figure and a common textbook value, there is considerable variation.

15. *Pumping water* The rectangular tank shown in the accompanying figure, with its top at ground level, is used to catch runoff water. Assume that the water weighs 62.4 lb/ft^3.

(a) How much work does it take to empty the tank by pumping the water back to ground level once the tank is full?

(b) If the water is pumped to ground level with a (5/11)-horsepower (hp) motor (work output 250 ft · lb/sec), how long will it take to empty the full tank (to the nearest minute)?

(c) Show that the pump in part (b) will lower the water level 10 ft (halfway) during the first 25 min of pumping.

(d) *The weight of water* What are the answers to parts (a) and (b) in a location where water weighs 62.26 lb/ft³? 62.59 lb/ft³?

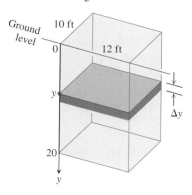

16. *Emptying a cistern* The rectangular cistern (storage tank for rainwater) shown below has its top 10 ft below ground level. The cistern, currently full, is to be emptied for inspection by pumping its contents to ground level.

(a) How much work will it take to empty the cistern?

(b) How long will it take a 1/2 hp pump, rated at 275 ft · lb/sec, to pump the tank dry?

(c) How long will it take the pump in part (b) to empty the tank halfway? (It will be less than half the time required to empty the tank completely.)

(d) *The weight of water* What are the answers to parts (a) through (c) in a location where water weighs 62.26 lb/ft³? 62.59 lb/ft³?

17. *Pumping water* How much work would it take to pump the water from the tank in Example 6 to the level of the top of the tank (instead of 4 m higher)?

18. *Pumping a half-full tank* Suppose that, instead of being full, the tank in Example 6 is only half full. How much work does it take to pump the remaining water to a level 4 m above the top of the tank?

19. *Emptying a tank* A vertical right circular cylindrical tank measures 30 ft high and 20 ft in diameter. It is full of kerosene weighing 51.2 lb/ft³. How much work does it take to pump the kerosene to the level of the top of the tank?

20. *Writing to Learn* The cylindrical tank shown here can be filled by pumping water from a lake 15 ft below the bottom of the tank. There are two ways to go about it. One is to pump the water through a hose attached to a valve in the bottom of the tank. The other is to attach the hose to the rim of the tank and let the water pour in. Which way will be faster? Give reasons for your answer.

21. (a) *Pumping milk* Suppose that the conical container in Example 7 contains milk (weighing 64.5 lb/ft³) instead of olive oil. How much work will it take to pump the contents to the rim?

(b) *Pumping oil* How much work will it take to pump the oil in Example 7 to a level 3 ft above the cone's rim?

22. *Pumping seawater* To design the interior surface of a huge stainless-steel tank, you revolve the curve $y = x^2$, $0 \le x \le 4$, about the y-axis. The container, with dimensions in meters, is to be filled with seawater, which weighs 10,000 N/m³. How much work will it take to empty the tank by pumping the water to the tank's top?

23. *Emptying a water reservoir* We model pumping from spherical containers the way we do from other containers, with the axis of integration along the vertical axis of the sphere. Use the figure here to find how much work it takes to empty a full hemispherical water reservoir of radius 5 m by pumping the water to a height of 4 m above the top of the reservoir. Water weighs 9800 N/m³.

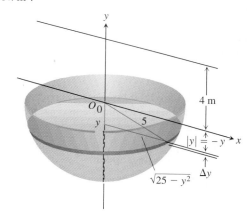

24. *Writing to Learn* You are in charge of the evacuation and repair of the storage tank shown here. The tank is a hemisphere of radius

10 ft and is full of benzene weighing 56 lb/ft^3. A firm you contacted says it can empty the tank for 1/2¢ per foot-pound of work. Find the work required to empty the tank by pumping the benzene to an outlet 2 ft above the top of the tank. If you have $5000 budgeted for the job, can you afford to hire the firm?

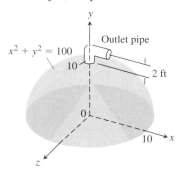

Work and Kinetic Energy

25. *Kinetic energy* If a variable force of magnitude $F(x)$ moves a body of mass m along the x-axis from x_1 to x_2, the body's velocity v can be written as dx/dt (where t represents time). Use Newton's Second Law of Motion $F = m(dv/dt)$ and the Chain Rule

$$\frac{dv}{dt} = \frac{dv}{dx}\frac{dx}{dt} = v\frac{dv}{dx}$$

to show that the net work done by the force in moving the body from x_1 to x_2 is

$$W = \int_{x_1}^{x_2} F(x)\,dx = \frac{1}{2}mv_2{}^2 - \frac{1}{2}mv_1{}^2,$$

where v_1 and v_2 are the body's velocities at x_1 and x_2. In physics, the expression $(1/2)mv^2$ is called the *kinetic energy* of the body moving with velocity v. Therefore, *the work done by the force equals the change in the body's kinetic energy*, and we can find the work by calculating this change.

In Exercises 26–32, use the result of Exercise 25.

26. *Tennis* A 2 oz tennis ball was served at 160 ft/sec (about 109 mph). How much work was done on the ball to make it go this fast? (To find the ball's mass from its weight, express the weight in pounds and divide by 32 ft/sec^2, the acceleration of gravity.)

27. *Baseball* How many foot-pounds of work does it take to throw a baseball 90 mph? A baseball weighs 5 oz, or 0.3125 lb.

28. *Golf* A 1.6 oz golf ball is driven off the tee at a speed of 280 ft/sec (about 191 mph). How many foot-pounds of work are done on the ball getting it into the air?

29. *Tennis* During the match in which Pete Sampras won the 1990 U.S. Open men's tennis championship, Sampras hit a serve that was clocked at a phenomenal 124 mph. How much work did Sampras have to do on the 2 oz ball to get it to that speed?

30. *Football* A quarterback threw a 14.5 oz football 88 ft/sec (60 mph). How many foot-pounds of work were done on the ball to get it to this speed?

31. *Softball* How much work has to be performed on a 6.5 oz softball to pitch it 132 ft/sec (90 mph)?

32. *A ball bearing* A 2 oz steel ball bearing is placed on a vertical spring whose force constant is $k = 18$ lb/ft. The spring is compressed 2 in. and released. About how high does the ball bearing go?

33. (*Continuation of Example 8*) *Pumping the funnel of the glory hole*

(a) Find the radius of the cross section (funnel portion) of the glory hole in Example 8 as a function of the height y above the floor of the dam (from $y = 325$ to $y = 375$).

(b) Find ΔV for the funnel section of the glory hole (from $y = 325$ to $y = 375$).

(c) Find the work necessary to pump out the funnel section by formulating and evaluating the appropriate definite integral.

34. (*Continuation of Exercise 33*) *Pumping water from a glory hole*

(a) Find the total work necessary to pump out the glory hole, by adding the work necessary to pump both the throat and funnel sections.

(b) Your answer to part (a) is in foot-pounds. A more useful form is horsepower-hours, since motors are rated in horsepower. To convert from foot-pounds to horsepower-hours, divide by 1.98×10^6. How many hours would it take a 1000 horsepower motor to pump out the glory hole, assuming that the motor was fully efficient?

Weight versus Mass
Weight is the force that results from gravity pulling on a mass. The two are related by the equation in Newton's second law,

$$\text{Weight} = \text{mass} \times \text{acceleration}.$$

Thus,

$$\text{Newtons} = \text{kilograms} \times \text{m/sec}^2,$$
$$\text{Pounds} = \text{slugs} \times \text{ft/sec}^2.$$

To convert mass to weight, multiply by the acceleration of gravity. To convert weight to mass, divide by the acceleration of gravity.

---

35. *Drinking a milkshake* The truncated conical container shown in the accompanying figure is full of strawberry milkshake that weighs 4/9 oz/in.^{3}. As you can see, the container is 7 in. deep, 2.5 in. across at the base, and 3.5 in. across at the top (a standard container size). The straw sticks up an inch above the top. About how much work does it take to suck up the milkshake through the straw (neglecting friction)? Answer in inch-ounces.

Dimensions in inches

36. *Water tower* Your town has decided to drill a well to increase its water supply. As the town engineer, you have determined that a water tower will be necessary to provide the pressure needed for

NOT TO SCALE

distribution, and you have designed the system shown here. The water is to be pumped from a 300 ft well through a vertical 4 in. pipe into the base of a cylindrical tank 20 ft in diameter and 25 ft high. The base of the tank will be 60 ft above ground. The pump is a 3 hp pump, rated at 1650 ft · lb/sec. To the nearest hour, how long will it take to fill the tank the first time? (Include the time it takes to fill the pipe.) Assume that water weighs 62.4 lb/ft³.

37. *Putting a satellite in orbit* The strength of Earth's gravitational field varies with the distance r from Earth's center, and the magnitude of the gravitational force experienced by a satellite of mass m during and after launch is

$$F(r) = \frac{mMG}{r^2}.$$

Here, $M = 5.975 \times 10^{24}$ kg is Earth's mass, $G = 6.6720 \times 10^{-11}$ N · m²kg⁻² is the universal gravitational constant, and r is measured in meters. The work it takes to lift a 1000 kg satellite from Earth's surface to a circular orbit 35,780 km above Earth's center is therefore given by the integral

$$\text{Work} = \int_{6,370,000}^{35,780,000} \frac{1000MG}{r^2} \, dr \text{ joules.}$$

Evaluate the integral. The lower limit of integration is Earth's radius in meters at the launch site. (This calculation does not take into account energy spent lifting the launch vehicle or energy spent bringing the satellite to orbit velocity.)

38. *Forcing electrons together* Two electrons r meters apart repel each other with a force of

$$F = \frac{23 \times 10^{-29}}{r^2} \text{ newtons.}$$

(a) Suppose that one electron is held fixed at the point $(1, 0)$ on the x-axis (units in meters). How much work does it take to move a second electron along the x-axis from the point $(-1, 0)$ to the origin?

(b) Suppose that an electron is held fixed at each of the points $(-1, 0)$ and $(1, 0)$. How much work does it take to move a third electron along the x-axis from $(5, 0)$ to $(3, 0)$?

5.6 Fluid Forces

The Constant-Depth Formula for Fluid Force • The Variable-Depth Formula

Engineers design dams to be thicker at the bottom than at the top (Figure 5.40) because the pressure against them increases with the depth. To generate hydroelectric power, gates (called *penstocks*) located near the base of the dam are opened to allow water under tremendous pressures to flow to turbine generators. Engineers must compute the total force against these gates at various depths to design the gates

FIGURE 5.40 To withstand the increasing pressure, dams are built thicker as they go down.

themselves and the hydraulic systems that open and close them. It is remarkable that the pressure at any point on a dam depends only on how far below the surface the point is and not on how much the surface of the dam happens to be tilted at that point. The pressure in pounds per square foot at a point h ft below the surface is always $62.4h$. The number 62.4 is the weight-density of water in pounds per cubic foot.

The formula, pressure $= 62.4h$, makes sense when you think of the units involved:

$$\frac{\text{lb}}{\text{ft}^2} = \frac{\text{lb}}{\text{ft}^3} \times \text{ft}.$$

As you can see, this equation depends only on units and not on the fluid involved. The pressure h feet below the surface of any fluid is the fluid's weight-density times h.

Weight–Density

A fluid's weight-density is its weight per unit volume. Typical values (in pounds per cubic feet) are

Gasoline	42
Mercury	849
Milk	64.5
Molasses	100
Olive oil	57
Seawater	64
Water	62.4

> **The Pressure-Depth Equation**
> In a fluid that is standing still, the pressure p at depth h is the fluid's weight-density w times h:
>
> $$p = wh. \tag{1}$$

In this section, we use the equation $p = wh$ to derive a formula for the total force exerted by a fluid against all or part of a vertical or horizontal containing wall.

The Constant-Depth Formula for Fluid Force

In a container of fluid with a flat horizontal base, the total force exerted by the fluid against the base can be calculated by multiplying the area of the base by the pressure at the base. We can do this because total force equals force per unit area (pressure) times area. (See Figure 5.41.) If F, p, and A are the total force, pressure, and area, then

$$F = \text{total force} = \text{force per unit area} \times \text{area}$$
$$= \text{pressure} \times \text{area} = pA$$
$$= whA. \qquad p = wh \text{ from Eq. (1)}$$

FIGURE 5.41 These containers are filled with water to the same depth and have the same base area. The total force is therefore the same on the bottom of each container. The containers' shapes do not matter here.

> **Fluid Force on a Constant-Depth Surface**
> $$F = pA = whA \tag{2}$$

Example 1 The Great Molasses Flood of 1919

FIGURE 5.42 Schematic drawing of the molasses tank in Example 1.

At 1:00 P.M. on January 15, 1919 (an unseasonably warm day), a 90-ft-high, 90-ft-diameter cylindrical metal tank in which the Puritan Distilling Company stored molasses at the corner of Foster and Commercial streets in Boston's North End exploded. Molasses flooded the streets 30 ft deep, trapping pedestrians and horses, knocking down buildings, and oozing into homes. It was eventually tracked all over town and even made its way into the suburbs via trolley cars and people's shoes. It took weeks to clean up. Given that the tank was full of molasses weighing 100 lb/ft³, what was the total force exerted by the molasses on the bottom of the tank at the time it ruptured (Figure 5.42)?

Solution At the bottom of the tank, the molasses exerted a constant pressure of

$$p = wh = \left(100 \frac{\text{lb}}{\text{ft}^3}\right)(90 \text{ ft}) = 9000 \frac{\text{lb}}{\text{ft}^2}.$$

Since the area of the base was $\pi(45)^2$, the total force on the base was

$$\left(9000 \frac{\text{lb}}{\text{ft}^2}\right)(2025 \pi \text{ ft}^2) \approx 57{,}255{,}526 \text{ lb}.$$

For a flat plate submerged *horizontally,* like the bottom of the molasses tank in Example 1, the downward force acting on its upper face due to liquid pressure is given by Equation (2). If the plate is submerged *vertically,* however, then the pressure against it will be different at different depths and Equation (2) no longer is usable in that form (because h varies). By dividing the plate into many narrow horizontal bands or strips, we can create a Riemann sum whose limit is the fluid force against the side of the submerged vertical plate. Here is the procedure.

The Variable-Depth Formula

Suppose we want to know the force exerted by a fluid against one side of a vertical plate submerged in a fluid of weight-density w. To find it, we model the plate as a region extending from $y = a$ to $y = b$ in the xy-plane (Figure 5.43). We partition $[a, b]$ in the usual way and imagine the region to be cut into thin horizontal strips by planes perpendicular to the y-axis at the partition points. The typical strip from y to $y + \Delta y$ is Δy units wide by $L(y)$ units long. We assume $L(y)$ to be a continuous function of y.

The pressure varies across the strip from top to bottom. If the strip is narrow enough, however, the pressure will remain close to its bottom-edge value of $w \times$ (strip depth). The force exerted by the fluid against one side of the strip will be about

$$\Delta F = \text{(pressure along bottom edge)} \times \text{(area)}$$
$$= w \times \text{(strip depth)} \times L(y)\, \Delta y.$$

The force against the entire plate will be about

$$\sum_a^b \Delta F = \sum_a^b (w \times \text{(strip depth)} \times L(y)\, \Delta y). \tag{3}$$

The sum in Equation (3) is a Riemann sum for a continuous function on $[a, b]$, and we expect the approximations to improve as the norm of the partition goes to zero. The force against the plate is the limit of these sums.

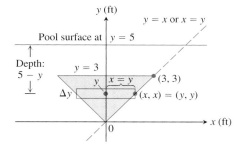

FIGURE 5.43 The force exerted by a fluid against one side of a thin horizontal strip is about $\Delta F = \text{pressure} \times \text{area} = w \times$ (strip depth) $\times\, L(y)\, \Delta y$.

The Integral for Fluid Force Against a Vertical Flat Plate
Suppose that a plate submerged vertically in fluid of weight-density w runs from $y = a$ to $y = b$ on the y-axis. Let $L(y)$ be the length of the horizontal strip measured from left to right along the surface of the plate at level y. Then the force exerted by the fluid against one side of the plate is

$$F = \int_a^b w \cdot \text{(strip depth)} \cdot L(y)\, dy. \tag{4}$$

Example 2   Applying the Integral for Fluid Force

A flat isosceles right triangular plate with base 6 ft and height 3 ft is submerged vertically, base up, 2 ft below the surface of a swimming pool. Find the force exerted by the water against one side of the plate.

Solution   We establish a coordinate system to work in by placing the origin at the plate's bottom vertex and running the y-axis upward along the plate's axis of symmetry (Figure 5.44). The surface of the pool lies along the line $y = 5$ and the plate's top edge along the line $y = 3$. The plate's right-hand edge lies along the line $y = x$, with the upper right vertex at $(3, 3)$. The length of a thin strip at level y is

$$L(y) = 2x = 2y.$$

FIGURE 5.44 To find the force on one side of the submerged plate in Example 2, we can use a coordinate system like the one here.

Historical Biography

William Thomson
(1824 — 1907)

The depth of the strip beneath the surface is $(5 - y)$. The force exerted by the water against one side of the plate is therefore

$$F = \int_a^b w \times \left(\begin{array}{c} \text{strip} \\ \text{depth} \end{array}\right) \times L(y)\, dy \qquad \text{Eq. (4)}$$

$$= \int_0^3 62.4(5 - y)\, 2y\, dy$$

$$= 124.8 \int_0^3 (5y - y^2)\, dy$$

$$= 124.8 \left[\frac{5}{2} y^2 - \frac{y^3}{3} \right]_0^3 = 1684.8 \text{ lb.}$$

How to Find Fluid Force

Whatever coordinate system you use, you can find the fluid force against one side of a submerged vertical plate or wall by taking these steps.

Step 1. *Find expressions* for the length and depth of a typical thin horizontal strip.

Step 2. *Multiply their product by the fluid's weight-density w and integrate* over the interval of depths occupied by the plate or wall.

Example 3   The Snake River Dam and Hydroelectric Power

Water from the Snake River Dam is used to generate hydroelectric power at an adjacent electrical plant. Water under extreme pressure flows through three pen-stock intake gates located as shown in Figure 5.45a. The lowest point on the gate is 101 ft above the base of the dam, whereas the highest point is 129 ft above the base of the dam. The width of the gate at its center is 16 ft. Each gate is inset to be vertical. To design and build the gate, engineers must know the maximum total force against the gate.

FIGURE 5.46 Dimensions and coordinates for calculating the maximum possible force against each penstock gate. (Example 3)

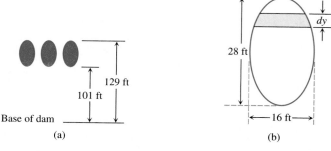

FIGURE 5.45 (a) Location of penstock gates from base of the dam. (b) Enlarged penstock gate subdivided into horizontal strips.

We now approximate the gate with horizontal rectangular strips of width dy and length $L(y)$. The force against the gate, from Equation (4), is

$$F = \int_a^b w \times \left(\frac{\text{strip}}{\text{depth}} \right) \times L(y) \, dy$$

$$= \int_{-14}^{14} 61.4 \, (274 - y)L(y) \, dy. \tag{5}$$

To complete the analysis, we use the equation for the gate's elliptical boundary to formulate $L(y)$ and perform the integration (Exercise 13). You will be astonished by the size of the force the gate must be designed to withstand. (Figure 5.46)

EXERCISES 5.6

The weight-densities of the fluids in the following exercises can be found in the table on page 465.

1. *Triangular plate* Calculate the fluid force on one side of the plate in Example 2 using the coordinate system shown here.

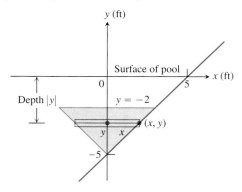

2. *Triangular plate* Calculate the fluid force on one side of the plate in Example 2 using the coordinate system shown here.

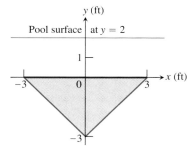

3. *Lowered triangular plate* The plate in Example 2 is lowered another 2 ft into the water. What is the fluid force on one side of the plate now?

4. *Raised triangular plate* The plate in Example 2 is raised to put its top edge at the surface of the pool. What is the fluid force on one side of the plate now?

5. *Triangular plate* The isosceles triangular plate shown here is submerged vertically 1 ft below the surface of a freshwater lake.

   (a) Find the fluid force against one face of the plate.

   (b) What would be the fluid force on one side of the plate if the water were seawater instead of freshwater?

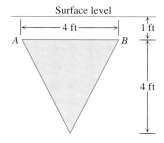

6. *Rotated triangular plate* The plate in Exercise 5 is revolved 180° about line *AB* so that part of the plate sticks out of the lake, as shown here. What force does the water exert on one face of the plate now?

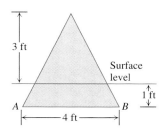

7. *New England Aquarium* The viewing portion of the rectangular glass window in a typical fish tank at the New England Aquarium in Boston is 63 in. wide and runs from 0.5 in. below the water's surface to 33.5 in. below the surface. Find the fluid force against this portion of the window. The weight-density of seawater is 64 lb/ft^3. (In case you were wondering, the glass is 3/4 in. thick and the tank walls extend 4 in. above the water to keep the fish from jumping out.)

8. *Fish tank* A horizontal rectangular freshwater fish tank with base 2×4 ft and height 2 ft (interior dimensions) is filled to within 2 in. of the top.

   (a) Find the fluid force against each side and end of the tank.

   (b) *Writing to Learn* If the tank is sealed and stood on end (without spilling), so that one of the square ends is the base, what does that do to the fluid forces on the rectangular sides?

9. *Semicircular plate* A semicircular plate 2 ft in diameter sticks straight down into fresh water with the diameter along the surface. Find the force exerted by the water on one side of the plate.

10. *Milk truck* A tank truck hauls milk in a 6-ft-diameter horizontal right circular cylindrical tank. How much force does the milk exert on each end of the tank when the tank is half full?

11. *Tank with parabolic gate* The cubical metal tank shown here has a parabolic gate, held in place by bolts and designed to withstand a fluid force of 160 lb without rupturing. The liquid you plan to store has a weight-density of 50 lb/ft^3.

   (a) What is the fluid force on the gate when the liquid is 2 ft deep?

   (b) What is the maximum height to which the container can be filled without exceeding its design limitation?

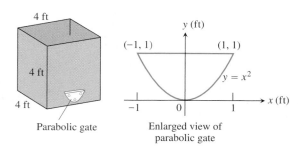

Parabolic gate   Enlarged view of parabolic gate

12. *Window in a tank* The rectangular tank shown here has a 1 ft $\times$ 1 ft square window 1 ft above the base. The window is designed to withstand a fluid force of 312 lb without cracking.

   (a) What fluid force will the window have to withstand if the tank is filled with water to a depth of 3 ft?

   (b) To what level can the tank be filled with water without exceeding the window's design limitation?

13. *Force against penstock gate (Example 3)*

   (a) Find an equation for the elliptical gate boundary in Figure 5.46.

   (b) Use the equation found in part (a) to write a formula for the strip length $L(y)$.

   T **(c)** Use the formula found in part (b) to complete the integral in Equation (5) and evaluate the integral numerically.

14. *Swimming pool drain plate* Water is running into the rectangular swimming pool shown here at the rate of 1000 ft^3/h.

   (a) Find the fluid force against the triangular drain plate after 9 h of filling.

   (b) The drain plate is designed to withstand a fluid force of 520 lb. How high can you fill the pool without exceeding this limitation?

Triangular drain plate

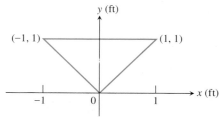

Enlarged view of drain plate

15. **(a)** *Average pressure* A vertical rectangular plate a units long by b units wide is submerged in a fluid of weight density w with its long edges parallel to the fluid's surface. Find the average value of the pressure along the vertical dimension of the plate.

(b) Show that the force exerted by the fluid on one side of the plate is the average value of the pressure times the area of the plate.

16. *Tank with a movable end* Water pours into the tank shown here at the rate of 4 ft³/min. The tank's cross sections are 4-ft-diameter semicircles. One end of the tank is movable, but moving it to increase the volume compresses a spring. The spring constant is $k = 100$ lb/ft. If the end of the tank moves 5 ft against the spring, the water will drain out of a safety hole in the bottom at the rate of 5 ft³/min. Will the movable end reach the hole before the tank overflows?

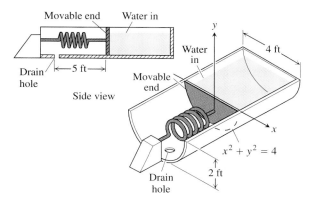

17. *Watering trough* The vertical ends of a watering trough are isosceles triangles like the one shown here (dimensions in feet).

(a) Find the fluid force against the ends when the trough is full.

(b) How many inches do you have to lower the water level in the trough to cut the fluid force on the ends in half? (Answer to the nearest half inch.)

(c) *Writing to Learn* Does it matter how long the trough is? Give reasons for your answer.

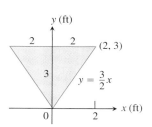

18. *Watering trough* The vertical ends of a watering trough are squares 3 ft on a side.

(a) Find the fluid force against the ends when the trough is full.

(b) How many inches do you have to lower the water level in the trough to reduce the fluid force by 25%?

19. *Milk carton* A rectangular milk carton measures 3.75×3.75 in. at the base and is 7.75 in. tall. Find the force of the milk on one side when the carton is full.

20. *Olive oil can* A standard olive oil can measures 5.75 by 3.5 in. at the base and is 10 in. tall. Find the fluid force against the base and each side when the can is full.

21. *Water in a trough* The end plates of the trough shown here were designed to withstand a fluid force of 6667 lb. How many cubic feet of water can the tank hold without exceeding this limitation? Round down to the nearest cubic foot.

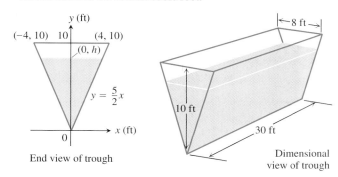

End view of trough   Dimensional view of trough

5.7   Moments and Centers of Mass

Masses Along a Line • Wires and Thin Rods • Masses Distributed over a Plane Region • Thin, Flat Plates • Centroids

Many structures and mechanical systems behave as if their masses were concentrated at a single point, called the center of mass (Figure 5.47). It is important to know how to locate this point, and doing so is basically a mathematical enterprise.

(a)

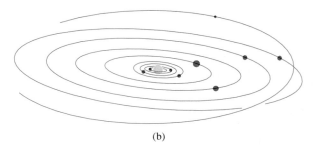

(b)

FIGURE 5.47 (a) The motion of this wrench gliding on ice seems haphazard until we notice that the wrench is simply turning about its center of mass as the center glides in a straight line. (b) The planets, asteroids, and comets of our solar system revolve about their collective center of mass. (It lies inside the sun.)

For the moment, we deal with one- and two-dimensional objects. Three-dimensional objects are best done with the multiple integrals of Chapter 12.*

Masses Along a Line

We develop our mathematical model in stages. The first stage is to imagine masses m_1, m_2, and m_3 on a rigid x-axis supported by a fulcrum at the origin.

The resulting system might balance, or it might not. It depends on how large the masses are and how they are arranged.

Each mass m_k exerts a downward force $m_k g$ equal to the magnitude of the mass times the acceleration of gravity. Each of these forces has a tendency to turn the axis about the origin, the way you turn a seesaw. This turning effect, called a **torque,** is measured by multiplying the force $m_k g$ by the signed distance x_k from the point of application to the origin. Masses to the left of the origin exert negative (counterclockwise) torque. Masses to the right of the origin exert positive (clockwise) torque.

The sum of the torques measures the tendency of a system to rotate about the origin. This sum is called the **system torque.**

$$\text{System torque} = m_1 g x_1 + m_2 g x_2 + m_3 g x_3 \tag{1}$$

The system will balance if and only if its torque is zero.

If we factor out the g in Equation (1), we see that the system torque is

$$\underbrace{g}_{\substack{\text{a feature of the} \\ \text{environment}}} \cdot \underbrace{(m_1 x_1 + m_2 x_2 + m_3 x_3)}_{\substack{\text{a feature of} \\ \text{the system}}}$$

Mass versus Weight

Weight is the force that results from gravity pulling on a mass. If an object of mass m is placed in a location where the acceleration of gravity is g, the object's weight there is

$$F = mg$$

(as in Newton's second law).

---

*The presentation in Chapter 12 is entirely self-contained. The present section may therefore be omitted, if desired. The essential ideas are repeated in Chapter 12.

Thus, the torque is the product of the gravitational acceleration g, which is a feature of the environment in which the system happens to reside, and the number ($m_1 x_1 + m_2 x_2 + m_3 x_3$), which is a feature of the system itself, a constant that stays the same no matter where the system is placed.

The number ($m_1 x_1 + m_2 x_2 + m_3 x_3$) is called the **moment of the system about the origin.** It is the sum of the **moments** $m_1 x_1, m_2 x_2, m_3 x_3$ of the individual masses.

$$M_0 = \text{Moment of system about origin} = \sum m_k x_k$$

(We shift to sigma notation here to allow for sums with more terms. For $\sum m_k x_k$, read "summation $m_k x_k$.")

We usually want to know where to place the fulcrum to make the system balance, that is, at what point $\bar{x}$ to place it to make the torques add to zero.

Special location
for balance

The torque of each mass about the fulcrum in this special location is

$$\text{Torque of } m_k \text{ about } \bar{x} = \begin{pmatrix} \text{signed distance} \\ \text{of } m_k \text{ from } \bar{x} \end{pmatrix} \begin{pmatrix} \text{downward} \\ \text{force} \end{pmatrix}$$

$$= (x_k - \bar{x}) m_k g.$$

When we write the equation that says that the sum of these torques is zero, we get an equation we can solve for $\bar{x}$:

$$\sum (x_k - \bar{x}) m_k g = 0 \qquad \text{Sum of the torques equals zero}$$

$$g \sum (x_k - \bar{x}) m_k = 0 \qquad \text{Constant Multiple Rule for Sums}$$

$$\sum (m_k x_k - \bar{x} m_k) = 0 \qquad g \text{ divided out, } m_k \text{ distributed}$$

$$\sum m_k x_k - \sum \bar{x} m_k = 0 \qquad \text{Difference Rule for Sums}$$

$$\sum m_k x_k = \bar{x} \sum m_k \qquad \begin{array}{l}\text{Rearranged, Constant} \\ \text{Multiple Rule again}\end{array}$$

$$\bar{x} = \frac{\sum m_k x_k}{\sum m_k}. \qquad \text{Solved for } \bar{x}$$

This last equation tells us to find $\bar{x}$ by dividing the system's moment about the origin by the system's total mass:

$$\bar{x} = \frac{\sum m_k x_k}{\sum m_k} = \frac{\text{system moment about origin}}{\text{system mass}}.$$

The point $\bar{x}$ is called the system's **center of mass.**

Wires and Thin Rods

In many applications, we want to know the center of mass of a rod or a thin strip of metal. In cases like these where we can model the distribution of mass with a continuous function, the summation signs in our formulas become integrals in a manner we now describe.

Imagine a long, thin strip lying along the x-axis from $x = a$ to $x = b$ and cut into small pieces of mass Δm_k by a partition of the interval $[a, b]$.

The kth piece is Δx_k units long and lies approximately x_k units from the origin. Now observe three things.

First, the strip's center of mass $\bar{x}$ is nearly the same as that of the system of point masses we would get by attaching each mass Δm_k to the point x_k:

$$\bar{x} \approx \frac{\text{system moment}}{\text{system mass}}.$$

Second, the moment of each piece of the strip about the origin is approximately $x_k \Delta m_k$, so the system moment is approximately the sum of the $x_k \Delta m_k$:

$$\text{System moment} \approx \sum x_k \Delta m_k.$$

Third, if the density of the strip at x_k is $\delta(x_k)$, expressed in terms of mass per unit length and if δ is continuous, then Δm_k is approximately equal to $\delta(x_k) \Delta x_k$ (mass per unit length times length):

$$\Delta m_k \approx \delta(x_k) \Delta x_k.$$

Combining these three observations gives

$$\bar{x} \approx \frac{\text{system moment}}{\text{system mass}} \approx \frac{\sum x_k \Delta m_k}{\sum \Delta m_k} \approx \frac{\sum x_k \delta(x_k) \Delta x_k}{\sum \delta(x_k) \Delta x_k}. \tag{2}$$

The sum in the last numerator in Equation (2) is a Riemann sum for the continuous function $x\delta(x)$ over the closed interval $[a, b]$. The sum in the denominator is a Riemann sum for the function $\delta(x)$ over this interval. We expect the approximations in Equation (2) to improve as the strip is partitioned more finely, and we are led to the equation

$$\bar{x} = \frac{\displaystyle\int_a^b x\delta(x)\,dx}{\displaystyle\int_a^b \delta(x)\,dx}.$$

This is the formula we use to find $\bar{x}$.

Density

A material's density is its mass per unit volume. In practice, however, we tend to use units we can conveniently measure. For wires, rods, and narrow strips we use mass per unit length. For flat sheets and plates we use mass per unit area.

To find a center of mass, divide moment by mass.

Moment, Mass, and Center of Mass of a Thin Rod or Strip Along the x-axis with Density Function $\delta(x)$

Moment about the origin: $M_0 = \displaystyle\int_a^b x\delta(x)\,dx$ (3a)

Mass: $M = \displaystyle\int_a^b \delta(x)\,dx$ (3b)

Center of mass: $\bar{x} = \dfrac{M_0}{M}$ (3c)

Example 1 Strips and Rods of Constant Density

Show that the center of mass of a straight, thin strip or rod of constant density lies halfway between its two ends.

Solution We model the strip as a portion of the x-axis from $x = a$ to $x = b$ (Figure 5.48). Our goal is to show that $\bar{x} = (a + b)/2$, the point halfway between a and b.

The key is the density's having a constant value. This enables us to regard the function $\delta(x)$ in the integrals in Equations (3) as a constant (call it δ), with the result that

$$M_0 = \int_a^b \delta x \, dx = \delta \int_a^b x \, dx = \delta \left[\frac{1}{2} x^2 \right]_a^b = \frac{\delta}{2} (b^2 - a^2)$$

$$M = \int_a^b \delta \, dx = \delta \int_a^b dx = \delta [x]_a^b = \delta(b - a)$$

$$\bar{x} = \frac{M_0}{M} = \frac{\frac{\delta}{2} (b^2 - a^2)}{\delta(b - a)}$$

$$= \frac{a + b}{2}. \qquad \text{The } \delta\text{'s cancel in the formula for } \bar{x}.$$

FIGURE 5.48 The center of mass of a straight, thin rod or strip of constant density lies halfway between its ends.

Example 2 Variable-Density Rod

The 10-m-long rod in Figure 5.49 thickens from left to right so that its density, instead of being constant, is $\delta(x) = 1 + (x/10)$ kg/m. Find the rod's center of mass.

Solution The rod's moment about the origin (Equation 3(a)) is

$$M_0 = \int_0^{10} x \delta(x) \, dx = \int_0^{10} x \left(1 + \frac{x}{10} \right) dx = \int_0^{10} \left(x + \frac{x^2}{10} \right) dx$$

$$= \left[\frac{x^2}{2} + \frac{x^3}{30} \right]_0^{10} = 50 + \frac{100}{3} = \frac{250}{3} \text{ kg} \cdot \text{m}. \qquad \text{The units of a moment are mass} \times \text{length.}$$

The rod's mass (Equation 3(b)) is

$$M = \int_0^{10} \delta(x) \, dx = \int_0^{10} \left(1 + \frac{x}{10} \right) dx = \left[x + \frac{x^2}{20} \right]_0^{10} = 10 + 5 = 15 \text{ kg}.$$

The center of mass (Equation 3(c)) is located at the point

$$\bar{x} = \frac{M_0}{M} = \frac{250}{3} \cdot \frac{1}{15} = \frac{50}{9} \approx 5.56 \text{ m}.$$

FIGURE 5.49 We can treat a rod of variable thickness as a rod of variable density. See Example 2.

Masses Distributed over a Plane Region

Suppose that we have a finite collection of masses located in the plane, with mass m_k at the point (x_k, y_k) (see Figure 5.50). The mass of the system is

$$\text{System mass:} \quad M = \sum m_k.$$

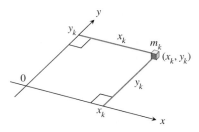

FIGURE 5.50 Each mass m_k has a moment about each axis.

Each mass m_k has a moment about each axis. Its moment about the x-axis is $m_k y_k$, and its moment about the y-axis is $m_k x_k$. The moments of the entire system about the two axes are

$$\text{Moment about } x\text{-axis:} \quad M_x = \sum m_k y_k,$$

$$\text{Moment about } y\text{-axis:} \quad M_y = \sum m_k x_k.$$

The x-coordinate of the system's center of mass is defined to be

$$\bar{x} = \frac{M_y}{M} = \frac{\sum m_k x_k}{\sum m_k}. \tag{4}$$

With this choice of $\bar{x}$, as in the one-dimensional case, the system balances about the line $x = \bar{x}$ (Figure 5.51).

The y-coordinate of the system's center of mass is defined to be

$$\bar{y} = \frac{M_x}{M} = \frac{\sum m_k y_k}{\sum m_k}. \tag{5}$$

With this choice of $\bar{y}$, the system balances about the line $y = \bar{y}$ as well. The torques exerted by the masses about the line $y = \bar{y}$ cancel out. Thus, as far as balance is concerned, the system behaves as if all its mass were at the single point $(\bar{x}, \bar{y})$. We call this point the system's *center of mass*.

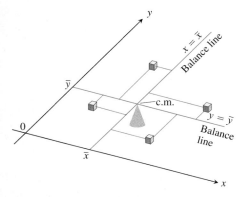

FIGURE 5.51 A two-dimensional array of masses balances on its center of mass.

Thin, Flat Plates

In many applications, we need to find the center of mass of a thin, flat plate: a disk of aluminum, say, or a triangular sheet of steel. In such cases, we assume the distribution of mass to be continuous, and the formulas we use to calculate $\bar{x}$ and $\bar{y}$ contain integrals instead of finite sums. The integrals arise in the following way.

Imagine the plate occupying a region in the xy-plane, cut into thin strips parallel to one of the axes (in Figure 5.52, the y-axis). The center of mass of a typical strip is $(\tilde{x}, \tilde{y})$. We treat the strip's mass Δm as if it were concentrated at $(\tilde{x}, \tilde{y})$. The moment of the strip about the y-axis is then $\tilde{x}\,\Delta m$. The moment of the strip about the x-axis is $\tilde{y}\,\Delta m$. Equations (4) and (5) then become

$$\bar{x} = \frac{M_y}{M} = \frac{\sum \tilde{x}\,\Delta m}{\sum \Delta m}, \qquad \bar{y} = \frac{M_x}{M} = \frac{\sum \tilde{y}\,\Delta m}{\sum \Delta m}.$$

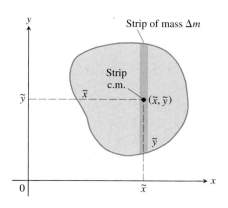

FIGURE 5.52 A plate cut into thin strips parallel to the y-axis. The moment exerted by a typical strip about each axis is the moment its mass Δm would exert if concentrated at the strip's center of mass $(\tilde{x}, \tilde{y})$.

As in the one-dimensional case, the sums are Riemann sums for integrals and approach these integrals as limiting values as the strips into which the plate is cut become narrower and narrower. We write these integrals symbolically as

$$\bar{x} = \frac{\int \tilde{x} \, dm}{\int dm} \qquad \text{and} \qquad \bar{y} = \frac{\int \tilde{y} \, dm}{\int dm}.$$

Moments, Mass, and Center of Mass of a Thin Plate Covering a Region in the xy-plane

Moment about the x-axis: $M_x = \displaystyle\int \tilde{y} \, dm$

Moment about the y-axis: $M_y = \displaystyle\int \tilde{x} \, dm$ (6)

Mass: $M = \displaystyle\int dm$

Center of mass: $\bar{x} = \dfrac{M_y}{M}, \qquad \bar{y} = \dfrac{M_x}{M}$

To evaluate these integrals, we picture the plate in the coordinate plane and sketch a strip of mass parallel to one of the coordinate axes. We then express the strip's mass dm and the coordinates $(\tilde{x}, \tilde{y})$ of the strip's center of mass in terms of x or y. Finally, we integrate $\tilde{y} \, dm$, $\tilde{x} \, dm$, and dm between limits of integration determined by the plate's location in the plane.

Example 3   Constant-Density Plate

The triangular plate shown in Figure 5.53 has a constant density of $\delta = 3 \text{ g/cm}^2$. Find

(a) the plate's moment M_y about the y-axis

(b) the plate's mass M

(c) the x-coordinate of the plate's center of mass (c.m.).

Solution

Method 1:   *Vertical strips* (Figure 5.54)

(a) The moment M_y: The typical vertical strip has

center of mass (c.m.): $(\tilde{x}, \tilde{y}) = (x, x)$

length: $2x$

width: dx

area: $dA = 2x \, dx$

mass: $dm = \delta \, dA = 3 \cdot 2x \, dx = 6x \, dx$

distance of c.m. from y-axis: $\tilde{x} = x.$

FIGURE 5.53 The plate in Example 3.

Units in centimeters

FIGURE 5.54 Modeling the plate in Example 3 with vertical strips.

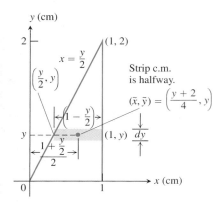

FIGURE 5.55 Modeling the plate in Example 3 with horizontal strips.

How to Find a Plate's Center of Mass

Step 1. Picture the plate in the xy-plane.

Step 2. Sketch a strip of mass parallel to one of the coordinate axes and find its dimensions.

Step 3. Find the strip's mass dm and center of mass $(\tilde{x}, \tilde{y})$.

Step 4. Integrate $\tilde{y}\,dm$, $\tilde{x}\,dm$, and dm to find M_x, M_y, and M.

Step 5. Divide the moments by the mass to calculate $\bar{x}$ and $\bar{y}$.

The moment of the strip about the y-axis is

$$\tilde{x}\,dm = x \cdot 6x\,dx = 6x^2\,dx.$$

The moment of the plate about the y-axis is therefore

$$M_y = \int \tilde{x}\,dm = \int_0^1 6x^2\,dx = 2x^3\Big]_0^1 = 2 \text{ g} \cdot \text{cm}.$$

(b) The plate's mass:

$$M = \int dm = \int_0^1 6x\,dx = 3x^2\Big]_0^1 = 3 \text{ g}.$$

(c) The x-coordinate of the plate's center of mass:

$$\bar{x} = \frac{M_y}{M} = \frac{2 \text{ g} \cdot \text{cm}}{3 \text{ g}} = \frac{2}{3} \text{ cm}.$$

By a similar computation, we could find M_x and $\bar{y} = M_x/M$.

Method 2:   *Horizontal strips* (Figure 5.55).

(a) The moment M_y: The y-coordinate of the center of mass of a typical horizontal strip is y (see the figure), so

$$\tilde{y} = y.$$

The x-coordinate is the x-coordinate of the point halfway across the triangle. This makes it the average of $y/2$ (the strip's left-hand x-value) and 1 (the strip's right-hand x-value):

$$\tilde{x} = \frac{(y/2) + 1}{2} = \frac{y}{4} + \frac{1}{2} = \frac{y+2}{4}.$$

We also have

length: $1 - \dfrac{y}{2} = \dfrac{2-y}{2}$

width: dy

area: $dA = \dfrac{2-y}{2}\,dy$

mass: $dm = \delta\,dA = 3 \cdot \dfrac{2-y}{2}\,dy$

distance of c.m. to y-axis: $\tilde{x} = \dfrac{y+2}{4}.$

The moment of the strip about the y-axis is

$$\tilde{x}\,dm = \frac{y+2}{4} \cdot 3 \cdot \frac{2-y}{2}\,dy = \frac{3}{8}(4 - y^2)\,dy.$$

The moment of the plate about the y-axis is

$$M_y = \int \tilde{x}\,dm = \int_0^2 \frac{3}{8}(4 - y^2)\,dy = \frac{3}{8}\left[4y - \frac{y^2}{3}\right]_0^2 = \frac{3}{8}\left(\frac{16}{3}\right) = 2 \text{ g} \cdot \text{cm}.$$

(b) The plate's mass:

$$M = \int dm = \int_0^2 \frac{3}{2}(2-y)\,dy = \frac{3}{2}\left[2y - \frac{y^2}{2}\right]_0^2 = \frac{3}{2}(4-2) = 3 \text{ g}.$$

(c) The x-coordinate of the plate's center of mass:

$$\bar{x} = \frac{M_y}{M} = \frac{2 \text{ g} \cdot \text{cm}}{3 \text{ g}} = \frac{2}{3} \text{ cm}.$$

By a similar computation, we could find M_x and $\bar{y}$.

If the distribution of mass in a thin, flat plate has an axis of symmetry, the center of mass will lie on this axis. If there are two axes of symmetry, the center of mass will lie at their intersection. These facts often help to simplify our work.

Example 4   Constant-Density Plate

Find the center of mass of a thin plate of constant density δ covering the region bounded above by the parabola $y = 4 - x^2$ and below by the x-axis (Figure 5.56).

Solution   Since the plate is symmetric about the y-axis and its density is constant, the distribution of mass is symmetric about the y-axis and the center of mass lies on the y-axis. Thus, $\bar{x} = 0$. It remains to find $\bar{y} = M_x / M$.

A trial calculation with horizontal strips (Figure 5.56a) leads to an inconvenient integration

$$M_x = \int_0^4 2\delta\, y\sqrt{4-y}\,dy.$$

We therefore model the distribution of mass with vertical strips instead (Figure 5.56b). The typical vertical strip has

center of mass (c.m.): $(\tilde{x}, \tilde{y}) = \left(x, \dfrac{4-x^2}{2}\right)$

length: $4 - x^2$

width: dx

area: $dA = (4 - x^2)\,dx$

mass: $dm = \delta\, dA = \delta(4 - x^2)\,dx$

distance from c.m. to x-axis: $\tilde{y} = \dfrac{4-x^2}{2}.$

The moment of the strip about the x-axis is

$$\tilde{y}\, dm = \frac{4-x^2}{2} \cdot \delta(4-x^2)\,dx = \frac{\delta}{2}(4-x^2)^2\,dx.$$

The moment of the plate about the x-axis is

$$M_x = \int \tilde{y}\, dm = \int_{-2}^2 \frac{\delta}{2}(4-x^2)^2\,dx$$

$$= \frac{\delta}{2}\int_{-2}^2 (16 - 8x^2 + x^4)\,dx = \frac{256}{15}\delta. \qquad (7)$$

(a)

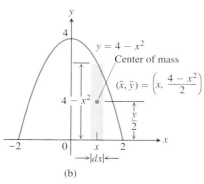

(b)

FIGURE 5.56 Modeling the plate in Example 4 with (a) horizontal strips leads to an inconvenient integration, so we model with (b) vertical strips instead.

The mass of the plate is

$$M = \int dm = \int_{-2}^{2} \delta(4 - x^2)\, dx = \frac{32}{3}\, \delta.$$ (8)

Therefore,

$$\bar{y} = \frac{M_x}{M} = \frac{(256/15)\delta}{(32/3)\delta} = \frac{8}{5}.$$

The plate's center of mass is the point

$$(\bar{x}, \bar{y}) = \left(0, \frac{8}{5}\right).$$

Example 5 Variable-Density Plate

Find the center of mass of the plate in Example 4 if the density at the point (x, y) is $\delta = 2x^2$, twice the square of the distance from the point to the y-axis.

Solution The mass distribution is still symmetric about the y-axis, so $\bar{x} = 0$. With $\delta = 2x^2$, Equations (7) and (8) become

$$M_x = \int \tilde{y}\, dm = \int_{-2}^{2} \frac{\delta}{2}(4 - x^2)^2\, dx = \int_{-2}^{2} x^2(4 - x^2)^2\, dx$$

$$= \int_{-2}^{2} (16x^2 - 8x^4 + x^6)\, dx = \frac{2048}{105}$$ (7′)

$$M = \int dm = \int_{-2}^{2} \delta(4 - x^2)\, dx = \int_{-2}^{2} 2x^2(4 - x^2)\, dx$$

$$= \int_{-2}^{2} (8x^2 - 2x^4)\, dx = \frac{256}{15}.$$ (8′)

Therefore,

$$\bar{y} = \frac{M_x}{M} = \frac{2048}{105} \cdot \frac{15}{256} = \frac{8}{7}.$$

The plate's new center of mass is

$$(\bar{x}, \bar{y}) = \left(0, \frac{8}{7}\right).$$

Example 6 Constant-Density Wire

Find the center of mass of a wire of constant density δ shaped like a semicircle of radius a.

Solution We model the wire with the semicircle $y = \sqrt{a^2 - x^2}$ (Figure 5.57). The distribution of mass is symmetric about the y-axis, so $\bar{x} = 0$. To find $\bar{y}$, we imagine the wire divided into short segments. The typical segment (Figure 5.57a) has

length: $ds = a\, d\theta$

mass: $dm = \delta\, ds = \delta a\, d\theta$ Mass per unit length times length

distance of c.m. to x-axis: $\tilde{y} = a \sin\theta.$

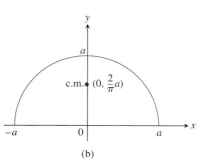

FIGURE 5.57 The semicircular wire in Example 6. (a) The dimensions and variables used in finding the center of mass. (b) The center of mass does not lie on the wire.

Hence,

$$\bar{y} = \frac{\int \tilde{y}\, dm}{\int dm} = \frac{\int_0^\pi a\,\sin\,\theta \cdot \delta a\, d\theta}{\int_0^\pi \delta a\, d\theta} = \frac{\delta a^2[-\cos\theta]_0^\pi}{\delta a\pi} = \frac{2}{\pi}\,a\,.$$

The center of mass lies on the axis of symmetry at the point $(0, 2a/\pi)$, about two-thirds of the way up from the origin (Figure 5.57b).

Centroids

When the density function is constant, it cancels out of the numerator and denominator of the formulas for $\bar{x}$ and $\bar{y}$. This happened in nearly every example in this section. As far as $\bar{x}$ and $\bar{y}$ were concerned, δ might as well have been 1. Thus, when the density is constant, the location of the center of mass is a feature of the geometry of the object and not of the material from which it is made. In such cases, engineers may call the center of mass the **centroid** of the shape, as in "Find the centroid of a triangle or a solid cone." To do so, just set δ equal to 1 and proceed to find $\bar{x}$ and $\bar{y}$ as before, by dividing moments by masses.

EXERCISES 5.7

Thin Rods

1. An 80 lb child and a 100 lb child are balancing on a seesaw. The 80 lb child is 5 ft from the fulcrum. How far from the fulcrum is the 100 lb child?

2. The ends of a log are placed on two scales. One scale reads 100 kg and the other 200 kg. Where is the log's center of mass?

3. The ends of two thin steel rods of equal length are welded together to make a right-angled frame. Locate the frame's center of mass. (*Hint:* Where is the center of mass of each rod?)

Right-angled weld

4. You weld the ends of two steel rods into a right-angled frame. One rod is twice the length of the other. Where is the frame's center of mass? (*Hint:* Where is the center of mass of each rod?)

Exercises 5–12 give density functions of thin rods lying along various intervals of the x-axis. Use Equations (3a) through (3c) to find each rod's moment about the origin, mass, and center of mass.

5. $\delta(x) = 4$, $\quad 0 \le x \le 2$

6. $\delta(x) = 4$, $\quad 1 \le x \le 3$

7. $\delta(x) = 1 + (x/3)$, $\quad 0 \le x \le 3$

8. $\delta(x) = 2 - (x/4)$, $\quad 0 \le x \le 4$

9. $\delta(x) = 1 + (1/\sqrt{x})$, $\quad 1 \le x \le 4$

10. $\delta(x) = 3(x^{-3/2} + x^{-5/2})$, $\quad 0.25 \le x \le 1$

11. $\delta(x) = \begin{cases} 2 - x, & 0 \le x < 1 \\ x, & 1 \le x \le 2 \end{cases}$

12. $\delta(x) = \begin{cases} x + 1, & 0 \le x < 1 \\ 2, & 1 \le x \le 2 \end{cases}$

Thin Plates with Constant Density

In Exercises 13–24, find the center of mass of a thin plate of constant density δ covering the given region.

13. The region bounded by the parabola $y = x^2$ and the line $y = 4$

14. The region bounded by the parabola $y = 25 - x^2$ and the x-axis

15. The region bounded by the parabola $y = x - x^2$ and the line $y = -x$

16. The region enclosed by the parabolas $y = x^2 - 3$ and $y = -2x^2$

17. The region bounded by the y-axis and the curve $x = y - y^3$, $0 \le y \le 1$

18. The region bounded by the parabola $x = y^2 - y$ and the line $y = x$

19. The region bounded by the x-axis and the curve $y = \cos x$, $-\pi/2 \le x \le \pi/2$

20. The region between the x-axis and the curve $y = \sec^2 x$, $-\pi/4 \le x \le \pi/4$

21. The region bounded by the parabolas $y = 2x^2 - 4x$ and $y = 2x - x^2$

22. (a) The region cut from the first quadrant by the circle $x^2 + y^2 = 9$

    (b) The region bounded by the x-axis and the semicircle $y = \sqrt{9 - x^2}$

    Compare your answer in part (b) with the answer in part (a).

23. The "triangular" region in the first quadrant between the circle $x^2 + y^2 = 9$ and the lines $x = 3$ and $y = 3$. (*Hint:* Use geometry to find the area.)

24. The region bounded above by the curve $y = 1/x^3$, below by the curve $y = -1/x^3$, and on the left and right by the lines $x = 1$ and $x = a > 1$. Also, find $\lim_{a \to \infty} \bar{x}$.

Thin Plates with Varying Density

25. Find the center of mass of a thin plate covering the region between the x-axis and the curve $y = 2/x^2$, $1 \le x \le 2$, if the plate's density at the point (x, y) is $\delta(x) = x^2$.

26. Find the center of a thin plate covering the region bounded below by the parabola $y = x^2$ and above by the line $y = x$ if the plate's density at the point (x, y) is $\delta(x) = 12x$.

27. The region bounded by the curves $y = \pm 4/\sqrt{x}$ and the lines $x = 1$ and $x = 4$ is revolved about the y-axis to generate a solid.

    (a) Find the volume of the solid.

    (b) Find the center of mass of a thin plate covering the region if the plate's density at the point (x, y) is $\delta(x) = 1/x$.

    (c) Sketch the plate and show the center of mass in your sketch.

28. The region between the curve $y = 2/x$ and the x-axis from $x = 1$ to $x = 4$ is revolved about the x-axis to generate a solid.

    (a) Find the volume of the solid.

    (b) Find the center of mass of a thin plate covering the region if the plate's density at the point (x, y) is $\delta(x) = \sqrt{x}$.

    (c) Sketch the plate and show the center of mass in your sketch.

Centroids of Triangles

29. *The centroid of a triangle lies at the intersection of the triangle's medians (Figure 5.58a)* You may recall that the point inside a triangle that lies one-third of the way from each side toward the opposite vertex is the point where the triangle's three medians intersect. Show that the centroid lies at the intersection of the medians by showing that it too lies one-third of the way from each side toward the opposite vertex. To do so, take the following steps.

(a)

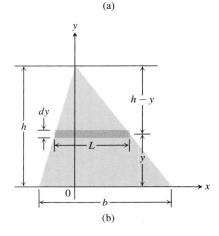

(b)

FIGURE 5.58 The triangle in Exercise 29. (a) **The centroid.** (b) **The dimensions and variables to use in locating the center** of mass.

    i. Stand one side of the triangle on the x-axis as in Figure 5.58b. Express dm in terms of L and dy.

    ii. Use similar triangles to show that $L = (b/h)(h - y)$. Substitute this expression for L in your formula for dm.

    iii. Show that $\bar{y} = h/3$.

    iv. Extend the argument to the other sides.

Use the result in Exercise 29 to find the centroids of the triangles whose vertices appear in Exercises 30–34. Assume $a, b > 0$.

30. $(-1, 0), (1, 0), (0, 3)$ 31. $(0, 0), (1, 0), (0, 1)$

32. $(0, 0), (a, 0), (0, a)$ 33. $(0, 0), (a, 0), (0, b)$

34. $(0, 0), (a, 0), (a/2, b)$

Thin Wires

35. *Constant density* Find the moment about the x-axis of a wire of constant density that lies along the curve $y = \sqrt{x}$ from $x = 0$ to $x = 2$.

36. *Constant density* Find the moment about the x-axis of a wire of constant density that lies along the curve $y = x^3$ from $x = 0$ to $x = 1$.

37. *Variable density* Suppose that the density of the wire in Example 6 is $\delta = k \sin \theta$ (k constant). Find the center of mass.

38. *Variable density* Suppose that the density of the wire in Example 6 is $\delta = 1 + k\,|\cos\theta|$ (k constant). Find the center of mass.

Engineering Formulas

Verify the statements and formulas in Exercises 39–42.

39. The coordinates of the centroid of a differentiable plane curve are

$$\bar{x} = \frac{\int x\,ds}{\text{length}}, \qquad \bar{y} = \frac{\int y\,ds}{\text{length}}.$$

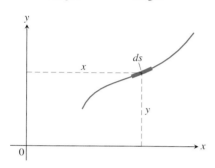

40. Whatever the value of $p > 0$ in the equation $y = x^2/(4p)$, the y-coordinate of the centroid of the parabolic segment shown here is $\bar{y} = (3/5)a$.

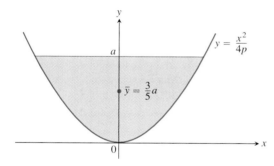

41. For wires and thin rods of constant density shaped like circular arcs centered at the origin and symmetric about the y-axis, the y-coordinate of the center of mass is

$$\bar{y} = \frac{a\,\sin\alpha}{\alpha} = \frac{ac}{s}.$$

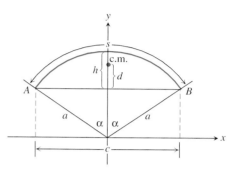

42. (*Continuation of Exercise 41*)

  (a) Show that when α is small, the distance d from the centroid to chord AB is about $2h/3$ (in the notation of the figure here) by taking the following steps.

  i. Show that

$$\frac{d}{h} = \frac{\sin\alpha - \alpha\cos\alpha}{\alpha - \alpha\cos\alpha}. \tag{9}$$

  ii. Graph

$$f(\alpha) = \frac{\sin\alpha - \alpha\cos\alpha}{\alpha - \alpha\cos\alpha}$$

  and use the trace feature to show that $\lim_{\alpha\to 0^+} f(\alpha) \approx 2/3$.

  (b) The error (difference between d and $2h/3$) is small even for angles greater than 45°. See for yourself by evaluating the right-hand side of Equation (9) for $\alpha = 0.2, 0.4, 0.6, 0.8,$ and 1.0 rad.

Questions to Guide Your Review

1. How do you define and calculate the volumes of solids by the method of slicing? Give an example.

2. How are the disk and washer methods for calculating volumes derived from the method of slicing? Give examples of volume calculations by these methods.

3. Describe the method of cylindrical shells. Give an example.

4. How do you define and calculate the length of the graph of a smooth function $y = f(x)$ over a closed interval? Give an example.

5. How do you find the length of a smooth parametrized curve $x = f(t), y = g(t), a \le t \le b$? What does smoothness have to do with length? What else do you need to know about the parametrization to find the curve's length? Give examples.

6. What is a first-order differential equation? When is a function a solution of such an equation?

7. How do you solve separable first-order differential equations?

8. What is the slope field of a differential equation $y' = f(x, y)$? What can we learn from such fields?

9. What is an initial value problem? How do you solve one? Give an example.

10. What is the law of exponential change? How can it be derived from an initial value problem? What are some of the applications of the law?

11. How do you define and calculate the work done by a variable force directed along a portion of the x-axis? How do you calculate the work it takes to pump a liquid from a tank? Give examples.

12. What is Hooke's law for springs? When might Hooke's law give poor results in modeling the behavior of a spring? Give examples.

13. How do you calculate the force exerted by a liquid against a portion of a flat vertical wall? Give an example.

14. What is a center of mass?

15. How do you locate the center of mass of a straight, narrow rod or strip of material? Give an example. If the density of the material is constant, you can tell right away where the center of mass is. Where is it?

16. How do you locate the center of mass of a thin flat plate of material? Give an example.

Practice Exercises

Volumes

Find the volumes of the solids in Exercises 1–16.

1. The solid lies between planes perpendicular to the x-axis at $x = 0$ and $x = 1$. The cross sections perpendicular to the x-axis between these planes are circular disks whose diameters run from the parabola $y = x^2$ to the parabola $y = \sqrt{x}$.

2. The base of the solid is the region in the first quadrant between the line $y = x$ and the parabola $y = 2\sqrt{x}$. The cross sections of the solid perpendicular to the x-axis are equilateral triangles whose bases stretch from the line to the curve.

3. The solid lies between planes perpendicular to the x-axis at $x = \pi/4$ and $x = 5\pi/4$. The cross sections between these planes are circular disks whose diameters run from the curve $y = 2 \cos x$ to the curve $y = 2 \sin x$.

4. The solid lies between planes perpendicular to the x-axis at $x = 0$ and $x = 6$. The cross sections between these planes are squares whose bases run from the x-axis up to the curve $x^{1/2} + y^{1/2} = \sqrt{6}$.

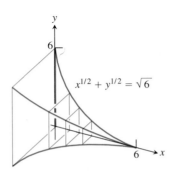

5. The solid lies between planes perpendicular to the x-axis at $x = 0$ and $x = 4$. The cross sections of the solid perpendicular to the x-axis between these planes are circular disks whose diameters run from the curve $x^2 = 4y$ to the curve $y^2 = 4x$.

6. The base of the solid is the region bounded by the parabola $y^2 = 4x$ and the line $x = 1$ in the xy-plane. Each cross section perpendicular to the x-axis is an equilateral triangle with one edge in the plane. (The triangles all lie on the same side of the plane.)

7. Find the volume of the solid generated by revolving the region bounded by the x-axis, the curve $y = 3x^4$, and the lines $x = 1$ and $x = -1$ about

(a) the x-axis

(b) the y-axis

(c) the line $x = 1$

(d) the line $y = 3$.

8. Find the volume of the solid generated by revolving the "triangular" region bounded by the curve $y = 4/x^3$ and the lines $x = 1$ and $y = 1/2$ about

(a) the x-axis

(b) the y-axis

(c) the line $x = 2$

(d) the line $y = 4$.

9. Find the volume of the solid generated by revolving the region bounded on the left by the parabola $x = y^2 + 1$ and on the right by the line $x = 5$ about

(a) the x-axis

(b) the y-axis

(c) the line $x = 5$.

10. Find the volume of the solid generated by revolving the region bounded by the parabola $y^2 = 4x$ and the line $y = x$ about

(a) the x-axis

(b) the y-axis

(c) the line $x = 4$

(d) the line $y = 4$.

11. Find the volume of the solid generated by revolving the region enclosed by the graphs of $y = e^{x/2}$, $y = 1$, and $x = \ln 3$ about the x-axis.

12. Find the volume of the solid generated by revolving the region bounded by the curve $y = \sin x$ and the lines $x = 0$, $x = \pi$, and $y = 2$ about the line $y = 2$.

13. Find the volume of the solid generated by revolving the region between the x-axis and the curve $y = x^2 - 2x$ about

   (a) the x-axis

   (b) the line $y = -1$

   (c) the line $x = 2$

   (d) the line $y = 2$.

14. Find the volume of the solid generated by revolving about the x-axis the region bounded by $y = 2 \tan x$, $y = 0$, $x = -\pi/4$, and $x = \pi/4$. (The region lies in the first and third quadrants and resembles a skewed bow tie.)

15. *Volume of a solid sphere with a hole* A round hole of radius $\sqrt{3}$ ft is bored through the center of a solid sphere of radius 2 ft. Find the volume of material removed from the sphere.

16. *Volume of a football* The profile of a football resembles the ellipse shown here. Find the football's volume to the nearest cubic inch.

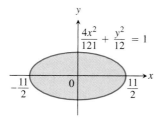

Lengths of Curves

Find the lengths of the curves in Exercises 17–22.

17. $y = x^{1/2} - (1/3)x^{3/2}$, $1 \le x \le 4$

18. $x = y^{2/3}$, $1 \le y \le 8$

19. $y = (5/12)x^{6/5} - (5/8)x^{4/5}$, $1 \le x \le 32$

20. $x = (y^3/12) + (1/y)$, $1 \le y \le 2$

21. $x = 5 \cos t - \cos 5t$, $y = 5 \sin t - \sin 5t$, $0 \le t \le \pi/2$

22. $x = t^2$, $y = 2t$, $0 \le t \le 1$

23. *Finding a function* Find a function f that has a continuous derivative on $(0, \infty)$ and that has both of the following properties.

   i. The graph of f goes through the point $(1, 1)$.

   ii. The length L of the curve from $(1, 1)$ to any point $(x, f(x))$ is given by the formula $L = \ln x + f(x) - 1$.

24. Find the length of the enclosed loop $x = t^2$, $y = (t^3/3) - t$ shown here. The loop starts at $t = -\sqrt{3}$ and ends at $t = \sqrt{3}$.

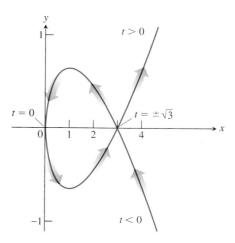

Work

25. *Lifting equipment* A rock climber is about to haul up 100 N (about 22.5 lb) of equipment that has been hanging beneath her on 40 m of rope that weighs 0.8 newton per meter. How much work will it take? (*Hint:* Solve for the rope and equipment separately, then add.)

26. *Leaky tank truck* You drove an 800 gal tank truck of water from the base of Mt. Washington to the summit and discovered on arrival that the tank was only half full. You started with a full tank, climbed at a steady rate, and accomplished the 4750 ft elevation change in 50 min. Assuming that the water leaked out at a steady rate, how much work was spent in carrying water to the top? Do not count the work done in getting yourself and the truck there. Water weighs 8 lb/U.S. gal.

27. *Stretching a spring* If a force of 20 lb is required to hold a spring 1 ft beyond its unstressed length, how much work does it take to stretch the spring this far? An additional foot?

28. *Garage door spring* A force of 200 N will stretch a garage door spring 0.8 m beyond its unstressed length. How far will a 300 N force stretch the spring? How much work does it take to stretch the spring this far from its unstressed length?

29. *Pumping a reservoir* A reservoir shaped like a right circular cone, point down, 20 ft across the top and 8 ft deep, is full of water. How much work does it take to pump the water to a level 6 ft above the top?

30. *Pumping a reservoir* (*Continuation of Exercise 29*) The reservoir is filled to a depth of 5 ft, and the water is to be pumped to the same level as the top. How much work does it take?

31. *Pumping a conical tank* A right circular conical tank, point down, with top radius 5 ft and height 10 ft is filled with a liquid whose weight-density is 60 lb/ft^3. How much work does it take to pump the liquid to a point 2 ft above the tank? If the pump is driven by a motor rated at 275 ft · lb/sec (1/2 hp), how long will it take to empty the tank?

32. *Pumping a cylindrical tank* A storage tank is a right circular cylinder 20 ft long and 8 ft in diameter with its axis horizontal. If the tank is half full of olive oil weighing 57 lb/ft^3, find the work done in emptying it through a pipe that runs from the bottom of the tank to an outlet that is 6 ft above the top of the tank.

Fluid Force

33. *Trough of water* The vertical triangular plate shown here is the end plate of a trough full of water ($w = 62.4$). What is the fluid force against the plate?

UNITS IN FEET

34. *Trough of maple syrup* The vertical trapezoidal plate shown here is the end plate of a trough full of maple syrup weighing 75 lb/ft^3. What is the force exerted by the syrup against the end plate of the trough when the syrup is 10 in. deep?

UNITS IN FEET

35. *Force on a parabolic gate* A flat vertical gate in the face of a dam is shaped like the parabolic region between the curve $y = 4x^2$ and the line $y = 4$, with measurements in feet. The top of the gate lies 5 ft below the surface of the water. Find the force exerted by the water against the gate ($w = 62.4$).

36. *Force on a trapezoid* The isosceles trapezoidal plate shown here is submerged vertically in water ($w = 62.4$) with its upper edge 4 ft below the surface. Find the fluid force on one side of the plate.

Dimensions in feet

First-Order Separable Equations

Solve the differential equations in Exercises 37–42.

37. $\dfrac{dy}{dx} = x^2\sqrt{y}, \quad y > 0$

38. $\dfrac{dy}{dx} = e^{2x+3y}$

39. $x\dfrac{dy}{dx} = y \ln x$

40. $\csc x \dfrac{dy}{dx} = e^{\cos x - y}$

41. $(\sec^2 \sqrt{x})\dfrac{dx}{dt} = \sqrt{x}$

42. $\sin t - (x \cos^2 t)\dfrac{dx}{dt} = 0, \quad -\pi/2 < t < \pi/2$

Centroids and Centers of Mass

43. Find the centroid of a thin, flat plate covering the region enclosed by the parabolas $y = 2x^2$ and $y = 3 - x^2$.

44. Find the centroid of a thin, flat plate covering the region enclosed by the x-axis, the lines $x = 2$ and $x = -2$, and the parabola $y = x^2$.

45. Find the centroid of a thin, flat plate covering the "triangular" region in the first quadrant bounded by the y-axis, the parabola $y = x^2/4$, and the line $y = 4$.

46. Find the centroid of a thin, flat plate covering the region enclosed by the parabola $y^2 = x$ and the line $x = 2y$.

47. *Variable density* Find the center of mass of a thin, flat plate covering the region enclosed by the parabola $y^2 = x$ and the line $x = 2y$ if the density function is $\delta(y) = 1 + y$. (Use horizontal strips.)

48. (a) *Constant density* Find the center of mass of a thin plate of constant density covering the region between the curve $y = 3/x^{3/2}$ and the x-axis from $x = 1$ to $x = 9$.

   (b) *Variable density* Find the plate's center of mass if, instead of being constant, the density is $\delta(x) = x$. (Use vertical strips.)

Additional Exercises: Theory, Examples, Applications

Volume and Length

1. A solid is generated by revolving about the x-axis the region bounded by the graph of the continuous function $y = f(x)$, the x-axis, and the lines $x = 0$ and $x = a$. Its volume, for all $a > 0$, is $a^2 + a$. Find $f(x)$.

2. Suppose that the function $f(x)$ is nonnegative and continuous for $x \geq 0$. Suppose also that, for every positive number b, revolving the region enclosed by the graph of f, the coordinate axes, and the line $x = b$ about the y-axis generates a solid of volume $2\pi b^3$. Find $f(x)$.

3. Suppose that the increasing function $f(x)$ is smooth for $x \geq 0$ and that $f(0) = a$. Let $s(x)$ denote the length of the graph of f from $(0, a)$ to $(x, f(x))$, $x > 0$. Find $f(x)$ if $s(x) = Cx$ for some constant C. What are the allowable values for C?

4. (a) Show that for $0 < \alpha \leq \pi/2$,

$$\int_0^\alpha \sqrt{1 + \cos^2\theta}\, d\theta > \sqrt{\alpha^2 + \sin^2\alpha}.$$

(b) Generalize the result in part (a).

Work and Fluid Force

5. *Work and kinetic energy* Suppose that a 1.6 oz golf ball is placed on a vertical spring with force constant $k = 2$ lb/in. The spring is compressed 6 in. and released. About how high does the ball go (measured from the spring's rest position)?

6. *Fluid force* A triangular plate ABC is submerged in water with its plane vertical. The side AB, 4 ft long, is 6 ft below the surface of the water, whereas the vertex C is 2 ft below the surface. Find the force exerted by the water on one side of the plate.

7. *Average pressure* A vertical rectangular plate is submerged in a fluid with its top edge parallel to the fluid's surface. Show that the force exerted by the fluid on one side of the plate equals the average value of the pressure up and down the plate times the area of the plate.

8. *Vertical square plate* The container profile shown here is filled with two nonmixing liquids of weight density w_1 and w_2. Find the fluid force on one side of the vertical square plate $ABCD$. The points B and D lie in the boundary layer and the square is $6\sqrt{2}$ ft on a side. (*Hint:* If y is measured downward from the fluid's surface, then the pressure is $p = w_1 y$ for $0 \leq y \leq 8$ and $p = 8w_1 + w_2(y - 8)$ for $y > 8$.)

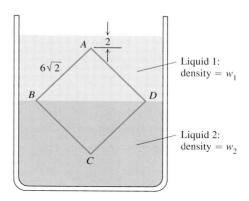

Liquid 1: density $= w_1$

Liquid 2: density $= w_2$

First-Order Differential Equations

9. *Radioactivity* A small amount A_0 of a radioactive substance is placed in a lead container. After 24 h, it is observed that 6/7 of the original amount is remaining. If the rate of decay is proportional to the amount of substance present at any time, what is the half-life of this substance? How long will it take for the substance to be reduced to 1/5 of its original amount?

10. *Gypsy moth population* The rate of increase of the gypsy moth caterpillar population is directly proportional to the number P of the caterpillars present at any time t. The results of a survey show that the caterpillars increased from 2 million in 1979 to 3 million in 1981. Predict the gypsy moth caterpillar population in 1985.

11. *Cooling broth* In a room at a constant temperature of 70°F, the temperature of a cup of broth changes from 200° at noon to 190° at 1 P.M. Predict the temperature of the broth at 3:30 P.M.

12. *A murder victim* The body of a murder victim was discovered at 11 P.M. one evening. The police doctor on call arrived at 11:30 P.M. and immediately took the temperature of the body, which was 94.6°F. He again took the temperature after 1 h, when it was 93.4°F, and he noted that the temperature of the room was a constant 70°F. Use Newton's Law of Cooling to estimate the time of death, assuming that the victim's normal body temperature was 98.6°F.

Moments and Centers of Mass

13. *Limiting position of a centroid* Find the centroid of the region bounded below by the x-axis and above by the curve $y = 1 - x^n$, n an even positive integer. What is the limiting position of the centroid as $n \to \infty$?

14. *Telephone pole* If you haul a telephone pole on a two-wheeled carriage behind a truck, you want the wheels to be 3 ft or so behind the pole's center of mass to provide an adequate "tongue" weight. NYNEX's class 1 40 ft wooden poles have a 27 in. circumference at the top and a 43.5 in. circumference at the base. About how far from the top is the center of mass?

15. *Constant density* Suppose that a thin metal plate of area A and constant density δ occupies a region R in the xy-plane, and let M_y be the plate's moment about the y-axis. Show that the plate's moment about the line $x = b$ is

(a) $M_y - b\delta A$ if the plate lies to the right of the line

(b) $b\delta A - M_y$ if the plate lies to the left of the line.

16. *Variable density* Find the center of mass of a thin plate covering the region bounded by the curve $y^2 = 4ax$ and the line $x = a$, $a = $ positive constant, if the density at (x, y) is directly proportional to (a) x, (b) $|y|$.

17. **(a)** *Concentric circles* Find the centroid of the region in the first quadrant bounded by two concentric circles and the coordinate axes, if the circles have radii a and b, $0 < a < b$, and their centers are at the origin.

(b) *Writing to Learn* Find the limits of the coordinates of the centroid as a approaches b and discuss the meaning of the result.

18. *Cutting a corner from a square* A triangular corner is cut from a square 1 ft on a side. The area of the triangle removed is 36 in². If the centroid of the remaining region is 7 in. from one side of the original square, how far is it from the remaining sides?

6 Transcendental Functions and Differential Equations

OVERVIEW   The functions $\ln x$ and e^x are probably the best-known function–inverse pair. In this chapter, we take a new view of the calculus of these functions and broaden our acquaintance with the amazing range of problems they solve. We also introduce the hyperbolic functions and their inverses, with applications to integration, skydiving, trucking, and hanging cables.

6.1 Logarithms

The Natural Logarithm Function   •   Derivative of $y = \ln x$ •   Derivative of $\log_a u$ •   The Integral $\int (1/u)\,du$ •   Integrals Involving $\log_a x$

In this section, we define the natural logarithm function as an integral, based on the Fundamental Theorem of Calculus. This is a different approach from starting with e^x and defining $\ln x$ as its inverse, as we did in the Preliminary chapter. We calculate the derivative of $\ln x$ based on its new definition, which differs from the way we made the calculation in Section 2.9. Of course, the results are exactly the same.

The importance of logarithms came at first from the improvement they brought to arithmetic. The revolutionary properties of logarithms made possible the calculations of the great seventeenth-century advances in offshore navigation and celestial mechanics. Nowadays we do complicated arithmetic with calculators, but the properties of logarithms remain as important as ever in calculus and modeling.

The Natural Logarithm Function

The natural logarithm of a positive number x, written as $\ln x$, is the value of an integral.

CD-ROM
WEBsite

Definition   The Natural Logarithm Function

$$\ln x = \int_1^x \frac{1}{t}\, dt, \qquad x > 0$$

489

If $x > 1$, then $\ln x$ is the area under the curve $y = 1/t$ from $t = 1$ to $t = x$ (Figure 6.1). For $0 < x < 1$, $\ln x$ gives the negative of the area under the curve from x to 1. The function is not defined for $x \leq 0$. We also have

$$\ln 1 = \int_1^1 \frac{1}{t}\, dt = 0. \qquad \text{Upper and lower limits equal}$$

Notice that we show the graph of $y = 1/x$ in Figure 6.1, but use $y = 1/t$ in the integral. Using x for everything would have us writing

$$\ln x = \int_1^x \frac{1}{x}\, dx,$$

with x meaning two different things. So we change the variable of integration to t.

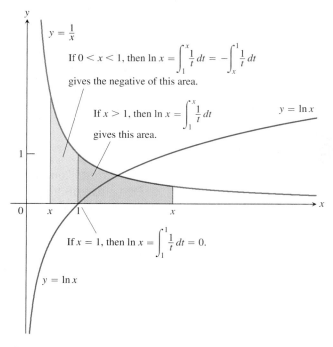

FIGURE 6.1 The graph of $y = \ln x$ and its relation to the function $y = 1/x$, $x > 0$. The graph of the logarithm rises above the x-axis as x moves from 1 to the right, and it falls below the axis as x moves from 1 to the left.

Derivative of $y = \ln x$

By the first part of the Fundamental Theorem of Calculus (in Section 4.5),

$$\frac{d}{dx}\ln x = \frac{d}{dx}\int_1^x \frac{1}{t}\, dt = \frac{1}{x}.$$

CD-ROM
WEBsite

For every positive value of x, therefore,

$$\frac{d}{dx}\ln x = \frac{1}{x}, \qquad x > 0$$

Since $x > 0$ in the definition of $\ln x$, we see that its derivative is always positive so $\ln x$ is increasing everywhere in its domain. The second derivative, $-1/x^2$, is negative, so the graph of $\ln x$ is everywhere concave down.

If u is a differentiable function of x whose values are positive, so that $\ln u$ is defined, then applying the Chain Rule

$$\frac{dy}{dx} = \frac{dy}{du}\frac{du}{dx}$$

to the function $y = \ln u$ gives

$$\frac{d}{dx}\ln u = \frac{d}{du}\ln u \cdot \frac{du}{dx} = \frac{1}{u}\frac{du}{dx}.$$

$$\frac{d}{dx}\ln u = \frac{1}{u}\frac{du}{dx}, \qquad u > 0 \qquad\qquad (1)$$

Example 1   Derivatives of Natural Logarithms

(a) $\dfrac{d}{dx}\ln 2x = \dfrac{1}{2x}\dfrac{d}{dx}(2x) = \dfrac{1}{2x}(2) = \dfrac{1}{x}$

(b) Equation (1) with $u = x^2 + 3$ gives

$$\frac{d}{dx}\ln (x^2 + 3) = \frac{1}{x^2 + 3} \cdot \frac{d}{dx}(x^2 + 3) = \frac{1}{x^2 + 3} \cdot 2x = \frac{2x}{x^2 + 3}.$$

Notice the remarkable occurrence in Example 1(a). The function $y = \ln 2x$ has the same derivative as the function $y = \ln x$. This is true of $y = \ln ax$ for any number a:

$$\frac{d}{dx}\ln ax = \frac{1}{ax} \cdot \frac{d}{dx}(ax) = \frac{1}{ax}(a) = \frac{1}{x}. \qquad\qquad (2)$$

Derivative of $\log_a u$

To find the derivative of a base a logarithm, we first convert it to a natural logarithm (see Preliminary Section 4). If u is a positive differentiable function of x, then

$$\frac{d}{dx}(\log_a u) = \frac{d}{dx}\left(\frac{\ln u}{\ln a}\right) = \frac{1}{\ln a}\frac{d}{dx}(\ln u) = \frac{1}{\ln a} \cdot \frac{1}{u}\frac{du}{dx}.$$

For $a > 0$ and $a \neq 1$,

$$\frac{d}{dx}(\log_a u) = \frac{1}{\ln a} \cdot \frac{1}{u}\frac{du}{dx}, \qquad u > 0 \qquad\qquad (3)$$

Equation (3) is the same formula we derived in Section 2.9.

Example 2 Differentiating Base a Logarithms

$$\frac{d}{dx} \log_{10}(3x+1) = \frac{1}{\ln 10} \cdot \frac{1}{3x+1}\frac{d}{dx}(3x+1) = \frac{3}{(\ln 10)(3x+1)}$$

Laws of Logarithms
For any numbers $a > 0$ and $x > 0$,

1. $\ln ax = \ln a + \ln x$

2. $\ln \frac{a}{x} = \ln a - \ln x$

3. $\ln x^n = n \ln x$

Proof that $\ln ax = \ln a + \ln x$ The argument is unusual, and elegant. It starts by observing that $\ln ax$ and $\ln x$ have the same derivative; see Equation (2). According to Corollary 1 of the Mean Value Theorem, then, the functions must differ by a constant, which means that

$$\ln ax = \ln x + C \tag{4}$$

for some C. With this much accomplished, it remains only to show that C equals $\ln a$.

Equation (4) holds for all positive values of x, so it must hold for $x = 1$. Hence,

$$\ln (a \cdot 1) = \ln 1 + C$$
$$\ln a = 0 + C \qquad \text{ln 1 = 0}$$
$$C = \ln a. \qquad \text{Rearranged}$$

Substituting $C = \ln a$ in Equation (4) gives the equation we wanted to prove:

$$\ln ax = \ln a + \ln x. \tag{5}$$

Proof that $\ln (a/x) = \ln a - \ln x$ We get this from Equation (5) in two stages. Equation (5) with a replaced by $1/x$ gives

$$\ln \frac{1}{x} + \ln x = \ln \left(\frac{1}{x} \cdot x\right)$$
$$= \ln 1 = 0,$$

so that

$$\ln \frac{1}{x} = -\ln x.$$

Equation (5) with x replaced by $1/x$ then gives

$$\ln \frac{a}{x} = \ln \left(a \cdot \frac{1}{x} \right) = \ln a + \ln \frac{1}{x}$$

$$= \ln a - \ln x.$$

Proof that ln xⁿ = n ln x (assuming n rational) We use the same-derivative argument again. For all positive values of x,

$$\frac{d}{dx} \ln x^n = \frac{1}{x^n} \frac{d}{dx} (x^n) \qquad \text{Eq. (1) with } u = x^n$$

$$= \frac{1}{x^n} n x^{n-1} \qquad \begin{array}{l} \text{Here is where we need } n \\ \text{to be rational, at least for} \\ \text{now. We have proved the} \\ \text{Power Rule only for} \\ \text{rational exponents.} \end{array}$$

$$= n \cdot \frac{1}{x} = \frac{d}{dx} (n \ln x).$$

Since $\ln x^n$ and $n \ln x$ have the same derivative,

$$\ln x^n = n \ln x + C$$

for some constant C. Taking x to be 1 identifies C as zero, and we're done. —

As for using the rule $\ln x^n = n \ln x$ for irrational values of n, go right ahead and do so. It does hold for all n, and there is no need to pretend otherwise. Be aware that the rule is far from proved.

The Integral ∫ (1/u) du

Equation (1) leads to the integral formula

$$\int \frac{1}{u} \, du = \ln u + C$$

when u is a positive differentiable function. As we saw in Section 4.2 this formula can be generalized to include negative functions as well.

Whether $u < 0$ or $u > 0$,

$$\int \frac{1}{u} \, du = \ln |u| + C. \tag{6}$$

We know that

$$\int u^n \, du = \frac{u^{n+1}}{n+1} + C, \qquad n \neq -1.$$

Equation (6) explains what to do when n equals -1.

Equation (6) says that integrals of a certain *form* lead to logarithms. That is,

$$\int \frac{f'(x)}{f(x)} dx = \ln |f(x)| + C$$

whenever $f(x)$ is a differentiable function that maintains a constant sign on the domain given for it.

Example 3   Using Substitution

$$\int_{-\pi/2}^{\pi/2} \frac{4 \cos \theta}{3 + 2 \sin \theta} d\theta = \int_{1}^{5} \frac{2}{u} du \quad \begin{array}{l} u = 3 + 2 \sin \theta, \, du = 2 \cos \theta \, d\theta, \\ u(-\pi/2) = 1, \, u(\pi/2) = 5 \end{array}$$

$$= 2 \ln |u| \Big]_{1}^{5} \quad \text{Eq. (6)}$$

$$= 2 \ln |5| - 2 \ln |1| = 2 \ln 5$$

Example 4   Integral of tan u

$$\int_{0}^{\pi/6} \tan 2x \, dx = \int_{0}^{\pi/3} \tan u \cdot \frac{du}{2} = \frac{1}{2} \int_{0}^{\pi/3} \tan u \, du \quad \begin{array}{l} \text{Substitute } u = 2x, \\ dx = du/2, \, u(0) = 0, \\ u(\pi/6) = \pi/3. \end{array}$$

$$= \frac{1}{2} \ln |\sec u| \Big]_{0}^{\pi/3} = \frac{1}{2} (\ln 2 - \ln 1) = \frac{1}{2} \ln 2$$

Integrals Involving $\log_a x$

To evaluate integrals involving base a logarithms, we convert them to natural logarithms.

Example 5   Using Substitution

$$\int \frac{\log_2 x}{x} dx = \frac{1}{\ln 2} \int \frac{\ln x}{x} dx \quad \log_2 x = \frac{\ln x}{\ln 2}$$

$$= \frac{1}{\ln 2} \int u \, du \quad u = \ln x, \, du = \frac{1}{x} dx$$

$$= \frac{1}{\ln 2} \frac{u^2}{2} + C = \frac{1}{\ln 2} \frac{(\ln x)^2}{2} + C = \frac{(\ln x)^2}{2 \ln 2} + C$$

EXERCISES 6.1

Derivatives of Logarithms

In Exercises 1–28, find the derivative of y with respect to x, t, or θ, as appropriate.

1. $y = \ln 3x$

2. $y = \ln(\theta + 1)$

3. $y = \ln x^3$

4. $y = (\ln x)^3$

5. $y = t(\ln t)^2$

6. $y = t\sqrt{\ln t}$

7. $y = \dfrac{x^4}{4} \ln x - \dfrac{x^4}{16}$

8. $y = \dfrac{1 + \ln t}{t}$

9. $y = \dfrac{\ln t}{t}$

10. $y = \dfrac{x \ln x}{1 + \ln x}$

11. $y = \dfrac{\ln x}{1 + \ln x}$

12. $y = \ln(\ln(\ln x))$

13. $y = \theta(\sin(\ln \theta) + \cos(\ln \theta))$

14. $y = \ln(\sec \theta + \tan \theta)$

15. $y = \ln \dfrac{1}{x\sqrt{x + 1}}$

16. $y = \sqrt{\ln \sqrt{t}}$

17. $y = \dfrac{1 + \ln t}{1 - \ln t}$

18. $y = \dfrac{1 + (\ln t)^2}{1 - (\ln t)^2}$

19. $y = \ln(\sec(\ln \theta))$

20. $y = \ln\left(\dfrac{(x^2 + 1)^5}{\sqrt{1 - x}}\right)$

21. $y = \log_2 5\theta$

22. $y = \log_4 x + \log_4 x^2$

23. $y = \log_2 r \cdot \log_4 r$

24. $y = \log_3\left(\left(\dfrac{x + 1}{x - 1}\right)^{\ln 3}\right)$

25. $y = \theta \sin(\log_7 \theta)$

26. $y = 3 \log_8 (\log_2 t)$

In Exercises 27 and 28, use Leibniz's Rule in the Chapter 4 Additional Exercises to find the derivative of y with respect to x.

27. $y = \displaystyle\int_{x^3/2}^{x^2} \ln \sqrt{t}\, dt$

28. $y = \displaystyle\int_{\sqrt{x}}^{\sqrt[3]{x}} \ln t\, dt$

Integration

Evaluate the integrals in Exercises 29–46.

29. $\displaystyle\int_{-3}^{-2} \dfrac{dx}{x}$

30. $\displaystyle\int_{-1}^{0} \dfrac{3\, dx}{3x - 2}$

31. $\displaystyle\int \dfrac{2y\, dy}{y^2 - 25}$

32. $\displaystyle\int \dfrac{8r\, dr}{4r^2 - 5}$

33. $\displaystyle\int_{0}^{\pi} \dfrac{\sin t}{2 - \cos t}\, dt$

34. $\displaystyle\int_{0}^{\pi/3} \dfrac{4 \sin \theta}{1 - 4 \cos \theta}\, d\theta$

35. $\displaystyle\int_{1}^{2} \dfrac{2 \ln x}{x}\, dx$

36. $\displaystyle\int_{2}^{4} \dfrac{dx}{x \ln x}$

37. $\displaystyle\int_{2}^{4} \dfrac{dx}{x(\ln x)^2}$

38. $\displaystyle\int_{2}^{16} \dfrac{dx}{2x\sqrt{\ln x}}$

39. $\displaystyle\int \dfrac{3 \sec^2 t}{6 + 3 \tan t}\, dt$

40. $\displaystyle\int \dfrac{\sec y \tan y}{2 + \sec y}\, dy$

41. $\displaystyle\int_{0}^{\pi/2} \tan \dfrac{x}{2}\, dx$

42. $\displaystyle\int_{\pi/4}^{\pi/2} \cot t\, dt$

43. $\displaystyle\int_{\pi/2}^{\pi} 2 \cot \dfrac{\theta}{3}\, d\theta$

44. $\displaystyle\int_{0}^{\pi/12} 6 \tan 3x\, dx$

45. $\displaystyle\int \dfrac{dx}{2\sqrt{x} + 2x}$

46. $\displaystyle\int \dfrac{\sec x\, dx}{\sqrt{\ln(\sec x + \tan x)}}$

Theory and Applications

47. *Absolute extrema* Locate and identify the absolute extreme values of

    (a) $\ln(\cos x)$ on $[-\pi/4, \pi/3]$

    (b) $\cos(\ln x)$ on $[1/2, 2]$.

48. *$\ln x < x$ if $x > 1$*

    (a) Prove that $f(x) = x - \ln x$ is increasing for $x > 1$.

    (b) Using part (a), show that $\ln x < x$ if $x > 1$.

49. *Area* Find the area between the curves $y = \ln x$ and $y = \ln 2x$ from $x = 1$ to $x = 5$.

50. *Area* find the area between the curve $y = \tan x$ and the x-axis from $x = -\pi/4$ to $x = \pi/3$.

Initial Value Problems

Solve the initial value problems in Exercises 51 and 52.

51. $\dfrac{dy}{dx} = 1 + \dfrac{1}{x}, \qquad y(1) = 3$

52. $\dfrac{d^2y}{dx^2} = \sec^2 x, \quad y(0) = 0 \quad \text{and} \quad y'(0) = 1$

Logarithms to Other Bases

Evaluate the integrals in Exercises 53–58.

53. $\displaystyle\int \frac{\log_{10} x}{x}\, dx$

54. $\displaystyle\int_1^4 \frac{\ln 2 \, \log_2 x}{x}\, dx$

55. $\displaystyle\int_0^2 \frac{\log_2 (x + 2)}{x + 2}\, dx$

56. $\displaystyle\int_0^9 \frac{2 \, \log_{10} (x + 1)}{x + 1}\, dx$

57. $\displaystyle\int \frac{dx}{x \, \log_{10} x}$

58. $\displaystyle\int \frac{dx}{x(\log_8 x)^2}$

59. *The linearization of* ln *(1 + x) at* x = 0 Instead of approximating ln x near $x = 1$, we approximate ln $(1 + x)$ near $x = 0$. We get a simpler formula this way.

 (a) Derive the linearization ln $(1 + x) \approx x$ at $x = 0$.

 (b) Estimate to 5 decimal places the error involved in replacing ln $(1 + x)$ by x on the interval $[0, 0.1]$.

 (c) Graph ln $(1 + x)$ and x together for $0 \le x \le 0.5$. Use different colors, if available. At what points does the approximation of ln $(1 + x)$ seem best? Least good? By reading coordinates from the graphs, find as good an upper bound for the error as your grapher will allow.

T **60.** *Estimating values of* ln *x with Simpson's Rule* Although linearizations are good for replacing the logarithmic function over short intervals, Simpson's Rule is better for estimating *particular* values of ln x.

 As a case in point, the values of ln (1.2) and ln (0.8) to 5 places are

$$\ln (1.2) = 0.18232, \qquad \ln (0.8) = -0.22314.$$

Estimate ln (1.2) and ln (0.8) first with the formula ln $(1 + x) \approx x$ and then use Simpson's Rule with $n = 2$. (Impressive, isn't it?)

6.2   Exponential Functions

The Inverse of ln x and the Number e • The Natural Exponential Function $y = e^x$ • Equations Involving ln x and e^x • The Derivative and Integral of e^x • The Number e Expressed as a Limit • The Derivative and Integral of a^u

---

Whenever we have a quantity y whose rate of change over time is proportional to the amount of y present, we have a function that satisfies the differential equation

$$\frac{dy}{dt} = ky.$$

If, in addition, $y = y_0$ when $t = 0$, the function is the exponential function $y = y_0 e^{kt}$. This section defines the exponential function as the inverse of ln x and explores the properties that account for the amazing frequency with which the function appears in mathematics and its applications. Our approach to the natural exponential function $y = e^x$ here is different from that taken in Section 2.9.

The Inverse of ln x and the Number e

The function ln x, being an increasing function of x with domain $(0, \infty)$ and range $(-\infty, \infty)$, has an inverse $\ln^{-1} x$ with domain $(-\infty, \infty)$ and range $(0, \infty)$.

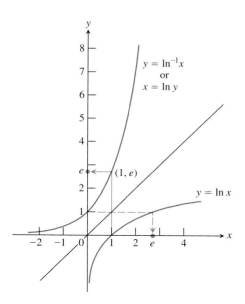

FIGURE 6.2 The graphs of $y = \ln x$ and $y = \ln^{-1} x$. The number e is $\ln^{-1} 1$.

The graph of $\ln^{-1} x$ is the graph of $\ln x$ reflected across the line $y = x$. As you can see from Figure 6.2,

$$\lim_{x \to \infty} \ln^{-1} x = \infty \qquad \text{and} \qquad \lim_{x \to -\infty} \ln^{-1} x = 0.$$

The number $\ln^{-1} 1$ is denoted by the letter e (Figure 6.2).

Definition The Number e

$$e = \ln^{-1} 1$$

Although e is not a rational number, later in this section we see one way to calculate it as a limit. In Equation (2) of Section 2.9, we used another limit to calculate e.

The Natural Exponential Function $y = e^x$

We can raise the number e to a rational power x in the usual way:

$$e^2 = e \cdot e, \qquad e^{-2} = \frac{1}{e^2}, \qquad e^{1/2} = \sqrt{e},$$

and so on. Since e is positive, e^x is positive too. Thus, e^x has a logarithm. When we take the logarithm, we find that

$$\ln e^x = x \ln e = x \cdot 1 = x. \tag{1}$$

Since $\ln x$ is one-to-one and $\ln (\ln^{-1} x) = x$, Equation (1) tells us that

$$e^x = \ln^{-1} x \qquad \text{for } x \text{ rational.} \tag{2}$$

Equation (2) provides a way to extend the definition of e^x to irrational values of x. The function $\ln^{-1} x$ is defined for all x, so we can use it to assign a value to e^x at every point where e^x had no previous value.

Definition Natural Exponential Function
For every real number x,

$$e^x = \ln^{-1} x.$$

Equations Involving $\ln x$ and e^x

Since $\ln x$ and e^x are inverses of one another, we have

Inverse Equations for e^x and $\ln x$

$$e^{\ln x} = x \qquad (\text{all } x > 0) \tag{3}$$

$$\ln (e^x) = x \qquad (\text{all } x) \tag{4}$$

The Derivative and Integral of e^x

We now calculate the derivative of e^x as the inverse of the natural logarithm. **This is a different method than we used in Section 2.9. The exponential function is differentiable because it is the inverse of a differentiable function whose derivative is never zero.** Starting with $y = e^x$, we have, in order,

$$y = e^x$$

$$\ln y = x \qquad \text{Logarithms of both sides}$$

$$\frac{1}{y}\frac{dy}{dx} = 1 \qquad \text{Derivative of both sides with respect to } x$$

$$\frac{dy}{dx} = y \qquad \text{Multiply both sides by } y$$

$$\frac{dy}{dx} = e^x \qquad y \text{ replaced by } e^x$$

As we saw in Section 2.9, the only functions that behave this way are constant multiples of e^x. The Chain Rule extends this result in the usual way to a more general form, just as we saw in Section 2.9. We repeat the rule here.

If u is any differentiable function of x, then

$$\frac{d}{dx}e^u = e^u \frac{du}{dx}. \tag{5}$$

The integral equivalent of Equation (5) is

$$\int e^u \, du = e^u + C.$$

Example 1   Integrating an Exponential

$$\int_0^{\ln 2} e^{3x} \, dx = \int_0^{\ln 8} e^u \cdot \frac{1}{3} \, du \qquad \begin{array}{l} u = 3x, \ \frac{1}{3}du = dx, \ u(0) = 0, \\ u(\ln 2) = 3\ln 2 = \ln 2^3 = \ln 8 \end{array}$$

$$= \frac{1}{3}\int_0^{\ln 8} e^u \, du$$

$$= \frac{1}{3} e^u \Big]_0^{\ln 8}$$

$$= \frac{1}{3}[8 - 1] = \frac{7}{3}$$

Example 2   Using Substitution

$$\int_0^{\pi/2} e^{\sin x} \cos x \, dx = \int_0^1 e^u \, du \qquad \begin{array}{l} u = \sin x, \ du = \cos x \, dx \\ u(0) = 0, \ u(\pi/2) = 1 \end{array}$$

$$= e^u \Big]_0^1$$

$$= e^1 - e^0 = e - 1$$

CD-ROM
WEBsite

Historical Biography

C. L. F. Lindemann
(1852 — 1939)

**Transcendental Numbers
and Transcendental Functions**

Numbers that are solutions of polynomial equations with rational coefficients are called **algebraic:** -2 is algebraic because it satisfies the equation $x + 2 = 0$, and $\sqrt{3}$ is algebraic because it satisfies the equation $x^2 - 3 = 0$. Numbers that are not algebraic are called **transcendental,** a term coined by Euler to describe numbers like e and π that appeared to "transcend the power of algebraic methods." Not until a hundred years after Euler's death (1873), however, did Charles Hermite prove the transcendence of e in the sense that we describe. A few years later (1882), C. L. F. Lindemann proved the transcendence of π.

Today, we call a function $y = f(x)$ algebraic if it satisfies an equation of the form

$$P_n y^n + \cdots + P_1 y + P_0 = 0$$

in which the P's are polynomials in x with rational coefficients. The function $y = 1/\sqrt{x+1}$ is algebraic because it satisfies the equation $(x + 1)y^2 - 1 = 0$. Here the polynomials are $P_2 = x + 1$, $P_1 = 0$, and $P_0 = -1$. Polynomials and rational functions with rational coefficients are algebraic, as are all sums, products, quotients, rational powers, and rational roots of algebraic functions.

Functions that are not algebraic are called transcendental. The six basic trigonometric functions are transcendental, as are the inverses of the trigonometric functions and the exponential and logarithmic functions that are the main subject of this chapter.

Example 3   Solving an Initial Value Problem

Solve the initial value problem

$$e^y \frac{dy}{dx} = 2x, \qquad x > \sqrt{3}, \qquad y(2) = 0.$$

Solution   We integrate both sides of the differential equation with respect to x to obtain

$$e^y = x^2 + C.$$

We use the initial condition to determine C:

$$C = e^0 - (2)^2$$
$$= 1 - 4 = -3.$$

This completes the formula for e^y:

$$e^y = x^2 - 3. \qquad (6)$$

To find y, we take logarithms of both sides:

$$\ln e^y = \ln (x^2 - 3)$$
$$y = \ln (x^2 - 3). \qquad (7)$$

Notice that the solution is valid for $x > \sqrt{3}$.

It is always a good idea to check a solution in the original equation. From Equations (6) and (7), we have

$$e^y \frac{dy}{dx} = e^y \frac{d}{dx} \ln (x^2 - 3) \qquad \text{Eq. (7)}$$
$$= e^y \frac{2x}{x^2 - 3}$$
$$= (x^2 - 3) \frac{2x}{x^2 - 3} \qquad \text{Eq. (6)}$$
$$= 2x.$$

The solution checks.

The Number e Expressed as a Limit

We have defined e as the number for which $\ln e = 1$. The next theorem shows one way to calculate e as a limit.

Theorem 1   The Number e as a Limit

$$\lim_{x \to 0} (1 + x)^{1/x} = e$$

Proof   If $f(x) = \ln x$, then $f'(x) = 1/x$ and $f'(1) = 1$. By definition of the derivative,

$$f'(1) = \lim_{h \to 0} \frac{f(1 + h) - f(1)}{h} = \lim_{x \to 0} \frac{f(1 + x) - f(1)}{x}$$

$$= \lim_{x \to 0} \frac{\ln (1 + x) - \ln 1}{x} = \lim_{x \to 0} \frac{1}{x} \ln (1 + x)$$

$$= \lim_{x \to 0} \ln (1 + x)^{1/x} = \ln \left[\lim_{x \to 0} (1 + x)^{1/x} \right] \qquad \ln \text{ is continuous.}$$

Because $f'(1) = 1$, we obtain

$$\ln\left[\lim_{x \to 0} (1 + x)^{1/x}\right] = 1$$

so that

$$\lim_{x \to 0} (1 + x)^{1/x} = e.$$

X	Y₁	
1.	2.	
0.1	2.59374	
0.01	2.70481	
0.001	2.71692	
0.0001	2.71815	
0.00001	2.71827	
0.000001	2.71828	

Y₁ = (1+ X)^(1/ X)

FIGURE 6.3 A table of values for $f(x) = (1 + x)^{1/x}$.

Using a calculator, we produced the table in Figure 6.3. To 15 places,

$$e = 2.7\ 1828\ 1828\ 45\ 90\ 45.$$

Since the derivative of the exponential function defined here as the inverse of $\ln x$ is the same as the derivative of the exponential function defined in Section 2.9, Corollary 2 of the Mean Value Theorem guarantees that the functions differ by a constant C. Both functions, however, have the same value at $x = 0$:

$$e^0 = 1.$$

Therefore, we obtain immediately that $C = 0$, and both functions are exactly the same. We simply took alternate ways in defining them.

The Derivative and Integral of a^u

We already know that the general exponential function is defined by

$$a^x = e^{x \ln a},$$

where $a > 0$ and $a \neq 1$. We also know (Section 2.9) that

$$\frac{d}{dx}(a^u) = a^u \ln a \frac{du}{dx}. \tag{8}$$

CD-ROM
WEBsite

Historical Biography

Simeon Denis Poisson
(1781 — 1840)

Since $a \neq 1$, so that $\ln a \neq 0$, we can divide both sides of Equation (8) by $\ln a$ to obtain

$$a^u \frac{du}{dx} = \frac{1}{\ln a} \frac{d}{dx}(a^u).$$

Integrating with respect to x then gives

$$\int a^u \frac{du}{dx} dx = \int \frac{1}{\ln a} \frac{d}{dx}(a^u)\, dx = \frac{1}{\ln a} \int \frac{d}{dx}(a^u)\, dx = \frac{1}{\ln a} a^u + C.$$

Writing the first integral in differential form gives

$$\int a^u \, du = \frac{a^u}{\ln a} + C. \tag{9}$$

Example 4 Integrating General Exponentials

(a) $\displaystyle\int 2^x \, dx = \frac{2^x}{\ln 2} + C$ Eq. (9) with $a = 2$, $u = x$

(b) $\displaystyle\int 2^{\sin x} \cos x \, dx$

$$= \int 2^u \, du = \frac{2^u}{\ln 2} + C$$

$$= \frac{2^{\sin x}}{\ln 2} + C \qquad u = \sin x \text{ in Eq. (9)}$$

EXERCISES 6.2

Derivatives of Natural Exponentials

In Exercises 1–14, find the derivative of y with respect to x, t, or θ, as appropriate.

1. $y = e^{-2x/3}$ **2.** $y = e^{5-7x}$

3. $y = e^{(4\sqrt{x}+x^2)}$ **4.** $y = (1 + 3x)e^{-x}$

5. $y = (x^2 - 2x + 2)e^x$ **6.** $y = e^{\theta}(\sin \theta + \cos \theta)$

7. $y = \ln(3\theta e^{-\theta})$ **8.** $y = \cos(e^{-\theta^2})$

9. $y = \ln(2e^{-t}\sin t)$ **10.** $y = \ln\left(\dfrac{e^{\theta}}{1 + e^{\theta}}\right)$

11. $y = \ln\left(\dfrac{\sqrt{\theta}}{1 + \sqrt{\theta}}\right)$ **12.** $y = e^{\sin t}(\ln t^2 + 1)$

13. $y = \displaystyle\int_0^{\ln x} \sin e^t \, dt$ **14.** $y = \displaystyle\int_{e^{4\sqrt{x}}}^{e^{2x}} \ln t \, dt$

Implicit Differentiation

In Exercises 15–18, find dy/dx.

15. $\ln y = e^y \sin x$ **16.** $\ln xy = e^{x+y}$

17. $e^{2x} = \sin(x + 3y)$ **18.** $\tan y = e^x + \ln x$

Integrals of Natural Exponentials

Evaluate the integrals in Exercises 19–32.

19. $\displaystyle\int (e^{3x} + 5e^{-x}) \, dx$ **20.** $\displaystyle\int_{\ln 2}^{\ln 3} e^x \, dx$

21. $\displaystyle\int 8e^{(x+1)} \, dx$ **22.** $\displaystyle\int_{\ln 4}^{\ln 9} e^{x/2} \, dx$

23. $\displaystyle\int \frac{e^{-\sqrt{r}}}{\sqrt{r}} \, dr$ **24.** $\displaystyle\int 2t \, e^{-t^2} \, dt$

25. $\displaystyle\int \frac{e^{1/x}}{x^2} \, dx$ **26.** $\displaystyle\int \frac{e^{-1/x^2}}{x^3} \, dx$

27. $\displaystyle\int_0^{\pi/4} (1 + e^{\tan \theta}) \sec^2 \theta \, d\theta$ **28.** $\displaystyle\int_{\pi/4}^{\pi/2} (1 + e^{\cot \theta}) \csc^2 \theta \, d\theta$

29. $\displaystyle\int e^{\sec \pi t} \sec \pi t \tan \pi t \, dt$ **30.** $\displaystyle\int_0^{\sqrt{\ln \pi}} 2xe^{x^2} \cos(e^{x^2}) \, dx$

31. $\displaystyle\int \frac{e^r}{1 + e^r} \, dr$ **32.** $\displaystyle\int \frac{dx}{1 + e^x}$

Derivatives of General Exponentials

In Exercises 33–44, find the derivative of y with respect to the given independent variable.

33. $y = 2^x$ **34.** $y = 5^{\sqrt{s}}$

35. $y = x^{\pi}$ **36.** $y = (\cos \theta)^{\sqrt{2}}$

37. $y = 7^{\sec \theta} \ln 7$ **38.** $y = 2^{\sin 3t}$

39. $y = t^{1-e}$ **40.** $y = (\ln \theta)^{\pi}$

41. $y = \log_3\left(\left(\dfrac{x + 1}{x - 1}\right)^{\ln 3}\right)$ **42.** $y = \log_5 \sqrt{\left(\dfrac{7x}{3x + 2}\right)^{\ln 5}}$

43. $y = \log_7\left(\dfrac{\sin \theta \cos \theta}{e^{\theta} 2^{\theta}}\right)$ **44.** $y = \log_2\left(\dfrac{x^2 e^2}{2\sqrt{x + 1}}\right)$

Logarithmic Differentiation

In Exercises 45–48, use logarithmic differentiation (Example 8, Section 2.9) to find the derivative of y with respect to the given independent variable.

45. $y = (x + 1)^x$ **46.** $y = t^{\sqrt{t}}$

47. $y = x^{\sin x}$ **48.** $y = (\ln x)^{\ln x}$

Integrals of General Exponentials

Evaluate the integrals in Exercises 49–56.

49. $\displaystyle\int_1^{\sqrt{2}} x 2^{(x^2)}\, dx$

50. $\displaystyle\int_0^{\pi/2} 7^{\cos t}\, \sin t\, dt$

51. $\displaystyle\int_1^2 \frac{2^{\ln x}}{x}\, dx$

52. $\displaystyle\int 3 x^{\sqrt{3}}\, dx$

53. $\displaystyle\int x^{\sqrt{2}-1}\, dx$

54. $\displaystyle\int_0^3 (\sqrt{2}+1)x^{\sqrt{2}}\, dx$

55. $\displaystyle\int_1^e x^{(\ln 2)-1}\, dx$

56. $\displaystyle\int_1^{e^x} \frac{3^{\ln t}}{t}\, dt$

Initial Value Problems

Solve the initial value problems in Exercises 57–60.

57. $\dfrac{dy}{dt} = e^t \sin (e^t - 2), \quad y(\ln 2) = 0$

58. $\dfrac{dy}{dt} = e^{-t} \sec^2 (\pi e^{-t}), \quad y(\ln 4) = 2/\pi$

59. $\dfrac{d^2 y}{dx^2} = 2e^{-x}, \quad y(0) = 1 \quad \text{and} \quad y'(0) = 0$

60. $\dfrac{d^2 y}{dt^2} = 1 - e^{2t}, \quad y(1) = -1 \quad \text{and} \quad y'(1) = 0$

Theory and Applications

61. *Absolute extrema* Find the absolute maximum and minimum values of $f(x) = e^x - 2x$ on $[0, 1]$.

62. *A periodic function* Where does the periodic function $f(x) = 2e^{\sin (x/2)}$ take on its extreme values, and what are these values?

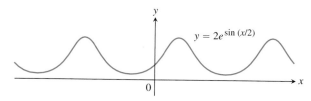

$$y = 2e^{\sin (x/2)}$$

63. *Absolute maximum* Find the absolute maximum value of $f(x) = x^2 \ln (1/x)$ and say where it is assumed.

64. *Area* Find the area of the "triangular" region in the first quadrant that is bounded above by the curve $y = e^{2x}$, below by the curve $y = e^x$, and on the right by the line $x = \ln 3$.

65. *Exponential limit* Show that $\lim_{k \to \infty} (1 + (r/k))^k = e^r$.

66. *Length of a curve* Find a curve through the origin in the xy-plane whose length from $x = 0$ to $x = 1$ is

$$L = \int_0^1 \sqrt{1 + \frac{1}{4}e^x}\, dx.$$

67. Show that for any number $a > 1$,

$$\int_1^a \ln x\, dx + \int_0^{\ln a} e^y\, dy = a \ln a.$$

(See the accompanying figure.)

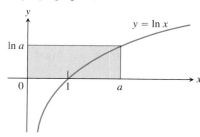

68. *The geometric, logarithmic, and arithmetic mean inequality*

(a) Show that the graph of e^x is concave up over every interval of x-values.

(b) Show, by reference to the accompanying figure, that if $0 < a < b$, then

$$e^{(\ln a + \ln b)/2} \cdot (\ln b - \ln a) < \int_{\ln a}^{\ln b} e^x\, dx <$$

$$\frac{e^{\ln a} + e^{\ln b}}{2} \cdot (\ln b - \ln a).$$

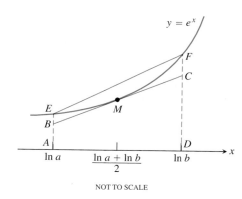

NOT TO SCALE

(c) Use the inequality in part (b) to conclude that

$$\sqrt{ab} < \frac{b - a}{\ln b - \ln a} < \frac{a + b}{2}.$$

This inequality says that the geometric mean of two positive numbers is less than their logarithmic mean, which in turn is less than their arithmetic mean.

(For more about this inequality, see "The Geometric, Logarithmic, and Arithmetic Mean Inequality" by Frank Burk, *American Mathematical Monthly*, Vol. 94, No. 6 (June–July 1987), pp. 527–528.)

T **69.** Graph $f(x) = (x - 3)^2 e^x$ and its first derivative together. Comment on the behavior of f in relation to the signs and values of f'. Identify significant points on the graphs with calculus, as necessary.

70. *The inverse relation between e^x and $\ln x$* Find out how good your calculator is at evaluating the composites

$$e^{\ln x} \quad \text{and} \quad \ln (e^x).$$

71. *The linearization of e^x at $x = 0$*

(a) Derive the linear approximation $e^x \approx 1 + x$ at $x = 0$.

T (b) Graph e^x and $1 + x$ together for $-2 \le x \le 2$. Use different colors, if available. On what intervals does the approximation appear to overestimate e^x? Underestimate e^x?

(c) Estimate to 5 decimal places the magnitude of the error involved in replacing e^x by $1 + x$ on the interval $[0, 0.2]$.

72. *A decimal representation of e* Find e to as many decimal places as your calculator allows by solving the equation $\ln x = 1$.

6.3 Linear First-Order Differential Equations

Linear First-Order Equations   •   Solving the Linear Equation   •   Mixture Problems   •   *RL* Circuits

In Section 5.4, we derived the law of exponential change, $y = y_0 e^{kt}$, as the solution of the initial value problem $dy/dt = ky$, $y(0) = y_0$. As we saw, this problem models radioactive decay, heat transfer, and a great many other phenomena. In this section, we study initial value problems based on the equation $dy/dx = f(x, y)$, in which f is a function of both the independent and dependent variables. The function f will have a particular form, called a *linear form*, and the associated differential equations have applications that are broader still.

Linear First-Order Equations

A first-order differential equation that can be written in the form

$$\frac{dy}{dx} + P(x)y = Q(x), \tag{1}$$

where P and Q are functions of x, is a **linear** first-order equation. Equation (1) is the equation's **standard form.**

Example 1   Finding the Standard Form

Put the following equation in standard form:

$$x \frac{dy}{dx} = x^2 + 3y, \qquad x > 0.$$

Solution

$$x \frac{dy}{dx} = x^2 + 3y$$

$$\frac{dy}{dx} = x + \frac{3}{x} y \qquad \text{Divide by } x.$$

$$\frac{dy}{dx} - \frac{3}{x} y = x \qquad \begin{array}{l}\text{Standard form with } P(x) = -3/x \\ \text{and } Q(x) = x\end{array}$$

Notice that $P(x)$ is $-3/x$, not $+3/x$. The standard form is $y' + P(x)y = Q(x)$, so the minus sign is part of the formula for $P(x)$.

Example 2   Linearity of the Exponential Growth Model

The equation

$$\frac{dy}{dx} = ky$$

with which we modeled resistance proportional to velocity and temperature change in Section 5.4 is a linear first-order equation. Its standard form is

$$\frac{dy}{dx} - ky = 0. \qquad \text{\scriptsize $P(x) = -k$ and $Q(x) = 0$}$$

Solving the Linear Equation

We solve the equation

$$\frac{dy}{dx} + P(x)y = Q(x) \tag{2}$$

by multiplying both sides by a *positive* function $v(x)$ that transforms the left-hand side into the derivative of the product $v(x) \cdot y$. We will show how to find v in a moment, but first we want to show how, once found, it provides the solution we seek.

Here is why multiplying by v works:

$$\frac{dy}{dx} + P(x)y = Q(x) \qquad \text{\scriptsize Original equation is in standard form.}$$

We call $v(x)$ an **integrating factor** for Equation (2) because its presence makes the equation integrable.

$$v(x)\frac{dy}{dx} + P(x)v(x)y = v(x)Q(x) \qquad \text{\scriptsize Multiply by positive $v(x)$.}$$

$$\frac{d}{dx}(v(x) \cdot y) = v(x)Q(x) \qquad \text{\scriptsize $v(x)$ is chosen to make $v\frac{dy}{dx} + Pvy = \frac{d}{dx}(v \cdot y)$.}$$

$$v(x) \cdot y = \int v(x)Q(x)\, dx \qquad \text{\scriptsize Integrate with respect to x.}$$

$$y = \frac{1}{v(x)} \int v(x)Q(x)\, dx. \qquad \text{\scriptsize Solve for y.} \tag{3}$$

Equation (3) expresses the solution of Equation (2) in terms of the functions $v(x)$ and $Q(x)$.

Why doesn't the formula for $P(x)$ appear in the solution as well? It does, but indirectly, in the construction of the positive function $v(x)$. We have

$$\frac{d}{dx}(vy) = v\frac{dy}{dx} + Pvy \qquad \text{\scriptsize Condition imposed on v}$$

$$v\frac{dy}{dx} + y\frac{dv}{dx} = v\frac{dy}{dx} + Pvy \qquad \text{\scriptsize Product Rule for derivatives}$$

$$y\frac{dv}{dx} = Pvy \qquad \text{\scriptsize The terms $v\frac{dy}{dx}$ cancel.}$$

This last equation will hold if

$$\frac{dv}{dx} = Pv \qquad \text{Variables separated}$$

$$\frac{dv}{v} = P\,dx \qquad \text{Integrate both sides.}$$

$$\int \frac{dv}{v} = \int P\,dx \qquad \text{Since } v > 0, \text{ we do not need absolute value signs in } \ln v.$$

$$\ln v = \int P\,dx \qquad \text{Exponentiate both sides to solve for } v.$$

$$e^{\ln v} = e^{\int P\,dx}$$

$$v = e^{\int P\,dx} \qquad\qquad (4)$$

From this, we see that any function v that satisfies Equation (4) will enable us to solve Equation (2) with the formula in Equation (3). We do not need the most general possible v, only one that will work. Therefore, it will do no harm to simplify our lives by choosing the simplest possible antiderivative of P for $\int P\,dx$. Notice too that any function v satisfying Equation (4) is positive.

CD-ROM
WEBsite

Historical Biography

Josiah Willard Gibbs
(1839 — 1903)

Theorem 2

The solution of the linear equation

$$\frac{dy}{dx} + P(x)\,y = Q(x)$$

is

$$y = \frac{1}{v(x)} \int v(x)\,Q(x)\,dx, \qquad\qquad (5)$$

where

$$v(x) = e^{\int P(x)\,dx}. \qquad\qquad (6)$$

In the formula for v, we do not need the most general antiderivative of $P(x)$. Any antiderivative will do.

Example 3   Applying Theorem 2

Solve the equation

$$x\frac{dy}{dx} = x^2 + 3y, \qquad x > 0.$$

Solution   We solve the equation in four steps.

Step 1: *Put the equation in standard form to identify P and Q.*

$$\frac{dy}{dx} - \frac{3}{x}y = x, \qquad P(x) = -\frac{3}{x}, \qquad Q(x) = x. \qquad \text{Example 1}$$

Step 2: *Find an antiderivative of P(x)* (any one will do).

$$\int P(x)\, dx = \int -\frac{3}{x}\, dx = -3\int \frac{1}{x}\, dx = -3\ln|x| = -3\ln x \qquad x > 0$$

Step 3: *Find the integrating factor v(x).*

$$v(x) = e^{\int P(x)\, dx} = e^{-3\ln x} = e^{\ln x^{-3}} = \frac{1}{x^3} \qquad \text{Eq. (6)}$$

Step 4: *Find the solution.*

$$y = \frac{1}{v(x)} \int v(x)Q(x)\, dx \qquad \text{Eq. (5)}$$

$$= \frac{1}{(1/x^3)} \int \left(\frac{1}{x^3}\right)(x)\, dx \qquad \text{Values from steps 1–3}$$

$$= x^3 \cdot \int \frac{1}{x^2}\, dx$$

$$= x^3\left(-\frac{1}{x} + C\right) \qquad \text{Don't forget the } C\ldots$$

$$= -x^2 + Cx^3 \qquad \begin{array}{l}\ldots \text{it provides part of}\\ \text{the answer.}\end{array}$$

The solution is $y = -x^2 + Cx^3$, $x > 0$.

How to Solve a Linear First-Order Equation

Step 1. Put it in standard form.

Step 2. Find an antiderivative of $P(x)$.

Step 3. Find $v(x) = e^{\int P(x)\, dx}$.

Step 4. Use Equation (5) to find y.

Example 4   Solving a Linear First-Order Initial Value Problem

Solve the equation

$$xy' = x^2 + 3y, \qquad x > 0,$$

given the initial condition $y(1) = 2$.

Solution   We first solve the differential equation (Example 3), obtaining

$$y = -x^2 + Cx^3, \qquad x > 0.$$

We then use the initial condition to find the right value for C:

$$y = -x^2 + Cx^3$$
$$2 = -(1)^2 + C(1)^3 \qquad y = 2 \text{ when } x = 1$$
$$C = 2 + (1)^2 = 3.$$

The solution of the initial value problem is the function $y = -x^2 + 3x^3$.

CD-ROM
WEBsite

Historical Biography

Adrien Marie Legendre
(1752 — 1833)

Mixture Problems

A chemical in a liquid solution (or dispersed in a gas) runs into a container holding the liquid (or the gas) with, possibly, a specified amount of the chemical dissolved as well. The mixture is kept uniform by stirring and flows out of the container at a known rate. In this process, it is often important to know the concentration of the chemical in the container at any given time. The differential equation describing the process is based on the formula

$$\begin{array}{c}\text{Rate of change}\\\text{of amount}\\\text{in container}\end{array} = \left(\begin{array}{c}\text{rate at which}\\\text{chemical}\\\text{arrives}\end{array}\right) - \left(\begin{array}{c}\text{rate at which}\\\text{chemical}\\\text{departs}.\end{array}\right) \tag{7}$$

If $y(t)$ is the amount of chemical in the container at time t and $V(t)$ is the total volume of liquid in the container at time t, then the departure rate of the chemical at time t is

$$\text{Departure rate} = \frac{y(t)}{V(t)} \cdot (\text{outflow rate})$$

$$= \left(\begin{array}{c}\text{concentration in}\\\text{container at time } t\end{array}\right) \cdot (\text{outflow rate}). \tag{8}$$

Accordingly, Equation (7) becomes

$$\frac{dy}{dt} = (\text{chemical's arrival rate}) - \frac{y(t)}{V(t)} \cdot (\text{outflow rate}). \tag{9}$$

If, say, y is measured in pounds, V in gallons, and t in minutes, the units in Equation (9) are

$$\frac{\text{pounds}}{\text{minutes}} = \frac{\text{pounds}}{\text{minutes}} - \frac{\text{pounds}}{\text{gallons}} \cdot \frac{\text{gallons}}{\text{minutes}}$$

Here is an example.

Example 5   Oil Refinery Storage Tank

In an oil refinery, a storage tank contains 2000 gal of gasoline that initially has 100 lb of an additive dissolved in it. In preparation for winter weather, gasoline containing 2 lb of additive per gallon is pumped into the tank at a rate of 40 gal/min. The well-mixed solution is pumped out at a rate of 45 gal/min. How much of the additive is in the tank 20 min after the pumping process begins (Figure 6.4)?

FIGURE 6.4 Storage tank in Example 5.

Solution

Differential Equation Model

Let y be the amount (in pounds) of additive in the tank at time t. We know that $y = 100$ when $t = 0$. The number of gallons of gasoline and additive in solution in the tank at any time t is

$$V(t) = 2000 \text{ gal} + \left(40 \frac{\text{gal}}{\text{min}} - 45 \frac{\text{gal}}{\text{min}}\right)(t \text{ min})$$

$$= (2000 - 5t) \text{ gal}.$$

Therefore,

$$\text{Rate out} = \frac{y(t)}{V(t)} \cdot \text{outflow rate} \qquad \text{Eq. (8)}$$

$$= \left(\frac{y}{2000 - 5t} \right) 45 \qquad \begin{array}{l} \text{Outflow rate is 45 gal/min.} \\ \text{and } V = 2000 - 5t. \end{array}$$

$$= \frac{45y}{2000 - 5t} \frac{\text{lb}}{\text{min}}.$$

Also,

$$\text{Rate in} = \left(2 \frac{\text{lb}}{\text{gal}} \right) \left(40 \frac{\text{gal}}{\text{min}} \right)$$

$$= 80 \frac{\text{lb}}{\text{min}}.$$

The differential equation modeling the mixture process is

$$\frac{dy}{dt} = 80 - \frac{45y}{2000 - 5t} \qquad \text{Eq. (9)}$$

in pounds per minute.

Analytical Solution

To solve this differential equation, we first write it in standard form:

$$\frac{dy}{dt} + \frac{45}{2000 - 5t} y = 80.$$

Thus, $P(t) = 45/(2000 - 5t)$ and $Q(t) = 80$.

An antiderivative for $P(t)$ is

$$\int P(t) \, dt = \int \frac{45}{2000 - 5t} \, dt$$

$$= -9 \ln (2000 - 5t). \qquad 2000 - 5t > 0$$

(Remember, any antiderivative for P will do.)

The integrating factor is

$$v(t) = e^{\int P \, dt}$$

$$= e^{-9 \ln (2000 - 5t)}$$

$$= (2000 - 5t)^{-9}.$$

The general solution to the differential equation is

$$y = \frac{1}{(2000 - 5t)^{-9}} \int (2000 - 5t)^{-9}(80) \, dt \quad \text{Eq. (5)}$$

$$= \frac{80}{(2000 - 5t)^{-9}} \left(\frac{(2000 - 5t)^{-8}}{(-8)(-5)} + C \right) \qquad \text{Don't forget the } C.$$

$$= 2(2000 - 5t) + C(2000 - 5t)^9. \qquad \text{Using } C \text{ again for } 80\,C$$

Because $y = 100$ when $t = 0$, we can determine the value of C:

$$100 = 2(2000 - 0) + C(2000 - 0)^9$$

$$C = -\frac{3900}{(2000)^9}.$$

The solution of the initial value problem is

$$y = 2(2000 - 5t) - \frac{3900}{(2000)^9}(2000 - 5t)^9.$$

Interpretation

The amount of additive 20 minutes after the pumping begins is

$$y(20) = 2[2000 - 5(20)] - \frac{3900}{(2000)^9}[2000 - 5(20)]^9 \approx 1342.03 \text{ lb}.$$

RL Circuits

The diagram in Figure 6.5 represents an electrical circuit whose total resistance is a constant R ohms and whose self-inductance, shown as a coil, is L henries, also a constant. There is a switch whose terminals at a and b can be closed to connect a constant electrical source of V volts.

Ohm's law, $V = RI$, has to be modified for such a circuit. The modified form is

$$L\frac{di}{dt} + Ri = V, \tag{10}$$

FIGURE 6.5 The RL circuit in Example 6.

where i is the intensity of the current in amperes and t is the time in seconds. By solving this equation, we can predict how the current will flow after the switch is closed.

Example 6   Electric Current Flow

The switch in the RL circuit in Figure 6.5 is closed at time $t = 0$. How will the current flow as a function of time?

Solution   Equation (10) is a linear first-order differential equation for i as a function of t. Its standard form is

$$\frac{di}{dt} + \frac{R}{L}i = \frac{V}{L}, \tag{11}$$

and the corresponding solution, from Theorem 2, given that $i = 0$ when $t = 0$, is

$$i = \frac{V}{R} - \frac{V}{R}e^{-(R/L)t} \tag{12}$$

(Exercise 32). Since R and L are positive, $-(R/L)$ is negative and $e^{-(R/L)t} \to 0$ as $t \to \infty$. Thus,

$$\lim_{t \to \infty} i = \lim_{t \to \infty}\left(\frac{V}{R} - \frac{V}{R}e^{-(R/L)t}\right) = \frac{V}{R} - \frac{V}{R} \cdot 0 = \frac{V}{R}.$$

At any given time, the current is theoretically less than V/R, but as time passes the current approaches the **steady-state value** V/R. According to the equation

$$L\frac{di}{dt} + Ri = V,$$

$I = V/R$ is the current that will flow in the circuit if either $L = 0$ (no inductance) or $di/dt = 0$ (steady current, $i =$ constant) (Figure 6.6).

Equation (12) expresses the solution of Equation (11) as the sum of two terms: a **steady-state solution** V/R and a **transient solution** $-(V/R)e^{-(R/L)t}$ that tends to zero as $t \to \infty$.

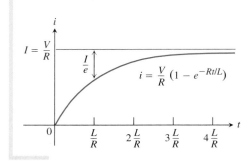

FIGURE 6.6 The growth of the current in the RL circuit in Example 6 is the current's steady-state value. The number $t = L/R$ is the time constant of the circuit. The current gets to within 5% of its steady-state value in 3 time constants. (Exercise 31)

EXERCISES 6.3

Linear First-Order Equations

Solve the differential equations in Exercises 1–14.

1. $x\dfrac{dy}{dx} + y = e^x, \quad x > 0$

2. $e^x\dfrac{dy}{dx} + 2e^x y = 1$

3. $xy' + 3y = \dfrac{\sin x}{x^2}, \quad x > 0$

4. $y' + (\tan x)y = \cos^2 x, \quad -\pi/2 < x < \pi/2$

5. $x\dfrac{dy}{dx} + 2y = 1 - \dfrac{1}{x}, \quad x > 0$

6. $(1 + x)y' + y = \sqrt{x}$

7. $2y' = e^{x/2} + y$

8. $e^{2x}y' + 2e^{2x}y = 2x$

9. $xy' - y = 2x \ln x$

10. $x\dfrac{dy}{dx} = \dfrac{\cos x}{x} - 2y, \quad x > 0$

11. $(t - 1)^3\dfrac{ds}{dt} + 4(t - 1)^2 s = t + 1, \quad t > 1$

12. $(t + 1)\dfrac{ds}{dt} + 2s = 3(t + 1) + \dfrac{1}{(t + 1)^2}, \quad t > -1$

13. $\sin \theta \dfrac{dr}{d\theta} + (\cos \theta)r = \tan \theta, \quad 0 < \theta < \pi/2$

14. $\tan \theta \dfrac{dr}{d\theta} + r = \sin^2 \theta, \quad 0 < \theta < \pi/2$

Solving Initial Value Problems

Solve the initial value problems in Exercises 15–20.

Differential equation	Initial condition
15. $\dfrac{dy}{dt} + 2y = 3$	$y(0) = 1$
16. $t\dfrac{dy}{dt} + 2y = t^3, \quad t > 0$	$y(2) = 1$
17. $\theta\dfrac{dy}{d\theta} + y = \sin \theta, \quad \theta > 0$	$y(\pi/2) = 1$
18. $\theta\dfrac{dy}{d\theta} - 2y = \theta^3 \sec \theta \tan \theta, \quad \theta > 0$	$y(\pi/3) = 2$
19. $(x + 1)\dfrac{dy}{dx} - 2(x^2 + x)y = \dfrac{e^{x^2}}{x + 1}, \quad x > -1$	$y(0) = 5$
20. $\dfrac{dy}{dx} + xy = x$	$y(0) = -6$

21. What do you get when you use Theorem 2 to solve the following initial value problem for y as a function of t?

$$\frac{dy}{dt} = ky \quad (k \text{ constant}), \quad y(0) = y_0$$

22. Use Theorem 2 to solve the following initial value problem for v as a function of t.

$$\frac{dv}{dt} + \frac{k}{m}v = 0 \quad (k \text{ and } m \text{ positive constants}), \quad v(0) = v_0$$

Theory and Examples

23. *Writing to Learn* Is either of the following equations correct? Give reasons for your answers.

(a) $x \int \frac{1}{x} \, dx = x \ln |x| + C$

(b) $x \int \frac{1}{x} \, dx = x \ln |x| + Cx$

24. *Continuous compounding* You have $1000 with which to open an account and plan to add $1000 per year. All funds in the account will earn 10% interest per year, compounded continuously. If the added deposits are also credited to your account continuously, the number of dollars x in your account at time t (years) will satisfy the initial value problem

$$\frac{dx}{dt} = 1000 + 0.10x, \qquad x(0) = 1000.$$

(a) Solve the initial value problem for x as a function of t.

(b) About how many years will it take for the amount in your account to reach $100,000?

25. *Salt mixture* A tank initially contains 100 gal of brine in which 50 lb of salt are dissolved. A brine containing 2 lb/gal of salt runs into the tank at the rate of 5 gal/min. The mixture is kept uniform by stirring and flows out of the tank at the rate of 4 gal/min.

(a) At what rate (pounds per minute) does salt enter the tank at time t?

(b) What is the volume of brine in the tank at time t?

(c) At what rate (pounds per minute) does salt leave the tank at time t?

(d) Write down and solve the initial value problem describing the mixing process.

(e) Find the concentration of salt in the tank 25 min after the process starts.

26. *Mixture problem* A 200 gallon tank is half full of distilled water. Beginning at time $t = 0$, a solution containing 0.5 lb/gal of concentrate enters the tank at the rate of 5 gal/min, and the well-stirred mixture is withdrawn at the rate of 3 gal/min.

(a) At what time will the tank be full?

(b) At the time the tank is full, how many pounds of concentrate will it contain?

27. *Fertilizer mixture* A tank contains 100 gal of fresh water. A solution containing 1 lb/gal of soluble lawn fertilizer runs into the tank at the rate of 1 gal/min, and the mixture is pumped out of the tank at the rate of 3 gal/min. Find the maximum amount of fertilizer in the tank and the time required to reach the maximum.

28. *Carbon monoxide pollution* An executive conference room of a corporation contains 4500 ft³ of air initially free of carbon monoxide. Starting at time $t = 0$, cigarette smoke containing 4% carbon monoxide is blown into the room at the rate of 0.3 ft³/min. A ceiling fan keeps the air in the room well circulated and the air leaves the room at the same rate of 0.3 ft³/min. Find the time when the concentration of carbon monoxide in the room reaches 0.01%.

29. *Current in a closed* RL *circuit* How many seconds after the switch in an *RL* circuit is closed will it take the current i to reach half of its steady-state value? Notice that the time depends on R and L and not on how much voltage is applied.

30. *Current in an open* RL *circuit* If the switch is thrown open after the current in an *RL* circuit has built up to its steady-state value $I = V/R$, the decaying current (graphed here) obeys the equation

$$L\frac{di}{dt} + Ri = 0,$$

which is Equation (10) with $V = 0$.

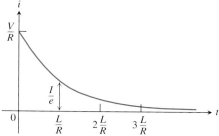

(a) Solve the equation above to express i as a function of t.

(b) How long after the switch is thrown will it take the current to fall to half its original value?

(c) Show that the value of the current when $t = L/R$ is I/e. (The significance of this time is explained in the next exercise.)

31. *Time constants* Engineers call the number L/R the *time constant* of the *RL* circuit in Figure 6.6. The significance of the time constant is that the current will reach 95% of its final value within 3 time constants of the time the switch is closed (Figure 6.6). Thus, the time constant gives a built-in measure of how rapidly an individual circuit will reach equilibrium.

(a) Find the value of i in Equation (12) that corresponds to $t = 3L/R$ and show that it is about 95% of the steady-state value $I = V/R$.

(b) Approximately what percentage of the steady-state current will be flowing in the circuit 2 time constants after the switch is closed (i.e., when $t = 2L/R$)?

32. *Derivation of Equation (12) in Example 6.*

(a) Use Theorem 2 to show that the solution of the equation

$$\frac{di}{dt} + \frac{R}{L}i = \frac{V}{L}$$

is

$$i = \frac{V}{R} + Ce^{-(R/L)t}.$$

(b) Then use the initial condition $i(0) = 0$ to determine the value of C. This will complete the derivation of Equation (12).

(c) Show that $i = V/R$ is a solution of Equation (11) and that $i = Ce^{-(R/L)t}$ satisfies the equation

$$\frac{di}{dt} + \frac{R}{L}i = 0.$$

6.4 Euler's Method: Population Models

Euler's Method • Graphical Solutions • Improved Euler's Method • Exponential Population Model • Logistic Population Model

If we do not require or cannot immediately find an *exact* solution for an initial value problem $y' = f(x, y)$, $y(x_0) = y_0$, we can probably use a computer to generate a table of approximate numerical values of y for values of x in an appropriate interval. Such a table is called a **numerical solution** of the problem and the method by which we generate the table is called a **numerical method.** Numerical methods are generally fast and accurate, and they are often the methods of choice when exact formulas are unnecessary, unavailable, or overly complicated. In this section, we study one such method, called Euler's method, upon which many other numerical methods are based.

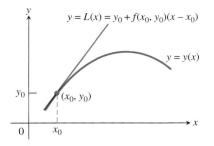

FIGURE 6.7 The linearization of $y = y(x)$ at $x = x_0$.

Euler's Method

Given a differential equation $dy/dx = f(x, y)$ and an initial condition $y(x_0) = y_0$, we can approximate the solution $y = y(x)$ by its linearization

$$L(x) = y(x_0) + y'(x_0)(x - x_0) \quad \text{or} \quad L(x) = y_0 + f(x_0, y_0)(x - x_0).$$

The function $L(x)$ gives a good approximation to the solution $y(x)$ in a short interval about x_0 (Figure 6.7). The basis of Euler's method is to patch together a string of linearizations to approximate the curve over a longer stretch. Here is how the method works.

We know the point (x_0, y_0) lies on the solution curve. Suppose that we specify a new value for the independent variable to be $x_1 = x_0 + dx$. If the increment dx is small, then

$$y_1 = L(x_1) = y_0 + f(x_0, y_0) \, dx$$

is a good approximation to the exact solution value $y = y(x_1)$. So from the point (x_0, y_0), which lies *exactly* on the solution curve, we obtain the point (x_1, y_1), which lies very close to the point $(x_1, y(x_1))$ on the solution curve (Figure 6.8).

Using the point (x_1, y_1) and the slope $f(x_1, y_1)$ of the solution curve through (x_1, y_1), we take a second step. Setting $x_2 = x_1 + dx$, we use the linearization of the solution curve through (x_1, y_1) to calculate

$$y_2 = y_1 + f(x_1, y_1) \, dx.$$

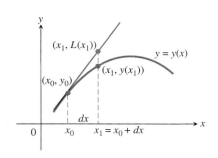

FIGURE 6.8 The first Euler step approximates $y(x_1)$ with $L(x_1)$.

This gives the next approximation (x_2, y_2) to values along the solution curve $y = y(x)$ (Figure 6.9). Continuing in this fashion, take a third step from the point (x_2, y_2) with slope $f(x_2, y_2)$ to obtain the third approximation

$$y_3 = y_2 + f(x_2, y_2) \, dx,$$

and so on. We are literally building an approximation to one of the solutions by following the direction of the slope field of the differential equation.

The steps in Figure 6.9 are drawn large to illustrate the construction process, so the approximation looks crude. In practice, dx would be small enough to make the red curve hug the blue one and give a good approximation throughout.

FIGURE 6.9 Three steps in the Euler approximation to the solution of the initial value problem $y' = f(x, y)$, $y(x_0) = y_0$. As we take more steps, the errors involved usually accumulate, but not in the exaggerated way shown here.

Example 1   Using Euler's Method

Find the first three approximations y_1, y_2, y_3 using Euler's method for the initial value problem

$$y' = 1 + y, \qquad y(0) = 1,$$

starting at $x_0 = 0$ with $dx = 0.1$.

Solution   We have $x_0 = 0$, $y_0 = 1$, $x_1 = x_0 + dx = 0.1$, $x_2 = x_0 + 2dx = 0.2$, and $x_3 = x_0 + 3dx = 0.3$.

$$\begin{aligned} \textit{First:} \quad y_1 &= y_0 + f(x_0, y_0)\, dx \\ &= y_0 + (1 + y_0)\, dx \\ &= 1 + (1 + 1)(0.1) = 1.2 \end{aligned}$$

$$\begin{aligned} \textit{Second:} \quad y_2 &= y_1 + f(x_1, y_1)\, dx \\ &= y_1 + (1 + y_1)\, dx \\ &= 1.2 + (1 + 1.2)(0.1) = 1.42 \end{aligned}$$

$$\begin{aligned} \textit{Third:} \quad y_3 &= y_2 + f(x_2, y_2)\, dx \\ &= y_2 + (1 + y_2)\, dx \\ &= 1.42 + (1 + 1.42)(0.1) = 1.662 \end{aligned}$$

The step-by-step process used in Example 1 can be continued easily. Using equally spaced values for the independent variable in the table and generating n of them, set

$$x_1 = x_0 + dx$$
$$x_2 = x_1 + dx$$
$$\cdot$$
$$\cdot$$
$$\cdot$$
$$x_n = x_{n-1} + dx.$$

Then calculate the approximations to the solution,

$$y_1 = y_0 + f(x_0, y_0)\, dx$$
$$y_2 = y_1 + f(x_1, y_1)\, dx$$
$$\cdot$$
$$\cdot$$
$$\cdot$$
$$y_n = y_{n-1} + f(x_{n-1}, y_{n-1})\, dx.$$

The number of steps n can be as large as we like, but errors can accumulate if n is too large.

Euler's method is easy to program on a computer or programmable calculator. A program generates a table of numerical solutions to an initial value problem, allowing us to input x_0 and y_0, the number of steps n, and the step size dx. It then calculates the approximate solution values y_1, y_2, $\ldots$, y_n in iterative fashion, as just described.

CD-ROM
WEBsite
Historical Biography
Leonhard Euler
(1703 — 1783)

In Exercise 13, you will show that the exact solution to the initial problem of Example 1 is $y = 2e^x - 1$. We use this information in Example 2.

Example 2   Investigating the Accuracy of Euler's Method

Use Euler's method to solve

$$y' = 1 + y, \qquad y(0) = 1,$$

on the interval $0 \le x \le 1$, starting at $x_0 = 0$ and taking

(a) $dx = 0.1$

(b) $dx = 0.05$.

Compare the approximations with the values of the exact solution $y = 2e^x - 1$.

Solution

(a) We used a programmable calculator to generate the approximate values in Table 6.1. The "error" column is obtained by subtracting the unrounded Euler values from the unrounded values found using the exact solution. All entries are then rounded to 4 decimal places.

  By the time we reach $x = 1$ (after 10 steps), the error is about 5.6% of the exact solution.

(b) One way to try to reduce the error is to decrease the step size. Table 6.2 shows the results and their comparisons with the exact solutions when we decrease the step size to 0.05, doubling the number of steps to 20. As in

Table 6.1 Euler solution of $y' = 1 + y$, $y(0) = 1$, step size $dx = 0.1$

x	y (Euler)	y (exact)	Error
0	1	1	0
0.1	1.2	1.2103	0.0103
0.2	1.42	1.4428	0.0228
0.3	1.662	1.6997	0.0377
0.4	1.9282	1.9836	0.0554
0.5	2.2210	2.2974	0.0764
0.6	2.5431	2.6442	0.1011
0.7	2.8974	3.0275	0.1301
0.8	3.2872	3.4511	0.1639
0.9	3.7159	3.9192	0.2033
1.0	4.1875	4.4366	0.2491

Table 6.1, all computations are performed before rounding. This time when we reach $x = 1$, the relative error is only about 2.9%.

Table 6.2 Euler solution of $y' = 1 + y$, $y(0) = 1$, step size $dx = 0.05$			
x	y (Euler)	y (exact)	Error
0	1	1	0
0.05	1.1	1.1025	0.0025
0.10	1.205	1.2103	0.0053
0.15	1.3153	1.3237	0.0084
0.20	1.4310	1.4428	0.0118
0.25	1.5526	1.5681	0.0155
0.30	1.6802	1.6997	0.0195
0.35	1.8142	1.8381	0.0239
0.40	1.9549	1.9836	0.0287
0.45	2.1027	2.1366	0.0340
0.50	2.2578	2.2974	0.0397
0.55	2.4207	2.4665	0.0458
0.60	2.5917	2.6442	0.0525
0.65	2.7713	2.8311	0.0598
0.70	2.9599	3.0275	0.0676
0.75	3.1579	3.2340	0.0761
0.80	3.3657	3.4511	0.0853
0.85	3.5840	3.6793	0.0953
0.90	3.8132	3.9192	0.1060
0.95	4.0539	4.1714	0.1175
1.00	4.3066	4.4366	0.1300

It might be tempting to reduce the step size even further in Example 2 to obtain greater accuracy. Each additional calculation, however, not only requires additional calculator time but more importantly adds to the buildup of round-off errors due to the approximate representations of numbers inside the grapher.

The analysis of error and the investigation of methods to reduce it when making numerical calculations are important but are appropriate for a more advanced course. There are numerical methods more accurate than Euler's method, as you will see in your further study of differential equations. We study one improvement here.

Graphical Solutions

Investigating a graph is usually more instructive than analyzing a large data table. If we plot the data pairs in a numerical solution of an initial value problem, we obtain a **graphical solution,** as shown in Example 3.

[–1, 2] by [–2, 6]

FIGURE 6.10 The graph of $y = 2e^x - 1$ superimposed on a scatter plot of the Euler approximations shown in Table 6.1. (Example 3)

CD-ROM
WEBsite

Historical Biography

Carl Runge
(1856 — 1927)

Example 3   Visualizing Euler's Approximations

Figure 6.10 gives a visualization of the numerical solution shown in Table 6.1 by superimposing the graph of the exact solution on a scatter plot of the data points in the table.

Improved Euler's Method

We can improve on Euler's method by taking an average of two slopes. We first estimate y_n as in the original Euler method, but denote it by z_n. We then take the average of $f(x_{n-1}, y_{n-1})$ and $f(x_n, z_n)$ in place of $f(x_{n-1}, y_{n-1})$ in the next step. Thus, we calculate the next approximation y_n using

$$z_n = y_{n-1} + f(x_{n-1}, y_{n-1}) \, dx$$

$$y_n = y_{n-1} + \left[\frac{f(x_{n-1}, y_{n-1}) + f(x_n, z_n)}{2} \right] dx.$$

Example 4   Investigating the Accuracy of the Improved Euler's Method

Use the improved Euler's method to solve

$$y' = 1 + y, \qquad y(0) = 1$$

on the interval $0 \le x \le 1$, starting at $x_0 = 0$ and taking $dx = 0.1$. Compare the approximations with the values of the exact solution $y = 2e^x - 1$.

Solution   We used a programmable calculator to generate the approximate values in Table 6.3. The "error" column is obtained by subtracting the unrounded improved Euler values from the unrounded values found using the exact solution. All entries are then rounded to 4 decimal places.

By the time we reach $x = 1$ (after 10 steps), the relative error is about 0.19%.

Table 6.3 Improved Euler solution of $y' = 1 + y$, $y(0) = 1$, step size $dx = 0.1$			
x	y (improved Euler)	y (exact)	Error
0	1	1	0
0.1	1.21	1.2103	0.0003
0.2	1.4421	1.4428	0.0008
0.3	1.6985	1.6997	0.0013
0.4	1.9818	1.9836	0.0018
0.5	2.2949	2.2974	0.0025
0.6	2.6409	2.6442	0.0034
0.7	3.0231	3.0275	0.0044
0.8	3.4456	3.4511	0.0055
0.9	3.9124	3.9192	0.0068
1.0	4.4282	4.4366	0.0084

By comparing Tables 6.1 and 6.3, we see that the improved Euler's method is considerably more accurate than the regular Euler's method, at least for the initial value problem $y' = 1 + y$, $y(0) = 1$.

Exponential Population Model

Strictly speaking, the number of individuals in a population is a discontinuous function of time because it takes on only whole number values. One common way to model a population, however, is with a differentiable function P growing at a rate proportional to the size of the population. Thus, for some constant k,

$$\frac{dP}{dt} = kP.$$

Notice that

$$\frac{dP/dt}{P} = k \tag{1}$$

is constant. This rate is called the **relative growth rate.** As we saw in Section 5.4, we can represent the population by the model $P = P_0 e^{kt}$, where P_0 is the size of the population at time $t = 0$.

Table 6.4 gives the world population at midyear for the years 1980 to 1989. Taking $dt = 1$ and $dP \approx \Delta P$, we see from the table that the relative growth rate in Equation (1) is approximately the constant 0.017.

Table 6.4 World population (midyear)		
Year	**Population (millions)**	**$\Delta P/P$**
1980	4454	$76/4454 \approx 0.0171$
1981	4530	$80/4530 \approx 0.0177$
1982	4610	$80/4610 \approx 0.0174$
1983	4690	$80/4690 \approx 0.0171$
1984	4770	$81/4770 \approx 0.0170$
1985	4851	$82/4851 \approx 0.0169$
1986	4933	$85/4933 \approx 0.0172$
1987	5018	$87/5018 \approx 0.0173$
1988	5105	$85/5105 \approx 0.0167$
1989	5190	

Source: U.S. Bureau of the Census (Sept., 1999): www.census.gov/ipc/www/worldpop.html.

Example 5   Predicting World Population

Find an initial value problem model for world population and use it to predict the population in midyear 1999. Graph the model and the data.

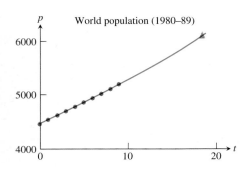

FIGURE 6.11 Notice that the value of the solution $P = 4454e^{0.017t}$ is 6152.16 when $t = 19$. (Example 5)

Solution We let $t = 0$ represent 1980, $t = 1$ represent 1981, and so forth. The year 1999 will be represented by $t = 19$. If we approximate the ratios in Table 6.4 by $k = 0.017$, we obtain the initial value problem

Differential equation: $\dfrac{dP}{dt} = 0.017P$

Initial condition: $P(0) = 4454.$

The solution to this initial value problem gives the population function $P = 4454e^{0.017t}$. Midyear 1999 is represented by $t = 19$, so

$$P(19) \approx 6152.$$

Figure 6.11 shows the graph of the model superimposed on a scatter plot of the data.

Interpret

This model predicts the world population in midyear 1999 to be about 6152 million, or 6.15 billion, which is more than the actual population of 5996 million given by the U.S. Bureau of the Census. In the next example, we examine more recent data to see if there has been a change in the growth rate.

Logistic Population Model

The exponential model for population growth assumes unlimited growth and assumes that the relative growth rate (Equation (1)) is constant. This may be reasonable for the years 1980 to 1989, but let's look at more recent data. Table 6.5 shows the world population for the years 1990 to 1999.

Table 6.5 Recent world population		
Year	**Population (millions)**	**$\Delta P/P$**
1990	5277	$82/5277 \approx 0.0155$
1991	5359	$83/5359 \approx 0.0155$
1992	5442	$81/5442 \approx 0.0149$
1993	5523	$80/5523 \approx 0.0145$
1994	5603	$79/5603 \approx 0.0141$
1995	5682	$79/5682 \approx 0.0139$
1996	5761	$79/5761 \approx 0.0137$
1997	5840	$79/5840 \approx 0.0135$
1998	5919	$77/5919 \approx 0.0130$
1999	5996	

Source: U.S. Bureau of the Census (Sept., 1999): www.census.gov/ipc/www/worldpop.html.

From Tables 6.4 and 6.5, we see that the relative growth rate is positive but decreases as the population increases due to environmental, economic, and other factors. On average, the growth rate decreases by about 0.00036 per year over the years 1990 to 1999. That is, the graph of k in Equation (1) is closer to being a line with a negative slope $-r = -0.00036$.

In Example 5 of Section 3.4, we proposed the **logistic growth model**

$$\frac{dP}{dt} = r(M - P)P, \tag{2}$$

where M is the maximum population, or **carrying capacity,** that the environment is capable of sustaining in the long run. Comparing Equation (2) with the exponential model we see that $k = r(M - P)$ is indeed a linearly decreasing function of the population. The graphical solution curves to the logistic model (2) were obtained in Section 3.4 and are displayed (again) in Figure 6.12. Notice from the graphs that if $P < M$, the population grows toward M; if $P > M$, the growth rate will be negative (as $r > 0$, $M > 0$) and the population decreasing.

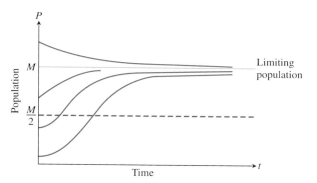

FIGURE 6.12 Solution curves to the logistic population model $dP/dt = r(M - P)P$.

CD-ROM
WEBsite

Example 6   Modeling a Bear Population

A national park is known to be capable of supporting 100 grizzly bears, but no more. Ten bears are in the park at present. We model the population with a logistic differential equation with $r = 0.001$.

(a) Draw and describe a slope field for the differential equation.

(b) Use Euler's method with step size $dt = 1$ to estimate the population size in 20 years.

(c) Find a logistic growth analytic solution $P(t)$ for the population and draw its graph.

(d) When will the bear population reach 50?

Solution

(a) *Slope field.* The carrying capacity is 100, so $M = 100$. The solution we seek is a solution to the following differential equation.

$$\frac{dP}{dt} = 0.001(100 - P)P$$

Figure 6.13 shows a slope field for this differential equation. There appears to be a horizontal asymptote at $P = 100$. The solution curves fall toward this level from above and rise toward it from below.

(b) *Euler's method.* With step size $dt = 1$, $t_0 = 0$, $P(0) = 10$, and

$$\frac{dP}{dt} = f(t, P) = 0.001(100 - P)P,$$

we obtain the approximations in Table 6.6, using the iteration formula

$$P_n = P_{n-1} + 0.001(100 - P_{n-1})P_{n-1}.$$

[0, 150] by [0, 150]

FIGURE 6.13 A slope field for the logistic differential equation

$$\frac{dP}{dt} = 0.001(100 - P)P.$$

(Example 6)

Table 6.6 Euler solution of $dP/dt = 0.001(100 - P)P$, $P(0) = 10$, step size $dt = 1$

t	P (Euler)	t	P (Euler)
0	10		
1	10.9	11	24.3629
2	11.8712	12	26.2056
3	12.9174	13	28.1395
4	14.0423	14	30.1616
5	15.2493	15	32.2680
6	16.5417	16	34.4536
7	17.9222	17	36.7119
8	19.3933	18	39.0353
9	20.9565	19	41.4151
10	22.6130	20	43.8414

There are approximately 44 grizzly bears after 20 years. Figure 6.14 shows a graph of the Euler approximation over the interval $0 \le t \le 150$ with step size $dt = 1$. It looks like the lower curves we sketched in Figure 6.12.

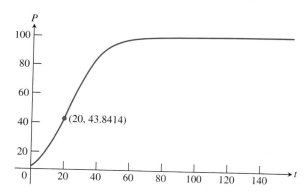

FIGURE 6.14 Euler approximations of the solution to $dP/dt = 0.001(100 - P)P$, $P(0) = 10$, step size $dt = 1$.

(c) *Analytic solution.* We can assume that $t = 0$ when the bear population is 10, so $P(0) = 10$. The logistic growth model we seek is the solution to the following initial value problem.

Differential equation: $\quad \dfrac{dP}{dt} = 0.001(100 - P)P$

Initial condition: $\quad P(0) = 10$

To prepare for integration, we rewrite the differential equation in the form

$$\frac{1}{P(100 - P)} \frac{dP}{dt} = 0.001.$$

The fraction $1/(P(100 - P))$ can be rewritten as

$$\frac{1}{P(100 - P)} = \frac{1}{100}\left(\frac{1}{P} + \frac{1}{100 - P}\right).$$

(You will learn a general method for obtaining such partial fraction decompositions in the next chapter.) Substituting this expression into the differential equation and multiplying both sides by 100 we obtain

$$\left(\frac{1}{P} + \frac{1}{100 - P}\right)\frac{dP}{dt} = 0.1$$

$$\ln |P| - \ln |100 - P| = 0.1t + C \qquad \text{Integrate with respect to } t.$$

$$\ln \left|\frac{P}{100 - P}\right| = 0.1t + C$$

$$\ln \left|\frac{100 - P}{P}\right| = -0.1t - C \qquad \ln \frac{a}{b} = -\ln \frac{b}{a}$$

$$\left|\frac{100 - P}{P}\right| = e^{-0.1t - C} \qquad \text{Exponentiate.}$$

$$\frac{100 - P}{P} = (\pm e^{-C})e^{-0.1t}$$

$$\frac{100}{P} - 1 = Ae^{-0.1t} \qquad \text{Let } A = \pm e^{-C}.$$

$$P = \frac{100}{1 + Ae^{-0.1t}}. \qquad \text{Solve for } P.$$

This is the general solution to the differential equation. When $t = 0$, $P = 10$, and we obtain

$$10 = \frac{100}{1 + Ae^0}$$

$$1 + A = 10$$

$$A = 9.$$

Thus, the logistic growth model is

$$P = \frac{100}{1 + 9e^{-0.1t}}.$$

Its graph (Figure 6.15) is superimposed on the slope field from Figure 6.13.

(d) *Interpretation.* When will the bear population reach 50? For this model,

$$50 = \frac{100}{1 + 9e^{-0.1t}}$$

$$1 + 9e^{-0.1t} = 2$$

$$e^{-0.1t} = \frac{1}{9}$$

$$e^{0.1t} = 9$$

$$t = \frac{\ln 9}{0.1} \approx 22 \text{ years}.$$

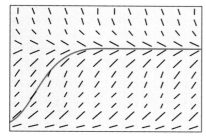

[0, 150] by [0, 150]

FIGURE 6.15 The graph of

$$P = \frac{100}{1 + 9e^{-0.1t}}$$

superimposed on a slope field for

$$\frac{dP}{dt} = 0.001(100 - P)P.$$

(Example 6)

The solution of the general logistic differential equation (Equation (2))

$$\frac{dP}{dt} = r(M - P)P$$

can be obtained as in Example 6. In Exercise 14, we ask you to show that the solution is

$$P = \frac{M}{1 + Ae^{-rMt}}.$$

The value of A is determined by an appropriate initial condition.

EXERCISES 6.4

Calculating Euler Approximations

In Exercises 1–4, use Euler's method to calculate the first three approximations to the given initial value problem for the specified increment size. Calculate the exact solution and compare the accuracy of your approximations with the exact solution. Round your results to 4 decimal places.

1. $y' = x(1 - y)$, $y(1) = 0$, $dx = 0.2$

2. $y' = 1 - \frac{y}{x}$, $y(2) = -1$, $dx = 0.5$

3. $y' = 2xy + 2y$, $y(0) = 3$, $dx = 0.2$

4. $y' = y^2(1 + 2x)$, $y(-1) = 1$, $dx = 0.5$

5. Use the Euler method with $dx = 0.2$ to estimate $y(1)$ if $y' = y$ and $y(0) = 1$. What is the exact value of $y(1)$?

6. Use the Euler method with $dx = 0.2$ to estimate $y(2)$ if $y' = y/x$ and $y(1) = 2$. What is the exact value of $y(2)$?

Improved Euler's Method

In Exercises 7 and 8, use the improved Euler's method to calculate the first three approximations to the given initial value problem. Compare the approximations with the values of the exact solution.

7. $y' = 2y(x + 1)$, $y(0) = 3$, $dx = 0.2$
(See Exercise 3 for the exact solution.)

8. $y' = x(1 - y)$, $y(1) = 0$, $dx = 0.2$
(See Exercise 1 for the exact solution.)

9. *Guppy population* A 2000 gal tank can support no more than 150 guppies. Six guppies are introduced into the tank. Assume that the rate of growth of the population is

$$\frac{dP}{dt} = 0.0015(150 - P)P,$$

where time t is in weeks.

(a) Find a formula for the guppy population in terms of t.

(b) How long will it take for the guppy population to be 100? 125?

10. *Gorilla population* A certain wild animal preserve can support no more than 250 lowland gorillas. Twenty-eight gorillas were known to be in the preserve in 1970. Assume that the rate of growth of the population is

$$\frac{dP}{dt} = 0.0004\,(250 - P)P,$$

where time t is in years.

(a) Find a formula for the gorilla population in terms of t.

(b) How long will it take for the gorilla population to reach the carrying capacity of the preserve?

11. *Pacific halibut fishery* The Pacific halibut fishery has been modeled by the logistic equation

$$\frac{dy}{dt} = r(M - y)y$$

where $y(t)$ is the total weight of the halibut population in kilograms at time t (measured in years), the carrying capacity is estimated to be $M = 8 \times 10^7$ kg, and $r = 0.08875 \times 10^{-7}$ per year.

(a) If $y(0) = 1.6 \times 10^7$ kg, what is the total weight of the halibut population after 1 year?

(b) When will the total weight in the halibut fishery reach 4×10^7 kg?

12. *Modified logistic model* Suppose that the logistic differential equation in Example 6 is modified to

$$\frac{dP}{dt} = 0.001(100 - P)P - c$$

for some constant c.

(a) *Writing to Learn* Explain the meaning of the constant c. What values for c might be realistic for the grizzly bear population?

T **(b)** Draw a direction field for the differential equation when $c = 1$. What are the equilibrium solutions (Section 3.4)?

(c) *Writing to Learn* Sketch several solution curves in your direction field from part (a). Describe what happens to the grizzly bear population for various initial populations.

13. *Exact solutions* Find the exact solutions to the following initial value problems.

(a) $y' = 1 + y$, $y(0) = 1$

(b) $y' = 0.5(400 - y)y$, $y(0) = 2$

14. *Logistic differential equation* Show that the solution of the differential equation

$$\frac{dP}{dt} = r(M - P)P$$

is

$$P = \frac{M}{1 + Ae^{-rMt}},$$

where A is an arbitrary constant.

15. *Catastrophic solution* Let k and P_0 be positive constants.

(a) Solve the initial value problem.

$$\frac{dP}{dt} = kP^2, \qquad P(0) = P_0$$

T **(b)** Show that the graph of the solution in part (a) has a vertical asymptote at a positive value of t. What is that value of t?

16. *Extinct populations* In Exercise 14, Section 3.4, we presented the population model

$$\frac{dP}{dt} = r(M - P)(P - m),$$

where $r > 0$, M is the maximum population, and m is the minimum population below which the species becomes extinct.

(a) Let $m = 100$ and $M = 1200$, and assume that $m < P < M$. Show that the differential equation can be rewritten in the form

$$\left[\frac{1}{1200 - P} + \frac{1}{P - 100} \right] \frac{dP}{dt} = 1100r.$$

Use a procedure similar to that used in Example 6 to solve this differential equation.

(b) Find the solution to part (a) that satisfies $P(0) = 300$.

(c) Solve the differential equation with the restriction $m < P < M$.

COMPUTER EXPLORATIONS

Euler's Method

In Exercises 17–20, use Euler's method with the specified step size to estimate the value of the solution at the given point x^*. Find the value of the exact solution at x^*.

17. $y' = 2xe^{x^2}$, $y(0) = 2$, $dx = 0.1$, $x^* = 1$

18. $y' = y + e^x - 2$, $y(0) = 2$, $dx = 0.5$, $x^* = 2$

19. $y' = y^2/\sqrt{x}$, $y(1) = -1$, $dx = 0.5$, $x^* = 5$

20. $y' = y - e^{2x}$, $y(0) = 1$, $dx = 1/3$, $x^* = 2$

In Exercises 21 and 22, **(a)** find the exact solution of the initial value problem. Then compare the accuracy of the approximation with $y(x^*)$ using Euler's method starting at x_0 with step size **(b)** 0.2, **(c)** 0.1, and **(d)** 0.05.

21. $y' = 2y^2(x - 1)$, $y(2) = -1/2$, $x_0 = 2$, $x^* = 3$

22. $y' = y - 1$, $y(0) = 3$, $x_0 = 0$, $x^* = 1$

Improved Euler's Method

In Exercises 23 and 24, compare the accuracy of the approximation with $y(x^*)$ using the improved Euler's method starting at x_0 with step size

(a) 0.2

(b) 0.1

(c) 0.05.

(d) *Writing to Learn* Describe what happens to the error as the step size decreases.

23. $y' = 2y^2(x - 1)$, $y(2) = -1/2$, $x_0 = 2$, $x^* = 3$

(See Exercise 21 for the exact solution.)

24. $y' = y - 1$, $y(0) = 3$, $x_0 = 0$, $x^* = 1$

(See Exercise 22 for the exact solution.)

Exploring Differential Equations Graphically

Use a CAS to explore graphically each of the differential equations in Exercises 25–28. Perform the following steps to help with your explorations.

(a) Plot a slope field for the differential equation in the given xy-window.

(b) Find the general solution for the differential equation using your CAS differential equation solver.

(c) Graph the solutions for the values of the arbitrary constant $C = -2, -1, 0, 1, 2$ superimposed on your slope field plot.

(d) Find and graph the solution that satisfies the specified initial condition over the interval $[0, b]$.

(e) Find the Euler numerical approximation to the solution of the initial value problem with 4 subintervals of the x-interval and plot the Euler approximation superimposed on the graph produced in part (d).

(f) Repeat part (e) for 8, 16, and 32 subintervals. Plot these three Euler approximations superimposed on the graph from part (e).

(g) Find the error (y (exact) $-$ y (Euler)) at the specified point $x = b$ for each of your four Euler approximations. Discuss the improvement in the percentage error.

25. $y' = x + y$, $y(0) = -7/10$; $-4 \le x \le 4$, $-4 \le y \le 4$; $b = 1$

26. $y' = -x/y$, $y(0) = 2$; $-3 \le x \le 3$, $-3 \le y \le 3$; $b = 2$

27. *A logistic equation* $y' = y(2 - y)$, $y(0) = 1/2$; $0 \le x \le 4$, $0 \le y \le 3$; $b = 3$

28. $y' = (\sin x)(\sin y), \quad y(0) = 2; \quad -6 \leq x \leq 6, \quad -6 \leq y \leq 6;$
$b = 3\pi/2$

Exercises 29 and 30 have no explicit solution in terms of elementary functions. Use a CAS to explore graphically each of the differential equations, performing as many of the steps (a) through (g) above as possible.

CD-ROM
WEBsite **29.** $y' = \cos(2x - y), \quad y(0) = 2; \quad 0 \leq x \leq 5, \quad 0 \leq y \leq 5; \quad y(2)$

CD-ROM
WEBsite **30.** *A Gompertz equation* $y' = y(1/2 - \ln y), \quad y(0) = 1/3;$
$0 \leq x \leq 4, \quad 0 \leq y \leq 3; \quad y(3)$

CD-ROM
WEBsite **31.** Use a CAS to find the solutions of $y' + y = f(x)$ subject to the initial condition $y(0) = 0$ if $f(x)$ is

(a) $2x$ (b) $\sin 2x$

(c) $3e^{x/2}$ (d) $2e^{-x/2}\cos 2x.$

Graph all four solutions over the interval $-2 \leq x \leq 6$ to compare the results.

CD-ROM
WEBsite **32. (a)** Use a CAS to plot the slope field of the differential equation

$$y' = \frac{3x^2 + 4x + 2}{2(y - 1)}$$

over the region $-3 \leq x \leq 3$ and $-3 \leq y \leq 3$.

(b) Separate the variables and use a CAS integrator to find the general solution in implicit form.

(c) Using a CAS implicit function grapher, plot solution curves for the arbitrary constant values $C = -6, -4, -2, 0, 2, 4, 6$.

(d) Find and graph the solution that satisfies the initial condition $y(0) = -1$.

6.5   Hyperbolic Functions

Definitions and Identities   •   Derivatives and Integrals   •   Inverse Hyperbolic Functions   •   Useful Identities   •   Derivatives and Integrals

Every function f that is defined on an interval centered at the origin can be written in a unique way as the sum of one even function and one odd function. The decomposition is

$$f(x) = \underbrace{\frac{f(x) + f(-x)}{2}}_{\text{even part}} + \underbrace{\frac{f(x) - f(-x)}{2}}_{\text{odd part}}.$$

If we write e^x this way, we get

$$e^x = \underbrace{\frac{e^x + e^{-x}}{2}}_{\text{even part}} + \underbrace{\frac{e^x - e^{-x}}{2}}_{\text{odd part}}.$$

The even and odd parts of e^x, called the hyperbolic cosine and hyperbolic sine of x, respectively, are useful in their own right. They describe the motions of waves in elastic solids, the shapes of hanging electric power lines, and the temperature distributions in metal cooling fins. The center line of the Gateway Arch to the West in St. Louis is a weighted hyperbolic cosine curve. In this section, we give a brief introduction to hyperbolic functions.

Definitions and Identities

The hyperbolic cosine and hyperbolic sine functions are defined by the first two equations in Table 6.7. The table also lists the definitions of the hyperbolic tangent, cotangent, secant, and cosecant. As we see, the hyperbolic functions bear a number of similarities to the trigonometric functions after which they are named. (See Exercise 84 as well.)

Hyperbolic functions satisfy the identities in Table 6.8. Except for differences in sign, these are identities we already know for trigonometric functions.

The notation $\cosh x$ is often read "kosh x," rhyming with "gosh x," and $\sinh x$ is pronounced as if spelled "cinch x."

Table 6.7 The six basic hyperbolic functions (see Figure 6.16 for graphs)

Hyperbolic cosine of x:	$\cosh x = \dfrac{e^x + e^{-x}}{2}$
Hyperbolic sine of x:	$\sinh x = \dfrac{e^x - e^{-x}}{2}$
Hyperbolic tangent:	$\tanh x = \dfrac{\sinh x}{\cosh x} = \dfrac{e^x - e^{-x}}{e^x + e^{-x}}$
Hyperbolic cotangent:	$\coth x = \dfrac{\cosh x}{\sinh x} = \dfrac{e^x + e^{-x}}{e^x - e^{-x}}$
Hyperbolic secant:	$\text{sech } x = \dfrac{1}{\cosh x} = \dfrac{2}{e^x + e^{-x}}$
Hyperbolic cosecant:	$\text{csch } x = \dfrac{1}{\sinh x} = \dfrac{2}{e^x - e^{-x}}$

Table 6.8 Identities for hyperbolic functions

$$\sinh 2x = 2 \sinh x \cosh x$$
$$\cosh 2x = \cosh^2 x + \sinh^2 x$$
$$\cosh^2 x = \frac{\cosh 2x + 1}{2}$$
$$\sinh^2 x = \frac{\cosh 2x - 1}{2}$$
$$\cosh^2 x - \sinh^2 x = 1$$
$$\tanh^2 x = 1 - \text{sech}^2 x$$
$$\coth^2 x = 1 + \text{csch}^2 x$$

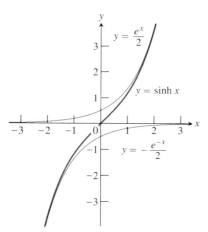

(a) The hyperbolic sine and its component exponentials

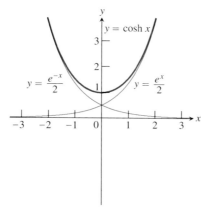

(b) The hyperbolic cosine and its component exponentials

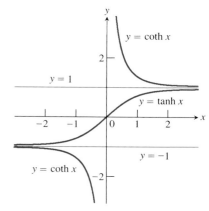

(c) The graphs of $y = \tanh x$ and $y = \coth x = 1/\tanh x$

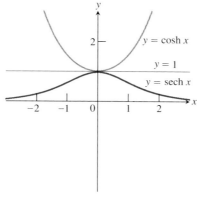

(d) The graphs of $y = \cosh x$ and $y = \text{sech } x = 1/\cosh x$

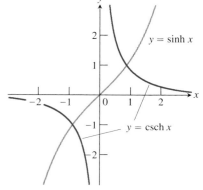

(e) The graphs of $y = \sinh x$ and $y = \text{csch } x = 1/\sinh x$

FIGURE 6.16 Graphs of the six basic hyperbolic functions.

Table 6.9 Derivatives of hyperbolic functions

$$\frac{d}{dx}(\sinh u) = \cosh u \frac{du}{dx}$$

$$\frac{d}{dx}(\cosh u) = \sinh u \frac{du}{dx}$$

$$\frac{d}{dx}(\tanh u) = \operatorname{sech}^2 u \frac{du}{dx}$$

$$\frac{d}{dx}(\coth u) = -\operatorname{csch}^2 u \frac{du}{dx}$$

$$\frac{d}{dx}(\operatorname{sech} u) = -\operatorname{sech} u \tanh u \frac{du}{dx}$$

$$\frac{d}{dx}(\operatorname{csch} u) = -\operatorname{csch} u \coth u \frac{du}{dx}$$

Evaluating Hyperbolic Functions

Like many standard functions, hyperbolic functions and their inverses are easily evaluated with calculators, which have special keys or keystroke sequences for that purpose.

Table 6.10 Integral formulas for hyperbolic functions

$$\int \sinh u \, du = \cosh u + C$$

$$\int \cosh u \, du = \sinh u + C$$

$$\int \operatorname{sech}^2 u \, du = \tanh u + C$$

$$\int \operatorname{csch}^2 u \, du = -\coth u + C$$

$$\int \operatorname{sech} u \tanh u \, du = -\operatorname{sech} u + C$$

$$\int \operatorname{csch} u \coth u \, du = -\operatorname{csch} u + C$$

Derivatives and Integrals

The six hyperbolic functions, being rational combinations of the differentiable functions e^x and e^{-x}, have derivatives at every point at which they are defined (Table 6.9).

Again, there are similarities with trigonometric functions. The derivative formulas in Table 6.9 lead to the integral formulas in Table 6.10.

Example 1   Finding Derivatives and Integrals

(a) $\dfrac{d}{dt}(\tanh \sqrt{1 + t^2}) = \operatorname{sech}^2 \sqrt{1 + t^2} \cdot \dfrac{d}{dt}(\sqrt{1 + t^2})$

$$= \frac{t}{\sqrt{1 + t^2}} \operatorname{sech}^2 \sqrt{1 + t^2}$$

(b) $\displaystyle\int \coth 5x \, dx = \int \frac{\cosh 5x}{\sinh 5x} \, dx = \frac{1}{5} \int \frac{du}{u}$ $u = \sinh 5x, \; du = 5 \cosh 5x \, dx$

$$= \frac{1}{5} \ln |u| + C = \frac{1}{5} \ln |\sinh 5x| + C$$

(c) $\displaystyle\int_0^1 \sinh^2 x \, dx = \int_0^1 \frac{\cosh 2x - 1}{2} \, dx$ Table 6.8

$$= \frac{1}{2} \int_0^1 (\cosh 2x - 1) \, dx = \frac{1}{2} \left[\frac{\sinh 2x}{2} - x \right]_0^1$$

$$= \frac{\sinh 2}{4} - \frac{1}{2} \approx 0.40672$$

(d) $\displaystyle\int_0^{\ln 2} 4e^x \sinh x \, dx = \int_0^{\ln 2} 4e^x \frac{e^x - e^{-x}}{2} \, dx = \int_0^{\ln 2} (2e^{2x} - 2) \, dx$

$$= [e^{2x} - 2x]_0^{\ln 2} = (e^{2 \ln 2} - 2 \ln 2) - (1 - 0)$$

$$= 4 - 2 \ln 2 - 1$$

$$\approx 1.6137$$

Inverse Hyperbolic Functions

We use the inverses of the six basic hyperbolic functions in integration. Since $d(\sinh x)/dx = \cosh x > 0$, the hyperbolic sine is an increasing function of x. We denote its inverse by

$$y = \sinh^{-1} x.$$

For every value of x in the interval $-\infty < x < \infty$, the value of $y = \sinh^{-1} x$ is the number whose hyperbolic sine is x. The graphs of $y = \sinh x$ and $y = \sinh^{-1} x$ are shown in Figure 6.17a.

The function $y = \cosh x$ is not one-to-one, as we can see from the graph in Figure 6.16b. The restricted function $y = \cosh x$, $x \geq 0$, however, is one-to-one and therefore has an inverse, denoted by

$$y = \cosh^{-1} x.$$

For every value of $x \geq 1$, $y = \cosh^{-1} x$ is the number in the interval $0 \leq y < \infty$ whose hyperbolic cosine is x. The graphs of $y = \cosh x$, $x \geq 0$, and $y = \cosh^{-1} x$ are shown in Figure 6.17b.

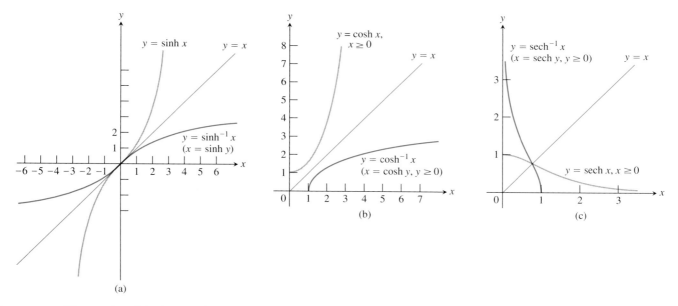

FIGURE 6.17 The graphs of the inverse hyperbolic sine, cosine, and secant of x. Notice the symmetries about the line $y = x$.

Like $y = \cosh x$, the function $y = \operatorname{sech} x = 1/\cosh x$ fails to be one-to-one, but its restriction to nonnegative values of x does have an inverse, denoted by

$$y = \operatorname{sech}^{-1} x.$$

For every value of x in the interval $(0, 1]$, $y = \operatorname{sech}^{-1} x$ is the nonnegative number whose hyperbolic secant is x. The graphs of $y = \operatorname{sech} x$, $x \geq 0$, and $y = \operatorname{sech}^{-1} x$ are shown in Figure 6.17c.

The hyperbolic tangent, cotangent, and cosecant are one-to-one on their domains and therefore have inverses, denoted by

$$y = \tanh^{-1} x, \qquad y = \coth^{-1} x, \qquad y = \operatorname{csch}^{-1} x.$$

These functions are graphed in Figure 6.18.

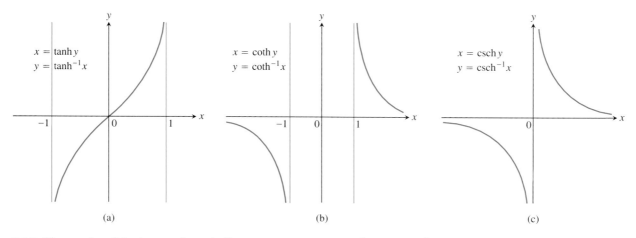

FIGURE 6.18 The graphs of the inverse hyperbolic tangent, cotangent, and cosecant of x.

Table 6.11 Identities for inverse hyperbolic functions

$$\text{sech}^{-1} x = \cosh^{-1} \frac{1}{x}$$

$$\text{csch}^{-1} x = \sinh^{-1} \frac{1}{x}$$

$$\coth^{-1} x = \tanh^{-1} \frac{1}{x}$$

Table 6.12 Derivatives of inverse hyperbolic functions

$$\frac{d(\sinh^{-1} u)}{dx} = \frac{1}{\sqrt{1 + u^2}} \frac{du}{dx}$$

$$\frac{d(\cosh^{-1} u)}{dx} = \frac{1}{\sqrt{u^2 - 1}} \frac{du}{dx}, \quad u > 1$$

$$\frac{d(\tanh^{-1} u)}{dx} = \frac{1}{1 - u^2} \frac{du}{dx}, \quad |u| < 1$$

$$\frac{d(\coth^{-1} u)}{dx} = \frac{1}{1 - u^2} \frac{du}{dx}, \quad |u| > 1$$

$$\frac{d(\text{sech}^{-1} u)}{dx} = \frac{-du/dx}{u\sqrt{1 - u^2}}, \quad 0 < u < 1$$

$$\frac{d(\text{csch}^{-1} u)}{dx} = \frac{-du/dx}{|u|\sqrt{1 + u^2}}, \quad u \neq 0$$

Useful Identities

We use the identities in Table 6.11 to calculate the values of $\text{sech}^{-1} x$, $\text{csch}^{-1} x$, and $\coth^{-1} x$ on calculators that give only $\cosh^{-1} x$, $\sinh^{-1} x$, and $\tanh^{-1} x$.

Derivatives and Integrals

The chief use of inverse hyperbolic functions lies in integrations that reverse the derivative formulas in Table 6.12.

The restrictions $|u| < 1$ and $|u| > 1$ on the derivative formulas for $\tanh^{-1} u$ and $\coth^{-1} u$ come from the natural restrictions on the values of these functions. (See Figure 6.18a and b.) The distinction between $|u| < 1$ and $|u| > 1$ becomes important when we convert the derivative formulas into integral formulas. If $|u| < 1$, the integral of $1/(1 - u^2)$ is $\tanh^{-1} u + C$. If $|u| > 1$, the appropriate integral is $\coth^{-1} u + C$.

Example 2   Derivative of the Inverse Hyperbolic Cosine

Show that if u is a differentiable function of x whose values are greater than 1, then

$$\frac{d}{dx}(\cosh^{-1} u) = \frac{1}{\sqrt{u^2 - 1}} \frac{du}{dx}.$$

Solution   First we find the derivative of $y = \cosh^{-1} x$ for $x > 1$:

$$y = \cosh^{-1} x$$

$$x = \cosh y \qquad \text{Equivalent equation}$$

$$1 = \sinh y \frac{dy}{dx} \qquad \text{Differentiation with respect to } x$$

$$\frac{dy}{dx} = \frac{1}{\sinh y} = \frac{1}{\sqrt{\cosh^2 y - 1}} \qquad \begin{array}{l}\text{Since } x > 1, \ y > 0 \text{ and} \\ \sinh y > 0.\end{array}$$

$$= \frac{1}{\sqrt{x^2 - 1}} \qquad \cosh y = x$$

In short,

$$\frac{d}{dx}(\cosh^{-1} x) = \frac{1}{\sqrt{x^2 - 1}}.$$

The Chain Rule gives the final result:

$$\frac{d}{dx}(\cosh^{-1} u) = \frac{1}{\sqrt{u^2 - 1}} \frac{du}{dx}.$$

With appropriate substitutions, the derivative formulas in Table 6.12 lead to the integration formulas in Table 6.13.

Example 3   Using Table 6.13

Evaluate

$$\int_0^1 \frac{2dx}{\sqrt{3 + 4x^2}}.$$

Table 6.13 Integrals leading to inverse hyperbolic functions

1. $\displaystyle\int \frac{du}{\sqrt{a^2 + u^2}} = \sinh^{-1}\left(\frac{u}{a}\right) + C, \qquad a > 0$

2. $\displaystyle\int \frac{du}{\sqrt{u^2 - a^2}} = \cosh^{-1}\left(\frac{u}{a}\right) + C, \qquad u > a > 0$

3. $\displaystyle\int \frac{du}{a^2 - u^2} = \begin{cases} \frac{1}{a}\tanh^{-1}\left(\frac{u}{a}\right) + C & \text{if } u^2 < a^2 \\ \frac{1}{a}\coth^{-1}\left(\frac{u}{a}\right) + C & \text{if } u^2 > a^2 \end{cases}$

4. $\displaystyle\int \frac{du}{u\sqrt{a^2 - u^2}} = -\frac{1}{a}\operatorname{sech}^{-1}\left(\frac{u}{a}\right) + C, \qquad 0 < u < a$

5. $\displaystyle\int \frac{du}{u\sqrt{a^2 + u^2}} = -\frac{1}{a}\operatorname{csch}^{-1}\left|\frac{u}{a}\right| + C, \qquad u \neq 0$

Solution The indefinite integral is

$$\int \frac{2\,dx}{\sqrt{3 + 4x^2}} = \int \frac{du}{\sqrt{a^2 + u^2}} \qquad u = 2x, \quad du = 2\,dx, \quad a = \sqrt{3}$$

$$= \sinh^{-1}\left(\frac{u}{a}\right) + C \qquad \text{Formula from Table 6.13}$$

$$= \sinh^{-1}\left(\frac{2x}{\sqrt{3}}\right) + C.$$

Therefore,

$$\int_0^1 \frac{2\,dx}{\sqrt{3 + 4x^2}} = \sinh^{-1}\left(\frac{2x}{\sqrt{3}}\right)\Big]_0^1 = \sinh^{-1}\left(\frac{2}{\sqrt{3}}\right) - \sinh^{-1}(0)$$

$$= \sinh^{-1}\left(\frac{2}{\sqrt{3}}\right) - 0 \approx 0.98665.$$

EXERCISES 6.5

Hyperbolic Function Values and Identities

Each of Exercises 1–4 gives a value of $\sinh x$ or $\cosh x$. Use the definitions and the identity $\cosh^2 x - \sinh^2 x = 1$ to find the values of the remaining five hyperbolic functions.

1. $\sinh x = -\frac{3}{4}$

2. $\sinh x = \frac{4}{3}$

3. $\cosh x = \frac{17}{15}, \quad x > 0$

4. $\cosh x = \frac{13}{5}, \quad x > 0$

Rewrite the expressions in Exercises 5–10 in terms of exponentials and simplify the results as much as you can.

5. $2\cosh(\ln x)$

6. $\sinh(2\ln x)$

7. $\cosh 5x + \sinh 5x$

8. $\cosh 3x - \sinh 3x$

9. $(\sinh x + \cosh x)^4$

10. $\ln(\cosh x + \sinh x) + \ln(\cosh x - \sinh x)$

11. Use the identities

$$\sinh(x + y) = \sinh x \cosh y + \cosh x \sinh y$$
$$\cosh(x + y) = \cosh x \cosh y + \sinh x \sinh y$$

to show that

(a) $\sinh 2x = 2 \sinh x \cosh x$

(b) $\cosh 2x = \cosh^2 x + \sinh^2 x$.

12. Use the definitions of $\cosh x$ and $\sinh x$ to show that

$$\cosh^2 x - \sinh^2 x = 1.$$

Derivatives

In Exercises 13–24, find the derivative of y with respect to the appropriate variable.

13. $y = 6 \sinh \dfrac{x}{3}$

14. $y = \dfrac{1}{2} \sinh (2x + 1)$

15. $y = 2\sqrt{t} \tanh \sqrt{t}$

16. $y = t^2 \tanh \dfrac{1}{t}$

17. $y = \ln(\sinh z)$

18. $y = \ln(\cosh z)$

19. $y = \operatorname{sech} \theta(1 - \ln \operatorname{sech} \theta)$

20. $y = \operatorname{csch} \theta(1 - \ln \operatorname{csch} \theta)$

21. $y = \ln \cosh v - \dfrac{1}{2} \tanh^2 v$

22. $y = \ln \sinh v - \dfrac{1}{2} \coth^2 v$

23. $y = (x^2 + 1) \operatorname{sech} (\ln x)$

  (*Hint:* Before differentiating, express in terms of exponentials and simplify.)

24. $y = (4x^2 - 1) \operatorname{csch} (\ln 2x)$

In Exercises 25–36, find the derivative of y with respect to the appropriate variable.

25. $y = \sinh^{-1} \sqrt{x}$

26. $y = \cosh^{-1} 2\sqrt{x + 1}$

27. $y = (1 - \theta) \tanh^{-1} \theta$

28. $y = (\theta^2 + 2\theta) \tanh^{-1} (\theta + 1)$

29. $y = (1 - t) \coth^{-1} \sqrt{t}$

30. $y = (1 - t^2) \coth^{-1} t$

31. $y = \cos^{-1} x - x \operatorname{sech}^{-1} x$

32. $y = \ln x + \sqrt{1 - x^2} \operatorname{sech}^{-1} x$

33. $y = \operatorname{csch}^{-1} \left(\dfrac{1}{2}\right)^{\theta}$

34. $y = \operatorname{csch}^{-1} 2^{\theta}$

35. $y = \sinh^{-1} (\tan x)$

36. $y = \cosh^{-1} (\sec x), \quad 0 < x < \pi/2$

Integration Formulas

Verify the integration formulas in Exercises 37–40.

37. (a) $\displaystyle \int \operatorname{sech} x \, dx = \tan^{-1} (\sinh x) + C$

(b) $\displaystyle \int \operatorname{sech} x \, dx = \sin^{-1} (\tanh x) + C$

38. $\displaystyle \int x \operatorname{sech}^{-1} x \, dx = \dfrac{x^2}{2} \operatorname{sech}^{-1} x - \dfrac{1}{2} \sqrt{1 - x^2} + C$

39. $\displaystyle \int x \coth^{-1} x \, dx = \dfrac{x^2 - 1}{2} \coth^{-1} x + \dfrac{x}{2} + C$

40. $\displaystyle \int \tanh^{-1} x \, dx = x \tanh^{-1} x + \dfrac{1}{2} \ln (1 - x^2) + C$

Indefinite Integrals

Evaluate the integrals in Exercises 41–50.

41. $\displaystyle \int \sinh 2x \, dx$

42. $\displaystyle \int \sinh \dfrac{x}{5} \, dx$

43. $\displaystyle \int 6 \cosh \left(\dfrac{x}{2} - \ln 3\right) dx$

44. $\displaystyle \int 4 \cosh (3x - \ln 2) \, dx$

45. $\displaystyle \int \tanh \dfrac{x}{7} \, dx$

46. $\displaystyle \int \coth \dfrac{\theta}{\sqrt{3}} \, d\theta$

47. $\displaystyle \int \operatorname{sech}^2 \left(x - \dfrac{1}{2}\right) dx$

48. $\displaystyle \int \operatorname{csch}^2 (5 - x) \, dx$

49. $\displaystyle \int \dfrac{\operatorname{sech} \sqrt{t} \tanh \sqrt{t} \, dt}{\sqrt{t}}$

50. $\displaystyle \int \dfrac{\operatorname{csch} (\ln t) \coth (\ln t) \, dt}{t}$

Definite Integrals

Evaluate the integrals in Exercises 51–60.

51. $\displaystyle \int_{\ln 2}^{\ln 4} \coth x \, dx$

52. $\displaystyle \int_{0}^{\ln 2} \tanh 2x \, dx$

53. $\displaystyle \int_{-\ln 4}^{-\ln 2} 2e^{\theta} \cosh \theta \, d\theta$

54. $\displaystyle \int_{0}^{\ln 2} 4e^{-\theta} \sinh \theta \, d\theta$

55. $\displaystyle \int_{-\pi/4}^{\pi/4} \cosh (\tan \theta) \sec^2 \theta \, d\theta$

56. $\displaystyle \int_{0}^{\pi/2} 2 \sinh (\sin \theta) \cos \theta \, d\theta$

57. $\displaystyle \int_{1}^{2} \dfrac{\cosh (\ln t)}{t} \, dt$

58. $\displaystyle \int_{1}^{4} \dfrac{8 \cosh \sqrt{x}}{\sqrt{x}} \, dx$

59. $\displaystyle \int_{-\ln 2}^{0} \cosh^2 \left(\dfrac{x}{2}\right) dx$

60. $\displaystyle \int_{0}^{\ln 10} 4 \sinh^2 \left(\dfrac{x}{2}\right) dx$

Evaluating Inverse Hyperbolic Functions and Related Integrals

When hyperbolic function keys are not available on a calculator, it is still possible to evaluate the inverse hyperbolic functions by expressing them as logarithms, as shown here.

$$\sinh^{-1} x = \ln \left(x + \sqrt{x^2 + 1}\right), \qquad -\infty < x < \infty$$

$$\cosh^{-1} x = \ln \left(x + \sqrt{x^2 - 1}\right), \qquad x \geq 1$$

$$\tanh^{-1} x = \dfrac{1}{2} \ln \dfrac{1 + x}{1 - x}, \qquad |x| < 1$$

$$\operatorname{sech}^{-1} x = \ln \left(\dfrac{1 + \sqrt{1 - x^2}}{x}\right), \qquad 0 < x \leq 1$$

$$\operatorname{csch}^{-1} x = \ln \left(\dfrac{1}{x} + \dfrac{\sqrt{1 + x^2}}{|x|}\right), \qquad x \neq 0$$

$$\coth^{-1} x = \dfrac{1}{2} \ln \dfrac{x + 1}{x - 1}, \qquad |x| > 1$$

Use the formulas in the table above to express the numbers in Exercises 61–66 in terms of natural logarithms.

61. $\sinh^{-1}(-5/12)$

62. $\cosh^{-1}(5/3)$

63. $\tanh^{-1}(-1/2)$

64. $\coth^{-1}(5/4)$

65. $\operatorname{sech}^{-1}(3/5)$

66. $\operatorname{csch}^{-1}(-1/\sqrt{3})$

Evaluate the integrals in Exercises 67–74 in terms of

    (a) inverse hyperbolic functions

    (b) natural logarithms.

67. $\displaystyle\int_{0}^{2\sqrt{3}} \frac{dx}{\sqrt{4+x^2}}$

68. $\displaystyle\int_{0}^{1/3} \frac{6\,dx}{\sqrt{1+9x^2}}$

69. $\displaystyle\int_{5/4}^{2} \frac{dx}{1-x^2}$

70. $\displaystyle\int_{0}^{1/2} \frac{dx}{1-x^2}$

71. $\displaystyle\int_{1/5}^{3/13} \frac{dx}{x\sqrt{1-16x^2}}$

72. $\displaystyle\int_{1}^{2} \frac{dx}{x\sqrt{4+x^2}}$

73. $\displaystyle\int_{0}^{\pi} \frac{\cos x\,dx}{\sqrt{1+\sin^2 x}}$

74. $\displaystyle\int_{1}^{e} \frac{dx}{x\sqrt{1+(\ln x)^2}}$

Applications and Theory

75. (a) Show that if a function f is defined on an interval symmetric about the origin (so that f is defined at $-x$ whenever it is defined at x), then

$$f(x) = \frac{f(x)+f(-x)}{2} + \frac{f(x)-f(-x)}{2}. \tag{1}$$

Then show that $(f(x)+f(-x))/2$ is even and that $(f(x)-f(-x))/2$ is odd.

(b) *Writing to Learn* Equation (1) simplifies considerably if f itself is (i) even or (ii) odd. What are the new equations? Give reasons for your answers.

76. *Writing to Learn* Derive the formula $\sinh^{-1} x = \ln(x + \sqrt{x^2+1})$, $-\infty < x < \infty$. Explain in your derivation why the plus sign is used with the square root instead of the minus sign.

77. *Skydiving* If a body of mass m falling from rest under the action of gravity encounters an air resistance proportional to the square of the velocity, then the body's velocity t sec into the fall satisfies the differential equation

$$m\frac{dv}{dt} = mg - kv^2,$$

where k is a constant that depends on the body's aerodynamic properties and the density of the air. (We assume that the fall is short enough so that the variation in the air's density will not affect the outcome significantly.)

(a) Show that

$$v = \sqrt{\frac{mg}{k}} \tanh\left(\sqrt{\frac{gk}{m}}\,t\right)$$

satisfies the differential equation and the initial condition that $v = 0$ when $t = 0$.

(b) Find the body's *limiting velocity*, $\lim_{t\to\infty} v$.

(c) For a 160 lb skydiver ($mg = 160$), with time in seconds and distance in feet, a typical value for k is 0.005. What is the diver's limiting velocity?

78. *Accelerations whose magnitudes are proportional to displacement* Suppose that the position of a body moving along a coordinate line at time t is

(a) $s = a\cos kt + b\sin kt$,

(b) $s = a\cosh kt + b\sinh kt$.

Show in both cases that the acceleration d^2s/dt^2 is proportional to s but that in the first case it is directed toward the origin, whereas in the second case it is directed away from the origin.

79. *Tractor trailers and the tractrix* When a tractor trailer turns into a cross street or driveway, its rear wheels follow a curve like the one shown here.

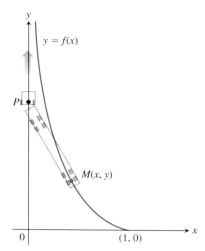

(This is why the rear wheels sometimes ride up over the curb.) We can find an equation for the curve if we picture the rear wheels as a mass M at the point $(1, 0)$ on the x-axis attached by a rod of unit length to a point P representing the cab at the origin. As the point P moves up the y-axis, it drags M along behind it. The curve traced by M—called a *tractrix* from the Latin word *tractum*, for "drag"—can be shown to be the graph of the function $y = f(x)$ that solves the initial value problem

Differential equation: $\dfrac{dy}{dx} = -\dfrac{1}{x\sqrt{1-x^2}} + \dfrac{x}{\sqrt{1-x^2}}$

  Initial condition: $y = 0$ when $x = 1$.

Solve the initial value problem to find an equation for the curve. (You need an inverse hyperbolic function.)

80. *Area* Show that the area of the region in the first quadrant enclosed by the curve $y = (1/a)\cosh ax$, the coordinate axes, and the line $x = b$ is the same as the area of a rectangle of height $1/a$

and length s, where s is the length of the curve from $x = 0$ to $x = b$.

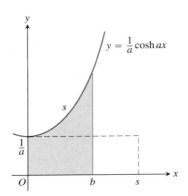

81. *Volume* A region in the first quadrant is bounded above by the curve $y = \cosh x$, below by the curve $y = \sinh x$, and on the left and right by the y-axis and the line $x = 2$, respectively. Find the volume of the solid generated by revolving the region about the x-axis.

82. *Volume* The region enclosed by the curve $y = \operatorname{sech} x$, the x-axis, and the lines $x = \pm\ln \sqrt{3}$ is revolved about the x-axis to generate a solid. Find the volume of the solid.

83. *Arc length* Find the length of the segment of the curve $y = (1/2) \cosh 2x$ from $x = 0$ to $x = \ln \sqrt{5}$.

84. *The hyperbolic in hyperbolic functions* In case you are wondering where the name *hyperbolic* comes from, here is the answer: Just as $x = \cos u$ and $y = \sin u$ are identified with points (x, y) on the unit circle, the functions $x = \cosh u$ and $y = \sinh u$ are identified with points (x, y) on the right-hand branch of the unit hyperbola, $x^2 - y^2 = 1$ (Figure 6.19).

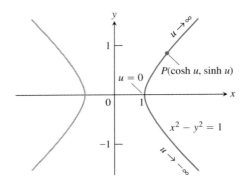

FIGURE 6.19 Since $\cosh^2 u - \sinh^2 u = 1$, the point $(\cosh u, \sinh u)$ lies on the right-hand branch of the hyperbola $x^2 - y^2 = 1$ for every value of u. (Exercise 84)

Another analogy between hyperbolic and circular functions is that the variable u in the coordinates $(\cosh u, \sinh u)$ for the points of the right-hand branch of the hyperbola $x^2 - y^2 = 1$ is

twice the area of the sector AOP pictured in Figure 6.20. To see why this is so, carry out the following steps.

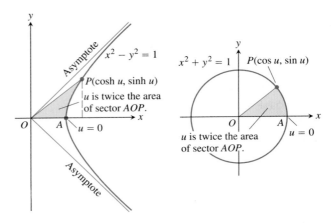

FIGURE 6.20 One of the analogies between hyperbolic and circular functions is revealed by these two diagrams. (Exercise 84)

(a) Show that the area $A(u)$ of sector AOP is

$$A(u) = \frac{1}{2} \cosh u \sinh u - \int_{1}^{\cosh u} \sqrt{x^2 - 1}\, dx.$$

(b) Differentiate both sides of the equation in part (a) with respect to u to show that

$$A'(u) = \frac{1}{2}.$$

(c) Solve this last equation for $A(u)$. What is the value of $A(0)$? What is the value of the constant of integration C in your solution? With C determined, what does your solution say about the relationship of u to $A(u)$?

Hanging Cables

85. Imagine a cable, like a telephone line or TV cable, strung from one support to another and hanging freely. The cable's weight per unit length is w and the horizontal tension at its lowest point is a vector of length H. If we choose a coordinate system for the plane of the cable in which the x-axis is horizontal, the force of gravity is straight down, the positive y-axis points straight up, and the lowest point of the cable lies at the point $y = H/w$ on the y-axis (Figure 6.21), then it can be shown that the cable lies along the graph of the hyperbolic cosine

$$y = \frac{H}{w} \cosh \frac{w}{H} x.$$

Such a curve is sometimes called a **chain curve** or a **catenary,** the latter deriving from the Latin *catena,* meaning "chain."

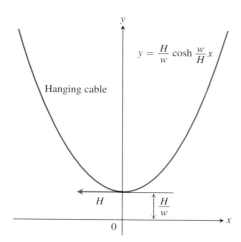

FIGURE 6.21 In a coordinate system chosen to match H and w in the manner shown, a hanging cable lies along the curve $y = (H/w)\cosh(wx/H)$.

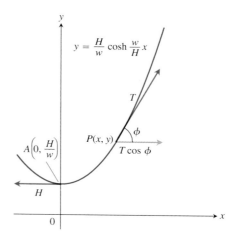

FIGURE 6.22 As discussed in Exercise 85, $T = wy$ in this coordinate system.

(a) Let $P(x, y)$ denote an arbitrary point on the cable. Figure 16.22 displays the tension at P as a vector of length (magnitude) T, as well as the tension H at the lowest point A. Show that the cable's slope at P is

$$\tan \phi = \frac{dy}{dx} = \sinh \frac{w}{H} x.$$

(b) Using the result from part (a) and the fact that the horizontal tension at P must equal H (the cable is not moving), show that $T = wy$. This means that the magnitude of the tension at $P(x, y)$ is exactly equal to the weight of y units of cable.

86. (*Continuation of Exercise 85*) The length of arc AP in Figure 6.22 is $s = (1/a)\sinh ax$, where $a = w/H$. Show that the coordinates of P may be expressed in terms of s as

$$x = \frac{1}{a}\sinh^{-1} as, \qquad y = \sqrt{s^2 + \frac{1}{a^2}}.$$

87. *The sag and horizontal tension in a cable* The ends of a cable 32 ft long and weighing 2 lb/ft are fastened at the same level to posts 30 ft apart.

(a) Model the cable with the equation

$$y = \frac{1}{a}\cosh ax, \qquad -15 \le x \le 15.$$

Use information from Exercise 86 to show that a satisfies the equation

$$16a = \sinh 15a. \qquad (2)$$

T (b) Solve Equation (2) graphically by estimating the coordinates of the points where the graphs of the equations $y = 16a$ and $y = \sinh 15a$ intersect in the ay-plane.

T (c) Solve Equation (2) for a numerically. Compare your solution with the value you found in part (b).

(d) Estimate the horizontal tension in the cable at the cable's lowest point.

T (e) Using the value found for a in part (c), graph the catenary

$$y = \frac{1}{a}\cosh ax$$

over the interval $-15 \le x \le 15$. Estimate the sag in the cable at its center.

Questions to Guide Your Review

1. What is the natural logarithm function in terms of an integral? What are its domain, range, and derivative? What arithmetic properties does it have? Comment on its graph.

2. What integrals lead to logarithms? Give examples.

3. How is the exponential function e^x defined in terms of the natural logarithm? What are its domain, range, and derivative? What laws of exponents does it obey? Comment on its graph.

4. How is the number e defined? How can it be expressed as a limit?

5. What is the slope field of a differential equation $y' = f(x, y)$? What can we learn from such fields?

6. How do you solve linear first-order differential equations? What applications are modeled by these equations?

7. Describe Euler's method for solving the initial value problem $y' = f(x, y)$, $y(x_0) = y_0$ numerically. Give an example. Comment on the method's accuracy. Why might you want to solve an initial value problem numerically?

8. Describe the improved Euler's method for solving the initial

value problem $y' = f(x, y)$, $y(x_0) = y_0$ numerically. How does it compare with Euler's method?

9. Why is the exponential model unrealistic for predicting long-term population growth? How does the logistic model correct for the deficiency in the exponential model for population growth? What is the logistic differential equation? What is the form of its solution? Describe the graph of the logistic solution.

10. What are the six basic hyperbolic functions? Comment on their domains, ranges, and graphs. What are some of the identities relating them?

11. What are the derivatives of the six basic hyperbolic functions? What are the corresponding integral formulas? What similarities do you see here with the six basic trigonometric functions?

12. How are the inverse hyperbolic functions defined? Comment on their domains, ranges, and graphs. How can you find values of $\text{sech}^{-1} x$, $\text{csch}^{-1} x$, and $\coth^{-1} x$ using a calculator's keys for $\cosh^{-1} x$, $\sinh^{-1} x$, and $\tanh^{-1} x$?

13. What integrals lead naturally to inverse hyperbolic functions?

Practice Exercises

Integration

Evaluate the integrals in Exercises 1–30.

1. $\displaystyle\int e^x \sin (e^x) \, dx$

2. $\displaystyle\int e^t \cos (3e^t - 2) \, dt$

3. $\displaystyle\int e^x \sec^2 (e^x - 7) \, dx$

4. $\displaystyle\int e^y \csc (e^y + 1) \cot (e^y + 1) \, dy$

5. $\displaystyle\int \sec^2 (x) e^{\tan x} \, dx$

6. $\displaystyle\int \csc^2 (x) \, e^{\cot x} \, dx$

7. $\displaystyle\int_{-1}^{1} \frac{dx}{3x - 4}$

8. $\displaystyle\int_{1}^{e} \frac{\sqrt{\ln x}}{x} \, dx$

9. $\displaystyle\int_{0}^{\pi} \tan \frac{x}{3} \, dx$

10. $\displaystyle\int_{1/6}^{1/4} 2 \cot \pi x \, dx$

11. $\displaystyle\int_{0}^{4} \frac{2t}{t^2 - 25} \, dt$

12. $\displaystyle\int_{-\pi/2}^{\pi/6} \frac{\cos t}{1 - \sin t} \, dt$

13. $\displaystyle\int \frac{\tan (\ln v)}{v} \, dv$

14. $\displaystyle\int \frac{dv}{v \ln v}$

15. $\displaystyle\int \frac{(\ln x)^{-3}}{x} \, dx$

16. $\displaystyle\int \frac{\ln (x - 5)}{x - 5} \, dx$

17. $\displaystyle\int \frac{1}{r} \csc^2 (1 + \ln r) \, dr$

18. $\displaystyle\int \frac{\cos (1 - \ln v)}{v} \, dv$

19. $\displaystyle\int x 3^{x^2} \, dx$

20. $\displaystyle\int 2^{\tan x} \sec^2 x \, dx$

21. $\displaystyle\int_{1}^{7} \frac{3}{x} \, dx$

22. $\displaystyle\int_{1}^{32} \frac{1}{5x} \, dx$

23. $\displaystyle\int_{-2}^{-1} e^{-(x+1)} \, dx$

24. $\displaystyle\int_{-\ln 2}^{0} e^{2w} \, dw$

25. $\displaystyle\int_{1}^{3} \frac{(\ln (v + 1))^2}{v + 1} \, dv$

26. $\displaystyle\int_{2}^{4} (1 + \ln t)t \ln t \, dt$

27. $\displaystyle\int_{1}^{8} \frac{\log_4 \theta}{\theta} \, d\theta$

28. $\displaystyle\int_{1}^{e} \frac{8 \ln 3 \, \log_3 \theta}{\theta} \, d\theta$

29. $\displaystyle\int \frac{2 \, dy}{\sqrt{1 + 25y^2}}$

30. $\displaystyle\int \frac{3 \, dy}{4 - 49y^2}$

Theory and Applications

In Exercises 31 and 32, find the absolute maximum and minimum values of each function on the given interval.

31. $y = x \ln 2x - x$, $\left[\dfrac{1}{2e}, \dfrac{e}{2}\right]$

32. $y = 10x(2 - \ln x)$, $(0, e^2]$

33. *Area* Find the area between the curve $y = 2(\ln x)/x$ and the x-axis from $x = 1$ to $x = e$.

34. (a) *Area* Show that the area between the curve $y = 1/x$ and the x-axis from $x = 10$ to $x = 20$ is the same as the area between the curve and the x-axis from $x = 1$ to $x = 2$.

(b) *Area* Show that the area between the curve $y = 1/x$ and the x-axis from ka to kb is the same as the area between the curve and the x-axis from $x = a$ to $x = b$ ($0 < a < b$, $k > 0$).

35. *Traveling particle* A particle is traveling upward and to the right along the curve $y = \ln x$. Its x-coordinate is increasing at the rate $(dx/dt) = \sqrt{x}$ m/sec. At what rate is the y-coordinate changing at the point $(e^2, 2)$?

36. *Volume* The region between the curve $y = 1/(2\sqrt{x})$ and the x-axis from $x = 1/4$ to $x = 4$ is revolved about the x-axis to generate a solid. Find the volume of the solid.

37. *Writing to Learn* The functions $f(x) = \ln 5x$ and $g(x) = \ln 3x$ differ by a constant. What constant? Give reasons for your answer.

38. *Writing to Learn*

(a) If $(\ln x)/x = (\ln 2)/2$, must $x = 2$?

(b) If $(\ln x)/x = -2 \ln 2$, must $x = 1/2$?

Give reasons for your answers.

39. *Escape velocity* The gravitational attraction F exerted by an airless moon on a body of mass m at a distance s from the moon's center is given by the equation $F = -mgR^2 s^{-2}$, where g is the acceleration of gravity at the moon's surface and R is the moon's radius (Figure 6.23). The force F is negative because it acts in the direction of decreasing s.

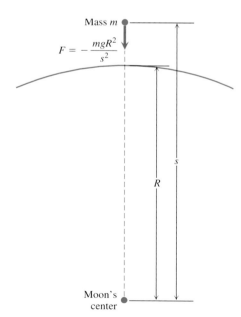

Mass m

$$F = -\frac{mgR^2}{s^2}$$

s

R

Moon's center

FIGURE 6.23 Diagram for Exercise 39.

If the body is projected vertically upward from the moon's surface with an initial velocity v_0 at time $t = 0$, use Newton's second law, $F = ma$, to show that the body's velocity at position s is given by the equation

$$v^2 = \frac{2gR^2}{s} + v_0^2 - 2gR.$$

Thus, the velocity remains positive as long as $v_0 \geq \sqrt{2gR}$. The velocity $v_0 = \sqrt{2gR}$ is the moon's **escape velocity**. A body projected upward with this velocity or a greater one will escape from the moon's gravitational pull.

40. (*Continuation of Exercise 39*) Show that if $v_0 = \sqrt{2gR}$, then

$$s = R\left(1 + \frac{3v_0}{2R}t\right)^{2/3}.$$

Initial Value Problems

Solve the initial value problems in Exercises 41–44.

Differential equation	Initial condition
41. $y' \cos x - y \sin x = \sin 2x$	$y(0) = 1$
42. $\dfrac{dy}{dx} = -\dfrac{y \ln y}{1 + x^2}$	$y(0) = e^2$
43. $(x + 1)\dfrac{dy}{dx} + 2y = x$, $\quad x > -1$	$y(0) = 1$
44. $x\dfrac{dy}{dx} + 2y = x^2 + 1$, $\quad x > 0$	$y(1) = 1$

T 45. *Extreme values* Graph the following functions and use what you see to locate and estimate the extreme values, identify the coordinates of the inflection points, and identify the intervals on which the graphs are concave up and concave down. Then confirm your estimates by working with the functions' derivatives.

(a) $y = (\ln x)/\sqrt{x}$ **(b)** $y = e^{-x^2}$

(c) $y = (1 + x)e^{-x}$

T 46. *Absolute minimum* Graph $f(x) = x \ln x$. Does the function appear to have an absolute minimum value? Confirm your answer with calculus.

47. *Age of charcoal* What is the age of a sample of charcoal in which 90% of the carbon-14 originally present has decayed?

48. *Cooling a pie* A deep-dish apple pie, whose internal temperature was 220°F when removed from the oven, was set out on a breezy 40°F porch to cool. Fifteen minutes later, the pie's internal temperature was 180°F. How long did it take the pie to cool from there to 70°F?

49. (a) Solve the differential equation

$$xy' + y = x \cos x, \qquad x > 0.$$

T (b) Graph several members of the family of solution curves. How does the curve vary as C changes?

50. *Confirming a solution* Show that

$$y = \int_0^x \sin (t^2)\, dt + x^3 + x + 2$$

is the solution of the initial value problem.

Differential equation: $\quad y'' = 2x \cos (x^2) + 6x$
Initial conditions: $\quad\quad y'(0) = 1, \quad y(0) = 2$

Slope Fields and Euler's Method

51. *Sketching solutions* Draw a possible graph for the function $y = f(x)$ with slope field given in the figure that satisfies the initial condition $y(0) = 0$.

[−10, 10] by [−10, 10]

In Exercises 52 and 53, use the stated method to solve the initial value problem on the given interval starting at x_0 with $dx = 0.1$.

T **52.** *Euler;* $y' = y + \cos x, \quad y(0) = 0; \quad 0 \le x \le 2; \quad x_0 = 0$

T **53.** *Improved Euler;* $y' = (2 - y)(2x + 3), \quad y(-3) = 1; \quad -3 \le x \le -1; \quad x_0 = -3$

In Exercises 54 and 55, use the stated method with $dx = 0.05$ to estimate $y(c)$ where y is the solution to the given initial value problem.

T **54.** *Improved Euler;* $c = 3; \quad \dfrac{dy}{dx} = \dfrac{x - 2y}{x + 1}, \quad y(0) = 1$

T **55.** *Euler;* $c = 4; \quad \dfrac{dy}{dx} = \dfrac{x^2 - 2y + 1}{x}, \quad y(1) = 1$

In Exercises 56 and 57, use the stated method to solve the initial value problem graphically, starting at $x_0 = 0$ with

(a) $dx = 0.1$ (b) $dx = -0.1$.

T **56.** *Euler;* $\dfrac{dy}{dx} = \dfrac{1}{e^{x+y+2}}, \quad y(0) = -2$

T **57.** *Improved Euler;* $\dfrac{dy}{dx} = -\dfrac{x^2 + y}{e^y + x}, \quad y(0) = 0$

58. (a) *Finding an exact solution* Use analytic methods to find the exact solution to

$$\frac{dP}{dt} = 0.002P\left(1 - \frac{P}{800}\right), \quad P(0) = 50.$$

T **(b)** *Numerical solution* Use Euler's method to solve the equation in part (a) for $0 \le t \le 20$ and $dt = 0.5$. Compare the approximation $P(20)$ with the exact solution value.

In Exercises 59–62, sketch part of the equation's slope field. Then add to your sketch the solution curve that passes through the point $P(1, -1)$. Use Euler's method with $x_0 = 1$ and $dx = 0.2$ to estimate $y(2)$. Round your answers to 4 decimal places. Find the exact value of $y(2)$ for comparison.

59. $y' = x$ **60.** $\quad y' = 1/x$

61. $y' = xy$ **62.** $\quad y' = 1/y$

CD-ROM
WEBsite

Additional Exercises: Theory, Examples, Applications

1. *Area between curves* Find the areas between the curves $y = 2(\log_2 x)/x$ and $y = 2(\log_4 x)/x$ and the x-axis from $x = 1$ to $x = e$. What is the ratio of the larger area to the smaller?

2. *Derivative of integral* Find $f'(2)$ if

$$f(x) = e^{g(x)} \quad \text{and} \quad g(x) = \int_2^x \frac{t}{1 + t^4}\, dt.$$

3. *Derivative of an integral*

(a) Find df/dx if

$$f(x) = \int_1^{e^x} \frac{2 \ln t}{t}\, dt.$$

(b) Find $f(0)$.

(c) *Writing to Learn* What can you conclude about the graph of f? Give reasons for your answer.

4. *Writing to Learn: The inequality* $\pi^e < e^\pi$

(a) Why does Figure 6.24 "prove" that $\pi^e < e^\pi$? (Source: "Proof without Words" by Fouad Nakhil, *Mathematics Magazine*, Vol. 60, No. 3 (June 1987), p. 165.)

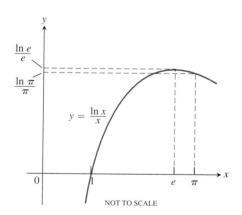

FIGURE 6.24 The figure for Exercise 4.

(b) Figure 6.24 assumes that $f(x) = (\ln x)/x$ has an absolute maximum value at $x = e$. How do you know it does?

5. *An unexpected equality* Use the accompanying figure to show that

$$\int_0^{\pi/2} \sin x \, dx = \frac{\pi}{2} - \int_0^1 \sin^{-1} x \, dx.$$

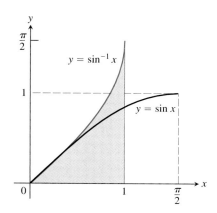

6. *Writing to Learn: Napier's inequality* Here are two pictorial proofs that

$$b > a > 0 \Rightarrow \frac{1}{b} < \frac{\ln b - \ln a}{b - a} < \frac{1}{a}.$$

Explain what is going on in each case.

(a)

(b)

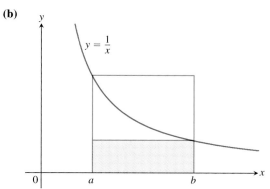

(*Source*: Roger B. Nelson, *College Mathematics Journal*, Vol. 24, No. 2 (March 1993), p. 165.)

7. *Supporting a solution* Give two ways to provide graphical support for the integral formula

$$\int x^2 \ln x \, dx = \frac{x^3}{3} \ln x - \frac{x^3}{9} + C.$$

8. *Writing to Learn* Graph $f(x) = (\sin x)^{\sin x}$ in the window $[0, 3\pi]$ by $[-2, 5]$. Explain what you see.

Applications

9. *Seasonal-growth model* In some population growth models, a periodic function of time is introduced as a multiplier of the population to account for seasonal variations in the rate of growth. For example, the population could be affected by seasonal changes in the food or water supplies.

   (a) Solve the seasonal-growth model

   T
   $$\frac{dP}{dt} = kP \cos(at - b), \qquad P(0) = P_0,$$

   where k, a, and b are positive constants.

   T **(b)** *Writing to Learn* Graph your solution for various values of k, a, and b. Explain the effects these values have on the solution. What happens to $P(t)$ in the long term (i.e., as $t \to \infty$)?

10. *Transport through a cell membrane* Under some conditions, the result of the movement of a dissolved substance across a cell's membrane is described by the equation

$$\frac{dy}{dt} = k \frac{A}{V}(c - y).$$

In this equation, y is the concentration of the substance inside the cell and dy/dt is the rate at which y changes over time. The letters k, A, V, and c stand for constants, k being the *permeability coefficient* (a property of the membrane), A the surface area of the membrane, V the cell's volume, and c the concentration of the substance outside the cell. The equation says that the rate at which the concentration changes within the cell is proportional to the difference between it and the outside concentration.

   (a) Solve the equation for $y(t)$, using y_0 to denote $y(0)$.

   (b) Find the steady state concentration, $\lim_{t \to \infty} y(t)$.
   (Based on *Some Mathematical Models in Biology* edited by R. M. Thrall, J. A. Mortimer, K. R. Rebman, R. F. Baum, rev. ed., December 1967, PB-202 364, pp. 101–103; distributed by N.T.I.S., U.S. Department of Commerce.)

T **11.** *Writing to Learn* Graph $f(x) = \tan^{-1} x + \tan^{-1}(1/x)$ for $-5 \le x \le 5$. Then use calculus to explain what you see. How would you expect f to behave beyond the interval $[-5, 5]$? Give reasons for your answer.

Integration Techniques, l'Hôpital's Rule, and Improper Integrals

7

OVERVIEW   We have seen how integrals arise in modeling real phenomena and in measuring objects in the world around us, and we know in theory how integrals are evaluated with antiderivatives. The more sophisticated our models become, however, the more involved our integrals become. We need to know how to change these more involved integrals into forms we can work with. One goal of this chapter is to show how to change unfamiliar integrals into integrals we can recognize, find in a table, or evaluate with a computer.

We already know two techniques for doing so: algebraic manipulation and substitution. Here, we carry these techniques a step further and introduce a powerful new technique called integration by parts. We also show how all rational functions can be integrated. Finally, we extend our ideas to integrals where one or both limits of integration are infinite and to integrals whose integrands become unbounded on the interval of integration. Just before doing so, we pause to present l'Hôpital's Rule for calculating limits of fractions whose numerators and denominators both approach zero. L'Hôpital's Rule was actually discovered by John Bernoulli but ended up being named for l'Hôpital after l'Hôpital popularized it in a calculus book he wrote.

7.1 Basic Integration Formulas

Algebraic Procedures

As we saw in Section 4.1, we evaluate an indefinite integral by finding an antiderivative of the integrand and adding an arbitrary constant. Table 7.1 shows the basic forms of the integrals we have evaluated so far. There is a more extensive table at the back of the book; we discuss it in Section 7.5.

Table 7.1 Basic integration formulas

1. $\displaystyle\int du = u + C$

2. $\displaystyle\int k\, du = ku + C$ (any number k)

3. $\displaystyle\int (du + dv) = \int du + \int dv$

4. $\displaystyle\int u^n\, du = \frac{u^{n+1}}{n+1} + C$ $(n \neq -1)$

5. $\displaystyle\int \frac{du}{u} = \ln |u| + C$

6. $\displaystyle\int \sin u\, du = -\cos u + C$

7. $\displaystyle\int \cos u\, du = \sin u + C$

8. $\displaystyle\int \sec^2 u\, du = \tan u + C$

9. $\displaystyle\int \csc^2 u\, du = -\cot u + C$

10. $\displaystyle\int \sec u \tan u\, du = \sec u + C$

11. $\displaystyle\int \csc u \cot u\, du = -\csc u + C$

12. $\displaystyle\int \tan u\, du = -\ln |\cos u| + C$
$\qquad\qquad\quad = \ln |\sec u| + C$

13. $\displaystyle\int \cot u\, du = \ln |\sin u| + C$
$\qquad\qquad\quad = -\ln |\csc u| + C$

14. $\displaystyle\int e^u\, du = e^u + C$

15. $\displaystyle\int a^u\, du = \frac{a^u}{\ln a} + C$ $(a > 0, a \neq 1)$

16. $\displaystyle\int \sinh u\, du = \cosh u + C$

17. $\displaystyle\int \cosh u\, du = \sinh u + C$

18. $\displaystyle\int \frac{du}{\sqrt{a^2 - u^2}} = \sin^{-1}\left(\frac{u}{a}\right) + C$

19. $\displaystyle\int \frac{du}{a^2 + u^2} = \frac{1}{a}\tan^{-1}\left(\frac{u}{a}\right) + C$

20. $\displaystyle\int \frac{du}{u\sqrt{u^2 - a^2}} = \frac{1}{a}\sec^{-1}\left|\frac{u}{a}\right| + C$

21. $\displaystyle\int \frac{du}{\sqrt{a^2 + u^2}} = \sinh^{-1}\left(\frac{u}{a}\right) + C$ $(a > 0)$

22. $\displaystyle\int \frac{du}{\sqrt{u^2 - a^2}} = \cosh^{-1}\left(\frac{u}{a}\right) + C$ $(u > a > 0)$

Algebraic Procedures

We often have to rewrite an integral to match it to a standard formula.

Example 1   Making a Simplifying Substitution

Evaluate

$$\int \frac{2x - 9}{\sqrt{x^2 - 9x + 1}}\, dx.$$

Solution

$$\int \frac{2x - 9}{\sqrt{x^2 - 9x + 1}}\, dx = \int \frac{du}{\sqrt{u}} \qquad \begin{array}{l} u = x^2 - 9x + 1, \\ du = (2x - 9)\, dx \end{array}$$

$$= \int u^{-1/2}\, du$$

$$= \frac{u^{(-1/2)+1}}{(-1/2) + 1} + C \qquad \begin{array}{l}\text{Table 7.1, Formula 4,}\\ \text{with } n = -1/2\end{array}$$

$$= 2u^{1/2} + C$$

$$= 2\sqrt{x^2 - 9x + 1} + C$$

Example 2   Completing the Square

Evaluate

$$\int \frac{dx}{\sqrt{8x - x^2}}.$$

Solution   We complete the square to write the radicand as

$$8x - x^2 = -(x^2 - 8x) = -(x^2 - 8x + 16 - 16)$$
$$= -(x^2 - 8x + 16) + 16 = 16 - (x - 4)^2.$$

Then

$$\int \frac{dx}{\sqrt{8x - x^2}} = \int \frac{dx}{\sqrt{16 - (x - 4)^2}}$$

$$= \int \frac{du}{\sqrt{a^2 - u^2}} \qquad \begin{array}{l} a = 4, u = (x - 4), \\ du = dx \end{array}$$

$$= \sin^{-1}\left(\frac{u}{a}\right) + C \qquad \text{Table 7.1, Formula 18}$$

$$= \sin^{-1}\left(\frac{x - 4}{4}\right) + C.$$

Example 3   Expanding a Power and Using a Trigonometric Identity

Evaluate

$$\int (\sec x + \tan x)^2 \, dx.$$

Solution   We expand the integrand and get

$$(\sec x + \tan x)^2 = \sec^2 x + 2 \sec x \tan x + \tan^2 x.$$

The first two terms on the right-hand side of this equation are old friends; we can integrate them at once. How about $\tan^2 x$? There is an identity that connects it with $\sec^2 x$:

$$\tan^2 x + 1 = \sec^2 x, \qquad \tan^2 x = \sec^2 x - 1.$$

We replace $\tan^2 x$ by $\sec^2 x - 1$ and get

$$\int (\sec x + \tan x)^2 \, dx = \int (\sec^2 x + 2 \sec x \tan x + \sec^2 x - 1) \, dx$$

$$= 2 \int \sec^2 x \, dx + 2 \int \sec x \tan x \, dx - \int 1 \, dx$$

$$= 2 \tan x + 2 \sec x - x + C.$$

Example 4   Eliminating a Square Root

Evaluate

$$\int_0^{\pi/4} \sqrt{1 + \cos 4x} \, dx.$$

Solution   We use the identity

$$\cos^2\theta = \frac{1 + \cos 2\theta}{2}, \qquad \text{or} \qquad 1 + \cos 2\theta = 2\cos^2\theta.$$

With $\theta = 2x$, this identity becomes

$$1 + \cos 4x = 2\cos^2 2x.$$

Hence,

$$\int_0^{\pi/4} \sqrt{1 + \cos 4x}\, dx = \int_0^{\pi/4} \sqrt{2}\,\sqrt{\cos^2 2x}\, dx$$

$$= \sqrt{2}\int_0^{\pi/4} |\cos 2x|\, dx \qquad \sqrt{u^2} = |u|$$

$$= \sqrt{2}\int_0^{\pi/4} \cos 2x\, dx \qquad \begin{array}{l}\text{On } [0, \pi/4], \cos 2x \geq 0,\\ \text{so } |\cos 2x| = \cos 2x.\end{array}$$

$$= \sqrt{2}\left[\frac{\sin 2x}{2}\right]_0^{\pi/4}$$

$$= \sqrt{2}\left[\frac{1}{2} - 0\right] = \frac{\sqrt{2}}{2}.$$

Example 5   Reducing an Improper Fraction

Evaluate

$$\int \frac{3x^2 - 7x}{3x + 2}\, dx.$$

Solution   The integrand is an improper fraction (degree of numerator greater than or equal to degree of denominator). To integrate it, we divide first, getting a quotient plus a remainder that is a proper fraction:

$$\frac{3x^2 - 7x}{3x + 2} = x - 3 + \frac{6}{3x + 2}.$$

$$\begin{array}{r} x - 3 \\ 3x + 2\overline{)3x^2 - 7x} \\ \underline{3x^2 + 2x} \\ -9x \\ \underline{-9x - 6} \\ +6 \end{array}$$

Therefore,

$$\int \frac{3x^2 - 7x}{3x + 2}\, dx = \int \left(x - 3 + \frac{6}{3x + 2}\right) dx = \frac{x^2}{2} - 3x + 2\ln|3x + 2| + C.$$

Reducing an improper fraction by long division (Example 5) does not always lead to an expression we can integrate directly. We see what to do about that in Section 7.3.

Example 6   Separating a Fraction

Evaluate

$$\int \frac{3x + 2}{\sqrt{1 - x^2}}\, dx.$$

Solution   We first separate the integrand to get

$$\int \frac{3x + 2}{\sqrt{1 - x^2}}\, dx = 3\int \frac{x\, dx}{\sqrt{1 - x^2}} + 2\int \frac{dx}{\sqrt{1 - x^2}}.$$

In the first of these new integrals, we substitute

$$u = 1 - x^2, \qquad du = -2x\,dx, \qquad \text{and} \qquad x\,dx = -\frac{1}{2}\,du.$$

$$3\int \frac{x\,dx}{\sqrt{1-x^2}} = 3\int \frac{(-1/2)\,du}{\sqrt{u}} = -\frac{3}{2}\int u^{-1/2}\,du$$

$$= -\frac{3}{2} \cdot \frac{u^{1/2}}{1/2} + C_1 = -3\sqrt{1-x^2} + C_1$$

The second of the new integrals is a standard form,

$$2\int \frac{dx}{\sqrt{1-x^2}} = 2\,\sin^{-1} x + C_2.$$

Combining these results and renaming $C_1 + C_2$ as C gives

$$\int \frac{3x+2}{\sqrt{1-x^2}}\,dx = -3\sqrt{1-x^2} + 2\,\sin^{-1} x + C.$$

Example 7 Multiplying by a Form of 1

Evaluate

$$\int \sec x\,dx.$$

Solution

$$\int \sec x\,dx = \int (\sec x)(1)\,dx = \int \sec x \cdot \frac{\sec x + \tan x}{\sec x + \tan x}\,dx$$

$$= \int \frac{\sec^2 x + \sec x \tan x}{\sec x + \tan x}\,dx$$

$$= \int \frac{du}{u} \qquad \begin{aligned} u &= \tan x + \sec x, \\ du &= (\sec^2 x + \sec x \tan x)\,dx \end{aligned}$$

$$= \ln |u| + C = \ln |\sec x + \tan x| + C.$$

With cosecants and cotangents in place of secants and tangents, the method of Example 7 leads to a companion formula for the integral of the cosecant (see Exercise 93).

Table 7.2 The secant and cosecant integrals
1. $\displaystyle\int \sec u\,du = \ln |\sec u + \tan u| + C$
2. $\displaystyle\int \csc u\,du = -\ln |\csc u + \cot u| + C$

Procedures for Matching Integrals to Basic Formulas

PROCEDURE	EXAMPLE
Making a simplifying substitution	$\dfrac{2x - 9}{\sqrt{x^2 - 9x + 1}}\, dx = \dfrac{du}{\sqrt{u}}$
Completing the square	$\sqrt{8x - x^2} = \sqrt{16 - (x - 4)^2}$
Using a trigonometric identity	$(\sec x + \tan x)^2 = \sec^2 x + 2 \sec x \tan x + \tan^2 x$
	$\qquad\qquad\quad = \sec^2 x + 2 \sec x \tan x + (\sec^2 x - 1)$
	$\qquad\qquad\quad = 2 \sec^2 x + 2 \sec x \tan x - 1$
Eliminating a square root	$\sqrt{1 + \cos 4x} = \sqrt{2 \cos^2 2x} = \sqrt{2}\,|\cos 2x|$
Reducing an improper fraction	$\dfrac{3x^2 - 7x}{3x + 2} = x - 3 + \dfrac{6}{3x + 2}$
Separating a fraction	$\dfrac{3x + 2}{\sqrt{1 - x^2}} = \dfrac{3x}{\sqrt{1 - x^2}} + \dfrac{2}{\sqrt{1 - x^2}}$
Multiplying by a form of 1	$\sec x = \sec x \cdot \dfrac{\sec x + \tan x}{\sec x + \tan x}$
	$\qquad = \dfrac{\sec^2 x + \sec x \tan x}{\sec x + \tan x}$

EXERCISES 7.1

Basic Substitutions

Evaluate each integral in Exercises 1–36 by using a substitution to reduce it to standard form.

1. $\displaystyle\int \frac{16x\, dx}{\sqrt{8x^2 + 1}}$

2. $\displaystyle\int \frac{3 \cos x\, dx}{\sqrt{1 + 3 \sin x}}$

3. $\displaystyle\int 3\sqrt{\sin v}\,\cos v\, dv$

4. $\displaystyle\int \cot^3 y\,\csc^2 y\, dy$

5. $\displaystyle\int_0^1 \frac{16x\, dx}{8x^2 + 2}$

6. $\displaystyle\int_{\pi/4}^{\pi/3} \frac{\sec^2 z}{\tan z}\, dz$

7. $\displaystyle\int \frac{dx}{\sqrt{x}\,(\sqrt{x} + 1)}$

8. $\displaystyle\int \frac{dx}{x - \sqrt{x}}$

9. $\displaystyle\int \cot (3 - 7x)\, dx$

10. $\displaystyle\int \csc (\pi x - 1)\, dx$

11. $\displaystyle\int e^\theta \csc (e^\theta + 1)\, d\theta$

12. $\displaystyle\int \frac{\cot (3 + \ln x)}{x}\, dx$

13. $\displaystyle\int \sec \frac{t}{3}\, dt$

14. $\displaystyle\int x \sec (x^2 - 5)\, dx$

15. $\displaystyle\int \csc (s - \pi)\, ds$

16. $\displaystyle\int \frac{1}{\theta^2} \csc \frac{1}{\theta}\, d\theta$

17. $\displaystyle\int_0^{\sqrt{\ln 2}} 2x\, e^{x^2}\, dx$

18. $\displaystyle\int_{\pi/2}^{\pi} (\sin y)\, e^{\cos y}\, dy$

19. $\displaystyle\int e^{\tan v} \sec^2 v\, dv$

20. $\displaystyle\int \frac{e^{\sqrt{t}}\, dt}{\sqrt{t}}$

21. $\displaystyle\int 3^{x+1}\, dx$

22. $\displaystyle\int \frac{2^{\ln x}}{x}\, dx$

23. $\displaystyle\int \frac{2^{\sqrt{w}}\, dw}{2\sqrt{w}}$

24. $\displaystyle\int 10^{2\theta}\, d\theta$

25. $\displaystyle\int \frac{9\, du}{1 + 9u^2}$

26. $\displaystyle\int \frac{4\, dx}{1 + (2x + 1)^2}$

27. $\displaystyle\int_0^{1/6} \frac{dx}{\sqrt{1 - 9x^2}}$

28. $\displaystyle\int_0^1 \frac{dt}{\sqrt{4 - t^2}}$

29. $\displaystyle\int \frac{2s\, ds}{\sqrt{1 - s^4}}$

30. $\displaystyle\int \frac{2\, dx}{x\sqrt{1 - 4 \ln^2 x}}$

31. $\displaystyle\int \frac{6\,dx}{x\sqrt{25x^2 - 1}}$

32. $\displaystyle\int \frac{dr}{r\sqrt{r^2 - 9}}$

33. $\displaystyle\int \frac{dx}{e^x + e^{-x}}$

34. $\displaystyle\int \frac{dy}{\sqrt{e^{2y} - 1}}$

35. $\displaystyle\int_1^{e^{\pi/3}} \frac{dx}{x \cos{(\ln x)}}$

36. $\displaystyle\int \frac{\ln x\,dx}{x + 4x \ln^2 x}$

Completing the Square

Evaluate each integral in Exercises 37–42 by completing the square and using a substitution to reduce it to standard form.

37. $\displaystyle\int_1^2 \frac{8\,dx}{x^2 - 2x + 2}$

38. $\displaystyle\int_2^4 \frac{2\,dx}{x^2 - 6x + 10}$

39. $\displaystyle\int \frac{dt}{\sqrt{-t^2 + 4t - 3}}$

40. $\displaystyle\int \frac{d\theta}{\sqrt{2\theta - \theta^2}}$

41. $\displaystyle\int \frac{dx}{(x + 1)\sqrt{x^2 + 2x}}$

42. $\displaystyle\int \frac{dx}{(x - 2)\sqrt{x^2 - 4x + 3}}$

Trigonometric Identities

Evaluate the integrals in Exercises 43–46 by using trigonometric identities and substitutions to reduce it to standard form.

43. $\displaystyle\int (\sec x + \cot x)^2\,dx$

44. $\displaystyle\int (\csc x - \tan x)^2\,dx$

45. $\displaystyle\int \csc x \sin 3x\,dx$

46. $\displaystyle\int (\sin 3x \cos 2x - \cos 3x \sin 2x)\,dx$

Improper Fractions

Evaluate each integral in Exercises 47–52 by reducing the improper fraction and using a substitution (if necessary) to reduce it to standard form.

47. $\displaystyle\int \frac{x}{x + 1}\,dx$

48. $\displaystyle\int \frac{x^2}{x^2 + 1}\,dx$

49. $\displaystyle\int_{\sqrt{2}}^3 \frac{2x^3}{x^2 - 1}\,dx$

50. $\displaystyle\int_{-1}^3 \frac{4x^2 - 7}{2x + 3}\,dx$

51. $\displaystyle\int \frac{4t^3 - t^2 + 16t}{t^2 + 4}\,dt$

52. $\displaystyle\int \frac{2\theta^3 - 7\theta^2 + 7\theta}{2\theta - 5}\,d\theta$

Separating Fractions

Evaluate each integral in Exercises 53–56 by separating the fraction and using a substitution (if necessary) to reduce it to standard form.

53. $\displaystyle\int \frac{1 - x}{\sqrt{1 - x^2}}\,dx$

54. $\displaystyle\int \frac{x + 2\sqrt{x - 1}}{2x\sqrt{x - 1}}\,dx$

55. $\displaystyle\int_0^{\pi/4} \frac{1 + \sin x}{\cos^2 x}\,dx$

56. $\displaystyle\int_0^{1/2} \frac{2 - 8x}{1 + 4x^2}\,dx$

Multiplying by a Form of 1

Evaluate each integral in Exercises 57–62 by multiplying by a form of 1 and using a substitution (if necessary) to reduce it to standard form.

57. $\displaystyle\int \frac{1}{1 + \sin x}\,dx$

58. $\displaystyle\int \frac{1}{1 + \cos x}\,dx$

59. $\displaystyle\int \frac{1}{\sec \theta + \tan \theta}\,d\theta$

60. $\displaystyle\int \frac{1}{\csc \theta + \cot \theta}\,d\theta$

61. $\displaystyle\int \frac{1}{1 - \sec x}\,dx$

62. $\displaystyle\int \frac{1}{1 - \csc x}\,dx$

Eliminating Square Roots

Evaluate each integral in Exercises 63–70 by eliminating the square root.

63. $\displaystyle\int_0^{2\pi} \sqrt{\frac{1 - \cos x}{2}}\,dx$

64. $\displaystyle\int_0^{\pi} \sqrt{1 - \cos 2x}\,dx$

65. $\displaystyle\int_{\pi/2}^{\pi} \sqrt{1 + \cos 2t}\,dt$

66. $\displaystyle\int_{-\pi}^0 \sqrt{1 + \cos t}\,dt$

67. $\displaystyle\int_{-\pi}^0 \sqrt{1 - \cos^2 \theta}\,d\theta$

68. $\displaystyle\int_{\pi/2}^{\pi} \sqrt{1 - \sin^2 \theta}\,d\theta$

69. $\displaystyle\int_{-\pi/4}^{\pi/4} \sqrt{1 + \tan^2 y}\,dy$

70. $\displaystyle\int_{-\pi/4}^0 \sqrt{\sec^2 y - 1}\,dy$

Assorted Integrations

Evaluate each integral in Exercises 71–82 by using any technique you think is appropriate.

71. $\displaystyle\int_{\pi/4}^{3\pi/4} (\csc x - \cot x)^2\,dx$

72. $\displaystyle\int_0^{\pi/4} (\sec x + 4 \cos x)^2\,dx$

73. $\displaystyle\int \cos \theta \csc (\sin \theta)\,d\theta$

74. $\displaystyle\int \left(1 + \frac{1}{x}\right) \cot (x + \ln x)\,dx$

75. $\displaystyle\int (\csc x - \sec x)(\sin x + \cos x)\,dx$

76. $\displaystyle\int 3 \sinh \left(\frac{x}{2} + \ln 5\right)\,dx$

77. $\displaystyle\int \frac{6\,dy}{\sqrt{y}\,(1 + y)}$

78. $\displaystyle\int \frac{dx}{x\sqrt{4x^2 - 1}}$

79. $\displaystyle\int \frac{7\,dx}{(x - 1)\sqrt{x^2 - 2x - 48}}$

80. $\displaystyle\int \frac{dx}{(2x + 1)\sqrt{4x^2 + 4x}}$

81. $\displaystyle\int \sec^2 t \tan (\tan t)\,dt$

82. $\displaystyle\int \frac{dx}{x\sqrt{3 + x^2}}$

Trigonometric Powers

83. **(a)** Evaluate $\int \cos^3 \theta\,d\theta$. (*Hint:* $\cos^2 \theta = 1 - \sin^2 \theta$.)

(b) Evaluate $\int \cos^5 \theta\,d\theta$.

(c) Without actually evaluating the integral, explain how you would evaluate $\int \cos^9 \theta\,d\theta$.

84. (a) Evaluate $\int \sin^3 \theta \, d\theta$. (*Hint:* $\sin^2 \theta = 1 - \cos^2 \theta$.)

(b) Evaluate $\int \sin^5 \theta \, d\theta$.

(c) Evaluate $\int \sin^7 \theta \, d\theta$.

(d) Without actually evaluating the integral, explain how you would evaluate $\int \sin^{13} \theta \, d\theta$.

85. (a) Express $\int \tan^3 \theta \, d\theta$ in terms of $\int \tan \theta \, d\theta$. Then evaluate $\int \tan^3 \theta \, d\theta$. (*Hint:* $\tan^2 \theta = \sec^2 \theta - 1$.)

(b) Express $\int \tan^5 \theta \, d\theta$ in terms of $\int \tan^3 \theta \, d\theta$.

(c) Express $\int \tan^7 \theta \, d\theta$ in terms of $\int \tan^5 \theta \, d\theta$.

(d) Express $\int \tan^{2k+1} \theta \, d\theta$, where k is a positive integer, in terms of $\int \tan^{2k-1} \theta \, d\theta$.

86. (a) Express $\int \cot^3 \theta \, d\theta$ in terms of $\int \cot \theta \, d\theta$. Then evaluate $\int \cot^3 \theta \, d\theta$. (*Hint:* $\cot^2 \theta = \csc^2 \theta - 1$.)

(b) Express $\int \cot^5 \theta \, d\theta$ in terms of $\int \cot^3 \theta \, d\theta$.

(c) Express $\int \cot^7 \theta \, d\theta$ in terms of $\int \cot^5 \theta \, d\theta$.

(d) Express $\int \cot^{2k+1} \theta \, d\theta$, where k is a positive integer, in terms of $\int \cot^{2k-1} \theta \, d\theta$.

Theory and Examples

87. *Area* Find the area of the region bounded above by $y = 2 \cos x$ and below by $y = \sec x$, $-\pi/4 \le x \le \pi/4$.

88. *Area* Find the area of the "triangular" region that is bounded from above and below by the curves $y = \csc x$ and $y = \sin x$, $\pi/6 \le x \le \pi/2$, and on the left by the line $x = \pi/6$.

89. *Volume* Find the volume of the solid generated by revolving the region in Exercise 87 about the x-axis.

90. *Volume* Find the volume of the solid generated by revolving the region in Exercise 88 about the x-axis.

91. *Arc length* Find the length of the curve $y = \ln (\cos x)$, $0 \le x \le \pi/3$.

92. *Arc length* Find the length of the curve $y = \ln (\sec x)$, $0 \le x \le \pi/4$.

93. *The integral of csc x* Repeat the derivation in Example 7, using cofunctions, to show that

$$\int \csc x \, dx = -\ln |\csc x + \cot x| + C.$$

94. *Using different substitutions* Show that the integral

$$\int ((x^2 - 1)(x + 1))^{-2/3} \, dx$$

can be evaluated with any of the following substitutions.

(a) $u = 1/(x + 1)$

(b) $u = ((x - 1)/(x + 1))^k$ for $k = 1, 1/2, 1/3, -1/3, -2/3$, and -1

(c) $u = \tan^{-1} x$

(d) $u = \tan^{-1} \sqrt{x}$ **(e)** $u = \tan^{-1} ((x - 1)/2)$

(f) $u = \cos^{-1} x$ **(g)** $u = \cosh^{-1} x$

What is the value of the integral? (*Source:* "Problems and Solutions," *College Mathematics Journal*, Vol. 21, No. 5 (Nov. 1990), pp. 425–426.)

7.2 Integration by Parts

Product Rule in Integral Form • Repeated Use • Solving for the Unknown Integral • Tabular Integration

Since

$$\int x \, dx = \frac{1}{2} x^2 + C$$

and

$$\int x^2 \, dx = \frac{1}{3} x^3 + C,$$

it is apparent that

$$\int x \cdot x \, dx \ne \int x \, dx \cdot \int x \, dx.$$

In other words, the integral of a product is generally *not* the product of the individual integrals:

$$\int f(x)g(x)\,dx \neq \int f(x)\,dx \cdot \int g(x)\,dx.$$

Integration by parts is a technique for simplifying integrals of the form

$$\int f(x)g(x)\,dx$$

in which f can be differentiated repeatedly and g can be integrated repeatedly without difficulty. The integral

$$\int xe^x\,dx$$

is such an integral because $f(x) = x$ can be differentiated twice to become zero and $g(x) = e^x$ can be integrated repeatedly without difficulty. Integration by parts also applies to integrals like

$$\int e^x \sin x\,dx$$

in which each part of the integrand appears again after repeated differentiation or integration.

In this section, we describe integration by parts and show how to apply it.

CD-ROM
WEBsite

Historical Biography

Charles Davies
(1798 — 1876)

Product Rule in Integral Form

When u and v are differentiable functions of x, the Product Rule for differentiation tells us that

$$\frac{d}{dx}(uv) = u\frac{dv}{dx} + v\frac{du}{dx}.$$

Integrating both sides with respect to x and rearranging leads to the integral equation

$$\int\left(u\frac{dv}{dx}\right)dx = \int\left(\frac{d}{dx}(uv)\right)dx - \int\left(v\frac{du}{dx}\right)dx$$

$$= uv - \int\left(v\frac{du}{dx}\right)dx.$$

When this equation is written in the simpler differential notation, we obtain the following formula.

Integration by Parts Formula

$$\int u\,dv = uv - \int v\,du \tag{1}$$

This formula expresses one integral, $\int u\,dv$, in terms of a second integral, $\int v\,du$. With a proper choice of u and v, the second integral may be easier to evaluate than the

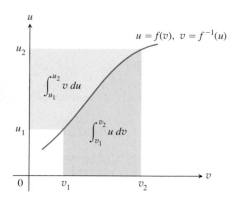

FIGURE 7.1 The area of the blue region, $\int_{v_1}^{v_2} u\,dv$, equals the area of the large rectangle, $u_2 v_2$, minus the areas of the small rectangle, $u_1 v_1$, and the sand-colored region, $\int_{u_1}^{u_2} v\,du$.

In symbols,

$$\int_{v_1}^{v_2} u\,dv = (u_2 v_2 - u_1 v_1) - \int_{u_1}^{u_2} v\,du.$$

first. This is the reason for the importance of the formula. When faced with an integral we cannot handle, we can replace it by one with which we might have more success.

The equivalent formula for definite integrals is

$$\int_{v_1}^{v_2} u\,dv = (u_2 v_2 - u_1 v_1) - \int_{u_1}^{u_2} v\,du. \tag{2}$$

Figure 7.1 shows how the different parts of the formula may be interpreted as areas.

Example 1   Using Integration by Parts

Evaluate

$$\int x \cos x\,dx.$$

Solution   We use the formula

$$\int u\,dv = uv - \int v\,du$$

with

$$u = x, \qquad dv = \cos x\,dx.$$

To complete the formula, we take the differential of u and find the simplest antiderivative of $\cos x$.

$$du = dx, \qquad v = \sin x$$

Then,

$$\int x \cos x\,dx = x \sin x - \int \sin x\,dx = x \sin x + \cos x + C.$$

Let's examine the choices available for u and v in Example 1.

Example 2   Investigating Integration by Parts

What are the choices for u and dv when we apply integration by parts to

$$\int x \cos x\,dx = \int u\,dv?$$

Which choices lead to a successful evaluation of the original integral?

Solution   There are four possible choices.

1. $u = 1$ and $dv = x \cos x\,dx$ **2.** $u = x$ and $dv = \cos x\,dx$

3. $u = x \cos x$ and $dv = dx$ **4.** $u = \cos x$ and $dv = x\,dx$

Choice 1 won't do because we still don't know how to integrate $dv = x \cos x\,dx$ to get v.

When and How to Use Integration by Parts

When: If substitution doesn't work, try integration by parts.

How: Start with an integral of the form

$$\int f(x)g(x)\,dx.$$

Match this with an integral of the form

$$\int u\,dv$$

by choosing dv to be part of the integrand including dx and possibly $f(x)$ or $g(x)$.

Guideline for choosing u and dv: The formula

$$\int u\,dv = uv - \int v\,du$$

gives a new integral on the right side of the equation. You must be able to readily integrate dv to get the right side. If the new integral is more complex than the original one, try a different choice for u and dv.

Choice 2 works well, as we saw in Example 1.

Choice 3 leads to

$$u = x\cos x, \qquad\qquad dv = dx,$$
$$du = (\cos x - x\sin x)\,dx, \qquad v = x,$$

and the new integral

$$\int v\,du = \int (x\cos x - x^2\sin x)\,dx.$$

This is worse than the integral we started with.

Choice 4 leads to

$$u = \cos x, \qquad\qquad dv = x\,dx,$$
$$du = -\sin x\,dx, \qquad v = \frac{x^2}{2},$$

and the new integral

$$\int v\,du = -\int \frac{x^2}{2}\sin x\,dx.$$

This, too, is worse.

The goal of integration by parts is to go from an integral $\int u\,dv$ that we don't see how to evaluate to an integral $\int v\,du$ that we can evaluate. Generally, you choose dv first to be as much of the integrand, including dx, as you can readily integrate; u is the leftover part. Keep in mind that integration by parts does not always work.

Example 3   Finding Area

Find the area of the region bounded by the curve $y = xe^{-x}$ and the x-axis from $x = 0$ to $x = 4$.

Solution   The region is shaded in Figure 7.2. Its area is

$$\int_0^4 xe^{-x}\,dx.$$

We use the formula $\int u\,dv = uv - \int v\,du$ with

$$u = x, \qquad dv = e^{-x}\,dx,$$
$$du = dx, \qquad v = -e^{-x}.$$

Then

$$\int xe^{-x}\,dx = -xe^{-x} - \int (-e^{-x})\,dx$$
$$= -xe^{-x} + \int e^{-x}\,dx$$
$$= -xe^{-x} - e^{-x} + C.$$

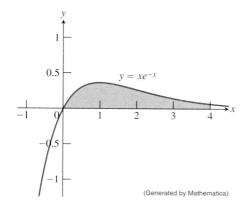

(Generated by Mathematica)

FIGURE 7.2 The region in Example 3.

Thus,

$$\int_0^4 xe^{-x} \, dx = [-xe^{-x} - e^{-x}]_0^4$$

$$= (-4e^{-4} - e^{-4}) - (-e^0) = 1 - 5e^{-4} \approx 0.91.$$

Example 4   Integral of the Natural Logarithm

Find

$$\int \ln x \, dx.$$

Solution   Since $\int \ln x \, dx$ can be written as $\int \ln x \cdot 1 \, dx$, we use the formula $\int u \, dv = uv - \int v \, du$ with

$u = \ln x$ Simplifies when differentiated $dv = dx$ Easy to integrate

$du = \dfrac{1}{x} \, dx,$ $v = x.$ Simplest antiderivative

Then

$$\int \ln x \, dx = x \ln x - \int x \cdot \frac{1}{x} \, dx = x \ln x - \int dx = x \ln x - x + C.$$

Repeated Use

Sometimes we have to use integration by parts more than once.

Example 5   Repeated Use of Integration by Parts

Evaluate

$$\int x^2 e^x \, dx.$$

Solution   With $u = x^2$, $dv = e^x \, dx$, $du = 2x \, dx$, and $v = e^x$, we have

$$\int x^2 e^x \, dx = x^2 e^x - 2 \int xe^x \, dx.$$

The new integral is less complicated than the original because the exponent on x is reduced by one. To evaluate the integral on the right, we integrate by parts again with $u = x$, $dv = e^x \, dx$. Then $du = dx$, $v = e^x$, and

$$\int xe^x \, dx = xe^x - \int e^x \, dx = xe^x - e^x + C.$$

Hence,

$$\int x^2 e^x \, dx = x^2 e^x - 2 \int xe^x \, dx$$

$$= x^2 e^x - 2xe^x + 2e^x + C.$$

The technique of Example 5 works for any integral $\int x^n e^x \, dx$ in which n is a positive integer, because differentiating x^n will eventually lead to zero and integrating e^x is easy. We say more on this later in this section when we discuss *tabular integration*.

Solving for the Unknown Integral

Integrals like the one in the next example occur in electrical engineering. Their evaluation requires two integrations by parts, followed by solving for the unknown integral.

Example 6   Solving for the Unknown Integral

Evaluate

$$\int e^x \cos x \, dx.$$

Solution   Let $u = e^x$ and $dv = \cos x \, dx$. Then $du = e^x \, dx$, $v = \sin x$, and

$$\int e^x \cos x \, dx = e^x \sin x - \int e^x \sin x \, dx.$$

The second integral is like the first except that it has $\sin x$ in place of $\cos x$. To evaluate it, we use integration by parts with

$$u = e^x, \qquad dv = \sin x \, dx, \qquad v = -\cos x, \qquad du = e^x \, dx.$$

Then

$$\int e^x \cos x \, dx = e^x \sin x - \left(-e^x \cos x - \int (-\cos x)(e^x \, dx) \right)$$

$$= e^x \sin x + e^x \cos x - \int e^x \cos x \, dx.$$

The unknown integral now appears on both sides of the equation. Adding the integral to both sides gives

$$2 \int e^x \cos x \, dx = e^x \sin x + e^x \cos x + C.$$

Dividing by 2 and renaming the constant of integration gives

$$\int e^x \cos x \, dx = \frac{e^x \sin x + e^x \cos x}{2} + C.$$

When making repeated use of integration by parts in circumstances like Example 6, once a choice for u and dv is made, it is usually not a good idea to switch choices in the second stage of the problem. Doing so may result in undoing the work. For example, if we had switched to the substitution $u = \sin x$, $dv = e^x \, dx$ in the second integration, we would have obtained

$$\int e^x \cos x \, dx = e^x \sin x - \left(e^x \sin x - \int e^x \cos x \, dx \right)$$

$$= \int e^x \cos x \, dx,$$

undoing the first integration by parts. The tabular integration technique introduced next prevents this mistake.

Tabular Integration

We have seen that integrals of the form $\int f(x)g(x)\,dx$, in which f can be differentiated repeatedly to become zero and g can be integrated repeatedly without difficulty, are natural candidates for integration by parts. If many repetitions are required, however, the calculations can be cumbersome. In situations like this, there is a way to organize the calculations that saves a great deal of work. It is **tabular integration,** as shown in Examples 7 and 8.

Example 7   Using Tabular Integration

Evaluate

$$\int x^2 e^x \, dx.$$

Solution   With $f(x) = x^2$ and $g(x) = e^x$, we list:

$f(x)$ and its derivatives		$g(x)$ and its integrals
x^2	$(+)$	e^x
$2x$	$(-)$	e^x
2	$(+)$	e^x
0		e^x

We combine the products of the functions connected by the arrows according to the operation signs above the arrows to obtain

$$\int x^2 e^x \, dx = x^2 e^x - 2xe^x + 2e^x + C.$$

Compare this with the result in Example 5.

Example 8   Using Tabular Integration

Evaluate

$$\int x^3 \sin x \, dx.$$

Solution   With $f(x) = x^3$ and $g(x) = \sin x$, we list:

$f(x)$ and its derivatives		$g(x)$ and its integrals
x^3	$(+)$	$\sin x$
$3x^2$	$(-)$	$-\cos x$
$6x$	$(+)$	$-\sin x$
6	$(-)$	$\cos x$
0		$\sin x$

For more about tabular integration, see the Additional Exercises at the end of this chapter.

Again we combine the products of the functions connected by the arrows according to the operation signs above the arrows to obtain

$$\int x^3 \sin x \, dx = -x^3 \cos x + 3x^2 \sin x + 6x \cos x - 6 \sin x + C.$$

EXERCISES 7.2

Integration by Parts

Evaluate the integrals in Exercises 1–24.

1. $\displaystyle\int x \sin \frac{x}{2} \, dx$

2. $\displaystyle\int \theta \cos \pi\theta \, d\theta$

3. $\displaystyle\int t^2 \cos t \, dt$

4. $\displaystyle\int x^2 \sin x \, dx$

5. $\displaystyle\int_1^2 x \ln x \, dx$

6. $\displaystyle\int_1^e x^3 \ln x \, dx$

7. $\displaystyle\int \tan^{-1} y \, dy$

8. $\displaystyle\int \sin^{-1} y \, dy$

9. $\displaystyle\int x \sec^2 x \, dx$

10. $\displaystyle\int 4x \sec^2 2x \, dx$

11. $\displaystyle\int x^3 e^x \, dx$

12. $\displaystyle\int p^4 e^{-p} \, dp$

13. $\displaystyle\int (x^2 - 5x)e^x \, dx$

14. $\displaystyle\int (r^2 + r + 1)e^r \, dr$

15. $\displaystyle\int x^5 e^x \, dx$

16. $\displaystyle\int t^2 e^{4t} \, dt$

17. $\displaystyle\int_0^{\pi/2} \theta^2 \sin 2\theta \, d\theta$

18. $\displaystyle\int_0^{\pi/2} x^3 \cos 2x \, dx$

19. $\displaystyle\int_{2/\sqrt{3}}^2 t \sec^{-1} t \, dt$

20. $\displaystyle\int_0^{1/\sqrt{2}} 2x \sin^{-1} (x^2) \, dx$

21. $\displaystyle\int e^\theta \sin \theta \, d\theta$

22. $\displaystyle\int e^{-y} \cos y \, dy$

23. $\displaystyle\int e^{2x} \cos 3x \, dx$

24. $\displaystyle\int e^{-2x} \sin 2x \, dx$

Substitution and Integration by Parts

Evaluate the integrals in Exercises 25–30 by using a substitution prior to integration by parts.

25. $\displaystyle\int e^{\sqrt{3s+9}} \, ds$

26. $\displaystyle\int_0^1 x\sqrt{1-x} \, dx$

27. $\displaystyle\int_0^{\pi/3} x \tan^2 x \, dx$

28. $\displaystyle\int \ln (x + x^2) \, dx$

29. $\displaystyle\int \sin (\ln x) \, dx$

30. $\displaystyle\int z (\ln z)^2 \, dz$

Differential Equations

In Exercises 31–34, solve the differential equation.

31. $\dfrac{dy}{dx} = x^2 e^{4x}$

32. $\dfrac{dy}{dx} = x^2 \ln x$

33. $\dfrac{dy}{d\theta} = \sin \sqrt{\theta}$

34. $\dfrac{dy}{d\theta} = \theta \sec \theta \tan \theta$

Theory and Examples

35. *Finding area* Find the area of the region enclosed by the curve $y = x \sin x$ and the x-axis (see the accompanying figure) for

   (a) $0 \le x \le \pi$

   (b) $\pi \le x \le 2\pi$

   (c) $2\pi \le x \le 3\pi$.

   (d) What pattern do you see here? What is the area between the curve and the x-axis for $n\pi \le x \le (n + 1)\pi$, n an arbitrary nonnegative integer? Give reasons for your answer.

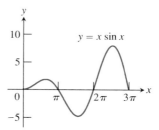

36. *Finding area* Find the area of the region enclosed by the curve $y = x \cos x$ and the x-axis (see the accompanying figure) for

   (a) $\pi/2 \le x \le 3\pi/2$

   (b) $3\pi/2 \le x \le 5\pi/2$

   (c) $5\pi/2 \le x \le 7\pi/2$.

   (d) What pattern do you see? What is the area between the curve and the x-axis for

$$\left(\frac{2n - 1}{2}\right)\pi \le x \le \left(\frac{2n + 1}{2}\right)\pi,$$

   n an arbitrary positive integer? Give reasons for your answer.

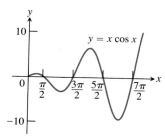

37. *Finding volume* Find the volume of the solid generated by revolving the region in the first quadrant bounded by the coordinate axes, the curve $y = e^x$, and the line $x = \ln 2$ about the line $x = \ln 2$.

38. *Finding volume* Find the volume of the solid generated by revolving the region in the first quadrant bounded by the coordinate axes, the curve $y = e^{-x}$, and the line $x = 1$

(a) about the y-axis,

(b) about the line $x = 1$.

39. *Finding volume* Find the volume of the solid generated by revolving the region in the first quadrant bounded by the coordinate axes and the curve $y = \cos x$, $0 \le x \le \pi/2$, about

(a) the y-axis

(b) the line $x = \pi/2$.

40. *Finding volume* Find the volume of the solid generated by revolving the region bounded by the x-axis and the curve $y = x \sin x$, $0 \le x \le \pi$, about

(a) the y-axis

(b) the line $x = \pi$.

(See Exercise 35 for a graph.)

41. *Average value* A retarding force, symbolized by the dashpot in the figure, slows the motion of the weighted spring so that the mass's position at time t is

$$y = 2e^{-t} \cos t, \qquad t \ge 0.$$

Find the average value of y over the interval $0 \le t \le 2\pi$.

Mass

Dashpot

42. *Average value* In a mass-spring-dashpot system like the one in Exercise 41, the mass's position at time t is

$$y = 4e^{-t}(\sin t - \cos t), \qquad t \ge 0.$$

Find the average value of y over the interval $0 \le t \le 2\pi$.

Reduction Formulas

In Exercises 43–46, use integration by parts to establish the *reduction formula*.

43. $\displaystyle \int x^n \cos x \, dx = x^n \sin x - n \int x^{n-1} \sin x \, dx$

44. $\displaystyle \int x^n \sin x \, dx = -x^n \cos x + n \int x^{n-1} \cos x \, dx$

45. $\displaystyle \int x^n e^{ax} \, dx = \frac{x^n e^{ax}}{a} - \frac{n}{a} \int x^{n-1} e^{ax} \, dx, \quad a \ne 0$

46. $\displaystyle \int (\ln x)^n \, dx = x (\ln x)^n - n \int (\ln x)^{n-1} \, dx$

Integrating Inverses of Functions

47. *Integrating inverse functions* Assume that the function f has an inverse.

(a) Show that

$$\int f^{-1}(x) \, dx = \int y f'(y) \, dy.$$

(*Hint:* Use the substitution $y = f^{-1}(x)$.)

(b) Use integration by parts on the second integral in part (a) to show that

$$\int f^{-1}(x) \, dx = \int y f'(y) \, dy = x f^{-1}(x) - \int f(y) \, dy.$$

48. *Integrating inverse functions* Assume that the function f has an inverse. Use integration by parts directly to show that

$$\int f^{-1}(x) \, dx = x f^{-1}(x) - \int x \left(\frac{d}{dx} f^{-1}(x) \right) dx.$$

In Exercises 49–52, evaluate the integral using

(a) the technique of Exercise 47

(b) the technique of Exercise 48.

(c) Show that the expressions (with $C = 0$) obtained in parts (a) and (b) are the same.

49. $\displaystyle \int \sin^{-1} x \, dx$

50. $\displaystyle \int \tan^{-1} x \, dx$

51. $\displaystyle \int \cos^{-1} x \, dx$

52. $\displaystyle \int \log_2 x \, dx$

7.3 Partial Fractions

Partial Fractions • General Description of the Method • The Heaviside "Cover-up" Method for Linear Factors • Other Ways to Determine the Coefficients

In studying population modeling in Example 6 (Section 6.4 or 6.6), we solved the logistic differential equation

$$\frac{dP}{dt} = 0.001P(100 - P)$$

by rewriting it as

$$\frac{100}{P(100 - P)} \, dP = 0.1 \, dt, \qquad \text{Variables separated}$$

expanding the fraction on the left into two basic fractions,

$$\frac{100}{P(100 - P)} = \frac{1}{P} + \frac{1}{100 - P},$$

and integrating both sides to find the solution

$$\ln |P| - \ln |100 - P| = 0.1t + C.$$

This expansion technique is the **method of partial fractions.** Any rational function can be written as a sum of basic fractions, called **partial fractions,** using the method of partial fractions. We can then integrate the rational function by integrating the sum of partial fractions instead.

Partial Fractions

We sum algebraic fractions by finding a common denominator, summing the resulting fractions, and finally simplifying. For instance,

$$\frac{2}{x + 1} + \frac{3}{x - 3} = \frac{2(x - 3)}{(x + 1)(x - 3)} + \frac{3(x + 1)}{(x - 3)(x + 1)}$$

$$= \frac{2x - 6 + 3x + 3}{x^2 - 2x - 3}$$

$$= \frac{5x - 3}{x^2 - 2x - 3}.$$

It is easy to find the integral

$$\int \frac{5x - 3}{x^2 - 2x - 3} \, dx$$

if we can "reverse" the process above and obtain

$$\int \frac{5x - 3}{x^2 - 2x - 3} \, dx = \int \frac{2}{x + 1} \, dx + \int \frac{3}{x - 3} \, dx$$

$$= 2 \ln |x + 1| + 3 \ln |x - 3| + C.$$

More generally, a theorem from advanced algebra (mentioned later in more detail) says that every rational function, no matter how complicated, can be rewritten as a sum of simpler fractions (that we can integrate with techniques we already know). Let's see how this simpler sum can be found by the method of partial fractions applied to our preceding example.

Example 1   Using Partial Fractions

Use the method of partial fractions to evaluate

$$\int \frac{5x - 3}{x^2 - 2x - 3}\, dx.$$

Undetermined Coefficients

The A and B in the partial fraction decomposition are referred to as **undetermined coefficients.**

Solution   First we factor the denominator: $x^2 - 2x - 3 = (x + 1)(x - 3)$. Then we determine the values of A and B so that

$$\frac{5x - 3}{x^2 - 2x - 3} = \frac{A}{x + 1} + \frac{B}{x - 3}.$$

We proceed by clearing the fractions:

$$5x - 3 = A(x - 3) + B(x + 1) \qquad \text{Multiply both sides of the equation by } (x + 1)(x - 3).$$
$$= (A + B)x - 3A + B. \qquad \text{Combine terms.}$$

Now equate the coefficients to obtain the following system of linear equations.

$$A + B = 5$$
$$-3A + B = -3$$

Solving these equations simultaneously yields $A = 2$ and $B = 3$. Therefore,

$$\int \frac{5x - 3}{x^2 - 2x - 3}\, dx = \int \frac{2}{x + 1}\, dx + \int \frac{3}{x - 3}\, dx$$
$$= 2 \ln |x + 1| + 3 \ln |x - 3| + C.$$

General Description of the Method

Success in writing a rational function $f(x)/g(x)$ as a sum of partial fractions depends on two things:

- *The degree of $f(x)$ must be less than the degree of $g(x)$.* That is, the fraction must be *proper*. If it isn't, divide $f(x)$ by $g(x)$ and work with the remainder term. See Example 4 of this section.

- *We must know the factors of $g(x)$.* In theory, any polynomial with real coefficients can be written as a product of real linear factors and real quadratic factors. In practice, the factors may be hard to find.

Here is how we find the partial fractions of a proper fraction $f(x)/g(x)$ when the factors of g are known.

Method of Partial Fractions (f(x)/g(x) Proper)

Step 1: Let $x - r$ be a linear factor of $g(x)$. Suppose that $(x - r)^m$ is the highest power of $x - r$ that divides $g(x)$. Then, to this factor, assign the sum of the m partial fractions:

$$\frac{A_1}{x - r} + \frac{A_2}{(x - r)^2} + \cdots + \frac{A_m}{(x - r)^m}.$$

Do this for each distinct linear factor of $g(x)$.

Step 2: Let $x^2 + px + q$ be a quadratic factor of $g(x)$. Suppose that $(x^2 + px + q)^n$ is the highest power of this factor that divides $g(x)$. Then, to this factor, assign the sum of the n partial fractions:

$$\frac{B_1 x + C_1}{x^2 + px + q} + \frac{B_2 x + C_2}{(x^2 + px + q)^2} + \cdots + \frac{B_n x + C_n}{(x^2 + px + q)^n}.$$

Do this for each distinct quadratic factor of $g(x)$ that cannot be factored into linear factors with real coefficients.

Step 3: Set the original fraction $f(x)/g(x)$ equal to the sum of all these partial fractions. Clear the resulting equation of fractions and arrange the terms in decreasing powers of x.

Step 4: Equate the coefficients of corresponding powers of x and solve the resulting equations for the undetermined coefficients.

Example 2   Using a Repeated Linear Factor

Express as a sum of partial fractions:

$$\frac{6x + 7}{(x + 2)^2}.$$

Solution   According to the description above, we must express the fraction as a sum of partial fractions with undetermined coefficients.

$$\frac{6x + 7}{(x + 2)^2} = \frac{A}{x + 2} + \frac{B}{(x + 2)^2}$$

$$6x + 7 = A(x + 2) + B \qquad \text{Multiply both sides by } (x + 2)^2.$$

$$= Ax + (2A + B) \qquad \text{Combine terms.}$$

Equating coefficients of corresponding powers of x gives

$$A = 6 \qquad \text{and} \qquad 2A + B = 12 + B = 7, \qquad \text{or}$$

$$A = 6 \qquad \text{and} \qquad B = -5.$$

Therefore,

$$\frac{6x + 7}{(x + 2)^2} = \frac{6}{x + 2} - \frac{5}{(x + 2)^2}.$$

Example 3 Using Partial Fractions

Evaluate

$$\int \frac{6x + 7}{(x + 2)^2} \, dx.$$

Solution

$$\int \frac{6x + 7}{(x + 2)^2} \, dx = \int \left(\frac{6}{x + 2} - \frac{5}{(x + 2)^2} \right) dx \qquad \text{Example 2}$$

$$= 6 \int \frac{dx}{x + 2} - 5 \int (x + 2)^{-2} \, dx$$

$$= 6 \ln |x + 2| + 5(x + 2)^{-1} + C$$

Example 4 Integrating an Improper Fraction

Evaluate

$$\int \frac{2x^3 - 4x^2 - x - 3}{x^2 - 2x - 3} \, dx.$$

Solution First we divide the denominator into the numerator to get a polynomial plus a proper fraction.

$$
\begin{array}{r}
2x \\
x^2 - 2x - 3 \overline{\smash{\big)}\ 2x^3 - 4x^2 - x - 3} \\
\underline{2x^3 - 4x^2 - 6x} \\
5x - 3
\end{array}
$$

Then we write the improper fraction as a polynomial plus a proper fraction.

$$\frac{2x^3 - 4x^2 - x - 3}{x^2 - 2x - 3} = 2x + \frac{5x - 3}{x^2 - 2x - 3}$$

Finally, using $\int 2x \, dx = x^2$ and Example 1 we obtain

$$\int \frac{2x^3 - 4x^2 - x - 3}{x^2 - 2x - 3} \, dx = \int 2x \, dx + \int \frac{5x - 3}{x^2 - 2x - 3} \, dx$$

$$= x^2 + 2 \ln |x + 1| + 3 \ln |x - 3| + C.$$

Example 5 Solving an Initial Value Problem

Find the solution to $dy/dx = 2xy \, (y^2 + 1)$ that satisfies $y(0) = 1$.

Solution Separating the variables, we rewrite the differential equation as

$$\frac{1}{y(y^2 + 1)} \, dy = 2x \, dx.$$

Integrating both sides gives

$$\int \frac{1}{y(y^2 + 1)} \, dy = \int 2x \, dx = x^2 + C_1.$$

We use partial fractions to rewrite the integrand on the left.

$$\frac{1}{y(y^2 + 1)} = \frac{A}{y} + \frac{By + C}{y^2 + 1}$$

Notice the numerator over $y^2 + 1$: For quadratic factors, we use first degree numerators, not constant numerators. Clearing the equations of fractions gives

$$1 = A(y^2 + 1) + (By + C)y \qquad \text{Multiply by } y(y^2 + 1).$$
$$= (A + B)y^2 + Cy + A$$

Equating coefficients of like terms gives $A + B = 0$, $C = 0$, and $A = 1$. Solving these equations simultaneously, we find $A = 1$, $B = -1$, and $C = 0$. Accordingly,

$$\int \frac{1}{y(y^2 + 1)} \, dy = \int \frac{1}{y} \, dy - \int \frac{y}{y^2 + 1} \, dy$$
$$= \ln |y| - \frac{1}{2} \ln (y^2 + 1) + C_2.$$

The solution to the differential equation is

$$\ln |y| - \frac{1}{2} \ln (y^2 + 1) = x^2 + C. \qquad C = C_1 - C_2$$

Substituting $x = 0$ and $y = 1$, we find

$$0 - \frac{1}{2} \ln 2 = C, \qquad \text{or} \qquad C = -\ln \sqrt{2}.$$

The solution to the initial value problem is

$$\ln |y| - \frac{1}{2} \ln (y^2 + 1) = x^2 - \ln \sqrt{2}.$$

Example 6   Integrating with an Irreducible Quadratic Factor in the Denominator

Evaluate

$$\int \frac{-2x + 4}{(x^2 + 1)(x - 1)^2} \, dx$$

using partial fractions.

A quadratic polynomial is **irreducible** if it cannot be written as the product of two linear factors with real coefficients.

Solution   The denominator has an irreducible quadratic factor as well as a repeated linear factor, so we write

$$\frac{-2x + 4}{(x^2 + 1)(x - 1)^2} = \frac{Ax + B}{x^2 + 1} + \frac{C}{x - 1} + \frac{D}{(x - 1)^2}. \qquad (1)$$

Clearing the equation of fractions gives

$$-2x + 4 = (Ax + B)(x - 1)^2 + C(x - 1)(x^2 + 1) + D(x^2 + 1)$$
$$= (A + C)x^3 + (-2A + B - C + D)x^2$$
$$+ (A - 2B + C)x + (B - C + D).$$

Equating coefficients of like terms gives

Coefficients of x^3:	$0 = A + C$
Coefficients of x^2:	$0 = -2A + B - C + D$
Coefficients of x^1:	$-2 = A - 2B + C$
Coefficients of x^0:	$4 = B - C + D$

We solve these equations simultaneously to find the values of A, B, C, and D:

$$-4 = -2A, \qquad A = 2 \qquad \text{Subtract fourth equation from second.}$$
$$C = -A = -2 \qquad \text{From the first equation}$$
$$B = 1 \qquad \text{$A = 2$ and $C = -2$ in third equation.}$$
$$D = 4 - B + C = 1. \qquad \text{From the fourth equation}$$

We substitute these values into Equation (1), obtaining

$$\frac{-2x + 4}{(x^2 + 1)(x - 1)^2} = \frac{2x + 1}{x^2 + 1} - \frac{2}{x - 1} + \frac{1}{(x - 1)^2} .$$

Finally, using the expansion above we can integrate:

$$\int \frac{-2x + 4}{(x^2 + 1)(x - 1)^2} \, dx = \int \left(\frac{2x + 1}{x^2 + 1} - \frac{2}{x - 1} + \frac{1}{(x - 1)^2} \right) dx$$

$$= \int \left(\frac{2x}{x^2 + 1} + \frac{1}{x^2 + 1} - \frac{2}{x - 1} + \frac{1}{(x - 1)^2} \right) dx$$

$$= \ln \, (x^2 + 1) + \tan^{-1} x - 2 \ln |x - 1| - \frac{1}{x - 1} + C.$$

The Heaviside "Cover-up" Method for Linear Factors

When the degree of the polynomial $f(x)$ is less than the degree of $g(x)$ and

$$g(x) = (x - r_1)(x - r_2) \cdots (x - r_n)$$

is a product of n distinct linear factors, each raised to the first power, there is a quick way to expand $f(x)/g(x)$ by partial fractions.

Example 7   Using the Heaviside Method

Find A, B, and C in the partial-fraction expansion

$$\frac{x^2 + 1}{(x - 1)(x - 2)(x - 3)} = \frac{A}{x - 1} + \frac{B}{x - 2} + \frac{C}{x - 3} . \qquad (2)$$

Solution   If we multiply both sides of Equation (2) by $(x - 1)$ to get

$$\frac{x^2 + 1}{(x - 2)(x - 3)} = A + \frac{B(x - 1)}{x - 2} + \frac{C(x - 1)}{x - 3}$$

and set $x = 1$, the resulting equation gives the value of A:

$$\frac{(1)^2 + 1}{(1 - 2)(1 - 3)} = A + 0 + 0,$$

$$A = 1.$$

Thus, the value of A is the number we would have obtained if we had covered the factor $(x - 1)$ in the denominator of the original fraction

$$\frac{x^2 + 1}{(x - 1)(x - 2)(x - 3)} \tag{3}$$

and evaluated the rest at $x = 1$:

$$A = \frac{(1)^2 + 1}{\boxed{(x - 1)} \ (1 - 2)(1 - 3)} = \frac{2}{(-1)(-2)} = 1.$$
$$\underset{\text{Cover}}{\Uparrow}$$

Similarly, we find the value of B in Equation (2) by covering the factor $(x - 2)$ in Equation (3) and evaluating the rest at $x = 2$:

$$B = \frac{(2)^2 + 1}{(2 - 1) \ \boxed{(x - 2)} \ (2 - 3)} = \frac{5}{(1)(-1)} = -5.$$
$$\underset{\text{Cover}}{\Uparrow}$$

Finally, C is found by covering the $(x - 3)$ in Equation (3) and evaluating the rest at $x = 3$:

$$C = \frac{(3)^2 + 1}{(3 - 1)(3 - 2) \ \boxed{(x - 3)}} = \frac{10}{(2)(1)} = 5.$$
$$\underset{\text{Cover}}{\Uparrow}$$

Heaviside Method

Step 1: *Write the quotient with g(x) factored:*

$$\frac{f(x)}{g(x)} = \frac{f(x)}{(x - r_1)(x - r_2) \cdots (x - r_n)}.$$

Step 2: *Cover the factors $(x - r_i)$ of g(x) one at a time,* each time replacing all the uncovered x's by the number r_i. This gives a number A_i for each root r_i:

$$A_1 = \frac{f(r_1)}{(r_1 - r_2) \cdots (r_1 - r_n)}$$

$$A_2 = \frac{f(r_2)}{(r_2 - r_1)(r_2 - r_3) \cdots (r_2 - r_n)}$$

$$\vdots$$

$$A_n = \frac{f(r_n)}{(r_n - r_1)(r_n - r_2) \cdots (r_n - r_{n-1})}.$$

Step 3: *Write the partial-fraction expansion of f(x)/g(x) as*

$$\frac{f(x)}{g(x)} = \frac{A_1}{(x - r_1)} + \frac{A_2}{(x - r_2)} + \cdots + \frac{A_n}{(x - r_n)}.$$

Example 8   Integrating with the Heaviside Method

Evaluate

$$\int \frac{x + 4}{x^3 + 3x^2 - 10x} \, dx.$$

Solution   The degree of $f(x) = x + 4$ is less than the degree of $g(x) = x^3 + 3x^2 - 10x$, and, with $g(x)$ factored,

$$\frac{x + 4}{x^3 + 3x^2 - 10x} = \frac{x + 4}{x(x - 2)(x + 5)}.$$

The roots of $g(x)$ are $r_1 = 0$, $r_2 = 2$, and $r_3 = -5$. We find

$$A_1 = \frac{0 + 4}{\boxed{x} \; (0 - 2)(0 + 5)} = \frac{4}{(-2)(5)} = -\frac{2}{5}$$

$$\underset{\text{Cover}}{\Uparrow}$$

$$A_2 = \frac{2 + 4}{2 \; \boxed{(x - 2)} \; (2 + 5)} = \frac{6}{(2)(7)} = \frac{3}{7}$$

$$\underset{\text{Cover}}{\Uparrow}$$

$$A_3 = \frac{-5 + 4}{(-5)(-5 - 2) \; \boxed{(x + 5)}} = \frac{-1}{(-5)(-7)} = -\frac{1}{35}.$$

$$\underset{\text{Cover}}{\Uparrow}$$

Therefore,

$$\frac{x + 4}{x(x - 2)(x + 5)} = -\frac{2}{5x} + \frac{3}{7(x - 2)} - \frac{1}{35(x + 5)},$$

and

$$\int \frac{x + 4}{x(x - 2)(x + 5)} \, dx = -\frac{2}{5} \ln |x| + \frac{3}{7} \ln |x - 2| - \frac{1}{35} \ln |x + 5| + C.$$

Other Ways to Determine the Coefficients

Another way to determine the constants that appear in partial fractions is to differentiate, as in the next example. Still another is to assign selected numerical values to x.

Example 9   Using Differentiation

Find A, B, and C in the equation

$$\frac{x - 1}{(x + 1)^3} = \frac{A}{x + 1} + \frac{B}{(x + 1)^2} + \frac{C}{(x + 1)^3}.$$

Solution   We first clear fractions:

$$x - 1 = A(x + 1)^2 + B(x + 1) + C.$$

Substituting $x = -1$ shows $C = -2$. We then differentiate both sides with respect to x, obtaining

$$1 = 2A(x + 1) + B.$$

Substituting $x = -1$ shows $B = 1$. We differentiate again to get $0 = 2A$, which shows $A = 0$. Hence,

$$\frac{x - 1}{(x + 1)^3} = \frac{1}{(x + 1)^2} - \frac{2}{(x + 1)^3}.$$

In some problems, assigning small values to x such as $x = 0, \pm 1, \pm 2$, to get equations in $A, B,$ and C provides a fast alternative to other methods.

Example 10 Assigning Numerical Values to x

Find $A, B,$ and C in

$$\frac{x^2 + 1}{(x - 1)(x - 2)(x - 3)} = \frac{A}{x - 1} + \frac{B}{x - 2} + \frac{C}{x - 3}.$$

Solution Clear fractions to get

$$x^2 + 1 = A(x - 2)(x - 3) + B(x - 1)(x - 3) + C(x - 1)(x - 2).$$

Then let $x = 1, 2, 3$ successively to find $A, B,$ and C:

$$
\begin{aligned}
x = 1: \quad & (1)^2 + 1 = A(-1)(-2) + B(0) + C(0) \\
& 2 = 2A \\
& A = 1 \\
x = 2: \quad & (2)^2 + 1 = A(0) + B(1)(-1) + C(0) \\
& 5 = -B \\
& B = -5 \\
x = 3: \quad & (3)^2 + 1 = A(0) + B(0) + C(2)(1) \\
& 10 = 2C \\
& C = 5.
\end{aligned}
$$

Conclusion:

$$\frac{x^2 + 1}{(x - 1)(x - 2)(x - 3)} = \frac{1}{x - 1} - \frac{5}{x - 2} + \frac{5}{x - 3}.$$

EXERCISES 7.3

Expanding Quotients into Partial Fractions

Expand the quotients in Exercises 1–8 by partial fractions.

1. $\dfrac{5x - 13}{(x - 3)(x - 2)}$

2. $\dfrac{5x - 7}{x^2 - 3x + 2}$

3. $\dfrac{x + 4}{(x + 1)^2}$

4. $\dfrac{2x + 2}{x^2 - 2x + 1}$

5. $\dfrac{z + 1}{z^2(z - 1)}$

6. $\dfrac{z}{z^3 - z^2 - 6z}$

7. $\dfrac{t^2 + 8}{t^2 - 5t + 6}$

8. $\dfrac{t^4 + 9}{t^4 + 9t^2}$

Nonrepeated Linear Factors

In Exercises 9–16, express the integrands as a sum of partial fractions and evaluate the integrals.

9. $\displaystyle\int \frac{dx}{1 - x^2}$

10. $\displaystyle\int \frac{dx}{x^2 + 2x}$

11. $\displaystyle\int \frac{x + 4}{x^2 + 5x - 6}\, dx$

12. $\displaystyle\int \frac{2x + 1}{x^2 - 7x + 12}\, dx$

13. $\displaystyle\int_4^8 \frac{y\, dy}{y^2 - 2y - 3}$

14. $\displaystyle\int_{1/2}^1 \frac{y + 4}{y^2 + y}\, dy$

15. $\displaystyle\int \frac{dt}{t^3 + t^2 - 2t}$

16. $\displaystyle\int \frac{x + 3}{2x^3 - 8x}\, dx$

Repeated Linear Factors

In Exercises 17–20, express the integrands as a sum of partial fractions and evaluate the integrals.

17. $\displaystyle\int_0^1 \frac{x^3\,dx}{x^2 + 2x + 1}$

18. $\displaystyle\int_{-1}^0 \frac{x^3\,dx}{x^2 - 2x + 1}$

19. $\displaystyle\int \frac{dx}{(x^2 - 1)^2}$

20. $\displaystyle\int \frac{x^2\,dx}{(x - 1)(x^2 + 2x + 1)}$

Irreducible Quadratic Factors

In Exercises 21–28, express the integrands as a sum of partial fractions and evaluate the integrals.

21. $\displaystyle\int_0^1 \frac{dx}{(x + 1)(x^2 + 1)}$

22. $\displaystyle\int_1^{\sqrt{3}} \frac{3t^2 + t + 4}{t^3 + t}\,dt$

23. $\displaystyle\int \frac{y^2 + 2y + 1}{(y^2 + 1)^2}\,dy$

24. $\displaystyle\int \frac{8x^2 + 8x + 2}{(4x^2 + 1)^2}\,dx$

25. $\displaystyle\int \frac{2s + 2}{(s^2 + 1)(s - 1)^3}\,ds$

26. $\displaystyle\int \frac{s^4 + 81}{s(s^2 + 9)^2}\,ds$

27. $\displaystyle\int \frac{2\theta^3 + 5\theta^2 + 8\theta + 4}{(\theta^2 + 2\theta + 2)^2}\,d\theta$

28. $\displaystyle\int \frac{\theta^4 - 4\theta^3 + 2\theta^2 - 3\theta + 1}{(\theta^2 + 1)^3}\,d\theta$

Improper Fractions

In Exercises 29–34, perform long division on the integrand, write the proper fraction as a sum of partial fractions, and then evaluate the integral.

29. $\displaystyle\int \frac{2x^3 - 2x^2 + 1}{x^2 - x}\,dx$

30. $\displaystyle\int \frac{x^4}{x^2 - 1}\,dx$

31. $\displaystyle\int \frac{9x^3 - 3x + 1}{x^3 - x^2}\,dx$

32. $\displaystyle\int \frac{16x^3}{4x^2 - 4x + 1}\,dx$

33. $\displaystyle\int \frac{y^4 + y^2 - 1}{y^3 + y}\,dy$

34. $\displaystyle\int \frac{2y^4}{y^3 - y^2 + y - 1}\,dy$

Evaluating Integrals

Evaluate the integrals in Exercises 35–40.

35. $\displaystyle\int \frac{e^t\,dt}{e^{2t} + 3e^t + 2}$

36. $\displaystyle\int \frac{e^{4t} + 2e^{2t} - e^t}{e^{2t} + 1}\,dt$

37. $\displaystyle\int \frac{\cos y\,dy}{\sin^2 y + \sin y - 6}$

38. $\displaystyle\int \frac{\sin\theta\,d\theta}{\cos^2\theta + \cos\theta - 2}$

39. $\displaystyle\int \frac{(x - 2)^2 \tan^{-1}(2x) - 12x^3 - 3x}{(4x^2 + 1)(x - 2)^2}\,dx$

40. $\displaystyle\int \frac{(x + 1)^2 \tan^{-1}(3x) + 9x^3 + x}{(9x^2 + 1)(x + 1)^2}\,dx$

Initial Value Problems

Solve the initial value problems in Exercises 41–48.

41. $(t^2 - 3t + 2)\dfrac{dx}{dt} = 1$ $(t > 2),$ $x(3) = 0$

42. $(3t^4 + 4t^2 + 1)\dfrac{dx}{dt} = 2\sqrt{3},$ $x(1) = -\pi\sqrt{3}/4$

43. $(t^2 + 2t)\dfrac{dx}{dt} = 2x + 2$ $(t, x > 0),$ $x(1) = 1$

44. $(t + 1)\dfrac{dx}{dt} = x^2 + 1$ $(t > -1),$ $x(0) = \pi/4$

45. $\dfrac{dy}{dx} = e^x(y^2 - y),$ $y(0) = 2$

46. $\dfrac{dy}{d\theta} = (y + 1)^2 \sin\theta$ $y(\pi/2) = 0$

47. $\dfrac{dy}{dx} = \dfrac{1}{x^2 - 3x + 2},$ $y(3) = 0$

48. $\dfrac{ds}{dt} = \dfrac{2s + 2}{t^2 + 2t},$ $s(1) = 1$

Applications and Examples

In Exercises 49 and 50, find the volume of the solid generated by revolving the shaded region about the indicated axis.

49. The x-axis

50. The y-axis

51. *Social diffusion* Sociologists sometimes use the phrase "social diffusion" to describe the way information spreads through a population. The information might be a rumor, a cultural fad, or news about a technical innovation. In a sufficiently large population, the number of people x who have the information is treated as a differentiable function of time t, and the rate of diffusion, dx/dt, is assumed to be proportional to the number of people who have the information times the number of people who do not. This leads to the equation

$$\frac{dx}{dt} = kx(N - x),$$

where N is the number of people in the population.

Suppose that t is in days, $k = 1/250$, and two people start a rumor at time $t = 0$ in a population of $N = 1000$ people.

(a) Find x as a function of t.

(b) When will half the population have heard the rumor? (This is when the rumor will be spreading the fastest.)

52. *Second-order chemical reactions* Many chemical reactions are the result of the interaction of two molecules that undergo a change to produce a new product. The rate of the reaction typically depends on the concentrations of the two kinds of molecules. If a is the amount of substance A and b is the amount of substance B at time $t = 0$, and if x is the amount of product at time t, then the rate of formation of x may be given by the differential equation

$$\frac{dx}{dt} = k(a - x)(b - x),$$

or

$$\frac{1}{(a - x)(b - x)} \frac{dx}{dt} = k,$$

where k is a constant for the reaction. Integrate both sides of this equation to obtain a relation between x and t

(a) if $a = b$

(b) if $a \neq b$.

Assume in each case that $x = 0$ when $t = 0$.

Trigonometric Substitutions

Three Basic Substitutions

Trigonometric substitutions enable us to replace the binomials $a^2 + x^2$, $a^2 - x^2$, and $x^2 - a^2$ by single squared terms and thereby transform a number of integrals containing square roots into integrals we can evaluate directly.

Three Basic Substitutions

The most common substitutions are $x = a \tan \theta$, $x = a \sin \theta$, and $x = a \sec \theta$. They come from the reference right triangles in Figure 7.3.

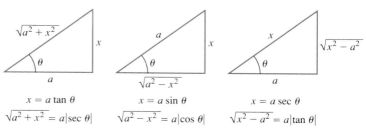

FIGURE 7.3 Reference triangles for trigonometric substitutions that change binomials into single squared terms.

With $x = a \tan \theta$,

$$a^2 + x^2 = a^2 + a^2 \tan^2 \theta = a^2(1 + \tan^2 \theta) = a^2 \sec^2 \theta.$$

With $x = a \sin \theta$,

$$a^2 - x^2 = a^2 - a^2 \sin^2 \theta = a^2(1 - \sin^2 \theta) = a^2 \cos^2 \theta.$$

With $x = a \sec \theta$,

$$x^2 - a^2 = a^2 \sec^2 \theta - a^2 = a^2(\sec^2 \theta - 1) = a^2 \tan^2 \theta.$$

Trigonometric Substitutions

1. $x = a \tan \theta$ replaces $a^2 + x^2$ by $a^2 \sec^2 \theta$.

2. $x = a \sin \theta$ replaces $a^2 - x^2$ by $a^2 \cos^2 \theta$.

3. $x = a \sec \theta$ replaces $x^2 - a^2$ by $a^2 \tan^2 \theta$.

We want any substitution we use in an integration to be reversible so that we can change back to the original variable afterward. For example, if $x = a \tan \theta$, we want to be able to set $\theta = \tan^{-1}(x/a)$ after the integration takes place. If $x = a \sin \theta$, we want to be able to set $\theta = \sin^{-1}(x/a)$ when we're done, and similarly for $x = a \sec \theta$.

For reversibility,

$$x = a \tan \theta \quad \text{requires} \quad \theta = \tan^{-1}\left(\frac{x}{a}\right) \quad \text{with} \quad -\frac{\pi}{2} < \theta < \frac{\pi}{2}$$

$$x = a \sin \theta \quad \text{requires} \quad \theta = \sin^{-1}\left(\frac{x}{a}\right) \quad \text{with} \quad -\frac{\pi}{2} \le \theta \le \frac{\pi}{2}$$

$$x = a \sec \theta \quad \text{requires} \quad \theta = \sec^{-1}\left(\frac{x}{a}\right) \quad \text{with} \quad \begin{cases} 0 \le \theta < \frac{\pi}{2}, & \frac{x}{a} \ge 1 \\ \frac{\pi}{2} < \theta \le \pi, & \frac{x}{a} \le -1. \end{cases}$$

Example 1   Using the Substitution $x = a \tan \theta$

Evaluate

$$\int \frac{dx}{\sqrt{4 + x^2}}.$$

Solution   We set

$$x = 2 \tan \theta, \qquad dx = 2 \sec^2 \theta\, d\theta, \qquad -\frac{\pi}{2} < \theta < \frac{\pi}{2},$$

$$4 + x^2 = 4 + 4 \tan^2 \theta = 4(1 + \tan^2 \theta) = 4 \sec^2 \theta.$$

Then

$$\int \frac{dx}{\sqrt{4 + x^2}} = \int \frac{2 \sec^2 \theta\, d\theta}{\sqrt{4 \sec^2 \theta}} = \int \frac{\sec^2 \theta\, d\theta}{|\sec \theta|} \qquad \sqrt{\sec^2 \theta} = |\sec \theta|$$

$$= \int \sec \theta\, d\theta \qquad \sec \theta > 0 \text{ for } -\frac{\pi}{2} < \theta < \frac{\pi}{2}$$

$$= \ln |\sec \theta + \tan \theta| + C$$

$$= \ln \left| \frac{\sqrt{4 + x^2}}{2} + \frac{x}{2} \right| + C \qquad \text{From Fig. 7.4}$$

$$= \ln |\sqrt{4 + x^2} + x| + C'. \qquad \text{Taking } C' = C - \ln 2$$

CD–ROM
WEBsite

Historical Biography

George Berkeley
(1685 — 1753)

FIGURE 7.4   Reference triangle for $x = 2 \tan \theta$ (Example 1):

$$\tan \theta = \frac{x}{2}$$

and

$$\sec \theta = \frac{\sqrt{4 + x^2}}{2}.$$

Notice how we expressed $\ln |\sec \theta + \tan \theta|$ in terms of x: We drew a reference triangle for the original substitution $x = 2 \tan \theta$ (Figure 7.4) and read the ratios from the triangle.

Example 2   Using the Substitution $x = a \sin \theta$

Evaluate

$$\int \frac{x^3 \, dx}{\sqrt{9 - x^2}}, \qquad -3 < x < 3.$$

Solution   We set

$$x = 3 \sin \theta, \qquad dx = 3 \cos \theta \, d\theta, \qquad -\frac{\pi}{2} < \theta < \frac{\pi}{2}$$

$$9 - x^2 = 9 - 9 \sin^2 \theta = 9(1 - \sin^2 \theta) = 9 \cos^2 \theta.$$

Then

$$\int \frac{x^3 \, dx}{\sqrt{9 - x^2}} = \int \frac{27 \sin^3 \theta \cdot 3 \cos \theta \, d\theta}{|3 \cos \theta|}$$

$$= 27 \int \sin^3 \theta \, d\theta \qquad \qquad \cos \theta > 0 \text{ for } -\frac{\pi}{2} < \theta < \frac{\pi}{2}$$

$$= 27 \int (1 - \cos^2 \theta) \sin \theta \, d\theta \qquad \sin^2 \theta = 1 - \cos^2 \theta$$

$$= -27 \cos \theta + 9 \cos^3 \theta + C$$

$$= -27 \cdot \frac{\sqrt{9 - x^2}}{3} + 9 \left(\frac{\sqrt{9 - x^2}}{3} \right)^3 + C \qquad \text{Figure 7.5, } a = 3$$

$$= -9\sqrt{9 - x^2} + \frac{(9 - x^2)^{3/2}}{3} + C.$$

FIGURE 7.5 Reference triangle for $x = 3 \sin \theta$ (Example 2):

$$\sin \theta = \frac{x}{3}$$

and

$$\cos \theta = \frac{\sqrt{9 - x^2}}{3}.$$

Example 3   Using the Substitution $x = a \sec \theta$

Evaluate

$$\int \frac{dx}{\sqrt{25x^2 - 4}}, \qquad x > \frac{2}{5}.$$

Solution   We first rewrite the radical as

$$\sqrt{25x^2 - 4} = \sqrt{25 \left(x^2 - \frac{4}{25} \right)}$$

$$= 5 \sqrt{x^2 - \left(\frac{2}{5} \right)^2}$$

FIGURE 7.6 If $x = (2/5)\sec\theta, 0 \le \theta < \pi/2$, then $\theta = \sec^{-1}(5x/2)$, and we can read the values of the other trigonometric functions of θ from this right triangle.

to put the radicand in the form $x^2 - a^2$. We then substitute

$$x = \frac{2}{5}\sec\theta, \qquad dx = \frac{2}{5}\sec\theta\tan\theta\,d\theta, \qquad 0 < \theta < \frac{\pi}{2}$$

$$x^2 - \left(\frac{2}{5}\right)^2 = \frac{4}{25}\sec^2\theta - \frac{4}{25}$$

$$= \frac{4}{25}(\sec^2\theta - 1) = \frac{4}{25}\tan^2\theta$$

$$\sqrt{x^2 - \left(\frac{2}{5}\right)^2} = \frac{2}{5}|\tan\theta| = \frac{2}{5}\tan\theta. \qquad \begin{matrix}\tan\theta > 0 \text{ for} \\ 0 < \theta < \pi/2\end{matrix}$$

With these substitutions, we have

$$\int \frac{dx}{\sqrt{25x^2 - 4}} = \int \frac{dx}{5\sqrt{x^2 - (4/25)}} = \int \frac{(2/5)\sec\theta\tan\theta\,d\theta}{5 \cdot (2/5)\tan\theta}$$

$$= \frac{1}{5}\int\sec\theta\,d\theta = \frac{1}{5}\ln|\sec\theta + \tan\theta| + C$$

$$= \frac{1}{5}\ln\left|\frac{5x}{2} + \frac{\sqrt{25x^2 - 4}}{2}\right| + C \qquad \text{Fig. 7.6}$$

A trigonometric substitution can sometimes help us to evaluate an integral containing an integer power of a quadratic binomial, as in the next example.

Example 4 Finding the Volume of a Solid of Revolution

Find the volume of the solid generated by revolving about the x-axis the region bounded by the curve $y = 4/(x^2 + 4)$, the x-axis, and the lines $x = 0$ and $x = 2$.

Solution   We sketch the region (Figure 7.7) and use the disk method:

$$V = \int_0^2 \pi[R(x)]^2\,dx = 16\pi\int_0^2 \frac{dx}{(x^2 + 4)^2}. \qquad R(x) = \frac{4}{x^2 + 4}$$

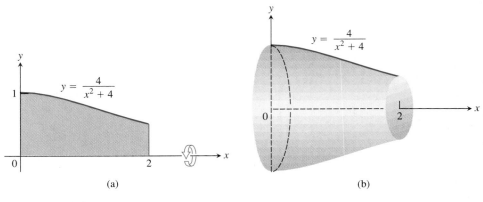

FIGURE 7.7 The region (a) and solid (b) in Example 4.

To evaluate the integral, we set

$$x = 2 \tan \theta, \qquad dx = 2 \sec^2 \theta \, d\theta, \qquad \theta = \tan^{-1} \frac{x}{2},$$

$$x^2 + 4 = 4 \tan^2 \theta + 4 = 4(\tan^2 \theta + 1) = 4 \sec^2 \theta$$

(Figure 7.8). With these substitutions,

$$V = 16\pi \int_0^2 \frac{dx}{(x^2 + 4)^2}$$

$$= 16\pi \int_0^{\pi/4} \frac{2 \sec^2 \theta \, d\theta}{(4 \sec^2 \theta)^2} \qquad\qquad \begin{array}{l} \theta = 0 \text{ when } x = 0; \\ \theta = \pi/4 \text{ when } x = 2 \end{array}$$

$$= 16\pi \int_0^{\pi/4} \frac{2 \sec^2 \theta \, d\theta}{16 \sec^4 \theta} = \pi \int_0^{\pi/4} 2 \cos^2 \theta \, d\theta$$

$$= \pi \int_0^{\pi/4} (1 + \cos 2\theta) \, d\theta = \pi \left[\theta + \frac{\sin 2\theta}{2} \right]_0^{\pi/4} \qquad 2 \cos^2 \theta = 1 + \cos 2\theta$$

$$= \pi \left[\frac{\pi}{4} + \frac{1}{2} \right] \approx 4.04.$$

FIGURE 7.8 Reference triangle for $x = 2 \tan \theta$. (Example 4)

EXERCISES 7.4

Basic Trigonometric Substitutions

Evaluate the integrals in Exercises 1–18.

1. $\displaystyle\int \frac{dy}{\sqrt{9 + y^2}}$

2. $\displaystyle\int \frac{3 \, dy}{\sqrt{1 + 9y^2}}$

3. $\displaystyle\int \sqrt{25 - t^2} \, dt$

4. $\displaystyle\int \sqrt{1 - 9t^2} \, dt$

5. $\displaystyle\int \frac{dx}{\sqrt{4x^2 - 49}}, \quad x > \frac{7}{2}$

6. $\displaystyle\int \frac{5 \, dx}{\sqrt{25x^2 - 9}}, \quad x > \frac{3}{5}$

7. $\displaystyle\int \frac{dx}{x^2 \sqrt{x^2 - 1}}, \quad x > 1$

8. $\displaystyle\int \frac{2 \, dx}{x^3 \sqrt{x^2 - 1}}, \quad x > 1$

9. $\displaystyle\int \frac{x^3 \, dx}{\sqrt{x^2 + 4}}$

10. $\displaystyle\int \frac{dx}{x^2 \sqrt{x^2 + 1}}$

11. $\displaystyle\int \frac{8 \, dw}{w^2 \sqrt{4 - w^2}}$

12. $\displaystyle\int \frac{\sqrt{9 - w^2}}{w^2} \, dw$

13. $\displaystyle\int \frac{dx}{(x^2 - 1)^{3/2}}, \quad x > 1$

14. $\displaystyle\int \frac{x^2 \, dx}{(x^2 - 1)^{5/2}}, \quad x > 1$

15. $\displaystyle\int \frac{(1 - x^2)^{3/2}}{x^6} \, dx$

16. $\displaystyle\int \frac{(1 - x^2)^{1/2}}{x^4} \, dx$

17. $\displaystyle\int \frac{8 \, dx}{(4x^2 + 1)^2}$

18. $\displaystyle\int \frac{6 \, dt}{(9t^2 + 1)^2}$

Combining Substitutions

In Exercises 19–26, use an appropriate substitution and then a trigonometric substitution to evaluate the integrals.

19. $\displaystyle\int_0^{\ln 4} \frac{e^t \, dt}{\sqrt{e^{2t} + 9}}$

20. $\displaystyle\int_{\ln (3/4)}^{\ln (4/3)} \frac{e^t \, dt}{(1 + e^{2t})^{3/2}}$

21. $\displaystyle\int_{1/12}^{1/4} \frac{2 \, dt}{\sqrt{t} + 4t\sqrt{t}}$

22. $\displaystyle\int_1^e \frac{dy}{y\sqrt{1 + (\ln y)^2}}$

23. $\displaystyle\int \frac{dx}{x\sqrt{x^2 - 1}}$

24. $\displaystyle\int \frac{dx}{1 + x^2}$

25. $\displaystyle\int \frac{x \, dx}{\sqrt{x^2 - 1}}$

26. $\displaystyle\int \frac{dx}{\sqrt{1 - x^2}}$

Initial Value Problems

Solve the initial value problems in Exercises 27–30 for y as a function of x.

27. $x \dfrac{dy}{dx} = \sqrt{x^2 - 4}, \quad x \geq 2, \quad y(2) = 0$

28. $\sqrt{x^2 - 9} \dfrac{dy}{dx} = 1, \quad x > 3, \quad y(5) = \ln 3$

29. $(x^2 + 4) \dfrac{dy}{dx} = 3, \quad y(2) = 0$

30. $(x^2 + 1)^2 \dfrac{dy}{dx} = \sqrt{x^2 + 1}, \quad y(0) = 1$

Applications

31. *Finding area* Find the area of the region in the first quadrant that is enclosed by the coordinate axes and the curve $y = \sqrt{9 - x^2}/3$.

32. *Finding volume* Find the volume of the solid generated by revolving about the x-axis the region in the first quadrant enclosd by the coordinate axes, the curve $y = 2/(1 + x^2)$, and the line $x = 1$.

The Substitution $z = \tan (x/2)$

33. The substitution

$$z = \tan \frac{x}{2}$$

(see figure) reduces the problem of integrating any rational function of $\sin x$ and $\cos x$ to a problem involving a rational function of z. This in turn can be solved by partial fractions. Show that each equation is true.

(a) $\tan \dfrac{x}{2} = \dfrac{\sin x}{1 + \cos x}$ **(b)** $\cos x = \dfrac{1 - z^2}{1 + z^2}$

(c) $\sin x = \dfrac{2z}{1 + z^2}$ **(d)** $dx = \dfrac{2\,dz}{1 + z^2}$

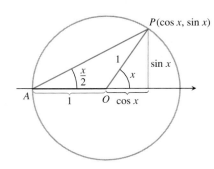

In Exercises 34–41, use the substitution $z = \tan (x/2)$ and the results in Exercise 33 to evaluate the integral.

34. $\displaystyle\int \frac{dx}{1 + \sin x}$ **35.** $\displaystyle\int \frac{dx}{1 - \cos x}$

36. $\displaystyle\int \frac{d\theta}{1 - \sin \theta}$ **37.** $\displaystyle\int \frac{dt}{1 + \sin t + \cos t}$

38. $\displaystyle\int_0^{\pi/2} \frac{d\theta}{2 + \cos \theta}$ **39.** $\displaystyle\int_{\pi/2}^{2\pi/3} \frac{\cos \theta\, d\theta}{\sin \theta \cos \theta + \sin \theta}$

40. $\displaystyle\int \frac{dt}{\sin t - \cos t}$ **41.** $\displaystyle\int \frac{\cos t\, dt}{1 - \cos t}$

7.5   Integral Tables, Computer Algebra Systems, and Monte Carlo Integration

Integral Tables • Integration with a CAS • Monte Carlo Numerical Integration

As you know, the basic techniques of integration are substitution and integration by parts. We apply these techniques to transform unfamiliar integrals into integrals whose forms we recognize or can find in a table. If a computer algebra system (CAS) is available, you can usually use it to evaluate an integral. Another technique for approximating the value of a definite integral is called *Monte Carlo Integration*. We investigate these methods for evaluating integrals in this section.

Integral Tables

A Brief Table of Integrals appears at the back of the book, after the index. The integration formulas are stated in terms of constants a, b, c, m, n, and so on. These constants can usually assume any real value and need not be integers. Occasional limitations on their values are stated with the formulas. Formula 5 requires $n \neq -1$, for example, and Formula 11 requires $n \neq -2$.

The formulas also assume that the constants do not take on values that require dividing by zero or taking even roots of negative numbers. For example, Formula 8 assumes that $a \neq 0$, and Formula 13(a) cannot be used unless b is negative.

Example 1   Using a Table of Integrals

Evaluate

$$\int \left(\frac{1}{x^2 + 1} + \frac{1}{x^2 - 2x + 5} \right) dx.$$

Solution

$$\int \left(\frac{1}{x^2 + 1} + \frac{1}{x^2 - 2x + 5} \right) dx = \int \frac{1}{x^2 + 1} \, dx + \int \frac{1}{x^2 - 2x + 5} \, dx \qquad (1)$$

Both integrals on the right-hand side of Equation 1 can be evaluated using Formula 16 from the Brief Table of Integrals.

$$\textbf{16.} \int \frac{dx}{a^2 + x^2} = \frac{1}{a} \tan^{-1} \frac{x}{a} + C$$

You probably remember that

$$\int \frac{dx}{1 + x^2} = \tan^{-1} x + C,$$

but if you don't, you can obtain the result by setting $a = 1$ in Formula 16.

To find the value of the second integral, we need to complete the square:

$$x^2 - 2x + 5 = x^2 - 2x + 1 + 4 = (x - 1)^2 + 4.$$

Then

$$\int \frac{dx}{x^2 - 2x + 5} = \int \frac{dx}{(x - 1)^2 + 4}$$

$$= \int \frac{du}{u^2 + 4} \qquad u = x - 1, \, du = dx$$

$$= \frac{1}{2} \tan^{-1} \frac{u}{2} + C \qquad \text{Formula 16 with } a = 2$$

$$= \frac{1}{2} \tan^{-1} \left(\frac{x - 1}{2} \right) + C.$$

Combining the two integrations gives

$$\int \left(\frac{1}{x^2 + 1} + \frac{1}{x^2 - 2x + 5} \right) dx = \tan^{-1} x + \frac{1}{2} \tan^{-1} \left(\frac{x - 1}{2} \right) + C.$$

The manipulation and substitution in Example 1 are typical of what is needed to use integral tables to evaluate integrals. Here's another example.

Example 2   Using a Table of Integrals

Find

$$\int \frac{dx}{x^2 \sqrt{2x - 4}}.$$

Solution   We begin with Formula 15:

$$15. \quad \int \frac{dx}{x^2\sqrt{ax+b}} = -\frac{\sqrt{ax+b}}{bx} - \frac{a}{2b} \int \frac{dx}{x\sqrt{ax+b}} + C.$$

With $a = 2$ and $b = -4$, we have

$$\int \frac{dx}{x^2\sqrt{2x-4}} = -\frac{\sqrt{2x-4}}{-4x} + \frac{2}{2\cdot 4} \int \frac{dx}{x\sqrt{2x-4}} + C.$$

We then use Formula 13(a) to evaluate the integral on the right:

$$13(a). \quad \int \frac{dx}{x\sqrt{ax-b}} = \frac{2}{\sqrt{b}} \tan^{-1} \sqrt{\frac{ax-b}{b}} + C.$$

With $a = 2$ and $b = 4$, we have

$$\int \frac{dx}{x\sqrt{2x-4}} = \frac{2}{\sqrt{4}} \tan^{-1} \sqrt{\frac{2x-4}{4}} + C = \tan^{-1} \sqrt{\frac{x-2}{2}} + C.$$

Combining the two integrations gives

$$\int \frac{dx}{x^2\sqrt{2x-4}} = \frac{\sqrt{2x-4}}{4x} + \frac{1}{4} \tan^{-1} \sqrt{\frac{x-2}{2}} + C.$$

Integration with a CAS

A powerful capability of computer algebra systems is their facility to integrate symbolically. This is performed with the **integrate command** specified by the particular system (e.g., **int** in Maple, **Integrate** in Mathematica).

Example 3   Using a CAS with a Named Function

Suppose that you want to evaluate the indefinite integral of the function

$$f(x) = x^2\sqrt{a^2+x^2}.$$

Using Maple, you first define or name the function:

$$> f := x\hat{}2 * sqrt (a\hat{}2 + x\hat{}2);$$

Then you use the integrate command on f, identifying the variable of integration:

$$> int(f, x);$$

Maple returns the answer

$$\frac{1}{4} x(a^2+x^2)^{3/2} - \frac{1}{8} a^2 x\sqrt{a^2+x^2} - \frac{1}{8} a^4 \ln{(x + \sqrt{a^2+x^2})}.$$

If you want to see if the answer can be simplified, enter

$$> simplify('');$$

Maple returns

$$\frac{1}{8} a^2 x\sqrt{a^2+x^2} + \frac{1}{4} x^3\sqrt{a^2+x^2} - \frac{1}{8} a^4 \ln{(x + \sqrt{a^2+x^2})}.$$

If you want the definite integral for $0 \le x \le \pi/2$, you can use the format

$$> \text{int}(f, x = 0..\text{Pi}/2);$$

Maple (Version 3.0) will return the expression

$$\frac{1}{64}(4a^2 + \pi^2)^{3/2}\pi - \frac{1}{8}a^4 \ln\left(\frac{1}{2}\pi + \frac{1}{2}\sqrt{4a^2 + \pi^2}\right)$$

$$-\frac{1}{32}a^2\sqrt{4a^2 + \pi^2}\,\pi + \frac{1}{8}a^4 \ln(\sqrt{a^2}).$$

You can also find the definite integral for a particular value of the constant a:

$$> a := 1;$$
$$> \text{int}(f, x = 0..1);$$

Maple returns the numerical answer

$$\frac{3}{8}\sqrt{2} - \frac{1}{8}\ln(1 + \sqrt{2}).$$

Example 4   Using a CAS Without Naming the Function

Use a CAS to find

$$\int \sin^2 x \cos^3 x \, dx.$$

Solution   With Maple, we have the entry

$$> \text{int}((\sin \,^2)(x)*(\cos \,^3)(x), x);$$

with the immediate return

$$-\frac{1}{5}\sin(x)\cos(x)^4 + \frac{1}{15}\cos(x)^2 \sin(x) + \frac{2}{15}\sin(x).$$

Example 5   A CAS May Not Return a Closed Form Solution

Use a CAS to find

$$\int (\cos^{-1} ax)^2 \, dx.$$

Solution   Using Maple, we enter

$$> \text{int}((\text{arccos}(a*x))^2, x);$$

and Maple returns the expression

$$\int \text{arccos}(ax)^2 \, dx,$$

indicating that it does not have a closed form solution. In the next chapter, you see how series expansion may help to evaluate such an integral.

Computer algebra systems vary in how they process integrations. We used Maple in Examples 3 through 5. Mathematica would have returned somewhat different results:

1. In Example 3, given

    In [1]: = Integrate [x^2 * Sqrt [a^2 + x^2], x]

    Mathematica returns

    $$\text{Out } [1] = \text{Sqrt } [a^2 + x^2]\left(\frac{a^2 x}{8} + \frac{x^3}{4}\right) - \frac{a^4 \text{ Log } [x + \text{Sqrt } [a^2 + x^2]]}{8}$$

    without having to simplify an intermediate result. The answer is close to Formula 22 in the integral tables.

2. The Mathematica answer to the integral

    In [2]: = Integrate [Sin [x]^2 * Cos [x]^3, x]

    in Example 4 is

    $$\text{Out } [2] = \frac{30 \text{ Sin } [x] - 5 \text{ Sin } [3x] - 3 \text{ Sin } [5x]}{240}$$

    differing from the Maple answer.

3. Mathematica does give a result for the integration

    In [3]: = Integrate [ArcCos [a * x]^2, x]

    in Example 5:

    $$\text{Out } [3] = -2x - \frac{2 \text{ Sqrt } [1 - a^2 x^2] \text{ ArcCos } [ax]}{a} + x\text{ArcCos } [ax]^2$$

Although a CAS is very powerful and can aid us in solving difficult problems, each CAS has its own limitations. There are even situations where a CAS may further complicate a problem (in the sense of producing an answer that is extremely difficult to use or interpret). On the other hand, a little mathematical thinking on your part may reduce the problem to one that is quite easy to handle. We provide an example in Exercise 49.

Monte Carlo Numerical Integration

In many applications, we encounter integrals such as $\int_1^{100} e^{x^2} dx$, which cannot be evaluated using the analytic techniques presented in this chapter. We can approximate a definite integral like that using numerical techniques such as the Trapezoidal Rule or Simpson's Rule (Section 4.7). Another numerical method for approximating definite integrals, called **Monte Carlo integration,** generalizes conveniently to multiple integrals (Chapter 12) as well. The method is typically accomplished with the aid of a computer. We illustrate how it works with a simple example.

Example 6 Monte Carlo Area Under a Nonnegative Curve

Imagine the area under the continuous curve $y = f(x)$, $0 \le f(x) \le M$, over the closed interval $a \le x \le b$ as a portion of the area of the total rectangle depicted in Figure 7.9. Randomly select a large number of points $P(x, y)$ within the rectangle and use them to estimate the area under the curve.

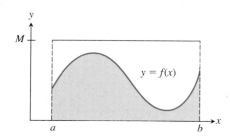

FIGURE 7.9 Imagine the rectangle as a dartboard.

Solution   To "randomly select" a point within the rectangle, we use a computer or calculator programmed with a random number generator. First we ask the computer to generate a random number x satisfying $a \le x \le b$. In theory, all numbers in the closed interval $[a, b]$ have an equal likelihood of being selected. Next we ask the computer to generate a second random number y satisfying $0 \le y \le M$. Again any number in $[0, M]$ has equal likelihood of being selected, at least in theory. The point $P(x, y)$ therefore lies somewhere inside the rectangle in Figure 7.9. Once the random point $P(x, y)$ is selected, ask yourself if it lies within the shaded region below the curve; that is, does the y-coordinate satisfy $0 \le y < f(x)$? If the answer is yes, then count the point P by adding one to some counter. Two counters will be necessary; one to count the total points generated and a second to count those points that lie below the curve (see Figure 7.10). We pick a very large number of points P using the method just described, remembering to count those that lie below the curve. We can then approximate the area under the curve from the formula

$$\frac{\text{Area under curve}}{\text{area of rectangle}} \approx \frac{\text{number of points counted below curve}}{\text{total number of random points}}. \tag{2}$$

For example, Table 7.3 gives estimates to the area below the curve $y = \cos x$ and above the x-axis over the interval $-\pi/2 \le x \le \pi/2$. We chose the number

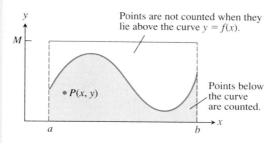

Points are not counted when they lie above the curve $y = f(x)$.

Points below the curve are counted.

FIGURE **7.10** The area under the nonnegative curve $y = f(x)$, $a \le x \le b$, is contained within the rectangle of height M and base $b - a$.

Table 7.3 Monte Carlo approximation to the area under the curve $y = \cos x$ over the interval $-\pi/2 \le x \le \pi/2$

Number of points	Approximation to area	Number of points	Approximation to area
100	2.07345	2,000	1.94465
200	2.13628	3,000	1.97711
300	2.01064	4,000	1.99962
400	2.12058	5,000	2.01429
500	2.04832	6,000	2.02319
600	2.09440	8,000	2.00669
700	2.02857	10,000	2.00873
800	1.99491	15,000	2.00978
900	1.99666	20,000	2.01093
1,000	1.96664	30,000	2.01186

$M = 2$ as the height of the rectangle, and we estimated the area after randomly selecting $100, 200, \ldots, 30{,}000$ points in the rectangle and applying the estimation formula (2).

The actual area under the curve $y = \cos x$ over the given interval is 2 square units. Note that even with the relatively large number of points generated the error is significant. For functions of one variable, Monte Carlo integration is generally not competitive with the techniques you learned in Section 4.7. The lack of an error bound and the potential difficulty in finding an upper bound M are disadvantages as well. Nevertheless, the Monte Carlo technique can be extended to functions of several variables and becomes more practical in that situation.

EXERCISES 7.5

Using Integral Tables

Use the table of integrals at the back of the book to evaluate the integrals in Exercises 1–20.

1. $\displaystyle\int \frac{dx}{x\sqrt{x-3}}$

2. $\displaystyle\int \frac{x\,dx}{\sqrt{x-2}}$

3. $\displaystyle\int x\sqrt{2x-3}\,dx$

4. $\displaystyle\int \frac{\sqrt{9-4x}}{x^2}\,dx$

5. $\displaystyle\int x\sqrt{4x-x^2}\,dx$

6. $\displaystyle\int \frac{dx}{x\sqrt{7+x^2}}$

7. $\displaystyle\int \frac{\sqrt{4-x^2}}{x}\,dx$

8. $\displaystyle\int \sqrt{25-p^2}\,dp$

9. $\displaystyle\int \frac{r^2}{\sqrt{4-r^2}}\,dr$

10. $\displaystyle\int \frac{d\theta}{5+4\sin 2\theta}$

11. $\displaystyle\int e^{2t}\cos 3t\,dt$

12. $\displaystyle\int x\cos^{-1}x\,dx$

13. $\displaystyle\int \frac{ds}{(9-s^2)^2}$

14. $\displaystyle\int \frac{\sqrt{4x+9}}{x^2}\,dx$

15. $\displaystyle\int \frac{\sqrt{3t-4}}{t}\,dt$

16. $\displaystyle\int x^2\tan^{-1}x\,dx$

17. $\displaystyle\int \sin 3x \cos 2x\,dx$

18. $\displaystyle\int 8\sin 4t \sin \frac{t}{2}\,dt$

19. $\displaystyle\int \cos \frac{\theta}{3}\cos \frac{\theta}{4}\,d\theta$

20. $\displaystyle\int \cos \frac{\theta}{2}\cos 7\theta\,d\theta$

Substitution and Integral Tables

In Exercises 21–32, use a substitution to change the integral into one you can find in the table. Then evaluate the integral.

21. $\displaystyle\int \frac{x^3+x+1}{(x^2+1)^2}\,dx$

22. $\displaystyle\int \frac{x^2+6x}{(x^2+3)^2}\,dx$

23. $\displaystyle\int \sin^{-1}\sqrt{x}\,dx$

24. $\displaystyle\int \frac{\cos^{-1}\sqrt{x}}{\sqrt{x}}\,dx$

25. $\displaystyle\int \cot t\sqrt{1-\sin^2 t}\,dt, \quad 0 < t < \pi/2$

26. $\displaystyle\int \frac{dt}{\tan t\sqrt{4-\sin^2 t}}$

27. $\displaystyle\int \frac{dy}{y\sqrt{3+(\ln y)^2}}$

28. $\displaystyle\int \frac{\cos \theta\,d\theta}{\sqrt{5+\sin^2 \theta}}$

29. $\displaystyle\int \frac{3\,dr}{\sqrt{9r^2-1}}$

30. $\displaystyle\int \frac{3\,dy}{\sqrt{1+9y^2}}$

31. $\displaystyle\int \cos^{-1}\sqrt{x}\,dx$

32. $\displaystyle\int \tan^{-1}\sqrt{y}\,dy$

Powers of x Times Exponentials

Evaluate the integrals in Exercises 33–36 using table Formulas 103–106. These integrals can also be evaluated using tabular integration (Section 7.2).

33. $\displaystyle\int xe^{3x}\,dx$

34. $\displaystyle\int x^3 e^{x/2}\,dx$

35. $\displaystyle\int x^2 2^x\,dx$

36. $\displaystyle\int x\pi^x\,dx$

Hyperbolic Functions

Use the integral tables to evaluate the integrals in Exercises 37–40.

37. $\displaystyle\int \frac{1}{8}\sinh^5 3x\,dx$

38. $\displaystyle\int \frac{\cosh^4 \sqrt{x}}{\sqrt{x}}\,dx$

39. $\displaystyle\int x^2\cosh 3x\,dx$

40. $\displaystyle\int x\sinh 5x\,dx$

Theory and Examples

Exercises 41–44 refer to formulas in the table of integrals at the back of the book.

41. Derive Formula 9 by using the substitution $u = ax + b$ to evaluate

$$\int \frac{x}{(ax+b)^2}\,dx.$$

42. Derive Formula 29 by using a trigonometric substitution to evaluate

$$\int \sqrt{a^2 - x^2}\, dx.$$

43. Derive Formula 110 by evaluating

$$\int x^n (\ln\, ax)^m \, dx$$

by integration by parts.

44. Derive Formula 99 by evaluating

$$\int x^n \sin^{-1} ax\, dx$$

by integration by parts.

45. *Finding volume* The head of your firm's accounting department has asked you to find a formula she can use in a computer program to calculate the year-end inventory of gasoline in the company's tanks. A typical tank is shaped like a right circular cylinder of radius r and length L, mounted horizontally, as shown here. The data come to the accounting office as depth measurements taken with a vertical measuring stick marked in centimeters.

(a) Show, in the notation of the figure here, that the volume of gasoline that fills the tank to a depth d is

$$V = 2L \int_{-r}^{-r+d} \sqrt{r^2 - y^2}\, dy.$$

(b) Evaluate the integral.

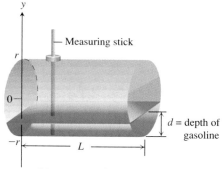

Measurements in centimeters

46. *Writing to Learn* What is the largest value

$$\int_a^b \sqrt{x - x^2}\, dx$$

can have for any a and b? Give reasons for your answer.

COMPUTER EXPLORATIONS

In Exercises 47 and 48, use a CAS to perform the integrations.

47. Evaluate the integrals in parts (a) through (c).

(a) $\displaystyle\int x \ln x\, dx$ **(b)** $\displaystyle\int x^2 \ln x\, dx$

(c) $\displaystyle\int x^3 \ln x\, dx$

(d) What pattern do you see? Predict a formula for $\int x^4 \ln x\, dx$ and then see if you are correct by evaluating it with a CAS.

(e) What is the formula for $\int x^n \ln x\, dx$, $n \geq 1$? Check your answer using a CAS.

48. Evaluate the integrals in parts (a) through (c).

(a) $\displaystyle\int \frac{\ln\, x}{x^2}\, dx$ **(b)** $\displaystyle\int \frac{\ln\, x}{x^3}\, dx$

(c) $\displaystyle\int \frac{\ln\, x}{x^4}\, dx$

(d) What pattern do you see? Predict the formula for

$$\int \frac{\ln\, x}{x^5}\, dx$$

and then see if you are correct by evaluating it with a CAS.

(e) What is the formula for

$$\int \frac{\ln x}{x^n}\, dx, \qquad n \geq 2?$$

Check your answer using a CAS.

49. (a) Use a CAS to evaluate

$$\int_0^{\pi/2} \frac{\sin^n x}{\sin^n x + \cos^n x}\, dx,$$

where n is an arbitrary positive integer. Does your CAS find the result?

(b) In succession, find the integral when $n = 1, 2, 3, 5, 7$. Comment on the complexity of the results.

(c) Now substitute $x = (\pi/2) - u$ and add the new and old integrals. What is the value of

$$\int_0^{\pi/2} \frac{\sin^n x}{\sin^n x + \cos^n x}\, dx?$$

This exercise illustrates how a little mathematical ingenuity solves a problem not immediately amenable to solution by a CAS.

Monte Carlo Integration

Use Monte Carlo integration to compute the integrals in Exercises 50–55. Compare your answers with exact answers obtained by using a computer algebra system or an analytic method previously presented.

50. $\displaystyle\int_0^1 x e^{-2x}\, dx$ **51.** $\displaystyle\int_{\pi/2}^{\pi} (\sin\, y) e^{\cos\, y}\, dy$

52. $\displaystyle\int_0^{1/\sqrt{2}} 2x \sin^{-1} (x^2)\, dx$ **53.** $\displaystyle\int_0^1 z\sqrt{1 - z}\, dz$

54. $\displaystyle\int_0^{1/2} \frac{t^3\, dt}{t^2 - 2t + 1}$ **55.** $\displaystyle\int_1^2 (\ln\, \theta)^3\, d\theta$

CD-ROM
WEBsite

7.6 L'Hôpital's Rule

Indeterminate Form 0/0 ● Indeterminate Forms ∞/∞, $\infty \cdot 0$, $\infty - \infty$ ● Indeterminate Forms 1^{∞}, 0^{0}, ∞^{0}

As noted earlier, John Bernoulli discovered a rule for calculating limits of fractions whose numerators and denominators both approach zero. The rule is known today as **l'Hôpital's Rule,** after Guillaume François Antoine de l'Hôpital, Marquis de St. Mesme, a French nobleman who wrote the first introductory differential calculus text, where the rule first appeared in print.

Indeterminate Form 0/0

If the continuous functions $f(x)$ and $g(x)$ are both zero at $x = a$, then

$$\lim_{x \to a} \frac{f(x)}{g(x)}$$

cannot be found by substituting $x = a$. The substitution produces 0/0, a meaningless expression known as an **indeterminate form.** Our experience so far has been that limits that lead to indeterminate forms may or may not be hard to find algebraically. It took a lot of analysis in Section 1.2 to find $\lim_{x \to 0} (\sin x)/x$. But we have had remarkable success with the limit

$$f'(a) = \lim_{x \to a} \frac{f(x) - f(a)}{x - a},$$

from which we calculate derivatives and which always produces the equivalent of 0/0 when we substitute $x = a$. L'Hôpital's Rule enables us to draw on our success with derivatives to evaluate limits that otherwise lead to indeterminate forms.

Theorem 1   L'Hôpital's Rule (First Form)
Suppose that $f(a) = g(a) = 0$, that $f'(a)$ and $g'(a)$ exist, and that $g'(a) \neq 0$. Then

$$\lim_{x \to a} \frac{f(x)}{g(x)} = \frac{f'(a)}{g'(a)}.$$

Caution

To apply l'Hôpital's Rule to f/g, divide the derivative of f by the derivative of g. Do not fall into the trap of taking the derivative of f/g. The quotient to use is f'/g', not $(f/g)'$.

Example 1   Using L'Hôpital's Rule

(a) $\lim_{x \to 0} \dfrac{3x - \sin x}{x} = \dfrac{3 - \cos x}{1}\Big|_{x=0} = 2$

(b) $\lim_{x \to 0} \dfrac{\sqrt{1 + x} - 1}{x} = \dfrac{\dfrac{1}{2\sqrt{1 + x}}}{1}\Big|_{x=0} = \dfrac{1}{2}$

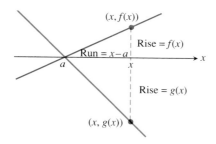

FIGURE 7.11 A zoom-in view of the graphs of the differentiable functions f and g at $x = a$. (Theorem 1)

Proof of Theorem 1

Graphical Argument

If we zoom in on the graphs of f and g at $(a, f(a)) = (a, g(a)) = (a, 0)$, the graphs (Figure 7.11) appear to be straight lines because differentiable functions are locally linear. Let m_1 and m_2 be the slopes of the lines for f and g, respectively. Then for x near a,

$$\frac{f(x)}{g(x)} = \frac{\dfrac{f(x)}{x - a}}{\dfrac{g(x)}{x - a}} = \frac{m_1}{m_2}.$$

As $x \to a$, m_1 and m_2 approach $f'(a)$ and $g'(a)$, respectively. Therefore,

$$\lim_{x \to a} \frac{f(x)}{g(x)} = \lim_{x \to a} \frac{m_1}{m_2} = \frac{f'(a)}{g'(a)}.$$

Confirm Analytically

Working backwards from $f'(a)$ and $g'(a)$, which are themselves limits, we have

$$\frac{f'(a)}{g'(a)} = \frac{\displaystyle\lim_{x \to a} \frac{f(x) - f(a)}{x - a}}{\displaystyle\lim_{x \to a} \frac{g(x) - g(a)}{x - a}} = \lim_{x \to a} \frac{\dfrac{f(x) - f(a)}{x - a}}{\dfrac{g(x) - g(a)}{x - a}}$$

$$= \lim_{x \to a} \frac{f(x) - f(a)}{g(x) - g(a)} = \lim_{x \to a} \frac{f(x) - 0}{g(x) - 0} = \lim_{x \to a} \frac{f(x)}{g(x)}.$$

Sometimes after differentiation, the new numerator and denominator both equal zero at $x = a$, as we see in Example 2. In these cases, we apply a stronger form of l'Hôpital's Rule.

Theorem 2   L'Hôpital's Rule (Stronger Form)

Suppose that $f(a) = g(a) = 0$, that f and g are differentiable on an open interval I containing a, and that $g'(x) \neq 0$ on I if $x \neq a$. Then

$$\lim_{x \to a} \frac{f(x)}{g(x)} = \lim_{x \to a} \frac{f'(x)}{g'(x)},$$

assuming that the limit on the right side exists.

A proof of the finite-limit case of Theorem 2 is given in Appendix 6.

Example 2   Applying Stronger Form of L'Hôpital's Rule

$$\lim_{x \to 0} \frac{\sqrt{1 + x} - 1 - x/2}{x^2} \qquad\qquad \frac{0}{0}$$

$$= \lim_{x \to 0} \frac{(1/2)(1 + x)^{-1/2} - 1/2}{2x} \qquad \text{Still } \frac{0}{0}; \text{differentiate again.}$$

$$= \lim_{x \to 0} \frac{-(1/4)(1 + x)^{-3/2}}{2} = -\frac{1}{8} \qquad \text{Not } \frac{0}{0}; \text{limit is found.}$$

When you apply l'Hôpital's Rule, look for a change from 0/0 to something else. This is where the limit is revealed.

Example 3 Incorrectly Applying Stronger Form of L'Hôpital's Rule

$$\lim_{x \to 0} \frac{1 - \cos x}{x + x^2} \qquad\qquad \frac{0}{0}$$

$$= \lim_{x \to 0} \frac{\sin x}{1 + 2x} = \frac{0}{1} = 0 \qquad \text{Not } \frac{0}{0}; \text{ limit is found.}$$

If we continue to differentiate in an attempt to apply l'Hôpital's Rule once more, we get

$$\lim_{x \to 0} \frac{1 - \cos x}{x + x^2} = \lim_{x \to 0} \frac{\sin x}{1 + 2x} = \lim_{x \to 0} \frac{\cos x}{2} = \frac{1}{2},$$

which is wrong.

L'Hôpital's Rule applies to one-sided limits as well.

Example 4 Using L'Hôpital's Rule with One-Sided Limits

(a) $\displaystyle \lim_{x \to 0^+} \frac{\sin x}{x^2}$ $\qquad\qquad \frac{0}{0}$

$\displaystyle = \lim_{x \to 0^+} \frac{\cos x}{2x} = \infty$ $\qquad\qquad \frac{1}{0}$

(b) $\displaystyle \lim_{x \to 0^-} \frac{\sin x}{x^2}$ $\qquad\qquad \frac{0}{0}$

$\displaystyle = \lim_{x \to 0^-} \frac{\cos x}{2x} = -\infty$ $\qquad\qquad \frac{-1}{0}$

When we reach a point where one of the derivatives approaches 0, as in Example 4, and the other does not, then the limit in question is 0 (if the numerator approaches 0) or infinity (if the denominator approaches 0).

Indeterminate Forms ∞/∞, $\infty \cdot 0$, $\infty - \infty$

A version of l'Hôpital's Rule also applies to quotients that lead to the indeterminant form ∞/∞. If $f(x)$ and $g(x)$ both approach infinity as $x \to a$, then

$$\lim_{x \to a} \frac{f(x)}{g(x)} = \lim_{x \to a} \frac{f'(x)}{g'(x)},$$

provided the latter limit exists. The a here (and in the indeterminate form 0/0) may itself be finite or infinite and may be an endpoint of the interval I of Theorem 2.

Example 5 Working with the Indeterminate Form ∞/∞

Find

(a) $\displaystyle \lim_{x \to \pi/2} \frac{\sec x}{1 + \tan x}$

(b) $\displaystyle \lim_{x \to \infty} \frac{\ln x}{2\sqrt{x}}.$

Solution

(a) The numerator and denominator are discontinuous at $x = \pi/2$, so we investigate the one-sided limits there. To apply l'Hôpital's Rule, we can choose I to be any open interval with $x = \pi/2$ as an endpoint.

$$\lim_{x \to (\pi/2)^-} \frac{\sec x}{1 + \tan x} \qquad \frac{\infty}{\infty} \text{ from the left}$$

$$= \lim_{x \to (\pi/2)^-} \frac{\sec x \tan x}{\sec^2 x} = \lim_{x \to (\pi/2)^-} \sin x = 1$$

The right-hand limit is 1 also, with $(-\infty)/(-\infty)$ as the indeterminate form. Therefore, the two-sided limit is equal to 1.

(b) $\lim\limits_{x \to \infty} \dfrac{\ln x}{2\sqrt{x}} = \lim\limits_{x \to \infty} \dfrac{1/x}{1/\sqrt{x}} = \lim\limits_{x \to \infty} \dfrac{1}{\sqrt{x}} = 0$

We can sometimes handle the indeterminate forms $\infty \cdot 0$ and $\infty - \infty$ by using algebra to get $0/0$ or ∞/∞ instead. Here again we do not mean to suggest that there is a number $\infty \cdot 0$ or $\infty - \infty$ any more than we mean to suggest that there is a number $0/0$ or ∞/∞. These forms are not numbers but are descriptions of function behavior.

Example 6 Working with the Indeterminate Form $\infty \cdot 0$

Find

(a) $\lim\limits_{x \to \infty} \left(x \sin \dfrac{1}{x} \right)$

(b) $\lim\limits_{x \to -\infty} \left(x \sin \dfrac{1}{x} \right)$.

Solution

(a)
$$\lim_{x \to \infty} \left(x \sin \frac{1}{x} \right) \qquad \infty \cdot 0$$

$$= \lim_{h \to 0^+} \left(\frac{1}{h} \sin h \right) \qquad \text{Let } h = 1/x.$$

$$= 1$$

(b) Similarly,

$$\lim_{x \to -\infty} \left(x \sin \frac{1}{x} \right) = 1.$$

Example 7 Working with the Indeterminate Form $\infty - \infty$

Find

$$\lim_{x \to 0} \left(\frac{1}{\sin x} - \frac{1}{x} \right).$$

Solution If $x \to 0^+$, then $\sin x \to 0^+$ and

$$\frac{1}{\sin x} - \frac{1}{x} \to \infty - \infty.$$

Similarly, if $x \to 0^-$, then $\sin x \to 0^-$ and

$$\frac{1}{\sin x} - \frac{1}{x} \to -\infty - (-\infty) = -\infty + \infty.$$

Neither form reveals what happens in the limit. To find out, we first combine the fractions:

$$\frac{1}{\sin x} - \frac{1}{x} = \frac{x - \sin x}{x \sin x} \qquad \text{Common denominator is } x \sin x.$$

Then apply l'Hôpital's Rule to the result:

$$\lim_{x \to 0} \left(\frac{1}{\sin x} - \frac{1}{x} \right) = \lim_{x \to 0} \frac{x - \sin x}{x \sin x} \qquad \frac{0}{0}$$

$$= \lim_{x \to 0} \frac{1 - \cos x}{\sin x + x \cos x} \qquad \text{Still } \frac{0}{0}$$

$$= \lim_{x \to 0} \frac{\sin x}{2 \cos x - x \sin x} = \frac{0}{2} = 0.$$

Indeterminate Forms 1^∞, 0^0, ∞^0

Limits that lead to the indeterminate forms 1^∞, 0^0, and ∞^0 can sometimes be handled by taking logarithms first. We use l'Hôpital's Rule to find the limit of the logarithm and then exponentiate to reveal the original function's behavior.

Since $b = e^{\ln b}$ for every positive number b, we can write $f(x)$ as

$$f(x) = e^{\ln f(x)}$$

for any positive function $f(x)$.

$$\lim_{x \to a} \ln f(x) = L \quad \Rightarrow \quad \lim_{x \to a} f(x) = \lim_{x \to a} e^{\ln f(x)} = e^{\lim_{x \to a} \ln f(x)} = e^L$$

Here a can be finite or infinite.

In Preliminary Section 3, we used graphs and tables to investigate the values of $f(x) = (1 + 1/x)^x$ as $x \to \infty$. Now we find this limit with l'Hôpital's Rule.

CD-ROM
WEBsite

Historical Biography

Augustin-Louis Cauchy
(1789 — 1857)

Example 8 Working with the Indeterminate Form 1^∞

Find

$$\lim_{x \to \infty} \left(1 + \frac{1}{x} \right)^x.$$

Solution Let $f(x) = (1 + 1/x)^x$. Then taking logarithms of both sides converts the indeterminate form 1^∞ to $0/0$, to which we can apply l'Hôpital's Rule.

$$\ln f(x) = \ln \left(1 + \frac{1}{x} \right)^x = x \ln \left(1 + \frac{1}{x} \right) = \frac{\ln \left(1 + \frac{1}{x} \right)}{\frac{1}{x}}$$

We apply l'Hôpital's Rule to the last expression above.

$$\lim_{x \to \infty} \ln f(x) = \lim_{x \to \infty} \frac{\ln\left(1 + \dfrac{1}{x}\right)}{\dfrac{1}{x}} \qquad \frac{0}{0}$$

$$= \lim_{x \to \infty} \frac{\dfrac{1}{1 + \dfrac{1}{x}}\left(-\dfrac{1}{x^2}\right)}{-\dfrac{1}{x^2}} \qquad \text{Differentiate numerator and denominator.}$$

$$= \lim_{x \to \infty} \frac{1}{1 + \dfrac{1}{x}} = 1$$

Therefore,

$$\lim_{x \to \infty}\left(1 + \frac{1}{x}\right)^x = \lim_{x \to \infty} f(x) = \lim_{x \to \infty} e^{\ln f(x)} = e^1 = e.$$

Example 9   Working with the Indeterminate Form 0^0

Determine whether $\lim_{x \to 0^+} x^x$ exists and find its value if it does.

Solution

The limit leads to the indeterminate form 0^0. To convert the problem to one involving $0/0$, we let $f(x) = x^x$ and take the logarithm of both sides.

$$\ln f(x) = x \ln x = \frac{\ln x}{1/x}$$

Applying l'Hôpital's Rule to $(\ln x)/(1/x)$, we obtain

$$\lim_{x \to 0^+} \ln f(x) = \lim_{x \to 0^+} \frac{\ln x}{1/x} \qquad \frac{-\infty}{\infty}$$

$$= \lim_{x \to 0^+} \frac{1/x}{-1/x^2} \qquad \text{Differentiate.}$$

$$= \lim_{x \to 0^+} (-x) = 0.$$

Therefore,

$$\lim_{x \to 0^+} x^x = \lim_{x \to 0^+} f(x) = \lim_{x \to 0^+} e^{\ln f(x)} = e^0 = 1.$$

Example 10   Working with the Indeterminate Form ∞^0

Find $\lim_{x \to \infty} x^{1/x}$.

Solution   Let $f(x) = x^{1/x}$. Then

$$\ln f(x) = \frac{\ln x}{x}.$$

Applying l'Hôpital's Rule to $\ln f(x)$, we obtain

$$\lim_{x\to\infty} \ln f(x) = \lim_{x\to\infty} \frac{\ln x}{x} \qquad \frac{\infty}{\infty}$$

$$= \lim_{x\to\infty} \frac{1/x}{1} \qquad \text{Differentiate.}$$

$$= \lim_{x\to\infty} \frac{1}{x} = 0.$$

Therefore,

$$\lim_{x\to\infty} x^{1/x} = \lim_{x\to\infty} f(x) = \lim_{x\to\infty} e^{\ln f(x)} = e^0 = 1.$$

EXERCISES 7.6

Finding Limits

In Exercises 1–6, use l'Hôpital's Rule to evaluate the limit. Then evaluate the limit using a method studied in Chapter 1.

1. $\displaystyle\lim_{x\to2} \frac{x-2}{x^2-4}$

2. $\displaystyle\lim_{x\to0} \frac{\sin 5x}{x}$

3. $\displaystyle\lim_{x\to\infty} \frac{5x^2-3x}{7x^2+1}$

4. $\displaystyle\lim_{x\to1} \frac{x^3-1}{4x^3-x-3}$

5. $\displaystyle\lim_{x\to0} \frac{1-\cos x}{x^2}$

6. $\displaystyle\lim_{x\to\infty} \frac{2x^2+3x}{x^3+x+1}$

Applying l'Hôpital's Rule

Use l'Hôpital's Rule to find the limits in Exercises 7–38.

7. $\displaystyle\lim_{\theta\to0} \frac{\sin\theta^2}{\theta}$

8. $\displaystyle\lim_{\theta\to\pi/2} \frac{1-\sin\theta}{1+\cos 2\theta}$

9. $\displaystyle\lim_{t\to0} \frac{\cos t-1}{e^t-t-1}$

10. $\displaystyle\lim_{t\to1} \frac{t-1}{\ln t-\sin\pi t}$

11. $\displaystyle\lim_{x\to\infty} \frac{\ln(x+1)}{\log_2 x}$

12. $\displaystyle\lim_{x\to\infty} \frac{\log_2 x}{\log_3(x+3)}$

13. $\displaystyle\lim_{y\to0^+} \frac{\ln(y^2+2y)}{\ln y}$

14. $\displaystyle\lim_{y\to\pi/2} \left(\frac{\pi}{2}-y\right)\tan y$

15. $\displaystyle\lim_{x\to0^+} x\ln x$

16. $\displaystyle\lim_{x\to\infty} x\tan\frac{1}{x}$

17. $\displaystyle\lim_{x\to0^+}(\csc x-\cot x+\cos x)$

18. $\displaystyle\lim_{x\to\infty}(\ln 2x-\ln(x+1))$

19. $\displaystyle\lim_{x\to0^+}(\ln x-\ln\sin x)$

20. $\displaystyle\lim_{x\to0^+}\left(\frac{1}{x}-\frac{1}{\sqrt{x}}\right)$

21. $\displaystyle\lim_{x\to0}(e^x+x)^{1/x}$

22. $\displaystyle\lim_{x\to0}\left(\frac{1}{x^2}\right)^x$

23. $\displaystyle\lim_{x\to\pm\infty} \frac{3x-5}{2x^2-x+2}$

24. $\displaystyle\lim_{x\to0} \frac{\sin 7x}{\tan 11x}$

25. $\displaystyle\lim_{x\to\infty}(\ln x)^{1/x}$

26. $\displaystyle\lim_{x\to\infty}(1+2x)^{1/(2\ln x)}$

27. $\displaystyle\lim_{x\to1}(x^2-2x+1)^{x-1}$

28. $\displaystyle\lim_{x\to(\pi/2)^-}(\cos x)^{\cos x}$

29. $\displaystyle\lim_{x\to0^+}(1+x)^{1/x}$

30. $\displaystyle\lim_{x\to1} x^{1/(x-1)}$

31. $\displaystyle\lim_{x\to0^+}(\sin x)^x$

32. $\displaystyle\lim_{x\to0^+}(\sin x)^{\tan x}$

33. $\displaystyle\lim_{x\to1^+} x^{1/(1-x)}$

34. $\displaystyle\lim_{x\to\infty} x^2 e^{-x}$

35. $\displaystyle\lim_{x\to\infty} \int_x^{2x} \frac{1}{t}\,dt$

36. $\displaystyle\lim_{x\to\infty} \frac{1}{x\ln x}\int_1^x \ln t\,dt$

37. $\displaystyle\lim_{\theta\to0} \frac{\cos\theta-1}{e^\theta-\theta-1}$

38. $\displaystyle\lim_{t\to\infty} \frac{e^t+t^2}{e^t-t}$

Theory and Applications

L'Hôpital's Rule does not help with the limits in Exercises 39–42. Try it; you just keep on cycling. Find the limits some other way.

39. $\displaystyle\lim_{x\to\infty} \frac{\sqrt{9x+1}}{\sqrt{x+1}}$

40. $\displaystyle\lim_{x\to0^+} \frac{\sqrt{x}}{\sqrt{\sin x}}$

41. $\displaystyle\lim_{x\to(\pi/2)^-} \frac{\sec x}{\tan x}$

42. $\displaystyle\lim_{x\to0^+} \frac{\cot x}{\csc x}$

43. *Writing to Learn* Which one is correct, and which one is wrong? Give reasons for your answers.

(a) $\displaystyle\lim_{x\to3} \frac{x-3}{x^2-3} = \lim_{x\to3} \frac{1}{2x} = \frac{1}{6}$

(b) $\displaystyle\lim_{x\to3} \frac{x-3}{x^2-3} = \frac{0}{6} = 0$

44. *∞/∞ form* Give an example of two differentiable functions f and g with $\lim_{x \to \infty} f(x) = \lim_{x \to \infty} g(x) = \infty$ that satisfy the following.

(a) $\lim_{x \to \infty} \dfrac{f(x)}{g(x)} = 3$ (b) $\lim_{x \to \infty} \dfrac{f(x)}{g(x)} = 0$

(c) $\lim_{x \to \infty} \dfrac{f(x)}{g(x)} = \infty$

45. *Writing to Learn: Continuous extension* Find a value of c that makes the function

$$f(x) = \begin{cases} \dfrac{9x - 3\sin 3x}{5x^3}, & x \neq 0 \\ c, & x = 0 \end{cases}$$

continuous at $x = 0$. Explain why your value of c works.

46. *L'Hôpital's Rule* Let

$$f(x) = \begin{cases} x + 2, & x \neq 0 \\ 0, & x = 0 \end{cases} \quad \text{and} \quad g(x) = \begin{cases} x + 1, & x \neq 0 \\ 0, & x = 0. \end{cases}$$

(a) Show that

$$\lim_{x \to 0} \frac{f'(x)}{g'(x)} = 1 \quad \text{but} \quad \lim_{x \to 0} \frac{f(x)}{g(x)} = 2.$$

(b) *Writing to Learn* Explain why this does not contradict l'Hôpital's Rule.

47. *Interest compounded continuously*

(a) Show that

$$\lim_{k \to \infty} A_0 \left(1 + \frac{r}{k} \right)^{kt} = A_0 e^{rt}.$$

(b) *Writing to Learn* Explain how the limit in part (a) connects interest compounded k times per year with interest compounded continuously.

T 48. *0/0 form* Estimate the value of

$$\lim_{x \to 1} \frac{2x^2 - (3x + 1)\sqrt{x} + 2}{x - 1}$$

by graphing. Then confirm your estimate with l'Hôpital's Rule.

T 49. *0/0 form*

(a) Estimate the value of

$$\lim_{x \to 1} \frac{(x - 1)^2}{x \ln x - x - \cos \pi x}$$

by graphing $f(x) = (x - 1)^2/(x \ln x - x - \cos \pi x)$ near $x = 1$. Then confirm your estimate with l'Hôpital's Rule.

(b) Graph f for $0 < x \leq 11$.

50. *Why 0^∞ and $0^{-\infty}$ are not indeterminate forms* Assume that $f(x)$ is nonnegative in an open interval containing c and $\lim_{x \to c} f(x) = 0$.

(a) If $\lim_{x \to c} g(x) = \infty$, show that $\lim_{x \to c} f(x)^{g(x)} = 0$.

(b) If $\lim_{x \to c} g(x) = -\infty$, show that $\lim_{x \to c} f(x)^{g(x)} = \infty$.

T 51. *Grapher precision* Let

$$f(x) = \frac{1 - \cos x^6}{x^{12}}.$$

Explain why some graphs of f may give false information about $\lim_{x \to 0} f(x)$. (*Hint:* Try the window $[-1, 1]$ by $[-0.5, 1]$.)

T 52. *∞ − ∞ form*

(a) Estimate the value of

$$\lim_{x \to \infty} (x - \sqrt{x^2 + x})$$

by graphing $f(x) = x - \sqrt{x^2 + x}$ over a suitably large interval of x-values.

(b) Now confirm your estimate by finding the limit with l'Hôpital's Rule. As the first step, multiply $f(x)$ by the fraction $(x + \sqrt{x^2 + x})/(x + \sqrt{x^2 + x})$ and simplify the new numerator.

53. *Exponential functions*

(a) Use the equation

$$a^x = e^{x \ln a}$$

to find the domain of

$$f(x) = \left(1 + \frac{1}{x} \right)^x.$$

(b) Find $\lim_{x \to -1^-} f(x)$.

(c) Find $\lim_{x \to -\infty} f(x)$.

54. *Extended exponentials* Given that $x > 0$, find the maximum value, if any, of

(a) $x^{1/x}$ (b) x^{1/x^2}

(c) x^{1/x^n} (n a positive integer)

(d) Show that $\lim_{x \to \infty} x^{1/x^n} = 1$ for every positive integer n.

T 55. *The place of ln x among the powers of x* The natural logarithm

$$\ln x = \int_1^x \frac{1}{t} \, dt$$

fills the gap in the set of formulas

$$\int t^{k-1} \, dt = \frac{t^k}{k} + C, \qquad k \neq 0,$$

but the formulas themselves do not reveal how well the logarithm fits in. We can see the nice fit graphically if we select the specific antiderivatives

$$\int_1^x t^{k-1}\, dt = \frac{x^k - 1}{k}, \qquad x > 0,$$

and compare their graphs with the graph of $\ln x$.

(a) Graph the functions $f(x) = (x^k - 1)/k$ together with $\ln x$ on the interval $0 \le x \le 50$ for $k = \pm 1,\ \pm 0.5,\ \pm 0.1$, and ± 0.05.

(b) Show that

$$\lim_{k \to 0} \frac{x^k - 1}{k} = \ln x.$$

(*Source:* "The Place of $\ln x$ Among the Powers of x" by Henry C. Finlayson, *American Mathematical Monthly*, Vol. 94, No. 5 (May 1987), p. 450.)

T **56.** *The continuous extension of* $(\sin x)^x$ *to* $[0, \pi]$

(a) Graph $f(x) = (\sin x)^x$ on the interval $0 \le x \le \pi$. What value would you assign to f to make it continuous at $x = 0$?

(b) Verify your conclusion in part (a) by finding $\lim_{x \to 0^+} f(x)$ with l'Hôpital's Rule.

(c) Returning to the graph, estimate the maximum value of f on $[0, \pi]$. About where is max f taken on?

(d) Sharpen your estimate in part (c) by graphing f' in the same window to see where its graph crosses the x-axis. To simplify your work, you might want to delete the exponential factor from the expression for f' and graph just the factor that has a zero.

(e) Sharpen your estimate of the location of max f further still by solving the equation $f' = 0$ numerically.

(f) Estimate max f by evaluating f at the locations you found in parts (c), (d), and (e). What is your best value for max f?

7.7 Improper Integrals

Infinite Limits of Integration • The Integral $\int_1^\infty dx/x^p$ • Integrands with Infinite Discontinuities • Tests for Convergence and Divergence • Computer Algebra Systems

CD-ROM
WEBsite

Up to now, we have required our definite integrals to have two properties: first, that the domain of integration, from a to b, be finite, and second, that the range of the integrand be finite on this domain. In practice, however, we frequently encounter problems that fail to meet one or both of these conditions. As an example of an infinite domain, we might consider the area under the curve $y = (\ln x)/x^2$ from $x = 1$ to $x = \infty$ (Figure 7.12a). As an example of an infinite range, we might

FIGURE **7.12** Are the areas under these infinite curves finite?

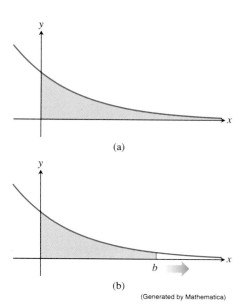

(a)

(b)

(Generated by Mathematica)

FIGURE 7.13 (a) The area in the first quadrant under the curve $y = e^{-x/2}$ is (b)

$$\lim_{b \to \infty} \int_0^b e^{-x/2} \, dx.$$

consider the area under the curve $y = 1/\sqrt{x}$ between $x = 0$ and $x = 1$ (Figure 7.12b). We treat both examples in the same reasonable way. We ask, "What is the integral when the domain is slightly less?" and examine the answer as the domain increases to the limit. We do the finite case and then see what happens as we approach infinity.

Infinite Limits of Integration

Consider the infinite region that lies under the curve $y = e^{-x/2}$ in the first quadrant (Figure 7.13a). You might think this region has infinite area, but we will see that the natural value to assign is finite. Here is how we assign a value to the area. First we find the area $A(b)$ of the portion of the region that is bounded on the right by $x = b$ (Figure 7.13b).

$$A(b) = \int_0^b e^{-x/2} \, dx = -2e^{-x/2} \Big]_0^b = -2e^{-b/2} + 2$$

Then we find the limit of $A(b)$ as $b \to \infty$.

$$\lim_{b \to \infty} A(b) = \lim_{b \to \infty} (-2e^{-b/2} + 2) = 2$$

The value we assign to the area under the curve from 0 to ∞ is

$$\int_0^\infty e^{-x/2} \, dx = \lim_{b \to \infty} \int_0^b e^{-x/2} \, dx = 2.$$

Definition   Improper Integrals with Infinite Integration Limits
Integrals with infinite limits of integration are **improper integrals.**

1. If $f(x)$ is continuous on $[a, \infty)$, then

$$\int_a^\infty f(x) \, dx = \lim_{b \to \infty} \int_a^b f(x) \, dx.$$

2. If $f(x)$ is continuous on $(-\infty, b]$, then

$$\int_{-\infty}^b f(x) \, dx = \lim_{a \to -\infty} \int_a^b f(x) \, dx.$$

3. If $f(x)$ is continuous on $(-\infty, \infty)$, then

$$\int_{-\infty}^\infty f(x) \, dx = \int_{-\infty}^c f(x) \, dx + \int_c^\infty f(x) \, dx,$$

where c is any real number.

In parts 1 and 2, if the limit is finite, the improper integral **converges** and the limit is the **value** of the improper integral. If the limit fails to exist, the improper integral **diverges**. In part 3, the integral on the left-hand side of the equation

converges if both improper integrals on the right-hand side converge; otherwise it **diverges** and has no value. It can be shown that the choice of c in part 3 is unimportant. We can evaluate or determine the convergence or divergence of $\int_{-\infty}^{\infty} f(x)\,dx$ with any convenient choice.

Example 1   Evaluating an Improper Integral on $[1, \infty)$

Is the area under the curve $y = (\ln x)/x^2$ from $x = 1$ to $x = \infty$ finite? If so, what is it?

Solution   We find the area under the curve from $x = 1$ to $x = b$ and examine the limit as $b \to \infty$. If the limit is finite, we take it to be the area under the infinite curve (Figure 7.14). The area from 1 to b is

$$\int_1^b \frac{\ln x}{x^2}\,dx = \left[(\ln x)\left(-\frac{1}{x}\right)\right]_1^b - \int_1^b \left(-\frac{1}{x}\right)\left(\frac{1}{x}\right)dx \qquad \begin{array}{l} \text{Integration by parts with}\\ u = \ln x,\ dv = dx/x^2,\\ du = dx/x,\ v = -1/x \end{array}$$

$$= -\frac{\ln b}{b} - \left[\frac{1}{x}\right]_1^b$$

$$= -\frac{\ln b}{b} - \frac{1}{b} + 1.$$

The limit of the area as $b \to \infty$ is

$$\int_1^{\infty} \frac{\ln x}{x^2}\,dx = \lim_{b \to \infty} \int_1^b \frac{\ln x}{x^2}\,dx$$

$$= \lim_{b \to \infty}\left[-\frac{\ln b}{b} - \frac{1}{b} + 1\right]$$

$$= -\left[\lim_{b \to \infty} \frac{\ln b}{b}\right] - 0 + 1$$

$$= -\left[\lim_{b \to \infty} \frac{1/b}{1}\right] + 1 = 0 + 1 = 1. \qquad \text{l'Hôpital's Rule}$$

Thus, the improper integral converges and the area has finite value 1.

Example 2   Evaluating an Integral on $(-\infty, \infty)$

Evaluate

$$\int_{-\infty}^{\infty} \frac{dx}{1 + x^2}.$$

Solution   According to the definition (part 3), we can write

$$\int_{-\infty}^{\infty} \frac{dx}{1 + x^2} = \int_{-\infty}^{0} \frac{dx}{1 + x^2} + \int_{0}^{\infty} \frac{dx}{1 + x^2}.$$

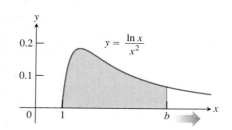

FIGURE 7.14 The area under this curve is

$$\lim_{b \to \infty} \int_1^b \frac{\ln x}{x^2}\,dx.$$

(Example 1)

Next we evaluate each improper integral on the right-hand side of the equation above.

$$\int_{-\infty}^{0} \frac{dx}{1+x^2} = \lim_{a \to -\infty} \int_{a}^{0} \frac{dx}{1+x^2}$$

$$= \lim_{a \to -\infty} \tan^{-1} x \Big]_{a}^{0}$$

$$= \lim_{a \to -\infty} (\tan^{-1} 0 - \tan^{-1} a) = 0 - \left(-\frac{\pi}{2}\right) = \frac{\pi}{2}$$

$$\int_{0}^{\infty} \frac{dx}{1+x^2} = \lim_{b \to \infty} \int_{0}^{b} \frac{dx}{1+x^2}$$

$$= \lim_{b \to \infty} \tan^{-1} x \Big]_{0}^{b}$$

$$= \lim_{b \to \infty} (\tan^{-1} b - \tan^{-1} 0) = \frac{\pi}{2} - 0 = \frac{\pi}{2}$$

Thus,

$$\int_{-\infty}^{\infty} \frac{dx}{1+x^2} = \frac{\pi}{2} + \frac{\pi}{2} = \pi.$$

The Integral $\displaystyle\int_{1}^{\infty} \frac{dx}{x^p}$

The function $y = 1/x$ is the boundary between the convergent and divergent improper integrals with integrands of the form $y = 1/x^p$. Example 3 explains.

Example 3   Determining Convergence

For what values of p does the integral $\int_{1}^{\infty} dx/x^p$ converge? When the integral does converge, what is its value?

Solution   If $p \neq 1$,

$$\int_{1}^{b} \frac{dx}{x^p} = \frac{x^{-p+1}}{-p+1} \Big]_{1}^{b} = \frac{1}{1-p}(b^{-p+1} - 1) = \frac{1}{1-p}\left(\frac{1}{b^{p-1}} - 1\right).$$

Thus,

$$\int_{1}^{\infty} \frac{dx}{x^p} = \lim_{b \to \infty} \int_{1}^{b} \frac{dx}{x^p}$$

$$= \lim_{b \to \infty} \left[\frac{1}{1-p}\left(\frac{1}{b^{p-1}} - 1\right)\right] = \begin{cases} \dfrac{1}{p-1}, & p > 1 \\ \infty, & p < 1 \end{cases}$$

because

$$\lim_{b \to \infty} \frac{1}{b^{p-1}} = \begin{cases} 0, & p > 1 \\ \infty, & p < 1. \end{cases}$$

Therefore, the integral converges to the value $1/(p - 1)$ if $p > 1$ and it diverges if $p < 1$.

If $p = 1$,

$$\int_1^\infty \frac{dx}{x^p} = \int_1^\infty \frac{dx}{x}$$

$$= \lim_{b \to \infty} \int_1^b \frac{dx}{x}$$

$$= \lim_{b \to \infty} \ln x \Big]_1^b$$

$$= \lim_{b \to \infty} (\ln b - \ln 1) = \infty$$

and the integral diverges.

Integrands with Infinite Discontinuities

Another type of improper integral arises when the integrand has a vertical asymptote—an infinite discontinuity—at a limit of integration or at some point between the limits of integration.

Consider the infinite region in the first quadrant that lies under the curve $y = 1/\sqrt{x}$ from $x = 0$ to $x = 1$ (Figure 7.12b). First we find the area of the portion from a to 1 (Figure 7.15).

$$\int_a^1 \frac{dx}{\sqrt{x}} = 2\sqrt{x} \Big]_a^1 = 2 - 2\sqrt{a}$$

Then we find the limit of this area as $a \to 0^+$.

$$\lim_{a \to 0^+} \int_a^1 \frac{dx}{\sqrt{x}} = \lim_{a \to 0^+} (2 - 2\sqrt{a}) = 2$$

The area under the curve from 0 to 1 is

$$\int_0^1 \frac{dx}{\sqrt{x}} = \lim_{a \to 0^+} \int_a^1 \frac{dx}{\sqrt{x}} = 2.$$

FIGURE 7.15 The area under this curve is

$$\lim_{a \to 0^+} \int_a^1 \left(\frac{1}{\sqrt{x}} \right) dx.$$

Definition   Improper Integrals with Infinite Discontinuities

Integrals of functions that become infinite at a point within the interval of integration are **improper integrals.**

1. If $f(x)$ is continuous on $(a, b]$, then

$$\int_a^b f(x)\, dx = \lim_{c \to a^+} \int_c^b f(x)\, dx.$$

2. If $f(x)$ is continuous on $[a, b)$, then

$$\int_a^b f(x)\, dx = \lim_{c \to b^-} \int_a^c f(x)\, dx.$$

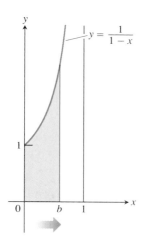

FIGURE 7.16 If the limit exists,

$$\int_0^1 \left(\frac{1}{1-x}\right) dx = \lim_{b \to 1^-} \int_0^b \frac{1}{1-x} dx.$$

(Example 4)

3. If $f(x)$ is continuous on $[a, c) \cup (c, b]$, then

$$\int_a^b f(x) \, dx = \int_a^c f(x) \, dx + \int_c^b f(x) \, dx.$$

In parts 1 and 2, if the limit is finite, the improper integral **converges** and the limit is the **value** of the improper integral. If the limit fails to exist, the improper integral **diverges**. In part 3, the integral on the left-hand side of the equation **converges** if both integrals on the right-hand side have values; otherwise it **diverges**.

Example 4   A Divergent Improper Integral

Investigate the convergence of

$$\int_0^1 \frac{1}{1-x} \, dx.$$

Solution   The integrand $f(x) = 1/(1-x)$ is continuous on $[0, 1)$ but becomes infinite as $x \to 1^-$ (Figure 7.16). We evaluate the integral as

$$\lim_{b \to 1^-} \int_0^b \frac{1}{1-x} \, dx = \lim_{b \to 1^-} \left[-\ln |1 - x| \right]_0^b$$

$$= \lim_{b \to 1^-} \left[-\ln (1 - b) + 0 \right] = \infty.$$

The limit is infinite, so the integral diverges.

Example 5   Infinite Discontinuity at an Interior Point

Evaluate

$$\int_0^3 \frac{dx}{(x-1)^{2/3}}.$$

Solution   The integrand has a vertical asymptote at $x = 1$ and is continuous on $[0, 1)$ and $(1, 3]$ (Figure 7.17). Thus, by part 3 of the definition above,

$$\int_0^3 \frac{dx}{(x-1)^{2/3}} = \int_0^1 \frac{dx}{(x-1)^{2/3}} + \int_1^3 \frac{dx}{(x-1)^{2/3}}.$$

Next, we evaluate each improper integral on the right-hand side of this equation.

$$\int_0^1 \frac{dx}{(x-1)^{2/3}} = \lim_{b \to 1^-} \int_0^b \frac{dx}{(x-1)^{2/3}}$$

$$= \lim_{b \to 1^-} 3(x-1)^{1/3} \bigg]_0^b$$

$$= \lim_{b \to 1^-} [3(b-1)^{1/3} + 3] = 3$$

$$\int_1^3 \frac{dx}{(x-1)^{2/3}} = \lim_{c \to 1^+} \int_c^3 \frac{dx}{(x-1)^{2/3}}$$

$$= \lim_{c \to 1^+} 3(x-1)^{1/3} \bigg]_c^3$$

$$= \lim_{c \to 1^+} [3(3-1)^{1/3} - 3(c-1)^{1/3}] = 3\sqrt[3]{2}$$

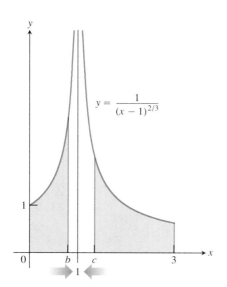

FIGURE 7.17 Example 5 investigates the convergence of

$$\int_0^3 \frac{1}{(x-1)^{2/3}} \, dx.$$

We conclude that

$$\int_0^3 \frac{dx}{(x-1)^{2/3}} = 3 + 3\sqrt[3]{2}.$$

Example 6   Finding the Volume of an Infinite Solid

The cross sections of the solid horn in Figure 7.18 perpendicular to the x-axis are circular disks with diameters reaching from the x-axis to the curve $y = e^x$, $-\infty < x \le \ln 2$. Find the volume of the horn.

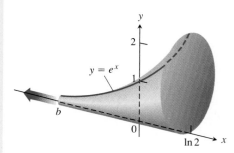

FIGURE 7.18 The calculation in Example 6 shows that this infinite horn has a finite volume.

Solution   The area of a typical cross section is

$$A(x) = \pi(\text{radius})^2 = \pi \left(\frac{1}{2} y\right)^2 = \frac{\pi}{4} e^{2x}.$$

We define the volume of the horn to be the limit as $b \to -\infty$ of the volume of the portion from b to $\ln 2$. As in Section 5.1 (the method of slicing), the volume of this portion is

$$V = \int_b^{\ln 2} A(x)\, dx = \int_b^{\ln 2} \frac{\pi}{4} e^{2x}\, dx = \frac{\pi}{8} e^{2x} \Big]_b^{\ln 2}$$

$$= \frac{\pi}{8} (e^{\ln 4} - e^{2b}) = \frac{\pi}{8} (4 - e^{2b}).$$

As $b \to -\infty$, $e^{2b} \to 0$ and $V \to (\pi/8)(4 - 0) = \pi/2$. The volume of the horn is $\pi/2$.

Example 7   Finding Circumference

Use the arc length formula (Section 5.3) to show that the circumference of the circle $x^2 + y^2 = 4$ is 4π.

Solution   One fourth of this circle is given by $y = \sqrt{4 - x^2}$, $0 \le x \le 2$. Its arc length is

$$L = \int_0^2 \sqrt{1 + (y')^2}\, dx, \qquad \text{where} \qquad y' = -\frac{x}{\sqrt{4 - x^2}}.$$

The integral is improper because y' is not defined at $x = 2$. We evaluate it as a limit.

$$
\begin{aligned}
L = \int_0^2 \sqrt{1 + (y')^2}\, dx &= \int_0^2 \sqrt{1 + \frac{x^2}{4 - x^2}}\, dx \\
&= \int_0^2 \sqrt{\frac{4}{4 - x^2}}\, dx \\
&= \lim_{b \to 2^-} \int_0^b \sqrt{\frac{4}{4 - x^2}}\, dx \\
&= \lim_{b \to 2^-} \int_0^b \sqrt{\frac{1}{1 - (x/2)^2}}\, dx \\
&= \lim_{b \to 2^-} 2\sin^{-1} \frac{x}{2} \Big]_0^b \\
&= \lim_{b \to 2^-} 2 \left[\sin^{-1} \frac{b}{2} - 0 \right] = \pi
\end{aligned}
$$

The circumference of the quarter circle is π; the circumference of the circle is 4π.

Bounded Monotonic Functions

It can be shown that a monotonic function $f(x)$ that is bounded on an infinite interval (a, ∞) must have a finite limit as $x \to \infty$. In Example 8, this is applied to the function

$$
f(b) = \int_1^b e^{-x^2}\, dx
$$

as $b \to \infty$.

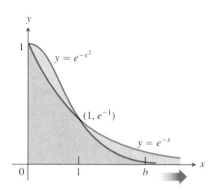

FIGURE 7.19 The graph of e^{-x^2} lies below the graph of e^{-x} for $x > 1$. (Example 8)

Tests for Convergence and Divergence

When we cannot evaluate an improper integral directly (often the case in practice), we first try to determine whether it converges or diverges. If the integral diverges, that's the end of the story. If it converges, we can then use numerical methods to approximate its value. The principal tests for convergence or divergence are the Direct Comparison Test and the Limit Comparison Test.

Example 8   Investigating Convergence

Does the integral $\int_1^\infty e^{-x^2}\, dx$ converge?

Solution   By definition,

$$
\int_1^\infty e^{-x^2}\, dx = \lim_{b \to \infty} \int_1^b e^{-x^2}\, dx.
$$

We cannot evaluate the latter integral directly because there is no simple formula for the antiderivative of e^{-x^2}. We must therefore determine its convergence or divergence some other way. Because $e^{-x^2} > 0$ for all x, $\int_1^b e^{-x^2}\, dx$ is an increasing function of b. Therefore, as $b \to \infty$, the integral either becomes infinite as $b \to \infty$ or it is bounded from above and is forced to converge (have a finite limit).

The two curves $y = e^{-x^2}$ and $y = e^{-x}$ intersect at $(1, e^{-1})$, and $0 < e^{-x^2} \le e^{-x}$ for $x \ge 1$ (Figure 7.19). Thus, for any $b > 1$,

$$
0 < \int_1^b e^{-x^2}\, dx \le \int_1^b e^{-x}\, dx = -e^{-b} + e^{-1} < e^{-1} \approx 0.368.
$$

Rounded up to be safe

As an increasing function of b bounded above by 0.368, the integral $\int_1^b e^{-x^2}\, dx$ must converge as $b \to \infty$. This does not tell us much about the value of the improper integral, however, except that it is positive and less than 0.368.

The comparison of e^{-x^2} and e^{-x} in Example 8 is a special case of the following test.

Theorem 3   Direct Comparison Test

Let f and g be continuous on $[a, \infty)$ with $0 \le f(x) \le g(x)$ for all $x \ge a$. Then

1. $\displaystyle\int_a^\infty f(x)\, dx$ converges if $\displaystyle\int_a^\infty g(x)\, dx$ converges

2. $\displaystyle\int_a^\infty g(x)\, dx$ diverges if $\displaystyle\int_a^\infty f(x)\, dx$ diverges.

Example 9   Using the Direct Comparison Test

(a) $\displaystyle\int_1^\infty \frac{\sin^2 x}{x^2}\, dx$ converges because

$$0 \le \frac{\sin^2 x}{x^2} \le \frac{1}{x^2} \quad \text{on} \quad [1, \infty) \quad \text{and} \quad \int_1^\infty \frac{1}{x^2}\, dx \text{ converges.} \qquad \text{Example 3}$$

(b) $\displaystyle\int_1^\infty \frac{1}{\sqrt{x^2 - 0.1}}\, dx$ diverges because

$$\frac{1}{\sqrt{x^2 - 0.1}} \ge \frac{1}{x} \quad \text{on} \quad [1, \infty) \quad \text{and} \quad \int_1^\infty \frac{1}{x}\, dx \text{ diverges.} \qquad \text{Example 3}$$

Theorem 4   Limit Comparison Test

If the positive functions f and g are continuous on $[a, \infty)$ and if

$$\lim_{x \to \infty} \frac{f(x)}{g(x)} = L, \qquad 0 < L < \infty,$$

then

$$\int_a^\infty f(x)\, dx \qquad \text{and} \qquad \int_a^\infty g(x)\, dx$$

both converge or both diverge.

A proof of Theorem 4 is given in advanced calculus.

Although the improper integrals of two functions from a to ∞ may both converge, this does not mean that their integrals necessarily have the same value, as the next example shows.

Example 10   Using the Limit Comparison Test

Show that

$$\int_1^\infty \frac{dx}{1+x^2}$$

converges by comparison with $\int_1^\infty (1/x^2)\,dx$. Find and compare the two integral values.

Solution   The functions $f(x) = 1/x^2$ and $g(x) = 1/(1+x^2)$ are positive and continuous on $[1, \infty)$. Also,

$$\lim_{x\to\infty} \frac{f(x)}{g(x)} = \lim_{x\to\infty} \frac{1/x^2}{1/(1+x^2)} = \lim_{x\to\infty} \frac{1+x^2}{x^2}$$

$$= \lim_{x\to\infty}\left(\frac{1}{x^2}+1\right) = 0+1 = 1,$$

a positive finite limit (Figure 7.20). Therefore, $\int_1^\infty \frac{dx}{1+x^2}$ converges because $\int_1^\infty \frac{dx}{1+x^2}$ converges.

The integrals converge to different values, however.

$$\int_1^\infty \frac{dx}{x^2} = \frac{1}{2-1} = 1 \qquad \text{Example 3}$$

and

$$\int_1^\infty \frac{dx}{1+x^2} = \lim_{b\to\infty}\int_1^b \frac{dx}{1+x^2}$$

$$= \lim_{b\to\infty}[\tan^{-1}b - \tan^{-1}1] = \frac{\pi}{2} - \frac{\pi}{4} = \frac{\pi}{4}$$

FIGURE 7.20 The functions in Example 10.

Example 11   Using the Limit Comparison Test

Show that

$$\int_1^\infty \frac{3}{e^x-5}\,dx$$

converges.

Solution   From Example 8, it is easy to see that $\int_1^\infty e^{-x}\,dx = \int_1^\infty (1/e^x)\,dx$ converges. Because

$$\lim_{x\to\infty}\frac{1/e^x}{3/(e^x-5)} = \lim_{x\to\infty}\frac{e^x-5}{3e^x} = \lim_{x\to\infty}\left(\frac{1}{3}-\frac{5}{3e^x}\right) = \frac{1}{3},$$

the integral

$$\int_1^\infty \frac{3}{e^x - 5}\, dx$$

also converges.

Computer Algebra Systems

Computer algebra systems can evaluate many convergent improper integrals.

Example 12   Using a CAS

Evaluate the integral

$$\int_2^\infty \frac{x + 3}{(x - 1)(x^2 + 1)}\, dx.$$

Solution   Using Maple, enter

> f: = (x + 3)/((x − 1)*(x^2 + 1));

Then use the integration command

> int(f, x=2..infinity);

Maple returns the answer

$$-\frac{1}{2}\pi + \ln(5) + \arctan(2).$$

To obtain a numerical result, use the evaluation command **evalf** and specify the number of digits, as follows:

> evalf(", 6);

The ditto symbol (") instructs the computer to evaluate the last expression on the screen, in this case $(-1/2)\pi + \ln(5) + \arctan(2)$. Maple returns 1.14579.
    Using Mathematica, entering

In [1]: = Integrate [(x + 3)/((x − 1)(x^2 + 1)), {x, 2, Infinity}]

returns

$$\text{Out } [1] = \frac{-\text{Pi}}{2} + \text{ArcTan}[2] + \text{Log}[5].$$

To obtain a numerical result with six digits, use the command "N[%, 6]"; it also yields 1.14579.

Types of Improper Integrals Discussed in This Section

INFINITE LIMITS OF INTEGRATION

1. Upper limit

$$\int_1^\infty \frac{\ln x}{x^2}\, dx = \lim_{b \to \infty} \int_1^b \frac{\ln x}{x^2}\, dx$$

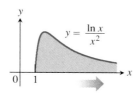

2. Lower limit

$$\int_{-\infty}^0 \frac{dx}{1 + x^2} = \lim_{a \to -\infty} \int_a^0 \frac{dx}{1 + x^2}$$

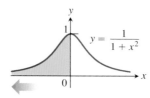

3. Both limits

$$\int_{-\infty}^\infty \frac{dx}{1 + x^2} = \lim_{b \to -\infty} \int_b^0 \frac{dx}{1 + x^2} + \lim_{c \to \infty} \int_0^c \frac{dx}{1 + x^2}$$

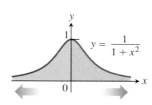

INTEGRAND BECOMES INFINITE

4. Upper endpoint

$$\int_0^1 \frac{dx}{(x - 1)^{2/3}} = \lim_{b \to 1^-} \int_0^b \frac{dx}{(x - 1)^{2/3}}$$

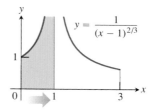

5. Lower endpoint

$$\int_1^3 \frac{dx}{(x - 1)^{2/3}} = \lim_{d \to 1^+} \int_d^3 \frac{dx}{(x - 1)^{2/3}}$$

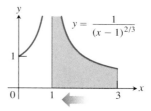

6. Interior point

$$\int_0^3 \frac{dx}{(x - 1)^{2/3}} = \int_0^1 \frac{dx}{(x - 1)^{2/3}} + \int_1^3 \frac{dx}{(x - 1)^{2/3}}$$

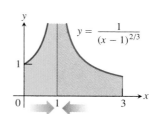

EXERCISES 7.7

Identifying Improper Integrals

In Exercises 1–6, do the following.

(a) State why the integral is improper or involves improper integrals.

(b) Determine whether the integral converges or diverges.

(c) Evaluate the integral if it converges.

1. $\displaystyle\int_0^\infty \frac{dx}{x^2 + 1}$ **2.** $\displaystyle\int_0^1 \frac{dx}{\sqrt{x}}$

3. $\displaystyle\int_{-8}^1 \frac{dx}{x^{1/3}}$ **4.** $\displaystyle\int_{-\infty}^\infty \frac{2x\, dx}{(x^2 + 1)^2}$

5. $\displaystyle\int_0^{\ln 2} x^{-2} e^{1/x}\, dx$ **6.** $\displaystyle\int_0^{\pi/2} \cot \theta\, d\theta$

Evaluating Improper Integrals

Evaluate each integral in Exercises 7–34 or state that it diverges.

7. $\displaystyle\int_1^\infty \frac{dx}{x^{1.001}}$ **8.** $\displaystyle\int_{-1}^1 \frac{dx}{x^{2/3}}$

9. $\displaystyle\int_0^4 \frac{dr}{\sqrt{4 - r}}$ **10.** $\displaystyle\int_0^1 \frac{dr}{r^{0.999}}$

11. $\displaystyle\int_0^1 \frac{dx}{\sqrt{1 - x^2}}$ **12.** $\displaystyle\int_{-\infty}^2 \frac{2\, dx}{x^2 + 4}$

13. $\displaystyle\int_{-\infty}^{-2} \frac{2\, dx}{x^2 - 1}$ **14.** $\displaystyle\int_2^\infty \frac{3\, dt}{t^2 - t}$

15. $\displaystyle\int_0^1 \frac{\theta + 1}{\sqrt{\theta^2 + 2\theta}}\, d\theta$ **16.** $\displaystyle\int_0^2 \frac{s + 1}{\sqrt{4 - s^2}}\, ds$

17. $\displaystyle\int_0^\infty \frac{dx}{(1 + x)\sqrt{x}}$ **18.** $\displaystyle\int_1^\infty \frac{dx}{x\sqrt{x^2 - 1}}$

19. $\displaystyle\int_1^2 \frac{ds}{s\sqrt{s^2 - 1}}$ **20.** $\displaystyle\int_{-1}^\infty \frac{d\theta}{\theta^2 + 5\theta + 6}$

21. $\displaystyle\int_2^\infty \frac{2}{v^2 - v}\, dv$ **22.** $\displaystyle\int_2^\infty \frac{2\, dt}{t^2 - 1}$

23. $\displaystyle\int_0^2 \frac{ds}{\sqrt{4 - s^2}}$ **24.** $\displaystyle\int_0^1 \frac{4r\, dr}{\sqrt{1 - r^4}}$

25. $\displaystyle\int_0^\infty \frac{dv}{(1 + v^2)(1 + \tan^{-1} v)}$ **26.** $\displaystyle\int_0^\infty \frac{16 \tan^{-1} x}{1 + x^2}\, dx$

27. $\displaystyle\int_{-1}^4 \frac{dx}{\sqrt{|x|}}$ **28.** $\displaystyle\int_0^2 \frac{dx}{\sqrt{|x - 1|}}$

29. $\displaystyle\int_{-\infty}^0 \theta e^\theta\, d\theta$ **30.** $\displaystyle\int_0^\infty 2e^{-\theta} \sin \theta\, d\theta$

31. $\displaystyle\int_{-\infty}^\infty e^{-|x|}\, dx$ **32.** $\displaystyle\int_{-\infty}^\infty 2xe^{-x^2}\, dx$

33. $\displaystyle\int_0^1 x \ln x\, dx$ **34.** $\displaystyle\int_0^1 (-\ln x)\, dx$

Testing for Convergence

In Exercises 35–64, use integration, the Direct Comparison Test, or the Limit Comparison Test to test the integrals for convergence. If more than one method applies, use whatever method you prefer.

35. $\displaystyle\int_0^{\pi/2} \tan \theta\, d\theta$ **36.** $\displaystyle\int_0^{\pi/2} \cot \theta\, d\theta$

37. $\displaystyle\int_0^\pi \frac{\sin \theta\, d\theta}{\sqrt{\pi - \theta}}$ **38.** $\displaystyle\int_{-\pi/2}^{\pi/2} \frac{\cos \theta\, d\theta}{(\pi - 2\theta)^{1/3}}$

39. $\displaystyle\int_0^{\ln 2} x^{-2} e^{-1/x}\, dx$ **40.** $\displaystyle\int_0^1 \frac{e^{-\sqrt{x}}}{\sqrt{x}}\, dx$

41. $\displaystyle\int_0^\pi \frac{dt}{\sqrt{t + \sin t}}$

42. $\displaystyle\int_0^1 \frac{dt}{t - \sin t}$ (*Hint:* $t \geq \sin t$ for $t \geq 0$)

43. $\displaystyle\int_0^2 \frac{dx}{1 - x^2}$ **44.** $\displaystyle\int_0^2 \frac{dx}{1 - x}$

45. $\displaystyle\int_{-1}^1 \ln |x|\, dx$ **46.** $\displaystyle\int_{-1}^1 -x \ln |x|\, dx$

47. $\displaystyle\int_1^\infty \frac{dx}{x^3 + 1}$ **48.** $\displaystyle\int_4^\infty \frac{dx}{\sqrt{x} - 1}$

49. $\displaystyle\int_2^\infty \frac{dv}{\sqrt{v} - 1}$ **50.** $\displaystyle\int_4^\infty \frac{2\, dt}{t^{3/2} - 1}$

51. $\displaystyle\int_0^\infty \frac{dx}{\sqrt{x^6 + 1}}$ **52.** $\displaystyle\int_2^\infty \frac{dx}{\sqrt{x^2 - 1}}$

53. $\displaystyle\int_1^\infty \frac{\sqrt{x + 1}}{x^2}\, dx$ **54.** $\displaystyle\int_2^\infty \frac{x\, dx}{\sqrt{x^4 - 1}}$

55. $\displaystyle\int_\pi^\infty \frac{2 + \cos x}{x}\, dx$ **56.** $\displaystyle\int_\pi^\infty \frac{1 + \sin x}{x^2}\, dx$

57. $\displaystyle\int_0^\infty \frac{d\theta}{1 + e^\theta}$ **58.** $\displaystyle\int_2^\infty \frac{1}{\ln x}\, dx$

59. $\displaystyle\int_1^\infty \frac{e^x}{x}\, dx$ **60.** $\displaystyle\int_{e^e}^\infty \ln (\ln x)\, dx$

61. $\displaystyle\int_1^\infty \frac{1}{\sqrt{e^x - x}}\, dx$ **62.** $\displaystyle\int_1^\infty \frac{1}{e^x - 2^x}\, dx$

63. $\displaystyle\int_{-\infty}^\infty \frac{dx}{\sqrt{x^4 + 1}}$ **64.** $\displaystyle\int_{-\infty}^\infty \frac{dx}{e^x + e^{-x}}$

Theory and Examples

65. Find the values of p for which each integral converges.

(a) $\displaystyle\int_1^2 \frac{dx}{x(\ln x)^p}$

(b) $\displaystyle\int_2^\infty \frac{dx}{x(\ln x)^p}$

66. $\int_{-\infty}^\infty f(x)\,dx$ *may not equal* $\lim_{b\to\infty}\int_{-b}^b f(x)\,dx$ Show that

$$\int_0^\infty \frac{2x\,dx}{x^2+1}$$

diverges and hence that

$$\int_{-\infty}^\infty \frac{2x\,dx}{x^2+1}$$

diverges. Then show that

$$\lim_{b\to\infty}\int_{-b}^b \frac{2x\,dx}{x^2+1} = 0.$$

Exercises 67–69 are about the infinite region in the first quadrant between the curve $y = e^{-x}$ and the x-axis.

67. *Area* Find the area of the region.

68. *Volume* Find the volume of the solid generated by revolving the region about the y-axis.

69. *Volume* Find the volume of the solid generated by revolving the region about the x-axis.

70. *Area* Find the area of the region that lies between the curves $y = \sec x$ and $y = \tan x$ from $x = 0$ to $x = \pi/2$.

71. *Writing to Learn* Here is an argument that $\ln 3$ equals $\infty - \infty$. Where does the argument go wrong? Give reasons for your answer.

$$\ln 3 = \ln 1 + \ln 3 = \ln 1 - \ln \frac{1}{3}$$

$$= \lim_{b\to\infty} \ln\left(\frac{b-2}{b}\right) - \ln\frac{1}{3}$$

$$= \lim_{b\to\infty} \left[\ln \frac{x-2}{x}\right]_3^b$$

$$= \lim_{b\to\infty} [\ln(x-2) - \ln x]_3^b$$

$$= \lim_{b\to\infty} \int_3^b \left(\frac{1}{x-2} - \frac{1}{x}\right) dx$$

$$= \int_3^\infty \left(\frac{1}{x-2} - \frac{1}{x}\right) dx$$

$$= \int_3^\infty \frac{1}{x-2}\,dx - \int_3^\infty \frac{1}{x}\,dx$$

$$= \lim_{b\to\infty} [\ln(x-2)]_3^b - \lim_{b\to\infty} [\ln x]_3^b$$

$$= \infty - \infty$$

72. *Comparing integrals* Show that if $f(x)$ is integrable on every interval of real numbers and a and b are real numbers with $a < b$, then

(a) $\int_{-\infty}^a f(x)\,dx$ and $\int_a^\infty f(x)\,dx$ both converge if and only if $\int_{-\infty}^b f(x)\,dx$ and $\int_b^\infty f(x)\,dx$ both converge

(b) $\int_{-\infty}^a f(x)\,dx + \int_a^\infty f(x)\,dx = \int_{-\infty}^b f(x)\,dx + \int_b^\infty f(x)\,dx$

when the integrals involved converge.

73. *Estimating the value of a convergent improper integral whose domain is infinite*

(a) Show that

$$\int_3^\infty e^{-3x}\,dx = \frac{1}{3}e^{-9} < 0.000042$$

and hence that $\int_3^\infty e^{-x^2}\,dx < 0.000042$. Explain why this means that $\int_0^\infty e^{-x^2}\,dx$ can be replaced by $\int_0^3 e^{-x^2}\,dx$ without introducing an error of magnitude greater than 0.000042.

T (b) *Numerical integrator* Evaluate $\int_0^3 e^{-x^2}\,dx$ numerically.

74. *Sine-integral function* The integral

$$\text{Si}(x) = \int_0^x \frac{\sin t}{t}\,dt,$$

called the **sine-integral function,** has important applications in optics.

T (a) Plot the integrand $(\sin t)/t$ for $t > 0$. Is the Si function everywhere increasing or decreasing? Do you think Si $(x) = 0$ for $x > 0$? Check your answers by graphing the function Si (x) for $0 \le x \le 25$.

(b) Explore the convergence of

$$\int_0^\infty \frac{\sin t}{t}\,dt.$$

If it converges, what is its value?

75. *Error function* The function

$$\text{erf}(x) = \int_0^x \frac{2e^{-t^2}}{\sqrt{\pi}}\,dt,$$

called the **error function,** has important applications in probability and statistics.

T (a) Plot the error function for $0 \le x \le 25$.

(b) Explore the convergence of

$$\int_0^\infty \frac{2e^{-t^2}}{\sqrt{\pi}}\,dt.$$

If it converges, what appears to be its value? You will see how to confirm your estimate in Section 12.3, Exercise 37.

76. *Normal probability distribution function* The function

$$f(x) = \frac{1}{\sigma\sqrt{2\pi}}\, e^{-\frac{1}{2}\left(\frac{x-\mu}{\sigma}\right)^2},$$

is called the **normal probability density function** with mean μ and standard deviation σ. The number μ tells where the distribution is centered, and σ measures the "scatter" around the mean.

From the theory of probability, it is known that

$$\int_{-\infty}^{\infty} f(x)\, dx = 1.$$

In what follows, let $\mu = 0$ and $\sigma = 1$.

T **(a)** Draw the graph of f. Find the intervals on which f is increasing, the intervals on which f is decreasing, and any local extreme values and where they occur.

(b) Evaluate

$$\int_{-n}^{n} f(x)\, dx$$

for $n = 1, 2, 3$.

(c) Give a convincing argument that

$$\int_{-\infty}^{\infty} f(x)\, dx = 1.$$

(*Hint:* Show that $0 < f(x) < e^{-x/2}$ for $x > 1$, and for $b > 1$,

$$\int_{b}^{\infty} e^{-x/2}\, dx \to 0 \qquad \text{as} \qquad b \to \infty.)$$

COMPUTER EXPLORATIONS

Exploring Integrals of $x^p \ln x$

In Exercises 77–80, use a CAS to explore the integrals for various values of p (include noninteger values). For what values of p does the integral converge? What is the value of the integral when it does converge? Plot the integrand for various values of p.

77. $\displaystyle\int_{0}^{e} x^p \ln x\, dx$ **78.** $\displaystyle\int_{e}^{\infty} x^p \ln x\, dx$

79. $\displaystyle\int_{0}^{\infty} x^p \ln x\, dx$ **80.** $\displaystyle\int_{-\infty}^{\infty} x^p \ln |x|\, dx$

Questions to Guide Your Review

1. What basic integration formulas do you know?

2. What procedures do you know for matching integrals to basic formulas?

3. What is the formula for integration by parts? Where does it come from? Why might you want to use it?

4. When applying the formula for integration by parts, how do you choose the u and dv? How can you apply integration by parts to an integral of the form $\int f(x)\, dx$?

5. What is tabular integration? Give an example.

6. What is the goal of the method of partial fractions?

7. When the degree of a polynomial $f(x)$ is less than the degree of a polynomial $g(x)$, how do you write $f(x)/g(x)$ as a sum of partial fractions if $g(x)$

   (a) is a product of distinct linear factors

   (b) consists of a repeated linear factor

   (c) contains an irreducible quadratic factor?

   What do you do if the degree of f is *not* less than the degree of g?

8. What substitutions are sometimes used to change quadratic binomials into single squared terms? Why might you want to make such a change?

9. What restrictions can you place on the variables involved in the three basic trigonometric substitutions to make sure the substitutions are reversible (have inverses)?

10. How are integral tables typically used? What do you do if a particular integral you want to evaluate is not listed in the table?

11. Describe Monte Carlo integration for finding the area beneath a nonnegative curve and above the x-axis.

12. Describe l'Hôpital's Rule. How do you know when to use the rule and when to stop? Give an example.

13. How can you sometimes handle limits that lead to indeterminate forms 1^∞, 0^0, and ∞^0? Give examples.

14. What is an improper integral? How are the values of various types of improper integrals defined? Give examples.

15. What tests are available for determining the convergence and divergence of improper integrals that cannot be evaluated directly? Give examples of their use.

Practice Exercises

Integration Using Substitutions

Evaluate the integrals in Exercises 1–46. To transform each inegral into a recognizable basic form, it may be necessary to use one or more of the techniques of algebraic substitution, completing the square, separating fractions, long division, or trigonometric substitution.

1. $\displaystyle \int x\sqrt{4x^2 - 9}\, dx$

2. $\displaystyle \int x(2x + 1)^{1/2}\, dx$

3. $\displaystyle \int \frac{x\, dx}{\sqrt{8x^2 + 1}}$

4. $\displaystyle \int \frac{y\, dy}{25 + y^2}$

5. $\displaystyle \int \frac{t^3\, dt}{\sqrt{9 - 4t^4}}$

6. $\displaystyle \int z^{2/3}(z^{5/3} + 1)^{2/3}\, dz$

7. $\displaystyle \int \frac{\sin 2\theta\, d\theta}{(1 - \cos 2\theta)^2}$

8. $\displaystyle \int \frac{\cos 2t}{1 + \sin 2t}\, dt$

9. $\displaystyle \int \sin 2x\, e^{\cos 2x}\, dx$

10. $\displaystyle \int e^{\theta} \sec^2 (e^{\theta})\, d\theta$

11. $\displaystyle \int 2^{x-1}\, dx$

12. $\displaystyle \int \frac{dv}{v \ln v}$

13. $\displaystyle \int \frac{dx}{(x^2 + 1)(2 + \tan^{-1} x)}$

14. $\displaystyle \int \frac{2\, dx}{\sqrt{1 - 4x^2}}$

15. $\displaystyle \int \frac{dt}{\sqrt{16 - 9t^2}}$

16. $\displaystyle \int \frac{dt}{9 + t^2}$

17. $\displaystyle \int \frac{4\, dx}{5x\sqrt{25x^2 - 16}}$

18. $\displaystyle \int \frac{dx}{\sqrt{4x - x^2 - 3}}$

19. $\displaystyle \int \frac{dy}{y^2 - 4y + 8}$

20. $\displaystyle \int \frac{dv}{(v + 1)\sqrt{v^2 + 2v}}$

21. $\displaystyle \int \cos^2 3x\, dx$

22. $\displaystyle \int \sin^3 \frac{\theta}{2}\, d\theta$

23. $\displaystyle \int \tan^3 2t\, dt$

24. $\displaystyle \int \frac{dx}{2 \sin x \cos x}$

25. $\displaystyle \int \frac{2\, dx}{\cos^2 x - \sin^2 x}$

26. $\displaystyle \int_{\pi/4}^{\pi/2} \sqrt{\csc^2 y - 1}\, dy$

27. $\displaystyle \int_{\pi/4}^{3\pi/4} \sqrt{\cot^2 t + 1}\, dt$

28. $\displaystyle \int_{0}^{2\pi} \sqrt{1 - \sin^2 \frac{x}{2}}\, dx$

29. $\displaystyle \int_{-\pi/2}^{\pi/2} \sqrt{1 - \cos 2t}\, dt$

30. $\displaystyle \int_{\pi}^{2\pi} \sqrt{1 + \cos 2t}\, dt$

31. $\displaystyle \int \frac{x^2}{x^2 + 4}\, dx$

32. $\displaystyle \int \frac{x^3}{9 + x^2}\, dx$

33. $\displaystyle \int \frac{2y - 1}{y^2 + 4}\, dy$

34. $\displaystyle \int \frac{y + 4}{y^2 + 1}\, dy$

35. $\displaystyle \int \frac{t + 2}{\sqrt{4 - t^2}}\, dt$

36. $\displaystyle \int \frac{2t^2 + \sqrt{1 - t^2}}{t\sqrt{1 - t^2}}\, dt$

37. $\displaystyle \int \frac{\tan x\, dx}{\tan x + \sec x}$

38. $\displaystyle \int x \csc (x^2 + 3)\, dx$

39. $\displaystyle \int \cot \left(\frac{x}{4}\right) dx$

40. $\displaystyle \int x\sqrt{1 - x}\, dx$

41. $\displaystyle \int (16 + z^2)^{-3/2}\, dz$

42. $\displaystyle \int \frac{dy}{\sqrt{25 + y^2}}$

43. $\displaystyle \int \frac{dx}{x^2\sqrt{1 - x^2}}$

44. $\displaystyle \int \frac{x^2\, dx}{\sqrt{1 - x^2}}$

45. $\displaystyle \int \frac{dx}{\sqrt{x^2 - 9}}$

46. $\displaystyle \int \frac{12\, dx}{(x^2 - 1)^{3/2}}$

Integration by Parts

Evaluate the integrals in Exercises 47–54 using integration by parts.

47. $\displaystyle \int \ln (x + 1)\, dx$

48. $\displaystyle \int x^2 \ln x\, dx$

49. $\displaystyle \int \tan^{-1} 3x\, dx$

50. $\displaystyle \int \cos^{-1} \left(\frac{x}{2}\right) dx$

51. $\displaystyle \int (x + 1)^2 e^x\, dx$

52. $\displaystyle \int x^2 \sin (1 - x)\, dx$

53. $\displaystyle \int e^x \cos 2x\, dx$

54. $\displaystyle \int e^{-2x} \sin 3x\, dx$

Partial Fractions

Evaluate the integrals in Exercises 55–66. It may be necessary to use a substitution first.

55. $\displaystyle \int \frac{x\, dx}{x^2 - 3x + 2}$

56. $\displaystyle \int \frac{dx}{x(x + 1)^2}$

57. $\displaystyle \int \frac{\sin \theta\, d\theta}{\cos^2 \theta + \cos \theta - 2}$

58. $\displaystyle \int \frac{3x^2 + 4x + 4}{x^3 + x}\, dx$

59. $\displaystyle \int \frac{v + 3}{2v^3 - 8v}\, dv$

60. $\displaystyle \int \frac{dt}{t^4 + 4t^2 + 3}$

61. $\displaystyle \int \frac{x^3 + x^2}{x^2 + x - 2}\, dx$

62. $\displaystyle \int \frac{x^3 + 4x^2}{x^2 + 4x + 3}\, dx$

63. $\displaystyle \int \frac{2x^3 + x^2 - 21x + 24}{x^2 + 2x - 8}\, dx$

64. $\displaystyle \int \frac{dx}{x(3\sqrt{x} + 1)}$

65. $\displaystyle \int \frac{ds}{e^s - 1}$

66. $\displaystyle \int \frac{ds}{\sqrt{e^s + 1}}$

Trigonometric Substitutions

Evaluate the integrals in Exercises 67–70 (a) without using a trigonometric substitution and then (b) using a trigonometric substitution.

67. $\int \dfrac{y\, dy}{\sqrt{16 - y^2}}$

68. $\int \dfrac{x\, dx}{\sqrt{4 + x^2}}$

69. $\int \dfrac{x\, dx}{4 - x^2}$

70. $\int \dfrac{t\, dt}{\sqrt{4t^2 - 1}}$

Quadratic Terms

Evaluate the integrals in Exercises 71–74.

71. $\int \dfrac{x\, dx}{9 - x^2}$

72. $\int \dfrac{dx}{x(9 - x^2)}$

73. $\int \dfrac{dx}{9 - x^2}$

74. $\int \dfrac{dx}{\sqrt{9 - x^2}}$

Assorted Integrations

Evaluate the integrals in Exercises 75–114. The integrals are listed in random order.

75. $\int \dfrac{x\, dx}{1 + \sqrt{x}}$

76. $\int \dfrac{dx}{x(x^2 + 1)^2}$

77. $\int \dfrac{\cos \sqrt{x}}{\sqrt{x}}\, dx$

78. $\int \dfrac{dx}{\sqrt{-2x - x^2}}$

79. $\int \dfrac{du}{\sqrt{1 + u^2}}$

80. $\int \dfrac{2 - \cos x + \sin x}{\sin^2 x}\, dx$

81. $\int \dfrac{9\, dv}{81 - v^4}$

82. $\int \theta \cos (2\theta + 1)\, d\theta$

83. $\int \dfrac{x^3\, dx}{x^2 - 2x + 1}$

84. $\int \dfrac{d\theta}{\sqrt{1 + \sqrt{\theta}}}$

85. $\int \dfrac{2 \sin \sqrt{x}\, dx}{\sqrt{x} \sec \sqrt{x}}$

86. $\int \dfrac{x^5\, dx}{x^4 - 16}$

87. $\int \dfrac{d\theta}{\theta^2 - 2\theta + 4}$

88. $\int \dfrac{dr}{(r + 1)\sqrt{r^2 + 2r}}$

89. $\int \dfrac{\sin 2\theta\, d\theta}{(1 + \cos 2\theta)^2}$

90. $\int \dfrac{dx}{(x^2 - 1)^2}$

91. $\int \dfrac{x\, dx}{\sqrt{2 - x}}$

92. $\int \dfrac{dy}{y^2 - 2y + 2}$

93. $\int \ln \sqrt{x - 1}\, dx$

94. $\int \dfrac{x\, dx}{\sqrt{8 - 2x^2 - x^4}}$

95. $\int \dfrac{z + 1}{z^2(z^2 + 4)}\, dz$

96. $\int x^3 e^{(x^2)}\, dx$

97. $\int \dfrac{\tan^{-1} x}{x^2}\, dx$

98. $\int \dfrac{e^t\, dt}{e^{2t} + 3e^t + 2}$

99. $\int \dfrac{1 - \cos 2x}{1 + \cos 2x}\, dx$

100. $\int \dfrac{\cos (\sin^{-1} x)}{\sqrt{1 - x^2}}\, dx$

101. $\int \dfrac{\cos x\, dx}{\sin^3 x - \sin x}$

102. $\int \dfrac{e^t\, dt}{1 + e^t}$

103. $\int_1^\infty \dfrac{\ln y}{y^3}\, dy$

104. $\int \dfrac{\cot v\, dv}{\ln \sin v}$

105. $\int \dfrac{dx}{(2x - 1)\sqrt{x^2 - x}}$

106. $\int e^{\ln \sqrt{x}}\, dx$

107. $\int e^\theta \sqrt{3 + 4e^\theta}\, d\theta$

108. $\int \dfrac{dv}{\sqrt{e^{2v} - 1}}$

109. $\int (27)^{3\theta + 1}\, d\theta$

110. $\int x^5 \sin x\, dx$

111. $\int \dfrac{dr}{1 + \sqrt{r}}$

112. $\int \dfrac{8\, dy}{y^3(y + 2)}$

113. $\int \dfrac{8\, dm}{m\sqrt{49m^2 - 4}}$

114. $\int \dfrac{dt}{t(1 + \ln t)\sqrt{(\ln t)(2 + \ln t)}}$

Limits

In Exercises 115–128, find the limit.

115. $\lim\limits_{t \to 0} \dfrac{t - \ln (1 + 2t)}{t^2}$

116. $\lim\limits_{t \to 0} \dfrac{\tan 3t}{\tan 5t}$

117. $\lim\limits_{x \to 0} \dfrac{x \sin x}{1 - \cos x}$

118. $\lim\limits_{x \to 1} x^{1/(1 - x)}$

119. $\lim\limits_{x \to \infty} x^{1/x}$

120. $\lim\limits_{x \to \infty} \left(1 + \dfrac{3}{x}\right)^x$

121. $\lim\limits_{r \to \infty} \dfrac{\cos r}{\ln r}$

122. $\lim\limits_{\theta \to \pi/2} \left(\theta - \dfrac{\pi}{2}\right) \sec \theta$

123. $\lim\limits_{x \to 1} \left(\dfrac{1}{x - 1} - \dfrac{1}{\ln x}\right)$

124. $\lim\limits_{x \to 0^+} \left(1 + \dfrac{1}{x}\right)^x$

125. $\lim\limits_{\theta \to 0^+} (\tan \theta)^\theta$

126. $\lim\limits_{\theta \to \infty} \theta^2 \sin \left(\dfrac{1}{\theta}\right)$

127. $\lim\limits_{x \to \infty} \dfrac{x^3 - 3x^2 + 1}{2x^2 + x - 3}$

128. $\lim\limits_{x \to \infty} \dfrac{3x^2 - x + 1}{x^4 - x^3 + 2}$

Improper Integrals

Evaluate the improper integrals in Exercises 129–138, or state that it diverges.

129. $\int_0^3 \dfrac{dx}{\sqrt{9 - x^2}}$

130. $\int_0^1 \ln x\, dx$

131. $\int_{-1}^1 \dfrac{dy}{y^{2/3}}$

132. $\int_{-2}^0 \dfrac{d\theta}{(\theta + 1)^{3/5}}$

133. $\int_3^\infty \dfrac{2\, du}{u^2 - 2u}$

134. $\int_1^\infty \dfrac{3v - 1}{4v^3 - v^2}\, dv$

135. $\displaystyle\int_0^\infty x^2 e^{-x}\, dx$

136. $\displaystyle\int_{-\infty}^0 x e^{3x}\, dx$

137. $\displaystyle\int_{-\infty}^\infty \frac{dx}{4x^2+9}$

138. $\displaystyle\int_{-\infty}^\infty \frac{4\, dx}{x^2+16}$

Convergence or Divergence

Which of the improper integrals in Exercises 139–144 converge, and which diverge? Give reasons for your answer.

139. $\displaystyle\int_6^\infty \frac{d\theta}{\sqrt{\theta^2+1}}$

140. $\displaystyle\int_0^\infty e^{-u}\cos u\, du$

141. $\displaystyle\int_1^\infty \frac{\ln z}{z}\, dz$

142. $\displaystyle\int_1^\infty \frac{e^{-t}}{\sqrt{t}}\, dt$

143. $\displaystyle\int_{-\infty}^\infty \frac{dx}{e^x+e^{-x}}$

144. $\displaystyle\int_{-\infty}^\infty \frac{dx}{x^2(1+e^x)}$

Initial Value Problems

In Exercises 145–148, solve the initial value problem.

145. $\dfrac{dy}{dx} = e^x(y^2 - y), \quad y(0) = 2$

146. $\dfrac{dy}{d\theta} = (y+1)^2 \sin\theta, \quad y(\pi/2) = 0$

147. $\dfrac{dy}{dx} = \dfrac{1}{x^2 - 3x + 2}, \quad y(3) = 0$

148. $\dfrac{ds}{dt} = \dfrac{2s+2}{t^2+2t}, \quad s(1) = 1$

Additional Exercises: Theory, Examples, Applications

Challenging Integrals

Evaluate the integrals in Exercises 1–10.

1. $\displaystyle\int (\sin^{-1} x)^2\, dx$

2. $\displaystyle\int \frac{dx}{x(x+1)(x+2)\cdots(x+m)}$

3. $\displaystyle\int x \sin^{-1} x\, dx$

4. $\displaystyle\int \sin^{-1}\sqrt{y}\, dy$

5. $\displaystyle\int \frac{d\theta}{1-\tan^2\theta}$

6. $\displaystyle\int \ln\left(\sqrt{x}+\sqrt{1+x}\right) dx$

7. $\displaystyle\int \frac{dt}{t-\sqrt{1-t^2}}$

8. $\displaystyle\int \frac{(2e^{2x}-e^x)\, dx}{\sqrt{3e^{2x}-6e^x-1}}$

9. $\displaystyle\int \frac{dx}{x^4+4}$

10. $\displaystyle\int \frac{dx}{x^6-1}$

Limits

Find the limits in Exercises 11–16.

11. $\displaystyle\lim_{b\to 1^-}\int_0^b \frac{dx}{\sqrt{1-x^2}}$

12. $\displaystyle\lim_{x\to\infty}\frac{1}{x}\int_0^x \tan^{-1} t\, dt$

13. $\displaystyle\lim_{x\to 0^+}(\cos\sqrt{x})^{1/x}$

14. $\displaystyle\lim_{x\to\infty}(x+e^x)^{2/x}$

15. $\displaystyle\lim_{x\to\infty}\int_{-x}^x \sin t\, dt$

16. $\displaystyle\lim_{x\to 0^+} x\int_x^1 \frac{\cos t}{t^2}\, dt$

Theory and Applications

17. *Finding arc length* Find the length of the curve
$$y = \int_0^x \sqrt{\cos 2t}\, dt, \qquad 0 \le x \le \pi/4.$$

18. *Finding arc length* Find the length of the curve $y = \ln(1-x^2)$, $0 \le x \le 1/2$.

19. *Finding volume* The region in the first quadrant that is enclosed by the x-axis and the curve $y = 3x\sqrt{1-x}$ is revolved about the y-axis to generate a solid. Find the volume of the solid.

20. *Finding volume* The region in the first quadrant that is enclosed by the x-axis, the curve $y = 5/(x\sqrt{5-x})$, and the lines $x = 1$ and $x = 4$ is revolved about the x-axis to generate a solid. Find the volume of the solid.

21. *Finding volume* The region in the first quadrant enclosed by the coordinate axes, the curve $y = e^x$, and the line $x = 1$ is revolved about the y-axis to generate a solid. Find the volume of the solid.

22. *Finding volume* The region in the first quadrant that is bounded above by the curve $y = e^x - 1$, below by the x-axis, and on the right by the line $x = \ln 2$ is revolved about the line $x = \ln 2$ to generate a solid. Find the volume of the solid.

23. *Finding volume* Let R be the "triangular" region in the first quadrant that is bounded above by the line $y = 1$, below by the curve $y = \ln x$, and on the left by the line $x = 1$. Find the volume of the solid generated by revolving R about

(a) the x-axis

(b) the line $y = 1$.

24. *Finding volume* (*Continuation of Exercise 23*) Find the volume of the solid generated by revolving the region R about

(a) the y-axis

(b) the line $x = 1$.

25. *Finding volume* The region between the curve

$$y = f(x) = \begin{cases} 0, & x = 0 \\ x \ln x, & 0 < x \le 2 \end{cases}$$

is revolved about the x-axis to generate the solid shown here.

(a) Show that f is continuous at $x = 0$.

(b) Find the volume of the solid.

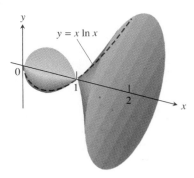

$y = x \ln x$

26. *Finding volume* The infinite region bounded by the coordinate axes and the curve $y = -\ln x$ in the first quadrant is revolved about the x-axis to generate a solid. Find the volume of the solid.

27. *Finding a limit* Find

$$\lim_{n \to \infty} \int_0^1 \frac{n \, y^{n-1}}{1 + y} \, dy.$$

28. *An integral formula* Derive the integral formula

$$\int x \left(\sqrt{x^2 - a^2} \right)^n dx = \frac{\left(\sqrt{x^2 - a^2} \right)^{n+2}}{n + 2} + C, \qquad n \ne -2.$$

29. *An inequality* Prove that

$$\frac{\pi}{6} < \int_0^1 \frac{dx}{\sqrt{4 - x^2 - x^3}} < \frac{\pi \sqrt{2}}{8}.$$

(*Hint:* Observe that for $0 < x < 1$, we have $4 - x^2 > 4 - x^2 - x^3 > 4 - 2x^2$, with the left-hand side becoming an equality for $x = 0$ and the right-hand side becoming an equality for $x = 1$.)

30. *Writing to Learn* For what value or values of a does

$$\int_1^\infty \left(\frac{ax}{x^2 + 1} - \frac{1}{2x} \right) dx$$

converge? Evaluate the corresponding integral(s).

31. *Evaluating an integral* Suppose that for a certain function f it is known that

$$f'(x) = \frac{\cos x}{x}, \qquad f(\pi/2) = a, \qquad \text{and} \qquad f(3\pi/2) = b.$$

Use integration by parts to evaluate

$$\int_{\pi/2}^{3\pi/2} f(x) \, dx.$$

32. *Finding equal areas* Find a positive number a satisfying

$$\int_0^a \frac{dx}{1 + x^2} = \int_a^\infty \frac{dx}{1 + x^2}.$$

33. *The length of an astroid* The graph of the equation $x^{2/3} + y^{2/3} = 1$ is one of a family of curves called *astroids* (not "asteroids") because of their starlike appearance (see accompanying figure). Find the length of this particular astroid.

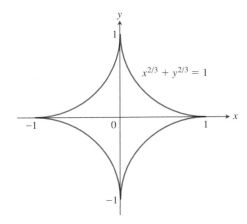

$x^{2/3} + y^{2/3} = 1$

34. Find a curve through the origin whose length is

$$\int_0^4 \sqrt{1 + \frac{1}{4x}} \, dx.$$

35. *A rational function* Find the second-degree polynomial $P(x)$ such that $P(0) = 1$, $P'(0) = 0$, and

$$\int \frac{P(x)}{x^3 (x - 1)^2} \, dx$$

is a rational function.

36. *Writing to Learn* Without evaluating either integral, explain why

$$2 \int_{-1}^1 \sqrt{1 - x^2} \, dx = \int_{-1}^1 \frac{dx}{\sqrt{1 - x^2}}.$$

(*Source:* Peter A. Lindstrom, *Mathematics Magazine*, Vol. 45, No. 1 (January 1972), p. 47.)

37. *Infinite area and finite volume* What values of p have the following property? The area of the region between the curve $y = x^{-p}$, $1 \le x < \infty$, and the x-axis is infinite, but the volume of the solid generated by revolving the region about the x-axis is finite.

38. *Infinite area and finite volume* What values of p have the following property? The area of the region in the first quadrant enclosed by the curve $y = x^{-p}$, the y-axis, the line $x = 1$, and the interval $[0, 1]$ on the x-axis is infinite, but the volume of the solid generated by revolving the region about the x-axis is finite.

39. *Finite area*

[T] **(a)** Graph the function $f(x) = e^{(x - e^x)}$, $-5 \le x \le 3$.

(b) Show that

$$\int_{-\infty}^{\infty} f(x)\, dx$$

converges and find its value.

40. *An integral connecting π to the approximation 22/7*

(a) Evaluate

$$\int_0^1 \frac{x^4(x - 1)^4}{x^2 + 1}\, dx.$$

(b) How good is the approximation $\pi \approx 22/7$? Find out by expressing $(\pi - 22/7)$ as a percentage of π.

[T] **(c)** Graph the function

$$y = \frac{x^4(x - 1)^4}{x^2 + 1}$$

for $0 \le x \le 1$. Experiment with the range on the y-axis set between 0 and 1, then between 0 and 0.5, and then decreasing the range until the graph can be seen. What do you conclude about the area under the curve?

Tabular Integration

The technique of tabular integration also applies to integrals of the form $\int f(x)g(x)\, dx$ when neither function can be differentiated repeatedly to become zero. For example, to evaluate

$$\int e^{2x} \cos x\, dx,$$

we begin as before with a table listing successive derivatives of e^{2x} and integrals of $\cos x$:

e^{2x} and its derivatives		$\cos x$ and its integrals	
e^{2x}	$+$	$\cos x$	
$2e^{2x}$	$-$	$\sin x$	
$4e^{2x}$	$+$	$-\cos x$	$\leftarrow$ *Stop here:* Row is same as first row except for multiplicative constants (4 on the left, -1 on the right).

We stop differentiating and integrating as soon as we reach a row that is the same as the first row except for multiplicative constants. We interpret the table as saying

$$\int e^{2x} \cos x\, dx = + (e^{2x} \sin x) - (2e^{2x}(-\cos x))$$

$$+ \int (4e^{2x})(-\cos x)\, dx.$$

We take signed products from the diagonal arrows and a signed integral for the last horizontal arrow. Transposing the integral on the right-hand side over to the left-hand side now gives

$$5 \int e^{2x} \cos x\, dx = e^{2x} \sin x + 2e^{2x} \cos x$$

or

$$\int e^{2x} \cos x\, dx = \frac{e^{2x} \sin x + 2e^{2x} \cos x}{5} + C,$$

after dividing by 5 and adding the constant of integration.

Use tabular integration to evaluate the integrals in Exercises 41–48.

41. $\displaystyle\int e^{2x} \cos 3x\, dx$

42. $\displaystyle\int e^{3x} \sin 4x\, dx$

43. $\displaystyle\int \sin 3x \sin x\, dx$

44. $\displaystyle\int \cos 5x \sin 4x\, dx$

45. $\displaystyle\int e^{ax} \sin bx\, dx$

46. $\displaystyle\int e^{ax} \cos bx\, dx$

47. $\displaystyle\int \ln (ax)\, dx$

48. $\displaystyle\int x^2 \ln (ax)\, dx$

The Gamma Function and Stirling's Formula

Euler's gamma function $\Gamma(x)$ ("gamma of x"; Γ is a Greek capital g) uses an integral to extend the factorial function from the nonnegative integers to other real values. The formula is

$$\Gamma(x) = \int_0^{\infty} t^{x-1} e^{-t}\, dt, \qquad x > 0.$$

For each positive x, the number $\Gamma(x)$ is the integral of $t^{x-1}e^{-t}$ with respect to t from 0 to ∞. Figure 7.21 shows the graph of Γ near the origin. You will see how to calculate $\Gamma(1/2)$ if you do Additional Exercise 31 in Chapter 12.

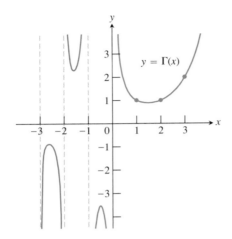

FIGURE 7.21 $\Gamma(x)$ is a continuous function of x whose value at each positive integer $n + 1$ is $n!$. The defining integral formula for Γ is valid only for $x > 0$, but we can extend Γ to negative noninteger values of x with the formula $\Gamma(x) = (\Gamma(x + 1))/x$, which is the subject of Exercise 49.

49. *If n is a nonnegative integer, $\Gamma(n + 1) = n!$*

(a) Show that $\Gamma(1) = 1$.

(b) Then apply integration by parts to the integral for $\Gamma(x + 1)$ to show that $\Gamma(x + 1) = x\Gamma(x)$. This gives

$$\Gamma(2) = 1\Gamma(1) = 1$$
$$\Gamma(3) = 2\Gamma(2) = 2$$
$$\Gamma(4) = 3\Gamma(3) = 6$$
$$\vdots$$
$$\Gamma(n + 1) = n\Gamma(n) = n! \qquad (1)$$

(c) Use mathematical induction to verify Equation (1) for every nonnegative integer n.

50. *Stirling's formula* Scottish mathematician James Stirling (1692–1770) showed that

$$\lim_{x \to \infty} \left(\frac{e}{x}\right)^x \sqrt{\frac{x}{2\pi}}\, \Gamma(x) = 1,$$

so for large x,

$$\Gamma(x) = \left(\frac{x}{e}\right)^x \sqrt{\frac{2\pi}{x}}\,(1 + \epsilon(x)), \qquad \epsilon(x) \to 0 \text{ as } x \to \infty. \qquad (2)$$

Dropping $\epsilon(x)$ leads to the approximation

$$\Gamma(x) \approx \left(\frac{x}{e}\right)^x \sqrt{\frac{2\pi}{x}} \qquad \textbf{(Stirling's formula).} \qquad (3)$$

(a) *Stirling's approximation for* n! Use Equation (3) and the fact that $n! = n\Gamma(n)$ to show that

$$n! \approx \left(\frac{n}{e}\right)^n \sqrt{2n\pi} \qquad \textbf{(Stirling's approximation).} \qquad (4)$$

As you will see if you do Exercise 64 in Section 8.1, Equation (4) leads to the approximation

$$\sqrt[n]{n!} \approx \frac{n}{e}. \qquad (5)$$

(b) Compare your calculator's value for $n!$ with the value given by Stirling's approximation for $n = 10, 20, 30, \ldots$, as far as your calculator can go.

(c) A refinement of Equation (2) gives

$$\Gamma(x) = \left(\frac{x}{e}\right)^x \sqrt{\frac{2\pi}{x}}\, e^{1/(12x)}(1 + \epsilon(x)),$$

or

$$\Gamma(x) \approx \left(\frac{x}{e}\right)^x \sqrt{\frac{2\pi}{x}}\, e^{1/(12x)},$$

which tells us that

$$n! \approx \left(\frac{n}{e}\right)^n \sqrt{2n\pi}\, e^{1/(12n)}. \qquad (6)$$

Compare the values given for 10! by your calculator, Stirling's approximation, and Equation (6).

8 Infinite Series

OVERVIEW   One infinite process that had puzzled mathematicians for centuries was the summing of infinite series. Sometimes an infinite series of terms added to a number, as in

$$\frac{1}{2} + \frac{1}{4} + \frac{1}{8} + \frac{1}{16} + \cdots = 1.$$

(You can see this by adding the areas in the "infinitely halved" unit square at the right.) Sometimes the infinite sum was infinite, however, as in

$$\frac{1}{1} + \frac{1}{2} + \frac{1}{3} + \frac{1}{4} + \frac{1}{5} + \cdots = \infty$$

(although this is far from obvious), and sometimes the infinite sum was impossible to pin down, as in

$$1 - 1 + 1 - 1 + 1 - 1 + \cdots .$$

(Is it 0? Is it 1? Is it neither?)

Nonetheless, mathematicians like Gauss and Euler successfully used infinite series to derive previously inaccessible results. Laplace used infinite series to prove the stability of the solar system (although that does not stop some people from worrying about it today when they feel that "too many" planets have swung to the same side of the sun). It was years later that careful analysts like Cauchy developed the theoretical foundation for series computations, sending many mathematicians (including Laplace) back to their desks to verify their results.

Infinite series form the basis for a remarkable formula that enables us to express many functions as "infinite polynomials" and at the same time tells how much error we incur if we truncate those polynomials to make them finite. In addition to providing effective polynomial approximations of differentiable functions, these infinite polynomials (called power series) have many other uses. We also see how to use infinite sums of trigonometric terms, called Fourier series, to represent important functions used in science and engineering applications. Infinite series provide an efficient way to evaluate nonelementary integrals, and they solve differential equations that give insight into heat flow, vibration, chemical diffusion, and signal transmission. What you learn here sets the stage for the roles played by series of functions of all kinds in science and mathematics.

Limits of Sequences of Numbers

Definitions and Notation • Convergence and Divergence •
Calculating Limits of Sequences • Using L'Hôpital's Rule • Limits That
Arise Frequently

Informally, a sequence is an ordered list of things, but in this chapter, the things will usually be numbers. We have seen sequences before, such as the sequence $x_0, x_1, \ldots, x_n, \ldots$ of numbers generated by Newton's method. Later we consider sequences involving powers of x and others involving trigonometric terms like $\sin x, \cos x, \sin 2x, \cos 2x, \ldots, \sin nx, \cos nx, \ldots$. A central question is whether a sequence has a limit or not.

CD-ROM
WEBsite

Historical Essay

Sequences and Series

Definitions and Notation

We can list the integer multiples of 3 by assigning each multiple a position:

$$
\begin{array}{lccccc}
\text{Domain:} & 1 & 2 & 3 & \ldots & n \ldots \\
& \downarrow & \downarrow & \downarrow & & \downarrow \\
\text{Range:} & 3 & 6 & 9 & & 3n
\end{array}
$$

The first number is 3, the second 6, the third 9, and so on. The assignment is a function that assigns $3n$ to the nth place. That is the basic idea for constructing sequences. There is a function placing each number in the range in its correct ordered position.

> **Definition** Sequence
> An infinite **sequence** of numbers is a function whose domain is the set of integers greater than or equal to some integer n_0.

Usually, n_0 is 1 and the domain of the sequence is the set of positive integers. Sometimes, however, we want to start sequences elsewhere. We take $n_0 = 0$ when we begin Newton's method. We might take $n_0 = 3$ if we were defining a sequence of n-sided polygons.

Sequences are defined the same way as other functions, some typical rules being

$$a(n) = \sqrt{n}, \qquad a(n) = (-1)^{n+1}\frac{1}{n}, \qquad a(n) = \frac{n-1}{n}$$

(Example 1 and Figure 8.1). To indicate that the domains are sets of integers, we use a letter like n from the middle of the alphabet for the independent variable, instead of the x, y, z, and t used widely in other contexts. The formulas in the defining rules, however, like those above, are often valid for domains larger than the set of positive integers. This can be an advantage, as we will see. The number

(a) The terms $a_n = \sqrt{n}$ eventually surpass every integer, so the sequence $\{a_n\}$ diverges, . . .

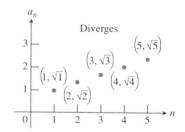

- -

(b) . . . but the terms $a_n = 1/n$ decrease steadily and get arbitrarily close to 0 as n increases, so the sequence $\{a_n\}$ converges to 0.

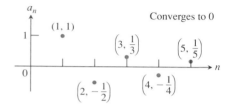

- -

(c) The terms $a_n = (-1)^{n+1}(1/n)$ alternate in sign but still converge to 0.

- -

(d) The terms $a_n = (n-1)/n$ approach 1 steadily and get arbitrarily close as n increases, so the sequence $\{a_n\}$ converges to 1.

- -

(e) The terms $a_n = (-1)^{n+1}[(n-1)/n]$ alternate in sign. The positive terms approach 1. But the negative terms approach -1 as n increases, so the sequence $\{a_n\}$ diverges.

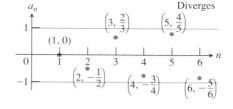

- -

(f) The terms in the sequence of constants $a_n = 3$ have the same value regardless of n, so the sequence $\{a_n\}$ converges to 3.

- -

FIGURE 8.1   The sequences of Example 1 are graphed here in two different ways: by plotting the numbers a_n on a horizontal axis and by plotting the points (n, a_n) in the coordinate plane.

$a(n)$ is the **nth term** of the sequence, or the **term with index n.** If $a(n) = (n-1)/n$, we have

First term	Second term	Third term		nth term
$a(1) = 0$	$a(2) = \dfrac{1}{2},$	$a(3) = \dfrac{2}{3},$	$\ldots,$	$a(n) = \dfrac{n-1}{n}.$

When we use the subscript notation a_n for $a(n)$, the sequence is written

$$a_1 = 0, \qquad a_2 = \frac{1}{2}, \qquad a_3 = \frac{2}{3}, \qquad \ldots, \qquad a_n = \frac{n-1}{n}.$$

To describe sequences, we often write the first few terms as well as a formula for the nth term.

Example 1   Describing Sequences

We write	For the sequence whose defining rule is
(a) $1, \sqrt{2}, \sqrt{3}, \sqrt{4}, \ldots, \sqrt{n}, \ldots$	$a_n = \sqrt{n}$
(b) $1, \dfrac{1}{2}, \dfrac{1}{3}, \ldots, \dfrac{1}{n}, \ldots$	$a_n = \dfrac{1}{n}$
(c) $1, -\dfrac{1}{2}, \dfrac{1}{3}, -\dfrac{1}{4}, \ldots, (-1)^{n+1}\dfrac{1}{n}, \ldots$	$a_n = (-1)^{n+1}\dfrac{1}{n}$
(d) $0, \dfrac{1}{2}, \dfrac{2}{3}, \dfrac{3}{4}, \ldots, \dfrac{n-1}{n}, \ldots$	$a_n = \dfrac{n-1}{n}$
(e) $0, -\dfrac{1}{2}, \dfrac{2}{3}, -\dfrac{3}{4}, \ldots, (-1)^{n+1}\left(\dfrac{n-1}{n}\right), \ldots$	$a_n = (-1)^{n+1}\left(\dfrac{n-1}{n}\right)$
(f) $3, 3, 3, \ldots, 3, \ldots$	$a_n = 3$

Notation   We refer to the sequence whose nth term is a_n with the notation $\{a_n\}$ ("the sequence a sub n"). The second sequence in Example 1 is $\{1/n\}$ ("the sequence 1 over n"); the last sequence is $\{3\}$ ("the constant sequence 3").

Convergence and Divergence

As Figure 8.1 shows, the sequences of Example 1 do not behave the same way. The sequences $\{1/n\}$, $\{(-1)^{n+1}(1/n)\}$, and $\{(n-1)/n\}$ each seem to approach a single limiting value as n increases, and $\{3\}$ is at a limiting value from the very first. On the other hand, terms of $\{(-1)^{n+1}(n-1)/n\}$ seem to accumulate near two different values, -1 and 1, whereas the terms of $\{\sqrt{n}\}$ become increasingly large and do not accumulate anywhere.

The following definition distinguishes those sequences that approach a unique limiting value L, as n increases, from those that do not.

Definitions Converges, Diverges, Limit

The sequence $\{a_n\}$ **converges** to the number L if to every positive number ϵ there corresponds an integer N such that for all n,

$$n > N \implies |a_n - L| < \epsilon.$$

If no such number L exists, we say that $\{a_n\}$ **diverges.**

If $\{a_n\}$ converges to L, we write $\lim_{n\to\infty} a_n = L$, or simply $a_n \to L$, and call L the **limit** of the sequence (Figure 8.2).

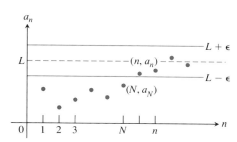

FIGURE 8.2 $a_n \to L$ if $y = L$ is a horizontal asymptote of the sequence of points $\{(n, a_n)\}$. In this figure, all the a_n's after a_N lie within ϵ of L.

Example 2 Testing the Definition

Show that

(a) $\lim_{n\to\infty} \dfrac{1}{n} = 0$

(b) $\lim_{n\to\infty} k = k$ (any constant k).

Solution

(a) Let $\epsilon > 0$ be given. We must show that there exists an integer N such that for all n,

$$n > N \implies \left| \frac{1}{n} - 0 \right| < \epsilon.$$

This implication will hold if $(1/n) < \epsilon$ or $n > 1/\epsilon$. If N is any integer greater than $1/\epsilon$, the implication will hold for all $n > N$. This proves $\lim_{n\to\infty} (1/n) = 0$.

(b) Let $\epsilon > 0$ be given. We must show that there exists an integer N such that for all n,

$$n > N \implies |k - k| < \epsilon.$$

Since $k - k = 0$, we can use any positive integer for N and the implication will hold. This proves that $\lim_{n\to\infty} k = k$ for any constant k.

Example 3   A Divergent Sequence

Show that $\{(-1)^{n+1}[(n-1)/n]\}$ diverges.

Solution   Take a positive ϵ smaller than 1 so that the bands shown in Figure 8.3 about the lines $y = 1$ and $y = -1$ do not overlap. Any $\epsilon < 1$ will do. Convergence to 1 would require every point of the graph beyond a certain index N to lie inside the upper band, but this will never happen. As soon as a point (n, a_n) lies in the upper band, every alternate point starting with $(n + 1, a_{n+1})$ will lie in the lower band. Hence, the sequence cannot converge to 1. Likewise, it cannot converge to -1. On the other hand, because the terms of the sequence get alternately closer to 1 and -1, they never accumulate near any other value. Therefore, the sequence diverges.

$$a_n = (-1)^{n+1}\left(\frac{n-1}{n}\right)$$

Neither the ϵ-interval about 1 nor the ϵ-interval about -1 contains all a_n satisfying $n \geq N$ for some N.

FIGURE 8.3 The sequence $\{(-1)^{n+1}[(n-1)/n]\}$ diverges.

The behavior of $\{(-1)^{n+1}[(n-1)/n]\}$ is qualitatively different from that of $\{\sqrt{n}\}$, which diverges because it outgrows every real number L. To describe the behavior of $\{\sqrt{n}\}$, we write

$$\lim_{n \to \infty} (\sqrt{n}) = \infty.$$

In speaking of infinity as a limit of a sequence $\{a_n\}$, we do not mean that the difference between a_n and infinity becomes small as n increases. We mean that a_n becomes numerically large as n increases.

Calculating Limits of Sequences

The study of limits would be cumbersome if we had to answer every question about convergence by applying the definition. Fortunately, three theorems make this largely unnecessary. The first theorem is not surprising, based on our previous work with limits. We omit the proofs.

Theorem 1   Limit Laws for Sequences

Let $\{a_n\}$ and $\{b_n\}$ be sequences of real numbers and let A and B be real numbers. The following rules hold if $\lim_{n\to\infty} a_n = A$ and $\lim_{n\to\infty} b_n = B$.

1. *Sum Rule:* $\lim_{n\to\infty} (a_n + b_n) = A + B$

2. *Difference Rule:* $\lim_{n\to\infty} (a_n - b_n) = A - B$

3. *Product Rule:* $\lim_{n\to\infty} (a_n \cdot b_n) = A \cdot B$

4. *Constant Multiple Rule:* $\lim_{n\to\infty} (k \cdot b_n) = k \cdot B$ (any number k)

5. *Quotient Rule:* $\lim_{n\to\infty} \dfrac{a_n}{b_n} = \dfrac{A}{B}$ if $B \neq 0$

Example 4   Applying the Limit Laws

By combining Theorem 1 with the limit results in Example 2, we have

(a) $\displaystyle\lim_{n\to\infty} \left(-\frac{1}{n}\right) = -1 \cdot \lim_{n\to\infty} \frac{1}{n} = -1 \cdot 0 = 0$

(b) $\displaystyle\lim_{n\to\infty} \left(\frac{n-1}{n}\right) = \lim_{n\to\infty} \left(1 - \frac{1}{n}\right) = \lim_{n\to\infty} 1 - \lim_{n\to\infty} \frac{1}{n} = 1 - 0 = 1$

(c) $\displaystyle\lim_{n\to\infty} \frac{5}{n^2} = 5 \cdot \lim_{n\to\infty} \frac{1}{n} \cdot \lim_{n\to\infty} \frac{1}{n} = 5 \cdot 0 \cdot 0 = 0$

(d) $\displaystyle\lim_{n\to\infty} \frac{4 - 7n^6}{n^6 + 3} = \lim_{n\to\infty} \frac{(4/n^6) - 7}{1 + (3/n^6)} = \frac{0 - 7}{1 + 0} = -7.$

Example 5   Constant Multiples of Divergent Sequences Diverge

Every nonzero multiple of a divergent sequence $\{a_n\}$ diverges. Suppose, to the contrary, that $\{ca_n\}$ converges for some number $c \neq 0$. Then, by taking $k = 1/c$ in the Constant Multiple Rule in Theorem 1, we see that the sequence

$$\left\{\frac{1}{c} \cdot ca_n\right\} = \{a_n\}$$

converges. Thus, $\{ca_n\}$ cannot converge unless $\{a_n\}$ also converges. If $\{a_n\}$ does not converge, then $\{ca_n\}$ does not converge.

You are asked to prove the next Theorem in Exercise 69.

Theorem 2   The Sandwich Theorem for Sequences

Let $\{a_n\}$, $\{b_n\}$, and $\{c_n\}$ be sequences of real numbers. If $a_n \leq b_n \leq c_n$ holds for all n beyond some index N and if $\lim_{n\to\infty} a_n = \lim_{n\to\infty} c_n = L$, then $\lim_{n\to\infty} b_n = L$ also.

An immediate consequence of Theorem 2 is that if $|b_n| \le c_n$ and $c_n \to 0$, then $b_n \to 0$ because $-c_n \le b_n \le c_n$. We use this fact in the next example.

Example 6   Using the Sandwich Theorem

Since $1/n \to 0$, we know that

(a) $\dfrac{\cos n}{n} \to 0$ because $\left| \dfrac{\cos n}{n} \right| = \dfrac{|\cos n|}{n} \le \dfrac{1}{n}$

(b) $\dfrac{1}{2^n} \to 0$ because $\dfrac{1}{2^n} \le \dfrac{1}{n}$

(c) $(-1)^n \dfrac{1}{n} \to 0$ because $\left| (-1)^n \dfrac{1}{n} \right| \le \dfrac{1}{n}$.

The application of Theorems 1 and 2 is broadened by a theorem stating that applying a continuous function to a convergent sequence produces a convergent sequence. We state the theorem without proof (Exercise 70).

Theorem 3   The Continuous Function Theorem for Sequences

Let $\{a_n\}$ be a sequence of real numbers. If $a_n \to L$ and if f is a function that is continuous at L and defined at all a_n, then $f(a_n) \to f(L)$.

Example 7   Applying Theorem 3

Show that $\sqrt{(n+1)/n} \to 1$.

Solution   We know that $(n+1)/n \to 1$. Taking $f(x) = \sqrt{x}$ and $L = 1$ in Theorem 3 gives $\sqrt{(n+1)/n} \to \sqrt{1} = 1$.

Example 8   The Sequence $\{2^{1/n}\}$

The sequence $\{1/n\}$ converges to 0. By taking $a_n = 1/n$, $f(x) = 2^x$, and $L = 0$ in Theorem 3, we see that $2^{1/n} = f(1/n) \to f(L) = 2^0 = 1$. The sequence $\{2^{1/n}\}$ converges to 1 (Figure 8.4).

Using L'Hôpital's Rule

The next theorem enables us to use l'Hôpital's Rule to find the limits of some sequences. It matches values of a (usually differentiable) function with the values of a given sequence.

Theorem 4

Suppose that $f(x)$ is a function defined for all $x \ge n_0$ and that $\{a_n\}$ is a sequence of real numbers such that $a_n = f(n)$ for $n \ge n_0$. Then

$$\lim_{x \to \infty} f(x) = L \;\Rightarrow\; \lim_{n \to \infty} a_n = L.$$

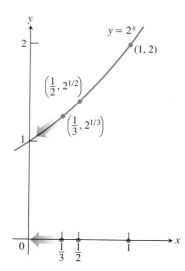

FIGURE 8.4 As $n \to \infty$, $1/n \to 0$ and $2^{1/n} \to 2^0$.

Example 9   Applying L'Hôpital's Rule

Show that

$$\lim_{n\to\infty} \frac{\ln n}{n} = 0.$$

Solution   The function $(\ln x)/x$ is defined for all $x \geq 1$ and agrees with the given sequence at positive integers. Therefore, by Theorem 4, $\lim_{n\to\infty} (\ln n)/n$ will equal $\lim_{x\to\infty} (\ln x)/x$ if the latter exists. A single application of l'Hôpital's Rule shows that

$$\lim_{x\to\infty} \frac{\ln x}{x} = \lim_{x\to\infty} \frac{1/x}{1} = \frac{0}{1} = 0.$$

We conclude that $\lim_{n\to\infty} (\ln n)/n = 0$.

When we use l'Hôpital's Rule to find the limit of a sequence, we often treat n as a continuous real variable and differentiate directly with respect to n. This saves us from having to rewrite the formula for a_n as we did in Example 9.

Example 10   Applying L'Hôpital's Rule

Find

$$\lim_{n\to\infty} \frac{2^n}{5n}.$$

Solution   By l'Hôpital's Rule (differentiating with respect to n),

$$\lim_{n\to\infty} \frac{2^n}{5n} = \lim_{n\to\infty} \frac{2^n \cdot \ln 2}{5}$$

$$= \infty.$$

Proof of Theorem 4   Suppose that $\lim_{x\to\infty} f(x) = L$. Then for each positive number ϵ there is a number M such that for all x,

$$x > M \implies |f(x) - L| < \epsilon.$$

Let N be an integer greater than M and greater than or equal to n_0. Then

$$n > N \implies a_n = f(n) \quad \text{and} \quad |a_n - L| = |f(n) - L| < \epsilon.$$

Example 11   Applying L'Hôpital's Rule to Determine Convergence

Does the sequence whose nth term is

$$a_n = \left(\frac{n+1}{n-1}\right)^n$$

converge? If so, find $\lim_{n\to\infty} a_n$.

Solution The limit leads to the indeterminate form 1^∞. We can apply l'Hôpital's Rule if we first change the form to $\infty \cdot 0$ by taking the natural logarithm of a_n:

$$\ln a_n = \ln \left(\frac{n+1}{n-1}\right)^n$$

$$= n \ln \left(\frac{n+1}{n-1}\right).$$

Then,

$$\lim_{n\to\infty} \ln a_n = \lim_{n\to\infty} n \ln \left(\frac{n+1}{n-1}\right) \qquad \infty \cdot 0$$

$$= \lim_{n\to\infty} \frac{\ln \left(\frac{n+1}{n-1}\right)}{1/n} \qquad \frac{0}{0}$$

$$= \lim_{n\to\infty} \frac{-2/(n^2-1)}{-1/n^2} \qquad \text{l'Hôpital's Rule}$$

$$= \lim_{n\to\infty} \frac{2n^2}{n^2-1} = 2.$$

Since $\ln a_n \to 2$ and $f(x) = e^x$ is continuous, Theorem 3 tells us that

$$a_n = e^{\ln a_n} \to e^2.$$

The sequence $\{a_n\}$ converges to e^2.

Limits That Arise Frequently

The limits in Table 8.1 arise frequently. The first limit is from Example 9. The next two can be proved by taking logarithms and applying Theorem 3 (Exercises 67 and 68). The remaining proofs can be found in Appendix 7.

Table 8.1

1. $\lim_{n\to\infty} \frac{\ln n}{n} = 0$

2. $\lim_{n\to\infty} \sqrt[n]{n} = 1$

3. $\lim_{n\to\infty} x^{1/n} = 1 \qquad (x > 0)$

4. $\lim_{n\to\infty} x^n = 0 \qquad (|x| < 1)$

5. $\lim_{n\to\infty} \left(1 + \frac{x}{n}\right)^n = e^x \qquad (\text{any } x)$

6. $\lim_{n\to\infty} \frac{x^n}{n!} = 0 \qquad (\text{any } x)$

In formulas (3) through (6), x remains fixed as $n \to \infty$.

Example 12 Limits from Table 8.1

(a) $\frac{\ln (n^2)}{n} = \frac{2 \ln n}{n} \to 2 \cdot 0 = 0$ Formula 1

(b) $\sqrt[n]{n^2} = n^{2/n} = (n^{1/n})^2 \to (1)^2 = 1$ Formula 2

(c) $\sqrt[n]{3n} = 3^{1/n}(n^{1/n}) \to 1 \cdot 1 = 1$ Formula 3 with $x = 3$ and Formula 2

(d) $\left(-\frac{1}{2}\right)^n \to 0$ Formula 4 with $x = -\frac{1}{2}$

(e) $\left(\frac{n-2}{n}\right)^n = \left(1 + \frac{-2}{n}\right)^n \to e^{-2}$ Formula 5 with $x = -2$

(f) $\frac{100^n}{n!} \to 0$ Formula 6 with $x = 100$

EXERCISES 8.1

Finding Terms of a Sequence

Each of Exercises 1–4 gives a formula for the nth term a_n of a sequence $\{a_n\}$. Find the values of a_1, a_2, a_3, and a_4.

1. $a_n = \dfrac{1 - n}{n^2}$

2. $a_n = \dfrac{1}{n!}$

3. $a_n = \dfrac{(-1)^{n+1}}{2n - 1}$

4. $a_n = \dfrac{2^n}{2^{n+1}}$

Finding Formulas for Sequences

In Exercises 5–12, find a formula for the nth term of the sequence.

5. The sequence $1, -1, 1, -1, 1, \ldots$ 1's with alternating signs

6. The sequence $1, -4, 9, -16, 25, \ldots$ Squares of the positive integers, with alternating signs

7. The sequence $0, 3, 8, 15, 24, \ldots$ Squares of the positive integers diminished by 1

8. The sequence $-3, -2, -1, 0, 1, \ldots$ Integers beginning with -3

9. The sequence $1, 5, 9, 13, 17, \ldots$ Every other odd positive integer

10. The sequence $2, 6, 10, 14, 18, \ldots$ Every other even positive integer

11. The sequence $1, 0, 1, 0, 1, \ldots$ Alternating 1's and 0's

12. The sequence $0, 1, 1, 2, 2, 3, 3, 4, \ldots$ Each positive integer repeated

Finding Limits

Which of the sequences $\{a_n\}$ in Exercises 13–56 converge, and which diverge? Find the limit of each convergent sequence.

13. $a_n = 2 + (0.1)^n$

14. $a_n = \dfrac{n + (-1)^n}{n}$

15. $a_n = \dfrac{1 - 2n}{1 + 2n}$

16. $a_n = \dfrac{1 - 5n^4}{n^4 + 8n^3}$

17. $a_n = \dfrac{n^2 - 2n + 1}{n - 1}$

18. $a_n = \dfrac{n + 3}{n^2 + 5n + 6}$

19. $a_n = 1 + (-1)^n$

20. $a_n = (-1)^n \left(1 - \dfrac{1}{n}\right)$

21. $a_n = \left(\dfrac{n + 1}{2n}\right)\left(1 - \dfrac{1}{n}\right)$

22. $a_n = \dfrac{(-1)^{n+1}}{2n - 1}$

23. $a_n = \sqrt{\dfrac{2n}{n + 1}}$

24. $a_n = \sin\left(\dfrac{\pi}{2} + \dfrac{1}{n}\right)$

25. $a_n = \dfrac{\sin n}{n}$

26. $a_n = \dfrac{\sin^2 n}{2^n}$

27. $a_n = \dfrac{n}{2^n}$

28. $a_n = \dfrac{\ln(n + 1)}{\sqrt{n}}$

29. $a_n = \dfrac{\ln n}{n^{1/n}}$

30. $a_n = \ln n - \ln(n + 1)$

31. $a_n = \left(1 + \dfrac{7}{n}\right)^n$

32. $a_n = \left(1 - \dfrac{1}{n}\right)^n$

33. $a_n = \sqrt[n]{10n}$

34. $a_n = \sqrt[n]{n^2}$

35. $a_n = \left(\dfrac{3}{n}\right)^{1/n}$

36. $a_n = (n + 4)^{1/(n+4)}$

37. $a_n = \sqrt[n]{4^n \, n}$

38. $a_n = \sqrt[n]{3^{2n+1}}$

39. $a_n = \dfrac{n!}{n^n}$ (*Hint*: Compare with $1/n$.)

40. $a_n = \dfrac{(-4)^n}{n!}$

41. $a_n = \dfrac{n!}{10^{6n}}$

42. $a_n = \dfrac{n!}{2^n \cdot 3^n}$

43. $a_n = \left(\dfrac{1}{n}\right)^{1/(\ln n)}$

44. $a_n = \ln\left(1 + \dfrac{1}{n}\right)^n$

45. $a_n = \left(\dfrac{3n + 1}{3n - 1}\right)^n$

46. $a_n = \left(\dfrac{n}{n + 1}\right)^n$

47. $a_n = \left(\dfrac{x^n}{2n + 1}\right)^{1/n}, \quad x > 0$

48. $a_n = \left(1 - \dfrac{1}{n^2}\right)^n$

49. $a_n = \dfrac{3^n \cdot 6^n}{2^{-n} \cdot n!}$

50. $a_n = \dfrac{n^2}{2n - 1} \sin \dfrac{1}{n}$

51. $a_n = \tan^{-1} n$

52. $a_n = \dfrac{1}{\sqrt{n}} \tan^{-1} n$

53. $a_n = \left(\dfrac{1}{3}\right)^n + \dfrac{1}{\sqrt{2^n}}$

54. $a_n = \sqrt[n]{n^2 + n}$

55. $a_n = \dfrac{(\ln n)^5}{\sqrt{n}}$

56. $a_n = n - \sqrt{n^2 - n}$

Calculator Explorations of Limits

In Exercises 57–60, experiment with a calculator to find a value of N that will make the inequality hold for all $n > N$. Assuming that the inequality is the one from the formal definition of the limit of a sequence, what sequence is being considered in each case, and what is its limit?

57. $\left|\sqrt[n]{0.5} - 1\right| < 10^{-3}$

58. $\left|\sqrt[n]{n} - 1\right| < 10^{-3}$

59. $(0.9)^n < 10^{-3}$

60. $(2^n/n!) < 10^{-7}$

Theory and Examples

61. A sequence of rational numbers is described as follows:

$$\frac{1}{1}, \frac{3}{2}, \frac{7}{5}, \frac{17}{12}, \ldots, \frac{a}{b}, \frac{a + 2b}{a + b}, \ldots.$$

Here the numerators form one sequence, the denominators form a second sequence, and their ratios form a third sequence. Let x_n and y_n be, respectively, the numerator and the denominator of the nth fraction $r_n = x_n/y_n$.

(a) Verify that $x_1^2 - 2y_1^2 = -1$, $x_2^2 - 2y_2^2 = +1$, and, more generally, that if $a^2 - 2b^2 = -1$ or $+1$, then

$$(a + 2b)^2 - 2(a + b)^2 = +1 \quad \text{or} \quad -1,$$

respectively.

(b) The fractions $r_n = x_n/y_n$ approach a limit as n increases. What is that limit? (*Hint:* Use part (a) to show that $r_n^2 - 2 = \pm(1/y_n)^2$ and that y_n is not less than n.)

62. (a) Suppose that $f(x)$ is differentiable for all x in [0, 1] and that $f(0) = 0$. Define the sequence $\{a_n\}$ by the rule $a_n = nf(1/n)$. Show that $\lim_{n\to\infty} a_n = f'(0)$.

Use the result in part (a) to find the limits of the following sequences $\{a_n\}$.

(b) $a_n = n \tan^{-1} \dfrac{1}{n}$

(c) $a_n = n(e^{1/n} - 1)$

(d) $a_n = n \ln \left(1 + \dfrac{2}{n}\right)$

63. *Pythagorean triples* A triple of positive integers a, b, and c is called a **Pythagorean triple** if $a^2 + b^2 = c^2$. Let a be an odd positive integer and let

$$b = \left\lfloor \frac{a^2}{2} \right\rfloor \quad \text{and} \quad c = \left\lceil \frac{a^2}{2} \right\rceil$$

be, respectively, the integer floor and ceiling for $a^2/2$.

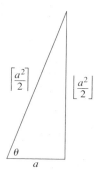

(a) Show that $a^2 + b^2 = c^2$. (*Hint:* Let $a = 2n + 1$ and express b and c in terms of n.)

(b) By direct calculation, or by appealing to the figure, find

$$\lim_{a\to\infty} \frac{\left\lfloor \dfrac{a^2}{2} \right\rfloor}{\left\lceil \dfrac{a^2}{2} \right\rceil}.$$

64. *The nth root of $n!$*

(a) Show that $\lim_{n\to\infty} (2n\pi)^{1/(2n)} = 1$ and hence, using Stirling's approximation (Chapter 7, Additional Exercise 50, part (a)), that

$$\sqrt[n]{n!} \approx \frac{n}{e} \quad \text{for large values of } n.$$

(b) Test the approximation in part (a) for $n = 40, 50, 60, \ldots$ as far as your calculator will allow.

65. (a) Assuming that $\lim_{n\to\infty} (1/n^c) = 0$ if c is any positive constant, show that

$$\lim_{n\to\infty} \frac{\ln n}{n^c} = 0$$

if c is any positive constant.

(b) Prove that $\lim_{n\to\infty} (1/n^c) = 0$ if c is any positive constant. (*Hint:* If $\epsilon = 0.001$ and $c = 0.04$, how large should N be to ensure that $|1/n^c - 0| < \epsilon$ if $n > N$?)

66. *The Zipper Theorem* Prove the "Zipper Theorem" for sequences: If $\{a_n\}$ and $\{b_n\}$ both converge to L, then the sequence

$$a_1, b_1, a_2, b_2, \ldots, a_n, b_n, \ldots$$

converges to L.

67. Prove that $\lim_{n\to\infty} \sqrt[n]{n} = 1$.

68. Prove that $\lim_{n\to\infty} x^{1/n} = 1 \ (x > 0)$.

69. Prove Theorem 2.

70. Prove Theorem 3.

71. *Terms become arbitrarily close in convergent sequences* Prove that if $\{a_n\}$ is a convergent sequence, then to every positive number ϵ there corresponds an integer N such that for all m and n,

$$m > N \quad \text{and} \quad n > N \Rightarrow |a_m - a_n| < \epsilon.$$

72. *Uniqueness of limits* Prove that limits of sequences are unique. That is, show that if L_1 and L_2 are numbers such that $a_n \to L_1$ and $a_n \to L_2$, then $L_1 = L_2$.

73. *Convergence and absolute value* Prove that a sequence $\{a_n\}$ converges to 0 if and only if the sequence of absolute values $\{|a_n|\}$ converges to 0.

74. *Improving automobile production* According to a front-page article in the December 15, 1992 issue of the *Wall Street Journal*, Ford Motor Company now uses about $7\frac{1}{4}$ h of labor to produce stampings for the average vehicle, down from an estimated 15 h in 1980. The Japanese need only about $3\frac{1}{2}$ h.

Ford's improvement since 1980 represents an average decrease of 6% per year. If that rate continues, then n years from now Ford will use about

$$S_n = 7.25(0.94)^n$$

hours of labor to produce stampings for the average vehicle. Assuming that the Japanese continue to spend $3\frac{1}{2}$ h per vehicle, how many more years will it take Ford to catch up? Find out two ways:

(a) Find the first term of the sequence $\{S_n\}$ that is less than or equal to 3.5.

T **(b)** Graph $f(x) = 7.25(0.94)^x$ and use Trace to find where the graph crosses the line $y = 3.5$.

COMPUTER EXPLORATIONS

Looking for Signs of Convergence and Divergence

Use a CAS to perform the following steps for the sequences in Exercises 75–84.

(a) Calculate and then plot the first 25 terms of the sequence. Does the sequence appear to converge or diverge? If it does converge, what is the limit L?

(b) If the sequence converges, find an integer N such that $|a_n - L| \le 0.01$ for $n \ge N$. How far in the sequence do you have to get for the terms to lie within 0.0001 of L?

75. $a_n = \sqrt[n]{n}$

76. $a_n = \left(1 + \dfrac{0.5}{n}\right)^n$

77. $a_n = \sin n$

78. $a_n = n \sin \dfrac{1}{n}$

79. $a_n = \dfrac{\sin n}{n}$

80. $a_n = \dfrac{\ln n}{n}$

81. $a_n = (0.9999)^n$

82. $a_n = 123456^{1/n}$

83. $a_n = \dfrac{8^n}{n!}$

84. $a_n = \dfrac{n^{41}}{19^n}$

8.2 Subsequences, Bounded Sequences, and Picard's Method

Subsequences • Monotonic and Bounded Sequences • Recursively Defined Sequences • Picard's Method for Finding Roots

This section continues our study of the convergence or divergence of a sequence.

Subsequences

If the terms of one sequence appear in another sequence in their given order, we call the first sequence a **subsequence** of the second.

> **Example 1** Subsequences of the Sequence of Positive Integers
>
> **(a)** The subsequence of even integers: $2, 4, 6, \ldots, 2n, \ldots$
>
> **(b)** The subsequence of odd integers: $1, 3, 5, \ldots, 2n - 1, \ldots$
>
> **(c)** The subsequence of primes: $2, 3, 5, 7, 11, \ldots$

Subsequences are important for two reasons:

1. If a sequence $\{a_n\}$ converges to L, then all its subsequences converge to L. If we know that a sequence converges, it may be quicker to find or estimate its limit by examining a particular subsequence.

2. If any subsequence of a sequence $\{a_n\}$ diverges or if two subsequences have different limits, then $\{a_n\}$ diverges. For example, the sequence $\{(-1)^n\}$ diverges because the subsequence $-1, -1, -1, \ldots$ of odd-numbered terms converge to -1, whereas the subsequence $1, 1, 1, \ldots$ of even-numbered terms converges to 1, a different limit.

The convergence or divergence of a sequence has nothing to do with how the sequence begins. It depends only on how the tails behave.

Subsequences also provide a new way to view convergence. A **tail** of a sequence is a subsequence that consists of all terms of the sequence from some index N on. In other words, a tail is one of the sets $\{a_n \mid n \geq N\}$. Another way to say that $a_n \to L$ is to say that every ϵ-interval about L contains a tail of the sequence.

Monotonic and Bounded Sequences

> **Definition**   Nondecreasing, Nonincreasing, Monotonic Sequence
>
> A sequence $\{a_n\}$ with the property that $a_n \leq a_{n+1}$ for all n is called a **nondecreasing sequence**; that is, $a_1 \leq a_2 \leq a_3 \leq \ldots$.
>
> It is called **nonincreasing** if $a_n \geq a_{n+1}$ for all n. A sequence is **monotonic** if it is either nondecreasing or nonincreasing.

CD-ROM
WEBsite

Historical Biography

Fibonacci
(1170 — 1240)

Example 2   Monotonic Sequences

(a) The sequence $1, 2, 3, \ldots, n, \ldots$ of natural numbers is nondecreasing.

(b) The sequence $\dfrac{1}{2}, \dfrac{2}{3}, \dfrac{3}{4}, \ldots, \dfrac{n}{n+1}, \ldots$ is nondecreasing.

(c) The sequence $\dfrac{3}{8}, \dfrac{3}{9}, \dfrac{3}{10}, \ldots, \dfrac{3}{n+7}, \ldots$ is nonincreasing.

(d) The constant sequence $\{3\}$ is both nondecreasing and nonincreasing.

Example 3   A Nondecreasing Sequence

Show that the sequence

$$a_n = \frac{n-1}{n+1}$$

is nondecreasing.

Solution

(a) We show that for all $n \geq 1$, $a_n \leq a_{n+1}$; that is,

$$\frac{n-1}{n+1} \leq \frac{(n+1)-1}{(n+1)+1}.$$

The inequality is equivalent to the one we get by cross multiplication:

$$\frac{n-1}{n+1} \leq \frac{(n+1)-1}{(n+1)+1} \Leftrightarrow \frac{n-1}{n+1} \leq \frac{n}{n+2}$$

$$\Leftrightarrow (n-1)(n+2) \leq n(n+1)$$

$$\Leftrightarrow n^2 + n - 2 \leq n^2 + n$$

$$\Leftrightarrow -2 \leq 0.$$

Since $-2 \leq 0$ is true, $a_n \leq a_{n+1}$ and the sequence $\{a_n\}$ is nondecreasing.

(b) Another way to show that $\{a_n\}$ is nondecreasing is by defining $f(n) = a_n$ and establishing that $f'(x) \geq 0$. In this example, $f(n) = (n-1)/(n+1)$ and

$$f'(x) = \frac{d}{dx}\left(\frac{x-1}{x+1}\right)$$

$$= \frac{(x+1)(1) - (x-1)(1)}{(x+1)^2} \qquad \text{Quotient Rule}$$

$$= \frac{2}{(x+1)^2} > 0.$$

Therefore, f is an increasing function, so $f(n+1) \geq f(n)$, or $a_{n+1} \geq a_n$.

Definition Bounded from Above, Upper Bound, Bounded from Below, Lower Bound, Bounded Sequence

A sequence $\{a_n\}$ is **bounded from above** if there exists a number M such that $a_n \leq M$ for all n. The number M is an **upper bound** for $\{a_n\}$. The sequence is **bounded from below** if there exists a number m such that $m \leq a_n$ for all n. The number m is a **lower bound** for $\{a_n\}$. If it is bounded from above and below, then $\{a_n\}$ is a **bounded sequence.**

Example 4 Applying The Definition For Boundedness

(a) The sequence $1, 2, 3, \dots, n, \dots$ has no upper bound, but it is bounded below by $m = 1$.

(b) The sequence $\frac{1}{2}, \frac{2}{3}, \frac{3}{4}, \dots, \frac{n}{n+1}, \dots$ is bounded above by $M = 1$ and below by $m = \frac{1}{2}$.

(c) The sequence $-1, 2, -3, 4, \dots, (-1)^n n, \dots$ is neither bounded from above nor from below.

We know that not every bounded sequence converges because the sequence $a_n = (-1)^n$ is bounded ($-1 \leq a_n \leq 1$) but is divergent. Also, not every monotonic sequence converges because the sequence $1, 2, 3, \dots, n, \dots$ of natural numbers is monotonic but diverges. If a sequence is *both* bounded and monotonic, however, then it must converge. This fact is summarized in the following theorem.

Theorem 5 Monotonic Sequence Theorem

Every bounded, monotonic sequence is convergent.

Although we do not prove Theorem 5, Figure 8.5 helps us understand why it is true for a nondecreasing sequence that is bounded from above. Since the sequence is nondecreasing and cannot go above M, the terms are forced to bunch together around some number $L \leq M$.

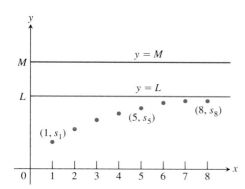

FIGURE 8.5 If the terms of a nondecreasing sequence have an upper bound M, they have a limit $L \leq M$.

Recursion formulas arise regularly in computer programs and numerical routines for solving differential equations, such as Euler's method.

Factorial Notation

The notation $n!$ ("n factorial") means the product $1 \cdot 2 \cdot 3 \cdots n$ of the integers from 1 to n. Notice that $(n + 1)! = (n + 1) \cdot n!$. Thus, $4! = 1 \cdot 2 \cdot 3 \cdot 4 = 24$ and $5! = 1 \cdot 2 \cdot 3 \cdot 4 \cdot 5 = 5 \cdot 4! = 120$. We define $0!$ to be 1. Factorials grow even faster than exponentials, as the following table suggests.

n	e^n(rounded)	$n!$
1	3	1
5	148	120
10	22,026	3,628,800
20	4.9×10^8	2.4×10^{18}

CD-ROM
WEBsite

Historical Biography

Charles Émile Picard
(1856 — 1941)

Example 5   Applying Theorem 5

(a) The nondecreasing sequence $\left\{\dfrac{n}{n + 1}\right\}$ is convergent because it is bounded above by $M = 1$. In fact,

$$\lim_{n\to\infty} \frac{n}{n + 1} = \lim_{n\to\infty} \frac{1}{1 + (1/n)}$$

$$= \frac{1}{1 + 0}$$

$$= 1,$$

and the sequence converges to $L = 1$.

(b) The nonincreasing sequence $\left\{\dfrac{1}{n + 1}\right\}$ is bounded below by $m = 0$ and is therefore convergent. It converges to $L = 0$.

Recursively Defined Sequences

So far, the sequences we have studied calculated each a_n directly from the value of n. Sequences, however, are often defined **recursively** by giving

1.   The value(s) of the initial term or terms and

2.   A rule, called a **recursion formula,** for calculating any later term from terms that precede it.

Example 6   Sequences Constructed Recursively

(a) The statements $a_1 = 1$ and $a_n = a_{n-1} + 1$ define the sequence $1, 2, 3, \ldots, n, \ldots$ of positive integers. With $a_1 = 1$, we have $a_2 = a_1 + 1 = 2$, $a_3 = a_2 + 1 = 3$, and so on.

(b) The statements $a_1 = 1$ and $a_n = n \cdot a_{n-1}$ define the sequence $1, 2, 6, 24, \ldots, n!, \ldots$ of factorials. With $a_1 = 1$, we have $a_2 = 2 \cdot a_1 = 2$, $a_3 = 3 \cdot a_2 = 6$, $a_4 = 4 \cdot a_3 = 24$, and so on.

(c) The statements $a_1 = 1$, $a_2 = 1$, and $a_{n+1} = a_n + a_{n-1}$ define the sequence $1, 1, 2, 3, 5, \ldots$ of **Fibonacci numbers.** With $a_1 = 1$ and $a_2 = 1$, we have $a_3 = 1 + 1 = 2$, $a_4 = 2 + 1 = 3$, $a_5 = 3 + 2 = 5$, and so on.

(d) As we can see by applying Newton's method, the statements $x_0 = 1$ and $x_{n+1} = x_n - [(\sin x_n - x_n^2)/(\cos x_n - 2x_n)]$ define a sequence that converges to a solution of the equation $\sin x - x^2 = 0$.

Picard's Method for Finding Roots

The problem of solving the equation

$$f(x) = 0 \tag{1}$$

is equivalent to that of solving the equation

$$g(x) = f(x) + x = x,$$

obtained by adding x to both sides of Equation (1). By this simple change, we cast Equation (1) into a form that may render it solvable on a computer by a powerful method called **Picard's method.**

If the domain of g contains the range of g, we can start with a point x_0 in the domain and apply g repeatedly to get

$$x_1 = g(x_0), \qquad x_2 = g(x_1), \qquad x_3 = g(x_2), \ldots.$$

Under simple restrictions that we describe shortly, the sequence generated by the recursion formula $x_{n+1} = g(x_n)$ will converge to a point x for which $g(x) = x$. This point solves the equation $f(x) = 0$ because

$$f(x) = g(x) - x = x - x = 0.$$

A point x for which $g(x) = x$ is a **fixed point** of g. We see from the last equation that the fixed points of g are precisely the roots of f.

Example 7   Testing Picard's Method

Solve the equation

$$\frac{1}{4}x + 3 = x.$$

Solution   By algebra, we know that the solution is $x = 4$. To apply Picard's method, we take

$$g(x) = \frac{1}{4}x + 3,$$

choose a starting point, say $x_0 = 1$, and calculate the initial terms of the sequence $x_{n+1} = g(x_n)$. Table 8.2 lists the results. In 10 steps, the solution of the original equation is found with an error of magnitude less than 3×10^{-6}.

Table 8.2 Successive iterates of $g(x) = (1/4)x + 3$, starting with $x_0 = 1$	
x_n	$x_{n+1} = g(x_n) = (1/4)\,x_n + 3$
$x_0 = 1$	$x_1 = g(x_0) = (1/4)(1) + 3 = 3.25$
$x_1 = 3.25$	$x_2 = g(x_1) = (1/4)(3.25) + 3 = 3.8125$
$x_2 = 3.8125$	$x_3 = g(x_2) = 3.9531\,25$
$x_3 = 3.9531\,25$	$x_4 = 3.9882\,8125$
$\cdot$	$x_5 = 3.9970\,70313$
$\cdot$	$x_6 = 3.9992\,67578$
$\cdot$	$x_7 = 3.9998\,16895$
	$x_8 = 3.9999\,54224$
	$x_9 = 3.9999\,88556$
	$x_{10} = 3.9999\,97139$
	$\vdots$

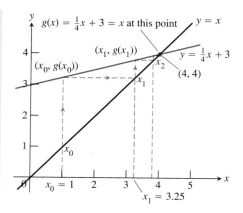

FIGURE 8.6 The Picard solution of the equation $g(x) = (1/4)x + 3 = x$. (Example 7)

Figure 8.6 shows the geometry of the solution. We start with $x_0 = 1$ and calculate the first value $g(x_0)$. This becomes the second x-value x_1. The second y-value $g(x_1)$ becomes the third x-value x_2, and so on. The process is shown as a path (called the *iteration path*) that starts at $x_0 = 1$, moves up to $(x_0, g(x_0)) = (x_0, x_1)$, over to (x_1, x_1), up to $(x_1, g(x_1))$, and so on. The path converges to the point where the graph of g meets the line $y = x$. This is the point where $g(x) = x$.

Example 8 Using Picard's Method

Solve the equation $\cos x = x$.

Solution We take $g(x) = \cos x$, choose $x_0 = 1$ as a starting value, and use the recursion formula $x_{n+1} = g(x_n)$ to find

$$x_0 = 1, \qquad x_1 = \cos 1, \qquad x_2 = \cos(x_1), \dots .$$

We can approximate the first 50 terms or so on a calculator in radian mode by entering 1 and taking the cosine repeatedly. The display stops changing when $\cos x = x$ to the number of decimal places in the display.

Try it for yourself. As you continue to take the cosine, the successive approximations lie alternately above and below the fixed point $x = 0.739085133\dots$.

Figure 8.7 shows that the values oscillate this way because the path of the procedure spirals around the fixed point.

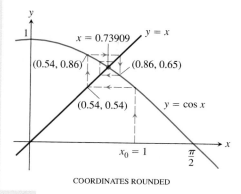

COORDINATES ROUNDED

FIGURE 8.7 The solution of $\cos x = x$ by Picard's method starting at $x_0 = 1$. (Example 8)

Example 9   Picard's Method May Fail to Solve an Equation

Picard's method will not solve the equation

$$g(x) = 4x - 12 = x.$$

As Figure 8.8 shows, any choice of x_0 except $x_0 = 4$, the solution itself, generates a divergent sequence that moves away from the solution.

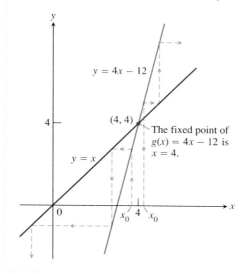

FIGURE 8.8 Applying the Picard method to $g(x) = 4x - 12$ will not find the fixed point unless x_0 is the fixed point 4 itself. (Example 9)

The difficulty in Example 9 can be traced to the slope of the line $y = 4x - 12$ exceeding 1, the slope of the line $y = x$. Conversely, the process worked in Example 7 because the slope of the line $y = (1/4)x + 3$ was numerically less than 1. A theorem from advanced calculus tells us that if $g'(x)$ is continuous on a closed interval I whose interior contains a solution of the equation $g(x) = x$ and if $|g'(x)| < 1$ on I, then any choice of x_0 in the interior of I will lead to the solution.

EXERCISES 8.2

Finding Terms of a Recursively Defined Sequence

Each of Exercises 1–6 gives the first term or two of a sequence along with a recursion formula for the remaining terms. Write out the first ten terms of the sequence.

1. $a_1 = 1$, $a_{n+1} = a_n + (1/2^n)$

2. $a_1 = 1$, $a_{n+1} = a_n/(n + 1)$

3. $a_1 = 2$, $a_{n+1} = (-1)^{n+1} a_n/2$

4. $a_1 = -2$, $a_{n+1} = na_n/(n + 1)$

5. $a_1 = a_2 = 1$, $a_{n+2} = a_{n+1} + a_n$

6. $a_1 = 2$, $a_2 = -1$, $a_{n+2} = a_{n+1}/a_n$

T 7. *Sequences generated by Newton's method* Newton's method, applied to a differentiable function $f(x)$, begins with a starting value x_0 and constructs from it a sequence of numbers $\{x_n\}$ that under favorable circumstances converges to a zero of f. The recursion formula for the sequence is

$$x_{n+1} = x_n - \frac{f(x_n)}{f'(x_n)}.$$

(a) Show that the recursion formula for $f(x) = x^2 - a$, $a > 0$, can be written as $x_{n+1} = (x_n + a/x_n)/2$.

(b) *Writing to Learn* Starting with $x_0 = 1$ and $a = 3$, calculate successive terms of the sequence until the display begins to repeat. What number is being approximated? Explain.

T **8.** (*Continuation of Exercise 7*) Repeat part (b) of Exercise 7 with $a = 2$ in place of $a = 3$.

T **9.** *Newton's method* The following sequences come from the recursion formula for Newton's method (see Exercise 7).

Do the sequences converge? If so, to what value? In each case, begin by identifying the function f that generates the sequence.

(a) $x_0 = 1$, $x_{n+1} = x_n - \dfrac{x_n^2 - 2}{2x_n} = \dfrac{x_n}{2} + \dfrac{1}{x_n}$

(b) $x_0 = 1$, $x_{n+1} = x_n - \dfrac{\tan x_n - 1}{\sec^2 x_n}$

(c) $x_0 = 1$, $x_{n+1} = x_n - 1$

T **10.** *A recursive definition of $\pi/2$* If you start with $x_1 = 1$ and define the subsequent terms of $\{x_n\}$ by the rule $x_n = x_{n-1} + \cos x_{n-1}$, you generate a sequence that converges rapidly to $\pi/2$.

(a) Try it.

(b) Use the accompanying figure to explain why the convergence is so rapid.

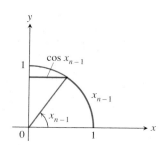

Theory and Examples

In Exercises 11–14, determine if the sequence is nondecreasing and if it is bounded from above.

11. $a_n = \dfrac{3n + 1}{n + 1}$

12. $a_n = \dfrac{(2n + 3)!}{(n + 1)!}$

13. $a_n = \dfrac{2^n 3^n}{n!}$

14. $a_n = 2 - \dfrac{2}{n} - \dfrac{1}{2^n}$

Which of the sequences in Exercises 15–24 converge, and which diverge? Give reasons for your answers.

15. $a_n = 1 - \dfrac{1}{n}$

16. $a_n = n - \dfrac{1}{n}$

17. $a_n = \dfrac{2^n - 1}{2^n}$

18. $a_n = \dfrac{2^n - 1}{3^n}$

19. $a_n = ((-1)^n + 1)\left(\dfrac{n + 1}{n}\right)$

20. The first term of a sequence is $x_1 = \cos(1)$. The next terms are $x_2 = x_1$ or $\cos(2)$, whichever is larger, and $x_3 = x_2$ or $\cos(3)$, whichever is larger (farther to the right). In general,

$$x_{n+1} = \max\{x_n, \cos(n+1)\}.$$

21. $a_n = \dfrac{n + 1}{n}$

22. $a_n = \dfrac{1 + \sqrt{2n}}{\sqrt{n}}$

23. $a_n = \dfrac{1 - 4^n}{2^n}$

24. $a_n = \dfrac{4^{n+1} + 3^n}{4^n}$

25. *Limits and subsequences* Prove that if two subsequences of a sequence $\{a_n\}$ have different limits $L_1 \neq L_2$, then $\{a_n\}$ diverges.

26. *Even and odd indices* For a sequence $\{a_n\}$, the terms of even index are denoted by a_{2k} and the terms of odd index by a_{2k+1}. Prove that if $a_{2k} \to L$ and $a_{2k+1} \to L$, then $a_n \to L$.

Picard's Method

Use Picard's method to solve the equations in Exercises 27–32.

27. $\sqrt{x} = x$

28. $x^2 = x$

29. $\cos x + x = 0$

30. $\cos x = x + 1$

31. $x - \sin x = 0.1$

32. $\sqrt{x} = 4 - \sqrt{1 + x}$ (*Hint:* Square both sides first.)

33. Solving the equation $\sqrt{x} = x$ by Picard's method finds the solution $x = 1$ but not the solution $x = 0$. Why? (*Hint:* Graph $y = x$ and $y = \sqrt{x}$ together.)

34. Solving the equation $x^2 = x$ by Picard's method with $|x_0| \neq 1$ can find the solution $x = 0$ but not the solution $x = 1$. Why? (*Hint:* Graph $y = x^2$ and $y = x$ together.)

COMPUTER EXPLORATIONS

Looking for Convergence of Recursively Defined Sequences

Use a CAS to perform the following steps for the sequences in Exercises 35 and 36.

(a) Calculate and then plot the first 25 terms of the sequence. Does the sequence appear to be bounded from above or below? Does it appear to converge or diverge? If it does converge, what is the limit L?

(b) If the sequence converges, find an integer N such that $|a_n - L| \leq 0.01$ for $n \geq N$. How far in the sequence do you have to get for the terms to lie within 0.0001 of L?

35. $a_1 = 1$, $a_{n+1} = a_n + \dfrac{1}{5^n}$

36. $a_1 = 1$, $a_{n+1} = a_n + (-2)^n$

37. *Compound interest, deposits, and withdrawals* If you invest an amount of money A_0 at a fixed annual interest rate r compounded m times per year and if the constant amount b is added to the account at the end of each compounding period (or taken from the account if $b < 0$), then the amount you have after $n + 1$ compounding periods is

$$A_{n+1} = \left(1 + \dfrac{r}{m}\right)A_n + b. \qquad (2)$$

(a) If $A_0 = 1000$, $r = 0.02015$, $m = 12$, and $b = 50$, calculate and plot the first 100 points (n, A_n). How much money is in your account at the end of 5 years? Does $\{A_n\}$ converge? Is $\{A_n\}$ bounded?

(b) Repeat part (a) with $A_0 = 5000$, $r = 0.0589$, $m = 12$, and $b = -50$.

(c) If you invest 5000 dollars in a certificate of deposit (CD) that pays 4.5% annually, compounded quarterly, and you make no further investments in the CD, approximately how many years will it take before you have 20,000 dollars? What if the CD earns 6.25%?

(d) It can be shown that for any $k \geq 0$, the sequence defined recursively by Equation (2) satisfies the relation

$$A_k = \left(1 + \frac{r}{m}\right)^k \left(A_0 + \frac{mb}{r}\right) - \frac{mb}{r}. \tag{3}$$

For the values of the constants A_0, r, m, and b given in part (a), validate this assertion by comparing the values of the first 50 terms of both sequences. Then show by direct substitution that the terms in Equation (3) satisfy the recursion formula (2).

38. *Logistic difference equation and bifurcation* The recursive relation

$$a_{n+1} = ra_n(1 - a_n)$$

is called the **logistic difference equation,** and when the initial value a_0 is given, the equation defines the **logistic sequence** $\{a_n\}$. Throughout this exercise, we choose a_0 in the interval $0 < a_0 < 1$, say $a_0 = 0.3$.

(a) Choose $r = 3/4$. Calculate and plot the points (n, a_n) for the first 100 terms in the sequence. Does it appear to converge? What do you guess is the limit? Does the limit seem to depend on your choice of a_0?

(b) Choose several values of r in the interval $1 < r < 3$ and repeat the procedures in part (a). Be sure to choose some points near the endpoints of the interval. Describe the behavior of the sequences you observe in your plots.

(c) Now examine the behavior of the sequence for values of r near the endpoints of the interval $3 < r < 3.45$. The transition value $r = 3$ is called a **bifurcation value,** and the new behavior of the sequence in the interval is called an **attracting 2-cycle.** Explain why this reasonably describes the behavior.

(d) Next explore the behavior for r values near the endpoints of each of the intervals $3.45 < r < 3.54$ and $3.54 < r < 3.55$. Plot the first 200 terms of the sequences. Describe in your own words the behavior observed in your plots for each interval. Among how many values does the sequence appear to oscillate for each interval? The values $r = 3.45$ and $r = 3.54$ (rounded to 2 decimal places) are also called bifurcation values because the behavior of the sequence changes as r crosses over those values.

(e) The situation gets even more interesting. There is actually an increasing sequence of bifurcation values $3 < 3.45 < 3.54 < \ldots < c_n < c_{n+1} \ldots$ such that for $c_n < r < c_{n+1}$, the logistic sequence $\{a_n\}$ eventually oscillates steadily among 2^n values, called an **attracting 2^n-cycle.** Moreover, the bifurcation sequence $\{c_n\}$ is bounded above by 3.57 (so it converges). If you choose a value of $r < 3.57$, you will observe a 2^n-cycle of some sort. Choose $r = 3.5695$ and plot 300 points.

(f) Let us see what happens when $r > 3.57$. Choose $r = 3.65$ and calculate and plot the first 300 terms of $\{a_n\}$. Observe how the terms wander around in an unpredictable, chaotic fashion. You cannot predict the value of a_{n+1} from the value of a_n.

(g) For $r = 3.65$, choose two starting values of a_0 that are close together, say $a_0 = 0.3$ and $a_0 = 0.301$. Calculate and plot the first 300 values of the sequences determined by each starting value. Compare the behaviors observed in your plots. How far out do you go before the corresponding terms of your two sequences appear to depart from each other? Repeat the exploration for $r = 3.75$. Can you see how the plots look different depending on your choice of a_0? We say that the logistic sequence is **sensitive to the initial condition** a_0.

8.3 Infinite Series

Series and Partial Sums • Geometric Series • Divergent Series • nth-Term Test for Divergence • Adding or Deleting Terms • Reindexing • Combining Series

In mathematics and science, we often write functions as infinite polynomials, such as

$$\frac{1}{1 - x} = 1 + x + x^2 + x^3 + \cdots + x^n + \cdots, \qquad |x| < 1$$

(we see the importance of doing so as the chapter continues). For any allowable value of x, we evaluate the polynomial as an infinite sum of constants, a sum we call an *infinite series*. The goal of this section is to familiarize ourselves with infinite series.

Series and Partial Sums

The first thing to get straight about an infinite series is that it is not simply an example of addition. Addition of real numbers is a *binary* operation, meaning that we really add numbers two at a time. The only reason that $1 + 2 + 3$ makes sense as "addition" is that we can *group* the numbers and then add them two at a time. The associative property of addition guarantees that we get the same sum no matter how we group them:

$$1 + (2 + 3) = 1 + 5 = 6 \quad \text{and} \quad (1 + 2) + 3 = 3 + 3 = 6.$$

In short, a *finite sum* of real numbers always produces a real number (the result of a finite number of binary additions), but an *infinite sum* of real numbers is something else entirely. That is why we need a careful definition of infinite series.

We begin by asking how to assign meaning to an expression like

$$1 + \frac{1}{2} + \frac{1}{4} + \frac{1}{8} + \frac{1}{16} + \cdots.$$

The way to do so is not to try to add all the terms at once (we cannot) but rather to add the terms one at a time from the beginning and look for a pattern in how these "partial sums" grow.

Partial sum		Value
First:	$s_1 = 1$	$2 - 1$
Second:	$s_2 = 1 + \frac{1}{2}$	$2 - \frac{1}{2}$
Third:	$s_3 = 1 + \frac{1}{2} + \frac{1}{4}$	$2 - \frac{1}{4}$
$\vdots$	$\vdots$	$\vdots$
nth	$s_n = 1 + \frac{1}{2} + \frac{1}{4} + \cdots + \frac{1}{2^{n-1}}$	$2 - \frac{1}{2^{n-1}}$

Indeed there is a pattern. The partial sums form a sequence whose nth term is

$$s_n = 2 - \frac{1}{2^{n-1}}.$$

(We see why momentarily.) This sequence converges to 2 because $\lim_{n \to \infty} (1/2^n) = 0$. We say,

"The sum of the infinite series $1 + \frac{1}{2} + \frac{1}{4} + \cdots + \frac{1}{2^{n-1}} + \cdots$ is 2."

Is the sum of any finite number of terms in this series equal to 2? No. Can we actually add an infinite number of terms one by one? No. We can, however, still define their sum by defining it to be the limit of the sequence of partial sums as $n \to \infty$, in this case 2 (Figure 8.9). Our knowledge of sequences and limits enables us to break away from the confines of finite sums to define this entirely new concept.

FIGURE 8.9 As the lengths $1, \frac{1}{2}, \frac{1}{4}, \frac{1}{8}, \ldots$ are added one by one, the sum approaches 2.

CD-ROM
WEBsite

Historical Biography

Blaise Pascal
(1623 — 1662)

Definition   Infinite Series

Given a sequence of numbers $\{a_n\}$, an expression of the form

$$a_1 + a_2 + a_3 + \cdots + a_n + \cdots$$

is an **infinite series**. The number a_n is the **nth term** of the series.

The **partial sums** of the series form a sequence

$$s_1 = a_1$$
$$s_2 = a_1 + a_2$$
$$s_3 = a_1 + a_2 + a_3$$
$$\vdots$$
$$s_n = \sum_{k=1}^{n} a_k$$
$$\vdots$$

of real numbers, each defined as a finite sum. If the sequence of partial sums has a limit S as $n \to \infty$, we say that the series **converges** to the sum S, and we write

$$a_1 + a_2 + a_3 + \cdots + a_n + \cdots = \sum_{k=1}^{\infty} a_k = S.$$

Otherwise, we say that the series **diverges.**

Example 1   Identifying a Convergent Series

Does the series

$$\frac{3}{10} + \frac{3}{100} + \frac{3}{1000} + \cdots + \frac{3}{10^n} + \cdots$$

converge?

Solution   Here is the sequence of partial sums, written in decimal form.

$$0.3, \ 0.33, \ 0.333, \ 0.3333, \ldots$$

This sequence has a limit $0.\overline{3}$, which we recognize as the fraction $1/3$. The series converges to the sum $1/3$.

When we begin to study a given series $a_1 + a_2 + \cdots + a_n + \cdots$, we might not know whether it converges or diverges. In either case, it is convenient to use sigma notation to write the series as

$$\sum_{n=1}^{\infty} a_n, \qquad \sum_{k=1}^{\infty} a_k, \qquad \text{or} \qquad \sum a_n.$$

A useful shorthand when summation from 1 to ∞ is understood

Geometric Series

The series in Example 1 is a **geometric series** because each term is obtained from its preceding term by multiplying by the same number r, in this case, $r = 1/10$. (The series of areas for the infinitely halved square at the beginning of this chapter is also geometric.) The convergence of geometric series is one of the few infinite processes with which mathematicians were reasonably comfortable prior to calculus. Let's see why.

Geometric series are series of the form

$$a + ar + ar^2 + \cdots + ar^{n-1} + \cdots = \sum_{n=1}^{\infty} ar^{n-1}$$

in which a and r are fixed real numbers and $a \neq 0$. The **ratio** r can be positive, as in

$$1 + \frac{1}{2} + \frac{1}{4} + \cdots + \left(\frac{1}{2}\right)^{n-1} + \cdots,$$

or negative, as in

$$1 - \frac{1}{3} + \frac{1}{9} - \cdots + \left(-\frac{1}{3}\right)^{n-1} + \cdots.$$

If $|r| \neq 1$, we can determine the convergence or divergence of the series in the following way, starting with the nth partial sum:

$$s_n = a + ar + ar^2 + \cdots + ar^{n-1}$$

$$rs_n = ar + ar^2 + \cdots + ar^{n-1} + ar^n \qquad \text{Multiply } s_n \text{ by } r.$$

$$s_n - rs_n = a - ar^n \qquad \text{Subtract } rs_n \text{ from } s_n. \text{ Most of the terms on the right cancel.}$$

$$s_n(1 - r) = a(1 - r^n) \qquad \text{Factor.}$$

$$s_n = \frac{a(1 - r^n)}{1 - r}, \qquad (r \neq 1). \qquad \text{We can solve for } s_n \text{ if } r \neq 1.$$

If $|r| < 1$, then $r^n \to 0$ as $n \to \infty$ (Table 8.1, Formula 4) and $s_n \to a/(1 - r)$. If $|r| > 1$, then $|r^n| \to \infty$ and the series diverges.

If $r = 1$, the nth partial sum of the geometric series is

$$s_n = a + a(1) + a(1)^2 + \cdots + a(1)^{n-1} = na,$$

and the series diverges because $\lim_{n \to \infty} s_n = \pm\infty$, depending on the sign of a. If $r = -1$, the series diverges because the nth partial sums alternate between a and 0. Let's summarize our results.

The equation

$$\sum_{n=1}^{\infty} ar^{n-1} = \frac{a}{1 - r}, \qquad |r| < 1$$

holds only if the summation begins with $n = 1$.

The geometric series

$$a + ar + ar^2 + ar^3 + \cdots + ar^{n-1} + \cdots = \sum_{n=1}^{\infty} ar^{n-1}$$

converges to the sum $a/(1 - r)$ if $|r| < 1$ and diverges if $|r| \geq 1$.

(a)

(b)

FIGURE 8.10 (a) Example 3 shows how to use a geometric series to calculate the total vertical distance traveled by a bouncing ball if the height of each rebound is reduced by the factor r. (b) A stroboscopic photo of a bouncing ball.

CD-ROM
WEBsite

This completely settles the issue for geometric series. We know which ones converge and which ones diverge, and for the convergent ones, we know what the sums must be. The interval $-1 < r < 1$ is the **interval of convergence.**

Example 2   Analyzing Geometric Series

Tell whether each series converges or diverges. If it converges, give its sum.

(a) $\displaystyle\sum_{n=1}^{\infty} 3\left(\frac{1}{2}\right)^{n-1}$

(b) $1 - \dfrac{1}{2} + \dfrac{1}{4} - \dfrac{1}{8} + \cdots + \left(-\dfrac{1}{2}\right)^{n-1} + \cdots$

(c) $\displaystyle\sum_{k=0}^{\infty} \left(\frac{3}{5}\right)^{k} = \sum_{k=1}^{\infty} \left(\frac{3}{5}\right)^{k-1}$

(d) $\dfrac{\pi}{2} + \dfrac{\pi^2}{4} + \dfrac{\pi^3}{8} + \cdots$

Solution

(a) First term is $a = 3$ and $r = 1/2$. The series converges to

$$\frac{3}{1 - (1/2)} = 6.$$

(b) First term is $a = 1$ and $r = -1/2$. The series converges to

$$\frac{1}{1 - (-1/2)} = \frac{2}{3}.$$

(c) First term is $a = (3/5)^0 = 1$ and $r = 3/5$. The series converges to

$$\frac{1}{1 - (3/5)} = \frac{5}{2}.$$

(d) In this series, $r = \pi/2 > 1$. The series diverges.

Example 3   A Bouncing Ball

You drop a ball from a meters above a flat surface. Each time the ball hits the surface after falling a distance h, it rebounds a distance rh, where r is positive but less than 1. Find the total distance the ball travels up and down (Figure 8.10).

Solution   The total distance is

$$s = a + \underbrace{2ar + 2ar^2 + 2ar^3 + \cdots}_{\text{This sum is } 2ar/(1-r).} = a + \frac{2ar}{1 - r} = a\frac{1 + r}{1 - r}.$$

If $a = 6$ m and $r = 2/3$, for instance, the distance is

$$s = 6\frac{1 + (2/3)}{1 - (2/3)} = 6\left(\frac{5/3}{1/3}\right) = 30 \text{ m}.$$

Example 4   Repeating Decimals

Express the repeating decimal 5.23 23 23 . . . as the ratio of two integers.

Solution

$$5.23\ 23\ 23\ldots = 5 + \frac{23}{100} + \frac{23}{(100)^2} + \frac{23}{(100)^3} + \cdots$$

$$= 5 + \frac{23}{100}\underbrace{\left(1 + \frac{1}{100} + \left(\frac{1}{100}\right)^2 + \cdots\right)}_{1/(1 - 0.01)} \qquad \begin{array}{l} a = 1, \\ r = 1/100 \end{array}$$

$$= 5 + \frac{23}{100}\left(\frac{1}{0.99}\right) = 5 + \frac{23}{99} = \frac{518}{99}$$

We have hardly begun our study of infinite series, but knowing everything there is to know about the convergence and divergence of an *entire class* of series (geometric) is an impressive start. Like the Renaissance mathematicians, we are ready to explore where this might lead.

Unfortunately, formulas like the one for the sum of a convergent geometric series are rare, and we usually have to settle for an estimate of a series' sum (more about this later). The next example, however, is another case in which we can find the sum exactly.

Example 5   A Nongeometric but Telescoping Series

Find the sum of the series

$$\sum_{n=1}^{\infty} \frac{1}{n(n + 1)}.$$

Solution   We look for a pattern in the sequence of partial sums that might lead to a formula for s_k. The key is partial fractions. The observation that

$$\frac{1}{k(k + 1)} = \frac{1}{k} - \frac{1}{k + 1}$$

permits us to write the partial sum

$$\sum_{n=1}^{k} \frac{1}{n(n + 1)} = \frac{1}{1 \cdot 2} + \frac{1}{2 \cdot 3} + \cdots + \frac{1}{k \cdot (k + 1)}$$

as

$$s_k = \left(\frac{1}{1} - \frac{1}{2}\right) + \left(\frac{1}{2} - \frac{1}{3}\right) + \cdots + \left(\frac{1}{k} - \frac{1}{k + 1}\right).$$

Removing parentheses and canceling the terms of opposite sign collapses the sum to

$$s_k = 1 - \frac{1}{k + 1}.$$

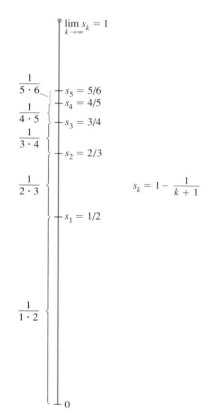

$\lim_{k \to \infty} s_k = 1$

$\frac{1}{5 \cdot 6}$ $s_5 = 5/6$

$\frac{1}{4 \cdot 5}$ $s_4 = 4/5$

  $s_3 = 3/4$

$\frac{1}{3 \cdot 4}$

  $s_2 = 2/3$

$\frac{1}{2 \cdot 3}$ $s_k = 1 - \frac{1}{k + 1}$

  $s_1 = 1/2$

$\frac{1}{1 \cdot 2}$

0

FIGURE 8.11 The partial sums of the series in Example 5.

We now see that $s_k \to 1$ as $k \to \infty$. The series converges, and its sum is 1 (Figure 8.11).

$$\sum_{n=1}^{\infty} \frac{1}{n(n + 1)} = 1.$$

Divergent Series

Geometric series with $|r| \geq 1$ are not the only series to diverge.

Example 6   Identifying a Divergent Series

Does the series $1 - 1 + 1 - 1 + 1 - 1 + \cdots$ converge?

Solution   You might be tempted to pair the terms as

$$(1 - 1) + (1 - 1) + (1 - 1) + \cdots.$$

That strategy, however, requires an *infinite* number of pairings, so it cannot be justified by the associative property of addition. This is an infinite series, not a finite sum, so if it has a sum it *has to be* the limit of its sequence of partial sums,

$$1, 0, 1, 0, 1, 0, 1, \ldots.$$

Since this sequence has no limit, the series has no sum. It diverges.

Example 7   Partial Sums Outgrowing Any Bound

(a) The series

$$\sum_{n=1}^{\infty} n^2 = 1 + 4 + 9 + \cdots + n^2 + \cdots$$

diverges because the partial sums grow beyond every number L. After $n = 1$, the partial sum $s_n = 1 + 4 + 9 + \cdots + n^2$ is greater than n^2.

(b) The series

$$\sum_{n=1}^{\infty} \frac{n + 1}{n} = \frac{2}{1} + \frac{3}{2} + \frac{4}{3} + \cdots + \frac{n + 1}{n} + \cdots$$

diverges because the partial sums eventually outgrow every preassigned number. Each term is greater than 1, so the sum of n terms is greater than n.

nth-Term Test for Divergence

Observe that $\lim_{n \to \infty} a_n$ must equal zero if the series $\sum_{n=1}^{\infty} a_n$ converges. To see why, let S represent the series' sum and $s_n = a_1 + a_2 + \cdots + a_n$ the nth partial sum. When n is large, both s_n and s_{n-1} are close to S, so their difference, a_n, is close to zero. More formally,

$$a_n = s_n - s_{n-1} \to S - S = 0. \qquad \text{Difference Rule for sequences}$$

CAUTION Theorem 6 *does not say* that $\sum_{n=1}^{\infty} a_n$ converges if $a_n \to 0$. It is possible for a series to diverge when $a_n \to 0$.

Theorem 6   Limit of the nth Term of a Convergent Series

If $\sum_{n=1}^{\infty} a_n$ converges, then $a_n \to 0$.

Theorem 6 leads to a test for detecting the kind of divergence that occurred in Examples 6 and 7.

---

nth-Term Test for Divergence

$\sum_{n=1}^{\infty} a_n$ diverges if $\lim_{n\to\infty} a_n$ fails to exist or is different from zero.

Example 8   Applying the nth-Term Test

(a) $\sum_{n=1}^{\infty} n^2$ diverges because $n^2 \to \infty$.

(b) $\sum_{n=1}^{\infty} \dfrac{n+1}{n}$ diverges because $\dfrac{n+1}{n} \to 1$.

(c) $\sum_{n=1}^{\infty} (-1)^{n+1}$ diverges because $\lim_{n\to\infty} (-1)^{n+1}$ does not exist.

(d) $\sum_{n=1}^{\infty} \dfrac{-n}{2n+5}$ diverges because $\lim_{n\to\infty} \left(\dfrac{-n}{2n+5} \right) = -\dfrac{1}{2} \neq 0$.

Example 9 $a_n \to 0$, but the Series Diverges

The series

$$1 + \underbrace{\frac{1}{2} + \frac{1}{2}}_{\text{2 terms}} + \underbrace{\frac{1}{4} + \frac{1}{4} + \frac{1}{4} + \frac{1}{4}}_{\text{4 terms}} + \cdots + \underbrace{\frac{1}{2^n} + \frac{1}{2^n} + \cdots + \frac{1}{2^n}}_{2^n \text{ terms}} + \cdots$$

$$= 1 + 1 + 1 + \cdots + 1 + \cdots$$

diverges even though its terms form a sequence that converges to 0.

Adding or Deleting Terms

We can always add a finite number of terms to a series or delete a finite number of terms without altering the series' convergence or divergence, although in the case of convergence, this will usually change the sum. If $\sum_{n=1}^{\infty} a_n$ converges, then $\sum_{n=k}^{\infty} a_n$ converges for any $k > 1$, and

$$\sum_{n=1}^{\infty} a_n = a_1 + a_2 + \cdots + a_{k-1} + \sum_{n=k}^{\infty} a_n.$$

Conversely, if $\sum_{n=k}^{\infty} a_n$ converges for any $k > 1$, then $\sum_{n=1}^{\infty} a_n$ converges. Thus,

$$\sum_{n=1}^{\infty} \frac{1}{5^n} = \frac{1}{5} + \frac{1}{25} + \frac{1}{125} + \sum_{n=4}^{\infty} \frac{1}{5^n}$$

and

$$\sum_{n=4}^{\infty} \frac{1}{5^n} = \left(\sum_{n=1}^{\infty} \frac{1}{5^n} \right) - \frac{1}{5} - \frac{1}{25} - \frac{1}{125}.$$

CD-ROM
WEBsite

Historical Biography

Richard Dedekind
(1831 — 1916)

Reindexing

As long as we preserve the order of its terms, we can reindex any series without altering its convergence (see Example 2c). To raise the starting value of the index h units, replace the n in the formula for a_n by $n - h$:

$$\sum_{n=1}^{\infty} a_n = \sum_{n=1+h}^{\infty} a_{n-h} = a_1 + a_2 + a_3 + \cdots.$$

To lower the starting value of the index h units, replace the n in the formula for a_n by $n + h$:

$$\sum_{n=1}^{\infty} a_n = \sum_{n=1-h}^{\infty} a_{n+h} = a_1 + a_2 + a_3 + \cdots.$$

It works like a horizontal shift.

Example 10   Reindexing A Geometric Series

We can write the geometric series that starts with

$$1 + \frac{1}{2} + \frac{1}{4} + \cdots$$

as

$$\sum_{n=0}^{\infty} \frac{1}{2^n}, \qquad \sum_{n=5}^{\infty} \frac{1}{2^{n-5}}, \qquad \text{or even} \qquad \sum_{n=-4}^{\infty} \frac{1}{2^{n+4}}.$$

The partial sums remain the same no matter what indexing we choose.

We usually give preference to indexings that lead to simple expressions.

Combining Series

Whenever we have two convergent series, we can add them term by term, subtract them term by term, or multiply them by constants to make new convergent series.

Theorem 7   Properties of Convergent Series

If $\Sigma a_n = A$ and $\Sigma b_n = B$ are convergent series, then

1. *Sum Rule:* $\Sigma (a_n + b_n) = \Sigma a_n + \Sigma b_n = A + B$

2. *Difference Rule:* $\Sigma (a_n - b_n) = \Sigma a_n - \Sigma b_n = A - B$

3. *Constant Multiple Rule:* $\Sigma k a_n = k \Sigma a_n = kA$ (any number k).

Example 11   Applying Theorem 7

Find the sums of the following series.

(a) $\displaystyle\sum_{n=1}^{\infty} \frac{3^{n-1} - 1}{6^{n-1}} = \sum_{n=1}^{\infty} \left(\frac{1}{2^{n-1}} - \frac{1}{6^{n-1}} \right)$

$\displaystyle\qquad = \sum_{n=1}^{\infty} \frac{1}{2^{n-1}} - \sum_{n=1}^{\infty} \frac{1}{6^{n-1}}$ Difference Rule

$\displaystyle\qquad = \frac{1}{1 - (1/2)} - \frac{1}{1 - (1/6)}$ Geometric series with $a = 1$ and $r =$ $1/2, 1/6$

$\displaystyle\qquad = 2 - \frac{6}{5}$

$\displaystyle\qquad = \frac{4}{5}$

(b) $\displaystyle\sum_{n=1}^{\infty} \frac{4}{2^{n-1}} = 4 \sum_{n=1}^{\infty} \frac{1}{2^{n-1}}$ Constant Multiple Rule

$\displaystyle\qquad = 4 \left(\frac{1}{1 - (1/2)} \right)$ Geometric series with $a = 1$, $r = 1/2$

$\displaystyle\qquad = 8$

Proof of Theorem 7   The three rules for series follow from the analogous rules for sequences in Theorem 1, Section 8.1. To prove the Sum Rule for series, let

$$A_n = a_1 + a_2 + \cdots + a_n, \qquad B_n = b_1 + b_2 + \cdots + b_n.$$

Then the partial sums of $\Sigma (a_n + b_n)$ are

$$S_n = (a_1 + b_1) + (a_2 + b_2) + \cdots + (a_n + b_n)$$
$$= (a_1 + \cdots + a_n) + (b_1 + \cdots + b_n)$$
$$= A_n + B_n.$$

Since $A_n \to A$ and $B_n \to B$, we have $S_n \to A + B$ by the Sum Rule for sequences. The proof of the Difference Rule is similar.

To prove the Constant Multiple Rule for series, observe that the partial sums of $\Sigma \, ka_n$ form the sequence

$$S_n = ka_1 + ka_2 + \cdots + ka_n = k(a_1 + a_2 + \cdots + a_n) = kA_n,$$

which converges to kA by the Constant Multiple Rule for sequences.   ▬

Interpreting Theorem 7 for Divergence

1. Every nonzero constant multiple of a divergent series diverges.

2. If $\Sigma \, a_n$ converges and $\Sigma \, b_n$ diverges, then $\Sigma \, (a_n + b_n)$ and $\Sigma \, (a_n - b_n)$ both diverge.

We omit the proofs.

EXERCISES 8.3

Finding nth Partial Sums

In Exercises 1–6, find a formula for the nth partial sum of each series and use it to find the series' sum if the series converges.

1. $2 + \dfrac{2}{3} + \dfrac{2}{9} + \dfrac{2}{27} + \cdots + \dfrac{2}{3^{n-1}} + \cdots$

2. $\dfrac{9}{100} + \dfrac{9}{100^2} + \dfrac{9}{100^3} + \cdots + \dfrac{9}{100^n} + \cdots$

3. $1 - \dfrac{1}{2} + \dfrac{1}{4} - \dfrac{1}{8} + \cdots + (-1)^{n-1}\dfrac{1}{2^{n-1}} + \cdots$

4. $1 - 2 + 4 - 8 + \cdots + (-1)^{n-1}2^{n-1} + \cdots$

5. $\dfrac{1}{2 \cdot 3} + \dfrac{1}{3 \cdot 4} + \dfrac{1}{4 \cdot 5} + \cdots + \dfrac{1}{(n+1)(n+2)} + \cdots$

6. $\dfrac{5}{1 \cdot 2} + \dfrac{5}{2 \cdot 3} + \dfrac{5}{3 \cdot 4} + \cdots + \dfrac{5}{n(n+1)} + \cdots$

Series with Geometric Terms

In Exercises 7–12, write out the first few terms of each series to show how the series starts. Then find the sum of the series.

7. $\displaystyle\sum_{n=0}^{\infty} \dfrac{(-1)^n}{4^n}$

8. $\displaystyle\sum_{n=1}^{\infty} \dfrac{7}{4^n}$

9. $\displaystyle\sum_{n=0}^{\infty} \left(\dfrac{5}{2^n} + \dfrac{1}{3^n}\right)$

10. $\displaystyle\sum_{n=0}^{\infty} \left(\dfrac{5}{2^n} - \dfrac{1}{3^n}\right)$

11. $\displaystyle\sum_{n=0}^{\infty} \left(\dfrac{1}{2^n} + \dfrac{(-1)^n}{5^n}\right)$

12. $\displaystyle\sum_{n=0}^{\infty} \left(\dfrac{2^{n+1}}{5^n}\right)$

Telescoping Series

Use partial fractions to find the sum of each series in Exercises 13–16.

13. $\displaystyle\sum_{n=1}^{\infty} \dfrac{4}{(4n-3)(4n+1)}$

14. $\displaystyle\sum_{n=1}^{\infty} \dfrac{6}{(2n-1)(2n+1)}$

15. $\displaystyle\sum_{n=1}^{\infty} \dfrac{40n}{(2n-1)^2(2n+1)^2}$

16. $\displaystyle\sum_{n=1}^{\infty} \dfrac{2n+1}{n^2(n+1)^2}$

Find the sums of the series in Exercises 17 and 18.

17. $\displaystyle\sum_{n=1}^{\infty} \left(\dfrac{1}{\sqrt{n}} - \dfrac{1}{\sqrt{n+1}}\right)$

18. $\displaystyle\sum_{n=1}^{\infty} \left(\dfrac{1}{\ln(n+2)} - \dfrac{1}{\ln(n+1)}\right)$

Convergence or Divergence

Which series in Exercises 19–32 converge, and which diverge? Give reasons for your answers. If a series converges, find its sum.

19. $\displaystyle\sum_{n=0}^{\infty} \left(\dfrac{1}{\sqrt{2}}\right)^n$

20. $\displaystyle\sum_{n=0}^{\infty} (\sqrt{2})^n$

21. $\displaystyle\sum_{n=1}^{\infty} (-1)^{n+1}\dfrac{3}{2^n}$

22. $\displaystyle\sum_{n=0}^{\infty} \dfrac{\cos n\pi}{5^n}$

23. $\displaystyle\sum_{n=0}^{\infty} e^{-2n}$

24. $\displaystyle\sum_{n=1}^{\infty} \ln\dfrac{1}{n}$

25. $\displaystyle\sum_{n=0}^{\infty} \dfrac{1}{x^n}, \quad |x| > 1$

26. $\displaystyle\sum_{n=0}^{\infty} \dfrac{2^n - 1}{3^n}$

27. $\displaystyle\sum_{n=1}^{\infty} \left(1 - \dfrac{1}{n}\right)^n$

28. $\displaystyle\sum_{n=0}^{\infty} \left(\dfrac{e}{\pi}\right)^n$

29. $\displaystyle\sum_{n=1}^{\infty} \ln\left(\dfrac{n}{n+1}\right)$

30. $\displaystyle\sum_{n=0}^{\infty} \dfrac{e^{n\pi}}{\pi^{ne}}$

31. $\displaystyle\sum_{n=0}^{\infty} \dfrac{n!}{1000^n}$

32. $\displaystyle\sum_{n=1}^{\infty} \dfrac{n^n}{n!}$

Geometric Series

In each of the geometric series in Exercises 33–36, write out the first few terms of the series to find a and r and find the sum of the series. Then express the inequality $|r| < 1$ in terms of x and find the values of x for which the inequality holds and the series converges.

33. $\displaystyle\sum_{n=0}^{\infty} (-1)^n x^n$

34. $\displaystyle\sum_{n=0}^{\infty} (-1)^n x^{2n}$

35. $\displaystyle\sum_{n=0}^{\infty} 3\left(\dfrac{x-1}{2}\right)^n$

36. $\displaystyle\sum_{n=0}^{\infty} \dfrac{(-1)^n}{2}\left(\dfrac{1}{3+\sin x}\right)^n$

In Exercises 37–40, find the values of x for which the given geometric series converges. Also, find the sum of the series (as a function of x) for those values of x.

37. $\displaystyle\sum_{n=0}^{\infty} 2^n x^n$

38. $\displaystyle\sum_{n=0}^{\infty} (-1)^n x^{-2n}$

39. $\displaystyle\sum_{n=0}^{\infty} \left(-\dfrac{1}{2}\right)^n (x-3)^n$

40. $\displaystyle\sum_{n=0}^{\infty} (\ln x)^n$

Repeating Decimals

Express each of the numbers in Exercises 41–46 as the ratio of two integers.

41. $0.\overline{23} = 0.23\ 23\ 23\ldots$

42. $0.\overline{234} = 0.234\ 234\ 234\ldots$

43. $0.\overline{7} = 0.7777\ldots$

44. $1.\overline{414} = 1.414\ 414\ 414\ldots$

45. $1.24\overline{123} = 1.24\ 123\ 123\ 123\ldots$

46. $3.\overline{142857} = 3.142857\ 142857\ldots$

Theory and Examples

CD-ROM
WEBsite

47. *Bouncing ball distance* A ball is dropped from a height of 4 m. Each time it strikes the pavement after falling from a height of h meters, it rebounds to a height of $0.75h$ m. Find the total distance the ball travels up and down.

CD-ROM
WEBsite

48. *Total time for bouncing* Find the total number of seconds the ball in Exercise 47 is traveling. (*Hint:* The formula $s = 4.9t^2$ gives $t = \sqrt{s/4.9}$.)

49. *Summing areas* The accompanying figure shows the first five of a sequence of squares. The outermost square has an area of 4 m². Each of the other squares is obtained by joining the midpoints of the sides of the squares before it. Find the sum of the areas of all the squares.

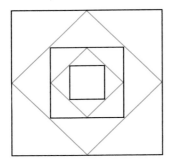

50. *Summing areas* The accompanying figure shows the first three rows and part of the fourth row of a sequence of rows of semicircles. There are 2^n semicircles in the nth row, each of radius $1/2^n$. Find the sum of the areas of all the semicircles.

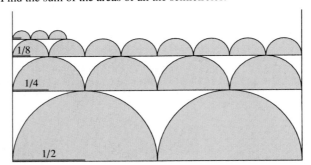

51. *Helge von Koch's snowflake curve* Start with an equilateral triangle whose sides have length 1, calling it Curve 1. On the middle third of each side, build an equilateral triangle pointing outward. Then erase the interiors of the old middle thirds. Call the expanded curve Curve 2. Now put equilateral triangles, again pointing outward, on the middle thirds of the sides of Curve 2. Erase the interiors of the old middle thirds to make Curve 3. Repeat the process, as shown, to define an infinite sequence of plane curves. The limit curve of the sequence is Koch's snowflake curve.

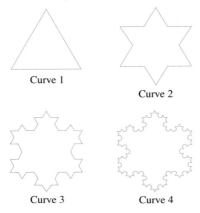

Here's how to show that the snowflake is a curve of infinite length enclosing a region of finite area.

(a) Find the length L_n of the nth curve C_n and show that $\lim_{n\to\infty} L_n = \infty$.

(b) Find the area A_n of the region enclosed by C_n and calculate $\lim_{n\to\infty} A_n$.

52. *Writing to Learn* The accompanying figure provides an informal proof that $\sum_{n=1}^{\infty} (1/n^2)$ is less than 2. Explain what is going on. (*Source:* "Convergence with Pictures" by P. J. Rippon, *American Mathematical Monthly*, Vol. 93, No. 6 (1986), pp. 476—478.)

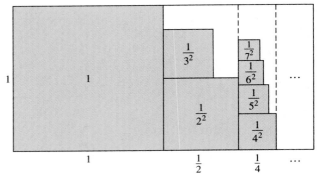

53. *Reindexing* The series in Exercise 5 can also be written as

$$\sum_{n=1}^{\infty} \frac{1}{(n+1)(n+2)} \qquad \text{and} \qquad \sum_{n=-1}^{\infty} \frac{1}{(n+3)(n+4)}.$$

Write it as a sum beginning with

(a) $n = -2$

(b) $n = 0$

(c) $n = 5$.

54. *Writing to Learn* Make up an infinite series of nonzero terms whose sum is

(a) 1

(b) -3

(c) 0.

Can you make an infinite series of nonzero terms that converges to any number you want? Explain.

55. *Geometric series* Find the value of b for which

$$1 + e^b + e^{2b} + e^{3b} + \cdots = 9.$$

56. *Modified geometric series* For what values of r does the infinite series

$$1 + 2r + r^2 + 2r^3 + r^4 + 2r^5 + r^6 + \cdots$$

converge? Find the sum of the series when it converges.

57. *Error using a partial sum* Show that the error $(L - s_n)$ obtained by replacing a convergent geometric series with one of its partial sums s_n is $ar^n/(1-r)$.

58. *Term-by-term product* Find convergent geometric series $A = \Sigma \, a_n$ and $B = \Sigma \, b_n$ that illustrate that $\Sigma \, a_n b_n$ may converge without being equal to AB.

59. *Term-by-term quotient* Show by example that $\Sigma \, (a_n/b_n)$ may converge to something other than A/B even when $A = \Sigma \, a_n$, $B = \Sigma \, b_n \neq 0$, and no b_n equals 0.

60. *Term-by-term quotient* Show by example that $\Sigma \, (a_n/b_n)$ may diverge even though $\Sigma \, a_n$ and $\Sigma \, b_n$ converge and no b_n equals 0.

61. *Term-by-term reciprocals* If $\Sigma \, a_n$ converges and $a_n > 0$ for all n, can anything be said about $\Sigma \, (1/a_n)$? Give reasons for your answer.

62. *Adding or deleting terms* What happens if you add a finite number of terms to a divergent series or delete a finite number of terms from a divergent series? Give reasons for your answer.

63. *Summing convergent and divergent series* If $\Sigma \, a_n$ converges and $\Sigma \, b_n$ diverges, can anything be said about their term-by-term sum $\Sigma \, (a_n + b_n)$? Give reasons for your answer.

8.4 Series of Nonnegative Terms

Integral Test   •   Harmonic Series and *p*-Series   •   Comparison Tests   •   Ratio and Root Tests

Given a series $\Sigma \, a_n$, we have two questions.

1.   Does the series converge?

2.   If it converges, what is its sum?

In this section, we study series that do not have negative terms. The reason for this restriction is that the partial sums of these series form nondecreasing sequences, and nondecreasing sequences that are bounded from above always converge. The partial sums are nondecreasing because $s_{n+1} = s_n + a_n$ and $a_n \geq 0$:

$$s_1 \leq s_2 \leq s_3 \leq \cdots \leq s_n \leq s_{n+1} \leq \cdots .$$

From the Monotonic Sequence Theorem (Theorem 5, Section 8.2), the series will converge if $\{s_n\}$ is bounded from above.

Corollary of Theorem 5

A series $\sum_{n=1}^{\infty} a_n$ of nonnegative terms converges if its partial sums are bounded from above.

This result is the basis for the tests to establish convergence we study in this section.

Integral Test

We introduce the Integral Test with an example.

Example 1   Applying Corollary of Theorem 5

Show that the series

$$\sum_{n=1}^{\infty} \frac{1}{n^2} = 1 + \frac{1}{4} + \frac{1}{9} + \frac{1}{16} + \cdots + \frac{1}{n^2} + \cdots$$

converges.

Solution   We determine the convergence of $\sum_{n=1}^{\infty}(1/n^2)$ by comparing it with $\int_1^{\infty}(1/x^2)\,dx$. To carry out the comparison, we think of the terms of the series as values of the function $f(x)=1/x^2$ and interpret these values as the areas of rectangles under the curve $y=1/x^2$.

As Figure 8.12 shows,

$$s_n = \frac{1}{1^2} + \frac{1}{2^2} + \frac{1}{3^2} + \cdots + \frac{1}{n^2}$$
$$= f(1) + f(2) + f(3) + \cdots + f(n)$$
$$< f(1) + \int_1^n \frac{1}{x^2}\,dx$$
$$< 1 + \int_1^{\infty} \frac{1}{x^2}\,dx$$
$$< 1 + 1 = 2. \qquad \text{As in Section 7.7. Example 3, with } p=2,\ \int_1^{\infty}(1/x^2)\,dx = 1.$$

Thus, the partial sums of $\sum_{n=1}^{\infty}(1/n^2)$ are bounded from above (by 2) and the series converges. The sum of the series is known to be $\pi^2/6 \approx 1.64493$.

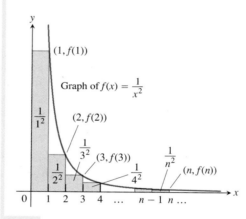

FIGURE 8.12 Figure for the area comparisons in Example 1.

The Integral Test

Let $\{a_n\}$ be a sequence of positive terms. Suppose that $a_n = f(n)$, where f is a continuous, positive, decreasing function of x for all $x \ge N$ (N a positive integer). Then the series $\sum_{n=N}^{\infty} a_n$ and the integral $\int_N^{\infty} f(x)\,dx$ both converge or both diverge.

Proof   We establish the test for the case $N=1$. The proof for general N is similar.

We start with the assumption that f is a decreasing function with $f(n)=a_n$ for every n. This leads us to observe that the rectangles in Figure 8.13a, which have areas $a_1, a_2, \ldots, a_n$, collectively enclose more area than that under the curve $y=f(x)$ from $x=1$ to $x=n+1$. That is,

$$\int_1^{n+1} f(x)\,dx \le a_1 + a_2 + \cdots + a_n.$$

CAUTION   The series and integral need not have the same value in the convergent case. In Example 1, $\sum_{n=1}^{\infty}(1/n^2) = \pi^2/6$, whereas $\int_1^{\infty}(1/x^2)\,dx = 1$.

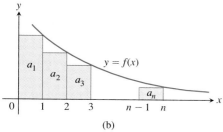

FIGURE 8.13 Subject to the conditions of the Integral Test, the series $\sum_{n=1}^{\infty} a_n$ and the integral $\int_1^{\infty} f(x)\, dx$ both converge or both diverge.

In Figure 8.13b the rectangles have been faced to the left instead of to the right. If we momentarily disregard the first rectangle, of area a_1, we see that

$$a_2 + a_3 + \cdots + a_n \leq \int_1^n f(x)\, dx.$$

If we include a_1, we have

$$a_1 + a_2 + \cdots + a_n \leq a_1 + \int_1^n f(x)\, dx.$$

Combining these results gives

$$\int_1^{n+1} f(x)\, dx \leq a_1 + a_2 + \cdots + a_n \leq a_1 + \int_1^n f(x)\, dx.$$

If $\int_1^{\infty} f(x)\, dx$ is finite, the right-hand inequality shows that $\sum a_n$ is finite. If $\int_1^{\infty} f(x)\, dx$ is infinite, the left-hand inequality shows that $\sum a_n$ is infinite.

Hence, the series and the integral are both finite or both infinite.

Example 2   Applying the Integral Test

Does $\displaystyle\sum_{n=1}^{\infty} \frac{1}{n\sqrt{n}}$ converge?

Solution   The Integral Test applies because

$$f(x) = \frac{1}{x\sqrt{x}}$$

is a continuous, positive, decreasing function of x for $x > 1$.

We have

$$\int_1^{\infty} \frac{1}{x\sqrt{x}}\, dx = \lim_{k \to \infty} \int_1^k x^{-3/2}\, dx$$

$$= \lim_{k \to \infty} \left[-2x^{-1/2} \right]_1^k$$

$$= \lim_{k \to \infty} \left(-\frac{2}{\sqrt{k}} + 2 \right)$$

$$= 2.$$

Since the integral converges, so must the series.

Harmonic Series and p-Series

The Integral Test can be used to settle the question of convergence for any series of the form $\sum_{n=1}^{\infty} (1/n^p)$, p a real constant. (The series in Example 2 had this form, with $p = 3/2$.) Such series are called **p-series.**

The p-Series

$$\sum_{n=1}^{\infty} \frac{1}{n^p} = \frac{1}{1^p} + \frac{1}{2^p} + \frac{1}{3^p} + \cdots + \frac{1}{n^p} + \cdots$$

(p a real constant) converges if $p > 1$ and diverges if $p \leq 1$.

Proof   From Example 3 in Section 7.7, the integral $\int_1^{\infty} dx/x^p$ converges if $p > 1$ and diverges if $p \leq 1$. From the Integral Test, the same holds true of the p-series $\sum_{n=1}^{\infty} (1/n^p)$: It converges if $p > 1$ and diverges if $p \leq 1$.   ▬

The p-series with $p = 1$ is the **harmonic series**, and it is probably the most famous divergent series in mathematics. The p-Series Test shows that the harmonic series is just *barely* divergent; if we increase p to 1.000000001, for instance, the series converges!

The slowness with which the partial sums of the harmonic series approaches infinity is most impressive. Consider the following example.

Example 3   The Slow Divergence of the Harmonic Series

Approximately how many terms of the harmonic series are required to form a partial sum larger than 20?

Solution   The graphs tell the story (Figure 8.14).

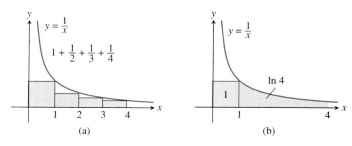

FIGURE 8.14 Finding an upper bound for one of the partial sums of the harmonic series. (Example 3)

Let H_n denote the nth partial sum of the harmonic series. Comparing the two graphs, we see that $H_4 < (1 + \ln 4)$ and (in general) that $H_n \leq (1 + \ln n)$. If we wish H_n to be greater than 20, then

$$1 + \ln n > H_n > 20$$
$$1 + \ln n > 20$$
$$\ln n > 19$$
$$n > e^{19}.$$

The exact value of e^{19} rounds up to 178,482,301. It will take *at least* that many terms of the harmonic series to move the partial sums beyond 20. It would take your calculator several weeks to compute a sum with this many terms. Nonetheless, the harmonic series really does diverge!

Comparison Tests

The p-Series Test tells everything there is to know about the convergence or divergence of series of the form $\Sigma (1/n^p)$. This is admittedly a rather narrow class of se-

---

What Is Harmonic About the Harmonic Series?

The terms in the harmonic series correspond to the nodes on a vibrating string that produce multiples of the fundamental frequency. For example, $1/2$ produces the harmonic that is twice the fundamental frequency, $1/3$ produces a frequency that is three times the fundamental frequency, and so on. The fundamental frequency is the lowest note or pitch we hear when a string is plucked.

ries, but we can test many other kinds (including those in which the nth term is any rational function of n) by *comparing* them with p-series.

The Direct Comparison Test

Let $\Sigma\, a_n$ be a series with no negative terms.

(a) $\Sigma\, a_n$ converges if there is a convergent series $\Sigma\, c_n$ with $a_n \le c_n$ for all $n > N$, for some integer N.

(b) $\Sigma\, a_n$ diverges if there is a divergent series of nonnegative terms $\Sigma\, d_n$ with $a_n \ge d_n$ for all $n > N$, for some integer N.

Proof In part (a), the partial sums of $\Sigma\, a_n$ are bounded above by

$$M = a_1 + a_2 + \cdots + a_n + \sum_{n=N+1}^{\infty} c_n.$$

They therefore form a nondecreasing sequence with a limit $L \le M$.

In part (b), the partial sums of $\Sigma\, a_n$ are not bounded from above. If they were, the partial sums for $\Sigma\, d_n$ would be bounded by

$$M^* = d_1 + d_2 + \cdots + d_N + \sum_{n=N+1}^{\infty} a_n$$

and $\Sigma\, d_n$ would have to converge instead of diverge. ⎯

To apply the Direct Comparison Test to a series, we need not include the early terms of the series. We can start the test with any index N provided that we include all the terms of the series being tested from there on.

Example 4 Applying the Direct Comparison Test

Does the following series converge?

$$5 + \frac{2}{3} + 1 + \frac{1}{7} + \frac{1}{2} + \frac{1}{3!} + \frac{1}{4!} + \cdots + \frac{1}{k!} + \cdots$$

Solution We ignore the first four terms and compare the remaining terms with those of the convergent geometric series $\sum_{n=1}^{\infty} (1/2^n)$. We see that

$$\frac{1}{2} + \frac{1}{3!} + \frac{1}{4!} + \cdots \le \frac{1}{2} + \frac{1}{4} + \frac{1}{8} + \cdots.$$

Therefore, the original series converges by the Direct Comparison Test.

To apply the Direct Comparison Test, we need to have on hand a list of series whose convergence or divergence we know. Here is what we know so far:

Convergent series	Divergent series
Geometric series with $\lvert r \rvert < 1$	Geometric series with $\lvert r \rvert \geq 1$
Telescoping series like $\sum\limits_{n=1}^{\infty} \dfrac{1}{n(n+1)}$	The harmonic series $\sum\limits_{n=1}^{\infty} \dfrac{1}{n}$
The series $\sum\limits_{n=0}^{\infty} \dfrac{1}{n!}$	Any series $\sum a_n$ for which $\lim_{n\to\infty} a_n$ does not exist or $\lim_{n\to\infty} a_n \neq 0$
Any p-series $\sum\limits_{n=1}^{\infty} \dfrac{1}{n^p}$ with $p > 1$	Any p-series $\sum\limits_{n=1}^{\infty} \dfrac{1}{n^p}$ with $p \leq 1$

The Direct Comparison Test is one method of comparison; the *Limit Comparison Test* is another.

The Limit Comparison Test

Suppose that $a_n > 0$ and $b_n > 0$ for all $n \geq N$ (N a positive integer).

1. If $\lim\limits_{n\to\infty} \dfrac{a_n}{b_n} = c$, $0 < c < \infty$, then $\Sigma\, a_n$ and $\Sigma\, b_n$ both converge or both diverge.

2. If $\lim\limits_{n\to\infty} \dfrac{a_n}{b_n} = 0$ and $\Sigma\, b_n$ converges, then $\Sigma\, a_n$ converges.

3. If $\lim\limits_{n\to\infty} \dfrac{a_n}{b_n} = \infty$ and $\Sigma\, b_n$ diverges, then $\Sigma\, a_n$ diverges.

Proof   We will prove part (1). Parts (2) and (3) are left as Exercise 67.

Since $c/2 > 0$, there exists an integer N such that for all n,

$$n > N \;\Rightarrow\; \left\lvert \frac{a_n}{b_n} - c \right\rvert < \frac{c}{2}. \qquad \text{\small Limit definition with } \epsilon = c/2,$$
$$\text{\small } L = c, \text{ and } a_n \text{ replaced by } a_n/b_n.$$

Thus, for $n > N$,

$$-\frac{c}{2} < \frac{a_n}{b_n} - c < \frac{c}{2},$$

$$\frac{c}{2} < \frac{a_n}{b_n} < \frac{3c}{2},$$

$$\left(\frac{c}{2}\right) b_n < a_n < \left(\frac{3c}{2}\right) b_n.$$

If $\Sigma\, b_n$ converges then $\Sigma\, (3c/2)b_n$ converges and $\Sigma\, a_n$ converges by the Direct Comparison Test. If $\Sigma\, b_n$ diverges, then $\Sigma\, (c/2)b_n$ diverges and $\Sigma\, a_n$ diverges by the Direct Comparison Test.

Example 5   Using the Limit Comparison Test

Determine whether the series converge or diverge.

(a) $\dfrac{3}{4} + \dfrac{5}{9} + \dfrac{7}{16} + \dfrac{9}{25} + \cdots = \displaystyle\sum_{n=1}^{\infty} \dfrac{2n+1}{(n+1)^2} = \sum_{n=1}^{\infty} \dfrac{2n+1}{n^2 + 2n + 1}$

(b) $\dfrac{1}{1} + \dfrac{1}{3} + \dfrac{1}{7} + \dfrac{1}{15} + \cdots = \displaystyle\sum_{n=1}^{\infty} \dfrac{1}{2^n - 1}$

(c) $\dfrac{1 + 2 \ln 2}{9} + \dfrac{1 + 3 \ln 3}{14} + \dfrac{1 + 4 \ln 4}{21} + \cdots = \displaystyle\sum_{n=2}^{\infty} \dfrac{1 + n \ln n}{n^2 + 5}$

Solution

We could just as well have taken $b_n = 2/n$, but $1/n$ is simpler.

(a) Let $a_n = (2n + 1)/(n^2 + 2n + 1)$. For n large, we expect a_n to behave like $2n/n^2 = 2/n$, so we let $b_n = 1/n$. Since

$$\sum_{n=1}^{\infty} b_n = \sum_{n=1}^{\infty} \frac{1}{n}$$

diverges and

$$\lim_{n \to \infty} \frac{a_n}{b_n} = \lim_{n \to \infty} \frac{2n^2 + n}{n^2 + 2n + 1} = 2,$$

$\Sigma\, a_n$ diverges by part 1 of the Limit Comparison Test.

(b) Let $a_n = 1/(2^n - 1)$. For n large, we expect a_n to behave like $1/2^n$, so we let $b_n = 1/2^n$. Since

$$\sum_{n=1}^{\infty} b_n = \sum_{n=1}^{\infty} \frac{1}{2^n}$$

converges and

$$\lim_{n \to \infty} \frac{a_n}{b_n} = \lim_{n \to \infty} \frac{2^n}{2^n - 1}$$

$$= \lim_{n \to \infty} \frac{1}{1 - (1/2^n)}$$

$$= 1,$$

$\Sigma\, a_n$ converges by part 1 of the Limit Comparison Test.

(c) Let $a_n = (1 + n \ln n)/(n^2 + 5)$. For n large, we expect a_n to behave like $(n \ln n)/n^2 = (\ln n)/n$, which is greater than $1/n$ for $n \geq 3$, so we take $b_n = 1/n$. Since

$$\sum_{n=2}^{\infty} b_n = \sum_{n=2}^{\infty} \frac{1}{n}$$

diverges and

$$\lim_{n \to \infty} \frac{a_n}{b_n} = \lim_{n \to \infty} \frac{n + n^2 \ln n}{n^2 + 5}$$

$$= \infty,$$

$\Sigma\, a_n$ diverges by part 3 of the Limit Comparison Test.

Ratio and Root Tests

The Ratio Test measures the rate of growth (or decline) of a series by examining the ratio a_{n+1}/a_n. For a geometric series $\Sigma\, ar^n$, this rate is a constant $((ar^{n+1})/(ar^n) = r)$, and the series converges if and only if its ratio is less than 1 in absolute value. The Ratio Test is a powerful rule extending that result.

The Ratio Test

Let $\Sigma\, a_n$ be a series with positive terms and suppose that

$$\lim_{n \to \infty} \frac{a_{n+1}}{a_n} = \rho.$$

Then

(a) the series *converges* if $\rho < 1$
(b) the series *diverges* if $\rho > 1$ or ρ is infinite
(c) the test is *inconclusive* if $\rho = 1$.

Proof

(a) $\boldsymbol{\rho < 1.}$ Let r be a number between ρ and 1. Then the number $\epsilon = r - \rho$ is positive. Since

$$\frac{a_{n+1}}{a_n} \to \rho,$$

a_{n+1}/a_n must lie within ϵ of ρ when n is large enough, say for all $n \geq N$. In particular

$$\frac{a_{n+1}}{a_n} < \rho + \epsilon = r, \qquad \text{when } n \geq N.$$

That is,

$$a_{N+1} < r\, a_N,$$
$$a_{N+2} < r\, a_{N+1} < r^2 a_N,$$
$$a_{N+3} < r\, a_{N+2} < r^3 a_N,$$
$$\vdots$$
$$a_{N+m} < r\, a_{N+m-1} < r^m a_N.$$

These inequalities show that the terms of our series, after the Nth term, approach zero more rapidly than the terms in a geometric series with ratio $r < 1$. More precisely, consider the series $\Sigma\, c_n$, where $c_n = a_n$ for $n = 1, 2, \ldots, N$ and $c_{N+1} = r a_N,\, c_{N+2} = r^2 a_N,\, \ldots,\, c_{N+m} = r^m a_N,\, \ldots$. Now $a_n \leq c_n$ for all n, and

$$\sum_{n=1}^{\infty} c_n = a_1 + a_2 + \cdots + a_{N-1} + a_N + r a_N + r^2 a_N + \cdots$$

$$= a_1 + a_2 + \cdots + a_{N-1} + a_N (1 + r + r^2 + \cdots).$$

The geometric series $1 + r + r^2 + \cdots$ converges because $|r| < 1$, so $\Sigma\, c_n$ converges. Since $a_n \le c_n$, $\Sigma\, a_n$ also converges.

(b) $1 < \rho \le \infty$. From some index M on,

$$\frac{a_{n+1}}{a_n} > 1 \qquad \text{and} \qquad a_M < a_{M+1} < a_{M+2} < \cdots.$$

The terms of the series do not approach zero as n becomes infinite, and the series diverges by the nth-Term Test.

(c) $\rho = 1$. The two series

$$\sum_{n=1}^{\infty} \frac{1}{n} \qquad \text{and} \qquad \sum_{n=1}^{\infty} \frac{1}{n^2}$$

show that some other test for convergence must be used when $\rho = 1$.

For $\displaystyle\sum_{n=1}^{\infty} \frac{1}{n}$: $\quad \dfrac{a_{n+1}}{a_n} = \dfrac{1/(n+1)}{1/n} = \dfrac{n}{n+1} \to 1.$

For $\displaystyle\sum_{n=1}^{\infty} \frac{1}{n^2}$: $\quad \dfrac{a_{n+1}}{a_n} = \dfrac{1/(n+1)^2}{1/n^2} = \left(\dfrac{n}{n+1}\right)^2 \to 1^2 = 1.$

In both cases, $\rho = 1$, yet the first series diverges, whereas the second converges.

Example 6 Applying the Ratio Test

Investigate the convergence of the following series.

The Ratio Test is often effective when the terms of a series contain factorials of expressions involving n or expressions raised to the nth power.

(a) $\displaystyle\sum_{n=0}^{\infty} \frac{2^n + 5}{3^n}$ $\qquad$ **(b)** $\displaystyle\sum_{n=1}^{\infty} \frac{(2n)!}{n!n!}$ $\qquad$ **(c)** $\displaystyle\sum_{n=1}^{\infty} \frac{4^n n! n!}{(2n)!}$

Solution

(a) For the series $\sum_{n=0}^{\infty} (2^n + 5)/3^n$,

$$\frac{a_{n+1}}{a_n} = \frac{(2^{n+1} + 5)/3^{n+1}}{(2^n + 5)/3^n} = \frac{1}{3} \cdot \frac{2^{n+1} + 5}{2^n + 5} = \frac{1}{3} \cdot \left(\frac{2 + 5 \cdot 2^{-n}}{1 + 5 \cdot 2^{-n}}\right) \to \frac{1}{3} \cdot \frac{2}{1} = \frac{2}{3}.$$

The series converges because $\rho = 2/3$ is less than 1. This does *not* mean that $2/3$ is the sum of the series. In fact,

$$\sum_{n=0}^{\infty} \frac{2^n + 5}{3^n} = \sum_{n=0}^{\infty} \left(\frac{2}{3}\right)^n + \sum_{n=0}^{\infty} \frac{5}{3^n} = \frac{1}{1 - (2/3)} + \frac{5}{1 - (1/3)} = \frac{21}{2}.$$

(b) If $a_n = \dfrac{(2n)!}{n!n!}$, then $a_{n+1} = \dfrac{(2n+2)!}{(n+1)!(n+1)!}$ and

$$\frac{a_{n+1}}{a_n} = \frac{n!n!(2n+2)(2n+1)(2n)!}{(n+1)!(n+1)!(2n)!}$$

$$= \frac{(2n+2)(2n+1)}{(n+1)(n+1)} = \frac{4n+2}{n+1} \to 4.$$

The series diverges because $\rho = 4$ is greater than 1.

(c) If $a_n = 4^n n! n!/(2n)!$, then

$$\frac{a_{n+1}}{a_n} = \frac{4^{n+1}(n+1)!(n+1)!}{(2n+2)(2n+1)(2n)!} \cdot \frac{(2n)!}{4^n n! n!}$$

$$= \frac{4(n+1)(n+1)}{(2n+2)(2n+1)} = \frac{2(n+1)}{2n+1} \to 1.$$

Because the limit is $\rho = 1$, we cannot decide from the Ratio Test whether the series converges. When we notice that $a_{n+1}/a_n = (2n+2)/(2n+1)$, we conclude that a_{n+1} is always greater than a_n because $(2n+2)/(2n+1)$ is always greater than 1. Therefore, all terms are greater than or equal to $a_1 = 2$, and the nth term does not approach zero as $n \to \infty$. The series diverges.

The nth-Root Test is another useful tool for answering the question of convergence for series with nonnegative terms. We state the result here without proof.

The nth-Root Test

Let $\Sigma \, a_n$ be a series with $a_n \geq 0$ for $n \geq N$ and suppose that

$$\lim_{n \to \infty} \sqrt[n]{a_n} = \rho.$$

Then

(a) the series *converges* if $\rho < 1$
(b) the series *diverges* if $\rho > 1$ or ρ is infinite
(c) the test is *inconclusive* if $\rho = 1$.

Example 7   Applying the nth-Root Test

Let

$$a_n = \begin{cases} n/2^n, & n \text{ odd} \\ 1/2^n, & n \text{ even}. \end{cases}$$

Does $\Sigma \, a_n$ converge?

Solution   We apply the nth-Root Test, finding that

$$\sqrt[n]{a_n} = \begin{cases} \sqrt[n]{n}/2, & n \text{ odd} \\ 1/2, & n \text{ even}. \end{cases}$$

Therefore,

$$\frac{1}{2} \leq \sqrt[n]{a_n} \leq \frac{\sqrt[n]{n}}{2}.$$

Since $\sqrt[n]{n} \to 1$ (Section 8.1, Table 8.1), we have $\lim_{n \to \infty} \sqrt[n]{a_n} = 1/2$ by the Sandwich Theorem. The limit is less than 1, so the series converges by the nth-Root Test.

Example 8   Applying the *n*th-Root Test

Which of the following series converge, and which diverge?

(a) $\displaystyle\sum_{n=1}^{\infty} \frac{n^2}{2^n}$ (b) $\displaystyle\sum_{n=1}^{\infty} \frac{2^n}{n^2}$

Solution

(a) $\displaystyle\sum_{n=1}^{\infty} \frac{n^2}{2^n}$ converges because $\sqrt[n]{\dfrac{n^2}{2^n}} = \dfrac{\sqrt[n]{n^2}}{\sqrt[n]{2^n}} = \dfrac{(\sqrt[n]{n})^2}{2} \to \dfrac{1}{2} < 1.$

(b) $\displaystyle\sum_{n=1}^{\infty} \frac{2^n}{n^2}$ diverges because $\sqrt[n]{\dfrac{2^n}{n^2}} = \dfrac{2}{(\sqrt[n]{n})^2} \to \dfrac{2}{1} > 1.$

EXERCISES 8.4

Integral Test

Use the Integral Test to determine which of the series in Exercises 1–8 converge and which diverge.

1. $\displaystyle\sum_{n=1}^{\infty} \frac{5}{n+1}$

2. $\displaystyle\sum_{n=1}^{\infty} \frac{1}{2n-1}$

3. $\displaystyle\sum_{n=2}^{\infty} \frac{\ln n}{n}$

4. $\displaystyle\sum_{n=2}^{\infty} \frac{\ln n}{\sqrt{n}}$

5. $\displaystyle\sum_{n=1}^{\infty} \frac{e^n}{1+e^{2n}}$

6. $\displaystyle\sum_{n=1}^{\infty} \frac{1}{\sqrt{n}(\sqrt{n}+1)}$

7. $\displaystyle\sum_{n=3}^{\infty} \frac{(1/n)}{(\ln n)\sqrt{\ln^2 n - 1}}$

8. $\displaystyle\sum_{n=1}^{\infty} \frac{1}{n(1 + \ln^2 n)}$

Direct Comparison Test

Use the Direct Comparison Test to determine which of the series in Exercises 9–14 converge and which diverge.

9. $\displaystyle\sum_{n=1}^{\infty} \frac{1}{2\sqrt{n}+\sqrt[3]{n}}$

10. $\displaystyle\sum_{n=1}^{\infty} \frac{3}{n+\sqrt{n}}$

11. $\displaystyle\sum_{n=1}^{\infty} \frac{\sin^2 n}{2^n}$

12. $\displaystyle\sum_{n=1}^{\infty} \frac{1 + \cos n}{n^2}$

13. $\displaystyle\sum_{n=1}^{\infty} \left(\frac{n}{3n+1}\right)^n$

14. $\displaystyle\sum_{n=3}^{\infty} \frac{1}{\ln(\ln n)}$

Limit Comparison Test

Use the Limit Comparison Test to determine which of the series in Exercises 15–20 converge and which diverge.

15. $\displaystyle\sum_{n=2}^{\infty} \frac{1}{(\ln n)^2}$

16. $\displaystyle\sum_{n=1}^{\infty} \frac{(\ln n)^2}{n^3}$

17. $\displaystyle\sum_{n=1}^{\infty} \frac{(\ln n)^3}{n^3}$

18. $\displaystyle\sum_{n=2}^{\infty} \frac{1}{\sqrt{n}\ln n}$

19. $\displaystyle\sum_{n=1}^{\infty} \frac{(\ln n)^2}{n^{3/2}}$

20. $\displaystyle\sum_{n=1}^{\infty} \frac{1}{1+\ln n}$

Ratio Test

Use the Ratio Test to determine which of the series in Exercises 21–28 converge and which diverge.

21. $\displaystyle\sum_{n=1}^{\infty} \frac{n^{\sqrt{2}}}{2^n}$

22. $\displaystyle\sum_{n=1}^{\infty} n^2 e^{-n}$

23. $\displaystyle\sum_{n=1}^{\infty} n!\, e^{-n}$

24. $\displaystyle\sum_{n=1}^{\infty} \frac{n!}{10^n}$

25. $\displaystyle\sum_{n=1}^{\infty} \frac{n^{10}}{10^n}$

26. $\displaystyle\sum_{n=1}^{\infty} \frac{n \ln n}{2^n}$

27. $\displaystyle\sum_{n=1}^{\infty} \frac{(n+1)(n+2)}{n!}$

28. $\displaystyle\sum_{n=1}^{\infty} e^{-n}(n^3)$

Root Test

Use the *n*th-Root Test to determine which of the series in Exercises 29–34 converge and which diverge.

29. $\displaystyle\sum_{n=1}^{\infty} \frac{(\ln n)^n}{n^n}$

30. $\displaystyle\sum_{n=1}^{\infty} \left(\frac{1}{n} - \frac{1}{n^2}\right)^n$

31. $\displaystyle\sum_{n=2}^{\infty} \frac{n}{(\ln n)^n}$

32. $\displaystyle\sum_{n=2}^{\infty} \frac{n}{(\ln n)^{(n/2)}}$

33. $\displaystyle\sum_{n=1}^{\infty} \frac{(n!)^n}{(n^n)^2}$

34. $\displaystyle\sum_{n=1}^{\infty} \frac{n^n}{(2^n)^2}$

Determining Convergence or Divergence

Which of the series in Exercises 35–60 converge, and which diverge? Give reasons for your answers. (When checking your answers, remember that there may be more than one way to determine a series' convergence or divergence.)

35. $\displaystyle\sum_{n=1}^{\infty} e^{-n}$

36. $\displaystyle\sum_{n=1}^{\infty} \frac{n}{n+1}$

37. $\displaystyle\sum_{n=1}^{\infty} \frac{3}{\sqrt{n}}$

38. $\displaystyle\sum_{n=1}^{\infty} \frac{-2}{n\sqrt{n}}$

39. $\displaystyle\sum_{n=1}^{\infty} \frac{1}{(1+\ln n)^2}$

40. $\displaystyle\sum_{n=2}^{\infty} \frac{\ln(n+1)}{n+1}$

41. $\displaystyle\sum_{n=2}^{\infty} \frac{1}{n\sqrt{n^2-1}}$

42. $\displaystyle\sum_{n=1}^{\infty} \left(\frac{n-2}{n}\right)^n$

43. $\displaystyle\sum_{n=1}^{\infty} \frac{(n+3)!}{3!n!3^n}$

44. $\displaystyle\sum_{n=1}^{\infty} \frac{n2^n(n+1)!}{3^n n!}$

45. $\displaystyle\sum_{n=1}^{\infty} \frac{n!}{(2n+1)!}$

46. $\displaystyle\sum_{n=1}^{\infty} \frac{n!}{n^n}$

47. $\displaystyle\sum_{n=1}^{\infty} \frac{8\tan^{-1}n}{1+n^2}$

48. $\displaystyle\sum_{n=1}^{\infty} \frac{n}{n^2+1}$

49. $\displaystyle\sum_{n=1}^{\infty} \operatorname{sech} n$

50. $\displaystyle\sum_{n=1}^{\infty} \operatorname{sech}^2 n$

51. $\displaystyle\sum_{n=1}^{\infty} \frac{2+(-1)^n}{1.25^n}$

52. $\displaystyle\sum_{n=1}^{\infty} \left(1-\frac{1}{3n}\right)^n$

53. $\displaystyle\sum_{n=1}^{\infty} \frac{\ln n}{n^3}$

54. $\displaystyle\sum_{n=1}^{\infty} \frac{\ln n}{n}$

55. $\displaystyle\sum_{n=1}^{\infty} \frac{10n+1}{n(n+1)(n+2)}$

56. $\displaystyle\sum_{n=3}^{\infty} \frac{5n^3-3n}{n^2(n-2)(n^2+5)}$

57. $\displaystyle\sum_{n=1}^{\infty} \frac{\tan^{-1}n}{n^{1.1}}$

58. $\displaystyle\sum_{n=1}^{\infty} \frac{\sec^{-1}n}{n^{1.3}}$

59. $\displaystyle\sum_{n=1}^{\infty} n\sin\frac{1}{n}$

60. $\displaystyle\sum_{n=1}^{\infty} \frac{2}{1+e^n}$

Terms Defined Recursively

Which of the series $\sum_{n=1}^{\infty} a_n$ defined by the formulas in Exercises 61–66 converge, and which diverge? Give reasons for your answers.

61. $a_1 = 2, \quad a_{n+1} = \dfrac{1+\sin n}{n} a_n$

62. $a_1 = 1, \quad a_{n+1} = \dfrac{1+\tan^{-1}n}{n} a_n$

63. $a_1 = \dfrac{1}{3}, \quad a_{n+1} = \dfrac{3n-1}{2n+5} a_n$

64. $a_1 = 3, \quad a_{n+1} = \dfrac{n}{n+1} a_n$

65. $a_1 = \dfrac{1}{3}, \quad a_{n+1} = \sqrt[n]{a_n}$

66. $a_1 = \dfrac{1}{2}, \quad a_{n+1} = (a_n)^{n+1}$

Theory and Examples

67. Prove

  (a) Part 2 of the Limit Comparison Test

  (b) Part 3 of the Limit Comparison Test.

68. *Writing to Learn* If $\sum_{n=1}^{\infty} a_n$ is a convergent series of nonnegative numbers, can anything be said about $\sum_{n=1}^{\infty} (a_n/n)$? Explain.

69. *Writing to Learn* Suppose that $a_n > 0$ and $b_n > 0$ for $n \geq N$ (N an integer). If $\lim_{n\to\infty} (a_n/b_n) = \infty$ and Σa_n converges, can anything be said about Σb_n? Give reasons for your answer.

70. *Term-by-term squaring* Prove that if Σa_n is a convergent series of nonnegative terms, then Σa_n^2 converges.

For what values of a, if any, do the series in Exercises 71 and 72 converge?

71. $\displaystyle\sum_{n=1}^{\infty} \left(\frac{a}{n+2} - \frac{1}{n+4}\right)$

72. $\displaystyle\sum_{n=3}^{\infty} \left(\frac{1}{n-1} - \frac{2a}{n+1}\right)$

73. *The Cauchy Condensation Test* The Cauchy Condensation Test says: Let $\{a_n\}$ be a nonincreasing sequence ($a_n \geq a_{n+1}$ for all n) of positive terms that converges to 0. Then Σa_n converges if and only if $\Sigma\, 2^n a_{2^n}$ converges. For example, $\Sigma\,(1/n)$ diverges because $\Sigma 2^n \cdot (1/2^n) = \Sigma 1$ diverges. Show why the test works.

74. Use the Cauchy Condensation Test from Exercise 73 to show that

  (a) $\displaystyle\sum_{n=2}^{\infty} \frac{1}{n\ln n}$ diverges

  (b) $\displaystyle\sum_{n=1}^{\infty} \frac{1}{n^p}$ converges if $p > 1$ and diverges if $p \leq 1$.

75. *Logarithmic p-series*

  (a) Show that

$$\int_2^{\infty} \frac{dx}{x(\ln x)^p} \qquad (p \text{ a positive constant})$$

  converges if and only if $p > 1$.

  (b) What implications does the fact in part (a) have for the convergence of the series

$$\sum_{n=2}^{\infty} \frac{1}{n(\ln n)^p}?$$

  Give reasons for your answer.

76. (Continuation of Exercise 75) Use the result in Exercise 75 to determine which of the following series converge and which diverge. Support your answer in each case.

  (a) $\displaystyle\sum_{n=2}^{\infty} \frac{1}{n(\ln n)}$

  (b) $\displaystyle\sum_{n=2}^{\infty} \frac{1}{n(\ln n)^{1.01}}$

  (c) $\displaystyle\sum_{n=2}^{\infty} \frac{1}{n\ln(n^3)}$

  (d) $\displaystyle\sum_{n=2}^{\infty} \frac{1}{n(\ln n)^3}$

77. *Another logarithmic p-series* Show that neither the Ratio Test nor the nth-Root Test provides information about the convergence of

$$\sum_{n=2}^{\infty} \frac{1}{(\ln n)^p} \qquad (p \text{ constant}).$$

78. Let

$$a_n = \begin{cases} n/2^n & \text{if } n \text{ is a prime number} \\ 1/2^n & \text{otherwise.} \end{cases}$$

Does $\Sigma \, a_n$ converge? Give reasons for your answer.

79. *p-Series* Neither the Ratio Test nor the nth-Root Test helps with p-series. Try these tests on

$$\sum_{n=1}^{\infty} \frac{1}{n^p}$$

and show that both fail to provide information about convergence.

COMPUTER EXPLORATION

A Current Mystery

80. It is not yet known whether the series

$$\sum_{n=1}^{\infty} \frac{1}{n^3 \sin^2 n}$$

converges or diverges. Use a CAS to explore the behavior of the series by performing the following steps.

(a) Define the sequence of partial sums

$$s_k = \sum_{n=1}^{k} \frac{1}{n^3 \sin^2 n}.$$

What happens when you try to find the limit of s_k as $k \to \infty$? Does your CAS find a closed-form answer for this limit?

(b) Plot the first 100 points (k, s_k) for the sequence of partial sums. Do they appear to converge? What would you estimate the limit to be?

(c) Next plot the first 200 points (k, s_k). Discuss the behavior in your own words.

(d) Plot the first 400 points (k, s_k). What happens when $k = 355$? Calculate the number $355/113$. Explain from your calculation what happened at $k = 355$. For what values of k would you guess this behavior might occur again?

   You will find an interesting discussion of this series in Chapter 72 of *Mazes for the Mind* by Clifford A. Pickover (New York: St. Martin's Press, 1992).

8.5 Alternating Series, Absolute and Conditional Convergence

Alternating Series • Absolute Convergence • Rearranging Series •
Procedure for Determining Convergence

The convergence tests investigated so far apply only to series with nonnegative terms. In this section, we learn how to deal with series that may have negative terms. An important example is the *alternating series*, whose terms alternate in sign. We also learn which convergent series can have their terms rearranged (that is, changing the order in which they appear) without changing their sum.

Alternating Series

A series in which the terms are alternately positive and negative is an **alternating series.**
   Here are three examples.

$$1 - \frac{1}{2} + \frac{1}{3} - \frac{1}{4} + \frac{1}{5} - \cdots + \frac{(-1)^{n+1}}{n} + \cdots \tag{1}$$

$$-2 + 1 - \frac{1}{2} + \frac{1}{4} - \frac{1}{8} + \cdots + \frac{(-1)^n 4}{2^n} + \cdots \tag{2}$$

$$1 - 2 + 3 - 4 + 5 - 6 + \cdots + (-1)^{n+1} n + \cdots \tag{3}$$

Series (1), called the **alternating harmonic series,** converges, as we see shortly. Series (2), a geometric series with $a = -2$ and $r = -1/2$, converges to $-2/[1 + (1/2)] = -4/3$. Series (3) diverges by the nth-Term Test.

We prove the convergence of the alternating harmonic series by applying the following test.

Theorem 8   The Alternating Series Test (Leibniz's Theorem)

The series

$$\sum_{n=1}^{\infty} (-1)^{n+1} u_n = u_1 - u_2 + u_3 - u_4 + \cdots$$

converges if all three of the following conditions are satisfied.

1. The u_n's are all positive.

2. $u_n \geq u_{n+1}$ for all $n \geq N$, for some integer N.

3. $u_n \to 0$.

Proof   If n is an even integer, say $n = 2m$, then the sum of the first n term is

$$s_{2m} = (u_1 - u_2) + (u_3 - u_4) + \cdots + (u_{2m-1} - u_{2m})$$
$$= u_1 - (u_2 - u_3) - (u_4 - u_5) - \cdots - (u_{2m-2} - u_{2m-1}) - u_{2m}.$$

The first equality shows that s_{2m} is the sum of m nonnegative terms, since each term in parentheses is positive or zero. Hence, $s_{2m+2} \geq s_{2m}$, and the sequence $\{s_{2m}\}$ is nondecreasing. The second equality shows that $s_{2m} \leq u_1$. Since $\{s_{2m}\}$ is nondecreasing and bounded from above, it has a limit, say

$$\lim_{m \to \infty} s_{2m} = L. \tag{4}$$

If n is an odd integer, say $n = 2m + 1$, then the sum of the first n terms is $s_{2m+1} = s_{2m} + u_{2m+1}$. Since $u_n \to 0$,

$$\lim_{m \to \infty} u_{2m+1} = 0$$

and, as $m \to \infty$,

$$s_{2m+1} = s_{2m} + u_{2m+1} \to L + 0 = L. \tag{5}$$

Combining the results of Equations (4) and (5) gives $\lim_{n \to \infty} s_n = L$ (Section 8.2, Exercise 26). Figure 8.15 illustrates the convergence of the partial sums to their limit L.

Figure 8.15 actually shows more than the *fact* of convergence; it also shows the *way* that an alternating series converges when it satisfies the conditions of the test. The partial sums keep "overshooting" the limit as they go back and forth on the number line, gradually closing in as the terms tend to zero. If we stop at the nth partial sum, we know that the next term ($\pm u_{n+1}$) will again cause us to overshoot the limit in the positive direction or negative direction, depending on the sign carried by u_{n+1}. This gives us a convenient bound for the **truncation error,** which we state as another theorem.

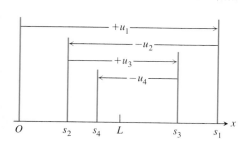

FIGURE 8.15 The partial sums of an alternating series that satisfies the hypotheses of Theorem 8 for $N = 1$ straddle the limit from the beginning.

> **Theorem 9** The Alternating Series Estimation Theorem
>
> If the alternating series $\sum_{n=1}^{\infty} (-1)^{n+1} u_n$ satisfies the conditions of Theorem 8, then the truncation error for the nth partial sum is less than u_{n+1} and has the same sign as the unused term.

Example 1 The Alternating Harmonic Series

Prove that the alternating harmonic series is convergent, but that the corresponding series of absolute values is not convergent. Find a bound for the truncation error after 99 terms.

Solution The terms are strictly alternating in sign and decrease in absolute value from the start:

$$1 > \frac{1}{2} > \frac{1}{3} > \cdots.$$

Also,

$$\frac{1}{n} \to 0.$$

By the Alternating Series Test,

$$\sum_{n=1}^{\infty} \frac{(-1)^{n+1}}{n}$$

converges.

On the other hand, the series $\sum_{n=1}^{\infty} (1/n)$ of absolute values is the harmonic series, which diverges.

The Alternating Series Estimation Theorem guarantees that the truncation error after 99 terms is less than $u_{99+1} = 1/(99 + 1) = 1/100$.

A Note on the Error Bound

Theorem 9 does not give a *formula* for the truncation error but a *bound* for the truncation error. The bound might be fairly conservative. For example, the first 99 terms of the alternating harmonic series add to about 0.6981721793, whereas the series itself has a sum of $\ln 2 \approx 0.6931471806$. That makes the actual truncation error very close to 0.005, about half the size of the bound of 0.01 given by Theorem 9.

Example 2 Applying The Estimation Theorem

We try Theorem 9 on a series whose sum we know:

$$\sum_{n=0}^{\infty} (-1)^n \frac{1}{2^n} = 1 - \frac{1}{2} + \frac{1}{4} - \frac{1}{8} + \frac{1}{16} - \frac{1}{32} + \frac{1}{64} - \frac{1}{128} \mid + \frac{1}{256} - \cdots.$$

The theorem says that if we truncate the series after the eighth term, we throw away a total that is positive and less than $1/256$. The sum of the first eight terms is 0.6640 625. The sum of the series is

$$\frac{1}{1 - (-1/2)} = \frac{1}{3/2} = \frac{2}{3}.$$

The difference, $(2/3) - 0.6640\ 625 = 0.0026\ 0416\ 6 \ldots$, is positive and less than $(1/256) = 0.0039\ 0625$.

Absolute Convergence

> **Definition** Absolute Convergence
>
> A series $\Sigma\, a_n$ **converges absolutely** (is **absolutely convergent**) if the corresponding series of absolute values, $\Sigma\,|a_n|$, converges.

The geometric series

$$1 - \frac{1}{2} + \frac{1}{4} - \frac{1}{8} + \cdots$$

converges absolutely because the corresponding series of absolute values

$$1 + \frac{1}{2} + \frac{1}{4} + \frac{1}{8} + \cdots$$

converges. The alternating harmonic series (Example 1) does not converge absolutely. The corresponding series of absolute values is the (divergent) harmonic series.

> **Definition** Conditional Convergence
>
> A series that converges but does not converge absolutely **converges conditionally.**

The alternating harmonic series converges conditionally.

Example 3 Absolute and Conditional Convergence

Determine which of the following series is absolutely or conditionally convergent, or divergent.

(a) $\displaystyle\sum_{n=1}^{\infty} (-1)^{n+1}\,\frac{1}{\sqrt{n}} = 1 - \frac{1}{\sqrt{2}} + \frac{1}{\sqrt{3}} - \frac{1}{\sqrt{4}} + \cdots$

(b) $\displaystyle\sum_{n=2}^{\infty} (-1)^n \left(1 - \frac{1}{n}\right)^n = \frac{1}{2} - \left(\frac{2}{3}\right)^3 + \left(\frac{3}{4}\right)^4 - \cdots$

(c) $\displaystyle\sum_{n=1}^{\infty} (-1)^{n(n+1)/2}\,\frac{1}{2^n} = -\frac{1}{2} - \frac{1}{4} + \frac{1}{8} + \frac{1}{16} - \cdots$

Solution

(a) The series converges by the Alternating Series Test since $(1/\sqrt{n}) > (1/\sqrt{n+1})$ and $(1/\sqrt{n}) \to 0$. The series $\sum_{n=1}^{\infty} (1/\sqrt{n})$ of absolute values diverges, however, because it is a p-series with $p = (1/2) < 1$. Therefore, the given series is *conditionally convergent*.

(b) The series *diverges* by the nth-Term Test since $\lim_{n\to\infty} (1 - (1/n))^n = e^{-1} \neq 0$ (Table 8.1, Formula 5).

(c) This is *not* an alternating series. However,

$$\sum_{n=1}^{\infty} \left| (-1)^{n(n+1)/2} \frac{1}{2^n} \right| = \sum_{n=1}^{\infty} \frac{1}{2^n}$$

is a convergent geometric series, and we conclude that the given series is *absolutely convergent*.

Absolute convergence is important for two reasons. First, we have good tests for convergence of series of positive terms. Second, if a series converges absolutely, then it converges. That is the thrust of the next theorem.

CAUTION   We can rephrase Theorem 10 to say that *every absolutely convergent series converges*. The converse statement is false, however: Many convergent series do not converge absolutely.

Theorem 10   The Absolute Convergence Test

If $\sum_{n=1}^{\infty} |a_n|$ converges, then $\sum_{n=1}^{\infty} a_n$ converges.

Proof   For each n,

$$-|a_n| \leq a_n \leq |a_n|, \qquad \text{so} \qquad 0 \leq a_n + |a_n| \leq 2|a_n|.$$

If $\sum_{n=1}^{\infty} |a_n|$ converges, then $\sum_{n=1}^{\infty} 2|a_n|$ converges and, by the Direct Comparison Test, the nonnegative series $\sum_{n=1}^{\infty} (a_n + |a_n|)$ converges. The equality $a_n = (a_n + |a_n|) - |a_n|$ now lets us express $\sum_{n=1}^{\infty} a_n$ as the difference of two convergent series:

$$\sum_{n=1}^{\infty} a_n = \sum_{n=1}^{\infty} (a_n + |a_n| - |a_n|) = \sum_{n=1}^{\infty} (a_n + |a_n|) - \sum_{n=1}^{\infty} |a_n|.$$

Therefore, $\sum_{n=1}^{\infty} a_n$ converges.   ▬

Example 4   Applying the Absolute Convergence Test

For

$$\sum_{n=1}^{\infty} (-1)^{n+1} \frac{1}{n^2} = 1 - \frac{1}{4} + \frac{1}{9} - \frac{1}{16} + \cdots,$$

the corresponding series of absolute values is the convergent series

$$\sum_{n=1}^{\infty} \frac{1}{n^2} = 1 + \frac{1}{4} + \frac{1}{9} + \frac{1}{16} + \cdots.$$

The original alternating series converges because it converges absolutely.

Example 5   Applying the Absolute Convergence Test

For

$$\sum_{n=1}^{\infty} \frac{\sin n}{n^2} = \frac{\sin 1}{1} + \frac{\sin 2}{4} + \frac{\sin 3}{9} + \cdots,$$

the corresponding series of absolute values is

$$\sum_{n=1}^{\infty} \left| \frac{\sin n}{n^2} \right| = \frac{|\sin 1|}{1} + \frac{|\sin 2|}{4} + \cdots,$$

which converges by comparison with $\sum_{n=1}^{\infty} (1/n^2)$ because $|\sin n| \leq 1$ for every n. The original series converges absolutely; therefore, it converges.

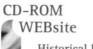

CD-ROM
WEBsite

Historical Biography

Georg Cantor
(1845 — 1918)

Example 6   Alternating *p*-Series

If p is a positive constant, the sequence $\{1/n^p\}$ is a decreasing sequence with limit zero. Therefore, the alternating *p*-series

$$\sum_{n=1}^{\infty} \frac{(-1)^{n-1}}{n^p} = 1 - \frac{1}{2^p} + \frac{1}{3^p} - \frac{1}{4^p} + \cdots, \qquad p > 0,$$

converges.

   If $p > 1$, the series converges absolutely. If $0 < p \leq 1$, the series converges conditionally.

   Conditional convergence: $1 - \dfrac{1}{\sqrt{2}} + \dfrac{1}{\sqrt{3}} - \dfrac{1}{\sqrt{4}} + \cdots$

   Absolute convergence: $1 - \dfrac{1}{2^{3/2}} + \dfrac{1}{3^{3/2}} - \dfrac{1}{4^{3/2}} + \cdots$

Rearranging Series

> **Theorem 11   The Rearrangement Theorem for Absolutely Convergent Series**
>
> If $\sum_{n=1}^{\infty} a_n$ converges absolutely and $b_1, b_2, \ldots, b_n, \ldots,$ is any arrangement of the sequence $\{a_n\}$, then $\sum b_n$ converges absolutely and
>
> $$\sum_{n=1}^{\infty} b_n = \sum_{n=1}^{\infty} a_n.$$

(For an outline of the proof, see Exercise 60.)

Example 7   Applying the Rearrangement Theorem

As we saw in Example 4, the series

$$1 - \frac{1}{4} + \frac{1}{9} - \frac{1}{16} + \cdots + (-1)^{n-1} \frac{1}{n^2} + \cdots$$

converges absolutely. A possible rearrangement of the terms of the series might start with a positive term, then two negative terms, then three positive terms, then four negative terms, and so on: After k terms of one sign, take $k + 1$ terms of the opposite sign. The first 10 terms of such a series look like this:

$$1 - \frac{1}{4} - \frac{1}{16} + \frac{1}{9} + \frac{1}{25} + \frac{1}{49} - \frac{1}{36} - \frac{1}{64} - \frac{1}{100} - \frac{1}{144} + \cdots.$$

The Rearrangement Theorem says that both series converge to the same value. In this example, if we had the second series to begin with, we would probably be glad to exchange it for the first, if we knew that we could. We can do even better: The sum of either series is also equal to

$$\sum_{n=1}^{\infty} \frac{1}{(2n-1)^2} - \sum_{n=1}^{\infty} \frac{1}{(2n)^2}.$$

(See Exercise 61.)

CAUTION If we rearrange infinitely many terms of a conditionally convergent series, we can get results that are far different from the sum of the original series.

The kind of behavior illustrated by this example is typical of what can happen with any conditionally convergent series. Moral: Add the terms of a conditionally convergent series in the order given.

Example 8 Rearranging the Alternating Harmonic Series

The alternating harmonic series

$$\frac{1}{1} - \frac{1}{2} + \frac{1}{3} - \frac{1}{4} + \frac{1}{5} - \frac{1}{6} + \frac{1}{7} - \frac{1}{8} + \frac{1}{9} - \frac{1}{10} + \frac{1}{11} - \cdots$$

can be rearranged to diverge or to reach any preassigned sum.

(a) *Rearranging $\sum_{n=1}^{\infty} (-1)^{n+1}/n$ to diverge.* The series of terms $\sum [1/(2n-1)]$ diverges to $+\infty$, and the series of terms $\sum (-1/2n)$ diverges to $-\infty$. No matter how far out in the sequence of odd-numbered terms we begin, we can always add enough positive terms to get an arbitrarily large sum. Similarly, with the negative terms, no matter how far out we start, we can add enough consecutive even-numbered terms to get a negative sum of arbitrarily large absolute value. If we wished to do so, we could start adding odd-numbered terms until we had a sum greater than $+3$, say, and then follow that with enough consecutive negative terms to make the new total less than -4. We could then add enough positive terms to make the total greater than $+5$ and follow with consecutive unused negative terms to make a new total less than -6, and so on. In this way, we could make the swings arbitrarily large in either direction.

(b) *Rearranging $\sum_{n=1}^{\infty} (-1)^{n+1}/n$ to converge to 1.* Another possibility is to focus on a particular limit. Suppose that we try to get sums that converge to 1. We start with the first term, $1/1$, and then subtract $1/2$. Next we add $1/3$ and $1/5$, which brings the total back to 1 or above. Then we add consecutive negative terms until the total is less than 1. We continue in this manner: When the sum is less than 1, add positive terms until the total is 1 or more, then subtract (add negative) terms until the total is again less than 1. This process can be continued indefinitely. Because both the odd-numbered terms and the even-numbered terms of the original series approach zero as $n \to \infty$, the amount by which our partial sums exceed 1 or fall below it approaches zero. So the new series converges to 1. The rearranged series starts like this:

$$\frac{1}{1} - \frac{1}{2} + \frac{1}{3} + \frac{1}{5} - \frac{1}{4} + \frac{1}{7} + \frac{1}{9} - \frac{1}{6} + \frac{1}{11} + \frac{1}{13} - \frac{1}{8} + \frac{1}{15} + \frac{1}{17} - \frac{1}{10}$$
$$+ \frac{1}{19} + \frac{1}{21} - \frac{1}{12} + \frac{1}{23} + \frac{1}{25} - \frac{1}{14} + \frac{1}{27} - \frac{1}{16} + \cdots.$$

Procedure For Determining Convergence

The following flowchart is often useful for deciding whether a given infinite series converges or diverges.

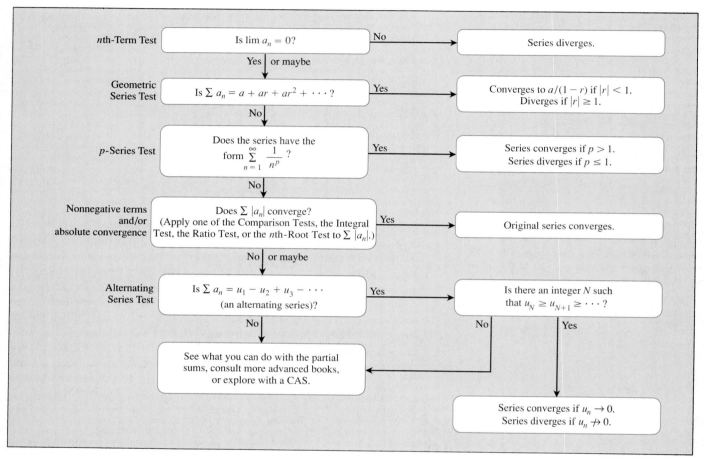

FLOWCHART 8.1 Procedure for determining convergence.

EXERCISES 8.5

Determining Convergence or Divergence

Which of the alternating series in Exercises 1–10 converge, and which diverge? Give reasons for your answers.

1. $\displaystyle\sum_{n=1}^{\infty} (-1)^{n+1} \frac{1}{n^2}$

2. $\displaystyle\sum_{n=1}^{\infty} (-1)^{n+1} \frac{1}{n^{3/2}}$

3. $\displaystyle\sum_{n=1}^{\infty} (-1)^{n+1} \left(\frac{n}{10}\right)^n$

4. $\displaystyle\sum_{n=1}^{\infty} (-1)^{n+1} \frac{10^n}{n^{10}}$

5. $\displaystyle\sum_{n=2}^{\infty} (-1)^{n+1} \frac{1}{\ln n}$

6. $\displaystyle\sum_{n=1}^{\infty} (-1)^{n+1} \frac{\ln n}{n}$

7. $\displaystyle\sum_{n=2}^{\infty} (-1)^{n+1} \frac{\ln n}{\ln n^2}$

8. $\displaystyle\sum_{n=1}^{\infty} (-1)^{n} \ln\left(1+\frac{1}{n}\right)$

9. $\displaystyle\sum_{n=1}^{\infty} (-1)^{n+1} \frac{\sqrt{n}+1}{n+1}$

10. $\displaystyle\sum_{n=1}^{\infty} (-1)^{n+1} \frac{3\sqrt{n}+1}{\sqrt{n}+1}$

Absolute Versus Conditional Convergence

Which of the series in Exercises 11–44 converge absolutely, which converge conditionally, and which diverge? Give reasons for your answers.

11. $\displaystyle\sum_{n=1}^{\infty} (-1)^{n+1} (0.1)^n$

12. $\displaystyle\sum_{n=1}^{\infty} (-1)^{n+1} \frac{(0.1)^n}{n}$

13. $\displaystyle\sum_{n=1}^{\infty} (-1)^n \frac{1}{\sqrt{n+1}}$

14. $\displaystyle\sum_{n=1}^{\infty} \frac{(-1)^n}{1+\sqrt{n}}$

15. $\displaystyle\sum_{n=1}^{\infty} (-1)^{n+1} \frac{n}{n^3+1}$

16. $\displaystyle\sum_{n=1}^{\infty} (-1)^{n+1} \frac{n!}{2^n}$

17. $\displaystyle\sum_{n=1}^{\infty} (-1)^n \frac{1}{n+3}$

18. $\displaystyle\sum_{n=1}^{\infty} (-1)^n \frac{\sin n}{n^2}$

19. $\displaystyle\sum_{n=1}^{\infty} (-1)^{n+1} \frac{3+n}{5+n}$

20. $\displaystyle\sum_{n=2}^{\infty} (-1)^n \frac{1}{\ln(n^3)}$

21. $\displaystyle\sum_{n=1}^{\infty} (-1)^{n+1} \frac{1+n}{n^2}$

22. $\displaystyle\sum_{n=1}^{\infty} \frac{(-2)^{n+1}}{n+5^n}$

23. $\displaystyle\sum_{n=1}^{\infty} (-1)^n n^2(2/3)^n$

24. $\displaystyle\sum_{n=1}^{\infty} (-1)^{n+1} (\sqrt[n]{10})$

25. $\displaystyle\sum_{n=1}^{\infty} (-1)^n \frac{\tan^{-1} n}{n^2+1}$

26. $\displaystyle\sum_{n=2}^{\infty} (-1)^{n+1} \frac{1}{n \ln n}$

27. $\displaystyle\sum_{n=1}^{\infty} (-1)^n \frac{n}{n+1}$

28. $\displaystyle\sum_{n=1}^{\infty} (-1)^n \frac{\ln n}{n - \ln n}$

29. $\displaystyle\sum_{n=1}^{\infty} \frac{(-100)^n}{n!}$

30. $\displaystyle\sum_{n=1}^{\infty} (-5)^{-n}$

31. $\displaystyle\sum_{n=1}^{\infty} \frac{(-1)^{n-1}}{n^2+2n+1}$

32. $\displaystyle\sum_{n=2}^{\infty} (-1)^n \left(\frac{\ln n}{\ln n^2}\right)^n$

33. $\displaystyle\sum_{n=1}^{\infty} \frac{\cos n\pi}{n\sqrt{n}}$

34. $\displaystyle\sum_{n=1}^{\infty} \frac{\cos n\pi}{n}$

35. $\displaystyle\sum_{n=1}^{\infty} \frac{(-1)^n (n+1)^n}{(2n)^n}$

36. $\displaystyle\sum_{n=1}^{\infty} \frac{(-1)^{n+1}(n!)^2}{(2n)!}$

37. $\displaystyle\sum_{n=1}^{\infty} (-1)^n \frac{(2n)!}{2^n n! \, n}$

38. $\displaystyle\sum_{n=1}^{\infty} (-1)^n \frac{(n!)^2 3^n}{(2n+1)!}$

39. $\displaystyle\sum_{n=1}^{\infty} (-1)^n (\sqrt{n+1} - \sqrt{n})$

40. $\displaystyle\sum_{n=1}^{\infty} (-1)^n (\sqrt{n^2+n} - n)$

41. $\displaystyle\sum_{n=1}^{\infty} (-1)^n (\sqrt{n+\sqrt{n}} - \sqrt{n})$

42. $\displaystyle\sum_{n=1}^{\infty} \frac{(-1)^n}{\sqrt{n} + \sqrt{n+1}}$

43. $\displaystyle\sum_{n=1}^{\infty} (-1)^n \operatorname{sech} n$

44. $\displaystyle\sum_{n=1}^{\infty} (-1)^n \operatorname{csch} n$

Error Estimation

In Exercises 45–48, estimate the magnitude of the error involved in using the sum of the first four terms to approximate the sum of the entire series.

45. $\displaystyle\sum_{n=1}^{\infty} (-1)^{n+1} \frac{1}{n}$ It can be shown that the sum is $\ln 2$.

46. $\displaystyle\sum_{n=1}^{\infty} (-1)^{n+1} \frac{1}{10^n}$

47. $\displaystyle\sum_{n=1}^{\infty} (-1)^{n+1} \frac{(0.01)^n}{n}$ As you will see in Section 8.6 the sum is $\ln(1.01)$.

48. $\displaystyle\frac{1}{1+t} = \sum_{n=0}^{\infty} (-1)^n t^n, \quad 0 < t < 1$

Approximate the sums in Exercises 49 and 50 with an error of magnitude less than 5×10^{-6}.

49. $\displaystyle\sum_{n=0}^{\infty} (-1)^n \frac{1}{(2n)!}$ As you will see in Section 8.7, the sum is cos 1, the cosine of 1 radian.

50. $\displaystyle\sum_{n=0}^{\infty} (-1)^n \frac{1}{n!}$ As you will see in Section 8.7, the sum is e^{-1}.

Theory and Examples

51. (a) *Writing to Learn* The series

$$\frac{1}{3} - \frac{1}{2} + \frac{1}{9} - \frac{1}{4} + \frac{1}{27} - \frac{1}{8} + \cdots + \frac{1}{3^n} - \frac{1}{2^n} + \cdots$$

does not meet one of the conditions of Theorem 8. Which one?

(b) Find the sum of the series in part (a).

52. The limit L of an alternating series that satisfies the conditions of Theorem 8 lies between the values of any two consecutive partial sums. This suggests using the average

$$\frac{s_n + s_{n+1}}{2} = s_n + \frac{1}{2}(-1)^{n+2} a_{n+1}$$

to estimate L. Compute

$$s_{20} + \frac{1}{2} \cdot \frac{1}{21}$$

as an approximation to the sum of the alternating harmonic series. The exact sum is $\ln 2 = 0.6931 \ldots$.

53. *The sign of the remainder of an alternating series that satisfies the conditions of Theorem 8* Prove the assertion in Theorem 9 that whenever an alternating series satisfying the conditions of Theorem 8 is approximated with one of its partial sums, then the remainder (sum of the unused terms) has the same sign as the first unused term. (*Hint:* Group the remainder's terms in consecutive pairs.)

54. *Writing to Learn* Show that the sum of the first $2n$ terms of the series

$$1 - \frac{1}{2} + \frac{1}{2} - \frac{1}{3} + \frac{1}{3} - \frac{1}{4} + \frac{1}{4} - \frac{1}{5} + \frac{1}{5} - \frac{1}{6} + \cdots$$

is the same as the sum of the first n terms of the series

$$\frac{1}{1 \cdot 2} + \frac{1}{2 \cdot 3} + \frac{1}{3 \cdot 4} + \frac{1}{4 \cdot 5} + \frac{1}{5 \cdot 6} + \cdots.$$

Do these series converge? What is the sum of the first $2n + 1$ terms of the first series? If the series converge, what is their sum?

55. *Divergence* Show that if $\sum_{n=1}^{\infty} a_n$ diverges, then $\sum_{n=1}^{\infty} |a_n|$ diverges.

56. Show that if $\sum_{n=1}^{\infty} a_n$ converges absolutely, then

$$\left| \sum_{n=1}^{\infty} a_n \right| \le \sum_{n=1}^{\infty} |a_n|.$$

57. *Rules of absolute convergence* Show that if $\sum_{n=1}^{\infty} a_n$ and $\sum_{n=1}^{\infty} b_n$ both converge absolutely, then so does

(a) $\displaystyle\sum_{n=1}^{\infty} (a_n + b_n)$ (b) $\displaystyle\sum_{n=1}^{\infty} (a_n - b_n)$

(c) $\displaystyle\sum_{n=1}^{\infty} ka_n$ (k any number)

58. *Term-by-term products* Show by example that $\sum_{n=1}^{\infty} a_n b_n$ may diverge even if $\sum_{n=1}^{\infty} a_n$ and $\sum_{n=1}^{\infty} b_n$ both converge.

T 59. *Rearrangement* In Example 8, suppose that the goal is to arrange the terms to get a new series that converges to $-1/2$. Start the new arrangement with the first negative term, which is $-1/2$. Whenever you have a sum that is less than or equal to $-1/2$, start introducing positive terms, taken in order, until the new total is greater than $-1/2$. Then add negative terms until the total is less than or equal to $-1/2$ again. Continue this process until your partial sums have been above the target at least three times and finish at or below it. If s_n is the sum of the first n terms of your new series, plot the points (n, s_n) to illustrate how the sums are behaving.

60. *Outline of the proof of the Rearrangement Theorem (Theorem 11)*

(a) Let ϵ be a positive real number, let $L = \sum_{n=1}^{\infty} a_n$, and let $s_k = \sum_{n=1}^{k} a_n$. Show that for some index N_1 and for some index $N_2 \geq N_1$,

$$\sum_{n=N_1}^{\infty} |a_n| < \frac{\epsilon}{2} \quad \text{and} \quad |s_{N_2} - L| \leq \frac{\epsilon}{2}.$$

Since all the terms $a_1, a_2, \ldots, a_{N_2}$ appear somewhere in the sequence $\{b_n\}$, there is an index $N_3 \geq N_2$ such that if $n \geq N_3$, then $\left(\sum_{k=1}^{k} b_k\right) - s_{N_2}$ is at most a sum of terms a_m with $m \geq N_1$. Therefore, if $n \geq N_3$,

$$\left|\sum_{k=1}^{n} b_k - L\right| \leq \left|\sum_{k=1}^{n} b_k - s_{N_2}\right| + |s_{N_2} - L|$$

$$\leq \sum_{k=N_1}^{\infty} |a_k| + |s_{N_2} - L| < \epsilon.$$

(b) The argument in part (a) shows that if $\sum_{n=1}^{\infty} a_n$ converges absolutely, then $\sum_{n=1}^{\infty} b_n$ converges and $\sum_{n=1}^{\infty} b_n = \sum_{n=1}^{\infty} a_n$. Now show that because $\sum_{n=1}^{\infty} |a_n|$ converges, $\sum_{n=1}^{\infty} |b_n|$ converges to $\sum_{n=1}^{\infty} |a_n|$.

61. *Unzipping absolutely convergent series*

(a) Show that if $\sum_{n=1}^{\infty} |a_n|$ converges and

$$b_n = \begin{cases} a_n & \text{if } a_n \geq 0 \\ 0 & \text{if } a_n < 0, \end{cases}$$

then $\sum_{n=1}^{\infty} b_n$ converges.

(b) Use the results in part (a) to show likewise that if $\sum_{n=1}^{\infty} |a_n|$ converges and

$$c_n = \begin{cases} 0 & \text{if } a_n \geq 0 \\ a_n & \text{if } a_n < 0, \end{cases}$$

then $\sum_{n=1}^{\infty} c_n$ converges.

In other words, if a series converges absolutely, its positive terms form a convergent series, and so do its negative terms. Furthermore,

$$\sum_{n=1}^{\infty} a_n = \sum_{n=1}^{\infty} b_n + \sum_{n=1}^{\infty} c_n$$

because $b_n = (a_n + |a_n|)/2$ and $c_n = (a_n - |a_n|)/2$.

62. *Alternating harmonic series revisited* What is wrong here: Multiply both sides of the alternating harmonic series.

$$S = 1 - \frac{1}{2} + \frac{1}{3} - \frac{1}{4} + \frac{1}{5} - \frac{1}{6} + \frac{1}{7} - \frac{1}{8} + \frac{1}{9} - \frac{1}{10} + \frac{1}{11} - \frac{1}{12} + \cdots$$

by 2 to get

$$2S = 2 - 1 + \frac{2}{3} - \frac{1}{2} + \frac{2}{5} - \frac{1}{3} + \frac{2}{7} - \frac{1}{4} + \frac{2}{9} - \frac{1}{5} + \frac{2}{11} - \frac{1}{6} + \cdots.$$

Collect terms with the same denominator, as the arrows indicate, to arrive at

$$2S = 1 - \frac{1}{2} + \frac{1}{3} - \frac{1}{4} + \frac{1}{5} - \frac{1}{6} + \cdots.$$

The series on the right-hand side of this equation is the series we started with. Therefore, $2S = S$, and dividing by S gives $2 = 1$. (*Source*: "Riemann's Rearrangement Theorem" by Stewart Galanor, *Mathematics Teacher*, Vol. 80, No. 8 (1987), pp. 675–681.)

63. Draw a figure similar to Figure 8.15 to illustrate the convergence of the series in Theorem 8 when $N > 1$.

8.6 Power Series

Power Series and Convergence • The Radius and Interval of Convergence
• Term-by-Term Differentiation • Term-by-Term Integration •
Multiplication of Power Series

If $|x| < 1$, then the geometric series formula assures us that

$$1 + x + x^2 + x^3 + \cdots + x^n + \cdots = \frac{1}{1 - x}.$$

Consider this statement for a moment. The expression on the right defines a function whose domain is the set of all numbers $x \neq 1$. The expression on the left defines a function whose domain is the interval of convergence, $|x| < 1$. The equality is understood to hold only on this latter domain, where both sides of the equation are defined. On this domain, the series *represents* the function $1/(1 - x)$.

In this section, we study the "infinite polynomials" like $\sum_{n=0}^{\infty} x^n$, and in the next section, we take up the question of representing a particular function with such an infinite polynomial (called a power series).

Power Series and Convergence

The expression $\sum_{n=0}^{\infty} c_n x^n$ is like a polynomial in that it is a sum of coefficients times powers of x, but polynomials have *finite* degrees and do not suffer from divergence for the wrong values of x. Just as an infinite series of numbers is not a mere sum, this series of powers of x is not a mere polynomial.

When we set $x = 0$ in the expression

$$\sum_{n=0}^{\infty} c_n x^n = c_0 + c_1 x + c_2 x^2 + \cdots + c_n x^n + \cdots ,$$

we get c_0 on the right but $c_0 \cdot 0^0$ on the left. Since 0^0 is not a number, this is a slight flaw in the notation, which we agree to overlook. The same situation arises when we set

$$x = a \quad \text{in} \quad \sum_{n=0}^{\infty} c_n (x - a)^n.$$

In either case, we agree that the expression will equal c_0. (It really *should* equal c_0, so we are not compromising the mathematics; we are clarifying the notation we use to convey the mathematics.)

Definition   Power Series

An expression of the form

$$\sum_{n=0}^{\infty} c_n x^n = c_0 + c_1 x + c_2 x^2 + \cdots + c_n x^n + \cdots$$

is a **power series centered at $x = 0$.** An expression of the form

$$\sum_{n=0}^{\infty} c_n (x - a)^n = c_0 + c_1 (x - a) + c_2 (x - a)^2 + \cdots + c_n (x - a)^n + \cdots$$

is a **power series centered at $x = a$.** The term $c_n (x - a)^n$ is the **nth term;** the number a is the **center.**

Example 1   The Geometric Series

The geometric series

$$\sum_{n=0}^{\infty} x^n = 1 + x + x^2 + \cdots + x^n + \cdots$$

is a power series centered at $x = 0$. It converges to $1/(1 - x)$ on the interval $-1 < x < 1$, also centered at $x = 0$. (Figure 8.16). This is typical behavior, as we soon see. A power series converges for all x, converges on a finite interval with the same center as the series, or converges only at the center itself.

Up to now, we have used the equation

$$\frac{1}{1 - x} = 1 + x + x^2 + \cdots + x^n + \cdots , \qquad -1 < x < 1$$

as a formula for the sum of the series on the right.

We now change the focus: We think of the partial sums of the series on the right as polynomials $P_n(x)$ that approximate the function on the left. For values of x near zero, we need take only a few terms of the series to get a good approximation. As

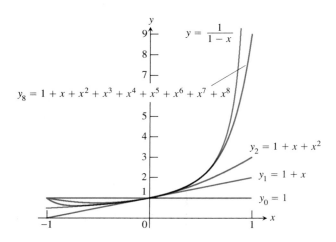

FIGURE 8.16 The graphs of $f(x) = 1/(1 - x)$ and four of its polynomial approximations. (Example 1)

we move toward $x = 1$, or -1, we must take more terms. Figure 8.16 shows the graphs of $f(x) = 1/(1 - x)$ and the approximating polynomials $y_n = P_n(x)$ for $n = 0, 1, 2$, and 8.

Example 2   Applying the Definition

The power series

$$1 - \frac{1}{2}(x - 2) + \frac{1}{4}(x - 2)^2 + \cdots + \left(-\frac{1}{2}\right)^n (x - 2)^n + \cdots \qquad (1)$$

is centered at $a = 2$ with coefficients $c_0 = 1$, $c_1 = -1/2$, $c_2 = 1/4, \ldots, c_n = (-1/2)^n$. This is a geometric series with first term 1 and ratio $r = -\dfrac{x - 2}{2}$. The series converges for $\left| \dfrac{x - 2}{2} \right| < 1$ or $0 < x < 4$. The sum is

$$\frac{1}{1 - r} = \frac{1}{1 + \dfrac{x - 2}{2}} = \frac{2}{x},$$

so

$$\frac{2}{x} = 1 - \frac{(x - 2)}{2} + \frac{(x - 2)^2}{4} - \cdots + \left(-\frac{1}{2}\right)^n (x - 2)^n + \cdots, \qquad 0 < x < 4.$$

Series (1) generates useful polynomial approximations of $f(x) = 2/x$ for values of x near 2:

$$P_0(x) = 1$$

$$P_1(x) = 1 - \frac{1}{2}(x - 2) = 2 - \frac{x}{2}$$

$$P_2(x) = 1 - \frac{1}{2}(x - 2) + \frac{1}{4}(x - 2)^2 = 3 - \frac{3x}{2} + \frac{x^2}{4},$$

and so on (Figure 8.17).

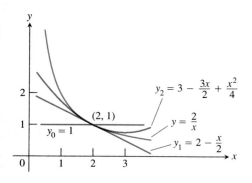

FIGURE 8.17 The graphs of $f(x) = 2/x$ and its first three polynomial approximations. (Example 2)

The Radius and Interval of Convergence

The power series in Examples 1 and 2 happen to be geometric, so we could find the intervals for which they converge. For nongeometric series, we begin by noting that any power series of the form $\sum_{n=0}^{\infty} c_n(x - a)^n$ always converges at $x = a$, thus assuring us of at least one coordinate on the real number line where the series must converge. We have encountered power series like the series in Examples 1 and 2 that converge only on a finite interval centered at a. Some power series converge for all real numbers. A useful fact about power series is that those are the only possibilities, as the following theorem attests.

Theorem 12   The Convergence Theorem for Power Series

There are three possibilities for $\sum_{n=0}^{\infty} c_n(x - a)^n$ with respect to convergence.

1. There is a positive number R such that the series diverges for $|x - a| > R$ but converges for $|x - a| < R$. The series may or may not converge at either of the endpoints $x = a - R$ and $x = a + R$.

2. The series converges for every x $(R = \infty)$.

3. The series converges at $x = a$ and diverges elsewhere $(R = 0)$.

The number R is the **radius of convergence,** and the set of all values of x for which the series converges is the **interval of convergence.** The radius of convergence completely determines the interval of convergence if R is either zero or infinite. For $0 < R < \infty$, however, there remains the question of what happens at the endpoints of the interval. The next example illustrates how to find the interval of convergence.

Example 3   Finding the Interval of Convergence Using the Ratio Test

For what values of x do the following power series converge?

(a) $\displaystyle\sum_{n=1}^{\infty} (-1)^{n-1} \frac{x^n}{n} = x - \frac{x^2}{2} + \frac{x^3}{3} - \cdots$

(b) $\displaystyle\sum_{n=1}^{\infty} (-1)^{n-1} \frac{x^{2n-1}}{2n - 1} = x - \frac{x^3}{3} + \frac{x^5}{5} - \cdots$

(c) $\displaystyle\sum_{n=0}^{\infty} \frac{x^n}{n!} = 1 + x + \frac{x^2}{2!} + \frac{x^3}{3!} + \cdots$

(d) $\displaystyle\sum_{n=0}^{\infty} n! \, x^n = 1 + x + 2! \, x^2 + 3! \, x^3 + \cdots$

Solution   Apply the Ratio Test to the series $\Sigma \, |u_n|$, where u_n is the nth term of the series in question.

(a) $\left| \dfrac{u_{n+1}}{u_n} \right| = \dfrac{n}{n + 1} \, |x| \to |x|.$

The series converges absolutely for $|x| < 1$. It diverges if $|x| > 1$ because the nth term does not converge to zero. At $x = 1$, we get the alternating harmonic series $1 - 1/2 + 1/3 - 1/4 + \cdots$, which converges. At $x = -1$, we get $-1 - 1/2 - 1/3 - 1/4 - \cdots$, the negative of the harmonic series; it diverges. Series (a) converges for $-1 < x \le 1$ and diverges elsewhere.

(b) $\left| \dfrac{u_{n+1}}{u_n} \right| = \dfrac{2n-1}{2n+1} x^2 \to x^2.$

The series converges absolutely for $x^2 < 1$. It diverges for $x^2 > 1$ because the nth term does not converge to zero. At $x = 1$, the series becomes $1 - 1/3 + 1/5 - 1/7 + \cdots$, which converges by the Alternating Series Theorem. It also converges at $x = -1$ because it is again an alternating series that satisfies the conditions for convergence. The value at $x = -1$ is the negative of the value at $x = 1$. Series (b) converges for $-1 \le x \le 1$ and diverges elsewhere.

(c) $\left| \dfrac{u_{n+1}}{u_n} \right| = \left| \dfrac{x^{n+1}}{(n+1)!} \cdot \dfrac{n!}{x^n} \right| = \dfrac{|x|}{n+1} \to 0$ for every x.

The series converges absolutely for all x.

(d) $\left| \dfrac{u_{n+1}}{u_n} \right| = \left| \dfrac{(n+1)! x^{n+1}}{n! x^n} \right| = (n+1)|x| \to \infty$ unless $x = 0$.

The series diverges for all values of x except $x = 0$.

Here's a summary of steps for finding the interval of convergence of a power series.

Finding the Interval of Convergence

Step 1: *Use the Ratio Test (or nth-Root Test) to find the interval where the series converges absolutely.* Ordinarily, this is an open interval

$$|x - a| < R \qquad \text{or} \qquad a - R < x < a + R.$$

Step 2: *If the interval of absolute convergence is finite, test for convergence or divergence at each endpoint,* as in Examples 3(a) and (b). Use a Comparison Test, the Integral Test, or the Alternating Series Test.

Step 3: *If the interval of absolute convergence is $a - R < x < a + R$, the series diverges for $|x - a| > R$* (it does not even converge conditionally), because the nth term does not approach zero for those values of x.

The convergence of a power series is absolute at every point in the interior of the interval. If a power series converges absolutely for all values of x, we say that its **radius of convergence is infinite.** If it converges only at $x = a$, the **radius of convergence is zero.**

Term-by-Term Differentiation

A theorem from advanced calculus says that a power series can be differentiated term by term at each interior point of its interval of convergence.

Theorem 13   The Term-by-Term Differentiation Theorem

If $\Sigma\, c_n(x - a)^n$ converges for $a - R < x < a + R$ for some $R > 0$, it defines a function f:

$$f(x) = \sum_{n=0}^{\infty} c_n (x - a)^n, \qquad a - R < x < a + R.$$

Such a function f has derivatives of all orders inside the interval of convergence. We can obtain the derivatives by differentiating the original series term by term:

$$f'(x) = \sum_{n=1}^{\infty} nc_n(x - a)^{n-1}$$

$$f''(x) = \sum_{n=2}^{\infty} n(n - 1)c_n(x - a)^{n-2},$$

and so on. Each of these derived series converges at every interior point of the interval of convergence of the original series.

Example 4   Applying Term-by-Term Differentiation

Find series for $f'(x)$ and $f''(x)$ if

$$f(x) = \frac{1}{1 - x} = 1 + x + x^2 + x^3 + x^4 + \cdots + x^n + \cdots$$

$$= \sum_{n=0}^{\infty} x^n, \qquad -1 < x < 1.$$

Solution

$$f'(x) = \frac{1}{(1 - x)^2} = 1 + 2x + 3x^2 + 4x^3 + \cdots + nx^{n-1} + \cdots$$

$$= \sum_{n=1}^{\infty} nx^{n-1}, \qquad -1 < x < 1$$

$$f''(x) = \frac{2}{(1 - x)^3} = 2 + 6x + 12x^2 + \cdots + n(n - 1)x^{n-2} + \cdots$$

$$= \sum_{n=2}^{\infty} n(n - 1)\, x^{n-2}, \qquad -1 < x < 1$$

Term-by-Term Integration

Another advanced theorem states that a power series can be integrated term by term throughout its interval of convergence.

Theorem 14   The Term-by-Term Integration Theorem

Suppose that

$$f(x) = \sum_{n=0}^{\infty} c_n(x - a)^n$$

converges for $a - R < x < a + R \ (R > 0)$. Then

$$\sum_{n=0}^{\infty} c_n \frac{(x - a)^{n+1}}{n + 1}$$

converges for $a - R < x < a + R$ and

$$\int f(x) \, dx = \sum_{n=0}^{\infty} c_n \frac{(x - a)^{n+1}}{n + 1} + C$$

for $a - R < x < a + R$.

Example 5   A Series for $\tan^{-1} x$, $-1 \le x \le 1$

Identify the function

$$f(x) = x - \frac{x^3}{3} + \frac{x^5}{5} - \cdots, \qquad -1 \le x \le 1.$$

Solution   We differentiate the original series term by term and get

$$f'(x) = 1 - x^2 + x^4 - x^6 + \cdots, \qquad -1 < x < 1.$$

This is a geometric series with first term 1 and ratio $-x^2$, so

$$f'(x) = \frac{1}{1 - (-x^2)} = \frac{1}{1 + x^2}.$$

We can now integrate $f'(x) = 1/(1 + x^2)$ to get

$$\int f'(x) \, dx = \int \frac{dx}{1 + x^2} = \tan^{-1} x + C.$$

> Notice that the original series in Example 5 converges at both endpoints of the original interval of convergence, but Theorem 13 can guarantee the convergence of the differentiated series only inside the interval.

The series for $f(x)$ is zero when $x = 0$, so $C = 0$. Hence,

$$f(x) = x - \frac{x^3}{3} + \frac{x^5}{5} - \frac{x^7}{7} + \cdots = \tan^{-1} x, \qquad -1 < x < 1.$$

In Section 8.8, we see that the series also converges to $\tan^{-1} x$ at $x = \pm 1$.

Example 6   A Series for $\ln (1 + x)$, $-1 < x \le 1$

The series

$$\frac{1}{1 + t} = 1 - t + t^2 - t^3 + \cdots$$

converges on the open interval $-1 < t < 1$. Therefore,

$$\ln{(1 + x)} = \int_0^x \frac{1}{1+t}\,dt = t - \frac{t^2}{2} + \frac{t^3}{3} - \frac{t^4}{4} + \cdots \Big]_0^x$$

$$= x - \frac{x^2}{2} + \frac{x^3}{3} - \frac{x^4}{4} + \cdots, \qquad -1 < x < 1.$$

It can also be shown that the series converges at $x = 1$ to the number $\ln 2$, but that was not guaranteed by the theorem.

Multiplication of Power Series

Still another advanced theorem states that absolutely converging power series can be multiplied the way we multiply polynomials to produce new absolutely convergent series.

Theorem 15   The Series Multiplication Theorem for Power Series

If $A(x) = \sum_{n=0}^{\infty} a_n x^n$ and $B(x) = \sum_{n=0}^{\infty} b_n x^n$ converge absolutely for $|x| < R$ and if

$$c_n = a_0 b_n + a_1 b_{n-1} + a_2 b_{n-2} + \cdots + a_{n-1} b_1 + a_n b_0 = \sum_{k=0}^{n} a_k b_{n-k},$$

then $\sum_{n=0}^{\infty} c_n x^n$ converges absolutely to $A(x)\,B(x)$ for $|x| < R$:

$$\left(\sum_{n=0}^{\infty} a_n x^n \right) \cdot \left(\sum_{n=0}^{\infty} b_n x^n \right) = \sum_{n=0}^{\infty} c_n x^n.$$

Example 7   Applying the Multiplication Theorem

Multiply the geometric series

$$\sum_{n=0}^{\infty} x^n = 1 + x + x^2 + \cdots + x^n + \cdots = \frac{1}{1-x} \qquad \text{for } |x| < 1,$$

by itself to get a power series for $1/(1-x)^2$, for $|x| < 1$.

Solution   Let

$$A(x) = \sum_{n=0}^{\infty} a_n x^n = 1 + x + x^2 + \cdots + x^n + \cdots = 1/(1-x)$$

$$B(x) = \sum_{n=0}^{\infty} b_n x^n = 1 + x + x^2 + \cdots + x^n + \cdots = 1/(1-x)$$

and

$$c_n = \underbrace{a_0 b_n + a_1 b_{n-1} + \cdots + a_k b_{n-k} + \cdots + a_n b_0}_{n+1 \text{ terms}}$$

$$= \underbrace{1 + 1 + \cdots + 1}_{n+1 \text{ ones}} = n + 1.$$

Then, by the Series Multiplication Theorem,

$$A(x) \cdot B(x) = \sum_{n=0}^{\infty} c_n x^n = \sum_{n=0}^{\infty} (n + 1)x^n$$

$$= 1 + 2x + 3x^2 + 4x^3 + \cdots + (n + 1)x^n + \cdots$$

is the series for $1/(1 - x)^2$. The series all converge absolutely for $|x| < 1$. Example 4 gives the same answer because

$$\frac{d}{dx}\left(\frac{1}{1 - x}\right) = \frac{1}{(1 - x)^2}.$$

EXERCISES 8.6

Intervals of Convergence

In Exercises 1–32, (a) find the series' radius and interval of convergence. For what values of x does the series converge (b) absolutely and (c) conditionally?

1. $\displaystyle\sum_{n=0}^{\infty} x^n$

2. $\displaystyle\sum_{n=0}^{\infty} (x + 5)^n$

3. $\displaystyle\sum_{n=0}^{\infty} (-1)^n(4x + 1)^n$

4. $\displaystyle\sum_{n=1}^{\infty} \frac{(3x - 2)^n}{n}$

5. $\displaystyle\sum_{n=0}^{\infty} \frac{(x - 2)^n}{10^n}$

6. $\displaystyle\sum_{n=0}^{\infty} (2x)^n$

7. $\displaystyle\sum_{n=0}^{\infty} \frac{nx^n}{n + 2}$

8. $\displaystyle\sum_{n=1}^{\infty} \frac{(-1)^n(x + 2)^n}{n}$

9. $\displaystyle\sum_{n=1}^{\infty} \frac{x^n}{n\sqrt{n}3^n}$

10. $\displaystyle\sum_{n=1}^{\infty} \frac{(x - 1)^n}{\sqrt{n}}$

11. $\displaystyle\sum_{n=0}^{\infty} \frac{(-1)^n x^n}{n!}$

12. $\displaystyle\sum_{n=0}^{\infty} \frac{3^n x^n}{n!}$

13. $\displaystyle\sum_{n=0}^{\infty} \frac{x^{2n+1}}{n!}$

14. $\displaystyle\sum_{n=0}^{\infty} \frac{(2x + 3)^{2n+1}}{n!}$

15. $\displaystyle\sum_{n=0}^{\infty} \frac{x^n}{\sqrt{n^2 + 3}}$

16. $\displaystyle\sum_{n=0}^{\infty} \frac{(-1)^n x^n}{\sqrt{n^2 + 3}}$

17. $\displaystyle\sum_{n=0}^{\infty} \frac{n(x + 3)^n}{5^n}$

18. $\displaystyle\sum_{n=0}^{\infty} \frac{nx^n}{4^n(n^2 + 1)}$

19. $\displaystyle\sum_{n=0}^{\infty} \frac{\sqrt{n}\,x^n}{3^n}$

20. $\displaystyle\sum_{n=1}^{\infty} \sqrt[n]{n}(2x + 5)^n$

21. $\displaystyle\sum_{n=1}^{\infty} \left(1 + \frac{1}{n}\right)^n x^n$

22. $\displaystyle\sum_{n=1}^{\infty} (\ln n) x^n$

23. $\displaystyle\sum_{n=1}^{\infty} n^n x^n$

24. $\displaystyle\sum_{n=0}^{\infty} n!\,(x - 4)^n$

25. $\displaystyle\sum_{n=1}^{\infty} \frac{(-1)^{n+1}(x + 2)^n}{n2^n}$

26. $\displaystyle\sum_{n=0}^{\infty} (-2)^n(n + 1)(x - 1)^n$

27. $\displaystyle\sum_{n=2}^{\infty} \frac{x^n}{n\,(\ln n)^2}$ (Get the information you need about $\sum 1/(n(\ln n)2)$ from Section 8.4, Exercise 75).

28. $\displaystyle\sum_{n=2}^{\infty} \frac{x^n}{n \ln n}$ (Get the information you need about $\sum 1/(n(\ln n)2)$ from Section 8.4, Exercise 75).

29. $\displaystyle\sum_{n=1}^{\infty} \frac{(4x - 5)^{2n+1}}{n^{3/2}}$

30. $\displaystyle\sum_{n=1}^{\infty} \frac{(3x + 1)^{n+1}}{2n + 2}$

31. $\displaystyle\sum_{n=1}^{\infty} \frac{(x + \pi)^n}{\sqrt{n}}$

32. $\displaystyle\sum_{n=0}^{\infty} \frac{(x - \sqrt{2})^{2n+1}}{2^n}$

Geometric Series in x

In Exercises 33–38, find the series' interval of convergence and, within this interval, the sum of the series as a function of x.

33. $\displaystyle\sum_{n=0}^{\infty} \frac{(x - 1)^{2n}}{4^n}$

34. $\displaystyle\sum_{n=0}^{\infty} \frac{(x + 1)^{2n}}{9^n}$

35. $\displaystyle\sum_{n=0}^{\infty} \left(\frac{\sqrt{x}}{2} - 1\right)^n$

36. $\displaystyle\sum_{n=0}^{\infty} (\ln x)^n$

37. $\displaystyle\sum_{n=0}^{\infty} \left(\frac{x^2 + 1}{3}\right)^n$

38. $\displaystyle\sum_{n=0}^{\infty} \left(\frac{x^2 - 1}{2}\right)^n$

Theory and Examples

39. *Term-by-term differentiation* For what values of x does the series

$$1 - \frac{1}{2}(x - 3) + \frac{1}{4}(x - 3)^2 + \cdots + \left(-\frac{1}{2}\right)^n (x - 3)^n + \cdots$$

converge? What is its sum? What series do you get if you differentiate the given series term by term? For what values of x does the new series converge? What is its sum?

40. *Term-by-term integration* If you integrate the series in Exercise 39 term by term, what new series do you get? For what values of x does the new series converge, and what is another name for its sum?

41. *Power series for* sin *x* The series

$$\sin\ x = x - \frac{x^3}{3!} + \frac{x^5}{5!} - \frac{x^7}{7!} + \frac{x^9}{9!} - \frac{x^{11}}{11!} + \cdots$$

converges to sin *x* for all *x*.

(a) Find the first six terms of a series for cos *x*. For what values of *x* should the series converge?

(b) By replacing *x* by 2*x* in the series for sin *x*, find a series that converges to sin 2*x* for all *x*.

(c) Using the result in part (a) and series multiplication, calculate the first six terms of a series for 2 sin *x* cos *x*. Compare your answer with the answer in part (b).

42. *Power series for* e^x The series

$$e^x = 1 + x + \frac{x^2}{2!} + \frac{x^3}{3!} + \frac{x^4}{4!} + \frac{x^5}{5!} + \cdots$$

converges to e^x for all *x*.

(a) Find a series for $(d/dx)e^x$. Do you get the series for e^x? Explain your answer.

(b) Find a series for $\int e^x\ dx$. Do you get the series for e^x? Explain your answer.

(c) Replace *x* by $-x$ in the series for e^x to find a series that converges to e^{-x} for all *x*. Then multiply the series for e^x and e^{-x} to find the first six terms of a series for $e^{-x} \cdot e^x$.

43. *Power series for* tan *x* The series

$$\tan\ x = x + \frac{x^3}{3} + \frac{2x^5}{15} + \frac{17x^7}{315} + \frac{62x^9}{2835} + \cdots$$

converges to tan *x* for $-\pi/2 < x < \pi/2$.

(a) Find the first five terms of the series for $\ln|\sec x|$. For what values of *x* should the series converge?

(b) Find the first five terms of the series for $\sec^2 x$. For what values of *x* should this series converge?

(c) Check your result in part (b) by squaring the series given for sec *x* in Exercise 44.

44. *Power series for* sec *x* The series for

$$\sec\ x = 1 + \frac{x^2}{2} + \frac{5}{24}x^4 + \frac{61}{720}x^6 + \frac{277}{8064}x^8 + \cdots$$

converges to sec *x* for $-\pi/2 < x < \pi/2$.

(a) Find the first five terms of a power series for the function $\ln|\sec x + \tan x|$. For what values of *x* should the series converge?

(b) Find the first four terms of a series for sec *x* tan *x*. For what values of *x* should the series converge?

(c) Check your result in part (b) by multiplying the series for sec *x* by the series given for tan *x* in Exercise 43.

45. *Uniqueness of convergent power series*

(a) Show that if two power series $\sum_{n=0}^{\infty} a_n x^n$ and $\sum_{n=0}^{\infty} b_n x^n$ are convergent and equal for all values of *x* in an open interval $(-c, c)$, then $a_n = b_n$ for every *n*. (*Hint:* Let $f(x) = \sum_{n=0}^{\infty} a_n x^n = \sum_{n=0}^{\infty} b_n x^n$. Differentiate term by term to show that a_n and b_n both equal $f^{(n)}(0)/(n!)$.)

(b) Show that if $\sum_{n=0}^{\infty} a_n x^n = 0$ for all *x* in an open interval $(-c, c)$, then $a_n = 0$ for every *n*.

46. *The sum of the series* $\sum_{n=0}^{\infty} (n^2/2^n)$ To find the sum of this series, express $1/(1 - x)$ as a geometric series, differentiate both sides of the resulting equation with respect to *x*, multiply both sides of the result by *x*, differentiate again, multiply by *x* again, and set *x* equal to 1/2. What do you get? (*Source:* David E. Dobbs's letter to the editor, *Illinois Mathematics Teacher*, Vol. 33, Issue 4 (1982), p. 27.)

47. *Convergence at endpoints* Show by examples that the convergence of a power series at an endpoint of its interval of convergence may be either conditional or absolute.

48. *Intervals of convergence* Make up a power series whose interval of convergence is

(a) $(-3, 3)$ **(b)** $(-2, 0)$ **(c)** $(1, 5)$.

8.7 Taylor and Maclaurin Series

Constructing a Series • Taylor and Maclaurin Series • Taylor Polynomials • Remainder of a Taylor Polynomial • Estimating the Remainder • Truncation Error • Table of Maclaurin Series • Combining Taylor Series

A comprehensive understanding of geometric series served us well in the last section, enabling us to find power series to represent certain functions and functions that are equivalent to certain power series (all these equivalencies being subject to

CD-ROM
WEBsite

the condition of convergence). In this section, we learn a more general technique for constructing power series, one that makes good use of the tools of calculus. In many cases, these series can provide useful polynomial approximations of the generating functions.

Constructing a Series

We know that within its interval of convergence, the sum of a power series is a continuous function with derivatives of all orders, but what about the other way around? If a function $f(x)$ has derivatives of all orders on an interval I, can it be expressed as a power series on I? If it can, what will its coefficients be?

We can answer the last question readily if we assume that $f(x)$ is the sum of a power series

$$f(x) = \sum_{n=0}^{\infty} a_n(x - a)^n$$

$$= a_0 + a_1(x - a) + a_2(x - a)^2 + \cdots + a_n(x - a)^n + \cdots$$

with a positive radius of convergence. By repeated term-by-term differentiation within the interval of convergence I, we obtain

$$f'(x) = a_1 + 2a_2(x - a) + 3a_3(x - a)^2 + \cdots + na_n(x - a)^{n-1} + \cdots$$
$$f''(x) = 1 \cdot 2a_2 + 2 \cdot 3a_3(x - a) + 3 \cdot 4a_4(x - a)^2 + \cdots$$
$$f'''(x) = 1 \cdot 2 \cdot 3a_3 + 2 \cdot 3 \cdot 4a_4(x - a) + 3 \cdot 4 \cdot 5a_5(x - a)^2 + \cdots,$$

with the nth derivative, for all n, being

$$f^{(n)}(x) = n!a_n + \text{a sum of terms with } (x - a) \text{ as a factor.}$$

Since these equations all hold at $x = a$, we have

$$f'(a) = a_1,$$
$$f''(a) = 1 \cdot 2a_2,$$
$$f'''(a) = 1 \cdot 2 \cdot 3a_3,$$

and, in general,

$$f^{(n)}(a) = n!a_n.$$

These formulas reveal a marvelous pattern in the coefficients of any power series $\sum_{n=0}^{\infty} a_n(x - a)^n$ that converges to the values of f on I ("represents f on I," we say). If there *is* such a series (still an open question), then there is only one such series and its nth coefficient is

$$a_n = \frac{f^{(n)}(a)}{n!}.$$

If f has a series representation, then the series must be

$$f(x) = f(a) + f'(a)(x - a) + \frac{f''(a)}{2!}(x - a)^2 + \cdots + \frac{f^{(n)}(a)}{n!}(x - a)^n + \cdots. \quad (1)$$

If we start with an arbitrary function f that is infinitely differentiable on an interval I centered at $x = a$ and use it to generate the series in Equation (1), however, will the series then converge to $f(x)$ at each x in the interior of I? The answer is maybe; for some functions it will, but for other functions it will not, as we will see.

Taylor and Maclaurin Series

CD-ROM
WEBsite

Historical Biographies

Brook Taylor
(1685 — 1731)
Colin Maclaurin
(1698 — 1746)

> **Definitions**   Taylor Series, Maclaurin Series
>
> Let f be a function with derivatives of all orders throughout some interval containing a as an interior point. Then the **Taylor series generated by f at $x = a$** is
>
> $$\sum_{k=0}^{\infty} \frac{f^{(k)}(a)}{k!}(x-a)^k = f(a) + f'(a)(x-a) + \frac{f''(a)}{2!}(x-a)^2 + \cdots +$$
>
> $$\frac{f^{(n)}(a)}{n!}(x-a)^n + \cdots.$$
>
> The **Maclaurin series generated by f** is
>
> $$\sum_{k=0}^{\infty} \frac{f^{(k)}(0)}{k!}x^k = f(0) + f'(0)x + \frac{f''(0)}{2!}x^2 + \cdots + \frac{f^{(n)}(0)}{n!}x^n + \cdots,$$
>
> the Taylor series generated by f at $x = 0$.

Example 1   Finding a Taylor Series

Find the Taylor series generated by $f(x) = 1/x$ at $a = 2$. Where, if anywhere, does the series converge to $1/x$?

Solution   We need to find $f(2), f'(2), f''(2), \ldots$. Taking derivatives, we get

$$f(x) = x^{-1}, \qquad\qquad f(2) = 2^{-1} = \frac{1}{2},$$

$$f'(x) = -x^{-2}, \qquad\qquad f'(2) = -\frac{1}{2^2},$$

$$f''(x) = 2!\,x^{-3}, \qquad\qquad \frac{f''(2)}{2!} = 2^{-3} = \frac{1}{2^3},$$

$$f'''(x) = -3!\,x^{-4}, \qquad\qquad \frac{f'''(2)}{3!} = -\frac{1}{2^4},$$

$$\vdots \qquad\qquad\qquad \vdots$$

$$f^{(n)}(x) = (-1)^n n!\,x^{-(n+1)}, \qquad \frac{f^{(n)}(2)}{n!} = \frac{(-1)^n}{2^{n+1}}.$$

The Taylor series is

$$f(2) + f'(2)(x-2) + \frac{f''(2)}{2!}(x-2)^2 + \cdots + \frac{f^{(n)}}{n!}(x-2)^n + \cdots$$

$$= \frac{1}{2} - \frac{(x-2)}{2^2} + \frac{(x-2)^2}{2^3} - \cdots + (-1)^n \frac{(x-2)^n}{2^{n+1}} + \cdots.$$

This is a geometric series with first term $1/2$ and ratio $r = -(x - 2)/2$. It converges absolutely for $|x - 2| < 2$, and its sum is

$$\frac{1/2}{1 + (x - 2)/2} = \frac{1}{2 + (x - 2)} = \frac{1}{x}.$$

In this example, the Taylor series generated by $f(x) = 1/x$ at $a = 2$ converges to $1/x$ for $|x - 2| < 2$ or $0 < x < 4$.

Taylor Polynomials

The linearization of a differentiable function f at a point a is the polynomial

$$P_1(x) = f(a) + f'(a)(x - a).$$

If f has derivatives of higher order at a, then it has higher-order polynomial approximations as well, one for each available derivative. These polynomials are called the Taylor polynomials of f.

We speak of a Taylor polynomial of *order n* rather than *degree n* because $f^{(n)}(a)$ may be zero. The first two Taylor polynomials of $\cos x$ at $x = 0$, for example, are $P_0(x) = 1$ and $P_1(x) = 1$. The first-order polynomial has degree zero, not one.

Definition Taylor Polynomial of Order n

Let f be a function with derivatives of order k for $k = 1, 2, \ldots, N$ in some interval containing a as an interior point. Then for any integer n from 0 through N, the **Taylor polynomial of order n** generated by f at $x = a$ is the polynomial

$$P_n(x) = f(a) + f'(a)(x - a) + \frac{f''(a)}{2!}(x - a)^2 + \cdots$$

$$+ \frac{f^{(k)}(a)}{k!}(x - a)^k + \cdots + \frac{f^{(n)}(a)}{n!}(x - a)^n.$$

Just as the linearization of f at $x = a$ provides the best linear approximation of f in the neighborhood of a, the higher-order Taylor polynomials provide the best polynomial approximations of their respective degrees. (See Exercise 58.)

CD-ROM
WEBsite

Example 2 Finding Taylor Polynomials for e^x

Find the Taylor series and the Taylor polynomials generated by $f(x) = e^x$ at $x = 0$.

Solution Since

$$f(x) = e^x, \qquad f'(x) = e^x, \qquad \ldots, \qquad f^{(n)}(x) = e^x, \ldots,$$

we have

$$f(0) = e^0 = 1, \qquad f'(0) = 1, \qquad \ldots, \qquad f^{(n)}(0) = 1, \ldots.$$

The Taylor series generated by f at $x = 0$ is

$$f(0) + f'(0)x + \frac{f''(0)}{2!}x^2 + \cdots + \frac{f^{(n)}(0)}{n!}x^n + \cdots = 1 + x + \frac{x^2}{2} + \cdots + \frac{x^n}{n!} + \cdots$$

$$= \sum_{k=0}^{\infty} \frac{x^k}{k!}.$$

By definition, this is also the Maclaurin series for e^x. We soon see that the series converges to e^x at every x.

The Taylor polynomial of order n at $x = 0$ is

$$P_n(x) = 1 + x + \frac{x^2}{2} + \cdots + \frac{x^n}{n!}.$$

See Figure 8.18.

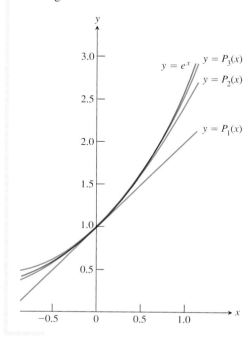

FIGURE 8.18 The graph of $f(x) = e^x$ and its Taylor polynomials

$$P_1(x) = 1 + x$$
$$P_2(x) = 1 + x + (x^2/2!)$$
$$P_3(x) = 1 + x + (x^2/2!) + (x^3/3!).$$

Notice the very close agreement near the center $x = 0$.

CD-ROM WEBsite

Example 3   Finding Taylor Polynomials For cos x

Find the Taylor series and Taylor polynomials generated by $f(x) = \cos x$ at $x = 0$.

Solution   The cosine and its derivatives are

$$f(x) = \cos x, \qquad f'(x) = -\sin x,$$
$$f''(x) = -\cos x, \qquad f^{(3)}(x) = \sin x,$$
$$\vdots \qquad\qquad \vdots$$
$$f^{(2n)}(x) = (-1)^n \cos x, \qquad f^{(2n+1)}(x) = (-1)^{n+1} \sin x.$$

At $x = 0$, the cosines are 1 and the sines are 0, so

$$f^{(2n)}(0) = (-1)^n, \qquad f^{(2n+1)}(0) = 0.$$

The Taylor series generated by f at 0 is

$$f(0) + f'(0)x + \frac{f''(0)}{2!}x^2 + \frac{f'''(0)}{3!}x^3 + \cdots + \frac{f^{(n)}(0)}{n!}x^n + \cdots$$
$$= 1 + 0 \cdot x - \frac{x^2}{2!} + 0 \cdot x^3 + \frac{x^4}{4!} + \cdots + (-1)^n \frac{x^{2n}}{(2n)!} + \cdots = \sum_{n=0}^{\infty} \frac{(-1)^n x^{2n}}{(2n)!}.$$

By definition, this is also the Maclaurin series for cos x. We see later that the series converges to cos x at every x.

Because $f^{(2n+1)}(0) = 0$, the Taylor polynomials of orders $2n$ and $2n + 1$ are identical:

$$P_{2n}(x) = P_{2n+1}(x) = 1 - \frac{x^2}{2!} + \frac{x^4}{4!} - \cdots + (-1)^n \frac{x^{2n}}{(2n)!}.$$

Figure 8.19 shows how well these polynomials approximate $f(x) = \cos x$ near $x = 0$. Only the right-hand portions of the graphs are given because the graphs are symmetric about the y-axis.

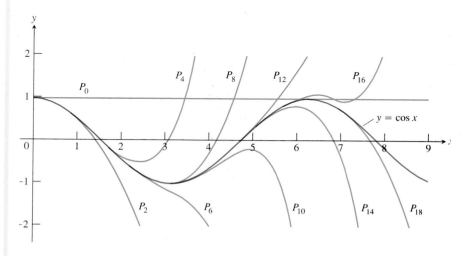

FIGURE 8.19 The polynomials

$$P_{2n}(x) = \sum_{k=0}^{n} \frac{(-1)^k x^{2k}}{(2k)!}$$

converge to $\cos x$ as $n \to \infty$. We can deduce the behavior of $\cos x$ arbitrarily far away solely from knowing the values of the cosine and its derivatives at $x = 0$.

Infinitely differentiable functions that are represented by their Taylor series only at isolated points are, in practice, quite rare.

Example 4   A Function f Whose Taylor Series Converges at Every x But Converges to $f(x)$ Only at $x = 0$

It can be shown (although not easily) that

$$f(x) = \begin{cases} 0, & x = 0 \\ e^{-1/x^2}, & x \neq 0 \end{cases}$$

(Figure 8.20) has derivatives of all orders at $x = 0$ and that $f^{(n)}(0) = 0$ for all n. Hence, the Taylor series generated by f at $x = 0$ is

$$f(0) + f'(0)x + \frac{f''(0)}{2!}x^2 + \cdots + \frac{f^{(n)}(0)}{n!}x^n + \cdots$$

$$= 0 + 0 \cdot x + 0 \cdot x^2 + \cdots + 0 \cdot x^n + \cdots$$

$$= 0 + 0 + \cdots + 0 + \cdots.$$

The series converges for every x (its sum is 0) but converges to $f(x)$ only at $x = 0$.

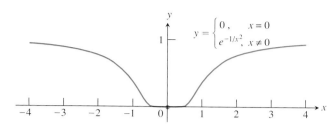

FIGURE 8.20 The graph of the continuous extension of $y = e^{-1/x^2}$ is so flat at the origin that all of its derivatives there are zero. (Example 4)

Two questions still remain.

1. For what values of x can we normally expect a Taylor series to converge to its generating function?

2. How accurately do a function's Taylor polynomials approximate the function on a given interval?

We answer these questions next.

Remainder of a Taylor Polynomial

We need a measure of the accuracy in approximating a function value $f(x)$ by its Taylor polynomial $P_n(x)$. We can use the idea of a **remainder** $R_n(x)$ defined by

$$f(x) = P_n(x) + R_n(x).$$

| Exact value | Approximate value | Remainder |

The absolute value $|R_n(x)| = |f(x) - P_n(x)|$ is called the **error** associated with the approximation.

The next theorem gives a way to estimate the remainder associated with a Taylor polynomial.

Theorem 16   Taylor's Theorem

If f is differentiable through order $n + 1$ in an open interval I containing a, then for each x in I, there exists a number c between x and a such that

$$f(x) = f(a) + f'(a)(x - a) + \frac{f''(a)}{2!}(x - a)^2 + \cdots + \frac{f^{(n)}(a)}{n!}(x - a)^n + R_n(x),$$

where

$$R_n(x) = \frac{f^{(n+1)}(c)}{(n + 1)!}(x - a)^{n+1}.$$

Taylor's Theorem is a generalization of the Mean Value Theorem (Exercise 49). The proof is lengthy and given in Appendix 8.

If $R_n(x) \to 0$ as $n \to \infty$ for all x in I, we say that the Taylor series generated by f at $x = a$ **converges** to f on I, and we write

$$f(x) = \sum_{k=0}^{\infty} \frac{f^{(k)}(a)}{k!} (x - a)^k.$$

Example 5   The Maclaurin Series for e^x Revisited

Show that the Taylor series generated by $f(x) = e^x$ at $x = 0$ converges to $f(x)$ for every real value of x.

Solution   The function has derivatives of all orders throughout the interval $I = (-\infty, \infty)$, and from Example 2,

$$e^x = 1 + x + \frac{x^2}{2!} + \cdots + \frac{x^n}{n!} + R_n(x),$$

where

$$R_n(x) = \frac{e^c}{(n + 1)!} x^{n+1} \qquad \text{for some } c \text{ between 0 and } x.$$

Since e^x is an increasing function of x, e^c lies between $e^0 = 1$ and e^x. When x is negative, so is c, and $e^c < 1$. When x is zero, $e^x = 1$ and $R_n(x) = 0$. When x is positive, so is c, and $e^c < e^x$. Thus,

$$|R_n(x)| \le \frac{|x|^{n+1}}{(n + 1)!} \qquad \text{when } x \le 0,$$

and

$$|R_n(x)| < e^x \frac{x^{n+1}}{(n + 1)!} \qquad \text{when } x > 0.$$

Finally, because

$$\lim_{n \to \infty} \frac{x^{n+1}}{(n + 1)!} = 0 \qquad \text{for every } x, \qquad \text{Table 8.1, Formula 6}$$

$\lim_{n \to \infty} R_n(x) = 0$, and the series converges to e^x for every x.

Estimating the Remainder

It is often possible to estimate $R_n(x)$ as we did in Example 5. This method of estimation is so convenient that we state it as a theorem for future reference.

Theorem 17   The Remainder Estimation Theorem
If there are positive constants M and r such that $|f^{(n+1)}(t)| \le Mr^{n+1}$ for all t between a and x, inclusive, then the remainder term $R_n(x)$ in Taylor's Theorem satisfies the inequality

$$|R_n(x)| \le M \frac{r^{n+1}|x - a|^{n+1}}{(n + 1)!}.$$

If these conditions hold for every n and all the other conditions of Taylor's Theorem are satisfied by f, then the series converges to $f(x)$.

In the simplest examples, we can take $r = 1$ provided f and all its derivatives are bounded in magnitude by some constant M. In other cases, we may need to consider r. For example, if $f(x) = 2 \cos(3x)$, each time we differentiate we get a factor of 3 and r needs to be greater than 1. In this particular case, we can take $r = 3$ along with $M = 2$.

We are now ready to look at some examples of how the Remainder Estimation Theorem and Taylor's Theorem can be used together to settle questions of convergence. As you will see, they can also be used to determine the accuracy with which a function is approximated by one of its Taylor polynomials.

Example 6 The Maclaurin Series for sin x

Show that the Maclaurin series for $\sin x$ converges to $\sin x$ for all x.

Solution The function and its derivatives are

$$
\begin{array}{ll}
f(x) = \sin x, & f'(x) = \cos x, \\
f''(x) = -\sin x, & f'''(x) = -\cos x, \\
\quad\vdots & \quad\vdots \\
f^{(2k)}(x) = (-1)^k \sin x, & f^{(2k+1)}(x) = (-1)^k \cos x,
\end{array}
$$

so

$$ f^{(2k)}(0) = 0 \quad \text{and} \quad f^{(2k+1)}(0) = (-1)^k. $$

The series has only odd-powered terms and, for $n = 2k + 1$, Taylor's Theorem gives

$$ \sin x = x - \frac{x^3}{3!} + \frac{x^5}{5!} - \cdots + \frac{(-1)^k x^{2k+1}}{(2k+1)!} + R_{2k+1}(x). $$

All the derivatives of $\sin x$ have absolute values less than or equal to 1, so we can apply the Remainder Estimation Theorem with $M = 1$ and $r = 1$ to obtain

$$ |R_{2k+1}(x)| \leq 1 \cdot \frac{|x|^{2k+2}}{(2k+2)!}. $$

Since $(|x|^{2k+2}/(2k+2)!) \to 0$ as $k \to \infty$, whatever the value of x, $R_{2k+1}(x) \to 0$, and the Maclaurin series for $\sin x$ converges to $\sin x$ for every x.

Example 7 The Maclaurin Series for cos x Revisited

Show that the Maclaurin series for $\cos x$ converges to $\cos x$ for every value of x.

Solution We add the remainder term to the Taylor polynomial for $\cos x$ in Example 3 to obtain Taylor's formula for $\cos x$ with $n = 2k$:

$$ \cos x = 1 - \frac{x^2}{2!} + \frac{x^4}{4!} - \cdots + (-1)^k \frac{x^{2k}}{(2k)!} + R_{2k}(x). $$

Because the derivatives of the cosine have absolute value less than or equal to 1, the Remainder Estimation Theorem with $M = 1$ and $r = 1$ gives

$$ |R_{2k}(x)| \leq 1 \cdot \frac{|x|^{2k+1}}{(2k+1)!}. $$

For every value of x, $R_{2k} \to 0$ as $k \to \infty$. Therefore, the series converges to $\cos x$ for every value of x.

Truncation Error

The Maclaurin series for e^x converges to e^x for all x, but we still need to decide how many terms to use to approximate e^x to a given degree of accuracy. We get this information from the Remainder Estimation Theorem.

Example 8   Calculating the Number e

Calculate e with an error of less than 10^{-6}.

Solution   We can use the result of Example 2 with $x = 1$ to write

$$e = 1 + 1 + \frac{1}{2!} + \cdots + \frac{1}{n!} + R_n(1),$$

with

$$R_n(1) = e^c \frac{1}{(n+1)!} \qquad \text{for some } c \text{ between 0 and 1.}$$

For the purposes of this example, we assume that we know that $e < 3$. Hence, we are certain that

$$\frac{1}{(n+1)!} < R_n(1) < \frac{3}{(n+1)!}$$

because $1 < e^c < 3$ for $0 < c < 1$.

By experiment, we find that $1/9! > 10^{-6}$, whereas $3/10! < 10^{-6}$. Thus, we should take $(n + 1)$ to be at least 10 or n to be at least 9. With an error of less than 10^{-6},

$$e = 1 + 1 + \frac{1}{2} + \frac{1}{3!} + \cdots + \frac{1}{9!} \approx 2.7182\,82.$$

Example 9   Sine Function as a Polynomial of Degree 3

For what values of x can we replace $\sin x$ by $x - (x^3/3!)$ with an error of magnitude no greater than 3×10^{-4}?

Solution   Using the result of Example 6, $x - (x^2/3!) = 0 + x + 0x^2 - (x^3/3!) + 0x^4$ is the Taylor polynomial of order 4 as well as of order 3 for $\sin x$. Then,

$$\sin x = x - \frac{x^3}{3!} + 0 + R_4,$$

and the Remainder Estimation Theorem with $M = r = 1$ gives

$$|R_4| \leq 1 \cdot \frac{|x|^5}{5!} = \frac{|x|^5}{120}.$$

Therefore, the error will be less than or equal to 3×10^{-4} if

$$\frac{|x|^5}{120} < 3 \times 10^{-4} \qquad \text{or} \qquad |x| < \sqrt[5]{360 \times 10^{-4}} \approx 0.514. \qquad \text{Rounded down, to be safe}$$

The Alternating Series Estimation Theorem tells us something that the Remainder Estimation Theorem does not: namely, that the estimate $x - (x^3/3!)$ for $\sin x$ is an underestimate when x is positive because then $x^5/120$ is positive.

Figure 8.21 shows the graph of $\sin x$, along with the graphs of a number of its approximating Taylor polynomials. The graph of $P_3(x) = x - (x^3/3!)$ is almost indistinguishable from the sine curve when $-1 \leq x \leq 1$.

FIGURE 8.21 The polynomials

$$P_{2n+1}(x) = \sum_{k=0}^{n} \frac{(-1)^k x^{2k+1}}{(2k+1)!}$$

converge to $\sin x$ as $n \to \infty$.

Table of Maclaurin Series

Here we list some of the most useful Maclaurin series, which have all been derived in one way or another in this chapter. The exercises will ask you to use these series as basic building blocks for constructing other series (e.g., $\tan^{-1} x^2$ or $7xe^x$). We also list the intervals of convergence.

Maclaurin series

1. $\dfrac{1}{1-x} = 1 + x + x^2 + \cdots + x^n + \cdots = \displaystyle\sum_{n=0}^{\infty} x^n \qquad (|x| < 1)$

2. $\dfrac{1}{1+x} = 1 - x + x^2 - \cdots + (-x)^n + \cdots = \displaystyle\sum_{n=0}^{\infty} (-1)^n x^n \qquad (|x| < 1)$

3. $e^x = 1 + x + \dfrac{x^2}{2!} + \cdots + \dfrac{x^n}{n!} + \cdots = \displaystyle\sum_{n=0}^{\infty} \dfrac{x^n}{n!} \qquad (\text{all real } x)$

4. $\sin x = x - \dfrac{x^3}{3!} + \dfrac{x^5}{5!} - \cdots + (-1)^n \dfrac{x^{2n+1}}{(2n+1)!} + \cdots \quad = \displaystyle\sum_{n=0}^{\infty} (-1)^n \dfrac{x^{2n+1}}{(2n+1)!} \ (\text{all real } x)$

5. $\cos x = 1 - \dfrac{x^2}{2!} + \dfrac{x^4}{4!} - \cdots + (-1)^n \dfrac{x^{2n}}{(2n)!} + \cdots \quad = \displaystyle\sum_{n=0}^{\infty} (-1)^n \dfrac{x^{2n}}{(2n)!} \qquad (\text{all real } x)$

6. $\ln(1+x) = x - \dfrac{x^2}{2} + \dfrac{x^3}{3} - \cdots + (-1)^{n-1} \dfrac{x^n}{n} + \cdots \quad = \displaystyle\sum_{n=1}^{\infty} (-1)^{n-1} \dfrac{x^n}{n} \quad (-1 < x \le 1)$

7. $\tan^{-1} x = x - \dfrac{x^3}{3} + \dfrac{x^5}{5} - \cdots + (-1)^n \dfrac{x^{2n+1}}{2n+1} + \cdots \quad = \displaystyle\sum_{n=0}^{\infty} (-1)^n \dfrac{x^{2n+1}}{2n+1} \quad (|x| \le 1)$

Combining Taylor Series

On the intersection of their intervals of convergence, Taylor series can be added, subtracted, and multiplied by constants and powers of x, and the results are once again Taylor series. The Taylor series for $f(x) + g(x)$ is the sum of the Taylor series for $f(x)$ and the Taylor series for $g(x)$ because the nth derivative of $f + g$ is $f^{(n)} + g^{(n)}$, and so on. We can obtain the Maclaurin series for $(1 + \cos 2x)/2$ by substituting $2x$ in the Maclaurin series for $\cos x$, adding 1, and dividing the result by 2. The Maclaurin series for $\sin x + \cos x$ is the term-by-term sum of the series for $\sin x$ and $\cos x$. We obtain the Maclaurin series for $x \sin x$ by multiplying all the terms of the Maclaurin series for $\sin x$ by x.

Example 10   Finding a Maclaurin Series by Substitution

Find the Maclaurin series for $\cos 2x$.

Solution   We can find the Maclaurin series for $\cos 2x$ by substituting $2x$ for x in the Maclaurin series for $\cos x$:

$$\cos 2x = \sum_{k=0}^{\infty} \frac{(-1)^k (2x)^{2k}}{(2k)!} = 1 - \frac{(2x)^2}{2!} + \frac{(2x)^4}{4!} - \frac{(2x)^6}{6!} + \cdots \qquad \text{Eq. (5)} \\ \text{with } 2x \\ \text{for } x$$

$$= 1 - \frac{2^2 x^2}{2!} + \frac{2^4 x^4}{4!} - \frac{2^6 x^6}{6!} + \cdots$$

$$= \sum_{k=0}^{\infty} (-1)^k \frac{2^{2k} x^{2k}}{(2k)!}.$$

Equation (5) holds for $-\infty < x < \infty$, implying that it holds for $-\infty < 2x < \infty$, so the newly created series converges for all x. Exercise 54 explains why the series is in fact the Maclaurin series for $\cos 2x$.

Example 11   Finding a Maclaurin Series by Multiplication

Find the Maclaurin series for $x \sin x$.

Solution   We can find the Maclaurin series for $x \sin x$ by multiplying the Maclaurin series for $\sin x$ (Equation 4) by x:

$$x \sin x = x \left(x - \frac{x^3}{3!} + \frac{x^5}{5!} - \frac{x^7}{7!} + \cdots \right)$$

$$= x^2 - \frac{x^4}{3!} + \frac{x^6}{5!} - \frac{x^8}{7!} + \cdots .$$

The new series converges for all x because the series for $\sin x$ converges for all x. Exercise 54 explains why the series is the Maclaurin series for $x \sin x$.

EXERCISES 8.7

Finding Taylor Polynomials

In Exercises 1–6 find the Taylor polynomials of orders 0, 1, 2, and 3 generated by f at a.

1. $f(x) = \ln x, \quad a = 1$

2. $f(x) = \ln(1 + x), \quad a = 0$

3. $f(x) = \dfrac{1}{(x + 2)}, \quad a = 0$

4. $f(x) = \sin x, \quad a = \pi/4$

5. $f(x) = \cos x, \quad a = \pi/4$

6. $f(x) = \sqrt{x}, \quad a = 4$

Finding Maclaurin Series

Find the Maclaurin series for the functions in Exercises 7–14.

7. e^{-x}

8. $\dfrac{1}{1 + x}$

9. $\sin 3x$

10. $7 \cos(-x)$

11. $\cosh x = \dfrac{e^x + e^{-x}}{2}$

12. $\sinh x = \dfrac{e^x - e^{-x}}{2}$

13. $x^4 - 2x^3 - 5x + 4$

14. $(x + 1)^2$

Finding Taylor Series

In Exercises 15–20, find the Taylor series generated by f at $x = a$.

15. $f(x) = x^3 - 2x + 4, \quad a = 2$

16. $f(x) = 3x^5 - x^4 + 2x^3 + x^2 - 2, \quad a = -1$

17. $f(x) = 1/x^2, \quad a = 1$

18. $f(x) = x/(1 - x), \quad a = 0$

19. $f(x) = e^x, \quad a = 2$

20. $f(x) = 2^x, \quad a = 1$

Maclaurin Series by Substitution

Use substitution as in Example 10 to find the Maclaurin series of the functions in Exercises 21–24.

21. e^{-5x}

22. $e^{-x/2}$

23. $\sin\left(\dfrac{\pi x}{2}\right)$

24. $\cos \sqrt{x}$

More Maclaurin Series

Using the series in the Maclaurin series table as basic building blocks, combine series expressions to find Maclaurin series for the functions in Exercises 25–34.

25. xe^x

26. $x^2 \sin x$

27. $\dfrac{x^2}{2} - 1 + \cos x$

28. $\sin x - x + \dfrac{x^3}{3!}$

29. $x \cos \pi x$

30. $\cos^2 x$ (*Hint*: $\cos^2 x = (1 + \cos 2x)/2$.)

31. $\sin^2 x$

32. $\dfrac{x^2}{1 - 2x}$

33. $x \ln(1 + 2x)$

34. $\dfrac{1}{(1 - x)^2}$

Error Estimates

35. *Writing to Learn* For approximately what values of x can you replace $\sin x$ by $x - (x^3/6)$ with an error of magnitude no greater than 5×10^{-4}? Give reasons for your answer.

36. *Writing to Learn* If $\cos x$ is replaced by $1 - (x^2/2)$ and $|x| < 0.5$, what estimate can be made of the error? Does $1 - (x^2/2)$ tend to be too large or too small? Give reasons for your answer.

37. *Linear approximation for* $\sin x$ How close is the approximation $\sin x = x$ when $|x| < 10^{-3}$? For which of these values of x is $x < \sin x$?

38. *Linear approximation for* $\sqrt{1 + x}$ The estimate $\sqrt{1 + x} = 1 + (x/2)$ is used when x is small. Estimate the error when $|x| < 0.01$.

39. *Quadratic approximation for* e^x

  (a) The approximation $e^x = 1 + x + (x^2/2)$ is used when x is small. Use the Remainder Estimation Theorem to estimate the error when $|x| < 0.1$.

  (b) When $x < 0$, the series for e^x is an alternating series. Use the Alternating Series Estimation Theorem to estimate the error that results from replacing e^x by $1 + x + (x^2/2)$ when $-0.1 < x < 0$. Compare your estimate with the one you obtained in part (a).

40. *Cubic approximation for* $\sinh x$ Estimate the error in the approximation $\sinh x = x + (x^3/3!)$ when $|x| < 0.5$. (*Hint*: Use R_4, not R_3.)

41. *Linear approximation for* e^h When $0 \le h \le 0.01$, show that e^h may be replaced by $1 + h$ with an error of magnitude no greater than 0.6% of h. Use $e^{0.01} = 1.01$.

42. *Approximating* $\ln(1 + x)$ *by* x For what positive values of x can you replace $\ln(1 + x)$ by x with an error of magnitude no greater than 1% of the value of x?

43. *Estimating $\pi/4$* You plan to estimate $\pi/4$ by evaluating the Maclaurin series for $\tan^{-1} x$ at $x = 1$. Use the Alternating Series Estimation Theorem to determine how many terms of the series you would have to add to be sure the estimate is good to 2 decimal places.

44. *Bounding* $y = (\sin x)/x$

(a) Use the Maclaurin series for $\sin x$ and the Alternating Series Estimation Theorem to show that

$$1 - \frac{x^2}{6} < \frac{\sin x}{x} < 1, \qquad x \neq 0.$$

T (b) *Writing to Learn* Graph $f(x) = (\sin x)/x$ together with the functions $y = 1 - (x^2/6)$ and $y = 1$ for $-5 \leq x \leq 5$. Comment on the relationships among the graphs.

Quadratic Approximations

)-ROM
WEBsite The Taylor polynomial of order 2 generated by a twice-differentiable function $f(x)$ at $x = a$ is called the **quadratic approximation** of f at $x = a$. In Exercises 45–48, find the

(a) linearization (Taylor polynomial of order 1) at $x = 0$

(b) quadratic approximation of f at $x = 0$.

45. $f(x) = \ln (\cos x)$ **46.** $f(x) = e^{\sin x}$

47. $f(x) = 1/\sqrt{1 - x^2}$ **48.** $f(x) = \cosh x$

Theory and Examples

49. *Taylor's Theorem and the Mean Value Theorem* Explain how the Mean Value Theorem (Section 3.2, Theorem 4) is a special case of Taylor's Theorem.

50. *Linearizations at inflection points* (*Continuation of Section 3.6, Exercise 49*) Show that if the graph of a twice-differentiable function $f(x)$ has an inflection point at $x = a$, then the linearization of f at $x = a$ is also the quadratic approximation of f at $x = a$. This explains why tangent lines fit so well at inflection points.

51. *The (Second) Second Derivative Test* Use the equation

$$f(x) = f(a) + f'(a)(x - a) + \frac{f''(c_2)}{2}(x - a)^2$$

to establish the following test.

    Let f have continuous first and second derivatives and suppose that $f'(a) = 0$. Then

(a) f has a local maximum at a if $f'' \leq 0$ throughout an interval whose interior contains a

(b) f has a local minimum at a if $f'' \geq 0$ throughout an interval whose interior contains a.

52. *A cubic approximation* Use Taylor's formula with $a = 0$ and $n = 3$ to find the standard cubic approximation of $f(x) = 1/(1 - x)$ at $x = 0$. Give an upper bound for the magnitude of the error in the approximation when $|x| \leq 0.1$.

53. *Improving approximations to π*

(a) Let P be an approximation of π accurate to n decimals. Show that $P + \sin P$ gives an approximation correct to $3n$ decimals. (*Hint*: Let $P = \pi + x$.)

(b) Try it with a calculator.

54. *The Maclaurin series generated by* $f(x) = \sum_{n=0}^{\infty} a_n x^n$ *is* $\sum_{n=0}^{\infty} a_n x^n$ A function defined by a power series $\sum_{n=0}^{\infty} a_n x^n$ with a radius of convergence $c > 0$ has a Maclaurin series that converges to the function at every point of $(-c, c)$. Show this by showing that the Maclaurin series generated by $f(x) = \sum_{n=0}^{\infty} a_n x^n$ is the series $\sum_{n=0}^{\infty} a_n x^n$ itself.

    An immediate consequence of this is that series like

$$x \sin x = x^2 - \frac{x^4}{3!} + \frac{x^6}{5!} - \frac{x^8}{7!} + \cdots$$

and

$$x^2 e^x = x^2 + x^3 + \frac{x^4}{2!} + \frac{x^5}{3!} + \cdots,$$

obtained by multiplying Maclaurin series by powers of x, as well as series obtained by integration and differentiation of convergent power series are themselves the Maclaurin series generated by the functions they represent.

55. *Maclaurin series for even functions and odd functions* Suppose that $f(x) = \sum_{n=0}^{\infty} a_n x^n$ converges for all x in an open interval $(-c, c)$.

(a) Show that if f is even, then $a_1 = a_3 = a_5 = \cdots = 0$; that is, the series for f contains only even powers of x.

(b) Show that if f is odd, then $a_0 = a_2 = a_4 = \cdots = 0$; that is, the series for f contains only odd powers of x.

56. *Taylor polynomials of periodic functions*

(a) Show that every continuous periodic function $f(x)$, $-\infty < x < \infty$, is bounded in magnitude by showing that there exists a positive constant M such that $|f(x)| \leq M$ for all x.

(b) Show that the graph of every Taylor polynomial of positive degree generated by $f(x) = \cos x$ must eventually move away from the graph of $\cos x$ as $|x|$ increases. You can see this in Figure 8.19. The Taylor polynomials of $\sin x$ behave in a similar way (Figure 8.21).

T **57.** (a) *Two graphs* Graph the curves $y = (1/3) - (x^2)/5$ and $y = (x - \tan^{-1} x)/x^3$ together with the line $y = 1/3$.

(b) Use a Maclaurin series to explain what you see. What is

$$\lim_{x \to 0} \frac{x - \tan^{-1} x}{x^3}?$$

58. *Of all polynomials of degree $\leq n$, the Taylor polynomial of n gives the best approximation* Suppose that $f(x)$ is differentiable on an interval centered at $x = a$ and that $g(x) = b_0 + b_1(x - a) + \cdots + b_n(x - a)^n$ is a polynomial of degree n with constant coefficients $b_0, \ldots, b_n$. Let $E(x) = f(x) - g(x)$. Show that if we impose on g the conditions

(a) $E(a) = 0$ The approximation error is zero at $x = a$.

(b) $\lim_{x \to a} \dfrac{E(x)}{(x-a)^n} = 0,$ The error is negligible when compared to $(x-a)^n$.

then

$$g(x) = f(a) + f'(a)(x-a) + \frac{f''(a)}{2!}(x-a)^2 + \cdots + \frac{f^{(n)}(a)}{n!}(x-a)^n.$$

Thus, the Taylor polynomial $P_n(x)$ is the only polynomial of degree less than or equal to n whose error is both zero at $x = a$ and negligible when compared with $(x-a)^n$.

COMPUTER EXPLORATIONS

Linear, Quadratic, and Cubic Approximations

Taylor's formula with $n = 1$ and $a = 0$ gives the linearization of a function at $x = 0$. With $n = 2$ and $n = 3$, we obtain the standard quadratic and cubic approximations. In these exercises, we explore the errors associated with these approximations. We seek answers to two questions:

(a) For what values of x can the function be replaced by each approximation with an error less than 10^{-2}?

(b) What is the maximum error we could expect if we replace the function by each approximation over the specified interval?

Using a CAS, perform the following steps to aid in answering questions (a) and (b) for the functions and intervals in Exercises 59–64.

Step 1: Plot the function over the specified interval.

Step 2: Find the Taylor polynomials $P_1(x)$, $P_2(x)$, and $P_3(x)$ at $x = 0$.

Step 3: Calculate the $(n + 1)$st derivative $f^{(n+1)}(c)$ associated with the remainder term for each Taylor polynomial. Plot the derivative as a function of c over the specified interval and estimate its maximum absolute value, M.

Step 4: Calculate the remainder $R_n(x)$ for each polynomial. Using the estimate M from step 3 in place of $f^{(n+1)}(c)$, plot $R_n(x)$ over the specified interval. Then estimate the values of x that answer question (a).

Step 5: Compare your estimated error with the actual error $E_n(x) = |f(x) - P_n(x)|$ by plotting $E_n(x)$ over the specified interval. This will help answer question (b).

Step 6: Graph the function and its three Taylor approximations together. Discuss the graphs in relation to the information discovered in steps 4 and 5.

59. $f(x) = \dfrac{1}{\sqrt{1+x}}$, $|x| \le \dfrac{3}{4}$

60. $f(x) = (1+x)^{3/2}$, $-\dfrac{1}{2} \le x \le 2$

61. $f(x) = \dfrac{x}{x^2 + 1}$, $|x| \le 2$

62. $f(x) = (\cos x)(\sin 2x)$, $|x| \le 2$

63. $f(x) = e^{-x} \cos 2x$, $|x| \le 1$

64. $f(x) = e^{x/3} \sin 2x$, $|x| \le 2$

8.8 Applications of Power Series

Binomial Series for Powers and Roots • Series Solutions of Differential Equations • Evaluating Indeterminate Forms • Arctangents

This section shows how power series are used by scientists and engineers in a variety of applications.

Binomial Series for Powers and Roots

The Maclaurin series generated by $f(x) = (1+x)^m$, when m is constant, is

$$1 + mx + \frac{m(m-1)}{2!}x^2 + \frac{m(m-1)(m-2)}{3!}x^3 + \cdots$$

$$+ \frac{m(m-1)(m-2)\cdots(m-k+1)}{k!}x^k + \cdots.$$

This series, called the **binomial series,** converges absolutely for $|x| < 1$. To derive the series, we first list the function and its derivatives:

$$f(x) = (1 + x)^m$$
$$f'(x) = m(1 + x)^{m-1}$$
$$f''(x) = m(m - 1)(1 + x)^{m-2}$$
$$f'''(x) = m(m - 1)(m - 2)(1 + x)^{m-3}$$

$$\vdots$$

$$f^{(k)}(x) = m(m - 1)(m - 2) \cdots (m - k + 1)(1 + x)^{m-k}.$$

We then evaluate these at $x = 0$ and substitute into the Maclaurin series formula to obtain the binomial series.

If m is an integer greater than or equal to zero, the series stops after $(m + 1)$ terms because the coefficients from $k = m + 1$ on are zero.

If m is not a positive integer or zero, the series is infinite and converges for $|x| < 1$. To see why, let u_k be the term involving x^k. Then apply the Ratio Test for absolute convergence to see that

$$\left| \frac{u_{k+1}}{u_k} \right| = \left| \frac{m - k}{k + 1} x \right| \to |x| \qquad \text{as } k \to \infty.$$

Our derivation of the binomial series shows only that it is generated by $(1 + x)^m$ and converges for $|x| < 1$. The derivation does not show that the series converges to $(1 + x)^m$. It does, but we assume that part without proof.

Binomial Series

For $-1 < x < 1$,

$$(1 + x)^m = 1 + \sum_{k=1}^{\infty} \binom{m}{k} x^k,$$

where we define

$$\binom{m}{1} = m, \qquad \binom{m}{2} = \frac{m(m - 1)}{2!},$$

and

$$\binom{m}{k} = \frac{m(m - 1)(m - 2) \cdots (m - k + 1)}{k!} \qquad \text{for } k \geq 3.$$

Example 1   Using the Binomial Series

If $m = -1$,

$$\binom{-1}{1} = -1, \qquad \binom{-1}{2} = \frac{-1(-2)}{2!} = 1,$$

and

$$\binom{-1}{k} = \frac{-1(-2)(-3) \cdots (-1 - k + 1)}{k!} = (-1)^k \left(\frac{k!}{k!} \right) = (-1)^k.$$

With these coefficient values, the binomial series formula gives the familiar geometric series

$$(1 + x)^{-1} = 1 + \sum_{k=1}^{\infty} (-1)^k x^k = 1 - x + x^2 - x^3 + \cdots + (-1)^k x^k + \cdots .$$

Example 2   Using the Binomial Series

We know from Section 3.6, Example 1, that $\sqrt{1 + x} \approx 1 + (x/2)$ for $|x|$ small. With $m = 1/2$, the binomial series gives quadratic and higher-order approximations as well, along with error estimates that come from the Alternating Series Estimation Theorem:

$$(1 + x)^{1/2} = 1 + \frac{x}{2} + \frac{\left(\dfrac{1}{2}\right)\left(-\dfrac{1}{2}\right)}{2!} x^2 + \frac{\left(\dfrac{1}{2}\right)\left(-\dfrac{1}{2}\right)\left(-\dfrac{3}{2}\right)}{3!} x^3$$

$$+ \frac{\left(\dfrac{1}{2}\right)\left(-\dfrac{1}{2}\right)\left(-\dfrac{3}{2}\right)\left(-\dfrac{5}{2}\right)}{4!} x^4 + \cdots$$

$$= 1 + \frac{x}{2} - \frac{x^2}{8} + \frac{x^3}{16} - \frac{5x^4}{128} + \cdots .$$

Substitution for x gives still other approximations. For example,

$$\sqrt{1 - x^2} \approx 1 - \frac{x^2}{2} - \frac{x^4}{8} \qquad \text{for } |x^2| \text{ small}$$

$$\sqrt{1 - \frac{1}{x}} \approx 1 - \frac{1}{2x} - \frac{1}{8x^2} \qquad \text{for } \left|\frac{1}{x}\right| \text{ small, that is, } |x| \text{ large.}$$

Series Solutions of Differential Equations

When we cannot find a relatively simple expression for the solution of an initial value problem or differential equation, we try to get information about the solution in other ways. One way is to try to find a power series representation for the solution. If we can do so, we immediately have a source of polynomial approximations of the solution, which may be all that we really need. The first example (Example 3) deals with a first-order linear differential equation that could be solved as a linear equation using the method studied earlier. The example shows how, not knowing this, we can solve the equation with power series. The second example (Example 4) deals with an equation that cannot be solved by previous methods.

Example 3   Series Solution of an Initial Value Problem

Solve the initial value problem

$$y' - y = x, \qquad y(0) = 1.$$

Solution   We assume that there is a solution of the form

$$y = a_0 + a_1 x + a_2 x^2 + \cdots + a_{n-1} x^{n-1} + a_n x^n + \cdots . \tag{1}$$

Our goal is to find values for the coefficients a_k that make the series and its first derivative

$$y' = a_1 + 2a_2x + 3a_3x^2 + \cdots + na_nx^{n-1} + \cdots \tag{2}$$

satisfy the given differential equation and initial condition. The series $y' - y$ is the difference of the series in Equations (1) and (2):

$$y' - y = (a_1 + a_0) + (2a_2 - a_1)x + (3a_3 - a_2)x^2 + \cdots$$
$$+ (na_n - a_{n-1})x^{n-1} + \cdots. \tag{3}$$

If y is to satisfy the equation $y' - y = x$, the series in Equation (3) must equal x. Since power series representations are unique, as you saw if you did Exercise 45 in Section 8.6, the coefficients in Equation (3) must satisfy the equations

$$a_1 - a_0 = 0 \qquad \text{Constant terms}$$
$$2a_2 - a_1 = 1 \qquad \text{Coefficients of } x$$
$$3a_3 - a_2 = 0 \qquad \text{Coefficients of } x^2$$
$$\vdots \qquad\qquad \vdots$$
$$na_n - a_{n-1} = 0 \qquad \text{Coefficients of } x^{n-1}$$
$$\vdots \qquad\qquad \vdots$$

CD-ROM
WEBsite

Historical Biography

John Van Neumann
(1903 — 1957)

We can also see from Equation (1) that $y = a_0$ when $x = 0$, so that $a_0 = 1$ (this being the initial condition). Putting it all together, we have

$$a_0 = 1, \qquad a_1 = a_0 = 1, \qquad a_2 = \frac{1 + a_1}{2} = \frac{1 + 1}{2} = \frac{2}{2},$$
$$a_3 = \frac{a_2}{3} = \frac{2}{3 \cdot 2} = \frac{2}{3!}, \qquad \cdots, \qquad a_n = \frac{a_{n-1}}{n} = \frac{2}{n!}, \cdots.$$

Substituting these coefficient values into the equation for y (Equation (1)) gives

$$y = 1 + x + 2 \cdot \frac{x^2}{2!} + 2 \cdot \frac{x^3}{3!} + \cdots + 2 \cdot \frac{x^n}{n!} + \cdots$$

$$= 1 + x + 2 \underbrace{\left(\frac{x^2}{2!} + \frac{x^3}{3!} + \cdots + \frac{x^n}{n!} + \cdots \right)}_{\text{The Maclaurin series for } e^x - 1 - x}$$

$$= 1 + x + 2(e^x - 1 - x) = 2e^x - 1 - x.$$

The solution of the initial value problem is $y = 2e^x - 1 - x$.
As a check, we see that

$$y(0) = 2e^0 - 1 - 0 = 2 - 1 = 1$$

and

$$y' - y = (2e^x - 1) - (2e^x - 1 - x) = x.$$

Example 4 Solving a Differential Equation

Find a power series solution for

$$y'' + x^2 y = 0. \tag{4}$$

Solution We assume that there is a solution of the form

$$y = a_0 + a_1 x + a_2 x^2 + \cdots + a_n x^n + \cdots \tag{5}$$

and find what the coefficients a_k have to be to make the series and its second derivative

$$y'' = 2a_2 + 3 \cdot 2a_3 x + \cdots + n(n-1)a_n x^{n-2} + \cdots \tag{6}$$

satisfy Equation (4). The series for $x^2 y$ is x^2 times the right-hand side of Equation (5):

$$x^2 y = a_0 x^2 + a_1 x^3 + a_2 x^4 + \cdots + a_n x^{n+2} + \cdots . \tag{7}$$

The series for $y'' + x^2 y$ is the sum of the series in Equations (6) and (7):

$$y'' + x^2 y = 2a_2 + 6a_3 x + (12a_4 + a_0)x^2 + (20a_5 + a_1)x^3$$
$$+ \cdots + (n(n-1)a_n + a_{n-4})x^{n-2} + \cdots . \tag{8}$$

Notice that the coefficient of x^{n-2} in Equation (7) is a_{n-4}. If y and its second derivative y'' are to satisfy Equation (4), the coefficients of the individual powers of x on the right-hand side of Equation (8) must all be zero:

$$2a_2 = 0, \quad 6a_3 = 0, \quad 12a_4 + a_0 = 0, \quad 20a_5 + a_1 = 0, \tag{9}$$

and for all $n \geq 4$,

$$n(n-1)a_n + a_{n-4} = 0. \tag{10}$$

We can see from Equation (5) that

$$a_0 = y(0), \qquad a_1 = y'(0).$$

In other words, the first two coefficients of the series are the values of y and y' at $x = 0$. The equations in Equation (9) and the recursion formula in Equation (10) enable us to evaluate all the other coefficients in terms of a_0 and a_1.

The first two of Equations (9) give

$$a_2 = 0, \qquad a_3 = 0.$$

Equation (10) shows that if $a_{n-4} = 0$, then $a_n = 0$, so we conclude that

$$a_6 = 0, \qquad a_7 = 0, \qquad a_{10} = 0, \qquad a_{11} = 0,$$

and whenever $n = 4k + 2$ or $4k + 3$, a_n is zero. For the other coefficients, we have

$$a_n = \frac{-a_{n-4}}{n(n-1)}$$

so that

$$a_4 = \frac{-a_0}{4 \cdot 3}, \qquad a_8 = \frac{-a_4}{8 \cdot 7} = \frac{a_0}{3 \cdot 4 \cdot 7 \cdot 8},$$

$$a_{12} = \frac{-a_8}{11 \cdot 12} = \frac{-a_0}{3 \cdot 4 \cdot 7 \cdot 8 \cdot 11 \cdot 12}$$

and

$$a_5 = \frac{-a_1}{5 \cdot 4}, \qquad a_9 = \frac{-a_5}{9 \cdot 8} = \frac{a_1}{4 \cdot 5 \cdot 8 \cdot 9},$$

$$a_{13} = \frac{-a_9}{12 \cdot 13} = \frac{-a_1}{4 \cdot 5 \cdot 8 \cdot 9 \cdot 12 \cdot 13}.$$

The answer is best expressed as the sum of two separate series, one multiplied by a_0, the other by a_1:

$$y = a_0 \left(1 - \frac{x^4}{3 \cdot 4} + \frac{x^8}{3 \cdot 4 \cdot 7 \cdot 8} - \frac{x^{12}}{3 \cdot 4 \cdot 7 \cdot 8 \cdot 11 \cdot 12} + \cdots \right)$$

$$+ a_1 \left(x - \frac{x^5}{4 \cdot 5} + \frac{x^9}{4 \cdot 5 \cdot 8 \cdot 9} - \frac{x^{13}}{4 \cdot 5 \cdot 8 \cdot 9 \cdot 12 \cdot 13} + \cdots \right).$$

Both series converge absolutely for all x, as is readily seen by the Ratio Test.

Evaluating Indeterminate Forms

We can sometimes evaluate indeterminate forms by expressing the functions involved as Taylor series.

Example 5 Limits Using Power Series

Evaluate

$$\lim_{x \to 0} \frac{\sin x - \tan x}{x^3}.$$

Solution The Maclaurin series for $\sin x$ and $\tan x$, to terms in x^5, are

$$\sin x = x - \frac{x^3}{3!} + \frac{x^5}{5!} - \cdots, \qquad \tan x = x + \frac{x^3}{3} + \frac{2x^5}{15} + \cdots.$$

Hence,

$$\sin x - \tan x = -\frac{x^3}{2} - \frac{x^5}{8} - \cdots = x^3 \left(-\frac{1}{2} - \frac{x^2}{8} - \cdots \right)$$

and

$$\lim_{x \to 0} \frac{\sin x - \tan x}{x^3} = \lim_{x \to 0} \left(-\frac{1}{2} - \frac{x^2}{8} - \cdots \right)$$

$$= -\frac{1}{2}.$$

If we apply series to calculate $\lim_{n \to 0} ((1/\sin x) - (1/x))$, we not only find the limit successfully but also discover an approximation formula for $\csc x$.

Example 6 Limits Using Power Series

Find

$$\lim_{x \to 0} \left(\frac{1}{\sin x} - \frac{1}{x} \right).$$

Solution

$$\frac{1}{\sin x} - \frac{1}{x} = \frac{x - \sin x}{x \sin x} = \frac{x - \left(x - \dfrac{x^3}{3!} + \dfrac{x^5}{5!} - \cdots\right)}{x \cdot \left(x - \dfrac{x^3}{3!} + \dfrac{x^5}{5!} - \cdots\right)}$$

$$= \frac{x^3\left(\dfrac{1}{3!} - \dfrac{x^2}{5!} + \cdots\right)}{x^2\left(1 - \dfrac{x^2}{3!} + \cdots\right)} = x\,\frac{\dfrac{1}{3!} - \dfrac{x^2}{5!} + \cdots}{1 - \dfrac{x^2}{3!} + \cdots}.$$

Therefore,

$$\lim_{x \to 0}\left(\frac{1}{\sin x} - \frac{1}{x}\right) = \lim_{x \to 0}\left(x\,\frac{\dfrac{1}{3!} - \dfrac{x^2}{5!} + \cdots}{1 - \dfrac{x^2}{3!} + \cdots}\right) = 0.$$

From the quotient on the right, we can see that if $|x|$ is small, then

$$\frac{1}{\sin x} - \frac{1}{x} \approx x \cdot \frac{1}{3!} = \frac{x}{6} \qquad \text{or} \qquad \csc x \approx \frac{1}{x} + \frac{x}{6}.$$

Arctangents

In Section 8.6, Example 5, we found a series for $\tan^{-1} x$ by differentiating to get

$$\frac{d}{dx} \tan^{-1} x = \frac{1}{1 + x^2} = 1 - x^2 + x^4 - x^6 + \cdots$$

and integrating to get

$$\tan^{-1} x = x - \frac{x^3}{3} + \frac{x^5}{5} - \frac{x^7}{7} + \cdots.$$

We did not, however, prove the term-by-term integration theorem on which this conclusion depended. We now derive the series again by integrating both sides of the finite formula

$$\frac{1}{1 + t^2} = 1 - t^2 + t^4 - t^6 + \cdots + (-1)^n t^{2n} + \frac{(-1)^{n+1} t^{2n+2}}{1 + t^2},$$

in which the last term comes from adding the remaining terms as a geometric series with first term $a = (-1)^{n+1} t^{2n+2}$ and ratio $r = -t^2$. Integrating both sides of the last equation from $t = 0$ to $t = x$ gives

$$\tan^{-1} x = x - \frac{x^3}{3} + \frac{x^5}{5} - \frac{x^7}{7} + \cdots + (-1)^n \frac{x^{2n+1}}{2n + 1} + R(n, x),$$

where

$$R(n, x) = \int_0^x \frac{(-1)^{n+1} t^{2n+2}}{1 + t^2}\, dt.$$

The denominator of the integrand is greater than or equal to 1; hence,

$$|R(n, x)| \le \int_0^{|x|} t^{2n+2} \, dt = \frac{|x|^{2n+3}}{2n+3}.$$

If $|x| \le 1$, the right side of this inequality approaches zero as $n \to \infty$. Therefore, $\lim_{n \to \infty} R(n, x) = 0$ if $|x| \le 1$ and

$$\tan^{-1} x = \sum_{n=0}^{\infty} \frac{(-1)^n x^{2n+1}}{2n+1}, \qquad |x| \le 1.$$

We take this route instead of finding the Maclaurin series directly because the formulas for the higher-order derivatives of $\tan^{-1} x$ are unmanageable.

When we put $x = 1$ in the series for $\tan^{-1} x$ we get **Leibniz's formula:**

$$\frac{\pi}{4} = 1 - \frac{1}{3} + \frac{1}{5} - \frac{1}{7} + \frac{1}{9} - \cdots + \frac{(-1)^n}{2n+1} + \cdots.$$

This series converges too slowly to be a useful source of decimal approximations of π. It is better to use a formula like

$$\pi = 48 \, \tan^{-1} \frac{1}{18} + 32 \, \tan^{-1} \frac{1}{57} - 20 \, \tan^{-1} \frac{1}{239},$$

which uses values of x closer to zero.

EXERCISES 8.8

Binomial Series

Find the first four terms of the binomial series for the functions in Exercises 1–10.

1. $(1 + x)^{1/2}$

2. $(1 + x)^{1/3}$

3. $(1 - x)^{-1/2}$

4. $(1 - 2x)^{1/2}$

5. $\left(1 + \frac{x}{2}\right)^{-2}$

6. $\left(1 - \frac{x}{2}\right)^{-2}$

7. $(1 + x^3)^{-1/2}$

8. $(1 + x^2)^{-1/3}$

9. $\left(1 + \frac{1}{x}\right)^{1/2}$

10. $\left(1 - \frac{2}{x}\right)^{1/3}$

Find the binomial series for the functions in Exercises 11–14.

11. $(1 + x)^4$

12. $(1 + x^2)^3$

13. $(1 - 2x)^3$

14. $\left(1 - \frac{x}{2}\right)^4$

Initial Value Problems

Find series solutions for the initial value problems in Exercises 15–32.

15. $y' + y = 0, \quad y(0) = 1$

16. $y' - 2y = 0, \quad y(0) = 1$

17. $y' - y = 1, \quad y(0) = 0$

18. $y' + y = 1, \quad y(0) = 2$

19. $y' - y = x, \quad y(0) = 0$

20. $y' + y = 2x, \quad y(0) = -1$

21. $y' - xy = 0, \quad y(0) = 1$

22. $y' - x^2 y = 0, \quad y(0) = 1$

23. $(1 - x)y' - y = 0, \quad y(0) = 2$

24. $(1 + x^2)y' + 2xy = 0, \quad y(0) = 3$

25. $y'' - y = 0, \quad y'(0) = 1$ and $y(0) = 0$

26. $y'' + y = 0, \quad y'(0) = 0$ and $y(0) = 1$

27. $y'' + y = x, \quad y'(0) = 1$ and $y(0) = 2$

28. $y'' - y = x, \quad y'(0) = 2$ and $y(0) = -1$

29. $y'' - y = -x, \quad y'(2) = -2$ and $y(2) = 0$

30. $y'' - x^2 y = 0, \quad y'(0) = b$ and $y(0) = a$

31. $y'' + x^2 y = x, \quad y'(0) = b$ and $y(0) = a$

32. $y'' - 2y' + y = 0, \quad y'(0) = 1$ and $y(0) = 0$

Approximating Integral Functions by Polynomials

In Exercises 33–36, find a polynomial that will approximate $F(x)$ throughout the given interval with an error of magnitude less than 10^{-3}.

33. $F(x) = \int_0^x \sin t^2 \, dt, \quad [0, 1]$

34. $F(x) = \int_0^x t^2 e^{-t^2} \, dt, \quad [0, 1]$

35. $F(x) = \int_0^x \tan^{-1} t \, dt, \quad$ (a) $[0, 0.5]$ (b) $[0, 1]$

36. $F(x) = \int_0^x \frac{\ln(1 + t)}{t} \, dt, \quad$ (a) $[0, 0.5]$ (b) $[0, 1]$

Indeterminate Forms

Use series to evaluate the limits in Exercises 37–42.

37. $\lim\limits_{x\to 0} \dfrac{e^x - (1 + x)}{x^2}$

38. $\lim\limits_{t\to 0} \dfrac{1 - \cos t - (t^2/2)}{t^4}$

39. $\lim\limits_{x\to\infty} x^2(e^{-1/x^2} - 1)$

40. $\lim\limits_{y\to 0} \dfrac{\tan^{-1} y - \sin y}{y^3 \cos y}$

41. $\lim\limits_{x\to 0} \dfrac{\ln (1 + x^2)}{1 - \cos x}$

42. $\lim\limits_{x\to\infty} (x + 1) \sin \dfrac{1}{x + 1}$

Theory and Examples

43. *Series for* $\ln (1 - x)$, $|x| < 1$ Replace x by $-x$ in the Maclaurin series for $\ln (1 + x)$ to obtain a series for $\ln (1 - x)$. Then subtract this from the Maclaurin series for $\ln (1 + x)$ to show that for $|x| < 1$,

$$\ln \frac{1 + x}{1 - x} = 2\left(x + \frac{x^3}{3} + \frac{x^5}{5} + \cdots\right).$$

44. *Writing to Learn* How many terms of the Maclaurin series for $\ln (1 + x)$ should you add to be sure of calculating $\ln (1.1)$ with an error of magnitude less than 10^{-8}? Give reasons for your answer.

45. *Writing to Learn* According to the Alternating Series Estimation Theorem, how many terms of the Maclaurin series for $\tan^{-1} 1$ would you have to add to be sure of finding $\pi/4$ with an error of magnitude less than 10^{-3}? Give reasons for your answer.

46. *Maclaurin series for* $\tan^{-1} x$ Show that the Maclaurin series for $f(x) = \tan^{-1} x$ diverges for $|x| > 1$.

47. *Taylor polynomial for* $\sin^{-1} x$

(a) Use the binomial series and that

$$\frac{d}{dx} \sin^{-1} x = (1 - x^2)^{-1/2}$$

to generate the first four nonzero terms of the Maclaurin series for $\sin^{-1} x$. What is the radius of convergence?

(b) *Taylor polynomial for* $\cos^{-1} x$ Use your result in part (a) to find the first five nonzero terms of the Maclaurin series for $\cos^{-1} x$.

48. *Maclaurin series for* $\sin^{-1} x$ Integrate the binomial series for the function $(1 - x^2)^{-1/2}$ to show that for $|x| < 1$,

$$\sin^{-1} x = x + \sum_{n=1}^{\infty} \frac{1 \cdot 3 \cdot 5 \cdot \cdots \cdot (2n - 1)}{2 \cdot 4 \cdot 6 \cdot \cdots \cdot (2n)} \frac{x^{2n+1}}{2n + 1}.$$

49. *Series for* $\tan^{-1} x$ *for* $|x| > 1$ Derive the series

$$\tan^{-1} x = \frac{\pi}{2} - \frac{1}{x} + \frac{1}{3x^3} - \frac{1}{5x^5} + \cdots, \quad x > 1$$

$$\tan^{-1} x = -\frac{\pi}{2} - \frac{1}{x} + \frac{1}{3x^3} - \frac{1}{5x^5} + \cdots, \quad x < -1$$

by integrating the series

$$\frac{1}{1 + t^2} = \frac{1}{t^2} \cdot \frac{1}{1 + (1/t^2)} = \frac{1}{t^2} - \frac{1}{t^4} + \frac{1}{t^6} - \frac{1}{t^8} + \cdots$$

in the first case from x to ∞ and in the second case from $-\infty$ to x.

50. *The value of* $\sum_{n=0}^{\infty} \tan^{-1}(2/n^2)$

(a) Use the formula for the tangent of the difference of two angles to show that

$$\tan (\tan^{-1} (n + 1) - \tan^{-1} (n - 1)) = \frac{2}{n^2}$$

and hence that

$$\tan^{-1} \frac{2}{n^2} = \tan^{-1} (n + 1) - \tan^{-1} (n - 1).$$

(b) Show that

$$\sum_{n=1}^{N} \tan^{-1} \frac{2}{n^2} = \tan^{-1} (N + 1) + \tan^{-1} N - \frac{\pi}{4}.$$

(c) Find the value of $\sum_{n=1}^{\infty} \tan^{-1} (2/n^2)$.

8.9   Fourier Series

Coefficients in the Fourier Series Expansion • Convergence of the Fourier Series • Periodic Extension

CD-ROM
WEBsite
Historical Biography
Jean-Baptiste
Joseph Fourier
(1766 — 1830)

When investigating the problem of heat conduction in a long thin insulated rod, French mathematician Jean-Baptiste Joseph Fourier needed to express a function $f(x)$ as a trigonometric series. Generally, if $f(x)$ is defined on the interval $-L < x < L$, we need to know the coefficients a_0, a_n, and b_n ($n \geq 1$) for which

$$f(x) = \frac{a_0}{2} + \sum_{n=1}^{\infty}\left(a_n \cos \frac{n\pi x}{L} + b_n \sin \frac{n\pi x}{L}\right). \tag{1}$$

Notice that the interval $-L < x < L$ is *symmetric* about the origin. Equation (1) is called a **Fourier series** for f on the interval $(-L, L)$. These series have a wide range of science and engineering applications in the study of heat conduction, wave phenomena, concentrations of chemicals and pollutants, and other models of the physical world. In this section, we introduce these important trigonometric series representations of a given function f.

Coefficients in the Fourier Series Expansion

Suppose that f is a function defined over the *symmetric* interval $-L < x < L$. Assume that f is expressible as the trigonometric series given by Equation (1). We want to find a way to calculate the coefficients $a_0, a_1, a_2, \ldots, b_1, b_2, \ldots$. The key to the calculations is the definite integral, based on the results in Table 8.3.

Table 8.3 Trigonometric Integrals

If m and n are positive integers, then

1. $\displaystyle\int_{-L}^{L} \cos \frac{n\pi x}{L}\, dx = 0$

2. $\displaystyle\int_{-L}^{L} \sin \frac{n\pi x}{L}\, dx = 0$

3. $\displaystyle\int_{-L}^{L} \cos \frac{n\pi x}{L} \cos \frac{m\pi x}{L}\, dx = \begin{cases} 0, & m \neq n, \\ L, & m = n \end{cases}$

4. $\displaystyle\int_{-L}^{L} \sin \frac{n\pi x}{L} \cos \frac{m\pi x}{L}\, dx = 0$

5. $\displaystyle\int_{-L}^{L} \sin \frac{n\pi x}{L} \sin \frac{m\pi x}{L}\, dx = \begin{cases} 0, & m \neq n, \\ L, & m = n. \end{cases}$

(We ask you to evaluate these trigonometric integrals in Exercises 17 through 21.)

Calculation of a_0 We integrate both sides of Equation (1) from $-L$ to L and assume that the operations for integration and summation can be interchanged to obtain

$$\int_{-L}^{L} f(x)\, dx = \frac{a_0}{2} \int_{-L}^{L} dx + \sum_{n=1}^{\infty} a_n \int_{-L}^{L} \cos \frac{n\pi x}{L}\, dx$$

$$+ \sum_{n=1}^{\infty} b_n \int_{-L}^{L} \sin \frac{n\pi x}{L}\, dx. \tag{2}$$

For every positive integer n, the last two integrals on the right-hand side of Equation (2) are zero (Formulas 1 and 2 in Table 8.3). Therefore,

$$\int_{-L}^{L} f(x)\, dx = \frac{a_0}{2} \int_{-L}^{L} dx = \frac{a_0 x}{2}\Big]_{-L}^{L} = La_0.$$

Solving for a_0 yields

$$a_0 = \frac{1}{L} \int_{-L}^{L} f(x)\, dx. \tag{3}$$

The term $a_0/2$ in Equation (1) keeps consistency with the formulas calculating the Fourier coefficients.

Calculation of a_m We multiply both sides of Equation (1) by $\cos(m\pi x/L)$, $m > 0$, and integrate the result from $-L$ to L:

$$\int_{-L}^{L} f(x) \cos \frac{m\pi x}{L} dx = \frac{a_0}{2} \int_{-L}^{L} \cos \frac{m\pi x}{L} dx$$

$$+ \sum_{n=1}^{\infty} a_n \int_{-L}^{L} \cos \frac{n\pi x}{L} \cos \frac{m\pi x}{L} dx \qquad (4)$$

$$+ \sum_{n=1}^{\infty} b_n \int_{-L}^{L} \sin \frac{n\pi x}{L} \cos \frac{m\pi x}{L} dx.$$

The first integral on the right-hand side of Equation (4) is zero (Formula 1 in Table 8.3). Formulas 3 and 4 in Table 8.3, further reduce the equation to

$$\int_{-L}^{L} f(x) \cos \frac{m\pi x}{L} dx = a_m \int_{-L}^{L} \cos \frac{m\pi x}{L} \cos \frac{m\pi x}{L} dx = La_m.$$

Therefore,

$$a_m = \frac{1}{L} \int_{-L}^{L} f(x) \cos \frac{m\pi x}{L} dx. \qquad (5)$$

Calculation of b_m We multiply both sides of Equation (1) by $\sin(m\pi x/L)$, $m > 0$, and integrate the result from $-L$ to L:

$$\int_{-L}^{L} f(x) \sin \frac{m\pi x}{L} dx = \frac{a_0}{2} \int_{-L}^{L} \sin \frac{m\pi x}{L} dx$$

$$+ \sum_{n=1}^{\infty} a_n \int_{-L}^{L} \cos \frac{n\pi x}{L} \sin \frac{m\pi x}{L} dx$$

$$+ \sum_{n=1}^{\infty} b_n \int_{-L}^{L} \sin \frac{n\pi x}{L} \sin \frac{m\pi x}{L} dx.$$

From Formulas 2, 4, and 5 in Table 8.3, we obtain

$$\int_{-L}^{L} f(x) \sin \frac{m\pi x}{L} dx = b_m \int_{-L}^{L} \sin \frac{m\pi x}{L} \sin \frac{m\pi x}{L} dx = Lb_m.$$

Therefore,

$$b_m = \frac{1}{L} \int_{-L}^{L} f(x) \sin \frac{m\pi x}{L} dx. \qquad (6)$$

The trigonometric series (1), whose coefficients a_0, a_n, b_n are determined by Equations (3), (5), and (6), respectively (with m replaced by n), is called the **Fourier series expansion** of the function f over the interval $-L < x < L$. The constants a_0, a_n, and b_n are the **Fourier coefficients** of f.

CD-ROM
WEBsite

Example 1   Finding a Fourier Series Expansion

Find the Fourier series expansion of the function

$$f(x) = \begin{cases} 1, & -\pi < x < 0, \\ x, & 0 < x < \pi, \end{cases}$$

(Figure 8.22).

FIGURE 8.22 The piecewise continuous function in Example 1.

Solution   Notice from Figure 8.22 that $L = \pi$. Thus, from Equation (3) we have

$$a_0 = \frac{1}{\pi} \int_{-\pi}^{\pi} f(x)\, dx$$

$$= \frac{1}{\pi} \int_{-\pi}^{0} dx + \frac{1}{\pi} \int_{0}^{\pi} x\, dx$$

$$= 1 + \frac{\pi}{2}.$$

To find a_n, we use Equation (5) with m replaced by n:

$$a_n = \frac{1}{\pi} \int_{-\pi}^{\pi} f(x)\, \cos\, nx\, dx$$

$$= \frac{1}{\pi} \int_{-\pi}^{0} \cos\, nx\, dx + \frac{1}{\pi} \int_{0}^{\pi} x \cos\, nx\, dx$$

$$= \frac{1}{n\pi} \sin\, nx \Big]_{-\pi}^{0} + \frac{1}{\pi} \left[\frac{x}{n} \sin\, nx \right]_{0}^{\pi} - \frac{1}{\pi n} \int_{0}^{\pi} \sin\, nx\, dx$$

$$= \frac{1}{\pi n^2} \cos\, nx \Big]_{0}^{\pi}$$

$$= \frac{1}{\pi n^2} (\cos\, n\pi - 1)$$

$$= \frac{(-1)^n - 1}{\pi n^2}. \qquad \cos\, n\pi = (-1)^n$$

In a similar manner, from Equation (6) with m replaced by n:

$$b_n = \frac{1}{\pi} \int_{-\pi}^{\pi} f(x)\, \sin\, nx\, dx$$

$$= \frac{1}{\pi} \int_{-\pi}^{0} \sin\, nx\, dx + \frac{1}{\pi} \int_{0}^{\pi} x \sin\, nx\, dx$$

$$= \frac{(-1)^n (1 - \pi) - 1}{n\pi}.$$

Therefore, the Fourier expansion is

$$f(x) = \frac{1}{2} + \frac{\pi}{4} + \sum_{n=1}^{\infty} \frac{(-1)^n - 1}{\pi n^2} \cos\, nx + \sum_{n=1}^{\infty} \frac{(-1)^n (1 - \pi) - 1}{\pi n} \sin\, nx.$$

A graph of the Fourier series approximations as n varies up to 1, 5, and 20 terms is given in Figure 8.23. Notice how the approximations get closer and closer to the graph of the function at all points of continuity as n increases. At the point $x = 0$, where f is discontinuous, the Fourier approximations approach the value

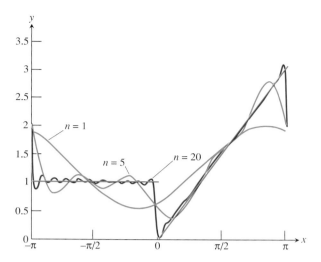

FIGURE 8.23 Fourier series approximations of the function in Example 1 as n varies up to 1, 5, and 20 terms in the infinite series. As n increases, the Fourier approximations approach the actual $f(x)$ values.

0.5 halfway between the jump. These results are consistent with the theorem on Fourier convergence stated below.

In calculating the coefficients a_0, a_n, and b_n, we assumed that f was integrable on the interval $(-L, L)$. We also assumed that the trigonometric series on the right side of Equation (1), as well as the series obtained when we multiply it by $\cos(m\pi x/L)$ or $\sin(m\pi x/L)$, converge in a way to permit term-by-term integration. These convergence issues are studied in advanced calculus, as is the question of when the Fourier series actually equals $f(x)$ for $-L < x < L$. Most of the functions you will encounter in applications guarantee both convergence of the series and its equality with f. We say a little more about this in a moment, but first we summarize our results.

Definition   Fourier Series

The **Fourier series** of a function $f(x)$ defined on the interval $-L < x < L$ is

$$f(x) = \frac{a_0}{2} + \sum_{n=1}^{\infty} \left[a_n \cos \frac{n\pi x}{L} + b_n \sin \frac{n\pi x}{L} \right], \tag{7}$$

where

$$a_0 = \frac{1}{L} \int_{-L}^{L} f(x)\, dx, \tag{8}$$

$$a_n = \frac{1}{L} \int_{-L}^{L} f(x) \cos \frac{n\pi x}{L}\, dx, \tag{9}$$

$$b_n = \frac{1}{L} \int_{-L}^{L} f(x) \, \sin \frac{n\pi x}{L}\, dx. \tag{10}$$

Convergence of the Fourier Series

We now state without proof the result concerning the convergence of the Fourier series expansion for a wide class of functions commonly encountered in simplified models of several physical behaviors. Recall that a function f is *piecewise continuous* over an interval I if both limits

$$\lim_{x \to c^+} f(x) = f(c^+) \qquad \text{and} \qquad \lim_{x \to c^-} f(x) = f(c^-)$$

exist at every interior point c in I and, moreover, the appropriate one-sided limits exist at the endpoints of I, and f has at most finitely many discontinuities in I. Notice that a piecewise continuous function over a closed interval must be bounded (so it cannot tend toward infinity).

Theorem 18   Convergence of Fourier Series

If the function f and its derivative f' are piecewise continuous over the interval $-L < x < L$, then f equals its Fourier series at all points of continuity. At a point c where a jump discontinuity occurs in f, the Fourier series converges to the average

$$\frac{f(c^+) + f(c^-)}{2},$$

where $f(c^+)$ and $f(c^-)$ denote the right and left limits of f at c, respectively.

Example 2   Convergence Values

The function in Example 1 satisfies the conditions of Theorem 18. For every $x \neq 0$ in the interval $-\pi < x < \pi$, the Fourier series converges to $f(x)$. At $x = 0$, the function has a jump discontinuity and the Fourier series converges to the average value

$$\frac{f(0^+) + f(0^-)}{2} = \frac{0 + 1}{2} = \frac{1}{2}$$

(Figure 8.23).

Periodic Extension

The trigonometric terms $\sin(n\pi x/L)$ and $\cos(n\pi x/L)$ in the Fourier series are periodic with period $2L$:

$$\sin \frac{n\pi(x + 2L)}{L} = \sin \frac{n\pi x}{L} \cos 2n\pi + \cos \frac{n\pi x}{L} \sin 2n\pi$$

$$= \sin \frac{n\pi x}{L}$$

and

$$\cos \frac{n\pi(x + 2L)}{L} = \cos \frac{n\pi x}{L} \cos 2n\pi - \sin \frac{n\pi x}{L} \sin 2n\pi$$

$$= \cos \frac{n\pi x}{L}.$$

It follows that the Fourier series is also periodic with period $2L$. Thus, the Fourier series not only represents the function f over the interval $-L < x < L$, but it also produces the **periodic extension** of f over the entire real-number line. From Theorem 18, the series converges to the average value $[f(L^-) + f(-L^+)]/2$ at the endpoints of the interval, as well as to this value extended periodically to $\pm 3L$, $\pm 5L$, $\pm 7L$, and so forth.

Example 3   Convergence and Periodic Extension

The Fourier series for $f(x) = x$ on $-\pi < x < \pi$ is

$$f(x) = \sum_{n=1}^{\infty} \frac{2(-1)^{n+1}}{n} \sin nx.$$

(You are asked to find this series in Exercise 3.) The series converges to the periodic extension of $f(x) = x$ on the entire x-axis. The solid dots in Figure 8.24 represent the value

$$\frac{f(\pi^+) + f(\pi^-)}{2} = \frac{\pi + (-\pi)}{2} = 0.$$

The series converges to 0 at the interval endpoints $\pm\pi$, $\pm 3\pi$, $\pm 5\pi$, ...

FIGURE 8.24 The Fourier series of $f(x) = x$ converges to f over the interval $-\pi < x < \pi$ and to its periodic extension along the real axis (Theorem 18).

EXERCISES 8.9

Finding Fourier Series

In Exercises 1–14, find the Fourier series expansion for the functions over the specified intervals.

1. $f(x) = 1,\quad -\pi < x < \pi$

2. $f(x) = \begin{cases} -1, & -\pi < x < 0 \\ 1, & 0 < x < \pi \end{cases}$

3. $f(x) = x,\quad -\pi < x < \pi$

4. $f(x) = 1 - x,\quad -\pi < x < \pi$

5. $f(x) = \dfrac{x^2}{4},\quad -\pi < x < \pi$

6. $f(x) = \begin{cases} 0, & -\pi < x < 0 \\ x^2, & 0 < x < \pi \end{cases}$

7. $f(x) = e^x,\quad -\pi < x < \pi$

8. $f(x) = \begin{cases} 0, & -\pi < x < 0 \\ e^x, & 0 < x < \pi \end{cases}$

9. $f(x) = \begin{cases} 0, & -\pi < x < 0 \\ \cos x, & 0 < x < \pi \end{cases}$

10. $f(x) = \begin{cases} -x, & -2 < x < 0 \\ 2, & 0 < x < 2 \end{cases}$

11. $f(x) = \begin{cases} 0, & -\pi < x < -\dfrac{\pi}{2} \\ 1, & -\dfrac{\pi}{2} < x < \dfrac{\pi}{2} \\ 0, & \dfrac{\pi}{2} < x < \pi \end{cases}$

12. $f(x) = |x|, \quad -1 < x < 1$

13. $f(x) = |2x - 1|, \quad -1 < x < 1$

14. $f(x) = x|x|, \quad -\pi < x < \pi$

Theory and Examples

15. Use the Fourier series in Exercise 5 to show that

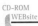

$$1 + \frac{1}{4} + \frac{1}{9} + \frac{1}{16} + \frac{1}{25} + \cdots = \frac{\pi^2}{6}.$$

16. Use the Fourier series in Exercise 6 to show that

$$1 - \frac{1}{4} + \frac{1}{9} - \frac{1}{16} + \cdots = \frac{\pi^2}{12}.$$

Establish the results in Exercises 17–21, where m and n are positive integers.

17. $\displaystyle\int_{-L}^{L} \cos \frac{m\pi x}{L}\, dx = 0 \quad$ for all m.

18. $\displaystyle\int_{-L}^{L} \sin \frac{m\pi x}{L}\, dx = 0 \quad$ for all m.

19. $\displaystyle\int_{-L}^{L} \cos \frac{n\pi x}{L} \cos \frac{m\pi x}{L}\, dx = \begin{cases} 0, & m \neq n \\ L, & m = n \end{cases}$

(*Hint:* $\cos A \cos B = (1/2)\,[\cos (A + B) + \cos (A - B)]$.)

20. $\displaystyle\int_{-L}^{L} \sin \frac{n\pi x}{L} \sin \frac{m\pi x}{L}\, dx = \begin{cases} 0, & m \neq n \\ L, & m = n \end{cases}$

(*Hint:* $\sin A \sin B = (1/2)\,[\cos (A - B) - \cos (A + B)]$.)

21. $\displaystyle\int_{-L}^{L} \sin \frac{n\pi x}{L} \cos \frac{m\pi x}{L}\, dx = 0 \quad$ for all m and n.

(*Hint:* $\sin A \cos B = (1/2)\,[\sin (A + B) + \sin (A - B)]$.)

22. *Writing to Learn: Fourier series of sums of functions* If $f(x)$ and $g(x)$ both satisfy the conditions of Theorem 18, is the Fourier series of $f(x) + g(x)$ on $(-L, L)$ the sum of the Fourier series of $f(x)$ and $g(x)$ on the interval? Give reasons for your answers.

23. *Term-by-term differentiation*

(a) Use Theorem 18 to verify that the Fourier series for $f(x) = x$ in Exercise 3 converges to $f(x)$ for $-\pi < x < \pi$.

(b) Although $f'(x) = 1$, show that the series obtained from term-by-term differentiation of the Fourier series in part (a) diverges.

(c) *Writing to Learn* What do you conclude from part (b)? Give reasons for your answer.

24. *Term-by-term integration* In advanced calculus, it is proved that the Fourier series of a piecewise continuous function on $[-L, L]$ can be integrated term by term. Use this fact to show that if $f(x)$ is piecewise continuous on $-\pi < x < \pi$, then

$$\int_{-\pi}^{\pi} f(s)\, ds = \frac{1}{2} a_0 (x + \pi)$$

$$+ \sum_{n=1}^{\infty} \frac{1}{n} (a_n \sin nx - b_n (\cos nx - \cos n\pi)) \quad \text{for } -\pi \leq x \leq \pi,$$

where a_0, a_n, and b_n are the Fourier coefficients of f.

8.10   Fourier Cosine and Sine Series

Integrals of Even and Odd Functions • Even Extension: Fourier Cosine Series • Odd Extension: Fourier Sine Series • Gibbs Phenomenon

In modeling heat conduction in a long, thin insulated rod or wire, we assume that the x-axis is aligned with the length L of the rod and that $0 < x < L$. The temperature $u(x, t)$ along the length of the rod generally varies with both position x and time t. (In Chapter 12, you will study functions like this, which depend on two or more independent variables for their values.) The problem is to determine $u(x, t)$ given the initial temperature $u(x, 0) = f(x)$ along the rod. For example, it might be hot at one end and cooler at the other end, so heat will flow from the hot to the cool end and we might want to know what the temperature distribution will look like in an hour's time. One method used to solve this problem requires the expansion

$$f(x) = \sum_{n=1}^{\infty} b_n \sin \frac{n\pi x}{L}$$

over the *nonsymmetric* interval $0 < x < L$. How then can we calculate a Fourier series expansion for f? To do so, we extend the function so that it is defined over the symmetric interval $-L < x < L$. How, though, do we define the extension of f for $-L < x < 0$? The answer is that we can define the extension to be *any function* over $-L < x < 0$ we choose as long as the extension and its derivative are piecewise continuous (in order to satisfy the hypothesis of Theorem 18). No matter what piecewise continuous function we define as the extension over $-L < x < 0$, the resulting Fourier series is guaranteed to equal $f(x)$ at all points of continuity over the original domain $0 < x < L$. Of course, the Fourier series also converges to whatever extension function we have chosen for $-L < x < 0$. Nevertheless, there are two special extensions that are particularly useful and whose Fourier coefficients are especially easy to calculate; these are the even and odd extensions of f.

Integrals of Even and Odd Functions

Recall (Preliminaries Chapter, Section 2) that a function $g(x)$ is an even function of x if $g(-x) = g(x)$ for all x in the domain of g. If $g(-x) = -g(x)$ instead, g is said to be an odd function of x. The function $\cos x$ is even, the function $\sin x$ is odd. The graph of an even function is symmetric about the y-axis, whereas the graph of an odd function is symmetric about the origin (Figure 8.25).

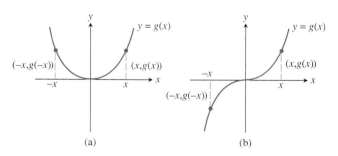

(a)   (b)

FIGURE 8.25 (a) The graph of an even function is symmetric about the y-axis. (b) The graph of an odd function is symmetric about the origin.

This observation can make the integrals of even and odd functions over intervals symmetric about the origin relatively easy to calculate. For instance, if we consider the "appropriately signed" portions of the graphs in Figure 8.25, we obtain the following:

Odd function:

$$\int_{-L}^{L} g(x)\, dx = 0. \qquad (1)$$

Even function:

$$\int_{-L}^{L} g(x)\, dx = 2\int_{0}^{L} g(x)\, dx. \qquad (2)$$

Because of rules (1) and (2), even and odd extensions of a function are convenient to use. The following results also hold for even and odd functions.

1. The product of two even functions is even.

2. The product of an even function with an odd function is odd.

3. The product of two odd functions is even.

CD-ROM
WEBsite
Historical Biography
John William
Strutt Rayleigh
(1842 — 1909)

Even Extension: Fourier Cosine Series

Suppose that the function $y = f(x)$ is specified for the interval $0 < x < L$. We define the **even extension of f** by requiring that

$$f(-x) = f(x), \qquad -L < x < L.$$

Graphically, we obtain the even extension by reflecting $y = f(x)$ about the y-axis. The even extension of a function is illustrated in Figure 8.26. Therefore, if we use the even extension for a function f, we obtain the Fourier coefficients

$$a_0 = \frac{1}{L}\int_{-L}^{L} f(x)\, dx = \frac{2}{L}\int_{0}^{L} f(x)\, dx,$$

$$a_n = \frac{1}{L}\int_{-L}^{L} \underbrace{f(x)\, \cos \frac{n\pi x}{L}}_{\text{Even}}\, dx = \frac{2}{L}\int_{0}^{L} f(x)\, \cos \frac{n\pi x}{L}\, dx,$$

$$b_n = \frac{1}{L}\int_{-L}^{L} \underbrace{f(x)\, \sin \frac{n\pi x}{L}}_{\text{Odd}}\, dx = 0.$$

The Fourier series of f is

$$f(x) = \frac{a_0}{2} + \sum_{n=1}^{\infty} a_n\, \cos \frac{n\pi x}{L}.$$

(a)

(b)

FIGURE 8.26 (a) The original piecewise continuous function f defined over nonsymmetric interval $0 < x < L$. (b) The even extension of f over $-L < x < L$.

Because the Fourier coefficients b_n are all zero, no sine terms appear in the Fourier series expansion, and the series is called the **Fourier cosine series** of the function f. It converges to the original function f over the interval $0 < x < L$ and to the even extension over the interval $-L < x < 0$ (assuming the piecewise continuity of f and f'). We summarize this result.

CD-ROM
WEBsite

Fourier Cosine Series

The Fourier series of an even function on the interval $-L < x < L$ is the **cosine series**

$$f(x) = \frac{a_0}{2} + \sum_{n=1}^{\infty} a_n \cos \frac{n\pi x}{L},$$ (3)

where

$$a_0 = \frac{2}{L} \int_0^L f(x)\, dx,$$ (4)

$$a_n = \frac{2}{L} \int_0^L f(x) \cos \frac{n\pi x}{L}\, dx.$$ (5)

FIGURE 8.27 The function given in Example 1.

Example 1   Finding a Fourier Cosine Series

Find the Fourier cosine series for the function

$$f(x) = \begin{cases} 1, & 0 < x < \dfrac{\pi}{2}, \\ 0, & \dfrac{\pi}{2} < x < \pi, \end{cases}$$

depicted in Figure 8.27.

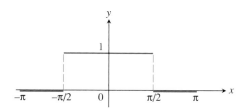

FIGURE 8.28 The even extension of the function in Example 1.

Solution   For the Fourier cosine series, we select the even extension of the function over $-\pi < x < \pi$ as shown in Figure 8.28. The Fourier coefficients are

$$a_0 = \frac{2}{\pi} \int_0^\pi f(x)\, dx \qquad \text{Eq. (4) with } L = \pi$$

$$= \frac{2}{\pi} \int_0^{\pi/2} dx \qquad \begin{matrix} f(x) = 0 \text{ for} \\ \pi/2 < x < \pi \end{matrix}$$

$$= \frac{2x}{\pi} \Bigg]_0^{\pi/2} = 1$$

$$a_n = \frac{2}{\pi} \int_0^\pi f(x) \cos \frac{n\pi x}{\pi}\, dx \qquad \text{Eq. (5) with } L = \pi$$

$$= \frac{2}{\pi} \int_0^{\pi/2} \cos nx\, dx$$

$$= \frac{2}{n\pi} \sin \frac{n\pi}{2}.$$

Therefore, we have the Fourier cosine expansion

$$f(x) = \frac{1}{2} + \sum_{n=1}^{\infty} \frac{2}{n\pi} \sin \frac{n\pi}{2} \cos nx.$$

The Fourier cosine series equals exactly the values of $f(x)$ for $x \neq \pi/2$; at the point $x = \pi/2$, the value of the Fourier cosine series is $1/2$. Plots of Fourier cosine approximations for $f(x)$ as n varies up to 1, 5, and 20 terms are given in Figure 8.29.

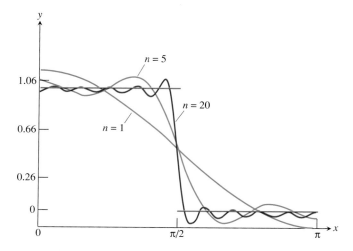

FIGURE 8.29 Fourier cosine series approximations of the function in Example 1 as n varies up to 1, 5, and 20 terms in the infinite series. As n increases, the Fourier cosine approximations approach the actual values of function $f(x)$. Each Fourier cosine approximation passes through the value $y = 0.5$, the midvalue of the jump, at the point of discontinuity $x = \pi/2$.

Odd Extension: Fourier Sine Series

Consider again a function $y = f(x)$ specified for the interval $0 < x < L$. We define the **odd extension of f** by requiring that

$$f(-x) = -f(x), \qquad -L < x < L.$$

Graphically, we obtain the odd extension by reflecting $y = f(x)$ about the origin. The odd extension of a function is illustrated in Figure 8.30. For the odd extension of f, we obtain

$$a_0 = \frac{1}{L} \int_{-L}^{L} f(x)\, dx = 0$$

$$a_n = \frac{1}{L} \int_{-L}^{L} \underbrace{f(x) \cos \frac{n\pi x}{L}}_{\text{Odd}}\, dx = 0$$

$$b_n = \frac{1}{L} \int_{-L}^{L} \underbrace{f(x) \sin \frac{n\pi x}{L}}_{\text{Even}}\, dx = \frac{2}{L} \int_{0}^{L} f(x) \sin \frac{n\pi x}{L}\, dx.$$

The Fourier series of f is

$$f(x) = \sum_{n=1}^{\infty} b_n \sin \frac{n\pi x}{L}.$$

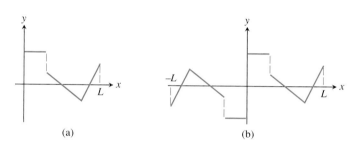

FIGURE 8.30 (a) The original piecewise continuous function f defined over nonsymmetric interval $0 < x < L$. (b) The odd extension of f over $-L < x < L$.

Because the Fourier coefficients a_0 and a_n are all zero, no cosine terms appear in the Fourier series expansion, and the series is called the **Fourier sine series** of the function f. This series converges to the original function f over the interval $0 < x < L$, and to the *odd* extension over the interval $-L < x < 0$ (assuming the piecewise continuity of f and f'). We summarize this result.

CD-ROM
WEBsite

Fourier Sine Series

The Fourier series of an odd function on the interval $-L < x < L$ is the **sine series**

$$f(x) = \sum_{n=1}^{\infty} b_n \sin \frac{n\pi x}{L}, \tag{6}$$

where

$$b_n = \frac{2}{L} \int_0^L f(x) \sin \frac{n\pi x}{L} \, dx. \tag{7}$$

Example 2   Finding a Fourier Sine Series

Find the Fourier sine series for the function

$$f(x) = \begin{cases} 1, & 0 < x < \dfrac{\pi}{2}, \\ 0, & \dfrac{\pi}{2} < x < \pi, \end{cases}$$

in Example 1.

Solution   We select the *odd* extension of the function $f(x)$. The Fourier coefficients are

$$b_n = \frac{2}{\pi} \int_0^\pi f(x) \sin \frac{n\pi x}{\pi} \, dx \qquad \text{Eq. (7) with } L = \pi$$

$$= \frac{2}{\pi} \int_0^{\pi/2} \sin nx \, dx \qquad \begin{aligned} &f(x) = 0 \text{ for} \\ &\pi/2 < x < \pi \end{aligned}$$

$$= -\frac{2}{n\pi} \cos nx \Big]_0^{\pi/2} = \frac{2}{n\pi}\left(1 - \cos \frac{n\pi}{2}\right).$$

Therefore, we have the Fourier sine expansion

$$f(x) = \sum_{n=1}^{\infty} \frac{2}{n\pi}\left(1 - \cos\frac{n\pi}{2}\right)\sin\,nx.$$

A graph of the Fourier sine approximations for $f(x)$ for $n = 1, 5$, and 20 terms is shown in Figure 8.31.

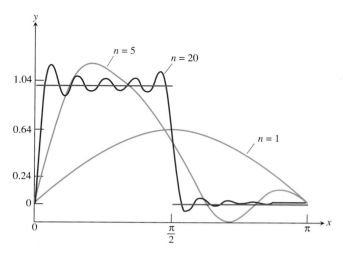

FIGURE 8.31 Fourier sine approximations of the function in Example 2 for $n = 1, 5$, and 20 terms in the infinite series. As n increases, the Fourier sine approximations approach the values of $f(x)$. The approximations converge to the midpoint of the jump at the point of discontinuity, $x = \pi/2$.

Gibbs Phenomenon

In Figures 8.29 and 8.31, the **overshoot** at $x = \pi/2^-$ and the **undershoot** at $x = \pi/2^+$ are characteristic of Fourier series expansions near points of discontinuity. Known as the **Gibbs phenomenon,** after American mathematical physicist Josiah Willard Gibbs, this characteristic persists even when a large number of terms are summed. The combined overshoot and undershoot amount to about 18% of the difference between the function's values at the point of discontinuity. Figure 8.32 reveals the phenomenon for

$$f(x) = \begin{cases} 1, & 0 \le x < 1 \\ 0, & x = 1 \end{cases}$$

with $n = 2, 4, 8, 16, 32$, and 64 terms. With more terms, the highest peak moves closer to the discontinuity at $x = 1$, but the overshoot remains approximately 1.09.

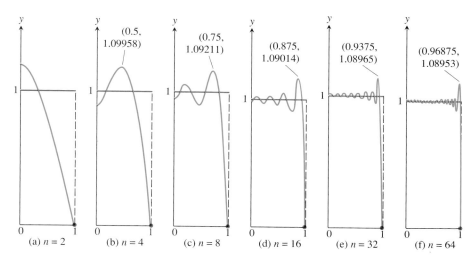

FIGURE 8.32 The Gibbs phenomenon for $n = 2, 4, 8, 16, 32$, and 64 terms. The highest peak moves from 0.5 to 0.75, then to 0.875, and so forth, getting closer to the discontinuity at $x = 1$. The overshoot is always close to 1.09, or about 9% of the distance between $y = 0$ and $y = 1$ at the discontinuity, $x = 1$.

EXERCISES 8.10

Finding Fourier Cosine Series

Each of Exercises 1–8 gives a function $f(x)$ defined on an interval $(0, L)$. Graph f and its even extension to $(-L, L)$. Then find the Fourier cosine series expansion for f.

1. $f(x) = x, \quad 0 < x < \pi$

2. $f(x) = \sin x, \quad 0 < x < \pi$

3. $f(x) = e^x, \quad 0 < x < 1$

4. $f(x) = \cos x, \quad 0 < x < \pi$

5. $f(x) = \begin{cases} 1, & 0 < x < 1 \\ -x, & 1 < x < 2 \end{cases}$

6. $f(x) = \begin{cases} -1, & 0 < x < 0.5 \\ 1, & 0.5 < x < 1 \end{cases}$

7. $f(x) = |2x - 1|, \quad 0 < x < 1$

8. $f(x) = |2x - \pi|, \quad 0 < x < \pi$

Finding Fourier Sine Series

Each of Exercises 9–16 gives a function $f(x)$ defined on an interval $(0, L)$. Graph f and its odd extension to $(-L, L)$. Then find the Fourier sine series expansion for f.

9. $f(x) = -x, \quad 0 < x < 1$

10. $f(x) = x^2, \quad 0 < x < \pi$

11. $f(x) = \cos x, \quad 0 < x < \pi$

12. $f(x) = e^x, \quad 0 < x < 1$

13. $f(x) = \sin x, \quad 0 < x < \pi$

14. $f(x) = \begin{cases} x, & 0 < x < 1 \\ 1, & 1 < x < 2 \end{cases}$

15. $f(x) = \begin{cases} 1 - x, & 0 < x < 1 \\ 0, & 1 < x < 2 \end{cases}$

16. $f(x) = |2x - \pi|, \quad 0 < x < \pi$

Theory and Examples

17. *A series for $\pi / 4$*

  (a) Find the Fourier sine series for

$$f(x) = \begin{cases} 1, & 0 < x < \pi \\ 0, & x = 0 \text{ and } x = \pi. \end{cases}$$

  (b) Use the result of part (a) to show that

$$\frac{\pi}{4} = 1 - \frac{1}{3} + \frac{1}{5} - \frac{1}{7} + \cdots.$$

18. *A function with a triangular graph*

    (a) Graph the triangular function

$$f(x) = \begin{cases} 1 - x, & 0 < x < 1 \\ x - 1, & 1 < x < 2 \end{cases}$$

    (b) Find a Fourier series expansion for $f(x)$.

    (c) Find a Fourier cosine series expansion for $f(x)$.

19. *Evaluating a series* Use the result of Exercise 2 to find the value of

$$\sum_{n=1}^{\infty} \frac{(-1)^n}{4n^2 - 1}.$$

20. *Fourier sine series* Given the function

$$f(x) = 2 - x, \qquad 0 < x < 2,$$

define a function whose Fourier sine series representation will converge to $f(x)$ for all values of x. (*Note*: The answer is not unique.)

Questions to Guide Your Review

1. What is an infinite sequence? What does it mean for such a sequence to converge? To diverge? Give examples.

2. What theorems are available for calculating limits of sequences? Give examples.

3. What theorem sometimes enables us to use l'Hôpital's Rule to calculate the limit of a sequence? Give an example.

4. What six sequence limits are likely to arise when you work with sequences and series?

5. What is a subsequence? Why are subsequences important? What uses can be found for subsequences? Give examples.

6. What is a nondecreasing sequence? A nonincreasing sequence? A monotonic sequence? Under what circumstances do these sequences have a limit? Give examples.

7. What is Picard's method for solving the equation $f(x) = 0$? Give an example.

8. What is an infinite series? What does it mean for such a series to converge? To diverge? Give examples.

9. What is a geometric series? When does such a series converge? Diverge? When it does converge, what is its sum? Give examples.

10. Besides geometric series, what other convergent and divergent series do you know?

11. What is the nth-Term Test for Divergence? What is the idea behind the test?

12. What can be said about term-by-term sums and differences of convergent series? About constant multiples of convergent and divergent series?

13. What happens if you add a finite number of terms to a convergent series? A divergent series? What happens if you delete a finite number of terms from a convergent series? A divergent series?

14. Under what circumstances will an infinite series of nonnegative terms converge? Diverge? Why study series of nonnegative terms?

15. What is the Integral Test? What is the reasoning behind it? Give an example of its use.

16. When do p-series converge? Diverge? How do you know? Give examples of convergent and divergent p-series.

17. What are the Direct Comparison Test and the Limit Comparison Test? What is the reasoning behind these tests? Give examples of their use.

18. What are the Ratio and Root Tests? Do they always give you the information you need to determine convergence or divergence? Give examples.

19. What is an alternating series? What theorem is available for determining the convergence of such a series?

20. How can you estimate the error involved in approximating the sum of an alternating series with one of the series' partial sums? What is the reasoning behind the estimate?

21. What is absolute convergence? Conditional convergence? How are the two related?

22. What do you know about rearranging the terms of an absolutely convergent series? Of a conditionally convergent series? Give examples.

23. What is a power series? How do you test a power series for convergence? What are the possible outcomes?

24. What are the basic facts about
    (a) term-by-term differentiation of power series?
    (b) term-by-term integration of power series?
    (c) multiplication of power series?
    Give examples.

25. What is the Taylor series generated by a function $f(x)$ at a point $x = a$? What information do you need about f to construct the series? Give an example.

26. What is a Maclaurin series?

27. Does a Taylor series always converge to its generating function? Explain.

28. What are Taylor polynomials? Of what use are they?

29. What is Taylor's Theorem? What does it say about the errors involved in using Taylor polynomials to approximate functions? In particular, what does the Remainder Estimation Theorem say about the error in a linearization? A quadratic approximation?

30. What is the binomial series? On what interval does it converge? How is it used?

31. What are the Maclaurin series for $1/(1 - x)$, $1/(1 + x)$, e^x, $\sin x$, $\cos x$, $\ln(1 + x)$, and $\tan^{-1} x$?

32. What is a Fourier series? How do you calculate the Fourier coefficients for a function $f(x)$ defined over the interval $-L < x < L$? Under what conditions does a Fourier series converge to its generating function? What happens at a point of discontinuity?

33. What is the periodic extension to the entire real line of a function $f(x)$ defined on $-L < x < L$?

34. What is the even extension to $-L < x < 0$ of a function $f(x)$ defined on $0 < x < L$? What is a Fourier cosine series? How do you calculate its coefficients?

35. What is the odd extension to $-L < x < 0$ of a function $f(x)$ defined on $0 < x < L$? What is a Fourier sine series? How do you calculate its coefficients?

36. What is the Gibbs phenomenon? How is it affected when you sum more and more terms in the Fourier series?

Practice Exercises

Convergent or Divergent Sequences

Which of the sequences whose nth terms appear in Exercises 1–18 converge, and which diverge? Find the limit of each convergent sequence.

1. $a_n = 1 + \dfrac{(-1)^n}{n}$

2. $a_n = \dfrac{1 - (-1)^n}{\sqrt{n}}$

3. $a_n = \dfrac{1 - 2^n}{2^n}$

4. $a_n = 1 + (0.9)^n$

5. $a_n = \sin \dfrac{n\pi}{2}$

6. $a_n = \sin n\pi$

7. $a_n = \dfrac{\ln (n^2)}{n}$

8. $a_n = \dfrac{\ln (2n + 1)}{n}$

9. $a_n = \dfrac{n + \ln n}{n}$

10. $a_n = \dfrac{\ln (2n^3 + 1)}{n}$

11. $a_n = \left(\dfrac{n - 5}{n}\right)^n$

12. $a_n = \left(1 + \dfrac{1}{n}\right)^{-n}$

13. $a_n = \sqrt[n]{\dfrac{3^n}{n}}$

14. $a_n = \left(\dfrac{3}{n}\right)^{1/n}$

15. $a_n = n(2^{1/n} - 1)$

16. $a_n = \sqrt[n]{2n + 1}$

17. $a_n = \dfrac{(n + 1)!}{n!}$

18. $a_n = \dfrac{(-4)^n}{n!}$

Convergent Series

Find the sums of the series in Exercises 19–24.

19. $\displaystyle\sum_{n=3}^{\infty} \dfrac{1}{(2n - 3)(2n - 1)}$

20. $\displaystyle\sum_{n=2}^{\infty} \dfrac{-2}{n(n + 1)}$

21. $\displaystyle\sum_{n=1}^{\infty} \dfrac{9}{(3n - 1)(3n + 2)}$

22. $\displaystyle\sum_{n=3}^{\infty} \dfrac{-8}{(4n - 3)(4n + 1)}$

23. $\displaystyle\sum_{n=0}^{\infty} e^{-n}$

24. $\displaystyle\sum_{n=1}^{\infty} (-1)^n \dfrac{3}{4^n}$

Convergent or Divergent Series

Which of the series in Exercises 25–40 converge absolutely, which converge conditionally, and which diverge? Give reasons for your answers.

25. $\displaystyle\sum_{n=1}^{\infty} \dfrac{1}{\sqrt{n}}$

26. $\displaystyle\sum_{n=1}^{\infty} \dfrac{-5}{n}$

27. $\displaystyle\sum_{n=1}^{\infty} \dfrac{(-1)^n}{\sqrt{n}}$

28. $\displaystyle\sum_{n=1}^{\infty} \dfrac{1}{2n^3}$

29. $\displaystyle\sum_{n=1}^{\infty} \dfrac{(-1)^n}{\ln (n + 1)}$

30. $\displaystyle\sum_{n=2}^{\infty} \dfrac{1}{n(\ln n)^2}$

31. $\displaystyle\sum_{n=1}^{\infty} \dfrac{\ln n}{n^3}$

32. $\displaystyle\sum_{n=3}^{\infty} \dfrac{\ln n}{\ln (\ln n)}$

33. $\displaystyle\sum_{n=1}^{\infty} \dfrac{(-1)^n}{n\sqrt{n^2 + 1}}$

34. $\displaystyle\sum_{n=1}^{\infty} \dfrac{(-1)^n 3n^2}{n^3 + 1}$

35. $\displaystyle\sum_{n=1}^{\infty} \dfrac{n + 1}{n!}$

36. $\displaystyle\sum_{n=1}^{\infty} \dfrac{(-1)^n(n^2 + 1)}{2n^2 + n - 1}$

37. $\displaystyle\sum_{n=1}^{\infty} \dfrac{(-3)^n}{n!}$

38. $\displaystyle\sum_{n=1}^{\infty} \dfrac{2^n 3^n}{n^n}$

39. $\displaystyle\sum_{n=1}^{\infty} \dfrac{1}{\sqrt{n(n + 1)(n + 2)}}$

40. $\displaystyle\sum_{n=2}^{\infty} \dfrac{1}{n\sqrt{n^2 - 1}}$

Power Series

In Exercises 41–50, (a) find the series' radius and interval of convergence. Then identify the values of x for which the series converges (b) absolutely and (c) conditionally.

41. $\displaystyle\sum_{n=1}^{\infty} \dfrac{(x + 4)^n}{n3^n}$

42. $\displaystyle\sum_{n=1}^{\infty} \dfrac{(x - 1)^{2n-2}}{(2n - 1)!}$

43. $\displaystyle\sum_{n=1}^{\infty} \dfrac{(-1)^{n-1}(3x - 1)^n}{n^2}$

44. $\displaystyle\sum_{n=0}^{\infty} \dfrac{(n + 1)(2x + 1)^n}{(2n + 1)2^n}$

45. $\displaystyle\sum_{n=1}^{\infty} \dfrac{x^n}{n^n}$

46. $\displaystyle\sum_{n=1}^{\infty} \dfrac{x^n}{\sqrt{n}}$

47. $\displaystyle\sum_{n=0}^{\infty} \dfrac{(n + 1) x^{2n-1}}{3^n}$

48. $\displaystyle\sum_{n=0}^{\infty} \dfrac{(-1)^n(x - 1)^{2n+1}}{2n + 1}$

49. $\displaystyle\sum_{n=1}^{\infty} (\operatorname{csch} n) x^n$

50. $\displaystyle\sum_{n=1}^{\infty} (\operatorname{coth} n) x^n$

Maclaurin Series

Each of the series in Exercises 51–56 is the value of the Maclaurin series of a function $f(x)$ at a particular point. What function and what point? What is the sum of the series?

51. $1 - \dfrac{1}{4} + \dfrac{1}{16} - \cdots + (-1)^n \dfrac{1}{4^n} + \cdots$

52. $\dfrac{2}{3} - \dfrac{4}{18} + \dfrac{8}{81} - \cdots + (-1)^{n-1} \dfrac{2^n}{n3^n} + \cdots$

53. $\pi - \dfrac{\pi^3}{3!} + \dfrac{\pi^5}{5!} - \cdots + (-1)^n \dfrac{\pi^{2n+1}}{(2n + 1)!} + \cdots$

54. $1 - \dfrac{\pi^2}{9 \cdot 2!} + \dfrac{\pi^4}{81 \cdot 4!} - \cdots + (-1)^n \dfrac{\pi^{2n}}{3^{2n}(2n)!} + \cdots$

55. $1 + \ln 2 + \dfrac{(\ln 2)^2}{2!} + \cdots + \dfrac{(\ln 2)^n}{n!} + \cdots$

56. $\dfrac{1}{\sqrt{3}} - \dfrac{1}{9\sqrt{3}} + \dfrac{1}{45\sqrt{3}} - \cdots + (-1)^{n-1} \dfrac{1}{(2n - 1)(\sqrt{3})^{2n-1}} + \cdots$

Find Maclaurin series for the functions in Exercises 57–64.

57. $\dfrac{1}{1-2x}$ **58.** $\dfrac{1}{1+x^3}$

59. $\sin \pi x$ **60.** $\sin \dfrac{2x}{3}$

61. $\cos (x^{5/2})$ **62.** $\cos \sqrt{5x}$

63. $e^{(\pi x/2)}$ **64.** e^{-x^2}

Taylor Series

In Exercises 65–68, find the first four nonzero terms of the Taylor series generated by f at $x = a$.

65. $f(x) = \sqrt{3 + x^2}$ at $x = -1$

66. $f(x) = 1/(1 - x)$ at $x = 2$

67. $f(x) = 1/(x + 1)$ at $x = 3$

68. $f(x) = 1/x$ at $x = a > 0$

Initial Value Problems

Use power series to solve the initial value problems in Exercises 69–76.

69. $y' + y = 0, \quad y(0) = -1$ **70.** $y' - y = 0, \quad y(0) = -3$

71. $y' + 2y = 0, \quad y(0) = 3$ **72.** $y' + y = 1, \quad y(0) = 0$

73. $y' - y = 3x, \quad y(0) = -1$ **74.** $y' + y = x, \quad y(0) = 0$

75. $y' - y = x, \quad y(0) = 1$ **76.** $y' - y = -x, \quad y(0) = 2$

Indeterminate Forms

In Exercises 77–82:

(a) Use power series to evaluate the limit.

T (b) Then use a grapher to support your calculation.

77. $\displaystyle\lim_{x\to 0} \dfrac{7 \sin x}{e^{2x} - 1}$ **78.** $\displaystyle\lim_{\theta\to 0} \dfrac{e^{\theta} - e^{-\theta} - 2\theta}{\theta - \sin \theta}$

79. $\displaystyle\lim_{t\to 0} \left(\dfrac{1}{2 - 2 \cos t} - \dfrac{1}{t^2} \right)$ **80.** $\displaystyle\lim_{h\to 0} \dfrac{(\sin h)/h - \cos h}{h^2}$

81. $\displaystyle\lim_{z\to 0} \dfrac{1 - \cos^2 z}{\ln (1 - z) + \sin z}$ **82.** $\displaystyle\lim_{y\to 0} \dfrac{y^2}{\cos y - \cosh y}$

83. Use a series representation of $\sin 3x$ to find values of r and s for which

$$\lim_{x\to 0} \left(\dfrac{\sin 3x}{x^3} + \dfrac{r}{x^2} + s \right) = 0.$$

84. (a) Show that the approximation $\csc x \approx 1/x + x/6$ in Section 8.8, Example 6, leads to the approximation $\sin x \approx 6x/(6 + x^2)$.

T (b) *Writing to Learn* Compare the accuracies of the approximations $\sin x \approx x$ and $\sin x \approx 6x/(6 + x^2)$ by comparing the graphs of $f(x) = \sin x - x$ and $g(x) = \sin x - (6x/(6 + x^2))$. Describe what you find.

Fourier Series

In Exercises 85–90, find the Fourier series of f on the given interval.

85. $f(x) = \begin{cases} -1, & -\pi < x < 0 \\ 2, & 0 < x < \pi \end{cases}$

86. $f(x) = \begin{cases} 0, & -1 < x < 0 \\ x, & 0 < x < 1 \end{cases}$

87. $f(x) = x + \pi, \quad -\pi < x < \pi$

88. $f(x) = \begin{cases} 0, & -\pi < x < 0 \\ \sin x, & 0 < x < \pi \end{cases}$

89. $f(x) = \begin{cases} 1, & -2 < x < 0 \\ 1 + x, & 0 < x < 2 \end{cases}$

90. $f(x) = \begin{cases} 0, & -2 < x < 0 \\ x, & 0 < x < 1 \\ 1, & 1 < x < 2 \end{cases}$

Fourier Cosine and Sine Series

In Exercises 91–96, find

(a) the Fourier cosine series

(b) the Fourier sine series of f on the given interval.

91. $f(x) = \begin{cases} 1, & 0 < x < 1/2 \\ 0, & 1/2 < x < 1 \end{cases}$

92. $f(x) = \begin{cases} 0, & 0 < x < 1 \\ x, & 1 < x < 2 \end{cases}$

93. $f(x) = \sin \pi x, \quad 0 < x < 1$

94. $f(x) = \cos x, \quad 0 < x < \pi/2$

95. $f(x) = 2x + x^2, \quad 0 < x < 3$

96. $f(x) = e^{-x}, \quad 0 < x < 2$

Theory and Examples

97. *A convergent series*

(a) Show that the series

$$\sum_{n=1}^{\infty} \left(\sin \dfrac{1}{2n} - \sin \dfrac{1}{2n + 1} \right)$$

converges.

(b) *Writing to Learn* Estimate the magnitude of the error involved in using the sum of the sines through $n = 20$ to approximate the sum of the series. Is the approximation too large, or too small? Give reasons for your answer.

98. (a) *A convergent series* Show that the series

$$\sum_{n=1}^{\infty} \left(\tan \frac{1}{2n} - \tan \frac{1}{2n+1} \right)$$

converges.

(b) *Writing to Learn* Estimate the magnitude of the error in using the sum of the tangents through $-\tan (1/41)$ to approximate the sum of the series. Is the approximation too large or too small? Give reasons for your answer.

99. *Radius of convergence* Find the radius of convergence of the series

$$\sum_{n=1}^{\infty} \frac{2 \cdot 5 \cdot 8 \cdot \cdots \cdot (3n-1)}{2 \cdot 4 \cdot 6 \cdot \cdots \cdot (2n)} x^n.$$

100. *Radius of convergence* Find the radius of convergence of the series

$$\sum_{n=1}^{\infty} \frac{3 \cdot 5 \cdot 7 \cdot \cdots \cdot (2n-1)}{4 \cdot 9 \cdot 14 \cdot \cdots \cdot (5n-1)} (x-1)^n.$$

101. *nth partial sum* Find a closed-form formula for the nth partial sum of the series $\sum_{n=2}^{\infty} \ln (1 - (1/n^2))$ and use it to determine the convergence or divergence of the series.

102. *nth partial sum* Evaluate $\sum_{k=2}^{\infty} (1/(k^2 - 1))$ by finding the limit as $n \to \infty$ of the series' nth partial sum.

103. (a) *Interval of convergence* Find the interval of convergence of the series

$$y = 1 + \frac{1}{6} x^3 + \frac{4}{720} x^6 + \cdots + \frac{1 \cdot 4 \cdot 7 \cdot \cdots \cdot (3n-2)}{(3n)!} x^{3n} + \cdots.$$

(b) *Differential equation* Show that the function defined by the series satisfies a differential equation of the form

$$\frac{d^2 y}{dx^2} = x^a y + b$$

and find the values of the constants a and b.

104. (a) *Maclaurin series* Find the Maclaurin series for the function $x^2/(1 + x)$.

(b) Does the series converge at $x = 1$? Explain.

105. *Writing to Learn* If $\sum_{n=1}^{\infty} a_n$ and $\sum_{n=1}^{\infty} b_n$ are convergent series of nonnegative numbers, can anything be said about $\sum_{n=1}^{\infty} a_n b_n$? Give reasons for your answer.

106. *Writing to Learn* If $\sum_{n=1}^{\infty} a_n$ and $\sum_{n=1}^{\infty} b_n$ are divergent series of nonnegative numbers, can anything be said about $\sum_{n=1}^{\infty} a_n b_n$? Give reasons for your answer.

107. *Sequence and series* Prove that the sequence $\{x_n\}$ and the series $\sum_{k=1}^{\infty} (x_{k+1} - x_k)$ both converge or both diverge.

108. *Convergence* Prove that $\sum_{n=1}^{\infty} (a_n/(1 + a_n))$ converges if $a_n > 0$ for all n and $\sum_{n=1}^{\infty} a_n$ converges.

109. (a) *Divergence* Suppose that $a_1, a_2, a_3, \ldots, a_n$ are positive numbers satisfying the following conditions:

  i. $a_1 \geq a_2 \geq a_3 \geq \cdots$

  ii. the series $a_2 + a_4 + a_8 + a_{16} + \cdots$ diverges.

Show that the series

$$\frac{a_1}{1} + \frac{a_2}{2} + \frac{a_3}{3} + \cdots$$

diverges.

(b) Use the result in part (a) to show that

$$1 + \sum_{n=2}^{\infty} \frac{1}{n \ln n}$$

diverges.

110. *Estimating an integral* Suppose that you wish to obtain a quick estimate for the value of $\int_0^1 x^2 e^x \, dx$. There are several ways to do this.

(a) Use the Trapezoidal Rule with $n = 2$ to estimate $\int_0^1 x^2 e^x \, dx$.

(b) Write out the first three nonzero terms of the Maclaurin series for $x^2 e^x$ to obtain the fourth Maclaurin polynomial $P(x)$ for $x^2 e^x$. Use $\int_0^1 P(x) \, dx$ to obtain another estimate for $\int_0^1 x^2 e^x \, dx$.

(c) *Writing to Learn* The second derivative of $f(x) = x^2 e^x$ is positive for all $x > 0$. Explain why this enables you to conclude that the trapezoidal rule estimate obtained in part (a) is too large.

(d) *Writing to Learn* All the derivatives of $f(x) = x^2 e^x$ are positive for $x > 0$. Explain why this enables you to conclude that all Maclaurin polynomial approximations to $f(x)$ for x in [0, 1] will be too small. (*Hint:* $f(x) = P_n(x) + R_n(x)$.)

(e) Use integration by parts to evaluate $\int_0^1 x^2 e^x \, dx$.

111. *Series for* $\tan^{-1} x$

(a) Integrate from $t = 0$ to $t = x$ both sides of the equation

$$\frac{1}{1 + t^2} = 1 - t^2 + t^4 - t^6 + \cdots + (-1)^n t^{2n} + \frac{(-1)^{n+1} t^{2n+2}}{1 + t^2}$$

in which the last term comes from adding the remaining terms as a geometric series with first term $a = (-1)^{n+1} t^{2n+2}$ and ratio $r = -t^2$.

(b) Show that the remainder term from part (a) is

$$R_n(x) = \int_0^x \frac{(-1)^{n+1} t^{2n+2}}{1 + t^2} \, dt$$

and find $\lim_{n \to \infty} R_n(x)$ if $|x| \leq 1$.

(c) Find a power series for $\tan^{-1} x$ based on the result in part (b).

(d) Set $x = 1$ in the series for $\tan^{-1} x$ to obtain **Leibniz's formula**

$$\frac{\pi}{4} = 1 - \frac{1}{3} + \frac{1}{5} - \frac{1}{7} + \frac{1}{9} - \cdots + \frac{(-1)^n}{2n+1} + \cdots.$$

112. *Evaluating nonelementary integrals* As you know, Maclaurin series can be used to express nonelementary integrals in terms of series.

(a) Express $\int_0^x \sin t^2 \, dt$ as a power series.

(b) According to the Alternating Series Estimation Theorem, how many terms of the series in part (a) should you use to estimate $\int_0^1 \sin x^2 \, dx$ with an error of less than 0.001?

113. *Picard's method for slope greater than 1* Example 9 in Section 8.2 showed that we cannot apply Picard's method to find a fixed point of $g(x) = 4x - 12$, but we can apply the method to find a fixed point of $g^{-1}(x) = (1/4)x + 3$ because the derivative of g^{-1} is $1/4$, whose value is less than 1 in magnitude on any interval.

In Example 7 of Section 8.2, we found the fixed point of g^{-1} to be $x = 4$. Now notice that 4 is also a fixed point of g, since

$$g(4) = 4(4) - 12 = 4.$$

In finding the fixed point of g^{-1}, we found the fixed point of g.

A function and its inverse always have the same fixed points. The graphs of the functions are symmetric about the line $y = x$ and therefore intersect the line at the same points.

We now see that the application of Picard's method is quite broad. Suppose g is one-to-one, with a continuous first derivative whose magnitude is greater than 1 on a closed interval I whose interior contains a fixed point of g. Then the derivative of g^{-1}, being the reciprocal of g', has magnitude less than 1 on I. Picard's method applied to g^{-1} on I will find the fixed point of g. As cases in point, find the fixed points of the following functions.

(a) $g(x) = 2x + 3$

(b) $g(x) = 1 - 4x$

Additional Exercises: Theory, Examples, Applications

Convergence or Divergence

Which of the series $\sum_{n=1}^{\infty} a_n$ defined by the formulas in Exercises 1–4 converge, and which diverge? Give reasons for your answers.

1. $\displaystyle\sum_{n=1}^{\infty} \frac{1}{(3n-2)^{n+(1/2)}}$

2. $\displaystyle\sum_{n=1}^{\infty} \frac{(\tan^{-1} n)^2}{n^2+1}$

3. $\displaystyle\sum_{n=1}^{\infty} (-1)^n \tanh n$

4. $\displaystyle\sum_{n=2}^{\infty} \frac{\log_n (n!)}{n^3}$

Which of the series $\sum_{n=1}^{\infty} a_n$ defined by the formulas in Exercises 5–8 converge, and which diverge? Give reasons for your answers.

5. $a_1 = 1, \quad a_{n+1} = \dfrac{n(n+1)}{(n+2)(n+3)} a_n$ (*Hint:* Write out several terms, see which factors cancel, and then generalize.)

6. $a_1 = a_2 = 7, \quad a_{n+1} = \dfrac{n}{(n-1)(n+1)} a_n$ if $n \geq 2$

7. $a_1 = a_2 = 1, \quad a_{n+1} = \dfrac{1}{1+a_n}$ if $n \geq 2$

8. $a_n = 1/3^n$ if n is odd, $a_n = n/3^n$ if n is even

Choosing Centers for Taylor Series

Taylor's formula

$$f(x) = f(a) + f'(a)(x-a) + \frac{f''(a)}{2!}(x-a)^2$$

$$+ \cdots + \frac{f^{(n)}(a)}{n!}(x-a)^n + \frac{f^{(n+1)}(c)}{(n+1)!}(x-a)^{n+1}$$

expresses the value of f at x in terms of the values of f and its derivatives at $x = a$. In numerical computations, we therefore need a to be a point where we know the values of f and its derivatives. We also need a to be close enough to the values of f we are interested in to make $(x-a)^{n+1}$ so small we can neglect the remainder.

In Exercises 9–14, what Taylor series would you choose to represent the function near the given value of x? (There may be more than one good answer.) Write out the first four nonzero terms of the series you choose.

9. $\cos x$ near $x = 1$

10. $\sin x$ near $x = 6.3$

11. e^x near $x = 0.4$

12. $\ln x$ near $x = 1.3$

13. $\cos x$ near $x = 69$

14. $\tan^{-1} x$ near $x = 2$

Theory and Examples

15. *nth root of* $a^n + b^n$ Let a and b be constants with $0 < a < b$. Does the sequence $\{(a^n + b^n)^{1/n}\}$ converge? If it does converge, what is the limit?

16. *Repeating decimal* Find the sum of the infinite series

$$1 + \frac{2}{10} + \frac{3}{10^2} + \frac{7}{10^3} + \frac{2}{10^4} + \frac{3}{10^5} + \frac{7}{10^6} + \frac{2}{10^7} + \frac{3}{10^8} + \frac{7}{10^9} + \cdots.$$

17. *Summing integrals* Evaluate

$$\sum_{n=0}^{\infty} \int_{n}^{n+1} \frac{1}{1+x^2}\, dx.$$

18. *Absolute convergence* Find all values of x for which

$$\sum_{n=1}^{\infty} \frac{nx^n}{(n+1)(2x+1)^n}$$

converges absolutely.

19. *Euler's constant* Graphs like those in Figure 8.13, suggest that as n increases, there is little change in the difference between the sum

$$1 + \frac{1}{2} + \cdots + \frac{1}{n}$$

and the integral

$$\ln n = \int_{1}^{n} \frac{1}{x}\, dx.$$

To explore this idea, carry out the following steps.

(a) By taking $f(x) = 1/x$ in Figure 8.13, show that

$$\ln (n+1) \le 1 + \frac{1}{2} + \cdots + \frac{1}{n} \le 1 + \ln n$$

or

$$0 < \ln (n+1) - \ln n \le 1 + \frac{1}{2} + \cdots + \frac{1}{n} - \ln n \le 1.$$

Thus, the sequence

$$a_n = 1 + \frac{1}{2} + \cdots + \frac{1}{n} - \ln n$$

is bounded from below and from above.

(b) Show that

$$\frac{1}{n+1} < \int_{n}^{n+1} \frac{1}{x}\, dx = \ln (n+1) - \ln n$$

and use this result to show that the sequence $\{a_n\}$ in part (a) is nonincreasing.

Since a nonincreasing sequence that is bounded from below converges, the numbers a_n defined in part (a) converge:

$$1 + \frac{1}{2} + \cdots + \frac{1}{n} - \ln n \to \gamma.$$

The number γ, whose value is $0.5772\ldots$, is called *Euler's constant*. In contrast to other special numbers like π and e, no other expression with a simple law of formulation has ever been found for γ.

20. *Generalizing Euler's constant* The figure below shows the graph of a positive twice-differentiable decreasing functizon f whose second derivative is positive on $(0, \infty)$. For each n, the number A_n is the area of the lunar region between the curve and the line segment joining the points $(n, f(n))$ and $(n+1, f(n+1))$.

(a) Use the figure to show that $\sum_{n=1}^{\infty} A_n < (1/2)(f(1) - f(2))$.

(b) Next show the existence of

$$\lim_{n\to\infty} \left[\sum_{k=1}^{n} f(k) - \frac{1}{2}(f(1) + f(n)) - \int_{1}^{n} f(x)\, dx \right].$$

(c) Then show the existence of

$$\lim_{n\to\infty} \left[\sum_{k=1}^{n} f(k) - \int_{1}^{n} f(x)\, dx \right].$$

If $f(x) = 1/x$, the limit in part (c) is Euler's constant. (*Source:* "Convergence with Pictures" by P. J. Rippon, *American Mathematical Monthly*, Vol. 93, No. 6 (1986), pp. 476–478.)

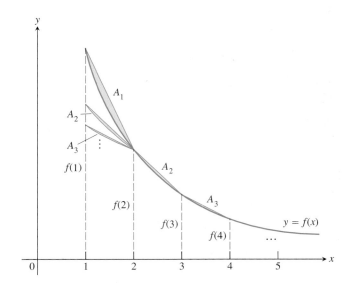

21. *Punching out triangles* This exercise refers to the "right side up" equilateral triangle with sides of length $2b$ in the accompanying figure. "Upside down" equilateral triangles are removed from the original triangle as the sequence of pictures suggests. The sum of the areas removed from the original triangle forms an infinite series.

(a) Find this infinite series.

(b) Find the sum of this infinite series and hence find the total area removed from the original triangle.

(c) Is every point on the original triangle removed? Explain why or why not.

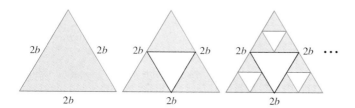

22. *A fast estimate of $\pi/2$* As you saw if you did Exercise 10 in Section 8.2, the sequence generated by starting with $x_0 = 1$ and applying the recursion formula $x_{n+1} = x_n + \cos x_n$ converges rapidly to $\pi/2$. To explain the speed of the convergence, let $\epsilon_n = (\pi/2) - x_n$. (See the accompanying figure.) Then

$$\epsilon_{n+1} = \frac{\pi}{2} - x_n - \cos x_n$$

$$= \epsilon_n - \cos\left(\frac{\pi}{2} - \epsilon_n\right)$$

$$= \epsilon_n - \sin \epsilon_n$$

$$= \frac{1}{3!}(\epsilon_n)^3 - \frac{1}{5!}(\epsilon_n)^5 + \cdots.$$

Use this equality to show that

$$0 < \epsilon_{n+1} < \frac{1}{6}(\epsilon_n)^3.$$

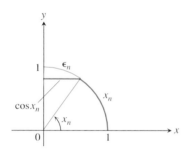

23. *Computer exploration*

(a) *Writing to Learn* Does the value of

$$\lim_{n\to\infty}\left(1 - \frac{\cos\,(a/n)}{n}\right)^n, \qquad a \text{ constant,}$$

appear to depend on the value of a? If so, how?

(b) *Writing to Learn* Does the value of

$$\lim_{n\to\infty}\left(1 - \frac{\cos\,(a/n)}{bn}\right)^n, \qquad a \text{ and } b \text{ constant, } b \neq 0,$$

appear to depend on the value of b? If so, how?

(c) Use calculus to confirm your findings in parts (a) and (b).

24. Show that if $\sum_{n=1}^{\infty} a_n$ converges, then

$$\sum_{n=1}^{\infty}\left(\frac{1 + \sin\,(a_n)}{2}\right)^n$$

converges.

25. *Radius of convergence* Find a value for the constant b that will make the radius of convergence of the power series

$$\sum_{n=2}^{\infty} \frac{b^n x^n}{\ln\,n}$$

equal to 5.

26. *Writing to Learn: Transcendental functions* How do you know that the functions $\sin x$, $\ln x$, and e^x are not polynomials? Give reasons for your answer.

27. *Raabe's (or Gauss's) Test* The following test, which we state without proof, is an extension of the Ratio Test.

Raabe's Test: If $\sum_{n=1}^{\infty} u_n$ is a series of positive constants and there exist constants C, K, and N such that

$$\frac{u_n}{u_{n+1}} = 1 + \frac{C}{n} + \frac{f(n)}{n^2},$$

where $|f(n)| < K$ for $n \geq N$, then $\sum_{n=1}^{\infty} u_n$ converges if $C > 1$ and diverges if $C \leq 1$.

Show that the results of Raabe's Test agree with what you know about the series $\sum_{n=1}^{\infty} (1/n^2)$ and $\sum_{n=1}^{\infty} (1/n)$.

28. *Using Raabe's Test* Suppose that the terms of $\sum_{n=1}^{\infty} u_n$ are defined recursively by the formulas

$$u_1 = 1, \qquad u_{n+1} = \frac{(2n-1)^2}{(2n)(2n+1)} u_n.$$

Apply Raabe's Test to determine whether the series converges.

29. Assume that $\sum_{n=1}^{\infty} a_n$ converges, $a_n \neq 1$, and $a_n > 0$ for all n.

(a) *Squaring terms* Show that $\sum_{n=1}^{\infty} a_n^2$ converges.

(b) *Writing to Learn* Does $\sum_{n=1}^{\infty} a_n/(1 - a_n)$ converge? Explain.

30. (*Continuation of Exercise 29*) If $\sum_{n=1}^{\infty} a_n$ converges and if $1 > a_n > 0$ for all n, show that $\sum_{n=1}^{\infty} \ln(1 - a_n)$ converges. (*Hint:* First show that $|\ln(1 - a_n)| \leq a_n/(1 - a_n)$.)

31. *Nicole Oresme's Theorem* Prove Nicole Oresme's Theorem that

$$1 + \frac{1}{2} \cdot 2 + \frac{1}{4} \cdot 3 + \cdots + \frac{n}{2^{n-1}} + \cdots = 4.$$

(*Hint:* differentiate both sides of the equation $1/(1 - x) = 1 + \sum_{n=1}^{\infty} x^n$.)

32. (a) *Term-by-term differentiation* Show that

$$\sum_{n=1}^{\infty} \frac{n(n + 1)}{x^n} = \frac{2x^2}{(x - 1)^3}$$

for $|x| > 1$ by differentiating the identity

$$\sum_{n=1}^{\infty} x^{n+1} = \frac{x^2}{1 - x}$$

twice, multiplying the result by x, and then replacing x by $1/x$.

(b) Use part (a) to find the real solution greater than 1 of the equation

$$x = \sum_{n=1}^{\infty} \frac{n(n + 1)}{x^n}.$$

33. *Summing exponential powers* Use the integral test to show that

$$\sum_{n=0}^{\infty} e^{-n^2}$$

converges.

34. *Writing to Learn* If $\sum_{n=1}^{\infty} a_n$ is a convergent series of positive numbers, can anything be said about the convergence of $\sum_{n=1}^{\infty} \ln(1 + a_n)$? Give reasons for your answer.

35. *Quality control*

(a) Differentiate the series

$$\frac{1}{1 - x} = 1 + x + x^2 + \cdots + x^n + \cdots$$

to obtain a series for $1/(1 - x)^2$.

(b) *Rolling dice* In one throw of two dice, the probability of getting a roll of 7 is $p = 1/6$. If you throw the dice repeatedly, the probability that a 7 will appear for the first time at the nth throw is $q^{n-1}p$, where $q = 1 - p = 5/6$. The expected number of throws until a 7 first appears is $\sum_{n=1}^{\infty} nq^{n-1}p$. Find the sum of this series.

(c) As an engineer applying statistical control to an industrial operation, you inspect items taken at random from the assembly line. You classify each sampled item as either "good" or "bad." If the probability of an item's being good is p and of an item's being bad is $q = 1 - p$, the probability that the first bad item found is the nth one inspected is $p^{n-1}q$.

The average number inspected up to and including the first bad item found is $\sum_{n=1}^{\infty} np^{n-1}q$. Evaluate this sum, assuming $0 < p < 1$.

36. *Expected value* Suppose that a random variable X may assume the values 1, 2, 3, $\ldots$, with probabilities $p_1, p_2, p_3, \ldots$, where p_k is the probability that X equals k ($k = 1, 2, 3, \ldots$). Suppose also that $p_k \geq 0$ and $\sum_{k=1}^{\infty} p_k = 1$. The **expected value** of X, denoted by $E(X)$, is the number $\sum_{k=1}^{\infty} kp_k$, provided the series converges. In each of the following cases, show that $\sum_{k=1}^{\infty} p_k = 1$ and find $E(X)$ if it exists. (*Hint:* See Exercise 35.)

(a) $p_k = 2^{-k}$ **(b)** $p_k = \dfrac{5^{k-1}}{6^k}$

(c) $p_k = \dfrac{1}{k(k + 1)} = \dfrac{1}{k} - \dfrac{1}{k + 1}$

37. *Safe and effective dosage* The concentration in the blood resulting from a single dose of a drug normally decreases with time as the drug is eliminated from the body. Doses may therefore need to be repeated periodically to keep the concentration from dropping below some particular level. One model for the effect of repeated doses gives the residual concentration just before the $(n + 1)$st dose as

$$R_n = C_0 e^{-kt_0} + C_0 e^{-2kt_0} + \cdots + C_0 e^{-nkt_0},$$

where $C_0 =$ the change in concentration achievable by a single dose (milligrams per milliliter), $k =$ the *elimination constant* (per hour), and $t_0 =$ time between doses (hours). See the accompanying figure.

(a) Write R_n in closed form as a single fraction and find $R = \lim_{n \to \infty} R_n$.

(b) Calculate R_1 and R_{10} for $C_0 = 1$ mg/mL, $k = 0.1$ h^{-1}, and $t_0 = 10$ h. How good an estimate of R is R_{10}?

(c) If $k = 0.01$ h^{-1} and $t_0 = 10$ h, find the smallest n such that $R_n > (1/2)R$.

(*Source: Prescribing Safe and Effective Dosage* by B. Horelick and S. Koont (Lexington, MA: COMAP, Inc., 1979).)

38. *Time between drug doses (continuation of Exercise 37)* If a drug is known to be ineffective below a concentration C_L and harmful above some higher concentration C_H, one needs to find values of

C_0 and t_0 that will produce a concentration that is safe (not above C_H) but effective (not below C_L). See the accompanying figure.

We therefore want to find values for C_0 and t_0 for which

$$R = C_L \quad \text{and} \quad C_0 + R = C_H.$$

Thus, $C_0 = C_H - C_L$. When these values are substituted in the equation for R obtained in part (a) of Exercise 37, the resulting equation simplifies to

$$t_0 = \frac{1}{k} \ln \frac{C_H}{C_L}.$$

To reach an effective level rapidly, one might administer a "loading" dose that would produce a concentration of C_H milligrams per milliliter. This could be followed every t_0 hours by a dose that raises the concentration by $C_0 = C_H - C_L$ milligrams per milliliter.

(a) Verify the preceding equation for t_0.

(b) If $k = 0.05$ h^{-1} and the highest safe concentration is e times the lowest effective concentration, find the length of time between doses that will assure safe and effective concentrations.

(c) *Writing to Learn* Given $C_H = 2$ mg/mL, $C_L = 0.5$ mg/mL, and $k = 0.02$ h^{-1}, determine a scheme for administering the drug.

(d) Suppose that $k = 0.2$ h^{-1} and that the smallest effective concentration is 0.03 mg/mL. A single dose that produces a concentration of 0.1 mg/mL is administered. About how long will the drug remain effective?

39. *An infinite product* The infinite product

$$\prod_{n=1}^{\infty} (1 + a_n) = (1 + a_1)(1 + a_2)(1 + a_3) \cdots$$

is said to converge if the series

$$\sum_{n=1}^{\infty} \ln (1 + a_n),$$

obtained by taking the natural logarithm of the product, converges. Prove that the product converges if $a_n > -1$ for every n and if $\sum_{n=1}^{\infty} |a_n|$ converges. (*Hint:* Show that

$$|\ln (1 + a_n)| \le \frac{|a_n|}{1 - |a_n|} \le 2 |a_n|$$

when $|a_n| < 1/2$.)

40. *Extended logarithmic p-series* If p is a constant, show that the series

$$1 + \sum_{n=3}^{\infty} \frac{1}{n \cdot \ln n \cdot [\ln (\ln n)]^p}$$

(a) converges if $p > 1$

(b) diverges if $p \le 1$.

In general, if $f_1(x) = x, f_{n+1}(x) = \ln (f_n(x))$ and n takes on the values 1, 2, 3, . . . , we find that $f_2(x) = \ln x, f_3(x) = \ln (\ln x)$, and so on. If $f_n(a) > 1$, then

$$\int_a^{\infty} \frac{dx}{f_1(x) f_2(x) \cdots f_n(x) (f_{n+1}(x))^p}$$

converges if $p > 1$ and diverges if $p \le 1$.

9

Vectors in the Plane and Polar Functions

CD-ROM
WEBsite

Historical Biography

Jakob Bernoulli
(1654 — 1705)

OVERVIEW   When a body travels in the *xy*-plane, the parametric equations $x = f(t)$ and $y = g(t)$ can be used to model the body's motion and path. In this chapter, we introduce the *vector* form of parametric equations, which allows us to track the positions of moving bodies with vectors, calculate the directions and magnitudes of their velocities and accelerations, and predict the effects of the forces we see working on them.

One of the principal applications of vector functions is the analysis of motion in space. Planetary motion is best described with polar coordinates (another of Newton's inventions, although Jakob Bernoulli usually gets the credit because he published first), so we investigate curves, derivatives, and integrals in this new coordinate system.

9.1 Vectors in the Plane

Component Form • Zero Vector and Unit Vectors • Vector Algebra Operations • Standard Unit Vectors • Length and Direction • Tangents and Normals

Some of the things we measure are determined by their magnitudes. To record mass, length, or time, for example, we need only write down a number and name an appropriate unit of measure. These are *scalar quantities*, and the associated real numbers are **scalars.** We need more information to describe a force, displacement, or velocity. To describe a force, we need to record the direction in which it acts as well as how large it is. To describe a body's displacement, we have to say in what direction it moved as well as how far. To describe a body's velocity, we have to know where the body is headed as well as how fast it is going.

Component Form

A quantity such as force, displacement, or velocity is represented by a **directed line segment** (Figure 9.1). The arrow points in the direction of the action and its length gives the magnitude of the action in terms of a suitably chosen unit. For example, a force vector points in the direction in which it is applied and its length is a measure of its strength; a velocity vector points in the direction of motion and its length is the speed of the moving object. (We say more about force and velocity in Sections 9.3 and 9.4.)

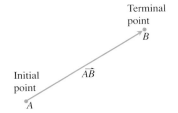

FIGURE 9.1 The directed line segment $\overrightarrow{AB}$.

717

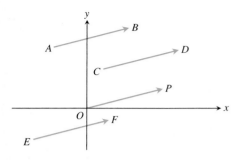

FIGURE 9.2 The four arrows (directed line segments) shown here have the same length and direction. They therefore represent the same vector, and we write $\overrightarrow{AB} = \overrightarrow{CD} = \overrightarrow{OP} = \overrightarrow{EF}$.

The directed line segment $\overrightarrow{AB}$ has **initial point** A and **terminal point** B; its **length** is denoted by $|\overrightarrow{AB}|$. Directed line segments that have the same length and direction are **equivalent.**

Definitions   Vector, Equal Vectors

A **vector** in the plane is a directed line segment. Two vectors are **equal** (or **the same**) if they have the same length and direction.

Thus, the arrows we use when we draw vectors are understood to represent the same vector if they have the same length, are parallel, and point in the same direction (Figure 9.2).

In textbooks, vectors are usually written in lowercase, boldface letters, for example **u**, **v**, and **w**. Sometimes we use uppercase boldface letters, such as **F,** to denote a force vector. In handwritten form, it is customary to draw small arrows above the letters, for example $\vec{u}$, $\vec{v}$, $\vec{w}$, and $\vec{F}$.

Example 1   Showing Vectors are Equal

Let $A = (0, 0)$, $B = (3, 4)$, $C = (-4, 2)$, and $D = (-1, 6)$. Show that the vectors $\mathbf{u} = \overrightarrow{AB}$ and $\mathbf{v} = \overrightarrow{CD}$ are equal.

Solution   We need to show that **u** and **v** have the same length and direction (Figure 9.3). We use the distance formula to find their lengths.

$$|\mathbf{u}| = |\overrightarrow{AB}| = \sqrt{(3 - 0)^2 + (4 - 0)^2} = 5$$

$$|\mathbf{v}| = |\overrightarrow{CD}| = \sqrt{(-1 - (-4))^2 + (6 - 2)^2} = 5$$

Next we calculate the slopes of the two line segments.

$$\text{Slope of } \overrightarrow{AB} = \frac{4 - 0}{3 - 0} = \frac{4}{3}, \qquad \text{slope of } \overrightarrow{CD} = \frac{6 - 2}{-1 - (-4)} = \frac{4}{3}$$

The line segments have the same direction because they are parallel and directed toward the upper right. Therefore, $\mathbf{u} = \mathbf{v}$ because they have the same length and direction.

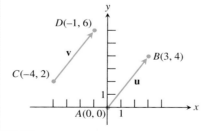

FIGURE 9.3 Two equal vectors. They have the same length and direction. (Example 1)

Let $\mathbf{v} = \overrightarrow{PQ}$. There is one directed line segment equivalent to $\overrightarrow{PQ}$ whose initial point is the origin (Figure 9.4). It is the representative of **v** in **standard position** and is the vector we normally use to represent **v**.

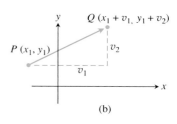

FIGURE 9.4 (a) The standard position of a vector is where the initial point is the origin. (b) The coordinates of Q satisfy $x_2 = x_1 + v_1$ and $y_2 = y_1 + v_2$.

Definition Component Form of a Vector

If $\mathbf{v}$ is a vector in the plane equal to the vector with initial point $(0, 0)$ and terminal point (v_1, v_2), then the **component form** of $\mathbf{v}$ is

$$\mathbf{v} = \langle v_1, v_2 \rangle.$$

Thus, a vector in the plane is also an ordered pair $\langle v_1, v_2 \rangle$ of real numbers. The numbers v_1 and v_2 are the **components** of $\mathbf{v}$. The vector $\langle v_1, v_2 \rangle$ is called the **position vector** of the point (v_1, v_2).

Observe that if $\mathbf{v} = \langle v_1, v_2 \rangle$ is represented by the directed line segment $\overrightarrow{PQ}$, where the initial point is $P(x_1, y_1)$ and the terminal point is $Q(x_2, y_2)$, then $x_1 + v_1 = x_2$ and $y_1 + v_2 = y_2$ (Figure 9.4) so that $v_1 = x_2 - x_1$ and $v_2 = y_2 - y_1$ are the components of $\overrightarrow{PQ}$. In summary,

Given the points $P(x_1, y_1)$ and $Q(x_2, y_2)$, the position vector $\mathbf{v} = \langle v_1, v_2 \rangle$ equivalent to $\overrightarrow{PQ}$ is

$$\mathbf{v} = \langle x_2 - x_1, y_2 - y_1 \rangle.$$

Two vectors $\langle a, b \rangle$ and $\langle c, d \rangle$ are equal if and only if $a = c$ and $b = d$, so that $\langle v_1, v_2 \rangle = \langle x_2 - x_1, y_2 - y_1 \rangle$.

The **magnitude** or **length** of the vector $\overrightarrow{PQ}$ is the length of any of its equivalent directed line segment representations. In particular, if $\mathbf{v} = \langle x_2 - x_1, y_2 - y_1 \rangle$ is the position vector for $\overrightarrow{PQ}$ (Figure 9.4), then the distance formula gives the magnitude or length of $\mathbf{v}$, denoted by the symbol $|\mathbf{v}|$ or $\|\mathbf{v}\|$.

The **magnitude** or **length** of the vector $\mathbf{v} = \overrightarrow{PQ}$ is

$$|\mathbf{v}| = \sqrt{v_1{}^2 + v_2{}^2} = \sqrt{(x_2 - x_1)^2 + (y_2 - y_1)^2}$$

(Figure 9.4).

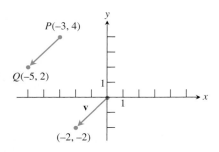

FIGURE 9.5 The vector $\overrightarrow{PQ}$ equals the position vector $v = \langle -2, -2 \rangle$. (Example 2)

Example 2 Finding the Component Form and Length of a Vector

Find the **(a)** component form and **(b)** length of the vector with initial point $P = (-3, 4)$ and terminal point $Q = (-5, 2)$.

Solution

(a) The position vector **v** representing $\overrightarrow{PQ}$ has components $v_1 = x_2 - x_1 = (-5) - (-3) = -2$ and $v_2 = y_2 - y_1 = 2 - 4 = -2$ (Figure 9.5). The component form of $\overrightarrow{PQ}$ is

$$\mathbf{v} = \langle -2, -2 \rangle.$$

(b) The length of $\mathbf{v} = \overrightarrow{PQ}$ is

$$|\mathbf{v}| = \sqrt{(-2)^2 + (-2)^2} = 2\sqrt{2}.$$

Example 3 Force Moving a Cart

A small cart is being pulled along a smooth horizontal floor with a 20 lb force **F** making a 45° angle to the floor (Figure 9.6). What is the effective force moving the cart forward?

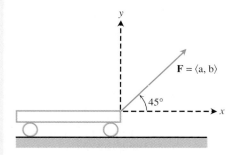

FIGURE 9.6 The force pulling the cart forward is represented by the vector **F** of length 20 (pounds) making an angle of 45° with the horizontal ground (positive x-axis). (Example 3)

Solution   The effective force is the horizontal component of $\mathbf{F} = \langle a, b \rangle$, given by

$$a = |\mathbf{F}| \cos 45° = (20)\left(\frac{\sqrt{2}}{2}\right) \approx 14.14 \text{ lb}.$$

Zero Vector and Unit Vectors

The only vector with length 0 is the **zero vector**

$$\mathbf{0} = \langle 0, 0 \rangle.$$

The zero vector is also the only vector with no specific direction.

Any vector **v** of length 1 is a **unit vector.** If $\mathbf{v} = \langle v_1, v_2 \rangle$ makes an angle θ with the positive x-axis, then

$$v_1 = |\mathbf{v}| \cos \theta = \cos \theta \qquad |v| = 1$$
$$v_2 = |\mathbf{v}| \sin \theta = \sin \theta$$

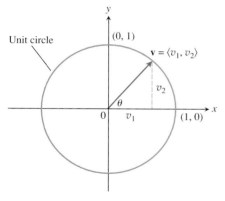

FIGURE 9.7 The unit vector $\mathbf{v} = \langle v_1, v_2 \rangle$ has length 1, so $v_1 = \cos \theta$ and $v_2 = \sin \theta$, where θ is the angle $\mathbf{v}$ makes with the positive x-axis. As θ varies from 0 to 2π, the terminal point of $\mathbf{v}$ traces the unit circle.

(Figure 9.7). In summary,

> A unit vector $\mathbf{v}$ in the plane having angle θ with the positive x-axis is represented by
>
> $$\mathbf{v} = \langle \cos \theta, \sin \theta \rangle.$$

As θ varies from 0 to 2π, the terminal point of a unit vector $\mathbf{v}$ traces the unit circle counterclockwise, taking into account all possible directions.

Vector Algebra Operations

Two principal operations involving vectors are *vector addition* and *scalar multiplication*.

Definitions   Vector Addition and Multiplication of a Vector by a Scalar

Let $\mathbf{u} = \langle u_1, u_2 \rangle$, $\mathbf{v} = \langle v_1, v_2 \rangle$ be vectors with k a scalar (real number).

Addition: $\qquad\qquad \mathbf{u} + \mathbf{v} = \langle u_1, u_2 \rangle + \langle v_1, v_2 \rangle = \langle u_1 + v_1, u_2 + v_2 \rangle$

Scalar multiplication: $\quad k\mathbf{u} = \langle ku_1, ku_2 \rangle$

The definition of vector addition is illustrated geometrically in Figure 9.8a, where the initial point of one vector is placed at the terminal point of the other. Another interpretation is shown in Figure 9.8b (called the **parallelogram law** of addition), where the sum, called the **resultant vector,** is the diagonal of the parallelogram. In physics, forces add vectorially as do velocities, accelerations, and so on.

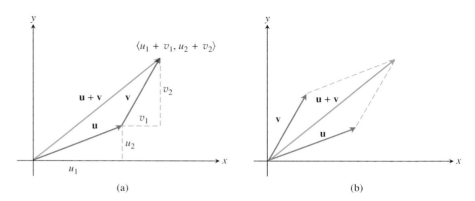

FIGURE 9.8 (a) Geometric interpretation of the vector sum. (b) The parallelogram law of vector addition.

A geometric interpretation of the product $k\mathbf{u}$ of the scalar k and vector $\mathbf{u}$ is displayed in Figure 9.9. First, if $k > 0$, then $k\mathbf{u}$ has the same direction as $\mathbf{u}$; if $k < 0$, then the direction of $k\mathbf{u}$ is opposite to that of $\mathbf{u}$. Comparing the lengths of $\mathbf{u} = \langle u_1, u_2 \rangle$ and $k\mathbf{u}$, we see that

$$|k\mathbf{u}| = \sqrt{(ku_1)^2 + (ku_2)^2} = \sqrt{k^2(u_1^2 + u_2^2)}$$
$$= \sqrt{k^2}\sqrt{u_1^2 + u_2^2} = |k|\,|\mathbf{u}|.$$

That is, the length of $k\mathbf{u}$ is the absolute value of the scalar k times the length of $\mathbf{u}$. In particular, the vector $(-1)\mathbf{u} = -\mathbf{u}$ has the same length as $\mathbf{u}$ but points in the opposite direction.

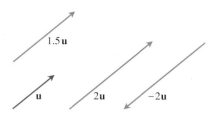

FIGURE 9.9 Scalar multiples of $\mathbf{u}$.

By the **difference** $\mathbf{u} - \mathbf{v}$ of two vectors, we mean

$$\mathbf{u} - \mathbf{v} = \mathbf{u} + (-\mathbf{v}).$$

If $\mathbf{u} = \langle u_1, u_2 \rangle$ and $\mathbf{v} = \langle v_1, v_2 \rangle$, then

$$\mathbf{u} - \mathbf{v} = \langle u_1 - v_1, u_2 - v_2 \rangle.$$

Note that $(\mathbf{u} - \mathbf{v}) + \mathbf{v} = \mathbf{u}$, so adding the vector $(\mathbf{u} - \mathbf{v})$ to $\mathbf{v}$ gives $\mathbf{u}$ (Figure 9.10a). Figure 9.10b shows the difference $\mathbf{u} - \mathbf{v}$ as the sum $\mathbf{u} + (-\mathbf{v})$.

(a)

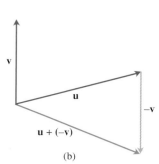

(b)

FIGURE 9.10 (a) The vector $\mathbf{u} - \mathbf{v}$, when added to $\mathbf{v}$, gives $\mathbf{u}$. (b) $\mathbf{u} - \mathbf{v} = \mathbf{u} + (-\mathbf{v})$.

Example 4   Performing Operations on Vectors

Let $\mathbf{u} = \langle -1, 3 \rangle$ and $\mathbf{v} = \langle 4, 7 \rangle$. Find

(a) $2\mathbf{u} + 3\mathbf{v}$ (b) $\mathbf{u} - \mathbf{v}$ (c) $\left|\dfrac{1}{2}\mathbf{u}\right|$.

Solution

(a) $2\mathbf{u} + 3\mathbf{v} = 2\langle -1, 3 \rangle + 3\langle 4, 7 \rangle$
$= \langle 2(-1) + 3(4),\ 2(3) + 3(7) \rangle = \langle 10, 27 \rangle$

(b) $\mathbf{u} - \mathbf{v} = \langle -1, 3 \rangle - \langle 4, 7 \rangle$
$= \langle -1 - 4,\ 3 - 7 \rangle = \langle -5, -4 \rangle$

(c) $\left|\dfrac{1}{2}\mathbf{u}\right| = \left|\left\langle -\dfrac{1}{2}, \dfrac{3}{2} \right\rangle\right| = \sqrt{\left(-\dfrac{1}{2}\right)^2 + \left(\dfrac{3}{2}\right)^2} = \dfrac{1}{2}\sqrt{10}$

Vector operations have many of the properties of ordinary arithmetic. These properties are readily verified using the definitions of vector addition and multiplication by a scalar.

> **Properties of Vector Operations**
> Let **u**, **v**, **w** be vectors and a, b be scalars.
>
> 1. $\mathbf{u} + \mathbf{v} = \mathbf{v} + \mathbf{u}$ 2. $(\mathbf{u} + \mathbf{v}) + \mathbf{w} = \mathbf{u} + (\mathbf{v} + \mathbf{w})$
>
> 3. $\mathbf{u} + \mathbf{0} = \mathbf{u}$ 4. $\mathbf{u} + (-\mathbf{u}) = \mathbf{0}$
>
> 5. $0\mathbf{u} = \mathbf{0}$ 6. $1\mathbf{u} = \mathbf{u}$
>
> 7. $a(b\mathbf{u}) = (ab)\mathbf{u}$ 8. $a(\mathbf{u} + \mathbf{v}) = a\mathbf{u} + a\mathbf{v}$
>
> 9. $(a + b)\mathbf{u} = a\mathbf{u} + b\mathbf{u}$

An important application of vectors occurs in navigation.

Example 5   Finding Ground Speed and Direction

A Boeing® 727® airplane, flying due east at 500 mph in still air, encounters a 70 mph tailwind acting in the direction 60° north of east. The airplane holds its compass heading due east but, because of the wind, acquires a new ground speed and direction. What are they?

Solution   If **u** = the velocity of the airplane alone and **v** = the velocity of the tailwind, then $|\mathbf{u}| = 500$ and $|\mathbf{v}| = 70$ (Figure 9.11). We need to find the magnitude and direction of the *resultant vector* **u** + **v**. If we let the positive x-axis represent east and the positive y-axis represent north, then the component forms of **u** and **v** are

$$\mathbf{u} = \langle 500, 0 \rangle \qquad \text{and} \qquad \mathbf{v} = \langle 70 \cos 60°, 70 \sin 60° \rangle = \langle 35, 35\sqrt{3} \rangle.$$

Therefore,

$$\mathbf{u} + \mathbf{v} = \langle 535, 35\sqrt{3} \rangle$$

$$|\mathbf{u} + \mathbf{v}| = \sqrt{535^2 + (35\sqrt{3})^2} \approx 538.4$$

and

$$\theta = \tan^{-1} \frac{35\sqrt{3}}{535} \approx 6.5°. \qquad \text{Fig. 9.11}$$

FIGURE 9.11 Vectors representing the velocities of the airplane and tailwind in Example 5.

NOT TO SCALE

Interpret

The new ground speed of the airplane is about 538.4 mph, and its new direction is about 6.5° north of east.

Standard Unit Vectors

Any vector $\mathbf{v} = \langle a, b \rangle$ in the plane can be written as a *linear combination* of the two **standard unit vectors**

$$\mathbf{i} = \langle 1, 0 \rangle \qquad \text{and} \qquad \mathbf{j} = \langle 0, 1 \rangle$$

as follows:

$$\mathbf{v} = \langle a, b \rangle = \langle a, 0 \rangle + \langle 0, b \rangle = a\langle 1, 0 \rangle + b\langle 0, 1 \rangle$$
$$= a\mathbf{i} + b\mathbf{j}.$$

The vector **v** is a **linear combination** of the vectors **i** and **j**; the scalar a is the **horizontal** or **i-component** of **v** and the scalar b is the **vertical** or **j-component** of **v**. The **slope** of a nonvertical vector $\mathbf{v} = \langle a, b \rangle$ is the slope shared by the lines parallel to it. Thus, if $a \neq 0$, the vector **v** has slope b/a (Figure 9.12).

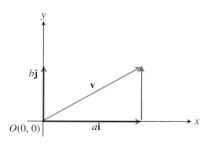

FIGURE 9.12 **v** is a linear combination of **i** and **j**.

Example 6   Expressing Vectors as Linear Combinations of i and j

Let $P = (-1, 5)$ and $Q = (3, 2)$. Write the vector $\mathbf{v} = \overrightarrow{PQ}$ as a linear combination of **i** and **j** and find its slope.

Solution   The component form of **v** is $\langle 3 - (-1), 2 - 5 \rangle = \langle 4, -3 \rangle$. Thus,

$$\mathbf{v} = \langle 4, -3 \rangle = 4\mathbf{i} + (-3)\mathbf{j} = 4\mathbf{i} - 3\mathbf{j}.$$

The slope of **v** is $-3/4$.

Length and Direction

In studying motion, we often want to know the direction an object is headed and how fast it is going.

Example 7   Expressing Velocity as Speed Times Direction

If $\mathbf{v} = 3\mathbf{i} - 4\mathbf{j}$ is a velocity vector, express **v** as a product of its speed times a unit vector in the direction of motion.

Solution   Speed is the magnitude (length) of **v**:

$$|\mathbf{v}| = \sqrt{(3)^2 + (-4)^2} = \sqrt{9 + 16} = 5.$$

The vector $\mathbf{v}/|\mathbf{v}|$ has the same direction as **v**:

$$\frac{\mathbf{v}}{|\mathbf{v}|} = \frac{3\mathbf{i} - 4\mathbf{j}}{5} = \frac{3}{5}\mathbf{i} - \frac{4}{5}\mathbf{j}.$$

Moreover, $\mathbf{v}/|\mathbf{v}|$ is a unit vector:

$$\left| \frac{\mathbf{v}}{|\mathbf{v}|} \right| = \sqrt{\left(\frac{3}{5}\right)^2 + \left(-\frac{4}{5}\right)^2} = \sqrt{\frac{9}{25} + \frac{6}{25}} = 1.$$

Thus,

$$\mathbf{v} = 3\mathbf{i} - 4\mathbf{j} = 5\left(\frac{3}{5}\mathbf{i} - \frac{4}{5}\mathbf{j}\right).$$

$\underset{\text{Length}}{\underbrace{}}$ $\underset{\text{Direction of motion}}{\underbrace{}}$

Length
(speed)   Direction of motion

Generally, if $\mathbf{v} \neq \mathbf{0}$, then its length $|\mathbf{v}|$ is not zero and

$$\left| \frac{1}{|\mathbf{v}|} \mathbf{v} \right| = \frac{1}{|\mathbf{v}|} |\mathbf{v}| = 1.$$

That is, $\mathbf{v}/|\mathbf{v}|$ is a unit vector in the direction of $\mathbf{v}$. We can therefore express $\mathbf{v}$ in terms of its two important features, length and direction, by writing $\mathbf{v} = |\mathbf{v}|(\mathbf{v}/|\mathbf{v}|)$.

If $\mathbf{v} \neq \mathbf{0}$, then

1. $\dfrac{\mathbf{v}}{|\mathbf{v}|}$ is a unit vector in the direction of $\mathbf{v}$;

2. the equation $\mathbf{v} = |\mathbf{v}| \dfrac{\mathbf{v}}{|\mathbf{v}|}$ expresses $\mathbf{v}$ in terms of its length and direction.

The unit vector $\mathbf{v}/|\mathbf{v}|$ is called the **direction** of $\mathbf{v}$. Thus, $\mathbf{v} = 5((3/5)\mathbf{i} - (4/5)\mathbf{j})$ expresses the velocity vector in Example 7 as a *product* of its length and direction.

Tangents and Normals

When an object is moving along a path in the plane (or in space), its velocity is a vector tangent to the path. Moreover, if the object is speeding up or slowing down, forces are acting in the tangent direction and perpendicular (or normal) to it. (We investigate motion along a path in the plane in Section 9.3.)

A vector is **tangent** or **normal** to a curve at a point P if it is parallel or normal, respectively, to the line that is tangent to the curve at P. Example 8 shows how to find such vectors for a differentiable curve $y = f(x)$ in the plane.

Example 8   Finding Vectors Tangent and Normal to a Curve

An object is moving along the curve

$$y = \frac{x^3}{2} + \frac{1}{2}.$$

Find unit vectors tangent and normal to the curve at the point $(1, 1)$.

Solution   We find the unit vectors that are parallel and normal to the curve's tangent line at $(1, 1)$ (Figure 9.13).

The slope of the line tangent to the curve at $(1, 1)$ is

$$y' = \frac{3x^2}{2} \bigg|_{x=1} = \frac{3}{2}.$$

We look for a unit vector with this slope. The vector $\mathbf{v} = 2\mathbf{i} + 3\mathbf{j}$ has slope $3/2$, as does every nonzero multiple of $\mathbf{v}$. To find a multiple of $\mathbf{v}$ that is a unit vector, we divide $\mathbf{v}$ by

$$|\mathbf{v}| = \sqrt{2^2 + 3^2} = \sqrt{13},$$

obtaining

$$\mathbf{u} = \frac{\mathbf{v}}{|\mathbf{v}|} = \frac{2}{\sqrt{13}} \mathbf{i} + \frac{3}{\sqrt{13}} \mathbf{j}.$$

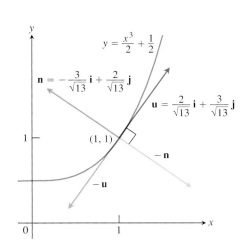

FIGURE 9.13 The unit tangent and normal vectors at the point $(1, 1)$ on the curve $y = (x^3/2) + 1/2$. (Example 8)

The vector **u** is tangent to the curve at $(1, 1)$ because it has the same direction as **v**. Of course,

$$-\mathbf{u} = -\frac{2}{\sqrt{13}}\mathbf{i} - \frac{3}{\sqrt{13}}\mathbf{j},$$

which points in the opposite direction, is also tangent to the curve at $(1, 1)$. Without some additional requirement (such as specifying the direction of motion), there is no reason to prefer one of these vectors to the other.

To find unit vectors normal to the curve at $(1, 1)$, we look for unit vectors whose slopes are the negative reciprocal of the slope of **u**. This is quickly done by interchanging the scalar components of **u** and changing the sign of one of them. We obtain

$$\mathbf{n} = -\frac{3}{\sqrt{13}}\mathbf{i} + \frac{2}{\sqrt{13}}\mathbf{j} \quad \text{and} \quad -\mathbf{n} = \frac{3}{\sqrt{13}}\mathbf{i} - \frac{2}{\sqrt{13}}\mathbf{j}.$$

Again, either one will do. The vectors have opposite directions but both are normal to the curve at $(1, 1)$. (See Figure 9.13.)

EXERCISES 9.1

Component Form

In Exercises 1–8, let $\mathbf{u} = \langle 3, -2 \rangle$ and $\mathbf{v} = \langle -2, 5 \rangle$. Find the **(a)** component form and **(b)** magnitude (length) of the vector.

1. $3\mathbf{u}$
2. $-2\mathbf{v}$
3. $\mathbf{u} + \mathbf{v}$
4. $\mathbf{u} - \mathbf{v}$
5. $2\mathbf{u} - 3\mathbf{v}$
6. $-2\mathbf{u} + 5\mathbf{v}$
7. $\frac{3}{5}\mathbf{u} + \frac{4}{5}\mathbf{v}$
8. $-\frac{5}{13}\mathbf{u} + \frac{12}{13}\mathbf{v}$

In Exercises 9–16, find the component form of the vector.

9. The vector $\overrightarrow{PQ}$, where $P = (1, 3)$ and $Q = (2, -1)$

10. The vector $\overrightarrow{OP}$ where O is the origin and P is the midpoint of segment RS, where $R = (2, -1)$ and $S = (-4, 3)$

11. The vector from the point $A = (2, 3)$ to the origin

12. The sum of $\overrightarrow{AB}$ and $\overrightarrow{CD}$, where $A = (1, -1)$, $B = (2, 0)$, $C = (-1, 3)$, and $D = (-2, 2)$

13. The unit vector that makes an angle $\theta = 2\pi/3$ with the positive x-axis

14. The unit vector that makes an angle $\theta = -3\pi/4$ with the positive x-axis

15. The unit vector obtained by rotating the vector $\langle 0, 1 \rangle$ $120°$ counterclockwise about the origin

16. The unit vector obtained by rotating the vector $\langle 1, 0 \rangle$ $135°$ counterclockwise about the origin

Geometry and Calculation

In Exercises 17 and 18, copy vectors **u**, **v**, and **w** head to tail as needed to sketch the indicated vector.

17.

(a) $\mathbf{u} + \mathbf{v}$ **(b)** $\mathbf{u} + \mathbf{v} + \mathbf{w}$

(c) $\mathbf{u} - \mathbf{v}$ **(d)** $\mathbf{u} - \mathbf{w}$

18.

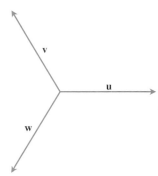

(a) $\mathbf{u} - \mathbf{v}$ **(b)** $\mathbf{u} - \mathbf{v} + \mathbf{w}$

(c) $2\mathbf{u} - \mathbf{v}$ **(d)** $\mathbf{u} + \mathbf{v} + \mathbf{w}$

Using Linear Combinations

Express the vectors in Exercises 19–24 in the form $a\mathbf{i} + b\mathbf{j}$ and sketch them as arrows in the coordinate plane beginning at the origin.

19. $\overrightarrow{P_1P_2}$ if P_1 is the point $(5, 7)$ and P_2 is the point $(2, 9)$

20. $\overrightarrow{P_1P_2}$ if P_1 is the point $(1, 2)$ and P_2 is the point $(-3, 5)$

21. $\overrightarrow{AB}$ if A is the point $(-5, 3)$ and B is the point $(-10, 8)$

22. $\overrightarrow{AB}$ if A is the point $(-7, -8)$ and B is the point $(6, 11)$

23. $\overrightarrow{P_1P_2}$ if P_1 is the point $(1, 3)$ and P_2 is the point $(2, -1)$

24. $\overrightarrow{P_3P_4}$ if P_3 is the point $(1, 3)$ and P_4 is the midpoint of the line segment P_1P_2 joining $P_1(2, -1)$ and $P_2(-4, 3)$

Unit Vectors

Sketch the vectors in Exercises 25–28 and express each vector in the form $a\mathbf{i} + b\mathbf{j}$.

25. The unit vectors $\mathbf{u} = (\cos\theta)\mathbf{i} + (\sin\theta)\mathbf{j}$ for $\theta = \pi/6$ and $\theta = 2\pi/3$. Include the circle $x^2 + y^2 = 1$ in your sketch.

26. The unit vectors $\mathbf{u} = (\cos\theta)\mathbf{i} + (\sin\theta)\mathbf{j}$ for $\theta = -\pi/4$ and $\theta = -3\pi/4$. Include the circle $x^2 + y^2 = 1$ in your sketch.

27. The unit vector obtained by rotating $\mathbf{j}$ counterclockwise $3\pi/4$ rad about the origin

28. The unit vector obtained by rotating $\mathbf{j}$ clockwise $2\pi/3$ rad about the origin

In Exercises 29–32, find a unit vector in the direction of the given vector.

29. $\langle 3, 4 \rangle$ **30.** $\langle 4, -3 \rangle$

31. $\langle -15, 8 \rangle$ **32.** $\langle -5, -2 \rangle$

For the vectors in Exercises 33 and 34, find unit vectors $\mathbf{u} = (\cos\theta)\mathbf{i} + (\sin\theta)\mathbf{j}$ in the same direction.

33. $6\mathbf{i} - 8\mathbf{j}$ **34.** $-\mathbf{i} + 3\mathbf{j}$

Length and Direction

In Exercises 35 and 36, express each vector as a product of its length and direction.

35. $5\mathbf{i} + 12\mathbf{j}$ **36.** $2\mathbf{i} - 3\mathbf{j}$

37. Find the unit vectors that are parallel to the vector $3\mathbf{i} - 4\mathbf{j}$ (two vectors in all).

38. Find a vector of length 2 whose direction is the opposite of the direction of the vector $-\mathbf{i} + 2\mathbf{j}$. How many such vectors are there?

Tangent and Normal Vectors

In Exercises 39–42, find the unit vectors that are tangent and normal to the curve at the given point (four vectors in all). Then sketch the vectors and curve together.

39. $y = x^2$, $(2, 4)$

40. $x^2 + 2y^2 = 6$, $(2, 1)$

41. $y = \tan^{-1}x$, $(1, \pi/4)$

42. $y = \sum_{n=0}^{\infty} \dfrac{x^n}{n!}$, $(0, 1)$

In Exercises 43–46, find the unit vectors that are tangent and normal to the curve at the given point (four vectors in all).

43. $3x^2 + 8xy + 2y^2 - 3 = 0$, $(1, 0)$

44. $x^2 - 6xy + 8y^2 - 2x - 1 = 0$, $(1, 1)$

45. $y = \int_0^x \sqrt{3 + t^4}\, dt$, $(0, 0)$

46. $y = \int_e^x \ln(\ln t)\, dt$, $(e, 0)$

Theory and Applications

47. *Linear combination* Let $\mathbf{u} = 2\mathbf{i} + \mathbf{j}$, $\mathbf{v} = \mathbf{i} + \mathbf{j}$, and $\mathbf{w} = \mathbf{i} - \mathbf{j}$. Find scalars a and b such that $\mathbf{u} = a\mathbf{v} + b\mathbf{w}$.

48. *Linear combination* Let $\mathbf{u} = \mathbf{i} - 2\mathbf{j}$, $\mathbf{v} = 2\mathbf{i} + 3\mathbf{j}$, and $\mathbf{w} = \mathbf{i} + \mathbf{j}$. Write $\mathbf{u} = \mathbf{u}_1 + \mathbf{u}_2$, where $\mathbf{u}_1$ is parallel to $\mathbf{v}$ and $\mathbf{u}_2$ is parallel to $\mathbf{w}$. (See Exercise 47.)

49. *Force vector* You are pulling on a suitcase with a force $\mathbf{F}$ (pictured here) whose magnitude is $|\mathbf{F}| = 10$ lb. Find the $\mathbf{i}$- and $\mathbf{j}$-components of $\mathbf{F}$.

50. *Force vector* A kite string exerts a 12 lb pull ($|\mathbf{F}| = 12$) on a kite and makes a 45° angle with the horizontal. Find the horizontal and vertical components of $\mathbf{F}$.

51. *Velocity* An airplane is flying in the direction 25° west of north at 800 km/h. Find the component form of the velocity of the airplane, assuming that the positive x-axis represents due east and the positive y-axis represents due north.

52. *Velocity* An airplane is flying in the direction 10° east of south at 600 km/h. Find the component form of the velocity of the airplane, assuming that the positive x-axis represents due east and the positive y-axis represents due north.

53. *Location* A bird flies from its nest 5 km in the direction 60° north of east, where it stops to rest on a tree. It then flies 10 km in the direction due southeast and lands atop a telephone pole. Place an *xy*-coordinate system so that the origin is the bird's nest, the *x*-axis points east, and the *y*-axis points north.

(a) At what point is the tree located?

(b) At what point is the telephone pole?

54. *Location* A bird flies from its nest 7 km in the direction northeast, where it stops to rest on a tree. It then flies 8 km in the direction 30° south of west and lands atop a telephone pole. Place an *xy*-coordinate system so that the origin is the bird's nest, the *x*-axis points east, and the *y*-axis points north.

(a) At what point is the tree located?

(b) At what point is the telephone pole?

Dot Products

Angle Between Vectors • Laws of the Dot Product • Perpendicular (Orthogonal) Vectors • Vector Projections • Work • Writing a Vector as a Sum of Orthogonal Vectors

If a force **F** is applied to a particle moving along a path, we often need to know the magnitude of the force in the direction of motion. If **v** is parallel to the tangent line to the path at the point where **F** is applied, then we want the magnitude of **F** in the direction of **v**. Figure 9.14 shows that the scalar quantity we seek is the length $|\mathbf{F}|\cos\theta$, where θ is the angle between the two vectors **F** and **v**.

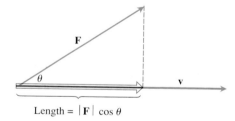

Length $= |\mathbf{F}|\cos\theta$

FIGURE 9.14 The magnitude of the force **F** in the direction of vector **v** is the length $|\mathbf{F}|\cos\theta$ of the projection of **F** onto **v**.

In this section, you learn how to calculate easily the angle between two vectors directly from their components. A key part of the calculation is an expression called the *dot product*. Dot products are also called *scalar* products because the product results in a scalar, not a vector. After investigating the dot product, we apply it to finding the projection of one vector onto another (as displayed in Figure 9.14) and to finding the work done by a constant force acting through a displacement.

Angle Between Vectors

When two nonzero vectors **u** and **v** are placed so their initial points coincide, they form an angle θ of measure $0 \leq \theta \leq \pi$ (Figure 9.15). This angle is the **angle between u and v.**

FIGURE 9.15 The angle between **u** and **v**.

Theorem 1 gives a formula we can use to determine the angle between two vectors.

> **Theorem 1**   Angle Between Two Vectors
>
> The angle θ between two nonzero vectors $\mathbf{u} = \langle u_1, u_2 \rangle$ and $\mathbf{v} = \langle v_1, v_2 \rangle$ is given by
>
> $$\theta = \cos^{-1} \frac{u_1 v_1 + u_2 v_2}{|\mathbf{u}\,||\,\mathbf{v}|}.$$

Before proving Theorem 1 (which is a consequence of the law of cosines), let's focus attention on the expression $u_1 v_1 + u_2 v_2$ in the calculation for θ.

> **Definition**   Dot Product (Inner Product)
>
> The **dot product** (or **inner product**) $\mathbf{u} \cdot \mathbf{v}$ ("**u** dot **v**") of vectors $\mathbf{u} = \langle u_1, u_2 \rangle$ and $\mathbf{v} = \langle v_1, v_2 \rangle$ is the number
>
> $$\mathbf{u} \cdot \mathbf{v} = u_1 v_1 + u_2 v_2.$$

Example 1   Finding Dot Products

(a) $\langle 1, -2 \rangle \cdot \langle -6, 2 \rangle = (1)(-6) + (-2)(2)$
$$= -6 - 4 = -10$$

(b) $\left(\frac{1}{2}\mathbf{i} + 3\mathbf{j} \right) \cdot (4\mathbf{i} - \mathbf{j}) = \left(\frac{1}{2} \right)(4) + (3)(-1)$
$$= 2 - 3 = -1$$

We can use the dot product to rewrite the formula in Theorem 1 for finding the angle between two vectors.

> **Corollary**   Angle Between Two Vectors
>
> The angle between nonzero vectors $\mathbf{u}$ and $\mathbf{v}$ is
>
> $$\theta = \cos^{-1} \left(\frac{\mathbf{u} \cdot \mathbf{v}}{|\mathbf{u}\,||\,\mathbf{v}|} \right).$$

Example 2   Finding an Angle of a Triangle

Find the angle θ in the triangle ABC determined by the vertices $A = (0, 0)$, $B = (3, 5)$, and $C = (5, 2)$ (Figure 9.16).

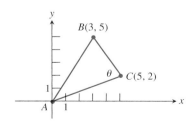

FIGURE 9.16 The triangle in Example 2.

Solution The angle θ is the angle between the vectors $\overrightarrow{CA}$ and $\overrightarrow{CB}$. The component forms of these two vectors are

$$\overrightarrow{CA} = \langle -5, -2 \rangle \qquad \text{and} \qquad \overrightarrow{CB} = \langle -2, 3 \rangle.$$

First we calculate the dot product and magnitudes of these two vectors.

$$\overrightarrow{CA} \cdot \overrightarrow{CB} = (-5)(-2) + (-2)(3) = 4$$
$$|\overrightarrow{CA}| = \sqrt{(-5)^2 + (-2)^2} = \sqrt{29}$$
$$|\overrightarrow{CB}| = \sqrt{(-2)^2 + (3)^2} = \sqrt{13}$$

Then applying the corollary to Theorem 1, we have

$$\theta = \cos^{-1}\left(\frac{\overrightarrow{CA} \cdot \overrightarrow{CB}}{|\overrightarrow{CA}||\overrightarrow{CB}|}\right)$$

$$= \cos^{-1}\left(\frac{4}{(\sqrt{29})(\sqrt{13})}\right)$$

$$\approx 78.1° \qquad \text{or} \qquad 1.36 \text{ radians.}$$

Proof of Theorem 1 Applying the law of cosines to the triangle in Figure 9.17, we find that

$$|\mathbf{w}|^2 = |\mathbf{u}|^2 + |\mathbf{v}|^2 - 2|\mathbf{u}||\mathbf{v}|\cos\theta$$
$$2|\mathbf{u}||\mathbf{v}|\cos\theta = |\mathbf{u}|^2 + |\mathbf{v}|^2 - |\mathbf{w}|^2.$$

Because $\mathbf{w} = \mathbf{u} - \mathbf{v}$, the component form of $\mathbf{w}$ is $\langle u_1 - v_1, u_2 - v_2 \rangle$. Thus,

$$|\mathbf{u}|^2 = (\sqrt{u_1^2 + u_2^2})^2 = u_1^2 + u_2^2$$
$$|\mathbf{v}|^2 = (\sqrt{v_1^2 + v_2^2})^2 = v_1^2 + v_2^2$$
$$|\mathbf{w}|^2 = (\sqrt{(u_1 - v_1)^2 + (u_2 - v_2)^2})^2 = (u_1 - v_1)^2 + (u_2 - v_2)^2$$
$$= (u_1^2 - 2u_1v_1 + v_1^2) + (u_2^2 - 2u_2v_2 + v_2^2)$$

and

$$|\mathbf{u}|^2 + |\mathbf{v}|^2 - |\mathbf{w}|^2 = 2(u_1v_1 + u_2v_2).$$

Therefore,

$$2|\mathbf{u}||\mathbf{v}|\cos\theta = |\mathbf{u}|^2 + |\mathbf{v}|^2 - |\mathbf{w}|^2 = 2(u_1v_1 + u_2v_2)$$
$$|\mathbf{u}||\mathbf{v}|\cos\theta = u_1v_1 + u_2v_2$$
$$\cos\theta = \frac{u_1v_1 + u_2v_2}{|\mathbf{u}||\mathbf{v}|}.$$

So

$$\theta = \cos^{-1}\left(\frac{u_1v_1 + u_2v_2}{|\mathbf{u}||\mathbf{v}|}\right).$$

FIGURE 9.17 The parallelogram law of addition of vectors gives $\mathbf{w} = \mathbf{u} - \mathbf{v}$.

Laws of the Dot Product

The dot product obeys many of the laws that hold for ordinary products of real numbers (scalars).

> **Properties of the Dot Product**
> If **u**, **v**, and **w** are any vectors and c is a scalar, then
>
> 1. $\mathbf{u} \cdot \mathbf{v} = \mathbf{v} \cdot \mathbf{u}$
>
> 2. $(c\mathbf{u}) \cdot \mathbf{v} = \mathbf{u} \cdot (c\mathbf{v}) = c(\mathbf{u} \cdot \mathbf{v})$
>
> 3. $\mathbf{u} \cdot (\mathbf{v} + \mathbf{w}) = \mathbf{u} \cdot \mathbf{v} + \mathbf{u} \cdot \mathbf{w}$
>
> 4. $\mathbf{u} \cdot \mathbf{u} = |\mathbf{u}|^2$
>
> 5. $\mathbf{0} \cdot \mathbf{u} = 0.$

CD-ROM
WEBsite

Historical Biography

Carl Friedrich Gauss
(1777 — 1855)

The properties are easy to prove using the definition. For instance, here are the proofs of Properties 1 and 3.

1. $\mathbf{u} \cdot \mathbf{v} = u_1 v_1 + u_2 v_2 = v_1 u_1 + v_2 u_2 = \mathbf{v} \cdot \mathbf{u}$

3. $\mathbf{u} \cdot (\mathbf{v} + \mathbf{w}) = \langle u_1, u_2 \rangle \cdot \langle v_1 + w_1, v_2 + w_2 \rangle$

$$= u_1 (v_1 + w_1) + u_2 (v_2 + w_2)$$
$$= u_1 v_1 + u_1 w_1 + u_2 v_2 + u_2 w_2$$
$$= (u_1 v_1 + u_2 v_2) + (u_1 w_1 + u_2 w_2)$$
$$= \mathbf{u} \cdot \mathbf{v} + \mathbf{u} \cdot \mathbf{w}$$

Perpendicular (Orthogonal) Vectors

Two nonzero vectors **u** and **v** are perpendicular or **orthogonal** if the angle between them is $\pi/2$. For such vectors, we automatically have $\mathbf{u} \cdot \mathbf{v} = 0$ because $\cos(\pi/2) = 0$. The converse is also true. If **u** and **v** are nonzero vectors with $\mathbf{u} \cdot \mathbf{v} = |\mathbf{u}||\mathbf{v}| \cos \theta = 0$, then $\cos \theta = 0$ and $\theta = \cos^{-1} 0 = \pi/2$.

> **Definition**   Orthogonal Vectors
> Vectors **u** and **v** are **orthogonal (perpendicular)** if and only if $\mathbf{u} \cdot \mathbf{v} = 0$.

Example 3   Applying the Definition of Orthogonality

(a) $\mathbf{u} = \langle 3, -2 \rangle$ and $\mathbf{v} = \langle 4, 6 \rangle$ are orthogonal because $\mathbf{u} \cdot \mathbf{v} = (3)(4) + (-2)(6) = 0.$

(b) $\mathbf{u} = \mathbf{i} + 2\mathbf{j}$ is orthogonal to $\mathbf{v} = -10\mathbf{i} + 5\mathbf{j}$ because $\mathbf{u} \cdot \mathbf{v} = (1)(-10) + (2)(5) = 0.$

(c) **0** is orthogonal to every vector **u** since $\mathbf{0} \cdot \mathbf{u} = 0$ from Property 5.

We now return to the problem of projecting one vector onto another, posed in the opening to this section.

Vector Projections

The **vector projection** of $\mathbf{u} = \overrightarrow{PQ}$ onto a nonzero vector $\mathbf{v} = \overrightarrow{PS}$ (Figure 9.18) is the vector $\overrightarrow{PR}$ determined by dropping a perpendicular from Q to the line PS. The notation for this vector is

$$\text{proj}_{\mathbf{v}} \, \mathbf{u} \qquad \text{(“the vector projection of } \mathbf{u} \text{ onto } \mathbf{v}\text{”)}.$$

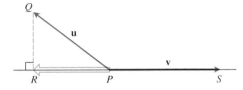

FIGURE 9.18 The vector projection of $\mathbf{u}$ onto $\mathbf{v}$.

If $\mathbf{u}$ represents a force, then $\text{proj}_{\mathbf{v}} \, \mathbf{u}$ represents the effective force in the direction of $\mathbf{v}$ (Figure 9.19).

If the angle θ between $\mathbf{u}$ and $\mathbf{v}$ is acute, $\text{proj}_{\mathbf{v}} \, \mathbf{u}$ has length $|\mathbf{u}| \cos \theta$ and direction $\mathbf{v}/|\mathbf{v}|$ (Figure 9.20). If θ is obtuse, $\cos \theta < 0$ and $\text{proj}_{\mathbf{v}} \, \mathbf{u}$ has length $-|\mathbf{u}| \cos \theta$ and direction $-\mathbf{v}/|\mathbf{v}|$. In any case,

$$\text{proj}_{\mathbf{v}} \, \mathbf{u} = (|\mathbf{u}| \cos \theta) \frac{\mathbf{v}}{|\mathbf{v}|}$$

$$= \left(\frac{\mathbf{u} \cdot \mathbf{v}}{|\mathbf{v}|} \right) \frac{\mathbf{v}}{|\mathbf{v}|} \qquad |\mathbf{u}| \cos \theta = \frac{|\mathbf{u}| |\mathbf{v}| \cos \theta}{|\mathbf{v}|}$$

$$= \left(\frac{\mathbf{u} \cdot \mathbf{v}}{|\mathbf{v}|^2} \right) \mathbf{v}. \qquad = \frac{\mathbf{u} \cdot \mathbf{v}}{|\mathbf{v}|}$$

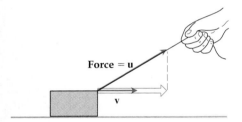

FIGURE 9.19 If we pull on the box with force $\mathbf{u}$, the effective force moving the box forward in the direction $\mathbf{v}$ is the projection of $\mathbf{u}$ onto $\mathbf{v}$.

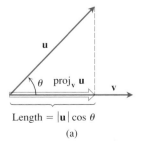

Length $= |\mathbf{u}| \cos \theta$

(a)

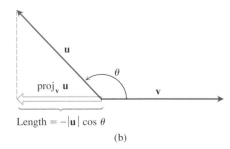

Length $= -|\mathbf{u}| \cos \theta$

(b)

FIGURE 9.20 The length of $\text{proj}_{\mathbf{v}} \, \mathbf{u}$ is (a) $|\mathbf{u}| \cos \theta$ if $\cos \theta \geq 0$ and (b) $-|\mathbf{u}| \cos \theta$ if $\cos \theta < 0$.

The number $|\mathbf{u}|\cos\theta$ is called the **scalar component of u in the direction of v.** To summarize,

---

Vector projection of **u** onto **v**:

$$\text{proj}_{\mathbf{v}}\,\mathbf{u} = \left(\frac{\mathbf{u}\cdot\mathbf{v}}{|\mathbf{v}|^2}\right)\mathbf{v}$$

Scalar component of **u** in the direction of **v**:

$$|\mathbf{u}|\cos\theta = \frac{\mathbf{u}\cdot\mathbf{v}}{|\mathbf{v}|} = \mathbf{u}\cdot\frac{\mathbf{v}}{|\mathbf{v}|}$$

Example 4   Finding Vector Projections and Scalar Components

Find the vector projection of a force $\mathbf{F} = 5\mathbf{i} + 2\mathbf{j}$ onto $\mathbf{v} = \mathbf{i} - 3\mathbf{j}$ and the scalar component of **F** in the direction of **v**.

Solution   The vector projection is

$$\begin{aligned}
\text{proj}_{\mathbf{v}}\,\mathbf{F} &= \left(\frac{\mathbf{F}\cdot\mathbf{v}}{|\mathbf{v}|^2}\right)\mathbf{v} \\
&= \frac{5-6}{1+9}(\mathbf{i}-3\mathbf{j}) = -\frac{1}{10}(\mathbf{i}-3\mathbf{j}) \\
&= -\frac{1}{10}\mathbf{i} + \frac{3}{10}\mathbf{j}.
\end{aligned}$$

The scalar component of **F** in the direction of **v** is

$$|\mathbf{F}|\cos\theta = \frac{\mathbf{F}\cdot\mathbf{v}}{|\mathbf{v}|} = \frac{5-6}{\sqrt{1+9}} = -\frac{1}{\sqrt{10}}.$$

Work

In Chapter 5, we calculated the work done by a constant force of magnitude F in moving an object through a distance d as $W = Fd$. That formula holds only if the force is directed along the line of motion. If a force **F** moving an object through a displacement $\mathbf{D} = \overrightarrow{PQ}$ has some other direction, the work is performed by the component of **F** in the direction of **D**. If θ is the angle between **F** and **D** (Figure 9.21), then

$$\begin{aligned}
\text{Work} &= \binom{\text{scalar component of }\mathbf{F}}{\text{in the direction of }\mathbf{D}}(\text{length of }\mathbf{D}) \\
&= (|\mathbf{F}|\cos\theta)|\mathbf{D}| \\
&= \mathbf{F}\cdot\mathbf{D}.
\end{aligned}$$

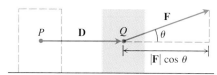

FIGURE 9.21 The work done by a constant force **F** during a displacement **D** is $(|\mathbf{F}|\cos\theta)|\mathbf{D}|$.

> **Definition**   Work Done by a Constant Force
>
> The **work** done by a constant force $\mathbf{F}$ acting through a displacement $\mathbf{D} = \overrightarrow{PQ}$ is
>
> $$W = \mathbf{F} \cdot \mathbf{D} = |\mathbf{F}| |\mathbf{D}| \cos \theta,$$
>
> where θ is the angle between $\mathbf{F}$ and $\mathbf{D}$.

Work

The standard units of work are the foot-pound and the newton-meter, both force-distance units. The newton-meter is usually called a *joule*.

Example 5   Applying the Definition of Work

If $|\mathbf{F}| = 40$ N (newtons), $|\mathbf{D}| = 3$ m, and $\theta = 60°$, the work done by $\mathbf{F}$ in acting from P to Q is

$$
\begin{aligned}
\text{Work} &= |\mathbf{F}| |\mathbf{D}| \cos \theta & \text{Definition}\\
&= (40)(3) \ \cos \ 60° & \text{Given values}\\
&= (120)(1/2)\\
&= 60 \text{ J (joules)}.
\end{aligned}
$$

We encounter more interesting work problems in Chapter 13 when we learn to find the work done by a variable force along a *path* in space.

Writing a Vector as a Sum of Orthogonal Vectors

We know one way to write a vector $\mathbf{u} = \langle a, b \rangle$ as a sum of orthogonal vectors:

$$\mathbf{u} = \langle a, b \rangle = a\mathbf{i} + b\mathbf{j}$$

(since $\mathbf{i} \cdot \mathbf{j} = 0$). Sometimes, however, it is more informative to express $\mathbf{u}$ as a different sum. In mechanics, for instance, we often need to write a vector $\mathbf{u}$ as a sum of a vector parallel to a given vector $\mathbf{v}$ and a vector orthogonal to $\mathbf{v}$. As an example, in studying the motion of a particle moving along a path in the plane (or space), it is desirable to know the components of the acceleration vector in the direction of the tangent to the path (at a point) and of the normal to the path. (These *tangential* and *normal components* of acceleration are investigated in Section 10.6.) The acceleration vector can then be expressed as the sum of its (vector) tangential and normal components (which reflect important geometric properties about the nature of the path itself, such as *curvature*). Velocity and acceleration vectors are studied in the next section.

Generally, for vectors $\mathbf{u}$ and $\mathbf{v}$, it is easy to see from Figure 9.22 that the vector

$$\mathbf{u} - \text{proj}_{\mathbf{v}} \mathbf{u}$$

is orthogonal to the projection vector $\text{proj}_{\mathbf{v}} \mathbf{u}$ (which has the same direction as $\mathbf{v}$). Thus,

$$\mathbf{u} = \text{proj}_{\mathbf{v}} \mathbf{u} + (\mathbf{u} - \text{proj}_{\mathbf{v}} \mathbf{u})$$

expresses $\mathbf{u}$ as a sum of orthogonal vectors.

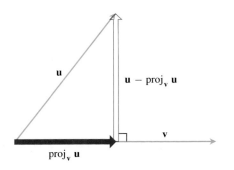

FIGURE 9.22 Writing $\mathbf{u}$ as the sum of vectors parallel and orthogonal to $\mathbf{v}$.

How to Write u as a Vector Parallel to v Plus a Vector Orthogonal to v

$$\mathbf{u} = \text{proj}_\mathbf{v}\, \mathbf{u} + (\mathbf{u} - \text{proj}_\mathbf{v}\, \mathbf{u})$$

$$= \underbrace{\left(\frac{\mathbf{u} \cdot \mathbf{v}}{|\mathbf{v}|^2} \right)\mathbf{v}}_{\text{Parallel to } v} + \underbrace{\left(\mathbf{u} - \left(\frac{\mathbf{u} \cdot \mathbf{v}}{|\mathbf{v}|^2} \right)\mathbf{v} \right)}_{\text{Orthogonal to } v}$$

Example 6   Sum of Orthogonal Vectors

In Example 8, Section 9.1, we found the vector $\mathbf{v} = 2\mathbf{i} + 3\mathbf{j}$ to be tangent to the path

$$y = \frac{x^3}{2} + \frac{1}{2}$$

at the point $(1, 1)$. If $\mathbf{u} = 4\mathbf{i} - \mathbf{j}$ is the acceleration at the point of a particle moving along the path, express $\mathbf{u}$ as the sum of a vector parallel to $\mathbf{v}$ and a vector orthogonal to $\mathbf{v}$.

Solution   With $\mathbf{u} \cdot \mathbf{v} = 8 - 3 = 5$ and $|\mathbf{v}|^2 = \mathbf{v} \cdot \mathbf{v} = 4 + 9 = 13$, we have

$$\mathbf{u} = \left(\frac{\mathbf{u} \cdot \mathbf{v}}{|\mathbf{v}|^2} \right)\mathbf{v} + \left(\mathbf{u} - \left(\frac{\mathbf{u} \cdot \mathbf{v}}{|\mathbf{v}|^2} \right)\mathbf{v} \right)$$

$$= \frac{5}{13}(2\mathbf{i} + 3\mathbf{j}) + \left(4\mathbf{i} - \mathbf{j} - \frac{5}{13}(2\mathbf{i} + 3\mathbf{j}) \right)$$

$$= \left(\frac{10}{13}\mathbf{i} + \frac{15}{13}\mathbf{j} \right) + \left(\frac{42}{13}\mathbf{i} - \frac{28}{13}\mathbf{j} \right).$$

Check: The first vector in the sum is parallel to $\mathbf{v}$ because it is $(5/13)\mathbf{v}$. The second vector is orthogonal to $\mathbf{v}$ because

$$\left(\frac{42}{13}\mathbf{i} - \frac{28}{13}\mathbf{j} \right) \cdot (2\mathbf{i} + 3\mathbf{j}) = \frac{84}{13} - \frac{84}{13} = 0.$$

In the next chapter, we learn how to write velocity and acceleration vectors as linear combinations of other mutually orthogonal vectors.

EXERCISES 9.2

Calculations

In Exercises 1–6, find

   (a) $\mathbf{v} \cdot \mathbf{u},\ |\mathbf{v}|,\ |\mathbf{u}|$

   (b) the cosine of the angle between $\mathbf{v}$ and $\mathbf{u}$

   (c) the scalar component of $\mathbf{u}$ in the direction of $\mathbf{v}$

   (d) the vector $\text{proj}_\mathbf{v}\, \mathbf{u}$.

1. $\mathbf{v} = 2\mathbf{i} - 4\mathbf{j}, \quad \mathbf{u} = 2\mathbf{i} + 4\mathbf{j}$

2. $\mathbf{v} = 2\mathbf{i} + 10\mathbf{j}, \quad \mathbf{u} = 2\mathbf{i} + 2\mathbf{j}$

3. $\mathbf{v} = -\mathbf{i} + \mathbf{j}, \quad \mathbf{u} = \sqrt{2}\mathbf{i} + \sqrt{3}\mathbf{j}$

4. $\mathbf{v} = 5\mathbf{i} + \mathbf{j}, \quad \mathbf{u} = 2\mathbf{i} + \sqrt{17}\mathbf{j}$

5. $v = \left\langle \dfrac{1}{\sqrt{2}}, \dfrac{1}{\sqrt{3}} \right\rangle$, $u = \left\langle \dfrac{1}{\sqrt{2}}, -\dfrac{1}{\sqrt{3}} \right\rangle$

6. $v = \left\langle \dfrac{1}{\sqrt{2}}, \dfrac{1}{\sqrt{2}} \right\rangle$, $u = \left\langle -\dfrac{1}{\sqrt{2}}, -\dfrac{1}{\sqrt{2}} \right\rangle$

Angles Between Vectors

Find the angles between the vectors in Exercises 7–10 to the nearest hundredth of a radian.

7. $v = 2i + j$, $u = i + 2j$

8. $v = 2i - 2j$, $u = 3i$

9. $v = \sqrt{3}i - 7j$, $u = \sqrt{3}i + j$

10. $v = i + \sqrt{2}j$, $u = -i + j$

11. *Triangle* Find the measures of the angles of the triangle whose vertices are $A = (-1, 0)$, $B = (2, 1)$, and $C = (1, -2)$.

12. *Rectangle* Find the measures of the angles between the diagonals of the rectangle whose vertices are $A = (1, 0)$, $B = (0, 3)$, $C = (3, 4)$, and $D = (4, 1)$.

Geometry and Examples

13. *Writing to Learn: Sums and differences* In the accompanying figure, it looks as if $v_1 + v_2$ and $v_1 - v_2$ are orthogonal. Is this mere coincidence, or are there circumstances under which we may expect the sum of two vectors to be orthogonal to their difference? Give reasons for your answer.

14. *Orthogonality on a circle* Suppose that AB is the diameter of a circle with center O and that C is a point on one of the two arcs joining A and B. Show that $\overrightarrow{CA}$ and $\overrightarrow{CB}$ are orthogonal.

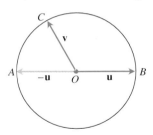

15. *Diagonals of a rhombus* Show that the diagonals of a rhombus (parallelogram with sides of equal length) are perpendicular.

16. *Perpendicular diagonals* Show that squares are the only rectangles with perpendicular diagonals.

17. *When parallelograms are rectangles* Prove that a parallelogram is a rectangle if and only if its diagonals are equal in length. (This fact is often exploited by carpenters.)

18. *Diagonal of parallelogram* Show that the indicated diagonal of the parallelogram determined by vectors u and v bisects the angle between u and v if $|u| = |v|$.

19. *Projectile motion* A gun with muzzle velocity of 1200 ft/sec is fired at an angle of 8° above the horizontal. Find the horizontal and vertical components of the velocity.

20. *Inclined plane* Suppose that a box is being towed up an inclined plane as shown in the figure. Find the force w needed to make the component of the force parallel to the inclined plane equal to 2.5 lb.

Theory and Examples

21. (a) *Cauchy–Schwartz inequality* Use the fact that $u \cdot v = |u||v| \cos \theta$ to show that the inequality $|u \cdot v| \leq |u||v|$ holds for any vectors u and v.

(b) *Writing to Learn* Under what circumstances, if any, does $|u \cdot v|$ equal $|u||v|$? Give reasons for your answer.

22. *Writing to Learn* Copy the axes and vector shown here. Then shade in the points (x, y) for which $(xi + yj) \cdot v \leq 0$. Justify your answer.

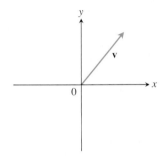

23. *Orthogonal unit vectors* If $\mathbf{u}_1$ and $\mathbf{u}_2$ are orthogonal unit vectors and $\mathbf{v} = a\mathbf{u}_1 + b\mathbf{u}_2$, find $\mathbf{v} \cdot \mathbf{u}_1$.

24. *Writing to Learn: Cancellation in dot products* In real-number multiplication, if $uv_1 = uv_2$ and $u \neq 0$, we can cancel the u and conclude that $v_1 = v_2$. Does the same rule hold for the dot product: If $\mathbf{u} \cdot \mathbf{v}_1 = \mathbf{u} \cdot \mathbf{v}_2$ and $\mathbf{u} \neq \mathbf{0}$, can you conclude that $\mathbf{v}_1 = \mathbf{v}_2$? Give reasons for your answer.

Equations for Lines in the Plane

25. *Line perpendicular to a vector* Show that the vector $\mathbf{v} = a\mathbf{i} + b\mathbf{j}$ is perpendicular to the line $ax + by = c$ by establishing that the slope of $\mathbf{v}$ is the negative reciprocal of the slope of the given line.

26. *Line parallel to a vector* Show that the vector $\mathbf{v} = a\mathbf{i} + b\mathbf{j}$ is parallel to the line $bx - ay = c$ by establishing that the slope of the line segment representing $\mathbf{v}$ is the same as the slope of the given line.

In Exercises 27–30, use the result of Exercise 25 to find an equation for the line through P perpendicular to $\mathbf{v}$. Then sketch the line. Include $\mathbf{v}$ in your sketch *as a vector starting at the origin.*

27. $P(2, 1)$, $\mathbf{v} = \mathbf{i} + 2\mathbf{j}$

28. $P(-1, 2)$, $\mathbf{v} = -2\mathbf{i} - \mathbf{j}$

29. $P(-2, -7)$, $\mathbf{v} = -2\mathbf{i} + \mathbf{j}$

30. $P(11, 10)$, $\mathbf{v} = 2\mathbf{i} - 3\mathbf{j}$

In Exercises 31–34, use the result of Exercise 26 to find an equation for the line through P parallel to $\mathbf{v}$. Then sketch the line. Include $\mathbf{v}$ in your sketch *as a vector starting at the origin.*

31. $P(-2, 1)$, $\mathbf{v} = \mathbf{i} - \mathbf{j}$

32. $P(0, -2)$, $\mathbf{v} = 2\mathbf{i} + 3\mathbf{j}$

33. $P(1, 2)$, $\mathbf{v} = -\mathbf{i} - 2\mathbf{j}$

34. $P(1, 3)$, $\mathbf{v} = 3\mathbf{i} - 2\mathbf{j}$

Work

35. *Work along a line* Find the work done by a force $\mathbf{F} = 5\,\mathbf{i}$ (magnitude 5 N) in moving an object along the line from the origin to the point $(1, 1)$ (distance in meters).

36. *Locomotive* The union Pacific's *Big Boy* locomotive could pull 6000-ton trains with a tractive effort (pull) of 602,148 N (135,375 lb). At this level of effort, about how much work did *Big Boy* do on the (approximately straight) 605 km journey from San Francisco to Los Angeles?

37. *Inclined plane* How much work does it take to slide a crate 20 m along a loading dock by pulling on it with a 200 N force at an angle of 30° from the horizontal?

38. *Sailboat* The wind passing over a boat's sail exerted a 1000 lb magnitude force $\mathbf{F}$ as shown here. How much work did the wind perform in moving the boat forward 1 mi? Answer in foot-pounds.

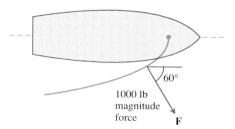

Angles Between Lines in the Plane

The acute angle between intersecting lines that do not cross at right angles is the same as the angle determined by vectors normal to the lines or by the vectors parallel to the lines.

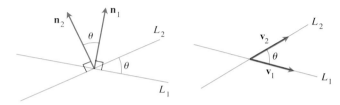

Use this fact and the results of Exercise 25 or 26 to find the acute angles between the lines in Exercises 39–44.

39. $3x + y = 5$, $2x - y = 4$

40. $y = \sqrt{3}x - 1$, $y = -\sqrt{3}x + 2$

41. $\sqrt{3}x - y = -2$, $x - \sqrt{3}y = 1$

42. $x + \sqrt{3}y = 1$, $(1 - \sqrt{3})x + (1 + \sqrt{3})y = 8$

43. $3x - 4y = 3$, $x - y = 7$

44. $12x + 5y = 1$, $2x - 2y = 3$

Angles Between Differentiable Curves

The angles between two differentiable curves at a point of intersection are the angles between the curves' tangent lines at these points. Find the angles between the curves in Exercises 45–48.

45. $y = (3/2) - x^2$, $y = x^2$ (two points of intersection)

46. $x = (3/4) - y^2$, $x = y^2 - (3/4)$ (two points of intersection)

47. $y = x^3$, $x = y^2$ (two points of intersection)

48. $y = -x^2$, $y = \sqrt[3]{x}$ (two points of intersection)

9.3   Vector-Valued Functions

Planar Curves • Limits and Continuity • Derivatives •
Motion • Integrals

In this section, we show how to use the calculus of vectors to study the paths, velocities, and accelerations of bodies moving in the plane. This allows us to answer questions later about the motions of projectiles.

Planar Curves

When a particle moves through the plane during a time interval I, we think of the particle's coordinates as functions defined on I:

$$x = f(t), \qquad y = g(t), \qquad t \in I. \tag{1}$$

The points $(x, y) = (f(t), g(t))$, $t \in I$, make up the curve in the plane that is the particle's **path.** The equations and interval in Equation (1) parametrize the curve. The vector

$$\mathbf{r}(t) = \overrightarrow{OP} = \langle f(t), g(t) \rangle = f(t)\mathbf{i} + g(t)\mathbf{j} \tag{2}$$

from the origin to the particle's **position** $P(f(t), g(t))$ at time t is the particle's **position vector.** The functions f and g are the **component functions (components)** of the position vector. We think of the particle's path as the **curve traced by r** during the time interval I (Figure 9.23).

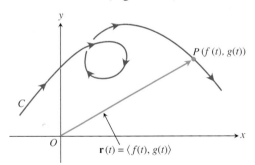

FIGURE 9.23 The path (curve) C is traced by the position vector $\mathbf{r}(t)$ during the time interval I.

Equation (2) defines $\mathbf{r}$ as a *vector function* of the real variable t on the interval I. More generally, a **vector function** or **vector-valued function** on a domain D is a rule that assigns a vector in the plane to each element in D. The curve traced by a vector function is its **graph.**

We refer to real-valued functions as **scalar functions** to distinguish them from vector functions. The components of $\mathbf{r}$ are scalar functions of t. When we define a vector-valued function by giving its component functions, we assume that the vector function's domain to be the common domain of the components.

Example 1   Graphing an Archimedes Spiral

Graph the vector function

$$\mathbf{r}(t) = (t \cos t)\mathbf{i} + (t \sin t)\mathbf{j}, \qquad t \geq 0.$$

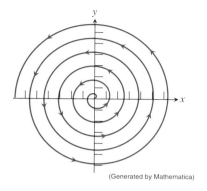

FIGURE 9.24 The graph of $\mathbf{r}(t) = (t\cos t)\mathbf{i} + (t\sin t)\mathbf{j}, t \geq 0$, is the curve $x = t\cos t, y = t\sin t, t \geq 0$. (Example 1)

(Generated by Mathematica)

Solution We can graph the vector function parametrically on a graphing calculator or computer using

$$x = t\cos t, \qquad y = t\sin t, \qquad t \geq 0.$$

As t increases from 0 to 2π, the point (x, y) starts at the origin $(0, 0)$ and then winds once around the origin, getting farther away from the origin as t increases. This spiral continues winding around the origin, getting farther and farther away as t increases beyond 2π. The spiral is shown in Figure 9.24.

Limits and Continuity

We define limits of vector functions in terms of their scalar components.

Definition Limit

Let $\mathbf{r}(t) = f(t)\mathbf{i} + g(t)\mathbf{j}$. If

$$\lim_{t\to c} f(t) = L_1 \qquad \text{and} \qquad \lim_{t\to c} g(t) = L_2,$$

then the **limit** of $\mathbf{r}(t)$ as t approaches c is

$$\lim_{t\to c} \mathbf{r}(t) = \mathbf{L} = L_1\mathbf{i} + L_2\mathbf{j}.$$

Example 2 Finding a Limit of a Vector Function

If $\mathbf{r}(t) = (\cos t)\mathbf{i} + (\sin t)\mathbf{j}$, then

$$\lim_{t\to\pi/4} \mathbf{r}(t) = \left(\lim_{t\to\pi/4} \cos t\right)\mathbf{i} + \left(\lim_{t\to\pi/4} \sin t\right)\mathbf{j} = \frac{\sqrt{2}}{2}\mathbf{i} + \frac{\sqrt{2}}{2}\mathbf{j}.$$

We define continuity for vector functions in the same way we define continuity for scalar functions.

Definition Continuity at a Point

A vector function $\mathbf{r}(t)$ is **continuous at a point** $t = c$ in its domain if

$$\lim_{t\to c} \mathbf{r}(t) = \mathbf{r}(c).$$

A vector function $\mathbf{r}(t)$ is **continuous** if it is continuous at every point in its domain. Since limits of vector functions are defined in terms of components, we have the following test for continuity.

Component Test for Continuity at a Point
The vector function $\mathbf{r}(t) = f(t)\mathbf{i} + g(t)\mathbf{j}$ is continuous at $t = c$ if and only if f and g are continuous at $t = c$.

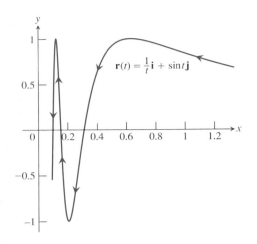

FIGURE 9.25 As $t > 0$ increases, the path of $\mathbf{r}(t)$ oscillates between $y = -1$ and $y = 1$, approaching the y-axis as the **i**-component approaches 0. As $t \to 0^+$, the **i**-component approaches ∞ and the **j**-component approaches 0 through positive values. (The portion of the graph to the left of the y-axis is not shown.)

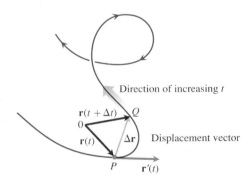

FIGURE 9.26 Between time t and time $t + \Delta t$, the particle moving along the path shown here undergoes the displacement $\overrightarrow{PQ} = \Delta\mathbf{r}$. The vector sum $\mathbf{r}(t) + \Delta\mathbf{r}$ gives the new position, $\mathbf{r}(t + \Delta t)$. As $\Delta t \to 0$, the point Q approaches P along the curve and the vector $\Delta\mathbf{r}/\Delta t$ approaches the limiting tangent position $\mathbf{r}'(t)$.

Example 3   Finding Points of Continuity and Discontinuity

(a) The vector function

$$\mathbf{r}(t) = (t \cos t)\mathbf{i} + (t \sin t)\mathbf{j}$$

is continuous everywhere because the component functions, $t \cos t$, and $t \sin t$ are continuous everywhere.

(b) The vector function

$$\mathbf{r}(t) = \frac{1}{t}\mathbf{i} + (\sin t)\mathbf{j}$$

is not continuous at $t = 0$ because the first component is not continuous at $t = 0$. It is a continuous vector function, however, because it is continuous on its domain, the set of all nonzero real numbers (Figure 9.25).

Derivatives

Suppose that $\mathbf{r}(t) = f(t)\mathbf{i} + g(t)\mathbf{j}$ is the position vector of a particle moving along a curve in the plane and that f and g are differentiable functions of t. Then (see Figure 9.26) the difference between the particle's positions at time $t + \Delta t$ and time t is $\Delta\mathbf{r} = \mathbf{r}(t + \Delta t) - \mathbf{r}(t)$. In terms of components,

$$\Delta\mathbf{r} = \mathbf{r}(t + \Delta t) - \mathbf{r}(t)$$
$$= [f(t + \Delta t)\mathbf{i} + g(t + \Delta t)\mathbf{j}] - [f(t)\mathbf{i} + g(t)\mathbf{j}]$$
$$= [f(t + \Delta t) - f(t)]\mathbf{i} + [g(t + \Delta t) - g(t)]\mathbf{j}.$$

As Δt approaches zero, three things seem to happen simultaneously. First, Q approaches P along the curve. Second, the secant line PQ seems to approach a limiting position tangent to the curve at P. Third, the quotient $\Delta\mathbf{r}/\Delta t$ approaches the limit

$$\lim_{\Delta t \to 0} \frac{\Delta\mathbf{r}}{\Delta t} = \left[\lim_{\Delta t \to 0} \frac{f(t + \Delta t) - f(t)}{\Delta t} \right]\mathbf{i} + \left[\lim_{\Delta t \to 0} \frac{g(t + \Delta t) - g(t)}{\Delta t} \right]\mathbf{j}$$

$$= \left[\frac{df}{dt} \right]\mathbf{i} + \left[\frac{dg}{dt} \right]\mathbf{j}.$$

We are therefore led by past experience to the following definition.

Definition   Derivative at a Point

The vector function $\mathbf{r}(t) = f(t)\mathbf{i} + g(t)\mathbf{j}$ has a **derivative (is differentiable)** at t if f and g have derivatives at t. The derivative is the vector function

$$\mathbf{r}'(t) = \frac{d\mathbf{r}}{dt} = \lim_{\Delta t \to 0} \frac{\mathbf{r}(t + \Delta t) - \mathbf{r}(t)}{\Delta t} = \frac{df}{dt}\mathbf{i} + \frac{dg}{dt}\mathbf{j}.$$

A vector function $\mathbf{r}$ is **differentiable** if it is differentiable at every point of its domain. The curve traced by $\mathbf{r}$ is **smooth** if $d\mathbf{r}/dt$ is continuous and never $\mathbf{0}$, that is,

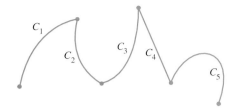

FIGURE 9.27 A piecewise smooth curve made by connecting five smooth curves end to end in continuous fashion.

if f and g have continuous first derivatives that are not simultaneously 0. On a smooth curve, there are no sharp corners or cusps.

The vector $d\mathbf{r}/dt$, when different from $\mathbf{0}$, is also a vector *tangent* to the curve at each point where it exists. The **tangent line** to the curve at a point $P = (f(a), g(a))$ is defined to be the line through P parallel to $d\mathbf{r}/dt$ at $t = a$ (Figure 9.26).

A curve that is made up of a finite number of smooth curves pieced together in a continuous fashion is **piecewise smooth** (Figure 9.27).

Example 4   Finding Derivatives

Find the derivative $d\mathbf{r}/dt$ of the vector function

$$\mathbf{r}(t) = (t \cos t)\mathbf{i} + (t \sin t)\mathbf{j}.$$

Solution

$$\mathbf{r}'(t) = \frac{d\mathbf{r}}{dt} = \frac{d}{dt}(t \cos t)\mathbf{i} + \frac{d}{dt}(t \sin t)\mathbf{j}$$

$$= (\cos t - t \sin t)\mathbf{i} + (\sin t + t \cos t)\mathbf{j}$$

Because the derivatives of vector functions are computed component by component, the rules for differentiating vector functions have the same form as the rules for differentiating scalar functions.

Differentiation Rules for Vector Functions
Let $\mathbf{u}$ and $\mathbf{v}$ be differentiable vector functions of t, $\mathbf{C}$ a constant vector, c any scalar, and f any differentiable scalar function.

1. *Constant Function Rule*: $\dfrac{d}{dt}\mathbf{C} = \mathbf{0}$

2. *Scalar Multiple Rules*: $\dfrac{d}{dt}[c\,\mathbf{u}(t)] = c\,\mathbf{u}'(t)$

    $\dfrac{d}{dt}[f(t)\,\mathbf{u}(t)] = f'(t)\,\mathbf{u}(t) + f(t)\,\mathbf{u}'(t)$

3. *Sum Rule*: $\dfrac{d}{dt}[\mathbf{u}(t) + \mathbf{v}(t)] = \mathbf{u}'(t) + \mathbf{v}'(t)$

4. *Difference Rule*: $\dfrac{d}{dt}[\mathbf{u}(t) - \mathbf{v}(t)] = \mathbf{u}'(t) - \mathbf{v}'(t)$

5. *Dot Product Rule*: $\dfrac{d}{dt}[\mathbf{u}(t) \cdot \mathbf{v}(t)] = \mathbf{u}'(t) \cdot \mathbf{v}(t) + \mathbf{u}(t) \cdot \mathbf{v}'(t)$

6. *Chain Rule*: $\dfrac{d}{dt}[\mathbf{u}(f(t))] = f'(t)\,\mathbf{u}'(f(t))$

We will prove the Dot Product Rule but leave the rest for Exercises 37–40.

Proof of Rule 5   Suppose that

$$\mathbf{u} = u_1(t)\mathbf{i} + u_2(t)\mathbf{j} \quad \text{and} \quad \mathbf{v} = v_1(t)\mathbf{i} + v_2(t)\mathbf{j}.$$

Then

$$\frac{d}{dt}(\mathbf{u} \cdot \mathbf{v}) = \frac{d}{dt}(u_1 v_1 + u_2 v_2)$$

$$= u_1' v_1 + u_2' v_2 + u_1 v_1' + u_2 v_2'$$

$$= \quad \mathbf{u}' \cdot \mathbf{v} \quad + \quad \mathbf{u} \cdot \mathbf{v}'.$$

Example 5   Applying the Differentiation Rules

For the functions given by $\mathbf{u}(t) = 2t^3\mathbf{i} - t^2\mathbf{j}$, $\mathbf{v}(t) = (1/t)\mathbf{i} + (\sin t)\mathbf{j}$, and $f(t) = e^{-t}$, find

(a) $\dfrac{d}{dt}[f(t)\,\mathbf{u}(t)]$ **(b)** $\dfrac{d}{dt}[\mathbf{u}(t) + \mathbf{v}(t)]$

(c) $\dfrac{d}{dt}[\mathbf{u}(t) \cdot \mathbf{v}(t)]$.

Solution

(a) Because $f'(t) = -e^{-t}$ and $\mathbf{u}'(t) = 6t^2\mathbf{i} - 2t\mathbf{j}$, we have

$$\frac{d}{dt}[f(t)\,\mathbf{u}(t)] = f'(t)\,\mathbf{u}(t) + f(t)\,\mathbf{u}'(t)$$

$$= (-e^{-t})(2t^3\mathbf{i} - t^2\mathbf{j}) + e^{-t}(6t^2\mathbf{i} - 2t\mathbf{j})$$

$$= e^{-t}(6t^2 - 2t^3)\mathbf{i} + e^{-t}(t^2 - 2t)\mathbf{j}$$

$$= 2t^2 e^{-t}(3 - t)\mathbf{i} + te^{-t}(t - 2)\mathbf{j}.$$

(b) $\mathbf{u}'(t) = 6t^2\mathbf{i} - 2t\mathbf{j}$ and $\mathbf{v}'(t) = -\dfrac{1}{t^2}\mathbf{i} + (\cos t)\mathbf{j}$ so that

$$\frac{d}{dt}[\mathbf{u}(t) + \mathbf{v}(t)] = \mathbf{u}'(t) + \mathbf{v}'(t)$$

$$= (6t^2\mathbf{i} - 2t\mathbf{j}) + \left(-\frac{1}{t^2}\mathbf{i} + (\cos t)\mathbf{j}\right)$$

$$= \left(6t^2 - \frac{1}{t^2}\right)\mathbf{i} + (\cos t - 2t)\mathbf{j}.$$

(c) Using the derivatives in part (b) and the Dot Product Rule, we have

$$\frac{d}{dt}[\mathbf{u}(t) \cdot \mathbf{v}(t)] = \mathbf{u}'(t) \cdot \mathbf{v}(t) + \mathbf{u}(t) \cdot \mathbf{v}'(t)$$

$$= (6t^2\mathbf{i} - 2t\mathbf{j}) \cdot \left(\frac{1}{t}\mathbf{i} + (\sin t)\mathbf{j}\right) + (2t^3\mathbf{i} - t^2\mathbf{j}) \cdot \left(-\frac{1}{t^2}\mathbf{i} + (\cos t)\mathbf{j}\right)$$

$$= (6t^2)\left(\frac{1}{t}\right) + (-2t)(\sin t) + (2t^3)\left(-\frac{1}{t^2}\right) + (-t^2)(\cos t)$$

$$= 6t - 2t \sin t - 2t - t^2 \cos t$$

$$= 4t - 2t \sin t - t^2 \cos t.$$

Observe that the derivative of the dot product of vector functions is a scalar function.

Check:

$$\frac{d}{dt}[\mathbf{u}(t) \cdot \mathbf{v}(t)] = \frac{d}{dt}(2t^2 - t^2 \sin\ t)$$

$$= 4t - 2t\ \sin\ t - t^2 \cos\ t.$$

CD-ROM
WEBsite

Historical Biography

Sir William Thomson
(1824 — 1907)

Motion

Look once again at Figure 9.26. We drew the figure for Δt positive, so $\Delta \mathbf{r}$ points forward, in the direction of the motion. The vector $\Delta \mathbf{r}/\Delta t$ (not shown), having the same direction as $\Delta \mathbf{r}$, points forward also. Had Δt been negative, $\Delta \mathbf{r}$ would have pointed backwards, against the direction of motion. The quotient $\Delta \mathbf{r}/\Delta t$, however, being a negative scalar multiple of $\Delta \mathbf{r}$, would have once again pointed forward. No matter how $\Delta \mathbf{r}$ points, $\Delta \mathbf{r}/\Delta t$ points forward and we expect the vector $d\mathbf{r}/dt = \lim_{\Delta t \to 0} \Delta \mathbf{r}/\Delta t$, when different from $\mathbf{0}$, to do the same. This means that the derivative $d\mathbf{r}/dt$ is just what we want for modeling a particle's velocity. It points in the direction of motion and gives the rate of change of position with respect to time. For a smooth curve, the velocity is never zero; the particle does not stop or reverse direction.

CD-ROM
WEBsite

Definitions   Velocity, Speed, Acceleration, Direction of Motion

If $\mathbf{r}$ is the position vector of a particle moving along a smooth curve in the plane, then at any time t,

1. $\mathbf{v}(t) = \dfrac{d\mathbf{r}}{dt}$ is the particle's **velocity vector** and is tangent to the curve

2. $|\mathbf{v}(t)|$, the magnitude of $\mathbf{v}$, is the particle's **speed**

3. $\mathbf{a}(t) = \dfrac{d\mathbf{v}}{dt} = \dfrac{d^2\mathbf{r}}{dt^2}$, the derivative of velocity and the second derivative of position, is the particle's **acceleration vector**

4. $\dfrac{\mathbf{v}}{|\mathbf{v}|}$, a unit vector, is the **direction of motion.**

We can express the velocity of a moving particle as the product of its speed and direction (Example 7, Section 9.1).

$$\text{Velocity} = |\mathbf{v}|\left(\frac{\mathbf{v}}{|\mathbf{v}|}\right) = (\text{speed})(\text{direction})$$

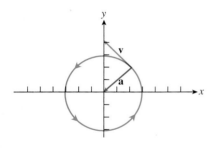

$\textsc{Figure}$ 9.28   At $t = \pi/4$, the velocity vector $-(3/\sqrt{2})\mathbf{i} + (3/\sqrt{2})\mathbf{j}$ is tangent to the circle and the acceleration vector $-(3/\sqrt{2})\mathbf{i} - (3/\sqrt{2})\mathbf{j}$ is perpendicular to the tangent, pointing toward the center of the circle. (Example 6)

Example 6   Studying Motion on a Circle

The vector $\mathbf{r}(t) = (3 \cos t)\mathbf{i} + (3 \sin t)\mathbf{j}$ gives the position of a particle at time t moving counterclockwise on the circle of radius 3 centered at the origin (Figure 9.28). Find

(a) the velocity and acceleration vectors

(b) the velocity, acceleration, speed, and direction of motion at $t = \pi/4$

(c) $\mathbf{v} \cdot \mathbf{a}$. Interpret this result geometrically.

Solution

(a) $\mathbf{v} = \dfrac{d\mathbf{r}}{dt} = (-3 \sin t)\mathbf{i} + (3 \cos t)\mathbf{j}$

$\qquad \mathbf{a} = \dfrac{d\mathbf{v}}{dt} = (-3 \cos t)\mathbf{i} - (3 \sin t)\mathbf{j}$

(b) At $t = \pi/4$, the particle's velocity and acceleration are

Velocity: $\mathbf{v}\left(\dfrac{\pi}{4}\right) = \left(-3 \sin \dfrac{\pi}{4}\right)\mathbf{i} + \left(3 \cos \dfrac{\pi}{4}\right)\mathbf{j} = -\dfrac{3}{\sqrt{2}}\mathbf{i} + \dfrac{3}{\sqrt{2}}\mathbf{j}$

Acceleration: $\mathbf{a}\left(\dfrac{\pi}{4}\right) = \left(-3 \cos \dfrac{\pi}{4}\right)\mathbf{i} - \left(3 \sin \dfrac{\pi}{4}\right)\mathbf{j} = -\dfrac{3}{\sqrt{2}}\mathbf{i} - \dfrac{3}{\sqrt{2}}\mathbf{j}.$

Its speed and direction are

Speed: $\left|\mathbf{v}\left(\dfrac{\pi}{4}\right)\right| = \sqrt{\left(\dfrac{-3}{\sqrt{2}}\right)^2 + \left(\dfrac{3}{\sqrt{2}}\right)^2} = 3$

Direction: $\dfrac{\mathbf{v}(\pi/4)}{|\mathbf{v}(\pi/4)|} = \dfrac{-3/\sqrt{2}}{3}\mathbf{i} + \dfrac{3/\sqrt{2}}{3}\mathbf{j} = -\dfrac{1}{\sqrt{2}}\mathbf{i} + \dfrac{1}{\sqrt{2}}\mathbf{j}.$

(c) $\mathbf{v} \cdot \mathbf{a} = 9 \sin t \cos t - 9 \sin t \cos t = 0$

Thus, in this example, $\mathbf{v}$ and $\mathbf{a}$ are perpendicular for all values of t.

Figure 9.28 shows the path and the velocity and acceleration vectors at $t = \pi/4$.

Example 7   Studying Motion

The vector $\mathbf{r}(t) = (2t^3 - 3t^2)\mathbf{i} + (t^3 - 12t)\mathbf{j}$ gives the position of a moving particle at time t.

(a) Write an equation for the line tangent to the path of the particle at the point where $t = -1$.

(b) Find the coordinates of each point on the path where the horizontal component of the velocity is 0.

Solution

(a) $\mathbf{v}(t) = \dfrac{d\mathbf{r}}{dt} = (6t^2 - 6t)\mathbf{i} + (3t^2 - 12)\mathbf{j}$

At $t = -1$, $\mathbf{r}(-1) = -5\mathbf{i} + 11\mathbf{j}$ and $\mathbf{v}(-1) = 12\mathbf{i} - 9\mathbf{j}$. Thus, we want the equation of the line through $(-5, 11)$ with slope $-9/12 = -3/4$.

$$y - 11 = -\frac{3}{4}(x + 5) \qquad \text{or} \qquad y = -\frac{3}{4}x + \frac{29}{4}$$

(b) The horizontal component of the velocity is $6t^2 - 6t$. It equals 0 when $t = 0$ and $t = 1$. The point corresponding to $t = 0$ is the origin $(0, 0)$; the point corresponding to $t = 1$ is $(-1, -11)$.

Integrals

A differentiable vector function $\mathbf{R}(t)$ is an **antiderivative** of a vector function $\mathbf{r}(t)$ on an interval I if $d\mathbf{R}/dt = \mathbf{r}$ at each point t of I. If $\mathbf{R}$ is an antiderivative of $\mathbf{r}$ on I, it can be shown, working one component at a time, that every antiderivative of $\mathbf{r}$ on I has the form $\mathbf{R} + \mathbf{C}$ for some constant vector $\mathbf{C}$ (Exercise 35).

Definition Indefinite Integral

The **indefinite integral** of $\mathbf{r}$ with respect to t is the set of all antiderivatives of $\mathbf{r}$, denoted by $\int \mathbf{r}(t)\,dt$. If $\mathbf{R}$ is any antiderivative of $\mathbf{r}$, then

$$\int \mathbf{r}(t)\,dt = \mathbf{R}(t) + \mathbf{C}.$$

The usual arithmetic rules for indefinite integrals apply.

Example 8 Finding Antiderivatives

$$\int ((\cos t)\mathbf{i} - 2t\mathbf{j})\,dt = \left(\int \cos t\,dt\right)\mathbf{i} - \left(\int 2t\,dt\right)\mathbf{j} \qquad (3)$$
$$= (\sin t + C_1)\mathbf{i} - (t^2 + C_2)\mathbf{j} \qquad (4)$$
$$= (\sin t)\mathbf{i} - t^2\mathbf{j} + \mathbf{C} \qquad \mathbf{C} = C_1\mathbf{i} + C_2\mathbf{j}$$

As with integration of scalar functions, we recommend that you skip the steps in Equations 3 and 4 and go directly to the final form. Find an antiderivative for each component and add a constant vector at the end.

As with derivatives and indefinite integrals, definite integrals of vector functions are calculated component by component.

Definition Definite Integral

If the components of $\mathbf{r}(t) = f(t)\mathbf{i} + g(t)\mathbf{j}$ are integrable on $[a, b]$, then so is $\mathbf{r}$, and the **definite integral** of $\mathbf{r}$ from a to b is

$$\int_a^b \mathbf{r}(t)\,dt = \left(\int_a^b f(t)\,dt\right)\mathbf{i} + \left(\int_a^b g(t)\,dt\right)\mathbf{j}.$$

Example 9   Evaluating Definite Integrals

$$\int_0^\pi ((\cos\ t)\mathbf{i} - 2t\mathbf{j})\ dt = \left(\int_0^\pi \cos\ t\ dt\right)\mathbf{i} - \left(\int_0^\pi 2t\ dt\right)\mathbf{j}$$

$$= \left(\sin\ t\Big]_0^\pi\right)\mathbf{i} - \left(t^2\Big]_0^\pi\right)\mathbf{j} = 0\mathbf{i} - \pi^2\mathbf{j} = -\pi^2\mathbf{j}$$

Example 10   Finding a Path

The velocity vector of a particle moving in the plane (scaled in meters) is

$$\frac{d\mathbf{r}}{dt} = \frac{1}{t+1}\mathbf{i} + 2t\mathbf{j}, \qquad t \ge 0.$$

(a) Find the particle's position as a vector function of t if $\mathbf{r} = (\ln 2)\mathbf{i}$ when $t = 1$.

(b) Find the distance the particle travels from $t = 0$ to $t = 2$.

Solution

(a) $\mathbf{r} = \left(\int \dfrac{dt}{t+1}\right)\mathbf{i} + \left(\int 2t\ dt\right)\mathbf{j} = (\ln\ (t+1))\mathbf{i} + t^2\mathbf{j} + \mathbf{C}$

$\mathbf{r}(1) = (\ln 2)\mathbf{i} + \mathbf{j} + \mathbf{C} = (\ln 2)\mathbf{i}$

Thus, $\mathbf{C} = -\mathbf{j}$ and

$$\mathbf{r} = (\ln\ (t+1))\mathbf{i} + (t^2 - 1)\mathbf{j}.$$

(b) The parametrization

$$x = \ln\ (t+1), \qquad y = t^2 - 1, \qquad 0 \le t \le 2$$

is smooth, and because x and y are increasing functions of t, the path is traversed exactly once as t increases from 0 to 2 (Figure 9.29). The length is

$$L = \int_0^2 \sqrt{\left(\frac{dx}{dt}\right)^2 + \left(\frac{dy}{dt}\right)^2}\ dt = \int_0^2 \sqrt{\left(\frac{1}{t+1}\right)^2 + (2t)^2}\ dt \approx 4.34\ \text{m}.$$

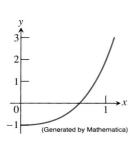

FIGURE 9.29 The path of the particle in Example 10 for $0 \le t \le 2$.

(Generated by Mathematica)

EXERCISES 9.3

Studying Motion

In Exercises 1–4, $\mathbf{r}(t)$ is the position vector of a particle in the plane at time t.

T **(a)** Draw the graph of the path of the particle.

(b) Find the velocity and acceleration vectors.

(c) Find the particle's speed and direction of motion at the given value of t.

(d) Write the particle's velocity at that time as the product of its speed and direction.

1. $\mathbf{r}(t) = (2\cos t)\mathbf{i} + (3\sin t)\mathbf{j}, \quad t = \pi/2$

2. $\mathbf{r}(t) = (\cos 2t)\mathbf{i} + (2\sin t)\mathbf{j}, \quad t = 0$

3. $\mathbf{r}(t) = (\sec t)\mathbf{i} + (\tan t)\mathbf{j}, \quad t = \pi/6$

4. $\mathbf{r}(t) = (2\ln (t+1))\mathbf{i} + (t^2)\mathbf{j}, \quad t = 1$

In Exercises 5–8, $\mathbf{r}(t)$ is the position vector of a particle in the plane at time t. Find the time, or times, in the given time interval when the velocity and acceleration vectors are perpendicular.

5. $\mathbf{r}(t) = (t - \sin t)\mathbf{i} + (1 - \cos t)\mathbf{j}, \quad 0 \le t \le 2\pi$

6. $\mathbf{r}(t) = (\sin t)\mathbf{i} + t\mathbf{j}, \quad t \ge 0$

7. $\mathbf{r}(t) = (3 \cos t)\mathbf{i} + (4 \sin t)\mathbf{j}, \quad t \geq 0$

8. $\mathbf{r}(t) = (5 \cos t)\mathbf{i} + (5 \sin t)\mathbf{j}, \quad t \geq 0$

In Exercises 9 and 10, $\mathbf{r}(t)$ is the position vector of a particle in the plane at time t. Find the angle between the velocity and acceleration vectors at the given value of t.

9. $\mathbf{r}(t) = (2 \cos t)\mathbf{i} + (\sin t)\mathbf{j}, \quad t = \pi/4$

10. $\mathbf{r}(t) = (3t + 1)\mathbf{i} + (t^2)\mathbf{j}, \quad t = 0$

Limits and Continuity

In Exercises 11 and 12, **(a)** evaluate the limit, and **(b)** find the values of t for which the vector function is continuous and **(c)** discontinuous.

11. $\lim\limits_{t \to 3} \left[t\mathbf{i} + \dfrac{t^2 - 9}{t^2 + 3t} \mathbf{j} \right]$

12. $\lim\limits_{t \to 0} \left[\dfrac{\sin 2t}{t} \mathbf{i} + (\ln \, (t + 1))\mathbf{j} \right]$

Tangents and Normals

In Exercises 13 and 14, find an equation for the line that is **(a)** tangent and **(b)** normal to the curve $\mathbf{r}(t)$ at the point determined by the given value of t.

13. $\mathbf{r}(t) = (\sin t)\mathbf{i} + (t^2 - \cos t)\mathbf{j}, \quad t = 0$

14. $\mathbf{r}(t) = (2 \cos t - 3)\mathbf{i} + (3 \sin t + 1)\mathbf{j}, \quad t = \pi/4$

Integration

In Exercises 15–18, evaluate the integral.

15. $\int_1^2 [(6 - 6t)\mathbf{i} + 3\sqrt{t}\,\mathbf{j}] \, dt$

16. $\int_{-\pi/4}^{\pi/4} [(\sin \, t)\mathbf{i} + (1 + \cos \, t)\mathbf{j}] \, dt$

17. $\int [(\sec \, t \tan \, t)\mathbf{i} + (\tan \, t)\mathbf{j}] \, dt$

18. $\int \left[\dfrac{1}{t} \mathbf{i} + \dfrac{1}{5 - t} \mathbf{j} \right] dt$

Initial Value Problems

In Exercises 19–22, solve the initial value problem for $\mathbf{r}$ as a vector function of t.

19. $\dfrac{d\mathbf{r}}{dt} = \dfrac{3}{2}(t + 1)^{1/2}\mathbf{i} + e^{-t}\mathbf{j}, \quad \mathbf{r}(0) = \mathbf{0}$

20. $\dfrac{d\mathbf{r}}{dt} = (t^3 + 4t)\mathbf{i} + t\mathbf{j}, \quad \mathbf{r}(0) = \mathbf{i} + \mathbf{j}$

21. $\dfrac{d^2\mathbf{r}}{dt^2} = -32\mathbf{j}, \quad \mathbf{r}(0) = 100\mathbf{i}, \quad \dfrac{d\mathbf{r}}{dt}\bigg|_{t=0} = 8\mathbf{i} + 8\mathbf{j}$

22. $\dfrac{d^2\mathbf{r}}{dt^2} = -\mathbf{i} - \mathbf{j}, \quad \mathbf{r}(0) = 10\mathbf{i} + 10\mathbf{j}, \quad \dfrac{d\mathbf{r}}{dt}\bigg|_{t=0} = \mathbf{0}$

Paths and Motion

23. *Finding distance traveled* The position of a particle in the plane at time t is given by
$$\mathbf{r}(t) = (1 - \cos t)\mathbf{i} + (t - \sin t)\mathbf{j}.$$
Find the distance the particle travels along the path from $t = 0$ to $t = 2\pi/3$.

24. *Length of a path* Let C be the path traced by
$$\mathbf{r}(t) = \left(\frac{1}{4} e^{4t} - t \right)\mathbf{i} + (e^{2t})\mathbf{j}, \quad 0 \leq t \leq 2.$$
(a) Find the initial and terminal points of C.

(b) Find the length of C.

25. *Velocity on a path* The position of a particle is given by
$$\mathbf{r}(t) = (\sin t)\mathbf{i} + (\cos 2t)\mathbf{j}.$$
(a) Find the velocity vector for the particle.

(b) For what values of t in the interval $0 \leq t \leq 2\pi$ is $d\mathbf{r}/dt$ equal to $\mathbf{0}$?

(c) *Writing to Learn* Find a Cartesian equation for a curve that contains the particle's path. What portion of the graph of the Cartesian equation is traced by the particle? Describe the motion as t increases from 0 to 2π.

26. *Revisiting Example 7* The position of a particle is given by $\mathbf{r}(t) = (2t^3 - 3t^2)\mathbf{i} + (t^3 - 12t)\mathbf{j}$.

(a) Find dy/dx in terms of t.

(b) *Writing to Learn* Find the x- and y-coordinates for each critical point of the path (point where dy/dx is zero or does not exist). Does the path have a vertical or horizontal tangent at the critical point? Explain.

27. *Finding a position vector* At time $t = 0$, a particle is located at the point $(1, 2)$. It travels in a straight line to the point $(4, 1)$, has speed 2 at $(1, 2)$, and constant acceleration $3\mathbf{i} - \mathbf{j}$. Find an equation for the position vector $\mathbf{r}(t)$ of the particle at time t.

28. *Studying a motion* The path of a particle for $t > 0$ is given by
$$\mathbf{r}(t) = \left(t + \frac{2}{t} \right)\mathbf{i} + (3t^2)\mathbf{j}.$$
(a) Find the coordinates of each point on the path where the horizontal component of the velocity of the particle is zero.

(b) Find dy/dx when $t = 1$.

(c) Find d^2y/dx^2 when $y = 12$.

29. *Motion on circular paths* Each of equations (a) through (e) describes the motion of a particle having the same path, namely, the unit circle $x^2 + y^2 = 1$. Although the path of each particle in (a) through (e) is the same, the behavior, or "dynamics," of each particle is different. For each particle, answer the following questions.

i. Does the particle have constant speed? If so, what is its constant speed?

ii. Is the particle's acceleration vector always orthogonal to its velocity vector?

iii. Does the particle move clockwise or counterclockwise around the circle?

iv. Does the particle begin at the point $(1, 0)$?

(a) $\mathbf{r}(t) = (\cos t)\mathbf{i} + (\sin t)\mathbf{j}, \quad t \geq 0$

(b) $\mathbf{r}(t) = (\cos 2t)\mathbf{i} + (\sin 2t)\mathbf{j}, \quad t \geq 0$

(c) $\mathbf{r}(t) = \cos (t - \pi/2)\mathbf{i} + \sin (t - \pi/2)\mathbf{j}, \quad t \geq 0$

(d) $\mathbf{r}(t) = (\cos t)\mathbf{i} - (\sin t)\mathbf{j}, \quad t \geq 0$

(e) $\mathbf{r}(t) = \cos (t^2)\mathbf{i} + \sin (t^2)\mathbf{j}, \quad t \geq 0$

30. *Motion on a parabola* A particle moves along the top of the parabola $y^2 = 2x$ from left to right at a constant speed of 5 units per second. Find the velocity of the particle as it moves through the point $(2, 2)$.

Applications

31. *Flying a kite* The position of a kite is given by

$$\mathbf{r}(t) = \frac{t}{8}\mathbf{i} - \frac{3}{64} t (t - 160)\mathbf{j},$$

where $t \geq 0$ is measured in seconds and distance is measured in meters.

(a) How long is the kite above ground?

(b) How high is the kite at $t = 40$ sec?

(c) At what rate is the kite's altitude increasing at $t = 40$ sec?

(d) At what time does the kite start to lose altitude?

32. *Colliding particles* The paths of two particles for $t \geq 0$ are given by

$$\mathbf{r}_1(t) = (t - 3)\mathbf{i} + (t - 3)^2\mathbf{j},$$

$$\mathbf{r}_2(t) = \left(\frac{3t}{2} - 4\right)\mathbf{i} + \left(\frac{3t}{2} - 2\right)\mathbf{j}.$$

(a) Determine the exact time(s) at which the particles collide.

(b) Find the direction of motion of each particle at the time(s) of collision.

33. *A satellite in circular orbit* A satellite of mass m is moving at a constant speed v around a planet of mass M in a circular orbit of radius r_0, as measured from the planet's center of mass. Determine the satellite's orbital period T (the time to complete one full orbit), as follows.

(a) Coordinatize the orbital plane by placing the origin at the planet's center of mass, with the satellite on the x-axis at $t = 0$ and moving counterclockwise, as in the accompanying figure.

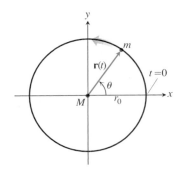

Let $\mathbf{r}(t)$ be the satellite's position vector at time t. Show that $\theta = vt/r_0$ and hence that

$$\mathbf{r}(t) = \left(r_0 \cos \frac{vt}{r_0}\right)\mathbf{i} + \left(r_0 \sin \frac{vt}{r_0}\right)\mathbf{j}.$$

(b) Find the acceleration of the satellite.

(c) According to Newton's law of gravitation, the gravitational force exerted on the satellite by the planet is directed toward the origin and is given by

$$\mathbf{F} = \left(-\frac{GmM}{r_0^2}\right)\frac{\mathbf{r}}{r_0},$$

where G is the universal constant of gravitation. Using Newton's second law, $\mathbf{F} = m\mathbf{a}$, show that $v^2 = GM/r_0$.

(d) Show that the orbital period T satisfies $vT = 2\pi r_0$.

(e) From parts (c) and (d), deduce that

$$T^2 = \frac{4\pi^2}{GM} r_0^3;$$

that is, the square of the period of a satellite in circular orbit is proportional to the cube of the radius from the orbital center.

34. *Rowing across a river* A straight river is 100 m wide. A rowboat leaves the far shore at time $t = 0$. The person in the boat rows at a rate of 20 m/min always toward the near shore. The velocity of the river at (x, y) is

$$\mathbf{v} = \left(-\frac{1}{250} (y - 50)^2 + 10\right)\mathbf{i} \quad \text{m/min}, \qquad 0 < y < 100.$$

(a) Given that $\mathbf{r}(0) = 0\mathbf{i} + 100\mathbf{j}$, what is the position of the boat at time t?

(b) How far downstream will the boat land on the near shore?

Theory and Examples

35. *Antiderivatives of vector functions*

(a) Use Corollary 2 in Section 3.2 (a consequence of the Mean Value Theorem for scalar functions) to show that two vector functions $\mathbf{R}_1(t)$ and $\mathbf{R}_2(t)$ that have identical derivatives on an interval I differ by a constant vector value throughout I.

(b) Use the result in part (a) to show that if $\mathbf{R}(t)$ is any antiderivative of $\mathbf{r}(t)$ on I, then every other antiderivative of $\mathbf{r}(t)$ on I equals $\mathbf{R}(t) + \mathbf{C}$ for some constant vector $\mathbf{C}$.

36. *Constant-length vector functions* Let $\mathbf{v}$ be a differentiable vector function of t. Show that if $\mathbf{v} \cdot (d\mathbf{v}/dt) = 0$ for all t, then $|\mathbf{v}|$ is constant.

37. *Constant function rule* Prove that if $\mathbf{u}$ is the vector function with the constant value $\mathbf{C}$, then $d\mathbf{u}/dt = \mathbf{0}$.

38. *Scalar multiple rules*

(a) Prove that if $\mathbf{u}$ is a differentiable function of t and c is any real number, then

$$\frac{d(c\mathbf{u})}{dt} = c\frac{d\mathbf{u}}{dt}.$$

(b) Prove that if $\mathbf{u}$ is a differentiable function of t and f is a differentiable scalar function of t, then

$$\frac{d(f\mathbf{u})}{dt} = \frac{df}{dt}\mathbf{u} + f\frac{d\mathbf{u}}{dt}.$$

39. *Sum and difference rules* Prove that if $\mathbf{u}$ and $\mathbf{v}$ are differentiable functions of t, then

(a) $\dfrac{d}{dt}(\mathbf{u} + \mathbf{v}) = \dfrac{d\mathbf{u}}{dt} + \dfrac{d\mathbf{v}}{dt}.$

(b) $\dfrac{d}{dt}(\mathbf{u} - \mathbf{v}) = \dfrac{d\mathbf{u}}{dt} - \dfrac{d\mathbf{v}}{dt}.$

40. *Chain rule* Prove that if $\mathbf{u}$ is a differentiable vector function of s and $s = f(t)$ is a differentiable scalar function of t, then

$$\frac{d}{dt}[\mathbf{u}(f(t))] = f'(t)\mathbf{u}'(f(t)).$$

41. *Differentiable vector functions are continuous* Show that if $\mathbf{r}(t) = f(t)\mathbf{i} + g(t)\mathbf{j}$ is differentiable at $t = c$, then $\mathbf{r}$ is continuous at c as well.

42. *Integration properties* Establish the following properties of integrable vector functions.

(a) *Constant Scalar Multiple Rule:*

$$\int_a^b k\mathbf{r}(t)\,dt = k\int_a^b \mathbf{r}(t)\,dt$$

for any scalar constant k.

(b) *Sum and Difference Rules:*

$$\int_a^b (\mathbf{r}_1(t) \pm \mathbf{r}_2(t))\,dt = \int_a^b \mathbf{r}_1(t)\,dt \pm \int_a^b \mathbf{r}_2(t)\,dt$$

(c) *Constant Vector Multiple Rule:*

$$\int_a^b \mathbf{C} \cdot \mathbf{r}(t)\,dt = \mathbf{C} \cdot \int_a^b \mathbf{r}(t)\,dt$$

for any constant vector $\mathbf{C}$.

43. *Fundamental theorem of calculus* The Fundamental Theorem of Calculus for scalar functions of a real variable holds for vector functions of a real variable as well.

(a) Prove this by using the theorem for scalar functions to show that if a vector function $\mathbf{r}(t)$ is continuous for $a \leq t \leq b$, then

$$\frac{d}{dt}\int_a^t \mathbf{r}(q)\,dq = \mathbf{r}(t)$$

at every point t of $[a, b]$.

(b) Use the conclusion in part (b) of Exercise 35 to show that if $\mathbf{R}$ is any antiderivative of $\mathbf{r}$ on $[a, b]$, then

$$\int_a^b \mathbf{r}(t)\,dt = \mathbf{R}(b) - \mathbf{R}(a).$$

9.4 Modeling Projectile Motion

CD-ROM
WEBsite

Ideal Projectile Motion • Height, Flight Time, and Range • Ideal Trajectories Are Parabolic • Firing from (x_0, y_0) • Projectile Motion with Wind Gusts

When we shoot a projectile into the air, we usually want to know beforehand how far it will go (will it reach the target?), how high it will rise (will it clear the hill?), and when it will land (when do we get results?) We get this information from the di-

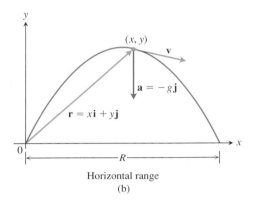

FIGURE 9.30 (a) Position, velocity, acceleration, and launch angle at $t = 0$. (b) Position, velocity, and acceleration at a later time t

rection and magnitude of the projectile's initial velocity vector, using Newton's second law of motion.

Ideal Projectile Motion

We are going to model *ideal* projectile motion. This assumes that the projectile behaves like a particle moving in a vertical coordinate plane and that the only force acting on the projectile during its flight (close to Earth's surface) is the constant force of gravity, always pointing straight down.

We assume that the projectile is launched from the origin at time $t = 0$ into the first quadrant with an initial velocity $\mathbf{v}_0$ (Figure 9.30). If $\mathbf{v}_0$ makes an angle α with the horizontal, then

$$\mathbf{v}_0 = (|\mathbf{v}_0| \cos \alpha)\mathbf{i} + (|\mathbf{v}_0| \sin \alpha)\mathbf{j}.$$

If we use the simpler notation v_0 for the initial speed $|\mathbf{v}_0|$, then

$$\mathbf{v}_0 = (v_0 \cos \alpha)\mathbf{i} + (v_0 \sin \alpha)\mathbf{j}. \tag{1}$$

The projectile's initial position is

$$\mathbf{r}_0 = 0\mathbf{i} + 0\mathbf{j} = \mathbf{0}. \tag{2}$$

Newton's second law of motion says that the force acting on the projectile is equal to the projectile's mass m times its acceleration, or $m(d^2\mathbf{r}/dt^2)$ if $\mathbf{r}$ is the projectile's position vector and t is time. If the force is solely the gravitational force $-mg\mathbf{j}$, then

$$m\frac{d^2\mathbf{r}}{dt^2} = -mg\mathbf{j} \quad \text{and} \quad \frac{d^2\mathbf{r}}{dt^2} = -g\mathbf{j}.$$

We find $\mathbf{r}$ as a function of t by solving the following initial value problem.

Differential equation: $\quad \dfrac{d^2\mathbf{r}}{dt^2} = -g\mathbf{j}$

Initial conditions: $\quad \mathbf{r} = \mathbf{r}_0 \quad \text{and} \quad \dfrac{d\mathbf{r}}{dt} = \mathbf{v}_0 \quad \text{when } t = 0$

The first integration gives

$$\frac{d\mathbf{r}}{dt} = -(gt)\mathbf{j} + \mathbf{v}_0.$$

A second integration gives

$$\mathbf{r} = -\frac{1}{2} gt^2\mathbf{j} + \mathbf{v}_0 t + \mathbf{r}_0.$$

Substituting the values of $\mathbf{v}_0$ and $\mathbf{r}_0$ from Equations 1 and 2 gives

$$\mathbf{r} = -\frac{1}{2} gt^2\mathbf{j} + \underbrace{(v_0 \cos \alpha)t\mathbf{i} + (v_0 \sin \alpha)t\mathbf{j}}_{\mathbf{v}_0 t} + \mathbf{0}$$

or

$$\mathbf{r} = (v_0 \cos \alpha)\, t\, \mathbf{i} + \left((v_0 \sin \alpha)\, t - \frac{1}{2} gt^2\right)\mathbf{j}. \tag{3}$$

Equation (3) is the **vector equation** for ideal projectile motion. The angle α is the projectile's **launch angle (firing angle, angle of elevation)**, and v_0, as we said before, is the projectile's **initial speed.**

The components of **r** give

$$x = (v_0 \cos \alpha)t \quad \text{and} \quad y = (v_0 \sin \alpha)t - \frac{1}{2}gt^2, \tag{4}$$

where x is the distance downrange and y is the height of the projectile at time $t \geq 0$.

Example 1   Firing an Ideal Projectile

A projectile is fired from the origin over horizontal ground at an initial speed of 500 m/sec and a launch angle of 60°. Where will the projectile be 10 sec later?

Solution   We use Equation (3) with $v_0 = 500$, $\alpha = 60°$, $g = 9.8$, and $t = 10$ to find the projectile's components 10 sec after firing.

$$\mathbf{r} = (v_0 \cos \alpha)t\,\mathbf{i} + \left((v_0 \sin \alpha)t - \frac{1}{2}gt^2\right)\mathbf{j}$$

$$= (500)\left(\frac{1}{2}\right)(10)\mathbf{i} + \left((500)\left(\frac{\sqrt{3}}{2}\right)10 - \left(\frac{1}{2}\right)(9.8)(100)\right)\mathbf{j}$$

$$\approx 2500\mathbf{i} + 3840\mathbf{j}.$$

Interpret

Ten seconds after firing, the projectile is about 3840 m in the air and 2500 m downrange.

Height, Flight Time, and Range

Equation (3) enables us to answer most questions about the ideal motion for a projectile fired from the origin.

The projectile reaches its highest point when its vertical velocity component is zero, that is, when

$$\frac{dy}{dt} = v_0 \sin \alpha - gt = 0, \quad \text{or} \quad t = \frac{v_0 \sin \alpha}{g}.$$

For this value of t, the value of y is

$$y_{\text{max}} = (v_0 \sin \alpha)\left(\frac{v_0 \sin \alpha}{g}\right) - \frac{1}{2}g\left(\frac{v_0 \sin \alpha}{g}\right)^2 = \frac{(v_0 \sin \alpha)^2}{2g}.$$

To find when the projectile lands when fired over horizontal ground, we set the vertical component equal to zero in Equation (3) and solve for t.

$$(v_0 \sin \alpha)t - \frac{1}{2}gt^2 = 0$$

$$t\left(v_0 \sin \alpha - \frac{1}{2}gt\right) = 0$$

$$t = 0, \quad t = \frac{2v_0 \sin \alpha}{g}$$

Since 0 is the time the projectile is fired, $(2v_0 \sin \alpha)/g$ must be the time when the projectile strikes the ground.

To find the projectile's **range** R, the distance from the origin to the point of impact on horizontal ground, we find the value of the horizontal component when $t = (2v_0 \sin \alpha)/g$.

$$x = (v_0 \cos \alpha)t$$

$$R = (v_0 \cos \alpha)\left(\frac{2v_0 \sin \alpha}{g}\right) = \frac{v_0^2}{g}(2 \sin \alpha \cos \alpha) = \frac{v_0^2}{g}\sin 2\alpha$$

The range is largest when $\sin 2\alpha = 1$ or $\alpha = 45°$.

Height, Flight Time, and Range for Ideal Projectile Motion

For ideal projectile motion when an object is launched from the origin over a horizontal surface with initial speed v_0 and launch angle α:

$$\text{Maximum height:} \qquad y_{max} = \frac{(v_0 \sin \alpha)^2}{2g}$$

$$\text{Flight time:} \qquad t = \frac{2v_0 \sin \alpha}{g}$$

$$\text{Range:} \qquad R = \frac{v_0^2}{g}\sin 2\alpha.$$

Example 2   Investigating Ideal Projectile Motion

Find the maximum height, flight time, and range of a projectile fired from the origin over horizontal ground at an initial speed of 500 m/sec and a launch angle of 60° (same projectile as Example 1).

Solution

Maximum height:
$$y_{max} = \frac{(v_0 \sin \alpha)^2}{2g}$$
$$= \frac{(500 \sin 60°)^2}{2(9.8)} \approx 9566.33 \text{ m}$$

Flight time:
$$t = \frac{2v_0 \sin \alpha}{g}$$
$$= \frac{2(500) \sin 60°}{9.8} \approx 88.37 \text{ sec}$$

Range:
$$R = \frac{v_0^2}{g}\sin 2\alpha$$
$$= \frac{(500)^2 \sin 120°}{9.8} \approx 22{,}092.48 \text{ m}$$

From Equation (3), the position vector of the projectile is

$$\mathbf{r} = (v_0 \cos \alpha)t\mathbf{i} + \left((v_0 \sin \alpha)t - \frac{1}{2}gt^2\right)\mathbf{j}$$

$$= (500 \cos 60°)t\mathbf{i} + \left((500 \sin 60°)t - \frac{1}{2}(9.8)t^2\right)\mathbf{j}$$

$$= 250t\mathbf{i} + ((250\sqrt{3})t - 4.9t^2)\mathbf{j}.$$

A graph of the projectile's path is shown in Figure 9.31.

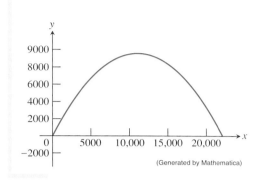

FIGURE 9.31 The graph of the parametric equations $x = 250t$, $y = (250\sqrt{3})t - 4.9t^2$ for $0 \leq t \leq 88.4$. (Example 2)

(Generated by Mathematica)

Ideal Trajectories Are Parabolic

It is often claimed that water from a hose traces a parabola in the air, but any one who looks closely enough will see this is not so. The air slows the water down, and its forward progress is too slow at the end to keep pace with the rate at which it falls.

What is really being claimed is that ideal projectiles move along parabolas, and this we can see from Equation (4). If we substitute $t = x/(v_0 \cos \alpha)$ from the first equation into the second, we obtain the Cartesian-coordinate equation

$$y = -\left(\frac{g}{2v_0{}^2 \cos^2 \alpha}\right)x^2 + (\tan \alpha)\,x.$$

This equation has the form $y = ax^2 + bx$, so its graph is a parabola.

Firing from (x_0, y_0)

If we fire our ideal projectile from the point (x_0, y_0) instead of the origin (Figure 9.32), the position vector for the path of motion is

$$\mathbf{v} = (x_0 + (v_0 \cos \alpha)\,t)\mathbf{i} + \left(y_0 + (v_0 \sin \alpha)\,t - \frac{1}{2}\,gt^2\right)\mathbf{j}, \qquad (5)$$

as you are asked to show in Exercise 19.

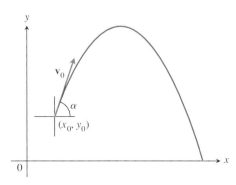

FIGURE 9.32 The path of a projectile fired from (x_0, y_0) with an initial velocity $\mathbf{v}_0$ at an angle of α degrees with the horizontal.

Example 3 Firing a Flaming Arrow

To open the 1992 Summer Olympics in Barcelona, bronze medalist archer Antonio Rebollo lit the Olympic torch with a flaming arrow (Figure 9.33). Suppose that Rebollo shot the arrow at a height of 6 ft above ground level 90 ft from the 70-ft-high cauldron, and he wanted the arrow to reach maximum height exactly 4 ft above the center of the cauldron (Figure 9.34).

FIGURE 9.33 Spanish archer Antonio Rebollo lights the Olympic torch in Barcelona with a flaming arrow.

(a) Express y_{max} in terms of the initial speed v_0 and firing angle α.

(b) Use $y_{max} = 74$ ft (Figure 9.34) and the result from part (a) to find the value of $v_0 \sin \alpha$.

(c) Find the value of $v_0 \cos \alpha$.

(d) Find the initial firing angle of the arrow.

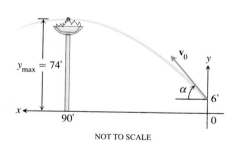

FIGURE 9.34 Ideal path of the arrow that lit the Olympic torch.

Solution

(a) We use a coordinate system in which the positive x-axis lies along the ground toward the left (to match the second photograph in Figure 9.33) and the coordinates of the flaming arrow at $t = 0$ are $x_0 = 0$ and $y_0 = 6$ (Figure 9.34). We have

$$y = y_0 + (v_0 \sin \alpha)t - \frac{1}{2}gt^2 \qquad \text{Equation (5), j-component}$$

$$= 6 + (v_0 \sin \alpha)t - \frac{1}{2}gt^2. \qquad y_0 = 6$$

We find the time when the arrow reaches its highest point by setting $dy/dt = 0$ and solving for t, obtaining

$$t = \frac{v_0 \sin \alpha}{g}.$$

For this value of t, the value of y is

$$y_{max} = 6 + (v_0 \sin \alpha) \left(\frac{v_0 \sin \alpha}{g} \right) - \frac{1}{2} g \left(\frac{v_0 \sin \alpha}{g} \right)^2$$

$$= 6 + \frac{(v_0 \sin \alpha)^2}{2g}.$$

(b) Using $y_{max} = 74$ and $g = 32$, we see from the preceeding equation in part (a) that

$$74 = 6 + \frac{(v_0 \sin \alpha)^2}{2(32)}$$

or

$$v_0 \sin \alpha = \sqrt{(68)(64)}.$$

(c) When the arrow reaches y_{max}, the horizontal distance traveled to the center of the cauldron is $x = 90$ ft. We substitute the time to reach y_{max} from part (a) and the horizontal distance $x = 90$ ft into the **i**-component of Equation (5) to obtain

$$x = x_0 + (v_0 \cos \alpha)t \qquad \text{Equation (5), i-component}$$
$$90 = 0 + (v_0 \cos \alpha)t \qquad x = 90, x_0 = 0$$
$$= (v_0 \cos \alpha) \left(\frac{v_0 \sin \alpha}{g} \right). \qquad t = (v_0 \sin \alpha)/g$$

Solving this equation for $v_0 \cos \alpha$ and using $g = 32$ and the result from part (b), we have

$$v_0 \cos \alpha = \frac{90g}{v_0 \sin \alpha} = \frac{(90)(32)}{\sqrt{(68)(64)}}.$$

(d) Parts (b) and (c) together tell us that

$$\tan \alpha = \frac{v_0 \sin \alpha}{v_0 \cos \alpha} = \frac{(\sqrt{(68)(64)})^2}{(90)(32)} = \frac{68}{45}$$

or

$$\alpha = \tan^{-1} \left(\frac{68}{45} \right) \approx 56.5°.$$

This is Rebollo's firing angle.

Projectile Motion with Wind Gusts

The next example shows how to account for another force acting on a projectile. We also assume that the path of the baseball in Example 4 lies in a vertical plane.

Example 4   Hitting a Baseball

A baseball is hit when it is 3 ft above the ground. It leaves the bat with initial speed of 152 ft/sec, making an angle of 20° with the horizontal. At the instant the ball is hit, an instantaneous gust of wind blows in the horizontal direction directly opposite the direction the ball is taking toward the outfield, adding a component of $-8.8\mathbf{i}$ (ft/sec) to the ball's initial velocity (8.8 ft/sec = 6 mph).

(a) Find a vector equation (position vector) for the path of the baseball.

(b) How high does the baseball go, and when does it reach maximum height?

(c) Assuming that the ball is not caught, find its range and flight time.

Solution

(a) Using Equation (1) and accounting for the gust of wind, the initial velocity of the baseball is

$$\mathbf{v}_0 = (v_0 \cos \alpha)\mathbf{i} + (v_0 \sin \alpha)\mathbf{j} - 8.8\mathbf{i}$$
$$= (152 \cos 20°)\mathbf{i} + (152 \sin 20°)\mathbf{j} - (8.8)\mathbf{i}$$
$$= (152 \cos 20° - 8.8)\mathbf{i} + (152 \sin 20°)\mathbf{j}.$$

The initial position is $\mathbf{r}_0 = 0\mathbf{i} + 3\mathbf{j}$. Integration of $d^2\mathbf{r}/dt^2 = -g\mathbf{j}$ gives

$$\frac{d\mathbf{r}}{dt} = (gt)\mathbf{j} + \mathbf{v}_0.$$

A second integration gives

$$\mathbf{r} = -\frac{1}{2} gt^2\mathbf{j} + \mathbf{v}_0 t + \mathbf{r}_0.$$

Substituting the values of $\mathbf{v}_0$ and $\mathbf{r}_0$ into the last equation gives the position vector of the baseball.

$$\mathbf{r} = -\frac{1}{2} gt^2\mathbf{j} + \mathbf{v}_0 t + \mathbf{r}_0$$
$$= -16t^2\mathbf{j} + (152 \cos 20° - 8.8)t\mathbf{i} + (152 \sin 20°)t\mathbf{j} + 3\mathbf{j}$$
$$= (152 \cos 20° - 8.8)t\mathbf{i} + (3 + (152 \sin 20°)t - 16t^2)\mathbf{j}.$$

(b) The baseball reaches its highest point when the vertical component of velocity is zero, or

$$\frac{dy}{dt} = 152 \sin 20° - 32t = 0.$$

Solving for t we find

$$t = \frac{152 \sin 20°}{32} \approx 1.62 \text{ sec.}$$

Substituting this time into the vertical component for $\mathbf{r}$ gives the maximum height

$$y_{max} = 3 + (152 \sin 20°)(1.62) - 16(1.62)^2$$
$$\approx 45.2 \text{ ft.}$$

That is, the maximum height of the baseball is about 45.2 ft, reached about 1.6 sec after leaving the bat.

(c) To find when the baseball lands, we set the vertical component for **r** equal to 0 and solve for t:

$$3 + (152 \, \sin \, 20°) \, t - 16t^2 = 0$$
$$3 + (51.99) \, t - 16t^2 = 0.$$

The solution values are about $t = 3.3$ sec and $t = -0.06$ sec. Substituting the positive time into the horizontal component for **r**, we find the range

$$R = (152 \, \cos \, 20° - 8.8)(3.3)$$
$$\approx 442.3 \text{ ft.}$$

Thus, the horizontal range is about 442.3 ft, and the flight time is about 3.3 sec.

In Exercises 29 through 31, we consider projectile motion when there is air resistance slowing down the flight.

EXERCISES 9.4

Projectile flights in the following exercises are to be treated as ideal unless stated otherwise. All launch angles are assumed to be measured from the horizontal. All projectiles are assumed to be launched from the origin over a horizontal surface unless stated otherwise.

1. *Travel time* A projectile is fired at a speed of 840 m/sec at an angle of 60°. How long will it take to get 21 km downrange?

2. *Finding muzzle speed* Find the muzzle speed of a gun whose maximum range is 24.5 km.

3. *Flight time and height* A projectile is fired with an initial speed of 500 m/sec at an angle of elevation of 45°.

   (a) When and how far away will the projectile strike?

   (b) How high overhead will the projectile be when it is 5 km downrange?

   (c) What is the greatest height reached by the projectile?

4. *Throwing a baseball* A baseball is thrown from the stands 32 ft above the field at an angle of 30° up from the horizontal. When and how far away will the ball strike the ground if its initial speed is 32 ft/sec?

5. *Shot put* An athlete puts a 16 lb shot at an angle of 45° to the horizontal from 6.5 ft above the ground at an initial speed of 44 ft/sec as suggested in the accompanying figure. How long after launch and how far from the inner edge of the stopboard does the shot land?

6. *(Continuation of Exercise 5)* Because of its initial elevation, the shot in Exercise 5 would have gone slightly farther if it had been launched at a 40° angle. How much farther? Answer in inches.

7. *Firing golf balls* A spring gun at ground level fires a golf ball at an angle of 45°. The ball lands 10 m away.

   (a) What was the ball's initial speed?

   (b) For the same initial speed, find the two firing angles that make the range 6 m.

8. *Beaming electrons* An electron in a TV tube is beamed horizontally at a speed of 5×10^6 m/sec toward the face of the tube 40 cm away. About how far will the electron drop before it hits?

9. *Finding golf ball speed* Laboratory tests designed to find how far golf balls of different hardness go when hit with a driver showed that a 100-compression ball hit with a club-head speed of 100 mph at a launch angle of 9° carried 248.8 yd. What was the launch speed of the ball? (It was more than 100 mph. At the same time the club head was moving forward, the compressed ball was kicking away from the club face, adding to the ball's forward speed.)

10. *Writing to Learn:* A *human cannonball* is to be fired with an initial speed of $v_0 = 80\sqrt{10/3}$ ft/sec. The circus performer (of the right caliber, naturally) hopes to land on a special cushion located 200 ft downrange at the same height as the muzzle of the cannon. The circus is being held in a large room with a flat ceiling 75 ft higher than the muzzle. Can the performer be fired to the cushion without striking the ceiling? If so, what should the cannon's angle of elevation be?

11. *Writing to Learn* A golf ball leaves the ground at a 30° angle at a speed of 90 ft/sec. Will it clear the top of a 30 ft tree that is in the way, 135 ft down the fairway? Explain.

12. *Elevated green* A golf ball is hit with an initial speed of 116 ft/sec at an angle of elevation of 45° from the tee to a green that is elevated 45 ft above the tee as shown in the diagram. Assuming that the pin, 369 ft downrange, does not get in the way, where will the ball land in relation to the pin?

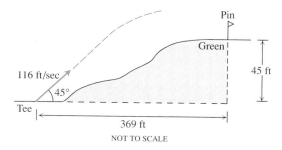

369 ft

NOT TO SCALE

13. *The Green Monster* A baseball hit by a Boston Red Sox player at a 20° angle from 3 ft above the ground just cleared the left end of the "Green Monster," the left-field wall in Fenway Park (retired after the 2002 season). This wall is 37 ft high and 315 ft from home plate (see the accompanying figure).

   (a) What was the initial speed of the ball?

   (b) How long did it take the ball to reach the wall?

14. *Equal-range firing angles* Show that a projectile fired at an angle of α degrees, $0 < \alpha < 90$, has the same range as a projectile fired at the same speed at an angle of $(90 - \alpha)$ degrees. (In models that take air resistance into account, this symmetry is lost.)

15. *Equal-range firing angles* What two angles of elevation will enable a projectile to reach a target 16 km downrange on the same level as the gun if the projectile's initial speed is 400 m/sec?

16. *Range and height versus speed*

   (a) Show that doubling a projectile's initial speed at a given launch angle multiplies its range by 4.

   (b) By about what percentage should you increase the initial speed to double the height and range?

17. *Shot put* In Moscow in 1987, Natalya Lisouskaya set a women's world record by putting an 8 lb 13 oz shot 73 ft 10 in. Assuming that she launched the shot at a 40° angle to the horizontal from 6.5 ft above the ground, what was the shot's initial speed?

18. *Height versus time* Show that a projectile attains three-quarters of its maximum height in half the time it takes to reach the maximum height.

19. *Firing from* (x_0, y_0) Derive the equations

$$x = x_0 + (v_0 \cos \alpha)t,$$

$$y = y_0 + (v_0 \sin \alpha)t - \frac{1}{2}gt^2,$$

(see Equation 5 in the text) by solving the following initial value problem for a vector **r** in the plane.

Differential equation: $\dfrac{d^2\mathbf{r}}{dt^2} = -g\mathbf{j}$

Initial conditions: $\mathbf{r}(0) = x_0\mathbf{i} + y_0\mathbf{j}$

$\dfrac{d\mathbf{r}}{dt}(0) = (v_0 \cos \alpha)\mathbf{i} + (v_0 \sin \alpha)\mathbf{j}$

20. *Flaming arrow* Using the firing angle found in Example 3, find the speed at which the flaming arrow left Rebollo's bow. See Figure 9.34.

21. *Flaming arrow* The cauldron in Example 3 is 12 ft in diameter. Using Equation 5 and Example 3c, find how long it takes the flaming arrow to cover the horizontal distance to the rim. How high is the arrow at this time?

22. *Writing to Learn* Describe the path of a projectile given by Equations 4 when $\alpha = 90°$.

23. *Model train* The accompanying multiflash photograph shows a model train engine moving at a constant speed on a straight horizontal track. As the engine moved along, a marble was fired into the air by a spring in the engine's smokestack. The marble, which continued to move with the same forward speed as the engine, rejoined the engine 1 sec after it was fired. Measure the angle the marble's path made with the horizontal and use the information to find how high the marble went and how fast the engine was moving.

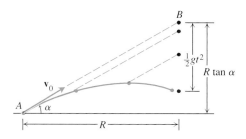

24. *Writing to Learn: Colliding marbles* The figure shows an experiment with two marbles. Marble A was launched toward marble B with launch angle α and initial speed v_0. At the same instant, marble B was released to fall from rest at $R \tan \alpha$ units directly above a spot R units downrange from A. The marbles were found to collide regardless of the value of v_0. Was this mere coincidence, or must this happen? Give reasons for your answer.

25. *Launching downhill* An ideal projectile is launched straight down an inclined plane as shown in the accompanying figure.

(a) Show that the greatest downhill range is achieved when the initial velocity vector bisects angle AOR.

(b) *Writing to Learn* If the projectile were fired uphill instead of down, what launch angle would maximize its range? Give reasons for your answer.

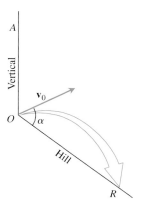

26. *Hitting a baseball under a wind gust* A baseball is hit when it is 2.5 ft above the ground. It leaves the bat with an initial velocity of 145 ft/sec at a launch angle of 23°. At the instant the ball is hit, an instantaneous gust of wind blows against the ball, adding a component of $-14\mathbf{i}$ (ft/sec) to the ball's initial velocity. A 15-ft-high fence lies 300 ft from home plate in the direction of the flight.

(a) Find a vector equation for the path of the baseball.

(b) How high does the baseball go, and when does it reach maximum height?

(c) Find the range and flight time of the baseball, assuming that the ball is not caught.

(d) When is the baseball 20 ft high? How far (ground distance) is the baseball from home plate at that height?

(e) *Writing to Learn* Has the batter hit a home run? Explain.

27. *Volleyball* A volleyball is hit when it is 4 ft above the ground and 12 ft from a 6-ft-high net. It leaves the point of impact with an initial velocity of 35 ft/sec at an angle of 27° and slips by the opposing team untouched.

(a) Find a vector equation for the path of the volleyball.

(b) How high does the volleyball go, and when does it reach maximum height?

(c) Find its range and flight time.

(d) When is the volleyball 7 ft above the ground? How far (ground distance) is the volleyball from where it will land?

(e) *Writing to Learn* Suppose that the net is raised to 8 ft. Does this change things? Explain.

28. *Where trajectories crest* For a projectile fired from the ground at launch angle α with initial speed v_0, consider α as a variable and v_0 as a fixed constant. For each α, $0 < \alpha < \pi/2$, we obtain a parabolic trajectory as shown in the accompanying figure. Show that the points in the plane that give the maximum heights of these parabolic trajectories all lie on the ellipse

$$x^2 + 4\left(y - \frac{v_0^{\,2}}{4g}\right)^2 = \frac{v_0^{\,4}}{4g^2},$$

where $x \geq 0$.

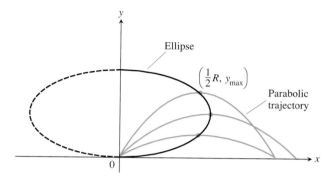

Projectile Motion with Linear Drag

The main force affecting the motion of a projectile, other than gravity, is air resistance. This slowing down force is **drag force,** and it acts in a direction *opposite* to the velocity of the projectile (see accompanying figure). For projectiles moving through the air at relatively low speeds, however, the drag force is (very nearly) proportional to the speed (to the first power) and so is called **linear.**

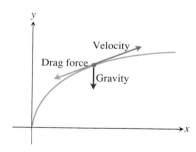

29. *Linear drag* Derive the equations

$$x = \frac{v_0}{k}(1 - e^{-kt})\cos\alpha$$

$$y = \frac{v_0}{k}(1 - e^{-kt})(\sin\alpha) + \frac{g}{k^2}(1 - kt - e^{-kt})$$

by solving the following initial value problem for a vector $\mathbf{r}$ in the plane.

Differential equation: $\dfrac{d^2\mathbf{r}}{dt^2} = -g\mathbf{j} - k\mathbf{v} = -g\mathbf{j} - k\dfrac{d\mathbf{r}}{dt}$

Initial conditions: $\mathbf{r}(0) = \mathbf{0}$

$$\left.\frac{d\mathbf{r}}{dt}\right|_{t=0} = \mathbf{v}_0 = (v_0\,\cos\,\alpha)\mathbf{i} + (v_0\,\sin\,\alpha)\mathbf{j}$$

The **drag coefficient** k is a positive constant representing resistance due to air density, v_0 and α are the projectile's initial speed and launch angle, and g is the acceleration of gravity.

30. *Hitting a baseball with linear drag* Consider the baseball problem in Example 4 when there is linear drag (see Exercise 29). Assume a drag coefficient $k = 0.12$, but no gust of wind.

   (a) From Exercise 29, find a vector form for the path of the baseball.

   (b) How high does the baseball go, and when does it reach maximum height?

   (c) Find the range and flight time of the baseball.

   (d) When is the baseball 30 ft high? How far (ground distance) is the baseball from home plate at that height?

   (e) *Writing to Learn* A 10-ft-high outfield fence is 340 ft from home plate in the direction of the flight of the baseball. The outfielder can jump and catch any ball up to 11 ft off the ground to stop it from going over the fence. Has the batter hit a home run?

31. *Hitting a baseball with linear drag under a wind gust* Consider again the baseball problem in Example 4. This time assume a drag coefficient of 0.08 *and* an instantaneous gust of wind that adds a component of $-17.6\mathbf{i}$ (ft/sec) to the initial velocity at the instant the baseball is hit.

   (a) Find a vector equation for the path of the baseball.

   (b) How high does the baseball go, and when does it reach maximum height?

   (c) Find the range and flight time of the baseball?

   (d) When is the baseball 35 ft high? How far (ground distance) is the baseball from home plate at that height?

   (e) *Writing to Learn* A 20-ft-high outfield fence is 380 ft from home plate in the direction of the flight of the baseball. Has the batter hit a home run? If "yes," what change in the horizontal component of the ball's initial velocity would have kept the ball in the park? If "no," what change would have allowed it to be a home run?

9.5   Polar Coordinates and Graphs

Polar Coordinates • Polar Graphing • Symmetry • Relating Polar and Cartesian Coordinates • Finding Points Where Polar Graphs Intersect

In radar tracking, an operator is interested in the bearing or angle the tracked object makes with some fixed ray (e.g., a directed line pointing due east) and how far away the object is currently located. In this section, we study a coordinate system devised by Newton, called the **polar coordinate system,** which is practical to use for such purposes.

Polar Coordinates

To define polar coordinates, we first choose a point in the plane called the **pole** (or **origin**) and labeled O. Then we draw an **initial ray** (or **polar axis**) starting at O. This ray is usually drawn horizontally and pointing to the right, corresponding to the positive x-axis in Cartesian coordinates (Figure 9.35). Then each point P can be located by assigning to it a **polar coordinate pair** (r, θ) in which r gives the directed distance from O to P and θ gives the directed angle from the initial ray to ray OP.

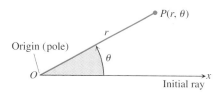

FIGURE 9.35 To define polar coordinates for the plane, we start with an origin, called the pole, and an initial ray.

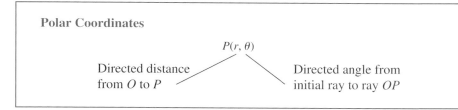

Polar Coordinates

$$P(r, \theta)$$

Directed distance   Directed angle from
from O to P initial ray to ray OP

As in trigonometry, θ is positive when measured counterclockwise and negative when measured clockwise. The angle associated with a given point is not unique. For instance, the point 2 units from the origin along the ray $\theta = \pi/6$ has polar coordinates $r = 2$, $\theta = \pi/6$. It also has coordinates $r = 2$, $\theta = -11\pi/6$ (Figure 9.36).

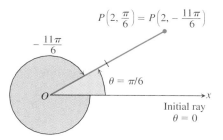

FIGURE 9.36 Polar coordinates are not unique.

There are also occasions when we wish to allow r to be negative. This is why we use *directed* distance in defining $P(r, \theta)$. The point $P(2, 7\pi/6)$ can be reached by turning $7\pi/6$ radians counterclockwise from the initial ray and going forward 2 units (Figure 9.37). It can also be reached by turning $\pi/6$ radians counterclockwise from the initial ray and going *backwards* 2 units. So the point also has polar coordinates $r = -2$, $\theta = \pi/6$.

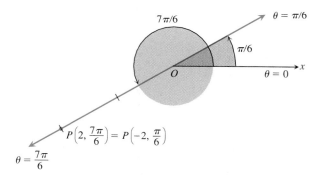

$$P\left(2, \frac{7\pi}{6}\right) = P\left(-2, \frac{\pi}{6}\right)$$

FIGURE 9.37 Polar coordinates can have negative r-values.

Polar Graphing

If we hold r fixed at a constant value $a \neq 0$, the point $P(r, \theta)$ will lie $|a|$ units from the origin O. As θ varies over any interval of length 2π, P traces a circle of radius $|a|$ centered at O (Figure 9.38).

If we hold θ fixed at a constant value $\theta = \alpha$ and let r vary between $-\infty$ and ∞, the point $P(r, \theta)$ traces the line through O that makes an angle of measure α with the initial ray.

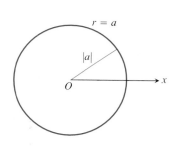

FIGURE 9.38 The polar equation for this circle is $r = a$.

Equation	Polar Graph
$r = a$	Circle of radius $|a|$ centered at O
$\theta = \alpha$	Line through O making an angle α with the initial ray

Example 1   Finding Polar Equations for Graphs

(a) $r = 1$ and $r = -1$ are equations for the circle of radius 1 centered at O.

(b) $\theta = \pi/6$, $\theta = 7\pi/6$, and $\theta = -5\pi/6$ are equations for the line in Figure 9.37.

Equations of the form $r = a$ and $\theta = \alpha$ can be combined to define regions, segments, and rays.

Example 2   Graphing Equations and Inequalities

Graph the set of points whose polar coordinates satisfy the given conditions.

(a) $1 \leq r \leq 2$ and $0 \leq \theta \leq \dfrac{\pi}{2}$

(b) $-3 \leq r \leq 2$ and $\theta = \dfrac{\pi}{4}$

(a)

(b)

(c)

(d)

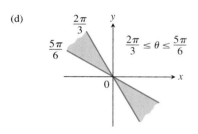

FIGURE 9.39 The graphs of typical inequalities in r and θ. (Example 2)

(c) $r \le 0$ and $\theta = \dfrac{\pi}{4}$

(d) $\dfrac{2\pi}{3} \le \theta \le \dfrac{5\pi}{6}$ (no restriction on r)

Solution   The graphs are shown in Figure 9.39.

Symmetry

Figure 9.40 illustrates the standard polar coordinate tests for symmetry.

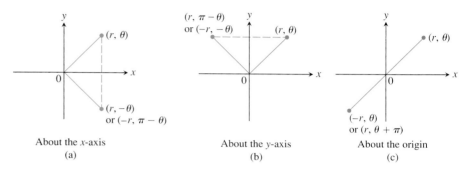

About the x-axis
(a)

About the y-axis
(b)

About the origin
(c)

FIGURE 9.40 Three tests for symmetry.

Symmetry Tests for Polar Graphs

1. **Symmetry about the x-axis:** If the point (r, θ) lies on the graph, the point $(r, -\theta)$ or $(-r, \pi - \theta)$ lies on the graph (Figure 9.40a).

2. **Symmetry about the y-axis:** If the point (r, θ) lies on the graph, the point $(r, \pi - \theta)$ or $(-r, -\theta)$ lies on the graph (Figure 9.40b).

3. **Symmetry about the origin:** If the point (r, θ) lies on the graph, the point $(-r, \theta)$ or $(r, \theta + \pi)$ lies on the graph (Figure 9.40c).

Example 3   A cardioid

Graph the curve $r = 1 - \cos \theta$.

Solution   The curve is symmetric about the x-axis because

$$(r, \theta) \text{ on the graph} \Rightarrow r = 1 - \cos \theta$$
$$\Rightarrow r = 1 - \cos (-\theta) \qquad \cos \theta = \cos (-\theta)$$
$$\Rightarrow (r, -\theta) \text{ on the graph.}$$

As θ increases from 0 to π, $\cos \theta$ decreases from 1 to -1 and $r = 1 - \cos \theta$ increases from a minimum value of 0 to a maximum value of 2. As θ continues on from π to 2π, $\cos \theta$ increases from -1 back to 1 and r decreases from 2 back to 0. The curve starts to repeat when $\theta = 2\pi$ because the cosine has period 2π.

The curve leaves the origin with slope $\tan (0) = 0$ and returns to the origin with slope $\tan (2\pi) = 0$.

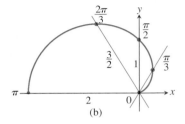

θ	$r = 1 - \cos\theta$
0	0
$\frac{\pi}{3}$	$\frac{1}{2}$
$\frac{\pi}{2}$	1
$\frac{2\pi}{3}$	$\frac{3}{2}$
π	2

(a)

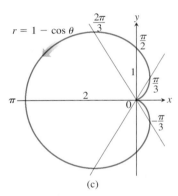

(b)

(c)

FIGURE 9.41 The steps in graphing the cardioid $r = 1 - \cos\theta$ (Example 3). The arrow shows the direction of increasing θ.

We make a table of values from $\theta = 0$ to $\theta = \pi$, plot the points, draw a smooth curve through them with a horizontal tangent at the origin, and reflect the curve across the x-axis to complete the graph (Figure 9.41). The curve is called a *cardioid* because of its heart shape. Cardioid shapes appear in the cams that direct the even layering of thread on bobbins and reels, and in the signal-strength patterns of certain radio antennae.

Example 4   Polar Graphing

Graph the curve $r^2 = 4\cos\theta$.

Solution   The equation $r^2 = 4\cos\theta$ requires $\cos\theta \geq 0$, so we get the entire graph by running θ from $-\pi/2$ to $\pi/2$. The curve is symmetric about the x-axis because

$$(r, \theta) \text{ on the graph} \Rightarrow r^2 = 4\cos\theta$$
$$\Rightarrow r^2 = 4\cos(-\theta) \qquad \cos\theta = \cos(-\theta)$$
$$\Rightarrow (r, -\theta) \text{ on the graph.}$$

The curve is also symmetric about the origin because

$$(r, \theta) \text{ on the graph} \Rightarrow r^2 = 4\cos\theta$$
$$\Rightarrow (-r)^2 = 4\cos\theta$$
$$\Rightarrow (-r, \theta) \text{ on the graph.}$$

Together, these two symmetries imply symmetry about the y-axis.

The curve passes through the origin when $\theta = -\pi/2$ and $\theta = \pi/2$. It has a vertical tangent both times because $\tan\theta$ is infinite.

For each value of θ in the interval between $-\pi/2$ and $\pi/2$, the formula $r^2 = 4\cos\theta$ gives two values of r:

$$r = \pm 2\sqrt{\cos\theta}.$$

We make a short table of values, plot the corresponding points, and use information about symmetry and tangents to guide us in connecting the points with a smooth curve (Figure 9.42).

θ	$\cos\theta$	$r = \pm 2\sqrt{\cos\theta}$
0	1	± 2
$\pm\frac{\pi}{6}$	$\frac{\sqrt{3}}{2}$	± 1.9
$\pm\frac{\pi}{4}$	$\frac{1}{\sqrt{2}}$	± 1.7
$\pm\frac{\pi}{3}$	$\frac{1}{2}$	± 1.4
$\pm\frac{\pi}{2}$	0	0

(a)

Loop for $r = -2\sqrt{\cos\theta}$,
$-\frac{\pi}{2} \leq \theta \leq \frac{\pi}{2}$

Loop for $r = 2\sqrt{\cos\theta}$,
$-\frac{\pi}{2} \leq \theta \leq \frac{\pi}{2}$

(b)

FIGURE 9.42 The graph of $r^2 = 4\cos\theta$. The arrows show the direction of increasing θ. The values of r in the table are rounded. (Example 4)

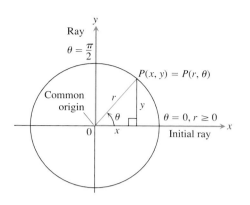

FIGURE 9.43 The usual way to relate polar and Cartesian coordinates.

Relating Polar and Cartesian Coordinates

When we use both polar and Cartesian coordinates in a plane, we place the two origins together and take the polar initial ray as the positive x-axis. The ray $\theta = \pi/2$, $r > 0$, becomes the positive y-axis (Figure 9.43). The two coordinate systems are then related by the following equations.

Equations Relating Polar and Cartesian Coordinates

$$x = r \cos \theta, \qquad y = r \sin \theta, \qquad x^2 + y^2 = r^2, \qquad \frac{y}{x} = \tan \theta$$

We use these equations and algebra (sometimes a lot of it!) to rewrite polar equations in Cartesian form and vice versa.

Example 5 Equivalent Equations

Polar equation	Cartesian equivalent
$r \cos \theta = 2$	$x = 2$
$r^2 \cos \theta \sin \theta = 4$	$xy = 4$
$r^2 \cos^2 \theta - r^2 \sin^2 \theta = 1$	$x^2 - y^2 = 1$
$r = 1 + 2r \cos \theta$	$y^2 - 3x^2 - 4x - 1 = 0$
$r = 1 - \cos \theta$	$x^4 + y^4 + 2x^2y^2 + 2x^3 + 2xy^2 - y^2 = 0$

With some curves, we are better off with polar coordinates; with others, we aren't.

Example 6 Converting Cartesian to Polar

Find a polar equation for the circle $x^2 + (y - 3)^2 = 9$ (Figure 9.44). Support graphically.

Solution

$$
\begin{aligned}
x^2 + y^2 - 6y + 9 &= 9 \qquad \text{Expand } (y - 3)^2. \\
x^2 + y^2 - 6y &= 0 \\
r^2 - 6r \sin \theta &= 0 \qquad x^2 + y^2 = r^2, \quad y = r \sin \theta \\
r = 0 \text{ or } r - 6 \sin \theta &= 0
\end{aligned}
$$

The equation $r = 6 \sin \theta$ includes the possibility that $r = 0$.

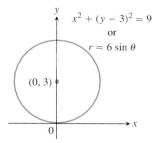

FIGURE 9.44 The circle in Example 6.

Example 7 Converting Polar to Cartesian

Find a Cartesian equivalent for the polar equation. Identify the graph.

(a) $r^2 = 4r \cos \theta$ **(b)** $r = \dfrac{4}{2 \cos \theta - \sin \theta}$

Solution

(a)

$$r^2 = 4r \cos \theta$$

$$x^2 + y^2 = 4x \qquad\qquad r^2 = x^2 + y^2, \ r \cos \theta = x$$

$$x^2 - 4x + y^2 = 0$$

$$x^2 - 4x + 4 + y^2 = 4 \qquad\qquad \text{Completing the square}$$

$$(x - 2)^2 + y^2 = 4$$

The graph of the equivalent Cartesian equation $(x - 2)^2 + y^2 = 4$ is a circle with radius 2 and center $(2, 0)$.

(b)

$$r = \frac{4}{2 \cos \theta - \sin \theta}$$

$$r (2 \cos \theta - \sin \theta) = 4$$

$$2r \cos \theta - r \sin \theta = 4$$

$$2x - y = 4 \qquad\qquad r \cos \theta = x, \ r \sin \theta = y$$

$$y = 2x - 4$$

The graph of the equivalent Cartesian equation $y = 2x - 4$ is a line with slope 2 and y-intercept -4.

Finding Points Where Polar Graphs Intersect

That we can represent a point in different ways in polar coordinates makes extra care necessary in deciding when a point lies on the graph of a polar equation and in determining the points in which polar graphs intersect. The problem is that a point of intersection may satisfy the equation of one curve with polar coordinates that are different from the ones with which it satisfies the equation of another curve. Thus, solving the equations of two curves simultaneously may not identify all their points of intersection. The only sure way to identify all the points of intersection is to graph the equations.

Example 8   Deceptive Coordinates

Show that the point $(2, \pi/2)$ lies on the curve $r = 2 \cos 2\theta$.

Solution   It may seem at first that the point $(2, \pi/2)$ does not lie on the curve because substituting the given coordinates into the equation gives

$$2 = 2 \cos 2 \left(\frac{\pi}{2} \right) = 2 \cos \pi = -2,$$

which is not a true equality. The magnitude is right, but the sign is wrong. This suggests looking for a pair of coordinates for the given point in which r is negative, for example, $(-2, -(\pi/2))$. If we try these in the equation $r = 2 \cos 2\theta$, we find

$$-2 = 2 \cos 2 \left(-\frac{\pi}{2} \right) = 2(-1) = -2,$$

and the equation is satisfied. The point $(2, \pi/2)$ does lie on the curve.

Example 9   Elusive Intersection Points

Find the points of intersection of the curves

$$r^2 = 4 \cos \theta \qquad \text{and} \qquad r = 1 - \cos \theta.$$

Solution   In Cartesian coordinates, we can always find the points where two curves cross by solving their equations simultaneously. In polar coordinates, the story is different. Simultaneous solution may reveal some intersection points without revealing others. In this example, simultaneous solution reveals only two of the four intersection points. The others are found by graphing. (Also, see Exercise 79.)

If we substitute $\cos \theta = r^2/4$ in the equation $r = 1 - \cos \theta$, we get

$$r = 1 - \cos \theta = 1 - \frac{r^2}{4}$$

$$4r = 4 - r^2$$

$$r^2 + 4r - 4 = 0$$

$$r = -2 \pm 2\sqrt{2}. \qquad \text{Quadratic formula}$$

The value $r = -2 - 2\sqrt{2}$ has too large an absolute value to belong to either curve. The values of θ corresponding to $r = -2 + 2\sqrt{2}$ are

$$\theta = \cos^{-1}(1 - r) \qquad \text{From } r = 1 - \cos \theta$$
$$= \cos^{-1}(1 - (2\sqrt{2} - 2)) \qquad \text{Set } r = 2\sqrt{2} - 2.$$
$$= \cos^{-1}(3 - 2\sqrt{2})$$
$$= \pm 80°. \qquad \text{Rounded to the nearest degree}$$

We have thus identified two intersection points: $(r, \theta) = (2\sqrt{2} - 2, \pm 80°)$.

If we graph the equations $r^2 = 4 \cos \theta$ and $r = 1 - \cos \theta$ together (Figure 9.45), as we can now do by combining the graphs in Figures 9.41 and 9.42, we see that the curves also intersect at the point $(2, \pi)$ and the origin. Why weren't the r-values of these points revealed by the simultaneous solution? The answer is that the points $(0, 0)$ and $(2, \pi)$ are not on the curves "simultaneously." They are not reached at the same value of θ. On the curve $r = 1 - \cos \theta$, the point $(2, \pi)$ is reached when $\theta = \pi$. On the curve $r^2 = 4 \cos \theta$, it is reached when $\theta = 0$, where it is identified not by the coordinates $(2, \pi)$, which do not satisfy the equation, but by the coordinates $(-2, 0)$, which do. Similarly, the cardioid reaches the origin when $\theta = 0$, but the curve $r^2 = 4 \cos \theta$ reaches the origin when $\theta = \pi/2$.

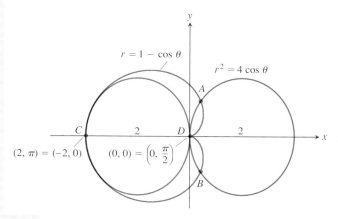

FIGURE 9.45 The four points of intersection of the curves $r = 1 - \cos \theta$ and $r^2 = 4 \cos \theta$ (Example 9). Only A and B were found by simultaneous solution. The other two were disclosed by graphing.

> **USING TECHNOLOGY**

Finding Intersections The *simultaneous mode* of a graphing utility gives new meaning to the *simultaneous solution* of a pair of polar coordinate equations. A simultaneous solution occurs only where the two graphs "collide" while they are being drawn simultaneously and not where one graph intersects the other at a point that had been illuminated earlier. The distinction is particularly important in the areas of traffic control or missile defense. For example, in traffic control, the only issue is whether two aircraft are in the same place at the same time. The question of whether the curves the craft follow intersect is unimportant.

To illustrate, graph the polar equations

$$r = \cos 2\theta \quad \text{and} \quad r = \sin 2\theta$$

in simultaneous mode with $0 \le \theta < 2\pi$, θ Step $= 0.1$ and view dimensions [xmin, xmax] $= [-1, 1]$ by [ymin, ymax] $= [-1, 1]$. *While the graphs are being drawn on the screen*, count the number of times the two graphs illuminate a single pixel simultaneously. Explain why these points of intersection of the two graphs correspond to simultaneous solutions of the equations. (You may find it helpful to slow down the graphing by making θ Step smaller, say 0.05, for example.) In how many points total do the graphs actually intersect?

EXERCISES 9.5

Polar Coordinate Pairs

In Exercises 1 and 2, determine which polar coordinate pairs name the same point.

1. (a) $(3, 0)$ **(b)** $(-3, 0)$ **(c)** $(2, 2\pi/3)$
   (d) $(2, 7\pi/3)$ **(e)** $(-3, \pi)$ **(f)** $(2, \pi/3)$
   (g) $(-3, 2\pi)$ **(h)** $(-2, -\pi/3)$

2. (a) $(-2, \pi/3)$ **(b)** $(2, -\pi/3)$ **(c)** (r, θ)
   (d) $(r, \theta + \pi)$ **(e)** $(-r, \theta)$ **(f)** $(2, -2\pi/3)$
   (g) $(-r, \theta + \pi)$ **(h)** $(-2, 2\pi/3)$

In Exercises 3 and 4, plot the points with the given polar coordinates and find their Cartesian coordinates.

3. (a) $(\sqrt{2}, \pi/4)$ **(b)** $(1, 0)$
   (c) $(0, \pi/2)$ **(d)** $(-\sqrt{2}, \pi/4)$

4. (a) $(-3, 5\pi/6)$ **(b)** $(5, \tan^{-1}(4/3))$
   (c) $(-1, 7\pi)$ **(d)** $(2\sqrt{3}, 2\pi/3)$

In Exercises 5 and 6, plot the points with the given Cartesian coordinates and find two sets of polar coordinates for each.

5. (a) $(-1, 1)$ **(b)** $(1, -\sqrt{3})$
   (c) $(0, 3)$ **(d)** $(-1, 0)$

6. (a) $(-\sqrt{3}, -1)$ **(b)** $(3, 4)$
   (c) $(0, -2)$ **(d)** $(2, 0)$

Graphing Polar Equations and Inequalities

In Exercises 7–18, graph the set of points whose polar coordinates satisfy the given equations and inequalities.

7. $r = 2$ **8.** $0 \le r \le 2$
9. $r \ge 1$ **10.** $0 \le \theta \le \pi/6, \quad r \ge 0$
11. $\theta = 2\pi/3, \quad r \le -2$ **12.** $\theta = \pi/3, \quad -1 \le r \le 3$
13. $0 \le \theta \le \pi, \quad r = 1$ **14.** $0 \le \theta \le \pi, \quad r = -1$
15. $\theta = \pi/2, \quad r \le 0$

16. $\pi/4 \le \theta \le 3\pi/4, \quad 0 \le r \le 1$

17. $-\pi/4 \le \theta \le \pi/4, \quad -1 \le r \le 1$

18. $0 \le \theta \le \pi/2, \quad 1 \le |r| \le 2$

Polar to Cartesian Equations

In Exercises 19–36, replace the polar equation by an equivalent Cartesian equation. Then identify or describe the graph.

19. $r \sin \theta = 0$

20. $r \cos \theta = 0$

21. $r = 4 \csc \theta$

22. $r = -3 \sec \theta$

23. $r \cos \theta + r \sin \theta = 1$

24. $r^2 = 1$

25. $r^2 = 4r \sin \theta$

26. $r = \dfrac{5}{\sin \theta - 2 \cos \theta}$

27. $r^2 \sin 2\theta = 2$

28. $r = \cot \theta \csc \theta$

29. $r = (\csc \theta) \, e^{r \cos \theta}$

30. $\cos^2 \theta = \sin^2 \theta$

31. $r \sin \theta = \ln r + \ln \cos \theta$

32. $r^2 + 2r^2 \cos \theta \sin \theta = 1$

33. $r^2 = -4r \cos \theta$

34. $r = 8 \sin \theta$

35. $r = 2 \cos \theta + 2 \sin \theta$

36. $r \sin \left(\theta + \dfrac{\pi}{6} \right) = 2$

Cartesian to Polar Equations

In Exercises 37–48, replace the Cartesian equation by an equivalent polar equation.

37. $x = 7$

38. $y = 1$

39. $x = y$

40. $x - y = 3$

41. $x^2 + y^2 = 4$

42. $x^2 - y^2 = 1$

43. $\dfrac{x^2}{9} + \dfrac{y^2}{4} = 1$

44. $xy = 2$

45. $y^2 = 4x$

46. $x^2 + xy + y^2 = 1$

47. $x^2 + (y - 2)^2 = 4$

48. $(x - 3)^2 + (y + 1)^2 = 4$

Symmetries and Polar Graphs

In Exercises 49–58, (a) graph the polar curve. (b) What is the shortest length a θ-interval can have and still produce the graph?

49. $r = 1 + \cos \theta$

50. $r = 2 - 2 \cos \theta$

51. $r^2 = -\sin 2\theta$

52. $r = 1 - \sin \theta$

53. $r = 1 - 2 \sin 3\theta$

54. $r = \sin (\theta/2)$

55. $r = \theta$

56. $r = 1 + \sin \theta$

57. $r = 2 \cos 3\theta$

58. $r = 1 + 2 \sin \theta$

In Exercises 59–62, determine the symmetries of the curve.

59. $r^2 = 4 \cos 2\theta$

60. $r^2 = 4 \sin 2\theta$

61. $r = 2 + \sin \theta$

62. $r^2 = -\cos 2\theta$

63. *Writing to Learn: Vertical and horizontal lines*

   (a) Explain why every vertical line in the plane has a polar equation of the form $r = a \sec \theta$.

   (b) Find an analogous polar equation for horizontal lines. Give reasons for your answer.

64. *Writing to Learn: Do two symmetries imply three?* If a curve has any two of the symmetries listed at the beginning of the section, can anything be said about its having or not having the third symmetry? Give reasons for your answer.

Intersections

65. Show that the point $(2, 3\pi/4)$ lies on the curve $r = 2 \sin 2\theta$.

66. Show that $(1/2, 3\pi/2)$ lies on the curve $r = -\sin (\theta/3)$.

Find the points of intersection of the pairs of curves in Exercises 67–70.

67. $r = 1 + \cos \theta, \quad r = 1 - \cos \theta$

68. $r = 2 \sin \theta, \quad r = 2 \sin 2\theta$

69. $r = \cos \theta, \quad r = 1 - \cos \theta$

70. $r = 1, \quad r^2 = 2 \sin 2\theta$

T Find the points of intersection of the pairs of curves in Exercises 71–74.

71. $r^2 = \sin 2\theta, \quad r^2 = \cos 2\theta$

72. $r = 1 + \cos \dfrac{\theta}{2}, \quad r = 1 - \sin \dfrac{\theta}{2}$

73. $r = 1, \quad r = 2 \sin 2\theta$

74. $r = 1, \quad r^2 = 2 \sin 2\theta$

T **75.** Which of the following has the same graph as $r = 1 - \cos \theta$?

   (a) $r = -1 - \cos \theta$

   (b) $r = 1 + \cos \theta$

   Confirm your answer with alegbra.

T **76.** *Rose curves* Let $r = 2 \sin n\theta$.

   (a) Graph $r = 2 \sin n\theta$ for $n = \pm2, \pm4, \pm6$. Describe the curves.

   (b) What is the smallest length a θ-interval can have and still produce the graphs in part (a)?

   (c) Based on your observations in part (a), describe the graph of $r = 2 \sin n\theta$ when n is a nonzero even integer.

   (d) Graph $r = 2 \sin n\theta$ for $n = \pm3, \pm5, \pm7$. Describe the curves.

   (e) What is the smallest length a θ-interval can have and still produce the graphs in part (d)?

   (f) Based on your observations in part (d), describe the graph of $r = 2 \sin n\theta$ when n is an odd integer different from ±1.

T **77.** *A rose within a rose* Graph the equation $r = 1 - 2 \sin 3\theta$.

78. *The nephroid of Freeth* Graph the nephroid of Freeth:

$$r = 1 + 2 \sin \frac{\theta}{2}.$$

Theory and Examples

79. (*Continuation of Example 9*) The simultaneous solution of the equations

$$r^2 = 4 \cos \theta \qquad (1)$$
$$r = 1 - \cos \theta \qquad (2)$$

in the text did not reveal the points $(0, 0)$ and $(2, \pi)$ in which their graphs intersected.

(a) We could have found the point $(2, \pi)$, however, by replacing the (r, θ) in Equation (1) by the equivalent $(-r, \theta + \pi)$ to obtain

$$r^2 = 4 \cos \theta$$
$$(-r)^2 = 4 \cos (\theta + \pi) \qquad (3)$$
$$r^2 = -4 \cos \theta.$$

Solve Equations (2) and (3) simultaneously to show that $(2, \pi)$ is a common solution. (This will still not reveal that the graphs intersect at $(0, 0)$.)

(b) The origin is still a special case. (It often is.) Here is one way to handle it: Set $r = 0$ in Equations (1) and (2) and solve each equation for a corresponding value of θ. Since $(0, \theta)$ is the origin for *any* θ, this will show that both curves pass through the origin even if they do so for different θ-values.

80. *Relating polar equations to parametric equations* Let $r = f(\theta)$ be a polar curve.

(a) *Writing to Learn* Explain why

$$x = f(t) \cos t, \qquad y = f(t) \sin t$$

are parametric equations for the curve.

(b) Use part (a) to write parametric equations for the circle $r = 3$. Support your answer by graphing the parametric equations.

(c) Repeat part (b) with $r = 1 - \cos \theta$.

(d) Repeat part (b) with $r = 3 \sin 2\theta$.

81. *Distance formula* Show that the distance between points (r_1, θ_1), and (r_2, θ_2) in polar coordinates is

$$d = \sqrt{r_1{}^2 + r_2{}^2 - 2r_1 r_2 \cos (\theta_1 - \theta_2)}.$$

82. *Height of a cardioid* Find the maximum height above the x-axis of the cardioid $r = 2(1 + \cos \theta)$.

9.6 Calculus of Polar Curves

Slope • Area in the Plane • Length of a Curve

In this section, you will see how to find slopes, areas, and lengths of polar curves $r = f(\theta)$.

Slope

The slope of a polar curve $r = f(\theta)$ is given by dy/dx, not by $r' = df/d\theta$. To see why, think of the graph of f as the graph of the parametric equations

$$x = r \cos \theta = f(\theta) \cos \theta, \qquad y = r \sin \theta = f(\theta) \sin \theta.$$

If f is a differentiable function of θ, then so are x and y, and when $dx/d\theta \neq 0$, we can calculate dy/dx from the parametric formula

$$\frac{dy}{dx} = \frac{dy/d\theta}{dx/d\theta} \qquad \text{Section 2.5, Equation (4) with } t = \theta$$

$$= \frac{\frac{d}{d\theta}(f(\theta) \sin \theta)}{\frac{d}{d\theta}(f(\theta) \cos \theta)}$$

$$= \frac{\frac{df}{d\theta} \sin \theta + f(\theta) \cos \theta}{\frac{df}{d\theta} \cos \theta - f(\theta) \sin \theta} \qquad \text{Product Rule for Derivatives}$$

> **Slope of the Polar Curve r = f(θ)**
>
> $$\left.\frac{dy}{dx}\right|_{(r,\theta)} = \frac{f'(\theta)\,\sin\,\theta + f(\theta)\,\cos\,\theta}{f'(\theta)\,\cos\,\theta - f(\theta)\,\sin\,\theta},$$ (1)
>
> provided $dx/d\theta \neq 0$ at (r,θ).

We can see from Equation 1 and its derivation that the curve $r = f(\theta)$ has a

1. Horizontal tangent at a point where $dy/d\theta = 0$ and $dx/d\theta \neq 0$

2. Vertical tangent at a point where $dx/d\theta = 0$ and $dy/d\theta \neq 0$.

If both derivatives are zero, no conclusion can be drawn without further investigation, as illustrated in Example 1.

Example 1   Finding Horizontal and Vertical Tangents

Find the horizontal and vertical tangents to the graph of the cardioid $r = 1 - \cos\,\theta$, $0 \le \theta \le 2\pi$.

Solution   The graph in Figure 9.46 suggests that there are at least two horizontal and three vertical tangents.

The parametric form of the equation is

$$x = r\,\cos\,\theta = (1 - \cos\,\theta)\,\cos\,\theta = \cos\,\theta - \cos^2\,\theta,$$
$$y = r\,\sin\,\theta = (1 - \cos\,\theta)\,\sin\,\theta = \sin\,\theta - \cos\,\theta\,\sin\,\theta.$$

We need to find the zeros of $dy/d\theta$ and $dx/d\theta$.

(a) Zeros of $dy/d\theta$ in $[0, 2\pi]$:

$$\frac{dy}{d\theta} = \cos\,\theta + \sin^2\,\theta - \cos^2\,\theta = \cos\,\theta + (1 - \cos^2\,\theta) - \cos^2\,\theta$$
$$= 1 + \cos\,\theta - 2\,\cos^2\,\theta = (1 + 2\,\cos\,\theta)(1 - \cos\,\theta).$$

Now,

$$1 - \cos\,\theta = 0 \Rightarrow \theta = 0, 2\pi$$
$$1 + 2\,\cos\,\theta = 0 \Rightarrow \theta = 2\pi/3, 4\pi/3.$$

Thus, $dy/d\theta = 0$ in $0 \le \theta \le 2\pi$ if $\theta = 0, 2\pi/3, 4\pi/3$, or 2π.

(b) Zeros of $dx/d\theta$ in $[0, 2\pi]$:

$$\frac{dx}{d\theta} = -\sin\,\theta + 2\,\cos\,\theta\,\sin\,\theta = (2\,\cos\,\theta - 1)\,\sin\,\theta.$$

Now,

$$2\,\cos\,\theta - 1 = 0 \Rightarrow \theta = \pi/3, 5\pi/3$$
$$\sin\,\theta = 0 \Rightarrow \theta = 0, \pi, 2\pi.$$

Thus, $dx/d\theta = 0$ in $0 \le \theta \le 2\pi$ if $\theta = 0, \pi/3, \pi, 5\pi/3$, or 2π.

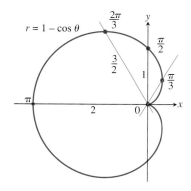

FIGURE 9.46 Where are the horizontal and vertical tangents to this cardioid? (Example 1)

We can now see that there are horizontal tangents ($dy/d\theta = 0$, $dx/d\theta \neq 0$) at the points where $\theta = 2\pi/3$ and $4\pi/3$, and vertical tangents ($dx/d\theta = 0$, $dy/d\theta \neq 0$) at the points where $\theta = \pi/3$, π, and $5\pi/3$.

At the points where $\theta = 0$ or 2π, the right side of Equation (1) takes the form $0/0$. We can use l'Hôpital's Rule (Section 7.6) to see that

$$\lim_{\theta \to 0,\, 2\pi} \frac{dy/d\theta}{dx/d\theta} = \lim_{\theta \to 0,2\pi} \frac{1 + \cos\theta - 2\cos^2\theta}{2\cos\theta\sin\theta - \sin\theta}$$

$$= \lim_{\theta \to 0,2\pi} \frac{-\sin\theta + 4\cos\theta\sin\theta}{2\cos^2\theta - 2\sin^2\theta - \cos\theta} = \frac{0}{1} = 0.$$

The curve has a horizontal tangent at the point where $\theta = 0$ or 2π. Summarizing, we have

Horizontal tangents at $\qquad (0, 0) = (0, 2\pi),\ (1.5, 2\pi/3),\ (1.5, 4\pi/3)$

Vertical tangents at $\qquad (0.5, \pi/3),\ (2, \pi),\ (0.5, 5\pi/3)$.

If the curve $r = f(\theta)$ passes through the origin at $\theta = \theta_0$, then $f(\theta_0) = 0$, and Equation (1) gives

$$\left. \frac{dy}{dx} \right|_{(0,\, \theta_0)} = \frac{f'(\theta_0)\sin\theta_0}{f'(\theta_0)\cos\theta_0} = \tan\theta_0,$$

provided $f'(\theta_0) \neq 0$ (not the case in Example 1). The reason we say "slope at $(0, \theta_0)$" and not just "slope at the origin" is that a polar curve may pass through the origin more than once, with different slopes at different θ-values.

Example 2   Finding Tangent Lines at the Pole (Origin)

Find the lines tangent to the rose curve

$$r = f(\theta) = 2\sin 3\theta, \qquad 0 \le \theta \le \pi,$$

at the pole.

Solution $f(\theta)$ is zero when $\theta = 0$, $\pi/3$, $2\pi/3$, and π. The derivative $f'(\theta) = 6\cos 3\theta$ is not zero at these four values of θ. Thus, this curve has tangent lines at the pole (Figure 9.47) with slopes $\tan 0 = \tan \pi = 0$, $\tan(\pi/3) = \sqrt{3}$, and $\tan(2\pi/3) = -\sqrt{3}$. The three corresponding tangent lines are $y = 0$, $y = \sqrt{3}x$, and $y = -\sqrt{3}x$.

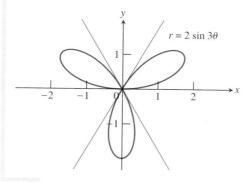

FIGURE 9.47 The three tangent lines to $r = f(\theta) = 2\sin 3\theta$, $0 \le \theta \le \pi$, are $y = 0$, $y = \sqrt{3}x$, and $y = -\sqrt{3}x$. (Example 2)

Area in the Plane

The region OTS in Figure 9.48 is bounded by the rays $\theta = \alpha$ and $\theta = \beta$ and the curve $r = f(\theta)$. We approximate the region with n nonoverlapping circular sectors based on a partition P of angle TOS. The typical sector has radius $r_k = f(\theta_k)$ and central angle of radian measure $\Delta\theta_k$. Its area is

$$A_k = \frac{1}{2} r_k{}^2 \, \Delta\theta_k = \frac{1}{2}(f(\theta_k))^2 \, \Delta\theta_k.$$

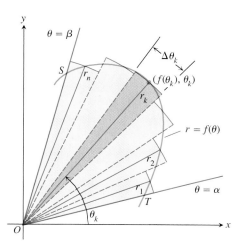

FIGURE 9.48 To derive a formula for the area of region OTS, we approximate the region with fan-shaped circular sectors.

The area of the region OTS is approximately

$$\sum_{k=1}^{n} A_k = \sum_{k=1}^{n} \frac{1}{2}(f(\theta_k))^2 \, \Delta\theta_k .$$

If f is continuous, we expect the approximations to improve as $\|P\| \to 0$, and we are led to the following formula for the region's area:

$$A = \lim_{\|P\| \to 0} \sum_{k=1}^{n} \frac{1}{2}(f(\theta_k))^2 \, \Delta\theta_k$$

$$= \int_{\alpha}^{\beta} \frac{1}{2}(f(\theta))^2 \, d\theta .$$

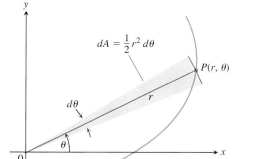

FIGURE 9.49 The area differential dA.

Area in Polar Coordinates

The **area** of the region **between the origin and the curve** $r = f(\theta)$, $\alpha \le \theta \le \beta$, is

$$A = \int_{\alpha}^{\beta} \frac{1}{2} r^2 \, d\theta.$$

This is the integral of the **area differential** (Figure 9.49),

$$dA = \frac{1}{2} r^2 \, d\theta .$$

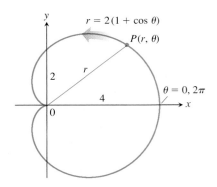

FIGURE 9.50 The cardioid in Example 3.

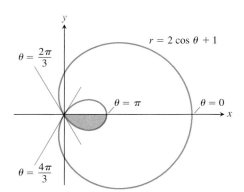

FIGURE 9.51 The limaçon in Example 4. Limaçon (pronounced LEE-ma-sahn) is an old French word for *snail*.

Example 3   Finding Area

Find the area of the region in the plane enclosed by the cardioid $r = 2(1 + \cos \theta)$.

Solution   We graph the cardioid (Figure 9.50) and determine that the *radius r* sweeps out the region exactly once as θ runs from 0 to 2π. The area is therefore

$$\int_{\theta=0}^{\theta=2\pi} \frac{1}{2} r^2 \, d\theta = \int_0^{2\pi} \frac{1}{2} \cdot 4(1 + \cos \theta)^2 \, d\theta$$

$$= \int_0^{2\pi} 2(1 + 2 \cos \theta + \cos^2 \theta) \, d\theta$$

$$= \int_0^{2\pi} \left(2 + 4 \cos \theta + 2 \frac{1 + \cos 2\theta}{2} \right) d\theta$$

$$= \int_0^{2\pi} (3 + 4 \cos \theta + \cos 2\theta) \, d\theta$$

$$= \left[3\theta + 4 \sin \theta + \frac{\sin 2\theta}{2} \right]_0^{2\pi} = 6\pi - 0 = 6\pi.$$

Example 4   Finding Area

Find the area inside the smaller loop of the limaçon $r = 2 \cos \theta + 1$.

Solution   After sketching the curve (Figure 9.51), we see that the smaller loop is traced out by the point (r, θ) as θ increases from $\theta = 2\pi/3$ to $\theta = 4\pi/3$. Since the curve is symmetric about the *x*-axis (the equation is unaltered when we replace θ by $-\theta$), we may calculate the area of the shaded half of the inner loop by integrating from $\theta = 2\pi/3$ to $\theta = \pi$. The area we seek will be twice the resulting integral:

$$A = 2 \int_{2\pi/3}^{\pi} \frac{1}{2} r^2 \, d\theta = \int_{2\pi/3}^{\pi} r^2 \, d\theta.$$

Since

$$r^2 = (2 \cos \theta + 1)^2 = 4 \cos^2 \theta + 4 \cos \theta + 1$$

$$= 4 \cdot \frac{1 + \cos 2\theta}{2} + 4 \cos \theta + 1$$

$$= 2 + 2 \cos 2\theta + 4 \cos \theta + 1$$

$$= 3 + 2 \cos 2\theta + 4 \cos \theta,$$

we have

$$A = \int_{2\pi/3}^{\pi} (3 + 2 \cos 2\theta + 4 \cos \theta) \, d\theta$$

$$= \left[3\theta + \sin 2\theta + 4 \sin \theta \right]_{2\pi/3}^{\pi}$$

$$= (3\pi) - \left(2\pi - \frac{\sqrt{3}}{2} + 4 \cdot \frac{\sqrt{3}}{2} \right)$$

$$= \pi - \frac{3\sqrt{3}}{2}.$$

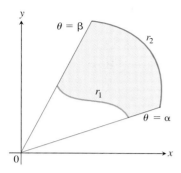

FIGURE 9.52 The area of the shaded region is calculated by subtracting the area of the region between r_1 and the origin from the area of the region between r_2 and the origin.

To find the area of a region like the one in Figure 9.52 which lies between two polar curves $r_1 = r_1(\theta)$ and $r_2 = r_2(\theta)$ from $\theta = \alpha$ to $\theta = \beta$, we subtract the integral of $(1/2)r_1^2$ from the integral of $(1/2)r_2^2$. This leads to the following formula.

> **Area Between Polar Curves**
>
> The area of the region $0 \le r_1(\theta) \le r_2(\theta)$, $\alpha \le \theta \le \beta$, is
>
> $$A = \int_\alpha^\beta \frac{1}{2} r_2^2 \, d\theta - \int_\alpha^\beta \frac{1}{2} r_1^2 \, d\theta = \int_\alpha^\beta \frac{1}{2}(r_2^2 - r_1^2) \, d\theta. \tag{2}$$

Example 5 Finding Area Between Curves

Find the area of the region that lies inside the circle $r = 1$ and outside the cardioid $r = 1 - \cos \theta$.

Solution The region is shown in Figure 9.53. The outer curve is $r_2 = 1$, the inner curve is $r_1 = 1 - \cos \theta$, and θ runs from $-\pi/2$ to $\pi/2$. The area, from Equation (2), is

$$A = \int_{-\pi/2}^{\pi/2} \frac{1}{2}(r_2^2 - r_1^2) \, d\theta$$

$$= 2 \int_0^{\pi/2} \frac{1}{2}(r_2^2 - r_1^2) \, d\theta \qquad \text{Symmetry}$$

$$= \int_0^{\pi/2} (1 - (1 - 2 \cos \theta + \cos^2 \theta)) \, d\theta$$

$$= \int_0^{\pi/2} (2 \cos \theta - \cos^2 \theta) \, d\theta = \int_0^{\pi/2} \left(2 \cos \theta - \frac{1 + \cos 2\theta}{2} \right) d\theta$$

$$= \left[2 \sin \theta - \frac{\theta}{2} - \frac{\sin 2\theta}{4} \right]_0^{\pi/2} = 2 - \frac{\pi}{4}.$$

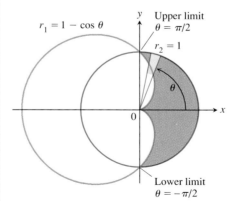

FIGURE 9.53 The region and limits of integration in Example 5.

Length of a Curve

We can obtain a polar coordinate formula for the length of a curve $r = f(\theta)$, $\alpha \le \theta \le \beta$, by parametrizing the curve as

$$x = r \cos \theta = f(\theta) \cos \theta, \qquad y = r \sin \theta = f(\theta) \sin \theta, \qquad \alpha \le \theta \le \beta. \tag{3}$$

The parametric length formula from Section 5.3 then gives the length as

$$L = \int_\alpha^\beta \sqrt{\left(\frac{dx}{d\theta}\right)^2 + \left(\frac{dy}{d\theta}\right)^2}\; d\theta.$$

This equation becomes

$$L = \int_\alpha^\beta \sqrt{r^2 + \left(\frac{dr}{d\theta}\right)^2}\; d\theta$$

when Equations (3) are substituted for x and y (Exercise 41).

CD-ROM
WEBsite

Historical Biography

Gaspard Monge
(1746 — 1818)

Length of a Polar Curve

If $r = f(\theta)$ has a continuous first derivative for $\alpha \le \theta \le \beta$ and if the point $P(r,\ \theta)$ traces the curve $r = f(\theta)$ exactly once as θ runs from α to β, then the length of the curve is

$$L = \int_\alpha^\beta \sqrt{r^2 + \left(\frac{dr}{d\theta}\right)^2}\; d\theta. \tag{4}$$

Example 6   Finding the Length of a Cardioid

Find the length of the cardioid $r = 1 - \cos\theta$.

Solution   The graph is shown in Figure 9.54. The point $P(r,\ \theta)$ traces the curve once counterclockwise as θ runs from 0 to 2π, so these are the values we take for the limits of integration.

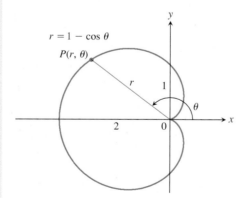

FIGURE 9.54 Example 6 calculates the length of this cardioid.

Since $r = 1 - \cos\theta$, $dr/d\theta = \sin\theta$, and we have

$$r^2 + \left(\frac{dr}{d\theta}\right)^2 = (1 - \cos\theta)^2 + (\sin\theta)^2$$

$$= 1 - 2\cos\theta + \underbrace{\cos^2\theta + \sin^2\theta}_{1} = 2 - 2\cos\theta.$$

Therefore,

$$L = \int_\alpha^\beta \sqrt{r^2 + \left(\frac{dr}{d\theta}\right)^2}\, d\theta \qquad \text{Equation (4)}$$

$$= \int_0^{2\pi} \sqrt{2 - 2\cos\theta}\, d\theta$$

$$= \int_0^{2\pi} \sqrt{4\sin^2\frac{\theta}{2}}\, d\theta \qquad 1 - \cos\theta = 2\sin^2\frac{\theta}{2}$$

$$= \int_0^{2\pi} \left|2\sin\frac{\theta}{2}\right|\, d\theta$$

$$= \int_0^{2\pi} 2\sin\frac{\theta}{2}\, d\theta \qquad \sin\frac{\theta}{2} \geq 0 \text{ for } 0 \leq \theta \leq 2\pi$$

$$= \left[-4\cos\frac{\theta}{2}\right]_0^{2\pi} = 4 + 4 = 8.$$

EXERCISES 9.6

Slopes of Polar Curves

In Exercises 1–4, find the slope of the curve at each indicated point.

1. $r = -1 + \sin\theta, \quad \theta = 0, \pi$

2. $r = \cos 2\theta, \quad \theta = 0, \pm\pi/2, \pi$

3. $r = 2 - 3\sin\theta$

4. $r = 3(1 - \cos\theta)$

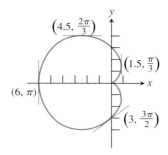

In Exercises 5–8, find the tangent lines at the pole.

5. $r = 3\cos\theta, \quad 0 \leq \theta \leq 2\pi$

6. $r = 2\cos 3\theta, \quad 0 \leq \theta \leq \pi$

7. $r = \sin 5\theta, \quad 0 \leq \theta \leq \pi$

8. $r = 2\sin 2\theta, \quad 0 \leq \theta \leq 2\pi$

In Exercises 9–12, find equations for the horizontal and vertical tangent lines to the curve.

9. $r = -1 + \sin\theta, \quad 0 \leq \theta \leq 2\pi$

10. $r = 1 + \cos\theta, \quad 0 \leq \theta \leq 2\pi$

11. $r = 2\sin\theta, \quad 0 \leq \theta \leq \pi$

12. $r = 3 - 4\cos\theta, \quad 0 \leq \theta \leq 2\pi$

Areas Inside Polar Curves

Find the areas of the regions in Exercises 13–18.

13. Inside the oval limaçon $r = 4 + 2\cos\theta$

14. Inside the cardioid $r = a(1 + \cos\theta), \quad a > 0$

15. Inside one leaf of the four-leaved rose $r = \cos 2\theta$

16. Inside the lemniscate $r^2 = 2a^2\cos 2\theta, \quad a > 0$

17. Inside one loop of the lemniscate $r^2 = 4\sin 2\theta$

18. Inside the six-leaved rose $r^2 = 2\sin 3\theta$

Areas Shared by Polar Regions

Find the areas of the regions in Exercises 19–28.

19. Shared by the circles $r = 2 \cos \theta$ and $r = 2 \sin \theta$

20. Shared by the circles $r = 1$ and $r = 2 \sin \theta$

21. Shared by the circle $r = 2$ and the cardioid $r = 2(1 - \cos \theta)$

22. Shared by the cardioids $r = 2(1 + \cos \theta)$ and $r = 2(1 - \cos \theta)$

23. Inside the lemniscate $r^2 = 6 \cos 2\theta$ and outside the circle $r = \sqrt{3}$

24. Inside the circle $r = 3a \cos \theta$ and outside the cardioid $r = a(1 + \cos \theta)$, $a > 0$

25. Inside the circle $r = -2 \cos \theta$ and outside the circle $r = 1$

26. **(a)** Inside the outer loop of the limaçon $r = 1 + 2 \cos \theta$ (See Figure 9.51.)

   (b) Inside the outer loop and outside the inner loop of the limaçon $r = 1 + 2 \cos \theta$

27. Inside the circle $r = 6$ above the line $r = 3 \csc \theta$

28. Inside the lemniscate $r^2 = 6 \cos 2\theta$ to the right of the line $r = (3/2) \sec \theta$

29. **(a)** Find the area of the shaded region in the accompanying figure.

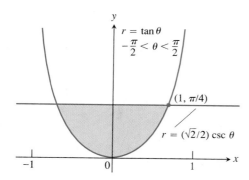

$r = \tan \theta$
$-\dfrac{\pi}{2} < \theta < \dfrac{\pi}{2}$

$(1, \pi/4)$

$r = (\sqrt{2}/2) \csc \theta$

   (b) *Writing to Learn* It looks as if the graph of $r = \tan \theta$, $-\pi/2 < \theta < \pi/2$, could be asymptotic to the lines $x = 1$ and $x = -1$. Is it? Give reasons for your answer.

30. *Writing to Learn* The area of the region that lies inside the cardioid curve $r = 1 + \cos \theta$ and outside the circle $r = \cos \theta$ is not

$$\frac{1}{2} \int_0^{2\pi} [(1 + \cos \theta)^2 - \cos^2 \theta] \, d\theta = \pi.$$

Why not? What *is* the area? Give reasons for your answers.

Lengths of Polar Curves

Find the lengths of the curves in Exercises 31–39.

31. The spiral $r = \theta^2$, $0 \le \theta \le \sqrt{5}$

32. The spiral $r = e^{\theta}/\sqrt{2}$, $0 \le \theta \le \pi$

33. The cardioid $r = 1 + \cos \theta$

34. The curve $r = a \sin^2 (\theta/2)$, $0 \le \theta \le \pi$, $a > 0$

35. The parabolic segment $r = 6/(1 + \cos \theta)$, $0 \le \theta \le \pi/2$

36. The parabolic segment $r = 2/(1 - \cos \theta)$, $\pi/2 \le \theta \le \pi$

37. The curve $r = \cos^3 (\theta/3)$, $0 \le \theta \le \pi/4$

38. The curve $r = \sqrt{1 + \sin 2\theta}$, $0 \le \theta \le \pi\sqrt{2}$

39. The curve $r = \sqrt{1 + \cos 2\theta}$, $0 \le \theta \le \pi\sqrt{2}$

40. *Circumferences of circles* As usual, when faced with a new formula, it is a good idea to try it on familiar objects to be sure it gives results consistent with past experience. Use the length formula in Equation (4) to calculate the circumferences of the following circles ($a > 0$):

   (a) $r = a$ **(b)** $r = a \cos \theta$ **(c)** $r = a \sin \theta$.

Theory and Examples

41. *Length of a polar curve* Assuming that the necessary derivatives are continuous, show how the substitutions

$$x = f(\theta) \cos \theta, \qquad y = f(\theta) \sin \theta$$

(Equations (3) in the text) transform

$$L = \int_\alpha^\beta \sqrt{\left(\frac{dx}{d\theta}\right)^2 + \left(\frac{dy}{d\theta}\right)^2} \, d\theta$$

into

$$L = \int_\alpha^\beta \sqrt{r^2 + \left(\frac{dr}{d\theta}\right)^2} \, d\theta.$$

42. *Average value* If f is continuous, the average value of the polar coordinate r over the curve $r = f(\theta)$, $\alpha \le \theta \le \beta$, with respect to θ is

$$r_{av} = \frac{1}{\beta - \alpha} \int_\alpha^\beta f(\theta) \, d\theta.$$

Use this formula to find the average value of r with respect to θ over the following curves ($a > 0$).

   (a) The cardioid $r = a(1 - \cos \theta)$

   (b) The circle $r = a$

   (c) The circle $r = a \cos \theta$, $-\pi/2 \le \theta \le \pi/2$

43. *Writing to Learn* Can anything be said about the relative lengths of the curves

$$r = f(\theta), \qquad \alpha \le \theta \le \beta,$$

and

$$r = 2f(\theta), \qquad \alpha \le \theta \le \beta?$$

Give reasons for your answer.

44. *Videocassette tape length* The length of a tape wound onto a take-up reel as shown in the figure is

$$L = \int_0^\alpha \sqrt{r^2 + \left(\frac{b}{2\pi}\right)^2}\, d\theta,$$

where b is the tape thickness and

$$r = r_0 + \left(\frac{\alpha}{2\pi}\right) b$$

is the radius of the tape on the take-up reel. The initial radius of the tape on the take-up reel is r_0 and α is the angle in radians through which the wheel has turned.

(a) Find a spiral that models the tape accumulating on the take-up reel using polar graphing with $r_0 = 1.75$ cm and $b = 0.06$ cm.

(b) Confirm the formula for L analytically.

(c) Determine the length of tape on the take-up reel if the reel has turned through an angle of 80π with $r_0 = 1.75$ cm and $b = 0.06$ cm.

(d) Assume that b is very small in comparison to r at any time. Show analytically that

$$L_a = \int_0^\alpha r\, d\theta$$

is an excellent approximation to the exact value of L.

(e) For the values given in part (c), compare L_a with L.

45. (*Continuation of Exercise 44*) Let n be the number of complete turns the take-up reel has made.

(a) Find a formula for n in terms of L, the tape length.

(b) When a VCR operates, the tape moves past the heads at a constant speed. Describe the speed of the take-up reel as time progresses.

(c) Suppose that the VCR tape counter is the number n of complete turns of the take-up reel. Describe the counter values as a function of time t.

Questions to Guide Your Review

1. When do directed line segments in the plane represent the same vector?

2. How are vectors added and subtracted geometrically? Algebraically?

3. How do you find a vector's magnitude and direction?

4. If a vector is multiplied by a positive scalar, how is the result related to the original vector? What if the scalar is zero? Negative?

5. Define the *dot product (scalar product)* of two vectors. Which algebraic laws are satisfied by dot products? Give examples. When is the dot product of two vectors equal to zero?

6. What geometric interpretation does the dot product have? Give examples.

7. What is the vector projection of a vector **u** onto a vector **v**? How do you write **u** as the sum of a vector parallel to **v** and a vector orthogonal to **v**?

8. State the rules for differentiating and integrating vector functions. Give examples.

9. How do you define and calculate the velocity, speed, direction of motion, and acceleration of a body moving along a sufficiently differentiable plane curve? Give examples.

10. What are the vector and parametric equations for ideal projectile motion? How do you find an ideal projectile's maximum height, flight time, and range? Give examples.

11. What are polar coordinates? What equations relate polar coordinates to Cartesian coordinates? Why might you want to change from one coordinate system to the other?

12. What consequence does the lack of uniqueness of polar coordinates have for graphing? Give an example.

13. How do you graph equations in polar coordinates? Include in your discussion symmetry and slope. Give examples.

14. How do you find the area of a region $0 \leq r_1(\theta) \leq r \leq r_2(\theta)$, $\alpha \leq \theta \leq \beta$, in the polar coordinate plane? Give examples.

15. Under what conditions can you find the length of a curve $r = f(\theta)$, $\alpha \leq \theta \leq \beta$, in the polar coordinate plane? Give an example of a typical calculation.

Practice Exercises

Vector Calculations

In Exercises 1–4, let $\mathbf{u} = \langle -3, 4 \rangle$ and $\mathbf{v} = \langle 2, -5 \rangle$. Find (a) the component form of the vector and (b) its magnitude.

1. $3\mathbf{u} - 4\mathbf{v}$

2. $\mathbf{u} + \mathbf{v}$

3. $-2\mathbf{u}$

4. $5\mathbf{v}$

In Exercises 5–8, find the component form of the vector.

5. The vector obtained by rotating $\langle 0, 1 \rangle$ through an angle of $2\pi/3$ radians

6. The unit vector that makes an angle of $\pi/6$ radian with the positive x-axis

7. The vector 2 units long in the direction $4\mathbf{i} - \mathbf{j}$

8. The vector 5 units long in the direction opposite to the direction of $(3/5)\mathbf{i} + (4/5)\mathbf{j}$

Length and Direction

Express the vectors in Exercises 9–12 in terms of their lengths and directions.

9. $\sqrt{2}\mathbf{i} + \sqrt{2}\mathbf{j}$

10. $-\mathbf{i} - \mathbf{j}$

11. Velocity vector to $\mathbf{r} = (2 \cos t)\mathbf{i} + (2 \sin t)\mathbf{j}$ at the point $(0, 2)$.

12. Velocity vector to $\mathbf{r} = (e^t \cos t)\mathbf{i} + (e^t \sin t)\mathbf{j}$ when $t = \ln 2$.

Tangent and Normal Vectors

In Exercises 13 and 14, find the unit vectors that are tangent and normal to the curve at point P.

13. $y = \tan x$, $P(\pi/4, 1)$

14. $x^2 + y^2 = 25$, $P(3, 4)$

Vector Projections

15. Copy the vectors $\mathbf{u}$ and $\mathbf{v}$ and sketch the vector projection of $\mathbf{v}$ onto $\mathbf{u}$.

16. Express vectors $\mathbf{a}$, $\mathbf{b}$, and $\mathbf{c}$ in terms of $\mathbf{u}$ and $\mathbf{v}$.

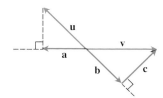

In Exercises 17 and 18, find $|\mathbf{v}|$, $|\mathbf{u}|$, $\mathbf{v} \cdot \mathbf{u}$, $\mathbf{u} \cdot \mathbf{v}$, the angle between $\mathbf{v}$ and $\mathbf{u}$, the scalar component of $\mathbf{u}$ in the direction of $\mathbf{v}$, and the vector projection of $\mathbf{u}$ onto $\mathbf{v}$.

17. $\mathbf{v} = \mathbf{i} + \mathbf{j}$

$\mathbf{u} = 2\mathbf{i} + \mathbf{j}$

18. $\mathbf{v} = \mathbf{i} + \mathbf{j}$

$\mathbf{u} = -\mathbf{i} - 3\mathbf{j}$

In Exercises 19 and 20, write $\mathbf{u}$ as the sum of a vector parallel to $\mathbf{v}$ and a vector orthogonal to $\mathbf{v}$.

19. $\mathbf{v} = 2\mathbf{i} - \mathbf{j}$

$\mathbf{u} = \mathbf{i} + \mathbf{j}$

20. $\mathbf{v} = \mathbf{i} - 2\mathbf{j}$

$\mathbf{u} = -\mathbf{i} + \mathbf{j}$

Velocity and Acceleration Vectors

In Exercises 21 and 22, $\mathbf{r}(t)$ is the position vector of a particle in the plane at time t.

(a) Find the velocity and acceleration vectors.

(b) Find the speed at the given value of t.

(c) Find the angle between the velocity and acceleration vectors at the given value of t.

21. $\mathbf{r}(t) = (4 \cos t)\mathbf{i} + (\sqrt{2} \sin t)\mathbf{j}$, $t = \pi/4$

22. $\mathbf{r}(t) = (\sqrt{3} \sec t)\mathbf{i} + (\sqrt{3} \tan t)\mathbf{j}$, $t = 0$

23. *Maximum speed* The position of a particle in the plane at time t is

$$\mathbf{r} = \frac{1}{\sqrt{1 + t^2}}\mathbf{i} + \frac{t}{\sqrt{1 + t^2}}\mathbf{j}.$$

Find the particle's greatest speed.

24. *Writing to Learn: Minimum speed* The position of a particle in the plane at time $t \geq 0$ is

$$\mathbf{r}(t) = (e^t \cos t)\mathbf{i} + (e^t \sin t)\mathbf{j}.$$

Find the particle's minimum speed. Does it have a maximum speed? Give reasons for your answer.

Integrals and Initial Value Problems

In Exercises 25 and 26, evaluate the integral.

25. $\int_0^1 [(3 + 6t)\mathbf{i} + (6\pi \cos \pi t)\mathbf{j}]\, dt$

26. $\int_e^{e^2} \left[\left(\frac{2 \ln t}{t}\right)\mathbf{i} + \left(\frac{1}{t \ln t}\right)\mathbf{j} \right] dt$

In Exercises 27–30, solve the initial value problem.

27. $\dfrac{d\mathbf{r}}{dt} = -(\sin t)\mathbf{i} + (\cos t)\mathbf{j}$, $\mathbf{r}(0) = \mathbf{j}$

28. $\dfrac{d\mathbf{r}}{dt} = \dfrac{1}{t^2 + 1}\mathbf{i} + \dfrac{t}{\sqrt{t^2 + 1}}\mathbf{j}$, $\mathbf{r}(0) = \mathbf{i} + \mathbf{j}$

29. $\dfrac{d^2\mathbf{r}}{dt^2} = 2\mathbf{j}$, $\left.\dfrac{d\mathbf{r}}{dt}\right|_{t=0} = \mathbf{0}$, $\mathbf{r}(0) = \mathbf{i}$

30. $\dfrac{d^2\mathbf{r}}{dt^2} = -2\mathbf{i} - 2\mathbf{j}$, $\left.\dfrac{d\mathbf{r}}{dt}\right|_{t=1} = 4\mathbf{i}$, $\mathbf{r}(1) = 3\mathbf{i} + 3\mathbf{j}$

Graphs in the Polar Plane

Sketch the regions defined by the polar coordinate inequalities in Exercises 31 and 32.

31. $0 \leq r \leq 6 \cos \theta$ **32.** $-4 \sin \theta \leq r \leq 0$

Match each graph in Exercises 33–40 with the appropriate equation (a) through (l). There are more equations than graphs, so some equations will not be matched.

(a) $r = \cos 2\theta$ (b) $r \cos \theta = 1$

(c) $r = \dfrac{6}{1 - 2 \cos \theta}$ (d) $r = \sin 2\theta$

(e) $r = \theta$ (f) $r^2 = \cos 2\theta$

(g) $r = 1 + \cos \theta$

(h) $r = 1 - \sin \theta$

(i) $r = \dfrac{2}{1 - \cos \theta}$

(j) $r^2 = \sin 2\theta$

(k) $r = -\sin \theta$

(l) $r = 2 \cos \theta + 1$

33. Four-leaved rose

34. Spiral

35. Limaçon

36. Lemniscate

37. Circle

38. Cardioid

39. Parabola

40. Lemniscate

In Exercises 41–44, **(a)** graph the polar curve. **(b)** What is the smallest length θ-interval that will produce the graph?

41. $r = \cos 2\theta$

42. $r \cos \theta = 1$

43. $r^2 = \sin 2\theta$

44. $r = -\sin \theta$

Tangent Lines to Polar Graphs

In Exercises 45 and 46, find the tangent lines at the pole.

45. $r = \cos 2\theta$, $0 \le \theta \le 2\pi$

46. $r = 1 + \cos 2\theta$, $0 \le \theta \le 2\pi$

In Exercises 47 and 48, find equations for the horizontal and vertical tangent lines to the curve.

47. $r = 1 - \cos (\theta/2)$, $0 \le \theta \le 4\pi$

48. $r = 2(1 - \sin \theta)$, $0 \le \theta \le 2\pi$

49. Find equations for the lines that are tangent to the tips of the petals of the four-leaved rose $r = \sin 2\theta$.

50. Find equations for the lines that are tangent to the graph of the cardioid $r = 1 + \sin \theta$ at the points where it crosses the x-axis.

Polar to Cartesian Equations

In Exercises 51–56, replace the polar equation by an equivalent Cartesian equation. Then identify or describe the graph.

51. $r \cos \theta = r \sin \theta$

52. $r = 3 \cos \theta$

53. $r = 4 \tan \theta \sec \theta$

54. $r \cos (\theta + \pi/3) = 2\sqrt{3}$

55. $r = 2 \sec \theta$

56. $r = -(3/2) \csc \theta$

Cartesian to Polar Equations

In Exercises 57–60, replace the Cartesian equation by an equivalent polar equation.

57. $x^2 + y^2 + 5y = 0$

58. $x^2 + y^2 - 2y = 0$

59. $x^2 + 4y^2 = 16$

60. $(x + 2)^2 + (y - 5)^2 = 16$

Area in the Polar Plane

Find the areas of the regions in the polar coordinate plane described in Exercises 61–64.

61. Enclosed by the limaçon $r = 2 - \cos \theta$

62. Enclosed by one leaf of the three-leaved rose $r = \sin 3\theta$

63. Inside the "figure eight" $r = 1 + \cos 2\theta$ and outside the circle $r = 1$

64. Inside the cardioid $r = 2(1 + \sin \theta)$ and outside the circle $r = 2 \sin \theta$

Lengths of Polar Curves

Find the lengths of the curves given by the polar coordinate equations in Exercises 65–68.

65. $r = -1 + \cos \theta$

66. $r = 2 \sin \theta + 2 \cos \theta$, $0 \le \theta \le \pi/2$

67. $r = 8 \sin^3 (\theta/3), \quad 0 \le \theta \le \pi/4$

68. $r = \sqrt{1 + \cos 2\theta}, \quad -\pi/2 \le \theta \le \pi/2$

Theory and Examples

69. *Navigation* An airplane, flying in the direction 80° east of north at 540 mph in still air, encounters a 55 mph tailwind acting in the direction 100° east of north. The airplane holds its compass heading but, because of the wind, acquires a different ground speed and direction. What are they?

70. *Combining forces* A force of 120 lb pulls up on an object at an angle of 20° with the horizontal. A second force of 300 lb pulls down on the object at an angle of −5°. Find the direction and length of the resultant force vector.

71. *Shot put* A shot leaves the thrower's hand 6.5 ft above the ground at a 45° angle at 44 ft/sec. Where is it 3 sec later?

72. *Javelin* A javelin leaves the thrower's hand 7 ft above the ground at a 45° angle at 80 ft/sec. How high does it go?

73. *Rolling wheel* A circular wheel with radius 1 ft and center C rolls to the right along the x-axis at a half-turn per second (see figure). At time t seconds, the position vector of the point P on the wheel's circumference is

$$\mathbf{r}(t) = (\pi t - \sin \pi t)\mathbf{i} + (1 - \cos \pi t)\mathbf{j}.$$

T **(a)** Graph the curve traced by P during the interval $0 \le t \le 3$.

(b) Find velocity and acceleration vectors $\mathbf{v}$ and $\mathbf{a}$ at $t = 0, 1, 2,$ and 3.

(c) *Writing to Learn* At any given time, what is the forward speed of the topmost point of the wheel? Of C? Give reasons for your answers.

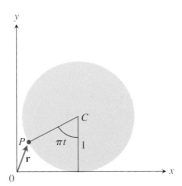

74. *The dictator* The Civil War mortar Dictator weighed so much (17,120 lb) that it had to be mounted on a railroad car. It had a 13 in. bore and used a 20 lb powder charge to fire a 200 lb shell. The mortar was made by Mr. Charles Knapp in his ironworks in Pittsburgh, Pennsylvania, and was used by the Union army in 1864 in the siege of Petersburg, Virginia. How far did it shoot? Here we have a difference of opinion. The ordnance manual claimed 4325 yd, whereas field officers claimed 4752 yd. Assuming a 45° firing angle, what muzzle speeds are involved here?

75. *World's record for popping a champagne cork*

(a) Until 1988, the world's record for popping a champagne cork, 109 ft 6 in., was held by Captain Michael Hill of the British Royal Artillery. Assuming that Captain Hill held the bottle at ground level at a 45° angle and the cork behaved like an ideal projectile, how fast was the cork going as it left the bottle?

(b) A new world record, 177 ft 9 in., was set on June 5, 1988, by Prof. Emeritus Heinrich of Rensselaer Polytechnic Institute, firing from 4 ft above ground level at a 45° angle at the Woodbury Vineyards Winery, New York. Assuming an ideal trajectory, what was the cork's initial speed?

76. *Javelin* In Potsdam in 1988, Petra Felke of (then) East Germany set a women's world record by throwing a javelin 262 ft 5 in.

(a) Assuming that Felke launched the javelin at a 40° angle to the horizontal from 6.5 ft above the ground, what was the javelin's initial speed?

(b) How high did the javelin go?

77. *Synchronous curves* By eliminating α from the ideal projectile equations

$$x = (v_0 \cos \alpha)t, \qquad y = (v_0 \sin \alpha)t - \frac{1}{2} gt^2,$$

show that $x^2 + (y + gt^2/2)^2 = v_0^2 t^2$. This shows that projectiles launched simultaneously from the origin at the same initial speed will, at any given instant, all lie on the circle of radius $v_0 t$ centered at $(0, -gt^2/2)$, regardless of their launch angle. These circles are the *synchronous curves* of the launching.

78. *Hitting a baseball under a wind gust* A baseball is hit when it is 4 ft above the ground. It leaves the bat with an intial velocity of 155 ft/sec, making an angle of 18° with the horizontal. At the instant the ball is hit, an instantaneous 11.7 ft/sec gust of wind blows in the horizontal direction against the ball, adding a component of

$-11.7\mathbf{i}$ to the ball's initial velocity. A 10-foot-high fence is 380 ft from home plate in the direction of the flight.

(a) Find vector and parametric equations for the path of the baseball.

(b) How high does the baseball go, and when does it reach maximum height?

(c) Find the range and flight time of the baseball.

(d) When is the baseball 25 ft high? How far (ground distance) is the baseball from home plate at that height?

(e) *Writing to Learn* Has the batter hit a home run? Explain.

79. *Linear drag* (*Continuation of Exercise 78*) Consider the baseball problem of Exercise 78 again. This time, assume a linear drag model with a drag coefficient of 0.09.

(a) Find vector and parametric equations for the path of the baseball.

(b) How high does the baseball go, and when does it reach maximum height?

(c) Find the range and flight time of the baseball.

(d) When is the baseball 30 ft high? How far (ground distance) is the baseball from home plate at that height?

(e) Has the batter hit a home run? If "yes," find a drag coefficient that would have prevented a home run. If "no," find a drag coefficient that would have allowed the hit to be a home run.

80. *Parallelogram* The accompanying figure shows parallelogram $ABCD$ and the midpoint P of diagonal BD.

(a) Express $\overrightarrow{BD}$ in terms of $\overrightarrow{AB}$ and $\overrightarrow{AD}$.

(b) Express $\overrightarrow{AP}$ in terms of $\overrightarrow{AB}$ and $\overrightarrow{AD}$.

(c) Prove that P is also the midpoint of diagonal AC.

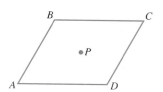

81. *Archimedes spirals* The graph of an equation of the form $r = a\theta$, where a is a nonzero constant, is called an Archimedes spiral. Is there anything special about the widths between the successive turns of such a spiral?

Additional Exercises: Theory, Examples, Applications

1. *Rowing across a river* A straight river is 20 m wide. The velocity of the river at (x, y) is

$$\mathbf{v} = -\frac{3x(20 - x)}{100}\mathbf{j} \quad \text{m/min}, \qquad 0 \le x \le 20.$$

A boat leaves the shore at $(0, 0)$ and travels through the water with a constant velocity. It arrives at the opposite shore at $(20, 0)$. The speed of the boat is always $\sqrt{20}$ m/min.

(a) Find the velocity of the boat.

(b) Find the location of the boat at time t.

(c) Sketch the path of the boat.

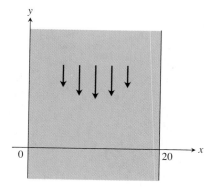

2. *Motion on a circle* A particle moves in the plane so that its velocity and position vectors are always orthogonal. Show that the particle moves in a circle centered at the origin.

3. *Angle between position and acceleration vectors* Suppose that $\mathbf{r}(t) = (e^t \cos t)\mathbf{i} + (e^t \sin t)\mathbf{j}$. Show that the angle between $\mathbf{r}$ and $\mathbf{a}$ never changes. What *is* the angle?

4. *Motion on a circle* A particle moves around the unit circle in the xy-plane. Its position at time t is $\mathbf{r} = x\mathbf{i} + y\mathbf{j}$, where x and y are differentiable functions of t. Find dy/dt if $\mathbf{v} \cdot \mathbf{i} = y$. Is the motion clockwise, or counterclockwise?

5. *Motion on a cubic* You send a message through a pneumatic tube that follows the curve $9y = x^3$ (distance in meters). At the point $(3, 3)$, $\mathbf{v} \cdot \mathbf{i} = 4$ and $\mathbf{a} \cdot \mathbf{i} = -2$. Find the values of $\mathbf{v} \cdot \mathbf{j}$ and $\mathbf{a} \cdot \mathbf{j}$ at $(3, 3)$.

6. *Angle bisection* Show that $\mathbf{w} = |\mathbf{v}|\mathbf{u} + |\mathbf{u}|\mathbf{v}$ bisects the angle between $\mathbf{u}$ and $\mathbf{v}$.

Polar Coordinates

7. (a) *Finding a polar equation* Find an equation in polar coordinates for the curve

$$x = e^{2t} \cos t, \qquad y = e^{2t} \sin t, \qquad -\infty < t < \infty.$$

(b) *Length of curve* Find the length of the curve from $t = 0$ to $t = 2\pi$.

8. *Length of curve* Find the length of the curve $r = 2 \sin^3 (\theta/3)$, $0 \le \theta \le 3\pi$, in the polar coordinate plane.

9. *Polar area* Sketch the regions enclosed by the curves $r = 2a \cos^2 (\theta/2)$ and $r = 2a \sin^2 (\theta/2)$, $a > 0$, in the polar coordinate plane and find the area of the portion of the plane they have in common.

T **10.** Graph the curve $r = \cos 5\theta + n \cos \theta$, $0 \le \theta \le \pi$ for integers $n = -5$ (heart) to $n = 5$ (bell). (*Source*: *The College Mathematics Journal*, Vol. 25, No. 1 (Jan. 1994).)

Appendices

A.1   Mathematical Induction

Many formulas, like

$$1 + 2 + \cdots + n = \frac{n(n + 1)}{2},$$

can be shown to hold for every positive integer n by applying an axiom called the *mathematical induction principle*. A proof that uses this axiom is called a *proof by mathematical induction* or a *proof by induction*.

The steps in proving a formula by induction are the following.

Step 1: Check that the formula holds for $n = 1$.

Step 2: Prove that if the formula holds for any positive integer $n = k$, then it also holds for the next integer, $n = k + 1$.

Once these steps are completed (the axiom says), we know that the formula holds for all positive integers n. By step 1, it holds for $n = 1$. By step 2, it holds for $n = 2$, and therefore by step 2 also for $n = 3$, and by step 2 again for $n = 4$, and so on. If the first domino falls, and the kth domino always knocks over the $(k + 1)$st when it falls, all the dominoes fall.

From another point of view, suppose we have a sequence of statements $S_1, S_2, \ldots, S_n, \ldots$, one for each positive integer. Suppose we can also show that assuming any one of the statements to be true implies the next statement in line is true. Finally, suppose we can show that S_1 is true. Then we may conclude that the statements are true from S_1 on.

Example 1   Sum of the First *n* Positive Integers

Prove that for every positive integer n,

$$1 + 2 + \cdots + n = \frac{n(n + 1)}{2}.$$

Solution   We accomplish the proof by carrying out the two steps above.

Step 1: The formula holds for $n = 1$ because

$$1 = \frac{1(1 + 1)}{2}.$$

1143

Step 2: If the formula holds for $n = k$, does it also hold for $n = k + 1$? The answer is yes, and here's why: If

$$1 + 2 + \cdots + k = \frac{k(k + 1)}{2},$$

then

$$1 + 2 + \cdots + k + (k + 1) = \frac{k(k + 1)}{2} + (k + 1) = \frac{k^2 + k + 2k + 2}{2}$$

$$= \frac{(k + 1)(k + 2)}{2} = \frac{(k + 1)((k + 1) + 1)}{2}.$$

The last expression in this string of equalities is the expression $n(n + 1)/2$ for $n = (k + 1)$.

The mathematical induction principle now guarantees the original formula for all positive integers n. Notice that all *we* have to do is carry out steps 1 and 2. The mathematical induction principle does the rest.

Example 2   Sum of Powers of 1/2

Prove that for all positive integers n,

$$\frac{1}{2^1} + \frac{1}{2^2} + \cdots + \frac{1}{2^n} = 1 - \frac{1}{2^n}.$$

Solution   We accomplish the proof by carrying out the two steps of mathematical induction.

Step 1: The formula holds for $n = 1$ because

$$\frac{1}{2^1} = 1 - \frac{1}{2^1}.$$

Step 2: If

$$\frac{1}{2^1} + \frac{1}{2^2} + \cdots + \frac{1}{2^k} = 1 - \frac{1}{2^k},$$

then

$$\frac{1}{2^1} + \frac{1}{2^2} + \cdots + \frac{1}{2^k} + \frac{1}{2^{k+1}} = 1 - \frac{1}{2^k} + \frac{1}{2^{k+1}} = 1 - \frac{1 \cdot 2}{2^k \cdot 2} + \frac{1}{2^{k+1}}$$

$$= 1 - \frac{2}{2^{k+1}} + \frac{1}{2^{k+1}} = 1 - \frac{1}{2^{k+1}}.$$

Thus, the original formula holds for $n = (k + 1)$ whenever it holds for $n = k$.

With these steps verified, the mathematical induction principle now guarantees the formula for every positive integer n.

Other Starting Integers

Instead of starting at $n = 1$, some induction arguments start at another integer. The steps for such an argument are as follows.

Step 1: Check that the formula holds for $n = n_1$ (the first appropriate integer).

Step 2: Prove that if the formula holds for any integer $n = k \geq n_1$, then it also holds for $n = (k + 1)$.

Once these steps are completed, the mathematical induction principle guarantees the formula for all $n \geq n_1$.

Example 3   Factorial Exceeding Exponential

Show that $n! > 3^n$ if n is large enough.

Solution   How large is large enough? We experiment:

n	1	2	3	4	5	6	7
$n!$	1	2	6	24	120	720	5040
3^n	3	9	27	81	243	729	2187

It looks as if $n! > 3^n$ for $n \geq 7$. To be sure, we apply mathematical induction. We take $n_1 = 7$ in step 1 and try for step 2.
   Suppose that $k! > 3^k$ for some $k \geq 7$. Then

$$(k + 1)! = (k + 1)(k!) > (k + 1) \, 3^k > 7 \cdot 3^k > 3^{k+1}.$$

Thus, for $k \geq 7$,

$$k! > 3^k \quad \Rightarrow \quad (k + 1)! > 3^{k+1}.$$

The mathematical induction principle now guarantees $n! \geq 3^n$ for all $n \geq 7$.

EXERCISES A.1

1. *General triangle inequality* Assuming that the triangle inequality $|a + b| \leq |a| + |b|$ holds for any two numbers a and b, show that

$$|x_1 + x_2 + \cdots + x_n| \leq |x_1| + |x_2| + \cdots + |x_n|$$

for any n numbers.

2. *Geometric sum* Show that if $r \neq 1$, then

$$1 + r + r^2 + \cdots + r^n = \frac{1 - r^{n+1}}{1 - r}$$

for every positive integer n.

3. *Positive integer power rule* Use the Product Rule,

$$\frac{d}{dx}(uv) = u\frac{dv}{dx} + v\frac{du}{dx},$$

and the equation

$$\frac{d}{dx}(x) = 1$$

to show that

$$\frac{d}{dx}(x^n) = nx^{n-1}$$

for every positive integer n.

4. *Products into sums* Suppose a function $f(x)$ has the property that $f(x_1 x_2) = f(x_1) + f(x_2)$ for any two positive numbers x_1 and x_2. Show that

$$f(x_1 x_2 \cdots x_n) = f(x_1) + f(x_2) + \cdots + f(x_n)$$

for the product of any n positive numbers $x_1, x_2 \ldots, x_n$.

5. *Geometric sum* Show that

$$\frac{2}{3^1} + \frac{2}{3^2} + \cdots + \frac{2}{3^n} = 1 - \frac{1}{3^n}$$

for all positive integers n.

6. Show that $n! > n^3$ if n is large enough.

7. Show that $2^n > n^2$ if n is large enough.

8. Show that $2^n \geq 1/8$ for $n \geq -3$.

9. *Sums of squares* Show that the sum of the squares of the first n positive integers is

$$\frac{n\left(n + \dfrac{1}{2}\right)(n + 1)}{3}.$$

10. *Sums of cubes* Show that the sum of the cubes of the first n positive integers is $(n(n + 1)/2)^2$.

11. *Rules for finite sums* Show that the following finite sum rules hold for every positive integer n.

(a) $\displaystyle\sum_{k=1}^{n} (a_k + b_k) = \sum_{k=1}^{n} a_k + \sum_{k=1}^{n} b_k$

(b) $\displaystyle\sum_{k=1}^{n} (a_k - b_k) = \sum_{k=1}^{n} a_k - \sum_{k=1}^{n} b_k$

(c) $\displaystyle\sum_{k=1}^{n} ca_k = c \cdot \sum_{k=1}^{n} a_k$ $\quad$ (any number c)

(d) $\displaystyle\sum_{k=1}^{n} a_k = n \cdot c$ $\quad$ (if a_k has the constant value c)

12. *Integer powers and absolute value* Show that $|x^n| = |x|^n$ for every positive integer n and every real number x.

A.2 Proofs of Limit Theorems in Section 1.2

This appendix proves Theorem 1 and Theorem 4 from Section 1.2.

Theorem 1 Properties of Limits

The following rules hold if $\lim_{x \to c} f(x) = L$ and $\lim_{x \to c} g(x) = M$ (L and M real numbers).

1. *Sum Rule:* $\qquad\qquad\qquad\qquad$ $\displaystyle\lim_{x \to c} [f(x) + g(x)] = L + M$

2. *Difference Rule:* $\qquad\qquad\quad$ $\displaystyle\lim_{x \to c} [f(x) - g(x)] = L - M$

3. *Product Rule:* $\qquad\qquad\qquad$ $\displaystyle\lim_{x \to c} f(x) \cdot g(x) = L \cdot M$

4. *Constant Multiple Rule:* $\qquad$ $\displaystyle\lim_{x \to c} k f(x) = k L$ $\quad$ (any number k)

5. *Quotient Rule:* $\qquad\qquad\qquad$ $\displaystyle\lim_{x \to c} \frac{f(x)}{g(x)} = \frac{L}{M}$, $\quad$ if $M \neq 0$

6. *Power Rule:* $\qquad\qquad\qquad\;$ If m and n are integers, then

$$\lim_{x \to c} [f(x)]^{m/n} = L^{m/n}$$

provided $L^{m/n}$ is a real number.

Proof of the Sum Rule Let $\epsilon > 0$ be given. We want to find a positive number δ such that for all x

$$0 < |x - c| < \delta \;\Rightarrow\; |f(x) + g(x) - (L + M)| < \epsilon.$$

Regrouping terms, we get

$$|f(x) + g(x) - (L + M)| = |(f(x) - L) + (g(x) - M)|$$

$$\leq |f(x) - L| + |g(x) - M|. \qquad \text{Triangle Inequality:}$$
$$\text{\scriptsize } |a + b| \leq |a| + |b|$$

Since $\lim_{x \to c} f(x) = L$, there exists a number $\delta_1 > 0$ such that for all x

$$0 < |x - c| < \delta_1 \implies |f(x) - L| < \epsilon/2.$$

Similarly, since $\lim_{x \to c} g(x) = M$, there exists a number $\delta_2 > 0$ such that for all x

$$0 < |x - c| < \delta_2 \implies |g(x) - M| < \epsilon/2.$$

Let $\delta = \min\{\delta_1, \delta_2\}$, the smaller of δ_1 and δ_2. If $0 < |x - c| < \delta$ then $|x - c| < \delta_1$, so $|f(x) - L| < \epsilon/2$, and $|x - c| < \delta_2$, so $|g(x) - M| < \epsilon/2$. Therefore,

$$|f(x) + g(x) - (L + M)| < \frac{\epsilon}{2} + \frac{\epsilon}{2} = \epsilon,$$

which shows that $\lim_{x \to c} (f(x) + g(x)) = L + M$.

CD-ROM
WEBsite

Historical Biography

John Wallis
(1616 — 1703)

The Difference Rule is obtained by replacing $g(x)$ by $-g(x)$ and M by $-M$ in the Sum Rule. The Constant Multiple Rule is the special case $g(x) = k$ of the Product Rule. The Power Rule is proved in more advanced texts, but we prove here the Product and Quotient Rules.

Proof of the Limit Product Rule   We show that for any $\epsilon > 0$ there exists a $\delta > 0$ such that for all x in the intersection D of the domains of f and g,

$$0 < |x - c| < \delta \implies |f(x)\, g(x) - LM| < \epsilon.$$

Suppose then that ϵ is a positive number and write $f(x)$ and $g(x)$ as

$$f(x) = L + (f(x) - L), \qquad g(x) = M + (g(x) - M).$$

Multiply these expressions together and subtract LM:

$$
\begin{aligned}
f(x) \cdot g(x) - LM &= (L + (f(x) - L))(M + (g(x) - M)) - LM \\
&= LM + L(g(x) - M) + M(f(x) - L) + \\
&\quad (f(x) - L)(g(x) - M) - LM \\
&= L(g(x) - M) + M(f(x) - L) + (f(x) - L)(g(x) - M).
\end{aligned}
\tag{1}
$$

Since f and g have limits L and M as $x \to c$, there exist positive numbers δ_1, δ_2, δ_3, and δ_4 such that for all x in D

$$
\begin{aligned}
0 < |x - c| < \delta_1 &\implies |f(x) - L| < \sqrt{\epsilon/3} \\
0 < |x - c| < \delta_2 &\implies |g(x) - M| < \sqrt{\epsilon/3} \\
0 < |x - c| < \delta_3 &\implies |f(x) - L| < \epsilon/(3(1 + |M|)) \\
0 < |x - c| < \delta_4 &\implies |g(x) - M| < \epsilon/(3(1 + |L|)).
\end{aligned}
\tag{2}
$$

If we take δ to be the smallest of the numbers δ_1 through δ_4, the inequalities on the right-hand side of (2) will hold simultaneously for $0 < |x - c| < \delta$. Therefore, for all x in D, $0 < |x - c| < \delta$ implies

$$|f(x) \cdot g(x) - LM|$$

$$\leq |L||g(x) - M| + |M||f(x) - L| + |f(x) - L||g(x) - M|$$

$$\leq (1 + |L|)|g(x) - M| + (1 + |M|)|f(x) - L| + |f(x) - L||g(x) - M|$$

$$\leq \frac{\epsilon}{3} + \frac{\epsilon}{3} + \sqrt{\frac{\epsilon}{3}}\sqrt{\frac{\epsilon}{3}} = \epsilon.$$

Triangle inequality applied to Eq. (1)

Values from Eq. (2)

This completes the proof of the Limit Product Rule.

Proof of the Limit Quotient Rule   We show that $\lim_{x \to c} (1/g(x)) = 1/M$. We can then conclude from the Limit Product Rule that

$$\lim_{x \to c} \frac{f(x)}{g(x)} = \lim_{x \to c} \left(f(x) \cdot \frac{1}{g(x)} \right) = \lim_{x \to c} f(x) \cdot \lim_{x \to c} g(x) = L \cdot \frac{1}{M} = \frac{L}{M}.$$

Let $\epsilon > 0$ be given. To show that $\lim_{x \to c} (1/g(x)) = 1/M,$ we need to show that there exists a $\delta > 0$ such that for all x

$$0 < |x - c| < \delta \implies \left| \frac{1}{g(x)} - \frac{1}{M} \right| < \epsilon.$$

Since $|M| > 0,$ there exists a positive number δ_1 such that for all x

$$0 < |x - c| < \delta_1 \implies |g(x) - M| < \left| \frac{M}{2} \right|. \tag{3}$$

CD-ROM
WEBsite

Historical Biography

Gilles Personne
de Roberval
(1602 — 1675)

For any numbers A and B it can be shown that $|A| - |B| \le |A - B|$ and $|B| - |A| \le |A - B|,$ from which it follows that $||A| - |B|| \le |A - B|.$ With $A = g(x)$ and $B = M,$ this becomes

$$||g(x)| - |M|| \le |g(x) - M|,$$

which can be combined with the inequality on the right in Eq. (3) to get, in turn,

$$||g(x)| - |M|| < \frac{|M|}{2}$$

$$-\frac{|M|}{2} < |g(x)| - |M| < \frac{|M|}{2}$$

$$\frac{|M|}{2} < |g(x)| < \frac{3|M|}{2} \tag{4}$$

$$|M| < 2|g(x)| < 3|M|$$

$$\frac{1}{|g(x)|} < \frac{2}{|M|} < \frac{3}{|g(x)|}.$$

Therefore, $0 < |x - c| < \delta_1$ implies that

$$\left| \frac{1}{g(x)} - \frac{1}{M} \right| = \left| \frac{M - g(x)}{Mg(x)} \right| \le \frac{1}{|M|} \cdot \frac{1}{|g(x)|} \cdot |M - g(x)|$$

$$< \frac{1}{|M|} \cdot \frac{2}{|M|} \cdot |M - g(x)|. \qquad \text{Inequality (4)} \tag{5}$$

Since $(1/2)|M|^2 \epsilon > 0,$ there exists a number $\delta_2 > 0$ such that for all x

$$0 < |x - c| < \delta_2 \implies |M - g(x)| < \frac{\epsilon}{2} |M|^2. \tag{6}$$

If we take δ to be the smaller of δ_1 and $\delta_2,$ the conclusions in Eqs. (5) and (6) both hold for all x such that $0 < |x - c| < \delta.$ Combining these conclusions gives

$$0 < |x - c| < \delta \implies \left| \frac{1}{g(x)} - \frac{1}{M} \right| < \epsilon.$$

This concludes the proof of the Limit Quotient Rule.

> ### Theorem 4   The Sandwich Theorem
> Suppose that $g(x) \leq f(x) \leq h(x)$ for all x in some open interval containing c, except possibly at $x = c$ itself. Suppose also that $\lim_{x \to c} g(x) = \lim_{x \to c} h(x) = L$. Then $\lim_{x \to c} f(x) = L$.

Proof for Right-hand Limits   Suppose that $\lim_{x \to c^+} g(x) = \lim_{x \to c^+} h(x) = L$. Then for any $\epsilon > 0$ there exists a $\delta > 0$ such that for all x the inequality $c < x < c + \delta$ implies

$$L - \epsilon < g(x) < L + \epsilon \qquad \text{and} \qquad L - \epsilon < h(x) < L + \epsilon.$$

These inequalities combine with $g(x) \leq f(x) \leq h(x)$ to give

$$L - \epsilon < g(x) \leq f(x) \leq h(x) < L + \epsilon,$$

$$L - \epsilon < f(x) < L + \epsilon,$$

$$-\epsilon < f(x) - L < \epsilon.$$

Therefore, for all x, the inequality $c < x < c + \delta$ implies $|f(x) - L| < \epsilon$.   ▬

Proof for Left-hand Limits   Suppose that $\lim_{x \to c^-} g(x) = \lim_{x \to c^-} h(x) = L$. Then for any $\epsilon > 0$ there exists a $\delta > 0$ such that for all x the inequality $c - \delta < x < c$ implies

$$L - \epsilon < g(x) < L + \epsilon \qquad \text{and} \qquad L - \epsilon < h(x) < L + \epsilon.$$

We conclude as before that for all x, $c - \delta < x < c$ implies $|f(x) - L| < \epsilon$.   ▬

Proof for Two-sided Limits   If $\lim_{x \to c} g(x) = \lim_{x \to c} h(x) = L$, then $g(x)$ and $h(x)$ both approach L as $x \to c^+$ and as $x \to c^-$; so $\lim_{x \to c^+} f(x) = L$ and $\lim_{x \to c^-} f(x) = L$. Hence, $\lim_{x \to c} f(x)$ exists and equals L.   ▬

EXERCISES A.2

1. *Generalized limit sum rule* Suppose that functions $f_1(x)$, $f_2(x)$, and $f_3(x)$ have limits L_1, L_2, and L_3, respectively, as $x \to c$. Show that their sum has limit $L_1 + L_2 + L_3$. Use mathematical induction (Appendix 1) to generalize this result to the sum of any finite number of functions.

2. *Generalized limit product rule* Use mathematical induction and the Limit Product Rule in Theorem 1 to show that if functions $f_1(x)$, $f_2(x), \dots, f_n(x)$ have limits $L_1, L_2, \dots, L_n$ as $x \to c$, then

$$\lim_{x \to c} f_1(x) f_2(x) \cdot \dots \cdot f_n(x) = L_1 \cdot L_2 \cdot \dots \cdot L_n.$$

3. *Positive integer power rule* Use the fact that $\lim_{x \to c} x = c$ and the result of Exercise 2 to show that $\lim_{x \to c} x^n = c^n$ for any integer $n > 1$.

4. *Limits of polynomials* Use the fact that $\lim_{x \to c} (k) = k$ for any number k together with the results of Exercises 1 and 3 to show that $\lim_{x \to c} f(x) = f(c)$ for any polynomial function

$$f(x) = a_n x^n + a_{n-1} x^{n-1} + \dots + a_1 x + a_0.$$

5. *Limits of rational functions* Use Theorem 1 and the result of Exercise 4 to show that if $f(x)$ and $g(x)$ are polynomial functions and $g(c) \neq 0$, then

$$\lim_{x \to c} \frac{f(x)}{g(x)} = \frac{f(c)}{g(c)}.$$

6. *Composites of continuous functions* Figure A.1 gives the diagram for a proof that the composite of two continuous functions is continuous. Reconstruct the proof from the diagram. The statement to be proved is this: If f is continuous at $x = c$ and g is continuous at $f(c)$, then $g \circ f$ is continuous at c.

Assume that c is an interior point of the domain of f and that $f(c)$ is an interior point of the domain of g. This will make the limits involved two-sided. (The arguments for the cases that involve one-sided limits are similar.)

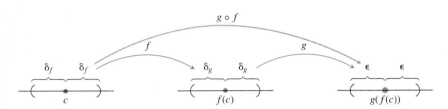

FIGURE A.1 A diagram for a proof that the composite of two continuous functions is continuous. The continuity of composites holds for any finite number of functions. The only requirement is that each function be continuous where it is applied. In the figure, f is to be continuous at c and g at $f(c)$.

A.3 Proof of the Chain Rule

This appendix proves the Chain Rule in Section 2.5 using ideas from Section 3.6.

> **Theorem 4** The Chain Rule
>
> If $f(u)$ is differentiable at the point $u = g(x)$, and $g(x)$ is differentiable at x, then the composite function $(f \circ g)(x) = f(g(x))$ is differentiable at x, and
>
> $$(f \circ g)'(x) = f'(g(x)) \cdot g'(x).$$
>
> In Leibniz's notation, if $y = f(u)$ and $u = g(x)$, then
>
> $$\frac{dy}{dx} = \frac{dy}{du} \cdot \frac{du}{dx},$$
>
> where dy/du is evaluated at $u = g(x)$.

Proof To be more precise, we show that if g is differentiable at x_0 and f is differentiable at $g(x_0)$, then the composite is differentiable at x_0 and

$$\left.\frac{dy}{dx}\right|_{x=x_0} = f'(g(x_0)) \cdot g'(x_0).$$

Let Δx be an increment in x and let Δu and Δy be the corresponding increments in u and y. As you can see in Figure A.2,

$$\left.\frac{dy}{dx}\right|_{x=x_0} = \lim_{\Delta x \to 0} \frac{\Delta y}{\Delta x},$$

so our goal is to show that this limit is $f'(g(x_0)) \cdot g'(x_0)$.

By Equation (3) in Section 3.6,

$$\Delta u = g'(x_0) \Delta x + \epsilon_1 \Delta x = (g'(x_0) + \epsilon_1) \Delta x,$$

where $\epsilon_1 \to 0$ as $\Delta x \to 0$. Similarly,

$$\Delta y = f'(u_0) \Delta u + \epsilon_2 \Delta u = (f'(u_0) + \epsilon_2) \Delta u,$$

FIGURE A.2 The graph of y as a function of x. The derivative of y with respect to x at $x = x_0$ is $\lim_{\Delta x \to 0} \Delta y / \Delta x$.

where $\epsilon_2 \to 0$ as $\Delta u \to 0$. Notice also that $\Delta u \to 0$ as $\Delta x \to 0$. Combining the equations for Δu and Δy gives

$$\Delta y = (f'(u_0) + \epsilon_2)(g'(x_0) + \epsilon_1)\, \Delta x,$$

so

$$\frac{\Delta y}{\Delta x} = f'(u_0)\, g'(x_0) + \epsilon_2\, g'(x_0) + f'(u_0)\, \epsilon_1 + \epsilon_2\, \epsilon_1.$$

Since ϵ_1 and ϵ_2 go to zero as Δx goes to zero, three of the four terms on the right vanish in the limit, leaving

$$\lim_{\Delta x \to 0} \frac{\Delta y}{\Delta x} = f'(u_0)\, g'(x_0) = f'(g(x_0)) \cdot g'(x_0).$$

This concludes the proof.

A.4 Complex Numbers

The Development of the Real Numbers • The Complex Number System • Argand Diagrams • Euler's Formula • Products • Quotients • Powers and De Moivre's Theorem • Roots • The Fundamental Theorem of Algebra

Complex numbers are expressions of the form $a + ib$, where a and b are real numbers and i is a symbol for $\sqrt{-1}$. Unfortunately, "real" and "imaginary" have connotations that somehow place $\sqrt{-1}$ in a less favorable position in our minds than $\sqrt{2}$. As a matter of fact, a good deal of imagination, in the sense of *inventiveness,* has been required to construct the *real* number system, which forms the basis of the calculus. In this appendix, we review the various stages of this invention. The further invention of a complex number system will then not seem so strange.

The Development of the Real Numbers

The earliest stage of number development was the recognition of the **counting numbers** $1, 2, 3, \ldots,$ which we now call the **natural numbers** or the **positive integers.** These numbers can be added or multiplied together without getting outside the system. That is, the system of positive integers is **closed** under the operations of addition and multiplication. If m and n are any positive integers, then

$$m + n = p \qquad \text{and} \qquad mn = q \tag{1}$$

are also positive integers.

Given the two positive integers on the left-hand side of either equation in Equations (1), we can find the corresponding positive integer on the right. More than this, we can sometimes specify the positive integers m and p and find a positive integer n such that $m + n = p$. For instance, $3 + n = 7$ can be solved when the only numbers we know are the positive integers. The equation $7 + n = 3$, however, cannot be solved unless the number system is enlarged.

The number zero and the negative integers were invented to solve equations like $7 + n = 3$. In a civilization that recognizes all the **integers**

$$\ldots, -3, -2, -1, 0, 1, 2, 3, \ldots, \tag{2}$$

an educated person can always find the missing integer that solves the equation $m + n = p$ when given the other two integers in the equation.

Suppose that our educated people also know how to multiply any two of the integers in Equation (2). If, in Equations (1), they are given m and q, they discover that sometimes they can find n and sometimes they cannot. If their imagination is still in good working order, they may be inspired to invent still more numbers and introduce fractions as ordered pairs m/n of integers m and n. The number zero has special properties that may bother them for a while, but they ultimately discover that it is handy to have all ratios of integers m/n, excluding only those having zero in the denominator. This system, called the set of **rational numbers,** is now rich enough for them to perform the so-called **rational operations** of arithmetic:

1. (a) addition
   (b) subtraction

2. (a) multiplication
   (b) division

on any two numbers in the system, *except that they cannot divide by zero.*

The geometry of the unit square (Figure A.3) and the Pythagorean Theorem showed that they could construct a geometric line segment that, in terms of some basic unit of length, has length equal to $\sqrt{2}$. Thus, they could solve the equation

$$x^2 = 2$$

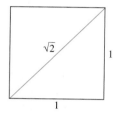

FIGURE A.3 With a straightedge and compass, it is possible to construct a line segment of irrational length.

by a geometric construction. Then they discovered, however, that the line segment representing $\sqrt{2}$ and the line segment representing the unit of length 1 were incommensurable quantities. This means that the ratio $\sqrt{2}/1$ cannot be expressed as the ratio of two *integer* multiples of some other, presumably more fundamental, unit of length. That is, our educated people could not find a rational number solution of the equation $x^2 = 2$.

They could not find it because there *is* no rational number whose square is 2. To see why, suppose that there were such a rational number. Then we could find integers p and q with no common factor other than 1 and such that

$$\left(\frac{p}{q}\right)^2 = 2$$

or

$$p^2 = 2q^2. \tag{3}$$

Since p and q are integers, p must be even; otherwise, its product with itself would be odd. In symbols, $p = 2p_1$, where p_1 is an integer. This step leads to $2p_1{}^2 = q^2$, which says q must be even, say $q = 2q_1$, where q_1 is an integer. This makes 2 a factor of both p and q, contrary to our choice of p and q as integers with no common factor other than 1. Hence, there is no rational number whose square is 2.

Although our educated people could not find a rational solution of the equation $x^2 = 2$, they could get a sequence of rational numbers

$$\frac{1}{1}, \quad \frac{7}{5}, \quad \frac{41}{29}, \quad \frac{239}{169}, \quad \cdots, \tag{4}$$

whose squares form a sequence

$$\frac{1}{1}, \quad \frac{49}{25}, \quad \frac{1681}{841}, \quad \frac{57{,}121}{28{,}561}, \quad \cdots \tag{5}$$

that converges to 2 as its limit. This time their imagination suggested that they needed a concept of a limit of a sequence of rational numbers. If we accept that an increasing sequence that is bounded from above always approaches a limit and observe that the sequence in Equation (4) has these properties, then we want it to have a limit L. This assumption would also mean, from Equation (5), that $L^2 = 2$, and hence L is *not* one of our rational numbers. If to the rational numbers we further add the limits of all bounded increasing sequences of rational numbers, we arrive at the system of all "real" numbers. The word *real* is placed in quotes because there is nothing that is either "more real" or "less real" about this system than there is about any other mathematical system.

The Complex Number System

Imagination was called upon at many stages during the development of the real number system. In fact, the art of invention was needed at least three times in constructing the systems we have discussed so far:

1. The *first invented* system: the set of *all integers* as constructed from the counting numbers

2. The *second invented* system: the set of *rational numbers* m/n as constructed from the integers

3. The *third invented* system: the set of all *real numbers* x as constructed from the rational numbers.

These invented systems form a hierarchy in which each system contains the previous system. Each system is also richer than its predecessor in that it permits additional operations to be performed without going outside the system:

1. In the system of all integers, we can solve all equations of the form

$$x + a = 0, \tag{6}$$

where a can be any integer.

2. In the system of all rational numbers, we can solve all equations of the form

$$ax + b = 0, \tag{7}$$

provided a and b are rational numbers and $a \neq 0$.

3. In the system of all real numbers, we can solve all the equations in Equations (6) and (7) and, in addition, all quadratic equations

$$ax^2 + bx + c = 0 \quad \text{having} \quad a \neq 0 \quad \text{and} \quad b^2 - 4ac \geq 0. \tag{8}$$

You are probably familiar with the formula that gives the solutions of Equation (8), namely,

$$x = \frac{-b \pm \sqrt{b^2 - 4ac}}{2a}, \tag{9}$$

and also know that when the discriminant, $d = b^2 - 4ac$, is negative, the solutions in Equation (9) do *not* belong to any of the systems discussed above. In fact, the very simple quadratic equation

$$x^2 + 1 = 0$$

is impossible to solve if the only number systems that can be used are the three invented systems mentioned so far.

Thus, we come to the *fourth invented* system, the set of all complex numbers $a + ib$. We could dispense entirely with the symbol i and use a notation like (a, b). We would then speak simply of a pair of real numbers a and b. Since, under algebraic operations, the numbers a and b are treated somewhat differently, it is essential to keep the *order* straight. We therefore might say that the **complex number system** consists of the set of all ordered pairs of real numbers (a, b), together with the rules by which they are to be equated, added, multiplied, and so on, listed below. We use both the (a, b) notation and the notation $a + ib$ in the discussion that follows. We call a the **real part** and b the **imaginary part** of the complex number (a, b).

We make the following definitions.

Equality

$a + ib = c + id$
if and only if
$a = c$ and $b = d$

Two complex numbers (a, b) and (c, d) are *equal* if and only if $a = c$ and $b = d$.

Addition

$(a + ib) + (c + id)$
$= (a + c) + i(b + d)$

The sum of the two complex numbers (a, b) and (c, d) is the complex number $(a + c, b + d)$.

Multiplication

$(a + ib)(c + id)$
$= (ac - bd) + i(ad + bc)$

The product of two complex numbers (a, b) and (c, d) is the complex number $(ac - bd, ad + bc)$.

$c(a + ib) = ac + i(bc)$

The product of a real number c and the complex number (a, b) is the complex number (ac, bc).

The set of all complex numbers (a, b) in which the second number b is zero has all the properties of the set of real numbers a. For example, addition and multiplication of $(a, 0)$ and $(c, 0)$ give

$$(a, 0) + (c, 0) = (a + c, 0)$$
$$(a, 0) \cdot (c, 0) = (ac, 0),$$

which are numbers of the same type with imaginary part equal to zero. Also, if we multiply a "real number" $(a, 0)$ and the complex number (c, d), we get

$$(a, 0) \cdot (c, d) = (ac, ad) = a(c, d).$$

In particular, the complex number $(0, 0)$ plays the role of zero in the complex number system, and the complex number $(1, 0)$ plays the role of unity.

The number pair $(0, 1)$, which has real part equal to zero and imaginary part equal to one, has the property that its square,

$$(0, 1)(0, 1) = (-1, 0),$$

has real part equal to minus one and imaginary part equal to zero. Therefore, in the system of complex numbers (a, b), there is a number $x = (0, 1)$ whose square can be added to unity $= (1, 0)$ to produce zero $= (0, 0)$; that is,

$$(0, 1)^2 + (1, 0) = (0, 0).$$

The equation

$$x^2 + 1 = 0$$

therefore has a solution $x = (0, 1)$ in this new number system.

You are probably more familiar with the $a + ib$ notation than you are with the notation (a, b). And since the laws of algebra for the ordered pairs enable us to write

$$(a, b) = (a, 0) + (0, b) = a(1, 0) + b(0, 1),$$

whereas $(1, 0)$ behaves like unity and $(0, 1)$ behaves like a square root of minus one, we need not hesitate to write $a + ib$ in place of (a, b). The i associated with b is like a tracer element that tags the imaginary part of $a + ib$. We can pass at will from the realm of ordered pairs (a, b) to the realm of expressions $a + ib$, and conversely. Yet there is nothing less "real" about the symbol $(0, 1) = i$ than there is about the symbol $(1, 0) = 1$, once we have learned the laws of algebra in the complex number system (a, b).

To reduce any rational combination of complex numbers to a single complex number, we apply the laws of elementary algebra, replacing i^2 wherever it appears by -1. Of course, we cannot divide by the complex number $(0, 0) = 0 + i0$. If $a + ib \neq 0$, however, we may carry out a division as follows:

$$\frac{c + id}{a + ib} = \frac{(c + id)(a - ib)}{(a + ib)(a - ib)} = \frac{(ac + bd) + i(ad - bc)}{a^2 + b^2}.$$

The result is a complex number $x + iy$ with

$$x = \frac{ac + bd}{a^2 + b^2}, \qquad y = \frac{ad - bc}{a^2 + b^2},$$

and $a^2 + b^2 \neq 0$, since $a + ib = (a, b) \neq (0, 0)$.

The number $a - ib$ that is used as multiplier to clear the i from the denominator is called the **complex conjugate** of $a + ib$. It is customary to use $\bar{z}$ (read "z bar") to denote the complex conjugate of z; thus,

$$z = a + ib, \qquad \bar{z} = a - ib.$$

Multiplying the numerator and denominator of the fraction $(c + id)/(a + ib)$ by the complex conjugate of the denominator will always replace the denominator by a real number.

Example 1   Operations with Complex Numbers

(a) $(2 + 3i) + (6 - 2i) = (2 + 6) + (3 - 2)i = 8 + i$

(b) $(2 + 3i) - (6 - 2i) = (2 - 6) + (3 - (-2))i = -4 + 5i$

(c) $(2 + 3i)(6 - 2i) = (2)(6) + (2)(-2i) + (3i)(6) + (3i)(-2i)$

$$= 12 - 4i + 18i - 6i^2 = 12 + 14i + 6 = 18 + 14i$$

(d) $\dfrac{2 + 3i}{6 - 2i} = \dfrac{2 + 3i}{6 - 2i} \dfrac{6 + 2i}{6 + 2i}$

$$= \frac{12 + 4i + 18i + 6i^2}{36 + 12i - 12i - 4i^2}$$

$$= \frac{6 + 22i}{40} = \frac{3}{20} + \frac{11}{20}i$$

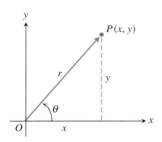

FIGURE A.4 This Argand diagram represents $z = x + iy$ both as a point $P(x, y)$ and as a vector $\overrightarrow{OP}$.

Argand Diagrams

There are two geometric representations of the complex number $z = x + iy$:

1.   as the point $P(x, y)$ in the xy-plane
2.   as the vector $\overrightarrow{OP}$ from the origin to P.

In each representation, the x-axis is called the **real axis** and the y-axis is the **imaginary axis**. Both representations are **Argand diagrams** for $x + iy$ (Figure A.4).

In terms of the polar coordinates of x and y, we have

$$x = r \cos \theta, \qquad y = r \sin \theta,$$

and

$$z = x + iy = r(\cos \theta + i \sin \theta). \tag{10}$$

We define the **absolute value** of a complex number $x + iy$ to be the length r of a vector $\overrightarrow{OP}$ from the origin to $P(x, y)$. We denote the absolute value by vertical bars; thus,

$$|x + iy| = \sqrt{x^2 + y^2}.$$

If we always choose the polar coordinates r and θ so that r is nonnegative, then

$$r = |x + iy|.$$

The polar angle θ is called the **argument** of z and is written $\theta = \arg z$. Of course, any integer multiple of 2π may be added to θ to produce another appropriate angle.

The following equation gives a useful formula connecting a complex number z, its conjugate $\bar{z}$, and its absolute value $|z|$, namely,

$$z \cdot \bar{z} = |z|^2.$$

Euler's Formula

The identity

$$e^{i\theta} = \cos \theta + i \sin \theta,$$

called **Euler's formula,** enables us to rewrite Equation (10) as

$$z = re^{i\theta}.$$

This formula, in turn, leads to the following rules for calculating products, quotients, powers, and roots of complex numbers. It also leads to Argand diagrams for $e^{i\theta}$. Since $\cos \theta + i \sin \theta$ is what we get from Equation (10) by taking $r = 1$, we can say that $e^{i\theta}$ is represented by a unit vector that makes an angle θ with the positive x-axis, as shown in Figure A.5.

Products

To multiply two complex numbers, we multiply their absolute values and add their angles. Let

$$z_1 = r_1 e^{i\theta_1}, \qquad z_2 = r_2 e^{i\theta_2}, \tag{11}$$

so that

$$|z_1| = r_1, \qquad \arg z_1 = \theta_1; \qquad |z_2| = r_2, \qquad \arg z_2 = \theta_2.$$

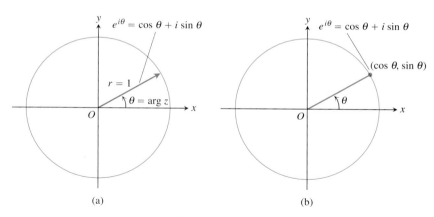

FIGURE A.5 Argand diagrams for $e^{i\theta} = \cos\theta + i\sin\theta$ (a) as a vector and (b) as a point.

Then

$$z_1 z_2 = r_1 e^{i\theta_1} \cdot r_2 e^{i\theta_2} = r_1 r_2 e^{i(\theta_1 + \theta_2)}$$

and hence

$$|z_1 z_2| = r_1 r_2 = |z_1| \cdot |z_2|$$
$$\arg(z_1 z_2) = \theta_1 + \theta_2 = \arg z_1 + \arg z_2. \tag{12}$$

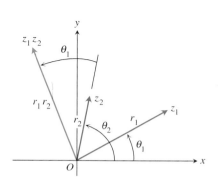

FIGURE A.6 When z_1 and z_2 are multiplied, $|z_1 z_2| = r_1 \cdot r_2$ and $\arg(z_1 z_2) = \theta_1 + \theta_2$.

Thus, the product of two complex numbers is represented by a vector whose length is the product of the lengths of the two factors and whose argument is the sum of their arguments (Figure A.6). In particular, a vector may be rotated counterclockwise through an angle θ by multiplying it by $e^{i\theta}$. Multiplication by i rotates 90°, by -1 rotates 180°, by $-i$ rotates 270°, and so on.

Example 2 Finding a Product of Complex Numbers

Let $z_1 = 1 + i$, $z_2 = \sqrt{3} - i$. We plot these complex numbers in an Argand diagram (Figure A.7) from which we read off the polar representations

$$z_1 = \sqrt{2} e^{i\pi/4}, \qquad z_2 = 2 e^{-i\pi/6}.$$

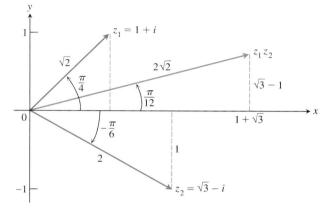

FIGURE A.7 To multiply two complex numbers, multiply their absolute values and add their arguments.

exp (A) stands for e^A.

Then

$$z_1 z_2 = 2\sqrt{2} \, \exp\left(\frac{i\pi}{4} - \frac{i\pi}{6}\right) = 2\sqrt{2} \, \exp\left(\frac{i\pi}{12}\right)$$

$$= 2\sqrt{2}\left(\cos\frac{\pi}{12} + i\sin\frac{\pi}{12}\right) \approx 2.73 + 0.73i.$$

Quotients

Suppose that $r_2 \neq 0$ in Equation (11). Then

$$\frac{z_1}{z_2} = \frac{r_1 e^{i\theta_1}}{r_2 e^{i\theta_2}} = \frac{r_1}{r_2} e^{i(\theta_1 - \theta_2)}.$$

Hence,

$$\left|\frac{z_1}{z_2}\right| = \frac{r_1}{r_2} = \frac{|z_1|}{|z_2|} \qquad \text{and} \qquad \arg\left(\frac{z_1}{z_2}\right) = \theta_1 - \theta_2 = \arg z_1 - \arg z_2.$$

That is, we divide lengths and subtract angles.

Example 3   Finding a Quotient of Complex Numbers

Let $z_1 = 1 + i$ and $z_2 = \sqrt{3} - i$, as in Example 2. Then

$$\frac{1 + i}{\sqrt{3} - i} = \frac{\sqrt{2}e^{i\pi/4}}{2e^{-i\pi/6}} = \frac{\sqrt{2}}{2} e^{5\pi i/12} \approx 0.707\left(\cos\frac{5\pi}{12} + i\sin\frac{5\pi}{12}\right)$$

$$\approx 0.183 + 0.683i.$$

Powers and De Moivre's Theorem

If n is a positive integer, we may apply the product formulas in Equations (12) to find

$$z^n = z \cdot z \cdot \cdots \cdot z. \qquad n \text{ factors}$$

With $z = re^{i\theta}$, we obtain

$$z^n = (re^{i\theta})^n = r^n e^{i(\theta + \theta + \cdots + \theta)} \qquad n \text{ summands}$$

$$= r^n e^{in\theta}. \tag{13}$$

The length $r = |z|$ is raised to the nth power and the angle $\theta = \arg z$ is multiplied by n.

If we take $r = 1$ in Equation (13), we obtain De Moivre's Theorem.

De Moivre's Theorem

$$(\cos\theta + i\sin\theta)^n = \cos n\theta + i\sin n\theta. \tag{14}$$

If we expand the left-hand side of De Moivre's equation (Equation 14) by the Binomial Theorem and reduce it to the form $a + ib$, we obtain formulas for $\cos n\theta$ and $\sin n\theta$ as polynomials of degree n in $\cos\theta$ and $\sin\theta$.

Example 4   Obtaining Formulas for cos 3θ and sin 3θ

Express $\cos 3\theta$ and $\sin 3\theta$ in terms of $\cos \theta$ and $\sin \theta$.

Solution   Taking $n = 3$ in De Moivre's Equation (Equation (14)), we have

$$(\cos \theta + i \sin \theta)^3 = \cos 3\theta + i \sin 3\theta.$$

The left-hand side of this equation is

$$\cos^3 \theta + 3i \cos^2 \theta \sin \theta - 3 \cos \theta \sin^2 \theta - i \sin^3 \theta.$$

The real part of this must equal $\cos 3\theta$ and the imaginary part must equal $\sin 3\theta$. Therefore,

$$\cos 3\theta = \cos^3 \theta - 3 \cos \theta \sin^2 \theta,$$
$$\sin 3\theta = 3 \cos^2 \theta \sin \theta - \sin^3 \theta.$$

Roots

If $z = re^{i\theta}$ is a complex number different from zero and n is a positive integer, then there are precisely n different complex numbers $w_0, w_1, \ldots, w_{n-1}$, that are nth roots of z. To see why, let $w = \rho e^{i\alpha}$ be an nth root of $z = re^{i\theta}$, so that

$$w^n = z$$

or

$$\rho^n e^{in\alpha} = re^{i\theta}.$$

Then

$$\rho = \sqrt[n]{r}$$

is the real, positive nth root of r. As regards the angle, although we cannot say that $n\alpha$ and θ must be equal, we can say that they may differ only by an integer multiple of 2π. That is,

$$n\alpha = \theta + 2k\pi, \qquad k = 0, \pm 1, \pm 2, \ldots.$$

Therefore,

$$\alpha = \frac{\theta}{n} + k\frac{2\pi}{n}.$$

Hence, all the nth roots of $z = re^{i\theta}$ are given by

$$\sqrt[n]{re^{i\theta}} = \sqrt[n]{r} \exp i\left(\frac{\theta}{n} + k\frac{2\pi}{n}\right), \qquad k = 0, \pm 1, \pm 2, \ldots. \tag{15}$$

There might appear to be infinitely many different answers corresponding to the infinitely many possible values of k, but $k = n + m$ gives the same answer as $k = m$ in Equation (15). Thus, we need only take n consecutive values for k to obtain all the different nth roots of z. For convenience, we take

$$k = 0, 1, 2, \ldots, n - 1.$$

All the nth roots of $re^{i\theta}$ lie on a circle centered at the origin and having radius equal to the real, positive nth root of r. One of them has argument $\alpha = \theta/n$. The others are uniformly spaced around the circle, each being separated from its neighbors by an angle equal to $2\pi/n$. Figure A.8 illustrates the placement of the three cube roots, w_0, w_1, w_2, of the complex number $z = re^{i\theta}$.

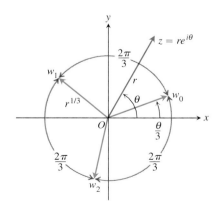

FIGURE A.8 The three cube roots of $z = re^{i\theta}$.

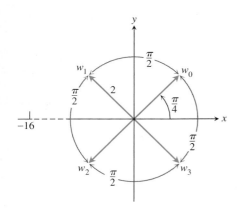

FIGURE A.9 The four fourth roots of −16.

Example 5   Finding Fourth Roots

Find the four fourth roots of −16.

Solution   As our first step, we plot the number −16 in an Argand diagram (Figure A.9) and determine its polar representation $re^{i\theta}$. Here, $z = -16$, $r = +16$, and $\theta = \pi$. One of the fourth roots of $16e^{i\pi}$ is $2e^{i\pi/4}$. We obtain others by successive additions of $2\pi/4 = \pi/2$ to the argument of this first one. Hence,

$$\sqrt[4]{16 \exp i\pi} = 2 \exp i \left(\frac{\pi}{4}, \frac{3\pi}{4}, \frac{5\pi}{4}, \frac{7\pi}{4} \right),$$

and the four roots are

$$w_0 = 2 \left[\cos \frac{\pi}{4} + i \sin \frac{\pi}{4} \right] = \sqrt{2}\,(1 + i)$$

$$w_1 = 2 \left[\cos \frac{3\pi}{4} + i \sin \frac{3\pi}{4} \right] = \sqrt{2}\,(-1 + i)$$

$$w_2 = 2 \left[\cos \frac{5\pi}{4} + i \sin \frac{5\pi}{4} \right] = \sqrt{2}\,(-1 - i)$$

$$w_3 = 2 \left[\cos \frac{7\pi}{4} + i \sin \frac{7\pi}{4} \right] = \sqrt{2}\,(1 - i).$$

The Fundamental Theorem of Algebra

One may well say that the invention of $\sqrt{-1}$ is all well and good and leads to a number system that is richer than the real number system alone, but where will this process end? Are we also going to invent still more systems so as to obtain $\sqrt[4]{-1}$, $\sqrt[6]{-1}$, and so on? By now it should be clear that is not necessary. These numbers are already expressible in terms of the complex number system $a + ib$. In fact, the Fundamental Theorem of Algebra says that with the introduction of the complex numbers, we have enough numbers to factor every polynomial into a product of linear factors and hence enough numbers to solve every possible polynomial equation.

The Fundamental Theorem of Algebra
Every polynomial equation of the form

$$a_n z^n + a_{n-1} z^{n-1} + a_{n-2} z^{n-2} + \cdots + a_1 z + a_0 = 0,$$

in which the coefficients $a_0, a_1, \ldots, a_n$ are any complex numbers, whose degree n is greater than or equal to one, and whose leading coefficient a_0 is not zero, has exactly n roots in the complex number system, provided each multiple root of multiplicity m is counted as m roots.

A proof of this theorem can be found in almost any text on the theory of functions of a complex variable.

EXERCISES A.4

Operations with Complex Numbers

1. *How computers multiply complex numbers* Find $(a, b) \cdot (c, d) = (ac - bd, ad + bc)$.

(a) $(2, 3) \cdot (4, -2)$ (b) $(2, -1) \cdot (-2, 3)$

(c) $(-1, -2) \cdot (2, 1)$

(This is how complex numbers are multiplied by computers.)

2. Solve the following equations for the real numbers, x and y.

(a) $(3 + 4i)^2 - 2(x - iy) = x + iy$

(b) $\left(\dfrac{1 + i}{1 - i}\right)^2 + \dfrac{1}{x + iy} = 1 + i$

(c) $(3 - 2i)(x + iy) = 2(x - 2iy) + 2i - 1$

Graphing and Geometry

3. How may the following complex numbers be obtained from $z = x + iy$ geometrically? Sketch.

(a) $\bar{z}$ (b) $\overline{(-z)}$

(c) $-z$ (d) $1/z$

4. Show that the distance between the two points z_1 and z_2 in an Argand diagram is $|z_1 - z_2|$.

In Exercises 5–10, graph the points $z = x + iy$ that satisfy the given conditions.

5. (a) $|z| = 2$ (b) $|z| < 2$ (c) $|z| > 2$

6. $|z - 1| = 2$ **7.** $|z + 1| = 1$

8. $|z + 1| = |z - 1|$ **9.** $|z + i| = |z - 1|$

10. $|z + 1| \geq |z|$

Express the complex numbers in Exercises 11–14 in the form $re^{i\theta}$, with $r \geq 0$ and $-\pi < \theta \leq \pi$. Draw an Argand diagram for each calculation.

11. $(1 + \sqrt{-3})^2$ **12.** $\dfrac{1 + i}{1 - i}$

13. $\dfrac{1 + i\sqrt{3}}{1 - i\sqrt{3}}$ **14.** $(2 + 3i)(1 - 2i)$

Powers and Roots

Use De Moivre's Theorem to express the trigonometric functions in Exercises 15 and 16 in terms of $\cos \theta$ and $\sin \theta$.

15. $\cos 4\theta$

16. $\sin 4\theta$

17. Find the three cube roots of 1.

18. Find the two square roots of i.

19. Find the three cube roots of $-8i$.

20. Find the six sixth roots of 64.

21. Find the four solutions of the equation $z^4 - 2z^2 + 4 = 0$.

22. Find the six solutions of the equation $z^6 + 2z^3 + 2 = 0$.

23. Find all solutions of the equation $x^4 + 4x^2 + 16 = 0$.

24. Solve the equation $x^4 + 1 = 0$.

Theory and Examples

25. *Complex numbers and vectors in the plane* Show with an Argand diagram that the law for adding complex numbers is the same as the parallelogram law for adding vectors.

26. *Complex arithmetic with conjugates* Show that the conjugate of the sum (product, or quotient) of two complex numbers, z_1 and z_2 is the same as the sum (product, or quotient) of their conjugates.

27. *Complex roots of polynomials with real coefficients come in complex-conjugate pairs*

(a) Extend the results of Exercise 26 to show that $f(\bar{z}) = \overline{f(z)}$ if

$$f(z) = a_n z^n + a_{n-1} z^{n-1} + \cdots + a_1 z + a_0$$

is a polynomial with real coefficients $a_0, \ldots, a_n$.

(b) If z is a root of the equation $f(z) = 0$, where $f(z)$ is a polynomial with real coefficients as in part (a), show that the conjugate $\bar{z}$ is also a root of the equation. (*Hint:* Let $f(z) = u + iv = 0$; then both u and v are zero. Use the fact that $f(\bar{z}) = \overline{f(z)} = u - iv$.)

28. *Absolute value of a conjugate* Show that $|\bar{z}| = |z|$.

29. *When $z = \bar{z}$* If z and $\bar{z}$ are equal, what can you say about the location of the point z in the complex plane?

30. *Real and imaginary parts* Let $Re(z)$ denote the real part of z and $Im(z)$ the imaginary part. Show that the following relations hold for any complex numbers z, z_1, and z_2.

(a) $z + \bar{z} = 2Re(z)$ (b) $z - \bar{z} = 2iIm(z)$

(c) $|Re(z)| \leq |z|$

(d) $|z_1 + z_2|^2 = |z_1|^2 + |z_2|^2 + 2Re(z_1\bar{z}_2)$

(e) $|z_1 + z_2| \leq |z_1| + |z_2|$

A.5   Simpson's One-Third Rule

Simpson's rule for approximating $\int_a^b f(x)\,dx$ is based on approximating the graph of f with parabolic arcs.

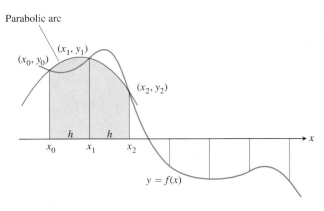

FIGURE A.10 Simpson's Rule approximates short stretches of curve with parabolic arcs.

The area of the shaded region under the parabola in Figure A.11 is

$$\text{Area} = \frac{h}{3}(y_0 + 4y_1 + y_2).$$

This formula is known as Simpson's One-Third Rule.

We can derive the formula as follows. To simplify the algebra, we use the coordinate system in Figure A.11. The area under the parabola is the same no matter where the y-axis is, as long as we preserve the vertical scale. The parabola has an equation of the form $y = Ax^2 + Bx + C$, so the area under it from $x = -h$ to $x = h$ is

$$\text{Area} = \int_{-h}^{h} (Ax^2 + Bx + C)\,dx = \left[\frac{Ax^3}{3} + \frac{Bx^2}{2} + Cx\right]_{-h}^{h}$$

$$= \frac{2Ah^3}{3} + 2Ch$$

$$= \frac{h}{3}(2Ah^2 + 6C).$$

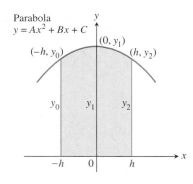

FIGURE A.11 By integrating from $-h$ to h, the shaded area is found to be

$$\frac{h}{3}(y_0 + 4y_1 + y_2).$$

Since the curve passes through $(-h, y_0)$, $(0, y_1)$, and (h, y_2), we also have

$$y_0 = Ah^2 - Bh + C, \qquad y_1 = C, \qquad y_2 = Ah^2 + Bh + C.$$

From these equations we obtain

$$C = y_1$$
$$Ah^2 - Bh = y_0 - y_1$$
$$Ah^2 + Bh = y_2 - y_1$$
$$2Ah^2 = y_0 + y_2 - 2y_1.$$

These substitutions for C and $2Ah^2$ give

$$\text{Area} = \frac{h}{3}(2Ah^2 + 6C) = \frac{h}{3}((y_0 + y_2 - 2y_1) + 6y_1) = \frac{h}{3}(y_0 + 4y_1 + y_2).$$

A.6 Cauchy's Mean Value Theorem and the Stronger Form of l'Hôpital's Rule

This appendix proves the finite-limit case of the stronger form of l'Hôpital's Rule (Section 7.6, Theorem 2) as stated here.

L'Hôpital's Rule (Stronger Form)

Suppose that

$$f(x_0) = g(x_0) = 0$$

and that the functions f and g are both differentiable on an open interval (a, b) that contains the point x_0. Suppose also that $g'(x) \neq 0$ at every point in (a, b) except possibly x_0. Then

$$\lim_{x \to x_0} \frac{f(x)}{g(x)} = \lim_{x \to x_0} \frac{f'(x)}{g'(x)}, \tag{1}$$

assuming the limit on the right exists.

The proof of the stronger form of l'Hôpital's Rule is based on Cauchy's Mean Value Theorem, a Mean Value Theorem that involves two functions instead of one. We prove Cauchy's Theorem first and then show how it leads to l'Hôpital's Rule.

Cauchy's Mean Value Theorem

Suppose functions f and g are continuous on $[a, b]$ and differentiable throughout (a, b) and also suppose $g'(x) \neq 0$ throughout (a, b). Then there exists a number c in (a, b) at which

$$\frac{f'(c)}{g'(c)} = \frac{f(b) - f(a)}{g(b) - g(a)}. \tag{2}$$

The ordinary Mean Value Theorem (Section 3.2, Theorem 4) is the case $g(x) = x$.

Proof of Cauchy's Mean Value Theorem We apply the Mean Value Theorem of Section 3.2 twice. First we use it to show that $g(a) \neq g(b)$. For if $g(b)$ did equal $g(a)$, then the Mean Value Theorem would give

$$g'(c) = \frac{g(b) - g(a)}{b - a} = 0$$

for some c between a and b, which cannot happen because $g'(x) \neq 0$ in (a, b).

We next apply the Mean Value Theorem to the function

$$F(x) = f(x) - f(a) - \frac{f(b) - f(a)}{g(b) - g(a)}[g(x) - g(a)].$$

This function is continuous and differentiable where f and g are, and $F(b) = F(a) = 0$. Therefore, there is a number c between a and b for which $F'(c) = 0$. When expressed in terms of f and g, this equation becomes

$$F'(c) = f'(c) - \frac{f(b) - f(a)}{g(b) - g(a)}[g'(c)] = 0$$

or

$$\frac{f'(c)}{g'(c)} = \frac{f(b) - f(a)}{g(b) - g(a)},$$

which is Equation (2).   —

Proof of the Stronger Form of l'Hôpital's Rule   We first establish Equation (1) for the case $x \to x_0^+$. The method needs almost no change to apply to $x \to x_0^-$, and the combination of these two cases establishes the result.

Suppose that x lies to the right of x_0. Then $g'(x) \neq 0$, and we can apply Cauchy's Mean Value Theorem to the closed interval from x_0 to x. This step produces a number c between x_0 and x such that

$$\frac{f'(c)}{g'(c)} = \frac{f(x) - f(x_0)}{g(x) - g(x_0)}.$$

But $f(x_0) = g(x_0) = 0$, so

$$\frac{f'(c)}{g'(c)} = \frac{f(x)}{g(x)}.$$

As x approaches x_0, c approaches x_0 because it lies between x and x_0. Therefore,

$$\lim_{x \to x_0^+} \frac{f(x)}{g(x)} = \lim_{c \to x_0^+} \frac{f'(c)}{g'(c)} = \lim_{x \to x_0^+} \frac{f'(x)}{g'(x)},$$

which establishes l'Hôpital's Rule for the case where x approaches x_0 from above. The case where x approaches x_0 from below is proved by applying Cauchy's Mean Value Theorem to the closed interval $[x, x_0]$, $x < x_0$.   —

A.7   Limits That Arise Frequently

This appendix verifies limits (4) through (6) in Section 8.1, Table 8.1.

Limit 4: If $|x| < 1$, $\lim_{n \to \infty} x^n = 0$ We need to show that to each $\epsilon > 0$ there corresponds an integer N so large that $|x^n| < \epsilon$ for all n greater than N. Since $\epsilon^{1/n} \to 1$, while $|x| < 1$, there exists an integer N for which $\epsilon^{1/N} > |x|$. In other words,

$$|x^N| = |x|^N < \epsilon. \tag{1}$$

This integer is the one we seek because, if $|x| < 1$, then

$$|x^n| < |x^N| \qquad \text{for all } n > N. \tag{2}$$

Combining Equations (1) and (2) produces $|x^n| < \epsilon$ for all $n > N$, concluding the proof.

Limit 5: For any number x, $\lim\limits_{n \to \infty} \left(1 + \dfrac{x}{n}\right)^n = e^x$ Let

$$a_n = \left(1 + \frac{x}{n}\right)^n.$$

Then

$$\ln a_n = \ln \left(1 + \frac{x}{n}\right)^n = n \ln \left(1 + \frac{x}{n}\right) \to x,$$

as we can see by the following application of l'Hôpital's Rule, in which we differentiate with respect to n:

$$\lim_{n \to \infty} n \ln \left(1 + \frac{x}{n}\right) = \lim_{n \to \infty} \frac{\ln (1 + x/n)}{1/n}$$

$$= \lim_{n \to \infty} \frac{\left(\dfrac{1}{1 + x/n}\right) \cdot \left(-\dfrac{x}{n^2}\right)}{-1/n^2} = \lim_{n \to \infty} \frac{x}{1 + x/n} = x.$$

Apply Theorem 3, Section 8.1, with $f(x) = e^x$ to conclude that

$$\left(1 + \frac{x}{n}\right)^n = a_n = e^{\ln a_n} \to e^x.$$

Limit 6: For any number x, $\lim\limits_{n \to \infty} \dfrac{x^n}{n!} = 0$ Since

$$-\frac{|x|^n}{n!} \le \frac{x^n}{n!} \le \frac{|x|^n}{n!},$$

all we need to show is that $|x|^n/n! \to 0$. We can then apply the Sandwich Theorem for Sequences (Section 8.1, Theorem 2) to conclude that $x^n/n! \to 0$.

The first step in showing that $|x|^n/n! \to 0$ is to choose an integer $M > |x|$, so that $(|x|/M) < 1$. By Limit 4, just proved, we then have $(|x|/M)^n \to 0$. We then restrict our attention to values of $n > M$. For these values of n, we can write

$$\frac{|x|^n}{n!} = \frac{|x|^n}{1 \cdot 2 \cdot \cdots \cdot M \cdot \underbrace{(M + 1)(M + 2) \cdot \cdots \cdot n}_{(n - M) \text{ factors}}}$$

$$\le \frac{|x|^n}{M! M^{n-M}} = \frac{|x|^n M^M}{M! M^n} = \frac{M^M}{M!} \left(\frac{|x|}{M}\right)^n.$$

Thus,

$$0 \le \frac{|x|^n}{n!} \le \frac{M^M}{M!} \left(\frac{|x|}{M}\right)^n.$$

Now, the constant $M^M/M!$ does not change as n increases. Thus, the Sandwich Theorem tells us that $|x|^n/n! \to 0$ because $(|x|/M)^n \to 0$.

A.8   Proof of Taylor's Theorem

This appendix proves Taylor's Theorem (Section 8.7, Theorem 16) in the following form.

> **Theorem 16**   Taylor's Theorem
>
> If f and its first n derivatives $f', f'', \dots, f^{(n)}$ are continuous on $[a, b]$ or on $[b, a]$, and $f^{(n)}$ is differentiable on (a, b) or on (b, a), then there exists a number c between a and b such that
>
> $$f(b) = f(a) + f'(a)(b - a) + \frac{f''(a)}{2!}(b - a)^2 + \cdots$$
>
> where
>
> $$+ \frac{f^{(n)}(a)}{n!}(b - a)^n + \frac{f^{(n+1)}(c)}{(n + 1)!}(b - a)^{n+1}.$$

Proof   We prove Taylor's Theorem assuming $a < b$. The proof for $a > b$ is nearly the same.

Then for any x in the interval $[a, b]$, the Taylor polynomial

$$P_n(x) = f(a) + f'(a)(x - a) + \frac{f''(a)}{2!}(x - a)^2 + \cdots + \frac{f^{(n)}(a)}{n!}(x - a)^n$$

and its first n derivatives match the function f and its first n derivatives at $x = a$. We do not disturb that matching if we add another term of the form $K(x - a)^{n+1}$, where K is any constant, because such a term and its first n derivatives are all equal to zero at $x = a$. The new function

$$\phi_n(x) = P_n(x) + K(x - a)^{n+1}$$

and its first n derivatives still agree with f and its first n derivatives at $x = a$.

We now choose the particular value of K that makes the curve $y = \phi_n(x)$ agree with the original curve $y = f(x)$ at $x = b$. In symbols,

$$f(b) = P_n(b) + K(b - a)^{n+1} \qquad \text{or} \qquad K = \frac{f(b) - P_n(b)}{(b - a)^{n+1}}. \tag{1}$$

With K defined by the last equation, the function

$$F(x) = f(x) - \phi_n(x)$$

measures the difference between the original function f and the approximating function ϕ_n for each x in $[a, b]$.

We now use Rolle's Theorem (Section 3.2). First, because $F(a) = F(b) = 0$ and both F and F' are continuous on $[a, b]$, we know that

$$F'(c_1) = 0 \qquad \text{for some } c_1 \text{ in } (a, b).$$

Next, because $F'(a) = F'(c_1) = 0$ and both F' and F'' are continuous on $[a, c_1]$, we know that

$$F''(c_2) = 0 \qquad \text{for some } c_2 \text{ in } (a, c_1).$$

Rolle's Theorem, applied successively to F'', F''', ..., $F^{(n-1)}$, implies the existence of

$$c_3 \text{ in } (a, c_2) \qquad \text{such that } F'''(c_3) = 0$$
$$c_4 \text{ in } (a, c_3) \qquad \text{such that } F^{(4)}(c_4) = 0$$
$$\vdots$$
$$c_n \text{ in } (a, c_{n-1}) \qquad \text{such that } F^{(n)}(c_n) = 0.$$

Finally, because $F^{(n)}$ is continuous on $[a, c_n]$ and differentiable on (a, c_n), and $F^{(n)}(a) = F^{(n)}(c_n) = 0$, Rolle's Theorem implies that there is a number c_{n+1} in (a, c_n) such that

$$F^{(n+1)}(c_{n+1}) = 0. \tag{2}$$

If we differentiate $F(x) = f(x) - P_n(x) - K(x - a)^{n+1}$ a total of $n + 1$ times, we get

$$F^{(n+1)}(x) = f^{(n+1)}(x) - 0 - (n + 1)!K. \tag{3}$$

Equations (2) and (3) together give

$$K = \frac{f^{(n+1)}(c)}{(n + 1)!} \qquad \text{for some number } c = c_{n+1} \text{ in } (a, b). \tag{4}$$

Equations (1) and (4) give

$$f(b) = P_n(b) + \frac{f^{(n+1)}(c)}{(n + 1)!} (b - a)^{n+1}. \tag{5}$$

This concludes the proof.

A.9 The Distributive Law for Vector Cross Products

In this appendix, we prove the distributive law

$$\mathbf{u} \times (\mathbf{v} + \mathbf{w}) = \mathbf{u} \times \mathbf{v} + \mathbf{u} \times \mathbf{w}$$

from Property 2 in Section 10.2.

Proof To derive the distributive law, we construct $\mathbf{u} \times \mathbf{v}$ a new way. We draw $\mathbf{u}$ and $\mathbf{v}$ from the common point O and construct a plane M perpendicular to $\mathbf{u}$ at O (Figure A.12). We then project $\mathbf{v}$ orthogonally onto M, yielding a vector $\mathbf{v}'$ with length $|\mathbf{v}| \sin \theta$. We rotate $\mathbf{v}'$ 90° about $\mathbf{u}$ in the positive sense to produce a vector $\mathbf{v}''$. Finally, we multiply $\mathbf{v}''$ by the length of $\mathbf{u}$. The resulting vector $|\mathbf{u}|\mathbf{v}''$

is equal to $\mathbf{u} \times \mathbf{v}$ since $\mathbf{v}''$ has the same direction as $\mathbf{u} \times \mathbf{v}$ by its construction (Figure A.12) and

$$|\mathbf{u}||\mathbf{v}''| = |\mathbf{u}||\mathbf{v}'| = |\mathbf{u}||\mathbf{v}| \sin \theta = |\mathbf{u} \times \mathbf{v}|.$$

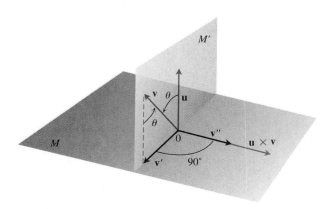

FIGURE A.12 As explained in the text, $\mathbf{u} \times \mathbf{v} = |\mathbf{u}|\mathbf{v}''$.

Now each of these three operations, namely,

1. projection onto M

2. rotation about $\mathbf{u}$ through 90°

3. multiplication by the scalar $|\mathbf{u}|$

when applied to a triangle whose plane is not parallel to $\mathbf{u}$, will produce another triangle. If we start with the triangle whose sides are $\mathbf{v}$, $\mathbf{w}$, and $\mathbf{v} + \mathbf{w}$ (Figure A.13) and apply these three steps, we successively obtain the following:

1. A triangle whose sides are $\mathbf{v}'$, $\mathbf{w}'$, and $(\mathbf{v} + \mathbf{w})'$ satisfying the vector equation

$$\mathbf{v}' + \mathbf{w}' = (\mathbf{v} + \mathbf{w})'$$

2. A triangle whose sides are $\mathbf{v}''$, $\mathbf{w}''$, and $(\mathbf{v} + \mathbf{w})''$ satisfying the vector equation

$$\mathbf{v}'' + \mathbf{w}'' = (\mathbf{v} + \mathbf{w})''$$

(the double prime on each vector has the same meaning as in Figure A.12)

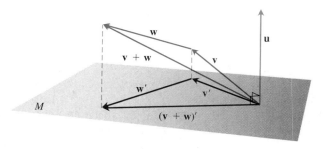

FIGURE A.13 The vectors, $\mathbf{v}$, $\mathbf{w}$, $\mathbf{v} + \mathbf{w}$, and their projections onto a plane perpendicular to $\mathbf{u}$.

3. A triangle whose sides are $|\mathbf{u}|\mathbf{v}''$, $|\mathbf{u}|\mathbf{w}''$, and $|\mathbf{u}|(\mathbf{v}+\mathbf{w})''$ satisfying the vector equation

$$|\mathbf{u}|\mathbf{v}'' + |\mathbf{u}|\mathbf{w}'' = |\mathbf{u}|(\mathbf{v}+\mathbf{w})''.$$

Substituting $|\mathbf{u}|\mathbf{v}'' = \mathbf{u}\times\mathbf{v}$, $|\mathbf{u}|\mathbf{w}'' = \mathbf{u}\times\mathbf{w}$, and $|\mathbf{u}|(\mathbf{v}+\mathbf{w})'' = \mathbf{u}\times(\mathbf{v}+\mathbf{w})$ from our discussion above into this last equation gives

$$\mathbf{u}\times\mathbf{v} + \mathbf{u}\times\mathbf{w} = \mathbf{u}\times(\mathbf{v}+\mathbf{w}),$$

which is the law we wanted to establish.

A.10   Determinants and Cramer's Rule

A rectangular array of numbers such as

$$A = \begin{bmatrix} 2 & 1 & 3 \\ 1 & 0 & -2 \end{bmatrix}$$

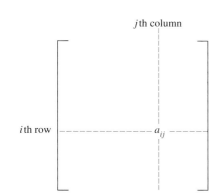

*j*th column

*i*th row .... a_{ij} ....

The vertical bars in the notation $|a_{ij}|$ do not mean absolute value.

---

is called a **matrix.** We call A a 2 by 3 matrix because it has two rows and three columns. An m by n matrix has m rows and n columns, and the **entry** or **element** (number) in the ith row and jth column is denoted by a_{ij}. The matrix

$$A = \begin{bmatrix} 2 & 1 & 3 \\ 1 & 0 & -2 \end{bmatrix}$$

has

$$a_{11} = 2, \qquad a_{12} = 1, \qquad a_{13} = 3,$$
$$a_{21} = 1, \qquad a_{22} = 0, \qquad a_{23} = -2.$$

A matrix with the same number of rows as columns is a **square matrix.** It is a **square matrix of order** n if the number of rows and columns is n.

With each square matrix A we associate a number $\det A$ or $|a_{ij}|$, called the **determinant** of A, calculated from the entries of A in the following way. For $n = 1$ and $n = 2$, we define

$$\det [a] = a \tag{1}$$

$$\det \begin{bmatrix} a_{11} & a_{12} \\ a_{21} & a_{22} \end{bmatrix} = a_{11}a_{22} - a_{21}a_{12}. \tag{2}$$

For a matrix of order 3, we write

$$\det A = \det \begin{bmatrix} a_{11} & a_{12} & a_{13} \\ a_{21} & a_{22} & a_{23} \\ a_{31} & a_{32} & a_{33} \end{bmatrix} = \begin{matrix} \text{sum of all signed products} \\ \text{of the form } \pm\, a_{1i}a_{2j}a_{3k}, \end{matrix} \tag{3}$$

where i, j, k is a permutation of 1, 2, 3 in some order. There are $3! = 6$ such permutations, so there are six terms in the sum. The sign is positive when the index of the permutation is even and negative when the index is odd.

Definition   Index of a Permutation

Given any permutation of the numbers $1, 2, 3, \ldots, n,$ denote the permutation by $i_1, i_2, i_3, \ldots, i_n.$ In this arrangement, some of the numbers following i_1 may be less than $i_1,$ and the number of these is called the **number of inversions** in the arrangement pertaining to $i_1.$ Likewise, there are a number of inversions pertaining to each of the other i's; it is the number of indices that come after that particular i in the arrangement and are less than it. The **index** of the permutation is the sum of all the numbers of inversions pertaining to the separate indices.

Example 1   Finding the Index of a Permutation

For $n = 5,$ the permutation

$$5 \quad 3 \quad 1 \quad 2 \quad 4$$

has four inversions pertaining to the first element, 5, two inversions pertaining to the second element, 3, and no further inversions, so the index is $4 + 2 = 6.$

The following table shows the permutations of 1, 2, 3, the index of each permutation, and the signed product in the determinant of Equation (3).

Permutation	Index	Signed product
1 2 3	0	$+a_{11}a_{22}a_{33}$
1 3 2	1	$-a_{11}a_{23}a_{32}$
2 1 3	1	$-a_{12}a_{21}a_{33}$
2 3 1	2	$+a_{12}a_{23}a_{31}$
3 1 2	2	$+a_{13}a_{21}a_{32}$
3 2 1	3	$-a_{13}a_{22}a_{31}$

The sum of the six signed products is

$$a_{11}(a_{22}a_{33} - a_{23}a_{32}) - a_{12}(a_{21}a_{33} - a_{23}a_{31}) + a_{13}(a_{21}a_{32} - a_{22}a_{31})$$

$$= a_{11}\begin{vmatrix} a_{22} & a_{23} \\ a_{32} & a_{33} \end{vmatrix} - a_{12}\begin{vmatrix} a_{21} & a_{23} \\ a_{31} & a_{33} \end{vmatrix} + a_{13}\begin{vmatrix} a_{21} & a_{22} \\ a_{31} & a_{32} \end{vmatrix} = \begin{vmatrix} a_{11} & a_{12} & a_{13} \\ a_{21} & a_{22} & a_{23} \\ a_{31} & a_{32} & a_{33} \end{vmatrix}.$$

The formula

$$\begin{vmatrix} a_{11} & a_{12} & a_{13} \\ a_{21} & a_{22} & a_{23} \\ a_{31} & a_{32} & a_{33} \end{vmatrix} = a_{11}\begin{vmatrix} a_{22} & a_{23} \\ a_{32} & a_{33} \end{vmatrix} - a_{12}\begin{vmatrix} a_{21} & a_{23} \\ a_{31} & a_{33} \end{vmatrix} + a_{13}\begin{vmatrix} a_{21} & a_{22} \\ a_{31} & a_{32} \end{vmatrix} \qquad (4)$$

reduces the calculation of a 3 by 3 determinant to the calculation of three 2 by 2 determinants.

Many people prefer to remember the following scheme for calculating the six signed products in the determinant of a 3 by 3 matrix:

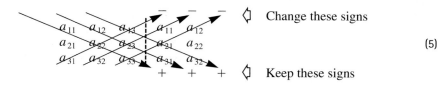

(5)

Minors and Cofactors

The second-order determinants on the right-hand side of Equation (4) are called the **minors** (short for "minor determinants") of the entries they multiply. Thus,

$$\begin{vmatrix} a_{22} & a_{23} \\ a_{32} & a_{33} \end{vmatrix} \text{ is the minor of } a_{11}, \qquad \begin{vmatrix} a_{21} & a_{23} \\ a_{31} & a_{33} \end{vmatrix} \text{ is the minor of } a_{12},$$

and so on. The minor of the element a_{ij} in a matrix A is the determinant of the matrix that remains after we delete the row and column containing a_{ij}.

$$\begin{vmatrix} a_{11} & a_{12} & a_{13} \\ \overline{a_{21}} & \overline{a_{22}} & \overline{a_{23}} \\ a_{31} & a_{32} & a_{33} \end{vmatrix}; \qquad \text{the minor of } a_{22} \text{ is } \begin{vmatrix} a_{11} & a_{13} \\ a_{31} & a_{33} \end{vmatrix}$$

$$\begin{vmatrix} a_{11} & a_{12} & a_{13} \\ \overline{a_{21}} & \overline{a_{22}} & \overline{a_{23}} \\ a_{31} & a_{32} & a_{33} \end{vmatrix}; \qquad \text{the minor of } a_{23} \text{ is } \begin{vmatrix} a_{11} & a_{12} \\ a_{31} & a_{32} \end{vmatrix}$$

The **cofactor** A_{ij} of a_{ij} is $(-1)^{i+j}$ times the minor of a_{ij}. Thus,

$$A_{22} = (-1)^{2+2} \begin{vmatrix} a_{11} & a_{13} \\ a_{31} & a_{33} \end{vmatrix} = \begin{vmatrix} a_{11} & a_{13} \\ a_{31} & a_{33} \end{vmatrix}$$

$$A_{23} = (-1)^{2+3} \begin{vmatrix} a_{11} & a_{12} \\ a_{31} & a_{32} \end{vmatrix} = -\begin{vmatrix} a_{11} & a_{12} \\ a_{31} & a_{32} \end{vmatrix}.$$

The factor $(-1)^{i+j}$ changes the sign of the minor when $i + j$ is odd. There is a checkerboard pattern for remembering these changes:

$$\begin{matrix} + & - & + \\ - & + & - \\ + & - & + \end{matrix}.$$

In the upper left corner, $i = 1$, $j = 1$ and $(-1)^{1+1} = +1$. In going from any cell to an adjacent cell in the same row or column, we change i by 1 or j by 1, but not both, so we change the exponent from even to odd or from odd to even, which changes the sign from + to − or from − to +.

When we rewrite Equation (4) in terms of cofactors we get

$$\det A = a_{11}A_{11} + a_{12}A_{12} + a_{13}A_{13}. \tag{6}$$

Example 2   Finding a Determinant Two Ways

Find the determinant of

$$A = \begin{bmatrix} 2 & 1 & 3 \\ 3 & -1 & -2 \\ 2 & 3 & 1 \end{bmatrix}.$$

Solution 1

Using Equation (6)

The cofactors are

$$A_{11} = (-1)^{1+1} \begin{vmatrix} -1 & -2 \\ 3 & 1 \end{vmatrix}, \qquad A_{12} = (-1)^{1+2} \begin{vmatrix} 3 & -2 \\ 2 & 1 \end{vmatrix},$$

$$A_{13} = (-1)^{1+3} \begin{vmatrix} 3 & -1 \\ 2 & 3 \end{vmatrix}.$$

To find $\det A$, we multiply each element of the first row of A by its cofactor and add:

$$\det A = 2 \begin{vmatrix} -1 & -2 \\ 3 & 1 \end{vmatrix} + (-1) \begin{vmatrix} 3 & -2 \\ 2 & 1 \end{vmatrix} + 3 \begin{vmatrix} 3 & -1 \\ 2 & 3 \end{vmatrix}$$

$$= 2(-1 + 6) - 1(3 + 4) + 3(9 + 2) = 10 - 7 + 33 = 36.$$

Solution 2

Using the Scheme (5)

We find

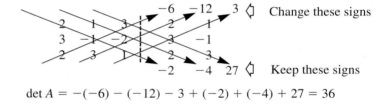

Change these signs

Keep these signs

$$\det A = -(-6) - (-12) - 3 + (-2) + (-4) + 27 = 36$$

Expanding by Columns or by Other Rows

The determinant of a square matrix can be calculated from the cofactors of any row or any column.

If we were to expand the determinant in Example 2 by cofactors according to elements of its third column, say, we would get

$$+3 \begin{vmatrix} 3 & -1 \\ 2 & 3 \end{vmatrix} - (-2) \begin{vmatrix} 2 & 1 \\ 2 & 3 \end{vmatrix} + 1 \begin{vmatrix} 2 & 1 \\ 3 & -1 \end{vmatrix}$$

$$= 3(9 + 2) + 2(6 - 2) + 1(-2 - 3) = 33 + 8 - 5 = 36.$$

Useful Facts About Determinants

Fact 1: If two rows (or columns) are identical, the determinant is zero.

Fact 2: Interchanging two rows (or columns) changes the sign of the determinant.

Fact 3: The determinant is the sum of the products of the elements of the ith row (or column) by their cofactors, for any i.

Fact 4: The determinant of the transpose of a matrix is the same as the determinant of the original matrix. (The **transpose** of a matrix is obtained by writing the rows as columns.)

Fact 5: Multiplying each element of some row (or column) by a constant c multiplies the determinant by c.

Fact 6: If all elements above the main diagonal (or all below it) are zero, the determinant is the product of the elements on the main diagonal. (The **main diagonal** is the diagonal from upper left to lower right.)

Example 3   Illustrating Fact 6

$$\begin{vmatrix} 3 & 4 & 7 \\ 0 & -2 & 5 \\ 0 & 0 & 5 \end{vmatrix} = (3)(-2)(5) = -30$$

Fact 7: If the elements of any row are multiplied by the cofactors of the corresponding elements of a different row and these products are summed, then the sum is zero.

Example 4   Illustrating Fact 7

If A_{11}, A_{12}, A_{13} are the cofactors of the elements of the first row of $A = (a_{ij})$, then the sums

$$a_{21}A_{11} + a_{22}A_{12} + a_{23}A_{13}$$

(elements of second row times cofactors of elements of first row) and

$$a_{31}A_{11} + a_{32}A_{12} + a_{33}A_{13}$$

are both zero.

Fact 8: If the elements of any column are multiplied by the cofactors of the corresponding elements of a different column and these products are summed, then the sum is zero.

Fact 9: If each element of a row is multiplied by a constant c and the results added to a different row, then the determinant is not changed. A similar result holds for columns.

Example 5   Adding a Multiple of One Row to Another Row

If we start with

$$A = \begin{bmatrix} 2 & 1 & 3 \\ 3 & -1 & -2 \\ 2 & 3 & 1 \end{bmatrix}$$

and add -2 times row 1 to row 2 (subtract 2 times row 1 from row 2), we get

$$B = \begin{bmatrix} 2 & 1 & 3 \\ -1 & -3 & -8 \\ 2 & 3 & 1 \end{bmatrix}.$$

Since det $A = 36$ (Example 2), we should find that det $B = 36$ as well. Indeed we do, as the following calculation shows:

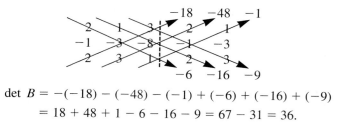

$$\det\ B = -(-18) - (-48) - (-1) + (-6) + (-16) + (-9)$$
$$= 18 + 48 + 1 - 6 - 16 - 9 = 67 - 31 = 36.$$

Example 6   Evaluating a Fourth-Order Determinant by Applying Fact 9

Evaluate the fourth order determinant

$$D = \begin{vmatrix} 1 & -2 & 3 & 1 \\ 2 & 1 & 0 & 2 \\ -1 & 2 & 1 & -2 \\ 0 & 1 & 2 & 1 \end{vmatrix}.$$

Solution   We subtract 2 times row 1 from row 2 and add row 1 to row 3 to get

$$D = \begin{vmatrix} 1 & -2 & 3 & 1 \\ 0 & 5 & -6 & 0 \\ 0 & 0 & 4 & -1 \\ 0 & 1 & 2 & 1 \end{vmatrix}.$$

We then multiply the elements of the first column by their cofactors to get

$$D = \begin{vmatrix} 5 & -6 & 0 \\ 0 & 4 & -1 \\ 1 & 2 & 1 \end{vmatrix} = 5(4 + 2) - (-6)(0 + 1) + 0 = 36.$$

Cramer's Rule

If the determinant $D = \det A = \begin{vmatrix} a_{11} & a_{12} \\ a_{21} & a_{22} \end{vmatrix} = 0,$ then the system

$$a_{11}x + a_{12}y = b_1$$
$$a_{21}x + a_{22}y = b_2 \tag{7}$$

has either infinitely many solutions or no solution at all. The system

$$x + y = 0$$
$$2x + 2y = 0$$

whose determinant is

$$D = \begin{vmatrix} 1 & 1 \\ 2 & 2 \end{vmatrix} = 2 - 2 = 0$$

has infinitely many solutions. We can find an x to match any given y. The system

$$x + y = 0$$
$$2x + 2y = 2$$

has no solution. If $x + y = 0,$ then $2x + 2y = 2(x + y)$ cannot be 2.

If $D \neq 0$, then the system (7) has a unique solution, and Cramer's Rule states that it may be found from the formulas

$$x = \frac{\begin{vmatrix} b_1 & a_{12} \\ b_2 & a_{22} \end{vmatrix}}{D}, \qquad y = \frac{\begin{vmatrix} a_{11} & b_1 \\ a_{21} & b_2 \end{vmatrix}}{D}. \tag{8}$$

The numerator in the formula for x comes from replacing the first column in A (the x-column) by the column of constants b_1 and b_2 (the b-column). Replacing the y-column by the b-column gives the numerator of the y-solution.

Example 7 Using Cramer's Rule

Solve the system

$$3x - y = 9$$
$$x + 2y = -4.$$

Solution   We use Equations (8). The determinant of the coefficient matrix is

$$D = \begin{vmatrix} 3 & -1 \\ 1 & 2 \end{vmatrix} = 6 + 1 = 7.$$

Hence,

$$x = \frac{\begin{vmatrix} 9 & -1 \\ -4 & 2 \end{vmatrix}}{D} = \frac{18 - 4}{7} = \frac{14}{7} = 2$$

$$y = \frac{\begin{vmatrix} 3 & 9 \\ 1 & -4 \end{vmatrix}}{D} = \frac{-12 - 9}{7} = \frac{-21}{7} = -3.$$

Systems of three equations in three unknowns work the same way. If

$$D = \det A = \begin{vmatrix} a_{11} & a_{12} & a_{13} \\ a_{21} & a_{22} & a_{23} \\ a_{31} & a_{32} & a_{33} \end{vmatrix} = 0,$$

then the system

$$a_{11}x + a_{12}y + a_{13}z = b_1$$
$$a_{21}x + a_{22}y + a_{23}z = b_2$$
$$a_{31}x + a_{32}y + a_{33}z = b_3$$

has either infinitely many solutions or no solution at all. If $D \neq 0$, then the system has a unique solution, given by Cramer's Rule:

$$x = \frac{1}{D} \begin{vmatrix} b_1 & a_{12} & a_{13} \\ b_2 & a_{22} & a_{23} \\ b_3 & a_{32} & a_{33} \end{vmatrix}, \qquad y = \frac{1}{D} \begin{vmatrix} a_{11} & b_1 & a_{13} \\ a_{21} & b_2 & a_{23} \\ a_{31} & b_3 & a_{33} \end{vmatrix}, \qquad z = \frac{1}{D} \begin{vmatrix} a_{11} & a_{12} & b_1 \\ a_{21} & a_{22} & b_2 \\ a_{31} & a_{32} & b_3 \end{vmatrix}.$$

The pattern continues in higher dimensions.

EXERCISES A.10

Evaluting Determinants

Evaluate the following determinants.

1. $\begin{vmatrix} 2 & 3 & 1 \\ 4 & 5 & 2 \\ 1 & 2 & 3 \end{vmatrix}$

2. $\begin{vmatrix} 2 & -1 & -2 \\ -1 & 2 & 1 \\ 3 & 0 & -3 \end{vmatrix}$

3. $\begin{vmatrix} 1 & 2 & 3 & 4 \\ 0 & 1 & 2 & 3 \\ 0 & 0 & 2 & 1 \\ 0 & 0 & 3 & 2 \end{vmatrix}$

4. $\begin{vmatrix} 1 & -1 & 2 & 3 \\ 2 & 1 & 2 & 6 \\ 1 & 0 & 2 & 3 \\ -2 & 2 & 0 & -5 \end{vmatrix}$

Evaluate the following determinants by expanding according to the cofactors of (a) the third row and (b) the second column.

5. $\begin{vmatrix} 2 & -1 & 2 \\ 1 & 0 & 3 \\ 0 & 2 & 1 \end{vmatrix}$

6. $\begin{vmatrix} 1 & 0 & -1 \\ 0 & 2 & -2 \\ 2 & 0 & 1 \end{vmatrix}$

7. $\begin{vmatrix} 1 & 1 & 0 & 0 \\ 0 & 0 & -2 & 1 \\ 0 & -1 & 0 & 7 \\ 3 & 0 & 2 & 1 \end{vmatrix}$

8. $\begin{vmatrix} 0 & 1 & 0 & 0 \\ 0 & 1 & 1 & 0 \\ 1 & 1 & 1 & 1 \\ 1 & 1 & 0 & 0 \end{vmatrix}$

Systems of Equations

Solve the following systems of equations by Cramer's Rule.

9. $\begin{aligned} x + 8y &= 4 \\ 3x - y &= -13 \end{aligned}$

10. $\begin{aligned} 2x + 3y &= 5 \\ 3x - y &= 2 \end{aligned}$

11. $\begin{aligned} 4x - 3y &= 6 \\ 3x - 2y &= 5 \end{aligned}$

12. $\begin{aligned} x + y + z &= 2 \\ 2x - y + z &= 0 \\ x + 2y - z &= 4 \end{aligned}$

13. $\begin{aligned} 2x + y - z &= 2 \\ x - y + z &= 7 \\ 2x + 2y + z &= 4 \end{aligned}$

14. $\begin{aligned} 2x - 4y &= 6 \\ x + y + z &= 1 \\ 5y + 7z &= 10 \end{aligned}$

15. $\begin{aligned} x - z &= 3 \\ 2y - 2z &= 2 \\ 2x + z &= 3 \end{aligned}$

16. $\begin{aligned} x_1 + x_2 - x_3 + x_4 &= 2 \\ x_1 - x_2 + x_3 + x_4 &= -1 \\ x_1 + x_2 + x_3 - x_4 &= 2 \\ x_1 + x_3 + x_4 &= -1 \end{aligned}$

Theory and Examples

17. *Infinitely many or no solutions* Find values of h and k for which the system

$$2x + hy = 8$$
$$x + 3y = k$$

has (a) infinitely many solutions and (b) no solution at all.

18. *A zero determinant* For what value of x will

$$\begin{vmatrix} x & x & 1 \\ 2 & 0 & 5 \\ 6 & 7 & 1 \end{vmatrix} = 0?$$

19. *A zero determinant* Suppose u, v, and w are twice-differentiable functions of x that satisfy the relation $au + bv + cw = 0$, where a, b, and c are constants, not all zero. Show that

$$\begin{vmatrix} u & v & w \\ u' & v' & w' \\ u'' & v'' & w'' \end{vmatrix} = 0.$$

20. *Partial fractions* Expanding the quotient

$$\frac{ax + b}{(x - r_1)(x - r_2)}$$

by partial fractions calls for finding the values of C and D that make the equation

$$\frac{ax + b}{(x - r_1)(x - r_2)} = \frac{C}{x - r_1} + \frac{D}{x - r_2}$$

hold for all x.

(a) Find a system of linear equations that determines C and D.

(b) *Writing to Learn* Under what circumstances does the system of equations in part (a) have a unique solution? That is, when is the determinant of the coefficient matrix of the system different from zero?

A.11 The Mixed Derivative Theorem and the Increment Theorem

This appendix derives the Mixed Derivative Theorem (Theorem 2, Section 11.3) and the Increment Theorem for Functions of Two Variables (Theorem 3, Section 11.3). Euler first published his theorem in 1734, in a series of papers he wrote on hydrodynamics.

Theorem 2   The Mixed Derivative Theorem

If $f(x, y)$ and its partial derivatives f_x, f_y, f_{xy}, and f_{yx} are defined throughout an open region containing a point (a, b) and are all continuous at (a, b), then $f_{xy}(a, b) = f_{yx}(a, b)$.

Proof   The equality of $f_{xy}(a, b)$ and $f_{yx}(a, b)$ can be established by four applications of the Mean Value Theorem (Theorem 4, Section 3.2). By hypothesis, the point (a, b) lies in the interior of a rectangle R in the xy-plane on which f, f_x, f_y, f_{xy}, and f_{yx} are all defined. We let h and k be the numbers such that the point $(a + h, b + k)$ also lies in R, and we consider the difference

$$\Delta = F(a + h) - F(a), \tag{1}$$

where

$$F(x) = f(x, b + k) - f(x, b). \tag{2}$$

We apply the Mean Value Theorem to F, which is continuous because it is differentiable, and Equation (1) becomes

$$\Delta = hF'(c_1), \tag{3}$$

where c_1 lies between a and $a + h$. From Equation (2),

$$F'(x) = f_x(x, b + k) - f_x(x, b),$$

so Equation (3) becomes

$$\Delta = h[f_x(c_1, b + k) - f_x(c_1, b)]. \tag{4}$$

Now we apply the Mean Value Theorem to the function $g(y) = f_x(c_1, y)$ and have

$$g(b + k) - g(b) = kg'(d_1),$$

or

$$f_x(c_1, b + k) - f_x(c_1, b) = kf_{xy}(c_1, d_1)$$

for some d_1 between b and $b + k$. By substituting this into Equation (4), we get

$$\Delta = hkf_{xy}(c_1, d_1) \tag{5}$$

for some point (c_1, d_1) in the rectangle R' whose vertices are the four points (a, b), $(a + h, b)$, $(a + h, b + k)$, and $(a, b + k)$. (See Figure A.14.)

By substituting from Equation (2) into Equation (1), we may also write

$$
\begin{aligned}
\Delta &= f(a + h, b + k) - f(a + h, b) - f(a, b + k) + f(a, b) \\
&= [f(a + h, b + k) - f(a, b + k)] - [f(a + h, b) - f(a, b)] \\
&= \phi(b + k) - \phi(b),
\end{aligned} \tag{6}
$$

where

$$\phi(y) = f(a + h, y) - f(a, y). \tag{7}$$

The Mean Value Theorem applied to Equation (6) now gives

$$\Delta = k\phi'(d_2) \tag{8}$$

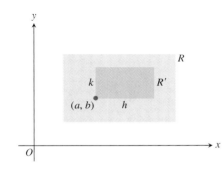

FIGURE A.14 The key to proving $f_{xy}(a, b) = f_{yx}(a, b)$ is that no matter how small R' is, f_{xy} and f_{yx} take on equal values somewhere inside R' (although not necessarily at the same point).

for some d_2 between b and $b + k$. By Equation (7),

$$\phi'(y) = f_y(a + h, y) - f_y(a, y). \tag{9}$$

Substituting from Equation (9) into Equation (8) gives

$$\Delta = k[f_y(a + h, d_2) - f_y(a, d_2)].$$

Finally, we apply the Mean Value Theorem to the expression in brackets and get

$$\Delta = kh f_{yx}(c_2, d_2) \tag{10}$$

for some c_2 between a and $a + h$.

Together, Equations (5) and (10) show that

$$f_{xy}(c_1, d_1) = f_{yx}(c_2, d_2), \tag{11}$$

where (c_1, d_1) and (c_2, d_2) both lie in the rectangle R' (Figure A.14). Equation (11) is not quite the result we want, since it says only that f_{xy} has the same value at (c_1, d_1) that f_{yx} has at (c_2, d_2). The numbers h and k in our discussion, however, may be made as small as we wish. The hypothesis that f_{xy} and f_{yx} are both continuous at (a, b) means that $f_{xy}(c_1, d_1) = f_{xy}(a, b) + \epsilon_1$ and $f_{yx}(c_2, d_2) = f_{yx}(a, b) + \epsilon_2$, where $\epsilon_1, \epsilon_2 \to 0$ as $h, k \to 0$. Hence, if we let h and $k \to 0$, we have $f_{xy}(a, b) = f_{yx}(a, b)$.

The equality of $f_{xy}(a, b)$ and $f_{yx}(a, b)$ can be proved with hypotheses weaker than the ones we assumed. For example, it is enough for f, f_x, and f_y to exist in R and for f_{xy} to be continuous at (a, b). Then f_{yx} will exist at (a, b) and will equal f_{xy} at that point.

The Increment Theorem for Functions of Two Variables

Suppose that the first partial derivatives of $z = f(x, y)$ are defined throughout an open region R containing the point (x_0, y_0) and that f_x and f_y are continuous at (x_0, y_0). Then the change $\Delta z = f(x_0 + \Delta x, y_0 + \Delta y) - f(x_0, y_0)$ in the value of f that results from moving from (x_0, y_0) to another point $(x_0 + \Delta x, y_0 + \Delta y)$ in R satisfies an equation of the form

$$\Delta z = f_x(x_0, y_0)\, \Delta x + f_y(x_0, y_0)\, \Delta y + \epsilon_1\, \Delta x + \epsilon_2\, \Delta y,$$

in which $\epsilon_1, \epsilon_2 \to 0$ as $\Delta x, \Delta y \to 0$.

FIGURE A.15 The rectangular region T in the proof of the Increment Theorem. The figure is drawn for Δx and Δy positive, but either increment might be zero or negative.

Proof We work within a rectangle T centered at $A(x_0, y_0)$ and lying within R, and we assume that Δx and Δy are already so small that the line segment joining A to $B(x_0 + \Delta x, y_0)$ and the line segment joining B to $C(x_0 + \Delta x, y_0 + \Delta y)$ lie in the interior of T (Figure A.15).

We may think of Δz as the sum $\Delta z = \Delta z_1 + \Delta z_2$ of two increments, where

$$\Delta z_1 = f(x_0 + \Delta x, y_0) - f(x_0, y_0)$$

is the change in the value of f from A to B and

$$\Delta z_2 = f(x_0 + \Delta x, y_0 + \Delta y) - f(x_0 + \Delta x, y_0)$$

is the change in the value of f from B to C (Figure A.16).

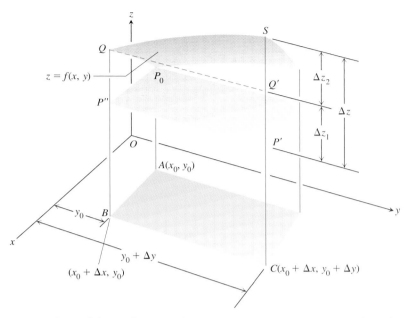

FIGURE A.16 Part of the surface $z = f(x, y)$ near $P_0(x_0, y_0, f(x_0, y_0))$. The points P_0, P', and P'' have the same height $z_0 = f(x_0, y_0)$ above the xy-plane. The change in z is $\Delta z = P'S$. The change

$$\Delta z_1 = f(x_0 + \Delta x, y_0) - f(x_0, y_0),$$

shown as $P''Q = P'Q'$, is caused by changing x from x_0 to $x_0 + \Delta x$ while holding y equal to y_0. Then, with x held equal to $x_0 + \Delta x$,

$$\Delta z_2 = f(x_0 + \Delta x, y_0 + \Delta y) - f(x_0 + \Delta x, y_0)$$

is the change in z caused by changing y from y_0 to $y_0 + \Delta y$, which is represented by $Q'S$. The total change in z is the sum of Δz_1 and Δz_2.

On the closed interval of x-values joining x_0 to $x_0 + \Delta x$, the function $F(x) = f(x, y_0)$ is a differentiable (and hence continuous) function of x, with derivative

$$F'(x) = f_x(x, y_0).$$

By the Mean Value Theorem (Theorem 4, Section 3.2), there is an x-value c between x_0 and $x_0 + \Delta x$ at which

$$F(x_0 + \Delta x) - F(x_0) = F'(c)\, \Delta x$$

or

$$f(x_0 + \Delta x, y_0) - f(x_0, y_0) = f_x(c, y_0)\, \Delta x$$

or

$$\Delta z_1 = f_x(c, y_0)\, \Delta x. \tag{12}$$

Similarly, $G(y) = f(x_0 + \Delta x, y)$ is a differentiable (and hence continuous) function of y on the closed y-interval joining y_0 and $y_0 + \Delta y$, with derivative

$$G'(y) = f_y(x_0 + \Delta x, y).$$

Hence, there is a y-value d between y_0 and $y_0 + \Delta y$ at which

$$G(y_0 + \Delta y) - G(y_0) = G'(d)\,\Delta y$$

or

$$f(x_0 + \Delta x, y_0 + \Delta y) - f(x_0 + \Delta x, y) = f_y(x_0 + \Delta x, d)\,\Delta y$$

or

$$\Delta z_2 = f_y(x_0 + \Delta x, d)\,\Delta y. \qquad (13)$$

Now, as Δx and $\Delta y \to 0$, we know that $c \to x_0$ and $d \to y_0$. Therefore, since f_x and f_y are continuous at (x_0, y_0), the quantities

$$\epsilon_1 = f_x(c, y_0) - f_x(x_0, y_0),$$

$$\epsilon_2 = f_y(x_0 + \Delta x, d) - f_y(x_0, y_0) \qquad (14)$$

both approach zero as Δx and $\Delta y \to 0$.

Finally,

$$
\begin{aligned}
\Delta z &= \Delta z_1 + \Delta z_2 \\
&= f_x(c, y_0)\,\Delta x + f_y(x_0 + \Delta x, d)\,\Delta y && \text{From (12) and (13)} \\
&= [f_x(x_0, y_0) + \epsilon_1]\,\Delta x + [f_y(x_0, y_0) + \epsilon_2]\,\Delta y && \text{From (14)} \\
&= f_x(x_0, y_0)\,\Delta x + f_y(x_0, y_0)\,\Delta y + \epsilon_1\,\Delta x + \epsilon_2\,\Delta y,
\end{aligned}
$$

where ϵ_1 and $\epsilon_2 \to 0$ as Δx and $\Delta y \to 0$, which is what we set out to prove. ▬

Analogous results hold for functions of any finite number of independent variables. Suppose that the first partial derivatives of $w = f(x, y, z)$ are defined throughout an open region containing the point (x_0, y_0, z_0) and that f_x, f_y, and f_z are continuous at (x_0, y_0, z_0). Then

$$
\begin{aligned}
\Delta w &= f(x_0 + \Delta x, y_0 + \Delta y, z_0 + \Delta z) - f(x_0, y_0, z_0) \\
&= f_x\,\Delta x + f_y\,\Delta y + f_z\,\Delta z + \epsilon_1\,\Delta x + \epsilon_2\,\Delta y + \epsilon_3\,\Delta z, \qquad (15)
\end{aligned}
$$

where $\epsilon_1, \epsilon_2, \epsilon_3 \to 0$ as $\Delta x, \Delta y,$ and $\Delta z \to 0$.

The partial derivatives f_x, f_y, f_z in Equation (15) are to be evaluated at the point (x_0, y_0, z_0).

Equation (15) can be proved by treating Δw as the sum of three increments,

$$\Delta w_1 = f(x_0 + \Delta x, y_0, z_0) - f(x_0, y_0, z_0) \qquad (16)$$

$$\Delta w_2 = f(x_0 + \Delta x, y_0 + \Delta y, z_0) - f(x_0 + \Delta x, y_0, z_0) \qquad (17)$$

$$\Delta w_3 = f(x_0 + \Delta x, y_0 + \Delta y, z_0 + \Delta z) - f(x_0 + \Delta x, y_0 + \Delta y, z_0), \qquad (18)$$

and applying the Mean Value Theorem to each of these separately. Two coordinates remain constant and only one varies in each of these partial increments Δw_1, Δw_2, Δw_3. In Equation (17), for example, only y varies, since x is held equal to $x_0 + \Delta x$ and z is held equal to z_0. Since $f(x_0 + \Delta x, y, z_0)$ is a continuous function of y with a derivative f_y, it is subject to the Mean Value Theorem, and we have

$$\Delta w_2 = f_y(x_0 + \Delta x, y_1, z_0)\,\Delta y$$

for some y_1 between y_0 and $y_0 + \Delta y$.

This appendix proves the result needed in Section 13.5 that $|(\mathbf{u} \times \mathbf{v}) \cdot \mathbf{p}|$ is the area of the projection of the parallelogram with sides determined by $\mathbf{u}$ and $\mathbf{v}$ onto any plane whose normal is $\mathbf{p}$. (See Figure A.17.)

Theorem

The area of the orthogonal projection of the parallelogram determined by two vectors $\mathbf{u}$ and $\mathbf{v}$ in space onto a plane with unit normal vector $\mathbf{p}$ is

$$\text{Area} = |(\mathbf{u} \times \mathbf{v}) \cdot \mathbf{p}|.$$

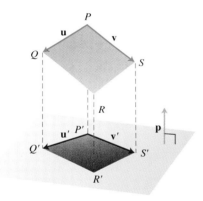

FIGURE A.17 The parallelogram determined by two vectors $\mathbf{u}$ and $\mathbf{v}$ in space and the orthogonal projection of the parallelogram onto a plane. The projection lines, orthogonal to the plane, lie parallel to the unit normal vector $\mathbf{p}$.

Proof In the notation of Figure A.17, which shows a typical parallelogram determined by vectors $\mathbf{u}$ and $\mathbf{v}$ and its orthogonal projection onto a plane with unit normal vector $\mathbf{p}$,

$$\begin{aligned}
\mathbf{u} &= \overrightarrow{PP'} + \mathbf{u'} + \overrightarrow{Q'Q} \\
&= \mathbf{u'} + \overrightarrow{PP'} - \overrightarrow{QQ'} \qquad (\overrightarrow{Q'Q} = -\overrightarrow{QQ'}) \\
&= \mathbf{u'} + s\mathbf{p}. \qquad \begin{array}{l}\text{(For some scalar } s \text{ because} \\ \overrightarrow{PP'} - \overrightarrow{QQ'}) \text{ is parallel to } \mathbf{p})\end{array}
\end{aligned}$$

Similarly,

$$\mathbf{v} = \mathbf{v'} + t\mathbf{p}$$

for some scalar t. Hence,

$$\begin{aligned}
\mathbf{u} \times \mathbf{v} &= (\mathbf{u'} + s\mathbf{p}) \times (\mathbf{v'} + t\mathbf{p}) \\
&= (\mathbf{u'} \times \mathbf{v'}) + s(\mathbf{p} \times \mathbf{v'}) + t(\mathbf{u'} \times \mathbf{p}) + \underbrace{st(\mathbf{p} \times \mathbf{p})}_{0}. \qquad (1)
\end{aligned}$$

The vectors $\mathbf{p} \times \mathbf{v'}$ and $\mathbf{u'} \times \mathbf{p}$ are both orthogonal to $\mathbf{p}$. Hence, when we dot both sides of Equation (1) with $\mathbf{p}$, the only surviving term on the right is $(\mathbf{u'} \times \mathbf{v'}) \cdot \mathbf{p}$. We are left with

$$(\mathbf{u} \times \mathbf{v}) \cdot \mathbf{p} = (\mathbf{u'} \times \mathbf{v'}) \cdot \mathbf{p}.$$

In particular,

$$|(\mathbf{u} \times \mathbf{v}) \cdot \mathbf{p}| = |(\mathbf{u'} \times \mathbf{v'}) \cdot \mathbf{p}|. \qquad (2)$$

The absolute value on the right is the volume of the box determined by $\mathbf{u'}$, $\mathbf{v'}$, and $\mathbf{p}$. The height of this particular box is $|\mathbf{p}| = 1$, so the box's volume is numerically the same as its base area, the area of parallelogram $P'Q'R'S'$. Combining this observation with Equation (2) gives

$$\text{Area of } P'Q'R'S' = |(\mathbf{u'} \times \mathbf{v'}) \cdot \mathbf{p}| = |(\mathbf{u} \times \mathbf{v}) \cdot \mathbf{p}|,$$

which says that the area of the orthogonal projection of the parallelogram determined by $\mathbf{u}$ and $\mathbf{v}$ onto a plane with unit normal vector $\mathbf{p}$ is $|(\mathbf{u} \times \mathbf{v}) \cdot \mathbf{p}|$, what we set out to prove.

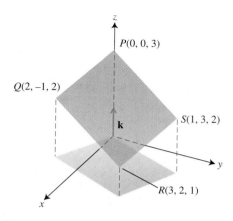

FIGURE A.18 Example 1 calculates the area of the orthogonal projection of parallelogram *PQRS* on the *xy*-plane.

Example 1   Finding the Area of a Projection

Find the area of the orthogonal projection onto the *xy*-plane of the parallelogram determined by the points $P(0, 0, 3)$, $Q(2, -1, 2)$, $R(3, 2, 1)$, and $S(1, 3, 2)$ (Figure A.18)

Solution   With

$$\mathbf{u} = \overrightarrow{PQ} = 2\mathbf{i} - \mathbf{j} - \mathbf{k}, \qquad \mathbf{v} = \overrightarrow{PS} = \mathbf{i} + 3\mathbf{j} - \mathbf{k}, \qquad \text{and} \qquad \mathbf{p} = \mathbf{k},$$

the area is

$$\text{Area} = (\mathbf{u} \times \mathbf{v}) \cdot \mathbf{p} = \begin{vmatrix} 2 & -1 & -1 \\ 1 & 3 & -1 \\ 0 & 0 & 1 \end{vmatrix} = \begin{vmatrix} 2 & -1 \\ 1 & 3 \end{vmatrix} = 7.$$

Answers

Section P.1, pp. 7–10

1. (a) $\Delta x = -2, \Delta y = -3$ **(b)** $\Delta x = 2, \Delta y = -4$

3. (a) $m = 3$ **(b)** $m = -\dfrac{1}{3}$

5. (a) $x = 2, y = 3$ **(b)** $x = -1, y = \dfrac{4}{3}$

7. (a) $y = 1(x - 1) + 1$ **(b)** $y = -1(x + 1) + 1$

9. (a) $3x - 2y = 0$ **(b)** $y = 1$

11. (a) $y = 3x - 2$ **(b)** $y = -x + 2$

13. $y = \dfrac{5}{2}x$

15. (a) (a) Slope: $-\dfrac{3}{4}$ **(b) (a)** Slope: -1

    (b) y-intercept: 3   **(b)** y-intercept: 2

    (c)   **(c)**

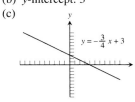

17. (a) $y = -x; y = x$

    (b) $y = -2x - 2; y = \dfrac{1}{2}x + 3$

19. $m = \dfrac{7}{2}, b = -\dfrac{3}{2}$ **21. (b)** $y = -1$

23. $y = 1(x - 3) + 4$
$y = x - 3 + 4$
$y = x + 1$, which is the same equation.

25. (a) $k = 2$ **(b)** $k = -2$

27. 5.97 atmospheres $(k = 0.0994)$

29. (a) Yes, $-40°F$ is the same as $-40°C$.

    (b) All three lines pass through the point $(-40, -40)$.

33. (a) $y = \dfrac{B}{A}(x - a) + b$

    (b) The coordinates of Q are $\left(\dfrac{B^2a + AC - ABb}{A^2 + B^2},\right.$

        $\left.\dfrac{A^2b + BC - ABa}{A^2 + B^2}\right)$

    (c) Distance $= \dfrac{|Aa + Bb - C|}{\sqrt{A^2 + B^2}}$

35. 40.25 ft

37. (a) $y = 0.680x + 9.013$

    (b) The slope is 0.68. It represents the approximate average weight gain in pounds per month.

    (c)   **(d)** 29 lb

39. (a) $y = 5632x - 11,080,280$

    (b) The rate at which the median price is increasing in dollars per year

    (c) $y = 2732x - 5,362,360$

    (d) In the Northeast

Section P.2, pp. 20–24

1. $A = \dfrac{\sqrt{3}}{4}x^2, p = 3x$

3. $x = \dfrac{d}{\sqrt{3}}, A = 2d^2, V = \dfrac{d^3}{3\sqrt{3}}$

5. (a) Not a function of x because some values of x have two values of y

    (b) A function of x because for every x there is only one possible y

7. (a) $D: (-\infty, \infty), R: [1, \infty)$
9. $D: [-2, 2], R: [0, 2]$
11. (a) Symmetric about the origin

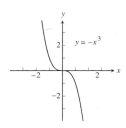

(b) Symmetric about the y-axis

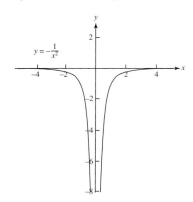

13. (a) For each positive value of x, there are two values of y.

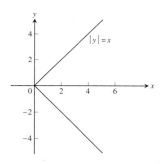

(b) For each value of $x \neq 0$, there are two values of y.

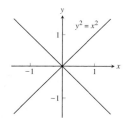

15. (a) Even   **(b)** Odd
17. (a) Odd   **(b)** Even
19. (a) Neither   **(b)** Even

21. (a) Domain = all reals,
range = $(-\infty, 2]$

(b) Domain = all reals,
range = $[-3, \infty)$

23. (a)

(b) All reals   **(c)** All reals
25. Because if the vertical line test holds, then for each x-coordinate there is at most one y-coordinate giving a point on the curve. This y-coordinate would correspond to the value assigned to the x-coordinate. Since there is only one y-coordinate, the assignment would be unique.

27. (a) $f(x) = \begin{cases} x, & 0 \le x \le 1 \\ -x + 2, & 1 < x \le 2 \end{cases}$

(b) $f(x) = \begin{cases} 2, & 0 \le x < 1 \\ 0, & 1 \le x < 2 \\ 2, & 2 \le x < 3 \\ 0, & 3 \le x \le 4 \end{cases}$

(c) $f(x) = \begin{cases} -x + 2, & 0 < x \le 2 \\ -\dfrac{1}{3}x + \dfrac{5}{3}, & 2 < x \le 5 \end{cases}$

(d) $f(x) = \begin{cases} -3x - 3, & -1 < x \le 0 \\ -2x + 3, & 0 < x \le 2 \end{cases}$

29. (a) Position 4   **(b)** Position 1
(c) Position 2   **(d)** Position 3
31. $(x + 2)^2 + (y + 3)^2 = 49$ **33.** $y + 1 = (x - 1)^{2/3}$

35. $y = \frac{1}{2}x$

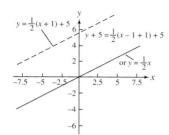

37. (a) 2 **(b)** 22 **(c)** $x^2 + 2$ **(d)** $x^2 + 10x + 22$
(e) 5 **(f)** -2 **(g)** $x + 10$ **(h)** $x^4 - 6x^2 + 6$

39. (a) $\frac{4}{x^2} - 5$ **(b)** $\frac{4}{x^2} - 5$ **(c)** $\left(\frac{4}{x} - 5\right)^2$

(d) $\left(\frac{1}{4x - 5}\right)^2$ **(e)** $\frac{1}{4x^2} - 5$ **(f)** $\frac{1}{(4x - 5)^2}$

41. (a) $f(g(x))$ **(b)** $j(g(x))$ **(c)** $g(g(x))$
(d) $j(j(x))$ **(e)** $g(h(f(x)))$ **(f)** $h(j(f(x)))$

43. (a) $g(x) = x^2$ **(b)** $g(x) = \frac{1}{x - 1}$ **(c)** $f(x) = \frac{1}{x}$

(d) $f(x) = x^2$ (Note that the domain of the composite is $[0, \infty)$.)

45. (a) *D:* [0, 2], *R:* [2, 3] **(b)** *D:* [0, 2], *R:* [−1, 0]

(c) *D:* [0, 2], *R:* [0, 2] **(d)** *D:* [0, 2], *R:* [−1, 0]

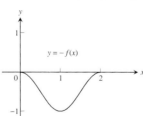

(e) *D:* [−2, 0], *R:* [0, 1] **(f)** *D:* [1, 3], *R:* [0, 1]

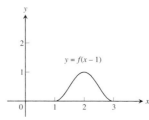

(g) *D:* [−2, 0], *R:* [0, 1] **(h)** *D:* [−1, 1], *R:* [0, 1]

47. (a) Because the circumference of the original circle was 8π and a piece of length x was removed

(b) $r = \frac{8\pi - x}{2\pi} = 4 - \frac{x}{2\pi}$

(c) $h = \sqrt{16 - r^2} = \frac{\sqrt{16\pi x - x^2}}{2\pi}$

(d) $V = \frac{1}{3}\pi r^2 h = \frac{(8\pi - x)^2 \sqrt{16\pi x - x^2}}{24\pi^2}$

49. (a) Yes. Since $(f \cdot g)(-x) = f(-x) \cdot g(-x) = f(x) \cdot g(x) = (f \cdot g)(x)$, the function $(f \cdot g)(x)$ will also be even.

(b) The product will be even, since $(f \cdot g)(-x) = (f \cdot g)(x)$.

53. (d) $(g \circ f)(x) = \sqrt{4 - x^2}$; $D(g \circ f) = [-2, 2]$; $R(g \circ f) = [0, 2]$
$(f \circ g)(x) = 4 - (\sqrt{x})^2 = 4 - x$ for $x \geq 0$; $D(f \circ g) = [0, \infty)$;
$R(f \circ g) = (-\infty, 4]$

55. (a) $y = 4.353951x^{0.518802}$

(b)

(c) 15 km/h

(d) $y = 0.917765x + 4.135356$; the power regression curve in part (a) better fits the data.

Section P.3, pp. 00–00

1. (a) **3. (e)** **5. (b)**

7. Domain: all reals, range: $(-\infty, 3)$, x-intercept: ≈ 1.585, y-intercept: 2

9. Domain: all reals, range: $(-2, \infty)$, x-intercept: ≈ 0.405, y-intercept: 1

11. 3^{4x}

13. 2^{-6x}

9. $D(f^{-1}) = [0, \infty); R(f^{-1}) = [0, \infty)$

15.

x	y	Δy
1	-1	
		2
2	1	
		2
3	3	
		2
4	5	

17.

x	y	Δy
1	1	
		3
2	4	
		5
3	9	
		7
4	16	

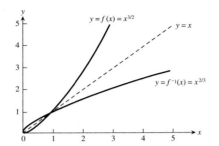

19. If the changes in x are constant for a linear function, say $\Delta x = c$, then the changes in y are also constant, specifically, $\Delta y = mc$.

21. $a = 3, k = 1.5$

23. $x \approx 2.3219$

25. $x \approx -0.6309$

27. 7609.7 million

29. After 19 years

31. (a) $A(t) = 6.6 \left(\dfrac{1}{2}\right)^{t/14}$

(b) About 38.1145 days later

33. ≈ 11.433 years

35. ≈ 11.090 years

37. ≈ 19.108 years

39. $2^{48} \approx 2.815 \times 10^{14}$

41. (a) Regression equation: $P(x) = 6.033(1.030)^x$, where $x = 0$ represents 1900

(b) Approximately 6.03 million, which is not very close to the actual population

(c) The annual rate of growth is approximately 3%.

Section P.4, pp. 00–00

1. One-to-one

3. Not one-to-one

5. One-to-one

7. $D: (0, 1], R: [0, \infty)$

11. $f^{-1}(x) = \sqrt{x - 1}$

13. $f^{-1}(x) = \sqrt[3]{x + 1}$

15. $f^{-1}(x) = \sqrt{x} - 1$

17. $f^{-1}(x) = \dfrac{x - 3}{2}$

19. $f^{-1}(x) = (x + 1)^{1/3}$ or $\sqrt[3]{x - 1}$

21. $f^{-1}(x) = -x^{1/2}$ or $-\sqrt{x}$

23. $f^{-1}(x) = 2 - (-x)^{1/2}$ or $2 - \sqrt{-x}$

25. $f^{-1}(x) = \dfrac{1}{x^{1/2}}$ or $\dfrac{1}{\sqrt{x}}$

27. $f^{-1}(x) = \dfrac{1 - 3x}{x - 2}$

29. $y = e^{x \ln 3} - 1$

(a) $D = (-\infty, \infty)$

(b) $R = (-1, \infty)$

31. $y = 1 - \ln x$

(a) $D = (0, \infty)$

(b) $R = (-\infty, \infty)$

(c)

33. $t = \dfrac{\ln 2}{\ln 1.045} \approx 15.75$

35. $x = \ln \left(\dfrac{3 \pm \sqrt{5}}{2}\right) \approx -0.96$ or 0.96

37. $y = e^{2t+4}$

39. (a) $f^{-1}(x) = \log_2 \left(\dfrac{x}{100 - x}\right)$

(b) $f^{-1}(x) = \log_{1.1} \left(\dfrac{x}{50 - x}\right)$

41. (a) Amount $= 8 \left(\dfrac{1}{2}\right)^{t/12}$

(b) 36 hours

43. ≈ 44.081 years

45. 10 db

47. $(4, 5)$

49. (a) $(1.58, 3)$

(b) No intersection

51. f and g are inverses of each other because $(f \circ g)(x) = (g \circ f)(x) = x.$
 (a)   **(b)** and **(c)**

53. f and g are inverses of each other because $(f \circ g)(x) = (g \circ f)(x) = x.$
 (a)   **(b)** and **(c)**

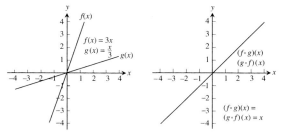

55. (a) The graphs of y, appear to be vertical translates of y_2.

(b)

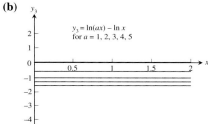

(c) $y_3 = y_1 - y_2 = \ln x - \ln ax = \ln x - (\ln a + \ln x) = -\ln a,$ a constant.

57. $x \approx -0.76666$

59. (a) $y(x) = -474.31 + 121.13 \ln x$; 59.48 million metric tons produced in 1982 and 83.51 million metric tons produced in 2000

(b)

(c) $y(82) \approx 59$ and $y(100) \approx 84$

Section P.5, pp. 00–00

1. (a) 8π m   **(b)** $\dfrac{55\pi}{9}$ m

3.

θ	$-\pi$	$-\dfrac{2\pi}{3}$	0	$\dfrac{\pi}{2}$	$\dfrac{3\pi}{4}$
$\sin\theta$	0	$-\dfrac{\sqrt{3}}{2}$	0	1	$\dfrac{1}{\sqrt{2}}$
$\cos\theta$	-1	$-\dfrac{1}{2}$	1	0	$-\dfrac{1}{\sqrt{2}}$
$\tan\theta$	0	$\sqrt{3}$	0	UND	-1
$\cot\theta$	UND	$\dfrac{1}{\sqrt{3}}$	UND	0	-1
$\sec\theta$	-1	-2	1	UND	$-\sqrt{2}$
$\csc\theta$	UND	$-\dfrac{2}{\sqrt{3}}$	UND	1	$\sqrt{2}$

5. (a) $\cos x = -\dfrac{4}{5}$, $\tan x = -\dfrac{3}{4}$

(b) $\sin x = -\dfrac{\sqrt{8}}{3}$, $\tan x = -\sqrt{8}$

7. (a) Period π **(b)** Period 2

9. (a) Period 2π

(b) Period 2π

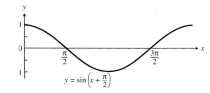

11. Period $\frac{\pi}{2}$, symmetric about the origin

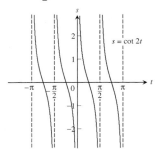

13. (a) $-\cos x$ **(b)** $-\sin x$

19. (a) $A = 2, B = 2\pi, C = -\pi, D = -1$

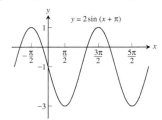

(b) $A = \frac{1}{2}, B = 2, C = 1, D = \frac{1}{2}$

21. (a) 37   **(b)** 365   **(c)** Right 101   **(d)** Up 25

23. (a) $\frac{\pi}{4}$ **(b)** $-\frac{\pi}{3}$ **(c)** $\frac{\pi}{6}$

25. (a) $\frac{\pi}{3}$ **(b)** $\frac{3\pi}{4}$ **(c)** $\frac{\pi}{6}$

37. (a) $c = \sqrt{7} \approx 2.646$ **(b)** $c \approx 1.951$

39. (a) Values of $\sin x$ approach values of x. At the origin, they are both equal to 0.
   (b) They both approach zero, but at any given x, the value of $\sin x$ is only a small percentage of the value of x.

41. (a) Domain: all real numbers except those having the form $\frac{\pi}{2} + k\pi$, where k is an integer; range: $-\frac{\pi}{2} < y < \frac{\pi}{2}$.
   (b) Domain: $-\infty < x < \infty$; range: $-\infty < y < \infty$

43. $x \approx 1.190$ and $x \approx 4.332$ **45.** $x \approx -1.911$ and $x \approx 1.911$

47. (a)

(b) Amplitude ≈ 1.414, period $= 2\pi$, horizontal shift ≈ -0.785 or 5.498, vertical shift: 0

49. (a) $p = 0.6 \sin (2479t - 2.801) + 0.265$

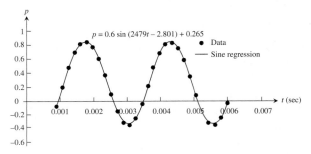

(b) ≈ 395 Hz

51. (a) $y = 3.0014 \sin (0.9996x + 2.0012) + 2.9999$
   (b) $y = 3 \sin (x + 2) + 3$

Section P.6, pp. 00–00

1.

3.

5.

7.

9.

$y = \sqrt{x}$

$t = 0$

11.

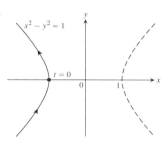

$x^2 - y^2 = 1$

$t = 0$

13.

$t > 0$

$y = -x + 2$

$t = 0$

$(1, 1)$

$t < 0$

15.

$t = 0$ $y = \sqrt{1 - x^2}$

$t = -1$

17.

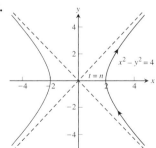

$x^2 - y^2 = 4$

$t = n$

19. **(a)** $x = a \cos t, y = -a \sin t, 0 \le t \le 2\pi$
   (b) $x = a \cos t, y = a \sin t, 0 \le t \le 2\pi$
   (c) $x = a \cos t, y = -a \sin t, 0 \le t \le 4\pi$
   (d) $x = a \cos t, y = a \sin t, 0 \le t \le 4\pi$
21. Possible answer: $x = -1 + 5t, y = -3 + 4t, 0 \le t \le 1$
23. Possible answer: $x = t^2 + 1, y = t, t \le 0$

25. Possible answer: $x = 2 - 3t, y = 3 - 4t, t \ge 0$
27. Graph (c). Window: $[-4, 4]$ by $[-3, 3], 0 \le t \le 2\pi$
29. Graph (d). Window: $[-10, 10]$ by $[-10, 10], 0 \le t \le 2\pi$

31.

$y_1 = e^x$
$y_2 = \ln x$
$y_3 = x$

33.

$y_1 = 2^{-x}$
$y_2 = -\log_2 x$
$y_3 = x$

35.

$y_1 = \ln x$
$y_2 = e^x$
$y_3 = x$

37.

$y_1 = \sin^{-1} x$
$y_2 = \sin x$
$y_3 = x$

39. $1 < t < 3$ **41.** $-5 \le t < -3$
49. $x = 2 \cot t, y = 2 \sin^2 t, 0 < t < \pi$
51. **(a)** The curve is traced from left to right and extends infinitely in
   both directions from the origin.

Section P.7, pp. 00–00

1. **(a)** The graph supports the assumption that y is proportional to x.
   The constant of proportionality is estimated from the slope of
   the regression line, which is 0.166.

$y = 0.166x$

● Data points
— Regression line

   (b) The graph supports the assumption that y is proportional to
   $x^{1/2}$. The constant of proportionality is estimated from the
   slope of the regression line, which is 2.03.

$y = 2.03x^{1/2}$

● Data points
— Regression line

3. Quadratic regression gives $y = 0.64555x^2 + 0.078422x + 4.88961$. The power regression fits the data well.

5. (a) Exponential regression gives $y = 0.5(0.69^x) = 0.5e^{-0.371x}$.

(b) The exponential function fits the data very well.

(c) The model predicts that after 12 hours, the amount of digoxin in the blood will be less than 0.006 mg.

7. (a) Exponential regression gives $C = 770(0.715^t) = 769e^{-0.3355t}$.

(b) The exponential function appears to capture a trend for these data.

(c) The model predicts that the blood concentration will fall below 10 ppm after 22 days and 23 hours.

9. The slope of the regression line is 0.008435, so the model that estimates the weight as a function of l is $w = 0.008435 l^3$.

11. Graph (c)   **13.** Graph (c)

15. (f) One possibility: Let y represent the number of people in your school who have the flu and let x represent the number of days that have elapsed after the first person gets sick. At first, the flu doesn't spread very quickly because there are only a few sick people to pass it on. As more people get sick, however, the disease spreads more rapidly. The most volatile mixture is when half the people are sick, because then there are a lot of sick people to spread the disease and a lot of uninfected people who can still catch it. As time continues and more people get sick, there are fewer and fewer people available to catch the flu and the spread of the disease begins to slow down. This behavior can be modeled with a function like the one represented by graph (f).

17.

$$h = \begin{cases} 6(1-t)(1+t) & \text{for } 0 \le t \le 1 \\ -4.5(t-1)(t-3) & \text{for } 1 < t \le 3 \\ -3.375(t-3)(t-5) & \text{for } 3 < t \le 5 \\ -2.531(t-5)(t-7) & \text{for } 5 < t \le 7 \\ -1.898(t-7)(t-9) & \text{for } 7 < t \le 9 \end{cases}$$

19. (a)

$$S = \begin{cases} 100(1 - e^{-0.175t}) & \text{for } 0 \le t \le 4 \\ 5 + 45.342e^{-0.5(t-4)} & \text{for } 4 < t \le 10 \end{cases}$$

(b)

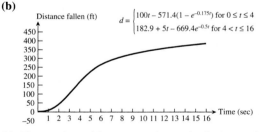

$$d = \begin{cases} 100t - 571.4(1 - e^{-0.175t}) & \text{for } 0 \le t \le 4 \\ 182.9 + 5t - 669.4e^{-0.5t} & \text{for } 4 < t \le 16 \end{cases}$$

21. (a) The graph could represent the angle that a pendulum makes with the vertical as it swings back and forth. The variable y represents the angle and x represents time. Because of friction, the amplitude of the oscillation decays, as depicted by the graph. When y is positive, the pendulum is on one side of the vertical, and when y is negative, the pendulum is on the other side.

(b) The graph could represent the angle the playground swing makes with the vertical as a child "pumps" on the swing to get it going. The variable y represents the angle and x represents time. Because the child puts energy into the system (swing + child), the amplitude of the oscillation grows with time, as depicted by the graph. When y is positive, the swing is on one side of the vertical, and when y is negative, the swing is on the other side.

Preliminary Chapter Practice Exercises, pp. 00–00

1. $y = 3x - 9$

3. $x = 0$

5. $y = 2$

7. $y = -3x + 3$

9. $y = -\dfrac{4}{3}x - \dfrac{20}{3}$

11. $y = \dfrac{2}{3}x + \dfrac{8}{3}$

13. $A \pi r^2,\ C = 2\pi r,\ A = \dfrac{C^2}{4\pi}$

15. $x = \tan\theta,\ y = \tan^2\theta$

17. Origin

19. Neither

21. Even

23. Even

25. Odd

27. Neither

29. (a) Domain: all reals
(b) Range: $[-2, \infty)$

31. (a) Domain: $[-4, 4]$
(b) Range: $[0, 4]$

33. (a) Domain: all reals
(b) Range: $(-3, \infty)$

35. (a) Domain: all reals
(b) Range: $[-3, 1]$

37. (a) Domain: $(3, \infty)$
(b) Range: all reals

39. (a) Domain: $[-4, 4]$
(b) Range: $[0, 2]$

41. $f(x) = \begin{cases} 1 - x, & 0 \le x < 1 \\ 2 - x, & 1 \le x \le 2 \end{cases}$

43. (a) 1

(b) $\dfrac{1}{\sqrt{2.5}}\left(=\sqrt{\dfrac{2}{5}}\right)$

(c) $x,\ x \ne 0$

(d) $\dfrac{1}{\sqrt{1/\sqrt{x+2}+2}}$

45. (a) $(f \circ g)(x) = -x,\ x \ge -2,\ (g \circ f)(x) = \sqrt{4 - x^2}$
(b) Domain $(f \circ g)$: $[-2, \infty)$, domain $(g \circ f)$: $[-2, 2]$
(c) Range $(f \circ g)$: $[-\infty, 2]$, range $(g \circ f)$: $[0, 2]$

47. Replace the portion for $x < 0$ with mirror image of portion for $x > 0$ to make the new graph symmetric with respect to the y-axis.

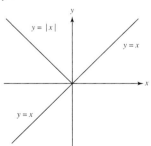

49. It does not change it.

51. Adds the mirror image of the portion for $x > 0$ to make the new graph symmetric with respect to the y-axis

53. Reflects the portion for $y < 0$ across the x-axis

55. Reflects the portion for $y < 0$ across the x-axis

57. (a) Symmetric about the line $y = x$

59. (a) $f^{-1}(x) = \dfrac{2 - x}{3}$

(b)

61. (a) $f(g(x)) = (\sqrt[3]{x})^3 = x,\ g(f(x)) = \sqrt[3]{x^3} = x$

(b)

63. (a) $f^{-1}(x) = x - 1$

(b) $f^{-1}(x) = x - b$. The graph of f^{-1} is a line parallel to the graph of f. The graphs of f and f^{-1} lie on opposite sides of the line $y = x$ and are equidistant from that line.
(c) Their graphs will be parallel to one another and lie on opposite sides of the line $y = x$ equidistant from that line.

65. 2.71828183

67. (a) 7.2

(b) $\dfrac{1}{x^2}$

(c) $\dfrac{x}{y}$

69. (a) 1

(b) 1

(c) $-x^2 - y^2$

71. ≈ 0.6435 radian or $36.8699°$

73. $\cos\theta = \dfrac{3}{7},\ \sin\theta = \dfrac{\sqrt{40}}{7},\ \tan\theta = \dfrac{\sqrt{40}}{3},\ \sec\theta = \dfrac{7}{3},$
$\csc\theta = \dfrac{7}{\sqrt{40}},\ \cot\theta = \dfrac{3}{\sqrt{40}}$

75. Period 4π

77.

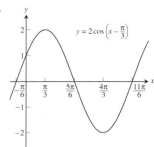

$$y = 2\cos\left(x - \frac{\pi}{3}\right)$$

79. (a) $a = 1; b = \sqrt{3}$
   (b) $c = 4/\sqrt{3}; a = 2/\sqrt{3}$

81. (a) $a = \dfrac{b}{\tan B}$ **(b)** $c = \dfrac{a}{\sin A}$

83. Since $\sin (x)$ has period 2π, $(\sin (x + 2\pi))^3 = (\sin (x))^3$. This function has period 2π. A graph shows that no smaller number works for the period.

87. $\dfrac{\sqrt{6} + \sqrt{2}}{4}$

89. (a) $\dfrac{\pi}{6}$ **91. (a)** $\dfrac{\pi}{4}$

   (b) $-\dfrac{\pi}{4}$ **(b)** $\dfrac{5\pi}{6}$

   (c) $\dfrac{\pi}{3}$ **(c)** $\dfrac{\pi}{3}$

93. 2
95. $-\dfrac{1}{2}$ **97.** $\sqrt{4x^2 + 1}$ **99.** $\dfrac{\sqrt{1 - x^2}}{x}$

101. (a) Defined; there is an angle whose tangent is 2.
   (b) Not defined; no angle has cosine 2.
103. (a) Not defined; no angle has secant 0.
   (b) Not defined; no angle has sine $\sqrt{2}$.
105. ≈ 16.98 m   **107. (b)** 4π
109. (a) $\left(\dfrac{x}{5}\right)^2 + \left(\dfrac{y}{2}\right)^2 = 1$; all

   (b) Initial point: $(5, 0)$, terminal point: $(5, 0)$

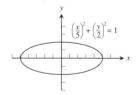

$$\left(\frac{x}{5}\right)^2 + \left(\frac{y}{2}\right)^2 = 1$$

111. (a) $y = 2x + 7$; from $(4, 15)$ to $(-2, 3)$
   (b) Initial point: $(4, 15)$, terminal point: $(-2, 3)$

$$y = 2x + 7$$
$$-2x \le x \le 4$$

113. Possible answer: $x = -2 + 6t, y = 5 - 2t, 0 \le t \le 1$
115. Possible answer: $x = 2 - 3t, y = 5 - 5t, 0 \le t$

Preliminary Chapter Additional Exercises, pp. 00–00

1. (a)

$$y = f(-x)$$

(b)

$$y = -f(x)$$

(c)

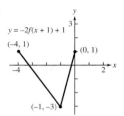

$$y = -2f(x + 1) + 1$$

(d)

$$y = 3f(x - 2) - 2$$

3. (a) $y = 100{,}000 - 10{,}000x, 0 \le x \le 10$
   (b) After 4.5 years

5. After $\dfrac{\ln (10/3)}{\ln 1.08} \approx 15.6439$ years. (If the bank only pays interest at the end of the year, it will take 16 years.)

11. (a) Yes
   (b) Not always. If f is odd, then h is odd.

13. Yes. For instance: $f(x) = \dfrac{1}{x}$ and $g(x) = \dfrac{1}{x}$, or $f(x) = 2x$ and $g(x) = \dfrac{x}{2}$, or $f(x) = e^x$ and $g(x) = \ln x$.

15. If $f(x)$ is odd, then $g(x) = f(x) - 2$ is not odd. Nor is $g(x)$ even, unless $f(x) = 0$ for all x. If f is even, then $g(x) = f(x) - 2$ is also even.

17.

$$|x| + |y| = 1 + x$$

21. If the graph of $f(x)$ passes the horizontal line test, so will the graph of $g(x) = -f(x)$ since it is the same graph reflected about the x-axis.

23. (a) Domain: all reals. Range: If $a > 0$, then (d, ∞); if $a < 0$, then $(-\infty, d)$.
   (b) Domain: (c, ∞), range: all reals

25. (a) The graph does not support the assumption that $y \propto x^2$.

● Data points
— Regression line
$$y = 3x^2$$

(b) The graph supports the assumption that $y \propto 4^x$. The constant of proportionality is estimated from the slope of the regression line, which is 0.6; therefore, $y = 0.6(4^x)$.

27. (a) Since the elongation of the spring is zero when the stress is $5(10^{-3})$(lb/in.²), the data should be adjusted by subtracting this amount from each of the stress data values. This gives the following table, where $\bar{s} = s - 5(10^{-3})$.

The slope of the graph is $\dfrac{(295 - 55)(10^5)}{(75 - 15)(10^{-3})} = 4.00(10^8)$ and the model is $e = 4(10^8)\bar{s}$ or $e = 4(10^8)(s - 5(10^{-3}))$.

$\bar{s} \times 10^{-3}$	0	5	15	25	35	45	55	65	75	85	95
$e \times 10^5$	0	19	57	94	134	173	216	256	297	343	390

(b) The model fits the data well.
(c) $c = 780(10^5)$(in./in.). Since $s = 200(10^{-3})$(lb/in.²) is well outside the range of the data used for the model, I would not feel comfortable with this prediction without further testing of the spring.

29. (a) $y = 20.627x + 338.622$

(b) Approximately 957
(c) Slope is 20.627. It represents the approximate annual increase in number of doctorates earned by Hispanic Americans per year.

31. (a) $f(x) = 2.000268 \sin(2.999187x - 1.000966) + 3.999881$

(b) $f(x) = 2\sin(3x - 1) + 4$
33. (a) $Q = 1.00(2.0138^x) = 1.00e^{0.7x}$
(b) The annual rate of increase during this time is 7.25%.

CHAPTER 1

Section 1.1, pp. 000–000

1. (a) 19
(b) 1

3. (a) $-\dfrac{4}{\pi}$
(b) $-\dfrac{3\sqrt{3}}{\pi}$

5. Graphs can shift during a press run, so your estimates may not completely agree with these.
(a)

PQ_1	PQ_2	PQ_3	PQ_4
43	46	49	50

The appropriate units are meters per second.

(b) ≈ 50 m/sec or 180 km/h

7. Lower bound: $a = 23.55$ ft/sec, upper bound: $b = 28.85$ ft/sec, $v(2) \approx \dfrac{a + b}{2} = 26.20$ ft/sec

9. (a) Does not exist. As x approaches 1 from the right, $g(x)$ approaches 0. As x approaches 1 from the left, $g(x)$ approaches 1. There is no single number L that all values $g(x)$ get arbitrarily close to as $x \to 1$.
(b) 1

11. (a) True   **(b)** True   **(c)** False
(d) False   **(e)** False   **(f)** True

13. As x approaches 0 from the left, $x/|x|$ approaches -1. As x approaches 0 from the right, $x/|x|$ approaches 1. There is no single number L that the function values all get arbitrarily close to as $x \to 0$.

15. Nothing can be said.
17. No
19. (a) $f(x) = (x^2 - 9)/(x + 3)$

x	-3.1	-3.01	-3.001	-3.0001	-3.00001	-3.000001
$f(x)$	-6.1	-6.01	-6.001	-6.0001	-6.00001	-6.000001

x	-2.9	-2.99	-2.999	-2.9999	-2.99999	-2.999999
$f(x)$	-5.9	-5.99	-5.999	-5.9999	-5.99999	-5.999999

(c) $\lim\limits_{x \to -3} f(x) = -6$

21. (a) $G(x) = (x + 6)/(x^2 + 4x - 12)$

x	-5.9	-5.99	-5.999
$G(x)$	-0.126582	-0.1251564	-0.1250156

x	-5.9999	-5.99999	-5.999999
$G(x)$	-0.1250016	-0.12500016	-0.12500002

x	-6.1	-6.01	-6.001
$G(x)$	-0.123457	-0.1248439	-0.1249844

x	-6.0001	-6.00001	-6.000001
$G(x)$	-0.1249984	-0.12499984	-0.12499998

(c) $\lim\limits_{x \to -6} G(x) = -\dfrac{1}{8} = -0.125$

23. $\delta = 0.1$

25. (a) $g(\theta) = (\sin \theta)/\theta$

θ	0.1	0.01	0.001
$g(\theta)$	0.998334	0.999983	0.999999

θ	0.0001	0.00001	0.000001
$g(\theta)$	0.999999	0.999999	0.999999

θ	-0.1	-0.01	-0.001
$g(\theta)$	0.998334	0.999983	0.999999

θ	-0.0001	-0.00001	-0.000001
$g(\theta)$	0.999999	0.999999	0.999999

$\lim\limits_{\theta \to 0} g(\theta) = 1$

27. (a) $f(x) = x^{1/(1-x)}$

x	0.9	0.99	0.999
$f(x)$	0.348678	0.366032	0.367695

θ	0.9999	0.99999	0.999999
$g(\theta)$	0.367861	0.367878	0.367879

x	1.1	1.01	1.001
$f(x)$	0.385543	0.369711	0.368063

θ	1.0001	1.00001	1.000001
$g(\theta)$	0.367898	0.367881	0.367880

$\lim\limits_{x \to 1} f(x) \approx 0.36788$

29. $\delta = \dfrac{7}{16}$

31. $(3.99, 4.01)$, $\delta = 0.01$

33. $(-0.19, 0.21)$, $\delta = 0.19$

35. $\left(\dfrac{10}{3}, 5\right)$, $\delta = \dfrac{2}{3}$

37. $[3.384, 3.387]$. To be safe, the left endpoint was rounded up and the right endpoint rounded down.

39. (a) $\lim\limits_{x \to 2} f(x) = 2$
(b) One possible answer: $a = 1.75$, $b = 2.28$
(c) One possible answer: $a = 1.99$, $b = 2.01$

41. (a) 14.7 m/sec
(b) 29.4 m/sec

43. (a)

x	-0.1	-0.01	-0.001	-0.0001
$f(x)$	-0.054402	-0.005064	-0.000827	-0.000031

(b)

x	0.1	0.01	0.001	0.0001
$f(x)$	-0.054402	-0.005064	-0.000827	-0.000031

The limit appears to be 0.

45. (a)

x	-0.1	-0.01	-0.001	-0.0001
$f(x)$	2.0567	2.2763	2.2999	2.3023

(b)

x	0.1	0.01	0.001	0.0001
$f(x)$	2.5893	2.3293	2.3052	2.3029

The limit appears to be approximately 2.3.

Section 1.2, pp. 108–111

1. (a) 3
(b) -2
(c) No limit
(d) 1
5. (a) 4
(c) No limit

3. (a) -4
(b) -4
(c) -4
(d) -4
(b) -3
(d) 4

7. (a) Quotient Rule
(b) Difference and Power Rules
(c) Sum and Constant Multiple Rules

9. (a) -10
(b) -20
(c) -1
(d) $\dfrac{5}{7}$

11. (a) -9
(b) -8
(c) $\dfrac{1}{5}$
(c) $\dfrac{3}{2}$

13. (a) -7
(c) 4

(b) $-\dfrac{1}{2}$
(d) $-\sin\left(\dfrac{1}{6}\right) \approx -0.1659$

15. (a) $\lim\limits_{x\to 0}\left(1 - \dfrac{x^2}{6}\right) = 1 - \dfrac{0}{6} = 1$ and $\lim\limits_{x\to 0} 1 = 1$;

by the Sandwich Theorem, $\lim\limits_{x\to 0}\dfrac{x \sin x}{2 - 2\cos x} = 1.$

(b) For $x \neq 0$, $y = (x \sin x)/(2 - 2\cos x)$ lies between the other two graphs in the figure, and the graphs converge as $x \to 0$.

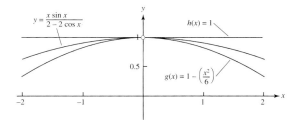

17. $\lim\limits_{h\to 0}\dfrac{(1+h)^2 - 1^2}{h} = \lim\limits_{h\to 0}(2 + h) = 2$

19. $\lim\limits_{h\to 0}\dfrac{\left(\dfrac{1}{-2+h}\right) - \left(\dfrac{1}{-2}\right)}{h} = \lim\limits_{h\to 0}\dfrac{-h}{h(4 - 2h)} = -\dfrac{1}{4}$

21. (a) False
(e) True
(i) False

(b) True
(f) True
(j) False

(c) False
(g) False
(k) True

(d) True
(h) False
(l) False

23. (a) No
(b) Yes, 0
(c) No

25. (a) D: $0 \le x \le 2$, R: $0 < y \le 1$ and $y = 2$
(b) $(0, 1) \cup (1, 2)$
(c) $x = 2$
(d) $x = 0$

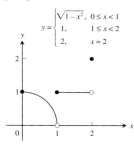

$$y = \begin{cases} \sqrt{1 - x^2}, & 0 \le x < 1 \\ 1, & 1 \le x < 2 \\ 2, & x = 2 \end{cases}$$

27. $\sqrt{3}$

29. $\dfrac{2}{\sqrt{5}}$

31. (a) 1
35. (a) 4
37. Yes
39. $\delta = \epsilon^2$, $\lim\limits_{x\to 5^+}\sqrt{x - 5} = 0$

(b) 21
(b) -2

Section 1.3, pp. 122–123

1. (a) π
3. (a) $-\dfrac{5}{3}$

(b) π
(b) $-\dfrac{5}{3}$

5. $-\dfrac{1}{x} \le \dfrac{\sin 2x}{x} \le \dfrac{1}{x} \Rightarrow \lim\limits_{x\to\infty}\dfrac{\sin 2x}{x} = 0$ by the Sandwich Theorem

7. (a) $\dfrac{2}{5}$
9. (a) $-\infty$
11. (a) ∞
13. (a) $-\dfrac{2}{3}$
15. 0
19. ∞

(b) $\dfrac{2}{5}$
(b) ∞
(b) $-\infty$
(b) $-\dfrac{2}{3}$
17. 1

21. Here is one possibility.

23. Here is one possibility.

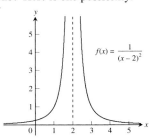

$f(x) = \dfrac{1}{(x-2)^2}$

25. $y = \dfrac{1}{x-1}$

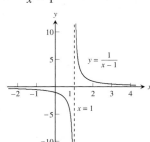

27. $y = \dfrac{2x^2 + x - 1}{x^2 - 1}$

29. $y = \dfrac{x^4 + 1}{x^2} = x^2 + \dfrac{1}{x^2}$

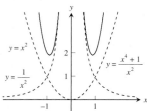

31. $y = \dfrac{x^2 - x + 1}{x - 1} = x + \dfrac{1}{x - 1}$

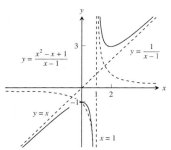

33. $y = \dfrac{8}{x^2 + 4}$

35. (a)

37. (d)

39. (a) e^x (b) $-2x$

41. (a) x (b) x

43. (a) $\dfrac{1}{2}$ (b) $\dfrac{1}{2}$

45. At most 2

47. $y = \dfrac{x}{\sqrt{4 - x^2}}$

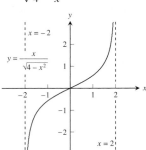

49. $y = x^{2/3} + \dfrac{1}{x^{1/3}}$

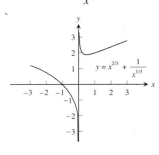

51. At ∞: ∞, at $-\infty$: 0

53. At ∞: 0, at $-\infty$: 0

55. 1

57. 3

59. $y = -\dfrac{x^2 - 4}{x + 1} = 1 - x + \dfrac{3}{x + 1}$

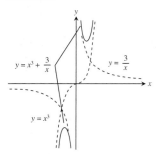

61. The graph of the function mimics each term as it becomes dominant.

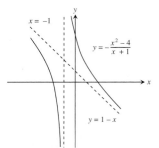

63. (a) $y \to \infty$ (see the accompanying graph)
  (b) $y \to \infty$ (see the accompanying graph)
  (c) Cusps at $x = \pm 1$ (see the accompanying graph)

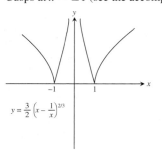

Section 1.4, pp. 132–134

1. No; discontinuous at $x = 2$; not defined at $x = 2$
3. Continuous
5. **(a)** Yes   **(b)** Yes   **(c)** Yes   **(d)** Yes
7. **(a)** No   **(b)** No
9. 0
13. Discontinuous when $x = 2$
15. Discontinuous when $t = 3$ or $t = 1$
17. Discontinuous at $\theta = 0$
19. Continuous on the interval $\left[-\dfrac{3}{2}, \infty \right)$
21. 0; continuous
23. 1; continuous
25. $f(x)$ is continuous on $[0, 1]$ and $f(0) < 0$, $f(1) > 0 \Rightarrow$ by the Intermediate Value Theorem $f(x)$ takes on every value between $f(0)$ and $f(1) \Rightarrow$ the equation $f(x) = 0$ has at least one solution between $x = 0$ and $x = 1$.

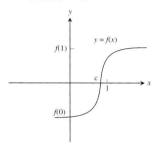

27. All five statements ask for the same information because of the Intermediate Value Property of continuous functions.
29. Answers may vary. For example, $f(x) = \dfrac{\sin (x - 2)}{x - 2}$ is discontinuous at $x = 2$ because it is not defined there. The discontinuity can be removed, however, because f has a limit (namely 1) as $x \to 2$.
31. $r_1 = \dfrac{1 - \sqrt{21}}{2} \approx -1.791$, $r_2 = r_3 = r_4 = 0$, and $r_5 = \dfrac{1 + \sqrt{21}}{2} \approx 2.791$
35. Yes, by the intermediate value property
39. **(b)** Continuous at all points in the domain $[0, 5)$ except at $t = 1$, 2, 3, 4

45. $x \approx 1.8794, -1.5321, -0.3473$ **47.** $x \approx 1.7549$
49. $x \approx 3.5156$
51. $x \approx 0.7391$

Section 1.5, pp. 139–141

1. P_1: $m_1 = 1$, P_2: $m_2 = 5$ **3.** P_1: $m_1 = \dfrac{5}{2}$, P_2: $m_2 = -\dfrac{1}{2}$
5. $y = 2x + 5$ **7.** $y = 12x + 16$
9. $y + 1 = -3(x - 1)$ **11.** $y - 3 = -2(u - 3)$
13. $-\dfrac{1}{4}$ **15.** $(-2, -5)$
17. If $x = 0$, $y = -(x + 1)$; if $x = 2$, $y = -(x - 3)$.
19. 19.6 m/sec   **21.** 6π
23. 3.72 m/sec   **25.** Yes   **27.** Yes
29. **(a)** $\dfrac{1 - e^{-2}}{2} \approx 0.432$ **(b)** $\dfrac{e^3 - e}{2} \approx 8.684$
31. **(a)** $-\dfrac{4}{\pi} \approx -1.273$ **(b)** $-\dfrac{3\sqrt{3}}{\pi} \approx -1.654$
33. **(a)** 0.3 billion dollars per year
(b) 0.5 billion dollars per year
(c) $y = 0.057x^2 - 0.1514x + 1.3943$

(d) 1994 to 1995: 0.31 billion dollars per year, 1995 to 1997: 0.53 billion dollars per year
(e) 0.65 billion dollars per year
35. **(a)** Nowhere   **37.** **(a)** At $x = 0$
39. **(a)** Nowhere   **41.** **(a)** At $x = 1$
43. **(a)** At $x = 0$

Chapter 1 Practice Exercises, pp. 142–143

1. At $x = -1$: $\lim\limits_{x \to -1^-} f(x) = \lim\limits_{x \to -1^+} f(x) = 1$, so $\lim\limits_{x \to -1} f(x) = 1 = f(-1)$; continuous at $x = -1$

At $x = 0$: $\lim\limits_{x \to 0^-} f(x) = \lim\limits_{x \to 0^+} f(x) = 0$, so $\lim\limits_{x \to 0} f(x) = 0$. However, $f(0) \neq 0$, so f is discontinuous at $x = 0$. The discontinuity can be removed by redefining $f(0)$ to be 0.

At $x = 1$: $\lim\limits_{x \to 1^-} f(x) = -1$ and $\lim\limits_{x \to 1^+} f(x) = 1$, so $\lim\limits_{x \to 1} f(x)$ does not exist. The function is discontinuous at $x = 1$, and the discontinuity is not removable.

3. **(a)** -21 **(b)** 49   **(c)** 0   **(d)** 1
(e) 1   **(f)** 7   **(g)** -7 **(h)** $-\dfrac{1}{7}$

5. 4

7. (a) $(-\infty, +\infty)$ **(b)** $[0, \infty)$
   (c) $(-\infty, 0)$ and $(0, \infty)$ **(d)** $(0, \infty)$

9. (a) Does not exist   **(b)** 0

11. $\frac{1}{2}$ **13.** $2x$ **15.** $-\frac{1}{4}$ **17.** $\frac{2}{5}$

19. 0   **21.** $-\infty$ **23.** 0   **25.** 1

27. 0   **29. (b)** 1.324717957

Chapter 1 Additional Exercises, pp. 143–145

3. 0; the left-hand limit was needed because the function is undefined for $v > c$.

5. $65 \le t \le 75$; within 5°F

7. (a) $\lim\limits_{a \to 0} r_+(a) = 0.5$, $\lim\limits_{a \to -1^+} r_+(a) = 1$
  (b) $\lim\limits_{a \to 0} r_-(a)$ does not exist, $\lim\limits_{a \to -1^+} r_-(a) = 1$

9. (a) True   **(b)** False   **(c)** True   **(d)** False

CHAPTER 2 DERIVATIVES

Section 2.1, pp. 157–160

1. $-2x, 6, 0,$ **3.** $-3t^2 - 2t, 5$

5. $\dfrac{3}{2\sqrt{3\theta}}, \sqrt{3}$ **7.** $2x + 1, 2$

9. $4x^2, 8x$

11. $y' = 2x^3 - 3x - 1 \Rightarrow y'' = 6x^2 - 3 \Rightarrow y''' = 12x \Rightarrow y^{(4)} = 12 \Rightarrow y^{(n)} = 0$ for all $n \ge 5$

13. (a) $\dfrac{dv}{dx} = 3x^2 - 4, y - 1 = 8(x - 2)$
  (b) $[-4, \infty)$
  (c) The equation of one such tangent line is found in part (a) when $x = 2$. The equation of the line tangent to the curve at the point $(-2, 1)$ is $y - 1 = 8(x - (-2))$.

15. (b)   **17. (d)**

19. (a) f' is not defined at $x = 0, 1, 4$. At these points, the left-hand and right-hand derivatives do not agree.
  (b)

21. $\lim\limits_{h \to 0^-} \dfrac{f(0 + h) - f(0)}{h} = 0$, $\lim\limits_{h \to 0^+} \dfrac{f(0 + h) - f(0)}{h} = 1 \Rightarrow$ the derivative $f'(0)$ does not exist.

23. (a) The function is differentiable on its domain $-2 \le x \le 3$.
  (b) None
  (c) None

25. (a) On $-1 \le x < 0$ and $0 < x \le 2$
  (b) At $x = 0$
  (c) None

27. (a) $y' = -2x$
  (b)

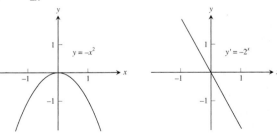

  (c) $x < 0, x = 0, x > 0$
  (d) $-\infty < x < 0, 0 < x < \infty$

29. (a) $y' = x^2$
  (b)

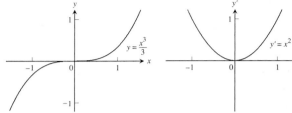

  (c) $x \ne 0, x = 0$, none
  (d) $-\infty < x < \infty$, none

31. $y' = 3x^2$ is never negative.

33. Yes, $y + 16 = -(x - 3)$ is tangent at $(3, -16)$.

35. No, the function $y = \lfloor x \rfloor$ does not satisfy the intermediate value property of derivatives.

37. Yes, $(-f)'(x) = -(f'(x))$

Section 2.2, pp. 169–172

1. (a) $\Delta s = -2m, v_{av} = -1$ m/sec
  (b) $|v(0)| = 3$ m/sec, $|v(2)| = 1$ m/sec, $a(0) = a(2) = 2$ m/sec^2
  (c) At $t = \dfrac{3}{2}$

3. (a) $\Delta s = -9m, v_{av} = -3$ m/sec
  (b) $|v(0)| = 3$ m/sec, $|v(3)| = 12$ m/sec, $a(0) = 6$ m/sec^2, and $a(3) = -12$ m/sec^2
  (c) Never changes direction

5. (a) $a(1) = -6$ m/sec^2 and $a(3) = 6$ m/sec^2
  (b) $|v(2)| = 3$ m/sec
  (c) Total distance $= |s(1) - s(0)| + |s(2) - s(1)| = 6$ m

7. $t \approx 7.5$ sec on Mars, $t \approx 1.2$ sec on Jupiter

9. $g_s = 0.75$ m/sec^2

11. (a) $v = -32t, |v| = 32t$ ft/sec, $a = -32$ ft/sec^2
  (b) $t \approx 3.3$ sec
  (c) $v \approx 107.0$ ft/sec

13. (a) $t = 2$, $t = 7$ **(b)** $3 \leq t \leq 6$

(c)   **(d)**

15. (a) 190 ft/sec
(b) 2 sec
(c) 8 sec, 0 ft/sec
(d) 10.8 sec, 90 ft/sec
(e) 2.8 sec
(f) Greatest acceleration happens 2 sec after launch
(g) Constant acceleration between 2 and 10.8 sec, -32 ft/sec^2

17. (a) 4/7 sec, 280 cm/sec
(b) 560 cm/sec, 980 cm/sec^2
(c) 29.75 flashes/sec

19. Graph C = position, graph B = velocity, graph A = acceleration
21. (a) \$2   **(b)** \$2   **(c)** 0
23. -8000 gal/min, $-10,000$ gal/min
25. (a) 16π ft^3/ft   **(b)** 3.2π ft^3
27. $80\sqrt{19}$ ft/sec, ≈ 238 mph

29. (a) $t = \dfrac{3}{2}$

(b) Left on $\left[0, \dfrac{3}{2}\right)$, right on $\left(\dfrac{3}{2}, 5\right]$

(c) $t = \dfrac{3}{2}$

(d) Speeds up on $\left(\dfrac{3}{2}, 5\right]$, slows down on $\left[0, \dfrac{3}{2}\right)$

(e) Fastest at $t = 5$, slowest at $t = 3/2$
(f) $t = 5$

31. (a) $t = \dfrac{6 \pm \sqrt{15}}{3}$

(b) Left on $\left[0, \dfrac{6 - \sqrt{15}}{3}\right) \cup \left(\dfrac{6 + \sqrt{15}}{3}, 4\right]$;

right on $\left(\dfrac{6 - \sqrt{15}}{3}, \dfrac{6 + \sqrt{15}}{3}\right)$

(c) $t = \dfrac{6 \pm \sqrt{15}}{3}$

(d) Speeds up on $\left(\dfrac{6 - \sqrt{15}}{3}, 2\right) \cup \left(\dfrac{6 + \sqrt{15}}{3}, 4\right]$,

slows down on $\left[0, \dfrac{6 - \sqrt{15}}{3}\right) \cup \left(2, \dfrac{6 + \sqrt{15}}{3}\right)$

(e) Fastest at $t = 0, 4$; slowest at $t = \dfrac{6 \pm \sqrt{15}}{3}$

(f) $t = \dfrac{6 + \sqrt{15}}{3}$

Section 2.3, pp. 178–179

1. $\dfrac{dy}{dx} = 12x - 10 + 10x^{-3}, \dfrac{d^2y}{dx^2} = 12 - 30x^{-4}$

3. $\dfrac{dr}{ds} = \dfrac{-2}{3s^3} + \dfrac{5}{2s^2}, \dfrac{d^2r}{ds^2} = \dfrac{2}{s^4} - \dfrac{5}{s^3}$

5. (a) $y' = (3 - x^2) \cdot \dfrac{d}{dx}(x^3 - x + 1) + (x^3 - x + 1) \cdot \dfrac{d}{dx}(3 - x^2) =$
$-5x^4 + 12x^2 - 2x - 3$
(b) $y = -x^5 + 4x^3 - x^2 - 3x + 3 \Rightarrow y' = -5x^4 + 12x^2 - 2x - 3$

7. $y' = -\dfrac{19}{(3x - 2)^2}$ **9.** $f'(t) = \dfrac{1}{(t + 2)^2}$

11. $f'(s) = \dfrac{1}{\sqrt{s}(\sqrt{s} + 1)^2}$

13. $\dfrac{dy}{dx} = \dfrac{-4x^3 - 3x^2 + 1}{(x^2 - 1)^2 (x^2 + x + 1)^2}$

15. $\dfrac{ds}{dt} = -5t^{-2} + 2t^{-3}, \dfrac{d^2s}{dt^2} = 10t^{-3} - 6t^{-4}$

17. $\dfrac{dw}{dz} = -z^{-2} - 1, \dfrac{d^2w}{dz^2} = 2z^{-3}$

19. (a) $\dfrac{d}{dx}(uv)\Big|_{x=0} = 7$ **(b)** $\dfrac{d}{dx}\left(\dfrac{u}{v}\right)\Big|_{x=0} = -13$

(c) $\dfrac{d}{dx}\left(\dfrac{v}{u}\right)\Big|_{x=0} = \dfrac{13}{25}$ **(d)** $\dfrac{d}{dx}(7v - 2u)\Big|_{x=0} = 8$

21. $y = 4x$ at $(0, 0)$, $y = 2$ at $(1, 2)$
23. $a = b = 1$ and $c = 0$, so $y = x^2 + x$

25. $\dfrac{d}{dx}(u \cdot c) = u \cdot \dfrac{dc}{dx} + c \cdot \dfrac{du}{dx} = u \cdot 0 + c\dfrac{du}{dx} = c\dfrac{du}{dx} \Rightarrow$ the
Constant Multiple Rule is a special case of the Product Rule

27. (a) $\dfrac{d}{dx}(uvw) = uvw' + uv'w + u'vw$

(b) $\dfrac{d}{dx}(u_1 u_2 u_3 u_4) = u_1 u_2 u_3 u'_4 + u_1 u_2 u'_3 u_4 + u_1 u'_2 u_3 u_4 +$
$u'_1 u_2 u_3 u_4$

(c) $\dfrac{d}{dx}(u_1 \cdots u_n) = u_1 u_2 \cdots u_{n-1} u'_n + u_1 u_2 \cdots u_{n-2} u'_{n-1} u_n +$
$\cdots + u'_1 u_2 \cdots u_n$

29. $\dfrac{dP}{dV} = -\dfrac{nRT}{(V - nb)^2} + \dfrac{2an^2}{V^3}$

Section 2.4, pp. 184–186

1. $-10 - 3\sin x$ **3.** $-\csc x \cot x - \dfrac{2}{\sqrt{x}}$

5. 0   **7.** $\dfrac{-\csc^2 x}{(1 + \cot x)^2}$

9. $4\tan x \sec x - \csc^2 x$ **11.** $x^2 \cos x$

13. $\sec^2 t - 1$ **15.** $\dfrac{-2\csc t \cot t}{(1 - \csc t)^2}$

17. $-\theta(\theta \cos \theta + 2\sin \theta)$
19. $\sec \theta \csc \theta(\tan \theta - \cot \theta) = \sec^2 \theta - \csc^2 \theta$

21. $\sec^2 q$

23. $\sec^2 q$

25. (a) $2 \csc^3 x - \csc x$

   (b) $2 \sec^3 x - \sec x$

27.

29.

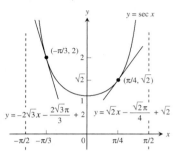

31. Yes, at $x = \pi$

33. No

35. $\left(-\dfrac{\pi}{4}, -1\right); \left(\dfrac{\pi}{4}, 1\right)$

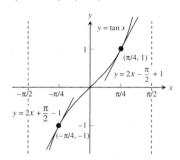

37. (a) $y = -x + \dfrac{\pi}{2} + 2$

   (b) $y = 4 - \sqrt{3}$

39. $-\sqrt{2}$ m/sec, $\sqrt{2}$ m/sec, $\sqrt{2}$ m/sec^2, $\sqrt{2}$ m/sec^3

41. $c = 9$

43. $\sin x$

Section 2.5, pp. 195–198

1. $12x^3$

3. $3 \cos (3x + 1)$

5. $10 \sec^2 (10x - 5)$

7. With $u = (4 - 3x)$, $y = u^9$: $\dfrac{dy}{dx} = \dfrac{dy}{du}\dfrac{du}{dx} = 9u^8 \cdot (-3) = -27(4 - 3x)^8$

9. With $u = \left(\dfrac{x^2}{8} + x - \dfrac{1}{x}\right)$, $y = u^4$: $\dfrac{dy}{dx} = \dfrac{dy}{du}\dfrac{du}{dx} =$

$4u^3 \cdot \left(\dfrac{x}{4} + 1 + \dfrac{1}{x^2}\right) = 4\left(\dfrac{x^2}{8} + x - \dfrac{1}{x}\right)^3 \left(\dfrac{x}{4} + 1 + \dfrac{1}{x^2}\right)$

11. With $u = \left(\pi - \dfrac{1}{x}\right)$, $y = \cot u$: $\dfrac{dy}{dx} = \dfrac{dy}{du}\dfrac{du}{dx} = (-\csc^2 u)\left(\dfrac{1}{x^2}\right) =$

$-\dfrac{1}{x^2} \csc^2 \left(\pi - \dfrac{1}{x}\right)$

13. $\dfrac{1 - r}{\sqrt{2r - r^2}}$

15. $\dfrac{\csc \theta}{\cot \theta + \csc \theta}$

17. $2x \sin^4 x + 4x^2 \sin^3 x \cos x + \cos^{-2} x + 2x \cos^{-3} x \sin x$

19. $(3x - 2)^6 - \dfrac{1}{x^3(4 - 1/2x^2)^2}$

21. $\sqrt{x} \sec^2 (2\sqrt{x}) + \tan (2\sqrt{x})$

23. $\dfrac{2 \sin \theta}{(1 + \cos \theta)^2}$

25. $\dfrac{dr}{d\theta} = (\sec \sqrt{\theta})\left[\dfrac{\tan \sqrt{\theta} \tan (1/\theta)}{2\sqrt{\theta}} - \dfrac{\sec^2 (1/\theta)}{\theta^2}\right]$

27. $2\pi \sin (\pi t - 2) \cos (\pi t - 2)$

29. $\dfrac{\csc^2 (t/2)}{[1 + \cot (1/2)]^3}$

31. $\left[1 + \tan^4\left(\dfrac{t}{12}\right)\right]^2 \left[\tan^3\left(\dfrac{t}{12}\right) \sec^2\left(\dfrac{t}{12}\right)\right]$

33. $y = -x + 2\sqrt{2}$, $\dfrac{d^2y}{dx^2}\Big|_{t=\pi/4} = -\sqrt{2}$

35. $y = x + \dfrac{1}{4}$, $\dfrac{d^2y}{dx^2}\Big|_{t=1/4} = -2$

37. $y = x - 4$, $\dfrac{d^2y}{dx^2}\Big|_{t=-1} = \dfrac{1}{2}$

39. $y = 2$, $\dfrac{d^2y}{dx^2}\Big|_{t=\pi/2} = -1$

41. $\dfrac{6}{x^3}\left(1 + \dfrac{1}{x}\right)\left(1 + \dfrac{2}{x}\right)$

43. $2 \csc^2 (3x - 1) \cot (3x - 1)$

45. 5/2

47. $-\pi/4$

49. 0

51. (a) 2/3

   (b) $2\pi + 5$

   (c) $15 - 8\pi$

   (d) 37/6

   (e) -1

   (f) $\sqrt{2}/24$

   (g) 5/32

   (h) $\dfrac{-5}{3\sqrt{17}}$

53. 5

55. (a) 1

   (b) 1

57. (a) $y = \pi x + 2 - \pi$

   (b) $\pi/2$

59. It multiplies the velocity, acceleration, and jerk by 2, 4, and 8, respectively.

61. $v(t) = 2/5$ m/sec, a(6) $= -4/125$ m/sec^2

71. $\left(\dfrac{\sqrt{2}}{2}, 1\right)$, $y = 2x$ at $t = 0$, $y = -2x$ at $t = \pi$

Section 2.6, pp. 204–206

1. $\dfrac{9}{4}x^{5/4}$

3. $\dfrac{7}{2(x + 6)^{1/2}}$

5. $\dfrac{2x^2 + 1}{(x^2 + 1)^{1/2}}$

7. $\dfrac{ds}{dt} = \dfrac{2}{7}t^{-5/7}$

9. $\dfrac{dy}{dt} = -\dfrac{4}{3}(2t + 5)^{-5/3} \cos [(2t + 5)^{-2/3}]$

11. $g'(x) = \frac{2}{3}(2x^{-1/2} + 1)^{-4/3}(x^{-3/2})$ **13.** $\dfrac{-2xy - y^2}{x^2 + 2xy}$

15. $\dfrac{y - 3x^2}{3y^2 - x}$ **17.** $\dfrac{1}{y(x + 1)^2}$

19. $\cos^2 y$

21. $\dfrac{-y^2}{y \sin\left(\frac{1}{y}\right) - \cos\left(\frac{1}{y}\right) + xy}$ **23.** $-\dfrac{\sqrt{r}}{\sqrt{\theta}}$

25. $\dfrac{-r}{\theta}$, $\cos(r\theta) \neq 0$

27. $y' = -\left(\dfrac{y}{x}\right)^{1/3}$, $y'' = \dfrac{y^{1/3}}{3x^{4/3}} + \dfrac{1}{3y^{1/3}x^{2/3}}$

29. $y' = \dfrac{\sqrt{y}}{\sqrt{y} + 1}$, $= y'' \dfrac{1}{2(\sqrt{y} + 1)^3}$

31. -2

33. 0

35. -6

37. $(-2, 1)$: $m = -1$, $(-2, -1)$: $m = 1$

39. (a) $y = \dfrac{7}{4}x - \dfrac{1}{2}$ **(b)** $y = -\dfrac{4}{7}x + \dfrac{29}{7}$

41. (a) $y = -x - 1$ **(b)** $y = x + 3$

43. (a) $y = -\dfrac{\pi}{2}x + \pi$ **(b)** $y = \dfrac{2}{\pi}x - \dfrac{2}{\pi} + \dfrac{\pi}{2}$

45. (a) $y = 2\pi x - 2\pi$ **(b)** $y = -\dfrac{x}{2\pi} + \dfrac{1}{2\pi}$

47. Points: $(-\sqrt{7}, 0)$ and $(\sqrt{7}, 0)$; slope: -2

49. $m = -1$ at $\left(\dfrac{\sqrt{3}}{4}, \dfrac{\sqrt{3}}{2}\right)$, $m = \sqrt{3}$ at $\left(\dfrac{\sqrt{3}}{4}, \dfrac{1}{2}\right)$

51. $(-3, 2)$: $m = -\dfrac{27}{8}$; $(-3, -2)$: $m = \dfrac{27}{8}$; $(3, 2)$:

$m = \dfrac{27}{8}$; $(3, -2)$: $m = -\dfrac{27}{8}$

53. (a) False   **(b)** True
    (c) True   **(d)** True

55. $(3, -1)$ **57.** $a = \dfrac{3}{4}$

59. $\dfrac{dy}{dx} = -\dfrac{y^3 + 2xy}{x^2 + 3xy^2}$, $\dfrac{dx}{dy} = -\dfrac{x^2 + 3xy^2}{y^3 + 2xy}$, $\dfrac{dx}{dy} = \dfrac{1}{dy/dx}$

Section 2.7, pp. 212–216

1. $\dfrac{dA}{dt} = 2\pi r \dfrac{dr}{dt}$

3. (a) $\dfrac{dV}{dt} = \pi r^2 \dfrac{dh}{dt}$

(b) $\dfrac{dV}{dt} = 2\pi h r \dfrac{dr}{dt}$

(c) $\dfrac{dV}{dt} = \pi r^2 \dfrac{dh}{dt} + 2\pi h r \dfrac{dr}{dt}$

5. (a) 1 volt/sec

(b) $-\dfrac{1}{3}$ amp/sec

(c) $\dfrac{dR}{dt} = \dfrac{1}{I}\left(\dfrac{dV}{dt} - \dfrac{V}{I}\dfrac{dI}{dt}\right)$

(d) $\dfrac{3}{2}$ ohms/sec, R is increasing

7. (a) $\dfrac{ds}{dt} = \dfrac{x}{\sqrt{x^2 + y^2}}\dfrac{dx}{dt}$

(b) $\dfrac{ds}{dt} = \dfrac{x}{\sqrt{x^2 + y^2}}\dfrac{dx}{dt} + \dfrac{y}{\sqrt{x^2 + y^2}}\dfrac{dy}{dt}$

(c) $\dfrac{dx}{dt} = -\dfrac{y}{x}\dfrac{dy}{dt}$

9. (a) $\dfrac{dA}{dt} = \dfrac{1}{2}ab \cos\theta \dfrac{d\theta}{dt}$

(b) $\dfrac{dA}{dt} = \dfrac{1}{2}ab \cos\theta \dfrac{d\theta}{dt} + \dfrac{1}{2}b \sin\theta \dfrac{da}{dt}$

(c) $\dfrac{dA}{dt} = \dfrac{1}{2}ab \cos\theta \dfrac{d\theta}{dt} + \dfrac{1}{2}b \sin\theta \dfrac{da}{dt} + \dfrac{1}{2}a \sin\theta \dfrac{db}{dt}$

11. (a) 14 cm²/sec, increasing
(b) 0 cm/sec, constant
(c) $-\dfrac{14}{13}$ cm/sec, decreasing

13. (a) -12 ft/sec   **(b)** -59.5 ft²/sec   **(c)** -1 rad/sec

15. 20 ft/sec

17. (a) $\dfrac{dh}{dt} = 11.19$ cm/min   **(b)** $\dfrac{dr}{dt} = 14.92$ cm/min

19. (a) $-\dfrac{1}{24\pi}$ m/min

(b) $r = \sqrt{26y - y^2}$ m

(c) $\dfrac{dr}{dt} = -\dfrac{5}{288\pi}$ m/min

21. 1 ft/min, 40π ft²/min   **23.** 11 ft/sec

25. Increasing at $\dfrac{466}{1681}$ L/min   **27.** 1 rad/sec

29. -5 m/sec   **31.** -1500 ft/sec

33. $\dfrac{5}{72\pi}$ in./min, $\dfrac{10}{3}$ in.²/min   **35.** 7.1 in./min

37. (a) $-32/\sqrt{13} \approx -8.875$ ft/sec

(b) $\dfrac{d\theta_1}{dt} = -\dfrac{8}{65}$ rad/sec, $\dfrac{d\theta_2}{dt} = \dfrac{8}{65}$ rad/sec

(c) $\dfrac{d\theta_1}{dt} = -\dfrac{1}{6}$ rad/sec, $\dfrac{d\theta_2}{dt} = \dfrac{1}{6}$ rad/sec

Section 2.8, pp. 221–222

1. $\dfrac{-2x}{\sqrt{1 - x^4}}$ **3.** $\dfrac{\sqrt{2}}{\sqrt{1 - 2t^2}}$

5. $\dfrac{1}{|2s + 1|\sqrt{s^2 + s}}$ **7.** $\dfrac{-2x}{(x^2 + 1)\sqrt{x^4 + 2x^2}}$

9. $\dfrac{-1}{\sqrt{1 - t^2}}$ **11.** $\dfrac{-1}{2\sqrt{t}(1 + t)}$

13. $\dfrac{-2s^2}{\sqrt{1-s^2}}$ **15.** 0

17. $\sin^{-1} x$

19. (a) $y = 2x - \dfrac{\pi}{2} + 1$ **(b)** $y = \dfrac{1}{2}x - \dfrac{1}{2} + \dfrac{\pi}{4}$

21. (a) $f'(x) = 3 - \sin x$ and $f'(x) \neq 0$. So, f has a differentiable inverse by Theorem 3.

(b) $f(0) = 1, f'(0) = 3$

(c) $f^{-1}(1) = 0, (f^{-1})'(1) = \dfrac{1}{3}$

23. $-\dfrac{1}{\sqrt{1-x^2}}$ **25.** $-\dfrac{1}{|x|\sqrt{x^2-1}}$

27. (a) $y = \dfrac{\pi}{2}$ **(b)** $y = -\dfrac{\pi}{2}$ **(c)** None

29. (a) $y = \dfrac{\pi}{2}$ **(b)** $y = \dfrac{\pi}{2}$ **(c)** None

31. (a) $f^{-1}(x) = \dfrac{x}{2} - \dfrac{3}{2}$

(b)   **(c)** 2, 1/2

33. (a) $f^{-1}(x) = -\dfrac{x}{4} + \dfrac{5}{4}$

(b)   **(c)** $-4, -1/4$

35. (b)

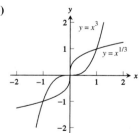

(c) Slope of f at $(1, 1)$: 3; slope of g at $(1, 1)$: 1/3; slope of f at $(-1, -1)$: 3; slope of g at $(-1, -1)$: 1/3

(d) $y = 0$ is tangent to $y = x^3$ at $x = 0$; $x = 0$ is tangent to $y = \sqrt[3]{x}$ at $x = 0$.

37. (a) $f^{-1}(x) = \dfrac{1}{m}x$

(b) The graph of f^{-1} is the line through the origin with slope $\dfrac{1}{m}$.

Section 2.9, pp. 231–232

1. $2e^x$ **3.** $-\dfrac{3}{2}e^{-3x/2}$

5. $\dfrac{2}{3}e^{2x/3}$ **7.** xe^x

9. $\dfrac{e^{\sqrt{x}}}{2\sqrt{x}}$ **11.** $\pi x^{\pi-1}$

13. $-\sqrt{2}x^{-\sqrt{2}-1}$ **15.** $8^x \ln 8$

17. $-3^{\csc x}(\ln 3)(\csc x \cot x)$ **19.** $\dfrac{e^x + 2}{e^{-2x} + 2e^{-x} + 1}$

21. $\dfrac{2}{x}$

23. $-\dfrac{1}{x}, x > 0$ **25.** $\dfrac{1}{x+2}, x > -2$

27. $\dfrac{\sin x}{2 - \cos x}$ **29.** $\dfrac{1}{x \ln x}$

31. $\dfrac{2}{x \ln 4} = \dfrac{1}{x \ln 2}$ **33.** $\dfrac{3}{(3x+1) \ln 2}, x > -\dfrac{1}{3}$

35. $-\dfrac{1}{x \ln 2}, x > 0$ **37.** $\dfrac{1}{x}, x > 0$

39. $\dfrac{1}{\ln 10}$ **41.** $(x^{\ln x})\left(\dfrac{\ln x^2}{x}\right)$

43. $(\sin x)^x [x \cot x + \ln(\sin x)]$

45. $\left(\dfrac{(x-3)^4(x^2+1)}{(2x+5)^3}\right)^{1/5}\left(\dfrac{4}{5(x-3)} + \dfrac{2x}{5(x^2+1)} - \dfrac{6}{5(2x+5)}\right)$

47. $y = ex$ **49.** Rate ≈ 0.098 gram/day

51. Answers will vary considerably with the number of decimal places retained. With (a) $k = 0.008754168326$, the population 8 years out is (b) 275,979,963 (nearest integer).

53. (a) $g'(0) = L$ **(b)** ≈ -0.6931

57. (a) $y = \dfrac{1}{e}x$

(b) Because the graph of $\ln x$ lies below the graph of the line for all positive $x \neq e$.

(c) Multiplying by e, $e(\ln x) < x$, or $\ln x^e < x$.

(d) Exponentiate both sides of the inequality in part (c).

(e) Let $x = \pi$ to see that $\pi^e < e^\pi$.

Chapter 2 Practice Exercises, pp. 233–238

1. $5x^4 - 0.25x + 0.25$ **3.** $7x^6 + \sqrt{7}$

5. $3(\theta^2 + \sec \theta + 1)^2(2\theta + \sec \theta \tan \theta)$

7. $-\dfrac{1}{2\sqrt{t}(\sqrt{t}-1)^2}$ **9.** $(-2 \csc x \cot x)(\csc x - 1)$

11. $\dfrac{6}{t^2} \cot^2\left(\dfrac{2}{t}\right) \csc^2\left(\dfrac{2}{5}\right)$ **13.** $\dfrac{\theta \cos \theta + \sin \theta}{\sqrt{2\theta \sin \theta}}$

15. $x \csc\left(\dfrac{2}{x}\right) + \csc\left(\dfrac{2}{x}\right) \cot\left(\dfrac{2}{x}\right)$ **17.** $-10x \csc^2(x^2)$

19. $\dfrac{-(t+1)}{8t^3}$ **21.** $\dfrac{6x + 5\sqrt{x}}{\sqrt{x + \sqrt{x}}}$

23. $\dfrac{3}{(3x-4)^{19/20}}$ **25.** xe^{4x}

27. $\dfrac{2 \sin \theta \cos \theta}{\sin^2 \theta} = 2 \cot \theta$

29. $\dfrac{3}{(\ln 5)(3x - 7)}$

31. $18x^{2.6}$

33. $(x + 2)^{x+2}(\ln(x + 2) + 1)$

35. $-\dfrac{1}{\sqrt{1 - u^2}}$

37. $\cos^{-1} z$

39. $2t \cot^{-1}(2t) + (1 + t^2)\left(\dfrac{-2}{1 + 4t^2}\right)$

41. -1

43. $-\dfrac{y + 2}{x + 3}$

45. $\dfrac{-3x^2 - 4y + 2}{4x - 4y^{1/3}}$

47. $-\dfrac{y}{x}$

49. $-1/2$

51. y/x

53. $-\dfrac{2e^{-\tan^{-1}x}}{1 + x^2}$

55. $\dfrac{dr}{ds} = (2r - 1)(\tan 2s)$

57. (a) $\dfrac{d^2y}{dx^2} = \dfrac{-2xy^3 - 2x^4}{y^5}$

(b) $\dfrac{d^2y}{dx^2} = \dfrac{-2xy^2 - 1}{x^4y^3}$

59. (a) 7

(b) -2

(c) $5/12$

(d) $\dfrac{1}{4}$

(e) 12

(f) $\dfrac{9}{2}$

(g) $\dfrac{3}{4}$

61. 0

63. $\dfrac{3\sqrt{2}e^{\sqrt{3/2}}}{4} \cos(e^{\sqrt{3/2}})$

65. $-1/2$

67. $\dfrac{-2}{(2t + 1)^2}$

69. (a)

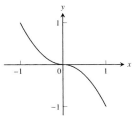

$f(x) = \begin{cases} x^2, & -1 \le x < 0 \\ -x^2, & 0 \le x \le 1 \end{cases}$

(b) Yes

(c) Yes

71. (a)

$f(x) = \begin{cases} x, & 0 \le x \le 1 \\ 2 - x, & 1 < x \le 2 \end{cases}$

(b) Yes

(c) No

73. $\left(\dfrac{5}{2}, \dfrac{9}{4}\right)$ and $\left(\dfrac{3}{2}, -\dfrac{1}{4}\right)$

75. $(0, 16)$ and $\left(-\dfrac{4}{3}, 0\right)$

79. $1/4$

81. Tangent: $y = -\dfrac{1}{4}x + \dfrac{9}{4}$; normal: $y = 4x - 2$

83. Tangent: $y = 2x - 4$; normal: $y = -\dfrac{1}{2}x + \dfrac{7}{2}$

85. Tangent: $y = -\dfrac{5}{4}x + 6$; normal: $y = \dfrac{4}{5}x - \dfrac{11}{5}$

87. $(1, 1)$: $m = -\dfrac{2}{3}$, $(1, -1)$: $m = 4$

89. $B = $ graph of f, $A = $ graph of f'

91.

93. (a) $0, 0$

(b) 1700 rabbits ≈ 1400 rabbits

95. (a) $\dfrac{dS}{dt} = (4\pi r + 2\pi h)\dfrac{dr}{dt}$

(b) $\dfrac{dS}{dt} = 2\pi r \dfrac{dh}{dt}$

(c) $\dfrac{dS}{dt} = (4\pi r + 2\pi h)\dfrac{dr}{dt} + 2\pi r \dfrac{dh}{dt}$

(d) $\dfrac{dr}{dt} = -\dfrac{r}{2r + h}\dfrac{dh}{dt}$

97. $40 \text{ m}^2/\text{sec}$

99. 0.02 ohm/sec

101. 2 m/sec

103. (a) $r = \dfrac{2}{5}h$

(b) $-\dfrac{125}{144\pi}$ ft/min

105. (a) $\dfrac{3}{5}$ km/sec or 600 m/sec

(b) $\dfrac{18}{\pi}$ rpm

107. $\dfrac{1}{e}$ m/sec

109. $\ln 5x - \ln 3x = \ln\left(\dfrac{5}{3}\right)$

111. $\dfrac{1}{2}$

113. $y'(r) = -\dfrac{1}{2r^2 l}\sqrt{\dfrac{T}{\pi d}}$, so increasing r decreases the frequency.

$y'(l) = -\dfrac{1}{2rl^2}\sqrt{\dfrac{T}{\pi d}}$, so increasing l decreases the frequency.

$y'(d) = -\dfrac{1}{4rl}\sqrt{\dfrac{T}{\pi d^3}}$, so increasing d decreases the frequency.

$y'(T) = \dfrac{1}{4rl\sqrt{\pi Td}}$, so increasing T increases the frequency.

115. (a) $x \neq k\frac{\pi}{4}$, where k is an odd integer

(b) $\left(-\frac{\pi}{2}, \frac{\pi}{2}\right)$

(c) Where it is not defined, at $x = k\frac{\pi}{4}$, k an odd integer

(d) It has period $\pi/2$ and continues to repeat the pattern seen in this window.

Chapter 2 Additional Exercises, pp. 238–240

1. (a) $\sin 2\theta = 2\sin\theta\cos\theta$; $2\cos 2\theta = 2\sin\theta(-\sin\theta) + \cos\theta(2\cos\theta)$; $2\cos 2\theta = -2\sin^2\theta + 2\cos^2\theta$; $\cos 2\theta = \cos^2\theta - \sin^2\theta$

(b) $\cos 2\theta = \cos^2\theta - \sin^2\theta$; $-2\sin 2\theta = 2\cos\theta(-\sin\theta) - 2\sin\theta(\cos\theta)$; $\sin 2\theta = \cos\theta\sin\theta + \sin\theta\cos\theta$; $\sin 2\theta = 2\sin\theta\cos\theta$

3. (a) $a = 1, b = 0, c = -1/2$

(b) $b = \cos a, c = \sin a$

5. $h = -4, k = \frac{9}{2}, a = \frac{5\sqrt{5}}{2}$

7. (a) $0.09y$

b) Increasing at 1% per year

9. Answers will vary. Here is one possibility.

11. (a) 2 sec, 64 ft/sec **(b)** ≈ 12.3 sec, ≈ 394 ft

13. $mv\dfrac{dv}{dt} = -kx\dfrac{dx}{dt} = -kxv \rightarrow m\dfrac{dv}{dt} = -kx$

15. (a) $m = -\dfrac{b}{\pi}$ **(b)** $m = -1, b = \pi$

17. (a) $a = 3/4, b = 9/4$ **19.** f odd $\Rightarrow f'$ is even

23. h' is defined but not continuous at $x = 0$; k' is defined *and* continuous at $x = 0$.

CHAPTER 3 APPLICATIONS OF DERIVATIVES

Section 3.1, pp. 251–254

1. Absolute minimum at $x = c_2$, absolute maximum at $x = b$
3. Absolute maximum at $x = c$, no absolute minimum
5. Absolute minimum at $x = a$, absolute maximum at $x = c$
7. Local minimum at $(-1, 0)$, local maximum at $(1, 0)$
9. Maximum at $(0, 5)$
11. (c) **13.** (d)

15. Absolute maximum: -3; absolute minimum: $-\dfrac{19}{3}$

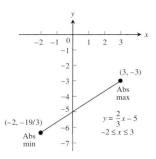

17. Absolute maximum: 4; absolute minimum: -5

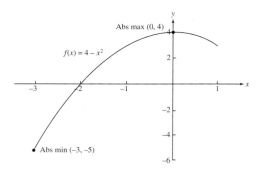

19. Maximum value is 1 at $x = \dfrac{\pi}{4}$; minimum value is -1 at $x = \dfrac{5\pi}{4}$; local minimum at $\left(0, \dfrac{1}{\sqrt{2}}\right)$; local maximum at $\left(\dfrac{7\pi}{4}, 0\right)$

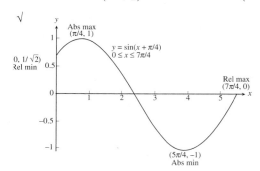

21. Absolute maximum: -0.25, absolute minimum: -4

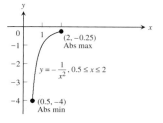

23. Maximum value is $1/4 + \ln 4$ at $x = 4$; minimum value is 1 at $x = 1$; local maximum at $(1/2, 2 - \ln 2)$

25. Maximum value is $\ln 4$ at $x = 3$; minimum value is 0 at $x = 0$.

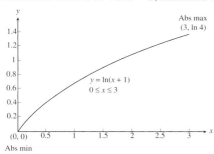

27. Minimum value is 1 at $x = 2$.
29. Minimum value is 1 at $x = 0$.
31. Maximum value is $1/e$ at $x = e$; there is no minimum value.
33. Maximum value is $1/2$ at $x = 1$; minimum value is $-1/2$ at $x = -1$.
35. Local maximum at $(-2, 17)$; local minimum at $\left(\dfrac{4}{3}, -\dfrac{41}{27}\right)$

37.

Critical point	Derivative	Extremum	Value
$x = -\dfrac{4}{5}$	0	Local max	$\dfrac{12}{25}\,10^{1/3} = 1.034$
$x = 0$	Undefined	Local min	0

39.

Critical point	Derivative	Extremum	Value
$x = -2$	Undefined	Local max	0
$x = -\sqrt{2}$	0	Minimum	-2
$x = \sqrt{2}$	0	Maximum	2
$x = 2$	Undefined	Local min	0

41.

Critical point	Derivative	Extremum	Value
$x = 1$	Undefined	Minimum	2

43.

Critical point	Derivative	Extremum	Value
$x = -1$	0	Maximum	5
$x = 1$	Undefined	Local min	1
$x = 3$	0	Maximum	5

45. (a) No
  (b) The derivative is defined and nonzero for $x \neq 2$. Also, $f(2) = 0$ and $f(x) > 0$ for all $x \neq 2$.
  (c) No, because $(-\infty, \infty)$ is not a closed interval.
  (d) The answers are the same as parts (a) and (b) with 2 replaced by a.
47. (a) $C(x) = 0.3\sqrt{16 + x^2} + 0.2(9 - x)$ million dollars, where $0 \leq x \leq 9$ mi. To minimize the cost of construction, the pipeline should be placed from the docking facility to point B, 3.58 mi along the shore from point A, and then along the shore from B to the refinery.
  (b) In theory, the underwater pipe cost per mile p would have to be infinite to justify running the pipe directly from the docking facility to point A (i.e., for x_c to be zero). For all values of $p > 0.218864$, there is always an x_c element of ϵ $(0, 9)$ that will give a minimum value for C. This is proved by looking at $C''(x_c) = \dfrac{16p}{(16 + x_c^2)^{3/2}}$, which is always positive for $p > 0$.

49. The length of pipeline is $L(x) = \sqrt{4 + x^2} + \sqrt{25 + (10 - x)^2}$ for $0 \leq x \leq 10$. Setting the derivative of $L(x)$ equal to zero gives $L'(x) = \dfrac{x}{\sqrt{4 + x^2}} - \dfrac{(10 - x)}{\sqrt{25 + (10 - x)^2}} = 0$. Note that $\dfrac{x}{\sqrt{4 + x^2}} = \cos\theta_A$ and $\dfrac{10 - x}{\sqrt{25 - (10 - x)^2}} = \cos\theta_B$; therefore, $L'(x) = 0$ when $\cos\theta_A = \cos\theta_B$, or $\theta_A = \theta_B$. Use simple proportions to determine x as follows: $\dfrac{x}{2} = \dfrac{10 - x}{5} \Rightarrow x = \dfrac{20}{7} \approx 2.857$ mi along the coast from town A to town B.

51. (a) Maximum value is 144 at $x = 2$.
  (b) The largest volume of the box is 144 cubic units, and it occurs when $x = 2$.
53. The largest possible area is $A\left(\dfrac{5}{\sqrt{2}}\right) = \dfrac{25}{4}$ cm^2.
55. $\dfrac{v_0^2}{2g} + s_0$
57. Yes
59. g assumes a local maximum at $-c$.

61. (a) $f'(x) = 3ax^2 + 2bx + c$ is a quadratic, so it can have 0, 1, or 2 zeros, which would be the critical points of f. Examples: The function $f(x) = x^3 - 3x$ has two critical points at $x = -1$ and $x = 1$.

$y = x^3 - 3x$

The function $f(x) = x^3 - 1$ has one critical point at $x = 0$.

$y = x^3 - 1$

The function $f(x) = x^3 + x$ has no critical points.

$y = x^3 + x$

(b) Two or none

63. Maximum value is 11 at $x = 5$; minimum value is 5 on the interval $[-3, 2]$; local maximum is at $(-5, 9)$.

$f(x) = |x - 2| + |x + 3|$

65. Maximum value is 5 on the interval $[3, \infty)$; minimum value is -5 on the interval $[-\infty, -2]$

$f(x) = |x - 2| + |x - 3|$

Section 3.2, pp. 261–263

1. (a) f is continuous on $[0, 1]$ and differentiable on $(0, 1)$.

(b) $c = \dfrac{1}{2}$

3. (a) f is continuous on $[-1, 1]$ and differentiable on $(-1, 1)$.

(b) $c = \pm\sqrt{1 - \dfrac{4}{\pi^2}} \approx \pm 0.771$

5. The function $f(x)$ is not continuous on $0 \le x \le 1$ because $\lim_{x \to 1^-} f(x) = \lim_{x \to 1^-} x = 1 \ne 0 = f(1)$; therefore, Rolle's Theorem does not apply because f is not continuous at $x = 1$.

7. By Corollary 1 of the Mean Value Theorem, $f'(x) = 0$ for all $x \Rightarrow f(x) = C$, where C is a constant. Since $f(-1) = 3$, we have $C = 3 \Rightarrow f(x) = 3$ for all x.

9. (a) $y = \dfrac{x^2}{2} + C$ **(b)** $y = \dfrac{x^3}{3} + C$ **(c)** $y = \dfrac{x^4}{4} + C$

11. (a) $y = \ln|\theta| + C$
(b) $y = \theta - \ln|\theta| + C$
(c) $y = 5\theta + \ln|\theta| + C$

13. $f(x) = x^2 - x$ **15.** $f(x) = 1 + \dfrac{e^{2x}}{2}$

17. $s = 4.9t^2 + 5t + 10$ **19.** $s = \dfrac{1 - \cos(\pi t)}{\pi}$

23. $s = \sin(2t) - 3$ **25.** 48 m/sec

27. 14 m/sec

29. (a) $v = 10t^{3/2} - 6t^{1/2}$ **(b)** $s = 4t^{5/2} - 4t^{3/2}$

31. If $T(t)$ is the temperature of the thermometer at time t, then $T(0) = -19°C$ and $T(14) = 100°C$. From the Mean Value Theorem, there exists a $0 < t_0 < 14$ such that $\dfrac{T(14) - T(0)}{14 - 0} = 8.5°C/sec = T'(t_0)$, the rate at which the temperature was changing at $t = t_0$ as measured by the rising mercury on the thermometer.

33. Because its average speed was approximately 7.667 knots, and by the Mean Value Theorem, it must have been going that speed at least once during the trip.

35. The conclusion of the Mean Value Theorem yields $\dfrac{1/b - 1/a}{b - a} = -\dfrac{1}{c^2} \Rightarrow c^2\left(\dfrac{a - b}{ab}\right) = a - b \Rightarrow c = \sqrt{ab}$.

39. $f(x)$ must be zero at least once between a and b by the Intermediate Value Theorem. Now suppose that $f(x)$ is zero twice between a and b. Then by the Mean Value Theorem, $f'(x)$ would have to be zero at least once between the two zeros of $f(x)$, but this cannot be true since we are given that $f'(x) \ne 0$ on this interval. Therefore, $f(x)$ is zero once and only once between a and b.

45. $1.09999 \le f(0.1) \le 1.1$

Section 3.3, pp. 272–276

1.

Loc max

Infl

P

Loc min

y'' y'

y

3.

5.

7. (a) Zero: $x = \pm 1$; positive: $(-\infty, -1)$ and $(1, \infty)$; negative: $(-1, 1)$
   (b) Zero: $x = 0$; positive: $(0, \infty)$; negative: $(-\infty, 0)$
9. (a) $(-\infty, -2]$ and $[0, 2]$
   (b) $[-2, 0]$ and $[2, \infty)$
   (c) Local maxima: $x = -2$ and $x = 2$; local minimum: $x = 0$
11. (a) $[0, 1]$, $[3, 4]$, and $[5.5, 6]$
   (b) $[1, 3]$ and $[4, 5.5]$
   (c) Local maxima: $x = 1$, $x = 4$ (if f is continuous at $x = 4$), and $x = 6$; local minima: $x = 0$, $x = 3$, and $x = 5.5$
13. (a) Critical points at -2 and 1
   (b) Increasing on $(-\infty, -2]$ and $[1, \infty)$, decreasing on $[-2, 1]$
   (c) Local maximum at $x = -2$ and a local minimum at $x = 1$
15. (a) Critical point at $x = 1$
   (b) Increasing on $[1, \infty)$, decreasing on $(-\infty, 1]$
   (c) Local (and absolute) minimum at $x = 1$
17. (a) $\left[\dfrac{1}{2}, \infty\right)$ **(b)** $\left(2\infty, \dfrac{1}{2}\right]$
   (c) $(-\infty, \infty)$ **(d)** Nowhere
19. (a) $[-1, 0]$ and $[1, \infty)$
   (b) $(-\infty, -1]$ and $[0, 1]$
   (c) $\left(-\infty, -\dfrac{1}{\sqrt{3}}\right)$ and $\left(\dfrac{1}{\sqrt{3}}, \infty\right)$
   (d) $\left(-\dfrac{1}{\sqrt{3}}, \dfrac{1}{\sqrt{3}}\right)$
   (e) Local maximum at $(0, 1)$; local (and absolute) minima at $(\pm 1, 1)$
   (f) $\left(\pm\dfrac{1}{\sqrt{3}}, -\dfrac{1}{9}\right)$
21. (a) $[-2, 2]$
   (b) $[-\sqrt{8}, -2]$ and $[2, \sqrt{8}]$
   (c) $(-\sqrt{8}, 0)$
   (d) $(0, \sqrt{8})$
   (e) Local maxima: $(-\sqrt{8}, 0)$ and $(2, 4)$; local minima: $(-2, -4)$ and $(\sqrt{8}, 0)$
   (f) $(0, 0)$

23. (a) $(-\infty, -2]$ and $\left[-\dfrac{3}{2}, \infty\right)$ **(b)** $\left[-2, -\dfrac{3}{2}\right]$
   (c) $\left(-\dfrac{7}{4}, \infty\right)$ **(d)** $\left(-\infty, -\dfrac{7}{4}\right)$
   (e) Local maximum: $(-2, -40)$; local minimum: $\left(-\dfrac{3}{2}, -\dfrac{161}{4}\right)$
   (f) $\left(-\dfrac{7}{4}, -\dfrac{321}{8}\right)$
25. (a) $(-\infty, \infty)$ **(b)** None   **(c)** $(-\infty, 0)$
   (d) $(0, \infty)$ **(e)** None   **(f)** $(0, 3)$
27. (a) $[1, \infty)$
   (b) $(-\infty, 1]$
   (c) $(-\infty, -2)$ and $(0, \infty)$
   (d) $(-2, 0)$
   (e) Local minimum: $(1, -3)$
   (f) $\approx(-2, 7.56)$ and $(0, 0)$
29. (a) $(-\infty, -\sqrt{2}]$ and $[\sqrt{2}, \infty)$
   (b) $[-\sqrt{2}, 0)$ and $(0, \sqrt{2}]$
   (c) $(0, \infty)$
   (d) $(-\infty, 0)$
   (e) Local maximum: $-\sqrt{2}, -\sqrt{2e} \approx (-1.41, -2.33)$; local minimum: $(\sqrt{2}, \sqrt{2e}) \approx (1.41, 2.33)$
   (f) None
31. (a) $[0, \infty)$ **(b)** None
   (c) $\left(\dfrac{9}{5}, \infty\right)$ **(d)** $\left(0, \dfrac{9}{5}\right)$
   (e) Local (and absolute) minimum: $(0, 0)$
   (f) $\left(\dfrac{9}{5}, \dfrac{24}{5} \cdot \sqrt[4]{\dfrac{9}{5}}\right) \approx (1.8, 5.56)$
33. (a) None
   (b) At $x = 2$
   (c) At $x = 1$ and $x = 5/3$
35. (a) Absolute maximum at $(1, 2)$; absolute minimum at $(3, -2)$
   (b) None
   (c) One possible answer:

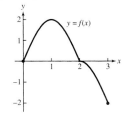

37. The zeros of y' are extrema of y. Inflection point at $x = 3$, local maximum at $(0, -240)$, local minimum at $(4, -496)$.

39. The zeros of $y' = 0$ and $y'' = 0$ are extrema and points of inflection, respectively. Inflection at $x = -\sqrt[3]{2}$, local maximum at $\left(-2, \dfrac{67}{5}\right)$, local minimum at $(0, -25)$.

43. (a) $v(t) = 2t - 4$
(b) $a(t) = 2$
(c) It begins at position 3 moving in a negative direction. It moves to position -1 when $t = 2$ and then changes direction, moving in a positive direction thereafter.

45. (a) $v(t) = 3t^2 - 3$
(b) $a(t) = 6t$
(c) It begins at position 3 moving in a negative direction. It moves to position 1 when $t = 1$ and then changes direction, moving in a positive direction thereafter.

47. (a) $t \approx 2.2, 6, 9.8$
(b) $t \approx 4, 8, 11$

49. No; f must have a horizontal tangent line at that point, but it could be increasing (or decreasing) on both sides of the point and there would be no local extremum.

51. One possible answer:

53. One possible answer:

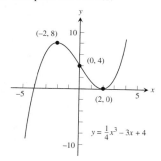

63. (b) $f'(x) = 3x^2 + k$; $-12k$; positive if $k < 0$, negative if $k > 0$, 0 if $k = 0$; f' has two zeros if $k < 0$, one zero if $k = 0$, no zeros if $k > 0$.

65. (a) $f'(x) = \dfrac{abce^{bx}}{(e^{bx} + a)^2}$, so the sign of $f'(x)$ is the same as the sign of the product abc.

(b) $f''(x) = \dfrac{ab^2ce^{bx}(e^{bx} - a)}{(e^{bx} + a)^3}$. Since $a > 0$, the equation changes sign when $x = \dfrac{\ln a}{b}$ due to the $e^{bx} - a$ factor in the numerator, and there is a point of inflection at that location.

Section 3.4, pp. 284–285

1. $y' = (y + 2)(y - 3)$
(a) $y = -2$ is a stable equilibrium value and $y = 3$ is an unstable equilibrium.
(b) $y'' = 2(y + 2)(y - 1/2)(y - 3)$

(c)

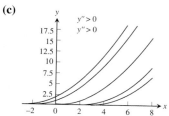

5. $y' = \sqrt{y}, y > 0$
(a) There are no equilibrium values.
(b) $y'' = \dfrac{1}{2}$

(c)

7. $y' = (y - 1)(y - 2)(y - 3)$

(a) $y = 1$ and $y = 3$ are unstable equilibria and $y = 2$ is a stable equilibrium.

(b) $y'' = (3y^2 - 12y + 11)(y - 1)(y - 2)(y - 3) =$

$$(y - 1)\left(y - \frac{6 - \sqrt{3}}{3}\right)(y - 2)\left(y - \frac{6 + \sqrt{3}}{3}\right)(y - 3)$$

(c)

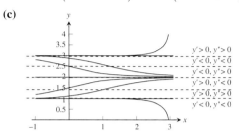

9. $\frac{dP}{dt} = 1 - 2P$ has a stable equilibrium at $P = \frac{1}{2}$; $\frac{d^2P}{dt^2} = -2\frac{dP}{dt} = -2(1 - 2P)$

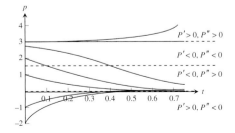

11. $\frac{dP}{dt} = 2P(P - 3)$ has a stable equilibrium at $P = 0$ and an unstable equilibrium at $P = 3$. $\frac{d^2P}{dt^2} = 2(2P - 3)\frac{dP}{dt} = 4P(2P - 3)(P - 3)$

13. Before the catastrophe, the population exhibits logistic growth and $P(t) \to M_0$, the stable equilibrium. After the catastrophe, the population declines logistically and $P(t) \to M_1$, the new stable equilibrium.

15. $\frac{dv}{dt} = g - \frac{k}{m}v^2$, $g, k, m > 0$ and $v(t) \geq 0$; equilibrium: $\frac{dv}{dt} = g - \frac{k}{m}v^2 = 0 \Rightarrow v = \sqrt{\frac{mg}{k}}$; concavity: $\frac{d^2v}{dt^2} = -2\left(\frac{k}{m}v\right)\frac{dv}{dt} = -2\left(\frac{k}{m}v\right)\left(g - \frac{k}{m}v^2\right)$

(a)

(b)

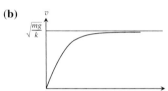

(c) $v_{\text{terminal}} = \sqrt{\frac{160}{0.005}} = 178.9$ ft/s $= 122$ mph

17. $F = F_p - F_r$; $ma = 50 - 5|v|$; $\frac{dv}{dt} = \frac{1}{m}(50 - 5|v|)$. The maximum velocity occurs when $\frac{dv}{dt} = 0$ or $v = 10$ ft/sec.

19. Phase line:

If the switch is closed at $t = 0$, then $i(0) = 0$, and the graph of the solution looks like this:

As $t \to \infty$, $i(t) \to i_{\text{steady state}} = V/R$.

Section 3.5, pp. 295–303

1. 16 in., 4 in. by 4 in.

3. **(a)** $(x, 1 - x)$
(b) $A(x) = 2x(1 - x)$
(c) 1/2 square units, $\frac{1}{2}$ in. $\times$ 1 in.

5. $\frac{2450}{27}$ in.^{3}, $\frac{14}{3} \times \frac{35}{3} \times \frac{5}{3}$ in.

7. 80,000 m^2, 400 m by 200 m

9. **(a)** The optimum dimensions of the tank are 10 ft on the base edges and 5 ft deep.
(b) Minimizing the surface area of the tank minimizes its weight for a given wall thickness. The thickness of the steel walls would likely be determined by other considerations such as structural requirements.

11. 9 $\times$ 18 in.

13. $\frac{\pi}{2}$

15. $h : r = 8 : \pi$

17. **(a)** $v(x) = 2x(24 - 2x)(18 - 2x)$
(b) Domain: $(0, 9)$

(c) Maximum volume $\approx$ 1309.95 in.^{3} when $x \approx 3.39$ in.
(d) $v'(x) = 24x^2 - 336x + 864$, so the critical point is at $x = 7 - \sqrt{13}$, which confirms the result in part (c).
(e) $x = 2$ in. or $x = 5$ in.

19. $r = \frac{10\sqrt{6}}{3}$, $h = \frac{20\sqrt{3}}{3}$, ≈ 2418.40 cm^3

21. **(a)** $h = 24$, $w = 18$
(b)

23. If r is the radius of the hemisphere, h the height of the cylinder, and V the volume, then $r = \left(\frac{3V}{8\pi}\right)^{1/3}$ and $h = \left(\frac{3V}{\pi}\right)^{1/3}$.

25. **(b)** $x = \frac{51}{8}$ **(c)** $L \approx 11$ in.

27. Radius $= \sqrt{2}$ m, height $= 1$ m, volume $\frac{2\pi}{3}$ m^3

29. $\frac{1}{\sqrt{2}}$ units long by $\frac{1}{\sqrt{e}}$ units high, $A = \frac{1}{\sqrt{2e}} \approx 0.43$ units^2

31. **(a)** $v(0) = 96$ ft/sec
(b) 256 ft at $t = 3$ sec
(c) Velocity when $s = 0$ is $v(7) = -128$ ft/sec

33. ≈ 46.87 ft **35.** 6 in. $\times 6\sqrt{3}$ in.

37. **(a)** $10\pi \approx 31.42$ cm/sec; when $t = 0.5$ sec, 1.5 sec, 2.5 sec, 3.5 sec; $s = 0$ acceleration is 0
(b) 10 cm from rest position; speed is 0

39. **(a)** $s = ((12 - 12t)^2 + 64t^2)^{1/2}$
(b) -12 knots, 8 knots
(c) No
(d) $4\sqrt{13}$. This limit is the square root of the sum of the squares of the individual speeds.

41. $x = \frac{a}{2}$, $v = \frac{ka^2}{4}$ **43.** $\frac{c}{2} + 50$

45. **(a)** $\sqrt{\frac{2km}{h}}$ **(b)** $\sqrt{\frac{2km}{h}}$

49. **(a)** The artisan should order px units of material to have enough until the next delivery.
(c) Average cost per day $= \frac{(d + (ps/2)x^2)}{x} = \frac{d}{x} + \frac{ps}{2} x$; $x^* = \sqrt{\frac{2d}{ps}}$, $px^* = \sqrt{\frac{2pd}{s}}$ gives a minimum
(d) The line and hyperbola intersect when $\frac{d}{x} = \frac{ps}{2} x$. For $x > 0$, $x_{\text{intersection}} = \sqrt{\frac{2d}{ps}} = x^*$. From this result, the average cost per day is minimized when the average daily cost of delivery is equal to the average daily cost of storage.

51. $M = \frac{C}{2}$

59. **(a)** The minimum distance is $\sqrt{5}/2$.
(b) The minimum distance is from the point $(3/2, 0)$ to the point $(1, 1)$ on the graph of $y = \sqrt{x}$, which occurs at the value $x = 1$ where $D(x)$, the distance squared, has its minimum value.

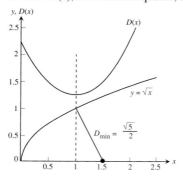

61. (a) $V(x) = \dfrac{\pi}{3}\left(\dfrac{2\pi a - x}{2\pi}\right)^2 \sqrt{a^2 - \left(\dfrac{2\pi a - x}{2\pi}\right)^2}$

(b) When $a = 4$: $r = \dfrac{4\sqrt{6}}{3}, h = \dfrac{4\sqrt{3}}{3}$; when $a = 5$: $r = \dfrac{5\sqrt{6}}{3}, h = \dfrac{5\sqrt{3}}{3}$; when $a = 6$: $r = 2\sqrt{6}, h = 2\sqrt{3}$; when $a = 8$: $r = \dfrac{8\sqrt{6}}{3}, h = \dfrac{8\sqrt{3}}{3}$

(c) Since $r = \dfrac{a\sqrt{6}}{3}$ and $h = \dfrac{a\sqrt{3}}{3}$, the relationship is $\dfrac{r}{h} = \sqrt{2}$.

Section 3.6, pp. 313–316

1. $10x - 13$ **3.** $L(x) = 2$
5. $L(x) = x - \pi$
7. $f(0) = 1$. Also, $f'(x) = k(1 + x)^{k-1}$, so $f'(0) = k$. This means that the linearization at $x = 0$ is $L(x) = 1 + kx$.
9. Center $= -1, L(x) = -5$
11. Center $= 1, L(x) = \dfrac{x}{4} + \dfrac{1}{4}$, or center $= 1.5, L(x) = \dfrac{4x}{25} + \dfrac{9}{25}$
13. (a) 1.01 **(b)** 1.003
15. $\left(3x^2 - \dfrac{3}{2\sqrt{x}}\right) dx$ **17. (a)** $(2x \ln x + x)\, dx$
19. $\dfrac{1 - y}{3\sqrt{y} + x}\, dx$ **21. (a)** $dy = (\cos x)e^{\sin x}\, dx$
23. $(1 + x)e^x\, dx$
25. (a) 0.21 **(b)** 0.2 **(c)** 0.01
27. (a) $-2/11$ **(b)** $-1/5$ **(c)** $1/55$
29. $4\pi a^2\, dr$ **31.** $3a^2\, dx$
33. (a) $0.08\pi\, \text{m}^2$ **(b)** 2%
35. $dV \approx 565.5 \text{ in.}^3$
37. $\dfrac{1}{3}\%$ **39.** 0.05%
41. The ratio equals 37.87, so a change in the acceleration of gravity on the moon has about 38 times the effect that a change of the same magnitude has on Earth.
47. (a) $L(x) = 1 + (\ln 2)x \approx 0.69x + 1$

Section 3.7, pp. 323–325

1. $x_2 = -\dfrac{5}{3}, \dfrac{13}{21}$ **3.** $x_2 = -\dfrac{51}{31}, \dfrac{5763}{4945}$
7. x, and later approximations will equal x.
9.

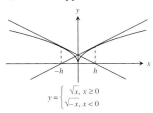

$y = \begin{cases} \sqrt{x}, & x \geq 0 \\ \sqrt{-x}, & x < 0 \end{cases}$

11. The points of intersection of $y = x^3$ and $y = 3x + 1$ or $y = x^3 - 3x$ and $y = 1$ have the same x-values as the roots of part (i) or the solutions of part (iv).
15. 1.165561185
17. (a) Two
(b) 0.35003501505249 and -1.0261731615301
19. $\pm 1.3065629648764, \pm 0.5411961001462$
21. $0, 0.53485$
23. The root is 1.17951.
25. (a) For $x_0 = -2$ or $x_0 = -0.8$, $x_i \to -1$ as i gets large.
(b) For $x_0 = -0.5$ or $x_0 = 0.25$, $x_i \to 0$ as i gets large.
(c) For $x_0 = 0.8$ or $x_0 = 2$, $x_i \to 1$ as i gets large.
(d) For $x_0 = -\sqrt{21}/7$ or $x_0 = \sqrt{21}/7$, Newton's method does not converge. The values of x_i alternate between $-\sqrt{21}/7$ and $\sqrt{21}/7$ as i increases.
27. Answers will vary with machine speed.
29. $2.45, 0.000245$

Chapter 3 Practice Exercises, pp. 326–329

1. Global minimum value of $\dfrac{1}{2}$ at $x = 2$
3. (a) $[-3, -2]$ and $[1, 2]$
(b) $[-2, 0)$ and $(0, 1]$
(c) Local maxima at $x = -2$ and $x = 2$; local minima at $x = -3$ and $x = 1$ (provided f is continuous at $x = 0$).
5. No
7. No minimum, absolute maximum: $f(1) = 16$, critical points: $x = 1$ and $\dfrac{11}{3}$
9. Yes   **11.** No
13. (b) One   **15. (b)** 0.8555996772
21.

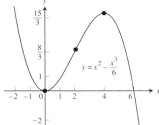

$y = x^2 - \dfrac{x^3}{6}$

23.

25.

$y = (x - 3)^2 e^x$

27. (a) Local maximum at $x = 4$, local minimum at $x = -4$, inflection point at $x = 0$

(b)

29. $x = \pm 1$ are the critical points; $y = 1$ is a horizontal asymptote in both directions; absolute minimum value of the function is $e^{-\sqrt{2}/2}$ at $x = -1$, and absolute maximum value is $e^{\sqrt{2}/2}$ at $x = 1$.

31. (a) $t = 0, 6, 12$
   (b) $t = 3, 9$
   (c) $6 < t < 12$
   (d) $0 < t < 6, 12 < t < 14$

33. (a) $v(t) = -3t^2 - 6t + 4$
   (b) $a(t) = -6t - 6$
   (c) The particle starts at position 3 moving in the positive direction, but decelerating. At approximately $t = 0.528$, it reaches position 4.128 and changes direction, beginning to move in the negative direction. After that, it continues to accelerate while moving in the negative direction.

35. $f(x) = -\dfrac{1}{4}x^{-4} - e^{-x} + C$

37. $f(x) = -\dfrac{2}{x} + \dfrac{1}{3}x^3 + x + C$ for $x > 0$

39. $s(t) = 4.9t^2 + 5t + 10$

41. (a) $y = -1$ is stable and $y = 1$ is unstable.
   (b) $\dfrac{d^2y}{dx^2} = 2y\dfrac{dy}{dx} = 2y(y^2 - 1)$

$$
\begin{array}{c|c|c|c|c}
 & y=-1 & & y=1 & \\
\hline
\frac{dy}{dx}>0 & \frac{dy}{dx}<0 & \frac{dy}{dx}<0 & \frac{dy}{dx}>0 \\
\frac{d^2y}{dx^2}<0 & \frac{d^2y}{dx^2}>0 & \frac{d^2y}{dx^2}<0 & \frac{d^2y}{dx^2}>0 \\
 & & y=0 & &
\end{array}
$$

(c)

43. $r = 25$ ft and $s = 50$ ft

45. Height $= 2$, radius $= \sqrt{2}$

47. $x = 5 - \sqrt{5}$ hundred ≈ 276 tires, $y = 2(5 - \sqrt{5})$ hundred ≈ 553 tires

49. Dimensions: base is 6 in. by 12 in., height $= 2$ in.; maximum volume $= 144$ in.^{3}

51. (a) $L(x) = 2x + \dfrac{\pi - 2}{2}$

$y = \tan x$
$y = 2x + (\pi - 2)/2$
$(-\pi/4, -1)$

(b) $L(x) = -\sqrt{2}x + \dfrac{\sqrt{2}(4 - \pi)}{4}$

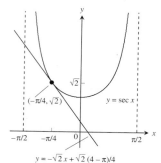

$(-\pi/4, \sqrt{2})$
$y = \sec x$
$y = -\sqrt{2}x + \sqrt{2}(4 - \pi)/4$

53. $L(x) = 2.5x - 0.1$

55. $dV = \dfrac{2}{3}\pi r_0 h\, dr$

57. (a) 4%   **(b)** 8%   **(c)** 12%

59. $x_5 = 1.732051$

Chapter 3 Additional Exercises, pp. 330–332

1. If M and m are the maximum and minimum values, respectively, then $m \le f(x) \le M$ for all $x \in I$. If $m = M$, then f is constant on I.

3. The extreme points will not be at the open end of an open interval.

5. (a) A local minimum at $x = -1$, points of inflection at $x = 0$ and $x = 2$
   (b) A local maximum at $x = 0$ and local minima at $x = -1$ and $x = 2$, points of inflection at $x = \dfrac{1 \pm \sqrt{7}}{3}$

11. $a = 1, b = 0, c = 1$ **13.** Yes

15. Drill the hole at $y = \dfrac{h}{2}$

17. $r = \dfrac{RH}{2(H - R)}$ for $H > 2R$, $r = R$ if $H \le 2R$

21. (a) 0.8156 ft
   (b) 0.00613 sec
   (c) It will lose about 8.83 min/day.

25. (b) 61°

CHAPTER 4 INTEGRATION

Section 4.1, pp. 340–343

1. (a) $3x^2$ (b) $\dfrac{x^8}{8}$ (c) $\dfrac{x^8}{8} - 3x^2 + 8x$

3. (a) $\dfrac{1}{x^2}$ (b) $-\dfrac{1}{4x^2}$ (c) $\dfrac{x^4}{4} + \dfrac{1}{2x^2}$

5. (a) $x^{2/3}$ (b) $x^{1/3}$ (c) $x^{-1/3}$

7. (a) $\tan x$ (b) $2\tan\left(\dfrac{x}{3}\right)$ (c) $-\dfrac{2}{3}\tan\left(\dfrac{3x}{2}\right)$

9. $\dfrac{x^2}{2} + x + C$

11. $\ln|x| - 5\tan^{-1} x + C$

13. $-e^{-x} + \dfrac{4^x}{\ln 4} + C$

15. $2\sin^{-1} y - \dfrac{4}{3}y^{3/4} + C$

17. $\dfrac{1}{7}y + \dfrac{4}{y^{1/4}} + C$

19. $2\sqrt{t} - \dfrac{2}{\sqrt{t}} + C$

21. $-21\cos\dfrac{\theta}{3} + C$

23. $\tan\theta + C$

25. $-\cos\theta + \theta + C$

31. (a) Wrong: $\dfrac{d}{dx}\left(\dfrac{x^2}{2}\sin x + C\right) = \dfrac{2x}{2}\sin x + \dfrac{x^2}{2}\cos x = x\sin x + \dfrac{x^2}{2}\cos x$

(b) Wrong: $\dfrac{d}{dx}(-x\cos x + C) = -\cos x + x\sin x$

(c) Right: $\dfrac{d}{dx}(-x\cos x + \sin x + C) = -\cos x + x\sin x + \cos x = x\sin x$

33. (b)

35. $y = x^2 - 7x + 10$

37. $y = 9x^{1/3} + 4$

39. $s = \sin t - \cos t$

41. $v = 3\sec^{-1} t - \pi$

43. $y = x^2 - x^3 + 4x + 1$

45. $y = x^3 - 4x^2 + 5$

47. $s = 4.9t^2 + 5t + 10$

49. $s = 16t^2 + 20t + 5$

51. $y = 2x^{3/2} - 50$

53. $y = x - x^{4/3} + \dfrac{1}{2}$

55. $y = -\sin x - \cos x - 2$

57. 48 m/sec

59. $t = \dfrac{88}{k}, k = 16$

61. (a) $v = 10t^{3/2} - 6t^{1/2}$
(b) $s = 4t^{5/2} - 4t^{3/2}$

65. (a) 1: 33.2 units; 2: 33.2 units; 3: 33.2 units
(b) True

Section 4.2, pp. 349–351

1. $-\dfrac{1}{4}\cos 2x^2 + C$

3. $-(7x - 2)^{-4} + C$

5. $-6(1 - r^3)^{1/2} + C$

7. $\dfrac{1}{3}(x^{3/2} - 1) - \dfrac{1}{6}\sin(2x^{3/2} - 2) + C$

9. (a) $-\dfrac{1}{4}(\cot^2 2\theta) + C$ (b) $-\dfrac{1}{4}(\csc^2 2\theta) + C$

11. $-\dfrac{1}{3}(3 - 2s)^{3/2} + C$

13. $\dfrac{3}{2 - x} + C$

15. $-\dfrac{1}{3}(7 - 3y^2)^{3/2} + C$

17. $4e^{\sqrt{x}} + C$

19. $\dfrac{1}{3}\tan(3x + 2) + C$

21. $-e^{1/t} + C$

23. $-\dfrac{2}{3}\cos(x^{3/2} + 1) + C$

25. $\dfrac{1}{2\cos(2t + 1)} + C$

27. $-\dfrac{2}{3}(\cot^3 y)^{1/2} + C$

29. $\dfrac{1}{2}\ln|2x - 1| + C$

31. $\ln|2 - \cos t| + C$

33. $\dfrac{1}{2}\sin^{-1}(2x) + C$

35. $\dfrac{1}{\sqrt{2}}\sec^{-1}\left|\dfrac{5x}{\sqrt{2}}\right| + C$

37. $\dfrac{1}{3}\tan^{-1}(3x + 1) + C$

39. $\tan^{-1}(e^x) + C$

41. (a) $-\dfrac{6}{2 + \tan^3 x} + C$ (b) $-\dfrac{6}{2 + \tan^3 x} + C$
(c) $-\dfrac{6}{2 + \tan^2 x} + C$

43. $\dfrac{1}{6}\sin\sqrt{3(2r - 1)^2 + 6} + C$

45. $s = \dfrac{1}{2}(3t^2 - 1)^4 - 5$

47. $s = 4t - 2\sin\left(2t + \dfrac{\pi}{6}\right) + 9$

49. $y = 1 - \cos(e^t - 2)$

51. $s = \sin\left(2t - \dfrac{\pi}{2}\right) + 100t + 1$

53. $y = 2(e^{-x} + x) - 1$

55. $y = \sin^{-1}(x)$

57. $y = \sec^{-1}(x) + \dfrac{2\pi}{3}, x > 1$

59. 6 m

63. (a) $\sin^{-1}\left(\dfrac{x}{3}\right) + C$

(b) $\dfrac{\sqrt{3}}{3}\tan^{-1}\left(\dfrac{\sqrt{3}\,x}{3}\right) + C$

Section 4.3, pp. 359–363

1. $\approx 44.8, 6.7$ L/m

3. (a) 87 in.   (b) 87 in.

5. (a) 3490 ft   (b) 3840 ft

7. (a) 80π (b) 6%

9. (a) $\dfrac{93\pi}{2}$, overestimate   (b) 9%

11. (a) 118.5π or ≈ 372.28 m³   (b) Error $\approx 11\%$

13. (a) 10π, underestimate   (b) 20%

15. (a) 74.65 ft/sec   (b) 45.28 ft/sec   (c) 146.59 ft

17. $\dfrac{31}{16}$ **19.** 1

21. (a) Upper = 758 gal, lower = 543 gal
(b) Upper = 2363 gal, lower = 1693 gal
(c) ≈ 31.4 h, ≈ 32.4 h

23. (a) 2   (b) $2\sqrt{2} \approx 2.828$
(c) $8\sin\left(\dfrac{\pi}{8}\right) \approx 3.061$
(d) Each area is less than the area of the circle, π. As n increases, the polygon area approaches π.

Section 4.4, pp. 372–374

1. $\dfrac{6(1)}{1+1} + \dfrac{6(2)}{2+1} = 7$

3. $\cos(1\pi) + \cos(2\pi) + \cos(3\pi) + \cos(4\pi) = 0$

5. $\sin\pi - \sin\dfrac{\pi}{2} + \sin\dfrac{\pi}{3} = \dfrac{\sqrt{3}-2}{2}$

7. (a)   **(b)**

(c)

9. (a)

(b)

(c)

11. $\displaystyle\int_{0}^{2} x^2\,dx$

13. $\displaystyle\int_{-7}^{5} (x^2 - 3x)\,dx$

15. $\displaystyle\int_{0}^{1} \sqrt{4 - x^2}\,dx$

17. Area = 21 square units

19. Area = 2.5 square units

21. Area = $\dfrac{b^2}{2}$ square units

23. $av(f) = 1/2$

25. $av(f) = \pi/4$

27. (a) 0   **(b)** -8 **(c)** -12
(d) 10   **(e)** -2 **(f)** 16

29. (a) 5   **(b)** $5\sqrt{3}$
(c) -5 **(d)** -5

31. (a) 4   **(b)** -4

33. $a = 0$ and $b = 1$ maximize the integral.

37. Upper bound = 1, lower bound = 1/2

39. 37.5 mph

Section 4.5, pp. 383–386

1. 6   **3.** 1

5. π **7.** 0

9. $\dfrac{2\pi^3}{3}$ **11.** $\dfrac{16\sqrt{2} - 17}{48}$

13. $\dfrac{e - 1}{2}$ **15.** $(\cos\sqrt{x})\left(\dfrac{1}{2\sqrt{x}}\right)$

17. $4t^5$ **19.** $\sqrt{1 + x^2}$

21. $-\dfrac{1}{2}x^{-1/2}\sin x$

23. $\dfrac{e^{x+1}}{3x^{2/3}}$

25. 0

27. $\dfrac{\pi}{2}+\sin 2$

29. $y=\displaystyle\int_{2}^{x}\sec t\,dt+3$

31. $y=1-\cos(e^{t}-2)$

33. $\dfrac{28}{3}$

35. 8

37. π

39. (a) $9.00 (b) $10.00

41. (a) $v=\dfrac{ds}{dt}=\dfrac{d}{dt}\displaystyle\int_{0}^{t}f(x)\,dx=f(t)\Rightarrow v(5)=f(5)=2$ m/sec

(b) $a=\dfrac{df}{dt}$ is negative since the slope of the tangent line at $t=5$ is negative.

(c) $s=\displaystyle\int_{0}^{3}f(x)\,dx=\dfrac{1}{2}(3)(3)=\dfrac{9}{2}$ m since the integral is the area of the triangle formed by $y=f(x)$, the x-axis, and $x=3$.

(d) $t=6$ since after $t=6$ to $t=9$, the region lies below the x-axis.

(e) At $t=4$ and $t=7$, since there are horizontal tangents there.

(f) Toward the origin between $t=6$ and $t=9$ since the velocity is negative on this interval. Away from the origin between $t=0$ and $t=6$ since the velocity is positive there.

(g) Right or positive side, because the integral of f from 0 to 9 is positive, there being more area above the x-axis than below.

43. $\displaystyle\int_{4}^{8}\pi(64-x^{2})\,dx=\dfrac{320\pi}{3}$

45. $2x-2$

47. $-3x+5$

49. (a) True; since f is continuous, g is differentiable by Part 1 of the Fundamental Theorem of Calculus.

(b) True: g is continuous because it is differentiable.

(c) True, since $g'(1)=f(1)=0$.

(d) False, since $g''(1)=f'(1)>0$.

(e) True, since $g'(1)=0$ and $g''(1)=f'(1)>0$.

(f) False: $g''(x)=f'(x)>0$, so g'' never changes sign.

(g) True, since $g'(1)=f(1)=0$ and $g'(x)=f(x)$ is an increasing function of x (because $f'(x)>0$).

51. (a) $\dfrac{125}{6}$ (b) $h=\dfrac{25}{4}$ (d) $\dfrac{2}{3}bh$

Section 4.6, pp. 393–394

1. (a) $\dfrac{14}{3}$ (b) $\dfrac{2}{3}$

3. (a) 2 (b) 2

5. (a) 0 (b) $\dfrac{1}{8}$

7. (a) 0 (b) 0

9. $\dfrac{1}{6}$

11. $\dfrac{1}{5}$

13. $e-1$

15. 1

17. $y(t)=\dfrac{1}{\pi}\left(3-\tan\left(\pi e^{-t}\right)\right)$

19. $\dfrac{\pi}{2}$

21. $\dfrac{128}{15}$

23. $\dfrac{38}{3}$

25. (a) 6 (b) $7\dfrac{1}{3}$

27. (a) 0 (b) $\dfrac{8}{3}$

29. $\dfrac{32}{3}$

31. $\dfrac{8}{3}$

33. 8

35. 4

37. $\dfrac{4-\pi}{\pi}$

39. 1

41. $\dfrac{32}{3}$

Section 4.7, pp. 403–407

1. I: (a) $1.5, 0$ (b) $1.5, 0$ (c) 0%
II: (a) $1.5, 0$ (b) $1.5, 0$ (c) 0%

3. I: (a) $2.75, 0.08$ (b) $2.67, 0.08$ (c) $0.0312\approx 3\%$
II: (a) $2.67, 0$ (b) $2.67, 0$ (c) 0%

5. I: (a) $6.25, 0.5$ (b) $6, 0.25$ (c) $0.0417\approx 4\%$
II: (a) $6, 0$ (b) $6, 0$ (c) 0%

7. I: (a) $0.509, 0.03125$
(b) $0.5, 0.009$
(c) $0.018\approx 2\%$
II: (a) $0.5004, 0.002604$
(b) $0.5, 0.0004$
(c) 0.08%

9. I: (a) $1.8961, 0.161$
(b) $2, 0.1039$
(c) $0.052\approx 5\%$
II: (a) $2.00456, 0.0066$
(b) $2, 0.0046$
(c) 0.23%

11. (a) 0.31929
(b) 0.32812
(c) $\dfrac{1}{3}, 0.01404, 0.00521$

13. (a) 1.95643
(b) 2.00421
(c) $2, 0.04357, -0.00421$

15. $15{,}990$ ft^{3}

17. 1.032 mi or 5443.5 ft

19. ≈ 10.63 ft

21. $4, 4$

23. (a) $S_{10}=0.8427$
(b) $|E_s|\le\dfrac{1-0}{180}(0.1)^{4}12\approx 6.7\times 10^{-6}$

25. (a) $f''(x) = 2\cos(x^2) - 4x^2\sin(x^2)$
(b)

(c) The graph shows that $-3 \le f''(x) \le 2$ for $-1 \le x \le 1$.
(d) $|E_T| \le \dfrac{1 - (-1)}{12}(h^2)(3) = \dfrac{h^2}{2}$

(e) $|E_T| \le \dfrac{h^2}{2} \le \dfrac{0.1^2}{2} < 0.01$

(f) $n \ge 20$
27. $S_{50} = 3.1379, S_{100} = 3.14029$
29. 1.37076
31. (a) $T_{10} \approx 1.983523538; T_{100} \approx 1.999835504, T_{1000} \approx$
1.999998355
(b)

n	$\|E_T\| = 2 - T_n$
10	$0.016476462 = 1.6476462 \times 10^{-2}$
100	1.64496×10^{-4}
1000	1.645×10^{-6}

(c) $|E_{T_{10n}}| \approx 10^{-2}|E_{T_n}|$

(d) $b - a = \pi; h^2 = \dfrac{\pi^2}{n^2}; M = 1$

$|E_{T_n}| \le \dfrac{\pi}{12}\left(\dfrac{\pi^2}{n^2}\right) = \dfrac{\pi^3}{12n^2}$

$|E_{T_{10n}}| \le \dfrac{\pi^3}{12(10n)^2} = 10^{-2}|E_{T_n}|$

Chapter 4 Practice Exercises, pp. 408–412

1. $\dfrac{x^4}{4} + \dfrac{5}{2}x^2 - 7x + C$ **3.** $2t^{3/2} - \dfrac{4}{t} + C$

5. $-\dfrac{1}{2(r^2 + 5)} + C$ **7.** $-(2 - \theta^2)^{3/2} + C$

9. $-\cos e^x + C$ **11.** $\tan(e^x - 7) + C$

13. $-[\ln|\cos(\ln v)|] + C$ **15.** $\dfrac{1}{2}x - \sin\dfrac{x}{2} + C$

17. $-\cot(1 + \ln r) + C$ **19.** $\dfrac{1}{2\ln 3}(3^{x^2}) + C$

21. (a) About 680 ft
(b)

23. $\displaystyle\int_1^5 (2x - 1)^{-1/2}\, dx = 2$ **25.** $\displaystyle\int_{-\pi}^0 \cos\dfrac{x}{2}\, dx = 2$

27. (a) 4   **(b)** 2   **(c)** -2
(d) -2π **(e)** 8/5
29. 16   **31.** $15/16 + \ln 2$
33. 1   **35.** 8
37. 1/6   **39.** $2\sqrt{2} - 2$
41. 9/14   **43.** $\pi/\sqrt{3}$
45. -1 **47.** $\pi/12$
49. 1   **51.** $\sqrt{3}\pi/4$
53. 8/3   **55.** 62
57. 1   **59.** 1/6
61. $\dfrac{\pi^2}{32} + \dfrac{\sqrt{2}}{2} - 1$ **63.** 4

65. Min: -4; max: 0; area: 27/4   **67.** $y = x - \dfrac{1}{x} - 1$

69. $r = 4t^{5/2} + 4t^{3/2} - 8t$ **73.** $y = \displaystyle\int_5^x \left(\dfrac{\sin t}{t}\right) dt - 3$

75. (a) b **(b)** b

79. $\sqrt{2 + \cos^3 x}$ **81.** $\dfrac{-6}{3 + x^4}$

83. $T = \pi, S = \pi$ **85.** $25°F$
87. $\cos \approx \$12{,}518.10$ (Trapezoidal Rule); no

89. Yes   **91.** $y = \displaystyle\int_5^x \dfrac{\sin t}{t}\, dt + 3$

93. (a) 0   **(b)** -1
(c) $-\pi$ **(d)** $x = 1$
(e) $y = 2x + 2 - \pi$ **(f)** $x = -1, x = 2$
(g) $[-2\pi, 0]$
95. 600, $18.00   **97.** 300, $6.00

Chapter 4 Additional Exercises, pp. 412–414

1. (a) Yes   **(b)** No
5. (a) 1/4   **(b)** $\sqrt[3]{12}$

7. $f(x) = \dfrac{x}{\sqrt{x^2 + 1}}$ **9.** $y = x^3 + 2x - 4$

11. 36/5   **13.** $\dfrac{1}{2} - \dfrac{2}{\pi}$

15. 13/3

17. 1/2

19. 2/x

21. $\dfrac{\sin 4y}{\sqrt{y}} - \dfrac{\sin y}{2\sqrt{y}}$

CHAPTER 5 APPLICATIONS OF INTEGRALS

Section 5.1, pp. 423–428

1. (a) $A(x) = \pi(1 - x^2)$ **(b)** $A(x) = 4(1 - x^2)$
(c) $A(x) = 2(1 - x^2)$ **(d)** $A(x) = \sqrt{3}(1 - x^2)$
3. 16
5. (a) $\dfrac{\pi^2}{2}$ **(b)** 2π
7. (a) $2\sqrt{3}$ **(b)** 8
9. 8π
11. (a) $s^2 h$ **(b)** $s^2 h$
13. $2\pi/3$ **15.** $4 - \pi$
17. $32\pi/5$ **19.** 36π
21. π **23.** $\pi\left(\dfrac{\pi}{2} + 2\sqrt{2} - \dfrac{11}{3}\right)$
25. 2π **27.** 2π
29. 3π **31.** $\pi^2 - 2\pi$
33. $2\pi/3$ **35.** $117\pi/5$
37. $\pi(\pi - 2)$ **39.** $4\pi/3$
41. 8π **43.** $7\pi/6$
45. (a) 8π **(b)** $32\pi/5$
(c) $8\pi/3$ **(d)** $224\pi/15$
47. (a) $16\pi/15$ **(b)** $56\pi/15$ **(c)** $64\pi/15$
49. $V = 2a^2 b\pi^2$
51. (a) $V = \dfrac{\pi h^2 (3a - h)}{3}$ **(b)** $\dfrac{1}{120\pi}$ m/sec
55. $V = 3308$ cm^3
57. (a) $c = 2\pi$ **(b)** $c = 0$
(c)

Section 5.2, pp. 433–435

1. 6π **3.** 2π
5. $14\pi/3$ **7.** 8π
9. $5\pi/6$ **11.** $\pi\left(1 - \dfrac{1}{3}\right)$
13. (b) 4π **15.** $\dfrac{16\pi}{15}(3\sqrt{2} + 5)$
17. $8\pi/3$ **19.** $4\pi/3$
21. $16\pi/3$
23. (a) $6\pi/5$ **(b)** $4\pi/5$
(c) 2π **(d)** 2π
25. (a) About the x-axis: $V = 2\pi/15$, about the y-axis: $V = \pi/6$
(b) About the x-axis: $V = 2\pi/15$, about the y-axis: $V = \pi/6$
27. (a) $5\pi/3$ **(b)** $4\pi/3$
(c) 2π **(d)** $2\pi/3$
29. (a) $4\pi/15$ **(b)** $7\pi/30$
31. (a) $24\pi/5$ **(b)** $48\pi/5$
33. (a) $9\pi/16$ **(b)** $9\pi/16$

Section 5.3, pp. 441–443

1. 12   **3.** 53/6
5. 123/32   **7.** 99/8
9. 2   **11.** $2\pi a$
13. $e^3 + 2$ **15.** $\sqrt{2}(e^\pi - 1)$
17. Yes, $f(x) = \pm x + C$, where C is any real number, from $(1, 1)$ to $(4, 2)$
19. (a) $y = \sqrt{x}$
(b) Only one. We know the derivative of the function and the value of the function at one value of x.
21. (a) $\displaystyle\int_{-1}^{2} \sqrt{1 + 4x^2}\, dx$ **(c)** ≈ 6.13
23. (a) $\displaystyle\int_{0}^{\pi} \sqrt{1 + \cos^2 y}\, dy$ **(c)** ≈ 3.82
25. (a) $\displaystyle\int_{-1}^{3} \sqrt{1 + (y + 1)^2}\, dy$ **(c)** ≈ 9.29
27. (a) $\displaystyle\int_{0}^{\pi/6} \sec x\, dx$ **(c)** ≈ 0.55
29. 21.07 in.

59. (a) 2.3, 1.6, 1.5, 2.1, 3.2, 4.8, 7.0, 9.3, 10.7, 10.7, 9.3, 6.4, 3.2
(b) $\dfrac{1}{4\pi}\displaystyle\int_{0}^{6} (C(y))^2\, dy$
(c) ≈ 34.7 in.^{3}
(d) $V \approx 34.75$ in.^{3} by Simpson's Rule. The Simpson's Rule estimate should be more accurate than the trapezoid estimate. The error in the Simpson's estimate is proportional to $h^4 = 0.0625$, whereas the error in the trapezoid estimate is proportional to $h^2 = 0.25$, a larger number when $h = 0.5$ in.

Section 5.4, pp. 451–453

5. $\frac{2}{3}y^{3/2} - x^{1/2} = C$

7. $e^y - e^x = C$

9. $-x + 2\tan\sqrt{y} = C$

11. $e^{-y} + 2e^{\sqrt{x}} = C$

13. $y = \sin(x^2 + C)$

15. (a) ≈ -0.121
   (b) ≈ 2.389 millibars
   (c) ≈ 0.977 km

17. ≈ 585.35 kg

19. ≈ 92.1 sec

21. (a) $\dfrac{dQ}{dt} = r - kQ$

   (b) $Q = \dfrac{r}{k} + \left(Q_0 - \dfrac{r}{K}\right)e^{-kt}$

   (c) $\dfrac{r}{k}$

23. 5°F

25. (a) $p(x) = 54.61e^{-0.01x}$ (in dollars)
   (b) $p(10) = \$49.41$ and $p(90) = \$22.20$

27. $\dfrac{dy}{dt} = -ky;\ y = y_0 e^{-kt}$

29. ≈ 864 years old

31. (a) 12,571 B.C.　(b) 12,101 B.C.　(c) 13,070 B.C.

33. (a) 550 ft　(b) ≈ 77.28 sec

35.

37.

39.

Section 5.5, pp. 460–464

1. 400 ft · lb

3. 780 J

5. 72,900 ft · lb

9. 400 N/m

11. 4 cm, 0.08 J

13. (a) 7238 lb/in.　(b) 905 in. · lb, 2714 in. · lb

15. (a) 1,497,600 ft · lb　(b) 1 h 40 min

   (c) $W = \displaystyle\int_0^{10} 62.4(120y)\,dy = 374{,}000$ ft · lbs $\to t = W/250 =$
   1948 sec ≈ 25 min

   (d) At 62.26 lb/ft³: part (a) 1,494,240 ft · lb, part (b) 1 h 40 min;
   at 62.59 lb/ft³: part (a) 1,502,160 ft · lb, part (b) 1 h 40 min

17. 38,484,510 J　**19.** 7,238,229.48 ft · lb

21. (a) 34,583 ft · lb　(b) 53,482 ft · lb

23. 15,073,100 J　**27.** ≈ 85.1 ft · lb

29. ≈ 64.6 ft · lb　**31.** ≈ 110.6 ft · lb

33. (a) $r(y) = 60 - \sqrt{50^2 - (y - 325)^2}$ for $325 \le y \le 375$ ft

   (b) $\Delta V \approx \pi[60 - \sqrt{2500 - (y - 325)^2}\,]^2\,\Delta y$

   (c) $W = 6.3358 \cdot 10^7$ ft · lb

35. 91.32 in. · oz　**37.** 5.144×10^{10} J

Section 5.6, pp. 469–471

3. 2808 lb

5. (a) 1164.8 lb　(b) 1194.7 lb

7. 1309 lb　**9.** 41.6 lb

11. (a) 93.33 lb　(b) 3 ft

13. (a) $\left(\dfrac{x}{8}\right)^2 + \left(\dfrac{y}{14}\right)^2 = 1$　(b) 3008 tons

15. (a) $\dfrac{wb}{2}$

17. (a) 374.4 lb　(b) 7.5 in.　(c) No

19. 4.2 lb　**21.** 1035 ft³

Section 5.7, pp. 481–483

1. 4 ft　**3.** $\left(\dfrac{L}{4}, \dfrac{L}{4}\right)$

5. $M_0 = 8, M = 8, \bar{x} = 1$

7. $M_0 = 15/2, M = 9/2, \bar{x} = 5/3$

9. $M_0 = 73/6, M = 5, \bar{x} = 73/30$

11. $M_0 = 3, M = 3, \bar{x} = 1$　**13.** $\bar{x} = 0, \bar{y} = 12/5$

15. $\bar{x} = 1, \bar{y} = -3/5$　**17.** $\bar{x} = 16/105, \bar{y} = 8/15$

19. $\bar{x} = 0, \bar{y} = \pi/8$　**21.** $\bar{x} = 1, \bar{y} = -2/5$

23. $\bar{x} = \bar{y} = \dfrac{2}{4 - \pi}$　**25.** $\bar{x} = 3/2, \bar{y} = 1/2$

27. (a) $224\pi/3$
   (b) $\bar{x} = 2, \bar{y} = 0$
   (c)
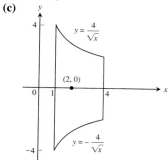

31. $\bar{x} = \bar{y} = 1/3$

33. $\bar{x} = a/3, \bar{y} = b/3$

35. $13\delta/6$

37. $\bar{x} = 0, \bar{y} = a\pi/4$

Chapter 5 Practice Exercises, pp. 484–486

1. $9\pi/280$

3. π^2

5. $72\pi/35$

7. (a) 2π

(b) π

(c) $\dfrac{12\pi}{5}$

(d) $26\pi/5$

9. (a) 8π

(b) $1088\pi/15$

(c) $512\pi/15$

11. $\pi(2 - \ln 3)$

13. (a) $16\pi/15$

(b) $8\pi/5$

(c) $8\pi/3$

(d) $32\pi/5$

15. $28\pi/3$

17. $10/3$

19. $285/8$

21. 10

23. $f(x) = \dfrac{x^2 - 2 \ln x + 3}{4}$

25. 4640 J

27. 10 ft · lb, 30 ft · lb

29. $418{,}208.81$ ft · lb

31. $22{,}500\pi$ ft · lb, 257 sec

33. 332.8 lb

35. 2196.48 lb

37. $y = \left(\dfrac{x^3}{6} + C\right)^2$

39. $y = Ce^{(\ln x)^2/2}$

41. $2 \tan \sqrt{x} = t + C$

43. $\bar{x} = 0, \bar{y} = 8/5$

45. $\bar{x} = 3/2, \bar{y} = 12/5$

47. $\bar{x} = 9/5, \bar{y} = 11/10$

Chapter 5 Additional Exercises, pp. 487–488

1. $f(x) = \sqrt{\dfrac{2x + 1}{\pi}}$

3. $f(x) = \sqrt{C^2 - 1}\,x + a$, where $C \geq 1$

5. 30 feet

9. 108.3042 h; 251.4747 h

11. $168.25°$F

13. $\bar{x} = 0, \bar{y} = \dfrac{n}{2n + 1}, \left(0, \dfrac{1}{2}\right)$

17. (a) $\bar{x} = \bar{y} = \dfrac{4(a^2 + ab + b^2)}{3\pi(a + b)}$

(b) $\bar{x} = \bar{y} = \dfrac{zb}{\pi}$

CHAPTER 6 TRANSCENDENTAL FUNCTIONS AND DIFFERENTIAL EQUATIONS

Section 6.1, pp. 495–496

1. $\dfrac{1}{x}$

3. $\dfrac{3}{x}$

5. $2(\ln t) + (\ln t)^2$

7. $x^3 \ln x$

9. $\dfrac{1 - \ln t}{t^2}$

11. $\dfrac{1}{x(1 + \ln x)^2}$

13. $2 \cos (\ln \theta)$

15. $-\dfrac{3x + 2}{2x(x + 1)}$

17. $\dfrac{2}{t(1 - \ln t)^2}$

19. $\dfrac{\tan (\ln \theta)}{\theta}$

21. $\dfrac{1}{\theta \ln 2}$

23. $\dfrac{2(\ln r)}{r(\ln 2)(\ln 4)}$

25. $\sin (\log_7 \theta) + \dfrac{1}{\ln 7} \cos (\log_7 \theta)$

27. $2x \ln |x| - x \ln \dfrac{|x|}{\sqrt{2}}$

29. $\ln\left(\dfrac{2}{3}\right)$

31. $\ln |y^2 - 25| + C$

33. $\ln 3$

35. $(\ln 2)^2$

37. $\dfrac{1}{\ln 4}$

39. $\ln |6 + 3 \tan t| + C$

41. $\ln 2$

43. $\ln 27$

25. $\ln(1 + \sqrt{x}) + C$

47. (a) Max $= 0$ at $x = 0$, min $= -\ln 2$ at $x = \pi/3$

(b) Max $= 1$ at $x = 1$, min $= \cos (\ln 2)$ at $x = 1/2$ and $x = 2$

49. $\ln 16$

51. $y = x + \ln |x| + 2$

53. $\dfrac{1}{\ln 10}\left(\dfrac{(\ln x)^2}{2}\right) + C$

55. $\dfrac{3 \ln 2}{2}$

57. $(\ln 10) \ln |\ln x| + C$

59. (b) 0.00469

Section 6.2, pp. 501–503

1. $-\dfrac{2}{3} e^{-2x/3}$

3. $\left(\dfrac{2}{\sqrt{x}} + 2x\right)e^{(4\sqrt{x}+x^2)}$

5. $x^2 e^x$

7. $\dfrac{1}{\theta} - 1$

9. $\dfrac{\cos t\, 2 \sin t}{\sin t}$

11. $\dfrac{1}{2\theta(1 + \theta^{1/2})}$

13. $\dfrac{\sin x}{x}$

15. $\dfrac{ye^y \cos x}{1 - ye^y \sin x}$

17. $\dfrac{2e^{2x} - \cos (x + 3y)}{3 \cos (x + 3y)}$

19. $\dfrac{1}{3} e^{3x} - 5e^{-s} + C$

21. $8e^{(x+1)} + C$

23. $-2e^{-\sqrt{r}} + C$

25. $-e^{1/x} + C$

27. e

29. $\dfrac{1}{\pi} e^{\sec \pi t} + C$

31. $\ln (1 + e^r) + C$

33. $2^x \ln 2$

35. $\pi x^{(\pi - 1)}$

37. $7^{\sec \theta}(\ln 7)^2 \sec \theta \tan \theta$

39. $(1 - e)t^{-e}$

41. $\dfrac{-2}{(x + 1)(x - 1)}$

43. $(\cot \theta - \tan \theta - 1 - \ln 2)\left(\dfrac{1}{\ln 7}\right)$

45. $(x + 1)^x\left(\dfrac{x}{x + 1} + \ln (x + 1)\right)$

47. $\dfrac{dy}{dx} = \dfrac{x \ln x \cos x + \sin x}{x^{\sin x + 1}}$

49. $\dfrac{1}{\ln 2}$

51. $\dfrac{2^{\ln 2} - 1}{\ln 2}$

53. $\dfrac{x^{\sqrt{2}}}{\sqrt{2}} + C$

55. $\dfrac{1}{\ln 2}$

57. $y = 1 - \cos(e^t - 2)$

59. $y = 2(e^{-x} + x) - 1$

61. Maximum: 1 at $x = 0$; minimum: $2 - 2 \ln 2$ at $x = \ln 2$

63. Absolute maximum of $\dfrac{1}{2e}$ assumed at $x = \dfrac{1}{\sqrt{e}}$

65. Let $x = \dfrac{r}{k} \Rightarrow k = \dfrac{r}{x}$ and as $k \to \infty$, $x \to 0 \Rightarrow \lim_{k \to \infty} \left(1 + \dfrac{r}{k}\right)^k =$

$\lim_{x \to 0} (1 + x)^{r/x} = \lim_{r \to 0} ((1 + x)^{1/x})^r = (\lim_{x \to 0} (1 + x)^{1/x})^r$, since u^r is

continuous. However, $\lim_{x \to 0} (1 + x)^{1/x} = e$ (by Theorem 2); there-

fore, $\lim_{k \to \infty} \left(1 + \dfrac{r}{k}\right)^k = e^r$.

71. (b) $|\text{Error}| \approx 0.02140$

Section 6.3, pp. 510–511

1. $y = \dfrac{e^x + C}{x}$

3. $y = \dfrac{C - \cos x}{x^3}, x > 0$

5. $y = \dfrac{1}{2} - \dfrac{1}{x} + \dfrac{C}{x^2}, x > 0$

7. $y = \dfrac{1}{2} xe^{x/2} + Ce^{x/2}$

9. $y = x(\ln x)^2 + Cx$

11. $s = \dfrac{t^3}{3(t-1)^4} - \dfrac{t}{(t-1)^4} + \dfrac{C}{(t-1)^4}$

13. $r = (\csc \theta)(\ln |\sec \theta| + C)$

15. $y = \dfrac{3}{2} - \dfrac{1}{2} e^{-2t}$

17. $y = -\dfrac{1}{\theta} \cos \theta + \dfrac{\pi}{2\theta}$

19. $y = 6e^{x^2} - \dfrac{e^{x^2}}{x+1}$

21. $y = y_0 e^{kt}$

23. Part (b) is correct, but part (a) is not.

25. (a) 10 lb/min

(b) $100 + t$ gal

(c) $4\left(\dfrac{y}{100 + t}\right)$ lb/min

(d) $\dfrac{dy}{dt} = 10 - \dfrac{4y}{100 + t}, y(0) = 50,$

$y = 2(100 + t) - \dfrac{150}{\left(1 + \dfrac{t}{100}\right)^4}$

(e) Concentration $= \dfrac{y(25)}{\text{amount brine in tank}} = \dfrac{188.6}{125} \approx 1.5$ lb/gal

27. $y(27.8) \approx 14.8$ lb, $t \approx 27.8$ min

29. $t = \dfrac{L}{R} \ln 2$ sec

31. (a) $i = \dfrac{V}{R} - \dfrac{V}{R} e^{-3} = \dfrac{V}{R}(1 - e^{-3}) \approx 0.95 \dfrac{V}{R}$ amp

(b) 86%

Section 6.4, pp. 522–524

1. $y(\text{exact}) = \dfrac{x}{2} - \dfrac{4}{x}$; $y_1 = -0.25, y_2 = 0.3, y_3 = 0.75$

3. $y(\text{exact}) = 3e^{x(x+2)}$; $y_1 = 4.2, y_2 = 6.216, y_3 = 9.697$

5. $y \approx 2.48832$, exact value is e

7.

x	z	y-approx.	y-exact	Error
0	—	3	3	0
0.2	4.2	4.608	4.658122	0.050122
0.4	6.81984	7.623475	7.835089	0.211614
0.6	11.89262	13.56369	14.27646	0.712777

9. (a) $P(t) = \dfrac{150}{1 + 24e^{-0.225t}}$

(b) About 17.21 weeks; 21.28 weeks

11. (a) $y(t) = \dfrac{8 \times 10^7}{1 + 4e^{-0.71t}} \Rightarrow y(1) \approx 2.697 \times 10^7$ kg

(b) $t \approx 1.95253$ years

13. (a) $y = 2e^t - 1$

(b) $y(t) = \dfrac{400}{1 + 199e^{-200t}}$

15. (a) $P(t) = \dfrac{P_0}{1 - kP_0 t}$

(b) Vertical asymptote at $t = \dfrac{1}{kP_0}$

17. $y \approx 3.45835$; exact value is $1 + e \approx 3.71828$.

19. $y \approx -0.2272$; exact value is $\dfrac{1}{1 - 2\sqrt{5}} \approx -0.2880$.

21. (a) $y = \dfrac{21}{x^2 - 2x + 2}, y(3) = -0.2$

(b) -0.1851, error ≈ 0.0149

(c) -0.1929, error ≈ 0.0071

(d) -0.1965, error ≈ 0.0035

23. The exact solution is $y = \dfrac{-1}{x^2 - 2x + 2}$, so $y(3) = -0.2$. To find

the approximation, let $z_n = y_{n+1} + 2_{n-1}{}^2(x_n - 1) \, dx$ and $y_n = y_{n-1}$

$+ (y_{n-1}{}^2(x_{n-1} - 1) + z_n{}^2(x_n{}^2 - 1)) \, dx$ with initial values $x_0 = 2$

and $y_0 = -\dfrac{1}{2}$. Use a spreadsheet, graphing calculator, or CAS as

indicated in parts (a) through (d).

(a) -0.2024, error ≈ 0.0024

(b) -0.2005, error ≈ 0.0005

(c) -0.2001, error ≈ 0.0001

(d) Each time the step size is cut in half, the error is reduced to ap-

proximately one-fourth of what it was for the larger step size.

Section 6.5, pp. 529–533

1. $\cosh x = \dfrac{5}{4}$, $\tanh x = -\dfrac{3}{5}$, $\coth x = -\dfrac{5}{3}$, $\operatorname{sech} x = \dfrac{4}{5}$, $\operatorname{csch} x = -\dfrac{4}{3}$

3. $\sinh x = \dfrac{8}{15}$, $\tanh x = \dfrac{8}{17}$, $\coth x = \dfrac{17}{8}$, $\operatorname{sech} x = \dfrac{15}{17}$, $\operatorname{csch} x = \dfrac{15}{8}$

5. $x + \dfrac{1}{x}$

7. e^{5x}

9. e^{4x}

13. $2 \cosh \dfrac{x}{3}$

15. $\operatorname{sech}^2 \sqrt{t} + \dfrac{\tanh \sqrt{t}}{\sqrt{t}}$

17. $\coth z$

19. $(\ln \text{sech }\theta)(\text{sech }\theta \tanh \theta)$

21. $\tanh^3 v$

23. 2

25. $\dfrac{1}{2\sqrt{x(1+x)}}$

27. $\dfrac{1}{1+\theta} - \tanh^{-1}\theta$

29. $\dfrac{1}{2\sqrt{t}} - \coth^{-1}\sqrt{t}$

31. $-\text{sech}^{-1} x$

33. $\dfrac{\ln 2}{\sqrt{1+\left(\dfrac{1}{2}\right)^{2\theta}}}$

35. $|\sec x|$

41. $\dfrac{\cosh 2x}{2} + C$

43. $12\sinh\left(\dfrac{x}{2} - \ln 3\right) + C$

45. $7\ln\left(e^{x/7} + e^{-x/7}\right) + C$

47. $\tanh\left(x - \dfrac{1}{2}\right) + C$

49. $-2\,\text{sech}\,\sqrt{t} + C$

51. $\ln\dfrac{5}{2}$

53. $\dfrac{3}{32} + \ln 2$

55. $e - e^{-1}$

57. $\dfrac{3}{4}$

59. $\dfrac{3}{8} + \ln\sqrt{2}$

61. $\ln\left(\dfrac{2}{3}\right)$

63. $-\dfrac{\ln 3}{2}$

65. $\ln 3$

67. (a) $\sinh^{-1}(\sqrt{3})$ **(b)** $\ln(\sqrt{3} + 2)$

69. (a) $\coth^{-1}(2) - \coth^{-1}\left(\dfrac{5}{4}\right)$ **(b)** $\left(\dfrac{1}{2}\right)\ln\left(\dfrac{1}{3}\right)$

71. (a) $-\text{sech}^{-1}\left(\dfrac{12}{13}\right) + \text{sech}^{-1}\left(\dfrac{4}{5}\right)$

(b) $-\ln\left(\dfrac{1 + \sqrt{1 - \left(\dfrac{12}{13}\right)^2}}{\left(\dfrac{12}{13}\right)}\right) + \ln\left(\dfrac{1 + \sqrt{1 - \left(\dfrac{4}{5}\right)^2}}{\left(\dfrac{4}{5}\right)}\right) =$

$-\ln\left(\dfrac{3}{2}\right) + \ln(2) = \ln\left(\dfrac{4}{3}\right)$

73. (a) 0 **(b)** 0

75. (b) (i) $f(x) = \dfrac{2f(x)}{2} + 0 = f(x)$

(ii) $f(x) = 0 + \dfrac{2f(x)}{2} = f(x)$

77. (b) $\sqrt{\dfrac{mg}{k}}$ **(c)** $80\sqrt{5} \approx 178.89$ ft/sec

79. $y = \text{sech}^{-1}(x) - \sqrt{1 - x^2}$ **81.** 2π

83. $\dfrac{6}{5}$

87. (c) $a \approx 0.0417525$ **(d)** ≈ 47.90 lb

Chapter 6 Practice Exercises, pp. 534–536

1. $-\cos e^x + C$

3. $\tan(e^x - 7) + C$

5. $e^{\tan x} + C$

7. $-\dfrac{\ln 7}{3}$

9. $\ln 8$

11. $\ln\left(\dfrac{9}{25}\right)$

13. $-[\ln|\cos(\ln v)|] + C$

15. $-\dfrac{1}{2}(\ln x)^{-2} + C$

17. $-\cot(1 + \ln r) + C$

19. $\dfrac{1}{2\ln 3}(3^{x^2}) + C$

21. $3\ln 7$

23. $e - 1$

25. $\dfrac{1}{3}[(\ln 4)^3 - (\ln 2)^3]$ or $\dfrac{7}{3}(\ln 2)^3$

27. $\dfrac{9\ln 2}{4}$

29. $\dfrac{2}{5}\sin^{-1}(5y) + C$

31. Absolute maximum $= 0$ at $x = e/2$, absolute minimum $= -0.5$ at $x = 0.5$

33. 1

35. $\dfrac{1}{e}$ m/sec

37. $\ln 5x - \ln 3x = \ln\left(\dfrac{5}{3}\right)$

41. $y = \dfrac{3 - \cos 2x}{2\cos x}$

45. (a) Absolute maximum of $\dfrac{2}{e}$ at $x = e^2$, inflection point $\left(e^{8/3}, \dfrac{8}{3}e^{-4/3}\right)$, concave up on $(e^{8/3}, \infty)$, concave down on $(0, e^{8/3})$

(b) Absolute maximum of 1 at $x = 0$, inflection points $\left(\pm\dfrac{1}{\sqrt{2}}, \dfrac{1}{\sqrt{e}}\right)$, concave up on $\left(-\infty, -\dfrac{1}{\sqrt{2}}\right) \cup \left(\dfrac{1}{\sqrt{2}}, \infty\right)$, concave down on $\left(-\dfrac{1}{\sqrt{2}}, \dfrac{1}{\sqrt{2}}\right)$,

(c) Absolute maximum of 1 at $x = 0$, inflection point $\left(1, \dfrac{2}{e}\right)$, concave up on $(1, \infty)$, concave down on $(-\infty, 1)$

47. 18,935 years

49. (a) $y = \sin x + \dfrac{C + \cos x}{x}$

(b) The graphs, from the bottom curve to the top, are for $C = -z, -1, 0, 1, z$, respectively.

51.

53. Let $z_n = y_{n-1} + ((2 - y_{n-1})(2x_{n-1} + 3))(0.1)$ and $y_n = y_{n-1} +$
$$\left(\frac{(2 - y_{n-1})(2x_{n-1} + 3) + (2 - z_n)(2x_n + 3)}{2}\right)(0.1) \text{ with initial}$$
values $x_0 = -3$, $y_0 = 1$, and 20 steps. Use a spreadsheet, graphing calculator, or CAS to obtain the values in the following table.

x	y	x	y
-3	1	-1.9	-5.9686
-2.9	0.6680	-1.8	-6.5456
-2.8	0.2599	-1.7	-6.9831
-2.7	-0.2294	-1.6	-7.2562
-2.6	-0.8011	-1.5	-7.3488
-2.5	-1.4509	-1.4	-7.2553
-2.4	-2.1687	-1.3	-6.9813
-2.3	-2.9374	-1.2	-6.5430
-2.2	-3.7333	-1.1	-5.9655
-2.1	-4.5268	-1.0	-5.2805
-2.0	-5.2840		

55. Let $y_n = y_{n-1} + \left(\dfrac{2x_{n-1}^2 - 2y_{n-1} + 1}{x_{n-1}}\right)(0.05)$ with initial values
$x_0 = 1$, $y_0 = 1$, and 60 steps. Use a spreadsheet, programmable calculator, or CAS to obtain $y(r) \approx 4.4974$.

57. Let $z_n = y_{n-1} - \left(\dfrac{x_{n-1}^2 + y_{n-1}}{e^{y_{n-1}} + x_{n-1}}\right)(dx)$ and $y_n = y_{n-1} +$
$\dfrac{1}{2}\left(\dfrac{x_{n-1}^2 + y_{n-1}}{e^{y_{n-1}} + x_{n-1}} + \dfrac{x_n^2 + z_n}{e^{z_n} + x_n}\right)(dx)$ with starting values $x_0 = 0$, $y_0 = 0$, and steps of 0.1 and -0.1. Use a spreadsheet, programmable calculator, or CAS to generate the following graphs.

(a)

(b)

59. $y(\text{exact}) = \dfrac{1}{2}x^2 - \dfrac{3}{2}$; $y \approx 0.4$; exact value is $1/2$.

61. $y(\text{exact}) = -e^{(x^2-1)/2}$; $y \approx -3.4192$; exact value is $-e^{3/2} \approx -4.4817$.

Chapter 6 Additional Exercises, pp. 536–537

1. $\dfrac{1}{\ln 2}, \dfrac{1}{2\ln 2}, 2:1$

3. (a) $2x$ (b) 0   (c) $f(x) = x^2$

9. (a) $P = P_0\, e^{\frac{k}{a}(\sin b + \sin (at - b))}$

CHAPTER 7

Section 7.1, pp. 544–546

1. $2\sqrt{8x^2 + 1} + C$

3. $2(\sin v)^{3/2} + C$

5. $\ln 5$

7. $2\ln(\sqrt{x} + 1) + C$

9. $-\dfrac{1}{7}\ln|\sin(3 - 7x)| + C$

11. $-\ln|\csc(e^\theta + 1) + \cot(e^\theta + 1)| + C$

13. $3\ln\left|\sec\dfrac{t}{3} + \tan\dfrac{t}{3}\right| + C$

15. $-\ln|\csc(s - \pi) + \cot(s - \pi)| + C$

17. 1

19. $e^{\tan v} + C$

21. $\dfrac{3^{x+1}}{\ln 3} + C$

23. $\dfrac{2^{\sqrt{w}}}{\ln 2} + C$

25. $3\tan^{-1} 3u + C$

27. $\dfrac{\pi}{18}$

29. $\sin^{-1} s^2 + C$

31. $6\sec^{-1}|5x| + C$

33. $\tan^{-1} e^x + C$

35. $\ln(2 + \sqrt{3})$

37. 2π

39. $\sin^{-1}(t - 2) + C$

41. $\sec^{-1}|x + 1| + C$, when $|x + 1| > 1$

43. $\tan x - 2\ln|\csc x + \cot x| - \cot x - x + C$

45. $x + \sin 2x + C$

47. $x - \ln|x + 1| + C$

49. $7 + \ln 8$

51. $2t^2 - t + 2\tan^{-1}\left(\dfrac{t}{2}\right) + C$

53. $\sin^{-1} x + \sqrt{1 - x^2} + C$

55. $\sqrt{2}$

57. $\tan x - \sec x + C$

59. $\ln|1 + \sin \theta| + C$

61. $\cot x + x + \csc x + C$

63. 4

65. $\sqrt{2}$

67. 2

69. $\ln|\sqrt{2} + 1| - \ln|\sqrt{2} - 1|$

71. $4 - \dfrac{\pi}{2}$

73. $-\ln|\csc(\sin \theta) + \cot(\sin \theta)| + C$

75. $\ln|\sin x| + \ln|\cos x| + C$

77. $12\tan^{-1}(\sqrt{y}) + C$

79. $\sec^{-1}\left|\dfrac{x - 1}{7}\right| + C$

81. $\ln|\sec(\tan t)| + C$

83. (a) $\sin \theta - \dfrac{1}{3}\sin^3 \theta + C$

    (b) $\sin \theta - \dfrac{2}{3}\sin^3 \theta + \dfrac{1}{5}\sin^5 \theta + C$

    (c) $\displaystyle\int \cos^9 \theta\, d\theta = \int \cos^8 \theta(\cos \theta)\, d\theta = \int (1 - \sin^2 \theta)^4 (\cos \theta)\, d\theta$

85. (a) $\displaystyle\int \tan^3 \theta\, d\theta = \dfrac{1}{2}\tan^2 \theta - \int \tan \theta\, d\theta = \dfrac{1}{2}\tan^2 \theta + \ln|\cos \theta| + C$

    (b) $\displaystyle\int \tan^5 \theta\, d\theta = \dfrac{1}{4}\tan^4 \theta - \int \tan^3 \theta\, d\theta$

    (c) $\displaystyle\int \tan^7 \theta\, d\theta = \dfrac{1}{6}\tan^6 \theta - \int \tan^5 \theta\, d\theta$

    (d) $\displaystyle\int \tan^{2k+1} \theta\, d\theta = \dfrac{1}{2k}\tan^{2k} \theta - \int \tan^{2k-1} \theta\, d\theta$

87. $2\sqrt{2} - \ln(3 + 2\sqrt{2})$

89. π^2

91. $\ln(2 + \sqrt{3})$

Section 7.2, pp. 553–554

1. $-2x \cos\left(\dfrac{x}{2}\right) + 4 \sin\left(\dfrac{x}{2}\right) + C$

3. $t^2 \sin t + 2t \cos t - 2 \sin t + C$

5. $\ln 4 - \dfrac{3}{4}$

7. $y \tan^{-1}(y) - \ln\sqrt{1 + y^2} + C$

9. $x \tan x + \ln|\cos x| + C$

11. $(x^3 - 3x^2 + 6x - 6)e^x + C$

13. $(x^2 - 7x + 7)e^x + C$

15. $(x^5 - 5x^4 + 20x^3 - 60x^2 + 120x - 120)e^x + C$

17. $\dfrac{\pi^2 - 4}{8}$ **19.** $\dfrac{5\pi - 3\sqrt{3}}{9}$

21. $\dfrac{1}{2}(-e^\theta \cos\theta + e^\theta \sin\theta) + C$

23. $\dfrac{e^{2x}}{13}(3 \sin 3x + 2 \cos 3x) + C$

25. $\dfrac{2}{3}(\sqrt{3s + 9}\, e^{\sqrt{3s+9}} - e^{\sqrt{3s+9}}) + C$

27. $\dfrac{\pi\sqrt{3}}{3} - \ln(2) - \dfrac{\pi^2}{18}$

29. $\dfrac{1}{2}[-x \cos(\ln x) + x \sin(\ln x)] + C$

31. $y = \left(\dfrac{x^2}{4} - \dfrac{x}{8} + \dfrac{1}{32}\right)e^{4x} + C$

33. $-2(\sqrt{\theta} \cos\sqrt{\theta} - \sin\sqrt{\theta}) + C$

35. (a) π (b) 3π
        (c) 5π (d) $(2n + 1)\pi$

37. $2\pi(1 - \ln 2)$

39. (a) $\pi(\pi - 2)$ (b) 2π

41. $\dfrac{1}{2\pi}(1 - e^{-2\pi})$

43. $u = x^n,\ dv = \cos x\, dx$ **45.** $u = x^n,\ dv = e^{ax}\, dx$

47. (a) Let $y = f^{-1}(x)$. Then $x = f(y)$, so $dx = f'(y)\, dy$. Substitute directly.
        (b) $u = y,\ dv = f'(y)\, dy$

49. (a) $\displaystyle\int \sin^{-1} x\, dx = x \sin^{-1} x + \cos(\sin^{-1} x) + C$
        (b) $\displaystyle\int \sin^{-1} x\, dx = x \sin^{-1} x + \sqrt{1 - x^2} + C$
        (c) $\cos(\sin^{-1} x) = \sqrt{1 - x^2}$

51. (a) $\displaystyle\int \cos^{-1} x\, dx = x \cos^{-1} x - \sin(\cos^{-1} x) + C$
        (b) $\displaystyle\int \cos^{-1} x\, dx = x \cos^{-1} x - \sqrt{1 - x^2} + C$
        (c) $\sin(\cos^{-1} x) = \sqrt{1 - x^2}$

Section 7.3, pp. 563–565

1. $\dfrac{2}{x - 3} + \dfrac{3}{x - 2}$ **3.** $\dfrac{1}{x + 1} + \dfrac{3}{(x + 1)^2}$

5. $\dfrac{-2}{z} + \dfrac{-1}{z^2} + \dfrac{2}{z - 1}$ **7.** $1 + \dfrac{17}{t - 3} + \dfrac{-12}{t - 2}$

9. $\dfrac{1}{2}[\ln|1 + x| - \ln|1 - x|] + C$

11. $\dfrac{1}{7} \ln|(x + 6)^2(x - 1)^5| + C$

13. $\dfrac{\ln 15}{2}$

15. $-\dfrac{1}{2} \ln|t| + \dfrac{1}{6} \ln|t + 2| + \dfrac{1}{3} \ln|t - 1| + C$

17. $3 \ln 2 - 2$

19. $\dfrac{1}{4} \ln\left|\dfrac{x + 1}{x - 1}\right| - \dfrac{x}{2(x^2 - 1)} + C$

21. $\dfrac{\pi + 2\ln 2}{8}$ **23.** $\tan^{-1} y - \dfrac{1}{y^2 + 1} + C$

25. $-(s - 1)^{-2} + (s - 1)^{-1} + \tan^{-1} s + C$

27. $\dfrac{-1}{\theta^2 + 2\theta + 2} + \ln|\theta^2 + 2\theta + 2| - \tan^{-1}(\theta + 1) + C$

29. $x^2 + \ln\left|\dfrac{x - 1}{x}\right| + C$

31. $9x + 2 \ln|x| + \dfrac{1}{x} + 7 \ln|x - 1| + C$

33. $\dfrac{y^2}{2} - \ln|y| + \dfrac{1}{2} \ln(1 + y^2) + C$

35. $\ln\left|\dfrac{e^t + 1}{e^t + 2}\right| + C$ **37.** $\dfrac{1}{5} \ln\left|\dfrac{\sin y - 2}{\sin y + 3}\right| + C$

39. $\dfrac{(\tan^{-1} 2x)^2}{4} - 3 \ln|x - 2| + \dfrac{6}{x - 2} + C$

41. $x = \ln|t - 2| - \ln|t - 1| + \ln 2$

43. $x = \dfrac{6t}{t + 2} - 1,\ t > 2/5$

45. $\ln|y - 1| - \ln|y| = e^x - 1 - \ln 2$

47. $y = \ln|x - 2| - \ln|x - 1| + \ln 2$

49. $3\pi \ln 25$

51. (a) $x = \dfrac{1000e^{4t}}{499 + e^{4t}}$ (b) 1.55 days

Section 7.4, pp. 569–570

1. $\ln|\sqrt{9 + y^2} + y| + C$

3. $\dfrac{25}{2} \sin^{-1}\left(\dfrac{t}{5}\right) + \dfrac{t\sqrt{25 - t^2}}{2} + C$

5. $\dfrac{1}{2} \ln\left|\dfrac{2x}{7} + \dfrac{\sqrt{4x^2 - 49}}{7}\right| + C$

7. $\dfrac{\sqrt{x^2 - 1}}{x} + C$

9. $\dfrac{1}{3}(x^2 + 4)^{3/2} - 4\sqrt{x^2 + 4} + C$ **11.** $\dfrac{-2\sqrt{4 - w^2}}{w} + C$

13. $-\dfrac{x}{\sqrt{x^2 - 1}} + C$ **15.** $-\dfrac{1}{5}\left(\dfrac{\sqrt{1 - x^2}}{x}\right)^5 + C$

17. $2 \tan^{-1} 2x + \dfrac{4x}{(4x^2 + 1)} + C$ **19.** $\ln 9 - \ln(1 + \sqrt{10})$

21. $\dfrac{\pi}{6}$ **23.** $\sec^{-1}|x| + C$

25. $\sqrt{x^2 - 1} + C$

27. $y = 2\left[\dfrac{\sqrt{x^2 - 4}}{2} - \sec^{-1}\left(\dfrac{x}{2}\right)\right]$

29. $y = \dfrac{3}{2}\tan^{-1}\left(\dfrac{x}{2}\right) - \dfrac{3\pi}{8}$ **31.** $\dfrac{3\pi}{4}$

33. (a) This can be seen geometrically in the figure.

  (b) Using part (a), substitute $z = \dfrac{\sin x}{1 + \cos x}$ and then obtain a trigonometric identity. Or, use the trigonometric identity $\dfrac{1 - \tan^2 \theta}{1 + \tan^2 \theta} = \cos 2\theta$ with $\theta = \dfrac{x}{2}$.

  (c) Using part (a), substitute $z = \dfrac{\sin x}{1 + \cos x}$ and then obtain a trigonometric identity. Or, use the trigonometric identity $\dfrac{2\tan \theta}{1 + \tan^2 \theta} = \sin 2\theta$ with $\theta = \dfrac{x}{2}$.

  (d) $dz = \left(\sec^2 \dfrac{x}{2}\right)\dfrac{1}{2}\,dx = \left(1 + \tan^2 \dfrac{x}{2}\right)\dfrac{1}{2}\,dx = (1 + z^2)\dfrac{1}{2}\,dx$, then solve for dx.

35. $-\dfrac{1}{\tan \frac{x}{2}} + C$ **37.** $\ln\left|1 + \tan \dfrac{t}{2}\right| + C$

39. $\dfrac{1}{2}(\ln \sqrt{3} - 1)$ **41.** $-\cot\left(\dfrac{t}{2}\right) - t + C$

Section 7.5, pp. 576–577

1. $\dfrac{2}{\sqrt{3}}\left(\tan^{-1}\sqrt{\dfrac{x-3}{3}}\right) + C$ **3.** $\dfrac{(2x-3)^{3/2}(x+1)}{5} + C$

5. $\dfrac{(x+2)(2x-6)\sqrt{4x - x^2}}{6} + 4\sin^{-1}\left(\dfrac{x-2}{2}\right) + C$

7. $\sqrt{4 - x^2} - 2\ln\left|\dfrac{2 + \sqrt{4 - x^2}}{x}\right| + C$

9. $2\sin^{-1}\dfrac{r}{2} - \dfrac{1}{2}r\sqrt{4 - r^2} + C$

11. $\dfrac{e^{2t}}{13}(2\cos 3t + 3\sin 3t) + C$

13. $\dfrac{s}{18(9 - s^2)} + \dfrac{1}{108}\ln\left|\dfrac{s+3}{s-3}\right| + C$

15. $2\sqrt{3t - 4} - 4\tan^{-1}\sqrt{\dfrac{3t - 4}{4}} + C$

17. $-\dfrac{\cos 5x}{10} - \dfrac{\cos x}{2} + C$

19. $6\sin\left(\dfrac{\theta}{12}\right) + \dfrac{6}{7}\sin\left(\dfrac{7\theta}{12}\right) + C$

21. $\dfrac{1}{2}\ln|x^2 + 1| + \dfrac{x}{2(1 + x^2)} + \dfrac{1}{2}\tan^{-1}x + C$

23. $\left(x - \dfrac{1}{2}\right)\sin^{-1}\sqrt{x} + \dfrac{1}{2}\sqrt{x - x^2} + C$

25. $\sqrt{1 - \sin^2 t} - \ln\left|\dfrac{1 + \sqrt{1 - \sin^2 t}}{\sin t}\right| + C$

27. $\ln|\ln y + \sqrt{3 + (\ln y)^2}| + C$

29. $\ln|3r + \sqrt{9r^2 - 1}| + C$

31. $x\cos^{-1}\sqrt{x} + \dfrac{1}{2}\sin^{-1}\sqrt{x} - \dfrac{1}{2}\sqrt{x - x^2} + C$

33. $\dfrac{e^{3x}}{9}(3x - 1) + C$

35. $\dfrac{x^2 2^x}{\ln 2} - \dfrac{2}{\ln 2}\left[\dfrac{x 2^x}{\ln 2} - \dfrac{2^x}{(\ln 2)^2}\right] + C$

37. $\dfrac{1}{120}\sinh^4 3x\cosh 3x - \dfrac{1}{90}\sinh^2 3x\cosh 3x + \dfrac{2}{90}\cosh 3x + C$

39. $\dfrac{x^2}{3}\sinh 3x - \dfrac{2x}{9}\cosh 3x + \dfrac{2}{27}\sinh 3x + C$

45. (b) Using Formula 29 in the text's integral table, $V =$
$$2L\left[\left(\dfrac{d-r}{2}\right)\sqrt{2rd - d^2} + \left(\dfrac{r^2}{2}\right)\left[\sin^{-1}\left(\dfrac{d-r}{r}\right) + \dfrac{\pi}{2}\right]\right]$$

49. (c) $\dfrac{\pi}{4}$ **51.** $1 - \dfrac{1}{e} \approx 0.632121$

53. $\dfrac{4}{15} \approx 0.266667$

55. $6 + 2[(\ln 2)^3 - 3(\ln 2)^2 + 6\ln 2 - 6] \approx 0.101097$

Section 7.6, pp. 584–586

1. $\dfrac{1}{4}$ **3.** $\dfrac{5}{7}$

5. $\dfrac{1}{2}$ **7.** 0

9. -1 **11.** $\ln 2$

13. 1   **15.** 0

17. 1   **19.** 0

21. e^2 **23.** 0

25. 1   **27.** 1

29. e **31.** 1

33. e^{-1} **35.** $\ln 2$

37. -1 **39.** 3

41. 1   **43. (b)** is correct, but **(a)** is not.

45. $c = \dfrac{27}{10}$

47. (a) $\ln\left(1 + \dfrac{r}{k}\right)^k = k\ln\left(1 + \dfrac{r}{k}\right)$. And, as $k \to \infty$,

$$\lim_{k\to\infty} k\ln\left(1 + \dfrac{r}{k}\right) = \lim_{k\to\infty}\dfrac{\ln\left(1 + \dfrac{r}{k}\right)}{\dfrac{1}{k}} = \lim_{k\to\infty}\dfrac{\dfrac{-r}{k^2}\Big/\left(1 + \dfrac{r}{k}\right)}{\dfrac{-1}{k^2}} =$$

$$\lim_{k\to\infty}\dfrac{r}{1 + \dfrac{r}{k}} = r. \text{ Therefore, } \lim_{k\to\infty}\left(1 + \dfrac{r}{k}\right)^k = e^r.$$

Hence, $\displaystyle\lim_{k\to\infty} A_0\left(1 + \dfrac{r}{k}\right)^{kt} = A_0 e^{rt}$.

(b) Part (a) shows that as the number of compoundings per year increases toward infinity, the limit of interest compounded k times per year is interest compounded continuously.

53. (a) $(-\infty, -1) \cup (0, \infty)$ **(b)** ∞ **(c)** e

Section 7.7, pp. 598–600

1. (a) Because of an infinite limit of integration
   (b) Converges
   (c) $\dfrac{\pi}{2}$

3. (a) Because the intergrand has an infinite discontinuity at $x = 0$
   (b) Converges
   (c) $-\dfrac{9}{2}$

5. (a) Because the integrand has an infinite discontinuity at $x = 0$
   (b) Diverges
   (c) No value

7. 1000   **9.** 4

11. $\dfrac{\pi}{2}$ **13.** $\ln 3$

15. $\sqrt{3}$ **17.** π

19. $\dfrac{\pi}{3}$ **21.** $\ln 4$

23. $\dfrac{\pi}{2}$ **25.** $\ln\left(1 + \dfrac{\pi}{2}\right)$

27. 6   **29.** -1

31. 2   **33.** $-\dfrac{1}{4}$

35. Diverges   **37.** Converges
39. Converges   **41.** Converges
43. Diverges   **45.** Converges
47. Converges   **49.** Diverges
51. Converges   **53.** Converges
55. Diverges   **57.** Converges
59. Diverges   **61.** Converges
63. Converges
65. (a) Converges when $p < 1$
   (b) Converges when $p > 1$

67. 1   **69.** $\dfrac{\pi}{2}$

73. (b) ≈ 0.88621 **75. (b)** 1

Chapter 7 Practice Exercises, pp. 601–602

1. $\dfrac{1}{12}(4x^2 - 9)^{3/2} + C$ **3.** $\dfrac{\sqrt{8x^2 + 1}}{8} + C$

5. $\dfrac{-\sqrt{9 - 4t^4}}{8} + C$ **7.** $-\dfrac{1}{2(1 - \cos 2\theta)} + C$

9. $-\dfrac{1}{2}e^{\cos 2x} + C$ **11.** $\dfrac{2^{x-1}}{\ln 2} + C$

13. $\ln\left| 2 + \tan^{-1} x \right| + C$ **15.** $\dfrac{1}{3}\sin^{-1}\left(\dfrac{3t}{4}\right) + C$

17. $\dfrac{1}{5}\sec^{-1}\left| \dfrac{5x}{4} \right| + C$ **19.** $\dfrac{1}{2}\tan^{-1}\left(\dfrac{y - 2}{2}\right) + C$

21. $\dfrac{x}{2} + \dfrac{\sin 6x}{12} + C$

23. $\dfrac{\tan^2 (2t)}{4} - \dfrac{1}{2}\ln\left| \sec 2t \right| + C$

25. $\ln\left| \sec 2x + \tan 2x \right| + C$

27. $\ln(3 + 2\sqrt{2})$ **29.** $2\sqrt{2}$

31. $x - 2\tan^{-1}\left(\dfrac{x}{2}\right) + C$

33. $\ln(y^2 + 4) - \dfrac{1}{2}\tan^{-1}\left(\dfrac{y}{2}\right) + C$

35. $-\sqrt{4 - t^2} + 2\sin^{-1}\left(\dfrac{t}{2}\right) + C$

37. $x - \tan x + \sec x + C$

39. $4\ln\left| \sin\left(\dfrac{x}{4}\right) \right| + C$

41. $\dfrac{z}{16(16 + z^2)^{1/2}} + C$

43. $\dfrac{-\sqrt{1 - x^2}}{x} + C$

45. $\ln\left| x + \sqrt{x^2 - 9} \right| + C$

47. $[(x + 1)(\ln(x + 1)) - (x + 1)] + C$

49. $x\tan^{-1}(3x) - \dfrac{1}{6}\ln(1 + 9x^2) + C$

51. $(x + 1)^2 e^x - 2(x + 1)e^x + 2e^x + C$

53. $\dfrac{2e^x \sin 2x}{5} + \dfrac{e^x \cos 2x}{5} + C$

55. $2\ln\left| x - 2 \right| - \ln\left| x - 1 \right| + C$

57. $-\dfrac{1}{3}\ln\left| \dfrac{\cos \theta - 1}{\cos \theta + 2} \right| + C$

59. $\dfrac{1}{16}\ln\left| \dfrac{(v - 2)^5 (v + 2)}{v^6} \right| + C$

61. $\dfrac{x^2}{2} + \dfrac{4}{3}\ln\left| x + 2 \right| + \dfrac{2}{3}\ln\left| x - 1 \right| + C$

63. $x^2 - 3x + \dfrac{2}{3}\ln\left| x + 4 \right| + \dfrac{1}{3}\ln\left| x - 2 \right| + C$

65. $\ln\left| 1 - e^{-s} \right| + C$ **67.** $-\sqrt{16 - y^2} + C$

69. $-\dfrac{1}{2}\ln\left| 4 - x^2 \right| + C$ **71.** $\ln \dfrac{1}{\sqrt{9 - x^2}} + C$

73. $\frac{1}{6} \ln \left| \frac{x+3}{x-3} \right| + C$

75. $\frac{2x^{3/2}}{3} - x + 2\sqrt{x} - 2 \ln (\sqrt{x} + 1) + C$

77. $2 \sin \sqrt{x} + C$

79. $\ln | u + \sqrt{1 + u^2} | + C$

81. $\frac{1}{12} \ln \left| \frac{3+v}{3-v} \right| + \frac{1}{6} \tan^{-1} \frac{v}{3} + C$

83. $\frac{x^2}{2} + 2x + 3 \ln |x - 1| - \frac{1}{x-1} + C$

85. $-\cos (2\sqrt{x}) + C$

87. $\frac{\sqrt{3}}{3} \tan^{-1} \left(\frac{\theta - 1}{\sqrt{3}} \right) + C$

89. $\frac{1}{4} \sec^2 \theta + C$

91. $-\frac{2}{3} (x + 4) \sqrt{2 - x} + C$

93. $\frac{1}{2} [x \ln |x - 1| - x - \ln |x - 1|] + C$

95. $\frac{1}{4} \ln |z| - \frac{1}{4z} - \frac{1}{4} \left[\frac{1}{2} \ln (z^2 + 4) + \frac{1}{2} \tan^{-1} \left(\frac{z}{2} \right) \right] + C$

97. $-\frac{\tan^{-1} x}{x} + \ln |x| - \ln \sqrt{1 + x^2} + C$

99. $\tan x - x + C$

101. $\ln | \csc (2x) + \cot (2x) | + C$ **103.** $\frac{1}{4}$

105. $\sec^{-1} |2x - 1| + C$ **107.** $\frac{1}{6} (3 + 4e^{\theta})^{3/2} + C$

109. $\frac{1}{3} \left(\frac{27^{3\theta + 1}}{\ln 27} \right) + C$

111. $2\sqrt{r} - 2 \ln (1 + \sqrt{r}) + C$

113. $4 \sec^{-1} \left| \frac{7m}{2} \right| + C$ **115.** The limit does not exist.

117. 2   **119.** 1

121. 0   **123.** $-\frac{1}{2}$

125. 1   **127.** ∞

129. $\frac{\pi}{2}$ **131.** 6

133. $\ln 3$ **135.** 2

137. $\frac{\pi}{6}$ **139.** Diverges

141. Diverges   **143.** Converges

145. $\ln |y - 1| - \ln |y| = e^x - 1 - \ln 2$

147. $y = \ln |x - 2| - \ln |x - 1| + \ln 2$

Chapter 7 Additional Exercises, pp. 603–606

1. $x(\sin^{-1} x)^2 + 2(\sin^{-1} x) \sqrt{1 - x^2} - 2x + C$

3. $\frac{x^2 \sin^{-1} x}{2} + \frac{x\sqrt{1 - x^2} - \sin^{-1} x}{4} + C$

5. $\frac{\ln |\sec 2\theta + \tan 2\theta| + 2\theta}{4} + C$

7. $\frac{1}{2} [\ln |t - \sqrt{1 - t^2}| - \sin^{-1} t] + C$

9. $\frac{1}{16} \ln \left| \frac{x^2 + 2x + 2}{x^2 - 2x + 2} \right| + \frac{1}{8}[\tan^{-1}(x + 1) + \tan^{-1}(x - 1)] + C$

11. $\frac{\pi}{2}$ **13.** $\frac{1}{\sqrt{e}}$

15. 0   **17.** 1

19. $\frac{32\pi}{35}$ **21.** 2π

23. (a) π **(b)** $\pi(2e - 5)$

25. (b) $\pi \left[\frac{8(\ln 2)^2}{3} - \frac{16(\ln 2)}{9} + \frac{16}{27} \right]$

27. $\frac{1}{2}$ **31.** $\frac{\pi}{2}(3b - a) + 2$

33. 6   **35.** $P(x) = -3x^2 + 1$

37. $\frac{1}{2} < p \le 1$ **39. (b)** 1

41. $\frac{e^{2x}}{13} (3 \sin 3x + 2 \cos 3x) + C$

43. $\frac{\cos x \sin 3x - 3 \sin x \cos 3x}{8} + C$

45. $\frac{e^{ax}}{a^2 + b^2} (a \sin bx - b \cos bx) + C$

47. $x \ln (ax) - x + C$

CHAPTER 8

Section 8.1, pp. 617–619

1. $a_1 = 0, a_2 = -\frac{1}{4}, a_3 = -\frac{2}{9}, a_4 = -\frac{3}{16}$

3. $a_1 = 1, a_2 = -\frac{1}{3}, a_3 = \frac{1}{5}, a_4 = -\frac{1}{7}$

5. $a_n = (-1)^{n+1}, n \ge 1$ **7.** $a_n = n^2 - 1, n \ge 1$

9. $a_n = 4n - 3, n \ge 1$ **11.** $a_n = \frac{1 + (-1)^{n+1}}{2}, n \ge 1$

13. Converges, 2   **15.** Converges, -1

17. Diverges   **19.** Diverges

21. Converges, $\frac{1}{2}$ **23.** Converges, $\sqrt{2}$

25. Converges, 0   **27.** Converges, 0

29. Diverges   **31.** Converges, e^7

33. Converges, 1   **35.** Converges, 1

37. Converges, 4   **39.** Converges, 0

41. Diverges   **43.** Converges, e^{-1}

45. Converges, $e^{2/3}$ **47.** Converges, $x \ (x > 0)$

49. Converges, 0   **51.** Converges, $\frac{\pi}{2}$

53. Converges, 0   **55.** Converges, 0

57. $N = 692, a_n = \sqrt[n]{0.5}, L = 1$ **59.** $N = 65, a_n = (0.9)^n, L = 0$

61. (b) $\sqrt{2}$ **63. (b)** 1

Section 8.2, pp. 625–627

1. $1, \dfrac{3}{2}, \dfrac{7}{4}, \dfrac{15}{8}, \dfrac{31}{16}, \dfrac{63}{32}, \dfrac{127}{64}, \dfrac{255}{128}, \dfrac{511}{256}, \dfrac{1023}{512}$

3. $2, 1, -\dfrac{1}{2}, -\dfrac{1}{4}, \dfrac{1}{8}, \dfrac{1}{16}, -\dfrac{1}{32}, -\dfrac{1}{64}, \dfrac{1}{128}, \dfrac{1}{256}$

5. 1, 1, 2, 3, 5, 8, 13, 21, 34, 55

7. (b) $\sqrt{3}$

9. (a) $f(x) = x^2 - 2,\ 1.414213562 \approx \sqrt{2}$

(b) $f(x) = \tan(x) - 1,\ 0.7853981635 \approx \dfrac{\pi}{4}$

(c) $f(x) = e^x$, diverges

11. Nondecreasing, bounded

13. Not nondecreasing, bounded

15. Converges, monotonic sequence theorem

17. Converges, monotonic sequence theorem

19. Diverges, definition of divergence

21. Converges, monotonic sequence theorem

23. Diverges, definition of divergence

27. 1   **29.** -0.73908513

31. 0.85375017

Section 8.3, pp. 637–639

1. $s_n = \dfrac{2\left[1 - \left(\frac{1}{3}\right)^n\right]}{1 - \left(\frac{1}{3}\right)}, 3$

3. $s_n = \dfrac{1 - \left(-\frac{1}{2}\right)^n}{1 - \left(-\frac{1}{2}\right)}, \dfrac{2}{3}$

5. $s_n = \dfrac{1}{2} - \dfrac{1}{n+2}, \dfrac{1}{2}$

7. $1 - \dfrac{1}{4} + \dfrac{1}{16} - \dfrac{1}{64} + \cdots, \dfrac{4}{5}$

9. $(5+1) + \left(\dfrac{5}{2} + \dfrac{1}{3}\right) + \left(\dfrac{5}{4} + \dfrac{1}{9}\right) + \left(\dfrac{5}{8} + \dfrac{1}{27}\right) + \cdots, \dfrac{23}{2}$

11. $(1+1) + \left(\dfrac{1}{2} - \dfrac{1}{6}\right) + \left(\dfrac{1}{4} + \dfrac{1}{25}\right) + \left(\dfrac{1}{8} - \dfrac{1}{125}\right) + \cdots, \dfrac{17}{6}$

13. 1   **15.** 5

17. 1   **19.** Converges, $2 + \sqrt{2}$

21. Converges, 1   **23.** Converges, $\dfrac{e^2}{e^2 - 1}$

25. Converges, $\dfrac{x}{x-1}$ **27.** Diverges

29. Diverges   **31.** Diverges

33. $a = 1, r = -x$; converges to $\dfrac{1}{1+x}$ for $|x| < 1$

35. $a = 3, r = \dfrac{x-1}{2}$, converges to $\dfrac{6}{3-x}$ for x in $(-1, 3)$

37. $|x| < \dfrac{1}{2}, \dfrac{1}{1-2x}$ **39.** $1 < x < 5, \dfrac{2}{x-1}$

41. $\dfrac{23}{99}$ **43.** $\dfrac{7}{9}$

45. $\dfrac{41,333}{33,300}$ **47.** 28 m

49. 8 m^2

51. (a) $3\left(\dfrac{4}{3}\right)^{n-1}$

(b) $A_n = A + \dfrac{1}{3}A + \dfrac{1}{3}\left(\dfrac{4}{9}\right)A + \cdots + \dfrac{1}{3}\left(\dfrac{4}{9}\right)^{n-2}A,\ \lim_{n\to\infty} A_n = \dfrac{2\sqrt{3}}{5}$

53. (a) $\displaystyle\sum_{n=-2}^{\infty} \dfrac{1}{(n+4)(n+5)}$

(b) $\displaystyle\sum_{n=0}^{\infty} \dfrac{1}{(n+2)(n+3)}$

(c) $\displaystyle\sum_{n=5}^{\infty} \dfrac{1}{(n-3)(n-2)}$

55. $\ln\left(\dfrac{8}{9}\right)$ **61.** It diverges.

Section 8.4, pp. 649–651

1. Diverges   **3.** Diverges

5. Converges   **7.** Converges

9. Diverges; $\dfrac{1}{2\sqrt{n} + \sqrt[3]{n}} \geq \dfrac{1}{2n+n} = \dfrac{1}{3n}$

11. Converges; $\dfrac{\sin^2 n}{2^n} \leq \dfrac{1}{2^n}$

13. Converges; $\left(\dfrac{n}{3n+1}\right)^n < \left(\dfrac{n}{3n}\right)^n = \left(\dfrac{1}{3}\right)^n$

15. Diverges; limit comparison with $\sum \dfrac{1}{n}$

17. Converges; limit comparison with $\sum \dfrac{1}{n^2}$

19. Converges; limit comparison with $\sum \dfrac{1}{n^{5/4}}$

21. Converges, $\rho = 1/2$ **23.** Diverges, $\rho = \infty$

25. Converges, $\rho = 1/10$ **27.** Converges, $\rho = 0$

29. Converges, $\rho = 0$ **31.** Converges, $\rho = 0$

33. Diverges, $\rho = \infty$

35. Converges; geometric series, $r = \dfrac{1}{e} < 1$

37. Diverges; p-series, $p < 1$

39. Diverges; limit comparison with $\sum \dfrac{1}{n}$

41. Converges; limit comparison with $\sum \dfrac{1}{n^{3/2}}$

43. Converges; Ratio Test   **45.** Converges; Ratio Test

47. Converges; Integral Test   **49.** Converges; Integral Test

51. Converges; compare with $\sum \dfrac{3}{(1.25)^n}$

53. Converges; compare with $\sum \dfrac{1}{n^2}$

55. Converges; compare with $\sum \dfrac{1}{n^2}$

57. Converges; $\dfrac{\tan^{-1} n}{n^{1.1}} < \dfrac{\left(\frac{\pi}{2}\right)}{n^{1.1}}$

59. Diverges; nth-Term Test

61. Converges; Ratio Test

63. Diverges; Ratio Test

65. Diverges; $a_n = \left(\dfrac{1}{3}\right)^{(1/n!)} \to 1$ **71.** $a = 1$

Section 8.5, pp. 658–660

1. Converges by Theorem 8

3. Diverges; $a_n \not\to 0$

5. Converges by Theorem 8

7. Diverges; $a_n \to \dfrac{1}{2} \neq 0$

9. Converges by Theorem 8

11. Converges absolutely. Series of absolute values is a convergent geometric series.

13. Converges conditionally. $\dfrac{1}{\sqrt{n}+1} \to 0$ but $\displaystyle\sum_{n=1}^{n} \dfrac{1}{\sqrt{n}+1}$ diverges.

15. Converges absolutely. Compare with $\displaystyle\sum_{n=1}^{\infty} \dfrac{1}{n^2}$.

17. Converges conditionally. $\dfrac{1}{n+3} \to 0$ but $\displaystyle\sum_{n=1}^{\infty} \dfrac{1}{n+3}$ diverges. Compare with $\displaystyle\sum_{n=1}^{\infty} \dfrac{1}{n}$.

19. Diverges; $\dfrac{3+n}{5+n} \to 1$

21. Converges conditionally; $\left(\dfrac{1}{n^2} + \dfrac{1}{n}\right) \to 0$ but $\dfrac{1+n}{n^2} > \dfrac{1}{n}$.

23. Converges absolutely; Ratio Test

25. Converges absolutely by Integral Test

27. Diverges; $a_n \not\to 0$

29. Converges absolutely by the Ratio Test

31. Converges absolutely; $\dfrac{1}{n^2 + 2n + 1} < \dfrac{1}{n^2}$

33. Converges absolutely since $\left| \dfrac{\cos n\pi}{n \sqrt{n}} \right| = \left| \dfrac{(-1)^{n+1}}{n^{3/2}} \right| = \dfrac{1}{n^{3/2}}$ (convergent p-series).

35. Converges absolutely n^{th} by Root Test

37. Diverges; $a_n \to \infty$

39. Converges conditionally; $\sqrt{n+1} - \sqrt{n} = \dfrac{1}{\sqrt{n} + \sqrt{n+1}} \to 0$, but series of absolute values diverges. Compare with $\Sigma \dfrac{1}{\sqrt{n}}$.

41. Diverges; $a_n \to \dfrac{1}{2} \neq 0$

43. Converges absolutely; sech $n = \dfrac{2}{e^n + e^{-n}} = \dfrac{2e^n}{e^{2n} + 1} < \dfrac{2e^n}{e^{2n}} = \dfrac{2}{e^n}$, a term from a convergent geometric series.

45. $|\,\text{Error}\,| < 0.2$ **47.** $|\,\text{Error}\,| < 2 \times 10^{-11}$

49. 0.54030

51. (a) $a_n \geq a_{n+1}$ fails

(b) $-\dfrac{1}{2}$

Section 8.6, pp. 668–669

1. (a) $1, -1 < x < 1$ (b) $-1 < x < 1$ (c) None

3. (a) $\dfrac{1}{4}, -\dfrac{1}{2} < x < 0$ (b) $-\dfrac{1}{2} < x < 0$ (c) None

5. (a) $10, -8 < x < 12$ (b) $-8 < x < 12$ (c) None

7. (a) $1, -1 < x < 1$ (b) $-1 < x < 1$ (c) None

9. (a) $3, [-3, 3]$ (b) $[-3, 3]$ (c) None

11. (a) ∞, for all x (b) For all x (c) None

13. (a) ∞, for all x (b) For all x (c) None

15. (a) $1, -1 \leq x < 1$ (b) $-1 < x < 1$ (c) $x = -1$

17. (a) $5, -8 < x < 2$ (b) $-8 < x < 2$ (c) None

19. (a) $3, -3 < x < 3$ (b) $-3 < x < 3$ (c) None

21. (a) $1, -1 < x < 1$ (b) $-1 < x < 1$ (c) None

23. (a) $0, x = 0$ (b) $x = 0$ (c) None

25. (a) $2, -4 < x \leq 0$ (b) $-4 < x < 0$ (c) $x = 0$

27. (a) $1, -1 \leq x \leq 1$ (b) $-1 \leq x \leq 1$ (c) None

29. (a) $\dfrac{1}{4}, 1 \leq x \leq \dfrac{3}{2}$ (b) $1 \leq x \leq \dfrac{3}{2}$ (c) None

31. (a) $1, (-1 - \pi) \leq x < (1 - \pi)$

(b) $(-1 - \pi) < x < (1 - \pi)$

(c) $x = -1 - \pi$

33. $-1 < x < 3, \dfrac{4}{3 + 2x - x^2}$

35. $0 < x < 16, \dfrac{2}{4 - \sqrt{x}}$

37. $-\sqrt{2} < x < \sqrt{2}, \dfrac{3}{2 - x^2}$

39. $1 < x < 5, \dfrac{2}{x - 1}, 1 < x < 5, \dfrac{-2}{(x - 1)^2}$

41. (a) $\cos x = 1 - \dfrac{x^2}{2!} + \dfrac{x^4}{4!} - \dfrac{x^6}{6!} + \dfrac{x^8}{8!} - \dfrac{x^{10}}{10!} + \cdots$; converges for all x

(b) and (c) $2x - \dfrac{2^3 x^3}{3!} + \dfrac{2^5 x^5}{5!} - \dfrac{2^7 x^7}{7!} + \dfrac{2^9 x^9}{9!} - \dfrac{2^{11} x^{11}}{11!} + \cdots$

43. (a) $\dfrac{x^2}{2} + \dfrac{x^4}{12} + \dfrac{x^6}{45} + \dfrac{17 x^8}{2520} + \dfrac{31 x^{10}}{14,175}, -\dfrac{\pi}{2} < x < \dfrac{\pi}{2}$

(b) $1 + x^2 + \dfrac{2x^4}{3} + \dfrac{17 x^6}{45} + \dfrac{62 x^8}{315} + \cdots, -\dfrac{\pi}{2} < x < \dfrac{\pi}{2}$

Section 8.7, pp. 681–683

1. $P_0(x) = 0, P_1(x) = x - 1, P_2(x) = (x - 1) - \dfrac{1}{2}(x - 1)^2,$

$P_3(x) = (x - 1) - \dfrac{1}{2}(x - 1)^2 + \dfrac{1}{3}(x - 1)^3$

3. $P_0(x) = \dfrac{1}{2}, P_1(x) = \dfrac{1}{2} - \dfrac{x}{4}, P_2(x) = \dfrac{1}{2} - \dfrac{x}{4} + \dfrac{x^2}{8},$

$P_3(x) = \dfrac{1}{2} - \dfrac{x}{4} + \dfrac{x^2}{8} - \dfrac{x^3}{16}$

5. $P_0(x) = \dfrac{1}{\sqrt{2}}$, $P_1(x) = \dfrac{1}{\sqrt{2}} - \dfrac{1}{\sqrt{2}}\left(x - \dfrac{\pi}{4}\right)$,

$P_2(x) = \dfrac{1}{\sqrt{2}} - \dfrac{1}{\sqrt{2}}\left(x - \dfrac{\pi}{4}\right) - \dfrac{1}{2\sqrt{2}}\left(x - \dfrac{\pi}{4}\right)^2$, $P_3(x) =$

$\dfrac{1}{\sqrt{2}} - \dfrac{1}{\sqrt{2}}\left(x - \dfrac{\pi}{4}\right) - \dfrac{1}{2\sqrt{2}}\left(x - \dfrac{\pi}{4}\right)^2 + \dfrac{1}{6\sqrt{2}}\left(x - \dfrac{\pi}{4}\right)^3$

7. $\displaystyle\sum_{n=0}^{\infty} \frac{(-x)^n}{n!} = 1 - x + \frac{x^2}{2!} - \frac{x^3}{3!} + \frac{x^4}{4!} - \cdots$

9. $\displaystyle\sum_{n=0}^{\infty} \frac{(-1)^n 3^{2n+1} x^{2n+1}}{(2n+1)!}$

11. $\displaystyle\sum_{n=0}^{\infty} \frac{x^{2n}}{(2n)!}$

13. $x^4 - 2x^3 - 5x + 4$

15. $8 + 10(x-2) + 6(x-2)^2 + (x-2)^3$

17. $\displaystyle\sum_{n=0}^{\infty} (-1)^n(n+1)(x-1)^n$

19. $\displaystyle\sum_{n=0}^{\infty} \frac{e^2}{n!}(x-2)^n$

21. $\displaystyle\sum_{n=0}^{\infty} \frac{(-5x)^n}{n!} = 1 - 5x + \frac{5^2 x^2}{2!} - \frac{5^3 x^3}{3!} + \cdots$

23. $\displaystyle\sum_{n=0}^{\infty} \frac{(-1)^n \left(\frac{\pi x}{2}\right)^{2n+1}}{(2n+1)!} = \frac{\pi x}{2} - \frac{\pi^3 x^3}{2^3 \cdot 3!} + \frac{\pi^5 x^5}{2^5 \cdot 5!} - \frac{\pi^7 x^7}{2^7 \cdot 7!} + \cdots$

25. $\displaystyle\sum_{n=0}^{\infty} \frac{x^{n+1}}{n!} = x + x^2 + \frac{x^3}{2!} + \frac{x^4}{3!} + \frac{x^5}{4!} + \cdots$

27. $\displaystyle\sum_{n=2}^{\infty} \frac{(-1)^n x^{2n}}{(2n)!} = \frac{x^4}{4!} - \frac{x^6}{6!} + \frac{x^8}{8!} - \frac{x^{10}}{10!} + \cdots$

29. $x - \dfrac{\pi^2 x^3}{2!} + \dfrac{\pi^4 x^5}{4!} - \dfrac{\pi^6 x^7}{6!} + \cdots = \displaystyle\sum_{n=0}^{\infty} \frac{(-1)^n \pi^{2n} x^{2n+1}}{(2n)!}$

31. $\displaystyle\sum_{n=1}^{\infty} \frac{(-1)^{n+1}(2x)^{2n}}{2 \cdot (2n)!} = \frac{(2x)^2}{2 \cdot 2!} - \frac{(2x)^4}{2 \cdot 4!} + \frac{(2x)^6}{2 \cdot 6!} - \frac{(2x)^8}{2 \cdot 8!} + \cdots$

33. $\displaystyle\sum_{n=1}^{\infty} \frac{(-1)^{n-1} 2^n x^{n+1}}{n} = 2x^2 - \frac{2^2 x^3}{2} + \frac{2^3 x^4}{3} - \frac{2^4 x^5}{4} + \cdots$

35. $|x| < (0.06)^{1/5} < 0.56968$

37. $|\,\text{Error}\,| < \dfrac{(10^{-3})^3}{6} < 1.67 \times 10^{-10}, \; -10^{-3} < x < 0$

39. (a) $|\,\text{Error}\,| < \dfrac{(3^{0.1})(0.1)^3}{6} < 1.87 \times 10^{-4}$

(b) $|\,\text{Error}\,| < \dfrac{(0.1)^3}{6} < 1.67 \times 10^{-4}$

45. (a) $L(x) = 0$ (b) $Q(x) = -\dfrac{x^2}{2}$

47. (a) $L(x) = 1$ (b) $Q(x) = 1 + \dfrac{x^2}{2}$

Section 8.8, pp. 690–691

1. $1 + \dfrac{x}{2} - \dfrac{x^2}{8} + \dfrac{x^3}{16}$

3. $1 + \dfrac{1}{2}x - \dfrac{3}{8}x^2 + \dfrac{5}{16}x^3 + \cdots$ **5.** $1 - x + \dfrac{3x^2}{4} - \dfrac{x^3}{2}$

7. $1 - \dfrac{x^3}{2} + \dfrac{3x^6}{8} - \dfrac{5x^9}{16}$ **9.** $1 + \dfrac{1}{2x} - \dfrac{1}{8x^2} + \dfrac{1}{16x^3}$

11. $(1 + x)^4 = 1 + 4x + 6x^2 + 4x^3 + x^4$

13. $(1 - 2x)^3 = 1 - 6x + 12x^2 - 8x^3$

15. $y = \displaystyle\sum_{n=0}^{\infty} \frac{(-1)^n}{n!} x^n = e^{-x}$

17. $y = \displaystyle\sum_{n=1}^{\infty} \frac{x^n}{n!} = e^x - 1$

19. $y = \displaystyle\sum_{n=2}^{\infty} \frac{x^n}{n!} = e^x - x - 1$

21. $y = \displaystyle\sum_{n=0}^{\infty} \frac{x^{2n}}{2^n n!} = e^{x^2/2}$

23. $y = \displaystyle\sum_{n=0}^{\infty} 2x^n = \frac{2}{1 - x}$

25. $y = \displaystyle\sum_{n=0}^{\infty} \frac{x^{2n+1}}{(2n + 1)!} = \sinh x$

27. $y = 2 + x - 2\displaystyle\sum_{n=1}^{\infty} \frac{(-1)^{n+1} x^{2n}}{(2n)!}$

29. $y = -2(x - 2) - \displaystyle\sum_{n=1}^{\infty} \left[\frac{2(x - 2)^{2n}}{(2n)!} + \frac{3(x - 2)^{2n+1}}{(2n + 1)!}\right]$

31. $y = a + bx + \dfrac{1}{6} x^3 - \dfrac{ax^4}{3 \cdot 4} - \dfrac{bx^5}{4 \cdot 5} - \dfrac{x^7}{6 \cdot 6 \cdot 7} + \dfrac{ax^8}{3 \cdot 4 \cdot 7 \cdot 8} +$

$\dfrac{bx^9}{4 \cdot 5 \cdot 8 \cdot 9} + \cdots$. For $n \geq 6$, $a_n = \dfrac{a_{n-4}}{n(n-1)}$

33. $\dfrac{x^3}{3} - \dfrac{x^7}{7 \cdot 3!} + \dfrac{x^{11}}{11 \cdot 5!}$

35. (a) $\dfrac{x^2}{2} - \dfrac{x^4}{12}$

(b) $\dfrac{x^2}{2} - \dfrac{x^4}{3 \cdot 4} + \dfrac{x^6}{5 \cdot 6} - \dfrac{x^8}{7 \cdot 8} + \cdots + (-1)^{15} \dfrac{x^{32}}{31 \cdot 32}$

37. $\dfrac{1}{2}$ **39.** -1

41. $2!$ **45.** 500 terms

47. (a) $x + \dfrac{x^3}{6} + \dfrac{3x^5}{40} + \dfrac{5x^7}{112}$, radius of convergence $= 1$

(b) $\dfrac{\pi}{2} - x - \dfrac{x^3}{6} - \dfrac{3x^5}{40} - \dfrac{5x^7}{112}$

Section 8.9, pp. 697–698

1. $f(x) = 1$ **3.** $f(x) = \displaystyle\sum_{n=1}^{\infty} \frac{2(-1)^{n+1}}{n} \sin nx$

5. $f(x) = \dfrac{\pi^2}{12} + \displaystyle\sum_{n=1}^{\infty} \frac{(-1)^n}{n^2} \cos nx$

7. $f(x) = \dfrac{2\sinh\pi}{\pi}\left[\dfrac{1}{2} + \displaystyle\sum_{n=1}^{\infty} \frac{(-1)^n}{n^2 + 1}(\cos nx - n \sin nx)\right]$

9. $f(x) = \dfrac{1}{2}\cos x + \dfrac{1}{\pi}\displaystyle\sum_{n=2}^{\infty} \frac{n(1 + (-1)^n)}{n^2 - 1}\sin nx$

11. $f(x) = \dfrac{1}{2} + \dfrac{2}{\pi}\displaystyle\sum_{k=0}^{\infty} \frac{(-1)^k}{2k + 1}\cos(2k + 1)x$

13. $f(x) = \dfrac{5}{4} + \dfrac{4}{\pi^2}\displaystyle\sum_{n=1}^{\infty} \frac{1}{n^2}\left[(-1)^n - \cos\frac{n\pi}{2}\right](\cos(n\pi x) +$

$\dfrac{2}{\pi}\displaystyle\sum_{n=1}^{\infty} \frac{1}{n}\left[(-1)^n - \frac{2}{n\pi}\sin\frac{n\pi}{2}\right]\sin(n\pi x)$

15. Set $x = \pi$, $\dfrac{\pi^2}{4} = \dfrac{\pi^2}{12} + \displaystyle\sum_{n=1}^{\infty} \frac{(-1)^n}{n^2}\cos n\pi$, or $\dfrac{\pi^2}{6} = \displaystyle\sum_{n=1}^{\infty} \frac{1}{n^2} =$

$1 + \dfrac{1}{4} + \dfrac{1}{9} + \dfrac{1}{16} + \cdots + \dfrac{1}{n^2} + \cdots$.

17. 0

19. 0 if $m \neq n$; L if $m = n$

21. 0 if $m \neq n$; 0 if $m = n$

Section 8.10, pp. 705–706

1. $f(x) = \dfrac{\pi}{2} + \dfrac{2}{\pi} \displaystyle\sum_{n=1}^{\infty} \dfrac{[(-1)^n - 1]}{n^2} \cos nx$

3. $f(x) = (e - 1) + 2 \displaystyle\sum_{n=1}^{\infty} \dfrac{[e(-1)^n - 1]}{1 + n^2\pi^2} \cos n\pi x$

5. $f(x) = -\dfrac{1}{4} + \dfrac{4}{\pi} \displaystyle\sum_{n=1}^{\infty} \left[\dfrac{1}{n} \sin \dfrac{n\pi}{2} + \dfrac{1}{\pi n^2}\left((-1)^{n+1} + \cos \dfrac{n\pi}{2} \right) \right]$
$\cos \dfrac{n\pi x}{2}$

7. $f(x) = \dfrac{1}{2} + \displaystyle\sum_{n=1}^{\infty} \dfrac{4}{n^2\pi^2}\left[1 + (-1)^n - 2\cos \dfrac{n\pi}{2} \right] \cos n\pi x$

9. $f(x) = 2 \displaystyle\sum_{n=1}^{\infty} \dfrac{(-1)^n}{n\pi} \sin n\pi x$

11. $f(x) = \dfrac{8}{\pi} \displaystyle\sum_{k=1}^{\infty} \dfrac{k}{4k^2 - 1} \sin 2kx$

13. $f(x) = \sin x$

15. $f(x) = \dfrac{2}{\pi} \displaystyle\sum_{n=1}^{\infty} \left[\dfrac{1}{n} - \dfrac{2}{n^2\pi} \sin \dfrac{n\pi}{2} \right] \sin \dfrac{n\pi x}{2}$

17. (a) $f(x) = \dfrac{4}{\pi}\left[\sin x + \dfrac{\sin 3x}{3} + \dfrac{\sin 5x}{5} + \dfrac{\sin 7x}{7} + \cdots \right]$

(b) Evaluate $f(x)$ at $x = \dfrac{\pi}{2} \Rightarrow \dfrac{\pi}{4} = 1 - \dfrac{1}{3} + \dfrac{1}{5} - \dfrac{1}{7} + \cdots$.

19. $\displaystyle\sum_{n=1}^{\infty} \dfrac{(-1)^n}{4n^2 - 1} = \dfrac{1}{2} - \dfrac{\pi}{4}$

Practice 8 Practice Exercises, pp. 708–711

1. Converges to 1

3. Converges to -1

5. Diverges

7. Converges to 0

9. Converges to 1

11. Converges to e^{-5}

13. Converges to 3

15. Converges to $\ln 2$

17. Diverges

19. $\dfrac{1}{6}$

21. $\dfrac{3}{2}$

23. $\dfrac{e}{e - 1}$

25. Diverges

27. Converges conditionally

29. Converges conditionally

31. Converges absolutely

33. Converges absolutely

35. Converges absolutely

37. Converges absolutely

39. Converges absolutely

41. (a) $3, -7 \leq x < -1$

(b) $-7 < x < -1$

(c) $x = -7$

43. (a) $\dfrac{1}{3}, 0 \leq x \leq \dfrac{2}{3}$ **(b)** $0 \leq x \leq \dfrac{2}{3}$ **(c)** None

45. (a) ∞, for all x **(b)** For all x **(c)** None

47. (a) $\sqrt{3}, -\sqrt{3} < x < \sqrt{3}$

(b) $-\sqrt{3} < x < \sqrt{3}$

(c) None

49. (a) $e, (-e, e)$ **(b)** $(-e, e)$ **(c)** $\{\ \}$

51. $\dfrac{1}{1 + x}, \dfrac{1}{4}, \dfrac{4}{5}$

53. $\sin x, \pi, 0$

55. $e^x, \ln 2, 2$

57. $\displaystyle\sum_{n=0}^{\infty} 2^n x^n$

59. $\displaystyle\sum_{n=0}^{\infty} \dfrac{(-1)^n \pi^{2n+1} x^{2n+1}}{(2n + 1)!}$

61. $\displaystyle\sum_{n=0}^{\infty} \dfrac{(-1)^n x^{5n}}{(2n)!}$

63. $\displaystyle\sum_{n=0}^{\infty} \dfrac{\left(\dfrac{\pi x}{2} \right)^n}{n!}$

65. $2 - \dfrac{(x + 1)}{2 \cdot 1!} + \dfrac{3(x + 1)^2}{2^3 \cdot 2!} + \dfrac{9(x + 1)^3}{2^5 \cdot 3!} + \cdots$

67. $\dfrac{1}{4} - \dfrac{1}{4^2}(x - 3) + \dfrac{1}{4^3}(x - 3)^2 - \dfrac{1}{4^4}(x - 3)^3 + \cdots$

69. $y = \displaystyle\sum_{n=0}^{\infty} \dfrac{(-1)^{n+1}}{n!} x^n = -e^{-x}$

71. $y = 3 \displaystyle\sum_{n=0}^{\infty} \dfrac{(-1)^n 2^n}{n!} x^n = 3e^{-2x}$

73. $y = -1 - x + 2 \displaystyle\sum_{n=2}^{\infty} \dfrac{x^n}{n!} = 2e^x - 3x - 3$

75. $y = -1 - x + 2 \displaystyle\sum_{n=0}^{\infty} \dfrac{x^n}{n!} = 2e^x - 1 - x$

77. (a) $\dfrac{7}{2}$

79. (a) $\dfrac{1}{12}$

81. (a) -2

83. $r = -3, s = \dfrac{9}{2}$

85. $f(x) = \dfrac{1}{2} + \dfrac{6}{\pi} \displaystyle\sum_{n=1}^{\infty} \dfrac{1}{2n - 1} \sin [(2n - 1)x]$

87. $f(x) = \pi - 2 \displaystyle\sum_{n=1}^{\infty} \dfrac{(-1)^n}{n} \sin nx$

89. $f(x) = \dfrac{3}{2} + \dfrac{2}{\pi^2} \displaystyle\sum_{n=1}^{\infty} \dfrac{(-1)^n - 1}{n^2} \cos \dfrac{n\pi x}{2} - \dfrac{2}{\pi} \displaystyle\sum_{n=1}^{\infty} \dfrac{(-1)^n}{n} \sin \dfrac{n\pi x}{2}$

91. (a) $f(x) = \dfrac{1}{2} + \dfrac{2}{\pi} \displaystyle\sum_{n=1}^{\infty} \dfrac{\sin \left(\dfrac{n\pi}{2} \right)}{n} \cos n\pi x$

(b) $f(x) = \dfrac{2}{\pi} \displaystyle\sum_{n=1}^{\infty} \dfrac{1}{n}\left(1 - \cos \dfrac{n\pi}{2} \right) \sin n\pi x$

93. (a) $f(x) = \dfrac{2}{\pi} + \dfrac{1}{\pi} \displaystyle\sum_{n=1}^{\infty} \left[\dfrac{1}{n + 1} - \dfrac{1}{n - 1} + \dfrac{\cos [(n - 1)\pi]}{n - 1} - \right.$
$\left. \dfrac{\cos [(n + 1)\pi]}{n + 1} \right] \cos n\pi x$

(b) $f(x) = \sin \pi x$

95. (a) $f(x) = 6 + \dfrac{12}{\pi^2} \displaystyle\sum_{n=1}^{\infty} \dfrac{4(-1)^n - 1}{n^2} \cos \dfrac{n\pi x}{3}$

(b) $f(x) = \dfrac{6}{\pi^3} \displaystyle\sum_{n=1}^{\infty} \dfrac{(6 - 5n^2\pi^2)(-1)^n - 6}{n^3} \sin \dfrac{n\pi x}{3}$

97. (b) $|\text{Error}| < \sin\left(\dfrac{1}{42} \right) < 0.02381$; an underestimate because the remainder is positive

99. $\dfrac{2}{3}$

101. $\ln\left(\dfrac{n+1}{2n}\right)$; the series converges to $\ln\left(\dfrac{1}{2}\right)$.

103. **(a)** ∞ **(b)** $a = 1, b = 0$

105. It converges.

113. **(a)** -3 is the fixed point.   **(b)** 0.2 is the fixed point.

Chapter 8 Additional Exercises, pp. 711–715

1. Converges; Direct Comparison Test

3. Diverges; nth-Term Test

5. Converges; Direct Comparison Test

7. Diverges; nth-Term Test

9. With $a = \dfrac{\pi}{3}$, $\cos x = \dfrac{1}{2} - \dfrac{\sqrt{3}}{2}\left(x - \dfrac{\pi}{3}\right) - \dfrac{1}{4}\left(x - \dfrac{\pi}{3}\right)^2 +$ $\dfrac{\sqrt{3}}{12}\left(x - \dfrac{\pi}{3}\right)^3 + \cdots$.

11. With $a = 0$, $e^x = 1 + x + \dfrac{x^2}{2!} + \dfrac{x^3}{3!} + \cdots$.

13. With $a = 22\pi$, $\cos x = 1 - \dfrac{1}{2}(x - 22\pi)^2 + \dfrac{1}{4!}(x - 22\pi)^4 -$ $\dfrac{1}{6!}(x - 22\pi)^6 + \cdots$.

15. Converges, limit $= b$ **17.** $\dfrac{\pi}{2}$

21. **(a)** $\dfrac{b^2\sqrt{3}}{4}\displaystyle\sum_{n=0}^{\infty}\dfrac{3^n}{4^n}$ **(b)** $\sqrt{3}b^2$

(c) No. For example, the three vertices of the original triangle are not removed. The set of points not removed has area 0.

23. **(a)** No, the limit does not appear to depend on the value of a.

(b) Yes, the limit depends on the value of b.

(c) $\displaystyle\lim_{n\to\infty}\left(1 - \dfrac{\cos\left(\dfrac{a}{n}\right)}{bn}\right)^n = e^{-1/b}$

25. $b = \pm\dfrac{1}{5}$ **29.** **(b)** Yes

35. **(a)** $\displaystyle\sum_{n=1}^{\infty} nx^{n-1}$ **(b)** 6   **(c)** $\dfrac{1}{q}$

37. **(a)** $R_n = \dfrac{C_0 e^{-kt_0}(1 - e^{-nkt_0})}{1 - e^{-kt_0}}$, $R = \dfrac{C_0(e^{-kt_0})}{1 - e^{-kt_0}} = \dfrac{C_0}{e^{kt_0} - 1}$

(b) $R_1 = \dfrac{1}{e} \approx 0.368$, $R_{10} = R(1 - e^{-10}) \approx R(0.9999546) \approx$ 0.58195; $R \approx 0.58198$; $0 < \dfrac{R - R_{10}}{R} < 0.0001$

(c) 7

CHAPTER 9

Section 9.1, pp. 726–728

1. **(a)** $\langle 9, -6\rangle$ **(b)** $3\sqrt{13}$

3. **(a)** $\langle 1, 3\rangle$ **(b)** $\sqrt{10}$

5. **(a)** $\langle 12, -19\rangle$ **(b)** $\sqrt{505}$

7. **(a)** $\left\langle \dfrac{1}{5}, \dfrac{14}{5}\right\rangle$ **(b)** $\dfrac{\sqrt{197}}{5}$

9. $\langle 1, -4\rangle$ **11.** $\langle -2, -3\rangle$

13. $\left\langle -\dfrac{1}{2}, \dfrac{\sqrt{3}}{2}\right\rangle$ **15.** $\left\langle -\dfrac{\sqrt{3}}{2}, \dfrac{1}{2}\right\rangle$

17. The vector $\mathbf{v}$ is horizontal and 1 in. long. The vectors $\mathbf{u}$ and $\mathbf{w}$ are $\dfrac{11}{16}$ in. long. $\mathbf{w}$ is vertical and $\mathbf{u}$ makes a $45°$ angle with the horizontal. All vectors must be drawn to scale.

(a)   **(b)**

(c)   **(d)**

19.

21.

23.

25.

27.

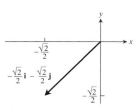

29. $\left\langle \dfrac{3}{5}, \dfrac{4}{5} \right\rangle$

31. $\left\langle -\dfrac{15}{17}, \dfrac{8}{17} \right\rangle$

33. $\dfrac{3}{5}\mathbf{i} - \dfrac{4}{5}\mathbf{j}$

35. $13\left(\dfrac{5}{13}\mathbf{i} + \dfrac{12}{13}\mathbf{j} \right)$

37. $\dfrac{3}{5}\mathbf{i} - \dfrac{4}{5}\mathbf{j}$ and $-\dfrac{3}{5}\mathbf{i} + \dfrac{4}{5}\mathbf{j}$

39. $\mathbf{u} = \dfrac{1}{\sqrt{17}}\mathbf{i} + \dfrac{4}{\sqrt{17}}\mathbf{j}, \ -\mathbf{u} = -\dfrac{1}{\sqrt{17}}\mathbf{i} - \dfrac{4}{\sqrt{17}}\mathbf{j},$

$\mathbf{n} = \dfrac{4}{\sqrt{17}}\mathbf{i} - \dfrac{1}{\sqrt{17}}\mathbf{j}, \ -\mathbf{n} = -\dfrac{4}{\sqrt{17}}\mathbf{i} + \dfrac{1}{\sqrt{17}}\mathbf{j}$

41. $\mathbf{u} = \dfrac{1}{\sqrt{5}}(2\mathbf{i} + \mathbf{j}), \ -\mathbf{u} = \dfrac{1}{\sqrt{5}}(-2\mathbf{i} - \mathbf{j}), \ \mathbf{n} = \dfrac{1}{\sqrt{5}}(-\mathbf{i} + 2\mathbf{j}), \ -$

$\mathbf{n} = \dfrac{1}{\sqrt{5}}(\mathbf{i} - 2\mathbf{j})$

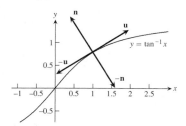

43. $\mathbf{u} = \dfrac{\pm 1}{5}(-4\mathbf{i} + 3\mathbf{j}), \ \mathbf{v} = \dfrac{\pm 1}{5}(3\mathbf{i} + 4\mathbf{j})$

45. $\mathbf{u} = \dfrac{\pm 1}{2}(\mathbf{i} + \sqrt{3}\mathbf{j}), \ \mathbf{v} = \dfrac{\pm 1}{2}(-\sqrt{3}\mathbf{i} + \mathbf{j})$

47. $a = \dfrac{3}{2}, b = \dfrac{1}{2}$

49. $5\sqrt{3}\,\mathbf{i}, 5\mathbf{j}$

51. $\approx \langle -338.095, 725.046 \rangle$

53. **(a)** $(5\cos 60°, 5\sin 60°) = \left(\dfrac{5}{2}, \dfrac{5\sqrt{3}}{2} \right)$

(b) $(5\cos 60° + 10\cos 315°, 5\sin 60° + 10\sin 315°) =$
$\left(\dfrac{5 + \sqrt{2}}{2}, \dfrac{5\sqrt{3} - 10\sqrt{2}}{2} \right)$

Section 9.2, pp. 735–737

	$\mathbf{v} \cdot \mathbf{u}$	$|\mathbf{v}|$	$|\mathbf{u}|$
1.	-20	$2\sqrt{5}$	$2\sqrt{5}$
	$\cos \theta$	$|\mathbf{u}|\cos\theta$	$\text{proj}_{\mathbf{v}}\,\mathbf{u}$
	-1	$-2\sqrt{5}$	$-2\mathbf{i} + 4\mathbf{j}$

	$\mathbf{v} \cdot \mathbf{u}$	$|\mathbf{v}|$	$|\mathbf{u}|$
3.	$\sqrt{3} - \sqrt{2}$	$\sqrt{2}$	$\sqrt{5}$
	$\cos \theta$	$|\mathbf{u}|\cos\theta$	$\text{proj}_{\mathbf{v}}\,\mathbf{u}$
	$\dfrac{\sqrt{30} - \sqrt{20}}{10}$	$\dfrac{\sqrt{6} - 2}{2}$	$\dfrac{\sqrt{3} - \sqrt{2}}{2}(-\mathbf{i} + \mathbf{j})$

	$\mathbf{v} \cdot \mathbf{u}$	$|\mathbf{v}|$	$|\mathbf{u}|$
5.	$\dfrac{1}{6}$	$\dfrac{\sqrt{30}}{6}$	$\dfrac{\sqrt{30}}{6}$
	$\cos \theta$	$|\mathbf{u}|\cos\theta$	$\text{proj}_{\mathbf{v}}\,\mathbf{u}$
	$\dfrac{1}{5}$	$\dfrac{1}{\sqrt{30}}$	$\dfrac{1}{5}\left\langle \dfrac{1}{\sqrt{2}}, \dfrac{1}{\sqrt{3}} \right\rangle$

7. ≈ 0.64 rad

9. ≈ 1.85

11. Angle at $A = \cos^{-1}\left(\dfrac{1}{\sqrt{5}} \right) \approx 63.435$ degrees, angle at $B =$

$\cos^{-1}\left(\dfrac{3}{5} \right) \approx 53.130$ degrees, angle at $C = \cos^{-1}\left(\dfrac{1}{\sqrt{5}} \right) \approx$

63.435 degrees.

13. The sum of two vectors of equal length is *always* orthogonal to their difference, as we can see from the equation $(\mathbf{v}_1 + \mathbf{v}_2) \cdot (\mathbf{v}_1 - \mathbf{v}_2) = \mathbf{v}_1 \cdot \mathbf{v}_1 + \mathbf{v}_2 \cdot \mathbf{v}_1 - \mathbf{v}_1 \cdot \mathbf{v}_2 - \mathbf{v}_2 \cdot \mathbf{v}_2 = |\mathbf{v}_1|^2 - |\mathbf{v}_2|^2 = 0$.

19. Horizontal component: ≈ 1188 ft/sec,
vertical component: ≈ 167 ft/sec

21. **(a)** Since $|\cos \theta| \leq 1$, we have $|\mathbf{u} \cdot \mathbf{v}| = |\mathbf{u}||\mathbf{v}||\cos \theta| \leq |\mathbf{u}||\mathbf{v}|(1) = |\mathbf{u}||\mathbf{v}|$.

(b) We have equality precisely when $|\cos \theta| = 1$ or when one or both of $\mathbf{u}$ and $\mathbf{v}$ are $\mathbf{0}$. In the case of nonzero vectors, we have equality when $\theta = 0$ or π, that is, when the vectors are parallel.

23. a

27. $x + 2y = 4$

29. $-2x + y = -3$

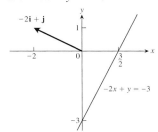

31. $x + y = -1$

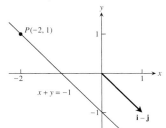

33. $2x - y = 0$

35. 5 J

37. 3464 J

39. $\dfrac{\pi}{4}$

41. $\dfrac{\pi}{6}$

43. 0.14

45. $\dfrac{\pi}{3}$ and $\dfrac{2\pi}{3}$ at each point

47. At $(0, 0)$: $\dfrac{\pi}{2}$; at $(1, 1)$: $\dfrac{\pi}{4}$ and $\dfrac{3\pi}{4}$

Section 9.3, pp. 746–749

1. (a)

(b) $\mathbf{v}(t) = (-2\sin t)\mathbf{i} + (3\cos t)\mathbf{j}$,
$\mathbf{a}(t) = (-2\cos t)\mathbf{i} + (-3\sin t)\mathbf{j}$
(c) Speed = 2, direction = $\langle -1, 0\rangle$
(d) Velocity = $2\langle -1, 0\rangle$

3. (a)

(b) $\mathbf{v} = (\sec t \tan t)\mathbf{i} + (\sec^2 t)\mathbf{j}$, $\mathbf{a}(t) = (\sec t \tan^2 t + \sec^3 t)\mathbf{i} + (2\sec^2 t \tan t)\mathbf{j}$

(c) Speed = $\dfrac{2\sqrt{5}}{3}$, direction = $\left\langle \dfrac{1}{\sqrt{5}}, \dfrac{2}{\sqrt{5}}\right\rangle$

(d) Velocity = $\left(\dfrac{2\sqrt{5}}{3}\right)\left\langle \dfrac{1}{\sqrt{5}}, \dfrac{2}{\sqrt{5}}\right\rangle$

5. $t = 0, \pi, 2\pi$

7. $t =$ all nonnegative integer multiples of $\dfrac{\pi}{2}$

9. $\cos^{-1}\left(\dfrac{3}{5}\right) \approx 53.130$ degrees

11. (a) $3\mathbf{i}$ **(b)** $t \neq 0, -3$ **(c)** $t = 0, -3$
13. (a) $y = -1$ **(b)** $x = 0$
15. $-3\mathbf{i} + (4\sqrt{2} - 2)\mathbf{j}$
17. $(\sec t)\mathbf{i} + (\ln|\sec t|)\mathbf{j} + \mathbf{C}$
19. $\mathbf{r}(t) = ((t + 1)^{3/2} - 1)\mathbf{i} - (e^{-t} - 1)\mathbf{j}$
21. $\mathbf{r}(t) = (8t + 100)\mathbf{i} + (-16t^2 + 8t)\mathbf{j}$
23. 2
25. (a) $\mathbf{v}(t) = (\cos t)\mathbf{i} - (2\sin 2t)\mathbf{j}$

(b) $t = \dfrac{\pi}{2}, \dfrac{3\pi}{2}$

(c) $y = 1 - 2x^2$, $-1 \le x \le 1$. The particle starts at $(0, 1)$, goes to $(1, -1)$, then goes to $(-1, -1)$, and then goes to $(0, 1)$, tracing the curve twice.

27. $\mathbf{r}(t) = \left(\dfrac{3}{2}t^2 + \dfrac{3\sqrt{10}}{5}t + 1\right)\mathbf{i} + \left(-\dfrac{1}{2}t^2 - \dfrac{\sqrt{10}}{5}t + 2\right)\mathbf{j}$

29. (a) i. Constant speed
    ii. Yes, orthogonal
    iii. Counterclockwise movement
    iv. Yes
  (b) i. Constant speed
    ii. Yes, orthogonal
    iii. Counterclockwise movement
    iv. Yes
  (c) i. Constant speed
    ii. Yes, orthogonal
    iii. Counterclockwise movement
    iv. No
  (d) i. Constant speed
    ii. Yes, orthogonal
    iii. Clockwise movement
    iv. Yes
  (e) i. Variable speed
    ii. Not orthogonal in general
    iii. Counterclockwise movement
    iv. Yes

31. (a) 160 sec   **(b)** 225 m   **(c)** $\dfrac{15}{4}$ m/sec

33. (a) Referring to the figure, look at the circular arc from the point where $t = 0$ to the point "m". On one hand, this arc has length given by $(r_0\theta)$, but it also has length given by (vt). Setting those two quantities equal gives the result.

(b) $\mathbf{a}(t) = -\dfrac{v^2}{r_0}\left[\left(\cos\dfrac{vt}{r_0}\right)\mathbf{i} + \left(\sin\dfrac{vt}{r_0}\right)\mathbf{j}\right]$

(c) From part (b), $\mathbf{a}(t) = -\left(\dfrac{v}{r_0}\right)^2\mathbf{r}(t)$. So, by Newton's second law, $\mathbf{F} = -m\left(\dfrac{v}{r_0}\right)^2\mathbf{r}$. Substituting for $\mathbf{F}$ in the law of gravitation gives the result.

(d) Set $\dfrac{vT}{r_0} = 2\pi$ and solve for vT.

(e) Substitute $\dfrac{2\pi r_0}{T}$ for v in $v^2 = \dfrac{GM}{r_0}$ and solve for T^2.

35. (a) Apply the corollary to each component separately.
(b) Follows immediately from part (a) since any two antiderivatives of $\mathbf{r}(t)$ must have identical derivatives, namely $\mathbf{r}(t)$.

37. Let $\mathbf{C} = \langle C_1, C_2\rangle$, $\dfrac{d\mathbf{C}}{dt} = \left\langle\dfrac{dC_1}{dt}, \dfrac{dC_2}{dt}\right\rangle = \langle 0, 0\rangle$.

39. $\mathbf{u} = \langle u_1, u_2\rangle$, $\mathbf{v} = \langle v_1, v_2\rangle$

(a) $\dfrac{d}{dt}(\mathbf{u}+\mathbf{v}) = \dfrac{d}{dt}(\langle u_1+v_1, u_2+v_2\rangle)$

$= \left\langle\dfrac{d}{dt}(u_1+v_1), \dfrac{d}{dt}(u_2+v_2)\right\rangle$

$= \langle u'_1+v'_1, u'_2+v'_2\rangle$

$= \langle u'_1, u'_2\rangle + \langle v'_1, v'_2\rangle = \dfrac{d\mathbf{u}}{dt} + \dfrac{d\mathbf{v}}{dt}$

(b) $\dfrac{d}{dt}(\mathbf{u}-\mathbf{v}) = \dfrac{d}{dt}(\langle u_1-v_1, u_2-v_2\rangle)$

$= \left\langle\dfrac{d}{dt}(u_1-v_1), \dfrac{d}{dt}(u_2-v_2)\right\rangle$

$= \langle u'_1-v'_1, u'_2-v'_2\rangle$

$= \langle u'_1, u'_2\rangle - \langle v'_1, v'_2\rangle = \dfrac{d\mathbf{u}}{dt} - \dfrac{d\mathbf{v}}{dt}$

41. $f(t)$ and $g(t)$ differentiable at $c \Rightarrow f(t)$ and $g(t)$ continuous at $c \Rightarrow$ $\mathbf{r}(t) = f(t)\mathbf{i} + g(t)\mathbf{j}$ is continuous at c.

43. (a) Let $\mathbf{r}(t) = f(t)\mathbf{i} + g(t)\mathbf{j}$. Then $\dfrac{d}{dt}\displaystyle\int_a^t \mathbf{r}(q)\,dq = \dfrac{d}{dt}\displaystyle\int_a^t [f(q)\mathbf{i} + g(q)\mathbf{j}]\,dq = \dfrac{d}{dt}\left[\left(\displaystyle\int_a^t f(q)\,dq\right)\mathbf{i} + \left(\displaystyle\int_a^t g(q)\,dq\right)\mathbf{j}\right] = \left(\dfrac{d}{dt}\displaystyle\int_a^t f(q)\,dq\right)\mathbf{i} + \left(\dfrac{d}{dt}\displaystyle\int_a^t g(q)\,dq\right)\mathbf{j} = f(t)\mathbf{i} + g(t)\mathbf{j} = \mathbf{r}(t)$.

(b) Let $\mathbf{S}(t) = \displaystyle\int_a^t \mathbf{r}(q)\,dq$. Then part (a) shows that $\mathbf{S}(t)$ is an antiderivative of $\mathbf{r}(t)$. Let $\mathbf{R}(t)$ be any antiderivative of $\mathbf{r}(t)$. Then, according to Exercise 35, part (b), $\mathbf{S}(t) = \mathbf{R}(t) + \mathbf{C}$. Letting $t = a$, we have $\mathbf{0} = \mathbf{S}(a) = \mathbf{R}(a) + \mathbf{C}$. Therefore,

$\mathbf{C} = -\mathbf{R}(a)$ and $\mathbf{S}(t) = \mathbf{R}(t) - \mathbf{R}(a)$. The result follows by letting $t = b$.

Section 9.4, pp. 757–760

1. 50 sec
3. (a) 72.2 sec, 25,510 m   **(b)** 4020 m   **(c)** 6378 m
5. $t \approx 2.135$ sec, $x \approx 66.42$ ft
7. (a) $v_0 \approx 9.9$ m/sec
(b) $\alpha \approx 18.4°$ or $71.6°$
9. 190 mph
11. The golf ball will clip the leaves at the top.
13. 149 ft/sec, 2.25 sec   **15.** $39.3°$ or $50.7°$
17. 46.6 ft/sec   **21.** 1.92 sec, 73.7 ft (approx.)
23. 4.00 ft, 7.80 ft/sec
25. (b) $\mathbf{v}_0$ would bisect $\angle AOR$
27. (a) (Assuming that "x" is zero at the point of impact.) $\mathbf{r}(t) = (x(t))\mathbf{i} + (y(t))\mathbf{j}$, where $x(t) = (35\cos 27°)t$ and $y(t) = 4 + (35\sin 27°)t - 16t^2$.
(b) At $t \approx 0.497$ sec, it reaches its maximum height of about 7.945 ft.
(c) Range $\approx 37,45$ ft, flight time ≈ 1.201 sec
(d) At $t \approx 0.254$ and $t \approx 0.740$ sec, when it is ≈ 29.554 and ≈ 14.396 ft from where it will land.
(e) Yes. It changes things because the ball won't clear the net.
31. (a) $\mathbf{r}(t) = (x(t))\mathbf{i} + (y(t))\mathbf{j}$, where $x(t) = \left(\dfrac{1}{0.08}\right)(1 - e^{-0.08t})$
$(152\cos 20° - 7.6)$ and $y(t) = 3 + \left(\dfrac{152}{0.08}\right)(1 - e^{-0.08t})$
$(\sin 20°) + \left(\dfrac{32}{0.08^2}\right)(1 - 0.08t - e^{-0.08t})$
(b) At $t \approx 1.527$ sec, it reaches its maximum height of about 41.893 ft.
(c) Range ≈ 351.734 ft, flight time ≈ 3.181 sec.
(d) At $t \approx 0.877$ and $t \approx 2.190$ sec, when it is about 106.028 and 251.530 ft from home plate.
(e) No. The wind gust would need to be greater than 12.846 ft/sec in the direction of the hit for the ball to clear the fence for a home run.

Section 9.5, pp. 768–770

1. (a) and **(e)** are the same.
(b) and **(g)** are the same.
(c) and **(h)** are the same.
(d) and **(f)** are the same.
3. (a) $(1, 1)$ **(b)** $(1, 0)$ **(c)** $(0, 0)$ **(d)** $(-1, -1)$

5. (a) $\left(\sqrt{2}, \dfrac{3\pi}{4}\right)$ or $\left(\sqrt{2}, -\dfrac{5\pi}{4}\right)$

(b) $\left(2, -\dfrac{\pi}{3}\right)$ or $\left(-2, \dfrac{2\pi}{3}\right)$

(c) $\left(3, \dfrac{\pi}{2}\right)$ or $\left(3, \dfrac{5\pi}{2}\right)$

(d) $(1, \pi)$ or $(-1, 0)$

7.

9.

11.

13.

15.

17.

19. $y = 0$, the x-axis

21. $y = 4$, a horizontal line

23. $x + y = 1$, a line (slope $= -1$, y-intercept $= 1$)

25. $x^2 + (y - 2)^2 = 4$, a circle (center $= (0, 2)$, radius $= 2$)

27. $xy = 1$ $\left(\text{or } y = \dfrac{1}{x}\right)$, a hyperbola

29. $y = e^x$, the exponential curve

31. $y = \ln x$, the logarithmic curve

33. $(x + 2)^2 + y^2 = 4$, a circle (center $= (-2, 0)$, radius $= 2$)

35. $(x - 1)^2 + (y - 1)^2 = 2$, a circle (center $= (1, 1)$, radius $= \sqrt{2}$)

37. $r \cos \theta = 7$

39. $\theta = \dfrac{\pi}{4}$

41. $r^2 = 4$ or $r = 2$

43. $r^2(4 \cos^2 \theta + 9 \sin^2 \theta) = 36$

45. $r \sin^2 \theta = 4 \cos \theta$

47. $r = 4 \sin \theta$

49. (a)

(b) Length of interval $= 2\pi$

51. (a)

(b) Length of interval $= \dfrac{\pi}{2}$

53. (a)

(b) Length of interval $= 2\pi$

55. (a)

(b) Required interval $= (-\infty, \infty)$

57. (a)

(b) Length of interval $= \pi$

59. x-axis, y-axis, origin   **61.** y-axis

63. (a) Because $r = a \sec \theta$ is equivalent to $r \cos \theta = a$, which is equivalent to the Cartesian equation $x = a$

(b) $r = a \csc \theta$ is equivalent to $y = a$.

67. $(0, 0), \left(1, \dfrac{\pi}{2}\right), \left(1, \dfrac{3\pi}{2}\right)$ **69.** $(0, 0), \left(\dfrac{1}{2}, \pm\dfrac{\pi}{3}\right)$

71. $(0, 0), \left(\pm\dfrac{1}{\sqrt[4]{2}}, \dfrac{\pi}{8}\right)$

73. $\left(1, \dfrac{\pi}{12}\right), \left(1, \dfrac{5\pi}{12}\right), \left(1, \dfrac{7\pi}{12}\right), \left(1, \dfrac{11\pi}{12}\right), \left(1, \dfrac{13\pi}{12}\right), \left(1, \dfrac{17\pi}{12}\right),$

$\left(1, \dfrac{19\pi}{12}\right), \left(1, \dfrac{23\pi}{12}\right)$

75. Part (a)

81. $d = [(x_2 - x_1)^2 + (y_2 - y_1)^2]^{1/2} = [(r_2 \cos \theta_2 - r_1 \cos \theta_1)^2 + (r_2 \sin \theta_2 - r_1 \sin \theta_1)^2]^{1/2}$, and then simplify using trigonometric identities.

Section 9.6, pp. 777–779

1. At $\theta = 0$: -1; at $\theta = \pi$: 1

3. At $(2, 0)$: $-\dfrac{2}{3}$; at $\left(-1, \dfrac{\pi}{2}\right)$: 0; at $(2, \pi)$: $\dfrac{2}{3}$; at $\left(5, \dfrac{3\pi}{2}\right)$: 0

5. $\theta = \dfrac{\pi}{2}\ [x = 0]$

7. $\theta = 0\ [y = 0]$, $\theta = \dfrac{\pi}{5}\left[y = \left(\tan \dfrac{\pi}{5}\right)x\right]$, $\theta = \dfrac{2\pi}{5}\left[y = \left(\tan \dfrac{2\pi}{5}\right)x\right]$,
$\theta = \dfrac{3\pi}{5}\left[y = \left(\tan \dfrac{3\pi}{5}\right)x\right]$, $\theta = \dfrac{4\pi}{5}\left[y = \left(\tan \dfrac{4\pi}{5}\right)x\right]$

9. Horizontal at $\left(-\dfrac{1}{2}, \dfrac{\pi}{6}\right)\left[y = -\dfrac{1}{4}\right]$, $\left(-\dfrac{1}{2}, \dfrac{5\pi}{6}\right)\left[y = -\dfrac{1}{4}\right]$,
$\left(-2, \dfrac{3\pi}{2}\right)[y = 2]$;

vertical at $\left(0, \dfrac{\pi}{2}\right)[x = 0]$, $\left(-\dfrac{3}{2}, \dfrac{7\pi}{6}\right)\left[x = \dfrac{3\sqrt{3}}{4}\right]$, $\left(-1.5, \dfrac{11\pi}{6}\right)$
$\left[x = -\dfrac{3\sqrt{3}}{4}\right]$

11. Horizontal at $(0, 0)\ [y = 0]$, $\left(2, \dfrac{\pi}{2}\right)[y = 2]$, $(0, \pi)\ [y = 0]$;

vertical at $\left(\sqrt{2}, \dfrac{\pi}{4}\right)[x = 1]$, $\left(\sqrt{2}, \dfrac{3\pi}{4}\right)[x = -1]$

13. 18π

15. $\dfrac{\pi}{8}$

17. 2

19. $\dfrac{\pi}{2} - 1$

21. $5\pi - 8$

23. $3\sqrt{3} - \pi$

25. $\dfrac{\pi}{3} + \dfrac{\sqrt{3}}{2}$

27. $12\pi - 9\sqrt{3}$

29. (a) $\dfrac{3}{2} - \dfrac{\pi}{4}$

31. $\dfrac{19}{3}$

33. 8

35. $3(\sqrt{2} + \ln(1 + \sqrt{2}))$

37. $\dfrac{\pi}{8} + \dfrac{3}{8}$

39. 2π

45. (a) Let $r = 1.75 + \dfrac{0.06\theta}{2\pi}$, $0 \le \theta \le 12\pi$.

(b) Since $\dfrac{dr}{d\theta} = \dfrac{b}{2\pi}$, this is just Equation (4) for the length of the curve.

(c) ≈ 741.420 cm, or ≈ 7.414 m

(d) $\left(r^2 + \left(\dfrac{b}{2\pi}\right)^2\right)^{1/2} = r\left(1 + \left(\dfrac{b}{2\pi r}\right)^2\right)^{1/2} \approx r$ since $\left(\dfrac{b}{2\pi r}\right)^2$ is a very small quantity squared

(e) $L \approx 741.420$ cm (from part (c)), $L_a \approx 741.416$ cm

Chapter 9 Practice Exercises, pp. 780–784

1. (a) $\langle -17, 32 \rangle$ (b) $\sqrt{1313}$

3. (a) $\langle 6, -8 \rangle$ (b) 10

5. $\left\langle -\dfrac{\sqrt{3}}{2}, -\dfrac{1}{2}\right\rangle$ [assuming counterclockwise]

7. $\left\langle \dfrac{8}{\sqrt{17}}, -\dfrac{2}{\sqrt{17}}\right\rangle$

9. Length $= 2$, direction is $\dfrac{1}{\sqrt{2}}\mathbf{i} + \dfrac{1}{\sqrt{2}}\mathbf{j}$.

11. $\left.\dfrac{d\mathbf{r}}{dt}\right|_{t=\pi/2} = 2(-\mathbf{i})$

13. Unit tangents $\pm\left(\dfrac{1}{\sqrt{5}}\mathbf{i} + \dfrac{2}{\sqrt{5}}\mathbf{j}\right)$, unit normals $\pm$
$\left(-\dfrac{2}{\sqrt{5}}\mathbf{i} + \dfrac{1}{\sqrt{5}}\mathbf{j}\right)$

15.

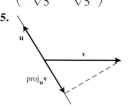

17. $|\mathbf{v}| = \sqrt{2}$, $|\mathbf{u}| = \sqrt{5}$, $\mathbf{u} \cdot \mathbf{v} = \mathbf{v} \cdot \mathbf{u} = 3$, $\theta = \cos^{-1}\left(\dfrac{3}{\sqrt{10}}\right) \approx$
0.32 rad, $|\mathbf{u}| \cos \theta = \dfrac{3\sqrt{2}}{2}$, $\text{proj}_v\, \mathbf{u} = \dfrac{3}{2}(\mathbf{i} + \mathbf{j})$

19. $\mathbf{u} = \left(\dfrac{2}{5}\mathbf{i} - \dfrac{1}{5}\mathbf{j}\right) + \left(\dfrac{3}{5}\mathbf{i} + \dfrac{6}{5}\mathbf{j}\right)$

21. (a) $\mathbf{v}(t) = (-4 \sin t)\mathbf{i} + (\sqrt{2} \cos t)\mathbf{j}$,
$\mathbf{a}(t) = (-4 \cos t)\mathbf{i} + (-\sqrt{2} \sin t)\mathbf{j}$

(b) 3

(c) $\cos^{-t}\dfrac{7}{9} \approx 38.942$ degrees

23. 1   **25.** 6i

27. $\mathbf{r}(t) = (\cos t - 1)\mathbf{i} + (\sin t + 1)\mathbf{j}$

29. $\mathbf{r}(t) = \mathbf{i} + t^2\mathbf{j}$

31.

33. (d)   **35.** (l)

37. (k)   **39.** (i)

41. (a)

(b) 2π

43. (a)

$r \cos \theta = 1$

(b) $\dfrac{\pi}{2}$

45. Tangent lines at $\theta = \dfrac{\pi}{4}, \dfrac{3\pi}{4}, \dfrac{5\pi}{4}$, and $\dfrac{7\pi}{4}$; Cartesian equations are
$y = \pm x$.

47. Horizontal: $y = 0$, $y \approx \pm 0.443$, $y \approx \pm 1.739$;
vertical: $x = 2$, $x \approx 0.067$, $x \approx -1.104$

49. $y = \pm x + \sqrt{2}$ and $y = \pm x - \sqrt{2}$

51. $x = y$, a line

53. $x^2 = 4y$, a parabola

55. $x = 2$, a vertical line

57. $r = -5 \sin \theta$

59. $r^2 \cos^2 \theta + 4r^2 \sin^2 \theta = 16$, or $r^2 = \dfrac{16}{\cos^2 \theta + 4 \sin^2 \theta}$

61. $\dfrac{9\pi}{2}$

63. $2 + \dfrac{\pi}{4}$

65. 8

67. $\pi - 3$

69. Speed ≈ 591.982 mph, direction ≈ 8.179 degrees north of east

71. It hits the ground ≈ 2.135 sec later, approximately 66.421 ft from where it left the thrower's hand. Assuming it does not bounce or roll, it will still be there 3 sec after it was thrown.

73. (a)

$\mathbf{r}(t) = (\pi t - \sin \pi t)\mathbf{i} + (1 - \cos \pi t)\mathbf{j}$

(b) $\mathbf{v}(0) = \langle 0, 0 \rangle$ $\mathbf{v}(1) = \langle 2\pi, 0 \rangle$
$\mathbf{a}(0) = \langle 0, \pi^2 \rangle$ $\mathbf{a}(1) = \langle 0, -\pi^2 \rangle$
$\mathbf{v}(2) = \langle 0, 0 \rangle$ $\mathbf{v}(3) = \langle 2\pi, 0 \rangle$
$\mathbf{a}(2) = \langle 0, \pi^2 \rangle$ $\mathbf{a}(3) = \langle 0, -\pi^2 \rangle$

(c) Topmost point: 2π ft/sec; center of wheel: π ft/sec. Reasons: Since the wheel rolls half a circumference, or π feet every second, the center of the wheel will move π feet every second. Since the rim of the wheel is turning at a rate of π ft/sec about the center, the velocity of the topmost point relative to the center is π ft/sec, giving it a total velocity of 2π ft/sec.

75. (a) ≈ 59.195 ft/sec
(b) ≈ 74.584 ft/sec

77. We have $x = (v_0 t) \cos \alpha$ and $y + \dfrac{gt^2}{2} = (v_0 t) \sin \alpha$. Squaring and adding gives $x^2 + \left(y + \dfrac{gt^2}{2}\right)^2 = (v_0 t)^2 (\cos^2 \alpha + \sin^2 \alpha) = v_0^2 t^2$.

79. (a) $\mathbf{r}(t) = \left[(155 \ \cos \ 18° - 11.7)\left(\dfrac{1}{0.09}\right)(1 - e^{-0.09t})\right]\mathbf{i} +$

$\left[4 + \left(\dfrac{155 \ \sin \ 18°}{0.09}\right)(1 - e^{-0.09t}) + \right.$

$\left. \dfrac{32}{0.09^2}(1 - 0.09t - e^{-0.09t})\right]\mathbf{j}$

$x(t) = (155 \cos 18° - 11.7)\left(\dfrac{1}{0.09}\right)(1 - e^{-0.09t})$

$y(t) = 4 + \left(\dfrac{155 \ \sin \ 18°}{0.09}\right)(1 - e^{-0.09t}) +$

$\dfrac{32}{0.09^2}(1 - 0.09t - e^{-0.09t})$

(b) At ≈ 1.404 sec, it reaches a maximum height of ≈ 36.921 ft.
(c) Range ≈ 352.52 ft, flight time ≈ 2.959 sec
(d) At times $t \approx 0.753$ and $t \approx 2.068$ sec, when it is ≈ 98.799 and ≈ 256.138 ft from home plate
(e) No, the batter has not hit a home run. If the drag coefficient k is less than ≈ 0.011, the hit will be a home run.

81. The widths between the successive turns are constant and are given by $2\pi a$.

Chapter 9 Additional Exercises, pp. 784–785

1. (a) $\mathbf{v} = 4\mathbf{i} + 2\mathbf{j}$
(b) $\mathbf{r}(t) = 4t\mathbf{i} + \left(2t + \dfrac{16t^3}{100} - \dfrac{120t^2}{100}\right)\mathbf{j}$, where $0 \le t \le 5$
(c)

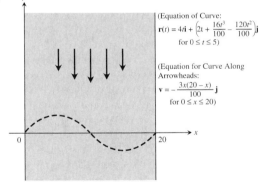

(Equation of Curve:
$\mathbf{r}(t) = 4t\mathbf{i} + \left(2t + \dfrac{16t^3}{100} - \dfrac{120t^2}{100}\right)\mathbf{j}$
for $0 \le t \le 5$)

(Equation for Curve Along Arrowheads:
$\mathbf{v} = -\dfrac{3x(20 - x)}{100}\mathbf{j}$
for $0 \le x \le 20$)

3. $\dfrac{\pi}{2}$ for all t

5. $\mathbf{v} \cdot \mathbf{j} = 12$, $\mathbf{a} \cdot \mathbf{j} = 26$

7. (a) $r = e^{2\theta}$

(b) $\dfrac{\sqrt{5}}{2}(e^{4\pi} - 1)$

9. $a^2\left(\dfrac{3\pi}{2} - 4\right)$

CHAPTER 10

Section 10.1, pp. 794–796

1. The line through the point $(2, 3, 0)$ parallel to the z-axis
3. The x-axis
5. The circle $x^2 + y^2 = 4$ in the plane $z = -2$
7. The circle $y^2 + z^2 = 1$ in the yz-plane
9. The circle $x^2 + y^2 = 16$ in the xy-plane
11. (a) The first quadrant of the xy-plane
　　(b) The fourth quadrant of the xy-plane
13. (a) The ball of radius 1 centered at the origin
　　(b) All points greater than 1 unit from the origin
15. (a) The upper hemisphere of radius 1 centered at the origin
　　(b) The solid upper hemisphere of radius 1 centered at the origin
17. (a) $x = 3$ 　　**(b)** $y = -1$ 　　**(c)** $z = -2$
19. (a) $z = 1$ 　　**(b)** $x = 3$ 　　**(c)** $y = -1$
21. (a) $x^2 + (y - 2)^2 = 4, z = 0$
　　(b) $(y - 2)^2 + z^2 = 4, x = 0$
　　(c) $x^2 + z^2 = 4, y = 2$
23. (a) $y = 3, z = -1$ 　　**(b)** $x = 1, z = -1$ 　　**(c)** $x = 1, y = 3$
25. $x^2 + y^2 + z^2 = 25, z = 3$
27. $0 \le z \le 1$ 　　　　　　**29.** $z \le 0$
31. (a) $(x - 1)^2 + (y - 1)^2 + (z - 1)^2 < 1$
　　(b) $(x - 1)^2 + (y - 1)^2 + (z - 1)^2 > 1$
33. $3\left(\dfrac{2}{3}\mathbf{i} + \dfrac{1}{3}\mathbf{j} - \dfrac{2}{3}\mathbf{k}\right)$
35. $5(\mathbf{k})$
37. $\sqrt{\dfrac{1}{2}}\left(\dfrac{1}{\sqrt{3}}\mathbf{i} - \dfrac{1}{\sqrt{3}}\mathbf{j} - \dfrac{1}{\sqrt{3}}\mathbf{k}\right)$
39. (a) $2\mathbf{i}$ 　　　　　　**(b)** $-\sqrt{3}\mathbf{k}$
　　(c) $\dfrac{3}{10}\mathbf{j} + \dfrac{2}{5}\mathbf{k}$ 　　**(d)** $6\mathbf{i} - 2\mathbf{j} + 3\mathbf{k}$
41. $\dfrac{7}{13}(12\mathbf{i} - 5\mathbf{k})$
43. (a) $5\sqrt{2}$
　　(b) $\dfrac{3}{5\sqrt{2}}\mathbf{i} + \dfrac{4}{5\sqrt{2}}\mathbf{j} - \dfrac{1}{\sqrt{2}}\mathbf{k}$
　　(c) $(1/2, 3, 5/2)$
45. (a) $\sqrt{3}$
　　(b) $-\dfrac{1}{\sqrt{3}}\mathbf{i} - \dfrac{1}{\sqrt{3}}\mathbf{j} - \dfrac{1}{\sqrt{3}}\mathbf{k}$
　　(c) $\left(\dfrac{5}{2}, \dfrac{7}{2}, \dfrac{9}{2}\right)$
47. $A(4, -3, 5)$
49. $(x - 1)^2 + (y - 2)^2 + (z - 3)^2 = 14$
51. $C(-2, 0, 2), a = 2\sqrt{2}$ 　　　**53.** $C(-2, 0, 2), a = 2\sqrt{2}$
55. $C\left(-\dfrac{1}{4}, -\dfrac{1}{4}, -\dfrac{1}{4}\right), a = \dfrac{5\sqrt{3}}{4}$
57. (a) $\sqrt{y^2 + z^2}$ 　　**(b)** $\sqrt{x^2 + z^2}$ 　　**(c)** $\sqrt{x^2 + y^2}$
59. (a) $\dfrac{3}{2}\mathbf{i} + \dfrac{3}{2}\mathbf{j} - 3\mathbf{k}$ 　**(b)** $\mathbf{i} + \mathbf{j} - 2\mathbf{k}$ 　**(c)** $(2, 2, 1)$

Section 10.2, pp. 805–807

1. (a) $-25, 5, 5$ 　　　　　**(b)** -1
　　(c) -5 　　　　　　　**(d)** $-2\mathbf{i} + 4\mathbf{j} - \sqrt{5}\mathbf{k}$
3. (a) $25, 15, 5$ 　　　　　**(b)** $\dfrac{1}{3}$
　　(c) $\dfrac{5}{3}$ 　　　　　　　**(d)** $\dfrac{1}{9}(10\mathbf{i} + 11\mathbf{j} - 2\mathbf{k})$
5. (a) $2, \sqrt{34}, \sqrt{3}$ 　　　**(b)** $\dfrac{2}{\sqrt{3}\sqrt{34}}$
　　(c) $\dfrac{2}{\sqrt{34}}$ 　　　　　**(d)** $\dfrac{1}{17}(5\mathbf{j} - 3\mathbf{k})$
7. $\left(\dfrac{3}{2}\mathbf{i} + \dfrac{3}{2}\mathbf{j}\right) + \left(-\dfrac{3}{2}\mathbf{i} + \dfrac{3}{2}\mathbf{j} + 4\mathbf{k}\right)$
9. $\left(\dfrac{14}{3}\mathbf{i} + \dfrac{28}{3}\mathbf{j} - \dfrac{14}{3}\mathbf{k}\right) + \left(\dfrac{10}{3}\mathbf{i} - \dfrac{16}{3}\mathbf{j} - \dfrac{22}{3}\mathbf{k}\right)$
11. 0.75 rad 　　　　　　**13.** 1.77 rad
17. $|\mathbf{u} \times \mathbf{v}| = 3$, direction is $\dfrac{2}{3}\mathbf{i} + \dfrac{1}{3}\mathbf{j} + \dfrac{2}{3}\mathbf{k}$; $|\mathbf{v} \times \mathbf{u}| = 3$, direction is $-\dfrac{2}{3}\mathbf{i} - \dfrac{1}{3}\mathbf{j} - \dfrac{2}{3}\mathbf{k}$
19. $|\mathbf{u} \times \mathbf{v}| = 0$, no direction; $|\mathbf{v} \times \mathbf{u}| = 0$, no direction
21. $|\mathbf{u} \times \mathbf{v}| = 6$, direction is $-\mathbf{k}$; $|\mathbf{v} \times \mathbf{u}| = 6$, direction is $\mathbf{k}$
23. $|\mathbf{u} \times \mathbf{v}| = 6\sqrt{5}$, direction is $\dfrac{1}{\sqrt{5}}\mathbf{i} - \dfrac{2}{\sqrt{5}}\mathbf{k}$; $|\mathbf{v} \times \mathbf{u}| = 6\sqrt{5}$, direction is $-\dfrac{1}{\sqrt{5}}\mathbf{i} + \dfrac{2}{\sqrt{5}}\mathbf{k}$
25. $\mathbf{u} \times \mathbf{v} = \mathbf{i} + \mathbf{k}$ 　　　　**27.** $\mathbf{u} \times \mathbf{v} = -2\mathbf{k}$

29. (a) $2\sqrt{6}$ 　　　　　**(b)** $\pm\dfrac{1}{\sqrt{6}}(2\mathbf{i} + \mathbf{j} + \mathbf{k})$
31. (a) $\dfrac{\sqrt{2}}{2}$ 　　　　　**(b)** $\pm\dfrac{1}{\sqrt{2}}(\mathbf{i} - \mathbf{j})$
33. 8 　　　　　　　**35.** 7
37. (a) None 　　　　**(b)** $\mathbf{u}$ and $\mathbf{w}$
39. $10\sqrt{3}$ ft · lb
41. (a) True 　　　　　**(b)** Not always true
　　(c) True 　　　　　**(d)** True
　　(e) Not always true 　**(f)** True
　　(g) True 　　　　　**(h)** True
43. (a) $\text{proj}_{\mathbf{v}}\,\mathbf{u} = \dfrac{\mathbf{u} \cdot \mathbf{v}}{\mathbf{v} \cdot \mathbf{v}}\mathbf{v}$ 　**(b)** $\pm\mathbf{u} \times \mathbf{v}$
　　(c) $\pm(\mathbf{u} \times \mathbf{v}) \times \mathbf{w}$ 　**(d)** $|(\mathbf{u} \times \mathbf{v}) \cdot \mathbf{w}|$
45. (a) Yes 　　**(b)** No 　　**(c)** Yes 　　**(d)** No
47. No, $\mathbf{v}$ need not equal $\mathbf{w}$. For example, $\mathbf{i} + \mathbf{j} \ne -\mathbf{i} + \mathbf{j}$, but $\mathbf{i} \times (\mathbf{i} + \mathbf{j}) = \mathbf{i} \times \mathbf{i} + \mathbf{i} \times \mathbf{j} = \mathbf{0} + \mathbf{k} = \mathbf{k}$ and $\mathbf{i} \times (-\mathbf{i} + \mathbf{j}) = -\mathbf{i} \times \mathbf{i} + \mathbf{i} \times \mathbf{j} = \mathbf{0} + \mathbf{k} = \mathbf{k}$.

Index

A Brief Table of Integrals

1. $\displaystyle\int u\,dv = uv - \int v\,du$

2. $\displaystyle\int a^u\,du = \frac{a^u}{\ln a} + C, \qquad a \neq 1, \qquad a > 0$

3. $\displaystyle\int \cos u\,du = \sin u + C$

4. $\displaystyle\int \sin u\,du = -\cos u + C$

5. $\displaystyle\int (ax+b)^n\,dx = \frac{(ax+b)^{n+1}}{a(n+1)} + C, \qquad n \neq -1$

6. $\displaystyle\int (ax+b)^{-1}\,dx = \frac{1}{a}\ln|ax+b| + C$

7. $\displaystyle\int x(ax+b)^n\,dx = \frac{(ax+b)^{n+1}}{a^2}\left[\frac{ax+b}{n+2} - \frac{b}{n+1}\right] + C, \qquad n \neq -1, -2$

8. $\displaystyle\int x(ax+b)^{-1}\,dx = \frac{x}{a} - \frac{b}{a^2}\ln|ax+b| + C$

9. $\displaystyle\int x(ax+b)^{-2}\,dx = \frac{1}{a^2}\left[\ln|ax+b| + \frac{b}{ax+b}\right] + C$

10. $\displaystyle\int \frac{dx}{x(ax+b)} = \frac{1}{b}\ln\left|\frac{x}{ax+b}\right| + C$

11. $\displaystyle\int (\sqrt{ax+b})^n\,dx = \frac{2}{a}\frac{(\sqrt{ax+b})^{n+2}}{n+2} + C, \qquad n \neq -2$

12. $\displaystyle\int \frac{\sqrt{ax+b}}{x}\,dx = 2\sqrt{ax+b} + b\int \frac{dx}{x\sqrt{ax+b}}$

13. (a) $\displaystyle\int \frac{dx}{x\sqrt{ax-b}} = \frac{2}{\sqrt{b}}\tan^{-1}\sqrt{\frac{ax-b}{b}} + C$

(b) $\displaystyle\int \frac{dx}{x\sqrt{ax+b}} = \frac{1}{\sqrt{b}}\ln\left|\frac{\sqrt{ax+b} - \sqrt{b}}{\sqrt{ax+b} + \sqrt{b}}\right| + C$

14. $\displaystyle\int \frac{\sqrt{ax+b}}{x^2}\,dx = -\frac{\sqrt{ax+b}}{x} + \frac{a}{2}\int \frac{dx}{x\sqrt{ax+b}} + C$

15. $\displaystyle\int \frac{dx}{x^2\sqrt{ax+b}} = -\frac{\sqrt{ax+b}}{bx} - \frac{a}{2b}\int \frac{dx}{x\sqrt{ax+b}} + C$

16. $\displaystyle\int \frac{dx}{a^2+x^2} = \frac{1}{a}\tan^{-1}\frac{x}{a} + C$

17. $\displaystyle\int \frac{dx}{(a^2+x^2)^2} = \frac{x}{2a^2(a^2+x^2)} + \frac{1}{2a^3}\tan^{-1}\frac{x}{a} + C$

18. $\displaystyle\int \frac{dx}{a^2-x^2} = \frac{1}{2a}\ln\left|\frac{x+a}{x-a}\right| + C$

19. $\displaystyle\int \frac{dx}{(a^2-x^2)^2} = \frac{x}{2a^2(a^2-x^2)} + \frac{1}{4a^3}\ln\left|\frac{x+a}{x-a}\right| + C$

20. $\displaystyle\int \frac{dx}{\sqrt{a^2+x^2}} = \sinh^{-1}\frac{x}{a} + C = \ln(x + \sqrt{a^2+x^2}) + C$

21. $\displaystyle\int \sqrt{a^2+x^2}\,dx = \frac{x}{2}\sqrt{a^2+x^2} + \frac{a^2}{2}\ln(x + \sqrt{a^2+x^2}) + C$

22. $\displaystyle\int x^2\sqrt{a^2+x^2}\,dx = \frac{x}{8}(a^2+2x^2)\sqrt{a^2+x^2} - \frac{a^4}{8}\ln(x + \sqrt{a^2+x^2}) + C$

23. $\displaystyle\int \frac{\sqrt{a^2+x^2}}{x}\,dx = \sqrt{a^2+x^2} - a\ln\left|\frac{a + \sqrt{a^2+x^2}}{x}\right| + C$

24. $\displaystyle\int \frac{\sqrt{a^2+x^2}}{x^2}\,dx = \ln(x + \sqrt{a^2+x^2}) - \frac{\sqrt{a^2+x^2}}{x} + C$

25. $\displaystyle\int \frac{x^2}{\sqrt{a^2+x^2}}\,dx = -\frac{a^2}{2}\ln(x + \sqrt{a^2+x^2}) + \frac{x\sqrt{a^2+x^2}}{2} + C$

26. $\displaystyle\int \frac{dx}{x\sqrt{a^2+x^2}} = -\frac{1}{a}\ln\left|\frac{a + \sqrt{a^2+x^2}}{x}\right| + C$

27. $\displaystyle\int \frac{dx}{x^2\sqrt{a^2+x^2}} = -\frac{\sqrt{a^2+x^2}}{a^2 x} + C$

28. $\displaystyle \int \frac{dx}{\sqrt{a^2 - x^2}} = \sin^{-1}\frac{x}{a} + C$

29. $\displaystyle \int \sqrt{a^2 - x^2}\, dx = \frac{x}{2}\sqrt{a^2 - x^2} + \frac{a^2}{2}\sin^{-1}\frac{x}{a} + C$

30. $\displaystyle \int x^2\sqrt{a^2 - x^2}\, dx = \frac{a^4}{8}\sin^{-1}\frac{x}{a} - \frac{1}{8}x\sqrt{a^2 - x^2}(a^2 - 2x^2) + C$

31. $\displaystyle \int \frac{\sqrt{a^2 - x^2}}{x}\, dx = \sqrt{a^2 - x^2} - a\ln\left|\frac{a + \sqrt{a^2 - x^2}}{x}\right| + C$

32. $\displaystyle \int \frac{\sqrt{a^2 - x^2}}{x^2}\, dx = -\sin^{-1}\frac{x}{a} - \frac{\sqrt{a^2 - x^2}}{x} + C$

33. $\displaystyle \int \frac{x^2}{\sqrt{a^2 - x^2}}\, dx = \frac{a^2}{2}\sin^{-1}\frac{x}{a} - \frac{1}{2}x\sqrt{a^2 - x^2} + C$

34. $\displaystyle \int \frac{dx}{x\sqrt{a^2 - x^2}} = -\frac{1}{a}\ln\left|\frac{a + \sqrt{a^2 - x^2}}{x}\right| + C$

35. $\displaystyle \int \frac{dx}{x^2\sqrt{a^2 - x^2}} = -\frac{\sqrt{a^2 - x^2}}{a^2 x} + C$

36. $\displaystyle \int \frac{dx}{\sqrt{x^2 - a^2}} = \cosh^{-1}\frac{x}{a} + C = \ln\left|x + \sqrt{x^2 - a^2}\right| + C$

37. $\displaystyle \int \sqrt{x^2 - a^2}\, dx = \frac{x}{2}\sqrt{x^2 - a^2} - \frac{a^2}{2}\ln\left|x + \sqrt{x^2 - a^2}\right| + C$

38. $\displaystyle \int (\sqrt{x^2 - a^2})^n\, dx = \frac{x(\sqrt{x^2 - a^2})^n}{n + 1} - \frac{na^2}{n + 1}\int (\sqrt{x^2 - a^2})^{n-2}\, dx, \qquad n \neq -1$

39. $\displaystyle \int \frac{dx}{(\sqrt{x^2 - a^2})^n} = \frac{x(\sqrt{x^2 - a^2})^{2-n}}{(2 - n)a^2} - \frac{n - 3}{(n - 2)a^2}\int \frac{dx}{(\sqrt{x^2 - a^2})^{n-2}}, \qquad n \neq 2$

40. $\displaystyle \int x(\sqrt{x^2 - a^2})^n\, dx = \frac{(\sqrt{x^2 - a^2})^{n+2}}{n + 2} + C, \qquad n \neq -2$

41. $\displaystyle \int x^2\sqrt{x^2 - a^2}\, dx = \frac{x}{8}(2x^2 - a^2)\sqrt{x^2 - a^2} - \frac{a^4}{8}\ln\left|x + \sqrt{x^2 - a^2}\right| + C$

42. $\displaystyle \int \frac{\sqrt{x^2 - a^2}}{x}\, dx = \sqrt{x^2 - a^2} - a\sec^{-1}\left|\frac{x}{a}\right| + C$

43. $\displaystyle \int \frac{\sqrt{x^2 - a^2}}{x^2}\, dx = \ln\left|x + \sqrt{x^2 - a^2}\right| - \frac{\sqrt{x^2 - a^2}}{x} + C$

44. $\displaystyle \int \frac{x^2}{\sqrt{x^2 - a^2}}\, dx = \frac{a^2}{2}\ln\left|x + \sqrt{x^2 - a^2}\right| + \frac{x}{2}\sqrt{x^2 - a^2} + C$

45. $\displaystyle \int \frac{dx}{x\sqrt{x^2 - a^2}} = \frac{1}{a}\sec^{-1}\left|\frac{x}{a}\right| + C = \frac{1}{a}\cos^{-1}\left|\frac{a}{x}\right| + C$

46. $\displaystyle \int \frac{dx}{x^2\sqrt{x^2 - a^2}} = \frac{\sqrt{x^2 - a^2}}{a^2 x} + C$

47. $\displaystyle \int \frac{dx}{\sqrt{2ax - x^2}} = \sin^{-1}\left(\frac{x - a}{a}\right) + C$

48. $\displaystyle \int \sqrt{2ax - x^2}\, dx = \frac{x - a}{2}\sqrt{2ax - x^2} + \frac{a^2}{2}\sin^{-1}\left(\frac{x - a}{a}\right) + C$

49. $\displaystyle \int (\sqrt{2ax - x^2})^n\, dx = \frac{(x - a)(\sqrt{2ax - x^2})^n}{n + 1} + \frac{na^2}{n + 1}\int (\sqrt{2ax - x^2})^{n-2}\, dx$

50. $\displaystyle \int \frac{dx}{(\sqrt{2ax - x^2})^n} = \frac{(x - a)(\sqrt{2ax - x^2})^{2-n}}{(n - 2)a^2} + \frac{n - 3}{(n - 2)a^2}\int \frac{dx}{(\sqrt{2ax - x^2})^{n-2}}$

51. $\displaystyle \int x\sqrt{2ax - x^2}\, dx = \frac{(x + a)(2x - 3a)\sqrt{2ax - x^2}}{6} + \frac{a^3}{2}\sin^{-1}\left(\frac{x - a}{a}\right) + C$

52. $\displaystyle \int \frac{\sqrt{2ax - x^2}}{x}\, dx = \sqrt{2ax - x^2} + a\sin^{-1}\left(\frac{x - a}{a}\right) + C$

53. $\displaystyle \int \frac{\sqrt{2ax - x^2}}{x^2}\, dx = -2\sqrt{\frac{2a - x}{x}} - \sin^{-1}\left(\frac{x - a}{a}\right) + C$

54. $\displaystyle \int \frac{x\, dx}{\sqrt{2ax - x^2}} = a\sin^{-1}\left(\frac{x - a}{a}\right) - \sqrt{2ax - x^2} + C$

55. $\displaystyle \int \frac{dx}{x\sqrt{2ax - x^2}} = -\frac{1}{a}\sqrt{\frac{2a - x}{x}} + C$

56. $\displaystyle \int \sin ax\, dx = -\frac{1}{a}\cos ax + C$

57. $\displaystyle \int \cos ax\, dx = \frac{1}{a}\sin ax + C$

58. $\displaystyle \int \sin^2 ax\, dx = \frac{x}{2} - \frac{\sin 2ax}{4a} + C$

59. $\displaystyle \int \cos^2 ax\, dx = \frac{x}{2} + \frac{\sin 2ax}{4a} + C$

60. $\displaystyle\int \sin^n ax \, dx = -\frac{\sin^{n-1} ax \cos ax}{na} + \frac{n-1}{n}\int \sin^{n-2} ax \, dx$

61. $\displaystyle\int \cos^n ax \, dx = \frac{\cos^{n-1} ax \sin ax}{na} + \frac{n-1}{n}\int \cos^{n-2} ax \, dx$

62. (a) $\displaystyle\int \sin ax \cos bx \, dx = -\frac{\cos(a+b)x}{2(a+b)} - \frac{\cos(a-b)x}{2(a-b)} + C, \qquad a^2 \neq b^2$

     (b) $\displaystyle\int \sin ax \sin bx \, dx = \frac{\sin(a-b)x}{2(a-b)} - \frac{\sin(a+b)x}{2(a+b)} + C, \qquad a^2 \neq b^2$

     (c) $\displaystyle\int \cos ax \cos bx \, dx = \frac{\sin(a-b)x}{2(a-b)} + \frac{\sin(a+b)x}{2(a+b)} + C, \qquad a^2 \neq b^2$

63. $\displaystyle\int \sin ax \cos ax \, dx = -\frac{\cos 2ax}{4a} + C$

64. $\displaystyle\int \sin^n ax \cos ax \, dx = \frac{\sin^{n+1} ax}{(n+1)a} + C, \qquad n \neq -1$

65. $\displaystyle\int \frac{\cos ax}{\sin ax} \, dx = \frac{1}{a} \ln |\sin ax| + C$

66. $\displaystyle\int \cos^n ax \sin ax \, dx = -\frac{\cos^{n+1} ax}{(n+1)a} + C, \qquad n \neq -1$

67. $\displaystyle\int \frac{\sin ax}{\cos ax} \, dx = -\frac{1}{a} \ln |\cos ax| + C$

68. $\displaystyle\int \sin^n ax \cos^m ax \, dx = -\frac{\sin^{n-1} ax \cos^{m+1} ax}{a(m+n)} + \frac{n-1}{m+n}\int \sin^{n-2} ax \cos^m ax \, dx, \qquad n \neq -m \qquad \text{(reduces } \sin^n ax\text{)}$

69. $\displaystyle\int \sin^n ax \cos^m ax \, dx = \frac{\sin^{n+1} ax \cos^{m-1} ax}{a(m+n)} + \frac{m-1}{m+n}\int \sin^n ax \cos^{m-2} ax \, dx, \qquad m \neq -n \qquad \text{(reduces } \cos^m ax\text{)}$

70. $\displaystyle\int \frac{dx}{b + c \sin ax} = \frac{-2}{a\sqrt{b^2-c^2}} \tan^{-1}\left[\sqrt{\frac{b-c}{b+c}} \tan\left(\frac{\pi}{4} - \frac{ax}{2}\right)\right] + C, \qquad b^2 > c^2$

71. $\displaystyle\int \frac{dx}{b + c \sin ax} = \frac{-1}{a\sqrt{c^2-b^2}} \ln\left|\frac{c + b \sin ax + \sqrt{c^2-b^2}\cos ax}{b + c \sin ax}\right| + C, \qquad b^2 < c^2$

72. $\displaystyle\int \frac{dx}{1 + \sin ax} = -\frac{1}{a} \tan\left(\frac{\pi}{4} - \frac{ax}{2}\right) + C$

73. $\displaystyle\int \frac{dx}{1 - \sin ax} = \frac{1}{a} \tan\left(\frac{\pi}{4} + \frac{ax}{2}\right) + C$

74. $\displaystyle\int \frac{dx}{b + c \cos ax} = \frac{2}{a\sqrt{b^2-c^2}} \tan^{-1}\left[\sqrt{\frac{b-c}{b+c}} \tan\frac{ax}{2}\right] + C, \qquad b^2 > c^2$

75. $\displaystyle\int \frac{dx}{b + c \cos ax} = \frac{1}{a\sqrt{c^2-b^2}} \ln\left|\frac{c + b \cos ax + \sqrt{c^2-b^2}\sin ax}{b + c \cos ax}\right| + C, \qquad b^2 < c^2$

76. $\displaystyle\int \frac{dx}{1 + \cos ax} = \frac{1}{a} \tan\frac{ax}{2} + C$

77. $\displaystyle\int \frac{dx}{1 - \cos ax} = -\frac{1}{a} \cot\frac{ax}{2} + C$

78. $\displaystyle\int x \sin ax \, dx = \frac{1}{a^2} \sin ax - \frac{x}{a} \cos ax + C$

79. $\displaystyle\int x \cos ax \, dx = \frac{1}{a^2} \cos ax + \frac{x}{a} \sin ax + C$

80. $\displaystyle\int x^n \sin ax \, dx = -\frac{x^n}{a} \cos ax + \frac{n}{a}\int x^{n-1} \cos ax \, dx$

81. $\displaystyle\int x^n \cos ax \, dx = \frac{x^n}{a} \sin ax - \frac{n}{a}\int x^{n-1} \sin ax \, dx$

82. $\displaystyle\int \tan ax \, dx = \frac{1}{a} \ln |\sec ax| + C$

83. $\displaystyle\int \cot ax \, dx = \frac{1}{a} \ln |\sin ax| + C$

84. $\displaystyle\int \tan^2 ax \, dx = \frac{1}{a} \tan ax - x + C$

85. $\displaystyle\int \cot^2 ax \, dx = -\frac{1}{a} \cot ax - x + C$

86. $\displaystyle\int \tan^n ax \, dx = \frac{\tan^{n-1} ax}{a(n-1)} - \int \tan^{n-2} ax \, dx, \qquad n \neq 1$

87. $\displaystyle\int \cot^n ax \, dx = -\frac{\cot^{n-1} ax}{a(n-1)} - \int \cot^{n-2} ax \, dx, \qquad n \neq 1$

88. $\displaystyle\int \sec ax \, dx = \frac{1}{a} \ln |\sec ax + \tan ax| + C$

89. $\displaystyle\int \csc ax \, dx = -\frac{1}{a} \ln |\csc ax + \cot ax| + C$

90. $\displaystyle\int \sec^2 ax \, dx = \frac{1}{a} \tan ax + C$

91. $\displaystyle\int \csc^2 ax \, dx = -\frac{1}{a} \cot ax + C$

92. $\displaystyle\int \sec^n ax \, dx = \frac{\sec^{n-2} ax \tan ax}{a(n-1)} + \frac{n-2}{n-1}\int \sec^{n-2} ax \, dx, \qquad n \neq 1$

93. $\displaystyle \int \csc^n ax \, dx = -\frac{\csc^{n-2} ax \cot ax}{a(n-1)} + \frac{n-2}{n-1} \int \csc^{n-2} ax \, dx, \qquad n \neq 1$

94. $\displaystyle \int \sec^n ax \tan ax \, dx = \frac{\sec^n ax}{na} + C, \qquad n \neq 0$

95. $\displaystyle \int \csc^n ax \cot ax \, dx = -\frac{\csc^n ax}{na} + C, \qquad n \neq 0$

96. $\displaystyle \int \sin^{-1} ax \, dx = x \sin^{-1} ax + \frac{1}{a}\sqrt{1 - a^2 x^2} + C$

97. $\displaystyle \int \cos^{-1} ax \, dx = x \cos^{-1} ax - \frac{1}{a}\sqrt{1 - a^2 x^2} + C$

98. $\displaystyle \int \tan^{-1} ax \, dx = x \tan^{-1} ax - \frac{1}{2a} \ln (1 + a^2 x^2) + C$

99. $\displaystyle \int x^n \sin^{-1} ax \, dx = \frac{x^{n+1}}{n+1} \sin^{-1} ax - \frac{a}{n+1} \int \frac{x^{n+1} \, dx}{\sqrt{1 - a^2 x^2}}, \qquad n \neq -1$

100. $\displaystyle \int x^n \cos^{-1} ax \, dx = \frac{x^{n+1}}{n+1} \cos^{-1} ax + \frac{a}{n+1} \int \frac{x^{n+1} \, dx}{\sqrt{1 - a^2 x^2}}, \qquad n \neq -1$

101. $\displaystyle \int x^n \tan^{-1} ax \, dx = \frac{x^{n+1}}{n+1} \tan^{-1} ax - \frac{a}{n+1} \int \frac{x^{n+1} \, dx}{\sqrt{1 + a^2 x^2}}, \qquad n \neq -1$

102. $\displaystyle \int e^{ax} \, dx = \frac{1}{a} e^{ax} + C$

103. $\displaystyle \int b^{ax} \, dx = \frac{1}{a} \frac{b^{ax}}{\ln b} + C, \qquad b > 0, \qquad b \neq 1$

104. $\displaystyle \int xe^{ax} \, dx = \frac{e^{ax}}{a^2} (ax - 1) + C$

105. $\displaystyle \int x^n e^{ax} \, dx = \frac{1}{a} x^n e^{ax} - \frac{n}{a} \int x^{n-1} e^{ax} \, dx$

106. $\displaystyle \int x^n b^{ax} \, dx = \frac{x^n b^{ax}}{a \ln b} - \frac{n}{a \ln b} \int x^{n-1} b^{ax} \, dx, \qquad b > 0, \qquad b \neq 1$

107. $\displaystyle \int e^{ax} \sin bx \, dx = \frac{e^{ax}}{a^2 + b^2} (a \sin bx - b \cos bx) + C$

108. $\displaystyle \int e^{ax} \cos bx \, dx = \frac{e^{ax}}{a^2 + b^2} (a \cos bx + b \sin bx) + C$

109. $\displaystyle \int \ln ax \, dx = x \ln ax - x + C$

110. $\displaystyle \int x^n (\ln ax)^m \, dx = \frac{x^{n+1} (\ln ax)^m}{n+1} - \frac{m}{n+1} \int x^n (\ln ax)^{m-1} \, dx, \qquad n \neq -1$

111. $\displaystyle \int x^{-1} (\ln ax)^m \, dx = \frac{(\ln ax)^{m+1}}{m+1} + C, \qquad m \neq -1$

112. $\displaystyle \int \frac{dx}{x \ln ax} = \ln |\ln ax| + C$

113. $\displaystyle \int \sinh ax \, dx = \frac{1}{a} \cosh ax + C$

114. $\displaystyle \int \cosh ax \, dx = \frac{1}{a} \sinh ax + C$

115. $\displaystyle \int \sinh^2 ax \, dx = \frac{\sinh 2ax}{4a} - \frac{x}{2} + C$

116. $\displaystyle \int \cosh^2 ax \, dx = \frac{\sinh 2ax}{4a} + \frac{x}{2} + C$

117. $\displaystyle \int \sinh^n ax \, dx = \frac{\sinh^{n-1} ax \cosh ax}{na} - \frac{n-1}{n} \int \sinh^{n-2} ax \, dx, \qquad n \neq 0$

118. $\displaystyle \int \cosh^n ax \, dx = \frac{\cosh^{n-1} ax \sinh ax}{na} + \frac{n-1}{n} \int \cosh^{n-2} ax \, dx, \qquad n \neq 0$

119. $\displaystyle \int x \sinh ax \, dx = \frac{x}{a} \cosh ax - \frac{1}{a^2} \sinh ax + C$

120. $\displaystyle \int x \cosh ax \, dx = \frac{x}{a} \sinh ax - \frac{1}{a^2} \cosh ax + C$

121. $\displaystyle \int x^n \sinh ax \, dx = \frac{x^n}{a} \cosh ax - \frac{n}{a} \int x^{n-1} \cosh ax \, dx$

122. $\displaystyle \int x^n \cosh ax \, dx = \frac{x^n}{a} \sinh ax - \frac{n}{a} \int x^{n-1} \sinh ax \, dx$

123. $\displaystyle \int \tanh ax \, dx = \frac{1}{a} \ln (\cosh ax) + C$

124. $\displaystyle \int \coth ax \, dx = \frac{1}{a} \ln |\sinh ax| + C$

125. $\displaystyle \int \tanh^2 ax \, dx = x - \frac{1}{a} \tanh ax + C$

126. $\displaystyle \int \coth^2 ax \, dx = x - \frac{1}{a} \coth ax + C$

127. $\displaystyle \int \tanh^n ax \, dx = -\frac{\tanh^{n-1} ax}{(n-1)a} + \int \tanh^{n-2} ax \, dx, \qquad n \neq 1$

128. $\displaystyle \int \coth^n ax \, dx = -\frac{\coth^{n-1} ax}{(n-1)a} + \int \coth^{n-2} ax \, dx, \qquad n \neq 1$

129. $\displaystyle \int \operatorname{sech} ax \, dx = \frac{1}{a} \sin^{-1} (\tanh ax) + C$

130. $\displaystyle \int \operatorname{csch} ax \, dx = \frac{1}{a} \ln \left| \tanh \frac{ax}{2} \right| + C$

131. $\displaystyle\int \text{sech}^2 \, ax \, dx = \frac{1}{a} \tanh ax + C$

132. $\displaystyle\int \text{csch}^2 \, ax \, dx = -\frac{1}{a} \coth ax + C$

133. $\displaystyle\int \text{sech}^n \, ax \, dx = \frac{\text{sech}^{n-2} \, ax \tanh ax}{(n-1)a} + \frac{n-2}{n-1} \int \text{sech}^{n-2} \, ax \, dx, \qquad n \neq 1$

134. $\displaystyle\int \text{csch}^n \, ax \, dx = -\frac{\text{csch}^{n-2} \, ax \coth ax}{(n-1)a} - \frac{n-2}{n-1} \int \text{csch}^{n-2} \, ax \, dx, \qquad n \neq 1$

135. $\displaystyle\int \text{sech}^n \, ax \tanh ax \, dx = -\frac{\text{sech}^n \, ax}{na} + C, \qquad n \neq 0$

136. $\displaystyle\int \text{csch}^n \, ax \coth ax \, dx = -\frac{\text{csch}^n \, ax}{na} + C, \qquad n \neq 0$

137. $\displaystyle\int e^{ax} \sinh bx \, dx = \frac{e^{ax}}{2}\left[\frac{e^{bx}}{a+b} - \frac{e^{-bx}}{a-b}\right] + C, \qquad a^2 \neq b^2$

138. $\displaystyle\int e^{ax} \cosh bx \, dx = \frac{e^{ax}}{2}\left[\frac{e^{bx}}{a+b} + \frac{e^{-bx}}{a-b}\right] + C, \qquad a^2 \neq b^2$

139. $\displaystyle\int_0^\infty x^{n-1} e^{-x} \, dx = \Gamma(n) = (n-1)!, \qquad n > 0$

140. $\displaystyle\int_0^\infty e^{-ax^2} \, dx = \frac{1}{2}\sqrt{\frac{\pi}{a}}, \qquad a > 0$

141. $\displaystyle\int_0^{\pi/2} \sin^n x \, dx = \int_0^{\pi/2} \cos^n x \, dx = \begin{cases} \dfrac{1 \cdot 3 \cdot 5 \cdots (n-1)}{2 \cdot 4 \cdot 6 \cdots n} \cdot \dfrac{\pi}{2}, & \text{if } n \text{ is an even integer} \geq 2 \\[2mm] \dfrac{2 \cdot 4 \cdot 6 \cdots (n-1)}{3 \cdot 5 \cdot 7 \cdots n}, & \text{if } n \text{ is an odd integer} \geq 3 \end{cases}$

Conic Sections

A **circle** is the set of points in a plane whose distance from a fixed point in the plane is constant. The fixed point is the **center** of the circle; the constant distance is the **radius**. An **ellipse** is the set of points in a plane whose distances from two fixed points in the plane have a constant sum. A **hyperbola** is the set of points in a plane whose distances from two fixed points in the plane have a constant difference. In each case, the fixed points are the **foci** of the conic section. A **parabola** is the set of points in a plane equidistant from a given fixed point and a given fixed line in the plane. The fixed point is the **focus** of the parabola; the line is the **directrix.**

Ellipses and Circle in Standard Position

For both ellipses:
a = semimajor axis
b = semiminor axis
$c = \sqrt{a^2 - b^2}$ = center-to-focus distance
Eccentricity: $e = c/a, 0 < e < 1$

Degenerate case:
    circle of radius a

Parabolas in Standard Position

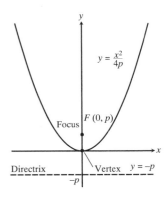

Hyperbolas in Standard Position

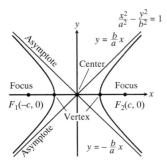

c = center-to-focus distance = $\sqrt{a^2 + b^2}$
Eccentricity: $e = c/a > 1$
Asymptotes: $y = \pm(b/a)x$

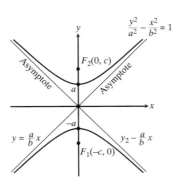

c = center-to-focus distance = $\sqrt{a^2 + b^2}$
Eccentricity: $e = c/a > 1$
Asymptotes: $y = \pm(a/b)x$

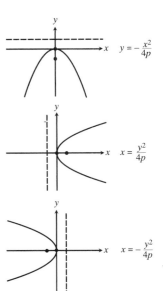

All parabolas have eccentricity $e = 1$

Vector Operator Formulas in Cartesian, Cylindrical, and Spherical Coordinates; Vector Identities

Formulas for Grad, Div, Curl, and the Laplacian

	Cartesian (x, y, z) **i, j,** and **k** are unit vectors in the directions of increasing x, y, and z. F_x, F_y, and F_z are the scalar components of $\mathbf{F}(x, y, z)$ in these directions.	**Cylindrical** (r, θ, z) $\mathbf{u}_r$, $\mathbf{u}_\theta$, and **k** are unit vectors in the directions of increasing r, θ, and z. F_r, F_θ, and F_z are the scalar components of $\mathbf{F}(r, \theta, z)$ in these directions.	**Spherical** (ρ, ϕ, θ) $\mathbf{u}_\rho$, $\mathbf{u}_\phi$, and $\mathbf{u}_\theta$ are unit vectors in the directions of increasing ρ, ϕ, and θ. F_ρ, F_ϕ, and F_θ are the scalar components of $\mathbf{F}(\rho, \phi, \theta)$ in these directions.
Gradient	$\nabla f = \dfrac{\partial f}{\partial x}\mathbf{i} + \dfrac{\partial f}{\partial y}\mathbf{j} + \dfrac{\partial f}{\partial z}\mathbf{k}$	$\nabla f = \dfrac{\partial f}{\partial r}\mathbf{u}_r + \dfrac{1}{r}\dfrac{\partial f}{\partial \theta}\mathbf{u}_\theta + \dfrac{\partial f}{\partial z}\mathbf{k}$	$\nabla f = \dfrac{\partial f}{\partial \rho}\mathbf{u}_\rho + \dfrac{1}{\rho}\dfrac{\partial f}{\partial \phi}\mathbf{u}_\phi + \dfrac{1}{\rho \sin \phi}\dfrac{\partial f}{\partial \theta}\mathbf{u}_\theta$
Divergence	$\nabla \cdot \mathbf{F} = \dfrac{\partial F_x}{\partial x} + \dfrac{\partial F_y}{\partial y} + \dfrac{\partial F_z}{\partial z}$	$\nabla \cdot \mathbf{F} = \dfrac{1}{r}\dfrac{\partial}{\partial r}(rF_r) + \dfrac{1}{r}\dfrac{\partial F_\theta}{\partial \theta} + \dfrac{\partial F_z}{\partial z}$	$\nabla \cdot \mathbf{F} = \dfrac{1}{\rho^2}\dfrac{\partial}{\partial \rho}(\rho^2 F_\rho)$ $+ \dfrac{1}{\rho \sin \phi}\dfrac{\partial}{\partial \phi}(F_\phi \sin \phi) + \dfrac{1}{\rho \sin \phi}\dfrac{\partial F_\theta}{\partial \theta}$
Curl	$\nabla \times \mathbf{F} = \begin{vmatrix} \mathbf{i} & \mathbf{j} & \mathbf{k} \\ \dfrac{\partial}{\partial x} & \dfrac{\partial}{\partial y} & \dfrac{\partial}{\partial z} \\ F_x & F_y & F_z \end{vmatrix}$	$\nabla \times \mathbf{F} = \begin{vmatrix} \dfrac{1}{r}\mathbf{u}_r & \mathbf{u}_\theta & \dfrac{1}{r}\mathbf{k} \\ \dfrac{\partial}{\partial r} & \dfrac{\partial}{\partial \theta} & \dfrac{\partial}{\partial z} \\ F_r & F_\theta & F_z \end{vmatrix}$	$\nabla \times \mathbf{F} = \begin{vmatrix} \dfrac{\mathbf{u}_\rho}{\rho^2 \sin \phi} & \dfrac{\mathbf{u}_\phi}{\rho \sin \phi} & \dfrac{\mathbf{u}_\theta}{\rho} \\ \dfrac{\partial}{\partial \rho} & \dfrac{\partial}{\partial \phi} & \dfrac{\partial}{\partial \theta} \\ F_\rho & \rho F_\phi & \rho \sin \phi\, F_\theta \end{vmatrix}$
Laplacian	$\nabla^2 f = \dfrac{\partial^2 f}{\partial x^2} + \dfrac{\partial^2 f}{\partial y^2} + \dfrac{\partial^2 f}{\partial z^2}$	$\nabla^2 f = \dfrac{1}{r}\dfrac{\partial}{\partial r}\left(r\dfrac{\partial f}{\partial r}\right) + \dfrac{1}{r^2}\dfrac{\partial^2 f}{\partial \theta^2} + \dfrac{\partial^2 f}{\partial z^2}$	$\nabla^2 f = \dfrac{1}{\rho^2}\dfrac{\partial}{\partial \rho}\left(\rho^2 \dfrac{\partial f}{\partial \rho}\right)$ $+ \dfrac{1}{\rho^2 \sin \phi}\dfrac{\partial}{\partial \phi}\left(\sin \phi \dfrac{\partial f}{\partial \rho}\right) + \dfrac{1}{\rho^2 \sin^2 \phi}\dfrac{\partial^2 f}{\partial \theta^2}$

Vector Triple Products

$(\mathbf{u} \times \mathbf{v}) \cdot \mathbf{w} = (\mathbf{v} \times \mathbf{w}) \cdot \mathbf{u} = (\mathbf{w} \times \mathbf{u}) \cdot \mathbf{v}$

$\mathbf{u} \times (\mathbf{v} \times \mathbf{w}) = (\mathbf{u} \cdot \mathbf{w})\mathbf{v} - (\mathbf{u} \cdot \mathbf{v})\mathbf{w}$

Vector Identities for the Cartesian Form of the Operator ∇

In the identities listed here, $f(x, y, z)$ and $g(x, y, z)$ are differentiable scalar functions and $\mathbf{u}(x, y, z)$ and $\mathbf{v}(x, y, z)$ are differentiable vector functions.

$\nabla \cdot f\mathbf{v} = f\nabla \cdot \mathbf{v} + \mathbf{v} \cdot \nabla f = f\nabla \cdot \mathbf{v} + (\mathbf{v} \cdot \nabla) f$

$\nabla \times f\mathbf{v} - f\nabla \times \mathbf{v} + \nabla f \times \mathbf{v}$

$\nabla \cdot (\nabla \times \mathbf{v}) = 0$

$\nabla \times (\nabla f) = \mathbf{0}$

$\nabla(fg) = f\nabla g + g\nabla f$

$\nabla(\mathbf{u} \cdot \mathbf{v}) = (\mathbf{u} \cdot \nabla)\mathbf{v} + (\mathbf{v} \cdot \nabla)\mathbf{u} + \mathbf{u} \times (\nabla \times \mathbf{v}) + \mathbf{v} \times (\nabla \times \mathbf{u})$

$\nabla \cdot (\mathbf{u} \times \mathbf{v}) = \mathbf{v} \cdot (\nabla \times \mathbf{u}) - \mathbf{u} \cdot (\nabla \times \mathbf{v})$

$\nabla \times (\mathbf{u} \times \mathbf{v}) = (\mathbf{v} \cdot \nabla)\mathbf{u} - (\mathbf{u} \cdot \nabla)\mathbf{v} + \mathbf{u}(\nabla \cdot \mathbf{v}) - \mathbf{v}(\nabla \cdot \mathbf{u})$

$\nabla \times (\nabla \times \mathbf{v}) = \nabla(\nabla \cdot \mathbf{v}) - (\nabla \cdot \nabla)\mathbf{v} = \nabla(\nabla \cdot \mathbf{v}) - \nabla^2 \mathbf{v}$

$(\nabla \times \mathbf{v}) \times \mathbf{v} = (\mathbf{v} \cdot \nabla)\mathbf{v} - \dfrac{1}{2}\nabla(\mathbf{v} \cdot \mathbf{v})$